HENLEY'S TWENTIETH CENTURY FORMULAS, RECIPES AND PROCESSES

Containing ten thousand selected household
and workshop formulas, recipes,
processes and moneymaking
methods for the practical
use of manufacturers,
mechanics,
housekeepers
and home workers

Edited by

Gardner Dexter Hiscox

PREFACE

In compiling this book of formulas, recipes and processes, the Editor has endeavored to meet the practical requirements of the home and workshop—the mechanic, the manufacturer, the artisan, the housewife, and the general home worker.

In addition to exercising the utmost care in selecting his materials from competent sources, the Editor has also modified formulas which were obviously ill adapted for his needs, but were valuable if altered. Processes of questionable merit he has discarded. By adhering to this plan the Editor trusts that he has succeeded in preparing a repository of useful knowledge representing the experience of experts in every branch of practical achievement. Much of the matter has been specially translated for this work from foreign technological periodicals and books. In this way the Editor has embodied much practical information otherwise inaccessible to most English-speaking people.

Each recipe is to be regarded as a basis of experiment, to be modified to suit the particular purpose in hand, or the peculiar conditions which may affect the experimenter. Chemicals are not always of uniform relative purity and strength; heat or cold may markedly influence the result obtained, and lack of skill in the handling of utensils and instruments may sometimes cause failure. Inasmuch as a particular formula may not always be applicable, the Editor has thought it advisable to give as many recipes as his space would allow under each heading. In some instances a series of formulas is given which apparently differ but slightly in their ingredients. This has been done on the principle that one or more may be chosen for the purpose in hand.

Recognizing the fact that works of a similar character are not unknown, the Editor has endeavored to present in these pages the most modern methods and formulas. Naturally, old recipes and so-called trade secrets which have proven their value by long use are also included, particularly where no noteworthy advance has been made; but the primary aim has been to modernize and bring the entire work up to the present date.

284534   THE EDITOR.

January, 1914.

PARTIAL LIST OF AUTHORITIES CONSULTED

Apothecary, The.
Berliner Drog. Zeitung.
Brass World.
British Journal of Photography.,
Chemical News.
Chemiker Zeitung Repertorium.
Chemisch Technische Fabrikant.
Chemische Zeitung.
Chemist-Druggist.
Comptes Rendus.
Cooley's Receipts.
Cosmos.
Dekorationsmaler, Der.
Deutsche Drog. Zeitung.
Deutsche Goldschmiede Zeitung.
Deutsche Handwerk.
Deutsche Maler Zeitung.
Deutsche Topfer und Ziefler Zeitung.
Dingler's Polytechnic Journal.
Drogisten Zeitung.
Druggists' Circular.
English Mechanic.
Farben Zeitung.
Gummi Zeitung.
Journal der Goldschmiedekunst.
Journal of Applied Microscopy.
Journal of the Franklin Institute.
Journal Society of Chemical Industry.
Journal Suisse d'Horlogerie.
Keramische Rundschau.
La Nature.
La Science en Famille.
La Vie Scientifique.
Lack und Farben Industrie.
Legierungen.
Le Genie Civil.
Le Praticien.
Leipziger Farber und Zeugdrucker Zeitung.

Maler Zeitung.
Metallarbeiter.
Mining and Scientific Press.
Neueste Erfindungen und Erfahrungen.
Nouvelles Scientifiques.
Oils, Colors, and Drysalteries.
Papier-Zeitung.
Parfumer, Der.
Pharmaceutische Zeitung.
Pharmaceutische Centralhalle.
Pharmaceutische Era.
Pharmaceutische Journal.
Pharmaceutische Journal Formulary.
Photo Times.
Polytech. Centralblatt.
Polyt. Notizblatt.
Popular Science News.
Pottery Gazette.
Practical Druggist.
Revue Chronometrique.
Revue de la Droguerie.
Revue des Produits Chimiques.
Revue Industrielle.
Science, Arts and Nature.
Science Pratique.
Seifensieder Zeitung, Der.
Seifenfabrikant, Der.
Spatula.
Stein der Weisen, Der.
Sudd. Apoth. Zeitung.
Technisches Centralblatt.
Technische Rundschau.
Uhland's Technische Rundschau.
Verzinnen Verzinken Vernickeln, Das.
Werkmeister Zeitung.
Wiener Drogisten Zeitung.
Wiener Gewerbe Zeitung.
Zeitschrift für die Gesammte Kohlensaure Industrie.

ABRASION REMEDY:
See Cosmetics and Ointments.

ABSINTHE:
See Wines and Liquors.

Acid-Proofing

An Acid-Proof Table Top.—

1.

Copper sulphate	1	part
Potassium chlorate	1	part
Water	8	parts

Boil until salts are dissolved.

2.

Aniline hydrochlorate	3	parts
Water	20	parts

Or, if more readily procurable:

Aniline	6	parts
Hydrochloric acid	9	parts
Water	50	parts

By the use of a brush two coats of solution No. 1 are applied while hot; the second coat as soon as the first is dry. Then two coats of solution No. 2, and the wood allowed to dry thoroughly. Later, a coat of raw linseed oil is to be applied, using a cloth instead of a brush, in order to get a thinner coat of the oil.

A writer in the *Journal of Applied Microscopy* states that he has used this method upon some old laboratory tables which had been finished in the usual way, the wood having been filled oiled, and varnished. After scraping off the varnish down to the wood, the solutions were applied, and the result was very satisfactory.

After some experimentations the formula was modified without materially affecting the cost, and apparently increasing the resistance of the wood to the action of strong acids and alkalies. The modified formula follows:

1.

Iron sulphate	4	parts
Copper sulphate	4	parts
Potassium permanganate	8	parts
Water, q. s.	100	parts

2.

Aniline	12	parts
Hydrochloric acid	18	parts
Water, q. s.	100	parts

Or:

Aniline hydrochlorate	15	parts
Water, q. s.	100	parts

Solution No. 2 has not been changed, except to arrange the parts per hundred. The method of application is the same, except that after solution No. 1 has dried the excess of the solution which has dried upon the surface of the wood is thoroughly rubbed off before the application of solution No. 2. The black color does not appear at once, but usually requires a few hours before becoming ebony black. The linseed oil may be diluted with turpentine without disadvantage, and after a few applications the surface will take on a dull and not displeasing polish. The table tops are easily cleaned by washing with water or suds after a course of work is completed, and the application of another coat of oil puts them in excellent order for another course of work. Strong acids or alkalies when spilled, if soon wiped off, have scarcely a perceptible effect.

A slate or tile top is expensive not only in its original cost, but also as a destroyer of glassware. Wood tops when painted, oiled, or paraffined have objectionable features, the latter especially in warm weather. Old table tops, after the paint or oil is scraped off down to the wood, take the above finish nearly as well as the new wood.

To Make Wood Acid- and Chlorine-Proof.—Take 6 pounds of wood tar and 12 pounds rosin, and melt them together in an iron kettle, after which stir in 8 pounds finely powdered brick dust. The damaged parts must be cleaned perfectly and dried, whereupon they may be painted over with the warm preparation or filled up and drawn off, leaving the film on the inside.

Protecting Cement Against Acid.—A paint to protect cement against acid is obtained by mixing pure asbestos, very finely powdered, with a thick solution of sodium silicate. The sodium silicate must be as alkaline as possible. The asbestos is first rubbed with a small quantity of the silicate, until a cake is obtained and then kept in well-closed vessels. For use this cake is simply thinned with a solution of the silicate, which furnishes a paint two or three applications of which protect the walls of reservoirs, etc., against any acid solid or liquid. This mass may also be employed for making a coating of sandstone.

To Make Corks Impermeable and Acid-Proof.—Choose your corks carefully. Then plunge them into a solution of gelatin or common glue, 15 parts, in 24 parts of glycerine and 500 parts of water, heated to 44° or 48° C. (112°–120° F.), and keep them there for several hours. On removing the corks, which should be weighted down in the solution, dry them in the shade until they are free from all surplus moisture. They are now perfectly tight, retaining at the same time the greater portion of their elasticity and suppleness. To render them acid-proof, they should be treated with a mixture of vaseline, 2 parts, and paraffine 7 parts, heated to about 105° F. This second operation may be avoided by adding to the gelatin solution a little ammonium dichromate and afterwards exposing the corks to the light.

Lining for Acid Receptacles.—Plates are formed of 1 part of brown slate, 2 of powdered glass, and 1 of Portland cement, the whole worked up with silicate of soda, molded and dried. Make a cement composed of ground slate and silicate of soda and smear the surface for the lining; then, while it is still plastic, apply the plates prepared as above described. Instead of these plates, slabs of glass or porcelain or similar substances may be employed with the same cement.

ACACIA, MUCILAGE OF:
See Adhesives under Mucilages.

ACID-PROOF GLASS:
See Glass.

ACID-RESISTING PAINT:
See Paint.

ACIDS, SOLDERING:
See Solders.

ACID STAINS FROM THE SKIN, TO REMOVE:
See Cleaning Preparations and Methods.

ACID TEST FOR VINEGAR:
See Vinegar.

Adhesives

GLUES:

Manufacture of Glue.—I.—The usual process of removing the phosphate of lime from bones for glue-making purposes by means of dilute hydrochloric acid has the disadvantage that the acid cannot be regenerated. Attempts to use sulphurous acid instead have so far proved unsuccessful, as, even with the large quantities used, the process is very slow. According to a German invention this difficulty with sulphurous acid can be avoided by using it in aqueous solution under pressure. The solution of the lime goes on very rapidly, it is claimed, and no troublesome precipitation of calcium sulphite takes place. Both phosphate of lime and sulphurous acid are regenerated from the lyes by simple distillation.

II.—Bones may be treated with successive quantities of combined sulphurous acid and water, from which the heat of combination has been previously dissipated, the solution being removed after each treatment, before the bone salts dissolved therein precipitate, and before the temperature rises above 74° F.— U. S. Pat. 783,784.

III.—A patent relating to the process for treating animal sinews, preparatory for the glue factory, has been granted to Florsheim, Chicago, and consists in immersing animal sinews successively in petroleum or benzine to remove the outer fleshy animal skin; in a hardening or preserving bath, as boric acid, or alum or copper sulphate; and in an alkaline bath to remove fatty matter from the fibrous part of the sinews. The sinews are afterwards tanned and disintegrated.

Test for Glue.—The more water the glue takes up, swelling it, the better it is. Four ounces of the glue to be examined are soaked for about 12 hours in a cool place in 4 pounds of cold water. If the glue has dissolved after this time, it is of bad quality and of little value; but if it is coherent, gelatinous, and weighing double, it is good; if it weighs up to 16 ounces, it is very good; if as much as 20 ounces, it may be called excellent.

To Prevent Glue from Cracking.—To prevent glue from cracking, which frequently occurs when glued articles are exposed to the heat of a stove, a little chloride of potassium is added. This prevents the glue from becoming dry enough to crack. Glue thus treated will adhere to glass, metals, etc., and may also be used for pasting on labels.

Preventing the Putrefaction of Strong Glues.—The fatty matter always existing in small quantity in sheets of ordinary glue affects the adhesive properties and facilitates the development of bacteria, and consequently putrefaction and decomposition. These inconveniences are remedied by adding a small quantity of caustic soda to the dissolved glue. The soda prevents decomposition absolutely; with the fatty matter it forms a hard soap which renders it harmless.

Liquid Glues.—

I.—Glue .... 3 ounces
Gelatin .... 3 ounces
Acetic acid .... 4 ounces
Water .... 2 ounces
Alum .... 30 grains

Heat together for 6 hours, skim, and add:

II.—Alcohol .... 1 fluidounce
Brown glue, No. 2.. 2 pounds
Sodium carbonate.. 11 ounces
Water............. 3½ pints
Oil of clove ....160 minims

Dissolve the soda in the water, pour the solution over the dry glue, let stand over night or till thoroughly soaked and swelled, then heat carefully on a water bath until dissolved. When nearly cold stir in the oil of cloves.

By using white glue, a finer article, fit for fancy work, may be made.

III.—Dissolve by heating 60 parts of borax in 420 parts of water, add 480 parts dextrin (pale yellow) and 50 parts of glucose and heat carefully with continued stirring, to complete solution; replace the evaporated water and pour through flannel.

The glue made in this way remains clear quite a long time, and possesses great adhesive power; it also dries very quickly, but upon careless and extended heating above 90° C. (194° F.), it is apt to turn brown and brittle.

IV.—Pour 50 parts of warm (not hot) water over 50 parts of Cologne glue and allow to soak over night. Next day the swelled glue is dissolved with moderate heat, and if still too thick, a little more water is added. When this is done, add from 2½ to 3 parts of crude nitric acid, stir well, and fill the liquid glue in well-corked bottles. This is a good liquid steam glue.

V.—Soak 1 pound of good glue in a quart of water for a few hours, then melt the glue by heating it, together with the unabsorbed water, then stir in ½ pound dry white lead, and when that is well mixed pour in 4 fluidounces of alcohol and continue the boiling 5 minutes longer.

VI.—Soak 1 pound of good glue in 1½ pints of cold water for 5 hours, then add 3 ounces of zinc sulphate and 2 fluidounces of hydrochloric acid, and keep the mixture heated for 10 or 12 hours at 175° to 190° F. The glue remains liquid and may be used for sticking a variety of materials.

VII.—A very inexpensive liquid glue may be prepared by first soaking and then dissolving gelatin in twice its own weight of water at a very gentle heat; then add glacial acetic acid in weight equal to the weight of the dry gelatin. It should be remembered, however, that all acid glues are not generally applicable.

VIII.—Glue .... 200 parts
Dilute acetic acid.. 400 parts

Dissolve by the aid of heat and add:

Alcohol .... 25 parts
Alum .... 5 parts

IX.—Glue............. 5 parts
Calcium chloride.. 1 part
Water .... 1 part

X.—Sugar of lead...... 1½ drachms
Alum............. 1½ drachms
Gum arabic....... 2½ drachms
Wheat flour....... 1 av. lb.
Water, q. s.

Dissolve the gum in 2 quarts of warm water; when cold mix in the flour, and add the sugar of lead and alum dissolved in water; heat the whole over a slow fire until it shows signs of ebullition. Let it cool, and add enough gum water to bring it to the proper consistence.

XI.—Dilute 1 part of official phosphoric acid with 2 parts of water and neutralize the solution with carbonate of ammonium. Add to the liquid an equal quantity of water, warm it on a water bath, and dissolve in it sufficient glue to form a thick syrupy liquid. Keep in well-stoppered bottles.

XII.—Dissolve 3 parts of glue in small pieces in 12 to 15 of saccharate of lime. By heating, the glue dissolves rapidly and remains liquid, when cold, without loss of adhesive power. Any desirable consistence can be secured by varying the amount of saccharate of lime. Thick glue retains its muddy color, while a thin solution becomes clear on standing.

The saccharate of lime is prepared by

dissolving 1 part of sugar in 3 parts of water, and after adding ¼ part of the weight of the sugar of slaked lime, heating the whole from 149° to 185° F., allowing it to macerate for several days, shaking it frequently. The solution, which has the properties of mucilage, is then decanted from the sediment.

XIII.—In a solution of borax in water soak a good quantity of glue until it has thoroughly imbibed the liquid. Pour off the surplus solution and then put on the water bath and melt the glue. Cool down until the glue begins to set, then add, drop by drop, with agitation, enough acetic acid to check the tendency to solidification. If, after becoming quite cold, there is still a tendency to solidification, add a few drops more of the acid. The liquid should be of the consistence of ordinary mucilage at all times.

XIV.—Gelatin............. 100 parts
Cabinetmakers' glue. 100 parts
Alcohol............ 25 parts
Alum.............. 2 parts
Acetic acid, 20 per
cent............. 800 parts

Soak the gelatin and glue with the acetic acid and heat on a water bath until fluid; then add the alum and alcohol.

XV.—Glue .... 10 parts
Water .... 15 parts
Sodium salicylate.... 1 part

XVI.—Soak 5 parts of Cologne glue in an aqueous calcium chloride solution (1:4) and heat on the water bath until dissolved, replacing the evaporating water; or slack 100 parts of lime with 150 parts of hot water, dissolve 60 parts of sugar in 180 parts of water, and add 15 parts of the slacked lime to the solution, heating the whole to 75° C. (167° F.). Place aside for a few days, shaking from time to time. In the clear sugar-lime solution collected by decanting soak 60 parts of glue and assist the solution by moderate heating.

XVII.—Molasses, 100 parts, dissolved in 300 parts of water, 25 parts of quicklime (slaked to powder), being then stirred in and the mixture heated to 167° F. on a water bath, with frequent stirrings. After settling for a few days a large portion of the lime will have dissolved, and the clear, white, thick solution, when decanted, behaves like rubber solution and makes a highly adherent coating.

XVIII.—Dissolve bone glue, 250 parts, by heating in 1,000 parts of water, and add to the solution barium peroxide 10 parts, sulphuric acid (66° B.) 5

parts, and water 15 parts. Heat for 48 hours on the water bath to 80° C. (176° F.). Thus a syrupy liquid is obtained, which is allowed to settle and is then decanted. This glue has no unpleasant odor, and does not mold.

XIX.—A glue possessing the adhesive qualities of ordinary joiners' glue, but constituting a pale yellow liquid which is ready for use without requiring heating and possesses great resistance to dampness, is produced by treating dry casein with a diluted borax solution or with enough ammonia solution to cause a faintly alkaline reaction. The preparation may be employed alone or mixed with liquid starch in any proportion.

Glue for Celluloid.—I.—Two parts shellac, 3 parts spirits of camphor, and 4 parts strong alcohol dissolved in a warm place, give an excellent gluing agent to fix wood, tin, and other bodies to celluloid. The glue must be kept well corked up.

II.—A collodion solution may be used, or an alcoholic solution of fine celluloid shavings.

Glue to Form Paper Pads.—

I.—Glue............ 3½ ounces
Glycerine........ 8 ounces
Water, a sufficient quantity.

Pour upon the glue more than enough water to cover it and let stand for several hours, then decant the greater portion of the water; apply heat until the glue is dissolved, and add the glycerin. If the mixture is too thick, add more water.

II.—Glue............ 6 ounces
Alum............ 30 grains
Acetic acid....... ½ ounce
Alcohol.......... 1½ ounces
Water.......... 6½ ounces

Mix all but the alcohol, digest on a water bath till the glue is dissolved, allow to cool and add the alcohol.

III.—Glue............ 5 ounces
Water............ 1 ounce
Calcium chloride.. 1 ounce

Dissolve the calcium chloride in the water, add the glue, macerate until it is thoroughly softened, and then heat until completely dissolved.

IV.—Glue............ 20 ounces
Glycerine........ 5 ounces
Syrupy glucose... 1 ounce
Tannin.......... 50 grains

Cover the glue with cold water, and let stand over night. In the morning pour off superfluous water, throw the glue on muslin, and manipulate so as to get rid of as much moisture as possible, then put in a water bath and melt. Add the glycerine and syrup, and stir well in. Finally, dissolve the tannin in the smallest quantity of water possible and add.

This mixture must be used hot.

V.—Glue............. 15 ounces
    Glycerine......... 5 ounces
    Linseed oil........ 2 ounces
    Sugar............. 1 ounce

Soak the glue as before, melt, add the sugar and glycerine, continuing the heat, and finally add the oil gradually under constant stirring.

This must be used hot.

Glue for Tablets.—

I.—Glue............. 3½ ounces
    Glycerine......... 8 ounces
    Water, a sufficient quantity.

Pour upon the glue more than enough water to cover it and let stand for several hours, then decant the greater portion of the water; apply heat until the glue is dissolved, and add the glycerine. If the mixture is too thick, add more water.

II.—Glue............. 6 ounces
    Alum............. 30 grains
    Acetic acid....... ½ ounce
    Alcohol.......... 1½ ounces
    Water........... 6½ ounces

Mix all but the alcohol, digest on a water bath till the glue is dissolved, allow to cool and add the alcohol.

III.—Glue............. 5 ounces
    Water........... 1 ounce
    Calcium chloride... 1 ounce

Dissolve the calcium chloride in the water, add the glue, macerate until it is thoroughly softened, and then apply heat until completely dissolved.

IV.—Glue, 1 pound; glycerine, 4 ounces; glucose syrup, 2 tablespoonfuls; tannin, $\frac{1}{16}$ ounce. Use warm, and give an hour to dry and set on the pads. This can be colored with any aniline dye.

Marine Glue.—Marine glue is a product consisting of shellac and caoutchouc, which is mixed differently according to the use for which it is required. The quantity of benzol used as solvent governs the hardness or softness of the glue.

I.—One part Pará caoutchouc is dissolved in 12 parts benzol; 20 parts powdered shellac are added to the solution, and the mixture is carefully heated.

II.—Stronger glue is obtained by dissolving 10 parts good crude caoutchouc in 120 parts benzine or naphtha which solution is poured slowly and in a fine stream into 20 parts asphaltum melted in a kettle, stirring constantly and heating. Pour the finished glue, after the solvent has almost evaporated and the mass has become quite uniform, into flat molds, in which it solidifies into very hard tablets of dark brown or black color. For use, these glue tablets are first soaked in boiling water and then heated over a free flame until the marine glue has become thinly liquid. The pieces to be glued are also warmed and a very durable union is obtained.

III.—Cut caoutchouc into small pieces and dissolve in coal naphtha by heat and agitation. Add to this solution powdered shellac, and heat the whole, constantly stirring until combination takes place, then pour it on metal plates to form sheets. When used it must be heated to 248° F., and applied with a brush.

Water-Proof Glues.—I.—The glue is put in water till it is soft, and subsequently melted in linseed oil at moderate heat. This glue is affected neither by water nor by vapors.

II.—Dissolve a small quantity of sandarac and mastic in a little alcohol, and add a little turpentine. The solution is boiled in a kettle over the fire, and an equal quantity of a strong hot solution of glue and isinglass is added. Then filter through a cloth while hot.

III.—Water-proof glue may also be produced by the simple addition of bichromate of potassium to the liquid glue solution, and subsequent exposure to the air.

IV.—Mix glue as usual, and then add linseed oil in the proportion of 1 part oil to 8 parts glue. If it is desired that the mixture remain liquid, ½ ounce of nitric acid should be added to every pound of glue. This will also prevent the glue from souring.

V.—In 1,000 parts of rectified alcohol dissolve 60 parts of sandarac and as much mastic whereupon add 60 parts of white oil of turpentine. Next, prepare a rather strong glue solution and add about the like quantity of isinglass, heating the solution until it commences to boil; then slowly add the hot glue solution till a thin paste forms, which can still be filtered through a cloth. Heat the solution before use and employ like ordinary glue. A connection effected with this glue is not dissolved by cold water and even resists hot water for a long time.

VI.—Soak 1,000 parts of Cologne glue in cold water for 12 hours and in another vessel for the same length of time 150 parts of isinglass in a mixture of lamp spirit and water. Then dissolve both masses together on the water bath in a suitable vessel, thinning, if necessary, with some hot water. Next add 100 parts of linseed oil varnish and filter hot through linen.

VII.—Ordinary glue is kept in water until it swells up without losing its shape. Thus softened it is placed in an iron crucible without adding water; then add linseed oil according to the quantity of the glue and leave this mixture to boil over a slow fire until a gelatinous mass results. Such glue unites materials in a very durable manner. It adheres firmly and hardens quickly. Its chief advantage, however, consists in that it neither absorbs water nor allows it to pass through, whereby the connecting places are often destroyed. A little borax will prevent putrefaction.

VIII.—Bichromate of potassium 40 parts (by weight); gelatin glue, 55 parts; alum, 5 parts. Dissolve the glue in a little water and add the bichromate of potassium and the alum.

IX.—This preparation permits an absolutely permanent gluing of pieces of cardboard, even when they are moistened by water. Melt together equal parts of good pitch and gutta-percha; of this take 9 parts, and add to it 3 parts of boiled linseed oil and 1½ parts of litharge. Place this over the fire and stir it till all the ingredients are intimately mixed. The mixture may be diluted with a little benzine or oil of turpentine, and must be warm when used.

Glue to Fasten Linoleum on Iron Stairs.—I.—Use a mixture of glue, isinglass, and dextrin which, dissolved in water and heated, is given an admixture of turpentine. The strips pasted down must be weighted with boards and brick on top until the adhesive agent has hardened.

II.—Soak 3 parts of glue in 8 parts water, add ½ part hydrochloric acid and ¾ part zinc vitriol and let this mixture boil several hours. Coat the floor and the back of the linoleum with this. Press the linoleum down uniformly and firmly and weight it for some time.

Glue for Attaching Gloss to Precious Metals.—Sandarac varnish, 15 parts; marine glue, 5 parts; drying oil, 5 parts; white lead, 5 parts; Spanish white, 5 parts; turpentine, 5 parts. Triturate all to form a rather homogeneous paste. This glue becomes very hard and resisting.

Elastic Glue.—Although elastic glue is less durable than rubber, and will not stand much heat, yet it is cheaper than rubber, and is not, like rubber affected by oil colors. Hence it is largely used for printing rollers and stamps. For

stamps, good glue is soaked for 24 hours in soft water. The water is poured off, and the swollen glue is melted and mixed with glycerine and a little salicylic acid and cast into molds. The durability is increased by painting the mass with a solution of tannin, or, better, of bichromate of potassium. Printing rollers require greater firmness and elasticity. The mass for them once consisted solely of glue and vinegar, and their manufacture was very difficult. The use of glycerine has remedied this, and gives great elasticity without adhesiveness, and has removed the liability of moldiness. Swollen glue, which has been superficially dried, is fused with glycerine and cast into oil molds. Similar mixtures are used for casting plaster ornaments, etc., and give very sharp casts. A mass consisting of glue and glycerine is poured over the model in a box. When the mold is removed, it is painted with plaster outside and with boiled oil inside, and can then be used many times for making reproductions of the model.

Glue for Paper and Metal.—A glue which will keep well and adhere tightly is obtained by diluting 1,000 parts by weight of potato starch in 1,200 parts by weight of water and adding 50 parts by weight of pure nitric acid. The mixture is kept in a hot place for 48 hours, taking care to stir frequently. It is afterwards boiled to a thick and transparent consistency, diluted with water if there is occasion, and then there are added in the form of a screened powder, 2 parts of sal ammoniac and 1 part of sulphur flowers.

Glue for Attaching Cloth Strips to Iron.—Soak 500 parts of Cologne glue in the evening with clean cold water in a clean vessel; in the morning pour off the water, place the softened glue without admixture of water into a clean copper or enamel receptacle, which is put on a moderate low fire (charcoal or steam apparatus). During the dissolution the mass must be continually stirred with a wooden trowel or spatula. If the glue is too thick, it is thinned with diluted spirit, but not with water. As soon as the glue has reached the boiling point, about 50 parts of linseed oil varnish (boiled oil) is added to the mass with constant stirring. When the latter has been stirred up well, add 50 parts of powdered colophony and shake it into the mass with stirring, subsequently removing the glue from the fire. In order to increase the binding qualities and to guard against moisture, it is well still to add about 50 parts of isinglass, which has been previously cut into narrow strips and placed, well beaten, in a vessel, into which enough spirit of wine has been poured to cover all. When dissolved, the last - named mass is added to the boiling glue with constant stirring. The adhesive agent is now ready for use and is employed hot, it being advisable to warm the iron also. Apply glue only to a surface equivalent to a single strip at a time. The strips are pressed down with a stiff brush or a wad of cloth.

Glue for Leather or Cardboard.—To attach leather to cardboard dissolve good glue (softened by swelling in water) with a little turpentine and enough water in an ordinary glue pot, and then having made a thick paste with starch in the proportion of 2 parts by weight, of starch powder for every 1 part, by weight, of dry glue, mix the compounds and allow the mixture to become cold before application to the cardboard.

For Wood, Glass, Cardboard, and all Articles of a Metallic or Mineral Character.—Take boiled linseed oil 20 parts, Flemish glue 20 parts, hydrated lime 15 parts, powdered turpentine 5 parts, alum 5 parts acetic acid 5 parts. Dissolve the glue with the acetic acid, add the alum, then the hydrated lime, and finally the turpentine and the boiled linseed oil. Triturate all well until it forms a homogeneous paste and keep in well-closed flasks. Use like any other glue.

Glue for Uniting Metals with Fabrics.—Cologne glue of good quality is soaked and boiled down to the consistency of that used by cabinetmakers. Then add, with constant stirring, sifted wood ashes until a moderately thick, homogeneous mass results. Use hot and press the pieces well together during the drying. For tinfoil about 2 per cent of boracic acid should be added instead of the wood ashes.

Glue or Paste for Making Paper Boxes.—

Chloral hydrate......	5 parts
Gelatin, white.......	8 parts
Gum arabic.........	2 parts
Boiling water........	30 parts

Mix the chloral, gelatin, and gum arabic in a porcelain container, pour the boiling water over the mixture and let stand for 1 day, giving it a vigorous stirring several times during the day. In cold weather this is apt to get hard and stiff, but this may be obviated by standing the container in warm water for a few minutes. This paste adheres to any surface whatever.

Natural Glue for Cementing Porcelain, Crystal Glass, etc.—The large shell snails which are found in vineyards have at the extremity of their body a small, whitish bladder filled with a substance of greasy and gelatinous aspect. If this substance extracted from the bladder is applied on the fragments of porcelain or any body whatever, which are juxtaposed by being made to touch at all parts, they acquire such adhesion that if one strives to separate them by a blow, they are more liable to break at another place than the cemented seam. It is necessary to give this glue sufficient time to dry perfectly, so as to permit it to acquire the highest degree of strength and tenacity.

Belt Glue.—A glue for belts can be prepared as follows: Soak 50 parts of gelatin in water, pour off the excess of water, and heat on the water bath. With good stirring add, first, 5 parts, by weight, of glycerine, then 10 parts, by weight, of turpentine, and 5 parts, by weight, of linseed oil varnish and thin with water as required. The ends of the belts to be glued are cut off obliquely and warmed; then the hot glue is applied, and the united parts are subjected to strong pressure, allowing them to dry thus for 24 hours before the belts are used.

Chromium Glue for Wood, Paper, and Cloth.—I.—(a) One-half pound strong glue (any glue if color is immaterial, white fish glue otherwise); soak 12 hours in 12 fluidounces of cold water. (b) One-quarter pound gelatin; soak 2 hours in 12 fluidounces cold water. (c) Two ounces bichromate of potassium dissolved in 8 fluidounces boiling water. Dissolve (a) after soaking, in a glue pot, and add (b). After (a) and (b) are mixed and dissolved, stir in (c). This glue is exceedingly strong, and if the article cemented be exposed to strong sunlight for 1 hour, the glue becomes perfectly waterproof. Of course, it is understood that the exposure to sunlight is to be made after the glue is thoroughly dry. The one objectionable feature of this cement is its color, which is a yellow-brown. By substituting chrome alum in place of the bichromate, an olive color is obtained.

II.—Use a moderately strong gelatin solution (containing 5 to 10 per cent of dry gelatin), to which about 1 part of acid chromate of potassium in solution is added to every 5 parts of gelatin. This mixture has the property of becoming insoluble by water through the action of sunlight under partial reduction of the chromic acid.

Fireproof Glue.—

Raw linseed oil	8 parts
Glue or gelatin	1 part
Quicklime	2 parts

Soak the glue or gelatin in the oil for 10 to 12 hours, and then melt it by gently heating the oil, and when perfectly fluid stir in the quicklime until the whole mass is homogeneous, then spread out in layers to dry gradually, out of the sun's rays. For use, reheat the glue in a glue pot in the ordinary way of melting glue.

CEMENTS.

Under this heading will be found only cements for causing one substance to adhere to another. Cements used primarily as fillers, such as dental cements, will be found under Cements, Putties, etc.

Cutlers' Cements for Fixing Knife Blades into Handles.—

I.—Rosin		4 pounds
Beeswax		1 pound
Plaster of Paris or brickdust		1 pound
II.—Pitch		5 pounds
Wood ashes		1 pound
Tallow		1 pound

III.—Rosin, 12; sulphur flowers, 3; iron filings, 5. Melt together, fill the handle while hot, and insert the instrument.

IV.—Plaster of Paris is ordinarily used for fastening loose handles. It is made into a moderately thick paste with water run into the hole in the head of the pestle, the handle inserted and held in place till the cement hardens. Some add sand to the paste, and claim to get better results.

V.—Boil together 1 part of caustic soda, 3 parts of rosin, and 5 parts of water till homogeneous and add 4 parts of plaster of Paris. The paste sets in half an hour and is but little affected by water.

VI.—Equal quantities of gutta percha and shellac are melted together and well stirred. This is best done in an iron capsule placed on a sandbath and heated over a gas furnace or on the top of a stove. The combination possesses both hardness and toughness, qualities that make it particularly desirable in mending mortars and pestles. In using, the articles to be cemented should be warmed to about the melting point of the mixture and retained in proper position until cool, when they are ready for use.

VII.—Rosin	600	⎫	Parts
Sulphur	150	⎬	by
Iron filings	250	⎭	weight.

Pour the mixture, hot, into the opening of the heated handle and shove in the knife likewise heated.

VIII.—Melt sufficient black rosin, and incorporate thoroughly with it one-fifth its weight of very fine silver sand. Make the pestle hot, pour in a little of the mixture, then force the handle well home, and set aside for a day before using.

IX.—Make a smooth, moderately soft paste with litharge and glycerine; fill the hole in the pestle with the cement, and firmly press the handle in place, keeping it under pressure for three or four days.

Cements for Stone.—I.—

An excellent cement for broken marble consists of 4 parts of gypsum and 1 part of finely powdered gum arabic. Mix intimately. Then with a cold solution of borax make into a mortarlike mass. Smear on each face of the parts to be joined, and fasten the bits of marble together. In the course of a few days the cement becomes very hard and holds very tenaciously. The object mended should not be touched for several days. In mending colored marbles the cement may be given the hue of the marble by adding the color to the borax solution.

II.—A cement which dries instantaneously, qualifying it for all sorts of repairing and only presenting the disadvantage of having to be freshly prepared each time, notwithstanding any subsequent heating, may be made as follows: In a metal vessel or iron spoon melt 4 to 5 parts of rosin (or preferably mastic) and 1 part of beeswax. This mixture must be applied rapidly, it being of advantage slightly to heat the surfaces to be united, which naturally must have been previously well cleaned.

III.—Slaked lime, 10 parts; chalk, 15 parts; kaolin, 5 parts; mix, and immediately before use stir with a corresponding amount of potash water glass.

IV.—Cement on Marble Slabs.—The whole marble slab is thoroughly warmed and laid face down upon a neatly cleaned planing bench upon which a woolen cloth is spread so as not to injure the polish of the slab. Next apply to the slab very hot, weak glue and quickly sift hot plaster of Paris on the glue in a thin even layer, stirring the plaster rapidly into the applied glue by means of a strong spatula, so that a uniform glue-plaster coating is formed on the warm slab. Before this has time to harden tip the respective piece of furniture on the slab. The frame, likewise warmed, will adhere very firmly to the slab after two days. Besides, this process has the advantage of great cleanliness.

V.—The following is a recipe used by marble workers, and which probably can be used to advantage: Flour of sulphur, 1 part; hydrochlorate of ammonia, 2 parts; iron filings, 16 parts. The above substances must be reduced to a powder, and securely preserved in closely stoppered vessels. When the cement is to be employed, take 20 parts very fine iron filings and 1 part of the above powder; mix them together with enough water to form a manageable paste. This paste solidifies in 20 days and becomes as hard as iron. A recipe for another cement useful for joining small pieces of marble or alabaster is as follows: Add ½ pint of vinegar to ½ pint skimmed milk; mix the curd with the whites of 5 eggs, well beaten, and sufficient powdered quicklime sifted in with constant stirring so as to form a paste. It resists water and a moderate degree of heat.

VI.—Cement for Iron and Marble.— For fastening iron to marble or stone a good cement is made as follows: Thirty parts plaster of Paris, 10 parts iron filings, ½ part sal ammoniac mixed with vinegar to a fluid paste fresh for use.

Cement for Sandstones.—One part sulphur and 1 part rosin are melted separately; the melted masses are mixed and 3 parts litharge and 2 parts ground glass stirred in. The latter ingredients must be perfectly dry, and have been well pulverized and mixed previously.

Equally good cement is obtained by melting together 1 part pitch and ⅒ part wax, and mixing with 2 parts brickdust.

The stones to be cemented, or between the joints of which the putty is to be poured, must be perfectly dry. If practicable, they should be warmed a little, and the surfaces to which the putty is to adhere painted with oil varnish once or twice. The above two formulæ are of especial value in case the stones are very much exposed to the heat of the sun in summer, as well as to cold, rain, and snow in winter. Experience has shown that in these instances the above-mentioned cements give better satisfaction than the other brands of cement.

Cements for Attaching Objects to Glass.—

    Rosin . . . . 1 part
    Yellow wax . . . . 2 parts

Melt together.

To Attach Copper to Glass.—Boil 1 part of caustic soda and 3 parts of colophony in 5 parts of water and mix with the like quantity of plaster of Paris.

This cement is not attacked by water, heat, and petroleum. If, in place of the plaster of Paris, zinc white, white lead, or slaked lime is used, the cement hardens more slowly.

To Fasten Brass upon Glass.—Boil together 1 part of caustic soda, 3 parts of rosin, 3 parts of gypsum, and 5 parts of water. The cement made in this way hardens in about half an hour, hence it must be applied quickly. During the preparation it should be stirred constantly. All the ingredients used must be in a finely powdered state.

Uniting Glass with Horn.—(1) A solution of 2 parts of gelatin in 20 parts water is evaporated up to one-sixth of its volume and ⅓ mastic dissolved in ½ spirit added and some zinc white stirred in. The putty is applied warm; it dries easily and can be kept a long time. (2) Mix gold size with the equal volume of water glass.

To Cement Glass to Iron.—

I.—Rosin . . . . 5 ounces
    Yellow wax . . . . 1 ounce
    Venetian red . . . . 1 ounce

Melt the wax and rosin on a water bath and add, under constant stirring, the Venetian red previously well dried. Stir until nearly cool, so as to prevent the Venetian red from settling to the bottom.

II.—Portland cement . . . . 2 ounces
    Prepared chalk . . . . . 1 ounce
    Fine sand . . . . 1 ounce
    Solution of sodium silicate enough to form a semi-liquid paste.

III.—Litharge . . . . 2 parts
    White lead . . . . 1 part

Work into a pasty condition by using 3 parts boiled linseed oil, 1 part copal varnish.

Celluloid Cements.—I.—To mend broken draughting triangles and other celluloid articles, use 3 parts alcohol and 4 parts ether mixed together and applied to the fracture with a brush until the edges become warm. The edges are then stuck together, and left to dry for at least 24 hours.

II.—Camphor, 1 part; alcohol, 4 parts. Dissolve and add equal quantity (by weight) of shellac to this solution.

III.—If firmness is desired in putting celluloid on wood, tin, etc., the following gluing agent is recommended, viz.: A compound of 2 parts shellac, 3 parts spirit of camphor, and 4 parts strong alcohol.

IV.—Shellac............ 2 ounces
    Spirits of camphor.. 2 ounces
    Alcohol, 90 per cent.. 6 to 8 ounces

V.—Make a moderately strong glue or solution of gelatin. In a dark place or a dark room mix with the above a small amount of concentrated solution of potassium dichromate. Coat the back of the label, which must be clean, with a thin layer of the mixture. Strongly press the label against the bottle and keep the two in close contact by tying with twine or otherwise. Expose to sunlight for some hours; this causes the cement to be insoluble even in hot water.

VI.—Lime............. av. oz. 1
    White of egg...... av. oz. 2½
    Plaster of Paris.... av. oz. 5½
    Water............ fl. oz. 1

Reduce the lime to a fine powder; mix it with the white of egg by trituration, forming a uniform paste. Dilute with water, rapidly incorporate the plaster of Paris, and use the cement immediately. The surfaces to be cemented must first be moistened with water so that the cement will readily adhere. The pieces must be firmly pressed together and kept in this position for about 12 hours.

Cementing Celluloid and Hard-Rubber Articles.—I.—Celluloid articles can be mended by making a mixture composed of 3 parts of alcohol and 4 parts of ether. This mixture should be kept in a well-corked bottle, and when celluloid articles are to be mended, the broken surfaces are painted over with the alcohol and ether mixture until the surfaces soften: then press together and bind and allow to dry for at least 24 hours.

II.—Dissolve 1 part of gum camphor in 4 parts of alcohol; dissolve an equal weight of shellac in such strong camphor solution. The cement is applied warm and the parts united must not be disturbed until the cement is hard. Hard-rubber articles are never mended to form a strong joint.

III.—Melt together equal parts of gutta percha and real asphaltum. The cement is applied hot, and the broken surfaces pressed together and held in place while cooling.

Sign-Letter Cements.—

I.—Copal varnish...... 15 parts
    Drying oil .... 5 parts
    Turpentine (spirits). 3 parts
    Oil of turpentine.... 2 parts
    Liquefied glue...... 5 parts

Melt all together on a water bath until well mixed, and then add 10 parts slaked lime.

II.—Mix 100 parts finely powdered white litharge with 50 parts dry white lead, knead together 3 parts linseed oil varnish and 1 part copal varnish into a firm dough. Coat the side to be attached with this, removing the superfluous cement. It will dry quickly and become very hard.

III.—Copal varnish...... 15 parts
    Linseed-oil varnish . 5 parts
    Raw turpentine..... 3 parts
    Oil of turpentine.... 2 parts
    Carpenters' glue, dis-
      solved in water... 5 parts
    Precipitated chalk.. 10 parts

IV.—Mastic gum........ 1 part
    Litharge, lead...... 2 parts
    White lead......... 1 part
    Linseed oil........ 3 parts

Melt together to a homogeneous mass. Apply hot. To make a thorough and reliable job, the letters should be heated to at least the temperature of the cement.

To Fix Gold Letters, etc., upon Glass. —I.—The glass must be entirely clean and polished, and the medium is prepared in the following manner: One ounce fish glue or isinglass is dissolved in water so that the latter covers the glue. When this is dissolved a quart of rectified spirit of wine is added, and enough water is poured in to make up one-quarter the whole. The substance must be kept well corked.

II.—Take ½ quart of the best rum and ¼ ounce fish glue, which is dissolved in the former at a moderate degree of heat. Then add ½ quart distilled water, and filter through a piece of old linen. The glass is laid upon a perfectly level table and is covered with this substance to the thickness of ⅛ inch, using a clean brush. Seize the gold leaf with a pointed object and place it smoothly upon the prepared mass, and it will be attracted by the glass at once. After 5 minutes hold the glass slightly slanting so that the superfluous mass can run off, and leave the plate in this position for 24 hours, when it will be perfectly dry. Now trace the letters or the design on a piece of paper, and perforate the lines with a thick needle, making the holes ⅕ inch apart. Then place the perforated paper upon the surface of the glass, and stamp the tracery on with powdered chalk. The paper pattern is then carefully removed, and the accurate design will remain upon the gold. The outlines are now filled out with an oily gold mass, mixed with a little chrome orange and diluted with boiled oil or turpentine. When all is dry the superfluous gold is washed off with water by means of a common rag. The back of the glass is then painted with a suitable color.

Attaching Enamel Letters to Glass.— To affix enamel letters to glass, first clean the surface of the glass perfectly, leaving no grease or sticky substance of any kind adhering to the surface. Then with a piece of soap sketch the outlines of the design. Make the proper division of the guide lines, and strike off accurately the position each letter is to occupy. Then to the back of the letters apply a cement made as follows: White lead ground in oil, 2 parts; dry white lead, 3 parts. Mix to a soft putty consistency with good copal varnish.

With a small knife or spatula apply the cement to the back of the letters, observing especial care in getting the mixture well and uniformly laid around the inside edges of the letter. In attaching the letters to the glass make sure to expel the air from beneath the characters, and to do this, work them up and down and sidewise. If the weather be at all warm, support the letters while drying by pressing tiny beads of sealing wax against the glass, close to the under side or bottom of the letters. With a putty knife, keenly sharpened on one edge, next remove all the surplus cement. Give the letters a hard, firm pressure against the glass around all edges to securely guard against the disruptive attacks of moisture.

The seepage of moisture beneath the surface of the letters is the main cause of their early detachment from the glass.

The removal of the letters from the glass may be effected by applying turpentine to the top of the characters, allowing it to soak down and through the cement. Oxalic acid applied in the same way will usually slick the letters off in a trice.

Cement for Porcelain Letters.—Slake 15 parts of fresh quicklime in 20 parts of water. Melt 50 parts of caoutchouc and 50 parts of linseed-oil varnish together, and bring the mixture to a boil. While boiling, pour the liquid on the slaked lime, little by little, under constant stirring. Pass the mixture, while still hot, through muslin, to remove any possible lumps, and let cool. It takes the cement 2 days to set completely, but when dry it makes a joint that will resist a great deal of strain. By thinning the mixture down with oil of turpentine, a brilliant, powerfully adhesive varnish is obtained.

Water - Glass Cements. — I. — Water glass (sodium of potassium silicate), which is frequently recommended for cementing glass, does not, as is often asserted, form a vitreous connection between the joined surfaces; and, in fact, some of the commercial varieties will not even dry, but merely form a thick paste, which has a strong affinity for moisture. Good 30° B. water glass is, however, suitable for mending articles that are exposed to heat, and is best applied to surfaces that have been gently warmed; when the pieces are put together they should be pressed warmly, to expel any superfluous cement, and then heated strongly.

To repair cracked glasses or bottles through which water will leak, water glasses may be used, the application being effected in the following easy manner: The vessel is warmed to induce rarefaction of the internal air, after which the mouth is closed, either by a cork in the case of bottles, or by a piece of parchment or bladder if a wide-mouthed vessel is under treatment. While still hot, the outside of the crack is covered with a little glass, and the vessel set aside to cool, whereupon the difference between the pressure of the external and internal air will force the cement into the fissure and close it completely. All that is then necessary is to take off the cover and leave the vessel to warm for a few hours. Subsequently rinse it out with lime water, followed by clean water, and it will then hold any liquid, acids and alkaline fluids alone excepted.

II.—When water glass is brought into contact with calcium chloride, a calcium silicate is at once formed which is insoluble in water. It seems possible that this reaction may be used in binding together masses of sand, etc. The process indicated has long been used in the preservation of stone which has become "weathered." The stone is first brushed with the water glass and afterwards with a solution of calcium chloride. The conditions here are of course different.

Calcium chloride must not be confounded with the so-called "chloride of lime" which is a mixture of calcium hypochlorite and other bodies.

To Fasten Paper Tickets to Glass —To attach paper tickets to glass, the employment of water glass is efficacious. Care should be taken to spread this product on the glass and not on the paper, and then to apply the paper dry, which should be done immediately. When the solution is dry the paper cannot be detached. The silicate should be somewhat diluted. It is spread on the glass with a rag or a small sponge.

JEWELERS' CEMENTS.

Jewelers and goldsmiths require, for the cementing of genuine and colored gems, as well as for the placing of colored folio under certain stones, very adhesive gluing agents, which must, however, be colorless. In this respect these are distinguished chiefly by the so-called diamond cement and the regular jewelers' cement. Diamond cement is much esteemed by jewelers for cementing precious stones and corals, but may also be employed with advantage for laying colored fluxes of glass on white glass. The diamond cement is of such a nature as to be able to remain for some time in contact with water without becoming soft. It adheres best between glass or between precious stones. It is composed as follows: Isinglass 8 parts, gum ammoniac 1 part, galbanum 1 part, spirit of wine 4 parts. Soak the isinglass in water with admixture of a little spirit of wine and add the solution of the gums in the remainder of the spirit of wine. Before use, heat the diamond cement a little so as to soften it. Jewelers' cement is used for similar purposes as is the diamond cement, and is prepared from: Isinglass (dry) 10 parts, mastic varnish 5 parts. Dissolve the isinglass in very little water, adding some strong spirit of wine. The mastic varnish is prepared by pouring a mixture of highly rectified spirit of wine and benzine over finely powdered mastic and dissolving it in the smallest possible quantity of liquid. The two solutions of isinglass and mastic are intimately ground together in a porcelain dish.

Armenian Cement.—The celebrated "Armenian" cement, so called formerly used by Turkish and Oriental jewelers generally, for setting precious stones, "facing diamonds," rubies, etc., is made as follows:

Mastic gum........	10 parts
Isinglass (fish glue) .	20 parts
Gum ammoniac....	5 parts
Alcohol absolute....	60 parts
Alcohol, 50 per cent..	35 parts
Water.............	100 parts

Dissolve the mastic in the absolute alcohol; dissolve, by the aid of gentle heat, on the water bath, the isinglass in the water, and add 10 parts of the dilute alcohol. Now dissolve the ammoniacum in the residue of the dilute alcohol. Add

the first solution to the second, mix thoroughly by agitation and then add the solution of gum ammoniac and stir well in. Finally put on the water bath, and keeping at a moderate heat, evaporate the whole down to 175 parts.

Cement for Enameled Dials.—The following is a good cement for enameled dials, plates, or other pieces: Grind into a fine powder 2¼ parts of dammar rosin and 2½ parts of copal, using colorless pieces if possible. Next add 2 parts of Venetian turpentine and enough spirit of wine so that the whole forms a thick paste. To this grind 3 parts of the finest zinc white. The mass now has the consistency of prepared oil paint. To remove the yellow tinge of the cement add a trifle of Berlin blue to the zinc white. Finally, the whole is heated until the spirit of wine is driven off and a molten mass remains, which is allowed to cool and is kept for use. Heat the parts to be cemented.

Watch-Lid Cement.—The hardest cement for fixing on watch lids is shellac. If the lids are exceedingly thin the engraving will always press through. Before cementing it on the inside of the lid, in order not to injure the polish, it is coated with chalk dissolved in alcohol, which is first allowed to dry. Next melt the shellac on the stick, heat the watch lid and put it on. After the engraving has been done, simply force the lid off and remove the remaining shellac from the latter by light tapping. If this does not remove it completely lay the lid in alcohol, leaving it therein until all the shellac has dissolved. All that remains to be done now is to wash out the watch lid.

Jewelers' Glue Cement.—Dissolve on a water bath 50 parts of fish glue in a little 95-per-cent alcohol, adding 4 parts, by weight, of gum ammoniac. On the other hand, dissolve 2 parts, by weight, of mastic in 10 parts, by weight, of alcohol. Mix these two solutions and preserve in a well-corked flask. For use it suffices to soften it on the water bath.

Casein Cements.—

I.—Borax.............	5 parts
Water.............	95 parts
Casein, sufficient quantity.	

Dissolve the borax in water and incorporate enough casein to produce a mass of the proper consistency.

II.—The casein is made feebly alkaline by means of soda or potash lye and then subjected for about 24 hours to a temperature of 140° F. Next follow the customary admixture, such as lime and water glass, and finally, to accomplish a quicker resinification, substances containing tannin are added. For tannic admixtures to the partially disintegrated casein, slight quantities—about 1 per cent—of gallic acid, cutch, or quercitannic acid are employed. The feebly alkaline casein cement containing tannic acid is used in the well-known manner for the gluing together of wood.

For Metals.—Make a paste with 16 ounces casein, 20 ounces slaked lime, and 20 ounces of sand, in water.

For Glass.—I.—Dissolve casein in a concentrated solution of borax.

II.—Make a paste of casein and water glass.

Pasteboard and Paper Cement.—I.—Let pure glue swell in cold water; pour and press off the excess; put on the water bath and melt. Paper or other material cemented with this is then immediately, before the cement dries, submitted to the action of formaldehyde and dried. The cement resists the action of water, even hot.

II.—Melt together equal parts of good pitch and gutta percha. To 9 parts of this mass add 3 parts of boiled linseed oil and ½ part litharge. The heat is kept up until, with constant stirring, an intimate union of all the ingredients has taken place. The mixture is diluted with a little benzine or oil of turpentine and applied while still warm. The cement is waterproof.

III.—The *National Druggist* says that experience with pasting or cementing parchment paper seems to show that about the best agent is casein cement, made by dissolving casein in a saturated aqueous solution of borax.

IV.—The following is recommended for paper boxes:

Chloral hydrate	5 parts
Gelatin, white	8 parts
Gum arabic	2 parts
Boiling water	30 parts

Mix the chloral, gelatin, and gum arabic in a porcelain container, pour the boiling water over the mixture and let stand for 1 day, giving it a vigorous stirring several times during the day. In cold weather this is apt to get hard and stiff, but this may be obviated by standing the container in warm water for a few minutes. This paste adheres to any surface whatever.

Waterproof Cements for Glass, Stoneware, and Metal.—I.—Make a paste of sulphur, sal ammoniac, iron filings, and boiled oil.

II.—Mix together dry: Whiting, 6 pounds; plaster of Paris, 3 pounds; sand, 3 pounds; litharge, 3 pounds; rosin, 1 pound. Make to a paste with copal varnish.

III.—Make a paste of boiled oil, 6 pounds; copal, 6 pounds; litharge, 2 pounds; white lead, 1 pound.

IV.—Make a paste with boiled oil, 3 pounds; brickdust 2 pounds; dry slaked lime, 1 pound.

V.—Dissolve 93 ounces of alum and 93 ounces of sugar of lead in water to concentration. Dissolve separately 152 ounces of gum arabic in 25 gallons of water, and then stir in 62½ pounds of flour. Then heat to a uniform paste with the metallic salts, but take care not to boil the mass.

VI.—For Iron and Marble to Stand in Heat.—In 3 pounds of water dissolve first, 1 pound water glass and then 1 pound of borax. With the solution make 2 pounds of clay and 1 pound of barytes, first mixed dry, to a paste.

VII.—Glue to Resist Boiling Water.—Dissolve separately in water 55 pounds of glue and a mixture of 40 pounds of bichromate and 5 pounds of alum. Mix as wanted.

VIII. (Chinese Glue).—Dissolve shellac in 10 times its weight of ammonia.

IX.—Make a paste of 40 ounces of dry slaked lime 10 ounces of alum, and 50 ounces of white of egg.

X.—Alcohol	1,000 parts
Sandarac	60 parts
Mastic	60 parts
Turpentine oil	60 parts

Dissolve the gums in the alcohol and add the oil and stir in. Now prepare a solution of equal parts of glue and isinglass, by soaking 125 parts of each in cold water until it becomes saturated, pouring and pressing off the residue, and melting on the water bath. This should produce a volume of glue nearly equal to that of the solution of gums. The latter should, in the meantime, have been cautiously raised to the boiling point on the water bath, and then mixed with the hot glue solution.

It is said that articles united with this substance will stand the strain of cold water for an unlimited time, and it takes hot water even a long time to affect it.

XI.—Burgundy pitch ..... . 6 parts
　　Gutta percha .... 1 part
　　Pumice stone, in fine
　　　　powder .... 3 parts

Melt the gutta percha very carefully, add the pumice stone, and lastly the pitch, and stir until homogeneous.

Use while still hot. This cement will withstand water and dilute mineral acids.

LEATHER AND RUBBER CEMENTS.

I.—Use a melted mixture of gutta percha and genuine asphalt, applied hot. The hard-rubber goods must be kept pressed together until the cement has cooled.

II.—A cement which is effective for cementing rubber to iron and which is especially valuable for fastening rubber bands to bandsaw wheels is made as follows: Powdered shellac, 1 part; strong water of ammonia, 10 parts. Put the shellac in the ammonia water and set it away in a tightly closed jar for 3 or 4 weeks. By that time the mixture will become a perfectly liquid transparent mass and is then ready for use. When applied to rubber the ammonia softens it, but it quickly evaporates, leaving the rubber in the same condition as before. The shellac clings to the iron and thus forms a firm bond between the iron and the rubber.

III.—Gutta percha. white. 1 drachm
　　Carbon disulphide.. 1 ounce

　Dissolve, filter, and add:

　　India rubber .... 15 grains

　Dissolve.

Cement for Metal on Hard Rubber.—I.—Soak good Cologne glue and boil down so as to give it the consistency of joiners' glue, and add with constant stirring, enough sifted wood ashes until a homogeneous, moderately thick mass results. Use warm and fit the pieces well together while drying.

How to Unite Rubber and Leather.—II.—Roughen both surfaces, the leather and the rubber, with a sharp glass edge; apply to both a diluted solution of gutta percha in carbon bisulphide and let this solution soak into the material. Then press upon each surface a skin of gutta percha ¹⁄₁₆ of an inch in thickness between rolls. The two surfaces are now united in a press, which should be warm but not hot. This method should answer in all cases in which it is applicable. The other prescription covers cases in which a press cannot be used. Cut 30 parts of rubber into small pieces, and dissolve

it in 140 parts of carbon bisulphide, the vessel being placed on a water bath of 30° C. (86° F.). Further, melt 10 parts of rubber with 15 of colophony, and add 35 parts of oil of turpentine. When the rubber has been completely dissolved, the two liquids may be mixed. The resulting cement must be kept well corked.

To Fasten Rubber to Wood.—I.—Make a cement by macerating virgin gum rubber, or as pure rubber as can be had, cut in small pieces, in just enough naphtha or gasoline to cover it. Let it stand in a very tightly corked or sealed jar for 14 days, or a sufficient time to become dissolved, shaking the mixture daily.

II.—Dissolve pulverized gum shellac, 1 ounce, in 9¼ ounces of strong ammonia. This of course must be kept tightly corked. It will not be as elastic as the first preparation.

III.—Fuse together shellac and gutta percha in equal weights.

IV.—India rubber .... 8 ounces
　　Gutta percha ...... 4 ounces
　　Isinglass .... 2 ounces
　　Bisulphide of carbon 32 ounces

V.—India rubber ...... 5 ounces
　　Gum mastic .... 1 ounce
　　Chloroform .... 3 ounces

VI.—Gutta percha ...... 16 ounces
　　India rubber ...... 4 ounces
　　Pitch .... 4 ounces
　　Shellac .... 1 ounce
　　Linseed oil .... 1 ounce

　Amalgamate by heat.

VII.—Mix 1 ounce of oil of turpentine with 10 ounces of bisulphide of carbon in which as much gutta percha as possible has been dissolved.

VIII.—Amalgamate by heat:

　　Gutta percha ...... 100 ounces
　　Venice turpentine. 80 ounces
　　Shellac .... 8 ounces
　　India rubber ..... 2 ounces
　　Liquid storax ..... 10 ounces

IX.—Amalgamate by heat:

　　India rubber ..... 100 ounces
　　Rosin .... 15 ounces
　　Shellac .... 10 ounces

　Then dissolve in bisulphide of carbon.

X.—Make the following solutions separately and mix:

　(a) India rubber ...... 5 ounces
　　　Chloroform ...... 140 ounces
　(b) India rubber ..... 5 ounces
　　　Rosin .... 2 ounces
　　　Venice turpentine. 1 ounce
　　　Oil of turpentine.. 20 ounces

Cement for Patching Rubber Boots and Shoes.—

I.—India rubber, finely
chopped......... 100 parts
Rosin .... 15 parts
Shellac............ 10 parts
Carbon disulphide,
q. s. to dissolve.

This will not only unite leather to leather, india rubber, etc., but will unite rubber to almost any substance.

II.—Caoutchouc, finely cut 4 parts
India rubber, finely
cut............... 1 part
Carbon disulphide... 32 parts

Dissolve the caoutchouc in the carbon disulphide, add the rubber, let macerate a few days, then mash with a palette knife to a smooth paste. The vessel in which the solution is made in both instances above must be kept tightly closed, and should have frequent agitations.

III.—Take 100 parts of crude rubber or caoutchouc, cut it up in small bits, and dissolve it in sufficient carbon bisulphide, add to it 15 parts of rosin and 10 parts of gum lac. The user must not overlook the great inflammability and exceedingly volatile nature of the carbon bisulphide.

Tire Cements.—

I.—India rubber..... 15 grams
Chloroform...... 2 ounces
Mastic.......... ½ ounce

Mix the india rubber and chloroform together, and when dissolved, the mastic is added in powder. It is then allowed to stand a week or two before using.

II.—The following is recommended as very good for cementing pneumatic tires to bicycle wheels:

Shellac.............. 1 ounce
Gutta percha .... 1 ounce
Sulphur............. 45 grains
Red lead............ 45 grains

Melt together the shellac and gutta percha, then add, with constant stirring, the sulphur and red lead. Use while hot.

III.—Raw gutta percha.. 16 ounces
Carbon bisulphide. 72 ounces
Eau de Cologne.... 2¾ ounces

This cement is the subject of an English patent and is recommended for patching cycle and motor tires, insulating electric wires, etc.

IV.—A good thick shellac varnish with which a small amount of castor oil has been mixed will be found a very excellent bicycle rim cement. The formula recommended by Edel is as follows:

Shellac............. 1 pound
Alcohol............. 1 pint

Mix and dissolve, then add:

Castor oil.......... ½ ounce

The castor oil prevents the cement from becoming hard and brittle.

A cement used to fasten bicycle tires may be made by melting together at a gentle heat equal parts of gutta percha and asphalt. Apply hot. Sometimes a small quantity each of sulphur and red lead is added (about 1 part of each to 20 parts of cement).

Cements for Leather.—

I.—Gutta percha ..,... 20 parts
Syrian asphalt, powdered............ 20 parts
Carbon disulphide.. 50 parts
Oil of turpentine... 10 parts

The gutta percha, shredded fine, is dissolved in the carbon disulphide and turpentine oil. To the solution add the asphalt and set away for several days, or until the asphalt is dissolved. The cement should have the consistency of honey. If the preparation is thinner than this let it stand, open, for a few days. Articles to be patched should first be washed with benzine.

II.—Glue .... 1 ounce
Starch paste......... 2 ounces
Turpentine .... 1 drachm
Water, a sufficient quantity.

Dissolve the glue in sufficient water with heat; mix the starch paste with water; add the turpentine, and finally mix with the glue while hot.

III.—Soak for one day 1 pound of common glue in enough water to cover, and 1 pound of isinglass in ale droppings. Then mix together and heat gently until boiling. At this point add a little pure tannin and keep boiling for an hour. If the glue and isinglass when mixed are too thick, add water. This cement should be used warm and the jointed leather pressed tightly together for 12 hours.

IV.—A waterproof cement for leather caoutchouc, or balata, is prepared by dissolving gutta percha, caoutchouc, benzoin, gum lac, mastic, etc., in some convenient solvent like carbon disulphide, chloroform, ether, or alcohol. The best solvent, however, in the case of gutta percha, is carbon disulphide and ether for mastic. The most favorable proportions are as follows: Gutta percha, 200 to 300 parts to 100 parts of the solvent, and 75 to 85 parts of mastic to 100 parts of ether. From 5 to 8 parts of the former solution are mixed with 1 part of the latter, and the mixture is then boiled on the water bath, or in a vessel fitted with a water jacket.

V.—Make a solution of 200 to 300 parts of caoutchouc, gutta percha india rubber, benzoin, or similar gum, in 1,000 parts of carbon disulphide, chloroform, ether, or alcohol, and of this add 5 to 8 parts to a solution of mastic (75 to 125 parts) in ether 100 parts, of equal volume and boil together. Use hot water as the boiling agent, or boil very cautiously on the water bath.

VI.—Forty parts of aluminum acetate, 10° B., 10 parts of glue, 10 parts of rye flour. These materials are either to be simultaneously mixed and boiled, or else the glue is to be dissolved in the aluminum acetate, and the flour stirred into the solution. This is an excellent cement for leather, and is used in so-called art work with leather, and with leather articles which are made of several pieces. It is to be applied warm.

Rubber Cement for Cloth.—The following formulas have been recommended:

I.—Caoutchouc, 5 parts; chloroform, 3 parts. Dissolve and add gum mastic (powder) 1 part.

II.—Gutta percha, 16 parts; india rubber, 4 parts; pitch, 2 parts; shellac, 1 part; linseed oil, 2 parts. Reduce the solids to small pieces, melt together with the oil and mix well.

III.—The following cement for mending rubber shoes and tires will answer similar purposes:

Caoutchouc in shavings..	10	Parts by weight.
Rosin................	4	
Gum turpentine........	40	
Oil turpentine, enough.		

Melt together first the caoutchouc and rosin, then add the gum turpentine, and when all is liquefied, add enough of oil of turpentine to preserve it liquid. A second solution is prepared by dissolving together:

Caoutchouc..........	10	Parts by weight.
Chloroform..........	280	

For use these two solutions are mixed. Wash the hole in the rubber shoe over with the cement, then a piece of linen dipped in it is placed over it; as soon as the linen adheres to the sole, the cement is then applied as thickly as required.

CEMENTS FOR METALS AND FOR ATTACHING VARIOUS SUBSTANCES TO METALS:

Cements for Iron.—I.—To make a good cement for iron on iron, make a thick paste, with water, of powdered iron, 60 parts; sal ammoniac, 2 parts, and sulphur flowers, 1 part. Use while fresh.

II.—Sulphur flowers, 6 parts; dry white lead 6 parts, and powdered borax, 1 part. Mix by sifting and keep as a dry powder in a closed tin box. To use, make into a thin paste with strong sulphuric acid and press together immediately. This cement will harden in 5 days.

III.—Graphite.........	50 pounds
Whiting.........	15 pounds
Litharge.........	15 pounds

Make to a paste with a boiled oil.

IV.—Make a paste of white lead and asbestos.

V.—Make a paste of litharge and glycerine. Red lead may be added. This also does for stone.

VI.—Make a paste of boiled oil of equal parts of white lead, pipe clay, and black oxide of manganese.

VII.—Make iron filings to a paste with water glass.

VIII.—Sal ammoniac....	4 ounces
Sulphur.........	2 ounces
Iron filings.......	32 ounces

Make as much as is to be used at once to a paste with a little water. This remark applies to both the following dry recipes:

IX.—Iron filings......	160 ounces
Lime...........	80 ounces
Red lead........	16 ounces
Alum...........	8 ounces
Sal ammoniac...	2 ounces

X.—Clay...........	10 ounces
Iron filings......	4 ounces
Salt...........	1 ounce
Borax.........	1 ounce
Black oxide of manganese ...	2 ounces

XI.—Mix:	
Iron filings......	180 ounces
Lime...........	45 ounces
Salt...........	8 ounces

XII.—Mix:	
Iron filings......	140 ounces
Hydraulic lime..	20 ounces
Sand...........	25 ounces
Sal ammoniac...	3 ounces

Either of these last two mixtures is made into a paste with strong vinegar just before use.

XIII.—Mix equal weights of zinc oxide and black oxide of manganese into a paste with water glass.

XIV.—Copal varnish, 15 parts; hydrated lime, 10 parts; glue *de nerfs* (of sinews), 5 parts; fat drying oil, 5 parts;

powdered turpentine, 3 parts; essence of turpentine, 2 parts. Dissolve the glue *de nerfs* on the water bath, add all the other substances, and triturate intimately.

XV.—Copal varnish, 15 parts; powdered turpentine, 3 parts; essence of turpentine, 2 parts; powdered fish glue, 3 parts; iron filings, 3 parts; ocher, 10 parts.

XVI.—To make a cement for cast iron, take 16 ounces cast-iron borings; 2 ounces sal ammoniac, and 1 ounce sulphur. Mix well and keep dry. When ready to use take 1 part of this powder to 20 parts of cast-iron borings and mix thoroughly into a stiff paste, adding a little water.

XVII.—Litharge............ 2 parts
Boiled linseed oil..... 2 parts
White lead.......... 1 part
Copal............. 1 part

Heat together until of a uniform consistence and apply warm.

XVIII.—A cement for iron which is said to be perfectly waterproof and fireproof is made by working up a mixture of equal weights of red lead and litharge with glycerine till the mass is perfectly homogeneous and has the consistency of a glazier's putty. This cement is said to answer well, even for very large iron vessels, and to be unsurpassable for stopping up cracks in large iron pans of steam pipes.

Cement for Metal, Glass, and Porcelain.—A soft alloy is prepared by mixing from 30 to 36 parts of copper precipitated in the form of a fine brown powder, with sulphuric acid of a specific gravity of 1.85 in a cast-iron or porcelain mortar and incorporating by stirring with 75 parts of mercury, the acid being afterwards removed by washing with water. In from 10 to 14 hours the amalgam becomes harder than tin, but when heated to 692° F., it can be kneaded like wax. In this condition it is applied to the surface to be cemented, and will fix them firmly together on cooling.

Dissolve 1 drachm of gum mastic in 3 drachms of spirits of wine. In a separate vessel containing water soak 3 drachms of isinglass. When thoroughly soaked take it out of the water and put it into 5 drachms of spirits of wine. Take a piece of gum ammoniacum the size of a large pea and grind it up finely with a little spirits of wine and isinglass until it has dissolved. Then mix the whole together with sufficient heat. It will be found most convenient to place the vessel on a hot-water bath. Keep this cement in a bottle closely stoppered, and when it is to be used, place it in hot water until dissolved.

Cements for Fastening Porcelain to Metal.—I.—Mix equal parts of alcohol (95 per cent) and water, and make a paste by incorporating the liquid with 300 parts of finely pulverized chalk and 250 parts of starch.

II.—Mix finely powdered burned lime, 300 parts, with powdered starch, 250 parts, and moisten the mixture with a compound of equal parts of water and alcohol of 95 per cent until a paste results.

III.—Cement or plaster can be used if the surfaces are sufficiently large; cement is the better article when the object may be exposed to moisture or subjected to much pressure. A process which can be recommended consists in mingling equal weights of chalk, brickdust, clay, and Romain cement. These materials, pulverized and sifted are incorporated with linseed oil in the proportion of half a kilo of oil to 3 kilos of the mingled powder. The Romain or Romanic cement is so designated from the district in France where the calcareous stone from which it is prepared is found in considerable quantity. Although its adhesive qualities are unquestioned, there are undoubtedly American cements equally as good.

IV.—Acetate of lead, 46½ parts by weight; alum, 46½ parts by weight; gum arabic, 76 parts by weight; flour, 500 parts by weight; water, 2,000 parts by weight. Dissolve the acetate of lead and the alum in a little water; on the other hand dissolve the gum arabic in water by pouring, for instance, the 2 liters of boiling water on the gum arabic reduced to powder. When the gum has dissolved, add the flower, put all on the fire, and stir well with a piece of wood; then add the solution of acetate of lead and the alum; agitate well so as to prevent any lumps from forming; retire from the fire before allowing to boil. This glue is used cold, does not peel off, and is excellent to make wood, glass, cardboard, etc. adhere to metals.

Cement for Leather and Iron.—To face a cast-iron pulley with leather apply acetic acid to the face of the pulley with a brush, which will roughen it by rusting, and then when dry apply a cement made of 1 pound of fish glue and ½ pound of common glue, melted in a mixture of alcohol and water. The leather should then be placed on the pulley and dried under pressure.

Amber Cements.—I.—To solder together two pieces of yellow amber, slightly heat the parts to be united and moisten them with a solution of caustic soda; then bring the two pieces together quickly.

II.—Dissolve in a closed bottle 75 parts of cut-up caoutchouc in 60 parts of chloroform. Add 15 parts of mastic and let the mixture stand in the cold until all has dissolved.

III.—Moisten the pieces to be joined with caustic potash and press them together when warm. The union is so perfect that no trace of the juncture is visible. A concentrated alcoholic solution of the rosin over the amber, soluble in alcohol, is also employed for this purpose. Another medium is a solution of hard and very finely powdered copal in pure sulphuric ether. Coat both fractures, previously well cleaned, with this solution and endeavor to combine them intimately by tying or pressing.

IV.—In 30 parts by weight of copal dissolve 30 parts by weight of alumina by means of a water bath. Bathe the surface to be cemented with this gelatinous liquid, but very slightly. Unite the fractures and press them together firmly until the mixture is dry.

Acid-Proof Cements for Stoneware and Glass.—I.—Mix with the aid of heat equal weights of pitch, rosin, and plaster of Paris.

II.—Mix silicate of soda to a paste with ground glass.

III.—Mix boiled oil to a paste with china clay.

IV.—Mix coal tar to a paste with pipe clay.

V.—Mix boiled oil to a paste with quicklime.

VI.—Mix with the aid of heat: Sulphur, 100 pounds; tallow, 2 pounds; rosin, 2 pounds. Thicken with ground glass.

VII.—Mix with the aid of heat: Rosin 2 pounds; sulphur, 2 pounds; brickdust, 4 pounds.

VIII.—Mix with the aid of heat 2 pounds of india rubber and 4 pounds of boiled oil. Thicken with 12 pounds of pipe clay.

IX.—Fuse 100 pounds of india rubber with 7 pounds of tallow. Then make to a paste with dry slaked lime and finally add 20 pounds of red lead.

X.—Mix with the aid of heat: Rosin, 24 pounds; red ocher, 8 pounds; boiled oil, 2 pounds; plaster of Paris, 4 pounds.

Acid-Proof Cement for Wood, Metals, etc.—

I.—Powdered asbestos... 2 parts
Ground baryta...... 1 part
Sodium water-glass solution .... 2 parts

Mix.

II.—To withstand hot nitric acid the following is used:

Sodium water-glass solution .... 2 parts
Sand............... 1 part
Asbestos............ 1 part

Mix.

III.—Asbestos .... 2 parts
Sulphate of barium... 3 parts
Silicate of sodium.... 2 parts

By mixing these ingredients a cement strong enough to resist the strongest nitric acid will be obtained.

IV.—If hot acids are dealt with, the following mixture will be found to possess still more resistant powers:

Silicate of sodium (50°
Baumé) .... 2 parts
Fine sand........... 1 part
Asbestos............ 1 part

Both these cements take a few hours to set. If the cement is wanted to set at once, use silicate of potassium, instead of silicate of sodium. This mixture will be instantly effective and possesses the same power of resistance as the other.

Directions for Repairing Broken Glass, Porcelain, Bric-à-Brac.—Broken glass, china, bric-à-brac, and picture frames, not to name casts, require each a different cement—in fact, several different cements. Glass may be beautifully mended to look at, but seldom so as to be safely used. For clear glass the best cement is isinglass dissolved in gin. Put 2 ounces of isinglass in a clean, wide-mouthed bottle, add half a pint of gin, and set in the sun until dissolved. Shake well every day, and before using strain through double lawn, squeezing very gently.

Spread a white cloth over the mending table and supply it with plenty of clean linen rags, strong rubber bands, and narrow white tape, also a basin of tepid water and a clean soft towel. Wash the broken glass very clean, especially along the break, but take care not to chip it further. Wet both broken edges well with the glue, using a camel's-hair pencil. Fit the break to a nicety, then slip on rubber bands length- and crosswise, every way they will hold. If they will not hold true as upon a stemmed thing, a vase or jug or scent bottle, string half a dozen bands of the same size and strength upon a bit of tape, and tie the tape about neck or base before beginning the gluing. After the parts are joined slip another tape through the same bands and tie it above the fracture; thus with all their strength the bands pull the break together. The bands can be used thus on casts of china—in fact, to hold anything mendable. In glass mending the greater the pressure the better—if only it stops short of the breaking point. Properly made the isinglass cement is as clear as water. When the pieces fit true one on the other the break should be hardly visible, if the pressure has been great enough to force out the tiny bubbles, which otherwise refract the light and make the line of cleavage distressingly apparent. Mended glass may be used to hold dry things—as rose leaves, sachets, violet powder, even candies and fruits. But it will not bear to have any sort of liquid left standing in it, nor to be washed beyond a quick rinsing in tepid water. In wiping always use a very soft towel, and pat the vessel dry with due regard to its infirmities.

Mend a lamp loose in the collar with sifted plaster of Paris mixed to a very soft paste with beaten white of egg. Have everything ready before wetting up the plaster, and work quickly so it may set in place. With several lamps to mend wet enough plaster for one at a time. It takes less than 5 minutes to set, and is utterly worthless if one tries working it over. Metal work apart from the glass needs the soldering iron. Dust the break well with powdered rosin, tie the parts firmly together, lay the stick of solder above the break, and fetch the iron down on it lightly but firmly. When the solder cools, remove the melted rosin with a cloth dipped in alcohol.

Since breakables have so unhappy a knack of fracturing themselves in such fashion they cannot possibly stand upright, one needs a sand box. It is only a box of handy size with 8 inches of clean, coarse sand in the bottom. Along with it there should be some small leaden weights, with rings cast in them, running from an ounce to a quarter pound. Two of each weight are needed. In use, tapes are tied to the rings, and the pair of weights swung outside the edges of the box, so as to press in place the upper part of a broken thing to which the tapes have been fastened.

Set broken platters on edge in the sand box with the break up. The sand will hold them firm, and the broken bit can be slapped on. It is the same with plates and saucers. None of these commonly requires weighting. But very fine pieces where an invisible seam is wanted should be held firm until partly set, then have the pair of heaviest weights accurately balanced across the broken piece. The weights are also very useful to prop and stay topheavy articles and balance them so they shall not get out of kilter. A cup broken, as is so common with cups, can have the tape passed around it, crossing inside the handle, then be set firmly in the sand, face down, and be held by the hanging weights pulling one against the other.

The most dependable cement for china is pure white lead, ground in linseed oil, so thick it will barely spread smoothly with a knife. Given time enough to harden (some 3 months), it makes a seam practically indestructible. The objection to it is that it always shows in a staring white line. A better cement for fine china is white of egg and plaster. Sift the plaster three times and tie a generous pinch of it loosely in mosquito netting. Then beat the egg until it will stick to the plaster. Have the broken egg very clean, cover both with the beaten egg, dust well with the plaster, fit together at once, tie, using rubber bands if possible, wrap loosely in very soft tissue paper, and bury head and ears in the sand box, taking care that the break lies so that the sand will hold it together. Leave in the box 24 hours. After a week the superfluous plaster may be gently scraped away.

General Formulas for Cements for Repairing Porcelain, Glassware, Crockery, Plaster, and Meerschaum.—I.— An excellent cement for joining broken crockery and similar small articles can be made by melting 4 or 5 parts of rosin (or, better still, gum mastic) with 1 part of beeswax in an iron spoon or similar vessel. Apply while hot. It will not stand great heat.

II.—An excellent cement for porcelain and stoneware is obtained by mixing 20 parts of fish glue with an equal weight of crystallizable acetic acid and evaporate the mixture carefully to a syrupy consistency so that it forms a gelatinous mass on cooling. For use the cement thus obtained is made liquid again by heating and applied to the fracture with a brush. The pieces should now be pressed firmly together, by winding a twine tightly around them, until the cement has hardened.

III.—For luting vessels made of glass,

porcelain, etc., which are to be used to hold strong acids, a mixture of asbestos powder, water glass, and an indifferent powder (permanent white, sand, etc.) is recommended. To begin with, asbestos powder is made into a pulp with three or four times the quantity (weight) of a solution of soda water glass (of 30° B.). The same is exceedingly fat and plastic, but is not very well suited for working, as it shrinks too much and cracks when drying. By an addition of fine writing sand of the same weight as the asbestos used, the mass can be made less fat, so as to obviate shrinking, without detracting from the plasticity. Small vessels were molded from it and dried in the air, to be tested afterwards. Put in water, the hardened mass becomes soft again and falls apart. Brought into contact, however, with very strong mineral acids, it becomes even firmer and withstands the liquid perfectly. Concentrated nitric acid was kept in such small vessels without the mass being visibly attacked or anything penetrating it. The action of the acid manifestly has the effect that silicic acid is set free from the water glass in excess, which clogs up the pores entirely and contributes to the lutation. Later on, the mass cannot be dissolved by pure water any more. The mass is also highly fireproof. One of the molded bodies can be kept glowing in a Bunsen gas flame for about half a day after treatment with acid, without slagging in the least. For many purposes it ought to be welcome to have such a mass at hand. It cannot be kept ready for use, however, as it hardens a few hours after being prepared; if potash water glass is used, instead of the soda composition, this induration takes place still more quickly.

IV.—Cement for Glass, Porcelain, etc.

Isinglass (fish glue) .. 50 parts
Gum ammoniac. . . . . 4 parts
Gum mastic. . . . . 2 parts
Alcohol, 95 per cent . . 10 parts
Water, q. s.

Soak the isinglass in cold water over night, or until it has become swollen and soft throughout. In the morning throw off any superfluous fluid and throw the isinglass on a clean towel or other coarse cloth, and hang it up in such a way that any free residual water will drain away. Upon doing this thoroughly depends, in a great measure, the strength of the cement. When the gelatin has become thoroughly drained put it into a flask or other container, place it in the water bath and heat carefully until it becomes

fluid, being careful not to let it come to a boil, as this injures its adhesive properties (the same may be said in regard to glues and gelatins of all kinds). Dissolve the gums in the alcohol and add the solution to the gelatin after removing the same from the water bath, and letting it cool down to about 160° F. Stir well together or mix by agitation.

The following precautions must be observed: 1. Both surfaces to be joined must be absolutely clean, free from dust, dirt, grease, etc. 2. Where the cement is one that requires the application of heat before use, the objects to be united should also *be heated to a point at least as high as the melting point of the cement.* Otherwise, the cement on application is chilled and consequently fails to make a lasting joint. 3. The thinner the layer of cement the stronger the joint; avoid, therefore, using too much of the binding material. Cover both surfaces to be united, coapt them exactly, and press together as closely as possible. In this manner the thinnest possible layer is secured. 4. Bind the parts securely together, and let remain without loosening or attempting to use the article for 2 or 3 days or longer. A liquid cement acquires its full strength only after evaporation of the fluids used as solvents, and this can occur only from the infinitesimal line of exposed surface.

V.—Liquid Porcelain Cement.—Fish glue, 20 parts; glass acetic acid, 20 parts; heat together until the mass gelatinizes on cooling.

VI.—Take 1 ounce of Russian isinglass, cut in small pieces, and bruise well; then add 6 ounces of warm water, and leave it in a warm place for from 24 to 48 hours. Evaporate the resulting solution to about 3 ounces. Next dissolve ½ ounce of mastic in 4 ounces of alcohol, and add the mastic solution to the isinglass in small quantities at a time, continuing the heat and stirring well. While still hot strain the liquid through muslin.

VII.—For optical glasses, Canada balsam is employed, the two pieces being firmly pressed together. After a while, especially by humidity, punctures will form, and the glass is separated by a mist of varying reflexes, while in certain climates the heat will melt the balsam. For all other glass articles which require only simple treatment, such as knobs of covers, plates, etc., silicate of potash is excellent.

VIII.—Glass Cement.—Dissolve in 150 parts of acetic acid of 96 per cent, 100 parts of gelatin by the use of heat, and add ammonium bichromate, 5 parts. This glue must be kept away from the light.

IX.—White glue......... 10 parts
Potassium bichromate 2 parts
Water............. 100 parts

The glue is dissolved in a portion of the water by the aid of heat, the bichromate in the remainder, and the liquids mixed, the mixing being done in a feebly lighted place, and the mixture is then kept in the dark. It is applied in feeble light, being reliquefied by gentle heat, and the glass, the fractured pieces being tightly clamped together, is then exposed to a strong light for some time. By this exposure the cement becomes insoluble. This is waterproof cement for glass.

X.—Diamond Glass Cement.—Dissolve 100 parts of fish glue in 150 parts of 90 per cent alcohol and add, with constant stirring, 200 parts of powdered rosin. This cement must be preserved in absolutely tight bottles, as it solidifies very quickly.

XI.—To unite objects of crystal dissolve 8 parts of caoutchouc and 100 parts of gum mastic in 600 parts of chloroform. Set aside, hermetically closed, for 8 days; then apply with a brush, cold.

XII.—To make a transparent cement for glass, digest together for a week in the cold 1 ounce of india rubber, 67 ounces of chloroform, and 40 ounces of mastic.

XIII.—A mixture of traumaticin, a solution of caoutchouc in chloroform, and a concentrated solution of water glass make a capital cement for uniting articles of glass. Not only is the joint very strong, but it is transparent. Neither changes of temperature nor moisture affect the cement.

XIV.—A transparent cement for porcelain is prepared by dissolving 75 parts of india rubber, cut into small pieces, in a bottle containing 60 parts chloroform; to this add 15 parts green mastic. Let the bottle stand in the cold until the ingredients have become thoroughly dissolved.

XV.—Some preparations resist the action of heat and moisture a short time, but generally yield very quickly. The following cement for glass has proven most resistant to liquids and heat:

Silver litharge .... 1,000 parts
White lead .... 50 parts
Boiled linseed oil.. 3 parts
Copal varnish .... 1 part

Mix the lead and litharge thoroughly, and the oil and copal in the same manner, and preserve separately. When needed for use, mix in the proportions indicated (150 parts of the powder to 4 parts of the liquid) and knead well together. Apply to the edges of the glass, bind the broken parts together, and let stand for from 24 to 48 hours.

XVI.—To reunite plaster articles dissolve small pieces of celluloid in ether; in a quarter of an hour decant, and use the pasty deposit which remains for smearing the edges of the articles. It dries rapidly and is insoluble in water.

XVII.—To Mend Wedgwood Mortars. —It is easy enough to mend mortars so that they may be used for making emulsions and other light work which does not tax their strength too much. But a mended mortar will hardly be able to stand the force required for powdering hard substances. A good cement for mending mortars is the following:

a.—Glass flour elutriated. 10 parts
Fluorspar, powdered
and elutriated..... 20 parts
Silicate of soda ...... 60 parts

Both glass and fluorspar must be in the finest possible condition, which is best done by shaking each in fine powder, with water allowing the coarser particles to deposit, and then to pour off the remainder, which holds the finest particles in suspension. The mixture must be made very rapidly by quick stirring, and when thoroughly mixed must be at once applied. This is said to yield an excellent cement.

b.—Freshly burnt plaster
of Paris............ 5 parts
Freshly burnt lime .... 1 part
White of egg, sufficient.

Reduce the first two ingredients to a very fine powder and mix them well; moisten the two surfaces to be united with a small quantity of white of egg to make them adhesive; then mix the powder very rapidly with the white of egg and apply the mixture to the broken surfaces. If they are large, two persons should do this, each applying the cement to one portion. The pieces are then firmly pressed together and left undisturbed for several days. The less cement is used the better will the articles hold together.

c.—If there is no objection to darkcolored cement, the very best that can be used is probably marine glue. This is made thus: Ten parts of caoutchouc or india rubber are dissolved in 120 parts of benzine or petroleum naphtha, with the aid of a gentle heat. When the solution is complete, which sometimes requires from 10 to 14 days, 20 parts of asphalt are melted in an iron vessel and the caoutchouc solution is poured in very slowly in a fine stream and under continued heating, until the mass has become homogeneous and nearly all the solvent has been driven off. It is then poured out and cast into greased tin molds. It forms dark brown or black cakes, which are very hard to break. This cement requires considerable heat to melt it; and to prevent it from being burnt it is best to heat a capsule containing a piece of it first on a water bath until the cake softens and begins to be liquid. It is then carefully wiped dry and heated over a naked flame, under constant stirring, up to about 300° F. The edges of the article to be mended should, if possible, also be heated to at least 212° F., so as to permit the cement to be applied at leisure and with care. The thinner the cement is applied the better it binds.

Meerschaum Cements.—I.—If the material is genuine (natural) meerschaum a lasting joint can be made between the parts by proceeding as follows: Clean a clove or two of garlic (the fresher the better) by removing all the outside hull of skin; throw into a little mortar and mash to a paste. Rub this paste over each surface to be united and join quickly. Bring the parts as closely together as possible and fasten in this position. Have ready some boiling fresh milk; place the article in it and continue the boiling for 30 minutes. Remove and let cool slowly. If properly done, this makes a joint that will stand any ordinary treatment, and is nearly invisible. For composition, use a cement made of quicklime, rubbed to a thick cream with egg albumen.

II.—Mix very fine meerschaum shavings with albumen or dissolve casein in water glass, stir finely powdered magnesia into the mass, and use the cement at once. This hardens quickly.

Asbestos Cement.—Ground asbestos may be made into a cement which will stand a high degree of heat by simply mixing it with a solution of sodium silicate. By subsequent treatment with a solution of calcium chloride the mass may be made insoluble, silicate of calcium being formed.

A cement said to stand a high degree of heat and to be suitable for cementing glass, porcelain, or other vessels intended to hold **corrosive acids**, is this one:

I.—Asbestos............. 2 parts
Barium sulphate..... 3 parts
Sodium silicate....... 2 parts

By mixing these ingredients a cement strong enough to resist the strongest nitric acid will be obtained. If hot acids are dealt with, the following mixture will be found to possess still more resistant powers:

II.—Sodium silicate....... 2 parts
Fine sand........... 1 part
Asbestos powder...... 1 part

Both these cements take a few hours to set. If the cement is wanted to set at once, use potassium silicate instead of sodium silicate. This mixture will be instantly effective, and possesses the same power of resistance as the other.

Parisian Cement.— Mix 1 part of finely ground glass powder, obtained by levigation, with 3 parts of finely powdered zinc oxide rendered perfectly free from carbonic acid by calcination. Besides prepare a solution of 1 part, by weight, of borax in a very small quantity of hot water and mix this with 50 parts of a highly concentrated zinc chloride solution of 1.5 to 1.6 specific gravity. As is well known the mixture of this powder with the liquid into a soft uniform paste is accomplished only immediately before use. The induration to a stonelike mass takes place within a few minutes, the admixture of borax retarding the solidification somewhat. The pure white color of the powder may be tinted with ocher, manganese, etc., according to the shade desired.

Strong Cement.— Pour over well-washed and cleaned casein 12½ parts of boiled linseed oil and the same amount of castor oil. Boil. Stir actively and add a small amount of a saturated aqueous solution of alum; remove from the fire and set aside. After a while a milky looking fluid will separate and rise. This should be poured off. To the residue add 120 parts of rock candy syrup and 6 parts of dextrin.

A Cheap and Excellent Cement.—A cheap and excellent cement, insoluble after drying in water, petroleum, oils, carbon disulphide, etc., very hard when dry and of very considerable tensile strength, is composed of casein and some tannic-acid compound, as, for instance, calcium tannate, and is prepared as follows:

First, a tannin solution is prepared either by dissolving a tannin salt, or by extraction from vegetable sources (as barks from certain trees, etc.), to which is added clear lime water (obtained by filtering milk of lime, or by letting the milk stand until the lime subsides) until no further precipitation occurs, and red litmus paper plunged in the fluid is turned blue. The liquid is now separated from its precipitate, either by decantation or otherwise, and the precipitate is dried. In operating with large quantities of the substance, this is done by passing a stream of atmospheric air through the same. The lime tannate obtained thus is then mixed with casein in proportions running from 1:1 up to 1:10, and the mixture, thoroughly dried, is milled into the consistency of the finest powder. This powder has now only to be mixed with water to be ready for use, the consistency of the preparation depending upon the use to which it is to be put.

Universal Cement.—Take gum arabic, 100 parts, by weight; starch, 75 parts, by weight; white sugar, 21 parts, by weight; camphor, 4 parts, by weight. Dissolve the gum arabic in a little water; also dissolve the starch in a little water. Mix and add the sugar and camphor. Boil on the water bath until a paste is formed which, on coating, will thicken.

Cement for Ivory.—Melt together equal parts of gutta percha and ordinary pitch. The pieces to be united have to be warmed.

Cement for Belts.—Mix 50 parts, by weight, of fish glue with equal parts of whey and acetic acid. Then add 50 parts, by weight, of garlic in paste form and boil the whole on the water bath. At the same time make a solution of 100 parts, by weight, of gelatin in the same quantity of whey, and mix both liquids. To the whole add, finally, 50 parts, by weight, of 90-per-cent alcohol and, after filtration, a cement is obtained which can be readily applied with a brush and possesses extraordinary binding qualities.

Cement for Chemical Apparatus.—Melt together 20 parts of gutta percha, 10 parts of yellow wax, and 30 parts of shellac.

Size Over Portland Cement. — The best size to use on Portland cement molding for wall paper would ordinarily be glue and alum size put on thin and warm, made in proportion of ¼ pound of glue and same weight of alum dissolved in separate pails, then poured together.

Aquarium Cements.—

I.—Litharge.......... 3 ounces
Fine white sand... 3 ounces
Plaster of Paris.... 3 ounces
Rosin, in fine powder............. 1 ounce
Linseed oil, enough.
Drier, enough.

Mix the first three ingredients, add sufficient linseed oil to make a homogeneous paste, and then add a small quantity of drier. This should stand a few hours before it is used. It is said that glass joined to iron with this cement will break before it will come loose.

II.—Litharge.......... 1 ounce
Fine white sand.... 1 ounce
Plaster of Paris.... 1 ounce
Manganese borate. 20 grains
Rosin, in fine powder............. 3½ pounds
Linseed varnish oil, enough.

III.—Take equal parts of flowers of sulphur, ammonium chloride, and iron filings, and mix thoroughly with boiled linseed oil. Finally, add enough white lead to form a thin paste.

IV.—Powdered graphite. 6 parts
Slaked lime....... 3 parts
Barium sulphate... 8 parts
Linseed varnish oil. 7 parts

V.—Simply mix equal parts of white and red lead with a little kettle-boiled linseed oil.

Substitute for Cement on Grinder Disks.—A good substitute in place of glue or various kinds of cement for fastening emery cloth to the disks of grinders of the Gardner type is to heat or warm the disk and apply a thin coating of beeswax; then put the emery cloth in place and allow to set and cool under pressure.

Knockenplombe.—If 1 part of thymol be mixed with 2 parts of iodoform we obtain a substance that retains its fluidity down to 72° C. (161.6° F.). If the temperature be carried down to 60° C. (140° F.) it suddenly becomes solid and hard. If, in its liquid condition, this substance be mixed intimately with an equal quantity of calcined bone, it forms a cement that can be molded or kneaded into any shape, that, at the temperature of the body (98° F.), becomes as hard as stone, a fact that suggests many useful purposes to which the mixture may be put.

Cement for General Use.—Take gum arabic, 100 parts, by weight; starch, 75 parts by weight; white sugar, 21 parts, by weight; camphor, 4 parts, by weight. Dissolve the gum arabic in a little water. On the other hand, dissolve the starch also in some water. When this is done add the sugar and the camphor and put in a water bath. Boil until a paste is formed, which must be rather thin, because it will thicken on cooling.

Strong Cement.—Pour over well-washed and cleaned casein 12½ parts of boiled linseed oil and the same amount of castor oil, put on the fire and bring to a boil; stir actively and add a small amount of a saturated aqueous solution of alum; remove from the fire and set aside. After standing a while a milky-looking fluid will separate at the bottom and rise to the top. This should be poured off and to the residue add 120 parts of rock-candy syrup and 6 parts of dextrine.

Syndeticon.—I.—Slake 100 parts of burnt lime with 50 parts of water, pour off the supernatant water; next, dissolve 60 parts of lump sugar in 160 parts of water, add to the solution 15 parts of the slaked lime, heat to 70° or 80° C. (158° to 176° F.), and set aside shaking frequently. Finally dissolve 50 to 60 parts of genuine Cologne glue in 250 parts of the clear solution.

II.—A solution of 10 parts gum arabic and 30 parts of sugar in 100 parts of soda water glass.

III.—A hot solution of 50 parts of Cologne glue in 60 parts of a 20-per-cent aqueous calcium-chloride solution.

IV.—A solution of 50 parts of Cologne glue in 60 parts of acetic acid.

V.—Soak isinglass (fish bladder) in acetic acid of 70 per cent until it swells up, then rub it up, adding a little water during the process.

"Shio Liao."—Under this name the Chinese manufacture an excellent cement which takes the place of glue, and with which gypsum, marble, porcelain, stone, and stoneware can be cemented. It consists of the following parts (by weight): Slaked powdered lime, 54 parts; powdered alum, 6 parts; and fresh, well-strained blood, 40 parts. These materials are stirred thoroughly until an intimately bound mass of the consistency of a more or less stiff salve is obtained. In paste form this mass is used as cement; in a liquid state it is employed for painting all sorts of articles which are to be rendered waterproof and durable. Cardboard covers, which are coated with it two or three times, become as hard as wood. The Chinese paint their houses with "shio liao" and glaze their barrels with it, in which they transport oil and other greasy substances.

LUTES.

Lutes always consist of a menstruum and dissolved or suspended solids, and they must not be attacked by the gases and liquids coming in contact with them. In some cases the constituents of the lute react to form a more strongly adhering mass.

The conditions of application are, in brief:

(a) Heating the composition to make it plastic until firmly fixed in place.

(b) Heating the surfaces.

(c) Applying the lute with water or a volatile solvent, which is allowed to volatilize.

(d) Moistening the surfaces with water, oil, etc. (the menstruum of the lute itself).

(e) Applying the lute in workable condition and the setting taking place by chemical reactions.

(f) Setting by hydration.

(g) Setting by oxidation.

These principles will be found to cover nearly all cases.

Joints should not be ill-fitting, depending upon the lute to do what the pipes or other parts of the apparatus should do. In most cases one part of the fitting should overlap the other, so as to make a small amount of the lute effective and to keep the parts of the apparatus rigid, as a luted joint is not supposed to be a particularly strong one, but rather one quickly applied, effective while in place and easily removed.

Very moderate amounts of the lute should be used, as large amounts are likely to develop cracks, be rubbed off, etc.

A classification may be given as follows:

(1) Plaster of Paris.

(2) Hydraulic cement.

(3) Clay.

(4) Lime.

(5) Asphalt and pitch.

(6) Rosin.

(7) Rubber.

(8) Linseed oil.

(9) Casein and albumen.

(10) Silicates of soda and oxychloride cements.

(11) Flour and starch.

(12) Miscellaneous, including core compounds.

I. Plaster of Paris is, of course, often used alone as a paste, which quickly solidines, for gas and wood distillation retorts, etc., and similar places where quickness of setting is requisite. It is more often, however, used with some fibrous material to give it greater strength. Asbestos is the most commonly used material of these, as it will stand a high temperature. When that is not so important, straw, plush trimmings, hair, etc., are used as binders, while broken stone, glass, and various mineral substances are used as fillers, but they do not add anything to the strength. These lutes seem to be particularly suitable for oil vapors and hydrocarbon gases.

Formulas:

(1) Plaster and water.
(2) Plaster (wet) and asbestos.
(3) Plaster (wet) and straw.
(4) Plaster (wet) and plush trimmings.
(5) Plaster (wet) and hair.
(6) Plaster (wet) and broken stone, etc.

II. Hydraulic Cement. — Cement is used either alone or with sand, asbestos, etc. These lutes are suitable for nitric acid. When used with substances such as rosin or sulphur, cement is probably employed because it is in such a fine state of division and used as a filler and not because of any powers of setting by hydration.

Formulas:

(1) Cement—neat.
(2) Cement and asbestos.
(3) Cement and sand.

III. Clay.—This most frequently enters into the composition of lutes as a filler, but even then the very finely divided condition of certain grades renders it valuable, as it gives body to a liquid, such as linseed oil, which, unless stiffened, would be pervious to a gas, the clay in all cases being neutral. Thus, for luting pipes carrying chlorine, a stiff paste of clay and molasses has been suggested by Theo. Köller in *Die Surrogate*, but it soon gives way.

Formulas:

(1) Clay and linseed oil.
(2) Same, using fire clay.
(3) Clay and molasses.

(1) Is suitable for steam, etc.; (2) for chlorine, and (3) for oil vapors.

IV. Lime is used in the old lute known as putty, which consists of caustic lime and linseed oil. Frequently the lime is replaced by chalk and china clay, but the lime should be, in part at least, caustic, so as to form a certain amount of lime soap. Lime is also used in silicate

and casein compositions, which are very strong and useful, but will be described elsewhere.

Formulas:

(1) Lime and boiled oil to stiff mass.
(2) Clay, etc., boiled oil to stiff mass.

V. Asphalt and Pitch. — These substances are used in lutes somewhat interchangeably. As a rule, pitch makes the stronger lutes. Tar is sometimes used, but, because of the light oils and, frequently, water contained, it is not so good as either of the others.

Asphalt dissolved in benzol is very useful for uniting glass for photographic, microscopical, and other uses. Also for coating wood, concrete, etc., where the melted asphalt would be too thick to cover well. Benzol is the cheapest solvent that is satisfactory for this purpose, as the only one that is cheaper would be a petroleum naphtha, which does not dissolve all the constituents of the asphalt. For waterproofing wood, brick, concrete, etc., melted asphalt alone is much used, but when a little paraffine is added, it improves its waterproofing qualities, and in particular cases boiled oil is also added to advantage.

Formulas:

1. Refined lake asphalt.

2. Asphalt .... 4 parts
   Paraffine .... 1 part

3. Asphalt .... 10 parts
   Paraffine .... 2 parts
   Boiled oil .... 1 part

Any of these may be thinned with hot benzol or toluol. Toluol is less volatile than benzol and about as cheap, if not cheaper, the straw-colored grades being about 24 cents per gallon.

Examples of so-called "stone cement" are:

4. Pitch .... 8 parts
   Rosin .... 6 parts
   Wax .... 1 part
   Plaster............¼ to ½ part
5. Pitch .... 8 parts
   Rosin .... 7 parts
   Sulphur .... 2 parts
   Stone powder .... 1 part

These compositions are used to unite slate slabs and stoneware for domestic, engineering, and chemical purposes. Various rosin and pitch mixtures are used for these purposes, and the proportions of these two ingredients are determined by the consistency desired. Sulphur and stone powder are added to prevent the formation of cracks, sulphur acting chemically and stone powder mechanically

Where the lute would come in contact with acid or vapors of the same, limestone should not be the powder used, otherwise it is about the best. Wax is a useful ingredient to keep the composition from getting brittle with age.

A class of lutes under this general grouping that are much used are so-called "marine glues" (q. v.). They must be tough and elastic. When used for calking on a vessel they must expand and contract with the temperature and not crack or come loose.

Formulas:

6. Pitch .... 3 parts
   Shellac .... 2 parts
   Pure crude rubber ... 1 part

7. Pitch .... 1 part
   Shellac .... 1 part
   Rubber substitute .... 1 part

These are used by melting over a burner.

VI. Rosin, Shellac, and Wax. — A strong cement, used as a stone cement, is:

1. Rosin .... 8 parts
   Wax .... 1 part
   Turpentine .... 1 part

It has little or no body, and is used in thin layers.

For nitric and hydrochloric acid vapors:

2. Rosin .... 1 part
   Sulphur .... 1 part
   Fire clay .... 2 parts

Sulphur gives great hardness and permanency to rosin lutes, but this composition is somewhat brittle.

Good waterproof lutes of this class are:

3. Rosin .... 1 part
   Wax .... 1 part
   Powdered stone ...... 2 parts

4. Shellac .... 5 parts
   Wax .... 1 part
   Turpentine .... 1 part
   Chalk, etc. .... 8 to 10 parts

For a soft air-tight paste for ground-glass surfaces:

5. Wax .... 1 part
   Vaseline .... 1 part

6. A strong cement, without body, for metals (other than copper or alloys of same), porcelain, and glass is made by letting 1 part of finely powdered shellac stand with 10 parts of ammonia water until solution is effected.

VII. Rubber. — Because of its toughness, elasticity, and resistance to alterative influences, rubber is a very useful con-

stituent in lutes, but its price makes its use very limited.

Leather Cement.

1. Asphalt .... 1 part
   Rosin .... 1 part
   Gutta percha .... 4 parts
   Carbon disulphide ... 20 parts

To stand acid vapors:

2. Rubber .... 1 part
   Linseed oil .... 3 parts
   Fire clay .... 3 parts

3. **Plain Rubber Cement.** — Cut the crude rubber in small pieces and then add the solvent. Carbon disulphide is the best, benzol good and much cheaper, but gasoline is probably most extensively used because of its cheapness.

4. To make corks and wood impervious to steam and water, soak in a rubber solution as above; if it is desired to protect them from oil vapors, use glue composition. (See Section IX.)

VIII. Linseed Oil. — This is one of the most generally useful substances we have for luting purposes, if absorbed by a porous substance that is inert.

Formulas: 1. China clay of general utility for aqueous vapors.

Linseed oil of general utility for aqueous vapors.

2. Lime forming the well-known putty.

Linseed oil forming the well-known putty.

3. Red or white lead and linseed oil.

These mixtures become very strong when set and are best diluted with powdered glass, clay, or graphite. There are almost an endless number of lutes using metallic oxides and linseed oil. A very good one, not getting as hard as those containing lead, is:

4. Oxide of iron and linseed oil.

IX. Casein, Albumen, and Glue. — These, if properly made, become very tough and tenacious; they stand moderate heat and oil vapors, but not acid vapors.

1. Finely powdered case-
   in .... 12 parts
   Slaked lime (fresh) ... 50 parts
   Fine sand .... 50 parts
   Water to thick mush.

A very strong cement which stands moderate heat is the following:

2. Casein in very fine
   powder .... 1 part
   Rubbed up with sili-
   cate of soda .... 3 parts

A strong lute for general purposes,

which must be used promptly when made:

3. White of egg made into a paste with slaked lime.

A composition for soaking corks, wood, packing, etc., to render them impervious to oil vapors, is:

Gelatine or good glue 2 parts
Glycerine.........½ to 1 part
Water.............. 6 parts
Oil of wintergreen,
    etc., to keep from
    spoiling.

X. Silicate of Oxychloride Cements.— For oil vapors, standing the highest heat:

1. A stiff paste of silicate of soda and asbestos.

Gaskets for superheated steam, retorts, furnaces, etc.:

2. Silicate of soda and powdered glass; dry the mixture and heat.

Not so strong, however, as the following:

3. Silicate of soda...... 50 parts
Asbestos .... 15 parts
Slaked lime .... 10 parts

Metal Cement:

4. Silicate of soda .... 1 part
Oxides of metal, such
    as zinc oxide; lith-
    arge, iron oxide,
    singly or mixed ..... 1 part

Very hard and extra strong compositions:

5. Zinc oxide .... 2 parts
Zinc chloride........ 1 part
Water to make a paste.

6. Magnesium oxide ... 2 parts
Magnesium chloride. 1 part
Water to make a paste.

XI. Flour and Starch Compositions.—
1. The well-known flaxseed poultice sets very tough, but does not stand water or condensed steam.

2. Flour and molasses, made by making a stiff composition of the two. This is an excellent lute to have at hand at all times for emergency use, etc.

3. Stiff paste of flour and strong zinc-chloride solution forms a more impervious lute, and is more permanent as a cement. This is good for most purposes, at ordinary temperature, where it would not be in contact with nitric-acid vapors or condensing steam.

4. A mixture of dextrine and fine sand makes a good composition, mainly used as core compound.

XII. Miscellaneous.—
1. Litharge.
Glycerine.

Mixed to form a stiff paste, sets and becomes very hard and strong, and is very useful for inserting glass tubes, etc., in iron or brass.

For a high heat:

2. Alumina............ 1 part
Sand................ 4 parts
Slaked lime......... 1 part
Borax.............. ½ part
Water sufficient.

A class of mixtures that can be classified only according to their intended use are core compounds.

I.—Dextrine, about...... 1 part
Sand, about......... 10 parts
With enough water to form a paste.

II.—Powdered anthracite coal, with molasses to form a stiff paste.

III.—Rosin, partly saponi-
    fied by soda lye.... 1 part
Flour.............. 2 parts
Sand (with sufficient
    water).......... 4 parts

(These proportions are approximate and the amount of sand can be increased for some purposes.)

IV.—Glue, powdered...... 1 part
Flour.............. 4 parts
Sand (with sufficient
    water).......... 6 parts

For some purposes the following mixture is used. It does not seem to be a gasket or a core compound:

V.—Oats (or wheat) ground 25 parts
Glue, powdered...... 6 parts
Sal ammoniac....... 1 part

Paper read by Samuel S. Sadtler before the Franklin Institute.

PASTES:

Dextrine Pastes.—

I.—Borax, powdered.... 60 parts
Dextrine, light yellow. 480 parts
Glucose........... 50 parts
Water............ 420 parts

By the aid of heat, dissolve the borax in the water and add the dextrine and glucose. Continue the heat, but do not let the mixture boil, and stir constantly until a homogeneous solution is obtained, from time to time renewing the water lost by evaporation with hot water. Finally, bring up to full weight (1,000 parts) by the addition of hot water, then strain through flannel. Prepared in this manner the paste remains bright and clear for a long time. It has extraordinary adhesive properties and dries very rapidly. If care is not taken to keep the cooking temperature below the boiling point of water, the paste is apt to become brown and to be very brittle on drying.

II.—Dissolve in hot water a sufficient quantity of dextrine to bring it to the consistency of honey. This forms a strong adhesive paste that will keep a long time unchanged, if the water is not allowed to evaporate. Sheets of paper may be prepared for extempore labels by coating one side with the paste and allowing it to dry; by slightly wetting the gummed side, the label will adhere to glass. This paste is very useful in the office or laboratory.

III.—Pour over 1,000 parts of dextrine 450 parts of soft water and stir the mixture for 10 minutes. After the dextrine has absorbed the water, put the mixture over the fire, or, preferably, on a water bath, and heat, with lively stirring for 5 minutes, or until it forms a light milk-like liquid, on the surface of which little bubbles begin to form and the liquid is apparently beginning to boil. Do not allow it to come to a boil. Remove from the fire and set in a bucket of cold water to cool off. When cold add to every 1,000 parts of the solution 51 parts glycerine and as much salicylic acid as will stand on the tip of a knife blade. If the solution is too thick, thin it with water that has been boiled and cooled off again. Do not add any more glycerine or the solution will never set.

IV.—Soften 175 parts of thick dextrine with cold water and 250 parts of boiling water added. Boil for 5 minutes and then add 30 parts of dilute acetic acid, 30 parts glycerine, and a drop or two of clove oil.

V.—Powder coarsely 400 parts dextrine and dissolve in 600 parts of water. Add 20 parts glycerine and 10 parts glucose and heat to 90° C. (195° F.).

VI. — Stir 400 parts of dextrine with water and thin the mass with 200 parts more water, 20 parts glucose, and 10 parts aluminum sulphate. Heat the whole to 90° C. (195° F.) in the water bath until the whole mass becomes clear and liquid.

VII. — Warm 2 parts of dextrine, 5 parts of water, 1 part of acetic acid, 1 part of alcohol together, with occasional stirring until a complete solution is attained.

VIII.—Dissolve by the aid of heat 100 parts of builders' glue in 200 parts of water add 2 parts of bleached shellac dissolved previously in 50 parts of alcohol. Dissolve by the aid of heat 50 parts of dextrine in 50 parts of water, and mix the two solutions by stirring the second slowly into the first. Strain the mixture through a cloth into a shallow dish and let it harden. When needed cut off a piece of sufficient size and warm until it becomes liquid and if necessary or advisable thin with water.

IX.—Stir up 10 parts of dextrine with sufficient water to make a thick broth. Then, over a light fire, heat and add 25 parts of sodium water glass.

X.—Dissolve 5 parts of dextrine in water and add 1 part of alum.

Fastening Cork to Metal.—In fastening cork to iron and brass, even when these are lacquered, a good sealing wax containing shellac will be found to serve the purpose nicely. Wax prepared with rosin is not suitable. The cork surface is painted with the melted sealing wax. The surface of the metal is heated with a spirit flame entirely free from soot, until the sealing wax melts when pressed upon the metallic surface. The wax is held in the flame until it burns, and it is then applied to the hot surface of the metal. The cork surface painted with sealing wax is now held in the flame, and as soon as the wax begins to melt the cork is pressed firmly on the metallic surface bearing the wax.

To Paste Celluloid on Wood, Tin, or Leather.—To attach celluloid to wood, tin, or leather, a mixture of 1 part of shellac, 1 part of spirit of camphor, 3 to 4 parts of alcohol and spirit of camphor (90°) is well adapted, in which 1 part of camphor is dissolved without heating in 7 parts of spirit of wine of 0.832 specific gravity, adding 2 parts of water.

To Paste Paper Signs on Metal or Cloth.—A piece of gutta percha of the same size as the label is laid under the latter and the whole is heated. If the heating cannot be accomplished by means of a spirit lamp the label should be ironed down under a protective cloth or paper in the same manner as woolen goods are pressed. This method is also very useful for attaching paper labels to minerals.

Paste for Fastening Leather, Oilcloth, or Similar Stuff to Table or Desk Tops, etc.—Use the same paste for leather as for oilcloth or other goods, but moisten the leather before applying the paste. Prepare the paste as follows: Mix 2¼ pounds of good wheat flour with 2 tablespoonfuls of pulverized gum arabic or powdered rosin and 2 tablespoonfuls of pulverized alum in a clean dish with water enough to make a uniformly thick batter; set it over a slow fire and stir continuously until the paste is uniform and free from lumps. When the mass has become so stout that the wooden spoon or stick will stand in it upright, it is taken from the fire and placed in another dish and covered so that no skin will form on top. When cold, the table or desk top, etc., is covered with a thin coat of the paste, the cloth, etc., carefully laid on and smoothed from the center toward the edges with a rolling pin. The trimming of edges is accomplished when the paste has dried. To smooth out the leather after pasting, a woolen cloth is of the best service.

To Paste Paper on Smooth Iron.— Over a water bath dissolve 200 parts, by weight, of gelatine in 150 parts, by weight, of water; while stirring add 50 parts, by weight, of acetic acid, 50 parts alcohol, and 50 parts, by weight, of pulverized alum. The spot upon which it is desired to attach the paper must first be rubbed with a bit of fine emery paper.

Paste for Affixing Cloth to Metal.—

Starch...............	20 parts
Sugar..............	10 parts
Zinc chloride......	1 part
Water.............	100 parts

Mix the ingredients and stir until a perfectly smooth liquid results entirely free from lumps, then warm gradually until the liquid thickens.

To Fix Paper upon Polished Metal.— Dissolve 400 parts, by weight, of dextrine in 600 parts, by weight, of water; add to this 10 parts, by weight, of glucose, and heat almost to boiling.

Albumen Paste.—Fresh egg albumen is recommended as a paste for affixing labels on bottles. It is said that labels put on with this substance, and well dried at the time, will not loosen even when bottles are put into water and left there for some time. Albumen, dry, is almost proof against mold or ferments. As to cost, it is but little if any higher than gum arabic, the white of one egg being sufficient to attach at least 100 medium-sized labels.

Paste for Parchment Paper.—The best agent is made by dissolving casein in a saturated aqueous solution of borax.

Medical Paste.—As an adhesive agent for medicinal purposes Professor Reihl, of Leipsic, recommends the viscous substance contained in the white mistletoe. It is largely present in the berries and the bark of the plant; it is called viscin, and can be produced at one-tenth the price of caoutchouc. Solutions in benzine may be used like those of caoutchouc without causing any irritation if applied mixed with medicinal remedies to the skin.

Paste That Will Not Mold.—Mix good white flour with cold water into a thick paste. Be sure to stir out all the lumps; then add boiling water, stirring all the time until thoroughly cooked. To 6 quarts of this add ½ pound light brown sugar and ¼ ounce corrosive sublimate, dissolved in a little hot water. When the paste is cool add 1 drachm oil of lavender. This paste will keep for a long time.

Pasting Wood and Cardboard on Metal.—In a little water dissolve 50 parts of lead acetate and 5 parts of alum. In another receptacle dissolve 75 parts of gum arabic in 2,000 parts of water. Into this gum-arabic solution pour 500 parts of flour, stirring constantly, and heat gradually to the boiling point. Mingle the solution first prepared with the second solution. It should be kept in mind that, owing to the lead acetate, this preparation is poisonous.

Agar Agar Paste.—The agar agar is broken up small, wetted with water, and exposed in an earthenware vessel to the action of ozone pumped under pressure into the vessel from the ozonizing apparatus. About an hour of this bleaches the agar agar and makes it freely soluble in boiling water, when solutions far more concentrated than has hitherto been possible can be prepared. On cooling, the solutions assume a milky appearance, but form no lumps and are readily reliquefied by heating. If the solution is completely evaporated, as of course happens when the adhesive is allowed to dry after use, it leaves a firmly holding mass which is insoluble in cold water. Among the uses to which the preparation can be applied are the dressing of textile fabrics and paper sizing, and the production of photographic papers, as well as the ordinary uses of an adhesive.

Strongly Adhesive Paste.—Four parts glue are soaked a few hours in 15 parts cold water, and moderately heated till the solution becomes perfectly clear, when 65 parts boiling water are added, while stirring. In another vessel 30 parts boiled starch are previously stirred together with 20 parts cold water, so that a thin, milky liquid without lumps results. The boiling glue solution is poured into this while stirring constantly, and the whole is kept boiling another 10 minutes.

Paste for Tissue Paper.—

(a) Pulverized gum arabic...............	2 ounces
White sugar........	4 drachms
Boiling water.......	3 fluidounces

(b) Common   laundry
     starch........... 1½ ounces
     Cold water........ 3   fluidounces
Make into a batter and pour into
     Boiling water......32 fluidounces
Mix (a) with (b), and keep in a wide-
mouthed bottle.

Waterproof and Acidproof Pastes.—

I.—Chromic acid......   2½ parts
     Stronger ammonia... 15   parts
     Sulphuric acid......   ½ part
     Cuprammonium so-
       lution........... 30   parts
     Fine white paper....   4   parts

II.—Isinglass, a sufficient
     quantity
     Acetic acid........   1   part
     Water.............   7   parts

Dissolve sufficient isinglass in the mixture of acetic acid and water to make a
thin mucilage.

One of the solutions is applied to the
surface of one sheet of paper and the other
to the other sheet, and they are then
pressed together.

III.—A fair knotting varnish free
from surplus oil is by far the best adhesive for fixing labels, especially on metal
surfaces. It dries instantly, insuring
a speedy job and immediate packing, if
needful, without fear of derangement.
It has great tenacity, and is not only
absolutely damp-proof itself, but is actually repellent of moisture, to which all
water pastes are subject. It costs more,
but the additional expense is often infinitesimal compared with the pleasure of a
satisfactory result.

Balkan Paste.—

     Pale glue...........   4   ounces
     White loaf sugar....   2   ounces
     Powdered starch....   1   ounce
     White dextrine......   ¼ pound
     Pure glycerine......   3   ounces
     Carbolic acid......   ¼ ounce
     Boiling water....... 32 ounces

Cut up the glue and steep it in ½ pint
boiling water; when softened melt in a
saucepan; add sugar, starch, and dextrine, and lastly the glycerine, in which
carbolic acid has been mixed; add remainder of water, and boil until it thickens. Pour into jars or bottles.

Permanent Paste.—

I.—Wheat flour........   1 pound
     Water, cold........   1 quart
     Nitric acid.........   4 fluidrachms
     Boric acid.......... 40 grains
     Oil of cloves....... 20 minims

Mix the flour, boric acid, and water,
then strain the mixture; add the nitric

acid, apply heat with constant stirring
until the mixture thickens; when nearly
cold add the oil of cloves. This paste
will have a pleasant smell, will not attract
flies, and can be thinned by the addition
of cold water as needed.

II.—Dissolve 4 ounces alum in 4
quarts hot water. When cool add as
much flour as will make it of the usual
consistency; then stir into it ½ ounce
powdered rosin; next add a little water
in which a dozen cloves have been
steeped; then boil it until thick as mush,
stirring from the bottom all the time.
Thin with warm water for use.

Preservatives for Paste.—Various antiseptics are employed for the preservation of flour paste, mucilage, etc. Boric
and salicylic acids, oil of cloves, oil of
sassafras, and solution of formaldehyde
are among those which have given best
service. A durable starch paste is produced by adding some borax to the
water used in making it. A paste from
10 parts (weight) starch to 100 parts
(weight) water with 1 per cent borax
added will keep many weeks, while without this addition it will sour after six
days. In the case of a gluing material
prepared from starch paste and joiners'
glue, borax has also demonstrated its preserving qualities. The solution is made
by mixing 10 parts (weight) starch into
a paste with water and adding 10 parts
(weight) glue soaked in water to the hot
solution; the addition of $\frac{1}{16}$ part (weight)
of borax to the solution will cause it to
keep for weeks. It is equal to the best
glue, but should be warmed and stirred
before use.

Board-Sizing.—A cheap sizing for
rough, weather-beaten boards may be
made by dissolving shellac in sal soda
and adding some heavy-bodied pigment.
This size will stick to grease spots. Linseed oil may be added if desired. Limewater and linseed oil make a good heavy
sizing, but hard to spread. They are
usually used half and half, though these
proportions may be varied somewhat.

Rice Paste.—Mix the rice flour with
cold water, and boil it over a gentle fire
until it thickens. This paste is quite
white and becomes transparent on drying. It is very adherent and of great use
for many purposes.

Casein Paste.—A solution of tannin,
prepared from a bark or from commercial tannin, is precipitated with limewater, the lime being added until the
solution just turns red litmus paper blue.
The supernatant liquid is then decanted,

and the precipitate is dried without artificial heat. The resulting calcium tannate is then mixed, according to the purpose for which the adhesive is intended, with from 1 to 10 times its weight of dry casein by grinding in a mill. The adhesive compound is soluble in water, petroleum, oils, and carbon bisulphide. It is very strong, and is applied in the form of a paste with water.

PASTES FOR PAPERHANGERS.

I.—Use a cheap grade of rye or wheat flour, mix thoroughly with cold water to about the consistency of dough, or a little thinner, being careful to remove all lumps; stir in a tablespoonful of powdered alum to a quart of flour, then pour in boiling water, stirring rapidly until the flour is thoroughly cooked. Let this cool before using, and thin with cold water.

II.—Venetian Paste.—
(a) 4 ounces white or fish glue
     8 fluidounces cold water
(b) 2 fluidounces Venice turpentine
(c) 1 pound rye flour
     16 fluidounces (1 pint) cold water
(d) 64 fluidounces (½ gallon) boiling water

Soak the 4 ounces of glue in the cold water for 4 hours; dissolve on a water bath (glue pot), and while hot stir in the Venice turpentine. Make up (c) into a batter free from lumps and pour into (d). Stir briskly, and finally add the glue solution. This makes a very strong paste, and it will adhere to a painted surface, owing to the Venice turpentine in its composition.

III.—Strong Adhesive Paste.—
(a) 4 pounds rye flour
     ½ gallon cold water
(b) 1½ gallons boiling water
(c) 2 ounces pulverized rosin

Make (a) into a batter free from lumps; then pour into (b). Boil if necessary, and while hot stir in the pulverized rosin a little at a time. This paste is exceedingly strong, and will stick heavy wall paper or thin leather. If the paste be too thick, thin with a little hot water; never thin paste with cold water.

IV.—Flour Paste.—
(a) 2 pounds wheat flour
     32 fluidounces (1 quart) cold water
(b) 1 ounce alum
     4 fluidounces hot water
(c) 96 fluidounces (½ gallon) boiling water

Work the wheat flour into a batter free from lumps with the cold water. Dissolve the alum as designated in (b).

Now stir in (a) and (c) and, if necessary, continue boiling until the paste thickens into a semitransparent mucilage, after which stir in solution (b). The above makes a very fine paste for wall paper.

V.—Elastic or Pliable Paste.—
(a) 4 ounces common starch
     2 ounces white dextrine
     10 fluidounces cold water
(b) 1 ounce borax
     3 fluidounces glycerine
     64 fluidounces (½ gallon) boiling water

Beat to a batter the ingredients of (a). Dissolve the borax in the boiling water; then add the glycerine, after which pour (a) into solution (b). Stir until it becomes translucent. This paste will not crack, and, being very pliable, is used for paper, cloth, leather, and other material where flexibility is required.

VI.—A paste with which wall paper can be attached to wood or masonry, adhering to it firmly in spite of dampness, is prepared, as usual, of rye flour, to which, however, are added, after the boiling, 8⅓ parts, by weight, of good linseed-oil varnish and 8⅓ parts, by weight, of turpentine to every 500 parts, by weight.

VII.—Paste for Wall Paper.—Soak 18 pounds of bolus (bole) in water, after it has been beaten into small fragments, and pour off the supernatant water. Boil 10 ounces of glue into glue water, mix it well with the softened bolus and 2 pounds plaster of Paris and strain through a sieve by means of a brush. Thin the mass with water to the consistency of a thin paste. The paste is now ready for use. It is not only much cheaper than other varieties, but has the advantage over them of adhering better to whitewashed walls, and especially such as have been repeatedly coated over the old coatings which were not thoroughly removed. For hanging fine wall paper this paste is less commendable, as it forms a white color, with which the paper might easily become soiled if great care is not exercised in applying it. If the fine wall paper is mounted on ground paper, however, it can be recommended for pasting the ground paper on the wall.

LABEL PASTES:

Pastes to Affix Labels to Tin.—Labels separate from tin because the paste becomes too dry. Some moisture is presumably always present; but more is required to cause continued adhesion in the case of tin than where the container is of glass. Paste may be kept moist by the addition of calcium chloride, which is strongly hygroscopic, or of glycerine.

The following formulas for pastes of the type indicated were proposed by Leo Eliel:

I.—Tragacanth........ 1 ounce
    Acacia............. 4 ounces
    Thymol .... 14 grains
    Glycerine.......... 4 ounces
    Water, sufficient to
      make............ 2 pints

Dissolve the gums in 1 pint of water, strain, and add the glycerine, in which the thymol is suspended; shake well and add sufficient water to make 2 pints. This separates on standing, but a single shake mixes it sufficiently for use.

II.—Rye flour.......... 8 ounces
    Powdered acacia.... 1 ounce
    Glycerine.......... 2 ounces
    Oil of cloves....... 40 drops

Rub the rye flour and acacia to a smooth paste with 8 ounces of cold water; strain through cheese cloth, and pour into 1 pint of boiling water, and continue the heat until as thick as desired. When nearly cold add the glycerine and oil of cloves.

III.—Rye flour.......... 5 parts
    Venice turpentine... 1 part
    Liquid glue, a sufficient quantity

Rub up the flour with the turpentine and then add sufficient freshly prepared glue (glue or gelatine dissolved in water) to make a stiff paste. This paste dries slowly.

IV.—Dextrine.......... 2 parts
    Acetic acid........ 1 part
    Water............. 5 parts
    Alcohol, 95 per cent . 1 part

Dissolve the dextrine and acetic acid in water by heating together in the water bath, and to the solution add the alcohol.

V.—Dextrine.......... 3 pounds
    Borax............. 2 ounces
    Glucose .... 5 drachms
    Water...........3 pints 2 ounces

Dissolve the borax in the water by warming, then add the dextrine and glucose, and continue to heat gently until dissolved.

Another variety is made by dissolving a cheap Ghatti gum in limewater, but it keeps badly.

VI.—Add tartaric acid to thick flour paste. The paste is to be boiled until quite thick, and the acid, previously dissolved in a little water, is added, the proportion being about 2 ounces to the pint of paste.

VII.—Gum arabic, 50 parts; glycerine, 10 parts; water, 30 parts; liq. Stibii chlorat., 2 parts.

VIII.—Boil rye flour and strong glue water into a mass to which are added, for 1,000 parts, good linseed-oil varnish 30 parts and oil of turpentine 30 parts. This mixture furnishes a gluing agent which, it is claimed, even renders the labels proof against being loosened by moisture.

IX.—Pour 140 parts of distilled cold water over 100 parts of gum arabic in a wide-necked bottle and dissolve by frequent shaking. To the solution, which is ready after standing for about 3 days, add 10 parts of glycerine; later, 20 parts of diluted acetic acid, and finally 6 parts of aluminum sulphate, then straining it through a fine-hair sieve.

X.—Good glue is said to be obtained by dissolving 1 part of powdered sugar in 4 parts of soda water glass.

XI.—A glue for bottle labels is prepared by dissolving borax in water; soak glue in this solution and dissolve the glue by boiling. Carefully drop as much acetic acid into the solution as will allow it to remain thin on cooling. Labels affixed with this agent adhere firmly and do not become moldy in damp cellars.

XII.—Dissolve some isinglass in acetic acid and brush the labels over with it. There will be no cause to complain of their coming off, nor of striking through the paper. Take a wide-mouthed bottle, fill about two-thirds with commercial acetic acid, and put in as much isinglass as the liquid will hold, and set aside in a warm place until completely dissolved. When cold it should form a jelly. To use it place the bottle in hot water. The cork should be well-fitting and smeared with vaseline or melted paraffine.

How to Paste Labels on Tin.—Brush over the entire back of the label with a flour paste, fold the label loosely by sticking both ends together without creasing the center, and throw to one side until this process has been gone through with the whole lot. Then unfold each label and place it on the can in the regular manner. The paste ought not to be thicker than maple syrup. When of this consistency it soaks through the label and makes it pliable and in a condition to be easily rubbed into position. If the paste is too thick it dries quickly, and does not soak through the label sufficiently. After the labels have been placed upon the cans the latter must be kept apart until dry. In putting the paste upon the labels in the first place, follow the method of placing the dry labels over one another, back sides up, with the edge of each just protruding over the edge of the one beneath it, so that the fingers may easily grasp the label after the pasting has been done.

Druggists' Label Paste.—This paste, when carefully made, is an admirable one for label use, and a very little will go a long way:

Wheat flour	4	ounces
Nitric acid	1	drachm
Boric acid	10	grains
Oil of cloves	5	drops
Carbolic acid	½	drachm

Stir flour and water together, mixing thoroughly, and add the other ingredients. After the stuff is well mixed, heat it, watching very carefully and removing the instant it stiffens.

To Attach Glass Labels to Bottles.—Melt together 1 part of rosin and 2 parts of yellow wax, and use while warm.

Photographic Mountants (see also Photography).—Owing to the nature of the different papers used for printing photographs, it is a matter of extreme importance to use a mountant that shall not set up decomposition in the coating of the print. For example, a mountant that exhibits acidity or alkalinity is injurious with most varieties of paper; and in photography the following formulas for pastes, mucilages, etc., have therefore been selected with regard to their absolute immunity from setting up decomposition in the print or changing its tone in any way. One of the usual mountants is rice starch or else rice water. The latter is boiled to a thick jelly, strained, and the strained mass used as an agglutinant for attaching photographic prints to the mounts. There is nothing of an injurious nature whatever in this mountant, neither is there in a mucilage made with gum dragon.

This gum (also called gum tragacanth) is usually in the form of curls (i.e., leaf gum), which take a long time to properly dissolve in water—several weeks, in fact—but during the past few years there has been put on the market a powdered gum dragon which does not occupy so many days in dissolving. To make a mucilage rom gum dragon a very large volume of water is required. For example, 1 ounce of the gum, either leaf or powder, will swell up and convert 1 gallon of water into a thickish mucilage in the course of 2 or 3 weeks.

Only cold water must be used, and before using the mucilage, all whitish lumps (which are particles of undissolved gum) should be picked out or else the mucilage strained. The time of solution can be considerably shortened (to a few hours) by acidifying the water in which the gum is placed with a little sulphuric or oxalic acid; but as the resultant mucilage would contain traces of their presence, such acids are not permissible when the gum-dragon mucilage is to be used for mounting photographs.

Glycerine and gum arabic make a very good adhesive of a fluid nature suited to mounting photographs; and although glycerine is hygroscopic by itself, such tendency to absorb moisture is checked by the reverse nature of the gum arabic; consequently an ideal fluid mucilage is produced. The proportions of the several ingredients are these:

Gum arabic, genuine (gum acacia, not Bassora gum)	4	ounces
Boiling water	12	ounces
Glycerine, pure	1	ounce

First dissolve the gum in the water, and then stir in the glycerine, and allow all *débris* from the gum to deposit before using. The following adhesive compound is also one that is free from chemical reactions, and is suited for photographic purposes:

Water	2	pints
Gum dragon, powdered	1	ounce
Gum arabic, genuine	4	ounces
Glycerine	4	ounces

Mix the gum arabic with half the water, and in the remainder of the water dissolve the gum dragon. When both solids are dissolved, mix them together, and then stir in the glycerine.

The following paste will be found a useful mountant:

Gum arabic, genuine	1	ounce
Rice starch	1	ounce
White sugar	4	ounces
Water, q. s.		

Dissolve the gum in just sufficient water to completely dissolve it, then add the sugar, and when that has completely dissolved stir in the starch paste, and then boil the mixture until the starch is properly cooked.

A very strong, stiff paste for fastening cardboard mounts to frames, wood, and other materials is prepared by making a bowl of starch paste in the usual way, and then adding 1 ounce of Venice turpentine per pound of paste, and boiling and stirring the mixture until the thick turpentine has become well incorporated. Venice turpentine stirred into flour paste and boiled will also be found a very adhesive cement for fastening cardboard, strawboard, leatherette, and skiver leather to wood or metal; but owing to the resinous nature of the Venice turpentine, such pastes are not suitable for mounting photographic prints. The following half-dozen compounds are suitable mountants to use with silver prints:

Alcohol, absolute.... 10 ounces
Gelatine, good...... 1 ounce
Glycerine........½ to 1 ounce

Soak the gelatine in water for an hour or two until it is completely softened; take the gelatine out of the water, and allow it to drain, and put it into a bottle and pour alcohol over it; add the glycerine (if the gelatine is soft, use only ½ ounce; if the gelatine is hard, use 1 ounce of the glycerine), then melt the gelatine by standing the bottle in a vessel of hot water, and shake up very well. For use, remelt by heat. The alcohol prevents the prints from stretching or cockling, as they are apt to, under the influence of the gelatine.

In the following compound, however, only sufficient alcohol is used to serve as an antiseptic, and prevent the agglutinant from decomposing: Dissolve 4 ounces of photographic gelatine in 16 ounces of water (first soaking the gelatine therein for an hour or two until it is completely softened), then remove the gelatine from the water, allow it to drain, and put it into the bottle, and pour the alcohol over it, and put in the glycerine (if the gelatine is soft, use only ½ ounce; if the gelatine is hard, use 1 ounce of the glycerine), then melt the gelatine by standing the bottle in a vessel of hot water, and shake up well and mix thoroughly. For use, remelt by heat. The alcohol prevents the print from stretching or cockling up under the influence of the gelatine.

The following paste agglutinant is one that is very permanent and useful for all purposes required in a photographic studio: Take 5 pints of water, 10 ounces of arrowroot, 1 ounce of gelatine, and a ½ pint (10 fluidounces) of alcohol, and proceed to combine them as follows: Make arrowroot into a thick cream with a little of the water, and in the remainder of the water soak the gelatine for a few hours, after which melt the gelatine in the water by heating it, add the arrowroot paste, and bring the mixture to the boil and allow to boil for 4 or 5 minutes, then allow to cool, and mix in the alcohol, adding a few drops of oil of cloves.

Perhaps one of the most useful compounds for photographic purposes is that prepared as follows: Soak 4 ounces of hard gelatine in 15 ounces of water for a few hours, then melt the gelatine by heating it in a glue pot until the solution is quite clear and free from lumps, stir in 65 fluidounces of cold water so that it is free from lumps, and pour in the boiling-hot solution of gelatine and continue stirring, and if the starch is not completely cooked, boil up the mixture for a few minutes until it "blows," being careful to keep it well stirred so as not to burn; when cold add a few drops of carbolic acid or some essential oil as an antiseptic to prevent the compound from decomposing or becoming sour.

A useful photographic mucilage, which is very liquid, is obtained by mixing equal bulks of gum-arabic and gum-dragon mucilages of the same consistence. The mixture of these mucilages will be considerably thinner than either of them when alone.

As an agglutinant for general use in the studio, the following is recommended: Dissolve 2 ounces of gum arabic in 5 ounces of water, and for every 250 parts of the mucilage add 20 parts of a solution of sulphate of aluminum, prepared by dissolving 1 part of the sulphate in 20 parts of water (common alum should not be used, only the pure aluminum sulphate, because common alum is a mixture of sulphates, and usually contaminated with iron salts). The addition of the sulphate solution to the gum mucilage renders the latter less hygroscopic, and practically waterproof, besides being very adhesive to any materials, particularly those exhibiting a smooth surface.

MUCILAGES:

For Affixing Labels to Glass and Other Objects.—I.—The mucilage is made by simply pouring over the gum enough water to a little more than cover it, and then, as the gum swells, adding more water from time to time in small portions, until the mucilage is brought to such consistency that it may be easily spread with the brush. The mucilage keeps fairly well without the addition of any antiseptic.

II.—Tragacanth....... 1 ounce
Acacia........... 4 ounces
Thymol......... 14 grains
Glycerine........ 4 ounces
Water, sufficient to
make.......... 2 pints

Dissolve the gums in 1 pint of water, strain and add the glycerine, in which the thymol is suspended; shake well and add sufficient water to make 2 pints. This separates on standing, but a single shake mixes it sufficiently for use.

III.—Rye flour....... 8 ounces
Powdered acacia. 1 ounce
Glycerine....... 2 ounces
Oil of cloves..... 40 drops
Water, a sufficient quantity.

Rub the rye flour and the acacia to a smooth paste with 8 ounces of cold water; strain through cheese cloth, and pour into 1 pint of boiling water and continue the heat until as thick as desired. When nearly cold add the glycerine and oil of cloves.

IV.—One part, by weight, of tragacanth, when mixed with 95-per-cent alcohol to form 4 fluidounces, forms a liquid in which a portion of the tragacanth is dissolved and the remainder suspended; this remains permanently fluid, never deteriorates, and can be used in place of the present mucilage; 4 to 8 minims to each ounce of mixture is sufficient to suspend any of the insoluble substances usually given in mixtures.

V.—To 250 parts of gum-arabic mucilage add 20 parts of water and 2 parts of sulphate of alumina and heat until dissolved.

VI.—Dissolve ½ pound gum tragacanth, powdered, ¼ pound gum arabic, powdered, cold water to the desired consistency, and add 40 drops carbolic acid.

Mucilage of Acacia.—Put the gum, which should be of the best kind, in a flask the size of which should be large enough to contain the mucilage with about one-fifth of its space to spare (i. e., the product should fill it about four-fifths full). Now tare, and wash the gum with distilled water, letting the latter drain away as much as possible before proceeding further. Add the requisite quantity of distilled water slowly, which, however, should first have added to it about 10 per cent of limewater. Now cork the flask, and lay it, without shaking, horizontally in a cool place and let it remain quietly for about 3 hours, then give it a half turn to the right without disturbing its horizontal position. Repeat this operation three or four times during the day, and keep it up until the gum is completely dissolved (which will not be until the fourth day probably), then strain through a thin cloth previously wet with distilled water, avoiding, in so doing, the formation of foam or bubbles. This precaution should also be observed in decantation of the percolate into smaller bottles provided with paraffine corks. The small amount of limewater, as will be understood, is added to the solvent water in order to prevent the action of free acid.

Commercial Mucilage. — Dissolve ½ pound white glue in equal parts water and strong vinegar, and add ¼ as much alcohol and ½ ounce alum dissolved in a little water. To proceed, first get good glue and soak in cold water until it swells and softens. Use pale vinegar. Pour off the cold water, then melt the glue to a thick paste in hot water, and add the vinegar hot. When a little cool add the alcohol and alum water.

To Render Gum Arabic More Adhesive.—I.—Add crystallized aluminum sulphate in the proportion of 2 dissolved in 20 parts of water to 250 parts of concentrated gum solution (75 parts of gum in 175 parts of water).

II.—Add to 250 parts of concentrated gum solution (2 parts of gum in 5 parts of water) 2 parts of crystallized aluminum sulphate dissolved in 20 parts of water. This mixture glues even unsized paper, pasteboard on pasteboard, wood on wood, glass, porcelain, and other substances on which labels frequently do not adhere well.

Envelope Gum.—The gum used by the United States Government on postage stamps is probably one of the best that could be used not only for envelopes but for labels as well. It will stick to almost any surface. Its composition is said to be the following:

Gum arabic......... 1 part
Starch.............. 1 part
Sugar.............. 4 parts
Water, sufficient to give the desired consistency.

The gum arabic is first dissolved in some water, the sugar added, then the starch, after which the mixture is boiled for a few minutes in order to dissolve the starch, after which it is thinned down to the desired consistency.

Cheaper envelope gums can be made by substituting dextrine for the gum arabic, glucose for the sugar, and adding boric acid to preserve and help stiffen it.

Mucilage to Make Wood and Pasteboard Adhere to Metals.—Dissolve 50 parts, by weight, of lead acetate together with 5 parts, by weight, of alum in a little water. Make a separate solution of 75 parts, by weight, of gum arabic in 2,000 parts, by weight, of water, stir in this 500 parts, by weight, of flour, and heat slowly to boiling, stirring the while. Let it cool somewhat, and mix with it the solution containing the lead acetate and alum, stirring them well together.

Preservation of Gum Solution.—Put a small piece of camphor in the mucilage bottle. Camphor vapors are generated which kill all the bacterial germs that have entered the bottle. The gum maintains its adhesiveness to the last drop.

ADULTERANTS IN FOODS:
See Foods.

ADUROL DEVÉLOPER:
See Photography.

ÆSCO-QUININE:
See Hórse Chestnut.

AGAR AGAR PASTE:
See Adhesives.

AGATE, BUTTONS OF ARTIFICIAL.

Prepare a mixture or frit of 33 parts of quartz sand, 65 parts calcium phosphate, and 2 parts of potash. The frit, which has been reduced by heat to the fusing point, is finely ground, intimately mingled with a small quantity of kaolin and pressed in molds which yield button-shaped masses. These masses, after having been fired, are given a transparent glaze by any of the well-known processes.

AGATE (IMITATION):
See Gems, Artificial.

AGING OF SILK:
See Silk.

AGING, SILVER AND GOLD:
See Plating.

AIR BATH.

This air bath is employed in cases in which, upon drying or heating substances, acid vapors arise because the walls of the bath are not attacked by them. For the production of the drying apparatus take a flask with the bottom burst off or a bell jar tubulated above. This is placed either upon a sand bath or upon asbestos paper, previously laid upon a piece of sheet iron. The sand bath or the sheet iron is put on a tripod, so that it can be heated by means of a burner placed underneath. The substance to be dried is placed in a glass or porcelain dish, which is put under the bell jar, and if desired the drying dish may be hung on the tripod. For regulating the temperature the tubulure of the jar is closed with a pierced cork,

through whose aperture the thermometer is thrust. In order to permit the vapors to escape, the cork is grooved lengthwise along the periphery.

AIR BUBBLES IN GELATINE:
See Gelatine.

AIR, EXCLUSION OF, FROM SOLUTIONS:
See Photography.

AIR-PURIFYING.

Ozonatine is a fragrant air-purifying preparation consisting of dextrogyrate turpentine oil scented with slight quantities of fragrant oils.

ALABASTER CLEANING:
See Cleaning Preparations and Methods.

ALBATA METAL:
See Alloys.

ALBUMEN IN URINE, DETECTION OF.

Patein (*Pharm. Zeit.*) recommends the following test for albumen in urine: Dissolve 250 grams of citric acid in a sufficient quantity of water, add enough ammonia to neutralize, then 50 grams of alcohol, and finally enough water to make 1 liter. To the acid (or acidulated) urine, one-tenth its volume of the ammonium-citrate solution made as above is added, and the whole heated in the usual manner. The appearance of the faintest turbidity is said to indicate with positive certainty the presence of albumen.

ALBUMEN PAPER:
See Photography.

ALBUMEN PASTE:
See Adhesives.

Alcohol

After the manuscript of this book was ready for the press, Congress passed the bill which has since become a law, whereby the prohibitive tax on industrial or denatured alcohol is removed. So important is this legislative measure that the Editor has deemed it wise to insert an article on the sources of alcohol and the manufacture of alcohol from farm products. Because the first portion of the book was in type when this step was decided upon, the Editor was compelled to relegate to a later page a monograph which should properly have appeared here. The reader will find the matter on alcohol referred to under the heading

"Spirit"; likewise methods of denaturing and a list of denaturants.

ALCOHOL, DILUTION OF:
See Tables.

Alcohol, Tests for Absolute.—The committee for the compilation of the German Arzneibuch established the following tests for the determination of absolute alcohol:

Absolute alcohol is a clear, colorless, volatile, readily imflammable liquid which burns with a faintly luminous flame. Absolute alcohol has a peculiar odor, a burning taste, and does not affect litmus paper. Boiling point, 78.50. Specific gravity, 0.795 to 0.797. One hundred parts contain 99.7 to 99.4 parts, by volume, or 99.6 to 99.0 parts, by weight, of alcohol.

Absolute alcohol should have no foreign smell and should mix with water without cloudiness.

After the admixture of 5 drops of silver-nitrate solution, 10 cubic centimeters of absolute alcohol should not become turbid or colored even on heating.

A mixture of 10 cubic centimeters of absolute alcohol and 0.2 cubic centimeter of potash lye evaporated down to 1 cubic centimeter should not exhibit an odor of fusel oil after supersaturation with dilute sulphuric acid.

Five cubic centimeters of sulphuric acid, carefully covered, in a test tube, with a stratum of 5 cubic centimeters of absolute alcohol, should not form a rose-colored zone at the surface of contact, even on standing for some time.

The red color of a mixture of 10 cubic centimeters of absolute alcohol and 1 cubic centimeter of potassium-permanganate solution should not pass into yellow before 20 minutes.

Absolute alcohol should not be dyed by hydrogen sulphide water or by aqueous ammonia.

Five cubic centimeters of absolute alcohol should not leave behind a weighable residue after evaporation on water bath.

Absolute Alcohol.—If gelatine be suspended in ordinary alcohol it will absorb the water, but as it is insoluble in alcohol, that substance will remain behind, and thus nearly absolute alcohol will be obtained without distillation.

Perfumed Denaturized Alcohol.—

East India lemon oil	1,250 parts
Mirbane oil	1,000 parts
Cassia oil	50 parts
Clove oil	75 parts
Lemon oil	100 parts
Amyl acetate	500 parts
Spirit (95 per cent)	7,000 parts

Dissolve the oils in the spirit and add the amyl acetate. The mixture serves for destroying the bad odor of denaturized spirit in distilling. Use 50 parts of the perfume per 1,000 parts of spirit.

Solid Alcohol.—I.—Heat 1,000 parts of denaturized alcohol (90 per cent) in a flask of double the capacity on the water bath to about 140° F., and then mix with 28 to 30 parts of well-dried, rasped Venetian soap and 2 parts of gum lac. After repeated shaking, complete dissolution will take place. The solution is put, while still warm, into metallic vessels, closing them up at once and allowing the mixture to cool therein. The admixture of gum lac effects a better preservation and also prevents the evaporation of the alcohol. On lighting the solid spirit the soap remains behind.

II.—Smaragdine is a trade name for solidified alcohol. It consists of alcohol and gun cotton, colored with malachite green. It appears in the market in the form of small cubes.

Alcohol in Fermented Beers.—Experience has shown that ¼ pound of sugar to 1 gallon of water yields about 2 per cent of proof spirit, or about 1 per cent of absolute alcohol. Beyond this amount it is not safe to go, if the legal limit is to be observed, yet a ginger beer brewed with ¼ pound per gallon of sugar would be a very wishy-washy compound, and there is little doubt that a much larger quantity is generally used. The more sugar that is used—up to 1½ or 1¼ pounds per gallon —the better the drink will be and the more customers will relish it; but it will be as "strong" as lager and contain perhaps 5 per cent of alcohol, which will make it anything but a "temperance" drink. Any maker who is using as much as even ½ pound of sugar per gallon is bound to get more spirit than the law allows. Meanwhile it is scarcely accurate to term ginger beers, etc., non-alcoholic.

Alcohol Deodorizer.—

Alcohol	160	ounces
Powdered quicklime	300	grains
Powdered alum	150	grains
Spirit of nitrous ether	1¼	drachms

Mix the lime and alum intimately by trituration; add the alcohol and shake well; then add the spirit of nitrous ether; set aside for 7 days and filter through animal charcoal.

Denaturized Alcohol.—There are two general classes or degrees of denaturizing, viz., the "complete" and the "incomplete," according to the purpose for which the alcohol so denaturized is to be ultimately used.

I.—Complete denaturization by the German system is accomplished by the addition to every 100 liters (equal to $26\frac{1}{2}$ gallons) of spirits:

(a) Two and one-half liters of the "standard" denaturizer, made of 4 parts of wood alcohol, 1 part of pyridine (a nitrogenous base obtained by distilling bone oil or coal tar), with the addition of 50 grams to each liter of oil of lavender or rosemary.

(b) One and one-fourth liters of the above "standard" and 2 liters of benzol with every 100 liters of alcohol.

II.—Incomplete denaturization—i. e., sufficient to prevent alcohol from being drunk, but not to disqualify it from use for various special purposes, for which the wholly denaturized spirits would be unavailable—is accomplished by several methods as follows, the quantity and nature of each substance given being the prescribed dose for each 100 liters ($26\frac{1}{2}$ gallons) of spirits:

(c) Five liters of wood alcohol or $\frac{1}{2}$ liter of pyridine.

(d) Twenty liters of solution of shellac, containing 1 part gum to 2 parts alcohol of 90-per-cent purity. Alcohol for the manufacture of celluloid and pegamoid is denaturized.

(e) By the addition of 1 kilogram of camphor or 2 liters oil of turpentine or $\frac{1}{4}$ liter benzol to each 100 liters of spirits. Alcohol to be used in the manufacture of ethers, aldehyde, agaricin, white lead, bromo-silver gelatines, photographic papers and plates, electrode plates, collodion, salicylic acid and salts, aniline chemistry, and a great number of other purposes, is denaturized by the addition of—

(f) Ten liters sulphuric ether, or 1 part of benzol, or $\frac{1}{2}$ part oil of turpentine, or 0.025 part of animal oil.

For the manufacture of varnishes and inks alcohol is denaturized by the addition of oil of turpentine or animal oil, and for the production of soda soaps by the addition of 1 kilogram of castor oil. Alcohol for the production of lanolin is prepared by adding 5 liters of benzine to each hectoliter of spirits.

ALE.

The ale of the modern brewer is manufactured in several varieties, which are determined by the wants of the consumer and the particular market for which it is intended. Thus, the finer kinds of Burton, East India, Bavarian, and other like ales, having undergone a thorough fermentation, contain only a small quantity of undecomposed sugar and gum, varying from 1 to 5 per cent. Some of these are highly "hopped" or "bittered," the further to promote their preservation during transit and change of temperature. Mild or sweet ales, on the contrary, are less accentuated by lengthened fermentation, and abound in saccharine and gummy matter. They are, therefore, more nutritious, though less intoxicating, than those previously referred to.

In brewing the finer kinds of ales, pale malt and the best hops of the current season's growth are always employed; and when it is desired to produce a liquor possessing little color, very great attention is paid to their selection. With the same object, the boiling is conducted with more than the usual precautions, and the fermentation is carried on at a somewhat lower temperature than that commonly allowed for other varieties of beer. For ordinary ale, intended for immediate use, the malt may be all pale; but, if the liquor be brewed for keeping, and in warm weather, when a slight color is not objectionable, one-fifth, or even one-fourth of amber malt may be advantageously employed. From $4\frac{1}{2}$ to 6 pounds of hops is the quantity commonly used to the one-fourth of malt, for ordinary ales; and 7 pounds to 10 pounds for "keeping" ales. The proportions, however, must greatly depend on the intended quality and description of the brewing and the period that will be allowed for its maturation.

The stronger varieties of ale usually contain from 6 to 8 per cent of "absolute alcohol"; ordinary strong ale, $4\frac{1}{2}$ to 6 per cent; mild ale, 3 to 4 per cent; and table ale, 1 to $1\frac{1}{2}$ per cent (each by volume); together with some undecomposed saccharine, gummy, and extractive matter, the bitter and narcotic principles of the hop, some acetic acid formed by the oxidation of the alcohol, and very small and variable quantities of mineral and saline matter.

Ordinary ale-wort (preferably pale), sufficient to produce 1 barrel, is slowly boiled with about 3 handfuls of hops, and 12 to 14 pounds of crushed groats, until the whole of the soluble matter of the latter is extracted. The resulting liquor, after being run through a coarse strainer and become lukewarm, is fermented with 2 or 3 pints of yeast; and, as soon as the fermentation is at its height, is either closely bunged up for draft or is at once put into strong stoneware bottles, which are then well corked and wired.

White ale is said to be very nutritious, though apt to prove laxative to those unaccustomed to its use. It is drunk in a state of effervescence or lively fermentation; the glass or cup containing it being kept in constant motion, when removed from the mouth, until the whole is consumed, in order that the thicker portion may not subside to the bottom.

ALE, GINGER:
See Beverages.

ALFENIDE METAL:
See Alloys.

ALKALI, HOW TO DETECT:
See Soaps.

ALKALOIDS, ANTIDOTES TO:
See Atropine.

Alloys

No general rules can be given for alloying metals. Alloys differing greatly in fusibility are commonly made by adding the more fusible ones, either in the melted state or in small portions at a time, to the other melted or heated to the lowest possible temperature at which a perfect union will take place between them. The mixture is usually effected under a flux, or some material that will promote liquefaction and prevent volatilization and unnecessary exposure to the air. Thus, in melting lead and tin together for solder, rosin or tallow is thrown upon the surface is rubbed with sal ammoniac; and in combining some metals, powdered charcoal is used for the same purpose. Mercury or quicksilver combines with many metals in the cold, forming AMALGAMS, or easily fusible alloys (q. v.).

Alloys generally possess characteristics unshared by their component metals. Thus, copper and zinc form brass, which has a different density, hardness, and color from either of its constituents. Whether the metals tend to unite in atomic proportions or in any definite ratio is still undetermined. The evidence afforded by the natural alloys of gold and silver, and by the phenomena accompanying the cooling of several alloys from the state of fusion, goes far to prove that such is the case (Rudberg). The subject is, however, one of considerable difficulty, as metals and metallic compounds are generally soluble in each other, and unite by simple fusion and contact. That they do not combine indifferently with each other, but exercise a species of elective affinity not dissimilar to other bodies, is clearly shown by the homogeneity and superior quality of many alloys in which the constituent metals are in atomic proportion. The variation of the specific gravity and melting points of alloys from the mean of those of their component metals also affords strong evidence of a chemical change having taken place. Thus, alloys generally melt at lower temperatures than their separate metals. They also usually possess more tenacity and hardness than the mean of their constituents.

Matthiessen found that when weights are suspended to spirals of hard-drawn wire made of copper, gold, or platinum, they become nearly straightened when stretched by a moderate weight; but wires of equal dimensions composed of copper-tin (12 per cent of tin), silver-platinum (36 per cent of platinum), and gold-copper (84 per cent of copper) scarcely undergo any permanent change in form when subjected to tension by the same weight.

The same chemist gives the following approximate results upon the tenacity of certain metals and wires hard-drawn through the same gauge (No. 23):

	Pounds
Copper, breaking strain.........	25–30
Tin, breaking strain.......under	7
Lead, breaking strain......under	7
Tin-lead (20% lead).......about	7
Tin-copper (12% copper)...about	7
Copper-tin (12% tin)......about	80–90
Gold (12% tin)................	20–25
Gold-copper (8.4% copper)......	70–75
Silver (8.4% copper)..........	45–50
Platinum (8.4% copper)........	45–50
Silver-platinum (30% platinum).	75–80

On the other hand, the malleability, ductility, and power of resisting oxygen of alloys is generally diminished. The alloy formed of two brittle metals is always brittle; that of a brittle and a ductile metal, generally so; and even two ductile metals sometimes unite to form a brittle compound. The alloys formed of metals having different fusing points are usually malleable while cold and brittle while hot. The action of the air on alloys is generally less than on their simple metals, unless the former are heated. A mixture of 1 part of tin and 3 parts of lead is scarcely acted on at common temperatures; but at a red heat it readily takes fire, and continues to burn for some time like a piece of bad turf. In like manner, a mixture of tin and zinc, when strongly heated, decomposes both moist air and steam with rapidity.

The specific gravity of alloys is rarely

the arithmetical mean of that of their constituents, as commonly taught; and in many cases considerable condensation or expansion occurs. When there is a strong affinity between two metals, the density of their alloy is generally greater than the calculated mean; and vice versa, as may be seen in the following table:

ALLOYS HAVING A DENSITY

Greater than the Mean of their Constituents:

Copper and bismuth,
Copper and palladium,
Copper and tin,
Copper and zinc,
Gold and antimony,
Gold and bismuth,
Gold and cobalt,
Gold and tin,
Gold and zinc,
Lead and antimony,
Palladium and bismuth,
Silver and antimony,
Silver and bismuth,
Silver and lead,
Silver and tin,
Silver and zinc.

Less than the Mean of their Constituents:

Gold and copper,
Gold and iridium,
Gold and iron,
Gold and lead,
Gold and nickel,
Gold and silver,
Iron and antimony,
Iron and bismuth,
Iron and lead,
Nickel and arsenic,
Silver and copper,
Tin and antimony,
Tin and lead,
Tin and palladium,
Zinc and antimony.

Compounding Alloys.—Considerable experience is necessary to insure success in compounding alloys, especially when the metals employed vary greatly in fusibility and volatility. The following are rules supplied by an experienced workman:

1. Melt the least fusible, oxidizable, and volatile first, and then add the others heated to their point of fusion or near it. Thus, if it is desired to make an alloy of exactly 1 part of copper and 3 of zinc, it will be impossible to do so by putting proportions of the metals in a crucible and exposing the whole to heat. Much of the zinc would fly off in vapor before the copper was melted. First, melt the copper and add the zinc, which has been melted in another crucible. The zinc should be in excess, as some of it will be lost anyway.

2. Some alloys, as copper and zinc, copper and arsenic, may be formed by exposing heated plates of the least fusible metal to the vapor of the other. In making brass in the large way, thin plates of copper are dissolved, as it were, in melted zinc until the proper proportions have been obtained.

3. The surface of all oxidizable metals should be covered with some protecting agent, as tallow for very fusible ones, rosin for lead and tin, charcoal for zinc, copper, etc.

4. Stir the metal before casting and if possible, when casting, with a whitewood stick; this is much better for the purpose than an iron rod.

5. If possible, add a small portion of old alloy to the new. If the alloy is required to make sharp castings and strength is not a very great object, the proportion of old alloy to the new should be increased. In all cases a new or thoroughly well-cleansed crucible should be used.

To obtain metals and metallic alloys from their compounds, such as oxides, sulphides, chlorides, etc., a process lately patented makes use of the reducing qualities of aluminum or its alloys with magnesium. The finely powdered material (e. g., chromic oxide) is placed in a crucible mixed with aluminum oxide. The mixture is set afire by means of a soldering pipe or a burning magnesium wire, and the desired reaction takes place. For igniting, one may also employ with advantage a special priming cartridge consisting of pulverized aluminum to which a little magnesium may be mixed, and peroxide of magnesia, which is shaped into balls and lighted with a magnesium wire. By suitable additions to the pulverized mixture, alloys containing aluminum, magnetism, chromium, manganese, copper, iron, boron, silicic acid, etc., are obtained.

ALUMINUM ALLOYS.

M. H. Pecheux has contributed to the *Comptes Rendus*, from time to time, the results of his investigations into the alloys of aluminum with soft metals, and the following constitutes a brief summary of his observations:

Lead.—When aluminum is melted and lead is added in proportion greater than 10 per cent, the metals separate on cooling into three layers—lead, aluminum, and between them an alloy containing from 90 to 97 per cent of aluminum.

The alloys with 93, 95, and 98 per cent have densities of 2.745, 2.674, and 2.600 respectively, and melting points near that of aluminum. Their color is like that of aluminum, but they are less lustrous. All are malleable, easily cut, softer than aluminum, and have a granular fracture. On remelting they become somewhat richer in lead, through a tendency to liquation. They do not oxidize in moist air, nor at their melting points. They are attacked in the cold by hydrochloric and by strong sulphuric acid, with evolution of hydrogen, and by strong nitric acid when hot; strong solution of potassium hydroxide also attacks them. They are without action on distilled water, whether cold or hot.

Zinc.—Well-defined alloys were obtained, corresponding to the formulas Zn_3Al, Zn_2Al, $ZnAl$, $ZnAl_2$, $ZnAl_3$, $ZnAl_4$, $ZnAl_6$, $ZnAl_{10}$, $ZnAl_{12}$. Their melting points and densities all lie between those of zinc and aluminum, and those containing most zinc are the hardest. They are all dissolved by cold hydrochloric acid and by hot dilute nitric acid. Cold concentrated nitric acid attacks the first three, and cold dilute acid the first five. The Zn_3Al, $ZnAl_6$, $ZnAl_{10}$, and $ZnAl_{12}$ are only slightly affected by cold potassium-hydroxide solution; the others are strongly attacked, potassium zincate and aluminate probably being formed.

Tin.—A filed rod of tin-aluminum alloy plunged in cold water gives off for some minutes bubbles of gas, composed of hydrogen and oxygen in explosive proportions. An unfiled rod, or a filed rod of either aluminum or tin, is without action, though the unfiled rod of alloy will act on boiling water. The filed rod of alloy, in faintly acid solution of copper or zinc sulphate, becomes covered with a deposit of copper or zinc, while bubbles of oxygen are given off. M. Pecheux believes that the metals are truly alloyed only at the surface, and that filing lays bare an almost infinitely numerous series of junctions of the two metals, which, heated by the filing, act as thermocouples.

Bismuth.—By the method used for lead, bismuth alloys were obtained containing 75, 85, 88, and 94 per cent of aluminum, with densities 2.86, 2.79, 2.78, and 2.74 respectively. They were sonorous, brittle, finely grained, and homogeneous, silver-white, and with melting points between those of their constituents, but nearer that of aluminum. They are not oxidized in air at the temperature of casting, but are readily attacked by acids, concentrated or dilute, and by potassium-hydroxide solution. The filed alloys behave like those of tin, but still more markedly.

Magnesium.—These were obtained with 66, 68, 73, 77, and 85 per cent of aluminum, and densities 2.24, 2.47, 2.32, 2.37, 2.47. They are brittle, with large granular fracture, silver-white, file well, take a good polish, and have melting points near that of aluminum. Being viscous when melted, they are difficult to cast, and when slowly cooled form a gray, spongy mass which cannot be remelted. They do not oxidize in air at the ordinary temperatures, but burn readily at a bright-red heat. They are attacked violently by acids and by potassium-hydroxide solution, decompose hydrogen peroxide, and slowly decompose water even in the cold.

Tin, Bismuth, and Magnesium.—The action of water on these alloys just referred to has been recently demonstrated on a larger scale, 5 to 6 cubic centimeters of hydrogen having been obtained in 20 minutes from 2 cubic centimeters of the filed tin alloy. The bismuth alloy yielded more hydrogen than the tin alloy, and the magnesium alloy more than the bismuth alloy. The oxygen of the decomposed water unites with the aluminum. Larger quantities of hydrogen are obtained from copper-sulphate solution, apart from the decomposition of this solution by precipitation of copper at the expense of the metal alloyed with the aluminum. The alloys of aluminum with zinc and lead do not decompose pure water, but do decompose the water of copper-sulphate solution, and, more slowly, that of zinc-sulphate solution.

Aluminum is a metal whose properties are very materially influenced by a proportionately small addition of copper. Alloys of 99 per cent aluminum and 1 per cent of copper are hard, brittle, and bluish in color; 95 per cent of aluminum and 5 per cent of copper give an alloy which can be hammered, but with 10 per cent of copper the metal can no longer be worked. With 80 per cent and upward of copper are obtained alloys of a beautiful yellow color, and these mixtures, containing from 5 to 10 per cent of aluminum and from 90 to 95 per cent of copper, are the genuine aluminum bronzes. The 10-per-cent alloys are of a pure golden-yellow color; with 5 per cent of aluminum they are reddish yellow, like gold heavily alloyed with copper, and a 2-per-cent admixture is of an almost pure copper red

As the proportion of copper increases, the brittleness is diminished, and alloys containing 10 per cent and less of aluminum can be used for industrial purposes, the best consisting of 90 per cent of copper and 10 of aluminum. The hardness of this alloy approaches that of the general bronzes, whence its name. It can be stretched out into thin sheets between rollers, worked under the hammer, and shaped as desired by beating or pressure, in powerful stamping presses. On account of its hardness it takes a fine polish, and its peculiar greenish-gold color resembles that of gold alloyed with copper and silver together.

Alloys with a still greater proportion of copper approach this metal more and more nearly in their character; the color of an alloy, for instance, composed of 95 per cent of copper and 5 per cent of aluminum, can be distinguished from pure gold only by direct comparison, and the metal is very hard, and also very malleable.

Electrical Conductivity of Aluminum Alloys.—During three years' exposure to the atmosphere, copper-aluminum alloys in one test gradually diminished in conductivity in proportion to the amount of copper they contained. The nickel-copper aluminum alloys, which show such remarkably increased tensile strength as compared with good commercial aluminum, considerably diminished in total conductivity. On the other hand, the manganese-copper aluminum alloys suffered comparatively little diminution in total conductivity, and one of them retained comparatively high tensile strength. It was thought that an examination of the structure of these alloys by aid of microphotography might throw some light on the great difference which exists between some of their physical properties. For instance, a nickel-copper aluminum alloy has 1.6 times the tensile strength of ordinary commercial aluminum. Under a magnification of 800 diameters practically no structure could be discovered. Considering the remarkable crystalline structure exhibited by ordinary commercial aluminum near the surface of an ingot, when allowed to solidify at an ordinary rate, the want of structure in these alloys must be attributed to the process of drawing down. The inference is that the great difference which exists between their tensile strengths and other qualities is not due to variation in structure.

Colored Alloys of Aluminum.—A purple scintillating composition is produced by an alloyage of 78 parts of gold and 22 parts aluminum. With platinum a gold-colored alloy is obtained; with palladium a copper-colored one; and with cobalt and nickel one of a yellow color. Easily fusible metals of the color of aluminum give white alloys. Metal difficult of fusion, such as iridium, osmium, titanium, etc., appear in abnormal tones of color through such alloyages.

Aluminum-Brass.—Aluminum, 1 per cent; specific gravity, 8.35; tensile strength, 40. Aluminum, 3 per cent; specific gravity, 8.33; tensile strength, 65. The last named is harder than the first.

Aluminum-Copper.— Minikin is principally aluminum with a small percentage of copper and nickel. It is alloyed by mixing the aluminum and copper, then adding the nickel. It resembles palladium and is very strong.

Aluminum-Silver.—I.—Silver, 3 per cent; aluminum, 97 per cent. A handsome color.

II.—A silver aluminum that is easily worked into various articles contains about one-fourth silver and three-fourths of aluminum.

Aluminum-Tin.—Bourbon metal is composed of equal parts of aluminum and tin; it solders readily.

Aluminum-Tungsten.—A new metal alloy consisting of aluminum and tungsten is used of late in France in the construction of conveyances, especially carriages, bicycles, and motor vehicles. The French call it partinium; the composition of the new alloy varies according to the purposes for which it is used. It is considerably cheaper than aluminum, almost as light, and has a greater resistance. The strength is stated at 32 to 37 kilograms per square millimeter.

Aluminum-Zinc. — Zinc, 3 per cent; aluminum, 97 per cent. Very ductile, white, and harder than aluminum.

AMALGAMS:
See Fusible Alloys.

Anti-Friction Bearing or Babbitt Metals.—These alloys are usually supported by bearings of brass, into which it is poured after they have been tinned, and heated and put together with an exact model of the axle, or other working piece, plastic clay being previously applied, in the usual manner, as a lute or outer mold. Soft gun metal is also excellent, and is much used for bearings. They all become less heated in working than the harder metals, and less grease or oil is consequently required when they are used.

I.—An anti-friction metal of excellent quality and one that has been used with success is made as follows : 17 parts zinc; 1 part copper; $1\frac{1}{2}$ parts antimony; prepared in the following way: Melt the copper in a small crucible, then add the antimony, and lastly the zinc, care being taken not to burn the zinc. Burning can be prevented by allowing the copper and antimony to cool slightly before adding the zinc. This metal is preferably cast into the shape desired and is not used as a lining metal because it requires too great a heat to pour. It machines nicely and takes a fine polish on bearing surfaces. It has the appearance of aluminum when finished. Use a lubricating oil made from any good grade of machine oil to which 3 parts of kerosene have been added.

II.—Copper, 6 parts; tin, 12 parts; lead, 150 parts; antimony, 30 parts; wrought iron, 1 part; cast iron, 1 part. For certain purposes the composition is modified as follows: Copper, 16 parts; tin, 40 parts; lead, 120 parts; antimony, 24 parts; wrought iron, 1 part; cast iron, 1 part. In both cases the wrought iron is cut up in small pieces, and in this state it will melt readily in fused copper and cast iron. After the mixture has been well stirred, the tin, lead, and antimony are added; these are previously melted in separate crucibles, and when mingled the whole mass is again stirred thoroughly. The product may then be run into ingots, to be employed when needed. When run into the molds the surface should be well skimmed, for in this state it oxidizes rapidly. The proportions may be varied without materially affecting the results.

III.—From tin, 16 to 20 parts; antimony, 2 parts; lead, 1 part; fused together, and then blended with copper, 80 parts. Used where there is much friction or high velocity.

IV.—Zinc, 6 parts; tin, 1 part; copper, 20 parts. Used when the metal is exposed to violent shocks.

V.—Lead, 1 part; tin, 2 parts; zinc, 4 parts; copper, 68 parts. Used when the metal is exposed to heat.

VI.—Tin, 48 to 50 parts; antimony, 5 parts; copper, 1 part.

VII.—(Fenton's.) Tin, with some zinc, and a little copper.

VIII.—(Ordinary.) Tin, or hard pewter, with or without a small portion of antimony or copper. Without the last it is apt to spread out under the weight of heavy machinery. Used for the bearings of locomotives, etc.

The following two compositions are for motor and dynamo shafts: 100 pounds tin; 10 pounds copper; 10 pounds antimony.

$83\frac{1}{2}$ pounds tin; $8\frac{1}{4}$ pounds antimony; $8\frac{1}{4}$ pounds copper.

IX.— Lead, 75 parts; antimony, 23 parts; tin, 2 parts.

X.—Magnolia Metal.—This is composed of 40 parts of lead, $7\frac{1}{2}$ parts of antimony, $2\frac{1}{2}$ of tin, $\frac{1}{8}$ of bismuth, $\frac{1}{8}$ of aluminum, and $\frac{1}{4}$ of graphite. It is used as an anti-friction metal, and takes its name from its manufacturer's mark, a magnolia flower.

ARGENTAN:
See German Silver, under this title.

BELL METAL.
The composition of bell metal varies considerably, as may be seen below:

I.—(Standard.) Copper, 78 parts; tin, 22 parts; fused together and cast. The most sonorous of all the alloys of copper and tin. It is easily fusible, and has a fine compact grain, and a vitreous conchoidal and yellowish-red fracture. According to Klaproth, the finest-toned Indian gongs have this composition.

II.—(Founder's Standard.) Copper, 77 parts; tin, 21 parts; antimony, 2 parts. Slightly paler and inferior to No. I.

III.—Copper, 80 parts; tin, 20 parts. Very deep-toned and sonorous. Used in China and India for the larger gongs, tam-tams, etc.

IV.—Copper, 78 to 80 parts; tin, 22 to 20 parts. Usual composition of Chinese cymbals, tam-tams, etc.

V.—Copper, 75 (= 3) parts; tin, 25 (=1) part. Somewhat brittle. In fracture, semivitreous and bluish-red. Used for church and other large bells.

VI.—Copper, 80 parts; tin, $10\frac{1}{4}$ parts; zinc, $5\frac{1}{2}$ parts; lead, $4\frac{1}{4}$ parts. English bell metal, according to Thomson. Inferior to the last; the lead being apt to form isolated drops, to the injury of the uniformity of the alloy.

VII.—Copper, 68 parts; tin, 32 parts. Brittle; fracture conchoidal and ashgray. Best proportions for house bells, hand bells, etc.; for which, however, 2 of copper and 1 of tin is commonly substituted by the founders.

VIII.—Copper, 72 parts; tin, $26\frac{1}{4}$ parts; iron, $1\frac{1}{2}$ parts. Used by the Paris houses for the bells of small clocks.

IX.—Copper, 72 parts; tin, 26 parts; zinc, 2 parts. Used, like the last, for very small bells.

X.—Copper, 70 parts; tin, 26 parts;

zinc, 2 parts. Used for the bells of repeating watches.

XI.—Melt together copper, 100 parts; tin, 25 parts. After being cast into the required object, it should be made red-hot, and then plunged immediately into cold water in order to impart to it the requisite degree of sonorousness. For cymbals and gongs.

XII.—Melt together copper, 80 parts; tin, 20 parts. When cold it has to be hammered out with frequent annealing.

XIII.—Copper, 78 parts; tin, 22 parts; This is superior to the former, and it can be rolled out. For tam-tams and gongs.

XIV.—Melt together copper, 72 parts; tin, 26 to 56 parts; iron $\frac{1}{44}$ part. Used in making the bells of ornamental French clocks.

Castings in bell metal are all more or less brittle; and, when recent, have a color varying from a dark ash-gray to grayish-white, which is darkest in the more cuprous varieties, in which it turns somewhat on the yellowish-red or bluish-red. The larger the proportion of copper in the alloy, the deeper and graver the tone of the bells formed of it. The addition of tin, iron, or zinc, causes them to give out their tones sharper. Bismuth and lead are also often used to modify the tone, which each metal affects differently. The addition of antimony and bismuth is frequently made by the founder to give a more crystalline grain to the alloy. All these conditions are, however, prejudicial to the sonorousness of bells, and of very doubtful utility. Rapid refrigeration increases the sonorousness of all these alloys. Hence M. D'Arcet recommends that the "pieces" be heated to a cherry-red after they are cast, and after having been suddenly plunged into cold water, that they be submitted to well-regulated pressure by skillful hammering, until they assume their proper form: after which they are to be again heated and allowed to cool slowly in the air. This is the method adopted by the Chinese with their gongs, etc., a casing of sheet iron being employed by them to support and protect the pieces during the exposure to heat. In a general way, however, bells are formed and completed by simple casting. This is necessarily the case with all very large bells. Where the quality of their tones is the chief object sought after, the greatest care should be taken to use commercially pure copper. The presence of a very little lead or any similar metal greatly lessens the sonorousness of this alloy; while that of silver increases it.

The specific gravity of a large bell is seldom uniform through its whole substance; nor can the specific gravity from any given portion of its constituent metals be exactly calculated owing to the many interfering circumstances. The nearer this uniformity is approached, or, in other words, chemical combination is complete, the more durable and finer-toned will be the bell. In general, it is found necessary to take about one-tenth more metal than the weight of the intended bell, or bells, in order to allow for waste and scorification during the operations of fusing and casting.

BISMUTH ALLOYS.

Bismuth possesses the unusual quality of expanding in cooling. It is, therefore, introduced in many alloys to reduce or check shrinkage in the mold.

For delicate castings, and for taking impressions from dies, medals, etc., various bismuth alloys are in use, whose composition corresponds to the following figures:

	I	II	III	IV
Bismuth	6	5	2	8
Tin	3	2	1	3
Lead	13	3	1	5

V.—Cliché Metal.—This alloy is composed of tin, 48 parts; lead, 32.5; bismuth, 9; and antimony, 10.5. It is especially well adapted to dabbing rollers for printing cotton goods, and as it possesses a considerable degree of hardness, it wears well.

VI.—For filling out defective places in metallic castings, an alloy of bismuth 1 part, antimony 3, lead 8, can be advantageously used.

VII.—For Cementing Glass.—Most of the cements in ordinary use are dissolved, or at least softened, by petroleum. An alloy of lead 3 parts, tin 2, bismuth 2.5, melting at 212° F., is not affected by petroleum, and is therefore very useful for cementing lamps made of metal and glass combined.

LIPOWITZ'S BISMUTH ALLOY:

See Cadmium Alloys.

BRASS.

In general brass is composed of two-thirds copper and one-third zinc, but a little lead or tin is sometimes advantageous, as the following:

I.—Red copper, 66 parts; zinc, 34 parts; lead, 1 part.

II.—Copper, 66 parts; zinc, 32 parts; tin, 1 part; lead, 1 part.

III.—Copper, 64.5 parts; zinc, 33.5 parts; lead, 1.5 parts; tin, 0.5 part.

Brass-Aluminum.—A small addition of aluminum to brass (1.5 to 8 per cent) greatly increases its hardness and elasticity, and this alloy is also easily worked for any purpose. Brass containing 8 per cent of aluminum has the valuable property of being but slightly affected by acids or gases. A larger percentage of aluminum makes the brass brittle. It is to be noted that aluminum brass decreases very materially in volume in casting, and the casts must be cooled slowly or they will be brittle. It is an alloy easily made, and its low price, combined with its excellent qualities, would seem to make it in many cases an advantageous substitute for the expensive phosphorous bronze.

Bristol Brass (Prince's Metal).—This alloy, which possesses properties similar to those of French brass, is prepared in the following proportions:

	I	II	III
Copper	75.7	67.2	60.8
Zinc	24.3	32.8	39.2

Particular care is required to prevent the zinc from evaporating during the fusing, and for this purpose it is customary to put only half of it into the first melting, and to add the remainder when the first mass is liquefied.

Brass-Iron (Aich's Metal).—This is a variety of brass with an admixture of iron, which gives it a considerable degree of tenacity. It is especially adapted for purposes which require a hard and, at the same time, tenacious metal. Analyses of the various kinds of this metal show considerable variation in the proportions. Even the amount of iron, to which the hardening effect must be attributed, may vary within wide limits without materially modifying the tenacity which is the essential characteristic of this alloy.

I.—The best variety of Aich's metal consists of copper, 60 parts; zinc, 38.2; iron, 1.8. The predominating quality of this alloy is its hardness, which is claimed to be not inferior to that of certain kinds of steel. It has a beautiful golden-yellow color, and is said not to oxidize easily, a valuable property for articles exposed to the action of air and water.

II.—Copper, 60.2 parts; zinc, 38.2; iron, 1.6. The permissible variations in the content of iron are from 0.4 to 3 per cent.

Sterro metal may properly be considered in connection with Aich's metal, since its constituents are the same and its properties very similar. The principal difference between the two metals is that sterro metal contains a much larger amount of iron. The composition of this alloy varies considerably with different manufacturers.

III.—Two varieties of excellent quality are the product of the Rosthorn factory, in Lower Austria—copper, 55.33 parts; zinc, 41.80; iron, 4.66. Also

IV.—English sterro metal (Gedge's alloy for ship sheathing), copper, 60 parts; zinc, 38.125; iron, 1.5.

The great value of this alloy lies in its strength, which is equaled only by that of the best steel. As an illustration of this, a wrought-iron pipe broke with a pressure of 267 atmospheres, while a similar pipe of sterro metal withstood the enormous pressure of 763 atmospheres without cracking. Besides its remarkable strength, it possesses a high degree of elasticity, and is, therefore, particularly suitable for purposes which require the combination of these two qualities, such as the construction of hydraulic cylinders. It is well known that these cylinders, at a certain pressure, begin to sweat, that is, the interior pressure is so great that the water permeates through the pores of the steel. With a sterro metal cylinder, the pressure can be considerably increased without any moisture being perceptible on the outside of the cylinder.

Sterro metal can be made even more hard and dense, if required for special purposes, but this is effected rather by mechanical manipulation than by any change in the chemical composition. If rolled or hammered in heat, its strength is increased, and it acquires, in addition, an exceedingly high degree of tenacity. Special care must be taken, however, in hammering not to overheat the metal, as in this case it would become brittle and might crack under the hammer. Sterro metal is especially suitable for all the purposes for which the so-called red metal has been in the past almost exclusively used. Axle bearings, for example, made of sterro metal have such excellent qualities that many machine factories are now using this material entirely for the purpose.

Cast Brass.—The various articles of bronze, so called, statuettes, clock cases, etc., made in France, where this industry has attained great perfection and extensive proportions, are not, in many cases genuine bronze, but fine cast brass. Following are the compositions of a few mixtures of metals most frequently used by French manufacturers:

	Copper	Zinc	Tin	Lead
I	63.70	33.55	2.50	0.25
II	64.45	32.44	0.25	2.86
III	70.90	24.05	2.00	3.05
IV	72.43	22.75	1.87	2.95

Their special advantage is that they can be readily cast, worked with file and chisel, and easily gilded.

To Cast Yellow Brass.—If good, clean, yellow brass sand castings are desired, the brass should not contain over 30 per cent of zinc. This will assure an alloy of good color and one which will run free and clean. Tin or lead may be added without affecting the property of casting clean. A mixture of 7 pounds of copper, 3 pounds of spelter, 4 ounces of tin, and 3 ounces of lead makes a good casting alloy and one which will cut free and is strong. If a stronger alloy be desired, more tin may be added, but 4 ounces is usually sufficient. If the alloy be too hard, reduce the proportion of tin.

Leaf Brass.—This alloy is also called Dutch gold, or imitation gold leaf. It is made of copper, 77.75 to 84.5 parts; zinc, 15.5 to 22.25. Its color is pale or bright yellow or greenish, according to the proportions of the metals. It has an unusual degree of ductility.

Malleable Brass.—This metal is affected less by sea water than pure copper, and was formerly much used for ship sheathing, and for making nails and rivets which were to come in contact with sea water. At the present day it has lost much of its importance, since all the larger ships are made of steel. It is usually composed of copper, 60 to 62 parts; and zinc, 40 to 38 parts. It is sometimes called yellow metal, or Müntz metal (called after its inventor), and is prepared with certain precautions, directed toward obtaining as fine a grain as possible, experience having shown that only a fine-grained alloy of uniform density can resist the action of the sea water evenly. A metal of uneven density will wear in holes. To obtain as uniform a grain as possible, small samples taken from the fused mass are cooled quickly and examined as to fracture. If they do not show the desired uniform grain, some zinc is added to the mass. After it has permeated the whole mass, a fresh sample is taken and tested, this being continued until the desired result is reached. It is scarcely necessary to remark that considerable experience is required to tell the correct composition of the alloy from the fracture. The mass is finally poured into molds and rolled cold. Malleable brass can be worked warm, like iron, being ductile in heat, a valuable quality.

Experiments with malleable brass show that all alloys containing up to 58.33 per cent of copper and up to 41.67 per cent of zinc are malleable. There is, in addition, a second group of such alloys, with 61.54 per cent of copper and 38.46 per cent of zinc, which are also malleable in heat.

The preparation of these alloys requires considerable experience, and is best accomplished by melting the metals together in the usual manner, and heating the fused mass as strongly as possible. It must be covered with a layer of charcoal dust to prevent oxidation of the zinc. The mass becomes thinly fluid, and an intimate mixture of the constituents is effected. Small pieces of the same alloy are thrown into the liquid mass until it no longer shows a reflecting surface, when it is cast into ingots in iron molds. The ingots are plunged into water while still red-hot, and acquire by this treatment a very high degree of ductility. The alloy, properly prepared, has a fibrous fracture and a reddish-yellow color.

Sheet Brass (For Sheet and Wire).—In the preparation of brass for the manufacture of wire, an especially pure quality of copper must be used; without this, all efforts to produce a suitable quality of brass will be in vain. That pure copper is indispensable to the manufacture of good, ductile brass may be seen from the great difference in the composition of the various kinds, all of which answer their purpose, but contain widely varying quantities of copper and zinc. The following table shows the composition of some excellent qualities of brass suitable for making sheet and wire:

Brass Sheet—Source	Copper	Zinc	Lead	Tin
Jemappes	64.6	33.7	1.4	0.2
Stolberg	64.8	32.8	2.0	0.4
Romilly	70.1	29.26	0.38	0.17
Rosthorn (Vienna)	68.1	31.9	....	....
Rosthorn (Vienna)	71.5	28.5	....	....
Rosthorn (Vienna)	71.1	27.6	1.3	....
Iserlohn & Romilly	70.1	29.9	....	....
Lüdenscheid	72.73	27.27	....	....
(Brittle)	63.66	33.02	2.52	....
Hegermühl	70.16	27.45	0.79	0.20
Oker	68.98	29.54	0.97	....
Brass Wire—				
England	70.29	29.26	0.28	0.17
Augsburg	71.89	27.63	0.85	....
Neustadt	70.16	27.45	0.2	0.79
Neustadt	71.36	28.15	....	....
Neustadt	71.5	28.5	....	....
Neustadt	71.0	27.6	....	....
(Good quality)	65.4	34.6	....	....
(Brittle)	65.5	32.4	2.1	....
For wire and sheet	67.0	32.0	0.5	0.5

As the above figures show, the percentage of zinc in the different kinds of brass lies between 27 and 34. Recently, alloys containing a somewhat larger quantity of zinc have been used, it having been found that the toughness and ductility of the brass are increased thereby, without injury to its tenacity. Alloys containing up to 37 per cent of zinc possess a high degree of ductility in the cold, and are well adapted for wire and sheet.

Gilders' Sheet Brass.—Copper, 1 part; zinc, 1 part; tin, $\frac{1}{10}$ part; lead, $\frac{1}{10}$ part. Very readily fusible and very dense.

White Brass.—Birmingham platina is an alloy of a pure white, almost silverwhite color, remaining unaffected by tolerably long exposure to the atmosphere. Unfortunately this alloy is so brittle that it can rarely be shaped except by casting. It is used only in the manufacture of buttons. The alloy is poured into molds giving rather sharp impressions and allowing the design on the button (letters or coat of arms) to stand out prominently with careful stamping. The composition of this alloy, also known by the name of platinum lead, is as follows:

	I	II
Copper	46.5	4
Zinc	53.5	16

III.—Zinc, 80 parts; copper, 10 parts; iron, 10 parts.

BRITANNIA METAL.

Britannia metal is an alloy consisting principally of tin and antimony. Many varieties contain only these two metals, and may be considered simply as tin hardened with antimony, while others contain, in addition, certain quantities of copper, sometimes lead, and occasionally, though rarely on account of its cost, bismuth. Britannia metal is always of a silvery-white color, with a bluish tinge, and its hardness makes it capable of taking a high polish, which is not lost through exposure to the air. Ninety per cent of tin and 10 per cent of antimony gives a composition which is the best for many purposes, especially for casting, as it fills out the molds well, and is readily fusible. In some cases, where articles made from it are to be subjected to constant wear, a harder alloy is required. In the proportions given above, the metal is indeed much harder than tin, but would still soon give way under usage.

A table is appended, giving the composition of some of the varieties of Britannia metal and their special names.

	Tin	Antimony	Copper	Zinc	Lead
English	81.90	16.25	1.84	...	....
English	90.62	7.81	1.46	...	....
English	90.1	6.3	3.1	0.5	....
English	85.4	9.66	0.81	3.06	....
Pewter	81.2	5.7	1.60	....	11.5
Pewter	89.3	7.6	1.8	....	1.8
Tutania	91.4	....	0.7	0.3	7.6
Queen's metal	88.5	7.1	3.5	0.9	....
German	72.0	24.0	4.0	...	....
German	84.0	9.0	2.0	5.0	....
German (for casting)	20.0	64.0	10.0	6.0	....
Malleable (for casting)	48.0	....	3.0	48.0	1.0

Britannia metal is prepared by melting the copper alone first, then adding a part of the tin and the whole of the antimony. The heat can then be quickly moderated, as the melting point of the new alloy is much lower than that of copper. Finally, the rest of the tin is added, and the mixture stirred constantly for some time to make it thoroughly homogeneous.

An alloy which bears a resemblance to Britannia metal is Ashberry metal, for which there are two formulas.

	I	II
Copper	2	3
Tin	8	79
Antimony	14	15
Zinc	1	2
Nickel	2	1

BRONZES.

The composition of bronze must be effected immediately before the casting, for bronze cannot be kept in store ready prepared. In forming the alloy, the refractory compound, copper, is first melted separately, the other metals, tin, zinc, etc., previously heated, being then added; the whole is then stirred and the casting carried out without loss of time. The process of forming the alloy must be effected quickly, so that there may be no loss of zinc, tin, or lead through oxidation, and also no interruption to the flow of metal, as metal added after an interval of time will not combine perfectly with the metal already poured in. It is important, therefore, to ascertain the specific weights of the metals, for the heavier metal will naturally tend to sink to the bottom and the lighter to collect at the top. Only in this way, and by vigorous stirring, can the complete blending of the two metals be secured. In adding the zinc, great care must be taken that the latter sinks at once to the level of the copper, otherwise a considerable portion will be volatilized before reaching the copper. When the castings are made, they must be cooled as quickly as possible, for the components of bronze have a tendency to form separate alloys of various composition, thus producing the so-called tin spots. This is much more likely to occur with a slow than with a sudden cooling of the mass.

Annealing Bronze.—This process is more particularly employed in the preparation of alloys used in the manufacture of cymbals, gongs, bells, etc. The alloy is naturally brittle, and acquires the properties essential to the purpose for which it is intended only after casting. The instruments are plunged into cold water while red-hot, hammered, reheated, and slowly cooled, when they become soft and sonorous. The alloy of copper and tin has the peculiar property that, whereas steel becomes hard through cooling, this mixture, when cooled suddenly, becomes noticeably soft and more malleable. The alloy is heated to a dark-red heat, or, in the case of thin articles, to the melting point of lead, and then plunged in cold water. The alloy may be hammered without splitting or breaking.

Aluminum Bronze.—This is prepared by melting the finest copper in a crucible, and adding the aluminum. The copper is cooled thereby to the thickly fluid point, but at the moment of the combination of the two metals, so much heat is released that the alloy becomes white hot and thinly fluid. Aluminum bronze thus prepared is usually brittle, and acquires its best qualities only after having been remelted several times. It may be remarked that, in order to obtain a bronze of the best quality, only the very purest copper must be used; with an inferior quality of copper, all labor is wasted. Aluminum bronze is not affected by exposure to the air; and its beautiful color makes it very suitable for manufacturing various ornamental articles, including clock cases, door knobs, etc.

Aluminum bronze wire is almost as strong as good steel wire, and castings made from it are almost as hard as steely iron; its resistance to bending or sagging is great.

I.—A good formula is 90 to 95 per cent of aluminum and 5 to 10 per cent of copper, of golden color, which keeps well in the air, without soon becoming dull and changing color like pure copper and its

alloys with tin and zinc (bronze, brass, etc.). It can be cast excellently, can be filed well and turned, possesses an extraordinary hardness and firmness, and attains a high degree of polish; it is malleable and forgeable. On the latter quality are founded applications which were formerly never thought of, viz.: forged works of art for decorative purposes. An alloy of 95 parts aluminum and 5 parts copper is used here. The technical working of bronze is not materially different from that of iron. The metal, especially in a hot condition, is worked like iron on the anvil, with hammer and chisel, only that the temperature to be maintained in forging lies between dark and light cherry red. If the articles are not forged in one piece and the putting together of the separate parts becomes necessary, riveting or soldering has to be resorted to. Besides forging, aluminum bronze is well suited for embossing, which is not surprising considering the high percentage of copper. After finishing the pieces, the metal can be toned in manifold ways by treatment with acid.

II.—Copper, 89 to 98 per cent; aluminum and nickel, 1 to 2 per cent. Aluminum and nickel change in the opposite way, that is to say, in increasing the percentage of nickel the amount of aluminum is decreased by the equal quantity. It should be borne in mind that the best ratio is aluminum, 9.5 per cent; nickel, 1 to 1.5 per cent at most. In preparing the alloy a deoxidizing agent is added, viz., phosphorus to 0.5 per cent; magnesium to 1.5 per cent. The phosphorus should always be added in the form of phosphorous copper or phosphor aluminum of exactly determined percentage. It is first added to the copper, then the aluminum and the nickel, and finally the magnesium, the last named at the moment of liquidity, are admixed.

III.—A gold bronze, containing 3 to 5 per cent aluminum; specific gravity, 8.37 to 8.15. Handsome golden color. This alloy oxidizes less on heating than copper and iron, and is therefore especially adapted for locomotive fireboxes and spindles, etc.

IV.—A steel bronze containing on an average 8.5 per cent aluminum (including 1 per cent silicium); specific gravity, 7.7. Very ductile and tough, but slightly elastic; hence its use is excluded where, with large demands upon tension and pressure, no permanent change of form must ensue. This is changed by working, such as rolling, drawing, etc. Especially useful where infrangibility is desired, as in machinery, ordnance, etc. At high temperature this bronze loses its elasticity again.

V.—This contains 8.5 per cent aluminum and 1¼ to 2 per cent silicium. Its use is advisable in cases where the metal is to possess a good elasticity, even in the cast state, and to retain it after being worked in red heat.

VI.—An acid bronze, containing 10 per cent aluminum; specific gravity, 7.65. Especially serviceable to resist oxidation and the action of acids.

VII.—Diamond bronze, containing 10 per cent aluminum and 2 per cent silicium. Specific gravity, 7.3. Very hard; of great firmness, but brittle.

Art Bronzes. (See also Aluminum Bronzes and Japanese Bronzes under this title.)—I.—Copper, 84 parts; zinc, 11 parts; tin, 5 parts.

II.—Copper, 90 parts; zinc, 6 parts; tin, 2 parts; lead, 2 parts.

III.—Copper, 65 parts; zinc, 30 parts; tin, 5 parts.

IV.—Copper, 90 parts; tin, 5 parts; zinc, 4 parts; lead, 1 part.

V.—Copper, 85 parts; zinc, 10 parts; tin, 3 parts; lead, 2 parts.

VI.—Copper, 72 parts; zinc, 23 parts; tin, 3 parts; lead, 2 parts.

Statuary Bronze.—Many of the antique statues were made of genuine bronze, which has advantages for this purpose, but has been superseded in modern times by mixtures of metals containing, besides copper and tin—the constituents of real bronze—a quantity of zinc, the alloy thus formed being really an intermediate product between bronze and brass. The reason for the use of such mixtures lies partly in the comparative cheapness of their production as compared with genuine bronze, and partly in the purpose for which the metal is to be used. A thoroughly good statuary bronze must become thinly fluid in fusing, fill the molds out sharply, allow of being easily worked with the file, and must take on the beautiful green coating called patina, after being exposed to the air for a short time.

Genuine bronze, however strongly heated, does not become thin enough to fill out the molds well, and it is also difficult to obtain homogeneous castings from it. Brass alone is also too thickly fluid, and not hard enough for the required fine chiseling or chasing of the finished object. Alloys containing zinc and tin, in addition to copper, can be prepared in such a manner that they will become very thinly fluid, and will give fine castings which can easily be worked with the file and chisel. The best proportions seem to be from 10 to 18 per cent of zinc and from 2 to 4 per cent of tin. In point of hardness, statuary bronze holds an intermediate position between genuine bronze and brass, being harder and tougher than the latter, but not so much so as the former.

Since statuary bronze is used principally for artistic purposes, much depends upon the color. This can be varied from pale yellow to orange yellow by slightly varying the content of tin or zinc, which must, of course, still be kept between the limits given above. Too much tin makes the alloy brittle and difficult to chisel; with too much zinc, on the other hand, the warm tone of color is lost, and the bronze does not acquire a fine patina.

The best proportions for statuary bronze are very definitely known at the present day; yet it sometimes happens that large castings have not the right character. They are either defective in color, or they do not take on a fine patina, or they are difficult to chisel. These phenomena may be due to the use of impure metals—containing oxides, iron, lead, etc.—or to improper treatment of the alloy in melting. With the most careful work possible, there is a considerable loss in melting—3 per cent at the very least, and sometimes as much as 10. This is due to the large proportion of zinc, and it is evident that, in consequence of it, the nature of the alloy will be different from what might be expected from the quantities of metals used in its manufacture.

It has been remarked that slight variations in composition quickly change the color of the alloy. The following table gives a series of alloys of different colors, suitable for statuary bronze:

	Copper	Zinc	Tin	Color
I...	84.42	11.28	4.30	Reddish yellow
II...	84.00	11.00	5.00	Orange red
III...	83.05	13.03	3.92	Orange red
IV...	83.00	12.00	5.00	Orange red
V...	81.05	15.32	3.63	Orange yellow
VI...	81.00	15.00	4.00	Orange yellow
VII...	78.09	18.47	3.44	Orange yellow
VIII...	73.58	23.27	3.15	Orange yellow
IX...	73.00	23.00	4.00	Pale orange
X...	70.36	26.88	2.76	Pale yellow
XI...	70.00	27.00	3.00	Pale yellow
XII...	65.95	31.56	2.49	Pale yellow

Perhaps the most satisfactory bronze metal is the alloy used in France for more than a century. It contains 91.60 per cent of copper, 5.33 per cent of zinc, 1.70 per cent of tin, and 1.37 per cent of lead. Somewhat more zinc is taken for articles to be gilded.

Bismuth Bronze.—Copper, 52 parts; nickel, 30 parts; zinc, 12 parts; lead, 5 parts; bismuth, 1 part. For metallic mirrors, lamp reflectors, etc.

Gun Bronze.—See Phosphor Bronze under this title.

Japanese Bronzes.—The formulas given below contain a large percentage of lead, which greatly improves the patina. The ingredients and the ratio of their parts for several sorts of modern Japanese bronze follow:

I.—Copper, 81.62 per cent; tin, 4.61 per cent; lead, 10.21 per cent.

II.—Copper, 76.60 per cent; tin, 4.38 per cent; lead, 11.88 per cent; zinc, 6.53 per cent.

III.—Copper, 88.55 per cent; tin, 2.42 per cent; lead, 4.72 per cent; zinc, 3.20 per cent.

Sometimes a little antimony is added just before casting, and such a composition would be represented more nearly by this formula:

IV.—Copper, 68.25 per cent; tin, 5.47 per cent; zinc, 8.88 per cent; lead, 17.06 per cent; antimony, 0.34 per cent.

For imitation Japanese bronze, see Plating under Bronzing.

Machine Bronze.—I.—Copper, 89 per cent; tin, 11 per cent.

II.—Copper, 80 per cent; tin, 16 per cent.

Phosphor Bronze.—Phosphor bronze is bronze containing varying amounts of phosphorus, from a few hundredths of 1 per cent to 1 or 2 per cent. Bronze containing simply copper and tin is very liable to be defective from the presence of oxygen, sulphur, or occluded gases. Oxygen causes the metal to be spongy and weak. Sulphur and occluded gases cause porosity. Oxygen gets into the metal by absorption from the air. It can be eliminated by adding to the metal something which combines with the oxygen and then fluxes off. Such deoxidizers are zinc, antimony, aluminum, manganese, silicon, and phosphorus. Sulphur and occluded gases can be eliminated by melting the metal, exposing it to the air, and letting it thus absorb some oxygen, which then burns the sulphur and gas. The oxygen can then be removed by adding one of the above-mentioned deoxidizers. The important use of phosphorus in bronze is, therefore, to remove oxygen and also indirectly to destroy occluded gas and sulphur.

A bronze is sometimes made with an extra high percentage of phosphorus, namely, 6 per cent. This alloy is made so as to have phosphorus in convenient form for use, and the process of manufacture is as follows: Ninety pounds of copper are melted under charcoal in a No. 70 crucible, which holds about 200 pounds of metal when full; 11 pounds of tin are added and the metal is allowed to become hot. The crucible is then removed from the furnace and 7 pounds of phosphorus are introduced in the following manner: A 3-gallon stone jar, half full of dilute solution of blue vitriol, is weighed. Then the weights are increased 7 pounds, and phosphorus in sticks about 4 inches long is added till the scales balance again. The phosphorus is left in this solution half an hour or longer, the phosphorus being given a coating of copper, so that it may be dried and exposed to the air without igniting. Have ready a pan about 30 inches square and 6 inches deep, containing about 2 inches of water. Over the water is a wire netting, which is laid loose on ledges or supports along the inner sides of the pan. On the netting is blotting paper, and on this the phosphorus is laid to dry when taken out of the blue-vitriol solution. The pan also has a lid which can be put down in case of ignition of the phosphorus.

The phosphorus is now ready for introduction into the metal. This is done by means of a cup-shaped instrument called a retort or phosphorizer. One man holds the retort on the rim of the crucible in a horizontal position. A second man takes about three pieces of phosphorus and throws them into the retort. The first man then immediately plunges the mouth of the retort below the surface of the metal before the phosphorus has a chance to fall or flow out. Of course the phosphorus immediately melts and also begins to volatilize. As the phosphorus comes in contact with the metal, it combines with it. This process is continued till all the 7 pounds of phosphorus has been put into the metal. The metal is then poured into slabs about 3 inches by 4 inches by 1 inch thick. The metal is so hard that a greater thickness would make it difficult to break it up. When finished, the metal contains, by analysis, 6 per cent of phosphorus. When phosphorus is to be added to metal, a little of this hardener is employed.

Copper is a soft, ductile metal, with its melting point at about 2,000° F. Molten copper has the marked property of absorbing various gases. It is for this reason that it is so difficult to make sound castings of clear copper. Molten copper combines readily with the oxygen of the air, forming oxide of copper, which dissolves in the copper and mixes homogeneously with it.

A casting made from such metal would be very spongy. The bad effect of oxygen is intended to be overcome by adding zinc to the extent of 1 per cent or more. This result can be much more effectively attained by the use of aluminum, manganese, or phosphorus. The action of these substances is to combine with the oxygen, and as the product formed separates and goes to the surface, the metal is left in a sound condition. Aluminum and manganese deoxidize copper and bronze very effectively, and the oxide formed goes to the surface as a scum. When a casting is made from such metal, the oxide or scum, instead of freeing itself from the casting perfectly, generally remains in the top part of the casting mixed with the metal, as a fractured surface will show. Phosphorus deoxidizes copper, and the oxide formed leaves the metal in the form of a gas, so that a casting made from such metal shows a clean fracture throughout, although the metal is not so dense as when aluminum or manganese is used.

Copper also has the property of absorbing or occluding carbon monoxide. But the carbonic oxide thus absorbed is in a different condition from the oxygen absorbed. When oxygen is absorbed by copper, the oxygen combines chemically with the copper and loses its own identity as a gas. But when coal gas is absorbed by the copper, it keeps its own physical identity and simply exists in the copper in a state of solution. All natural waters, such as lake water, river water, spring water, etc., contain air in solution or occlusion. When such water is cooled and frozen, just at the time of changing from the liquid to the solid state, the dissolved gas separates and forms air bubbles, which remain entangled in the ice. The carbonic oxide which is dissolved or occluded in copper acts in exactly the same way.

Hydrogen acts in exactly the same manner as carbonic oxide. Sulphur also has a bad effect upon copper and bronze. Sulphur combines with copper and other metals, forming sulphide of copper, etc. When molten copper or bronze containing sulphur comes in contact with air it absorbs some oxygen, and this in turn combines with the sulphur present,

forming sulphur dioxide, which is a gas which remains occluded in the metal.

Tin is a soft, white metal, melting at 440° F. Toward gases it acts something like copper, but not in so marked a degree. Although copper and tin are both soft, yet when mixed they make a harder metal. When bronze cools from the molten state, the copper and the copper-tin alloy tend to crystallize by themselves. The quicker the cooling occurs the less separation will there be, and also the fracture will be more homogeneous in appearance.

Gun bronze contains copper and tin in the proportion of 9 or 10 parts of copper to 1 of tin. This is the metal used when an ordinary bronze casting is wanted. A harder bronze is copper and tin in the ratio of 6 to 1. This is often used as a bearing metal. When either of these metals is to be turned in the machine shop, they should contain about 3 per cent of lead, which will make them work very much better, but it also decreases their tensile strength. Bearing metal now generally contains about 10 per cent of lead, with copper and tin in varying ratios. The large percentage of lead is put in that the metal may wear away slower. Lead, although a metal having properties similar to tin, acts entirely different toward copper. Copper and tin have a good deal of affinity for each other, but copper and lead show no attraction at all for each other. Copper and tin mix in all proportions, but copper and lead mix only to a very limited extent. About 3 per cent of lead can be mixed with copper. With bronze about 15 per cent to 20 per cent of lead can be mixed. In bearing bronze the lead keeps its own physical properties, so that the constituent lead melts long before the metal attains a red heat. It sometimes happens when a bearing runs warm that the lead actually sweats out and forms pimples on the metal. Or, sometimes, in remelting a bearing bronze casting the lead may be seen to drop out while the metal is warming up. All of these metals, however, should contain something to flux or deoxidize them, such as zinc, manganese, aluminum, silicon, antimony, or phosphorus.

The phosphor bronze bearing metal in vogue has the following composition: Copper, 79.7 per cent; tin, 10 per cent; lead, 10 per cent; and phosphorus, 0.3 per cent.

Melt 140 pounds of copper in a No. 70 pot, covering with charcoal. When copper is all melted, add 17½ pounds of tin to 17½ pounds of lead, and allow the metal to become sufficiently warm, but not any hotter than is needed. Then add 10 pounds of "hardener" (made as previously described) and stir well. Remove from furnace, skim off the charcoal, cool the metal with gates to as low a temperature as is consistent with getting a good casting, stir well again, and pour. The molds for this kind of work are faced with plumbago.

There are several firms that make phosphor-bronze bearings with a composition similar to the above one, and most of them, or perhaps all, make it by melting the metals and then charging with phosphorus to the extent of 0.7 to 1 per cent. But some metal from all brands contains occluded gas. So that after such metal is cast (in about two minutes or so) the metal will ooze or sweat out through the gate, and such a casting will be found to be porous. But not one such experience with metal made as described above has yet been found.

This practical point should be heeded, viz., that pig phosphor bronze should be brought to the specifications that the metal should have shrunk in the ingot mold in cooling, as shown by the concave surface of the upper side, and that it should make a casting in a sand mold without rising in the gate after being poured.

In bearing metal, occluded gas is very objectionable, because the gas, in trying to free itself, shoves the very hard copper-tin compound (which has a low melting point and remains liquid after the copper has begun to set) into spots, and thus causes hard spots in the metal.

Phosphorus is very dangerous to handle, and there is great risk from fire with it, so that many would not care to handle the phosphorus itself. But phosphor copper containing 5 per cent of phosphorus, and phosphor tin containing 2 to 7 per cent of phosphorus, and several other such alloys can be obtained in the market. It may be suggested to those who wish to make phosphor bronze, but do not want to handle phosphorus itself, to make it by using the proper amounts of one of these high phosphorus alloys. In using phosphorus it is only necessary to use enough to thoroughly deoxidize the metal, say 0.3 per cent. More than this will make the metal harder, but not any sounder.

Phosphor bronze is not a special kind of alloy, but any bronze can be made into phosphor bronze; it is, in fact, simply a deoxidized bronze, produced under treatment with phosphorus compounds.

Although the effect of phosphorus in improving the quality of bronze has been known for more than fifty years, it is only of late that the mode for preparing phosphor bronze has been perfected. It is now manufactured in many localities. Besides its action in reducing the oxides dissolved in the alloy, the phosphorus exerts another very material influence upon the properties of the bronze. The ordinary bronzes consist of mixtures in which the copper is really the only crystallized constituent, since the tin crystallizes with great difficulty. As a consequence of this dissimilarity in the nature of the two metals, the alloy is not so solid as it would be if both were crystallized. The phosphorus causes the tin to crystallize, and the result is a more homogeneous mixture of the two metals.

If enough phosphorus is added, so that its presence can be detected in the finished bronze, the latter may be considered an alloy of crystallized phosphor tin with copper. If the content of phosphor is still more increased, a part of the copper combines with the phosphorus, and the bronze then contains, besides copper and tin, compounds of crystallized copper phosphide with phosphide of tin. The strength and tenacity of the bronze are not lessened by a larger amount of phosphorus, and its hardness is considerably increased. Most phosphor bronzes are equal in this respect to the best steel, and some even surpass it in general properties.

The phosphorus is added to the bronze in the form of copper phosphide or phosphide of tin, the two being sometimes used together. They must be specially prepared for this purpose, and the best methods will be here given. Copper phosphide is prepared by heating a mixture of 4 parts of superphosphate of lime, 2 parts of granulated copper, and 1 part of finely pulverized coal in a crucible at a temperature not too high. The melted copper phosphide, containing 14 per cent of phosphorus, separates on the bottom of the crucible.

Tin phosphide is prepared as follows: Place a bar of zinc in an aqueous solution of tin chloride. The tin will be separated in the form of a sponge-like mass. Collect it, and put it into a crucible, upon the bottom of which sticks of phosphorus have been placed. Press the tin tightly into the crucible, and expose to a gentle heat. Continue the heating until flames of burning phosphorus are no longer observed on the crucible. The pure tin phosphide, in the form of a coarsely crystalline mass, tin-white in color, will be found on the bottom of the crucible.

To prepare the phosphor bronze, the

alloy to be treated is melted in the usual way, and small pieces of the copper phosphide and tin phosphide are added.

Phosphor bronze, properly prepared, has nearly the same melting point as that of ordinary bronze. In cooling, however, it has the peculiarity of passing directly from the liquid to the solid state, without first becoming thickly fluid. In a melted state it retains a perfectly bright surface, while ordinary bronze in this condition is always covered with a thin film of oxide.

If phosphor bronze is kept for a long time at the melting point, there is not any loss of tin, but the amount of phosphorus is slightly diminished.

The most valuable properties of phosphor bronze are its extraordinary tenacity and strength. It can be rolled, hammered, and stretched cold, and its strength is nearly double that of the best ordinary bronze. It is principally used in cases where great strength and power of resistance to outward influences are required, as, for instance, in objects which are to be exposed to the action of sea water.

Phosphor bronze containing about 4 per cent of tin is excellently well adapted for sheet bronze. With not more than 5 per cent of tin, it can be used, forged, for firearms. Seven to 10 per cent of tin gives the greatest hardness, and such bronze is especially suited to the manufacture of axle bearings, cylinders for steam fire engines, cogwheels, and, in general, for parts of machines where great strength and hardness are required. Phosphor bronze, if exposed to the air, soon becomes covered with a beautiful, closely adhering patina, and is therefore well adapted to purposes of art. The amount of phosphorus added varies from 0.25 to 2.5 per cent, according to the purpose of the bronze. The composition of a number of kinds of phosphor bronze is given below:

	Copper	Tin	Zinc	Lead	Iron	Phosphorus
I.	85.55	9.85	3.77	0.62	trs.	0.05
II.	.....	4–15	.....	4–15	....	0.5–3
III.	....	4–15	8–20	4–15	....	.25–2
IV.	77.85	11.00	7.65	....	....	....
V.	72.50	8.00	17.00	....	....	....
VI.	73.50	6.00	19.00	....	....	....
VII.	74.50	11.00	11.00	....	....	....
VIII.	83.50	8.00	3.00	....	....	....
IX.	90.34	8.90	....	....	....	0.76
X.	90.86	8.56	....	....	....	0.196
XI.	94.71	4.39	....	....	....	0.053

I for axle bearings, II and III for harder and softer axle bearings, IV to VIII for railroad purposes, IV especially for valves of locomotives, V and VI axle bearings for wagons, VII for connecting rods, VIII for piston rods in hydraulic presses.

Steel Bronze.—Copper, 60; ferromanganese (containing 70 to 80 per cent manganese), 40; zinc, 15.

Silicon Bronze.—Silicon, similarly to phosphorus, acts as a deoxidizing agent, and the bronzes produced under its influence are very ductile and elastic, do not rust, and are very strong. On account of these qualities silicon bronze is much used for telegraph and telephone wires. The process of manufacture is similar to that of phosphor bronze; the silicon is used in the form of copper silicide. Some good silicon bronzes are as follows:

	I	II
Copper	97.12	97.37
Tin	1.14	1.32
Zinc	1.10	1.27
Silicon	0.05	0.07

Sun Bronze.—The alloy called sun bronze contains 10 parts of aluminum, 30 to 50 parts of copper, and 40 to 60 parts of cobalt. The mixture known by the name of metalline has 25 per cent of aluminum, 30 of copper, 10 of iron, and 35 of cobalt. These alloys melt at a point approaching the melting point of copper, are tenacious, ductile, and very hard.

Tobin Bronze.—This alloy is nearly similar in composition and properties to Delta metal.

	I	II	III	IV
Copper	61.203	59.00	61.20	82.67
Zinc	27.440	38.40	37.14	3.23
Tin	0.906	2.16	0.90	12.40
Iron	0.180	0.11	0.18	0.10
Lead	0.359	0.31	0.35	2.14
Silver	....	....	....	0.07
Phosphorus	....	....	....	0.005

The alloy marked IV is sometimes called deoxidized bronze.

Violet-colored bronze is 50 parts copper and 50 parts antimony.

CADMIUM ALLOYS:

See also Fusible Alloys.

Lipowitz's Alloy.—I.—This alloy is composed of cadmium, 3 parts; tin, 4; bismuth, 15; and lead, 8. The simplest method of preparation is to heat the metals, in small pieces, in a crucible, stirring constantly, as soon as fusion begins, with a stick of hard wood. The stirring is important, in order to prevent the metals, whose specific gravity varies considerably, from being deposited in layers. The alloy softens at 140° F. and melts completely at 158° F. The color is silvery white, with a luster like polished silver, and the metal can be bent, hammered, and turned. These properties would make it valuable for many purposes where a beautiful appearance is of special importance, but on account of the considerable amount of cadmium and bismuth which it contains, it is rather expensive, and therefore limited in use. Casts of small animals, insects, lizards, etc., have been prepared from it, which were equal in sharpness to the best galvanoplastic work. Plaster of Paris is poured over the animal to be cast, and after sharp drying, the animal is removed and the mold filled up with Lipowitz's metal. The mold is placed in a vessel of water, and by heating to the boiling point the metal is melted and deposited in the finest impressions of the mold.

This alloy is most excellent for soldering tin, lead, Britannia metal, and nickel, being especially adapted to the last two metals on account of its silver-white color. But here again its costliness prevents its general use, and cheaper alloys possessing the same properties have been sought. In cases where the silver-white color and the low melting point are not of the first importance, the alloys given below may very well be used in the place of it.

II.—Cadmium alloy (melting point, 170° F.): Cadmium, 2 parts; tin, 3; lead, 11; bismuth, 16.

III.—Cadmium alloy (melting point, 167° F.): Cadmium, 10 parts; tin, 3; lead, 8; bismuth, 8.

Cadmium alloys (melting point, 203° F.):

	IV	V	VI
Cadmium	1	1	1 parts
Tin	2	3	1 "
Bismuth	3	5	2 "

VII.—A very fusible alloy, melting at 150° F., is composed of tin, 1 or 2 parts; lead, 2 or 3; bismuth, 4 or 15; cadmium, 1 or 2.

VIII.—Wood's alloy melts between 140° and 161.5° F. It is composed of lead, 4 parts; tin, 2; bismuth, 5 to 8; cadmium, 1 to 2. In color it resembles platinum, and is malleable to a certain extent.

IX.—Cadmium alloy (melting point, 179.5° F.): Cadmium, 1 part; lead, 6

parts; bismuth, 7. This, like the preceding, can be used for soldering in hot water.

X.—Cadmium alloy (melting point, 300° F.): Cadmium, 2 parts; tin, 4; lead, 2. This is an excellent soft solder, with a melting point about 86 degrees below that of lead and tin alone.

Cadmium Alloys with Gold, Silver, and Copper.—I.—Gold, 750 parts; silver, 166 parts; cadmium, 84 parts. A malleable and ductile alloy of green color.

II.—Gold, 750 parts; silver, 125 parts; and cadmium, 125 parts. Malleable and ductile alloy of yellowish-green hue.

III. — Gold, 746 parts; silver, 114 parts; copper, 97 parts; and cadmium, 43 parts. Likewise a malleable and ductile alloy of a peculiar green shade. All these alloys are suitable for plating. As regards their production, each must be carefully melted together from its ingredients in a covered crucible lined with coal dust, or in a graphite crucible. Next, the alloy has to be remelted in a graphite crucible with charcoal (or rosin powder) and borax. If, in spite thereof, a considerable portion of the cadmium should have evaporated, the alloy must be re-fused once more with an addition of cadmium.

ALLOYS FOR CASTING COINS, MEDALLIONS, ETC.

Alloys which fulfill the requirements of the medalist, and capable, therefore, of reproducing all details, are the following:

	I	II
Tin	3	6 parts
Lead	13	8 "
Bismuth	6	14 "

III.—A soft alloy suitable to take impressions of woodcuts, coins, metals, engravings, etc., and which must melt at a low degree of heat, is made out of bismuth, 3 parts; tin, 1½ parts; lead, 2½ parts; and worn-out type, 1 part.

Acid-proof Alloy.—This alloy is characterized by its power of resisting the action of acids, and is therefore especially adapted to making cocks, pipes, etc., which are to come in contact with acid fluids. It is composed of copper, zinc, lead, tin, iron, nickel, cobalt, and antimony, in the following proportions:

Copper	74.75 parts
Zinc	0.61 "
Lead	16.35 "
Tin	0.91 "
Iron	0.43 "
Nickel } Cobalt }	0.24 "
Antimony	6.78 "

Albata Metal.—Copper, 40 parts; zinc, 32 parts; and nickel, 8 parts.

Alfenide Metal.—Copper, 60 parts; zinc, 30; nickel, 10; traces of iron.

Bath Metal.—This alloy is used especially in England for the manufacture of teapots, and is very popular owing to the fine white color it possesses. It takes a high polish, and articles made from this alloy acquire in the course of time, upon only being rubbed with a white cloth, a permanent silver luster. The composition of Bath metal is copper, 55 parts; zinc, 45 parts.

Baudoin Metal.—This is composed of 72 parts of copper, 16.6 of nickel, 1.8 of cobalt, 1 of zinc; ½ per cent of aluminum may be added.

CASTING COPPER:

Macht's Yellow Metal.—I.—This alloy consists of 33 parts of copper and 25 of zinc. It has a dark golden-yellow color, great tenacity, and can be forged at a red heat, properties which make it especially suitable for fine castings.
II.—Yellow.—Copper, 67 to 70 parts; zinc, 33 to 30 parts.
III.—Red.—Copper, 82 parts; zinc, 18 parts.

Copper Arsenic.—Arsenic imparts to copper a very fine white color, and makes it very hard and brittle. Before German silver was known, these alloys were sometimes used for the manufacture of such cast articles as were not to come in contact with iron. When exposed to the air, they soon lose their whiteness and take on a brownish shade. On account of this, as well as the poisonous character of the arsenic, they are very little used at the present time. Alloys of copper and arsenic are best prepared by pressing firmly into a crucible a mixture of 70 parts of copper and 30 of arsenic (the copper to be used in the form of fine shavings) and fusing this mixture in a furnace with a good draught, under a cover of glass.

Copper Iron.—The alloys of copper and iron are little used in the industries of the present day, but it would seem that in earlier times they were frequently prepared for the purpose of giving a considerable degree of hardness to copper; for in antique casts, consisting principally of copper, we regularly find large quantities of iron, which leads to the supposition that they were added intentionally.
These alloys, when of a certain composition, have considerable strength and hardness. With an increase in the quantity of the iron the hardness increases, but the solidity is lessened. A copper and iron alloy of considerable strength, and at the same time very hard, is made of copper, 66 parts; iron, 34. These alloys acquire, on exposure to air, an ugly color inclining toward black, and are therefore not adapted for articles of art.

Copper Nickel.—A. Morrell, of New York, has obtained a patent on a nickel-copper alloy which he claims is valuable on account of its noncorrosive qualities, therefore making it desirable for ships, boiler tubes, and other uses where the metal comes much in contact with water. The process of making the metal is by smelting ore containing sulphide of nickel and copper, and besemerizing the resultant matter. This is calcined in order to obtain the nickel and copper in the form of oxides. The latter are reduced in reverberating furnace with carbon, or the like, so as to produce an alloy which preferably contains 2 parts of nickel and 1 part of copper.

Delta Metal.—An alloy widely used for making parts of machinery, and also for artistic purposes, is the so-called Delta metal. This is a variety of brass hardened with iron; some manufacturers add small quantities of tin and lead; also, in some cases, nickel. The following analysis of Delta metal (from the factory at Düsseldorf) will show its usual composition:

	I	II	III	IV	V
Copper....	55.94	55.80	55.82	54.22	58.65
Zinc.......	41.61	40.07	41.41	42.25	38.95
Lead......	0.72	1.82	0.76	1.10	0.67
Iron.......	0.87	1.28	0.86	0.99	1.62
Manganese	0.81	0.96	1.38	1.09	....
Nickel.....	traces.	traces.	0.06	0.16	0.11
Phosphorus	0.013	0.011	traces.	0.02	....

I is cast, II hammered, III rolled, and IV hot-stamped metal. Delta metal is produced by heating zinc very strongly in crucibles (to about 1600° F.), and adding ferromanganese or "spiegeleisen," producing an alloy of 95 per cent zinc and 5 per cent of iron. Copper and brass and a very small amount of copper phosphate are also added.

Gong Metal.—A sonorous metal for cymbals, gongs, and tam-tams consists of 100 parts of copper with 25 parts tin. Ignite the piece after it is cast and plunge it into cold water immediately.

Production of Minargent.—This alloy consists of copper, 500 parts; nickel, 350; tungsten, 25, and aluminum, 5. The metal obtained possesses a handsome white color and greatly resembles silver.

Minofor.—The so-called Minofor metal is composed of copper, tin, antimony, zinc, and iron in the following proportions:

	I	II
Copper	3.26	4
Tin	67.53	66
Antimony	17.00	20
Zinc	8.94	9
Iron	....	1

Minargent and Minofor are sometimes used in England for purposes in which the ordinary Britannia metal, 2 parts tin and 1 part antimony, might equally well be employed; the latter surpasses both of them in beauty of color, but they are, on the other hand, harder.

Retz Alloy.—This alloy, which resists the corrosive action of alkalies and acids, is composed of 15 parts of copper, 2.34 of tin, 1.82 of lead, and 1 of antimony. It can be utilized in the manufacture of receivers, for which porcelain and ebonite are usually employed.

Ruoltz Metal.—This comprises 20 parts of silver, 50 of copper, 30 of nickel. These proportions may, however, vary.

Tissier's Metal.—This alloy contains arsenic, is of a beautiful tombac red color, and very hard. Its composition varies a great deal, but the peculiar alloy which gives the name is composed of copper, 97 parts; zinc, 2 parts; arsenic, 1 or 2. It may be considered a brass with a very high percentage of copper, and hardened by the addition of arsenic. It is sometimes used for axle bearings, but other alloys are equally suitable for this purpose, and are to be preferred on account of the absence of arsenic, which is always dangerous.

FILE ALLOYS.—Many copper-tin alloys are employed for the making of files which, in distinction from the steel files, are designated composition files. Such alloys have the following compositions:

Geneva Composition Files.—

	I	II
Copper	64.4	62
Tin	18.0	20
Zinc	10.0	10
Lead	7.6	8

Vogel's Composition Files.—

	III	IV	V
Copper	57.0	61.5	73.0
Tin	28.5	31.0	19.0
Zinc	78.0	....	8.0
Lead	7.0	8.5	8.0

VI.—Another alloy for composition files is copper, 8 parts; tin, 2; zinc, 1, and lead, 1—fused under a cover of borax.

EASILY FUSIBLE OR PLASTIC ALLOYS.

(These have a fusing point usually below 300° F.)

(See also Solders.)

I. Rose's Alloy. — Bismuth, 2 parts; lead, 1 part; tin, 1 part. Melting point, 200° F.

II. Darcet Alloy.—This is composed of 8 parts of bismuth, 5 of lead, and 3 of tin. It melts at 176° F. To impart greater fusibility, $\frac{1}{16}$ part of mercury is added; the fusing is then lowered to 149° F.

III.—Newton alloy melts at 212° F., and is composed of 5 parts of bismuth, 2 of lead, and 3 of tin.

IV.—Wood's Metal.—

Tin	2 parts
Lead	4 parts
Bismuth	5 to 8 parts

This silvery, fine-grained alloy fuses between 151° and 162° F., and is excellently adapted to soldering.

V.—Bismuth, 7 parts; lead, 6 parts; cadmium, 1 part. Melting point, 180° F.

VI.—Bismuth, 7 to 8 parts; lead, 4; tin, 2; cadmium, 1 to 2. Melting point, 149° to 160° F.

Other easily fusible alloys:

	VII	VIII	IX
Lead	1	2	3
Tin	1	2	3
Bismuth	1	1	1
Melting Point	258° F.	283°	311°

Fusible Alloys for Electric Installations.—These alloys are employed in electric installations as current interrupters. Serving as conductors on a short length of circuit, they melt as soon as the current becomes too strong. Following is the composition of some of these alloys.

	Fusing temperature	Lead	Tin	Bismuth	Cadmium
I	203° F.	250	500	500	...
II	193° F.	397	...	532	71
III	168° F.	344	94	500	62
IV	153° F.	260	148	522	70
V	150° F.	249	142	501	108
VI	145° F.	267	136	500	100

These alloys are prepared by melting the lead in a stearine bath and adding successively, and during the cooling, first, the cadmium; second, the bismuth; third, the tin. It is absolutely necessary to proceed in this manner, since these metals fuse at temperatures ranging from 850° F. (for lead), to 551° F. (for tin).

Fusible Safety Alloys for Steam Boilers.—

	Bismuth	Lead	Zinc	Melting point	Atmos. pressure
I....	8	5	3	212° F.	1
II....	8	8	4	235° F.	1.5
III...	8	8	3	253° F.	2
IV....	8	10	8	266° F.	2.5
V....	8	12	8	270° F.	3
VI....	8	16	14	280° F.	3.5
VII...	8	16	12	285° F.	4
VIII..	8	22	24	309° F.	5
IX....	8	32	36	320° F.	6
X....	8	32	28	330° F.	7
XI....	8	30	24	340° F.	8

Lipowitz Metal.—This amalgam is prepared as follows: Melt in a dish, cadmium, 3 parts, by weight; tin, 4 parts; bismuth, 15 parts; and lead, 8 parts, adding to the alloy, while still in fusion, 2 parts of quicksilver previously heated to about 212° F. The amalgamation proceeds easily and smoothly. The liquid mass in the dish, which should be taken from the fire immediately upon the introduction of the mercury, is stirred until the contents solidify. While Lipowitz alloy softens already at 140° F. and fuses perfectly at 158°, the amalgam has a still lower fusing point, which lies around 143⅗° F.

This amalgam is excellently adapted for the production of impressions of various objects of nature, direct impressions of leaves, and other delicate parts of plants having been made with its aid which, in point of sharpness, are equal to the best plaster casts and have a very pleasing appearance. The amalgam has a silver-white color and a fine gloss. It is perfectly constant to atmospheric influences. This amalgam has also been used with good success for the making of small statuettes and busts, which are hollow and can be readily gilt or bronzed by electro-deposition. The production of small statues is successfully carried out by making a hollow gypsum mold of the articles to be cast and heating the mold evenly to about 140° F. A corresponding quantity of the molten amalgam is then poured in and the mold moved rapidly to and fro, so that the alloy is thrown against the sides all over. The shaking should be continued until it is certain that the amalgam has solidified. When the mold has cooled off it is taken apart and the seams removed by means of a sharp knife. If the operation is carried on correctly, a chasing of the cast mass becomes unnecessary, since the alloy fills out the finest depressions of the mold with the greatest sharpness.

Amalgam for Plaster.—Tin, 1 part; bismuth, 1 part; mercury, 1 part. Melt the bismuth and the tin together, and when the two metals are in fusion add the mercury while stirring. For use, rub up the amalgam with a little white of egg and brush like a varnish on the plaster articles.

Plastic Metal Composition.—I. Copper oxide is reduced by means of hydrogen or copper sulphate by boiling a solution of the same in water with some zinc filings in order to obtain entirely pure copper. Of the copper powder obtained in this manner, 20, 30, or 36 parts, by weight, according to the degree of hardness desired for the composition (the greater the quantity of copper used the harder will the composition become), are thoroughly moistened in a cast-iron or porcelain mortar with sulphuric acid of 1.85 specific gravity; 70 parts, by weight, of mercury are then added to this paste, the whole being constantly stirred. When all the copper has been thoroughly amalgamated with the mercury, the sulphuric acid is washed out again with boiling water, and in 12 hours after it has become cold the composition will be so hard that it can be polished. It is impervious to the action of dilute acids, alcohol, ether, and boiling water. It contains the same specific gravity, alike in the soft or the hard condition. When used as a cement, it can at any time be rendered soft and plastic in the following manner: If applied while hot and plastic to the deoxidized surfaces of two pieces of metal, these latter will unite so firmly that in about 10 or 12 hours the metal may be subjected to any mechanical process. The properties of this composition render it very useful for various purposes, and it forms a most effective cement for fine metal articles which cannot be soldered in fire.

II.—Bismuth, 5.5 parts; lead, 3; tin, 1.5.

III. Alloy d'Homburg. — Bismuth,

3 parts; lead, 3; tin, 3. This alloy is fusible at 251° F., and is of a silvery white. It is employed for reproductions of medals.

IV. Alloy Valentine Rose.—Bismuth, 4 to 6 parts; lead, 2 parts; tin, 2 to 3 parts. This alloy fuses at 212° to 250° F.

V. Alloy Rose père. — Bismuth, 2 parts; lead, 2; tin, 2. This alloy fuses at 199° F.

The remainder are plastic alloys for reproducing cuts, medals, coins, etc.:

VI.—Bismuth, 4 parts; lead, 2 parts; tin, 1 part.

VII.—Bismuth, 3 parts; lead, 3 parts; tin, 2 parts.

VIII.—Bismuth, 4 parts; lead, 2 parts; tin, 2 parts.

IX.—Bismuth, 5 parts; lead, 2 parts; tin, 3 parts.

X.—Bismuth, 2 parts; lead, 2 parts; tin, 2 parts.

Quick - Water. — That the amalgam may easily take hold of bronze objects and remain there, it is customary to cover the perfectly cleansed and shining article with a thin coat of mercury, which is usually accomplished by dipping it into a so-called quick-water bath.

In the form of minute globules the mercury immediately separates itself from the solution and clings to the bronze object, which thereupon presents the appearance of being plated with silver. After it has been well rinsed in clean water, the amalgam may be evenly and without difficulty applied with the scratch brush.

This quick-water (in reality a solution of mercurous nitrate), is made in the simplest manner by taking 10 parts of mercury and pouring over it 11 parts of nitric acid of a specific gravity equal to 1.33; now let it stand until every part of the mercury is dissolved; then, while stirring vigorously, add 540 parts of water. This solution must be kept in closed flasks or bottles to prevent impurities, such as dust, etc., from falling into it.

The preparatory work on the object to be gilded consists mainly in cleansing it from every trace of oxidation. First, it must be well annealed by placing it in a bed of glowing coal, care being exercised that the heating be uniform. When cooled, this piece is plunged into a highly diluted sulphuric-acid bath in order to dissolve in a measure the oxide. Next it is dipped in a 36° nitric-acid bath, of a specific gravity equal to 1.33, and brushed off with a long brush; it is now dipped into nitric acid into which a little

lampblack and table salt have been thrown. It is now ready for washing in clean water and drying in unsoiled sawdust. It is of the greatest importance that the surface to be gilded should appear of a pale yellow tint all over. If it be too smooth the gold will not take hold easily, and if it be too dull it will require too much gold to cover it.

GOLD ALLOYS:

Colored Gold Alloys.—The alloys of gold with copper have a reddish tinge; those of gold with silver are whiter, and an alloy of gold, silver, and copper together is distinguished by a greenish tone. Manufacturers of gold ware make use of these different colors, one piece being frequently composed of several pieces of varying color. Below are given some of these alloys, with their colors:

	Gold	Silver	Copper	Steel	Cadmium
I..	2.6	1.0	....	....	....
II..	75.0	16.6	....	....	8.4
III..	74.6	11.4	9.7	....	4.3
IV..	75.0	12.6	....	....	12.5
V..	1.0	2.0	....	....	....
VI..	4.0	3.0	1.0	....	....
VII..	14.7	7.0	6.0	....	....
VIII..	14.7	9.0	4.0	....	....
IX..	3.0	1.0	1.0	....	....
X..	10.0	1.0	4.0	....	....
XI..	1.0	....	1.0	....	....
XII..	1.0	....	2.0	....	....
XIII..	30.0	3.0	....	2.0	....
XIV..	4.0	....	....	1.0	....
XV..	29.0	11.0	....	....	....
XVI..	1.3	....	....	1.0	....

Nos. I, II, III, and IV are green gold; No. V is pale yellow; Nos. VI, VII, and VIII bright yellow; Nos. IX and X pale red; Nos. XI and XII bright red; Nos. XIII, XIV, and XV gray; while No. XVI exhibits a bluish tint. The finished gold ware, before being put upon the market, is subjected to a special treatment, consisting either in the simple pickling or in the so-called coloring, which operation is conducted especially with alloys of low degree of fineness, the object being to give the layers a superficial layer of pure gold.

The presence of silver considerably modifies the color of gold, and the jeweler makes use of this property to obtain alloys of various shades. The following proportions are to be observed, viz.:

Color of Gold	Gold per 1,000	Silver per 1,000	Copper per 1,000
I. Green.........	750	250	...
II. Dead leaves....	700	300	...
III. Sea green......	600	400	...
IV. Pink..........	750	200	50
V. English yellow..	750	125	125
VI. English white...	750	150	100
VII. Whiter........	750	170	80
VIII. Less white......	750	190	60
IX. Red...........	750	...	250

Other colored gold alloys are the following:

X. Blue.—Fine gold, 75; iron, 25.

XI. Dark Gray. — Fine gold, 94; iron, 6.

XII. Pale Gray. — Fine gold, 191; iron, 9.

XIII. Cassel Yellow. — Fine gold, 75; fine silver, 12½; rose copper, 12½.

The above figures are understood to be by weight.

The gold solders, known in France under the names of *soudures au quart* (13½ carat), *au tiers* (12 carat), and *au deux* (9 carat), are composed of 3, 2, or 1 part of gold respectively, with 1 part of an alloy consisting of two-thirds silver and one-third copper. Gold also forms with aluminum a series of alloys of greatly varying coloration, the most curious of them, composed of 22 parts of aluminum for 88 parts of gold, possessing a pretty purple shade. But all these alloys, of a highly crystalline base, are very brittle and cannot be worked, for which reason their handsome colorings have not yet been capable of being utilized.

Enameling Alloys.— I. Transparent. —This alloy should possess the property of transmitting rays of light so as to give the highest possible effect to the enamel. The alloy of gold for transparent green should be pale; a red or copper alloy does not do for green enamel, the copper has a tendency to darken the color and thus take away a part of its brilliancy. The following alloy for transparent green possesses about the nearest print, in color, to the enamel—which should represent, as near as possible, the color and brilliancy of the emerald—that can be arrived at:

	ozs.	dwts.	grs.
Fine gold........	0	18	8
Fine silver.......	0	1	6
Fine copper......	0	0	10

No borax must be used in the melting of this alloy, it being of a more fusible nature than the ordinary alloy, and will not take so high a heat in enameling.

II. Red Enamel.—The enamel which forms this color being of a higher fusing point, if proper care be not taken, the gold will melt first, and the work become ruined. In the preparation of red enamel, the coloring matter is usually an oxide of gold, and this so raises the temperature at which it melts that, in order to prevent any mishap, the gold to be enameled on should be what is called a 22-carat red, that is, it should contain a preponderance of copper in the alloying mixture so as to raise the fusing point of the gold. The formula is:

	ozs.	dwts.	grs.
Fine gold........	0	18	8
Fine silver.......	0	0	10
Fine copper......	0	1	6

Gold-leaf Alloys.—All gold made into leaf is more or less alloyed. The gold used by the goldbeater is alloyed according to the variety of color required. Fine gold is commonly supposed to be incapable of being reduced to thin leaves. This, however, is not the case, although its use for ordinary purposes is undesirable on account of its greater cost. It also adheres by contact of one leaf with another, thus causing spoiled material and wasted labor; but for work exposed to the weather it is much preferable, as it is more durable and does not tarnish or change color.

The following is a list of the principal classes of leaf recognized and ordinarily prepared by beaters with the proportion of alloy they contain:

	Gold grs.	Silver grs.	Copper grs.
I. Red gold...	456–460	...	20–24
II. Pale red...	464	...	16
III. Extra deep.	456	12	12
IV. Deep......	444	24	12
V. Citron.....	440	30	10
VI. Yellow....	408	72	....
VII. Pale yellow	384	96	....
VIII. Lemon....	360	120	
IX. Green or pale	312	168	....
X. White.....	240	240	....

Gold-Plate Alloys.— Gold, 92 parts; copper, 8 parts.

II.—Gold, 84 parts; copper, 16 parts.

III.—Gold, 75 parts; copper, 25 parts.

IMITATION GOLD.

I.—One hundred parts, by weight, of copper of the purest quality; 14 of zinc or tin; 6 of magnesia; ¾ of sal ammoniac, limestone, and cream of tartar. The copper is first melted, then the magnesia, sal ammoniac, limestone, and cream of tartar in powder are added separately and gradually. The whole mass is kept stirred for a half hour, the zinc or tin being dropped in piece by piece, the stirring being kept up till they melt. Finally the crucible is covered and the mass is kept in fusion 35 minutes and, the same being removed, the metal is poured into molds, and is then ready for use. The alloy thus made is said to be fine-grained, malleable, takes a high polish, and does not easily oxidize.

II.—An invention, patented in Germany, covers a metallic alloy, to take the place of gold, which, even if exposed for some time to the action of ammoniacal and acid vapors, does not oxidize or lose its gold color. It can be rolled and worked like gold and has the appearance of genuine gold without containing the slightest admixture of that metal. The alloy consists of copper and antimony in the approximate ratio of 100 to 6, and is produced by adding to molten copper, as soon as it has reached a certain degree of heat, the said percentage of antimony. When the antimony has likewise melted and entered into intimate union with the copper, some charcoal ashes, magnesium, and lime spar are added to the mass when the latter is still in the crucible.

III. **Aluminum Gold.** — This alloy, called Nuremberg gold, is used for making cheap gold ware, and is excellent for this purpose, as its color is exactly that of pure gold, and does not change in the air. Articles made of Nuremberg gold need no gilding, and retain their color under the hardest usage; even the fracture of this alloy shows the pure gold color. The composition is usually 90 parts of copper, 2.5 of gold, and 7.5 of aluminum.

IV.—Imitation gold, capable of being worked and drawn into wire, consists of 950 parts copper, 45 aluminum, and 2 to 5 of silver.

V.—Chrysochalk is similar in composition to Mannheim gold:

	I	II
Copper	90.5	58.68
Zinc	7.9	40.22
Lead	1.6	1.90

In color it resembles gold, but quickly loses its beauty if exposed to the air, on account of the oxidation of the copper. It can, however, be kept bright for a long time by a coating of colorless varnish, which excludes the air and prevents oxidation. Chrysochalk is used for most of the ordinary imitations of gold. Cheap watch chains and jewelry are manufactured from it, and it is widely used by the manufacturers of imitation bronze ornaments.

Mannheim Gold or Similor.—Mannheim gold is composed of copper, zinc, and tin, in proportions about as follows:

	I	II
Copper	83.7	89.8
Zinc	9.3	9.9
Tin	7.0	0.6

It has a fine yellow color, and was formerly much used in making buttons and pressed articles resembling gold. Later alloys, however, surpass it in color, and it has fallen somewhat into disuse. One variety of Mannheim gold, so called, contains 1.40 parts of brass (composition 3 Cu_2 1 Zn) to 10 of copper and 0.1 of zinc.

Mosaic Gold.—This is an alloy composed—with slight deviations—of 100 parts of copper and 50 to 55 of zinc. It has a beautiful color, closely resembling that of gold, and is distinguished by a very fine grain, which makes it especially suitable for the manufacture of castings which are afterwards to be gilded. The best method of obtaining a thoroughly homogeneous mixture of the two metals is first to put into the crucible one-half of the zinc to be used, place the cover upon it, and fuse the mixture under a cover of borax at as low a temperature as possible. Have ready the other half of the zinc, cut into small pieces and heated almost to melting, and when the contents of the crucible are liquid throw it in, a small portion at a time, stirring constantly to effect as intimate a mixture of the metals as possible.

Oreïde or Oroïde (French Gold).—The so-called French gold, when polished, so closely resembles genuine gold in color that it can scarcely be distinguished from it. Besides its beautiful color, it has the valuable properties of being very ductile and tenacious, so that it can easily be stamped into any desired shape; it also takes a high polish. It is frequently used for the manufacture of spoons, forks, etc., but is unsuitable for this purpose on account of the large amount of copper contained in it, rendering it injurious to health. The directions for preparing this alloy vary greatly. The products of some Paris factories show the following composition:

	I	II	III
Copper	90	80.5	86.21
Zinc	10	14.5	31.52
Tin	..	...	0.48
Iron	..	...	0.24

A special receipt for oreïde is the following:

IV.—Melt 100 parts of copper and add, with constant stirring, 6 parts of magnesia, 3.6 of sal ammoniac, 1.8 of lime, and 9 of crude tartar. Stir again thoroughly, and add 17 parts of granulated zinc, and after mixing it with the copper by vigorous stirring keep the alloy liquid for one hour. Then carefully remove the scum and pour off the alloy.

Pinchbeck.—This was first manufactured in England. Its dark gold color is the best imitation of gold alloyed with copper. Being very ductile, it can easily be rolled out into thin plates, which can be given any desired shape by stamping. It does not readily oxidize, and thus fulfills all the requirements for making cheap jewelry, which is its principal use.

Copper	88.8	93.6
Zinc	11.2	6.4

Or

Copper	2.1	1.28
Zinc	....	0.7
Brass	1.0	0.7

Palladium Gold.—Alloys of gold, copper, silver, and palladium have a brownish-red color and are nearly as hard as iron. They are sometimes (although rarely) used for the bearings for the axles of the wheels of fine watches, as they invite little friction and do not rust in the air. The composition used in the Swiss and English watch factories consists usually of gold 18 parts, copper 13 parts, silver 11, and palladium 6.

Talmi Gold.—The name of talmi gold was first applied to articles of jewelry, chains, earrings, bracelets, etc., brought from Paris, and distinguished by beautiful workmanship, a low price, and great durability. Later, when this alloy had acquired a considerable reputation, articles were introduced under the same name, but which were really made of other metals, and which retained their beautiful gold color only as long as they were not used. The fine varieties of talmi gold are manufactured from brass, copper, or tombac, covered with a thin plate of gold, combined with the base by rolling, under strong pressure. The plates are then rolled out by passing through rollers, and the coating not only acquires considerable density, but adheres so closely to the base that the metal will keep its beautiful appearance for years. Of late, many articles of talmi gold have been introduced whose gold coating is produced by electroplating, and is in many cases so thin that hard rubbing will bring through the color of the base. Such articles, of course, are not durable. In genuine talmi gold, the coating, even though it may be thin, adheres very closely to the base, for the reason that the two metals are actually welded by the rolling, and also because alloyed gold is always used, which is much harder than pure gold. The pure gold of electroplating is very soft. The composition of some varieties of talmi gold are here given. It will be seen that the content of gold varies greatly, and the durability of the alloy will, of course, correspond to this. The alloys I, II, III are genuine Paris talmi gold; IV, V, VI are electroplated imitations; and VII is an alloy of a wrong composition, to which the gold does not adhere firmly:

	Copper	Zinc	Tin	Iron	Gold
I.	89.9	9.3	....	....	1.3
II.	90.8	8.3	....	....	0.9
III.	90.0	8.9	....	....	0.9
IV. {	90.7	89.0 }	....	....	0.5
{	88.2	11.4 }			
V. {	87.5	12.4 }	....	....	0.3
{	83.1	17.0 }			
VI. {	93.5	6.6 }	....	....	0.05
{	84.5	15.8 }			
VII.	86.0	12.0	1.1	0.3	...

Japanese Alloys.—In Japan some specialties in metallic alloys are in use of which the composition is as follows:

Shadke consists of copper with from 1 to 10 per cent of gold. Articles made from this alloy are laid in a pickle of blue vitriol, alum, and verdigris, until they acquire a bluish-black color.

Gui-shi-bu-ichi is an alloy of copper containing 30 to 50 per cent of silver. It possesses a peculiar gray shade.

Mokume consists of several compositions. Thus, about 30 gold foils (genuine) are welded together with shadke, copper, silver, and gui-shi-bu-ichi and pierced. The pierced holes are, after firmly hammering together the plates, filled up with the above-named pickle.

The finest Japanese brass consists of 10 parts copper and 8 parts zinc, and is called siachu. The bell metal kara kane is composed of copper 10 parts, tin 10 parts, iron 0.5 part, and zinc 1.5 parts. The copper is first fused, then the remaining metals are added in rotation.

GERMAN SILVER OR ARGENTAN.

The composition of this alloy varies considerably, but from the adjoined figures an average may be found, which will represent, approximately, the normal composition:

Copper	50 to 66 parts
Zinc	19 to 31 parts
Nickel	13 to 18 parts

The properties of the different kinds, such as their color, ductility, fusibility, etc., vary with the proportions of the single metals. For making spoons, forks, cups, candlesticks, etc., the most suitable proportions are 50 parts of copper, 25 of zinc, and 25 of nickel. This metal has a beautiful blue-white color, and does not tarnish easily.

German silver is sometimes so brittle that a spoon, if allowed to fall upon the floor, will break; this, of course, indicates faulty composition. But the following table will show how the character of the alloy changes with the varying percentage of the metals composing it:

	Copper	Zinc	Nickel	Quality
I.	8	3.5	4	Finest quality.
II.	8	3.5	6	Beautiful, but refractory.
III.	8	6.5	3	Ordinary, readily fusible.
IV.	52	26.0	22	First quality.
V.	59	30.0	11	Second quality.
VI.	63	31.0	6	Third quality.

The following analyses give further particulars in regard to different kinds of German silver:

For sheet	Copper	Zinc	Nickel	Lead	Iron
(French)	50.0	31.3	18.7	...	...
(French)	50.0	30.0	20.0	...	...
(French)	58.3	25.0	16.7	...	...
Vienna	50.0	25.0	25.0	...	...
Vienna	55.6	22.0	22.0	...	...
Vienna	60.0	20.0	20.0	...	...
Berlin	54.0	28.0	18.0	...	...
Berlin	55.5	29.1	17.5	...	...
English	63.34	17.01	19.13	...	...
English	62.40	22.15	15.05	...	...
English	62.63	26.05	10.85	...	...
English	57.40	25.	13.0	...	3.0
Chinese	26.3	36.8	36.8	...	...
Chinese	43.8	40.6	15.6	...	...
Chinese	45.7	36.9	17.9	...	...
Chinese	40.4	25.4	31.6	...	2.6
Castings	48.5	24.3	24.3	2.9	...
Castings	54.5	21.8	21.8	1.9	...
Castings	58.3	19.4	19.4	2.9	...
Castings	57.8	27.1	14.3	0.8	...
Castings	57.	20.0	20.0	3.0	...

In some kinds of German silver are found varying quantities of iron, manganese, tin, and very frequently lead, added for the purpose of changing the properties of the alloy or cheapening the cost of production. But all these metals have a detrimental rather than a beneficial effect upon the general character of the alloy, and especially lessen its power of resistance to the action of dilute acids, one of its most valuable properties. Lead makes it more fusible; tin acts somewhat as in bronze, making it denser and more resonant, and enabling it to take a higher polish. With iron or manganese the alloy is whiter, but it becomes at the same time more refractory and its tendency toward brittleness is increased.

SUBSTITUTES FOR GERMAN SILVER.

There are many formulas for alloys which claim to be substitutes for German silver; but no one of them has yet become an article of general commerce. It will be sufficient to note these materials briefly, giving the composition of the most important.

Nickel Bronze.—This is prepared by fusing together very highly purified nickel (99.5 per cent) with copper, tin, and zinc. A bronze is produced containing 20 per cent of nickel, light-colored and very hard.

Bismuth Bronze.—

	I	II	III	IV
Copper	25.0	45.0	69.0	47.0
Nickel	24.0	32.5	10.0	30.9
Antimony	50.0	...	...	...
Bismuth	1.0	1.0	1.0	0.1
Tin	...	16.0	15.0	1.0
Zinc	...	21.5	20.0	21.0
Aluminum	...	...	1.0	...

I is hard and very lustrous, suitable for lamp reflectors and axle bearings; II is hard, resonant, and not affected by sea water, for parts of ships, pipes, telegraph wires, and piano strings; III and IV are for cups, spoons, etc.

Manganese Argentan.—

Copper	52 to 50 parts
Nickel	17 to 15 "
Zinc	5 to 10 "
Manganese	1 to 5 "
Copper, with 15 per cent phosphorus.	3 to 5 "

Readily cast for objects of art.

Aphtite.—

Iron	66 parts
Nickel	23 "
Tungsten	4 "
Copper	5 "

Arguzoid.—

Copper	55.78 parts
Zinc	23.198 "
Nickel	13.406 "
Tin	4.035 "
Lead	3.544 "

Silver white, almost ductile, suited for artistic purposes.

Ferro-Argentan.—

Copper	70.0 parts
Nickel	20.0 "
Zinc	5.5 "
Cadmium	4.5 "

Resembles silver; worked like German silver.

Silver Bronze.—Manganese, 18 per cent; aluminum, 1.2 per cent; silicium, 5 per cent; zinc, 13 per cent; copper, 67.5 per cent. The electric resistance of silver bronze is greater than that of German silver, hence it ought to be highly suitable for rheostats.

Instrument Alloys. — The following are suitable for physical and optical instruments, metallic mirrors, telescopes, etc.:

I.—Copper, 62 parts; tin, 33 parts; lead, 5 parts.

II.—Copper, 80; antimony, 11; lead, 9.

III.—Copper, 10; tin, 10; antimony, 10; lead, 40.

IV.—Copper, 30; tin, 50; silver, 2; arsenic, 1.

V.—Copper, 66; tin, 33.

VI.—Copper, 64; tin, 26.

VII.—Steel, 90; nickel, 10.

VIII.—Platinum, 60; copper, 40.

IX.—Platinum, 45; steel, 55.

X.—Platinum, 55; iron, 45.

XI.—Platinum, 15; steel, 85.

XII.—Platinum, 20; copper, 79; arsenic, 1.

XIII.—Platinum, 62; iron, 28; gold, 10.

XIV.—Gold, 48; zinc, 52.

XV.—Steel, 50; rhodium, 50.

XVI.—Platinum, 12; iridium, 88.

XVII.—Copper, 89.5; tin, 8.5; zinc, 2.

LEAD ALLOYS.

The following alloys, principally lead, are used for various purposes:

Bibra Alloy.—This contains 8 parts of bismuth, 9 of tin, and 38 to 40 of lead.

Metallic Coffins.—Tin, 40 parts; lead, 45 parts; copper, 15 parts.

Plates for Engraving.—I.—Lead, 84 parts; antimony, 16 parts.

II.—Lead, 86 parts; antimony, 14 parts.

III.—Lead, 87 parts; antimony, 12 parts; copper, 1 part.

IV.—Lead, 81 parts; antimony, 14 parts; tin, 5 parts.

V.—Lead, 73 parts; antimony, 17 parts; zinc, 10 parts.

VI.—Tin, 53 parts; lead, 43 parts; antimony, 4 parts.

Hard lead is made of lead, 84 parts; antimony, 16 parts.

Sheet Metal Alloy.—

Tin	35 parts
Lead	250 parts
Copper	2.5 parts
Zinc	0.5 part

This alloy has a fine white color, and can be readily rolled into thin sheets. For that reason it is well adapted for lining tea chests and for the production of tobacco and chocolate wrappers. The copper and zinc are used in the form of fine shavings. The alloy should be immediately cast into thin plates, which can then be passed through rolls.

MAGNETIC ALLOYS.

Alloys which can be magnetized most strongly are composed of copper, manganese, and aluminum, the quantities of manganese and aluminum being proportional to their atomic weights (55.0 to 27.1, or about 2 to 1). The maximum magnetization increases rapidly with increase of manganese, but alloys containing much manganese are exceedingly brittle and cannot be wrought. The highest practicable proportion of manganese at present is 24 per cent.

These magnetic alloys were studied by Hensler, Haupt, and Starck, and Gumlich has recently examined them at the Physikalisch - technische Reichsanstalt, with very remarkable and interesting results.

The two alloys examined were composed as follows:

Alloy I.—Copper, 61.5 per cent; manganese, 23.5 per cent; aluminum, 15 per cent; lead, 0.1 per cent, with traces of iron and silicon.

Alloy II.—Copper, 67.7 per cent; manganese, 20.5 per cent; aluminum, 10.7 per cent; lead, 1.2 per cent, with traces of iron and silicon.

Alloy II could be worked without difficulty, but alloy I was so brittle that it broke under the hammer. A bar 7 inches long and ¼ inch thick was obtained by grinding. This broke in two during the measurements, but, fortunately, without invalidating them. Such a material is evidently unsuited to practical uses.

The behavior of magnetic alloys at high temperatures is very peculiar. Alloy I is indifferent to temperature changes, which scarcely affect its magnetic properties, but the behavior of alloy II is very different. Prolonged heating to 230° F. produces a great increase in its capability of magnetization, which, after 544 hours' heating, rises from 1.9 to 3.2 kilogauss, approaching the strength of alloy I. But when alloy II is heated to 329° F., its capability of magnetization fails again and the material suffers permanent injury, which can be partly, but not wholly, cured by prolonged heating.

Another singular phenomenon was exhibited by both of these alloys. When a bar of iron is magnetized by an electric current, it acquires its full magnetic strength almost instantaneously on the closure of the circuit. The magnetic alloys, on the contrary, do not attain their full magnetization for several minutes. In some of the experiments a gradual increase was observed even after the current had been flowing five minutes.

In magnetic strength alloy I proved far superior to alloy II, which contained smaller proportions of manganese and aluminum. Alloy I showed magnetic strengths up to 4.5 kilogauss, while the highest magnetization obtained with alloy II was only 1.9 kilogauss. But even alloy II may be called strongly magnetic, for its maximum magnetization is about one-tenth that of good wrought iron (18 to 20 kilogauss), or one-sixth that of cast iron (10 to 12 kilogauss). Alloy I is nearly equal in magnetic properties to nickel, which can be magnetized up to about 5 kilogauss.

MANGANESE ALLOYS:

Manganese bronze is a bronze deprived of its oxide by an admixture of manganese. The manganese is used as copper manganese containing 10 to 30 per cent manganese and added to the bronze to the amount of 0.5 to 2 per cent.

Manganese Copper.—The alloys of copper with manganese have a beautiful silvery color, considerable ductility, great hardness and tenacity, and are more readily fusible than ordinary bronze. A special characteristic is that they exactly fill out the molds, without the formation of blowholes, and present no difficulties in casting.

Cupromanganese is suitable for many purposes for which nothing else but bronze can advantageously be used, and the cost of its production is no greater than that of genuine bronze. In preparing the alloy, the copper is used in the form of fine grains, obtained by pouring melted copper into cold water. These copper grains are mixed with the dry oxide of manganese, and the mixture put into a crucible holding about 66 pounds. Enough space must be left in the crucible to allow a thick cover of charcoal, as the manganese oxidizes easily. The crucible is placed in a well-drawing

wind furnace and subjected to a strong white heat. The oxide of manganese is completely reduced to manganese, which at once combines with the copper to form an alloy. In order to prevent, as far as possible, the access of air to the fusing mass, it is advisable to cover the crucible with a lid which has an aperture in the center for the escape of the carbonic oxide formed during the reduction.

When the reduction is complete and the metals fused, the lid is removed and the contents of the crucible stirred with an iron rod, in order to make the alloy as homogeneous as possible. By repeated remelting of the cupromanganese a considerable quantity of the manganese is reconverted into oxide; it is, therefore, advisable to make the casts directly from the crucible. When poured out, the alloy rapidly solidifies, and resembles in appearance good German silver. Another reason for avoiding remelting is that the crucible is strongly attacked by the cupromanganese, and can be used but a few times.

The best kinds of cupromanganese contain between 10 and 30 per cent of manganese. They have a beautiful white color, are hard, tougher than copper, and can be worked under the hammer or with rolls. Some varieties of cupromanganese which are especially valuable for technical purposes are given below:

	I	II	III	IV
Copper.....	75	60	65	60
Manganese.	25	25	20	20
Zinc........	..	15	5	..
Tin........	..	..	..	10
Nickel.....	..	..	10	10

Manganin.—This is an alloy of copper, nickel, and manganese for electric resistances.

MIRROR ALLOYS:

Amalgams for Mirrors.—I.—Tin, 70 parts; mercury, 30 parts.

II.—For curved mirrors. Tin, 1 part; lead, 1 part; bismuth, 1 part; mercury, 9 parts.

III.—For glass balls. Tin, 80 parts; mercury, 20 parts.

IV.—Metallic cement. Copper, 30 parts; mercury, 70 parts.

V.—Mirror metal.—Copper, 100 parts; tin, 50 parts; Chinese copper, 8 parts; lead, 1 part; antimony, 1 part.

Reflector Metals. — I. — (Cooper's.) Copper, 35 parts; platinum, 6; zinc, 2; tin, 16.5; arsenic, 1. On account of the hardness of this alloy, it takes a very high polish; it is impervious to the effects of the weather, and is therefore remarkably well adapted to the manufacture of mirrors for fine optical instruments.

II.—(Duppler's.) Zinc, 20 parts; silver, 80 parts.

III.—Copper, 66.22 parts; tin, 33.11 parts; arsenic, 0.67 part.

IV.—Copper, 64 parts; tin, 32 parts; arsenic, 4 parts.

V.—Copper, 82.18 parts; lead, 9.22 parts; antimony, 8.60 parts.

VI.—(Little's.) Copper, 69.01 parts; tin, 30.82 parts; zinc, 2.44 parts; arsenic, 1.83 parts.

Speculum Metal.—Alloys consisting of 2 parts of copper and 1 of tin can be very brilliantly polished, and will serve for mirrors. Good speculum metal should have a very fine-grained fracture, should be white and very hard, the highest degree of polish depending upon these qualities. A composition to meet these requirements must contain at least 35 to 36 per cent of copper. Attempts have frequently been made to increase the hardness of speculum metal by additions of nickel, antimony, and arsenic. With the exception of nickel, these substances have the effect of causing the metal to lose its high luster easily, any considerable quantity of arsenic in particular having this effect.

The real speculum metal seems to be a combination of the formula Cu_4Sn, composed of copper 68.21 per cent, tin 31.7. An alloy of this nature is sometimes separated from ordnance bronze by incorrect treatment, causing the so-called tin spots; but this has not the pure white color which distinguishes the speculum metal containing 31.5 per cent of tin. By increasing the percentage of copper the color gradually shades into yellow; with a larger amount of tin into blue. It is dangerous to increase the tin too much, as this changes the other properties of the alloy, and it becomes too brittle to be worked. Below is a table showing different compositions of speculum metal. The standard alloy is undoubtedly the best.

	Copper	Tin	Zinc	Arsenic	Silver
Standard alloy....	68.21	31.7	....	....	....
Otto's alloy....	68.5	31.5	....	....	....
Richardson's alloy	65.3	30.0	0.7	2.	2.
Sollit's alloy......	64.6	31.3	4.1	Nickel	....
Chinese speculum metal...	80.83	....	....	8.5	Antimony
Old Roman	63.39	19.05	....	17.29	Lead

PALLADIUM ALLOYS.

I.—An alloy of palladium 24 parts, gold 80, is white, hard as steel, unchangeable in the air, and can, like the other alloys of palladium, be used for dental purposes.

II.—Palladium 6 parts, gold 18, silver 11, and copper 13, gives a reddish-brown, hard, and very fine-grained alloy, suitable for the bearings of pivots in clock works.

The alloys of most of the other platinum metals, so called, are little used on account of their rarity and costliness. Iridium and rhodium give great hardness to steel, but the commercial rhodium and iridium steel, so called, frequently contains not a trace of either. The alloy of iridium with osmium has great hardness and resistance and is recommended for pivots, fine instruments, and points of ship compasses.

Palladium Silver.—This alloy, composed of 9 parts of palladium and 1 of silver, is used almost exclusively for dental purposes, and is well suited to the manufacture of artificial teeth, as it does not oxidize. An alloy even more frequently used than this consists of platinum 10 parts, palladium 8, and gold 6.

Palladium Bearing Metal.—This alloy is extremely hard, and is used instead of jewel bearings in watches. It is composed of palladium 24 parts, gold 72, silver 44, copper 92.

PLATINUM ALLOYS.

Platinum has usually been alloyed with silver in goldsmith's work, 2 parts silver to 1 of platinum being taken to form the favorite "platinum silver." The object has been to produce an alloy having a white appearance, which can be polished, and at the same time has a low melting point. In addition to this platinum alloy the following are well known:

I.—A mixture of 7 parts platinum with 3 parts iridium. This gives to platinum the hardness of steel, which can be still further increased by taking 4 parts of iridium.

II.—An alloy of 9 parts platinum and 1 part iridium is used by the French in the manufacture of measuring instruments of great resisting power.

Compounds of copper, nickel, cadmium, and tungsten are also used in the construction of parts of watches; the latter acquire considerable hardness without becoming magnetic or rusting like steel.

III.—For this purpose a compound of

62.75 parts platinum, 18 parts copper, 1.25 parts cadmium, and 18 parts nickel is much recommended.

IV.—Very ductile platinum-copper alloys have also been made, e. g., the so-called Cooper gold, consisting of 3 parts platinum and 13 parts copper, which is almost equal to 18-carat gold in regard to color, finish, and ductility. If 4 per cent of platinum is taken, these latter alloys acquire a rose-red color, while a golden-yellow color can be produced by further adding from 1 to 2 per cent (in all 5 to 6 per cent) of platinum. The last-named alloy is extensively used for ornaments, likewise alloy V.

V.—Ten parts platinum, 60 parts nickel, and 220 parts brass, or 2 parts platinum, 1 part nickel and silver respectively, 2 parts brass, and 5 parts copper; this also gives a golden-yellow color.

VI.—For table utensils a favorite alloy is composed of 1 part platinum, 100 parts nickel, and 10 parts tin. Articles made of the latter alloy are impervious to atmospheric action and keep their polish for a long time. Pure white platinum alloys have for some time been used in dental work, and they have also proved serviceable for jewelry.

VII.—A mixture of 30 parts platinum, 10 parts gold, and 3 parts silver, or 7 parts platinum, 2 parts gold, and 3 parts silver.

VIII.—For enameled articles: Platinum, 35 parts; silver, 65 parts. First fuse the silver, then add the platinum in the spongy form. A good solder for this is platinum 80 parts, copper 20 parts.

IX.—For pens: Platinum, 4 parts; silver, 3 parts; copper, 1 part.

Platinum Gold.—Small quantities of platinum change the characteristics of gold in many respects. With a small percentage the color is noticeably lighter than that of pure gold, and the alloys are extremely elastic; alloys containing more than 20 per cent of platinum, however, almost entirely lose their elasticity. The melting point of the platinum-gold alloy is high, and alloys containing 70 per cent of platinum can be fused only in the flame of oxyhydrogen gas, like platinum itself. Alloys with a smaller percentage of platinum can be prepared in furnaces, but require the strongest white heat. In order to avoid the chance of an imperfect alloy from too low a temperature, it is always safer to fuse them with the oxyhydrogen flame. The alloys of platinum and gold have a somewhat limited application. Those which contain from 5 to 10 per cent of platinum are used for sheet and wire in the manufacture of artificial teeth.

Platinum-Gold Alloys for Dental Purposes.—

	I	II	III
Platinum........	6	14	10
Gold............	2	4	6
Silver..........	1	6	..
Palladium.......	..	..	8

Platinum Silver.—An addition of platinum to silver makes it harder, but also more brittle, and changes the white color to gray. An alloy which contains only a very small percentage of platinum is noticeably darker in color than pure silver. Such alloys are prepared under the name of *platine au titre*, containing between 17 and 35 per cent of platinum. They are almost exclusively used for dental purposes.

Imitation Platinum.—I.—Brass, 100 parts; zinc, 65 parts.

II.—Brass, 120 parts; zinc, 75 parts.

III.—Copper, 5 parts; nickel, 4 parts; zinc, 1½ parts; antimony, 1 part; lead, 1 part; iron, 1 part; tin, 1 part.

Cooper's Pen Metal.—This alloy is especially well adapted to the manufacture of pens, on account of its great hardness, elasticity, and power of resistance to atmospheric influences, and would certainly have superseded steel if it were possible to produce it more cheaply than is the case. The compositions most frequently used for pen metal are copper 1 part, platinum 4, and silver 3; or, copper 21, platinum 50, and silver 36.

Pens have been manufactured, consisting of several sections, each of a different alloy, suited to the special purpose of the part. Thus, for instance, the sides of the pen are made of the elastic composition just described; the upper part is of an alloy of silver and platinum; and the point is made either of minute cut rubies or of an extremely hard alloy of osmium and iridium, joined to the body of the pen by melting in the flame of the oxyhydrogen blowpipe. The price of such pens, made of expensive materials and at the cost of great labor, is of course exceedingly high, but their excellent qualities repay the extra expense. They are not in the least affected by any kind of ink, are most durable, and can be used constantly for years without showing any signs of wear.

The great hardness and resistance to the atmosphere of Cooper's alloys make them very suitable for manufacturing mathematical instruments where great precision is required. It can scarcely be calculated how long a chronometer, for instance, whose wheels are constructed of this alloy, will run before showing any irregularities due to wear. In the construction of such instruments, the price of the material is not to be taken into account, since the cost of the labor in their manufacture so far exceeds this.

PEWTER.

This is an alloy of tin and lead only, or of tin with antimony and copper. The first is properly called pewter. Three varieties are known in trade:

I (Plate Pewter).—From tin, 79 per cent; antimony, 7 per cent; bismuth and copper, of each 2 per cent; fused together. Used to make plates, teapots, etc. Takes a fine polish.

II (Triple Pewter).—From tin, 79 per cent; antimony, 15 per cent; lead, 6 per cent; as the last. Used for minor articles, syringes, toys, etc.

III (Ley Pewter).—From tin, 80 per cent; lead, 20 per cent. Used for measures, inkstands, etc.

According to the report of a French commission, pewter containing more than 18 parts of lead to 82 parts of tin is unsafe for measures for wine and similar liquors, and, indeed, for any other utensils exposed to contact with food or beverages. The legal specific gravity of pewter in France is 7.764; if it be greater, it contains an excess of lead, and is liable to prove poisonous. The proportions of these metals may be approximately determined from the specific gravity; but correctly only by an assay for the purpose.

SILVER ALLOYS:

Aluminum Silver.—Aluminum and silver form beautiful white alloys which are considerably harder than pure aluminum, and take a very high polish. They have the advantage over copper alloys of being unchanged by exposure to the air, and of retaining their white color.

The properties of aluminum and silver alloys vary considerably according to the percentage of aluminum.

I.—An alloy of 100 parts of aluminum and 5 parts of silver is very similar to pure aluminum, but is harder and takes a finer polish.

II.—One hundred and sixty-nine parts of aluminum and 5 of silver make an elastic alloy, recommended for watch springs and dessert knives.

III.—An alloy of equal parts of silver and aluminum is as hard as bronze.

IV.—Five parts of aluminum and 1 part of silver make an alloy that is easily worked.

V.—Also aluminum, 3 parts, and silver, 1 part.

VI. Tiers-Argent.—This alloy is prepared chiefly in Paris, and used for the manufacture of various utensils. As indicated by its name (one-third silver), it consists of 33.33 parts of silver and 66.66 parts of aluminum. Its advantages over silver consist in its lower price and greater hardness; it can also be stamped and engraved more easily than the alloys of copper and silver.

VII.—This is a hard alloy which has been found very useful for the operating levers of certain machines, such as the spacing lever of a typewriter. The metal now generally used for this purpose by the various typewriter companies is "aluminum silver," or "silver metal." The proportions are given as follows:

Copper	57.00
Nickel	20.00
Zinc	20.00
Aluminum	3.00

This alloy when used on typewriting machines is nickel-plated for the sake of the first appearance, but so far as corrosion is concerned, nickeling is unnecessary. The alloy is stiff and strong and cannot be bent to any extent without breaking, especially if the percentage of aluminum is increased to 3.5 per cent; it casts free from pinholes and blowholes; the liquid metal completely fills the mold, giving sharp, clean castings, true to pattern; its cost is not greater than brass; its color is silver white, and its hardness makes it susceptible to a high polish.

Arsenic.—Alloys which contain small quantities of arsenic are very ductile, have a beautiful white color, and were formerly used in England in the manufacture of tableware. They are not, however, suitable for this purpose, on account of the poisonous character of the arsenic. They are composed usually of 49 parts of silver, 49 of copper, and 2 of arsenic.

China Silver.—Copper, 65.24 per cent; tin, 19.52 per cent; nickel, 13.00 per cent; silver, 2.05 per cent.

Copper-Silver.—When silver is alloyed with copper only one proportion is known which will give a uniform casting. The proportion is 72 per cent silver to 28 per cent copper. With more silver than 72 per cent the center of a cast bar will be richer than the outside, which chills first; while with a less percentage than 72 per cent the center of the bar will be poorer and the outside richer than the average. This characteristic of silver-copper alloys is known to metallurgists as "segregation."

When nickel is added to the silver and copper, several good alloys may be formed, as the following French compositions:

	I	II	III
Silver....	33	40	20
Copper....	37–42	30–40	45–55
Nickel....	25–30	20–30	25–35

The whitening of alloys of silver and copper is best accomplished by annealing the alloy until it turns black on the surface. Cool in a mixture of 20 parts, by weight, of concentrated sulphuric acid to 1,000 parts of distilled water and leave therein for some time. In place of the sulphuric acid, 40 parts of potassium bisulphate may be used per 1,000 parts of liquid. Repeat the process if necessary.

Copper, Silver, and Cadmium Alloys. —Cadmium added to silver alloys gives great flexibility and ductility, without affecting the white color; these properties are valuable in the manufacture of silver-plated ware and wire. The proportions of the metals vary in these alloys. Some of the most important varieties are given below.

	Silver	Copper	Cadmium
I.....	980	15	5
II.....	950	15	35
III.....	900	18	82
IV.....	860	20	180
V.....	666	25	309
VI.....	667	50	284
VII.....	500	50	450

In preparing these alloys, the great volatility of cadmium must be taken into account. It is customary to prepare first the alloy of silver and copper, and add the cadmium, which, as in the case of the alloys of silver and zinc, must be wrapped in paper. After putting it in, the mass is quickly stirred, and the alloy poured immediately into the molds. This is the surest way to prevent the volatilization of the cadmium.

Silver, Copper, Nickel, and Zinc Alloys. —These alloys, from the metals contained in them, may be characterized as argentan or German silver with a certain percentage of silver. They have been used for making small coins, as in the older coins of Switzerland. Being quite hard, they have the advantage of wearing well, but soon lose their beautiful white color and take on a disagreeable shade of yellow, like poor brass. The silver contained in them can be regained only by a laborious process, which is a great drawback to their use in coinage. The composition of the Swiss fractional coins is as follows:

	20 centimes	10 centimes	5 centimes
Silver........	15	10	5
Copper.......	50	55	60
Nickel........	25	25	25
Zinc..........	10	10	10

Mousset's Alloy.—Copper, 59.06; silver, 27.56; zinc, 9.57; nickel, 3.42. This alloy is yellowish with a reddish tinge, but white on the fractured surface. It ranks next after Argent-Ruolz, which also contains sometimes certain quantities of zinc, and in this case may be classed together with the alloy just described. The following alloys can be rolled into sheet or drawn into wire:

	I	II	III
Silver.......	33.3	34	40.0
Copper......	41.8	42	44.6
Nickel.......	8.6	8	4.6
Zinc........	16.3	16	10.8

Japanese (Gray) Silver.—An alloy is prepared in Japan which consists of equal parts of copper and silver, and which is given a beautiful gray color by boiling in a solution of alum, to which copper sulphate and verdigris are added. The so-called "mokum," also a Japanese alloy, is prepared by placing thin plates of gold, silver, copper, and the alloy just described over each other and stretching them under the hammer. The cross sections of the thin plates obtained in this way show the colors of the different metals, which give them a peculiar striped appearance. Mokum is principally used for decorations upon gold and silver articles.

Silver-Zinc.—Silver and zinc have great affinity for each other, and alloys of these two metals are therefore easily made. The required quantity of zinc, wrapped in paper, is thrown into the melted and strongly heated silver, the mass is thoroughly stirred with an iron rod, and at once poured out into molds. Alloys of silver and zinc can be obtained which are both ductile and flexible. An alloy consisting of 2 parts of zinc and 1 of silver closely resembles silver in color, and is quite ductile. With a larger proportion of zinc the alloy becomes brittle. In preparing the alloy, a somewhat larger quantity of zinc must be taken than the finished alloy is intended to contain, as a small amount always volatilizes.

Imitation Silver Alloys.—There are a number of alloys, composed of different metals, which resemble silver, and may be briefly mentioned here.

I.—Warne's metal is composed of tin 10 parts, bismuth 7, and cobalt 3. It is white, fine-grained, but quite difficult to fuse.

II.—Tonca's metal contains coppei 5 parts, nickel 4, tin 1, lead 1, iron 1, zinc 1, antimony 1. It is hard, difficult to fuse, not very ductile, and cannot be recommended.

III.—Trabuk metal contains tin 87.5, nickel 5.5, antimony 5, bismuth 5.

IV.—Tourun-Leonard's metal is composed of 500 parts of tin and 64 of bell metal.

V.—Silveroid is an alloy of copper, nickel, tin, zinc, and lead.

VI.—Minargent. Copper, 100 parts; nickel, 70 parts; tungsten, 5 parts; aluminum, 1 part.

VII.—Nickel, 23 parts; aluminum, 5 parts; copper, 5 parts; iron, 65 parts; tungsten, 4 parts.

VIII.—Argasoid. Tin, 4.035; lead, 3.544; copper, 55.780; nickel, 13.406; zinc, 23.198; iron, trace.

SOLDERS:

See Solders.

STEEL ALLOYS:

See also Steel.

For Locomotive Cylinders.—This mixture consists of 20 per cent steel castings, old steel springs, etc.; 20 per cent No. 2 coke iron, and 60 per cent scrap. From this it is stated a good solid metal can be obtained, the castings being free from honeycombing, and finishing better than the ordinary cast-iron mixture, over which it has the advantage of 24 per cent greater strength. Its constituents are: Silicon, 1.51; manganese, 0.33; phosphorus, 0.65; sulphur, 0.068; combined carbon, 0.62; graphite, 2.45.

Nickel steel is composed of nickel 36 per cent, steel 64 per cent.

Tungsten steel is crucible steel with 5 to 12 per cent tungsten.

STEREOTYPE METAL.

Lead..................	2 parts
Tin..................	3 parts
Bismuth.............	5 parts

The melting point of this alloy is 196° F. The alloy is rather costly because of the amount of bismuth which it contains. The following mixtures are cheaper:

	I	II	III	IV
Tin........	1	3	1	2
Lead.......	1	5	1.5	2
Bismuth....	2	8	3	5
Antimony...	..	..	..	1

TIN ALLOYS:

Alloys for Dentists' Molds and Dies.
—I.—Very hard. Tin, 16 parts; antimony, 1 part; zinc, 1 part.

II.—Softer than the former. Tin, 8 parts; zinc, 1 part; antimony, 1 part.

III.—Very hard. Tin, 12 parts; antimony, 2 parts; copper, 1 part.

Cadmium Alloy, about the Hardness of Zinc.—Tin, 10 parts; antimony, 1 part; cadmium, 1 part.

Tin-Lead.—Tin is one of those metals which is not at all susceptible to the action of acids, while lead, on the other hand, is very easily attacked by them. In such alloys, consequently, used for cooking utensils, the amount of lead must be limited, and should properly not exceed 10 or 15 per cent; but cases have been known in which the so-called tin contained a third part, by weight, of lead.

Alloys containing from 10 to 15 per cent of lead have a beautiful white color, are considerably harder than pure tin, and much cheaper. Many alloys of tin and lead are very lustrous, and are used for stage jewelry and mirrors for reflecting the light of lamps, etc. An especially brilliant alloy is called "Fahlun brilliants." It is used for stage jewelry, and consists of 29 parts of tin and 19 of lead. It is poured into molds faceted in the same way as diamonds, and when seen by artificial light, the effect is that of diamonds. Other alloys of tin and lead are employed in the manufacture of toys. These must fill the molds well, and must also be cheap, and therefore as much as 50 per cent of lead is used. Toys can also be made from type metal, which is even cheaper than the alloys of tin and lead, but has the disadvantage of readily breaking if the articles are sharply bent. The alloys of tin and lead give very good castings, if sharp iron or brass molds are used.

Lead...............	19 parts
Tin................	29 parts

This alloy is very bright and possesses a permanent sheen. It is well adapted for the making of artificial gems for stage use. It is customary in carrying out the process to start with two parts of tin and one part of lead. Tin is added until a sample drop which is allowed to fall upon an iron plate forms a mirror. The artificial gems are produced by dipping into the molten alloy pieces of glass cut to the proper shape. The tin coating of metal which adheres to the glass cools rapidly and adheres tenaciously. Outwardly these artificial gems appear rough and gray, but inwardly they are highly reflective and quite deceptive when seen in artificial light.

If the reflective surfaces be coated with red, blue, or green aniline, various colored effects can be obtained. Instead of fragile glass the gems may be produced by means of well-polished pieces of steel or bronze.

Other Tin-Lead Alloys.— Percentage of lead and specific gravity.

P. C.	S. G.	P. C.	S. G.
0	7.290	28	8.105
1	7.316	29	8.137
2	7.342	30	8.169
3	7.369	31	8.202
4	7.396	32	8.235
5	7.423	33	8.268
6	7.450	34	8.302
7	7.477	35	8.336
8	7.505	36	8.379
9	7.533	37	8.405
10	7.562	38	8.440
11	7.590	39	8.476
12	7.619	40	8.512
13	7.648	41	8.548
14	7.677	42	8.584
15	7.706	43	8.621
16	7.735	44	8.658
17	7.764	45	8.695
18	7.794	46	8.732
19	7.824	47	8.770
20	7.854	48	8.808
21	7.885	49	8.846
22	7.916	50	8.884
23	7.947	60	9.299
24	7.978	70	9.736
25	8.009	80	10.225
26	8.041	90	10.767
27	8.073	100	11.370

Tin Statuettes, Buttons, etc.—

1.—Tin.................. 4 parts
Lead................ 3 parts

This is a very soft solder which sharply reproduces all details.

Another easily fusible alloy but somewhat harder, is the following:

II.—Tin.................. 8 parts
Lead.................. 6 parts
Antimony........ 0.5 part

Miscellaneous Tin Alloys.—I.—Alger Metal.—Tin, 90 parts; antimony, 10 parts. This alloy is suitable as a protector.

II. Argentine Metal.—Tin, 85.5 per cent; antimony, 14.5 per cent.

III.—Ashberry metal is composed of 78 to 82 parts of tin, 16 to 20 of antimony, 2 to 3 of copper.

IV. Quen's Metal.—Tin, 9 parts; lead, 1 part; antimony, 1 part; bismuth, 1 part.

Type Metal.—An alloy which is to serve for type metal must be readily cast, fill out the molds sharply, and be as hard as possible. It is difficult to satisfy all these requirements, but an alloy of antimony and lead answers the purpose best. At the present day there are a great many formulas for type metal in which other metals besides lead and antimony are used, either to make the alloy more readily fusible, as in the case of additions of bismuth, or to give it greater power of resistance, the latter being of especial importance for types that are subjected to constant use. Copper and iron have been recommended for this purpose, but the fusibility of the alloys is greatly impaired by these, and the manufacture of the types is consequently more difficult than with an alloy of lead and antimony alone. In the following table some alloys suitable for casting type are given:

	Lead	Anti-mony	Cop-per	Bis-muth	Zinc	Tin	Nickel
I	3	1	..	..	..	..	..
II	5	1	..	..	..	..	..
III	10	1	..	..	..	..	..
IV	10	2	..	1	..	..	..
V	70	18	2	..	..	10	..
VI	60	20	..	..	..	20	..
VII	55	25	..	..	..	20	..
VIII	55	30	..	..	..	15	..
IX	100	30	8	2	..	20	8
X	6	..	4	..	90	..	..

The French and English types contain a certain amount of tin, as shown by the following analyses:

	English Types			French Types
	I	II	III	
Lead	69.2	61.3	55.0	55
Antimony	19.5	18.8	22.7	30
Tin	9.1	20.2	22.1	15
Copper	1.7	....	....	..

Ledebur gives the composition of type metal as follows:

	I	II	III	IV
Lead	75	60	80	82
Antimony	23	25	20	14.8
Tin	22	15	..	3.2

WATCHMAKERS' ALLOYS:
See Watchmakers' Formulas.

WHITE METALS.
The so-called white metals are employed almost exclusively for bearings. (See Anti-friction Metals under Alloys.) In the technology of mechanics an accurate distinction is made between the different kinds of metals for bearings; and they may be classed in two groups, red brass and white metal. The red-

brass bearings are characterized by great hardness and power of resistance, and are principally used for bearings of heavily loaded and rapidly revolving axles. For the axles of large and heavy flywheels, revolving at great speed, bearings of red brass are preferable to white metal, though more expensive.

In recent years many machinists have found it advantageous to substitute for the soft alloys generally in use for bearings a metal almost as hard as the axle itself. Phosphor bronze (q. v.) is frequently employed for this purpose, as it can easily be made as hard as wrought or cast steel. In this case the metal is used in a thin layer, and serves only, as it were, to fill out the small interstices caused by wear on the axle and bearing, the latter being usually made of some rather easily fusible alloy of lead and tin. Such bearings are very durable, but expensive, and can only be used for large machines. For small machines, running gently and uniformly, white-metal bearings are preferred, and do excellent work, if the axle is not too heavily loaded. For axles which have a high rate of revolution, bearings made of quite hard metals are chosen, and with proper care —which, indeed, must be given to bearings of any material—they will last for a long time without needing repair.

White Metal for Bearing.

No.		Tin	Antimony	Zinc	Iron	Lead	Copper
I	German, light loads	85.00	10.00				5.00
II	German, light loads	82.00	11.00				7.00
III	German, light loads	80.00	12.00				8.00
IV	German, light loads	76.00	17.00				7.00
V	German, light loads	90.00	8.00				2.00
VII	German, heavy loads	86.81	7.62				5.57
VIII	German, heavy loads	76.70	15.50				7.80
IX	English, heavy loads	17.47		76.14			5.62
X	English, medium loads	72.00	26.00				2.00
XI	English, medium loads	72.70	18.20				9.10
XII	For mills	15.00	1.00	40.00		42.00	2.00
XIII	For mills	38.00	6.00	47.00		5.00	4.00
XIV	Heavy axles	55.00		10.00		2.00	3.00
XVI	Rapidly revolving axles	3.00	1.00		70.00	4.00	1.00
XVII	Very hard metal	17.00	77.00				6.00
XVIII	Very hard metal	12.00	82.00	2.00			2.50
XIX	Cheap metal	2.00	2.00	88.00			8.00
XX	Cheap metal	1.50	1.50	90.00			7.00

Other white bearing metals are:

XXI.—Tin, 8.5; antimony, 10; copper, 5 parts.

XXII.—Tin, 42; antimony, 16; lead, 42 parts.

XXIII.—Tin, 72; antimony, 26; copper, 2 parts.

XXIV.—Tin, 81; antimony, 12.5; copper, 6.5 parts.

White Metals Based on Copper.—
I.—Copper, 65 parts; arsenic, 55 parts.

II.—Copper, 64 parts; arsenic, 50 parts.

III.—Copper, 10 parts; zinc, 20 parts; nickel, 30 parts.

IV.—Nickel, 70 parts; copper, 30 parts; zinc, 20 parts.

V.—Nickel, 60 parts; copper, 30 parts; zinc, 30 parts.

VI.—Copper, 8 parts; nickel, 4 parts; zinc, 4 parts.

VII.—Copper, 10 parts; nickel, 5 parts; zinc, 5 parts.

VIII.—Copper, 8 parts; nickel, 3 parts; zinc, 4 parts.

IX.—Copper, 50 parts; nickel, 25 parts; zinc, 25 parts.

X.—Copper, 55 parts; nickel, 24 parts; zinc, 21 parts.

XI.—Copper, 55 parts; nickel, 24 parts; zinc, 16 parts; iron, 2 parts; tin, 3 parts.

IX, X, and XI are suitable for tableware.

XII.—Copper, 67 parts, and arsenic, 53 parts.

XIII.—Copper, 63 parts, and arsenic, 57 parts.

XII and XIII are bright gray, unaffected by the temperature of boiling water; they are fusible at red heat.

White Metals Based on Platinum.—
I.—Platinum, 1 part; copper, 4 parts; or platinum, 1½ parts; copper, 3½ parts.

II.—Platinum, 10 parts; tin, 90 parts; or platinum, 8 parts; tin, 92 parts.

III.—Platinum, 7 parts; copper, 13 parts; tin, 80 parts.

IV.—Platinum, 2 parts; steel, 98 parts.

V.—Platinum, 2.5 parts; steel, 97.5 parts.

IV and V are for gun metal.

Miscellaneous White-Metal Alloys.—
I.—For lining cross-head slides: Lead, 65 parts; antimony, 25 parts; copper, 10 parts. Some object to white metal containing lead or zinc. It has been found, however, that lead and zinc have properties of great use in these alloys.

II.—Tin, 85 parts; antimony, 7½ parts; copper, 7½ parts.

III.—Tin, 90 parts; copper, 3 parts; antimony, 7 parts.

ZINC ALLOYS:

Bidery Metal. — This is sometimes composed of 31 parts of zinc, 2 parts of copper, and 2 parts of lead; the whole is melted on a layer of rosin or wax to avoid oxidation. This metal is very resistive; it does not oxidize in air or moisture. It takes its name from the town of Bider, near Hyderabad (India), where it was prepared for the first time industrially for the manufacture of different utensils.

Other compositions of Indian Bidery metal (frequently imitated in England) are about as follows:

	P.C.	P.C.	P.C.
Copper...	3.5	11.4	16
Zinc.....	93.4	84.3	112
Tin......	....	1.4	2
Lead.....	3.1	2.9	4

Erhardt recommends the following as being both ductile and hard:

Zinc	89 to 93
Tin	9 to 6
Lead	2 to 4
Copper	2 to 4

The tin is first melted, and the lead, zinc, and copper added successively.

Zinc-Nickel. — Zinc, 90 parts; nickel, 10 parts. Used in powder form for painting and cloth printing purposes.

Platine for Dress Buttons. — Copper, 43 parts; zinc, 57 parts.

UNCLASSIFIED ALLOYS:

Alloys for Drawing Colors on Steel. — Alloys of various composition are successfully used for drawing colors on steel. To draw to a straw color use 2 parts of lead and 1 part of tin, and melt in an iron ladle. Hold the steel piece to be drawn in the alloy as it melts and it will turn to straw color. This mixture melts at a temperature of about 437° F. For darker yellow use 9 parts of lead to 4 parts of tin, which melts at 458° F. For purple, use 3 parts of lead to 1 part of tin, the melting temperature being 482° F. For violet, use 9 parts of lead to 2 parts of tin, which melts at 494° F. Lead without any alloy will draw steel to a dark blue. The above apply to steel only since iron requires a somewhat greater heat and is more or less uncertain in handling.

Alloy for Pattern Letters and Figures. — A good alloy for casting pattern letters and figures and similar small parts of brass, iron, or plaster molds, is made of lead 80 parts, and antimony 20 parts. A better alloy will be lead 70 parts, an-timony and bismuth each 15 parts. To insure perfect work the molds should be quite hot by placing them over a Bunsen burner.

Alloy for Caliper and Gage-Rod Castings. — A mixture of 30 parts zinc to 70 parts aluminum gives a light and durable alloy for gage rods and caliper legs; the gage rods must be steel tipped, for the alloy is soft and wears away too rapidly for gage points.

Alloys for Small Casting Molds. — Tin, 75 parts, and lead, 22 parts; or 75 parts of zinc and 25 parts of tin; or 30 parts of tin and 70 parts of lead; or 60 parts of lead and 40 parts of bismuth.

ALLOYS FOR METAL FOIL:
See Metal Foil.

ALMOND COLD CREAM:
See Cosmetics.

ALMOND LIQUEURS:
See Wines and Liquors.

ALTARS, TO CLEAN:
See Cleaning Preparations and Methods.

ALUM:

Burnt Alum. — I. — Heat the alum in a porcelain dish or other suitable vessel till it liquefies, then raise and continue the heat, not allowing it to exceed 400°, till aqueous vapor ceases to be disengaged, and the salt has lost 47 per cent of its weight. Reduce the residue to powder, and preserve it in a well-stoppered bottle. — *Cooley.*

II. — Heat ordinary alum (alumina alum) with constant stirring in an iron pan in which it will first melt quietly, and then commence to form blisters. Continue heating until a dry white mass of a loose character remains, which is powdered and kept in well-closed glasses.

ALUM BATH:
See Photography.

Aluminum and its Treatment

HOW TO COLOR ALUMINUM:

Blanching of Aluminum. — Aluminum is one of the metals most inalterable by air; nevertheless, the objects of aluminum tarnish quickly enough without being

...tered. They may be restored to their mat whiteness in the following manner: Immerse the aluminum articles in a boiling bath of caustic potash; next plunge them quickly into nitric acid, rinse and let dry. It must be understood that this method is applicable only to pieces entirely of aluminum.

Decolorized Aluminum.—Gray or unsightly aluminum may be restored to its white color by washing with a mixture of 30 parts of borax dissolved in 1,000 parts of water, with a few drops of ammonia added.

Mat Aluminum.—In order to impart to aluminum the appearance of mat silver, plunge the article into a hot bath composed of a 10-per-cent solution of caustic soda saturated with kitchen salt. Leave it in the bath for 15 to 20 seconds, then wash and brush; put back into the bath for half a minute, wash anew and dry in sawdust.

To Blacken Aluminum.—I.—The surface of the sheet to be colored is polished with very fine emery powder or finest emery cloth. After polishing pour a thin layer of olive oil over the surface and heat slowly over an alcohol flame. Large sheets must, of course, be heated in the drying oven. After a short while pour on oil again, in order to obtain absolute uniformity of the coating, and heat the plate once more. Under the action of the heat the plate turns first brown, then black, according to the degrees of heat. When the desired coloration has been attained, the plate is polished over again, after cooling, with a woolen rag or soft leather.

II.—White arsenic ...... 1 ounce
Sulphate of iron .... 1 ounce
Hydrochloric acid .. 12 ounces
Water............. 12 ounces

When the arsenic and iron are dissolved by the acid add the water. The aluminum to be blackened should be well cleaned with fine emery powder and washed before immersing in the blackening solution. When the deposit of black is deep enough dry off with fine sawdust and lacquer.

Decorating Aluminum.—A process for decorating aluminum, patented in Germany, prescribes that the objects be first corroded, which is usually done with caustic soda lye, or, better still, by a new method which consists in heating 3 parts of sulphuric acid with 1 part of water to 140° to 158° F., in an enameled vessel. Into this liquid dip the aluminum arti-

cles, rinsing them off clean and then drying them well. The corroded articles are now placed in a bath consisting of 1,000 parts of alcohol (90 per cent), 1.50 parts of antimony, 250 parts of chemically pure hydrochloric acid, 100 parts of manganous nitrate, and 20 parts of purified and finally elutriated graphite. In this bath, which is heated to 86°–95° F., the objects are left until fumes develop around them, which takes place in a few seconds. Now they are put over a coal fire or similar arrangement until the alcohol is burned up and there is no more smoke. After they are somewhat cooled off, they are laid into cold water and worked with a brush, then rinsed with water and well dried. The pieces are now provided with a gray metallic coating, consisting mainly of antimony, manganese, and graphite. This metallic layer renders them capable of receiving a lacquer which is best prepared from 1,000 parts of alcohol (90 per cent), 50 parts of sandarac, 100 parts of shellac, and 100 parts of nigrosine (black aniline color). Then the articles are quickly but thoroughly rinsed off, dried in warmed air for a few minutes, and baked in ovens or over a moderate coal fire until they do not smoke any more and no more gloss can be seen. Finally they are rubbed with a cotton rag saturated with thin linseed-oil varnish, and the objects thus treated now appear dull black, like velvet. The covering withstands all action of the weather, so that cooking vessels coated with this varnish on the outside can be placed on the fire without injury to the coating. If the articles are engraved, the aluminum appears almost glossy white under the black layer at the engraved places. When the pieces have been provided with the gray metallic coating, colored lacquer may also be applied with the brush. In this manner paintings, etc., may be done on aluminum, while not possible on unprepared aluminum surfaces, which will not retain them.

Making Castings in Aluminum.—The method adopted in preparing molds and cores for aluminum work is necessarily somewhat the same as for brass, but there are particular points which need attention to insure successful work. Both in the sand and the making of the molds there are some small differences which make considerable variation in the results, and the temperature at which the metal is poured is a consideration of some importance.

In selecting the sand, which should

not have been previously used, that of a fine grain should be chosen, but it should not have any excess of aluminous matter, or it will not permit of the free escape of gases and air, this being an important matter. Besides this, the sand must be used as dry as possible consistent with its holding against the flow of the metal, and having only moderate compression in ramming.

In making the molds it is necessary to remember that aluminum has a large contraction in cooling, and also that at certain temperatures it is very weak and tears readily, while all metals shrink away from the mold when this is wholly outside the casting, but they shrink on to cores or portions of the mold partly inclosed by metal. Thus, if casting a plate or bar of metal, it will shrink away from the mold in all directions; but if casting a square frame, it shrinks away from the outside only, while it shrinks on to the central part or core. With brass, or iron, or such metals, this is not of much importance, but with some others, including aluminum, it is of great importance, because if the core or inclosed sand will not give somewhat with the contraction of the metal, torn or fractured castings will be the result. Both for outside and inside molds, and with cores used with aluminum, the sand should be compressed as little as possible, and hard ramming must in every case be avoided, particularly where the metal surrounds the sand. The molds must be very freely vented, and not only at the joint of the mold, but by using the vent wire freely through the body of the mold itself; in fact, for brass the venting would be considered excessive. With aluminum it is, however, necessary to get the air off as rapidly as possible, because the metal soon gets sluggish in the mold, and unless it runs up quickly it runs faint at the edges. The ingates should be wide and of fair area, but need careful making to prevent their drawing where they enter the casting, the method of doing this being known to most molders.

If it is considered desirable to use a specially made-up facing sand for the molds where the metal is of some thickness, the use of a little pea or bean meal will be all that is necessary. To use this, first dry as much sand as may be required and pass through a 20-mesh sieve, and to each bushel of the fine sand rub in about 4 quarts of meal, afterwards again passing through the sieve to insure regular mixing. This sand should then be damped as required, being careful

that all parts are equally moist, rubbing on a board being a good way to get it tough, and in good condition, with the minimum of moisture.

The molds should not be sleeked with tools, but they may be dusted over with plumbago or steatite, smoothing with a camel's-hair brush, in cases in which a very smooth face is required on the castings. Preferably, however, the use of the brush even should be avoided. Patterns for aluminum should be kept smooth and well varnished.

In melting the metal it is necessary to use a plumbago crucible which is clean and which has not been used for other metals. Clay or silica crucibles are not good for this metal, especially silica, on account of the metal absorbing silicon and becoming hard under some conditions of melting. A steady fire is necessary, and the fuel should reach only about halfway up the crucible, as it is not desirable to overheat the crucible or metal. The metal absorbs heat for some time and then fuses with some rapidity, hence the desirability of a steady heat; and as the metal should be poured when of a claret color under the film of oxide which forms on the surface, too rapid a heating is not advisable. The molding should always be well in advance of the pouring, because the metal should be used as soon as it is ready; for not only is waste caused, but the metal loses condition if kept in a molten state for long periods. The metal should be poured rapidly, but steadily, and when cast up there should not be a large head of metal left on top of the runner. In fact, it is rather a disadvantage to leave a large head, as this tends to draw rather than to feed the casting.

With properly prepared molds, and careful melting, fluxes are not required, but ground cryolite—a fluoride of sodium and aluminum—is sometimes used to increase the fluidity of the metal. In using this, a few ounces according to the bulk of metal to be treated is put into the molten metal before it is taken from the furnace, and well stirred in, and as soon as the reaction apparently ceases the pot is lifted and the metal at once skimmed and poured. The use of sodium in any form with aluminum is very undesirable, however, and should be avoided, and the same remark applies to tin, but there is no objection to alloying with zinc, when the metal thus produced is sold as an alloy.

Aluminum also casts very well in molds of plaster of Paris and crushed bath brick when such molds are perfectly dry and well vented, smoothness being secured by brushing over with dry steatite or plumbago. When casting in metal molds, these should be well brushed out with steatite or plumbago, and made fairly hot before pouring, as in cold molds the metal curdles and becomes sluggish, with the result that the castings run up faint.

To Increase the Toughness, Density, and Tenacity of Aluminum.—For the purpose of improving aluminum, without increasing its specific gravity, the aluminum is mixed with 4 to 7 per cent of phosphorus, whereby the density, tenacity, and especially the toughness are said to be enhanced.

WORKING OF SHEET ALUMINUM:

The great secret, if there is any, in working aluminum, either pure or alloyed, consists in the proper lubricant and the shape of the tool. Another great disadvantage in the proper working of the metal is that, when a manufacturer desires to make up an article, he will procure the pure metal in order to make his samples, which, of course, is harder to work than the alloy. But the different grades of aluminum sheet which are on the market are so numerous for different classes of work that it might be advisable to consider them for a moment before passing to the method of working them.

The pure metal, to begin with, can be purchased of all degrees of hardness, from the annealed, or what is known as the "dead soft" stock, to the pure aluminum hard rolled. Then comes a harder grade of alloys, running from "dead soft" metal, which will draw up hard, to the same metal hard rolled; and, still again, another set of alloys which, perhaps, are a little harder still when hard rolled, and will, when starting with the "dead soft," spin up into a utensil which, when finished, will probably be as stiff as brass. These latter alloys are finding a large sale for replacing brass used in all classes of manufactured articles.

To start with lathe work on aluminum, probably more difficulty has been found here, especially in working pure metal, and more complaints are heard from this source than from any other. As stated before, however, these difficulties can all be readily overcome, if the proper tools and the proper lubricants are used, as automatic screw machines are now made so that they can be operated when working aluminum just as readily as when they are working brass, and in some cases more readily. To start with

the question of the tool, this should be made as what is known as a "shearing tool," that is, instead of a short, stubby point, such as would be used in turning brass, the point should be lengthened out and a lot of clearance provided on the inside of the tool, so as to give the chips of the metal a good chance to free themselves and not cause a clogging around the point of the tool—a similar tool, for instance, to what would be used for turning wood.

The best lubricant to be used would be coal oil or water, and plenty of it. The latter is almost as good as coal oil if enough of it is used, and with either of these lubricants and a tool properly made, there should be no difficulty whatsoever in the rapid working of aluminum, either on the lathe or on automatic screw machines.

To go from the lathe to the drawing press, the same tools here would be used in drawing up shapes of aluminum as are used for drawing up brass or other metals; the only precaution necessary in this instance being to use a proper lubricant, which in this case is a cheap grade of vaseline, or in some cases lard oil, but in the majority of instances better results will be secured by the use of vaseline. Aluminum is probably susceptible of deeper drawing with less occasion to anneal than any of the other commercial metals. It requires but one-third or one-fourth of as much annealing as brass or copper. For instance, an article which is now manufactured in brass, requiring, say, three or four operations before the article is finished, would probably have to be annealed after every operation. With aluminum, however, if the proper grade is used, it is generally possible to perform these three operations without annealing the metal at all, and at the same time to produce a finished article which, to all intents and purposes, is as stiff as an article made of sheet brass.

Too much stress cannot be laid on the fact of starting with the proper grade of metal, for either through ignorance or by not observing this point is the foundation of the majority of the complaints that aluminum "has been tried and found wanting." If, however, it should be found necessary to anneal aluminum, this can be readily accomplished by heating it in an ordinary muffle, being careful that the temperature shall not be too high—about 650° or 700° F. The best test as to when the metal has reached the proper temperature is to take a soft pine stick and draw it across the metal. If it chars the stick and leaves a black mark on the metal, it is sufficiently annealed and is in a proper condition to proceed with further operation.

Next taking up the question of spinning aluminum, success again depends particularly on starting with the proper metal. The most satisfactory speed for articles from 5 to 8 inches in diameter is about 2,600 revolutions a minute, and for larger or smaller diameters the speed should be so regulated as to give the same velocity at the circumference. Aluminum is a very easy metal to spin and no difficulty should be found at all in spinning the proper grades of sheets. Several factories that are using large quantities of aluminum now, both for spinning and stamping, are paying their men by the piece the same amount that they formerly paid on brass and tin work, and it is stated that the men working on this basis make anywhere from 10 to 20 per cent more wages by working aluminum.

After aluminum has been manufactured into the shape of an article, the next process is the finishing of it. The best polish can be obtained by first cutting down the metal with an ordinary rag buff on which use tripoli, and then finish it with a dry red rouge which comes in the lump form, or that which is known as "White Diamond Rouge." One point, however, that it is necessary to observe carefully is that both the tripoli and the rouge should be procured ground as fine as it is possible to grind them; for, if this is not done, the metal will have little fine scratches all over it, and will not appear as bright and as handsome as it otherwise would.

If it is desired to put on a frosted appearance, this can either be done by scratch brushing or sand blasting. A brass wire scratch brush, made of crimped wire of No. 32 to No. 36 B. & S. gage, with three or four rows of bristles, will probably give the best results. This work of scratch brushing can be somewhat lessened, however, if, before applying the scratch brush to the surface of the aluminum, the article is first cut down by the use of a porpoise-hide wheel and fine Connecticut sand, placing the sand between the surface of the aluminum and the wheel, so that the skin and the irregularities on the surface are removed, and then putting the article on a buffing wheel before attempting to scratch brush it. This method, however, is probably more advantageous in the treating of aluminum castings than for articles manufactured out of the sheet metal, as in the majority of cases it is simply necessary before scratch brushing to cut down the article with tripoli, and then polish it with rouge as already described, before putting on the scratch brush; in this way the brush seems to take hold quicker and better, and to produce a more uniform polish.

An effect similar to the scratch-brush finish can be got by sand blasting, and by first sand blasting and then scratch brushing the sheets, a good finish is obtained with very much less labor than by scratch brushing alone. Another very pretty frosted effect is procured by first sand blasting and then treated as hereinafter described by "dipping" and "frosting," and many variations in the finish of aluminum can be got by varying the treatment, either by cutting down with tripoli and polishing, scratch brushing, sand blasting, dipping, and frosting, and by combinations of those treatments. A very pretty mottled effect is secured on aluminum by first polishing and then scratch brushing and then holding the aluminum against a soft pine wheel, run at a high rate of speed on a lathe, and by careful manipulation, quite regular forms of a mottled appearance can be obtained.

The dipping and frosting of aluminum sheet is probably the cheapest way of producing a nice finish. First remove all grease and dirt from the article by dipping in benzine, then dip into water in order that the benzine adhering to the article may be removed, so as not to affect the strength of the solution into which it is next dipped. After they have been taken out of the water and well shaken, the articles should be plunged in a strong solution of caustic soda or caustic potash, and left there a sufficient length of time until the aluminum starts to turn black. Then they should be removed, dipped in water again, and then into a solution of concentrated nitric and sulphuric acid, composed of 24 parts of nitric acid to 1 part of sulphuric acid. After being removed, the article should be washed thoroughly in water and dried in hot sawdust in the usual way. This finish can also be varied somewhat by making the solution of caustic soda of varying degrees of strength, or by adding a small amount of common salt to the solution.

In burnishing the metal use a bloodstone or a steel burnisher. In burnishing use a mixture of melted vaseline and coal oil, or a solution composed of 2 tablespoonfuls of ground borax dissolved in about a quart of hot water, with a few drops of ammonia added. In engraving, which adds materially to the appearance of finished castings, book covers, picture frames, and similar articles made of sheet, probably the best lubricant to use on an engraver's tool in order to obtain a clean cut, which is bright, is naphtha or coal oil, or a mixture of coal oil and vaseline. The naphtha, however, is preferred, owing to the fact that it does not destroy the satin finish in the neighborhood of the cut, as the other lubricants are very apt to do. There is, however, as much skill required in using and making a tool in order to give a bright, clean cut as there is in the choice of the lubricant to be used. The tool should be made somewhat on the same plan as the lathe tools already outlined. That is, they should be brought to a sharp point and be "cut back" rather far, so as to give plenty of clearance.

There has been one class of work in aluminum that has been developed lately and only to a certain extent, in which there are great possibilities, and that is in drop forging the metal. Some very superior bicycle parts have been manufactured by drop forging. This can be accomplished probably more readily with aluminum than with other metals, for the reason that it is not necessary with all the alloys to work them hot; consequently, they can be worked and handled more rapidly.

ALUMINUM, TO CLEAN:
See Cleaning Preparations and Methods.

ALUMINUM ALLOYS:
See Alloys.

ALUMINUM BRONZE:
See Alloys under Bronzes.

ALUMINUM CASTINGS:
See Casting.

ALUMINUM PAPER:
See Paper.

ALUMINUM PLATING:
See Plating.

ALUMINUM POLISHES:
See Polishes.

Amalgams

See also Easily Fusible Alloys under Alloys.

The name amalgam is given to alloys of metals containing mercury. The term comes to us from the alchemists. It signifies softening, because an excess of mercury dissolves a large number of metals.

Preparation of Amalgams.—Mercury forms amalgams with most metals. It unites directly and readily, either cold or hot, with potassium, sodium, barium, strontium, calcium, magnesium, zinc, cadmium, tin, antimony, lead, bismuth, silver, and gold; directly, but more difficultly, with aluminum, copper, and palladium. This combination takes place oftenest at the ordinary temperature; certain metals, however, like aluminum and antimony, combine only when heated in presence of quicksilver.

Quicksilver has no direct action on metals of high fusing points: manganese, iron, nickel, cobalt, uranium, platinum, and their congeners. Still, amalgams of these metals can be obtained of butyrous consistency, either by electrolysis of their saline solutions, employing quicksilver as the negative electrode, or by the action of an alkaline amalgam (potassium or sodium), on their concentrated and neutral saline solutions. These same refractory metals are also amalgamated superficially when immersed in the amalgam of sodium or of ammonium in presence of water.

Processes for preparing amalgams by double decomposition between an alkaline amalgam and a metallic salt, or by electrolysis of saline solutions, with employment of mercury as the negative electrode, apply *a fortiori* to metals capable of combining directly with the quicksilver. The latter of these methods is especially utilized for the preparation of alkaline earthy metals by electrolytic decomposition of the solutions of their salts or hydrated oxides with quicksilver as a cathode.

General Properties of Amalgams.—Amalgams are liquid when the quicksilver is in great excess; solid, but readily fusible, when the alloyed metal predominates.

They have a metallic luster, and a metallic structure which renders them brittle. They even form crystallized metallic combinations of constant proportions, dissolved in an excess of quicksilver, when the excess is separated by compression in a chamois skin, or by filtration in a glass funnel of slender stem, terminating with an orifice almost capillary.

According as the fusing heat of a metal is less or greater than its combination heat with quicksilver, the amalgamation of this metal produces an elevation or a lowering of temperature. Thus potassium, sodium, and cadmium, in alloy with quicksilver, disengage heat; while zinc, antimony, tin, bismuth, lead, and silver combine with mercury with absorption of heat. The amalgamation of 162 parts of quicksilver with 21 parts of lead, 12 parts of tin or of antimony, and 28.5 parts of bismuth, lowers the temperature of the mixture 79° F.

Amalgams formed with disengagement of heat are electro-negative with reference to the metals alloyed with the quicksilver. The products with absorption of heat are electro-negative with reference to the metals combined with the quicksilver; consequently, in a battery of elements of pure cadmium and amalgamated cadmium, the cadmium will be the negative pole; in case of zinc and amalgamated zinc, the zinc will be the positive pole.

Heat decomposes all amalgams, vaporizing the mercury and leaving the metal alloys as a residue.

Water is decomposed by the amalgams of potassium and sodium, because the heat of formation of these amalgams, although considerable, is even less than the heat disengaged by potassium and sodium, on decomposing water. The alkaline amalgams may, therefore, serve as a source of nascent hydrogen in presence of water, giving rise to an action less energetic, and often more advantageous, than that of the alkaline metals alone. Thus is caused the frequent employment of sodium amalgam for hydrogenizing a large number of bodies. As a consequence of their action on water, the alkaline amalgams are changed by moist air, with production of free alkali or alkaline carbonate.

Applications of Potassium Amalgams. —I.—They furnish a process for preparing potassium by the decomposition of potash by the electric current, by employing quicksilver as the cathode, and vaporizing the quicksilver of the amalgam formed by heating this in a current of dry hydrogen.

II.—They can serve for the preparation of the amalgams of the metals, other than those of the alkaline group, by decomposing the salts of these metals, with formation of a salt of potash and of the amalgam of the metal corresponding to the original salt.

III.—They can be employed as a source of nascent hydrogen in presence of water for hydrogenizing many substances.

Applications of Sodium Amalgams. — These are nearly the same as those of the potassium amalgams, but the sodium amalgams are employed almost exclusively, because sodium is easier to handle than potassium, and is cheaper. These employments are the following:

I.—Sodium amalgam furnishes a process for the preparation of sodium when soda is decomposed by means of the electric current, employing quicksilver as the cathode, and afterwards vaporizing the quicksilver of the amalgam formed by heating this in a current of dry hydrogen.

II.—Amalgams of sodium serve for the preparation of amalgams of the other metals, particularly alkaline earthy metals and metals of high fusing points, by decomposing the salts of these metals, with formation of a salt of soda and of the amalgam of the metal corresponding to the original salt.

III.—They serve for amalgamating superficially the metals of high fusing point, called "refractory," such as iron and platinum, when a well-cleaned plate of these metals is immersed in sodium amalgam in presence of water.

IV.—An amalgam of 2 or 3 per cent of sodium is employed in the processes of extraction of gold by amalgamation. It has the property of rendering quicksilver more brilliant, and consequently more energetic, by acting as a deoxidant on the pellicle of oxide formed on its surface in presence of certain ores, which, by keeping it separated from the particles of gold, destroy its activity. Sodium amalgam of 3 per cent is utilized with success for the amalgamated plates employed in crushers and other apparatus for treating the ores of gold. If a few drops of this amalgam are spread on a plate of copper, of tin, or of zinc, a brilliant coating of an amalgam of tin, copper, or zinc is immediately formed.

V.—Amalgams of from 2 to 8 per cent of sodium serve frequently in laboratories for reducing or hydrogenizing organic combinations, without running the risk of a partial destruction of these compounds by too intense action, as may occur by employing free sodium instead of its amalgam.

Applications of Barium Amalgams.— These can, by distillation, furnish barium. It is one of the processes for preparing this metal, which, when thus obtained, almost always retains a little sodium.

Applications of Strontium Amalgams. —These amalgams, washed and dried rapidly immediately after their preparation, and then heated to a nascent red in a current of dry hydrogen, yield a fused mass of strontium.

Applications of Cadmium Amalgams.—Amalgams of cadmium, formed of equal weights of cadmium and quicksilver, have much power of cohesion and are quite malleable; the case is the same with an amalgam formed of 1 part of cadmium and 2 parts of quicksilver. They are used as dental cements for plugging teeth; for the same purpose an amalgam of 2 parts of quicksilver, 1 part of cadmium, and 2 parts of tin may be used.

Applications of Zinc Amalgams.—The principal employment of zinc amalgams is their use as a cathode or negative electrode in the batteries of Munson, Daniels, and Lechanché. This combination is designed to render the zinc non-attackable by the exciting liquid of the battery with open circuit. The action of the mercury is to prevent the zinc from forming a large number of small voltaic elements when foreign bodies are mingled with the metal; in a word, the giving to ordinary zinc the properties of pure zinc, and consequently of causing a great saving in expense.

For amalgamating a zinc plate it is plunged for a few seconds into water in which there is one-sixteenth in volume of sulphuric acid, then rubbing with a copper-wire brush which has been dipped in the quicksilver. The mercury takes more readily on the zinc when, after the zinc has been cleaned with water sharpened with sulphuric acid, it is moistened with a solution of corrosive sublimate, which is reduced and furnishes a first very thin coat of amalgam, on which the quicksilver is immediately fixed by simple immersion without rubbing.

The zinc of a battery may be amalgamated by putting at the bottom of the compartment containing each element, a little quicksilver in such a way that the zinc touches the liquid. The amalgamation is effected under the influence of the current, but this process applies only on condition that the zinc alone touches the bottom of the vessel containing the quicksilver.

Applications of Manganese Amalgams.—These may serve for the preparation of manganese. For this purpose it is sufficient to distil in a current of pure hydrogen. The manganese remains in the form of a grayish powder.

Applications of Tin Amalgams.—I.—Tinning of glass. This operation is accomplished in the following manner:

On a cast-iron table, quite horizontal, a sheet of tin of the dimensions of the glass is spread out and covered with a layer of quicksilver, 5 or 6 millimeters in thickness. The glass is made to slide on the sheet of tin in such a way as to drive off the excess of quicksilver; when the two surfaces are covered without interposition of air, weights are placed on the glass. In a few days, the glass may be removed, having been covered with an adhering pellicle of amalgam of 4 parts of tin and 1 part of quicksilver. (See also Mirrors.)

II.—An amalgam consisting of 2 parts of zinc and 1 part tin may be used for covering the cushions of frictional electric machines. This amalgam is prepared by first melting the zinc and tin in a crucible and adding the quicksilver previously heated.

III.—Mention has been made of the cadmium amalgam employed for plugging teeth, an amalgam of 2 parts of quicksilver, 2 parts of tin, and 1 part of cadmium. For the same purpose an amalgam of tin, silver, and gold is employed. (See also Cements, Dental.)

Applications of Copper Amalgams.—I.—An amalgam of 30 per cent of copper has been employed for filling teeth. This use has been abandoned on account of the inconvenience occasioned by the great changeableness of the product.

II.—The amalgam of 30 per cent of copper, designated by the name of "metallic mastic," is an excellent cement for repairing objects and utensils of porcelain. For this employment, the broken surfaces are heated to 662° F., and a little of the amalgam, previously heated to the consistency of melted wax, is applied.

III.—Copper amalgam, of 30 to 45 per cent of copper, rendered plastic by heating and grinding, may serve for obtaining with slight compression copies of delicate objects, which may, after hardening of the amalgam, be reproduced, either in wax or by galvanic process.

IV.—According to Debray, when a medal, obtained with an amalgam of 45 per cent of copper, by compression in the soft state, in molds of gutta percha, is heated progressively to redness in an atmosphere of hydrogen, the quicksilver is volatilized gradually, and the particles of copper come together without fusion in such a way as to produce a faithful reproduction, formed exclusively of metallic copper, of the original medal.

V.—In the metallurgy of gold the crushers are furnished with amalgamated plates of copper for retaining the gold. The preparation of these plates, which are at least 0.128 inches in thickness, is delicate, requiring about two weeks. They are freed from greasy matter by rubbing with ashes, or, better, with a little sand and caustic soda, or if more rapid action is desired, with a cloth dipped in dilute nitric acid; they are washed with water, then with a solution of potassium cyanide, and finally brushed with a mixture of sal ammoniac and a little quicksilver, until the surface is completely amalgamated. They are finally made to absorb as much quicksilver as possible. But the plates thus treated are useful for only a few days when they are sufficiently covered with a layer of gold amalgam; in the meantime they occasion loss of time and of gold. So it is preferable to cover them artificially with a little gold amalgam, which is prepared by dissolving gold in quicksilver. Sometimes the amalgam of gold is replaced by an amalgam of silver, which is readily poured and more economical.

Another method giving better results consists in silvering copper slabs by electroplating and covering them with a layer of silver. Then it is only necessary to apply a little quicksilver, which adheres quite rapidly, so that they are ready for use almost immediately, and are quite active at the outset.

These amalgamation slabs ought to be cleaned before each operation. Potassium cyanide removes fatty matter, and sal ammoniac the oxides of the low metals.

Applications of Lead Amalgams.— These meet with an interesting employment for the autogenous soldering of lead. After the surfaces to be soldered have been well cleaned, a layer of lead amalgam is applied. It is afterwards sufficient to pass along the line of junction a soldering iron heated to redness, in order that the heat should cause the volatilization of the quicksilver, and that the lead, liberated in a state of fine division, should be melted and cause the adherence of the two surfaces. The only precaution necessary is to avoid breathing the mercurial vapor, which is quite poisonous.

Applications of Bismuth Amalgams.— The amalgam formed of 1 per cent of bismuth and 4 parts of quicksilver will cause the strong adherence of glass. It is employed with advantage in the tinning of glass globes. For this operation it is poured into a dry hot receiver, and then passed over the whole surface of the glass; it solidifies on cooling. For the purpose of economizing the bismuth, the price of which is high, the preceding amalgam is replaced by another composed of 2 parts of quicksilver, 1 part of bismuth, 1 part of lead, and 1 part of tin. The bismuth, broken into small fragments, is added to the tin and lead, previously melted in the crucible, and when the mixture of the three metals becomes fluid, the quicksilver is poured in, while stirring with an iron rod. The impurities floating on the surface are removed, and when the temperature is sufficiently lowered this amalgam is slowly poured into the vessels to be tinned, which have been previously well cleaned and slightly heated. M. Ditte recommends for the same employment, as a very strong adherent to the glass, an amalgam obtained by dissolving hot 2 parts of bismuth and 1 part of lead in a solution of 1 part of tin in 10 parts of quicksilver. By causing a quantity of this amalgam to move around the inside of a receiver, clean, dry, and slightly heated, the surface will be covered with a thin, brilliant layer, which hardens quite rapidly.

For the injection of anatomical pieces an amalgam formed of 10 parts of quicksilver, 50 parts of bismuth, 31 parts of lead, and 18 parts of tin, fusible at 77.5° and solidifiable at 60° C., is made use of; or, again, an amalgam composed of 9 parts of Darcet alloy and 1 part of quicksilver fusible at 127½° F., and pasty at a still lower temperature. This last amalgam may also be used for filling carious teeth. The Darcet alloy, as known, contains 2 parts of bismuth, 1 part of lead, and 1 part of tin, and melts at 199½° F. The addition of 1 part of quicksilver lowers the fusing point to 104° F.

Applications of Silver Amalgams.—I.— In the silvering of mirrors by the Petitjean method, which has almost universally replaced tinning, the property of silver in readily amalgamating is taken advantage of, by substituting the glass after silvering to the action of a dilute solution of double cyanide of mercury and potassium in such a manner as to form an amalgam of white and brilliant silver adhering strongly to the glass. To facilitate the operation and utilize all the silver, while economizing the double cyanide, M. Lenoir has recommended the following: Sprinkle the glass at the time when it is covered with the mercurial solution with very fine zinc powder, which precipitates the quicksilver and regulates the amalgamation.

II.—The metallurgy of silver also takes advantage of the property of this metal in combining cold with quicksilver; this for the treatment of poor silver ores.

In the Saxon or Freiburg process for treating silver ores, recourse is had to quicksilver in the case of amalgam in amalgamating casks, in which the ore, after grinding, is shaken with disks of iron, and with mercury and water. The amalgam, collected and filtered under strong pressure, contains from 30 to 33 per cent of silver. It is distilled either in cylindrical retorts of cast iron, furnished with an exit tube immersed in the water for condensing the mercurial vapors, or on plates of iron, arranged over each other along a vertical iron stem, supported by a tripod at the bottom of a tank filled with water, and covered with an iron receiver, which is itself surrounded with ignited charcoal. It should be remarked that the last portions of quicksilver in a silver amalgam submitted to distillation are volatilized only under the action of a high and prolonged temperature.

Applications of Gold Amalgams.—I.—

Gilding with quicksilver. This process of gilding, much employed formerly, is now but little used. It can be applied only to metals slightly fusible and capable of amalgamation, like silver, copper, bronze, and brass. Iron can also be gilded by this method, provided it is previously covered with a coating of copper. To perform this gilding the surface is well cleaned, and the gold amalgam, consisting of 2 parts of gold and 1 part of quicksilver, prepared as mentioned before, is applied. The piece is afterwards heated to about the red, so as to volatilize the mercury. The gold remains, superficially alloyed with the metal, and forms an extremely solid layer of deadened gold, which can be afterwards polished. The volatilization should be effected under a chimney having strong draught, in order to avoid the poisonous action of the mercurial vapors.

II.—The amalgamation of gold finds its principal applications in the treatment of auriferous ores. The extraction of small spangles of gold scattered in gold-bearing sands is based on the ready dissolution of gold in quicksilver, and on the formation of an amalgam of solid gold by compression and filtering through a chamois skin, in a state more or less liquid. The spangles of gold are shaken with about their weight of quicksilver, collected in the cavities of sluices and mixed with a small quantity of sand. The gold is dissolved and the sand re-

mains. The amalgam thus obtained is compressed in a chamois skin, so as to separate the excess of mercury which passes through the pores of the skin; or, yet again, it is filtered through a glass funnel having a very slender stem, with almost capillary termination. In both cases an amalgam of solid gold remains, which is submitted to the action of heat in a crucible or cast-iron retort, communicating with a bent-iron tube, of which the extremity, surrounded with a cloth immersed in water, is arranged above a receiver half full of water. The quicksilver is vaporized and condensed in the water. The gold remains in the retort.

The property of gold of combining readily with quicksilver is also used in many kinds of amalgamating apparatus for extraction and in the metallurgy of gold.

In various operations it is essential to keep the quicksilver active by preserving its limpidity. For this purpose potassium cyanide and ammonium chloride are especially employed; sometimes wood ashes, carbonate of soda, hyposulphite of soda, nitrate of potash, cupric sulphate, sea salt, and lime; the latter for precipitating the soluble sulphates proceeding from the decomposition of pyrites.

The amalgamation of gold is favored by a temperature of 38° to 45° C. (100° to 113° F.), and still more by the employment of quicksilver in the nascent state. This last property is the base of the Designol process, which consists in treating auriferous or auro-argentiferous ores, first ground with sea salt, in revolving cylinders of cast iron, with iron and mercury bichloride, in such a way that the mercury precipitated collects the gold and eventually the silver more efficaciously.

Gold Amalgam.—

Eight parts of gold and 1 of mercury are formed into an amalgam for plating by rendering the gold into thin plates, making it red hot, and then putting it into the mercury while the latter is also heated to ebullition. The gold immediately disappears in combination with the mercury, after which the mixture may be turned into water to cool. It is then ready for use.

Zinc Amalgam for Electric Batteries.

—Dissolve 2 parts of mercury in 1 part of aqua regia. This accomplished, add 5 parts of hydrochloric acid. This solution is made warm. It suffices to dip the zinc to be amalgamated into this liquid only for a few seconds.

Amalgam for Cementing Glass, Porcelain, Etc.—Take tin 2 parts, and cadmium 1 part. Fuse in an iron spoon or some vessel of the same material. When the two materials are in fusion add a little mercury, previously heated. Place all in an iron crucible and boil, agitating the mass with a pestle. This amalgam is soft and can be kneaded between the fingers. It may be employed for luting glass or porcelain vessels, as well as for filling teeth. It hardens in a short while.

Amalgam for Silvering Glass Balls.—Lead, 25 parts; tin, 25 parts; bismuth, 25 parts; mercury, 25 parts; or, lead, 20 parts; tin, 20 parts; bismuth, 20 parts; mercury, 40 parts. Melt the lead and the tin, then add the bismuth; skim several times and add the mercury, stirring the composition vigorously.

(See also Mirror-Silvering).

Copper Amalgam.—Copper amalgam, or so-called Viennese metal cement, crystallizes with the greatest readiness and acquires such hardness on solidifying that it can be polished like gold. The amalgam may also be worked under the hammer or between rollers; it can also be stamped, and retains its metallic luster for a long time in the air. In air containing hydrogen sulphide, however, it quickly tarnishes and turns black. A very special property of copper amalgam consists in that it becomes very soft when laid in water, and attains such pliancy that it can be employed for modeling the most delicate objects. After a few hours the amalgam congeals again into a very fine-grained, rather malleable mass. An important application of copper amalgam is that for cementing metals. All that is necessary for this purpose is to heat the metals, which must be bright, to 80–90° C. (176–194° F.), to apply the amalgam and to press the metal pieces together. They will cohere as firmly as though soldered together.

Copper amalgam may be prepared in the following manner:

Place strips of zinc in a solution of blue vitriol and agitate the solution thoroughly. The copper thus obtained in the form of a very fine powder is washed and, while still moist, treated in a mortar with a solution of mercury nitrate. The copper powder thereby amalgamates more readily with the quicksilver. Next, hot water is poured over the copper, the mortar is kept hot, and the mercury added. Knead with the pestle of the mortar until the copper, pulverulent in the beginning, has united with the mercury into a very plastic mass. The longer the kneading is continued the more uniform will be the mass. As soon as the amalgam has acquired the suitable character—for its production 3 parts of copper and 7 parts of mercury are used—the water is poured off and the amalgam still soft is given the shape in which it is to be kept.

For cementing purposes, the amalgam is rolled out into small cylinders, whose diameter is about 0.16 to 0.2 inches, with a length of a few inches. In order to produce with this amalgam impressions of castings, which are made after woodcuts, the amalgam is rolled out hot into a thin plate and pressed firmly onto the likewise heated plaster cast. After the amalgam has hardened the thin plate of it may be reinforced by pouring on molten type metal.

Silver Amalgam.—Silver amalgam can easily be made with the help of finely powdered silver. The mercury need only be heated to 250° to 300° C. (482° to 572° F.); silver powder is then sprinkled on it, and mixed with it by stirring. The vessel is heated for several minutes and then allowed to cool, the excess of mercury being removed from the granulated crystalline amalgam by pressing in a leather bag. Silver amalgam can also easily be made by dissolving silver in nitric acid, evaporating the solution till the excess of free acid is eliminated, diluting with distilled water, and adding mercury to the fluid in the proportion of 4 parts, by weight, of mercury to 1 of the silver originally used. The mercury precipitates the silver in a metallic state, and immediately forms an amalgam with it; the fluid standing above after a time contains no more silver, but consists of a solution of mercury nitrate mixed with whatever copper was contained in the dissolved silver in the form of copper nitrate. The absence of a white precipitate, if a few drops of hydrochloric acid are added to a sample of the fluid in a test tube, shows that all the silver has been eliminated from the solution and is present in the form of amalgam.

Amalgam for the Rubber of Electric Machines.—Mercury, 100 parts; zinc, 50 parts; tin, 50 parts. This amalgam reduced to powder and incorporated with grease can be applied to the rubber of electric machines.

AMALGAM GOLD PLATING:

See Gilding under Plating.

AMBER:

Imitation Amber.—Melt carefully together pine rosin, 1; lacca in tabulis, 2; white colophony, 15 parts.

AMBER CEMENT:
See Adhesives under Cements.

AMBER VARNISH:
See Varnishes.

AMBROSIA POWDER:
See Salts (Effervescent).

AMIDOL DEVELOPER:
See Photography.

AMETHYST (IMITATION):
See Gems, Artificial.

AMMON-CARBONITE:
See Explosives.

Ammonia

Household Ammonia.—(See also Household Formulas.)—Household ammonia is simply diluted ammonia water to which borax and soap have been added. To make it cloudy add potassium nitrate or methylated spirit. The following are good formulas:

I.—Ammonia water .... 16 parts
Yellow soap .... 64 parts
Potassium nitrate... 1 part
Soft water, sufficient
to make........ 200 parts

Shave up the soap and dissolve it in the water by heating, add the potassium nitrate and dissolve. Cool, strain, skim off any suds or bubbles, add the ammonia, mix, and bottle at once.

II.—Yellow soap....... 10 grains
Borax............ 1 drachm
Lavender water.... 20 minims
Stronger ammonia
water.......... 6 ounces
Water, enough to
make.......... 20 ounces

Dissolve the soap and borax in 5 ounces of boiling water; when cold add the lavender water and ammonia, and make up to a pint with water.

III.—Methylated spirit... 1 gallon
Soft water........ 1 gallon
Stronger ammonia
water.......... 1 gallon

IV.—Ammonia water.... 5 pints
Distilled water..... 5 pints
Soap............ 100 grains
Olive oil......... 5 drachms

Cut the soap in shavings, boil with the oil and water, cool, add the ammonia water, and bottle. For use in laundries, baths, and for general household purposes add one tablespoonful to one gallon of water.

V.—The best quality:
Alcohol, 94 per cent.. 4 ounces
Soft water.......... 4 gallons
Oil of rosemary...... 4 drachms
Oil of citronella...... 3 drachms

Dissolve the oils in the alcohol and add to the water. To the mixture add 4 ounces of talc (or fuller's earth will answer), mix thoroughly, strain through canvas, and to the colate add 1, 2, or 3 gallons of ammonia water, according to the strength desired, in which has been dissolved 1, 2, or 3 ounces of white curd, or soft soap.

Liquor Ammonii Anisatus.—
Oil of anise, by weight........ 1 part
Alcohol, by weight............ 24 parts
Water of ammonia, by weight.. 5 parts

Dissolve the oil in the alcohol and add the water of ammonia.

It should be a clear, yellowish liquid.

Violet Color for Ammonia.—A purple-blue color may be given to ammonia water by adding an aqueous solution of litmus. The shade, when pale enough, will probably meet all views as to a violet color.

Perfumed Ammonia Water.—The following are typical formulas:

I.—Stronger water of am-
monia............ 6 ounces
Lavender water..... 1 ounce
Soft soap.......... 10 grains
Water, enough to
make............ 16 ounces

II.—Soft soap.......... 1 ounce
Borax............ 2 drachms
Cologne water....... ½ ounce
Stronger water of am-
monia............ 5½ ounces
Water, enough to
make............ 12 ounces

Rub up the soap and borax with water until dissolved, strain and add the other ingredients. The perfumes may be varied to suit the price.

AMMONIA FOR FIXING PRINTS:
See Photography.

ANGOSTURA BITTERS:
See Wines and Liquors.

ANILINE:
See Dyes.

ANILINE IN PIGMENTS, TESTS FOR:
See Pigments.

ANILINE STAINS, TO REMOVE:
See Cleaning Preparations and Methods.

ANISE CORDIAL:
See Wines and Liquors.

ANKARA:
See Butter.

ANNEALING OF STEEL, TOOLS, WIRE, AND SPRINGS:
See Steel.

ANODYNES:
See Pain Killers.

ANT DESTROYERS:
See Insecticides.

Antidotes for Poisons

POISON, SYMPTOMS AND ANTIDOTES.

When a person has taken poison the first thing to do is to compel the patient to vomit, and for that purpose give any emetic that can be most readily and quickly obtained, and which is prompt and energetic, but safe in its action. For this purpose there is, perhaps, nothing better than a large teaspoonful of ground mustard in a tumblerful of warm water, and it has the advantage of being almost always at hand. If the dry mustard is not to be had use mixed mustard from the mustard pot. Its operation may generally be facilitated by the addition of a like quantity of common table salt. If the mustard is not at hand, give two or three teaspoonfuls of powdered alum in syrup or molasses, and give freely of warm water to drink; or give 10 to 20 grains of sulphate of zinc (white vitriol), or 20 to 30 grains of ipecac, with 1 or 2 grains of tartar emetic, in a large cup of warm water, and repeat every ten minutes until three or four doses are given, unless free vomiting is sooner produced. After vomiting has taken place large draughts of warm water should be given, so that the vomiting will continue until the poisonous substances have been thoroughly evacuated, and then suitable antidotes should be given. If vomiting cannot be produced the stomach pump should be used. When it is known what particular kind of poison has been swallowed, then the proper antidote for that poison should be given; but when this cannot be ascertained, as is often the case, give freely of equal parts of calcined magnesia, pulverized charcoal, and sesquioxide of iron, in a sufficient quantity of water. This is a very harmless mixture and is likely to be of great benefit, as the ingredients, though very simple, are antidotes for the most common and active poisons. In case this mixture cannot be obtained, the stomach should be soothed and protected by the free administration of demulcent, mucilaginous, or oleaginous drinks, such as the whites of eggs, milk, mucilage of gum arabic, or slippery-elm bark, flaxseed tea, starch, wheat flour, or arrowroot mixed in water, linseed or olive oil, or melted butter or lard. Subsequently the bowels should be moved by some gentle laxative, as a tablespoonful or two of castor oil, or a teaspoonful of calcined magnesia; and pain or other evidence of inflammation must be relieved by the administration of a few drops of laudanum, and the repeated application of hot poultices, fomentations, and mustard plasters.

The following are the names of the substances that may give rise to poisoning, most commonly used, and their antidotes:

Mineral Acids—Sulphuric Acid (Oil of Vitriol), Nitric Acid (Aqua Fortis), Muriatic Acid (Spirits of Salts).—Symptoms: Acid, burning taste in the mouth, acute pain in the throat, stomach, and bowels; frequent vomiting, generally bloody; mouth and lips excoriated, shriveled, white or yellow; hiccough, copious stools, more or less bloody, with great tenderness in the abdomen; difficult breathing, irregular pulse, excessive thirst, while drink increases the pain and rarely remains in the stomach; frequent but vain efforts to urinate; cold sweats, altered countenance; convulsions, generally preceding death. Nitric acid causes yellow stains; sulphuric acid, black ones. Treatment: Mix calcined magnesia in milk or water to the consistence of cream, and give freely to drink a glassful every couple of minutes, if it can be swallowed. Common soap (hard or soft), chalk, whiting, or even mortar from the wall mixed in water may be given, until magnesia can be obtained. Promote vomiting by tickling the throat, if necessary, and when the poison is got rid of, flaxseed or slippery-elm tea, gruel, or other mild drinks. The inflammation which always follows needs good treatment to save the patient's life.

Vegetable Acids—Acetic, Citric, Oxalic, Tartaric.—Symptoms: Intense burning pain of mouth, throat, and stomach; vomiting blood which is highly acid, violent purging, collapse, stupor, death.

Oxalic acid is frequently taken in

mistake for Epsom salts, to which in shops it often bears a strong resemblance. Treatment: Give chalk or magnesia in a large quantity of water, or large draughts of limewater. If these are not at hand, scrape the wall or ceiling, and give the scrapings mixed with water.

Prussic or Hydrocyanic Acid—Laurel Water, Cyanide of Potassium, Bitter Almond Oil, Etc.—Symptoms: In large doses almost invariably instantaneously fatal; when not immediately fatal, sudden loss of sense and control of the voluntary muscles. The odor of the poison generally noticeable on the breath. Treatment: Chlorine, in the form of chlorine water, in doses of from 1 to 4 fluidrachms, diluted. Weak solution of chloride lime of soda; water of ammonia (spirits of hartshorn), largely diluted, may be given, and the vapor of it cautiously inhaled. Cold affusion, and chloroform in half to teaspoonful doses in glycerine or mucilage, repeated every few minutes, until the symptoms are ameliorated. Artificial respiration.

Aconite — Monkshood, Wolfsbane.— Symptoms: Numbness and tingling in the mouth and throat, and afterwards in other portions of the body, with sore throat, pain over the stomach, and vomiting; dimness of vision, dizziness, great prostration, loss of sensibility, and delirium. Treatment: An emetic and then brandy in tablespoonful doses, in ice water, every half hour; spirits of ammonia in half-teaspoonful doses in like manner; the cold douche over the head and chest, warmth to the extremities, etc.

Alkalis and Their Salts—Concentrated Lye, Wood-ash Lye, Caustic Potash, Ammonia, Hartshorn.—Symptoms: Caustic, acrid taste, excessive heat in the throat, stomach, and intestines; vomiting of bloody matter, cold sweats, hiccough, purging of bloody stools. Treatment: The common vegetable acids. Common vinegar, being always at hand, is most frequently used. The fixed oils, as castor, flaxseed, almond, and olive oils form soaps with the alkalis and thus also destroy their caustic effect. They should be given in large quantity.

Antimony and Its Preparations—Tartar Emetic, Antimonial Wine, Kerme's Mineral. — Symptoms: Faintness and nausea, soon followed by painful and continued vomiting, severe diarrhea, constriction and burning sensation in the throat, cramps, or spasmodic twitchings, with symptoms of nervous derangement, and great prostration of strength, often terminating in death. Treatment: If vomiting has not been produced, it should be effected by tickling the fauces, and administering copious draughts of warm water. Astringent infusions, such as of gall, oak bark, Peruvian bark, act as antidotes, and should be given promptly. Powdered yellow bark may be used until the infusion is prepared, or very strong green tea should be given. To stop the vomiting, should it continue, blister over the stomach by applying a cloth wet with strong spirits of hartshorn, and then sprinkle on one-eighth to one-fourth of a grain of morphia.

Arsenic and Its Preparations—Ratsbane, Fowler's Solution, Etc.— Symptoms: Generally within an hour pain and heat are felt in the stomach, soon followed by vomiting, with a burning dryness of the throat and great thirst; the matters vomited are generally colored either green yellow, or brown, and are sometimes bloody. Diarrhea or dysentery ensues, while the pulse becomes small and rapid, yet irregular. Breathing much oppressed; difficulty in vomiting may occur, while cramps, convulsions, or even paralysis often precede death, which sometimes takes place within five or six hours after arsenic has been taken. Treatment: Give a prompt emetic, and then hydrate of peroxide of iron (recently prepared) in tablespoonful doses every 10 or 15 minutes until the urgent symptoms are relieved. In the absence of this, or while it is being prepared, give large draughts of new milk and raw eggs, limewater and oil, melted butter, magnesia in a large quantity of water, or even if nothing else is at hand, flour and water, always, however, giving an emetic the first thing, or causing vomiting by tickling the throat with a feather, etc. The inflammation of the stomach which follows must be treated by blisters, hot fomentations, mucilaginous drinks, and the like.

Belladonna, or Deadly Nightshade.— Symptoms: Dryness of the mouth and throat, great thirst, difficulty of swallowing, nausea, dimness, confusion or loss of vision, great enlargement of the pupils, dizziness, delirium, and coma. Treatment: There is no known antidote. Give a prompt emetic and then reliance must be placed on continual stimulation with brandy, whisky, etc., and to necessary artificial respiration. Opium and its preparations, as morphia, laudanum, etc., are thought by some to counteract the effect of belladonna, and may be given in small and repeated doses, as also strong black coffee and green tea.

Blue Vitriol, or Blue Stone.—See Copper.

Cantharides (Spanish or Blistering Fly) and Modern Potato Bug.—Symptoms: Sickening odor of the breath, sour taste, with burning heat in the throat, stomach, and bowels; frequent vomiting, often bloody; copious bloody stools, great pain in the stomach, with burning sensation in the bladder and difficulty to urinate followed with terrible convulsions, delirium, and death. Treatment: Excite vomiting by drinking plentifully of sweet oil or other wholesome oils, sugar and water, milk, or slippery-elm tea; give injections of castor oil and starch, or warm milk. The inflammatory symptoms which generally follow must be treated by a physician. Camphorated oil or camphorated spirits should be rubbed over the bowels, stomach, and thighs.

Caustic Potash.—See Alkalis under this title.

Cobalt, or Fly Powder.—Symptoms: Heat and pain in the throat and stomach, violent retching and vomiting, cold and clammy skin, small and feeble pulse, hurried and difficult breathing, diarrhea, etc. Treatment: An emetic, followed by the free administration of milk, eggs, wheat flour and water, and mucilaginous drinks.

Copper—Blue Vitriol, Verdigris or Pickles or Food Cooked in Copper Vessels.—Symptoms: General inflammation of the alimentary canal, suppression of urine; hiccough, a disagreeable metallic taste, vomiting, violent colic, excessive thirst, sense of tightness of the throat, anxiety; faintness, giddiness, and cramps and convulsions generally precede death. Treatment: Large doses of simple syrup as warm as can be swallowed, until the stomach rejects the amount it contains. The whites of eggs and large quantities of milk. Hydrated peroxide of iron.

Creosote—Carbolic Acid.—Symptoms: Burning pain, acrid, pungent taste, thirst, vomiting, purging, etc. Treatment: An emetic and the free administration of albumen, as the whites of eggs, or, in the absence of these, milk, or flour and water.

Corrosive Sublimate.—See Mercury under this title.

Deadly Nightshade.—See Belladonna under this title.

Foxglove, or Digitalis.—Symptoms: Loss of strength, feeble, fluttering pulse, faintness, nausea and vomiting and stupor; cold perspiration, dilated pupils, sighing, irregular breathing, and sometimes convulsions. Treatment: After vomiting, give brandy and ammonia in frequently repeated doses, apply warmth to the extremities, and if necessary resort to artificial respiration.

Gases—Carbonic Acid, Chlorine, Cyanogen, Hydrosulphuric Acid, Etc.—Symptoms: Great drowsiness, difficult respiration, features swollen, face blue as in strangulation. Treatment: Artificial respiration, cold douche, friction with stimulating substances to the surface of the body. Inhalation of steam containing preparations of ammonia. Cupping from nape of neck. Internal use of chloroform.

Hellebore, or Indian Poke.—Symptoms: Violent vomiting and purging, bloody stools, great anxiety, tremors, vertigo, fainting, sinking of the pulse, cold sweats, and convulsions. Treatment: Excite speedy vomiting by large draughts of warm water, molasses and water, tickling the throat with the finger or a feather, and emetics; give oily and mucilaginous drinks, oily purgatives, and clysters, acids, strong coffee, camphor, and opium.

Hemlock (Conium).—Symptoms: Dryness of the throat, tremors, dizziness, difficulty of swallowing, prostration, and faintness, limbs powerless or paralyzed, pupils dilated, pulse rapid and feeble; insensibility and convulsions sometimes precede death. Treatment: Empty the stomach and give brandy in tablespoonful doses, with half teaspoonful of spirits of ammonia, frequently repeated, and if much pain and vomiting, give bromide of ammonium in 5-grain doses every half hour. Artificial respiration may be required.

Henbane, or Hyoscyamus.—Symptoms: Muscular twitching, inability to articulate plainly, dimness of vision and stupor; later, vomiting and purging, small intermittent pulse, convulsive movement of the extremities, and coma. Treatment: Similar to opium poisoning, which see.

Iodine.—Symptoms: Burning pain in throat, lacerating pain in the stomach, fruitless effort to vomit, excessive tenderness of the epigastrium. Treatment:

Free emesis, prompt administration of starch, wheat flour, or arrowroot, beaten up in water.

Lead—Acetate of Lead, Sugar of Lead, Dry White Lead, Red Lead, Litharge, or Pickles, Wine, or Vinegar Sweetened by Lead.—Symptoms: When taken in large doses, a sweet but astringent metallic taste exists, with constriction in the throat, pain in the region of the stomach, painful, obstinate, and frequently bloody vomitings, hiccough, convulsions or spasms, and death. When taken in small but long-continued doses it produces colic, called painters' colic; great pain, obstinate constipation, and in extreme cases paralytic symptoms, especially wrist-drop, with a blue line along the edge of the gums. Treatment: To counteract the poison give alum in water 1½ ounce to a quart; or, better still, Epsom salts or Glauber's salts, an ounce of either in a quart of water; or dilute sulphuric acid, a teaspoonful to a quart of water. If a large quantity of sugar of lead has been recently taken, empty the stomach by an emetic of sulphate of zinc (1 drachm in a quart of water), giving one-fourth to commence, and repeating smaller doses until free vomiting is produced; castor oil should be given to clear the bowels and injections of oil and starch freely administered. If the body is cold use the warm bath.

Meadow Saffron.—See Belladonna.

Laudanum.—See Opium.

Lobelia — Indian Poke. — Symptoms: Excessive vomiting and purging, pains in the bowels, contraction of the pupils, delirium, coma, and convulsions. Treatment: Mustard over the stomach, and brandy and ammonia.

Mercury—Corrosive Sublimate (bug poisons frequently contain this poison), **Red Precipitate, Chinese or English Vermilion.**—Symptoms: Acrid, metallic taste in the mouth, immediate constriction and burning in the throat, with anxiety and tearing pains in both stomach and bowels, sickness, and vomiting of various-colored fluids, and sometimes bloody and profuse diarrhea, with difficulty and pain in urinating; pulse quick, small, and hard; faint sensations, great debility, difficult breathing, cramps, cold sweats, syncope, and convulsions. Treatment: If vomiting does not already exist, emetics must be given immediately—white of eggs in continuous large doses, and infusion of catechu afterwards, sweet milk, mixtures of flour and

water in successive cupfuls, and to check excessive salivation put a half ounce of chlorate of potash in a tumbler of water, and use freely as a gargle, and swallow a tablespoonful every hour or two.

Morphine.—See Opium.

Nitrate of Silver (Lunar Caustic).—Symptoms: Intense pain and vomiting, and purging of blood, mucus, and shreds of mucous membranes; and if these stand they become dark. Treatment: Give freely of a solution of common salt in water, which decomposes the poison, and afterwards flaxseed or slippery-elm-bark tea, and after a while a dose of castor oil.

Opium and All Its Compounds—Morphine, Laudanum, Paregoric, Etc.—Symptoms: Giddiness, drowsiness, increasing to stupor, and insensibility; pulse usually, at first, quick and irregular, and breathing hurried, and afterwards pulse slow and feeble, and respiration slow and noisy; the pupils are contracted and the eyes and face congested, and later, as death approaches, the extremities become cold, the surface is covered with cold, clammy perspiration, and the sphincters relax. The effects of opium and its preparations, in poisonous doses, appear in from a half to two hours from its administration. Treatment: Empty the stomach immediately with an emetic or with the stomach pump. Then give very strong coffee without milk; put mustard plasters on the wrists and ankles; douche the head and chest with cold water, and if the patient is cold and sinking, give brandy, or whisky and ammonia. Belladonna is thought by many to counteract the poisonous effects of opium, and may be given in doses of half to a teaspoonful of the tincture, or 2 grains of the extract, every 20 minutes, until some effect is observed in causing the pupils to expand. Use warmth and friction, and if possible prevent sleep for some hours, for which purpose the patient should be walked about between two persons. Finally, as a last resort, use artificial respiration, persistence in which will sometimes be rewarded with success in apparently hopeless cases. Electricity should also be tried.

Cooley advises as follows: Vomiting must be induced as soon as possible, by means of a strong emetic and tickling the fauces. If this does not succeed, the stomach pump should be applied. The emetic may consist of a half drachm of sulphate of zinc dissolved in a half pint of warm water, of which one-third should be taken at once, and the remainder at the rate of a wineglassful every 5 or 10 minutes, until vomiting commences. When there is much drowsiness or stupor 1 or 2 fluidrachms of tincture of capsicum will be found a useful addition; or one of the formulas for emetic draughts may be taken instead. Infusion of galls, cinchona, or oak bark should be freely administered before the emetic, and water soured with vinegar and lemon juice, after the stomach has been well cleared out. To rouse the system spirit and water or strong coffee may be given. To keep the sufferer awake, rough friction should be applied to the skin, an upright posture preserved, and walking exercise enforced, if necessary. When this is ineffectual cold water may be dashed over the chest, head, and spine, or mild shocks of electricity may be had recourse to. To allow the sufferer to sleep is to abandon him to destruction. Bleeding may be subsequently necessary in plethoric habits, or in threatened congestion. The costiveness that accompanies convalescence may be best met by aromatic aperients; and the general tone of the habit restored by stimulating tonics and the shower bath. The smallest fatal dose of opium in the case of an adult within our recollection was 4½ grains. Children are much more susceptible to the action of opium than of other medicines, and hence the dose of it for them must be diminished considerably below that indicated by the common method of calculation depending on the age.

Oxalic Acid.—See Acids.

Phosphorus—Found in Lucifer Matches and Some Rat Poisons.—Symptoms: Symptoms of irritant poisoning; pain in the stomach and bowels; vomiting, diarrhea; tenderness and tension of the abdomen. Treatment: An emetic is to be promptly given; copious draughts containing magnesia in suspension; mucilaginous drinks. General treatment for inflammatory symptoms.

Poisonous Mushrooms.—Symptoms: Nausea, heat and pains in the stomach and bowels; vomiting and purging, thirst, convulsions, and faintings; pulse small and frequent, dilated pupil and stupor, cold sweats and death. Treatment: The stomach and bowels are to be cleared by an emetic of ground mustard or sulphate of zinc, followed by frequent doses of Glauber's or of Epsom salts, and large stimulating clysters. After the poison is evacuated, either may be given with small quantities of brandy and water. But if inflammatory symptoms manifest themselves such stimuli should be avoided, and these symptoms appropriately treated. A hypodermic injection of $\frac{1}{32}$ grain of atropine is the latest discovered antidote.

Potash.—See Alkali.

Prussic or Hydrocyanic Acid.—See Acids.

Poison Ivy.—Symptoms: Contact with, and with many persons the near approach to, the vine gives rise to violent erysipelatous inflammation, especially of the face and hands, attended with itching, redness, burning, and swelling, with watery blisters. Treatment: Give saline laxatives, and apply weak sugar of lead and laudanum, or limewater and sweet oil, or bathe the parts freely with spirits of niter. Anointing with oil will prevent poisoning from it.

Saltpeter (Nitrate of Potash).—Symptoms: Only poisonous in large quantities, and then causes nausea, painful vomiting, purging, convulsions, faintness, feeble pulse, cold feet and hands, with tearing pains in stomach and bowels. Treatment: Treat as is directed for arsenic, for there is no antidote known, and emptying the stomach and bowels with mild drinks must be relied on.

Savine.—Symptoms: Sharp pains in the bowels, hot skin, rapid pulse, violent vomiting and sometimes purging, with great prostration. Treatment: Mustard and hot fomentations over the stomach and bowels and ice allowed in the stomach only until the inflammation ceases. If prostration comes on, food and stimulants must be given by injection.

Stramonium, Thorn Apple, or Jamestown Weed.—Symptoms: Vertigo, headache, perversion of vision, slight delirium, sense of suffocation, disposition to sleep, bowels relaxed, and all secretions augmented. Treatment: Same as for belladonna.

Snake Bites, Cure for.—The Inspector of Police in the Bengal Government reports that of 939 cases in which ammonia was freely administered, 207 victims have recovered, and in the cured instances the remedy was not administered till about 3½ hours after the attack; on the average of the fatal cases the corresponding duration of time was 4½ hours.

Strychnine or Nux Vomica.—The characteristic symptom is the special influence exerted upon the nervous system, which is manifested by a general contraction of all the muscles of the body, with rigidity of the spinal column. A profound calm soon succeeds, which is followed by a new tetanic seizure, longer than the first, during which the respiration is suspended. These symptoms then cease, the breathing becomes easy, and there is stupor, followed by another general contraction. In fatal cases these attacks are renewed, at intervals, with increasing violence, until death ensues. One phenomenon which is found only in poisonings by substances containing strychnine is that touching any part of the body, or even threatening to do so, instantly produces the tetanic spasm. Antidote: The stomach should be immediately cleared by means of an emetic, tickling the fauces, etc. To counteract the asphyxia from tetanus, etc., artificial respiration should be practiced with diligence and care. "If the poison has been applied externally, we ought immediately to cauterize the part, and apply a ligature tightly above the wound. If the poison has been swallowed for some time we should give a purgative clyster, and administer draughts containing sulphuric ether or oil of turpentine, which in most cases produce a salutary effect. Lastly, injections of chlorine and decoction of tannin are of value."

According to Ch. Gunther the greatest reliance may be placed on full doses of opium, assisted by venesection, in cases of poisoning by strychnia or nux vomica. His plan is to administer this drug in the form of solution or mixture, in combination with a saline aperient.

Another treatment is to give, if obtainable, 1 ounce or more of bone charcoal mixed with water, and follow with an active emetic; then to give chloroform in teaspoonful doses, in flour and water or glycerine, every few minutes while the spasms last, and afterwards brandy and stimulants, and warmth of the extremities if necessary. Recoveries have followed the free and prompt administration of oils or melted butter or lard. In all cases empty the stomach if possible.

Sulphate of Zinc—White Vitriol.—See Zinc.

Tin—Chloride of Tin, Solution of Tin (used by dyers), **Oxide of Tin, or Putty Powder.**—Symptoms: Vomiting, pains in the stomach, anxiety, restlessness, frequent pulse, delirium, etc. Treatment: Empty the stomach, and give whites of eggs in water, milk in large quantities,

or flour beaten up in water, with magnesia or chalk.

Tartar Emetic.—See Antimony.

Tobacco.—Symptoms: Vertigo, stupor, fainting, nausea, vomiting, sudden nervous debility, cold sweat, tremors, and at times fatal prostration. Treatment: After the stomach is empty apply mustard to the abdomen and to the extremities, and give strong coffee, with brandy and other stimulants, with warmth to the extremities.

Zinc—Oxide of Zinc, Sulphate of Zinc, White Vitriol, Acetate of Zinc.—Symptoms: Violent vomiting, astringent taste, burning pain in the stomach, pale countenance, cold extremities, dull eyes, fluttering pulse. Death seldom ensues, in consequence of the emetic effect. Treatment: The vomiting may be relieved by copious draughts of warm water. Carbonate of soda, administered in solution, will decompose the sulphate of zinc. Milk and albumen will also act as antidotes. General principles to be observed in the subsequent treatment.

Woorara.—Symptoms: When taken into the stomach it is inert; when absorbed through a wound it causes sudden stupor and insensibility, frothing at the mouth, and speedy death. Treatment: Suck the wound immediately, or cut it out and tie a cord around the limb between the wound and the heart. Apply iodine, or iodide of potassium, and give it internally, and try artificial respiration.

ANTIFERMENTS.

The following are tried and useful formulas:

I.—Sulphite (not sulphate) of lime, in fine powder, 1 part; marble dust, ground oyster shells, or chalk, 7 parts; mix, and pack tight, so as to exclude the air.

II.—Sulphite (not sulphate) of potassa, 1 part; new black-mustard seed (ground in a pepper mill), 7 parts; mix, and pack so as to exclude air and moisture perfectly. Dose (of either), $\frac{1}{2}$ ounce to $1\frac{1}{2}$ ounces per hogshead.

III.—Mustard seed, 14 pounds; cloves and capsicum, of each, $1\frac{1}{4}$ pounds; mix, and grind them to powder in a pepper mill. Dose, $\frac{1}{4}$ to $\frac{1}{2}$ pound per hogshead.

A portion of any one of these compounds added to cider, or the like, soon allays fermentation, when excessive, or when it has been renewed. The first formula is preferred when there is a tendency to acidity. The second and third may be advantageously used for wine and beer, as well as for cider. The third compound greatly improves the flavor and the apparent strength of the liquor, and also improves its keeping qualities.

Anchovy Preparations

Extemporaneous Anchovy Sauce.—

Anchovies, chopped small	3 or 4
Butter	3 ounces
Water	2 ounces
Vinegar	1 ounce
Flour	1 ounce

Mix, place over the fire, and stir until the mixture thickens. Then rub through a coarse sieve.

Essence of Anchovies.—Remove the bones from 1 pound of anchovies, reduce the remaining portions of the fish to a pulp in a Wedgewood mortar, and pass through a clean hair or brass sieve. Boil the bones and other portions which will not pass through the sieve in 1 pint of water for 15 minutes, and strain. To the strained liquor add 2½ ounces of salt and 2¼ ounces of flour, and the pulped anchovies. Let the whole simmer over the fire for three or four minutes; remove from the fire, and when the mixture has cooled a little add 4 ounces of strong vinegar. The product (nearly 3 pounds) may be then bottled, and the corks tied over with bladder, and either waxed or capsuled.

Anchovy Paste.—

Anchovies	7	pounds
Water	9	pints
Salt	1	pound
Flour	1	pound
Capsicum	¼	ounce
Grated lemon peel	1	
Mushroom catsup	4	ounces

Anchovy Butter.—

Anchovies, boned and beaten to a paste	1 part
Butter	2 parts
Spice	enough

ANTIFOULING COMPOSITIONS:
See Paints.

ANTIFREEZING SOLUTION:
See Freezing Preventives.

ANTIFRICTION METAL:
See Alloys, under Phosphor Bronze and Antifriction Metals.

ANTIQUES, TO PRESERVE.

The best process for the preservation of antique metallic articles consists in a retransformation of the metallic oxides into metal by the electrolytic method. For this purpose a zinc strip is wound around the article and the latter is laid in a soda-lye solution of 5 per cent, or suspended as the negative pole of a small battery in a potassium cyanide solution of 2 per cent. Where this method does not seem practicable it is advisable to edulcorate the objects in running water, in which operation fragile or easily destroyed articles may be protected by winding with gauze; next, they should be carefully dried, first in the air, then with moderate heat, and finally protected from further destruction by immersion in melted paraffine. A dry place is required for storing the articles, since paraffine is not perfectly impermeable to water in the shape of steam.

ANTIRUST COMPOSITIONS:
See Rust Preventives.

Antiseptics

Antiseptic Powders.—

I.—Borax	3	ounces
Dried alum	3	ounces
Thymol	22	grains
Eucalyptol	20	drops
Menthol	1½	grains
Phenol	15	grains
Oil of gaultheria	4	drops
Carmine to give a pink tint.		

II.—Alum, powdered	50	
Borax, powdered	50	
Carbolic acid, crystals	5	Parts by weight
Oil of eucalyptus	5	
Oil of wintergreen	5	
Menthol	5	
Thymol	5	

III.—Boracic acid	10	ounces
Sodium biborate	4	ounces
Alum	1	ounce
Zinc sulphocarbolate	1	ounce
Thymic acid	1	drachm.

Mix thoroughly. For an antiseptic wash dissolve 1 or 2 drachms in a quart of warm water.

IV.—Ektogan is a new dusting powder which is a mixture of zinc hydroxide and dioxide. It is equivalent to about 8 per cent of active oxygen. It is a yellowish-white odorless and tasteless powder, insoluble in water. It is used externally in wounds and in skin diseases as a moist dressing mixed with citric, tartaric, or tannic acid, which causes the liberation of oxygen. With iodides it liberates iodine. It is stated to be strongly antiseptic; it is used in the form of a powder, a gauze, and a plaster.

Antiseptic Pencils.—

I.—Tannin . . . . q. s.
    Alcohol, q. s . . . . 1 part
    Ether, q. s . . . . 3 parts

Make into a mass, using as an excipient the alcohol and ether previously mixed. Roll into pencils of the desired length and thickness. Then coat with collodion, roll in pure silver leaf, and finally coat with the following solution of gelatine and set aside to dry:

Gelatine . . . . 1 drachm
Water . . . . 1 pint

Dissolve by the aid of a gentle heat.

When wanted for use, shave away a portion of the covering, dip the pencil into tepid water and apply.

II.—Pencils for stopping bleeding are prepared by mixing:

Purified alum	480	
Borax	24	Parts by weight
Oxide zinc	2½	
Thymol	8	
Formalin	4	

Melting carefully in a water bath, adding some perfume, and forming mixture into pencils or cones.

A very convenient way to form into pencils where no mold need be made is to take a small glass tube, roll a piece of oil paper around the tube, remove the glass tube, crimp the paper tube thus formed on one end and stand it on end or in a bottle, and pour the melted solution in it and leave until cool, then remove the paper.

Antiseptic Paste (Poison) for Organic Specimens.—

(a) Wheat flour . . . . 16 ounces
    Beat to a batter with
      cold water . . . . . . 16 fluidounces
    Then pour into boil-
      ing water . . . . 32 fluidounces
(b) Pulverized gum ar-
      abic . . . . 2 ounces
    Dissolve in boil-
      ing water . . . . 4 fluidounces
(c) Pulverized alum . . . 2 ounces
    Dissolve in boil-
      ing water . . . . 4 fluidounces
(d) Acetate of lead . . . . 2 ounces
    Dissolve in boil-
      ing water . . . . 4 fluidounces
(e) Corrosive sublimate 10 grains

Mix (a) and (b) while hot and continue to simmer; meanwhile stir in (c) and mix thoroughly; then add (d). Stir briskly, and pour in the dry corrosive sublimate. This paste is very poisonous. It is used for anatomical work and for pasting organic tissue, labels on skeletons, etc.

Mouth Antiseptics.—I.—Thymic acid, 25 centigrams (3¼ grains); benzoic acid, 3 grams (45 grains); essence of peppermint, 75 centigrams (10 minims); tincture of eucalyptus, 15 grams (4¼ drachms); alcohol, 100 grams (3 ounces). Put sufficient in a glass of water to render latter milky.

II.—Tannin, 12 grams (3 drachms); menthol, 8 grams (2 drachms); thymol, 1 gram (15 grains); tincture benzoin, 6 grams (90 minims); alcohol, 100 grams (3 ounces). Ten drops in a half-glassful of tepid water.

See also Dentifrices for Mouth Washes.

Antiseptic Paste.—Difficulty is often experienced in applying an antiseptic dressing to moist surfaces, such as the lips after operation for harelip. A paste for this purpose is described by its originator, Socin. The composition is: Zinc oxide, 50 parts; zinc chloride, 5 parts; distilled water, 50 parts. The paste is applied to the wound, previously dried by means of a brush or spatula, allowed to dry on, and to remain in place five or six days. It may then be removed and a fresh application made.

Potassium bicarbonate	32.0 grams
Sodium benzoate . .	32.0 grams
Sodium borate	8.0 grams
Thymol	0.2 gram
Eucalyptol	2.0 c. cent.
Oil of peppermint . .	0.2 c. cent.
Oil of wintergreen . .	0.4 c. cent.
Tincture of cudbear	15.0 c. cent.
Alcohol	60.0 c. cent.
Glycerine	250.0 c. cent.
Water, enough to make	1,000.0 c. centimeters

Dissolve the salts in 650 cubic centimeters of water, and the thymol, eucalyptol, and oils in the alcohol. Mix the alcoholic solution with the glycerine and add the aqueous liquid, then the tincture of cudbear, and lastly enough water to make 1,000 cubic centimeters. Allow to stand a few days, then filter, adding a little magnesium carbonate to the filter, if necessary, to get a brilliant filtrate.

This is from the Formulary of the Bournemouth Pharmaceutical Association, as reported in the Canadian Pharmaceutical Association:

Alkaline Glycerine of Thymol.—

Sodium bicarbonate..	100 grains
Sodium biborate.....	200 grains
Sodium benzoate.....	80 grains
Sodium salicylate....	40 grains
Menthol...........	2 grains
Pumilio pine oil......	4 minims
Wintergreen oil......	2 minims
Thymol...........	4 grains
Eucalyptol.........	12 minims

Compound Solution of Thymol.—

A

Benzoic acid........	64 grains
Borax.............	64 grains
Boric acid..........	128 grains
Distilled water.......	6 ounces

Dissolve.

B

Thymol...........	20 grains
Menthol...........	6 grains
Eucalyptol.........	4 minims
Oil of wintergreen....	4 minims
Oil of peppermint....	2 minims
Oil of thyme........	1 minim
Alcohol (90 per cent).	3 ounces

Dissolve.

Mix solutions A and B, make up to 20 fluidounces with distilled water, and filter.

Oil of Cinnamon as an Antiseptic.—

Oil of cinnamon in a 9-per-cent emulsion, when used upon the hands, completely sterilizes them. A 7-to 8-per-cent emulsion is equal to a 1-per-cent solution of corrosive sublimate and is certainly far more agreeable to use. Oil of thyme in an 11-per-cent solution is equal to a 7-per-cent solution of cinnamon oil.

Green Coloring for Antiseptic Solutions.—

The safest coloring substance for use in a preparation intended either for internal administration or for application to the skin is the coloring matter of leaves, chlorophyll. A tincture of spinach or of grass made by macerating 2 ounces of the freshly cut leaves in a pint of alcohol for five days will be found to give good results. If the pure coloring substance is wanted the solvent should be evaporated off.

Antiseptic Bromine Solution.—

Bromine...........	1 ounce
Sodium chloride.....	8 ounces
Water.............	8 pints

Dissolve the sodium chloride in the water and add the bromine. This solution is to be diluted, when applied to broken skin surfaces, 1 part with 15 parts of water.

Substitute for Rubber Gloves.—Mur-

phy has found that a 4-, 6-, or 8-per-cent solution of gutta-percha in benzine, when applied to the hands of the surgeon or the skin of the patient, will seal these surfaces with an insoluble, impervious, and practically imperceptible coating— a coating that will not allow the secretions of the skin to escape, and will not admit secretions, blood, or pus into the crevices of the skin. At the same time it does not impair the sense of touch nor the pliability of the skin. A similar solution in acetone also meets most of the requirements.

Murphy's routine method of hand preparation is as follows: First, five to seven minutes' scrubbing with spirits of green soap and running hot water; second, three minutes' washing with alcohol; third, when the hands are thoroughly dried, the gutta-percha solution is poured over the hands and forearms, care being taken to fill in around and beneath the nails. The hands must be kept exposed to the air with the fingers separated until thoroughly dry. The coating is very thin and can be recognized only by its glazed appearance. It will resist soap and water, but is easily removed by washing in benzine. The hands can be washed in bichloride or any of the antiseptic solutions without interfering with the coating or affecting the skin. If the operations be many, or prolonged, the coating wears away from the tips of the fingers, but is easily renewed. For the remaining portion of the hands one application is sufficient for a whole morning's work.

The 4-per-cent solution of rubber wears better on the tips of the fingers, in handling instruments, sponges, and tissues than the acetone solution.

For the abdomen the acetone solution has the advantage, and it dries in three to four seconds after its application, while the benzine solution takes from three to four and a half minutes to make a dry, firm coating.

The preparation of the patient's skin consists in five minutes' scrubbing with spirits of green soap, washing with ether, followed by alcohol. The surface is then swabbed over thoroughly with the benzine or acetone solution.

The gutta-percha solution is prepared by dissolving the pure gutta-percha chips in sterile benzine or acetone. These solutions do not stand boiling, as this impairs the adhesiveness and elasticity of the coating.

ANTISEPTICS FOR CAGED BIRDS:

See Veterinary Formulas.

APOLLINARIS:
See Waters.

APPLE SYRUP:
See Essences and Extracts.

AQUA FORTIS FOR BRIGHT LUSTER:
See Castings.

AQUA FORTIS FOR THE TOUCHSTONE:
See Gold.

AQUARIUM CEMENTS:
See Adhesives.

AQUARIUM PUTTY:
See Putty.

ARGENTAN:
See Alloys.

ARMENIAN CEMENT:
See Adhesives under Jewelers' Cements.

ARMS, OIL FOR:
See Lubricants.

ARNICA SALVE:
See Ointments.

ARSENIC ALLOYS:
See Alloys.

ASBESTOS CEMENT:
See Adhesives.

ASBESTOS FABRIC:
See Fireproofing.

ASPHALT AS AN INGREDIENT OF INDIA RUBBER:
See Rubber.

ASPHALT IN PAINTING:
See Paint.

ASPHALT VARNISHES:
See Varnishes.

ASSAYING:
See Gold.

ASTHMA CURES.—*Asthma Papers.*—
I.—Impregnate bibulous paper with the following: Extract of stramonium, 10; potassium nitrate, 17; sugar, 20; warm water, 200 parts. Dry.

II.—Blotting or gray filter paper, 120; potassium nitrate, 60; powdered belladonna leaves, 5; powdered stramonium leaves, 5; powdered digitalis leaves, 5; powdered lobelia, 5; myrrh, 10; olibanum, 10; phellandrium fruits, 5 parts.

Stramonium Candle.—Powdered stramonium leaves, 120; potassium nitrate, 72; Peruvian balsam, 3; powdered sugar, 1; powdered tragacanth, 4 parts. (Water, q. s. to mass; roll into suitable shapes and dry.)

Cleary's Asthma Fumigating Powder.—Powdered stramonium, 15; powdered belladonna leaves, 15; powdered opium, 2; potassium nitrate, 5.

Asthma Fumigating Powders.—I.—Powdered stramonium leaves, 4; powdered aniseed, 2; potassium nitrate, 2 parts.

II.—Powdered stramonium, 30; potassium nitrate, 5; powdered tea, 15; powdered eucalyptus leaves, 15; powdered Indian hemp, 15; powdered lobelia, 15; powdered aniseed, 2; distilled water, 45 parts. (All the herbal ingredients in coarse powder; moisten with the water in which the potassium nitrate has been previously dissolved, and dry.)

Schiffmann's Asthma Powder.—Potassium nitrate, 25; stramonium, 70; belladonna leaves, 5 parts.

Neumeyer's Asthma Powder.—Potassium nitrate, 6 parts; sugar, 4; stramonium, 6; powdered lobelia, 1.

Fischer's Asthma Powder.—Stramonium, 5 parts; potassium nitrate, 1; powdered *Achillea millefolium* leaves, 1.

Vorlaender's Asthma Powder.—Stramonium, 150; lobelia, 80; arnica flowers, 80; potassium nitrate, 30; potassium iodide, 3; naphthol, 1,100 parts.

Asthma Cigarettes.—I.—Belladonna leaves, 5 parts; stramonium leaves, 5 parts; digitalis leaves, 5 parts; sage leaves, 5 parts; potassium nitrate, 75 parts; tincture of benzoin, 40 parts; boiling water, 1,000 parts. Extract the leaves with the boiling water, filter, and in the filtrate dissolve the salts. Immerse in the fluid sheets of bibulous paper (Swedish filter paper will answer) and let remain for 24 hours. At the end of this time remove, dry, cut into pieces about 2¼ by 4 inches, and roll into cigarettes.

II.—Sodium arseniate, 3 grains; extract of belladonna, 8 grains; extract of stramonium, 8 grains. Dissolve the arseniate of sodium in a small quantity of water, and rub it with the two extracts. Then soak up the whole mixture with fine blotting paper, which is dried and cut into 24 equal parts. Each part is rolled up in a piece of cigarette paper. Four or five inhalations are generally sufficient as a dose.

ASTHMA IN CANARIES:
See Veterinary Formulas.

ASTRINGENT FOR HORSES:
See Veterinary Formulas.

ATOMIC WEIGHTS:
See Weights and Measures.

ATROPINE, ANTIDOTE TO.

The usual physiological antidotes to the mydriatic alkaloids from belladonna, stramonium, and hyoscyamus are morphine or eserine. Strong tea, coffee, or brandy are usually administered as stimulants. Chief reliance has usually been placed upon a stomach siphon and plenty of water to wash out the contents of the stomach. The best antidote ever reported was that of muscarine extracted by alcohol from the mushroom, *Amanita muscaria*, but the difficulty of securing the same has caused it to be overlooked and almost forgotten. Experiments with this antidote showed it to be an almost perfect opposite of atropine in its effects upon the animal body and that it neutralized poisonous doses.

AQUA AROMATICA.—

Cort. cinnam. chinens.	3 parts
Flor. lavandulæ	5 parts
Fol. Menth. pip......	5 parts
Fol. rosmarini.......	5 parts
Fol. salviæ	10 parts
Fruct. fœniculi	3 parts
Spiritus.......... ..	70 parts
Aqua..............	300 parts

Macerate the drugs in the mixed alcohol and water for 24 hours and distill 200 parts.

AQUA REGIA.—Aqua regia consists in principle of 2 parts of hydrochloric acid and 1 part of nitric acid. But this quantity varies according to the shop where it is used for gilding or jewelry, and sometimes the proportion is brought to 4 parts of hydrochloric acid to 1 of nitric acid.

AUTOMOBILES, ANTIFREEZING SOLUTION FOR:
See Freezing Preventives.

AXLE GREASE:
See Lubricants.

BABBITT METAL:
See Alloys.

Baking Powders

I.—Tartaric acid, 3 parts; sodium bicarbonate, 1 part; starch, 0.75 part. Of this baking powder the required amount for 500 parts of flour is about 20 parts for rich cake, and 15 parts for lean cake.

The substances employed must be dry, each having been previously sifted by itself, so that no coarse pieces are present; the starch is mixed with the sodium bicarbonate before the acid is added. When large quantities are prepared the mixing is done by machine; smaller quantities are best mixed together in a spacious mortar, and then passed repeatedly through a sieve. Instead of starch, flour may be used, but starch is preferable, because it interferes with the action of the acid on the alkali.

II.—A formula proposed by Crampton, of the United States Department of Agriculture, as the result of an investigation of the leading baking powders of the market, is:

Potassium bitartrate...	2 parts
Sodium bicarbonate...	1 part
Cornstarch..........	1 part

The addition of the starch serves the double purpose of a "filler" to increase the weight of the powder and as a preservative. A mixture of the chemicals alone does not keep well.

The stability of the preparation is increased by drying each ingredient separately by exposure to a gentle heat, mixing at once, and immediately placing in bottles or cans and excluding access of air and consequently of moisture.

This is not a cheap powder; but it is the best that can be made, as to healthfulness.

III.—Sodium acid phosphate		20 parts
Calcium acid phosphate		20 parts
Sodium bicarbonate		25 parts
Starch		35 parts

Caution as to drying the ingredients and keeping them dry must be observed. Even the mixing should be done in a room free from excessive humidity.

IV.—Alum Baking Powder.—

Ammonium alum, anhydrous		15 parts
Sodium bicarbonate		18 parts
Cornstarch, q. s. to make 100 parts.		

Mix. The available carbon dioxide yielded is $7\frac{1}{2}$ per cent or 8 per cent.

BALANCE SPRING:
See Watchmakers' Formulas.

BALDNESS:
See Hair Preparations.

BALL BLUE:
See Laundry Preparations.

BALSAMS:
See also Ointments.

Wild-Cherry Balsam.—

Wild-cherry bark..	1 ounce
Licorice root......	1 ounce
Ipecac.............	1 ounce
Bloodroot........	1 drachm
Sassafras.........	1 drachm
Compound tincture of opium........	1 fluidounce
Fluid extract of cubeb..........	4 fluidrachms

Moisten the ground drugs with the fluid extract and tincture and enough menstruum consisting of 25 per cent alcohol, and after six or eight hours pack in a percolator, and pour on menstruum until percolation begins. Then cork the orifice, cover the percolator, and allow to macerate for 24 hours. Then percolate to 10 fluidounces, pouring back the first portion of percolate until it comes through clear. In the percolate dissolve ½ ounce of ammonium chloride and ¼ pound of sugar by cold percolation, adding simple syrup to make 16 fluidounces. Finally add 1 fluidrachm of chloroform.

Balsam Spray Solution.—

Oil of Scotch pine...	30 minims
Oil of eucalyptus....	1 drachm
Oil of cinnamon....	30 minims
Menthol crystals....	q. s.
Fluid extract of balm-of-Gilead buds...	1 drachm
Tincture of benzoin, enough to make..	4 ounces

This formula can, of course, be modified to suit your requirements. The oils of eucalyptus and cinnamon can be omitted and such quantities of tincture of tolu and tincture of myrrh incorporated as may be desired.

Birch Balsam.—

	Parts by weight
Alcohol...........	30,000
Birch juice.........	3,000
Glycerine..........	1,000
Bergamot oil........	90
Vanillin............	10
Geranium oil........	50
Water.............	14,000

BALSAM STAINS, TO REMOVE:
See Cleaning Preparations and Methods.

BANANA BRONZING SOLUTION:
See Plating.

BANANA SYRUP:
See Essences and Extracts.

BANANA TRICK, THE BURNING:
See Pyrotechnics.

BANJO SOUR:
See Beverages under Lemonade.

BAR POLISHES:
See Polishes.

BARBERS'-ITCH CURE:
See Ointments.

BARBERS' POWDER:
See Cosmetics.

BAROMETERS (PAPER):
See Hygrometers and Hygroscopes.

BATH, AIR:
See Air Bath.

BATH METAL:
See Alloys.

BATH POWDER:
See Cosmetics.

BATH TABLETS, EFFERVESCENT.

Tartaric acid........	10 parts
Sodium bicarbonate..	9 parts
Rice flour..........	6 parts

A few spoonfuls of this, when stirred into a bathtubful of water, causes a copious liberation of carbon dioxide, which is refreshing. This mixture can be made into tablets by compression, moistening, if necessary, with alcohol. Water, of course, cannot be used in making them, as its presence causes the decomposition referred to. Perfume may be added to this powder, essential oils being a good form. Oil of lavender would be a suitable addition, in the proportion of a fluidrachm or more to the pound of powder. A better but more expensive perfume may be obtained by mixing 1 part of oil of rose geranium with 6 parts of oil of lavender. A perfume still more desirable may be had by adding a mixture of the oils from which Cologne water is made. For an ordinary quality the following will suffice:

Oil of lavender..	4 fluidrachms
Oil of rosemary..	4 fluidrachms
Oil of bergamot..	1 fluidounce
Oil of lemon.....	2 fluidounces
Oil of clove......	30 minims

For the first quality the following may be taken:

Oil of neroli.....	6 fluidrachms
Oil of rosemary..	3 fluidrachms
Oil of bergamot..	3 fluidrachms
Oil of cedrat.....	7 fluidrachms
Oil of orange peel	7 fluidrachms

A fluidrachm or more of either of these mixtures may be used to the pound, as in the case of lavender.

These mixtures may also be used in the preparation of a bath powder (non-effervescent) made by mixing equal parts of powdered soap and powdered borax.

BATH-TUB ENAMEL:
See Varnishes.

BATH-TUB PAINTS:
See Paint.

BATTERY FILLERS AND SOLUTIONS.

I.—In the so-called dry batteries the exciting substance is a paste instead of a fluid; moisture is necessary to cause the reaction. These pastes are generally secret preparations. One of the earlier "dry" batteries is that of Gassner. The apparatus consists of a containing vessel of zinc, which forms the positive element; the negative one is a cylinder of carbon, and the space between is filled with a paste, the recipe for which is:

Oxide of zinc.........	1 part
Sal ammoniac........	1 part
Plaster.............	3 parts
Chloride of zinc.......	1 part
Water...............	2 parts

The usual form of chloride-of-silver battery consists of a sealed cell containing a zinc electrode, the two being generally separated by some form of porous septum. Around the platinum or silver electrode is cast a quantity of silver chloride. This is melted and generally poured into molds surrounding the metallic electrode. The exciting fluid is either a solution of ammonium chloride, caustic potassa, or soda, or zinc sulphate. As ordinarily constructed, these cells contain a paste of the electrolyte, and are sealed up hermetically in glass or hard-rubber receptacles.

II.—The following formula is said to yield a serviceable filling for dry batteries:

Charcoal...........	3 ounces
Graphite...........	1 ounce
Manganese dioxide...	3 ounces
Calcium hydrate.....	1 ounce
Arsenic acid.........	1 ounce
Glucose mixed with dextrine or starch..	1 ounce

Intimately mix, and then work into a paste of proper consistency with a saturated solution of sodium and ammonium chlorides containing one-tenth of its volume of a mercury-bichloride solution and an equal volume of hydrochloric acid. Add the fluid gradually, and well work up the mass.

III.—

Calcium chloride, crystallized......	30 parts
Calcium chloride, granulated.......	30 parts
Ammonium sulphate	15 parts
Zinc sulphate......	25 parts

Solutions for Batteries.—The almost exclusively employed solution of sal ammoniac (ammonium chloride) presents the drawback that the zinc rods, glasses, etc., after a short use, become covered with a fine, yellow, very difficultly soluble, basic zinc salt, whereby the generation of the electric current is impaired, and finally arrested altogether. This evil may be remedied by an admixture of cane sugar. For a battery of ordinary size about 20 to 25 grams of sugar, dissolved in warm water, is sufficient per 50 to 60 grams of sal ammoniac. After prolonged use only large crystals (of a zinc saccharate) form, which, however, become attached only to the zinc rod in a few places, having very little disadvantageous effect upon the action of the batteries and being easy to remove, owing to their ready solubility.

BAUDOIN METAL:
See Alloys.

BAY RUM.

I.—

Oil of bay.......	1 drachm
Alcohol........	18 ounces
Water..........	18 ounces

Mix and filter through magnesia.

II.—

Bay-leaf otto........	½ ounce
Magnesium carbonate.	½ ounce
Jamaica rum........	2 pints
Alcohol.............	3 pints
Water..............	3 pints

Triturate the otto with the magnesium carbonate, gradually adding the other ingredients, previously mixed, and filter. If the rum employed contains sufficient sugar or mucilaginous matter to cause any stickiness to be felt on the skin, rectification will be necessary.

BEAR FAT:
See Fats.

BEARING LUBRICANT:
See Lubricants.

BEARING METAL:
See Babbitt Metal, Bearing Metal, and Phosphor Bronze, under Alloys.

BEDBUG DESTROYERS:
See Insecticides.

BEEF, IRON, AND WINE.

Extract of beef....	512 grains
Detannated sherry wine...........	26 ounces
Alcohol...........	4 ounces
Citrate of iron and ammonia.......	256 grains
Simple sirup......	12 ounces

Tincture of orange.   2 ounces
Tincture of carda-
   mom co......... 1 ounce
Citric acid........ 10 grains
Water, enough to make 4 pints

Let stand 24 hours, agitate frequently, and filter. See that the orange is fresh.

BEEF PEPTONOIDS:
See Peptonoids.

BEEF PRESERVATIVES:
See Foods.

BEEF TEA:
See Beverages.

BEERS, ALCOHOL IN:
See Alcohol.

BEER, GINGER, HOP-BITTER, SCOTCH, AND SPRUCE:
See Beverages.

BEER, RESTORATION OF SPOILED.

I.—Powdered chalk is poured into the cask and allowed to remain in the beer until completely precipitated.

II.—The liquor of boiled raisins may be poured into the beer, with the result that the sour taste of the beer is disguised.

III.—A small quantity of a solution of potash will remove the sour taste of beer. Too much potash must not be added; otherwise the stomach will suffer. Beer thus restored will not keep long.

IV.—If the beer is not completely spoiled it may be restored by the addition of coarsely powdered charcoal.

V.—If the addition of any of the above-mentioned substances should affect the taste of the beer, a little powdered zingiber may be used to advantage. Syrup or molasses may also be employed.

BEES, FOUL BROOD IN.

"Foul brood" is a contagious disease to which bees are subject. It is caused by bacteria and its presence may be known by the bees becoming languid. Dark, stringy, and elastic masses are found in the bottom of the cells, while the caps are sunken or irregularly punctured. Frequently the disease is said to be accompanied by a peculiar offensive odor. Prompt removal of diseased colonies, their transfer to clean and thoroughly disinfected hives, and feeding on antiseptically treated honey or syrup are the means taken for the prevention and cure of the disease. The antiseptics used are salicylic acid, carbolic acid, or formic acid. Spraying the brood with any one of these remedies in a solution and feeding with a honey or syrup medicated with them will usually be all that is required by way of treatment. It is also said that access to salt water is important for the health of bees.

BEETLE POWDER:
See Insecticides.

BELL METAL:
See Alloys.

BELLADONNA, ANTIDOTES TO:
See Antidotes and Atropine.

BELT PASTES FOR INCREASING ADHESION.

I.—Tallow............ 50 parts
    Castor oil, crude.... 20 parts
    Fish oil............ 20 parts
    Colophony......... 10 parts

Melt on a moderate fire and stir until the mass cools.

II.—Melt 250 parts of gum elastic with 250 parts of oil of turpentine in an iron, well-closed crucible at 122° F. (caution!) and mix well with 200 parts of colophony. After further melting add 200 parts of yellow wax and stir carefully. Melt in 750 parts of heated train oil, 250 parts of tallow, and to this add, with constant stirring, the first mixture when the latter is still warm, and let cool slowly with stirring. This grease is intended for cotton belts.

III.—Gutta-percha ...... 40 parts
     Rosin............. 10 parts
     Asphalt........... 15 parts
     Petroleum........ 60 parts

Heat in a glass vessel on the water bath for a few hours, until a uniform solution is obtained. Let cool and add 15 parts of carbon disulphide and allow the mixture to stand, shaking it frequently.

Directions for Use.—The leather belts to be cemented should first be roughened at the joints, and after the cement has been applied they should be subjected to a strong pressure between warm rollers, whereupon they will adhere together with much tenacity.

Preservation of Belts.—In a well-covered iron vessel heat at a temperature of 50° C. (152° F.) 1 part by weight of caoutchouc, cut in small pieces, with 1 part by weight of rectified turpentine. When the caoutchouc is dissolved add 0.8 part of colophony, stir until this is dissolved, and add to the mixture 0.1 part of yellow wax. Into another vessel of suitable size pour 3 parts of fish oil, add 1 part of tallow, and heat the mixture until the tallow is melted; then pour on the contents of the first vessel, constantly stirring—an operation to be continued until the matter is cooled and congealed. This grease is to be rubbed on the inside of the belts from time to time, while they are in use. The belts run easily and do not slip. The grease may also serve for improving old belts. For this purpose the grease should be rubbed on both sides in a warm place. A first layer is allowed to soak in, and another applied.

To Make a Belt Pull.—Hold a piece of tar soap on the inside of the belt while it is running.

BELT CEMENT:
See Adhesives.

BELT GLUE:
See Adhesives.

BELT LUBRICANT:
See Lubricants.

BÉNÉDICTINE:
See Wines and Liquors.

Benzine

Benzine, to Color Green.—Probably the simplest and cheapest as well as the best method of coloring benzine green is to dissolve in it sufficient oil soluble aniline green of the desired tint to give the required shade.

Purification of Benzine.—Ill-smelling benzine, mixed with about 1 to 2 per cent of its weight of free fatty acid, will dissolve therein. One-fourth per cent of tannin is added and all is mixed well. Enough potash or soda lye, or even lime milk, is added until the fatty acids are saponified, and the tannic acid is neutralized, shaking repeatedly. After a while the milky liquid separates into two layers, viz., a salty, soapy, mud-sediment and clear, colorless, and almost odorless benzine above. This benzine, filtered, may be employed for many technical purposes, but gives an excellent, pure product upon a second distillation.

Fatty acid from tallow, olive oil, or other fats may be used, but care should be taken that they have as slight an odor of rancid fat as possible. The so-called elaine or olein—more correctly oleic acid —of the candle factories may likewise be employed, but it should first be agitated with a $\frac{1}{10}$-per-cent soda solution to get rid of the bad-smelling fatty acids, especially the butyric acid.

The Prevention of the Inflammability of Benzine.—A mixture of 9 volumes tetrachloride and 1 volume of benzine is practicably inflammable. The flame is soon extinguished by itself.

Substitute for Benzine as a Cleansing Agent.—

I.—Chloroform......... 75 parts
    Ether............... 75 parts
    Alcohol............ 600 parts
    Decoction of quillaya
      bark......... 22,500 parts
Mix.

II.—Acetic ether, technically pure....... 10 parts
    Amyl acetate........ 10 parts
    Ammonia water..... 10 parts
    Alcohol dilute....... 70 parts
Mix.

III.—Acetone............. 1 part
    Ammonia water...... 1 part
    Alcohol dilute........ 1 part
Mix.

Deodorizing Benzine.—

I.—Benzine......... 20 ounces
    Oil of lavender... 1 fluidrachm
    Potassium dichromate......... 1 ounce
    Sulphuric acid... 1 fluidounce
    Water.......... 20 fluidounces

Dissolve the dichromate in the water, add the acid and, when the solution is cold, the benzine. Shake every hour during the day, allow to stand all night, decant the benzine, wash with a pint of water and again decant, then add the oil of lavender.

II.—First add to the benzine 1 to 2 per cent of oleic acid, which dissolves. Then about a quarter of 1 per cent of tannin is incorporated by shaking. A sufficient quantity of caustic potassa solution, or milk of lime, to combine with the acids is then well shaken into the mixture, and the whole allowed to stand. The benzine rises to the top of the watery fluid, sufficiently deodorized and decolorized for practical purposes.

III.—To 1,750 parts of water add 250 parts of sulphuric acid, and when it has cooled down add 30 parts of potassium permanganate and let dissolve. Add this solution to 4,500 parts of benzine, stir well together, and set aside for 24 hours. Now decant the benzine and to it add a solution of 7½ parts of potassium permanganate and 15 parts of sodium hydrate in 1,000 parts of water, and agitate the substances well together. Let stand until the benzine separates, then draw off.

IV.—Dissolve 3 parts of litharge and 18 parts of sodium hydrate in 40 parts of water. Add this to 200–250 parts of benzine and agitate well together for two minutes, then let settle and draw off the benzine. Rinse the latter by agitating it with plenty of clear water, let settle, draw off the benzine, and, if necessary, repeat the operation.

BENZINE, CLEANING WITH:
See Cleaning Preparations and Methods, under Miscellaneous Methods.

BENZOIC ACID IN FOOD:
See Food.

BENZOIN SOAP:
See Soap.

BENZOPARAL:
A neutral, bland, oily preparation of benzoin, useful for applying various antiseptics by the aid of an atomizer, nebulizer, or vaporizer. Can be used plain or in combination with other easily dissolved medicinals.

Paraffine, liquid	16 ounces
Gum benzoin	1 ounce

Digest on a sand bath for a half hour and filter.

Beverages

GINGER ALE AND GINGER BEER:

Old-Fashioned Ginger Beer.—

Lemons, large and sound	6	only
Ginger, bruised	3	ounces
Sugar	6	cups
Yeast, compressed	¼	cake
Boiling water	4	gallons
Water		enough

Slice the lemons into a large earthenware vessel, removing the seed. Add the ginger, sugar, and water. When the mixture has cooled to lukewarmness, add the yeast, first diffused in a little water. Cover the vessel with a piece of cheese cloth, and let the beer stand 24 hours. At the end of that time strain and bottle it. Cork securely, but not so tightly that the bottles would break before the corks would fly out, and keep in a cool place.

Ginger Beer.—
Honey gives the beverage a peculiar softness and, from not having fermented with yeast, is the less violent in its action when opened. Ingredients: White sugar, ¼ pound; honey, ¼ pound; bruised ginger, 5 ounces; juice of sufficient lemons to suit the taste; water, 4½ gallons. Boil the ginger in 3 quarts of the water for half an hour, then add the ginger, lemon juice, and honey, with the remainder of the water; then strain through a cloth; when cold, add the quarter of the white of an egg and a teaspoonful of essence of lemon. Let the whole stand for four days before bottling. This quantity will make a hundred bottles.

Ginger Beer without Yeast.—

Ginger, bruised	1½	pounds
Sugar	20	pounds
Lemons	1	dozen
Honey	1	pound
Water		enough

Boil the ginger in 3 gallons of water for half an hour; add the sugar, the lemons (bruised and sliced), the honey, and 17 gallons of water. Strain and, after three or four days, bottle.

Package Pop.—

Cream of tartar	3	ounces
Ginger, bruised	1	ounce
Sugar	24	ounces
Citric acid	2	drachms

Put up in a package, and direct that it be shaken in 1½ gallons of boiling water, strained when cooled, fermented with 1 ounce of yeast, and bottled.

Ginger-Ale Extract.—

I.—Jamaica ginger, coarse powder	4	ounces
Mace, powder	½	ounce
Canada snakeroot, coarse powder	60	grains
Oil of lemon	1	fluidrachm
Alcohol	12	fluidounces
Water	4	fluidounces
Magnesium carbonate or purified talcum	1	av. ounce

Mix the first four ingredients, and make 16 fluidounces of tincture with the alcohol and water, by percolation. Dissolve the oil of lemon in a small quantity of alcohol, rub with magnesia or talcum, add gradually with constant trituration the tincture, and filter. The extract may be fortified by adding 4 avoirdupois ounces of powdered grains of paradise to the ginger, etc., of the above before extraction with alcohol and water.

II.—Capsicum, coarse powder	8	ounces
Water	6	pints
Essence of ginger	8	fluidounces
Diluted alcohol	7	fluidounces
Vanilla extract	2	fluidounces
Oil of lemon	20	drops
Caramel	1	fluidounce

Boil the capsicum with water for three hours, occasionally replacing the water lost by evaporation; filter, concentrate the filtrate on a hot water bath to the consistency of a thin extract, add the remaining ingredients, and filter.

III.—Jamaica ginger,
ground........ 12 ounces
Lemon peel, fresh,
cut fine........ 2 ounces
Capsicum, powder 1 ounce
Calcined magne-
sia........·... 1 ounce
Alcohol } of each. sufficient
Water }

Extract the mixed ginger and capsicum by percolation so as to obtain 16 fluidounces of water, set the mixture aside for 24 hours, shaking vigorously from time to time, then filter, and pass through the filter enough of a mixture of 2 volumes of alcohol and 1 of water to make the filtrate measure 32 fluidounces. In the latter macerate the lemon peel for 7 days, and again filter.

Ginger Beer.—

Brown sugar........ 2 pounds
Boiling water........ 2 gallons
Cream of tartar...... 1 ounce
Bruised ginger root... 2 ounces

Infuse the ginger in the boiling water, add the sugar and cream of tartar; when lukewarm strain; then add half pint good yeast. Let it stand all night, then bottle; one lemon and the white of an egg may be added to fine it.

Lemon Beer.—

Boiling water........ 1 gallon
Lemon, sliced....... 1
Ginger, bruised...... 1 ounce
Yeast............. 1 teacupful
Sugar............. 1 pound

Let it stand 12 to 20 hours, and it is ready to be bottled.

Hop Beer.—

Water............. 5 quarts
Hops............. 6 ounces

Boil 3 hours, strain the liquor, add:

Water............. 5 quarts
Bruised ginger....... 4 ounces

and boil a little longer, strain, and add 4 pounds of sugar, and when milk-warm, 1 pint of yeast. Let it ferment; in 24 hours it is ready for bottling.

Œnanthic Ether as a Flavoring for Ginger Ale.—A fruity, vinous bouquet and delightful flavor are produced by the presence of œnanthic ether or brandy flavor in ginger ale. This ether throws off a rich, pungent, vinous odor, and gives a smoothness very agreeable to any liquor or beverage of which it forms a part. It is a favorite with "brandy sophisticators." Add a few drops of the ether (previously dissolved in eight times its bulk of Cologne spirit) to the ginger-ale syrup just before bottling.

Soluble Extract of Ginger Ale.—Of the following three formulas the first is intended for soda-fountain use, the second is a "cheap" extract for the bottlers who want a one-ounce-to-the-gallon extract, and the third is a bottlers' extract to be used in the proportion of three ounces to a gallon of syrup. This latter is a most satisfactory extract and has been sold with most creditable results, both as to clearness of the finished ginger ale and delicacy of flavor.

It will be noted that in these formulas oleoresin of ginger is used in addition to the powdered root. Those who do not mind the additional expense might use one-fourth of the same quantity of volatile oil of ginger instead. This should develop an excellent flavor, since the oil is approximately sixteen times as strong as the oleoresin, and has the additional advantage of being free from resinous extractive.

The following are the formulas:

I.—(To be used in the proportion of 4 ounces of extract to 1 gallon of syrup.)

Jamaica ginger, in
fine powder....... 8 pounds
Capsicum, in fine pow-
der............. 6 ounces
Alcohol, a sufficient quantity.

Mix the powders intimately, moisten them with a sufficient quantity of alcohol, and set aside for 4 hours. Pack in a cylindrical percolator and percolate with alcohol until 10 pints of percolate have resulted. Place the percolate in a bottle of the capacity of 16 pints, and add to it 2 fluidrachms of oleoresin of ginger; shake, add 2½ pounds of finely powdered pumice stone, and agitate thoroughly at intervals of one-half hour for 12 hours. Then add 14 pints of water in quantities of 1 pint at each addition, shaking briskly meanwhile. This part of the operation is most important. Set the mixture aside for 24 hours, agitating it strongly every hour or so during that period. Then take

Oil of lemon....... 1½ fluidounces
Oil of rose (or ge-
ranium)........ 3 fluidrachms
Oil of bergamot.... 2 fluidrachms

Oil of cinnamon.... 3 fluidrachms
Magnesium carbon-
ate .... 3 fluidounces

Rub the oils with the magnesia in a large mortar and add 9 ounces of the clear portion of the ginger mixture to which have been previously added 2 ounces of alcohol, and continue trituration, rinsing out the mortar with the ginger mixture. Pass the ginger mixture through a double filter and add through the filter the mixture of oils and magnesia; finally pass enough water through the filter to make the resulting product measure 24 pints, or 3 gallons. If the operator should desire an extract of more or less pungency, he may obtain his desired effect by increasing or decreasing the quantity of powdered capsicum in the formula.

II.—(To be used in the proportion of 1 ounce to 1 gallon of syrup.)

Ginger, in moderately
fine powder...... 6 pounds
Capsicum, in fine pow-
der.............. 2½ pounds
Alcohol, a sufficient quantity.

Mix, moisten the powder with 3 pints of alcohol, and set aside in a suitable vessel for 4 hours. Then pack the powder firmly in a cylindrical percolator, and percolate until 6 pints of extract are obtained. Set this mixture aside and label Percolate No. 1, and continue the percolation with 1½ pints of alcohol mixed with 1½ pints of water. Set the resultant tincture aside, and label Percolate No. 2.

Take oleoresin ginger 5 fluid ounces and add to Percolate No. 1. Then take:

Oil of lemon....... 1½ fluidounces
Oil of cinnamon... 1 fluidounce
Oil of geranium.... ½ fluidounce
Magnesium carbon-
ate .... 8 ounces

Triturate the oils with the magnesia, add gradually Percolate No. 2, and set aside. Then place Percolate No. 1 in a large bottle, add 3¼ pounds of finely powdered pumice stone, and shake at intervals of half an hour for six hours. This being completed, add the mixture of oils, and later 10 pints of water, in quantities of ½ a pint at a time, shaking vigorously after each solution. Let the mixture stand for 24 hours, shaking it at intervals, and then pass it through a double filter. Finally add enough water through the filter to make the product measure 24 pints, or 3 gallons.

III.—(To be used in proportion of 3 ounces to 1 gallon of syrup.)

Ginger, in moderately
fine powder....... 8 pounds
Capsicum, in moder-
ately fine powder .. 2 pounds
Alcohol, q. s.

Mix, moisten with alcohol, and set aside as in the preceding formula; then percolate with alcohol until 10 pints of extract are obtained. To this add oleoresin of ginger 3 drachms, and place in a large bottle. Add 2½ pounds of powdered pumice stone, and shake as directed for formula No. 1. Then add 14 pints of water, in quantities of 1 pint at a time, shaking vigorously after each addition. Set the mixture aside for 24 hours, shaking at intervals. Then take:

Oil of lemon....... 1½ fluidounces
Oil of geranium ... ½ fluidounce
Oil of cinnamon ... 3 fluidrachms
Magnesia carbonate 3 ounces

Rub these in a mortar with the magnesia, and add 9 ounces of the clear portion of the ginger mixture mixed with 2 ounces of alcohol, rubbing the mixture until it becomes smooth. Prepare a double filter, and filter the ginger mixture, adding through the filter the mixture of oils and magnesia. Finally add enough water through the filter to make the final product measure 24 pints, or 3 gallons.

If these formulas are properly manipulated the extracts should keep for a reasonable length of time without a precipitate. If, however, a precipitate occur after the extract has stood for a week, it should be refiltered.

LEMONADES:

Lemonade Preparations for the Sick.— I.—Strawberry Lemonade: Citric acid, 6 parts; water, 100 parts; sugar, 450 parts; strawberry syrup, 600 parts; cherry syrup, 300 parts; claret, 450 parts; aromatic tincture, ad lib.

II.—Lemonade Powder: Sodium bicarbonate, 65; tartaric acid, 60; sugar, 125; lemon oil, 12 drops.

III.—Lemonade juice: Sugar syrup, 200; tartaric acid, 15; distilled water, 100; lemon oil, 3; tincture of vanilla, 6 drops.

IV.—Lemonade Lozenges: Tartaric acid, 10; sugar, 30; gum arabic, 2; powdered starch, 0.5; lemon oil, 6 drops; tincture of vanilla, 25 drops; and sufficient diluted spirit of wine so that 30 lozenges can be made with it.

Lemonade for Diabetics.—The following is said to be useful for assuaging the thirst of diabetics:

Citric acid.........	1 part
Glycerine...........	50 parts
Cognac.............	50 parts
Distilled water......	500 parts

Hot Lemonade.—Take 2 large, fresh lemons, and wash them clean with cold water. Roll them until soft; then divide each into halves, and use a lemon-squeezer or reamer to express the juice into a small pitcher. Remove all the seeds from the juice, to which add 4 or more tablespoonfuls of white sugar, according to taste. A pint of boiling water is now added, and the mixture stirred until the sugar is dissolved. The beverage is very effective in producing perspiration, and should be drunk while hot. The same formula may be used for making cold lemonade, by substituting ice water for the hot water, and adding a piece of lemon peel. If desired, a weaker lemonade may be made by using more water.

Lemonades, Lemon and Sour Drinks for Soda-Water Fountains.—**Plain Lemonade.**—Juice of 1 lemon; pulverized sugar, 2 teaspoonfuls; filtered water, sufficient; shaved ice, sufficient.

Mix and shake well. Garnish with fruit, and serve with both spoon and straws.

Huyler's Lemonade.—Juice of 1 lemon; simple syrup, 2 ounces; soda water, sufficient. Dress with sliced pineapple, and serve with straws. In mixing, do not shake, but stir with a spoon.

Pineapple Lemonade.—Juice of 1 lemon; pineapple syrup, 2 ounces; soda water, sufficient. Dress with fruit. Serve with straws.

Seltzer Lemonade.—Juice of 1 lemon; pulverized sugar, 2 teaspoonfuls. Fill with seltzer. Dress with sliced lemon.

Apollinaris Lemonade.—The same as seltzer, substituting apollinaris water for seltzer.

Limeade.—Juice of 1 lime; pulverized sugar, 2 teaspoonfuls; water, sufficient. Where fresh limes are not obtainable, use bottled lime juice.

Orangeade.—Juice of 1 orange; pulverized sugar, 2 teaspoonfuls; water, sufficient; shaved ice, sufficient. Dress with sliced orange and cherries. Serve with straws.

Seltzer and Lemon.—Juice of 1 lemon; seltzer, sufficient. Serve in a small glass.

Claret Lemonade.—Juice of 1 lemon; pulverized sugar, 3 teaspoonfuls. Make lemonade, pour into a glass containing shaved ice until the glass lacks about one inch of being full. Pour in sufficient claret to fill the glass. Dress with cherries and sliced pineapple.

Claret Punch.—Juice of 1 lemon; pulverized sugar, 3 teaspoonfuls; claret wine, 2 ounces; shaved ice, sufficient. Serve in small glass. Dress with sliced lemon, and fruit in season. Bright red cherries and plums make attractive garnishings.

Raspberry Lemonade.—I.—Juice of 1 lemon; 3 teaspoonfuls powdered sugar; 1 tablespoonful raspberry juice; shaved ice; plain water; shake.

II.—Juice of 1 lemon; 2 teaspoonfuls powdered sugar; ½ ounce raspberry syrup; shaved ice; water; shake.

Banjo Sour.—Pare a lemon, cut it in two, add a large tablespoonful of sugar, then thoroughly muddle it; add the white of an egg; an ounce of sloe gin; 3 or 4 dashes of abricotine; shake well; strain into a goblet or fizz glass, and fill balance with soda; decorate with a slice of pineapple and cherry.

Orgeat Punch.—Orgeat syrup, 12 drachms; brandy, 1 ounce; juice of 1 lemon.

Granola.—Orange syrup, 1 ounce; grape syrup, 1 ounce; juice of ½ lemon; shaved ice, q. s. Serve with straws. Dress with sliced lemon or pineapple.

American Lemonade.—One ounce orange syrup; 1 ounce lemon syrup; 1 teaspoonful powdered sugar; 1 dash acidphosphate solution; ⅓ glass shaved ice. Fill with coarse stream. Add slice of orange, and run two straws through it.

Old-Fashioned Lemonade.—Put in a freezer and freeze almost hard, then add the fruits, and freeze very hard. Serve in a silver sherbet cup.

"Ping Pong" Frappé.—Grape juice, unfermented, 1 quart; port wine (California), ½ pint; lemon syrup, 12 ounces; pineapple syrup, 2 ounces; orange syrup, 4 ounces; Bénédictine cordial, 4 ounces; sugar, 1 pound.

Dissolve sugar in grape juice and put in wine; add the syrup and cordial; serve from a punch bowl, with ladle, into 12-ounce narrow lemonade glass and fill with solid stream; garnish with slice of orange and pineapple, and serve with straw.

Orange Frappé.—Glass half full of fine ice; tablespoonful powdered sugar; ½ ounce orange syrup; 2 dashes lemon syrup; dash prepared raspberry; ¼ ounce acid-phosphate solution. Fill with soda and stir well; strain into a mineral glass and serve.

Hot Lemonades.—

I.—Lemon essence.. 4 fluidrachms
Solution of citric
acid........ 1 fluidounce
Syrup, enough to
make....... 32 fluidounces

In serving, draw 2½ fluidounces of the syrup into an 8-ounce mug, fill with hot water, and serve with a spoon.

II.—Lemon......... 1
Alcohol........ 1 fluidounce
Solution of citric
acid .... 2 fluidrachms
Sugar .... 20 av. ounces
Water .... 20 fluidounces
White of....... 1 egg

Grate the peel of the lemon, macerate with the alcohol for a day; express; also express the lemon, mix the two, add the sugar and water, dissolve by agitation, and add the solution of citric acid and the white of egg, the latter first beaten to a froth. Serve like the preceding.

Egg Lemonade.—I.—Break 1 egg into a soda glass, add 1¼ ounces lemon syrup, a drachm of lemon juice, and a little shaved ice; then draw carbonated water to fill the glass, stirring well.

II.—Shaved ice..... ½ tumblerful
Powdered sugar 4 tablespoonfuls
Juice of....... 1 lemon
Yolk of........ 1 egg

Shake well, and add carbonated water to fill the glass.

HOT SODA-WATER DRINKS:

Chocolate.—I.—This may be prepared in two ways, from the powdered cocoa or from a syrup. To prepare the cocoa for use, dry mix with an equal quantity of pulverized sugar and use a heaping teaspoonful to a mug. To prepare a syrup, take 12 ounces of cocoa, 5 pints of water, and 4 pounds of sugar. Reduce the cocoa to a smooth paste with a little warm water. Put on the fire. When the water becomes hot add the paste, and then allow to boil for 3 or 4 minutes; remove from fire and add the sugar; stir carefully while heating, to prevent scorching; when cold add 3 drachms of vanilla; ½ to ¾ ounce will suffice for a cup of chocolate; top off with whipped cream.

II.—Baker's fountain choc-
olate............. 1 pound
Syrup .... 1 gallon
Extract vanilla...... enough

Shave the chocolate into a gallon porcelained evaporating dish and melt with a gentle heat, stirring with a thin-bladed spatula. When melted remove from the fire and add 1 ounce of cold water, mixing well. Add gradually 1 gallon of hot syrup and strain; flavor to suit. Use 1 ounce to a mug.

III.—Hot Egg Chocolate.—Break a fresh egg into a soda tumbler; add 1½ ounces chocolate syrup and 1 ounce cream; shake thoroughly, add hot soda slowly into the shaker, stirring meanwhile; strain carefully into mug; top off with whipped cream and serve.

IV.—Hot Chocolate and Milk.—

Chocolate syrup.... 1 ounce
Hot milk.......... 4 ounces

Stir well, fill mug with hot soda and serve.

V.—Hot Egg Chocolate.—One egg, 1½ ounces chocolate syrup, 1 teaspoonful sweet cream; shake, strain, add 1 cup hot soda, and 1 tablespoonful whipped cream.

Coffee.—I.—Make an extract by macerating 1 pound of the best Mocha and Java with 8 ounces of water for 20 minutes, then add hot water enough to percolate 1 pint. One or 2 drachms of this extract will make a delicious cup of coffee. Serve either with or without cream, and let customer sweeten to taste.

II.—Pack ½ pound of pulverized coffee in a percolator. Percolate with 2 quarts of boiling water, letting it run through twice. Add to this 2 quarts of milk; keep hot in an urn and draw as a finished drink. Add a lump of sugar and top off with whipped cream.

III.—Coffee syrup may be made by adding boiling water from the apparatus to 1 pound of coffee, placed in a suitable filter or coffeepot, until 2 quarts of the infusion are obtained. Add to this 3 pounds of sugar. In dispensing, first put sufficient cream in the cup, add the coffee, then sweeten, if necessary, and mix with the stream from the draught tube.

IV.—Mocha coffee (ground
fine)............. 4 ounces
Java coffee (ground
fine).............. 4 ounces
Granulated sugar.... 6 pounds
Hot water.......... q. s.

Percolate the coffee with hot water until the percolate measures 72 ounces. Dissolve the sugar in the percolate by agitation without heat and strain.

Hot Egg Orangeade.—One egg; juice

of ½ orange; 2 teaspoonfuls powdered sugar. Shake, strain, add 1 cup of hot water. Stir, serve with nutmeg.

Hot Egg Bouillon. — One-half ounce liquid extract beef; 1 egg; salt and pepper; hot water to fill 8-ounce mug. Stir extract, egg, and seasoning together; add water, still stirring; strain and serve.

Hot Celery Punch. — One - quarter ounce of clam juice; ¼ ounce beef extract; 1 ounce of cream; 4 dashes of celery essence. Stir while adding hot water, and serve with spices.

Chicken Bouillon. — Two ounces concentrated chicken; ½ ounce sweet cream and spice. Stir while adding hot water.

Ginger. —

Fluid extract of ginger   2½ ounces
Sugar............   40   ounces
Water, to..........   2½ pints

Take 10 ounces of the sugar and mix with the fluid extract of ginger; heat on the water bath until the alcohol is evaporated. Then mix with 20 ounces of water and shake till dissolved. Filter and add the balance of the water and the sugar. Dissolve by agitation.

Cocoa Syrup. —

I.—Cocoa, light, soluble.   4 ounces
Granulated sugar....   2 pounds
Boiling hot water....   1 quart
Extract vanilla......   1 ounce

Dissolve the cocoa in the hot water, by stirring, then add the sugar and dissolve. Strain, and when cold add the vanilla extract.

II.—Cocoa syrup........   2 ounces
Cream.............   1 ounce

Turn on the hot water stream and stir while filling. Top off with whipped cream.

Hot Soda Toddy. —

Lemon juice........   2 fluidrachms
Lemon syrup.......   1 fluidounce
Aromatic bitters....   1 fluidrachm
Hot water, enough to fill an 8-ounce mug.

Sprinkle with nutmeg or cinnamon.

Hot Orange Phosphate. —

Orange syrup......   1 fluidounce
Solution of acid phosphate.......   1 fluidrachm
Hot water, enough to fill an 8-ounce mug.

It is prepared more acceptably by mixing the juice of half an orange with acid phosphate, sugar, and hot water.

Pepsin Phosphate. — One teaspoonful of liquid pepsin; 2 dashes of acid phosphate; 1 ounce of lemon syrup; 1 cup hot water.

Cream Beef Tea. — Use 1 teaspoonful of liquid beef extract in a mug of hot water, season with salt and pepper, then stir in a tablespoonful of rich cream. Put a teaspoonful of whipped cream on top and serve with flakes.

Cherry Phosphate. — Cherry-phosphate syrup, 1½ ounces; hot water to make 8 ounces.

Cherry-phosphate syrup is made as follows: Cherry juice, 3 pints; sugar, 6 pounds; water, 1 pint; acid phosphate, 4 ounces. Bring to a boil, and when cool add the acid phosphate.

Celery Clam Punch. — Clam juice, 2 drachms; beef extract, 1 drachm; cream, 1 ounce; essence of celery, 5 drops; hot water to make 8 ounces.

Claret Punch. — Claret wine, 2 ounces; sugar, 3 teaspoonfuls; juice of ½ lemon; hot water to make 8 ounces.

Ginger. — Extract of ginger, 2 drachms; sugar, 2 drachms; lemon juice, 2 dashes; hot water to make 8 ounces.

Lemon Juice, Plain. — Fresh lemon juice, 2½ drachms; lemon syrup, 1 ounce; hot water, q. s. to make 8 ounces.

Lime Juice. — Lime juice, ¾ drachm; lemon syrup, 1 ounce; hot water to make 8 ounces. Mix. Eberle remarks that lemon juice or lime juice enters into many combinations. In plain soda it may be combined with ginger and other flavors, as, for instance, chocolate and coffee.

Lemonade. — Juice of 1 lemon; powdered sugar, 2 teaspoonfuls; hot water to make 8 ounces. A small piece of fresh lemon peel twisted over the cup lends an added flavor.

Hot Malt. — Extract of malt, 1 ounce; cherry syrup, 1 ounce; hot water, sufficient to make 8 ounces. Mix.

Malted Milk. — Horlick's malted milk, 2 tablespoonfuls; hot water, quantity sufficient to make 8 ounces; flavoring to suit. Mix. Essence of coffee, chocolate, etc., and many of the fruit syrups go well with malted milk.

Hot Malted Milk Coffee (or Chocolate). — Malted milk, 2 teaspoonfuls; coffee (or chocolate) syrup, 1 ounce; hot water, quantity sufficient to make 8 ounces.

Hot Beef Tea. — I. — Best beef extract, 1 tablespoonful; sweet cream, 1 ounce; hot water, 7 ounces; pepper, salt, etc., quantity sufficient. Mix.

II.—Extract beef bouillon, 1 teaspoonful; extract aromatic soup herbs (see Condiments), 10 drops; hot soda, 1 cupful. Mix.

III.—Extract of beef..... 1 teaspoonful
Hot water... ..... q. s.
Pepper, salt, and celery salt.
Mix.

Hot Bouillon.—

Beef extract......... 1 ounce
Hot water, q. s. to
make............. 8 ounces
Pepper, salt, etc...... q. s.
Mix.

Clam Bouillon.—

I.—Clam juice........... 12 drachms
Cream.............. 2 ounces
Hot water, q. s. to make 8 ounces
Mix.

II.—Extract clam bouillon 2 ounces
Prepared milk....... 2 drachms
Extract of aromatic
soup herbs......... 5 drops
Extract white pepper.. 5 drops
Hot soda........... 1 cupful
Mix.

III.—Clam juice may be served with hot water, salt and pepper added. Adding butter makes this bouillon a broth.

It may also be served with milk or cream, lemon juice, tomato catsup, etc. Hot oyster juice may be served in the same way.

Hot Tea.—

I.—Tea syrup.......... sufficient
Hot water, q. s. to
make............ 1 cupful

II.—Loaf sugar.......... 4 cubes
Extract of Oolong
tea, about...... 1 dessertsp'ful
Prepared milk, about 1 dessertsp'ful
Hot soda......... 1 cupful
Whipped cream... 1 tablespoonful

Mix the tea extract, sugar, and prepared milk, pour on water, and dissolve. Top off with whipped cream.

Hot Egg Drinks.—I.—One-half to 1 ounce liquid extract of beef, 1 egg, salt and pepper to season, hot water to fill an 8-ounce mug. Stir the extract, egg, and seasoning together with a spoon, to get well mixed, add the water, stirring briskly meanwhile; then strain, and serve. Or shake the egg and extract in a shaker, add the water, and mix by pouring back and forth several times, from shaker to mug.

II.—Hot Egg Chocolate.—One to 1½ ounces chocolate syrup, 1 egg, ½ ounce cream, hot water sufficient to fill an 8-ounce mug.

Mix the syrup, egg, and cream together in an egg-shaker; shake as in making cold drinks; add the hot water, and mix all by pouring back and forth several times, from shaker to mug. Or, prepare by beating the egg with a spoon, add the syrup and cream, mix all quickly with the spoon, and add hot water, stirring constantly, and strain.

III.—Hot Egg Coffee.—One egg, 1 dessertspoonful extract of coffee, 1 teaspoonful sweet cream, 1 ounce syrup. Shake well, strain, and add 1 cupful hot water and top with whipped cream.

IV.—Hot Egg Lemonade.—One egg, juice of 1 lemon, 3 teaspoonfuls powdered sugar. Beat the egg with lemon juice and sugar thoroughly. Mix while adding the water. Serve grated nutmeg and cinnamon. The amount of lemon juice and sugar may be varied to suit different tastes.

V.—Hot Egg Milk.—Two teaspoonfuls sugar, 1 ounce cream, 1 egg, hot milk to fill an 8-ounce mug. Prepare as in hot egg chocolate, top with whipped cream, and sprinkle with nutmeg. If there are no facilities for keeping hot milk, use about 2 ounces of cream, and fill mug with hot water.

VI.—Hot Egg Nogg.—Plain syrup, ¾ ounce; brandy, ½ ounce; Angostura bitters, 3 drops; 1 egg. Put in shaker and beat well. Strain in 10-ounce mug, and fill with hot milk; finish with whipped cream and nutmeg.

VII.—Hot Egg Phosphate.—Two ounces lemon syrup, 1 egg, ½ ounce solution of acid phosphate. Mix in a glass, and shake together thoroughly; pour into another glass, heated previously, and slowly draw full of hot water; season with nutmeg.

VIII.—Hot Egg Phosphate.—Break fresh egg into shaker and add ½ ounce pineapple syrup, ½ ounce orange syrup, 1 dash phosphate. Shake, without ice, and pour into bouillon cup. Draw cupful of hot water, sprinkle a touch of cinnamon, and serve with wafers.

FANCY SODA DRINKS:

Coffee Cream Soda.—Serve in a 12-ounce glass. Draw 1½ ounces of syrup and 1 ounce of cream. Into the shaker draw 8 ounces of carbonated water, pour into the glass sufficient to fill it to within

1 inch of the top; pour from glass to shaker and back, once or twice, to mix thoroughly; give the drink a rich, creamy appearance, and make it cream sufficiently to fill the glass.

Iced Coffee.—Serve in a 10-ounce glass. Draw 1 ounce into glass, fill nearly full with ice-cold milk, and mix by stirring.

Egg Malted Milk Coffee.—Prepare same as malted milk coffee, with the exception of adding the egg before shaking, and top off with a little nutmeg, if desired. This drink is sometimes called coffee light lunch.

Coffee Frappé.—Serve in a 12-ounce glass. Coffee syrup, 1½ ounces; white of 1 egg; 1 to 1½ ounces of pure, rich, sweet cream; a small portion of fine shaved ice; shake thoroughly to beat the white of the egg light, and then remove the glass, leaving the contents in the shaker. Now fill the shaker two-thirds full, using the fine stream only. Draw as quickly as possible that the drink may be nice and light. Now pour into glass and back, and then strain into a clean glass. Serve at once, and without straws. This should be drunk at once, else it will settle, and lose its lightness and richness.

Coffee Nogg.—

Coffee syrup	2	ounces
Brandy	4	drachms
Cream	2	ounces
One egg.		

Coffee Cocktail.—

Coffee syrup	1	ounce
One egg.		
Port wine	1	ounce
Brandy	2	drachms

Shake, strain into a small glass, and add soda. Mace on top.

Chocolate and Milk.—

Chocolate syrup	2	ounces
Sweet milk, sufficient.		

Fill a glass half full of shaved ice, put in the syrup, and add milk until the glass is almost full. Shake well, and serve without straining. Put whipped cream on top and serve with straws.

Chocolate Frappé.—

Frozen whipped cream, sufficient.
Shaved ice, sufficient.

Fill a glass half full of frozen whipped cream, fill with shaved ice nearly to the top, and pour in chocolate syrup. Other syrups may be used, if desired.

Royal Frappé.—This drink consists of 3 parts black coffee and 1 part of brandy, frozen in a cooler, and served while in a semifrozen state.

Mint Julep.—One-half tumbler shaved ice, teaspoonful powdered sugar, dash lemon juice, 2 or 3 sprigs of fresh mint. Crush the mint against side of the glass to get the flavor. Then add claret syrup, ½ ounce; raspberry syrup, 1½ ounces; and draw carbonated water nearly to fill glass. Insert bunch of mint and fill glass, leaving full of shaved ice. Serve with straws, and decorate with fruits of the season.

Grape Glacé.—Beat thoroughly the whites of 4 eggs and stir in 1 pound of powdered sugar, then add 1 pint grape juice, 1 pint water, and 1 pound more of powdered sugar. Stir well until sugar is dissolved, and serve from a pitcher or glass dish, with ladle.

"Golf Goblet."—Serve in a 12-ounce glass; fill two-thirds full of cracked ice, add ½ ounce pineapple juice, 1 teaspoonful lemon juice, 1 teaspoonful raspberry vinegar. Put spoon in glass, and fill to within one-half inch of top with carbonated water; add shaved ice, heaping full. Put strawberry or cherry on top, and stick slice of orange down side of glass. Serve with spoon and straws.

Goldenade.—Shaved ice, ¼ tumblerful; powdered sugar; juice of 1 lemon; yolk of 1 egg. Shake well, add soda water from large stream, turn from tumbler to shaker, and vice versa, several times, and strain through julep strainer into a 12-ounce tumbler.

Lunar Blend.—Take two mixing glasses, break an egg, putting the yolk in one glass, the white into the other; into the glass with the yolk add 1 ounce cherry syrup and some cracked ice; shake, add small quantity soda, and strain into a 12-ounce glass. Into the other mixing glass add 1 ounce plain sweet cream, and beat with bar spoons until well whipped; add ½ ounce lemon syrup, then transfer it into the shaker, and add soda from fine stream only, and float on top of the one containing the yolk and sherry. Serve with two straws.

Egg Chocolate.—

Chocolate syrup	2	ounces
Cream	4	ounces
White of one egg.		

Egg Crême de Menthe.—

Mint syrup	12 drachms
Cream	3 ounces
White of one egg.	
Whisky	4 drachms

Egg Sherbet.—

Sherry syrup	4 drachms
Pineapple syrup	4 drachms
Raspberry syrup	4 drachms
One egg.	
Cream.	

Egg Claret.—

Claret syrup	2 ounces
Cream	3 ounces
One egg.	

Royal Mist.—

Orange syrup	1 ounce
Catawba syrup	1 ounce
Cream	2 ounces
One egg.	

Banana Cream.—

Banana syrup	12 drachms
Cream	4 ounces
One egg.	

Egg Coffee.—

Coffee syrup	2 ounces
Cream	3 ounces
One egg.	
Shaved ice.	

Cocoa Mint.—

Chocolate syrup	1 ounce
Peppermint syrup	1 ounce
White of one egg.	
Cream	2 ounces

The peppermint syrup is made as follows:

Oil of peppermint	30 minims
Syrup simplex	1 gallon
Soda foam	1 ounce

Egg Lemonade.—

Juice of one lemon.	
Pulverized sugar	3 teasp'fuls
One egg.	
Water, q. s.	

Shake well, using plenty of ice, and serve in a small glass.

Nadjy.—

Raspberry juice	1 ounce
Pineapple syrup	1 ounce
One egg.	
Cream	2 ounces

Siberian Flip.—

Orange syrup	1 ounce
Pineapple syrup	1 ounce
One egg.	
Cream	2 ounces

Egg Orgeat.—

Orgeat syrup	12 drachms
Cream	3 ounces
One egg.	

Normona.—

Peach syrup	1 ounce
Grape syrup	1 ounce
Cream	3 ounces
Brandy	2 drachms
One egg.	

Silver Fizz.—

Catawba syrup	2 ounces
Holland gin	2 drachms
Lemon juice	8 dashes
White of one egg.	

Golden Fizz.—

Claret syrup	2 ounces
Holland gin	¼ ounce
Lemon juice	8 dashes
Yolk of one egg.	

Rose Cream.—

Rose syrup	12 drachms
Cream	4 ounces
White of one egg.	

Violet Cream.—

Violet syrup	12 drachms
Cream	4 ounces
White of one egg.	

Rose Mint.—

Rose syrup	6 drachms
Mint syrup	6 drachms
Cream	3 ounces
White of one egg.	

Currant Cream.—

Red-currant syrup	2 ounces
Cream	3 ounces
One egg.	

Quince Flip.—

Quince syrup	2 ounces
Cream	3 ounces
One egg.	
Shaved ice.	

Coffee Nogg.—

Coffee syrup	2 ounces
Brandy	4 drachms
Cream	2 ounces
One egg.	

Egg Sour.—

Juice of one lemon.	
Simple syrup	12 drachms
One egg.	

Shake, strain, and fill with soda. Mace on top.

Lemon Sour.—

    Lemon syrup....... 12 drachms
    Juice of one lemon.
    One egg.

Raspberry Sour.—

    Raspberry syrup.... 12 drachms
    One egg.
    Juice of one lemon.

Yama.—

    One egg.
    Cream 2   ounces
    Sugar.......... 2   teaspoonfuls
    Jamaica rum..... ½ ounce

Shake well, put into cup, and add hot water. Serve with whipped cream, and sprinkle mace on top.

Prairie Oyster.—

    Cider vinegar........ 2 ounces
    One egg.

Put vinegar into glass, and break into it the egg. Season with salt and pepper. Serve without mixing.

Fruit Frappé.—

    Granulated gelatin... 1 ounce
    Juice of six lemons.
    Beaten whites of two eggs.
    Water.............. 5 quarts
    Syrup.............. 1 quart
    Maraschino cherries.. 8 ounces
    Sliced peach......... 4 ounces
    Sliced pineapple..... 4 ounces
    Whole strawberries... 4 ounces
    Sliced orange........ 4 ounces

Dissolve the gelatin in 1 quart boiling hot water; add the syrup and the balance of the water; add the whites of the eggs and lemon juice.

KOUMISS.

The original koumiss is the Russian, made from mare's milk, while that produced in this country and other parts of Europe is usually, probably always, made from cow's milk. For this reason there is a difference in the preparation which may or may not be of consequence. It has been asserted that the ferment used in Russia differs from ordinary yeast, but this has not been established.

In an article on this subject, contributed by D. H. Davies to the *Pharmaceutical Journal and Transactions*, it is pointed out that mare's milk contains less casein and fatty matter than cow's milk, and he states that it is "therefore far more easy of digestion." He thinks that cow's milk yields a better preparation when diluted with water to reduce the percentage of casein, etc. He proposes the following formula:

    Fresh milk...... .. 12 ounces
    Water............. 4 ounces
    Brown sugar.......150 grains
    Compressed yeast... 24 grains
    Milk sugar......... 3 drachms

Dissolve the milk sugar in the water, add to the milk, rub the yeast and brown sugar down in a mortar with a little of the mixture, then strain into the other portion.

Strong bottles are very essential, champagne bottles being frequently used, and the corks should fit tightly; in fact, it is almost necessary to use a bottling machine for the purpose, and once the cork is properly fixed it should be wired down. Many failures have resulted because the corks did not fit properly, the result being that the carbon dioxide escaped as formed and left a worthless preparation. It is further necessary to keep the preparation at a moderate temperature, and to be sure that the article is properly finished the operator should gently shake the bottles each day for about 10 minutes to prevent the clotting of the casein. It is well to take the precaution of rolling a cloth around the bottle during the shaking process, as the amount of gas generated is great, and should the bottle be weak it might explode.

Kogelman says that if 1 volume of buttermilk be mixed with 1 or 2 volumes of sweet milk, in a short time lively fermentation sets in, and in about 3 days the work is completed. This, according to the author, produces a wine-scented fluid, rich in alcohol, carbon dioxide, lactic acid, and casein, which, according to all investigations yet made, is identical with koumiss. The following practical hints are given for the production of a good article: The sweet milk used should not be entirely freed from cream; the bottles should be of strong glass; the fermenting milk must be industriously shaken by the operator at least 3 times a day, and then the cork put in firmly, so that the fluid will become well charged with carbon-dioxide gas; the bottles must be daily opened and at least twice each day brought nearly to a horizontal position, in order to allow the carbon dioxide to escape and air to enter; otherwise fermentation rapidly ceases. If a drink is desired strong in carbonic acid, the bottles, toward the end of fermentation, should be placed with the necks down. In order to ferment a fresh quantity of milk, simply add ⅓ of its volume of either actively fermenting or freshly fermented milk. The temperature should be from 50° to 60° F., about 60° being the most favorable.

Here are some miscellaneous formulas:

I.—Fill a quart champagne bottle up to the neck with pure milk; add 2 tablespoonfuls of white sugar, after dissolving the same in a little water over a hot fire; add also a quarter of a 2-cent cake of compressed yeast. Then tie the cork in the bottle securely, and shake the mixture well; place it in a room of the temperature of 50° to 95° F. for 6 hours, and finally in the ice box over night. Handle wrapped in a towel as protection if the bottle should burst. Be sure that the milk is pure, that the bottle is sound, that the yeast is fresh, to open the mixture in the morning with great care, on account of its effervescent properties; and be sure not to drink it at all if there is any curdle or thickening part resembling cheese, as this indicates that the fermentation has been prolonged beyond the proper time.

II.—Dilute the milk with $\frac{1}{8}$ part of hot water, and while still tepid add $\frac{1}{8}$ of very sour (but otherwise good) buttermilk. Put it into a wide jug, cover with a clean cloth, and let stand in a warmish place (about 75° F.) for 24 hours; stir up well, and leave for another 24 hours. Then beat thoroughly together, and pour from jug to jug till perfectly smooth and creamy. It is now "still" koumiss, and may be drunk at once. To make it sparkling, which is generally preferred, put it into champagne or soda-water bottles; do not quite fill them, secure the corks well, and lay them in a cool cellar. It will then keep for 6 or 8 weeks, though it becomes increasingly acid. To mature some for drinking quickly, it is as well to keep a bottle or two to start with in some warmer place, and from time to time shake vigorously. With this treatment it should, in about 3 days, become sufficiently effervescent to spurt freely through a champagne tap, which must be used for drawing it off as required. Later on, when very frothy and acid it is more pleasant to drink if a little sweetened water (or milk and water) is first put into the glass. Shake the bottle, and hold it inverted well into the tumbler before turning the tap. Having made one lot of koumiss as above you can use some of that instead of buttermilk as a ferment for a second lot, and so on 5 or 6 times in succession; after which it will be found advisable to begin again as at first. Mare's milk is the best for koumiss; then ass's milk. Cow's milk may be made more like them by adding a little sugar of milk (or even loaf sugar) with the hot water before fermenting. But perhaps the chief drawback to cow's milk is that the cream separates permanently, whereas that of mare's milk will remix. Hence use partially skimmed milk; for if there is much cream it only forms little lumps of butter, which are apt to clog the tap, or are left behind in the bottle.

Kwass.—Kwass is a popular drink among the Russian population of Kunzews, prepared as follows: In a big kettle put from 13 to 15 quarts of water, and bring to a boil, and when in active ebullition pour in 500 grams of malt. Let boil for 20 minutes, remove from the fire, let cool down, and strain off. The liquid is now put into a clean keg or barrel, 30 grams (about an ounce) of best compressed yeast added along with about 600 grams (20 ounces) of sugar, and the cask is put in a warm place to ferment. As soon as bubbles of carbonic gas are detected on the surface of the liquid, it is a signal that the latter is ready for bottling. In each of the bottles, which should be strong and clean, put one big raisin, fill, cork, and wire down. The bottles should be placed on the side, and in the coolest place available—best, on ice. The liquor is ready for drinking in from 2 to 3 days, and is said to be most palatable.

"Braga."—Braga is a liquid of milky turbidity, resembling *café au lait* in color, and forming a considerable precipitate if left alone. When shaken it sparkles and a little gas escapes. Its taste is more or less acid, possessing a pleasant flavor.

About 35 parts of crushed millet, to which a little wheat flour is added, are placed in a large kettle. On this about 400 parts of water are poured. The mixture is stirred well and boiled for 3 hours. After settling for 1 hour the lost water is renewed and the boiling continued for another 10 hours. A viscous mass remains in the kettle, which substance is spread upon large tables to cool. After it is perfectly cool, it is stirred with water in a wooden trough and left to ferment for 8 hours. This pulp is sifted, mixed with a little water, and after an hour the braga is ready for sale. The taste is a little sweetish at first, but becomes more and more sourish in time. Fermentation begins only in the trough.

WINTER BEVERAGES:

Campchello.—Thoroughly beat the yolks of 12 fresh eggs with $2\frac{1}{4}$ pounds finely powdered, refined sugar, the juice of 3 lemons and 2 oranges, and 3 bottles of Grâves or other white wine, over the fire, until rising. Remove, and slowly beat 1 bottle of Jamaica rum with it.

Egg Wine.—Vigorously beat 4 whole eggs and the yolks of 4 with ½ pound of fine sugar; next add 2 quarts of white wine and beat over a moderate fire until rising.

Bavaroise au Cognac.—Beat up the yolks of 8 eggs in 1 quart of good milk over the fire, until boiling, then quickly add 5 ounces of sugar and ⅓ quart of fine cognac.

Bavaroise au Café.—Heat 1 pint of strong coffee and 1 pint of milk, 5 ounces of sugar, and the yolks of 8 eggs, until boiling, then add 1/16 quart of Jamaica rum.

Carbonated Pineapple Champagne.—

Plain syrup, 42°.....	10	gallons
Essence of pineapple	8	drachms
Tincture of lemon...	5	ounces
Carbonate of magnesia..............	1	ounce
Liquid saffron......	2½	ounces
Citric-acid solution..	30	ounces
Caramel...........	2½	ounces

Filter before adding the citric-acid solution and limejuice. Use 2 ounces to each bottle.

A German Drink.—To 100 parts of water add from 10 to 15 parts of sugar, dissolve and add to the syrup thus formed an aqueous extract of 0.8 parts of green or black tea. Add fresh beer or brewers' yeast, put in a warm place and let ferment. When fermentation has progressed to a certain point the liquid is cleared, and then bottled, corked, and the corks tied down. The drink is said to be very pleasant.

Limejuice Cordial.—Limejuice cordial that will keep good for any length of time may be made as follows: Sugar, 6 pounds; water, 4 pints; citric acid, 4 ounces; boric acid, ⅓ ounce. Dissolve by the aid of a gentle heat, and when cold add refined limejuice, 60 ounces; tincture of lemon peel, 4 ounces; water to make up to 2 gallons, and color with caramel.

Summer Drink.—

Chopped ice......	2	tablespoonfuls
Chocolate syrup ..	2	tablespoonfuls
Whipped cream ...	3	tablespoonfuls
Milk............	½	cup
Carbonated water.	¼	cup

Shake or stir well before drinking. A tablespoonful of vanilla ice cream is a desirable addition. A plainer drink is made by combining the syrup, ¾ cup of milk, and the ice, and shaking well.

American Champagne.—Good cider (crab-apple cider is the best), 7 gallons; best fourth-proof brandy, 1 quart; genuine champagne wine, 5 pints; milk, 1 gallon; bitartrate of potassa, 2 ounces. Mix, let stand a short time; bottle while fermenting. An excellent imitation.

British Champagne.—Loaf sugar, 56 pounds; brown sugar (pale), 48 pounds; water (warm), 45 gallons; white tartar, 4 ounces; mix, and at a proper temperature add yeast, 1 quart; and afterwards sweet cider, 5 gallons; bruised wild cherries, 14 or 15 ounces; pale spirits, 1 gallon; orris powder, ½ ounce. Bottle while fermenting.

Champagne Cider.—Good pale cider, 1 hogshead; spirits, 3 gallons; sugar, 20 pounds; mix, and let it stand one fortnight; then fine with skimmed milk, ½ gallon; this will be very pale, and a similar article, when properly bottled and labeled, opens so briskly that even good judges have mistaken it for genuine champagne.

BEER:

Scotch Beer.—Add 1 peck malt to 4 gallons of boiling water and let it mash for 8 hours, and then strain, and in the strained liquor boil:

Hops..............	4 ounces
Coriander seeds....:.	1 ounce
Honey...... ,......	1 pound
Orange peel.........	2 ounces
Bruised ginger.......	1 ounce

Boil for half an hour, then strain and ferment in the usual way.

Hop Bitter Beer.—

Coriander seeds......	2 ounces
Orange peel.........	4 ounces
Ginger.............	1 ounce
Gentian root........	½ ounce

Boil in 5 gallons of water for half an hour, then strain and put into the liquor 4 ounces hops and 3 pounds of sugar, and simmer for 15 minutes, then add sufficient yeast, and bottle when ready.

Sarsaparilla Beer.—I.—Compound extract of sarsaparilla, 1½ ounces; hot water, 1 pint; dissolve, and when cold, add of good pale or East India ale, 7 pints.

II.—Sarsaparilla (sliced), 1 pound; guaiacum bark (bruised small), ¼ pound; guaiacum wood (rasped) and licorice root (sliced), of each, 2 ounces; aniseed (bruised), 1½ ounces; mezereon root-bark, 1 ounce; cloves (cut small), $\frac{1}{4}$ ounce; moist sugar, $3\frac{1}{2}$ pounds; hot water (not boiling), 9 quarts; mix in a clean stone jar, and keep it in a moderately warm room (shaking it twice or thrice daily) until active fermentation sets in, then let it repose for about a week, when it will be ready for use. This is said to be superior to the other preparations of sarsaparilla as an alterative or purifier of the blood, particularly in old affections. That usually made has generally only $\frac{1}{2}$ of the above quantity of sugar, for which molasses is often substituted; but in either case it will not keep well; whereas, with proper caution, the products of the above formulas may be kept for 1 or even 2 years. No yeast must be used. Dose: A small tumblerful 3 or 4 times a day, or oftener.

Spruce Beer.—I.—Sugar, 1 pound; essence of spruce, $\frac{1}{2}$ ounce; boiling water, 1 gallon; mix well, and when nearly cold add of yeast $\frac{1}{2}$ wineglassful; and the next day bottle like ginger beer.

II.—Essence of spruce, $\frac{1}{2}$ pint; pimento and ginger (bruised), of each, 5 ounces; hops, $\frac{1}{2}$ pound; water, 3 gallons; boil the whole for 10 minutes, then add of moist sugar, 12 pounds (or good molasses, 14 pounds); warm water, 11 gallons; mix well, and, when only lukewarm, further add of yeast, 1 pint; after the liquid has fermented for about 24 hours, bottle it.

This is diuretic and antiscorbutic. It is regarded as an agreeable summer drink, and often found useful during long sea voyages. When made with lump sugar it is called White Spruce Beer; when with moist sugar or treacle, Brown Spruce Beer. An inferior sort is made by using less sugar or more water.

Treacle Beer.—I.—From treacle or molasses, $\frac{3}{4}$ to 2 pounds per gallon (according to the desired strength); hops, $\frac{1}{4}$ to $\frac{3}{4}$ ounce; yeast, a tablespoonful; water, q. s.; treated as below.

II.—Hops, $1\frac{1}{2}$ pounds; corianders, 1 ounce; capsicum pods (cut small), $\frac{1}{2}$ ounce; water, 8 gallons; boil for 10 or 15 minutes, and strain the liquor through a coarse sieve into a barrel containing treacle, 28 pounds; then throw back the hops, etc., into the copper and reboil them, for 10 minutes, with a second 8 gallons of water, which must be strained into the barrel, as before; next "rummage" the whole well with a stout stick, add of cold water 21 gallons (sufficient to make the whole measure 37 gallons), and, again after mixing, stir in $\frac{1}{2}$ pint of good fresh yeast; lastly, let it remain for 24 hours in a moderately warm place, after which it may be put into the cellar, and in 2 or 3 days bottled or tapped on draught. In a week it will be fit to drink. For a stronger beer, 36 pounds, or even half a hundredweight of molasses may be used. It will then keep good for a twelvemonth. This is a wholesome drink, but apt to prove laxative when taken in large quantities.

Weiss Beer.—This differs from the ordinary lager beer in that it contains wheat malt. The proportions are $\frac{2}{3}$ wheat to $\frac{1}{3}$ barley malt, 1 pound hops being used with a peck of the combined malt to each 20 gallons of water. A good deal depends on the yeast, which must be of a special kind, the best grades being imported from Germany.

Yellow Coloring for Beverages.—The coloring agents employed are fustic, saffron, turmeric, quercitron, and the various aniline dyes. Here are some formulas:

I.—Saffron.......... 1 ounce
    Deodorized alcohol.......... 4 fluidounces
    Distilled water... 4 fluidounces

Mix alcohol and water, and then add the saffron. Allow the mixture to stand in a warm place for several days, shaking occasionally; then filter. The tincture thus prepared has a deep orange color, and when diluted or used in small quantities gives a beautiful yellow tint to syrups, etc.

II.—Ground fustic wood.......... $1\frac{1}{2}$ ounces
    Deodorized alcohol.......... 4 fluidounces
    Distilled water... 4 fluidounces

This color may be made in the same manner as the liquid saffron, and is a fine coloring for many purposes.

III.—Turmeric powder.... 2 ounces
    Alcohol, dilute..... 16 ounces

Macerate for several days, agitating frequently, and filter. For some beverages the addition of this tincture is not to be recommended, as it possesses a very spicy taste.

The nonpoisonous aniline dyes recommended for coloring confectionery, beverages, liquors, essences, etc., yellow are those known as acid yellow R and tropæolin 000 (orange I).

BICYCLE-TIRE CEMENT:
See Adhesives, under Rubber Cements.

BICYCLE VARNISHES:
See Varnishes.

BIDERY METAL:
See Alloys.

BILLIARD BALLS:
See Ivory and Casein.

BIRCH BALSAM:
See Balsam.

BIRCH WATER:
See Hair Preparations.

BIRD DISEASES AND THEIR REMEDIES:
See Veterinary Formulas.

BIRD FOODS:
See also Veterinary Formulas.

Mixed Birdseed.—

Canary seed	6 parts
Rape seed	2 parts
Maw seed	1 part
Millet seed	2 parts

Mocking-Bird Food.—

Cayenne pepper	2 ounces
Rape seed	8 ounces
Hemp seed	16 ounces
Corn meal	2 ounces
Rice	2 ounces
Cracker	8 ounces
Lard oil	2 ounces

Mix the solids, grinding to a coarse powder, and incorporate the oil.

Food for Redbirds.—

Sunflower seed	8 ounces
Hemp seed	16 ounces
Canary seed	10 ounces
Wheat	8 ounces
Rice	6 ounces

Mix and grind to coarse powder.

BIRD LIME:
See Lime.

BIRD PASTE:
See Canary-Bird Paste.

BISCHOFF:
See Wines and Liquors.

BISCUIT, DOG:
See Dog Biscuit.

BISMUTH ALLOYS:
See Alloys.

BISMUTH, PURIFICATION OF:
See Gold.

BITTERS:
See Wines and Liquors.

BITTER WATER:
See Waters.

BLACKING FOR HARNESS:
See Leather.

BLACKING FOR SHOES:
See Shoedressings.

BLACKING, STOVE:
See Stove Blackings and Polishes.

BLACKBERRY CORDIAL AND BLACKBERRY MIXTURE AS A CHOLERA REMEDY:
See Cholera Remedy.

BLACKBOARD PAINT AND VARNISH:
See Paint and Varnish.

BLACKHEAD REMEDIES:
See Cosmetics.

BLANKET WASHING:
See Household Formulas.

BLASTING POWDER:
See Explosives.

Bleaching

Linen.—Mix common bleaching powder in the proportion of 1 pound to a gallon of water; stir it occasionally for 3 days, let it settle, and pour it off clear. Then make a lye of 1 pound of soda to 1 gallon of boiling water, in which soak the linen for 12 hours, and boil it half an hour; next soak it in the bleaching liquor, made as above; and lastly, wash it in the usual manner. Discolored linen or muslin may be restored by putting a portion of bleaching liquor into the tub wherein the articles are soaking.

Straw.—I.—Dip the straw in a solution of oxygenated muriatic acid, saturated with potash. (Oxygenated muriate of lime is much cheaper.) The straw is thus rendered very white, and its flexibility is increased.

II.—Straw is bleached by simply exposing it in a closed chamber to the fumes of burning sulphur. An old flour barrel is the apparatus most used for the purpose by milliners, a flat stone being laid on the ground, the sulphur ignited thereon, and the barrel containing the goods to be bleached turned over it. The goods should be previously washed in pure water.

Wool, Silk, or Straw.—Mix together 4 pounds of oxalic acid, 4 pounds of table salt, water 50 gallons. The goods are laid in this mixture for 1 hour; they are then generally well bleached, and only require to be thoroughly rinsed and worked. For bleaching straw it is best to soak the goods in caustic soda, and afterwards to make use of chloride of lime or Javelle water. The excess of chlorine is afterwards removed by hyposulphite of soda.

Feathers.—Place the feathers from 3 to 4 hours in a tepid dilute solution of bichromate of potassa, to which, cautiously, some nitric acid has been added (a small quantity only). To remove a greenish hue induced by this solution, place them in a dilute solution of sulphuric acid, in water, whereby the feathers become perfectly white and bleached.

Bleaching Solution.—Aluminum hypochloride, or Wilson's bleaching liquid, is produced by adding to a clear solution of lime chloride a solution of aluminum sulphate (alumina, alum) as long as a precipitate keeps forming. By mutual decomposition aluminum chloride results, which remains in solution, and lime sulphate (gypsum), which separates out in the form of an insoluble salt.

BLIGHT REMEDIES.

I.—Soft soap	40	parts
Amyl alcohol	50	parts
Methylated spirit	20	parts
Water	1,000	parts
II.—Soft soap	30	parts
Sulphureted potash	2	parts
Amyl alcohol	32	parts
Water	1,000	parts
III.—Soft soap	15	parts
Sulphureted potash	29	parts
Water	1,000	parts

BLEACHING SOLUTIONS FOR THE LAUNDRY:
See Laundry Preparations.

BLEACHING SOLUTION FOR PHOTOGRAPHS:
See Photography.

BLEEDING, LOCAL:
See Styptics.

BLISTER CURE:
See Turpentine.

BLISTERS, FOR HORSES:
See Veterinary Formulas.

BLOCK, HOLLOW CONCRETE BUILDING:
See Stone, Artificial.

BLOCK FOR SOLDERING:
See Soldering.

BLOTTING PAPER:
See Paper.

BLUE FROM GREEN AT NIGHT, TO DISTINGUISH:
To distinguish blue from green at night, use either the light of a magnesium wire for this purpose or take a number of Swedish (parlor) matches, light them, and as soon as they flash up, observe the 2 colors, when the difference can be easily told.

BLUE (BALL):
See Dyes.

BLUING:
See Laundry Preparations.

BLUING OF STEEL:
See Steel.

BLUE PRINTS, TO MAKE CHANGES AND CORRECTIONS ON:
Use a solution of sodium carbonate and water, with a little red ink mixed in. This gives a very pleasing pink color to the changes which, at the same time, is very noticeable. The amount of sodium carbonate used depends upon the surface of the blue-print paper, as some coarse-grained papers will look better if less soda is used and *vice versa*. However, the amount of powdered soda held on a small coin dissolved in a bottle of water gives good results.

BLUE-PRINT PAPER MAKING:
See Photography.

BLUE PRINTS, TO TURN BROWN:
See Photography, under Toning.

BOIL REMEDY.
Take a piece of soft linen or borated gauze, rub some vaseline upon one side of it, quickly pour upon it some chloroform, apply it to the unopened boil or carbuncle, and place a bandage over all. It smarts a little at first, but this is soon succeeded by a pleasing, cool sensation. The patient is given a bottle of the remedy, and directed to change the cloth often. In from 2 hours to 1 day the boil (no matter how indurated) softens and opens.

Boiler Compounds

There are three chemicals which are known to attack boiler scale. These are caustic soda, soda ash, and tannic-acid compounds, the last being derived from sumac, catechu, and the exhausted bark liquor from tanneries.

Caustic soda in large excess is injurious to boiler fittings, gaskets, valves, etc. That it is injurious, in reasonable excess, to the boiler tubes themselves is yet to be proved. Foaming and priming may be caused through excess of caustic soda or soda ash, as is well known by every practical engineer. Tannic acid is to be condemned and the use of its salts is not to be recommended. It may unite with the organic matter, present in the form of albuminoids, and with calcium and magnesium carbonates. That it removes scale is an assured fact; that it removes iron with the scale is also assured, as tannic acid corrodes an iron surface rapidly.

Compounds of vegetable origin are widely advertised, but they often contain dextrine and gum, both of which are dangerous, as they coat the tubes with a compact scale, not permitting the water to reach the iron. Molasses is acid and should not be used in the boiler. Starch substances generally should be avoided. Kerosene must be dangerous, as it is very volatile and must soon leave the boiler and pass over and through the engine.

There are two materials the use of which in boilers is not prohibited through action upon the metal itself or on account of price. These are soda ash and caustic soda. Sodium triphosphate and sodium fluoride have both been used with success, but their cost is several hundred per cent greater than soda ash. If prescribed as per analysis, in slight excess, there should be no injurious results through the use of caustic soda and soda ash. It would be practicable to manufacture an intimate mixture of caustic soda and carbonate of soda, containing enough of each to soften the average water of a given district.

There is a great deal of fraud in connection with boiler compounds generally. The better class of venders advertise to prepare a special compound for special water. This is expensive, save on a large scale, in reference to a particular water, for it would mean a score or more of tanks with men to make up the mixtures. The less honest of the boiler-compound guild consign each sample of water to the sewer and send the regular goods. Others have a stock analysis which is sent to customers of a given locality, whether it contains iron, lime, or magnesium sulphates or carbonates.

Any expense for softening water in excess of 3 cents per 1,000 gallons is for the privilege of using a ready-made softener. Every superintendent in charge of a plant should insist that the compound used be pronounced by competent authority free from injurious materials, and that it be adapted to the water in use.

Boiler compounds should contain only such ingredients as will neutralize the scale-forming salts present. They should be used only by prescription, so many gallons per 1,000 gallons of feed water. A properly proportioned mixture of soda ought to answer the demands of all plants depending upon that method of softening water in limestone and shale regions.

The honest boiler compounds are, however, useful for small isolated plants, because of the simplicity of their action. For plants of from 75 to 150 horse power two 24-hour settling tanks will answer the purpose of a softening system. Each of these, capable of holding a day's supply, provided with a soda tank in common, and with sludge valves, has paddles for stirring the contents. Large plants are operated on this principle, serving boilers of many thousand horse power. Such a system has an advantage over a continuous system, in that the exact amount of chemical solutions required for softening the particular water can be applied. For some variations of such a system, several companies have secured patents. The fundamental principles, however, have been used for many years and are not patentable.

Prevention of Boiler Scale.—The lime contained in the feed water, either as bicarbonate or as sulphate, is precipitated in the shape of a light mud, but the walls of the boiler remain perfectly bright without being attacked in any manner. While under ordinary atmospheric pressure calcium chromate in solution is precipitated by soda or Glauber's salt as calcium carbonate or as calcium sulphate; the latter is separated under higher pressure by chromates as calcium chromate. An excess of chromates or chromic acid does not exercise any deleterious action upon the metal, nor upon the materials used for packing. By the slight admixture of chromates, two pounds are sufficient for a small boiler for weeks; no injurious ingredients are carried in by the wet steam, the injection water, on the contrary, having been found to be chemically pure.

Protecting Boiler Plates from Scale.— I.—For a 5-horse-power boiler, fed with water which contains calcic sulphate, take catechu, 2 pounds; dextrine, 1 pound; crystallized soda, 2 pounds; potash, $\frac{1}{2}$ pound; cane sugar, $\frac{1}{2}$ pound; alum, $\frac{1}{2}$ pound; gum arabic, $\frac{1}{2}$ pound.

II.—For a boiler of the same size, fed with water which contains lime: Turmeric, 2 pounds; dextrine, 1 pound; sodium bicarbonate, 2 pounds; potash, $\frac{1}{2}$ pound; alum, $\frac{1}{2}$ pound; molasses, $\frac{1}{2}$ pound.

III.—For a boiler of the same size, fed with water which contains iron: Gamboge, 2 pounds; soda, 2 pounds; dextrine, 1 pound; potash, $\frac{1}{2}$ pound; sugar, $\frac{1}{2}$ pound; alum, $\frac{1}{2}$ pound; gum arabic, $\frac{1}{2}$ pound.

IV.—For a boiler of the same size, fed with sea water: Catechu, 2 pounds; Glauber's salt, 2 pounds; dextrine, 2 pounds; alum, $\frac{1}{2}$ pound; gum arabic, $\frac{1}{2}$ pound.

When these preparations are used add 1 quart of water, and in ordinary cases charge the boiler every month; but if the incrustation is very bad, charge every two weeks.

V.—Place within the boiler of 100 horse power 1 bucketful of washing soda; put in 2 gallons of kerosene oil (after closing the blow-off cock), and fill the boiler with water. Feed in at least 1 quart of kerosene oil every day through a sight-feed oil cup attached to the feed pipe near the boiler—i. e., between the heater and the boiler—so that the oil is not entrapped within the heater. If it is inconvenient to open the boiler, then dissolve the washing soda in hot water and feed it in with the pump or through a tallow cock (attached between the ejector and the valve in the suction pipe) when the ejector is working.

VI.—A paint for protecting boiler plates from scale, and patented in Germany, is composed of 10 pounds each of train oil, horse fat, paraffine, and of finely ground zinc white. To this mixture is added 40 pounds of graphite and 10 pounds of soot made together into a paste with $1\frac{1}{2}$ gallons of water, and about a pound of carbolic acid. The horse fat and the zinc oxide make a soap difficult to fuse, which adheres strongly to the plates, and binds the graphite and the soot. The paraffine prevents the water from penetrating the coats. The scale which forms on this application can be detached, it is said, with a wooden mallet, without injuring the paint.

VII.—M. E. Asselin, of Paris, recommends the use of glycerine as a preventive. It increases the solubility of combinations of lime, and especially of the sulphate. It forms with these combinations soluble compounds. When the quantity of lime becomes so great that it can no longer be dissolved, nor form soluble combinations, it is deposited in a gelatinous substance, which never adheres to the surface of the iron plates. The gelatinous substances thus formed are not carried with the steam into the cylinder of the engine. M. Asselin advises the employment of 1 pound of glycerine for every 300 pounds or 400 pounds of coal burnt.

Prevention of Electrolysis.—In order to prevent the eating away of the sheets and tubes by electrolytic action, it has long been the practice of marine engineers to suspend slabs of zinc in their boilers. The zinc, being more susceptible to the electrolytic action than the iron, is eaten away, while the iron remains unimpaired. The use of zinc in this way has been found also to reduce the trouble from boiler scale. Whether it be due to the formation of hydrogen bubbles between the heating surfaces and incipient scale, to the presence in the water of the zinc salts resulting from the dissolution of the zinc, or to whatever cause, it appears to be a general conclusion among those who have used it that the zinc helps the scale, as well as the corrosion. Nobody has ever claimed for it that it prevented the attachment of scale altogether, but the consensus of opinion is that it "helps some."

BOILER PRESSURE.

It hardly pays to reduce pressure on boilers, except in very extreme cases, but if it can be done by throttling before the steam reaches the cylinder of the engine it would be an advantage, because this retains the heat units due to the higher pressure in the steam, and the throttling has a slight superheating effect. As a matter of fact, tests go to show that for light loads and high pressure a throttling engine may do better than an automatic cut-off. The ideal arrangement is to throttle the steam for light loads; for heavier loads, allow the variable cut-off to come into play. This practice has been carried into effect by the design of Mr. E. J. Armstrong, in which he arranges the shaft governor so that there is negative lead up to nearly one-quarter cut-off, after which the lead becomes positive, and this has the effect of throttling the steam for the earlier loads and undoubtedly gives better economy, in addition to making the engine run more quietly.

BONE BLACK:

Bone or Ivory Black.—All bones (and ivory is bone in a sense) consist of a framework of crystallized matter or bone earth, in the interstices of which organic matter is embedded. Hence if bones are heated red-hot in a closed vessel, the organic matter is destroyed, leaving carbon, in a finely divided state, lodged in the bony framework. If the heat is applied gradually the bone retains its shape, but is quite black and of much less weight than at first. This bone black or animal charcoal is a substance which has great power of absorbing coloring matter from liquids, so that it is largely used for bleaching such liquids. For example, in the vast industry of beet-sugar manufacture the solutions first made are very dark in color, but after filtration through animal charcoal will give colorless crystals on evaporation. Chemical trades require such large quantities of bone charcoal that its production is a large industry in itself. As in breaking up the charred bones a considerable amount of waste is produced, in the form of dust and small grains which cannot be used for bleaching purposes, this waste should be worked up into a pigment. This is done by dissolving out the mineral with hydrochloric acid, and then rinsing and drying the carbon.

The mineral basis of bones consists mainly of the phosphates of lime and magnesia, salts soluble in not too dilute hydrochloric acid. A vat is half filled with the above-mentioned waste, which is then just covered with a mixture of equal volumes of commercial hydrochloric acid and water. As the mineral matter also contains carbonates, a lively effervescence at once ensues, and small quantities of hydrofluoric acid are also formed from the decomposition of calcium fluoride in the bones. Now hydrofluoric acid is a very dangerous substance, as air containing even traces of it is very injurious to the lungs. Hence the addition of hydrochloric acid should be done in the open air, and the vat should be left by itself until the evolution of fumes ceases. A plug is then pulled out at the bottom and the carbon is thoroughly drained. It is then stirred up with water and again drained, when it has fully settled to the bottom. This rinsing with clear water is repeated till all the hydrochloric acid is washed away and only pure carbon remains in the vat. As for pigment-making purposes it is essential that the carbon should be as finely divided as possible, it is as well to grind the washed carbon in an ordinary color mill. Very little power is required for this purpose, as when once the bone earth is removed the carbon particles have little cohesion. The properly ground mass forms a deep-black mud, which can be left to dry or be dried by artificial heat. When dry, the purified bone black is of a pure black and makes a most excellent pigment.

Bone black is put upon the market under all sorts of names, such as ivory black, *ebur ustum*, Frankfort black, neutral black, etc. All these consist of finely ground bone black purified from mineral matter. If leather scraps or dried blood are to be worked up, iron tubes are employed, closed at one end, and with a well-fitting lid with a small hole in it at the other. As these bodies give off large volumes of combustible gas during the charring, it is a good plan to lead the vapors from the hole by a bent tube so that they can be burnt and help to supply the heat required and so save fuel. Leather or blood gives a charcoal which hardly requires treatment with hydrochloric acid, for the amount of mineral salts present is so small that its removal appears superfluous.

BONES, A TEST FOR BROKEN.

Place a stethoscope on one side of the supposed fracture, and a tuning fork on the other. When the latter is vibrated, and there is no breakage, the sound will be heard distinctly through bone and stethoscope. Should any doubt exist, comparison should be made with the same bone on the other side of the body. This test shows the difference in the power of conducting sound possessed by bone and soft tissue.

BONE BLEACHES:
See Ivory.

BONE FAT:
See Fats.

BONE FAT, PURIFICATION AND BLEACHING OF:
See Soap.

BONE POLISHES:
See Polishes.

BONE FERTILIZERS:
See Fertilizers.

BONES, TREATMENT OF, IN MANUFACTURING GLUE:
See Adhesives.

BONE, UNITING GLASS WITH:
See Adhesives.

BOOKS, THEIR HANDLING AND PRESERVATION:

The Preservation of Books in Hot Climates.—Books in hot climates quickly deteriorate unless carefully guarded. There are three destructive agencies: (1) damp, (2) a small black insect, (3) cockroaches.

(1) Books which are kept in a damp atmosphere deteriorate on account of molds and fungi that grow rapidly when the conditions are favorable. Books are best kept on open, airy, well-lighted shelves. When there has been a prolonged spell of moist weather their covers should be wiped, and they should be placed in the sun or before a fire for a few hours. Damp also causes the bindings and leaves of some books to separate.

(2) A small black insect, one-eighth of an inch long and a sixteenth of an inch broad, somewhat resembling a beetle, is very destructive, and books will be found, if left untouched, after a few months to have numerous holes in the covers and leaves. If this insect be allowed plenty of time for its ravages it will make so many holes that bindings originally strong can be easily torn to pieces. All damage may be prevented by coating the covers of books with the varnish described under (3). When books are found to contain the insects they should be well wrapped and placed in the sun before varnishing.

(3) The appearance of a fine binding may be destroyed in a single night by cockroaches. The lettering of the binding may, in two or three days, be completely obliterated.

The following varnishes have been found to prevent effectually the ravages of cockroaches and of all insects that feed upon books:

I.—Dammar resin....... 2 ounces
Mastic............. 2 ounces
Canada balsam...... 1 ounce
Creosote........... ½ ounce
Spirit of wine....... 20 fl. ounces

Macerate with occasional shaking for a few days if wanted at once, but for a longer time when possible, as a better varnish will result after a maceration of several months.

II.—Corrosive sublimate, 1 ounce; carbolic acid, 1 ounce; methylated or rum spirit, 1 quart.

Where it is necessary to keep books or paper of any description in boxes, cupboards, or closed bookcases, some naphthalene balls or camphor should be always present with them. If camphor be used it is best to wrap it in paper, otherwise it volatilizes more quickly than is necessary. In dry weather the doors of closed bookcases should be left open occasionally, as a damp, still atmosphere is most favorable for deterioration.

How to Open a Book.—Never force the back of the book. Hold the book with its back on a smooth or covered table; let the front board down, then the other, holding the leaves in one hand while you open a few leaves at the back, then a few at the front, and so on, alternately opening back and front, gently pressing open the sections till you reach the center of the volume. Do this two or three times and you will obtain the best results. Open the volume violently or carelessly in any one place and you will probably break the back or cause a start in the leaves.

BOOK DISINFECTANT:
See Disinfectants.

BOOKS, TO REMOVE FINGER-MARKS FROM:
See Cleaning Preparations and Methods.

BOOKBINDERS' VARNISH:
See Varnishes.

BOOKWORMS:
See Insecticides.

BOOT DRESSINGS:
See Shoe Dressings.

BOOT LUBRICANT:
See Lubricant.

BOOTS, WATERPROOFING:
See Waterproofing.

BORAX FOR SPRINKLING.

I.—Sprinkling borax is not only cheaper, but also dissolves less in soldering than pure borax.

The borax is heated in a metal vessel until it has lost its water of crystallization and mixed with calcined cooking salt and potash—borax, 8 parts; cooking salt, 3 parts; potash, 3 parts. Next i. is pounded in a mortar into a fine powder, constituting the sprinkling borax.

II.—Another kind of sprinkling borax is prepared by substituting glass-gall for the potash. Glass-gall is the froth floating on the melted glass, which can be skimmed off.

The borax is either dusted on in powder form from a sprinkling box or stirred with water before use into a thin paste.

BORAX AND BORIC ACID IN FOOD:
See Food.

BORDEAUX MIXTURE:
See Insecticides.

BOROTONIC:
See Dentifrices.

BOTTLE-CAP LACQUER:
See Lacquer.

BOTTLE CLEANERS:
See Cleaning Preparations and Methods, under Miscellaneous Methods.

BOTTLE STOPPERS:
See Stoppers.

BOTTLE VARNISH:
See Varnishes.

BOTTLE WAX:
See Photography.

BOUILLON:
See Beverages.

BOURBON METALS:
See Alloys.

BOWLS OF FIRE TRICK:
See Pyrotechnics.

BOX GLUE:
See Adhesives.

BRAGA:
See Beverages.

BRAN, SAWDUST IN.

For the detection of sawdust in bran use a solution of 1 part of phloroglucin in 15 parts of alcohol, 15 parts of water, and 10 parts of syrupy phosphoric acid. Place 2 parts of the solution in a small porcelain dish, add a knifepointful of the bran and heat moderately. Sawdust is dyed red while bran parts only seldom acquire a faint red color. By a microscopic examination of the reddish parts, sawdust will be readily recognized.

Bottles

Magic Bottles.—

The mystery of the "wonderful bottle," from which can be poured in succession port wine, sherry, claret, water, champagne, or ink, at the will of the operator, is easily explained. The materials consist of an ordinary dark-colored pint wine bottle, seven wine glasses of different patterns, and the chemicals described below:

Solution A: A mixture of tincture of ferric chloride, drachms vi; hydrochloric acid, drachms ii.

Solution B: Saturated solution of ammonium sulphocyanide, drachm i.

Solution C: Strong solution of ferric chloride, drachm i.

Solution D: A weak solution of ammonium sulphocyanide.

Solution E: Concentrated solution of lead acetate.

Solution F: Solution of ammonium sulphide, drachm i; or pyrogallic acid, drachm i.

Package G: Pulverized potassium bicarbonate, drachm iss.

Having poured two teaspoonfuls of solution A into the wine bottle, treat the wine glasses with the different solutions, noting and remembering into which glasses the several solutions are placed. Into No. 1 wine glass pour one or two drops of solution B; into No. 2 glass pour one or two drops of solution C; into No. 3 one or two drops of Solution D; leave No. 4 glass empty; into No. 5 glass pour a few drops of Solution E; into No. 6 glass place a few grains of Package G; into No. 7 glass pour a little of solution F.

Request some one to bring you some cold drinking water, and to guarantee that it is pure show that your wine bottle is (practically) empty. Fill it up from the carafe, and having asked the audience whether you shall produce wine or water, milk or ink, etc., you may obtain any of these by pouring a little of the water from the bottle into the prepared glass. Thus No. 1 glass gives a port-wine color; No. 2 gives a sherry color; No. 3 gives a claret color; No. 4 is left empty to prove that the solution in the bottle is colorless; No. 5 produces milk; No. 6, effervescing champagne; No. 7, ink.

Bottle-Capping Mixtures.—

I.—Soak 7 pounds of good gelatin in 10 ounces of glycerine and 60 ounces of water, and heat over a water bath until dissolved, and add any desired color. Pigments may be used, and various tints can be obtained by the use of aniline colors. The resulting compound should be stored in jars. To apply liquefy the mass and dip the cork and portion of the neck of the bottle into the liquid; it sets very quickly.

II.—Gelatin	1 ounce
Gum arabic	1 ounce
Boric acid	20 grains
Starch	1 ounce
Water	16 fluidounces

Mix the gelatin, gum arabic, and boric acid with 14 fluidounces of cold water, stir occasionally until the gum is dissolved, heat the mixture to boiling, remove the scum, and strain. Also mix the starch intimately with the remainder of the water, and stir this mixture into the hot gelatin mixture until a uniform product results. As noted above, the composition may be tinted with any suitable dye. Before using, it must be softened by the application of heat.

III.—Shellac.......... 3 ounces
    Venice turpentine 1½ ounces
    Boric acid....... 72 grains
    Powdered talcum. 3 ounces
    Ether........... 6 fluidrams
    Alcohol........ 12½ fluidounces

Dissolve the shellac, turpentine, and boric acid in the mixed alcohol and ether, color with a spirit-soluble dye, and add the talcum. During use the mixture must be agitated frequently.

Show Bottles.—

I.—Place in a cylindrical bottle the following liquids in the order named:

First, sulphuric acid, tinted blue with indigo; second, chloroform; third, glycerine, slightly tinted with caramel; fourth, castor oil, colored with alkanet root; fifth, 40-per-cent alcohol, slightly tinted with aniline green; sixth, codliver oil, containing 1 per cent of oil of turpentine. The liquids are held in place by force of gravity, and alternate with fluids which are not miscible, so that the strata of layers are clearly defined and do not mingle by diffusion.

II.—Chromic acid...... 1 drachm
    Commercial "muri-
      atic" acid........ 2 ounces
    Nitric acid......... 2 ounces
    Water, enough to
      make.......... 3 gallons

The color is magenta.

The following makes a fine pink for show carboys:

III.—Cobalt oxide........ 2 parts
    Nitric acid, c. p...... 1 part
    Hydrochloric acid.... 1 part

Mix and dissolve, and to the solution add:

    Strongest water of
      ammonia......... 6 parts
    Sulphuric acid...... 1 part
    Water, distilled, q. s.
      to make......... 400 parts

This should be left standing in a dark, cool place for at least a month before putting in the window.

IV.—Green.—Copper sulphate, 300 parts, by weight; hydrochloric acid, 450 parts, by weight; distilled water, to 4,500 parts, by weight.

V.—Blue.—Copper sulphate, 480 parts, by weight; sulphuric acid, 60 parts, by weight; distilled water, to 450 parts, by weight.

VI.—Yellowish Brown.—Potassium dichromate, 120 parts, by weight; nitric acid, 150 parts, by weight; distilled water, to 4,500 parts, by weight.

VII.—Yellow.—Potassium dichromate, 30 parts, by weight; sodium bicarbon-ate, 225 parts, by weight; distilled water, to 4,500 parts, by weight.

VIII.—Red.—Liquid ferric chloride, officinal, 60 parts, by weight; concentrated ammonium-acetate solution, 120 parts, by weight; acetic acid, 30 per cent, 30 parts, by weight; distilled water, to 9,000 parts, by weight.

IX.—Crimson.—Potassium iodide, 7.5 parts, by weight; iodine, 7.5 parts, by weight; hydrochloric acid, 60 parts, by weight; distilled water, to 4,500 parts, by weight.

All the solutions IV to IX should be filtered. If distilled water be used these solutions should keep for five to ten years. In order to prevent them from freezing, either add 10 per cent of alcohol, or reduce the quantity of water by 10 per cent.

A Cheap and Excellent Warming Bottle.—Mix sodium acetate and sodium hyposulphate in the proportion of 1 part of the former to 9 parts of the latter, and with the mixture fill an earthenware bottle about three-quarters full. Close the vessel well with a cork and place it either in hot water or in the oven, and let remain until the salts within melt. For at least a half day the jug will radiate its heat, and need only be well shaken from time to time to renew its heat-giving energy.

Bottle Deodorizer.—Powdered black mustard seed is successfully employed. Pour a little of it with some lukewarm water into the receptacle, rinsing it afterwards with water. If necessary, repeat the process.

BRANDY AND BRANDY BITTERS
See Wines and Liquors.

Brass

Formulas for the making of Brass will be found under Alloys.

Colors for Polished Brass.—The brass objects are put into boiling solutions composed of different salts, and the intensity of the shade obtained is dependent upon the duration of the immersion. With a solution composed of

    Sulphate of copper.... 120 grains
    Hydrochlorate of am-
      monia............ 30 grains
    Water.............. 1 quart

greenish shades are obtained. With the following solution all the shades of brown from orange brown to cinnamon are obtained:

Chlorate of potash...	150 grains
Sulphate of copper...	150 grains
Water..............	1 quart

The following solution gives the brass first a rosy tint and then colors it violet and blue:

Sulphate of copper...	435 grains
Hyposulphite of soda	300 grains
Cream of tartar......	150 grains
Water..............	1 pint

Upon adding to the last solution

| Ammoniacal sulphate of iron............ | 300 grains |
| Hyposulphite of soda | 300 grains |

there are obtained, according to the duration of the immersion, yellowish, orange, rosy, then bluish shades. Upon polarizing the ebullition the blue tint gives way to yellow, and finally to a pretty gray. Silver, under the same circumstances, becomes very beautifully colored. After a long ebullition in the following solution we obtain a yellow-brown shade, and then a remarkable fire red:

Chlorate of potash...	75 grains
Carbonate of nickel..	30 grains
Salt of nickel........	75 grains
Water..............	16 ounces

The following solution gives a beautiful, dark-brown color:

Chlorate of potash..	75 grains
Salt of nickel.......	150 grains
Water..............	10 ounces

The following gives, in the first place, a red, which passes to blue, then to pale lilac, and finally to white:

Orpiment.............	75 grains
Crystallized sal sodæ	150 grains
Water.............	10 ounces

The following gives a yellow brown:

Salt of nickel.......	75 grains
Sulphate of copper..	75 grains
Chlorate of potash..	75 grains
Water..............	10 ounces

On mixing the following solutions, sulphur separates and the brass becomes covered with iridescent crystallizations:

I.—Cream of tartar.....	75 grains
Sulphate of copper..	75 grains
Water.............	10 ounces

| II.—Hyposulphite of soda | 225 grains |
| Water............. | 5 ounces |

Upon leaving the brass objects immersed in the following mixture contained in corked vessels they at length acquire a very beautiful blue color:

Hepar of sulphur....	15 grains
Ammonia...........	75 grains
Water..............	4 ounces

Miscellaneous Coloring of Brass.—Yellow to bright red: Dissolve 2 parts native copper carbonate with 1 part caustic soda in 10 parts water. Dip for a few minutes into the liquor, the various shades desired being obtained according to the length of time of the immersion. Green: Dissolve 1 part copper acetate (verdigris), 1 part blue vitriol, and 1 part alum in 10 parts of water and boil the articles therein. Black: For optical articles, photographic apparatus, plates, rings, screws, etc., dissolve 45 parts of malachite (native copper carbonate) in 1,000 parts of sal ammoniac. For use clean and remove the grease from the article by pickling and dip it into the bath until the coating is strong enough. The bath operates better and quicker if heated. Should the oxidation be a failure it should be removed by dipping into the brass pickle.

A verdigris color on brass is produced by treating the articles with dilute acids, acetic acid, or sulphuric acid, and drying.

Brown in all varieties of shades is obtained by immersing the metal in solutions of nitrates or ferric chloride after it has been corroded with dilute nitric acid, cleaned with sand and water, and dried. The strength of the solutions governs the deepness of the resulting color.

Violet is caused by immersing the thoroughly cleaned objects in a solution of ammonium chloride.

Chocolate color results if red ferric oxide is strewn on and burned off, followed by polishing with a small quantity of galena.

Olive green is produced by blackening the surface with a solution of iron in hydrochloric acid, polishing with galena, and coating hot with a lacquer composed of 1 part varnish, 4 parts cincuma, and 1 part gamboge.

A steel-blue coloring is obtained by means of a dilute boiling solution of chloride of arsenic, and a blue one by a treatment with strong hyposulphite of soda. Another formula for bluing brass is: Dissolve 10 parts of antimony chloride in 200 parts of water, and add 30 parts of pure hydrochloric acid. Dip the article until it is well blued, then wash and dry in sawdust.

Black is much used for optical brass articles and is produced by coating with a solution of platinum or auric chloride mixed with nitrate of tin.

Coloring Unpolished Brass.—A yellow color of handsome effect is obtained on unpolished brass by means of antimony-chloride solution. This is produced by finely powdering gray antimony and boiling it with hydrochloric acid. With formation of hydrogen sulphide a solution of antimony results, which must not be diluted with water, since a white precipitate of antimony oxychloride is immediately formed upon admixture of water. For dilution, completely saturated cooking-salt solution is employed, using for 1 part of antimony chloride 2 parts of salt solution.

Coloring Fluid for Brass.—Caustic soda, 33 parts; water, 24 parts; hydrated carbonate of copper, 5.5 parts.

Dissolve the salt in water and dip the metal in the solution obtained. The intensity of the color will be proportional to the time of immersion. After removing the object from the liquid, rinse with water and dry in sawdust.

Black Color on Brass.—A black or oxidized surface on brass is produced by a solution of carbonate of copper in ammonia. The work is immersed and allowed to remain until the required tint is observed. The carbonate of copper is best used in a plastic condition, as it is then much more easily dissolved. Plastic carbonate of copper may be mixed as follows: Make a solution of blue vitriol (sulphate of copper) in hot water, and add a strong solution of common washing soda to it as long as any precipitate forms. The precipitate is allowed to settle, and the clear liquid is poured off. Hot water is added, and the mass stirred and again allowed to settle. This operation is repeated six or eight times to remove the impurities. After the water has been removed during the last pouring, and nothing is left but an emulsion of the thick plastic carbonate in a small quantity of water, liquid ammonia is added until everything is dissolved and a clear, deep-blue liquid is produced. If too strong, water may be added, but a strong solution is better than a weak one. If it is desired to make the solution from commercial plastic carbonate of copper the following directions may be followed: Dissolve 1 pound of the plastic carbonate of copper in 2 gallons of strong ammonia. This gives the required strength of solution.

The brass which it is desired to blacken is first boiled in a strong potash solution to remove grease and oil, then well rinsed and dipped in the copper solution, which has previously been heated to from 150° to 175° F. This solution, if heated too hot, gives off all the ammonia.

The brass is left in the solution until the required tint is produced. The color produced is uniform, black, and tenacious. The brass is rinsed and dried in sawdust. A great variety of effects may be produced by first finishing the brass before blackening, as the oxidizing process does not injure the texture of the metal. A satisfactory finish is produced by first rendering the surface of the brass matt, either by scratch-brush or similar methods, as the black finish thus produced by the copper solution is dead—one of the most pleasing effects of an oxidized surface. Various effects may also be produced by coloring the entire article and then buffing the exposed portions.

The best results in the use of this solution are obtained by the use of the so-called red metals—i. e., those in which the copper predominates. The reason for this is obvious. Ordinary sheet brass consists of about 2 parts of copper and 1 part of zinc, so that the large quantity of the latter somewhat hinders the production of a deep-black surface. Yellow brass is colored black by the solution, but it is well to use some metal having a reddish tint, indicating the presence of a large amount of copper. The varieties of sheet brass known as gilding or bronze work well. Copper also gives excellent results. Where the best results are desired on yellow brass a very light electroplate of copper before the oxidizing works well and gives an excellent black. With the usual articles made of yellow brass this is rarely done, but the oxidation carried out directly.

Black Finish for Brass.—I.—A handsome black finish may be put on brass by the following process: Dissolve in 1,000 parts of ammonia water 45 parts of natural malachite, and in the solution put the object to be blackened, after first having carefully and thoroughly cleaned the same. After letting it stand a short time gradually warm the mixture, examining the article from time to time to ascertain if the color is deep enough. Rinse and let dry.

II.—The blacking of brass may be accomplished by immersing it in the following solution and then heating over a Bunsen burner or a spirit flame: Add a saturated solution of ammonium carbonate to a saturated copper-sulphate solution, until the precipitate resulting in the beginning has almost entirely dissolved. The immersion and heating are repeated until the brass turns dark; then it is brushed and dipped in negative varnish or dull varnish.

To Give a Brown Color to Brass.—I.— In 1.000 parts of rain or distilled water dissolve 5 parts each of verdigris (copper acetate) and ammonium chloride. Let the solution stand 4 hours, then add 1,500 parts of water. Remove the brass to be browned from its attachment to the fixtures and make the surface perfectly bright and smooth and free from grease. Place it over a charcoal fire and heat until it "sizzes" when touched with the dampened finger. The solution is then painted over the surface with a brush or swabbed on with a rag. If one swabbing does not produce a sufficient depth of color, repeat the heating and the application of the liquid until a fine durable brown is produced. For door plates, knobs, and ornamental fixtures generally, this is one of the handsomest as well as the most durable surfaces, and is easily applied.

II.—A very handsome brown may be produced on brass castings by immersing the thoroughly cleaned and dried articles in a warm solution of 15 parts of sodium hydrate and 5 parts of cupric carbonate in 100 parts of water. The metal turns dark yellow, light brown, and finally dark brown, with a greenish shimmer, and, when the desired shade is reached, is taken out of the bath, rinsed, and dried.

III.—Paint the cleaned and dried surface uniformly with a dilute solution of ammonium sulphide. When this coating is dry, it is rubbed over, and then painted with a dilute ammoniacal solution of arsenic sulphide, until the required depth of color is attained. If the results are not satisfactory the painting can be repeated after washing over with ammonia. Prolonged immersion in the second solution produces a grayish-green film, which looks well, and acquires luster when polished with a cloth.

Refinishing Gas Fixtures.—Gas fixtures which have become dirty or tarnished from use may be improved in appearance by painting with bronze paint and then, if a still better finish is required, varnishing after the paint is thoroughly dry with some light-colored varnish that will give a hard and brilliant coating.

If the bronze paint is made up with ordinary varnish it is liable to become discolored from acid which may be present in the varnish. One method proposed for obviating this is to mix the varnish with about five times its volume of spirit of turpentine, add to the mixture dried slaked lime in the proportion of about 40 grains to the pint, agitate well, repeating the agitation several times, and finally allowing the suspended matter to settle and decanting the clear liquid. The object of this is to neutralize any acid which may be present. To determine how effectively this has been done the varnish may be chemically tested.

Steel Blue and Old Silver on Brass.—For the former dissolve 100 parts of carbonic carbonate in 750 parts of ammonia and dilute this solution with distilled water, whereupon the cleaned articles are dipped into the liquid by means of a brass wire. After two to three minutes take them out, rinse in clean water, and dry in sawdust. Old silver on brass is produced as follows: The articles are first silvered and next painted with a thin paste consisting of graphite, 6 parts; pulverized hematite, 1 part; and turpentine. Use a soft brush and dry well; then brush off the powder. Oxidized silver is obtained by dipping the silvered goods into a heated solution of liver of sulphur, 5 parts; ammonia carbonate, 10 parts; and water, 10,000 parts. Only substantially silvered objects are suited for oxidation, as a weak silvering is taken off by this solution. Unsatisfactory coloring is removed with potassium-cyanide solution. It is advisable to lay the articles in hydrogen sulphide-ammonia solution diluted with water, wherein they acquire a blue to a deep-black shade.

Tombac Color on Brass.—This is produced by immersion in a mixture of copper carbonate, 10 parts; caustic soda, 30 parts; water, 200 parts. This layer will only endure wiping with a cloth, not vigorous scouring with sand.

Graining of Brass.—Brass parts of timepieces are frequently provided with a dead grained surface. For this purpose they are fastened with flat-headed pins on cork disks and brushed with a paste of water and finest powdered pumice stone. Next they are thoroughly washed and placed in a solution of 10 quarts of water, 30 grains of mercuric nitrate, and 60 grains of sulphuric acid. In this amalgamating solution the objects become at once covered with a layer of mercury, which forms an amalgam with the copper, while the zinc passes into solution. After the articles have again been washed they are treated with graining powder, which consists of silver powder, tartar, and cooking salt. These substances must be pure, dry, and very finely pulverized. The mixing is done with moderate heat. According to whether a coarser or finer grain is desired, more cooking salt or more tartar must be contained in the powder. The ordinary proportions are:

Silver powder..	28	28	28 parts
Tartar........	283	110–140	85 parts
Cooking salt...	900	370	900 parts

This powder is moistened with water and applied to the object. Place the article with the cork support in a flat dish and rub on the paste with a stiff brush while turning the dish incessantly. Gradually fresh portions of graining powder are put on until the desired grain is obtained. These turn out the rounder the more the dish and brush are turned. When the right grain is attained, rinse off with water, and treat the object with a scratch brush, with employment of a decoction of saponaria. The brushes must be moved around in a circle in brushing with the pumice stone, as well as in rubbing on the graining powder and in using the scratch brush. The required silver powder is produced by precipitating a diluted solution of silver nitrate with some strips of sheet copper. The precipitated silver powder is washed out on a paper filter and dried at moderate heat.

The Dead, or Matt, Dip for Brass.— The dead dip is used to impart a satiny or crystalline finish to the surface. The bright dip gives a smooth, shiny, and perfectly even surface, but the dead dip is the most pleasing of any dip finish, and can be used as a base for many secondary finishes.

The dead dip is a mixture of oil of vitriol (sulphuric acid) and aqua fortis (nitric acid) in which there is enough sulphate of zinc (white vitriol) to saturate the solution. It is in the presence of the sulphate of zinc that the essential difference between the bright and the dead dip exists. Without it the dead or matt surface cannot be obtained.

The method generally practiced is to add the sulphate of zinc to the mixed acids (sulphuric and nitric), so that some remains undissolved in the bottom of the vessel. It is found that the sulphate of zinc occurs in small crystals having the appearance of very coarse granulated sugar. These crystals readily settle to the bottom of the vessel and do not do the work of matting properly. If they are finely pulverized the dip is slightly improved, but it is impossible to pulverize such material to a fineness that will do the desired work. The use of sulphate of zinc, then, leaves much to be desired. The most modern method of making

up the dead dip is to produce the sulphate of zinc directly in the solution and in the precipitated form. It is well known that the most finely divided materials are those which are produced by precipitation, and in the dead dip it is very important that the sulphate of zinc shall be finely divided so that it will not immediately settle to the bottom. Therefore it should be precipitated so that when it is mixed with the acids it will not settle immediately. The method of making the sulphate of zinc directly in the solution is as follows:

Take 1 gallon of yellow aqua fortis (38° F.) and place in a stone crock which is surrounded with cold water. The cold water is to keep the heat, formed by the reaction, from evaporating the acid. Add metallic zinc in small pieces until the acid will dissolve no more. The zinc may be in any convenient form—sheet clippings, lumps, granulated, etc., that may be added little by little. If all is added at once it will boil over. When the acid will dissolve no more zinc it will be found that some of the acid has evaporated by the heat, and it will be necessary to add enough fresh acid to make up to the original gallon. When this is done add 1 gallon of strong oil of vitriol. The mixture should be stirred with a wooden paddle while the oil of vitriol is being added.

As the sulphuric acid is being added the solution begins to grow milky, and finally the whole has the consistency of thick cream. This is caused by the sulphuric acid (oil of vitriol) precipitating out the sulphate of zinc. Thus the very finely divided precipitate of sulphate of zinc is formed. If one desires to use known quantities of acid and zinc the following amounts may be taken: Oil of vitriol, 1 gallon; aqua fortis (38° F.), 1 gallon; metallic zinc, 6 ounces.

In dissolving the zinc in the aqua fortis it is necessary to be sure that none remains undissolved in the bottom.

The dead or matt dip is used hot, and, therefore, is kept in a stone crock surrounded with hot water. The articles to be matted are polished and cleaned, and the dip thoroughly stirred with a wooden paddle, so as to bring up the sulphate of zinc which has settled. Dip the work in the solution and allow it to remain until the matt is obtained. This is a point which can be learned only by experience. When the brass article is first introduced there is a rapid action on the surface, but in a few seconds this slows down. Remove the article and rinse and immediately dip into the usual bright dip. This is necessary for the reason that the dead dip produces a dark coating upon the surface, which, were it left on, would not show the real effect or the color of the metal. The bright dip, however, removes this and exposes the true dead surface.

The usual rule for making up the dead dip is to use equal parts of oil of vitriol and aqua fortis; but these may be altered to suit the case. More oil of vitriol gives a finer matt, while a larger quantity of aqua fortis will give a coarser matt. When the dip becomes old it is unnecessary to add more zinc, as a little goes into the solution each time anything is dipped. After a while, however, the solution becomes loaded with copper salts, and should be thrown away.

A new dip does not work well, and will not give good results when used at once. It is usual to allow it to remain over night, when it will be found to be in a better working condition in the morning. A new dip will frequently refuse to work, and the addition of a little water will often start it. The water must be used sparingly, however, and only when necessary. Water, as a usual thing, spoils a dead dip, and must be avoided. After a while it may be necessary to add a little more aqua fortis, and this may be introduced as desired. Much care is needed in working the dead dip, and it requires constant watching and experience. The chief difficulty in working the dead dip is to match a given article. The only way that it can be done is to "cut and try," and add aqua fortis or oil of vitriol as the case requires.

The dead or matt dip can be obtained only upon brass or German silver; in other words, only on alloys which contain zinc. The best results are obtained upon yellow brass high in zinc.

To Improve Deadened Brass Parts.— Clock parts matted with oilstone and oil, such as the hour wheels, minute wheels, etc., obtain, by mere grinding, a somewhat dull appearance, with a sensitive surface which readily takes spots. This may be improved by preparing the following powder, rubbing a little of it on a buff stick, and treating the deadened parts, which have been cleansed with benzine, by rubbing with slight pressure on cork. This imparts to the articles a handsome, permanent, metallic matt luster. The smoothing powder consists of 2 parts of jewelers' red and 8 parts of lime carbonate, levigated in water, and well dried. Jewelers' red alone may be employed, but this requires some practice and care, especially in the treatment of wheels, because rays are liable to form from the teeth toward the center.

Pickle for Brass.—Stir 10 parts (by weight) of shining soot or snuff, 10 parts of cooking salt, and 10 parts of red tartar with 250 parts of nitric acid, and afterwards add 250 parts of sulphuric acid; or else mix 7 parts of aqua fortis (nitric acid) with 10 parts of English sulphuric acid. For the mixing ratio of the acid, the kind and alloy of the metal should be the guidance, and it is best found out by practical trials. The better the alloy and the less the percentage of zinc or lead, the handsomer will be the color. Genuine bronze, for instance, acquires a golden shade. In order to give brass the appearance of handsome gilding it is often coated with gold varnish by applying same thinly with a brush or sponge and immediately heating the metal over a coal fire.

Pickling Brass to Look Like Gold.— To pickle brass so as to make it resemble gold allow a mixture of 6 parts of chemically pure nitric acid and 1 part of English sulphuric acid to act for some hours upon the surface of the brass; then wash with a warm solution, 20 parts of tartar in 50 parts of water, and rub off neatly with dry sawdust. Then coat the article with the proper varnish.

Pickle for Dipping Brass.—To improve the appearance of brass, tombac, and copper goods, they are usually dipped. For this purpose they are first immersed in diluted oil of vitriol (brown sulphuric acid), proportion, 1 to 10; next in a mixture of 10 parts of red tartar; 10 parts of cooking salt; 250 parts of English sulphuric acid, as well as 250 parts of aqua fortis (only for a moment), rinsing off well in water and drying in sawdust. For obtaining a handsome matt gold color $\frac{1}{10}$ part of zinc vitriol (zinc sulphate) is still added to the pickle.

Restoration of Brass Articles.—The brass articles are first freed from adhering dirt by the use of hot soda lye; if bronzed they are dipped in a highly dilute solution of sulphuric acid and rinsed in clean water. Next they are yellowed in a mixture of nitric acid, 75 parts; sulphuric acid, 100 parts; shining lampblack, 2 parts; cooking salt, 1 part; then rinsed and polished and, to prevent oxidation, coated with a colorless spirit varnish, a celluloid varnish being best for this purpose.

Tempering Brass.—If hammered too brittle brass can be tempered and made of a more even hardness throughout by warming it, as in tempering steel; but the heat must not be nearly so great. Brass, heated to the blue heat of steel, is almost soft again. To soften brass, heat it nearly to a dull red and allow it to cool, or, if time is an object, it may be cooled by plunging into water.

Drawing Temper from Brass.—Brass is rendered hard by hammering or rolling, therefore when a brass object requires to be tempered the material must be prepared before the article is shaped. Temper may be drawn from brass by heating it to a cherry red and then simply plunging it into water, the same as though steel were to be tempered.

BRASS, FASTENING PORCELAIN TO:
See Adhesives.

BRASS POLISHES:
See Polishes.

BRASS SOLDERS:
See Solders.

BRASS BRONZING:
See Plating.

BRASS CLEANERS:
See Cleaning Preparations and Methods.

BRASS PLATINIZING:
See Plating.

BRASS, SAND HOLES IN:
See Castings.

BRASSING:
See Plating.

BREAD, DOG:
See Dog Biscuit.

BREATH PERFUMES:
See also Dentifrices.

Remedies for Fetid Breath.—Fetid breath may be due to the expelled air (i. e., to disease of the respirational tract), to gases thrown off from the digestive tract, or to a diseased mouth. In the first two cases medication must be directed to the causative diseases, with the last, antisepsis principally and the neutralization of the saliva, also the removal of all residual food of dental caries.

I.—Potassium perman-
ganate.......... 1 part
Distilled water.... 10 parts

Mix and dissolve. Add from 5 to 8 drops of this solution to a glass of water and with it gargle the mouth.

II.—Infusion of salvia 250 parts
Glycerine....... 30 parts
Tincture of myrrh 12 parts
Tincture of laven-
der........... 12 parts
Labarraque's so-
lution........ 30 parts

Mix. Rinse the mouth frequently with this mixture.

III.—Decoction of cham-
omile.......... 30 parts
Glycerine........ 80 parts
Chlorinated water. 15 parts

Mix. Use as a gargle and mouth wash.

IV.—Peppermint water 500 parts
Cherry-laurel wa-
ter........... 60 parts
Borax.......... 25 parts

Mix and dissolve. Use as gargle and mouth wash.

V.—Thymol......... 3 parts
Spirit of cochlea-
ria........... 300 parts
Tincture of rhat-
any.......... 100 parts
Oil of peppermint 15 parts
Oil of cloves..... 10 parts

Mix. Gargle and wash mouth well with 10 drops in a glass of water.

VI.—Salol........... 5 parts
Alcohol........ 1,000 parts
Tincture of white
canella....... 30 parts
Oil of pepper-
mint........ 1 part

Mix. Use as a dentifrice.

VII.—Hydrogen perox-
ide.......... 25 parts
Distilled water... 100 parts

Mix. Gargle the mouth twice daily with 2 tablespoonfuls of the mixture in a glass of water.

VIII.—Sodium bicarbon-
ate........... 2 parts
Distilled water.... 70 parts
Spirit of cochlearia 30 parts

Mix a half-teaspoonful in a wine-glassful of water. Wash mouth two or three times daily.

BRICK STAIN.

To stain brick flat the color of brownstone, add black to Venetian red until the desired shade is obtained. If color ground in oil is used, thin with turpentine, using a little japan as a drier. If necessary to get the desired shade add yellow ocher to the mixture of red and black. If the work is part old and part new, rub the wall down, using a brick for a rubber, until the surface is uniform, and keep it well wet while rubbing with cement water, made by stirring Portland cement into water until the water looks the color of the cement. This operation fills the pores of the brick and makes a smooth, uniform surface to paint on. Tinge the wash with a little dry Venetian red and lampblack. This will help bring the brick to a uniform color, so that an even color can be obtained with one coat of stain.

BRICKS:
See Ceramics.

BRICKS OF SAND-LIME:
See Stone, Artificial.

BRICK POLISHES:
See Polishes.

BRICK WALLS, TO CLEAN:
See Cleaning Preparations and Methods and Household Formulas.

BRICK WATERPROOFING:
See Waterproofing.

BRICKMAKERS' NOTES:
See Ceramics.

BRIDGE PAINT:
See Paint.

BRILLIANTINE:
See Hair Preparations.

BRIMSTONE (BURNING):
See Pyrotechnics.

BRIONY ROOTS: THEIR PRESERVATION:
See Roots.

BRITANNIA METAL:
See Alloys.

BRITANNIA METAL, TO CLEAN:
See Cleaning Preparations and Methods.

BRITANNIA, SILVERPLATING:
See Plating.

BROMINE, ANTISEPTIC:
See Antiseptics.

BROMOFORM.
Bromoform is insoluble in dilute alcohol, but may be dissolved by the aid of glycerine. The following formula has been devised:

Bromoform.........	1	part
Alcohol.............	2	parts
Compound tincture of cardamon.........	2	parts
Glycerine..........	1½	parts

Some other formulas are:

Syrup of Bromoform.—Bromoform, 5 parts; alcohol (95 per cent), 45 parts; glycerine, 150 parts; syrup, 800 parts. Mix in the order given and place the container in warm water until the syrup becomes perfectly clear.

Emulsion of Bromoform.—Add 3 parts of bromoform to 20 parts of expressed oil of almond; emulsify this mixture in the usual manner with 2 parts of powdered tragacanth, 4 parts of powdered acacia, and sufficient water, using for the completed emulsion a total of 120 parts of water, and add, finally, 4 parts of cherry-laurel water.

Bromoform Rum.—Bromoform, 1.2 parts; chloroform, 0.8 parts; rum, sufficient to make 120 parts. Claimed to be an effective remedy in the treatment of whooping cough.

BRONZES:
See Alloys.

BRONZE CASTING:
See Casting.

BRONZE, IMITATION:
See Plaster.

BRONZE POLISHES:
See Polishes.

BRONZE, RENOVATION OF:
See Cleaning Compounds.

Bronze Powders, Liquid Bronzes, Bronze Substitutes, and Bronzing

BRONZE POWDERS.
Gold bronze is a mixture of equal parts of oxide of tin and sulphur, which are heated for some time in an earthen retort. Silver bronze is a mixture of equal parts of bismuth, tin, and mercury, which are fused in a crucible, adding the mercury only when the tin and the bismuth are in fusion. Next reduce to a very fine powder. To apply these bronzes, white of egg, gum arabic, or varnish is used. It is preferable to apply them dry upon one of the above-named mediums serving as size, than to mix them with the liquids themselves, for in the latter case their luster is impaired.

Simple Coloring of Bronze Powder.—In order to impart different colors to bronze powders, such as pale yellow, dark yellow to copper red, the powder is heated with constant stirring in flat iron pans until through the oxidation of the copper—the bronzes consist of the brass powder of an alloy from which the so-called Dutch gold is produced—the desired shade of color is reached. As a rule a very small quantity of fat, wax, or even paraffine is added in this operation. The bronze powders are employed to produce coatings or certain finishes on metals themselves or to give articles of wood, stone, pasteboard, etc., a metallic appearance.

General Directions for Bronzing.—The choice of bronze powders is determined by the degree of brilliancy to be obtained. The powder is mixed with strong gum water or isinglass, and laid on with a brush or pencil, almost but not absolutely dry. A piece of soft leather, wrapped around the finger, is dipped into the powder and rubbed over the work; when all this has been covered with the bronze it must be left to dry, and the loose powder is then cleared away with a hair pencil.

LIQUID BRONZES.

Liquid Bronzes.—I.—For the production of liquid bronze, acid-free varnish should be used, as bronze ground with ordinary varnish will form verdigris. For the deacidification of dammar rosin pour 1,000 parts of petroleum benzine over 350 parts of finely ground dammar rosin, and dissolve by repeated shaking. Next add to the solution 250 parts of a 10-per-cent aqueous solution of caustic soda and shake up well for 10 minutes. After standing for a short time two strata will have formed, the upper one consisting of benzine-rosin solution and the lower, aqueous one containing the resinic acid dissolved as soda salts. Pour off the benzine layers and agitate again assiduously with 250 parts of the 10-per-cent caustic-soda solution. Now set aside for a complete classification and separation of the two liquids. The dammar solution siphoned off will be perfectly free from acid. To obtain gold-bronze varnish add to the deacidified dammar solution about 250 parts of bronze or brocade per liter.

II.—Or else carefully mix 100 parts of finely ground dammar rosin with 50 parts of calcined soda and heat to fusion, in which state it is maintained 2 or 3 hours with frequent stirring. Let cool, grind the turbid mass obtained, and pour a little coal benzine or petroleum benzine over

it in a flask. By repeated shaking of the flask the soluble portion of the molten mass is dissolved; filter after allowing to settle: into the filtrate put 300 to 400 parts of bronze powder of any desired shade, the brocades being especially well adapted for this purpose. If the metallic powder remains distributed over the mass for a long time it is of the right consistency; if it deposits quickly it is too thin and a part of the solvent must be evaporated before stirring in the bronze powder.

III.—A liquid bronze, which, while it contains no metallic constituent, yet possesses a metallic luster and a bronze appearance, and answers excellently for many purposes, is made as follows: Dissolve by the aid of gentle heat 10 parts of aniline red and 5 parts of aniline purple in 100 parts of alcohol. When solution is complete, add 5 parts of benzoic acid, raise the heat, and let boil from 5 to 10 minutes, or until the greenish color of the mixture passes over to a clear bronze brown. For "marbling" or bronzing paper articles, this answers particularly well.

Incombustible Bronze Tincture.—Finely pulverize 5 parts, by weight, of prime Dammar rosin and 1.5 parts of ammonia soda. Heat gently, and stir frequently, until no more carbonic acid bubbles up. Cool and pulverize again. Put the powder into a glass carboy, and pour over it 50 parts of carbon tetrachloride; let this stand for 2 days, stirring frequently. Then filter. Ten parts of the fluid are mixed with 5 parts of metallic bronze of any desired shade, and put into bottles. Shake well before using.

General Formulas for Bronzing Preparations.—I.—Take 240 parts subacetate of copper, 120 parts oxide of zinc in powder form, 60 parts borax, 60 parts saltpeter, and 3.5 parts corrosive sublimate. Prepare a paste from it with oil, stir together, and continue working with boiled linseed oil and turpentine.

II.—Dissolve 120 parts sulphate of copper and add 120 parts chipping of tin; stir well and gather the precipitating copper. After complete drying, grind very finely in boiled linseed oil and turpentine.

III.—Melt in a crucible 60 parts sulphur and 60 parts stannic acid; stir with a clay tube until the mixture takes on the appearance of Dutch gold and pour out. When cold mix the color with boiled linseed oil and turpentine, adding a small quantity of drier. These three bronzes must be covered with a pale, resistant lacquer, otherwise they will soon tarnish in rooms where gas is burned.

Florentine Bronzes.—I.—To produce a Florentine bronzing, apply to the articles, which must have previously been dipped, a varnish composed of cherry gum lac dissolved in alcohol. This varnish is put on with a brush, and after that the bronzed piece is passed through the stove.

II.—If the article is of brass it must be given a coat of copper by means of the battery. Next dip a brush in olive oil and brush the piece uniformly; let dry for 5 or 6 hours and place in sawdust. Then heat the article on a moderate charcoal dust fire.

Preparation of French Bronze.— French bronze may be prepared by reducing to a powder hematite, 5 parts, and plumbago, 8 parts, and mixing into a paste with spirit of wine. Apply the composition with a soft brush to the article to be bronzed and set it aside for some hours. By polishing with a tolerably hard brush the article will assume the beautiful appearance of real bronze. The desired tint may be regulated by the proportions of the ingredients.

How to Bronze Metals.—Prepare a solution of 1½ ounces of sodium hyposulphite in 1 pint of water and add to the same a solution of 1½ ounces of lead acetate dissolved in 1 pint of water.

If, instead of lead acetate, an equal weight of sulphuric acid (1½ ounces) is added to the sodium hyposulphite and the process carried on as before, the brass becomes coated with a very beautiful red, which changes to green, and finally a splendid brown with a green and red iridescence. This last is a very durable coating and may be especially recommended. It is very difficult to obtain exact shades by this process without some experience. The thorough cleansing of all articles from grease by boiling in potash is absolutely necessary to success. By substituting other metal salts for the lead acetate many changes in tints and quality of the coatings can also be effected.

When this mixture is heated to a temperature a little below the boiling point it precipitates sulphide of lead in a state of fine division. If some metal is present some of the lead is precipitated on the surface and, according to the thickness of the layer, different colors are produced. To produce an even color the articles must be evenly heated. By immersion of brass articles for 5 minutes the same may be coated with colors varying from gold to copper red, then to carmine, dark red, and from light blue to blue white, and at last a reddish white, depending on the time the metal remains in the solution and the temperature used. Iron objects treated in this solution take a steel-blue color, zinc a brown color. In the case of copper objects a golden yellow cannot be obtained.

New Bronzing Liquid.—Dissolve 10 parts of fuchsine and 5 parts of aniline purple in 100 parts of alcohol (95 per cent) and add to the solution 5 parts of benzoic acid. Boil the whole for 10 minutes until the color turns bronze brown. This liquid can be applied to all metals and dries quickly.

A Bronze for Brass.—Immerse the articles, freed from dirt and grease, in a cold solution of 10 parts of potassium permanganate, 50 parts of iron sulphate, 5 parts of hydrochloric acid in 1,000 parts of water. Let remain 30 seconds, then withdraw, rinse, and let dry in fine, soft sawdust. If the articles have become too dark, or if a reddish-brown color be desired, immerse for about 1 minute in a warm (140° F.) solution of chromic acid, 10 parts; hydrochloric acid, 10 parts; potassium permanganate, 10 parts; iron sulphate, 50 parts; water, 1,000 parts. Treat as before. If the latter solution alone be used the product will be a brighter dark-yellow or reddish-brown color. By heating in a drying oven the tone of the colors is improved.

To Bronze Copper.—This process is analogous to the one practiced at the Mint of Paris for bronzing medals.

Spread on the copper object a solution composed of:

Acetate or chlorhydrate of ammonia..	30 parts
Sea salt.............	10 parts
Cream of tartar......	10 parts
Acetate of copper....	10 parts
Diluted acetic acid...	100 parts

Let dry for 24 to 48 hours at an ordinary temperature. The surface of the metal will become covered with a series of varying tints. Brush with a waxed brush. The green portions soaked with chlorhydrate of ammonia will assume a blue coloring, and those treated with carbonate will be thick and darkened.

Bronzing and Patinizing of Small Zinc Articles.—Coatings of bronze tones and patina shades may be produced on zinc by means of various liquids, but the articles, before being worked upon, should be rubbed down with very fine glass or emery paper, to make them not only perfectly metallic, but also somewhat rough, as a consequence of which the bronze or patina coatings will adhere much better. The best bronze or patina effects on bronze are obtained by electroplating the article with a fairly thick deposit of brass rich in copper and then treating it like genuine bronze. The solutions used, however, must always be highly diluted, otherwise they may eat entirely through the thin metallic coating.

Bronzing of Zinc.—Mix thoroughly 30 parts of sal ammoniac, 10 parts of oxalate of potash, and 1,000 parts of vinegar. Apply with a brush or a rag several times, until the desired tint is produced.

Bronze Gilding on Smooth Moldings.—A perfect substitute for dead gilding cannot be obtained by bronzing, because of the radically different reflection of the light, for the matt gilding presents to the light a perfectly smooth surface, while in bronzing every little scale of bronze reflects the light in a different direction. In consequence of this diffusion of light, all bronzing, even the best executed, is somewhat darker and dimmer than leaf gilding. This dimness, it is true, extends over the whole surface, and therefore is not perceptible to the layman, and cannot be called an evil, as the genuine leaf gold is so spotted that a bronzed surface is cleaner than a gilt one. The following process is the best known at present: Choose only the best bronze, which is first prepared thick with pure spirit. Next add a quantity of water and stir again. After the precipitation, which occurs promptly, the water is poured off and renewed repeatedly by fresh water. When the spirit has been washed out again in this manner, the remaining deposit, i. e., the bronze, is thinned with clean, good gold size. The bronze must be thin enough just to cover. The moldings are coated twice, the second time commencing at the opposite end. Under no circumstances should the dry, dead gilding give off color when grasping it firmly. If it does that, either the size is inferior or the solution too weak or the mixture too thick.

Incombustible Bronze Tincture.—Five parts of prime dammar rosin and 1.5 parts of ammonia soda, very finely pulverized. Heat gently, with frequent stirring, until the evolution of carbonic acid ceases. Then take from the fire, and when cool pulverize again. Put the powder into a glass carboy, and pour over it 50 parts of carbon tetrachloride; let this stand for 2 days, stirring frequently, then filter. Ten parts of the fluid are to be mixed with each 5 parts of metallic bronze of any desired shade, and put into bottles. Shake the tincture well before using.

Bronzing Engraved Ornaments.—Take bronze and stir with it pale copal varnish diluted one-half with turpentine. With this paint the ornaments neatly. In ½ hour the bronze will have dried. The places from which the bronze is to be removed, i. e., where the bronze has overrun the polished surface, are dabbed with a small rag soaked with kerosene, taking care that it is not too wet, so as to prevent the kerosene from running into the ornament. After a short while the bronze will have dissolved and can be wiped off with a soft rag. If this does not remove it entirely, dab and wipe again. Finally finish wiping with an especially soft, clean rag. Kerosene does not attack polish on wood. The bronze must become dull and yet adhere firmly, under which condition it has a hardened color. If it does not become dull the varnish is too strong and should be diluted with turpentine.

Durable Bronze on Banners.—To render bronzes durable on banners, etc., the ground must be primed with gum arabic and a little glycerine. Then apply the bronze solution, prepared with dammar and one-tenth varnish. Instead of gum arabic with glycerine, gelatine glue may also be employed as an underlay.

BRONZE SUBSTITUTES.

The following recipe is used in making imitation gold bronzes:

Sandarac..........	50 parts
Mastic............	10 parts
Venice turpentine...	5 parts
Alcohol...........	135 parts

In the above dissolve:

Metanil yellow and gold orange......	0.4 parts

and add

Aluminum, finely powdered........	20 parts

and shake.

If a deeper shade is desired it is well to use ethyl orange and gold orange in the same proportion, instead of the dyes.

For the production of imitation copper bronze take the above-mentioned rosin mixture and dissolve therein only **gold** orange 0.8 parts, and add aluminum 20 parts, whereby a handsome copper color is produced. Metanil yellow 0.4 parts without gold orange gives with the same amount of lacquer a greenish tone of bronze. The pigments must not be made use of in larger quantities, because the luster of the bronze is materially affected. Only pigments of certain properties, such as solubility in alcohol, relative constancy. to reductive agents, are suitable; unsuitable are, for instance, naphthol yellow, phenylene-diamin, etc. Likewise only a lacquer of certain composition is fit for use, other lacquers of commerce, such as zapon (celluloid) lacquer being unsuitable. The bronzes prepared in this manner excel in luster and color effect; the cost is very low. They are suitable for bronzing low-priced articles, as tinware, toys, etc. Under the action of sun and moisture the articles lose some of their luster, but objects kept indoors such as figures of plaster of Paris, inkstands, wooden boxes, etc., retain their brilliancy for years.

Some use powdered aluminum and yellow organic dyestuffs, such as gold orange. These are employed together with a varnish of certain composition, which imparts the necessary gloss to the mixture.

BRONZE COLORING:

To Color Bronze.—Bronze articles acquire handsome tempering colors by heating. In order to impart an old appearance to new objects of bronze, they may be heated over a flame and rubbed with a woolen rag dipped in finely powdered graphite, until the desired shade is attained. Or else a paste is applied on the article, consisting of graphite 5 parts and bloodstone 15 parts, with a sufficient quantity of alcohol. After 24 hours brush off the dry powder. A hot solution composed of sal ammoniac 4 parts, sorrel salt 1 part, vinegar 200 parts, may also be brushed on. Another way is to dip the pieces into a boiling solution of cupric acetate 20 parts, and sal ammoniac 10 parts, dissolved in 60 to 100 parts of vinegar.

Patent bronzes (products colored by means of aniline dyes) have hitherto been used in the manufacture of toys and *de luxe* or fancy paper, but makers of wall or stained paper have recently given their attention to these products. Wall —or *moiré*—paper prepared with these dyes furnishes covers or prints of silken gloss with a peculiar double-color effect in which the metallic brilliancy characteristic of bronze combines with the shades of the tar pigments used. Very beautiful reliefs, giving rise to the most charming play of colors in perpendicular or laterally reflected light, are produced by pressing the paper lengths or web painted with aniline-bronze dyes. The brass brocade and tin bronzes serve as bases for the aniline dyes; of the tar pigments only basic aniline dyes soluble in alcohol are used. In coloring the pulverized bronze care must be taken that the latter is as free as possible from organic fats. Tar dyes should be dissolved in as concentrated a form as possible in alcohol and stirred with the bronze, the pigment being then fixed on the vehicle with an alcoholic solution of tannin. The patent bronze is then dried by allowing the alcohol to evaporate. This method of coloring is purely mechanical, as the tar dyes do not combine with the metallic bronze, as is the case with pigments in which hydrate of alumina is used. A coating of aniline bronze of this kind is therefore very sensitive to moisture, unless spread over the paper surface with a suitable protective binding medium, or protected by a transparent coat of varnish, which of course must not interfere with the special color effect.

Pickle for Bronzes.—Sulphuric acid, 1,000 parts; nitric acid, 500 parts; soot, 10 parts; sea salt, 5 parts.

Imitation Japanese Bronze.—When the copper or coppered article is perfectly dry and the copper or copper coating made brilliant, which is produced by rubbing with a soft brush, put graphite over the piece to be bronzed so that the copper is simply dyed. Wipe off the raised portions with a damp cloth, so that the copper makes its appearance. Next put on a thin coat of Japanese varnish; wipe the relief again and let dry. Apply 1 or 2 coats after the first is perfectly dry. Handsome smoked hues may be obtained by holding the bronze either over the dust of lighted peat or powdered rosin thrown on lighted coal, so as to obtain a smoke which will change the color of the varnish employed. The varnish must be liquid enough to be worked easily, for this style of bronzing is only applicable to brass.

Green Bronze on Iron.—Abietate of silver, 1 part; essence of lavender, 19 parts. Dissolve the abietate of silver in the essence of lavender. After the articles have been well pickled apply the abietate-of-silver solution with a brush; next place the objects in a stove and let the temperature attain about 150° C.

Blue Bronze.—Blue bronze is produced by the wet process by coloring white bronze (silver composition) with aniline blue. A blue-bronze color can be produced in the ordinary way from white-bronze color, the product of pure English tin, and with an alum solution consisting of 20 parts of alum in 4,500 parts of water boiled for 5 hours and washed clean and dried. The bronze prepared in this manner is placed in a porcelain dish, mixed with a solution of 15 parts of aniline blue in 1,500 parts of alcohol, stirring the bronze powder and liquid until the alcohol has evaporated entirely and the bronze color becomes dry. This manipulation must be repeated 6 or 8 times, until the desired blue shade is reached. When the bronze is dark enough it is washed out in warm water, and before entirely dry 1 tablespoonful of petroleum is poured on 2 pounds of bronze, which is intimately mixed and spread out into a thin layer, exposed to the air, whereby the smell is caused to disappear in a few days.

Bronzing with Soluble Glass.—To bronze wood, porcelain, glass, and metal by means of a water-glass solution, coat the article with potash water-glass of 30° Bé. and sprinkle on the respective bronze powder.

Brown Oxidation on Bronze.—Genuine bronze can be beautifully oxidized by painting it with a solution of 4 parts of sal ammoniac and 1 part of oxalium (oxalate of potash) in 200 parts of vinegar, allowing it to dry, and repeating the operation several times. These articles, protected against rain, soon lose the unpleasant glaring metallic luster and assume instead a soft brown tint, which bronze articles otherwise acquire only after several years' exposure to the atmosphere. A beautiful bronze color which will remain unaffected by heat can be imparted to bronze articles by the following process: The object is first washed in a solution of 1 part of crystallized verdigris and 2 parts of sal ammoniac in 260 parts of water, and then dried before an open fire till the green color begins to disappear. The operation is repeated 10 to 20 times, but with a solution of 1 part of verdigris crystals and 2 parts of sal ammoniac in 600 parts of water. The color of the article, olive green at first, gradually turns to brown, which will remain unaltered even when exposed to strong heat.

BRONZE POWDERS:
See also Plating for general methods of bronzing, and Varnishes.

Gold and Silver Bronze Powders.—Genuine gold bronze is produced from the waste and parings obtained in gold beating. The parings, etc., are ground with honey or a gum solution, upon a glass plate or under hard granite stones, into a very fine powder, which is repeatedly washed out with water and dried. There are various shades of gold bronze, viz., red, reddish, deep yellow, pale yellow, as well as greenish. These tints are caused by the various percentages of gold or the various mixtures of the gold with silver and copper.

By the use of various salt solutions or acidulated substances other shades can be imparted to bronze. In water containing sulphuric acid, nitric acid, or hydrochloric acid, it turns a bright yellow; by treatment with a solution of crystallized verdigris or blue vitriol in water it assumes more of a reddish hue; other tints are obtained with the aid of cooking salt, tartar, green vitriol, or saltpeter in water.

Gold bronze is also obtained by dissolving gold in aqua regia and mixing with a solution of green vitriol in water, whereupon the gold falls down as a metallic powder which may be treated in different ways. The green vitriol, however, must be dissolved in boiling water and mixed in a glass, drop by drop, with sulphuric acid and stirred until the basic iron sulphate separating in flakes has redissolved. Another way of producing gold bronze is by dissolving gold in aqua regia and evaporating the solution in a porcelain dish. When it is almost dry add a little pure hydrochloric acid and repeat this to drive out all the free chlorine and to produce a pure hydrochlorate of gold. The gold salt is dissolved in distilled water, taking ½ liter per ducat (3½ grams fine gold); into this solution drop, while stirring by means of a glass rod, an 8° solution (by Beaumé) of antimony chloride, as long as a precipitate forms. This deposit is gold bronze, which, dried after removal of all liquids, is chiefly employed in painting, for bronzing, and for china and glass decoration.

Metallic gold powder is, furthermore, obtained by dissolving pure and alloyed gold in aqua regia and precipitating it again by an electro-positive metal, such as iron or zinc, which is placed in the liquid in the form of rods. The gold is completely separated thereby. The rods must be perfectly clean and polished bright. The color of the gold bronze depends upon the proportions of the gold. In order to further increase the brilliancy the dried substance may still be ground.

Mosaic Gold.—Mosaic gold, generally a compound of tin, 64.63 parts, and sulphur, 35.37 parts, is odorless and tasteless, and dissolves only in chlorine solution, aqua regia, and boiling potash lye. It is employed principally for bronzing plaster-of-Paris figures, copper, and brass, by mixing it with 6 parts of bone ashes, rubbing it on wet, or applying it with varnish or white of egg in the preparation of gold paper or for gilding cardboard and wood. Mosaic gold of golden-yellow color is produced by heating 6 parts of sulphur and 16 parts of tin amalgam with equal parts of mercury and 4 parts of sulphur; 8 parts of precipitate from stannic muriate (stannic acid) and 4 parts of sulphur also give a handsome mosaic gold.

The handsomest, purest, and most gold-like mosaic gold is obtained by melting 12 parts of pure tin, free from lead, and mixing with 6 parts of mercury to an amalgam. This is mixed with 7 parts of flowers of sulphur and 6 parts of sal ammoniac, whereupon the mass is subjected for several hours to a heat which at first does not attain redness, but eventually when no more fumes are generated is increased to dark-red heat. This operation is conducted either in a glass retort or in an earthenware crucible. The sal ammoniac escapes first on heating, next vermilion sublimates and some stannic chloride, while the mosaic gold remains on the bottom, the upper layer, consisting of lustrous, golden, delicately translucent leaflets, being the handsomest mosaic gold.

Genuine Silver Bronze.—This is obtained by the finely ground waste from beating leaf silver or by dissolving silver in aqua fortis. This solution is then diluted with water and brightly scoured copper plates are put in, whereby the silver precipitates as a metallic powder.

Imitation Silver Bronze.—This is obtained through the waste in beating imitation leaf silver, which, finely ground, is then washed and dried. In order to increase the luster it is ground again in a dry condition.

Mosaic Silver.—Mosaic silver is an amalgam of equal parts of mercury, bismuth, and tin. One may also melt 50 parts of good tin in a crucible, and as soon as it becomes liquid add 50 parts of bismuth, stirring all with an iron wire until the bismuth is fused as well. As soon as this occurs the crucible must be removed from the fire; then stir in, as long as the contents are still liquid, 25 parts of mercury and mix the whole mass evenly until it can be ground on a stone slab.

BRONZE VARNISHES:
See Varnishes.

BRONZING SOLUTIONS FOR PAINTS:
See Paints.

BRONZING OF WOOD:
See Wood.

BROOCHES, PHOTOGRAPHS ON:
See Photography.

BROWN OINTMENT:
See Ointments.

BROWNING OF STEEL:
See Plating.

BROWNSTONE, IMITATION:
See Brick Stain.

BRUNETTE POWDER:
See Cosmetics.

Brushes

HOW TO TAKE CARE OF PAINT AND VARNISH BRUSHES.

It is a good plan to fill the varnish brush before putting it in the keeper.

Whitewash or kalsomine brushes should not be put into newly slaked lime or hot kalsomine.

Cement-set brushes should never be put in any alcohol mixture, such as shellacs and spirit stains.

Varnish brushes should be selected with a view to their possessing the following qualities: 1st, excellence of material; 2d, excellence of make, which includes fullness of hair or bristles and permanency of binding; 3d, life and spring, or elasticity sufficient to enable the varnisher to spread the varnish without reducing it with turpentine; and 4th, springing, when in use, to a true chisel edge.

Temperature for Brushes.—The bristles of every brush are held in place by the handle. It passes through the shank of the brush and is kiln-dried to fit perfectly. If it shrinks, however, its outward tension is lost and the bristles loosened. For this reason the first principle in brush care is to keep the tool, when it is new or not soaking, in a cool place, out of hot rooms, and any temperature that would tend to shrink the wood of the handle.

Cleaning Paint Brushes.—No new brush should be dipped in the paint and put to work without first being cleaned. By working it with a brisk movement back and forth through the hand most of the dust and loose hairs will be taken out. A paint brush, when thus thoroughly dry cleaned, should be placed in water for a few minutes, not long enough to soak or swell it, but only until wet through, and then swung and shaken dry. It is then ready to dip in the paint, and although some of the hairs may still be loose, most of them will come out in the first few minutes' working and can be easily picked from the surface.

Cleaning Varnish Brushes.—Varnish brushes, and brushes used in varnish stain, buggy paint, and all color in varnish require different handling than paint brushes. They should be more thoroughly dry cleaned, in order that all loose hairs may be worked out. After working them through the hand it is a good thing to pass the brush back and forth over a sheet of sandpaper. This rough surface will pull out the loose bristles and smooth down the rough ends of the chisel point. The brush should then be washed by working it for a few minutes in clean turpentine and swinging it dry. It should never be put in water. For carriage work and fine varnishing the brush should be broken in on the rubbing coat in order to work out all the dust particles before it is used on the finishing coats.

Setting the Paint-Brush Bristles.—For the first 2 or 3 days new brushes require special care while at rest. They should be dipped in raw oil or the paint itself and smoothed out carefully, then laid on their sides over night. The chisel-pointed brushes should be set at an incline, the handle supported just enough to allow the brush to lie along the point. This is done to prevent twisting of the bristles, and to keep the shape of the brush. It is necessary to do this only 2 or 3 times before the shape becomes set.

Paint Brushes at Rest.—An important principle in brush care is never to leave the brush on end while at rest. Even for temporary rest during a job the brush should never stand on end. At night it should always be placed in a "brush-keeper"—a water-tight box, or a paint keg, with nails driven through the sides on which the brushes can be suspended in water. Holes are bored in the handles so the brush will hang free of the bottom, but with the bristles entirely under water. Before placing them in water the brushes should be wiped so as not to be too full of paint, but not cleaned.

Varnish Brushes at Rest.—Varnish brushes should be kept at rest in turpentine and varnish, or better, in some of the varnish that the brush is used for. They should preferably not be kept in turpentine, as that makes the brush "lousy"—roughening the bristles.

Washing Brushes.—All brushes should be washed in benzine or turpentine and shaken dry—not whipped—when it is desired to change from one color to another, or from one varnish to another.

To Restore Brushes.—A good remedy to restore lettering brushes which have lost their elasticity and do not keep a point, is as follows:

Put the pencil in oil and brush it several times over a hot iron in such a manner that the hairs touch the iron from each side; then dip the pencil quickly in cold water.

A Removable Binding.—The bristle bunch of brushes is bound with rope so as to keep them together for use. Instead of the twine, a covering of rubber may be employed, which is easily slipped over the bristles and can be conveniently removed again. The cleaning of the brush is much facilitated thereby, and the breadth of the stripe to be drawn with the brush can be accurately regulated, according to how far the covering is slipped over the brush.

See also Cleaning Preparations and Methods.

BUBBLES IN GELATIN:
See Gelatin.

BUBBLE (SOAP) LIQUID:
See Soap Bubble Liquid.

BUBBLES.

Bubbles of air often adhere to molds immersed in depositing solutions. They may be prevented by previously dipping the object into spirits of wine, or be removed by the aid of a soft brush, or by directing a powerful current of the liquid against them by means of a vulcanized india-rubber bladder, with a long and curved glass tube attached to it; but the liquid should be free from sediment.

BUG KILLERS:
See Insecticides.

BUNIONS:
See Corn Cures.

BURNS:

See also Ointments and Turpentine.

Mixture for Burns.—I.—A mixture of castor oil with the white of egg is recommended for burns. The eggs are broken into a bowl and the castor oil slowly poured in while the eggs are beaten. Enough oil is added to make a thick, creamy paste, which is applied to the burn. The applications are repeated often enough to prevent their becoming dry or sticky. Leave the surface uncovered.

II.—Put 27 parts, by measure, of menthol into 44 parts, by measure, of witch hazel (distillate) and apply freely. A good plan is to bandage the parts and wet the wrappings with this mixture.

III.—A very efficacious remedy for burns is a solution of cooking salt in water. It is best to immerse fingers, hands, and arms in the solution, which must be tolerably strong. For burns in the face and other parts of the body, salt water poultices are applied.

Butter

(See also Foods.)

Butter Color.—Orlean, 80 parts, by weight; curcuma root (turmeric), 80 parts, by weight; olive oil, 240 parts, by weight; saffron, 1 part, by weight; alcohol, 5 parts, by weight. The orlean and turmeric are macerated with olive oil and expressed. The weight of the filtered liquid is made up again to 240 parts, by weight, with olive oil, next the filtered saffron-alcohol extract is added, and the alcohol is expelled again by heating the mixture.

Artificial Butter.—I.—Carefully washed beef suet furnishes a basis for the manufactures of an edible substitute for natural butter. The thoroughly washed and finely chopped suet is rendered in a steam-heated tank; 1,000 parts of fat, 300 parts of water, 1 part of potassium carbonate, and 2 stomachs of pigs or sheep, are taken. The temperature of the mixture is raised to 113° F. After 2 hours, under the influence of the pepsin in the stomachs, the membranes are dissolved and the fat is melted and rises to the top of the mixture. After the addition of a little salt the melted fat is drawn off, stood to cool so as to allow the stearine and palmitin to separate, and then pressed in bags in a hydraulic press. Forty to 50 per cent of solid stearine remains, while 50 to 60 per cent

of fluid oleopalmitin (so-called "oleomargarine") is pressed out. The "oleo oil" is then mixed with 10 per cent of its weight of milk and a little butter color and churned. The product is then worked, salted, and constituted the "oleomargarine," or butter substitute. Leaf lard can be worked in the same way as beef suet, and will yield an oleopalmitin suitable for churning up into a butter substitute.

II.—Fat from freshly slaughtered cattle after thorough washing is placed in clean water and surrounded with ice, where it is allowed to remain until all animal heat has been removed. It is then cut into small pieces by machinery and cooked at a temperature of about 150° F. (65.6° C.) until the fat in liquid form has separated from the tissue, then settled until it is perfectly clear. Then it is drawn into the graining vats and allowed to stand for a day, when it is ready for the presses. The pressing extracts the stearine, leaving a product commercially known as oleo oil which, when churned with cream or milk, or both, and with usually a proportion of creamery butter, the whole being properly salted, gives the new food product, oleomargarine.

III.—In making butterine use neutral lard, which is made from selected leaf lard in a very similar manner to oleo oil, excepting that no stearine is extracted. This neutral lard is cured in salt brine for from 48 to 70 hours at an ice-water temperature. It is then taken and, with the desired proportion of oleo oil and fine butter, is churned with cream and milk, producing an article which when properly salted and packed is ready for the market. In both cases coloring matter is used, which is the same as that used by dairymen to color their butter. At certain seasons of the year—viz., in cold weather, a small quantity of sesame oil or salad oil made from cottonseed oil is used to soften the texture of the product.

IV.—"Ankara" is a substance which in general appearance resembles a good article of butter, being rather firmer at ordinary temperatures than that substance, approaching the consistency of cocoa butter. It is quite odorless, but in taste it resembles that of a fair article of butter and, what is more, its behavior under heat is very similar to that of butter—it browns and forms a sort of spume like that of fat. Ankara consists of a base of cocoa butter, carrying about 10 per cent of milk, colored with yolk of egg. While not derived from milk, on the one hand, nor does it come from a single vegetable or animal fat on the other, ankara may be considered as belonging to the category of the margarines. Ankara is obtained in the market in the form of cakes or tablets of 2 pounds in weight.

V.—Fresh butter, 150 parts, by weight; animal fat, 80 parts, by weight; sunflower oil, 40 parts, by weight; cocoanut oil, 30 parts, by weight.

VI.—Fresh butter, 100 parts, by weight; animal fat, 100 parts, by weight; sunflower oil, 80 parts, by weight; cocoanut oil, 20 parts, by weight.

VII.—Fresh butter, 50 parts, by weight; animal fat, 150 parts, by weight; sunflower oil, 80 parts, by weight; cocoanut oil, 20 parts, by weight.

It is seen that these three varieties contain respectively 50, 33, and about 16 per cent of cow's butter. The appearance of the mixture is nearly perfect.

Formulas V to VII are for a Russian artificial butter called " Perepusk."

To Impart the Aroma and Taste of Natural Butter to Margarine.—In order to give margarine the aroma and flavor of cow butter, add to it a fatty acid product, which is obtained by saponification of butter, decomposition of the soap, and distillation in the vacuum at about 140° F. The addition of the product is made upon emulsification of the fats with milk. The margarine will keep for months.

Harmless Butter Color.—Alum, pulverized finely, 30 parts; extract of turmeric, 1 part. With the extract dampen the powder as evenly as possible, then spread out and dry over some hot surface. When dry, again pulverize thoroughly. Protect the product from the light. As much of the powder as will lie on the point of a penknife is added to a churnful of milk, or cream, before churning, and it gives a beautiful golden color, entirely harmless. To make the extract of turmeric add 1 part of powdered turmeric to 5 parts of alcohol, and let macerate together for fully a week.

To Sweeten Rancid Butter.—I.—Wash the butter first with fresh milk and afterwards with spring water, carefully working out the residual water.

II.—Add 25 to 30 drops of lime chloride to every 2 pounds of butter, work the mass up thoroughly, then wash in plenty of fresh, cold water, and work out the residual water.

III.—Melt the butter in a water bath, along with some freshly burned animal charcoal, coarsely powdered and carefully sifted to free it from dust. After this has remained in contact for a few minutes, the butter is strained through a clean flannel. If the rancid odor is not completely removed, complete the process.

An English Margarine.—A mixture of edible fats of suitable consistency, e. g., oleo oil, 5 parts; neutral lard, 7 parts; and butter, 1 part; is mixed with albuminous "batter," 4 parts, with the addition of 1 part of salt as a preservative. If the albuminous constituent be composed of the whites and yolks of eggs beaten to a foam the product will have the consistency and color of butter. The molten fats are added to the egg batter and the whole is stirred at a temperature sufficient to produce coagulation of the albumen (150–200° F.). The mass is then cooled gradually with continuous stirring, and the salt is worked in.

Olive-Oil Paste.—If an ounce of peeled garlic be rubbed up into a pulp, in a clean Wedgwood mortar, and to this be added from 3 to 4 ounces of good olive oil, with constant rubbing up with the pestle, the oil becomes converted into a pasty mass, like butter. It is possible that the mucilage obtainable from other bulbs of the *Lilium* tribe would prove equally efficient in conferring semi-solidity on the oil, without imparting any strong smell. The above composition is largely used by the Spanish peasantry, instead of butter, which runs liquid in the Spanish summer. It is known as "aleoli." The more easily solidified portion of olive oil is stearine, and this may be cheaply prepared from mutton fat. If added, in certain proportions, to olive oil, it would certainly raise its melting point.

BUTTERMILK, ARTIFICIAL.

Buttermilk powder, 10 parts; vinegar, 1 part; syrup of buckthorn, 1 part. Dissolve the powder in the water and add the vinegar and syrup. The powder is prepared as follows: Sodium chloride, 50 parts; milk sugar, 100 parts; potassium nitrate, 5 parts; alum, 5 parts. Mix.

BUTTER, ARTIFICIAL: TESTS FOR:
See Foods.

BUTTER COLORANT:
See Foods.

BUTTONS OF ARTIFICIAL AGATE:
See Agate.

CADMIUM ALLOYS:
See Alloys.

CALCIUM CARBIDE:

Preservation and Use of Calcium Carbide.—Calcium carbide is readily attacked by the air and the moisture contained in the generators and consequently decomposes during the storing, with formation of acetylene gas. Aside from the loss, this decomposition is also attended with dangers. One of the oldest methods of preservation is the saturation of the carbide with petroleum. In using such carbide a layer of petroleum forms on the surface of the water in the generator, which prevents the water from evaporating, thus limiting the subsequent generation of acetylene from the remaining carbide. Instead of petroleum many other substances have been proposed which answer the purpose equally well, e. g., toluol, oils, solid bodies, which previously have to be liquefied, such as stearine, paraffine, rosin, etc.

Of a different nature is a medium offered by Létang of Paris. He employs sugar or saccharine bodies to which he adds, if necessary, a little petroleum, turpentine, vaseline, or varnish of any kind, as well as chalk, limestone, talc, sulphur, or sand. The carbide is coated with this mixture. The saccharine substances dissolve in the generating water, and also have a dissolving action on the slaked lime, which is formed by the decomposition of the carbide which admits of its easy removal.

According to another process carbide is put on the market in such a shape that, without weighing, merely by counting or measuring one is in a position to use equivalent quantities for every charge. Gearing casts molten carbide in the shape of bars, and pours a layer of gelatin, glue, and water soluble varnish over the carbide bars. Others make shells containing a certain quantity of reduced carbide. For this ordinary and varnished pasteboard, wax paper, tinfoil, thin sheet zinc, and similar substances may be used which ward off atmospheric moisture, thus protecting the carbide from premature decomposition. Before use, the cartridge-like shell is pierced or cut open, so that the water can get at the contents. The more or less reduced carbide is filled in the shell, either without any admixture or united into a compact mass by a binding agent, such as colophony, pitch, tar, sand, etc.

Deodorization of Calcium Carbide.—Calcium carbide is known to possess a very unpleasant odor because it constantly develops small quantities of impure acetylene in contact with the moisture of the air. Le Roy, of Rouen, proposes for portable—especially bicycle—lamps, in which the evil is more noticeable than in large plants, simply to pour some petroleum over the carbide and to pour off the remainder not absorbed. The petroleum, to which it is well to add some nitro-benzol (mirbane essence), prevents the access of air to the carbide, but permits a very satisfactory generation of gas on admission of water.

CALCIUM SULPHIDE (LUMINOUS):
See Paints.

CALFSKIN:
See Leather.

CAMERA RENOVATION:
See Photography.

CAMPHOR PREPARATIONS:

Fragrant Naphthalene Camphor.—

Naphthalene white, in scales	3,000 parts
Camphor	1,000 parts

Melt on the steam bath and add to the hot mass:

Coumarin	2 parts
Mirbane oil	10 parts

Cast in plates or compressed tablets. The preparation is employed as a moth preventive.

Powdered Camphor in Permanent Form.—I.—Powder the camphor in the usual manner, with the addition of a little alcohol. When it is nearly reduced to the proper degree of fineness add a few drops of fluid petrolatum and immediately triturate again. In this manner a powder as fine as flour is obtained, which does not cake together. This powdered camphor may be used for all purposes except for solution in alcohol, as it will impart to the latter a faint opalescence, owing to the insolubility of the petrolatum.

II.—Take equal parts of strong ether and alcohol to reduce the camphor to powder. It is claimed for this method that it only takes one-half of the time required when alcohol alone is used, and that the camphor dries more quickly. Before sifting add 1 per cent of white vaseline and 5 per cent of sugar of milk. Triturate fairly dry, spread out in the air, say 15 minutes, then pass through a moderately fine wire sieve, using a stubby shaving brush to assist in working it through.

Camphor Pomade —

Oil of bitter almonds.	1	drachm
Oil of cloves........	20	drops
Camphor..........	1½	ounces
White wax.........	4	ounces
Lard, prepared.....	1	pound

Melt the wax and lard together, then add the camphor in saturated solution in spirit; put in the oils when nearly cold.

Camphor Ice.—

I.—

White wax.........	16 parts
Benzoated suet.....	48 parts
Camphor, powdered.	8 parts
Essential oil, to perfume.	

Melt the wax and suet together. When nearly cold, add the camphor and perfume, mix well, and pour into molds.

II.—

Oil of almond.......	16 parts
White wax.........	4 parts
Spermaceti........	4 parts
Paraffine..........	8 parts
Camphor, powdered.	1 part
Perfume, quantity sufficient.	

Dissolve the camphor in the oil by the aid of a gentle heat. Melt the solids together, remove, and let cool, but before the mixture begins to set add the camphorated oil and the perfume, mix, and pour into molds.

III.—

Stearine (stearic acid)	8 pounds
Lard...............	10 pounds
White wax.........	5 pounds
Spermaceti.........	5 pounds

Melt on a water bath in an earthen or porcelain dish; strain into a similar vessel; add a solution of 2 ounces powdered borax in 1 pound of glycerine, previously warmed, to the melted substance when at the point of cooling; stir well; add camphor, 2 pounds, powdered by means of alcohol, 3 ounces; stir well and pour into molds.

CAMPHOR SUBSTITUTES IN THE PREPARATION OF CELLULOID:
See Celluloid.

CAMPHOR AND RHUBARB AS A REMEDY FOR CHOLERA:
See Cholera Remedies.

CAN VARNISH:
See Varnishes.

CANARY-BIRD PASTE.

The following is a formula much used by German canary-bird raisers:

Sweet almonds, blanched........	16 parts
Pea meal..........	32 parts
Butter, fresh (unsalted).........	3 parts
Honey, quantity sufficient to make a stiff paste.	

The ingredients are worked into a stiff paste, which is pressed through a colander or large sieve to granulate the mass. Some add to every 5 pounds, 10 or 15 grains of saffron and the yolks of 2 eggs.

CANARY BIRDS AND THEIR DISEASES:
See Veterinary Formulas.

CANDLES:

Coloring Ceresine Candles for the Christmas Tree.—For coloring these candles only dye stuffs soluble in oil can be employed. Blue: 23–24 lavender blue, pale or dark, 100–120 parts per 5,000 parts of ceresine. Violet: 26 fast violet R, 150 parts per 5,000 parts of ceresine. Silver gray: 29 silver gray, 150 parts per 5,000 parts of ceresine. Yellow and orange: 30 wax yellow, medium, 200 parts per 5,000 parts of ceresine; 61 old gold, 200 parts per 5,000 parts of ceresine. Pink and red: 27 peach-pink, or 29 chamois, about 100 parts per 5,000 parts of ceresine. Green: 16–17 brilliant green, 33 May green, 41 May green, 200–250 parts per 5,000 parts of ceresine. The above-named colors should be ground in oil and the ceresine tinted with them afterwards.

Manufacture of Composite Paraffine Candles.—Three parts of hydroxystearic acid are dissolved in 1 part of a suitable solvent (e. g., stearic acid), and the solution is mixed with paraffine wax to form a stock for the manufacture of composite candles.

Transparent Candles.—The following are two recipes given in a German patent specification. The figures denote parts by weight:

I.—Paraffine wax, 70; stearine, 15; petroleum, 15.

II.—Paraffine wax, 90; stearine, 5; petroleum, 5. Recipe I of course gives candles more transparent than does recipe II. The 15 per cent may be regarded as the extreme limit consistent with proper solidity of the candles.

To Prevent the Trickling of Burning Candles.—Dip the candles in the following mixture:

Magnesium sulphate	15 parts
Dextrin............	15 parts
Water.............	100 parts

The solution dries quickly and does not affect the burning of the candle,

Candle Coloring.—Candles are colored either throughout or they sometimes consist of a white body that is covered with a colored layer of paraffine wax. According to the material from which candles are made (stearine, paraffine, or ozokerite), the process of coloring varies.

Stearine, owing to its acid character, dissolves the coal-tar colors much more readily than do the perfectly neutral paraffine and ozokerite waxes. For coloring stearine the necessary quantity of the color is added to the melted mass and well stirred in; if the solution effected happens to be incomplete, a small addition of alcohol will prove an effective remedy. It is also an advantage to dissolve the colors previously in alcohol and add the concentrated solution to the melted stearine. The alcohol soon evaporates, and has no injurious effect on the quality of the stearine.

For a number of years there have been on the market so-called "fat colors," formed by making concentrated solutions of the color, and also special preparations of the colors in stearine. They are more easily applied, and are, therefore, preferred to the powdered aniline colors, which are apt to cause trouble by being accidentally distributed in soluble particles, where they are not wanted. Since paraffine and ozokerite dissolve comparatively little, they will not become colored, and so must be colored indirectly. One way is to dissolve the color in oleic acid or in stearine acid and add the solution to the wax to be colored. Turpentine may be employed for the same purpose. Concerning the colors suitable for candles, there are the eosine colors previously mentioned, and also chroline yellow, auramine, taniline blue, tartrazine, brilliant green, etc. The latter, however, bleaches so rapidly that it can hardly be recommended. An interesting phenomenon is the change some colors undergo in a warm temperature; for instance, some blues turn red at a moderate degree of heat (120° F.) and return to blue only when completely cooled off; this will be noticed while the candle mixture is being melted previous to molding into candles.

CANDLES (FUMIGATING):
See Fumigants.

CANDY COLORS AND FLAVORS:
See Confectionery.

CANDY:
See Confectionery.

CANVAS WATERPROOFING:
See Waterproofing.

CAOUTCHOUC:
See Rubber.

CAOUTCHOUC SOLUTION FOR PAINTS:
See Paint.

CAPPING MIXTURES FOR BOTTLES:
See Bottle-Capping Mixtures.

CAPSULE VARNISH:
See Varnishes.

CARAMEL:

Cloudless Caramel Coloring.—I.—When it is perfectly understood that in the manufacture of caramel, sugar is to be deprived of the one molecule of its water of constitution, it will be apparent that heat must not be carried on to the point of carbonization. Cloudy caramel is due to the fact that part of the sugar has been dissociated and reduced to carbon, which is insoluble in water. Hence the cloudiness. Caramel may be made on a small scale in the following manner: Place 4 or 5 ounces of granulated sugar in a shallow porcelain-lined evaporating dish and apply either a direct heat or that of an oil bath, continuing the heat until caramelization takes place or until tumescence ceases and the mass has assumed a dark-brown color. Then carefully add sufficient water to bring the viscid mass to the consistence of a heavy syrup. Extreme *care* must be taken and the face and hands protected during the addition of the water, owing to the intensity of the heat of the mass, and consequent sputtering.

II.—The ordinary sugar coloring material is made from sugar or glucose by heating it, while being constantly stirred, up to a temperature of about 405° F. A metal pan capable of holding nearly ten times as much as the sugar used, is necessary so as to retain the mass in its swollen condition. As soon as it froths up so as nearly to fill the pan, an action which occurs suddenly, the fire must instantly be extinguished or removed. The finished product will be insoluble if more than about 15 per cent of its weight is driven off by the heat.

CARAMEL IN FOOD:
See Food.

CARAMELS:
See Confectionery.

CARBOLIC ACID.

Perfumed Carbolic Acid.—

I.—Carbolic acid (cryst.). 1 ounce
Alcohol............ 1 ounce
Oil bergamot......... 10 minims
Oil eucalyptus...... 10 minims
Oil citronella........ 3 minims
Tincture cudbear.... 10 minims
Water, to make...... 10 ounces

Set aside for several days, and then filter through fuller's earth.

II.—Carbolic acid (cryst.) 4 drachms
Cologne water....... 4 drachms
Dilute acetic acid.... 9 ounces

Keep in a cool place for a few days, and filter.

Treatment of Carbolic-Acid Burns.—
Thoroughly wash the hands with alcohol, and the burning and tingling will almost immediately cease. Unless employed immediately, however, the alcohol has no effect. When the time elapsed since the burning is too great for alcohol to be of value, brush the burns with a saturated solution of picric acid in water.

Decolorization of Carbolic Acid.—
To decolorize the acid the following simple method is recommended. For purifying carbolic acid which has already become quite brown-red on account of having been kept in a tin vessel, the receptacle is exposed for a short time to a temperature of 25° C. (77° F.), thus causing only a part of the contents to melt. In this state the acid is put into glass funnels and left to stand for 10 to 12 days in a room which is likewise kept at the above temperature. Clear white crystals form from the drippings, which remained unchanged, protected from air and light, while by repeating the same process more clear crystals are obtained from the solidified dark colored mother lye. In this manner 75 to 80 per cent of clear product is obtained altogether.

Disguising Odor of Carbolic Acid.—
Any stronger smelling substance will disguise the odor of carbolic acid, to an extent at least, but it is a difficult odor to disguise on account of its persistence. Camphor and some of the volatile oils, such as peppermint, cajeput, caraway, clove, and wintergreen may be used.

To Restore Reddened Carbolic Acid.
—Demont's method consists in melting the acid on the water bath, adding 12 per cent of alcohol of 95 per cent, letting cool down and, after the greater part of the substance has crystallized out, decanting the liquid residue. The crystals obtained in this manner are snowy white, and on being melted yield a nearly colorless liquid. The alcohol may be recovered by redistillation at a low temperature. This is a rather costly procedure.

CARBOLIC SOAP:
See Soap.

CARBOLINEUM:
See also Paints and Wood.

Preparation of Carbolineum.—I.—
Melt together 50 parts of American rosin (F) and 150 parts of pale paraffine oil (yellow oil), and add, with stirring, 20 parts of rosin oil (rectified).

II.—Sixty parts, by weight, of black coal tar oil of a specific gravity higher than 1.10; 25 parts, by weight, of creosote oil; 25 parts, by weight, of beechwood tar oil of a higher specific weight than 0.9. Mix together and heat to about 347° F., or until the fumes given off begin to deposit soot. The resulting carbolineum is brown, and of somewhat thick consistency; when cool it is ready for use and is packed in casks. This improved carbolineum is applied to wood or masonry with a brush; the surfaces treated dry quickly, very soon loose the odor of the carbolineum, and are effectively protected from dampness and formation of fungi.

CARBON PRINTING:
See Photography.

CARBON PROCESS IN PHOTOGRAPHY:
See Photography.

CARBONYLE:
See Wood.

CARBUNCLE REMEDIES:
See Boil Remedy.

CARDS (PLAYING), TO CLEAN:
See Cleaning Preparations and Methods.

CARDBOARD, WATERPROOF GLUE FOR:
See Adhesives under Cements and Waterproof Glues.

CARDBOARD, WATERPROOFING:
See Waterproofing.

CARMINATIVES:
See Pain Killers.

CARPET PRESERVATION:
See Household Formulas.

CARPET SOAP:
See Soap.

CARRIAGE-TOP DRESSING:
See Leather.

CARRON OIL:
See Cosmetics.

CASE HARDENING:
See Steel.

Casein

Dried Casein, its Manufacture and Uses.—For the production of casein, skimmed milk or buttermilk is used, articles of slight value, as they cannot be employed for feeding hogs or for making cheese, except of a very inferior sort, of little or no alimentive qualities. This milk is heated to from 70° to 90° C. (175°–195° F.), and sulphuric or hydrochloric acid is added until it no longer causes precipitation. The precipitate is washed to free it from residual lactose, redissolved in a sodium carbonate solution, and again precipitated, this time by lactic acid. It is again washed, dried, and pulverized. It takes 8 gallons of skimmed milk to make 1 pound of dry casein.

In the manufacture of fancy papers, or papers that are made to imitate the appearance of various cloths, laces, and silks, casein is very widely used. It is also largely used in waterproofing tissues, for preparation of waterproof products, and various articles prepared from agglomeration of cork (packing boards, etc.). With lime water casein makes a glue that resists heat, steam, etc. It also enters into the manufacture of the various articles made from artificial ivory (billiard balls, combs, toilet boxes, etc.), imitation of celluloid, meerschaum, etc., and is finding new uses every day.

Casein, as known, may act the part of an acid and combine with bases to form caseinates or caseates; among these compounds, caseinates of potash, of soda, and of ammonia are the only ones soluble in water; all the others are insoluble and may be readily prepared by double decomposition. Thus, for example, to obtain caseinate of alumina it is sufficient to add to a solution of casein in caustic soda, a solution of sulphate of alumina; an insoluble precipitate of casein, or caseinate of alumina, is instantly formed.

This precipitate ought to be freed from the sulphate of soda (formed by double decomposition), by means of prolonged washing. Pure, ordinary cellulose may be incorporated with it by this process, producing a new compound, cheaper than pure cellulose, although possessing the same properties, and capable of replacing it in all its applications.

According to the results desired, in transparency, color, hardness, etc., the most suitable caseinate should be selected. Thus, if a translucent compound is to be obtained, the caseinate of alumina yields the best. If a white compound is desired, the caseinate of zinc, or of magnesia, should be chosen; and for colored products the caseinates of iron, copper, and nickel will give varied tints.

The process employed for the new products, with a base of celluloid and caseinate, is as follows: On one hand casein is dissolved in a solution of caustic soda (100 parts of water for 10 to 25 parts of soda), and this liquid is filtered to separate the matters not dissolved and the impurities. On the other hand, a salt of the base of which the caseinate is desired is dissolved, and the solution filtered. It is well not to operate on too concentrated a solution. The two solutions are mixed in a receptacle provided with a mechanical stirrer, in order to obtain the insoluble caseinate precipitate in as finely divided a state as possible. This precipitate should be washed thoroughly, so as to free it from the soda salt formed by double decomposition, but on account of its gummy or pasty state, this washing presents certain difficulties, and should be done carefully. After the washing the mass is freed from the greater part of water contained, by draining, followed by drying, or energetic pressing; then it is washed in alcohol, dried or pressed again, and is ready to be incorporated in the plastic mass of the celluloid.

For the latter immersion and washing it has been found that an addition of 1 to 5 per cent of borax is advantageous, for it renders the mass more plastic, and facilitates the operation of mixing. This may be conducted in a mixing apparatus; but, in practice, it is found preferable to effect it with a rolling mill, operating as follows:

The nitro-cellulose is introduced in the plastic state, and moistened with a solution of camphor in alcohol (40 to 50 parts of camphor in 50 to 70 of alcohol for 100 of nitro-cellulose) as it is practiced in celluloid factories.

This plastic mass of nitro-cellulose is placed in a rolling mill, the cylinders of which are slightly heated at the same time as the caseinate, prepared as above; then the whole mass is worked by the cylinders until the mixture of the two is perfectly homogeneous, and the final mass is sufficiently hard to be drawn out in leaves in the same way as practiced for pure celluloid.

These leaves are placed in hydraulic presses, where they are compressed, first hot, then cold, and the block thus formed is afterwards cut into leaves of the thickness desired. These leaves are dried in an apparatus in the same way as ordinary celluloid. The product resembles celluloid, and has all its properties. At 90° to 100° C. (194° to 212° F.), it becomes quite plastic, and is easily molded. It may be sawed, filed, turned, and carved without difficulty, and takes on a superb polish. It burns less readily than celluloid, and its combustibility diminishes in proportion as the percentage of caseinate increases; finally, the cost price is less than that of celluloid, and by using a large proportion of caseinate, products may be manufactured at an extremely low cost.

Phosphate of Casein and its Production.—The process is designed to produce a strongly acid compound of phosphoric acid and casein, practically stable and not hydroscopic, which may be employed as an acid ingredient in bakers' yeast and for other purposes.

The phosphoric acid may be obtained by any convenient method; for example, by decomposing dicalcic or monocalcic phosphate with sulphuric acid. The commercial phosphoric acid may also be employed.

The casein may be precipitated from the skimmed milk by means of a suitable acid, and should be washed with cold water to remove impurities. A caseinate may also be employed, such as a compound of casein and an alkali or an alkaline earth.

The new compound is produced in the following way: A sufficient quantity of phosphoric acid is incorporated with the casein or a caseinate in such a way as to insure sufficient acidity in the resulting compound. The employment of 23 to 25 parts by weight of phosphoric acid with 75 to 77 parts of casein constitutes a good proportion.

An aqueous solution of phosphoric acid is made, and the casein introduced in the proportion of 25 to 50 per cent of the weight of the phosphoric acid present. The mixture is then heated till the curdled form of the casein disappears, and it assumes a uniform fluid form. Then the mixture is concentrated to a syrupy consistency. The remainder of the casein or of the caseinate is added and mixed with the solution until it is intimately incorporated and the mass becomes uniform. The compound is dried in a current of hot air, or in any other way that will not discolor it, and it is ground to a fine powder. The intimate union of the phosphoric acid and casein during the gradual concentration of the mixture and during the grinding and drying, removes the hydroscopic property of the phosphoric acid, and produces a dry and stable product, which may be regarded as a hyperphosphate of casein. When it is mixed with water, it swells and dissolves slowly. When this compound is mingled with its equivalent of sodium bicarbonate it yields about 17 per cent of gas.

CASEIN CEMENTS:
See Adhesives.

CASEIN VARNISH:
See Varnishes.

CASKS:
To Render Shrunken Wooden Casks Watertight.—When a wooden receptacle has dried up it naturally cannot hold the water poured into it for the purpose of swelling it, and the pouring has to be repeated many times before the desired end is reached. A much quicker way is to stuff the receptacle full of straw or bad hay, laying a stone on top and then filling the vessel with water. Although the water runs off again, the moistened straw remains behind and greatly assists the swelling up of the wood.

CASSIUS, PURPLE OF:
See Gold.

CASKET TRIMMINGS:
See Castings.

CASTS (PLASTER), PRESERVATION OF:
See Plaster.

CASTS, REPAIRING OF BROKEN:
See Adhesives and Lutes.

CASTS FROM WAX MODELS:
See Modeling.

Casting

Castings Out of Various Metals.—Until recent years metal castings were all made in sand molds; that is, the patterns were used for the impressions in the sand, the same as iron castings are produced to-day. Nearly all of the softer metals are now cast in brass, copper, zinc, or iron molds, and only the silver and German silver articles, like wire real bronze, are cast the old way, in sand. Aluminum can be readily cast in iron molds, especially if the molds have been previously heated to nearly the same temperature as the molten aluminum, and after the molds are full the metal is cooled gradually and the casting taken out as soon as cooled enough to prevent breaking from the shrinkage. Large bicycle frames have been successfully cast in this manner.

The French bronzes, which are imitations, are cast in copper or brass molds. The material used is principally zinc and tin, and an unlimited number of castings can be made in the mold, but if a real bronze piece is to be produced it must be out of copper and the mold made in sand. To make the castings hollow, with sand, a core is required. This fills the inside of the figure so that the molten copper runs around it, and as the core is made out of sand, the same can be afterwards washed out. If the casting is to be hollow and is to be cast in a metal mold, then the process is very simple. The mold is filled with molten metal, and when the operator thinks the desired thickness has cooled next to the walls, he pours out the balance. An experienced man can make hollow castings in this way, and make the walls of any thickness.

Casket hardware trimmings, which are so extensively used on coffins, especially the handles, are nearly all cast out of tin and antimony, and in brass molds. The metal used is brittle, and requires strengthening at the weak portions, and this is mostly done with wood filling or with iron rods, which are secured in the molds before the metal is poured in.

Aluminum castings, which one has procured at the foundries, are usually alloyed with zinc. This has a close affinity with aluminum, and alloys readily; but this mixture is a detriment and causes much trouble afterwards. While this alloy assists the molder to produce his castings easily, on the other hand it will not polish well and will corrode in a short time. Those difficulties may be avoided if pure aluminum is used.

Plaster of Paris molds are the easiest made for pieces where only a few castings are wanted. The only difficulty is that it requires a few days to dry the plaster thoroughly, and that is absolutely necessary to use them successfully. Not only can the softer metals be run into plaster molds, but gold and silver can be run into them. A plaster mold should be well smoked over a gaslight, or until well covered with a layer of soot, and the metal should be poured in as cool a state as it will run.

To Prevent the Adhesion of Modeling Sand to Castings.—Use a mixture of finely ground coke and graphite. Although the former material is highly porous, possessing this quality even as a fine powder, and the fine pulverization is a difficult operation, still the invention attains its purpose of producing an absolutely smooth surface. This is accomplished by mixing both substances intimately and adding melted rosin, whereupon the whole mass is exposed to heat, so that the rosin decomposes, its carbon residue filling up the finest pores of the coke. The rosin, in melting, carries the fine graphite particles along into the pores. After cooling the mass is first ground in edge mills, then again in a suitable manner and sifted. Surprising results are obtained with this material. It is advisable to take proportionately little graphite, as the different co-efficients of expansion of the two substances may easily exercise a disturbing action. One-fifth of graphite, in respect to the whole mass, gives the best results, but it is advisable to add plenty of rosin. The liquid mixture must, before burning, possess the consistency of mortar.

Sand Holes in Cast-Brass Work.— Cast-brass work, when it presents numerous and deep sand holes, should be well dipped into the dipping acid before being polished, in order thoroughly to clean these objectionable cavities; and the polishing should be pushed to an extent sufficient to obliterate the smaller sand holes, if possible, as this class of work looks very unsightly, when plated and finished, if pitted all over with minute hollows. The larger holes cannot, without considerable labor, be obliterated; indeed, it not infrequently happens that in endeavoring to work out such cavities they become enlarged, as they often extend deep into the body of the metal. An experienced hand knows how far he dare go in polishing work of this awkward character.

Black Wash for Casting Molds.— Gumlac, 1 part; wood spirit, 2 parts; lampblack, in sufficient quantity to color.

How to Make a Plaster Cast of a Coin or Medal.—The most exact observance of any written or printed directions is no guarantee of success. Practice alone can give expertness in this work.

The composition of the mold is of the most varied, but the materials most generally used are plaster of Paris and brick dust, in the proportion of 2 parts of the first to 1 of the second, stirred in water, with the addition of a little sal ammoniac. The best quality of plaster for this purpose is the so-called alabaster, and the brick dust should be as finely powdered as possible. The addition of clay, dried and very finely powdered, is recommended. With very delicate objects the proportion of plaster may be slightly increased. The dry material should be thoroughly mixed before the addition of water.

As the geometrically exact contour of the coin or medal is often the cause of breaking of the edges, the operator sometimes uses wax to make the edges appear half round and it also allows the casting to be more easily removed from the second half of the mold. Each half of the mold should be about the thickness of the finger. The keys, so called, of every plaster casting must not be forgotten. In the first casting some little half-spherical cavities should be scooped out, which will appear in the second half-round knobs, and which, by engaging with the depressions, will ensure exactness in the finished mold.

After the plaster has set, cut a canal for the flow of the molten casting material, then dry the mold thoroughly in an oven strongly heated. The halves are now ready to be bound together with a light wire. When bound heat the mold gradually and slowly and let the mouth of the canal remain underneath while the heating is in progress, in order to prevent the possible entry of dirt or foreign matter. The heating should be continued as long as there is a suspicion of remaining moisture. When finally assured of this fact, take out the mold, open it, and blow it out, to make sure of absolute cleanness. Close and bind again and place on a hearth of fine, hot sand. The mold should still be glowing when the casting is made. The ladle should contain plenty of metal, so as to hold the heat while the casting is being made. The presence of a little zinc in the metal ensures a sharp casting. Finally, to ensure success, it is always better to provide two molds in case of accident. Even the most practiced metal molders take this precaution, especially when casting delicate objects.

How to Make Castings of Insects.— The object—a dead beetle, for example —is first arranged in a natural position, and the feet are connected with an oval rim of wax. It is then fixed in the center of a paper or wooden box by means of pieces of fine wire, so that it is perfectly free, and thicker wires are run from the sides of the box to the object, which subsequently serve to form air channels in the mold by their removal. A wooden stick, tapering toward the bottom, is placed upon the back of the insect to produce a runner for casting. The box is then filled up with a paste with 3 parts of plaster of Paris and 1 of brick dust, made up with a solution of alum and sal ammoniac. It is also well first to brush the object with this paste to prevent the formation of air bubbles. After the mold thus formed has set, the object is removed from the interior by first reducing it to ashes. It is, therefore, allowed to dry, very slowly at first, by leaving in the shade at a normal temperature (as in India this is much higher than in our zone, it will be necessary to place the mold in a moderately warm place), and afterwards heating gradually to a red heat. This incinerates the object, and melts the waxen base upon which it is placed. The latter escapes, and is burned as it does so, and the object, reduced to fine ashes, is removed through the wire holes as suggested above. The casting is then made in the ordinary manner.

Casting of Soft Metal Castings.—I.—It is often difficult to form flat back or half castings out of the softer metals so that they will run full, owing mostly to the thin edges and frail connections. Instead of using solid metal backs for the molds it is better to use cardboard, or heavy, smooth paper, fastened to a wooden board fitted to the back of the other half of the mold. By this means very thin castings may be produced that would be more difficult with a solid metal back.

II.—To obtain a full casting in brass molds for soft metal two important points should be observed. One is to have the deep recesses vented so the air will escape, and the other is to have the mold properly blued. The bluing is best done by dipping the mold in sulphuric acid, then placing it on a gas stove until the mold is a dark color. Unless this bluing is done it will be impossible to obtain a sharp casting.

Drosses.—All the softer grades of metal throw off considerable dross, which is usually skimmed off; especially with tin and its composition. Should much of this gather on the top of the molten metal, the drosses should all be saved, and melted down when there is enough for a kettle full. Dross may be remelted five or six times before all the good metal is out.

Fuel.—Where a good soft coal can be had at a low price, as in the middle West, this is perhaps the cheapest and easiest fuel to use; and, besides, it has some advantages over gas, which is so much used in the East. A soft-coal fire can be regulated to keep the metal at an even temperature, and it is especially handy to keep the metal in a molten state during the noon hour. This refers particularly to the gas furnaces that are operated from the power plant in the shop; when this power shuts down during the noon hour the metal becomes chilled, and much time is lost by the remelting after one o'clock, or at the beginning in the morning.

Molds.—I.— Brass molds for the casting of soft metal ornaments out of britannia, pewter, spelter, etc., should be made out of brass that contains enough zinc to produce a light-colored brass. While this hard brass is more difficult for the mold maker to cut, the superiority over the dark red copper-colored brass is that it will stand more heat and rougher usage and thereby offset the extra labor of cutting the hard brass. The mold should be heavy enough to retain sufficient heat while the worker is removing a finished casting from the mold so that the next pouring will come full. If the mold is too light it cools more quickly, and consequently the castings are chilled and will not run full. Where the molds are heavy enough they will admit the use of a swab and water after each pouring. This chills the casting so that it can be removed easily with the plyers.

II.—Molds for the use of soft metal castings may be made out of soft metal. This is done with articles that are not numerous, or not often used; and may be looked upon as temporary. The molds are made in part the same as when of brass, and out of tin that contains as much hardening as possible. The hardening consists of antimony and copper. This metal mold must be painted over several times with Spanish red, which tends to prevent the metal from melting. The metal must not be used too hot, otherwise it will melt the mold. By a little careful manipulation many pieces can be cast with these molds.

III.—New iron or brass molds must be blued before they can be used for casting purposes. This is done by placing the mold face downward on a charcoal fire, or by swabbing with sulphuric acid, then placing over a gas flame or charcoal fire until the mold is perfectly oxidized.

IV.—A good substantial mold for small castings of soft metal is made of brass. The expense of making the cast mold is considerable, however, and, on that account, some manufacturers are making their molds by electro-deposition. This produces a much cheaper mold, which can be made very quickly. The electro-deposited mold, however, is very frail in comparison with a brass casting, and consequently must be handled very carefully to keep its shape. The electro-deposited ones are made out of copper, and the backs filled in with a softer metal. The handles are secured with screws.

Plaster Molds.—Castings of any metal can be done in a plaster mold, provided the mold has dried, at a moderate heat, for several days. Smoke the mold well with a brand of rosin to insure a full cast. Where there are only one or two ornaments or figures to cast, it may be done in a mold made out of dental plaster. After the mold is made and set enough so that it can be taken apart, it should be placed in a warm place and left to dry for a day or two. When ready to use the inside should be well smoked over a gaslight; the mold should be well warmed and the metal must not be too hot. Very good castings may be obtained this way; the only objection being the length of time needed for a thorough drying of the mold.

Temperature of Metal.—Metals for casting purposes should not be overheated. If any of the softer metals show blue colors after cooling it is an indication that the metal is too hot. The metal should be heated enough so that it can be poured, and the finished casting have a bright, clean appearance. The mold may be very warm, then the metal need not be so hot for bright, clean castings. Some of the metals will not stand reheating too often, as this will cause them to run sluggish. Britannia metal should not be skimmed or stirred too much, otherwise there will be too much loss in the dross.

CASTING IN WAX:
See Modeling.

CASTINGS, TO SOFTEN IRON:
See Iron.

CASTOR OIL:

Purifying Rancid Castor Oil.—To clean rancid castor oil mix 100 parts of the oil at 95° F. with a mixture of 1 part of alcohol (96 per cent) and 1 part of sulphuric acid. Allow to settle for 24 hours and then carefully decant from the precipitate. Now wash with warm water, boiling for ½ hour; allow to settle for 24 hours in well closed vessels, after which time the purified oil may be taken off.

How to Pour Out Castor Oil.—Any one who has tried to pour castor oil from a square, 5-gallon can, when it is full, knows how difficult it is to avoid a mess. This, however, may be avoided by having a hole punched in the cap which screws onto the can, and a tube, 2 inches long and ¾ of an inch in diameter, soldered on. With a wire nail a hole is punched in the top of the can between the screw cap and the edge of the can. This will admit air while pouring. Resting the can on a table, with the screw-cap tube to the rear, the can is carefully tilted forward with one hand and the shop bottle held in the other. In this way the bottle may be filled without spilling any of the oil and that, too, without a funnel. It is preferable to rest the can on a table when pouring from a 1- or 2-gallon square varnish can, when filling shop bottles. With the opening to the rear, the can is likewise tilted forward slowly so as to allow the surface of the liquid to become "at rest." Even mobile liquids, such as spirits of turpentine, may be poured into shop bottles without a funnel. Of course, the main thing is that the can be lowered slowly, otherwise the first portion may spurt out over the bottle. With 5-gallon round cans it is possible to fill shop bottles in the same manner by resting the can on a box or counter. When a funnel is used for non-greasy liquids, the funnel may be slightly raised with the thumb and little finger from the neck of the bottle, while holding the bottle by the neck between the middle and ring fingers, to allow egress of air.

Tasteless Castor Oil.—

I.—Pure castor oil .. 1 pint
Cologne spirit .. 3 fluidounces
Oil of winter-
green........ 40 minims
Oil of sassafras. 20 minims
Oil of anise..... 15 minims
Saccharine..... 5 grains
Hot water, a sufficient quantity.

Place the castor oil in a gallon bottle.

Add a pint of hot water and shake vigorously for about 15 minutes. Then pour the mixture into a vessel with a stopcock at its base, and allow the mixture to stand for 12 hours. Draw off the oil, excepting the last portion, which must be rejected. Dissolve the essential oils and saccharine in the cologne spirit and add to the washed castor oil.

II.—First prepare an aromatic solution of saccharine as follows:

Refined saccharine.. 25 parts
Vanillin............ 5 parts
Absolute alcohol.... 950 parts
Oil of cinnamon .... 20 parts

Dissolve the saccharine and vanillin in the alcohol, then add the cinnamon oil, agitate well and filter. Of this liquid add 20 parts to 980 parts of castor oil and mix by agitation. Castor oil, like cod-liver oil, may be rendered nearly tasteless, it is claimed, by treating it as follows: Into a matrass of suitable size put 50 parts of freshly roasted coffee, ground as fine as possible, and 25 parts of purified and freshly prepared bone or ivory black. Pour over the mass 1,000 parts of the oil to be deodorized and rendered tasteless, and mix. Cork the container tightly, put on a water bath, and raise the temperature to about 140° F. Keep at this heat from 15 to 20 minutes, then let cool down, slowly, to 90°, at which temperature let stand for 3 hours. Finally filter, and put up in small, well-stoppered bottles.

III.—Vanillin......... 3 grains
Garantose........ 4 grains
Ol. menth. pip.... 8 minims
Alcoholis......... 3 drachms
Ol. ricinus....... 12 ounces
Ol. olivæ (im-
ported), quan-
tity sufficient... 1 pint

M. ft. sol.
Mix vanillin, garantose, ol. menth. pip. with alcohol and add castor oil and olive oil.

Dose: One drachm to 2 fluidounces.

IV.—The following keeps well:

Castor oil........ 24 parts
Glycerine........ 24 parts
Tincture of orange
peel ..:........ 8 parts
Tincture of senega 2 parts
Cinnamon water
enough to make. 100 parts

Mix and make an emulsion. Dose is 1 tablespoonful.

V.—One part of common cooking molasses to 2 of castor oil is the best disguise for the taste of the oil that can be used.

VI.—

Castor oil	1½ ounces
Powdered acacia	2 drachms
Sugar	2 drachms
Peppermint water	4 ounces

Triturate the sugar and acacia, adding the oil gradually; when these have been thoroughly incorporated add the peppermint water in small portions, triturating the mixture until an emulsion is formed.

VII.—This formula for an emulsion is said to yield a fairly satisfactory product:

Castor oil	500 c.c.
Mucilage of acacia	125 c.c.
Spirit of gaultheria	10 grams
Sugar	1 gram
Sodium bicarbonate	1 gram

VIII.—

Castor oil	1 ounce
Compound tincture of cardamom	4 drachms
Oil of wintergreen	3 drops
Powdered acacia	3 drachms
Sugar	2 drachms
Cinnamon water	enough to make 4 ounces.

IX.—

Castor oil	12 ounces
Vanillin	3 grains
Saccharine	4 grains
Oil of peppermint	8 minims
Alcohol	3 drachms
Olive oil enough to make 1 pint.	

In any case, use only a fresh oil.

How to Take Castor Oil.—The disgust for castor oil is due to the odor, not to the taste. If the patient grips the nostrils firmly before pouring out the dose, drinks the oil complacently, and then thoroughly cleanses the mouth, lips, larynx, etc., with water, removing the last vestige of the oil before removing the fingers, he will not get the least taste from the oil, which is bland and tasteless. It all depends upon preventing any oil from entering the nose during the time while there is any oil present.

Castor-Oil Chocolate Lozenges.—

Cacao, free from oil	250 parts
Castor oil	250 parts
Sugar, pulverized	500 parts
Vanillin sugar	5 parts

Mix the chocolate and oil and heat in the water, both under constant stirring. Have the sugar well dried and add, stirring constantly, to the molten mass. Continue the heat for 30 minutes, then pour out and divide into lozenges in the usual way.

CAT DISEASES AND THEIR REMEDIES: See Insecticides and Veterinary Formulas.

CATATYPY.

It is a well-known fact that the reactions of the compounds of silver, platinum, and chromium in photographic processes are generally voluntary ones and that the light really acts only as an accelerator, that is to say the chemical properties of the preparations also change in the dark, though a longer time is required. When these preparations are exposed to the light under a negative, the modification of their chemical properties is accelerated in such a way that, through the gradations of the tone-values in the negative, the positive print is formed. Now it has been found that we also have such accelerators in material substances that can be used in the light, the process being termed catalysis. It is remarkable that these substances, called catalyzers, apparently do not take part in the process, but bring about merely by their presence, decomposition or combination of other bodies during or upon contact. Hence, catalysis may be defined, in short, as the act of changing or accelerating the speed of a chemical reaction by means of agents which appear to remain stable.

Professor Ostwald and Dr. O. Gros, of the Leipsic University, have given the name of "catatypy" to the new copying process. The use of light is entirely done away with, except that for the sake of convenience the manipulations are executed in the light. All that is necessary is to bring paper and negative into contact, no matter whether in the light or in the dark. Hence the negative (if necessary a positive may also be employed) need not even be transparent, for the ascending and descending action of the tone values in the positive picture is produced only by the quantity in the varying density of the silver powder contained in the negative. Hence no photographic (light) picture, but a catatypic picture (produced by contact) is created, but the final result is the same.

Catatypy is carried out as follows: Pour dioxide of hydrogen over the negative, which can be done without any damage to the latter, and lay a piece of paper on (sized or unsized, rough or smooth, according to the effect desired); by a contact lasting a few seconds the paper receives the picture, dioxide of hydrogen being destroyed. From a single application several prints can be made. The acquired picture—still invisible—may now in the further course of the process, have a reducing or oxydizing action. As picture-producing bodies, the large group of iron salts are above all eminently adapted, but other substances, such as chromium, manganese, etc., as well as pigments with glue solutions may also be employed. The development takes place as follows: When the paper which has been in contact with the negative is drawn through a solution of ferrous oxide, the protoxide is transformed into oxide by the peroxide, hence a yellow positive picture, consisting of iron oxide, results, which can be readily changed into other compounds, so that the most varying tones of color can be obtained. With the use of pigments, in conjunction with a glue solution, the action is as follows: In the places where the picture is, the layer with the pigments becomes insoluble and all other dye stuffs can be washed off with water.

The chemical inks and reductions, as well as color pigments, of which the pictures consist, have been carefully tested and are composed of such as are known to possess unlimited durability.

After a short contact, simply immerse the picture in the respective solution, wash out, and a permanent picture is obtained.

CATERPILLAR DESTROYERS:
See Insecticides.

CATGUT:
Preparation of Catgut Sutures.—The catgut is stretched tightly over a glass plate tanned in 5 per cent watery extract of quebracho, washed for a short time in water, subjected to the action of a 4 per cent formalin solution for 24 to 48 hours, washed in running water for 24 hours, boiled in water for 10 to 15 minutes, and stored in a mixture of absolute alcohol with 5 per cent glycerine and 4 per cent carbolic acid. In experiments on dogs, this suture material in aseptic wounds remained intact for 65 days, and was absorbed after 83 days. In infected wounds it was absorbed after 32 days.

CATSUP (ADULTERATED):
See Foods.

CATTLE DIPS AND APPLICATIONS:
See Disinfectants and Insecticides.

CEILING CLEANERS:
See Cleaning Preparations and Methods, and also Household Formulas.

CELERY COMPOUND.

Celery (seed ground).	25	parts
Coca leaves (ground).	25	parts
Black haw (ground)..	25	parts
Hyoscyamus leaves (ground)..........	12½	parts
Podophyllum (powdered)...........	10	parts
Orange peel (ground)	6	parts
Sugar (granulated)...	100	parts
Alcohol.............	150	parts
Water, q. s. ad.......	400	parts

Mix the alcohol with 150 parts of water and macerate drugs for 24 hours; pack in percolator and pour on menstruum till 340 parts is obtained; dissolve sugar in it and strain.

CELLS, SOLUTIONS AND FILLERS FOR BATTERY:
See Battery Solutions and Fillers.

CELLARS, WATERPROOF:
See Household Formulas.

CELLOIDIN PAPER:
See Paper.

Celluloid

New Celluloid.—M. Ortmann has ascertained that turpentine produced by the *Pinus larix*, generally denominated Venice turpentine, in combination with acetone (dimethyl ketone), yields the best results; but other turpentines, such as the American from the *Pinus australis*, the Canada turpentine from the *Pinus balsamea*, the French turpentine from the *Pinus maritima*, and ketones, such as the ketone of methyl-ethyl, the ketone of dinaphthyl, the ketone of methyloxynaphthyl, and the ketone of dioxynaphthyl, may be employed.

To put this process in practice, 1,000 parts of pyroxyline is prepared in the usual manner, and mixed with 65 parts of turpentine, or 250 parts of ketone and 250 parts of ether; 500 parts or 750 parts of methyl alcohol is added, and a colorant, such as desired. Instead of turpentine, rosins derived from it may be employed. If the employment of camphor is desired to a certain extent, it may be added to the mixture. The whole is shaken and left at rest for about 12 hours. It is then passed between hot rollers, and finally pressed, cut, and dried, like ordinary celluloid.

The product thus obtained is without odor, when camphor is not employed; and in appearance and properties it cannot be distinguished from ordinary celluloid, while the expense of production is considerably reduced.

Formol Albumen for Preparation of Celluloid.—Formol has the property of forming combinations with most albuminoid substances. These are not identical with reference to plasticity, and the use which may be derived from them for the manufacture of plastic substances. This difference explains why albumen should not be confounded with gelatin or casein. With this in view, the Société Anonyme l'Oyonnaxienne has originated the following processes:

I.—The albumen may be that of the egg or that of the blood, which are readily found in trade. The formolizing may be effected in the moist state or in the dry state. The dry or moist albumen is brought into contact with the solution of commercial formol diluted to 5 or 10 per cent for an hour. Care must be taken to pulverize the albumen, if it is dry. The formol penetrates rapidly into the albuminoid matter, and is filtered or decanted and washed with water until all the formol in excess has completely disappeared; this it is easy to ascertain by means of aniline water, which produces a turbid white as long as a trace of formic aldehyde remains.

The formol albumen is afterwards dried at low temperature by submitting it to the action of a current of dry air at a temperature not exceeding 107° F. Thus obtained, the product appears as a transparent corneous substance. On pulverizing, it becomes opaque and loses its transparency. It is completely insoluble in water, but swells in this liquid.

II.—The formol albumen is reduced to a perfecily homogeneous powder, and mixed intimately with the plastic matter before rolling. This cannot be considered an adequate means for effecting the mixture. It is necessary to introduce the formol albumen, in the course of the moistening, either by making an emulsion with camphor alcohol, or by mixing it thoroughly with nitro-cellulose, or by making simultaneously a thorough mixture of the three substances. When the mixture is accomplished, the paste is rolled according to the usual operation. The quantity of formol albumen to add is variable, being diminished according to the quantity of camphor.

Instead of adding the desiccated formol albumen, it may previously be swollen in water in order to render it more malleable.

Instead of simple water, alkalinized or acidified water may be taken for this purpose, or even alcoholized water. The albumen, then, should be pressed between paper or cloth, in order to remove the excess of moisture.

Plastic Substances of Nitro-Cellulose Base.—To manufacture plastic substances the Compagnie Française du Celluloid commences by submitting casein to a special operation. It is soaked with a solution of acetate of urea in alcohol; for 100 parts of casein 5 parts of acetate of urea and 50 parts of alcohol are employed. The mass swells, and in 48 hours the casein is thoroughly penetrated. It is then ready to be incorporated with the camphored nitro-cellulose. The nitro-cellulose, having received the addition of camphor, is soaked in the alcohol, and the mass is well mixed. The casein prepared as described is introduced into the mass. The whole is mixed and left at rest for 2 days.

The plastic pulp thus obtained is rolled, cut, and dried like ordinary cellulose, and by the same processes and apparatus. The pulp may also be converted into tubes and other forms, like ordinary celluloid.

It is advisable to subject the improved plastic pulp to a treatment with formaldehyde for the purpose of rendering insoluble the casein incorporated in the celluloid. The plastic product of nitro-cellulose base, thus obtained, presents in employment the same general properties as ordinary celluloid. It may be applied to the various manufacturing processes in use for the preparation of articles of all kinds, and its cost price diminishes more or less according to the proportion of casein associated with the ordinary celluloid. In this plastic product various colorants may be incorporated, and the appearance of shell, pearl, wood, marble, or ivory may also be imparted.

Improved Celluloid.—This product is obtained by mingling with celluloid, under suitable conditions, gelatin or strong glue of gelatin base. Iti s clear that the replacement of part of the celluloid by the gelatin, of which the cost is much less, lowers materially the cost of the final product. The result is obtained without detriment to the qualities of the objects. These are said to be of superior properties, having more firmness than those of celluloid. And the new material is worked more readily than the celluloid employed alone.

The new product may be prepared in open air or in a closed vessel under pressure. When operated in the air, the gelatin is first immersed cold (in any form, and in a state more or less pure) in alcohol marking about 140° F., with the addition of a certain quantity (for example, 5 to 10 per cent) of crystallizable acetic acid. In a few hours the material has swollen considerably, and it is then introduced in alcohol of about 90 per cent, and at the same time the celluloid pulp (camphor and gun cotton), taking care to add a little acetone. The proportion of celluloid in the mixture may be 50 to 75 per cent of the weight of the gelatin, more or less, according to the result desired. After heating the mixture slightly, it is worked, cold, by the rollers ordinarily employed for celluloid and other similar pastes, or by any other suitable methods.

The preparation in a closed vessel does not differ from that which has been described, except for the introduction of the mixture of gelatin, celluloid, alcohol, and acetone, at the moment when the heating is to be accomplished in an autoclave heated with steam, capable of supporting a pressure of 2 to 5 pounds, and furnished with a mechanical agitator. This method of proceeding abridges the operation considerably ; the paste comes from the autoclave well mingled, and is then submitted to the action of rollers. There is but little work in distilling the alcohol and acetic acid in the autoclave. These may be recovered, and on account of their evaporation the mass presents the desired consistency when it reaches the rollers. Whichever of the two methods of preparation may be employed, the substance may be rolled as in the ordinary process, if a boiler with agitator is made use of; the mass may be produced in any form.

Preparation of Uninflammable Celluloid.—The operation of this process by Woodward is the following: In a receiver of glass or porcelain, liquefied fish glue and gum arabic are introduced and allowed to swell for 24 hours in a very dry position, allowing the air to circulate freely. The receiver is not covered. Afterwards it is heated on a water bath, and the contents stirred (for example, by means of a porcelain spatula) until the gum is completely liquefied. The heating of the mass should not exceed 77° F. Then the gelatin is added in

such a way that there are no solid pieces. The receiver is removed from the water bath and colza oil added, while agitating anew. When the mixture is complete it is left to repose for 24 hours.

Before cooling, the mixture is passed through a sieve in order to retain the pieces which may not have been dissolved. After swelling, and the dissolution and purification by means of the sieve, it is allowed to rest still in the same position, with access of air. The films formed while cooling may be removed. The treatment of celluloid necessitates employing a solution completely colorless and clear. The celluloid to be treated while it is still in the pasty state should be in a receiver of glass, porcelain, or similar material.

The mass containing the fish glue is poured in, drop by drop, while stirring carefully, taking care to pour it in the middle of the celluloid and to increase the surface of contact.

When the mixture is complete, the celluloid is ready to be employed and does not produce flame when exposed.

The solution of fish glue may be prepared by allowing 200 parts of it to swell for 48 hours in 1,000 parts of cold distilled water. It is then passed through the sieve, and the pieces which may remain are broken up, in order to mingle them thoroughly with the water. Ten parts of kitchen salt are then added, and the whole mass passed through the sieve.

This product may be utilized for the preparation of photographic films or for those used for cinematographs, or for replacing hard caoutchouc for the insulation of electric conductors, and for the preparation of plastic objects.

Substitute for Camphor in the Preparation of Celluloid and Applicable to Other Purposes.—In this process commercial oil of turpentine, after being rectified by distillation over caustic soda, is subjected to the action of gaseous chlorhydric acid, in order to produce the solid monochlorhydrate of turpentine. After having, by means of the press, extracted the liquid monochlorhydrate, and after several washings with cold water, the solid matter is desiccated and introduced into an autoclave apparatus capable of resisting a pressure of 6 atmospheres. Fifty per cent of caustic soda, calculated on the weight of the monochlorhydrate, and mingled with an equal quantity of alcohol, is added in the form of a thick solution. The apparatus is closed and heated for several hours at the temperature of 284° to 302° F. The material is washed several times for freeing it from the mingled sodium chloride and sodium hydrate, and the camphor resulting from this operation is treated in the following manner:

In an autoclave constructed for the purpose, camphene and water strongly mixed with sulphuric acid are introduced and heated so as to attain 9 pounds of pressure. Then an electric current is applied, capable of producing the decomposition of water. The mass is constantly stirred, either mechanically or more simply by allowing a little of the steam to escape by a tap. In an hour, at least, the material is drawn from the apparatus, washed and dried, sublimed according to need, and is then suitable for replacing camphor in its industrial employments, for the camphene is converted entirely or in greater part into camphor, either right-hand camphor, or a product optically inactive, according to the origin of the oil of turpentine made use of.

In the electrolytic oxidation of the camphene, instead of using acidulated water, whatever is capable of furnishing, under the influence of the electric current, the oxygen necessary for the reaction, such as oxygenized water, barium bioxide, and the permanganates, may be employed.

Plastic and Elastic Composition.—Formaldehyde has the property, as known, of removing from gelatin its solubility and its fusibility, but it has also another property, prejudicial in certain applications, of rendering the composition hard and friable. In order to remedy this prejudicial action M. Deborda adds to the gelatin treated by means of formaldehyde, oil of turpentine, or a mixture of oil of turpentine and German turpentine or Venice turpentine. The addition removes from the composition its friability and hardness, imparting to it great softness and elasticity. The effect is accomplished by a slight proportion, 5 to 10 per cent.

Production of Substances Resembling Celluloid.—Most of the substitutes for camphor in the preparation of celluloid are attended with inconveniences limiting their employment and sometimes causing their rejection. Thus, in one case the celluloid does not allow of the preparation of transparent bodies; in another it occasions too much softness in the products manufactured; and in still another it does not allow of pressing, folding, or other operations, because the mass

is too brittle; in still others combinations are produced which in time are affected unfavorably by the coloring substances employed.

Callenberg has found that the halogenous derivatives of etherized oils, principally oil of turpentine, and especially the solid chloride of turpentine, which is of a snowy and brilliant white, and of agreeable odor, are suitable for yielding, either alone or mixed with camphor or one of its substitutes, and combined by ordinary means with nitrated cellulose, or other ethers of cellulose, treated with acetic ether, a celluloidic product, which, it is said, is not inferior to ordinary celluloid and has the advantage of reduced cost.

Elastic Substitute for Celluloid.—Acetic cellulose, like nitro-cellulose, can be converted into an elastic corneous compound. The substances particularly suitable for the operation are organic substances containing one or more hydroxy, aldehydic, amide, or ketonic groups, as well as the acid amides. Probably a bond is formed when these combinations act on the acetate of cellulose, but the bond cannot well be defined, considering the complex nature of the molecule of cellulose. According to the mode of preparation, the substances obtained form a hard mass, more or less flexible. In the soft state, copies of engraved designs can be reproduced in their finest details. When hardened, they can be cut and polished. In certain respects they resemble celluloid, without its inflammability, and they can be employed in the same manner. They can be produced by the following methods—the Lederer process:

I.—Melt together 1 part of acetate of cellulose and 1½ parts of phenol at about the temperature of 104° to 122° F. When a clear solution is obtained place the mass of reaction on plates of glass or metal slightly heated and allow it to cool gradually. After a rest of several days the mass, which at the outset is similar to caoutchouc, is hard and forms flexible plates, which can be worked like celluloid.

II.—Compress an intimate mixture of equal parts of acetic cellulose and hydrate of chloride or of aniline, at a temperature of 122° to 140° F., and proceed as in the previous case.

In the same way a ketone may be employed, as acetophenone, or an acid amide, as acetamide.

III.—A transparent, celluloid-like substance which is useful for the production of plates, tubes, and other articles, but especially as an underlay for sensitive films in photography, is produced by dissolving 1.8 parts, by weight, of nitrocellulose in 16 parts of glacial acetic acid, with heating and stirring and addition of 5 parts of gelatin. After this has swelled up, add 7.5 parts, by weight, of alcohol (96 per cent), stirring constantly. The syrupy product may be pressed into molds or poured, after further dilution with the said solvents in the stated proportion, upon glass plates to form thin layers. The dried articles are well washed with water, which may contain a trace of soda lye, and dried again. Photographic foundations produced in this manner do not change, nor attack the layers sensitive to light, nor do they become electric, and in developing they remain flat.

IV.—Viscose is the name of a new product of the class of substances like celluloid, pegamoid, etc., substances having most varied and valuable applications. It is obtained directly from cellulose by mascerating this substance in a 1 per cent dilution of hydrochloric acid. The maceration is allowed to continue for several hours, and at its close the liquid is decanted and the residue is pressed off and washed thoroughly. The mass (of which we will suppose there is 100 grams) is then treated with a 20 per cent aqueous solution of sodium hydrate, which dissolves it. The solution is allowed to stand for 3 days in a tightly closed vessel; 100 grams carbon disulphide are then added, the vessel closed and allowed to stand for 12 hours longer, when it is ready for purification. Viscose thus formed is soluble in water, cold or tepid, and yields a solution of a pale brownish color, from which it is precipitated by alcohol and sodium chloride, which purifies it, but at the expense of much of its solubility. A solution of the precipitated article is colorless, or of a slightly pale yellow. Under the action of heat, long continued, viscose is decomposed, yielding cellulose, caustic soda, and carbon disulphide.

See also Casein for Celluloid Substitutes.

Celluloid of Reduced Inflammability.

—I.—A practicable method consists in incorporating silica, which does not harm the essential properties of the celluloid. The material is divided by the usual methods, and dissolved by means of the usual solvents, to which silica has been added, either in the state of amylic, ethylic, or methylic silicate, or in the state of any ether derivative of silicic acid. The suitable proportions vary according to the degree of inflammability desired, and according to the proportion of silica in the ether derivative employed; but sufficient freedom from inflammability for practical purposes is attained by the following proportions: Fifty-five to 65 parts in volume of the solvent of the celluloid, and 35 to 45 parts of the derivative of silicic acid.

When the ether derivative is in the solid form, such, for instance, as ethyl disilicate, it is brought to the liquid state by means of any of the solvents. The union of the solvent and of the derivative is accomplished by mixing the two liquids and shaking out the air as much as possible. The incorporation of this mixture with the celluloid, previously divided or reduced to the state of chips, is effected by pouring the mixture on the chips, or inversely, shaking or stirring as free from the air as possible. The usual methods are employed for the desiccation of the mass. A good result is obtained by drying very slowly, preferably at a temperature not above 10° C. (50° F.). The resulting residue is a new product scarcely distinguished from ordinary celluloid, except that the inherent inflammability is considerably reduced. It is not important to employ any individual silicate or derivative. A mixture of the silicates or derivatives mentioned will accomplish the same results.

II.—Any ignited body is extinguished in a gaseous medium which is unsuitable for combustion; the attempt has therefore been made to find products capable of producing an uninflammable gas; and products have been selected that yield chlorine, and others producing bromine; it is also necessary that these bodies should be soluble in a solvent of celluloid; therefore, among chlorated products, ferric chloride has been taken; this is soluble in the ether-alcohol mixture.

This is the process: An ether-alcohol solution of celluloid is made; then an ether-alcohol solution of ferric perchloride. The two solutions are mingled, and a clear, syrupy liquid of yellow color, yielding no precipitate, is obtained. The liquid is poured into a cup or any suitable vessel; it is left for spontaneous evaporation, and a substance of shell-color is produced, which, after washing and drying, effects the desired result. The celluloid thus treated loses none of its properties in pliability and transparency, and is not only uninflammable, but also incombustible.

Of bromated compounds, calcium bromide has been selected, which produces nearly the same result; the product obtained fuses in the flame; outside, it is extinguished, without the power of ignition.

It may be objected that ferric perchloride and calcium bromide, being soluble in water, may present to the celluloid a surface capable of being affected by moist air; but the mass of celluloid, not being liable to penetration by water, fixes the chlorinated or brominated product. Still, as the celluloid undergoes a slight decomposition, on exposure to the light, allowing small quantities of camphor to evaporate, the surface of the perchlorinated celluloid may be fixed by immersion in albuminous water, after previous treatment with a solution of oxalic acid, if a light yellow product is desired.

For preventing the calcium bromide from eventually oozing on the surface of the celluloid, by reason of its deliquescence, it may be fixed by immersing the celluloid in water acidulated with sulphuric acid. For industrial products, such as toilet articles, celluloid with ferric perchloride may be employed.

Another method of preparing an uninflammable celluloid, based on the principle above mentioned, consists in mixing bromide of camphor with cotton powder, adding castor oil to soften the product, in order that it may be less brittle. The latter product is not incombustible, but it is uninflammable, and its facility of preparation reduces at least one-half the apparatus ordinarily made use of in the manufacture of celluloid. The manufacture of this product is not at all dangerous, for the camphor bromide is strictly uninflammable, and may be melted without any danger of dissolving the gun cotton.

III.—Dissolve 25 parts of ordinary celluloidin in 250 parts of acetone and add a solution of 50 parts of magnesium chloride in 150 parts of alcohol, until a paste results, which occurs with a proportion of about 100 parts of the former solution to 20 parts of the latter solution. This paste is carefully mixed and worked through, then dried, and gives an absolutely incombustible material.

IV.—Glass-like plates which are impervious to acids, salts, and alkalies, flexible, odorless, and infrangible, and still possess a transparency similar to ordinary glass, are said to be obtained by dissolving 4 to 8 per cent of collodion wool (soluble pyroxylin) in 1 per cent of ether or alcohol and mixing the solution with 2 to 4 per cent of castor oil, or a similar non-resinifying oil, and with 4 to 6 per cent of Canada balsam. The inflammability of these plates is claimed to be much less than with others of collodion, and may be almost entirely obviated by admixture of magnesium chloride. An addition of zinc white produces the appearance of ivory.

Solvents for Celluloid.—Celluloid dissolves in acetone, sulphuric ether, alcohol, oil of turpentine, benzine, amyl acetate, etc., alone, or in various combinations of these agents. The following are some proportions for solutions of celluloid:

I.—Celluloid .... 5 parts
Amyl acetate ...... 10 parts
Acetone .... 16 parts
Sulphuric ether .... 16 parts

II.—Celluloid .... 10 parts
Sulphuric ether .... 30 parts
Acetone .... 30 parts
Amyl acetate ...... 30 parts
Camphor .... 3 parts

III.—Celluloid .... 5 parts
Alcohol .... 50 parts
Camphor .... 5 parts

IV.—Celluloid .... 5 parts
Amyl acetate ...... 50 parts

V.—Celluloid .... 5 parts
Amyl acetate ...... 25 parts
Acetone .... 25 parts

Softening and Cementing Celluloid.—If celluloid is to be warmed only sufficiently to be able to bend it, a bath in boiling water will answer. In steam at 120° C. (248° F.), however, it becomes so soft that it may be easily kneaded like dough, so that one may even imbed in it metal, wood, or any similar material. If it be intended to soften it to solubility, the celluloid must then be scraped fine and macerated in 90 per cent alcohol, whereupon it takes on the character of cement and may be used to join broken pieces of celluloid together. Solutions of celluloid may be prepared: 1. With 5 parts, by weight, of celluloid in 16 parts, by weight, each of amyl acetate, acetone, and sulphuric ether. 2. With 10 parts, by weight, of celluloid in 30 parts, by weight, each of sulphuric ether, acetone, amyl acetate, and 4 parts, by weight, camphor. 3. With 5 parts, by weight, celluloid in 50 parts, by weight, alcohol and 5 parts, by weight, camphor. 4. With 5 parts, by weight, celluloid in 50 parts, by weight, amyl acetate. 5. With 5 parts, by weight, celluloid in 25 parts, by weight, amyl acetate and 25 parts, by weight, acetone.

It is often desirable to soften celluloid so that it will not break when hammered. Dipping it in water warmed to 40° C. (104° F.) will suffice for this.

Mending Celluloid.—Celluloid dishes which show cracks are easily repaired by brushing the surface repeatedly with alcohol, 3 parts, and ether, 4 parts, until the mass turns soft and can be readily squeezed together. The pressure must be maintained for about one day. By putting only 1 part of ether in 3 parts of alcohol and adding a little shellac, a cement for celluloid is obtained, which, applied warm, produces quicker results. Another very useful gluing agent for celluloid receptacles is concentrated acetic acid. The celluloid fragments dabbed with it stick together almost instantaneously.

See also Adhesives for Methods of Mending Celluloid.

Printing on Celluloid.—Printing on celluloid may be done in the usual way. Make ready the form so as to be perfectly level on the impression—that is, uniform to impressional touch on the face. The tympan should be hard. Bring up the form squarely, allowing for about a 3- or 4-sheet cardboard to be withdrawn from the tympan when about to proceed with printing on the celluloid; this is to allow for the thickness of the sheet of celluloid. Use live but dry and well-seasoned rollers. Special inks of different colors are made for this kind of presswork; in black a good card-job quality will be found about right, if a few drops of copal varnish are mixed with the ink before beginning to print.

Colored Celluloid.—

Black: First dip into pure water, then into a solution of nitrate of silver; let dry in the light.

Yellow: First immerse in a solution of nitrate of lead, then in a concentrated solution of chromate of potash.

Brown: Dip into a solution of permanganate of potash made strongly alkaline by the addition of soda.

Blue: Dip into a solution of indigo neutralized by the addition of soda.

Red: First dip into a diluted bath of nitric acid; then into an ammoniacal solution of carmine.

Green: Dip into a solution of verdigris.

Aniline colors may also be employed but they are less permanent.

Bleaching Celluloid.—If the celluloid has become discolored throughout, its whiteness can hardly be restored, but if merely superficially discolored, wipe with a woolen rag wet with absolute alcohol and ether mixed in equal proportions. This dissolves and removes a minute superficial layer and lays bare a new surface. To restore the polish rub briskly first with a woolen cloth and finish with silk or fine chamois. A little jeweler's rouge or putzpomade greatly facilitates matters. Ink marks may be removed in the same manner. Printer's ink may be removed from celluloid by rubbing first with oil of turpentine and afterwards with alcohol and ether.

Process of Impregnating Fabrics with Celluloid.—The fabric is first saturated with a dilute celluloid solution of the consistency of olive oil, which solution penetrates deeply into the tissue; dry quickly in a heating chamber and saturate with a more concentrated celluloid solution, about as viscous as molasses. If oil be added to the celluloid solution, the quantity should be small in the first solution, e. g., 1 to 2 per cent, in the following ones 5 to 8 per cent, while the outer layer contains very little or no oil. A fabric impregnated in this manner possesses a very flexible surface, because the outer layer may be very thin, while the interior consists of many flexible fibers surrounded by celluloid.

CELLULOID CEMENTS AND GLUES: See Adhesives.

CELLULOID LACQUER: See Lacquer.

CELLULOID PUTTY: See Cements.

Cements

(See also Putties.)

For Adhesive Cements intended for repairing broken articles, see Adhesives.

Putty for Celluloid.—To fasten celluloid to wood, tin, etc., use a compound of 2 parts shellac, 3 parts spirit of camphor, and 4 parts strong alcohol.

Plumbers' Cement.—A plumbers' cement consists of 1 part black rosin, melted, and 2 parts of brickdust, thoroughly powdered and dried.

Cement for Steam and Water Pipes.—A cement for pipe joints is made as follows: Ten pounds fine yellow ocher; 4

pounds ground litharge; 4 pounds whiting, and $\frac{1}{2}$ pound of hemp, cut up fine. Mix together thoroughly with linseed oil to about the consistency of putty.

Gutter Cement.—Stir sand and fine lime into boiled paint skins while hot and thick. Use hot.

Cement for Pipe Joints.—A good cement for making tight joints in pumps, pipes, etc., is made of a mixture of 15 parts of slaked lime, 30 parts of graphite, and 40 parts of barium sulphate. The ingredients are powdered, well mixed together, and stirred up with 15 parts of boiled oil. A stiffer preparation can be made by increasing the proportions of graphite and barium sulphate to 30 and 40 parts respectively, and omitting the lime. Another cement for the same purpose consists of 15 parts of chalk and 50 of graphite, ground, washed, mixed, and reground to fine powder. To this mixture is added 20 parts of ground litharge, and the whole mixed to a stiff paste with about 15 parts of boiled oil. This last preparation possesses the advantage of remaining plastic for a long time when stored in a cool place. Finally, a good and simple mixture for tightening screw connections is made from powdered shellac dissolved in 10 per cent ammonia. The mucinous mass is painted over the screw threads, after the latter have been thoroughly cleaned, and the fitting is screwed home. The ammonia soon volatilizes, leaving behind a mass which hardens quickly, makes a tight joint, and is impervious to hot and cold water.

Protection for Cement Work.—A coating of soluble glass will impart to cement surfaces exposed to ammonia not only a protective covering, but also increased solidness.

Cemented surfaces can be protected from the action of the weather by repeated coats of a green vitriol solution consisting of 1 part of green vitriol and 3 parts of water. Two coatings of 5 per cent soap water are said to render the cement waterproof; after drying and rubbing with a cloth or brush, this coating will become glossy like oil paint. This application is especially recommended for sick rooms, since the walls can be readily cleaned by washing with soapy water. The coating is rendered more and more waterproof thereby. The green vitriol solution is likewise commendable for application on old and new plastering, since it produces thereon waterproof coatings. From old plastering the loose particles have first to be removed by washing.

Puncture Cement.—A patented preparation for automatically repairing punctures in bicycle tires consists of glycerine holding gelatinous silica or aluminum hydrate in suspension. Three volumes of glycerine are mixed with 1 volume of liquid water glass, and an acid is stirred in. The resulting jelly is diluted with 3 additional volumes of glycerine, and from 4 to 6 ounces of this fluid are placed in each tire. In case of puncture, the internal pressure of the air forces the fluid into the hole, which it closes.

To Fix Iron in Stone.—Of the quickly hardening cements, lead and sulphur, the latter is popularly employed. It can be rendered still more suitable for purposes of pouring by the admixture of Portland cement, which is stirred into the molten sulphur in the ratio of 1 to 3 parts by weight. The strength of the latter is increased by this addition, since the formation of so coarse a crystalline structure as that of solidifying pure sulphur is disturbed by the powder added.

White Portland Cement.—Mix together feldspar, 40–100 parts, by weight; kaolin, 100 parts; limestone, 700 parts; magnesite, 20–40 parts; and sodium chloride, 2.5–5 parts, all as pure as possible, and heat to 1430° to 1500° C. (2606° to 2732° F.), until the whole has become sintered together, and forms a nice, white cement-like mass.

Cement for Closing Cracks in Stoves.—Make a putty of reduced iron (iron by hydrogen) and a solution of sodium or potassium silicate, and force it into the crack. If the crack be a very narrow one, make the iron and silicate into paste instead of putty. This material grows firmer and harder the longer the mended article is used.

Cement for Waterpipe. I.—Mix together 11 parts, by weight, Portland cement; 4 parts, by weight, lead white; 1 part, by weight, litharge; and make to a paste with boiled oil in which 3 per cent of its weight of colophony has been dissolved.

II.—Mix 1 part, by weight, torn-up wadding; 1 part, by weight, of quicklime, and 3 parts, by weight, of boiled oil. This cement must be used as soon as made.

Cement for Pallet Stones.—Place small pieces of shellac around the stone when in position and subject it to heat. Often the lac spreads unevenly or swells up; and this, in addition to being unsightly, is apt to displace the stone. This can be avoided as follows: The pallets are held in long sliding tongs. Take a piece of shellac, heat it and roll it into a cylinder between the fingers; again heat the extremity and draw it out into a fine thread. This thread will break off, leaving a point at the end of the lac. Now heat the tongs at a little distance from the pallets, testing the degree of heat by touching the tongs with the shellac. When it melts easily, lightly touch the two sides of the notch with it; a very thin layer can thus be spread over them, and the pallet stone can then be placed in position and held until cold enough. The tongs will not lose the heat suddenly, so that the stone can easily be raised or lowered as required. The projecting particles of cement can be removed by a brass wire filed to an angle and forming a scraper. To cement a ruby pin, or the like, one may also use shellac dissolved in spirit, applied in the consistency of syrup, and liquefied again by means of a hot pincette, by seizing the stone with it.

DENTAL CEMENTS:

Fairthorne's Cement.—Powdered glass, 5 parts; powdered borax, 4 parts; silicic acid, 8 parts; zinc oxide, 200 parts. Powder very finely and mix; then tint with a small quantity of golden ocher or manganese. The compound, mixed before use with concentrated syrupy zinc-chloride solution, soon becomes as hard as marble and constitutes a very durable tooth cement.

Huebner's Cement.—Zinc oxide, 500.0 parts; powdered manganese, 1.5 parts; yellow ocher, powdered, 1.5-4.0 parts; powdered borax, 10.0 parts; powdered glass, 100.0 parts.

As a binding liquid it is well to use acid-free zinc chloride, which can be prepared by dissolving pure zinc, free from iron, in concentrated, pure, hydrochloric acid, in such a manner that zinc is always in excess. When no more hydrogen is evolved the zinc in excess is still left in the solution for some time. The latter is filtered and boiled down to the consistency of syrup.

Commercial zinc oxide cannot be employed without previous treatment, because it is too loose; the denser it is the better is it adapted for dental cements, and the harder the latter will be. For this reason it is well, in order to obtain a dense product, to stir the commercial pure zinc oxide into a stiff paste with water to which 2 per cent of nitric acid has been added; the paste is dried and heated for some time at white heat in a Hessian crucible.

After cooling, the zinc oxide, thus obtained, is very finely powdered and kept in hermetically sealed vessels, so that it cannot absorb carbonic acid. The dental cement prepared with such zinc oxide turns very hard and solidifies with the concentrated zinc-chloride solution in a few minutes.

Phosphate Cement.—Concentrate pure phosphoric acid till semi-solid, and mix aluminum phosphate with it by heating. For use, mix with zinc oxide to the consistency of putty. The cement is said to set in 2 minutes.

Zinc Amalgam, or Dentists' Zinc.—This consists of pure zinc filings combined with twice their weight of mercury, a gentle heat being employed to render the union more complete. It is best applied as soon as made. Its color is gray, and it is said to be effective and durable.

Sorel's Cement.—Mix zinc oxide with half its bulk of fine sand, add a solution of zinc chloride of 1.260 specific gravity, and rub the whole thoroughly together in a mortar. The mixture must be applied at once, as it hardens very quickly.

Metallic Cement.—Pure tin, with a small proportion of cadmium and sufficient mercury, forms the most lasting and, for all practical purposes, the least objectionable amalgam. Melt 2 parts of tin with 1 of cadmium, run it into ingots, and reduce it to filings. Form these into a fluid amalgam with mercury, and squeeze out the excess of the latter through leather. Work up the solid residue in the hand, and press it into the tooth. Or melt some beeswax in a pipkin, throw in 5 parts of cadmium, and when melted add 7 or 8 parts of tin in small pieces. Pour the melted metals into an iron or wooden box, and shake them until cold, so as to obtain the alloy in a powder. This is mixed with 2½ to 3 times its weight of mercury in the palm of the hand, and used as above described.

CEMENT COLORS:
See Stone.

CEMENT, MORDANT FOR:
See Mordants.

CEMENT, PAINTS FOR:
See Paint.

CEMENT, PROTECTION OF, AGAINST ACID:
See Acid-Proofing.

CHAIN OF FIRE:

See Pyrotechnics.

CHAINS (WATCH), TO CLEAN:

See Cleaning Preparations and Methods.

CHALK FOR TAILORS.

Knead together ordinary pipe clay, moistened with ultramarine blue for blue, finely ground ocher for yellow, etc., until they are uniformly mixed, roll out into thin sheets, cut and press into wooden or metallic molds, well oiled to prevent sticking, and allow to dry slowly at ordinary temperature or at a very gentle heat.

CHAPPED HANDS:

See Cosmetics.

CHARTA SINAPIS:

See Mustard Paper.

CHARTREUSE:

See Wines and Liquors.

Ceramics

GROUND CERAMICS—LAYING OIL FOR:

See Oil.

Notes for Potters, Glass-, and Brickmakers.—It is of the highest importance in selecting oxides, minerals, etc., for manufacturing different articles, for potters' use, to secure pure goods, especially in the purchase of the following: Lead, manganese, oxide of zinc, borax, whiting, oxide of iron, and oxide of cobalt. The different ingredients comprising any given color or glaze should be thoroughly mixed before being calcined, otherwise the mass will be of a streaky or variegated kind. Calcination requires care, especially in the manufacture of enamel colors. Over-firing, particularly of colors or enamels composed in part of lead, borax, antimony, or litharge, causes a dullness of shade, or film, that reduces their value for decorative purposes, where clearness and brilliancy are of the first importance.

To arrest the unsightly defect of "crazing," the following have been the most successful methods employed, in the order given:

I.—Flux made of 10 parts tincal; 4 parts oxide of zinc; 1 part soda.

II.—A calcination of 5 parts oxide of zinc; 1 part pearl ash.

III.—Addition of raw oxide of zinc, 6 pounds to each hundredweight of glaze.

To glazed brick and tile makers, whose chief difficulty appears to be the production of a slip to suit the contraction of their clay, and adhere strongly to either a clay or a burnt brick or tile, the following method may be recommended:

Mix together:

Ball clay	10	parts
Cornwall stone	10	parts
China clay	7	parts
Flint	6½	parts

To be mixed and lawned one week before use.

To Cut Pottery.—Pottery or any soft or even hard stone substance can be cut without chipping by a disk of soft iron, the edge of which has been charged with emery, diamond, or other grinding powder, that can be obtained at any tool agency. The cutting has to be done with a liberal supply of water fed continually to the revolving disk and the substance to be cut.

BRICK AND TILEMAKERS' GLAZED BRICKS:

White.—When the brick or tile leaves the press, with a very soft brush cover the part to be glazed with No. 1 Slip; afterwards dip the face in the same mixture.

No. 1 Slip.—

Same clay as brick	9	parts
Flint	1	part
Ball clay	5	parts
China	4	parts

Allow the brick to remain slowly drying for 8 to 10 hours, then when moist dip in the white body.

White Body.—

China clay	24	parts
Ball clay	8	parts
Feldspar	8	parts
Flint	4	parts

The brick should now be dried slowly but thoroughly, and when perfectly dry dip the face in clean cold water, and immediately afterwards in glaze.

Hard Glaze.—

Feldspar	18	parts
Cornwall stone	3½	parts
Whiting	1½	parts
Oxide of zinc	1½	parts
Plaster of Paris	¾	part

Soft Glaze.—

White lead	13	parts
Feldspar	20	parts
Oxide of zinc	3	parts
Plaster of Paris	1	part
Flint glass	13	parts
Cornwall stone	3½	parts
Paris white	1¼	parts

Where clay is used that will stand a very high fire, the white lead and glass may be left out. A wire brush should now be used to remove all superfluous glaze, etc., from the sides and ends of the brick, which is then ready for the kiln. In placing, set the bricks face to face, about an inch space being left between the two glazed faces. All the mixtures, after being mixed with water to the consistency of cream, must be passed 2 or 3 times through a very fine lawn. The kiln must not be opened till perfectly cold.

Process for Colored Glazes.—Use color, 1 part, to white body, 7 parts. Use color, 1 part, to glaze, 9 parts.

Preparation of Colors.—The specified ingredients should all be obtained finely ground, and after being mixed in the proportions given should, in a saggar or some clay vessel, be fired in the brick kiln and afterwards ground for use. In firing the ingredients the highest heat attainable is necessary.

Turquoise.—

Oxide of zinc	8	parts
Oxide of cobalt	1¼	parts

Grass Green.—

Oxide of chrome	6	parts
Flint	1	part
Oxide of copper	½	part

Royal Blue.—

Pure alumina	20	parts
Oxide of zinc	8	parts
Oxide of cobalt	4	parts

Mazarine Blue.—

Oxide of cobalt	10	parts
Paris white	9	parts
Sulphate barytes	1	part

Red Brown.—

Oxide of zinc	40	parts
Crocus of martis	6	parts
Oxide of chrome	6	parts
Red lead	5	parts
Boracic acid	5	parts
Red oxide of iron	1	part

Orange.—

Pure alumina	5	parts
Oxide of zinc	2	parts
Bichromate of potash	1	part
Iron scale	½	part

Claret Brown.—

Bichromate of potash	2	parts
Flint	2	parts
Oxide of zinc	1	part
Iron scale	1	part

Blue Green.—

Oxide of chrome	6	parts
Flint	2	parts
Oxide of cobalt	¾	part

Sky Blue.—

Flint	9	parts
Oxide of zinc	13	parts
Cobalt	2½	parts
Phosphate soda	1	part

Chrome Green.—

Oxide of chrome	3	parts
Oxide of copper	1	part
Carbonate of cobalt	1	part
Oxide of cobalt	2	parts

Olive.—

Oxide of chrome	3	parts
Oxide of zinc	2	parts
Flint	5	parts
Oxide of cobalt	1	part

Blood Red.—

Oxide of zinc	30	parts
Crocus martis	7	parts
Oxide of chrome	7	parts
Litharge	5	parts
Borax	5	parts
Red oxide of iron	2	parts

Black.—

Chromate of iron	24	parts
Oxide of nickel	2	parts
Oxide of tin	2	parts
Oxide of cobalt	5	parts

Imperial Blue.—

Oxide of cobalt	10	parts
Black color	1½	parts
Paris white	7½	parts
Flint	2½	parts
Carbonate of soda	1	part

Mahogany.—

Chromate of iron	30	parts
Oxide of manganese	20	parts
Oxide of zinc	12	parts
Oxide of tin	4	parts
Crocus martis	2	parts

Gordon Green.—

Oxide of chrome	12	parts
Paris white	8	parts
Bichromate of potash	4½	parts
Oxide of cobalt	¾	part

Violet.—

Oxide of cobalt	2½	parts
Oxide of manganese	4	parts
Oxide of zinc	8	parts
Cornwall stone	8	parts

Lavender.—

Calcined oxide of zinc	5	parts
Carbonate of cobalt..	¾	part
Oxide of nickel	¼	part
Paris white	1	part

Brown.—

Manganese	4	parts
Oxide of chrome	2	parts
Oxide of zinc	4	parts
Sulphate barytes	2	parts

Dove.—

Oxide of nickel	7	parts
Oxide of cobalt	2	parts
Oxide of chrome	1	part
Oxide of flint	18	parts
Paris white	3	parts

Yellow Green.—

Flint	6	parts
Paris white	4	parts
Bichromate of potash.	4½	parts
Red lead	2	parts
Fluorspar	2	parts
Plaster of Paris	1½	parts
Oxide of copper	½	part

BODIES REQUIRING NO STAIN:

Ivory.—

Cane marl	16	parts
Ball clay	12	parts
Feldspar	8	parts
China clay	6	parts
Flint	4	parts

Cream.—

Ball clay	22	parts
China clay	5½	parts
Flint	5	parts
Feldspar	3½	parts
Cane marl	12	parts

Black.—

Ball clay	120	parts
Ground ocher	120	parts
Ground manganese.	35	parts

Buff.—

Ball clay	12	parts
China clay	10	parts
Feldspar	8	parts
Bull fire clay	16	parts
Yellow ocher	3	parts

Drab.—

Cane marl	30	parts
Ball clay	10	parts
Stone	7	parts
Feldspar	4	parts

Brown.—

Red marl	50	parts
China clay	7	parts
Ground manganese..	6	parts
Feldspar	3	parts

In making mazarine blue glazed bricks use the white body and stain the glaze only.

Mazarine blue	1	part
Glaze	7	parts

For royal blue use 1 part stain to 6 parts white body, and glaze unstained.

Blood-Red Stain.—Numerous brick manufacturers possess beds of clay from which good and sound bricks or tiles can be made, the only drawback being that the clay does not burn a good color. In many cases this arises from the fact that the clay contains more or less sulphur or other impurity, which spoils the external appearance of the finished article. The following stain will convert clay of any color into a rich, deep red, mixed in proportions of stain, 1 part, to clay, 60 parts.

Stain.—

Crocus martis	20	parts
Yellow ocher	4	parts
Sulphate of iron	10	parts
Red oxide of iron	2	parts

A still cheaper method is to put a slip or external coating upon the goods. The slip being quite opaque, effectively hides the natural color of the brick or tile upon which it may be used.
The process is to mix:

Blood-red stain	1	part
Good red clay	6	parts

Add water until the mixture becomes about the consistency of cream, then with a sponge force the liquid two or three times through a very fine brass wire lawn, No. 80, and dip the goods in the liquid as soon as they are pressed or molded.

Blue Paviors.—Blue paving bricks may be produced with almost any kind of clay that will stand a fair amount of heat, by adopting the same methods as in the former case of blood-red bricks, that is, the clay may be stained throughout, or an outside coating may be applied.

Stain for Blue Paviors.—

Ground ironstone	20	parts
Chromate of iron	5	parts
Manganese	6	parts
Oxide of nickel	1	part

Use 1 part clay and 1 part stain for coating, and 50 or 60 parts clay and 1 part stain for staining through.
Fire blue paviors very hard.

Buff Terra-Cotta Slip.—

Buff fire clay	16	parts
China clay	6	parts

Yellow ocher........ 3 parts
Ball clay........... 10 parts
Flint. .... 4 parts

Add water to the materials after mixing well, pass through the fine lawn, and dip the goods when soft in the liquid.

Transparent Glaze.—

Ground flint glass..... 4 parts
Ground white lead.... 4 parts
Ground oxide of zinc. ¼ part

This glaze is suitable for bricks or tiles made of very good red clay, the natural color of the clay showing through the glaze. The goods must first be fired sufficiently hard to make them durable, afterwards glazed, and fired again. The glaze being comparatively soft will fuse at about half the heat required for the first burning. The glaze may be stained, if desired, with any of the colors given in glazed-brick recipes, in the following proportions: Stain, 1 part; glaze, 1 part.

SPECIAL RECIPES FOR POTTERY AND BRICK AND TILE WORKS:

Vitrifiable Bodies.—The following mixtures will flux only at a very high heat. They require no glaze when a proper heat is attained, and they are admirably adapted for stoneware glazes.

I.—Cornwall stone.... 20 parts
Feldspar .... 12 parts
China clay........ 3 parts
Whiting........... 2 parts
Plaster of Paris ... 1½ parts

II.—Feldspar .... 30 parts
Flint............. 9 parts
Stone............. 8 parts
China clay........ 3 parts

III.—Feldspar .... 20 parts
Stone............. 5 parts
Oxide of zinc...... 3 parts
Whiting........... 2 parts
Plaster of Paris ... 1 part
Soda crystals, dissolved.......... 1 part

Special Glazes for Bricks or Pottery at One Burning.—To run these glazes intense heat is required.

I.—Cornwall stone..... 40 parts
Flint............. 7 parts
Paris white........ 4 parts
Ball clay........... 15 parts
Oxide of zinc...... 6 parts
White lead........ 15 parts

II.—Feldspar........... 20 parts
Cornwall stone.... 5 parts
Oxide of zinc...... 3 parts
Flint............. 3 parts
Lynn sand........ 1½ parts
Sulphate barytes... 1½ parts

III.—Feldspar .... 25 parts
Cornwall stone..... 6 parts
Oxide of zinc...... 2 parts
China clay........ 2 parts

IV.—Cornwall stone.....118 parts
Feldspar .... 40 parts
Paris white........ 28 parts
Flint............. 4 parts

V.—Feldspar .... 16 parts
China clay........ 4 parts
Stone............. 4 parts
Oxide of zinc...... 2 parts
Plaster of Paris .... 1 part

VI.—Feldspar........... 10 parts
Stone............. 5 parts
Flint............. 2 parts
Plaster............ ½ part

The following glaze is excellent for bricks in the biscuit and pottery, which require an easy firing:

White.—

White lead........ 20 parts
Stone............. 9 parts
Flint............. 9 parts
Borax............. 4 parts
Oxide of zinc...... 2 parts
Feldspar.......... 3 parts

These materials should be procured finely ground, and after being thoroughly mixed should be placed in a fire-clay crucible, and be fired for 5 or 6 hours, sharply, or until the material runs down into a liquid, then with a pair of iron tongs draw the crucible from the kiln and pour the liquid into a bucket of cold water, grind the flux to an extremely fine powder, and spread a coating upon the plate to be enameled, previously brushing a little gum thereon. The plate must then be fired until a sufficient heat is attained to run or fuse the powder.

POTTERY BODIES AND GLAZES:

Ordinary.—

I.—China clay........ 2½ parts
Stone............. 1½ parts
Bone.............. 3 parts

II.—China clay........ 5 parts
Stone............. 2½ parts
Bone............. 7 parts
Barytes........... 3 parts

III.—Chain clay...... 5 parts
Stone............. 3 parts
Flint............. ¼ part
Barytes........... 8 parts

Superior.—

I.—China clay........ 35 parts
Cornwall stone..... 23 parts
Bone.............. 40 parts
Flint............. 2 parts

II.—China clay . . . .   35 parts
Cornwall stone . . . . .   8 parts
Bone . . . .   50 parts
Flint . . . .   3 parts
Blue clay . . . .   4 parts

III.—China clay . . . .   8 parts
Cornwall stone . . . . .   40 parts
Bone . . . .   29 parts
Flint . . . .   5 parts
Blue clay . . . .   18 parts

IV.—China clay . . . .   32 parts
Cornwall stone . . . . .   23 parts
Bone . . . .   34 parts
Flint . . . .   6 parts
Blue clay . . . .   5 parts

V.—China clay . . . .   7 parts
Stone . . . .   40 parts
Bone . . . .   28 parts
Flint . . . .   5 parts
Blue clay . . . .   20 parts

Finest China Bodies.—

I.—China clay . . . .   20 parts
Bone . . . .   60 parts
Feldspar . . . .   20 parts

II.—China clay . . . .   30 parts
Bone . . . .   40 parts
Feldspar . . . .   30 parts

III.—China clay . . . .   25 parts
Stone . . . .   10 parts
Bone . . . .   45 parts
Feldspar . . . .   20 parts

IV.—China clay . . . .   30 parts
Stone . . . .   15 parts
Bone . . . .   35 parts
Feldspar . . . .   20 parts

Earthenware Bodies.—

I.—Ball clay   13   parts
China clay . . . .   9½ parts
Flint . . . .   5½ parts
Cornwall stone   4   parts

II.—Ball clay . . . .   12½ parts
China clay   8   parts
Flint . . . .   5½ parts
Cornwall stone . . .   2½ parts
One pint of cobalt
stain to 1 ton of
glaze.

III.—Ball clay . . . .   13¼ parts
China clay   11   parts
Flint   4   parts
Cornwall stone   5   parts
Feldspar   4   parts
Stain as required.

IV.—Ball clay . . . .   18½ parts
China clay . . . .   13½ parts
Flint . . . .   8½ parts
Stone   4   parts
Blue stain, 2 pints to ton.

V.—Ball clay . . . .   15 parts
China clay . . . .   12 parts
Flint . . . .   6 parts
Stone . . . .   4 parts
Feldspar . . . .   4 parts
Blue stain, 2 pints to ton.

VI. (Parian).—
Stone . . . .   11 parts
Feldspar . . . .   10 parts
China clay . . . .   8 parts

COLORED BODIES:

Ivory Body.—
Ball clay   22   parts
China . . . .   5½ parts
Flint   5   parts
Stone . . . .   3½ parts

Dark Drab Body.—
Cane marl . . . .   30 parts
Ball clay . . . .   10 parts
Cornwall stone . . . . .   7 parts
Feldspar . . . .   4 parts

Black Body.—
Ball clay . . . .   120 parts
Ocher . . . .   120 parts
Manganese . . . .   35 parts
Cobalt carbonate . .   2 parts
Grind the three last mentioned ingre-
dients first.

Caledonia Body.—
Yellow clay . . . .   32 parts
China clay . . . .   10 parts
Flint . . . .   4 parts

Brown Body.—
Red clay   50   parts
Common clay . . . . . .   7½ parts
Manganese   1   part
Flint   1   part

Jasper Body.—
Cawk clay   10   parts
Blue clay   10   parts
Bone   5   parts
Flint   2   parts
Cobalt . . . .   ¼ part

Stone Body.—
Stone . . . .   48 parts
Blue clay . . . .   25 parts
China clay . . . .   24 parts
Cobalt . . . .   10 parts

Egyptian Black.—
Blue clay . . . .   235 parts
Calcined ocher . . . .   225 parts
Manganese . . . .   45 parts
China clay . . . . . .   15 parts

Ironstone Body.—
Stone . . . .   200 parts
Cornwall clay . . . . .   150 parts

Blue clay......... 200 parts
Flint............ 100 parts
Calx............. 1 part

Cream Body.—

Blue clay.......... 1½ parts
Brown clay........ 1½ parts
Black clay......... 1 part
Cornish clay....... 1 part
Common ball clay.. ¼ part
Buff color......... ¼ part

Light Drab.—

Cane marl........ 30 parts
Ball clay.......... 24 parts
Feldspar .... 7 parts

Sage Body.—

Cane marl........ 15 parts
Ball clay.......... 15 parts
China clay........ 5 parts
Stained with turquoise stain.

COLORED GLAZES FOR POTTERY:

Blue.—

White glaze...... 100 parts
Oxide of cobalt... 3 parts
Red lead......... 10 parts
Flowing blue..... 3 parts
Enamel blue...... 3 parts

Grind.

Pink.—

White glaze...... 100 parts
Red lead......... 8 parts
Marone pink U. G. 8 parts
Enamel red....... 3 parts

Grind.

Buff.—

White glaze...... 100 parts
Red lead......... 10 parts
Buff color........ 8 parts

Grind.

Ivory.—

White glaze...... 100 parts
Red lead......... 8 parts
Enamel amber.... 8 parts
Yellow underglaze 2 parts

Grind.

Turquoise.—

White glaze...... 100 parts
Red lead......... 10 parts
Carbonate of soda. 5 parts
Enamel blue...... 4 parts
Malachite, 110.... 4 parts

Grind.

Yellow.—

I.—White glaze...... 100 parts
Red lead........ 10 parts
Oxide of uranium. 8 parts

Grind.

II.—Dried flint.......... 5 parts
Cornwall stone...... 15 parts
Litharge............ 50 parts
Yellow underglaze... 4 parts

Grind.

Green.—

I.—Oxide of copper..... 8 parts
Flint of glass........ 3 parts
Flint................ 1 part
Red lead............ 6 parts

Grind, then take:

Of above........... 1 part
White glaze......... 6 parts

Or stronger as required.

II.—Red lead..........60 parts
Stone.............24 parts
Flint.............12 parts
Flint glass........12 parts
China clay......... 3 parts
Calcined oxide of
copper...........14 parts
Oxide of cobalt..... ¼ part

Grind only.

Green Glaze, Best.—

III.—Stone.............80 parts
Flint............. 8 parts
Soda crystals....... 4 parts
Borax............. 3½ parts
Niter............. 2 parts
Whiting........... 2 parts
Oxide of cobalt..... ¼ part

Glost fire, then take:

Above frit.........60 parts
Red lead..........57 parts
Calcined oxide of
copper.......... 5¼ parts

Black.—

Red lead.......... 24 parts
Raddle............ 4 parts
Manganese........ 4 parts
Flint............. 2 parts
Oxide of cobalt..... 2 parts
Carbonate of cobalt. 2 parts

Glost fire.

WHITE GLAZES:

China.—Frit:

I.—Stone............. 6 parts
Niter............. 2 parts
Borax............. 12 parts
Flint............. 4 parts
Pearl ash.......... 2 parts

To mill:

Frit...............24 parts
Stone.............15½ parts
Flint.............6½ parts
White lead........31 parts

II.—Frit:

Stone	24	parts
Borax	53	parts
Lynn sand	40	parts
Feldspar	32	parts
Paris white	16	parts

To mill:

Frit	90	parts
Stone	30	parts
White lead	90	parts
Flint	4	parts
Glass	2	parts

III.—Frit:

Stone	50	parts
Borax	40	parts
Flint	30	parts
Flint glass	30	parts
Pearl barytes	10	parts

To mill:

Frit	160	parts
Red lead	30	parts
Enamel blue	$\frac{1}{2}$	part
Flint glass	2	parts

IV.—Frit:

Borax	100	parts
China clay	55	parts
Whiting	60	parts
Feldspar	75	parts

To mill:

Frit	200	parts
China clay	16	parts
White clay	$3\frac{1}{2}$	parts
Stone	3	parts
Flint	2	parts

V.—Frit:

Stone	40	parts
Flint	25	parts
Niter	10	parts
Borax	20	parts
White lead	10	parts
Flint glass	40	parts

To mill:

Frit	145	parts
Stone	50	parts
Borax	16	parts
Flint	15	parts
Red lead	60	parts
Flint glass	8	parts

Earthenware.—Frit:

I.—

Flint	108	parts
China clay	45	parts
Paris white	60	parts
Borax	80	parts
Soda crystals	30	parts

To mill:

Frit	270	parts
Flint	20	parts

Paris white	15	parts
Stone	80	parts
White lead	65	parts

II.—Frit:

Flint	62	parts
China clay	30	parts
Paris white	38	parts
Boracic acid	48	parts
Soda crystals	26	parts

To mill:

Frit	230	parts
Stone	160	parts
Flint	60	parts
Lead	120	parts

III.—Frit:

Stone	56	parts
Paris white	55	parts
Flint	60	parts
China clay	20	parts
Borax	120	parts
Soda crystals	15	parts

To mill:

Frit	212	parts
Stone	130	parts
Flint	50	parts
Lead	110	parts

Stain as required.

IV.—Frit:

Stone	100	parts
Flint	44	parts
Paris white	46	parts
Borax	70	parts
Niter	10	parts

To mill:

Frit	200	parts
Stone	60	parts
Lead	80	parts

Pearl White Glaze.—Frit:

Flint	50	parts
Stone	100	parts
Paris white	20	parts
Borax	60	parts
Soda crystals	20	parts

To mill:

Frit	178	pounds
Lead	55	pounds
Stain	3	ounces

Opaque Glaze.—Frit:

Borax	74	parts
Stone	94	parts
Flint	30	parts
China clay	22	parts
Pearl ash	$5\frac{1}{2}$	parts

To mill:

Frit	175	parts
Lead	46	parts

Flint.............. 10 parts
Oxide of tin........ 12 parts
Flint glass......... 12 parts

Glaze for Granite.—Frit:

I.—Stone............ 100 parts
Flint............. 80 parts
China clay........ 30 parts
Paris white....... 30 parts
Feldspar.......... 40 parts
Soda crystals...... 40 parts
Borax............. 80 parts

To mill:

Frit............. 360 parts
Flint............. 50 parts
Stone............. 50 parts
Lead............. 80 parts

II.—Frit:

Borax............ 100 parts
Stone............. 50 parts
Flint............. 50 parts
Paris white....... 40 parts
China clay........ 20 parts

To mill:

Frit............. 210 parts
Stone............. 104 parts
Flint............. 64 parts
Lead............. 95 parts

Raw Glazes.—White:

I.—White lead......., 160 parts
Borax............ 32 parts
Stone............. 48 parts
Flint............. 52 parts

Stain with blue and grind.

II.—White lead........ 80 parts
Litharge.......... 60 parts
Boracic acid...... 40 parts
Stone............. 45 parts
Flint............. 50 parts

Treat as foregoing.

III.—White lead....... 100 parts
Borax............ 4 parts
Flint............. 11 parts
Cornwall stone.... 50 parts

IV.—Red lead........ 80 parts
Litharge.......... 60 parts
Tincal........... 40 parts
Stone............. 40 parts
Flint............. 52 parts

ROCKINGHAM GLAZES.

I.—Litharge.......... 50 parts
Stone............. 7½ parts
Red marl.......... 3 parts
Oxide of manganese 5 parts
Red oxide of iron... 1 part

II.—White lead........ 30 parts
Stone............. 3 parts
Flint............. 9 parts
Red marl.......... 3 parts
Manganese........ 5 parts

III.—Red lead.......... 20 parts
Stone............. 3 parts
Flint............. 2 parts
China clay........ 2 parts
Manganese........ 3 parts
Red oxide of iron... 1 part

Stoneware Bodies.—

Ball clay........,.... 14 parts
China clay........ 10 parts
Stone............. 8 parts

Ball clay........:.. 8 parts
China clay........ 5 parts
Flint............. 3 parts
Stone............. 4 parts

Ball clay......... 14 parts
China clay........ 11 parts
Flint............. 4 parts
Stone............. 5 parts
Feldspar.......... 4 parts

Cane marl......... 16 parts
China clay........ 10 parts
Stone............. 9 parts
Flint............. 5 parts

Glazes.—Hard glaze:

Stone............. 10 parts
Flint............. 5 parts
Whiting........... 1½ parts
Red lead.......... 10 parts

Hard glaze:

Feldspar.......... 25 parts
Flint............. 5 parts
Red lead.......... 15 parts
Plaster........... 1 part

Softer:

White lead........ 13 parts
Flint glass........ 10 parts
Feldspar.......... 18 parts
Stone............. 3 parts
Whiting........... 1½ parts

Best:

Feldspar.......... 20 parts
Flint glass........ 14 parts
White lead........ 14 parts
Stone............. 3 parts
Oxide of zinc...... 3 parts
Whiting........... 1½ parts
Plaster........... 1 part

Rockingham Bodies.—

Ball clay.......... 20 parts
China clay........ 13 parts
Flint............. 7 parts
Stone............. 1 part

Cane marl......... 22 parts
China clay........ 15 parts
Flint............. 8 parts
Feldspar.......... 1 part

Glazes.—

I.—Red lead . . . .   60 parts
Stone . . . .   8 parts
Red clay . . . .   3 parts
Best manganese . . .   5 parts

II.—White lead . . . .   60 parts
Feldspar . . . .   6 parts
Flint . . . .   16 parts
Red clay . . . .   6 parts
Manganese . . . .   12 parts

III.—Red lead . . . . 100 parts
Stone . . . .   15 parts
Flint . . . .   10 parts
China clay . . . .   10 parts
Manganese . . . .   40 parts
Crocus martis . . . . .   2 parts

IV.—Litharge . . . . 100 parts
Feldspar . . . .   14 parts
China clay . . . .   20 parts
Manganese . . . .   40 parts
Oxide of iron . . . . . .   2 parts

Jet.—Procure some first-class red marl, add water, and, by passing through a fine lawn, make it into a slip, and dip the ware therein.

When fired use the following:

Glaze.—

Stone 60   parts
Flint 30   parts
Paris white . . . .   7½   parts
Red lead 140   parts

One part mazarine blue stain to 10 parts glaze.

Mazarine Blue Stain.—

Oxide of cobalt . . . . . 10 parts
Paris white . . . .   9 parts
Sulphate barytes . . . .   1 part

Calcine.

Another Process Body.—

Ball clay . . . . 16 parts
China clay . . . . 12 parts
Flint clay . . . .   9 parts
Stone clay . . . .   6 parts
Black stain . . . .   7 parts

Glaze.—

Litharge . . . . 70 parts
Paris white . . . .   3 parts
Flint . . . . 12 parts
Stone . . . . 30 parts
Black stain . . . . 20 parts

Black Stain.—

Chromate of iron . . . 12 parts
Oxide of nickel . . . . .   2 parts
Oxide of tin . . . .   2 parts
Carbonate of cobalt.   5 parts
Oxide of manganese .   2 parts

Calcine and grind.

Blue Stains.—

I.—Oxide of cobalt . . . . .   2½ parts
Oxide of zinc . . . . . .   7½ parts
Stone . . . .   7½ parts

Fire this very hard.

II.—Zinc . . . . 6 pounds
Flint . . . . 4 pounds
China clay . . . . 4 pounds
Oxide of cobalt . . . . . 5 ounces

Hard fire.

III.—Whiting . . . . 3¾ parts
Flint . . . . 3¾ parts
Oxide of cobalt . . . . . 2½ parts

Glost fire.

Turquoise Stain.—

Prepared cobalt . . . . 1½ parts
Oxide of zinc . . . . 6 parts
China clay . . . . 6 parts
Carbonate of soda . . 1 part

Hard fire.

MATERIALS:

Tin Ash.—

Old lead . . . . 4 parts
Grain tin . . . . 2 parts

Melt in an iron ladle, and pour out in water, then spread on a dish, and calcine in glost oven with plenty of air.

Oxide of Tin.—

Granulated tin . . . . . 5 pounds
Niter . . . . ½ pound

Put on saucers and fire in glost oven.

Oxide of Chrome is made by mixing powdered bichromate of potash with sulphur as follows:

Potash . . . . 6 parts
Flowers of sulphur . . 1 part

Put in saggar, inside kiln, so that fumes are carried away, and place 4 or 5 pieces of red-hot iron on the top so as to ignite it. Leave about 12 hours, then pound very fine, and put in saggar again. Calcine in hard place of biscuit oven. Wash this until the water is quite clear, and dry for use.

Production of Luster Colors on Porcelain and Glazed Pottery.—The luster colors are readily decomposed by acids and atmospheric influences, because they do not contain, in consequence of the low baking temperature, enough silicic acid to form resistive compounds. In order to attain this, G. Alefeld has patented a process according to which such compounds are added to the luster preparations as leave behind after the burning an acid which transforms the luster preparation into more resisting compounds. In this connection the admixture of such bodies has been found advantageous, as they form phosphides with the metallic oxides of the lusters after the burning. These phosphides are especially fitted for the production of saturated resisting compounds, not only on account of their insolubility in water, but also on account of their colorings. Similarly titanic, molybdic, tungstic, and vanadic compounds may be produced. The metallic phosphates produced by the burning give a luster coating which, as regards gloss, is not inferior to the nonsaturated metallic oxides, while it materially excels them in power of resistance. Since the lusters to be applied are used dissolved in essential oils, it is necessary to make the admixture of phosphoric substance also in a form soluble in essential oils. For the production of this admixture the respective chlorides, preeminently phosphoric chloride, are suitable. They are mixed with oil of lavender in the ratio of 1 to 5, and the resulting reaction product is added to the commercial metallic oxide luster, singly or in conjunction with precious metal preparations (glossy gold, silver, platinum, etc.) in the approximate proportion of 5 to 1. Then proceed as usual. Instead of the chlorides, nitrates and acetates, as well as any readily destructible organic compounds, may also be employed, which are entered into fusing rosin or rosinous liquids.

Metallic Luster on Pottery.—According to a process patented in Germany, a mixture is prepared from various natural or artificial varieties of ocher, to which 25–50 per cent of finely powdered more or less metalliferous or sulphurous coal is added. The mass treated in this manner is brought together in saggars with finely divided organic substances, such as sawdust, shavings, wood-wool, cut straw, etc., and subjected to feeble red heat. After the heating the material is taken out. The glazings now exhibit that thin but stable metallic color which is governed by the substances used. Besides coal, salts and oxides of silver, cobalt, cadmium, chrome iron, nickel, manganese, copper, or zinc may be employed. The color-giving layer is removed by washing or brushing, while the desired color is burned in and remains. In this manner handsome shades can be produced.

Metallic Glazes on Enamels.—The formulas used by the Arabs and their Italian successors are partly disclosed in manuscripts in the British and South Kensington Museums; two are given below:

	Arab	Italian
Copper sulphide	26.87	24.74
Silver sulphide	1.15	1.03
Mercury sulphide	....	24.74
Red ocher	71.98	49.49

These were ground with vinegar and applied with the brush to the already baked enamel. A great variety of iridescent and metallic tones can be obtained by one or the other, or a mixture of the following formulas:

	I	II	III	IV	V	VI
Copper carbonate	30	..	..	28	..	95
Copper oxalate	..	..	..	..	5	..
Copper sulphide	..	..	20	..	..	..
Silver carbonate	..	3	..	2	1	5
Bismuth subnitrate	..	12	..	..	10	..
Stannous oxide	..	..	25	..	..	..
Red ocher	70	85	55	70	84	..

Silver chloride and yellow ocher may be respectively substituted for silver carbonate and red ocher. The ingredients, ground with a little gum tragacanth and water, are applied with a brush to enamels melting about 1814° F., and are furnaced at 1202° F. in a reducing atmosphere. After cooling the ferruginous deposit is rubbed off, and the colors thus brought out.

Sulphur, free or combined, is not necessary, cinnabar has no action, ocher may be dispensed with, and any organic gummy matter may be used instead of vinegar, and broom is not needed in the furnace. The intensity and tone of the iridescence depend on the duration of the reduction, and the nature of the enamel. Enamels containing a coloring base—copper, iron, antimony, nickel—especially in presence of tin, give the best results.

To Toughen China.—To toughen china or glass place the new article in cold water, bring to boil gradually, boil for 4 hours, and leave standing in the water till cool. Glass or china toughened in this way will never crack with hot water.

How to Tell Pottery and Porcelain.—The following simple test will serve: Hold the piece up to the light, and if it can be seen through—that is, if it is translucent—it is porcelain. Pottery is opaque, and not so hard and white as porcelain. The main differences in the manufacture of stoneware, earthenware, and porcelain are due to the ingredients used, to the way they are mixed, and to the degree of heat to which they are subjected in firing. Most of the old English wares found in this country are pottery or semichina, although the term china is commonly applied to them all.

Cheese

Manufacture.—The process of cheese making is one which is eminently interesting and scientific, and which, in every gradation, depends on principles which chemistry has developed and illustrated. When a vegetable or mineral acid is added to milk, and heat applied, a coagulum is formed, which, when separated from the liquid portion, constitutes cheese. Neutral salts, earthy and metallic salts, sugar, and gum arabic, as well as some other substances, also produce the same effect; but that which answers the purpose best, and which is almost exclusively used by dairy farmers, is rennet, or the mucous membrane of the last stomach of the calf. Alkalies dissolve this curd at a boiling heat, and acids again precipitate it. The solubility of casein in milk is occasioned by the presence of the phosphates and other salts of the alkalies. In fresh milk these substances may be readily detected by the property it possesses of restoring the color of reddened litmus paper. The addition of an acid neutralizes the alkali, and so precipitates the curd in an insoluble state. The philosophy of cheese making is thus expounded by Liebig:

"The acid indispensable to the coagulation of milk is not added to the milk in the preparation of cheese, but it is formed in the milk at the expense of the milk-sugar present. A small quantity of water is left in contact with a small quantity of a calf's stomach for a few hours, or for a night; the water absorbs so minute a portion of the mucous membrane as to be scarcely ponderable; this is mixed with milk; its state of transformation is communicated (and this is a most important circumstance) not to the cheese, but to the milk-sugar, the elements of which transpose themselves into lactic acid, which neutralizes the alkalies, and thus causes the separation of the cheese. By means of litmus paper the process may be followed and observed through all its stages; the alkaline reaction of the milk ceases as soon as the coagulation begins. If the cheese is not immediately separated from the whey, the formation of lactic acid continues, the fluid turns acid, and the cheese itself passes into a state of decomposition.

"When cheese-curd is kept in a cool place a series of transformation takes place, in consequence of which it assumes entirely new properties; it gradually becomes semi-transparent, and more or less soft, throughout the whole mass; it exhibits a feebly acid reaction, and develops the characteristic caseous odor. Fresh cheese is very sparingly soluble in water, but after having been left to itself for two or three years it becomes (especially if all the fat be previously removed) almost completely soluble in cold water, forming with it a solution which, like milk, is coagulated by the addition of the acetic or any mineral acid. The cheese, which whilst fresh is insoluble, returns during the maturation, or ripening, as it is called, to a state similar to that in which it originally existed in the milk. In those English, Dutch, and Swiss cheeses which are nearly inodorous, and in the superior kinds of French cheese, the casein of the milk is present in its unaltered state.

"The odor and flavor of the cheese is due to the decomposition of the butter; the non-volatile acids, the margaric and oleic acids, and the volatile butyric acid, capric and caproic acids are liberated in consequence of the decomposition of glycerine. Butyric acid imparts to cheese its characteristic caseous odor, and the differences in its pungency or aromatic flavor depend upon the proportion of free butyric, capric, and caproic acids present. In the cheese of certain dairies and districts, valerianic acid has been detected along with the other acids just referred to. Messrs Jljenjo and Laskowski found this acid in the cheese of Limbourg, and M. Bolard in that of Roquefort.

"The transition of the insoluble into soluble casein depends upon the decomposition of the phosphate of lime by the margaric acid of the butter; margarate of lime is formed, whilst the phosphoric acid combines with the casein, forming a compound soluble in water.

"The bad smell of inferior kinds of cheese, especially those called meager or poor cheeses, is caused by certain fetid products containing sulphur, and which are formed by the decomposition or putrefaction of the casein. The alteration which the butter undergoes (that is, in becoming rancid), or which occurs in the milk-sugar still present, being transmitted to the casein, changes both the composition of the latter substance and its nutritive qualities.

"The principal conditions for the preparation of the superior kinds of cheese (other obvious circumstances being of course duly regarded) are a careful removal of the whey, which holds the milk-sugar in solution, and a low temperature during the maturation or ripening of the cheese."

Cheese differs vastly in quality and flavor according to the method employed in its manufacture and the richness of the milk of which it is made. Much depends upon the quantity of cream it contains, and, consequently, when a superior quality of cheese is desired cream is frequently added to the curd. This plan is adopted in the manufacture of Stilton cheese and others of a like description. The addition of a pound or two of butter to the curd for a middling size cheese also vastly improves the quality of the product. To insure the richness of the milk, not only should the cows be properly fed, but certain breeds chosen. Those of Alderney, Cheddar, Cheshire, etc., have been widely preferred.

The materials employed in making cheese are milk and rennet. Rennet is used either fresh or salted and dried; generally in the latter state. The milk may be of any kind, according to the quality of the cheese required. Cows' milk is that generally employed, but occasionally ewes' milk is used; and sometimes, though more rarely, that from goats.

In preparing his cheese the dairy farmer puts the greater portion of the milk into a large tub, to which he adds the remainder, sufficiently heated to raise the temperature to that of new milk. The whole is then whisked together, the rennet or rennet liquor added, and the tub covered over. It is now allowed to stand until completely "turned," when the curd is gently struck down several times with the skimming dish, after which it is allowed to subside. The vat, covered with cheese cloth, is next placed on a "horse" or "ladder" over the tub, and filled with curd by means of the skimmer, care being taken to allow as little as possible of the oily particles or butter to run back with the whey. The curd is pressed down with the hands, and more added as it sinks. This process is repeated until the curd rises to about two inches above the edge. The newly formed cheese, thus partially separated from the whey, is now placed in a clean tub, and a proper quantity of salt, as well as of annotta, added when that coloring is used, after which a board is placed over and under it, and pressure applied for about 2 or 3 hours. The cheese is next turned out and surrounded by a fresh cheese cloth, and then again submitted to pressure in the cheese press for 8 or 10 hours, after which it is commonly removed from the press, salted all over, and again pressed for 15 to 20 hours. The quality of the cheese especially depends on this part of the process, as if any of the whey is left in the cheese it rapidly becomes bad-flavored. Before placing it in the press the last time the common practice is to pare the edges smooth and sightly. It now only remains to wash the outside of the cheese in warm whey or water, to wipe it dry, and to color it with annotta or reddle, as is usually done.

The storing of the newly made cheese is the next point that engages the attention of the maker and wholesale dealer. The same principles which influence the maturation or ripening of fermented liquors also operate here. A cool cellar, neither damp nor dry, and which is uninfluenced by change of weather or season, is commonly regarded as the best for the purpose. If possible, the temperature should on no account be permitted to exceed 50° or 52° F. at any portion of the year. An average of about 45° F. is preferable when it can be procured. A place exposed to sudden changes of temperature is as unfit for storing cheese as it is for storing beer. "The quality of Roquefort cheese, which is prepared from sheep's milk, and is very excellent, depends exclusively upon the places where the cheeses are kept after pressing and during maturation. These are cellars, communicating with mountain grottoes and caverns which are kept constantly cool, at about 41° to 42° F., by currents of air from clefts in the mountains. The value of these cellars as storehouses varies with their property of maintaining an equable and low temperature."

It will thus be seen that very slight differences in the materials, in the preparation, or in storing of the cheese, materially influence the quality and flavor of this article. The richness of the milk; the addition to or subtraction of cream from the milk; the separation of the curd from the whey with or without compression; the salting of the curd; the collection of the curd, either whole or broken, before pressing; the addition of coloring matter, as annotta or saffron, or of flavoring; the place and method of storing; and the length of time allowed for maturation, all tend to alter the taste and odor of the cheese in some or other particular, and that in a way readily perceptible to the palate of the connoisseur. No other alimentary substance appears to be so seriously affected by slight variations in the quality of the materials from which it is made, or by such apparently trifling differences in the methods of preparing.

The varieties of cheese met with in commerce are very numerous, and differ greatly from each other in richness, color, and flavor. These are commonly distinguished by names indicative of the places in which they have been manufactured, or of the quality of the materials from which they have been prepared. Thus we have Dutch, Gloucester, Stilton, skimmed milk, raw milk, cream, and other cheeses; names which explain themselves. The following are the principal varieties:

American Factory.—Same as Cheddar.

Brickbat.—Named from its form; made, in Wiltshire, of new milk and cream.

Brie.—A soft, white, cream cheese of French origin.

Cheddar.—A fine, spongy kind of cheese, the eyes or vesicles of which contain a rich oil; made up into round, thick cheeses of considerable size (150 to 200 pounds).

Cheshire.—From new milk, without skimming, the morning's milk being mixed with that of the preceding evening's, previously warmed, so that the whole may be brought to the heat of new milk. To this the rennet is added, in less quantity than is commonly used for other kinds of cheese. On this point much of the flavor and mildness of the cheese is said to depend. A piece of dried rennet, of the size of a half-dollar put into a pint of water over night, and allowed to stand until the next morning, is sufficient for 18 or 20 gallons of milk; in large, round, thick cheeses (100 to 200 pounds each). They are generally solid, homogeneous, and dry, and friable rather than viscid.

Cottenham.—A rich kind of cheese, in flavor and consistence not unlike Stilton, from which, however, it differs in shape, being flatter and broader than the latter.

Cream.—From the "strippings" (the last of the milk drawn from the cow at each milking), from a mixture of milk and cream, or from raw cream only, according to the quality desired. It is usually made in small oblong, square, or rounded cakes, a general pressure only (that of a 2- or 4-pound weight) being applied to press out the whey. After 12 hours it is placed upon a board or wooden trencher, and turned every day until dry. It ripens in about 3 weeks. A little salt is generally added, and frequently a little powdered lump sugar.

Damson.—Prepared from damsons boiled with a little water, the pulp passed through a sieve, and then boiled with about one-fourth the weight of sugar, until the mixture solidifies on cooling; it is next poured into small tin molds previously dusted out with sugar. Cherry cheese, gooseberry cheese, plum cheese, etc., are prepared in the same way, using the respective kinds of fruit. They are all very agreeable candies or confections.

Derbyshire.—A small, white, rich variety, very similar to Dunlop cheese.

Dunlop.—Rich, white, and buttery; in round forms, weighing from 30 to 60 pounds.

Dutch (Holland).—Of a globular form, 5 to 14 pounds each. Those from Edam are very highly salted; those from Gouda less so.

Emmenthaler.—Same as Gruyère.

Gloucester.—Single Gloucester, from milk deprived of part of its cream; double Gloucester, from milk retaining the whole of the cream. Mild tasted, semibuttery consistence, without being friable; in large, round, flattish forms.

Green or Sage.—From milk mixed with the juice of an infusion or decoction of sage leaves, to which marigold flowers and parsley are frequently added.

Gruyère.—A fine kind of cheese made in Switzerland, and largely consumed on the Continent. It is firm and dry, and exhibits numerous cells of considerable magnitude.

Holland.—Same as Dutch.

Leguminous.—The Chinese prepare an actual cheese from peas, called taofoo, which they sell in the streets of Canton. The paste from steeped ground peas is boiled, which causes the starch to dissolve with the casein; after straining the liquid it is coagulated by a solution of gypsum; this coagulum is worked up like sour milk, salted, and pressed into molds.

Limburger.—A strong variety of cheese, soft and well ripened.

Lincoln.—From new milk and cream; in pieces about 2 inches thick. Soft, and will not keep over 2 or 3 months.

Neufchâtel.—A much-esteemed variety of Swiss cheese; made of cream, and weighs about 5 or 6 ounces.

Norfolk.—Dyed yellow with annotta or saffron; good, but not superior; in cheeses of 30 to 50 pounds.

Parmesan.—From the curd of skimmed milk, hardened by a gentle heat. The rennet is added at about 120°, and an hour afterwards the curdling milk is set on a slow fire until heated to about 150° F., during which the curd separates in small lumps. A few pinches of saffron are then thrown in. About a fortnight after making the outer crust is cut off, and the new surface varnished with linseed oil, and one side colored red.

Roquefort.—From ewes' milk; the best prepared in France. It greatly resembles Stilton, but is scarcely of equal richness or quality, and possesses a peculiar pungency and flavor.

Roquefort, Imitation.—The gluten of wheat is kneaded with a little salt and a small portion of a solution of starch, and made up into cheeses. It is said that this mixture soon acquires the taste, smell, and unctuosity of cheese, and when kept a certain time is not to be distinguished from the celebrated Roquefort cheese, of which it possesses all the peculiar pungency. By slightly varying the process other kinds of cheese may be imitated.

Sage.—Same as green cheese.

Slipcoat or Soft.—A very rich, white cheese, somewhat resembling butter; for present use only.

Stilton.—The richest and finest cheese made in England. From raw milk to which cream taken from other milk is added; in cheeses generally twice as high as they are broad. Like wine, this cheese is vastly improved by age, and is therefore seldom eaten before it is 2 years old. A spurious appearance of age is sometimes given to it by placing it in a warm, damp cellar, or by surrounding it with masses of fermenting straw or dung.

Suffolk.—From skimmed milk; in round, flat forms, from 24 to 30 pounds each. Very hard and horny.

Swiss.—The principal cheeses made in Switzerland are the Gruyère, the Neufchâtel, and the Schabzieger or green cheese. The latter is flavored with melitot.

Westphalian.—Made in small balls or rolls of about 1 pound each. It derives its peculiar flavor from the curd being allowed to become partially putrid before being pressed. In small balls or rolls of about 1 pound each.

Wiltshire.—Resembles Cheshire or Gloucester. The outside is painted with reddle or red ocher or whey.

York.—From cream. It will not keep.

We give below the composition of some of the principal varieties of cheese:

	Cheddar	Double Gloucester	Skim
Water	36.64	35.61	43.64
Casein	23.38	21.76	45.64
Fatty matter	35.44	38.16	5.76
Mineral matter	4.54	4.47	4.96
	100.00	100.00	100.00

	Stilton	Cotherstone
Water	32.18	38.28
Butter	37.36	30.89
Casein	24.31	23.93
Milk, sugar, and extractive matters	2.22	3.70
Mineral matter	3.93	3.20
	100.00	100.00

	Gruyère (Swiss)	Ordinary Dutch
Water	40.00	36.10
Casein	31.50	29.40
Fatty matter	24.00	27.50
Salts	3.00	.90
Non-nitrogenous organic matter and loss	1.50	6.10
	100.00	100.00

When a whole cheese is cut, and the consumption small, it is generally found to become unpleasantly dry, and to lose flavor before it is consumed. This is best prevented by cutting a sufficient quantity for a few days' consumption from the cheese, and keeping the remainder in a cool place, rather damp than dry, spreading a thin film of butter over the fresh surface, and covering it with a cloth or pan to keep off the dirt. This removes the objection existing in small families against purchasing a whole cheese at a time. The common practice of buying small quantities of cheese should be avoided, as not only a higher price is paid for any given quality, but there is little likelihood of obtaining exactly the same flavor twice running. Should cheese become too dry to be agreeable, it may be used for stewing, or for making grated cheese, or Welsh rarebits.

Goats' Milk Cheese.—Goats' milk cheese is made as follows: Warm 20 quarts of milk and coagulate it with rennet, either the powder or extract. Separate the curds from the whey in a colander. After a few days the dry curd may be shaped into larger or smaller cheeses, the former only salted, the latter containing salt and caraway seed. The cheeses must be turned every day, and sprinkled with salt, and any mold removed. After a few days they may be put away on shelves to ripen, and left for several weeks. Pure goat's milk cheese should be firm and solid all the way through. Twenty quarts of milk will make about 4 pounds of cheese.

CHEESE COLORANT:
See Food.

CHEMICAL GARDENS:
See Gardens, Chemical.

CHERRY BALSAM:
See Balsam.

CHERRY CORDIAL:
See Wines and Liquors.

Chewing Gums

Manufacture.—The making of chewing gum is by no means the simple operation which it seems to be. Much experience in manipulation is necessary to succeed, and the published formulas can at best serve as a guide rather than as something to be absolutely and blindly followed. Thus, if the mass is either too hard or soft, change the proportions until it is right; often it will be found that different purchases of the same article will vary in their characteristics when worked up. But given a basis, the manufacturer can flavor and alter to suit himself. The most successful manufacturers attribute their success to the employment of the most approved machinery and the greatest attention to details. The working formulas and the processes of these manufacturers are guarded as trade secrets, and aside from publishing general formulas, little information can be given.

Chicle gum is purified by boiling with water and separating the foreign matter. Flavorings, pepsin, sugar, etc., are worked in under pressure by suitable machinery. Formula:

I.—Gum chicle........ 1 pound
　　Sugar............. 2 pounds
　　Glucose............ 1 pound
　　Caramel butter..... 1 pound

First mash and soften the gum at a gentle heat. Place the sugar and glucose in a small copper pan; add enough water to dissolve the sugar; set on a fire and cook to 244° F.; lift off the fire; add the caramel butter and lastly the gum; mix well into a smooth paste; roll out on a smooth marble, dusting with finely powdered sugar, run through sizing machine to the proper thickness, cut into strips, and again into thin slices.

II.—Chicle............ 6 ounces
　　Paraffine.......... 2 ounces
　　Balsam of Tolu.... 2 drachms
　　Balsam of Peru... 1 drachm
　　Sugar............ 20 ounces
　　Glucose.......... 8 ounces
　　Water............ 6 ounces
　　Flavoring, enough.

Triturate the chicle and balsams in water, take out and add the paraffine, first heated. Boil the sugar, glucose, and water together to what is known to confectioners as "crack" heat, pour the syrup over the oil slab and turn into it the gum mixture, which will make it tough and plastic. Add any desired flavor.

III.—Gum chicle....... 122 parts
　　Paraffine......... 42 parts
　　Balsam of Tolu.... 4 parts
　　Sugar............ 384 parts
　　Water............ 48 parts

Dissolve the sugar in the water by the aid of heat and pour the resultant syrup on an oiled slab. Melt the gum, balsam, and paraffine together and pour on top of the syrup, and work the whole up together.

IV.—Gum chicle....... 240 parts
　　White wax........ 64 parts
　　Sugar............ 640 parts
　　Glucose.......... 128 parts
　　Water............ 192 parts
　　Balsam of Peru... 4 parts
　　Flavoring matter, enough.

Proceed as indicated in II.

V.—Balsam of Tolu...... 4 parts
　　Benzoin........... 1 part
　　White wax......... 1 part
　　Paraffine.......... 1 part
　　Powdered sugar. ... 1 part

Melt together, mix well, and roll into sticks of the usual dimensions.

Mix, and, when sufficiently cool, roll out into sticks or any other desirable form.

Spruce Chewing Gum.—

Spruce gum	20 parts
Chicle	20 parts
Sugar, powdered	60 parts

Melt the gums separately, mix while hot, and immediately add the sugar, a small portion at a time, kneading it thoroughly on a hot slab. When completely incorporated remove to a cold slab, previously dusted with powdered sugar, roll out at once into sheets, and cut into sticks. Any desired flavor or color may be added to or incorporated with the sugar.

CHICKEN-COOP APPLICATION:
See Insecticides.

CHICKEN DISEASES AND THEIR REMEDIES:
See Veterinary Formulas.

CHICORY, TESTS FOR:
See Foods.

CHILBLAINS:
See Ointments.

CHILBLAIN SOAP:
See Soap.

CHILDREN, DOSES FOR:
See Doses.

CHILLS, BITTERS FOR:
See Wines and Liquors.

CHINA CEMENTS:
See Adhesives and Lutes.

CHINA:
See Ceramics.

CHINA, TO REMOVE BURNED LETTERS FROM:
See Cleaning Preparations and Methods, under Miscellaneous Methods.

CHINA REPAIRING:
See Porcelain.

CHINA RIVETING.

China riveting is best left to practical men, but it can be done with a drill made from a splinter of a diamond fixed on a handle. If this is not to be had, get a small three-cornered file, harden it by placing it in the fire till red hot, and then plunging it in cold water. Next grind the point on a grindstone and finish on an oilstone. With the point pick out the place to be bored, taking care to do it gently for fear of breaking the article. In a little while a piece will break off, then the hole can easily be made by working the point round. The wire may then be passed through and fastened. A good cement may be made from 1 ounce of grated cheese, $\frac{1}{2}$ ounce of finely powdered quicklime, and white of egg sufficient to make a paste. The less cement applied the better, using a feather to spread it over the broken edge.

CHLORIDES, PLATT'S:
See Disinfectants.

CHLORINE-PROOFING:
See Acid-Proofing.

CHOCOLATE.

Prepare 1,000 parts of finished cacao and 30 parts of fresh cacao oil, in a warmed, polished, iron mortar, into a liquid substance, add to it 800 parts of finely powdered sugar, and, after a good consistency has been reached, 60 parts of powdered iron lactate and 60 parts of sugar syrup, finely rubbed together. Scent with 40 parts of vanilla sugar. Of this mass weigh out tablets of 125 parts into the molds.

Coating Tablets with Chocolate.—If a chocolate which is free from sugar be placed in a dish over a water bath, it will melt into a fluid of proper consistence for coating tablets. No water must be added. The coating is formed by dipping the tablets. When they are sufficiently hardened they are laid on oiled paper to dry.

CHOCOLATE CASTOR-OIL LOZENGES:
See Castor Oil.

CHOCOLATE CORDIAL:
See Wines and Liquors.

CHOCOLATE EXTRACTS:
See Essences and Extracts.

CHOCOLATE SODA WATER:
See Beverages.

CHOKING IN CATTLE:
See Veterinary Formulas.

CHOLERA REMEDIES:

Sun Cholera Mixture.—

Tincture of opium	1 part
Tincture of capsicum	1 part
Tincture of rhubarb	1 part
Spirit of camphor	1 part
Spirit of peppermint	1 part

Squibb's Diarrhea Mixture.—

Tincture opium	40 parts
Tincture capsicum	40 parts
Spirit camphor	40 parts
Chloroform	15 parts
Alcohol	65 parts

Aromatic Rhubarb.—

Cinnamon, ground..	8 parts
Rhubarb..........	8 parts
Calumba..........	4 parts
Saffron...........	1 part
Powdered opium....	2 parts
Oil peppermint.....	5 parts
Alcohol, q. s. ad....	100 parts

Macerate the ground drugs with 75 parts alcohol in a closely covered percolator for several days, then allow percolation to proceed, using sufficient alcohol to obtain 95 parts of percolate. In percolate dissolve the oil of peppermint.

Rhubarb and Camphor.—

Tincture capsicum...	2 ounces
Tincture opium......	2 ounces
Tincture camphor....	3 ounces
Tincture catechu.....	4 ounces
Tincture rhubarb....	4 ounces
Spirit peppermint....	4 ounces

Blackberry Mixture.—

Fluid extract blackberry root........	2	pints
Fluid ginger, soluble.	5½	ounces
Fluid catechu.......	5⅓	ounces
Fluid opium for tincture.............	160	minims
Brandy............	8	ounces
Sugar.............	4	pounds
Essence cloves......	256	minims
Essence cinnamon..	256	minims
Chloroform........	128	minims
Alcohol (25 per cent), q. s. ad..........	1	gallon

CHOWCHOW:
See Condiments.

CHROME YELLOW, TEST FOR:
See Pigments.

CHROMIUM GLUE:
See Adhesives.

CHROMO MAKING.

The production of chromo pictures requires a little skill. Practice is necessary. The glass plate to be used should be washed off with warm water, and then laid in a 10 per cent solution of nitric acid. After one hour, wash with clean, cold water, dry with a towel, and polish the plate with good alcohol on the inside—hollow side—until no finger marks or streaks are visible. This is best ascertained by breathing on the glass; the breath should show an even blue surface on the glass.

Coat the unmounted photograph to be colored with benzine by means of wad-

ding, but without pressure, so that the retouching of the picture is not disturbed. Place 2 tablets of ordinary kitchen gelatin in 8½ ounces of distilled or pure rain water, soak for an hour, and then heat until the gelatin has completely dissolved. Pour this warm solution over the polished side of the glass, so that the liquid is evenly distributed. The best way is to pour the solution on the upper right-hand corner, allowing it to flow into the left-hand corner, from there to the left below and right below, finally letting the superfluous liquid run off. Take the photograph, which has been previously slightly moistened on the back, lay it with the picture side on the gelatin-covered plate, centering it nicely, and squeeze out the excess gelatin solution gently, preferably by means of a rubber squeegee. Care must be taken, however, not to displace the picture in this manipulation, as it is easily spoiled.

The solution must never be allowed to boil, since this would render the gelatin brittle and would result in the picture, after having been finished, cracking off from the glass in a short time. When the picture has been attached to the glass plate without blisters (which is best observed from the back), the edge of the glass is cleansed of gelatin, preferably by means of a small sponge and lukewarm water, and the plate is allowed to dry over night.

When the picture and the gelatin are perfectly dry, coat the back of the picture a few times with castor oil until it is perfectly transparent; carefully remove the oil without rubbing, and proceed with the painting, which is best accomplished with good, not over-thick oil colors. The coloring must be observed from the glass side, and for this reason the small details, such as eyes, lips, beard, and hair, should first be sketched in. When the first coat is dry the dress and the flesh tints are painted. The whole surface may be painted over, and it is not necessary to paint shadows, as these are already present in the picture, and consequently show the color through in varying strength.

When the coloring has dried, a second glass plate should be laid on for protection, pasting the two edges together with narrow strips of linen.

Cider

To Make Cider.—Pick the apples off the tree by hand. Every apple before going into the press should be carefully wiped. As soon as a charge of apples is ground, remove the pomace and put in a cask with a false bottom and a strainer beneath it, and a vessel to catch the drainage from pomace. As fast as the juice runs from the press place it in clean, sweet, open tubs or casks with the heads out and provide with a faucet, put in about two inches above bottom. The juice should be closely watched and as soon as the least sign of fermentation appears (bubbles on top, etc.) it should be run off into casks prepared for this purpose and placed in a moderately cool room. The barrels should be entirely filled, or as near to the bunghole as possible. After fermentation is well under way the spume or foam should be scraped off with a spoon several times a day. When fermentation has ceased the cider is racked off into clean casks, filled to the bunghole, and the bung driven in tightly. It is now ready for use or for bottling.

Champagne Cider.—I.—To convert ordinary cider into champagne cider, proceed as follows: To 100 gallons of good cider add 3 gallons of strained honey (or 24 pounds of white sugar will answer), stir in well, tightly bung, and let alone for a week. Clarify the cider by adding a half gallon of skimmed milk, or 4 ounces of gelatin dissolved in sufficient hot water and add 4 gallons of proof spirit. Let stand 3 days longer, then syphon off, bottle, cork, and tie or wire down. Bunging the cask tightly is done in order to induce a slow fermentation, and thus retain in the cider as much carbonic acid as possible.

II.—Put 10 gallons of old and clean cider in a strong and iron-bound cask, pitched within (a sound beer cask is the very thing), and add and stir in well 40 ounces of simple syrup. Add 5 ounces of tartaric acid, let dissolve, then add $7\frac{1}{2}$ ounces sodium bicarbonate in powder. Have the bung ready and the moment the soda is added put it in and drive it home. The cider will be ready for use in a few hours.

Cider Preservative.—I.—The addition of 154 grains of bismuth subnitrate to 22 gallons of cider prevents, or materially retards, the hardening of the beverage on exposure to air; moreover, the bismuth salt renders alcoholic fermentation more complete.

II.—Calcium sulphite (sulphite of lime) is largely used to prevent fermentation in cider. About $\frac{1}{8}$ to $\frac{1}{4}$ of an ounce of the sulphite is required for 1 gallon of cider. It should first be dissolved in a small quantity of cider, then added to the bulk, and the whole agitated until thoroughly mixed. The barrel should then be bunged and allowed to stand for several days, until the action of the sulphite is exerted. It will preserve the sweetness of cider perfectly, but care should be taken not to add too much, as that would impart a slight sulphurous taste.

Artificial Ciders.—To 25 gallons of soft water add 2 pounds of tartaric acid, 25 or 30 pounds of sugar, and a pint of yeast; put in a warm place, and let ferment for 15 days, then add the flavoring matter to suit taste. The various fruit ethers are for sale at any wholesale drug house.

Bottling Sweet Cider.—Champagne quarts are generally used for bottling cider, as they are strong and will stand pressure, besides being a convenient size for consumers. In making cider champagne the liquor should be clarified and bottled in the sweet condition, that is to say, before the greater part of the sugar which it contains has been converted into alcohol by fermentation. The fermentation continues, to a certain extent, in the bottle, transforming more of the sugar into alcohol, and the carbonic acid, being unable to escape, is dissolved in the cider and produces sparkling.

The greater the quantity of sugar contained in the liquor, when it is bottled, the more complete is its carbonation by the carbonic-acid gas, and consequently the more sparkling it is when poured out. But this is true only within certain limits, for if the production of sugar is too high the fermentation will be arrested.

To make the most sparkling cider the liquor is allowed to stand for three, four, five, or six weeks, during which fermentation proceeds. The time varies according to the nature of the apples, and also to the temperature; when it is very warm the first fermentation is usually completed in 7 days.

Before bottling, the liquid must be fined, and this is best done with catechu dissolved in cold cider, 2 ounces of catechu to the barrel of cider. This is well stirred and left to settle for a few days.

The cider at this stage is still sweet, and it is a point of considerable nicety not to carry the first fermentation too far. The bottle should not be quite filled, so as to allow more freedom for the carbonic-acid gas which forms.

When the bottles have been filled,

corked, and wired down, they should be placed in a good cellar, which should be dry, or else the cider will taste of the cork. The bottles should not be laid for four or five weeks, or breakage will ensue. When they are being laid they should be placed on laths of wood or on dry sand; they should never be allowed on cold or damp floors.

Should the cider be relatively poor in sugar, or if it has been fermented too far, about 1 ounce of powdered loaf sugar can be added to each bottle, or else a measure of sugar syrup before pouring in the cider.

Imitation Cider.—

I.—A formula for an imitation cider is as follows:

Rain water.........	100 gallons
Honey, unstrained..	6 gallons
Catechu, powdered.	3 ounces
Alum, powdered....	5 ounces
Yeast (brewer's preferably).........	2 pints

Mix and put in a warm place to ferment. Let ferment for about 15 days; then add the following, stirring well in:

Bitter almonds, crushed	8 ounces
Cloves	8 ounces

Let stand 24 hours, add two or three gallons of good whiskey, and rack off into clean casks. Bung tightly, let stand 48 hours, then bottle. If a higher color is desired use caramel sufficient to produce the correct tinge. If honey is not obtainable, use sugar-house molasses instead, but honey is preferable.

II.—The following, when properly prepared, makes a passable substitute for cider, and a very pleasant drink:

Catechu, powdered.	3 parts
Alum, powdered...	5 parts
Honey............	640 parts
Water............	12,800 parts
Yeast............	32 parts

Dissolve the catechu, alum, and honey in the water, add the yeast, and put in some warm place to ferment. The container should be filled to the square opening, made by sawing out five or six inches of the center of a stave, and the spume skimmed off daily as it arises. In cooler weather from 2 weeks to 18 days will be required for thorough fermentation. In warmer weather from 12 to 13 days will be sufficient. When fermentation is complete add the following solution:

Oil of bitter almonds	1 part
Oil of clover........	1 part
Caramel...........	32 parts
Alcohol...........	192 parts

The alcohol may be replaced by twice its volume of good bourbon whiskey. A much cheaper, but correspondingly poor substitute for the above may be made as follows:

Twenty-five gallons of soft water, 2 pounds tartaric acid, 25 pounds of brown sugar, and 1 pint of yeast are allowed to stand in a warm place, in a clean cask with the bung out, for 24 hours. Then bung up the cask, after adding 3 gallons of whiskey, and let stand for 48 hours, after which the liquor is ready for use.

CIDER VINEGAR:

See Vinegar.

Cigars

Cigar Sizes and Colors.—Cigars are named according to their color and shape. A dead-black cigar, for instance, is an "Oscuro," a very dark-brown one is a "Colorado," a medium brown is a "Colorado Claro," and a yellowish light brown is a "Claro." Most smokers know the names of the shades from "Claro" to "Colorado," and that is as far as most of them need to know. As to the shapes, a "Napoleon" is the biggest of all cigars—being 7 inches long; a "Perfecto" swells in the middle and tapers down to a very small head at the lighting end; a "Panatela" is a thin, straight, up-and-down cigar without the graceful curve of the "Perfecto"; a "Conchas" is very short and fat, and a "Londres" is shaped like a "Perfecto" except that it does not taper to so small a head at the lighting end. A "Reina Victoria" is a "Londres" that comes packed in a ribbon-tied bundle of 50 pieces, instead of in the usual four layers of 13, 12, 13 and 12.

How to Keep Cigars.—Cigars kept in a case are influenced every time the case is opened. Whatever of taint there may be in the atmosphere rushes into the case, and is finally taken up by the cigars. Even though the cigars have the appearance of freshness, it is not the original freshness in which they were received from the factory. They have been dry, or comparatively so, and have absorbed more moisture than has been put in the case, and it matters not what that moisture may be, it can never restore the flavor that was lost during the drying-out process.

After all, it is a comparatively simple matter to take good care of cigars. All that is necessary is a comparatively air-tight, zinc-lined chest. This should be behind the counter in a place where the temperature is even. When a customer calls for a cigar the dealer takes the box out of the chest, serves his customer, and then puts the box back again. The box being opened for a moment the cigars are not perceptibly affected. The cigars in the close, heavy chest are always safe from atmospheric influences, as the boxes are closed, and the chest is open but a moment, while the dealer is taking out a box from which to serve his customer.

Some of the best dealers have either a large chest or a cool vault in which they keep their stock, taking out from time to time whatever they need for use. Some have a number of small chests, in which they keep different brands, so as to avoid opening and closing one particular chest so often.

It may be said that it is only the higher priced cigars that need special care in handling, although the cheaper grades are not to be handled carelessly. The Havana cigars are more susceptible to change, for there is a delicacy of flavor to be preserved that is never present in the cheaper grades of cigars.

Every dealer must, of course, make a display in his show case, but he need not serve his patrons with these cigars. The shrinkage in value of the cigars in the case is merely a business proposition of profit and loss.

Cigar Flavoring. —I.— Macerate 2 ounces of cinnamon and 4 ounces of tonka beans, ground fine, in 1 quart of rum.

II.—Moisten ordinary cigars with a strong tincture of cascarilla, to which a little gum benzoin and storax may be added. Some persons add a small quantity of camphor or oil of cloves or cassia.

III.—
Tincture of valerian.	4 drachms
Butyric aldehyde...	4 drachms
Nitrous ether......	1 drachm
Tincture vanilla....	2 drachms
Alcohol...........	5 ounces
Water enough to make...........	16 ounces

IV.—
Extract vanilla.....	4 ounces
Alcohol...........	½ gallon
Jamaica rum.......	½ gallon
Tincture valerian...	8 ounces
Caraway seed......	2 ounces
English valerian root	2 ounces
Bitter orange peel...	2 ounces
Tonka beans.......	4 drachms
Myrrh...........	16 ounces

Soak the myrrh for 3 days in 6 quarts of water, add the alcohol, tincture valerian, and extract of vanilla, and after grinding the other ingredients to a coarse powder, put all together in a jug and macerate for 2 weeks, occasionally shaking; lastly, strain.

V.—Into a bottle filled with ½ pint of French brandy put 1½ ounces of cascarilla bark and 1¼ ounces of vanilla previously ground with ¼ pound of sugar; carefully close up the flask and distil in a warm place. After 3 days pour off the liquid, and add ¼ pint of mastic extract. The finished cigars are moistened with this liquid, packed in boxes, and preserved from air by a well-closed lid. They are said to acquire a pleasant flavor and mild strength through this treatment.

Cigar Spots.—The speckled appearance of certain wrappers is due to the work of a species of fungus that attacks the growing tobacco. In a certain district of Sumatra, which produces an exceptionally fine tobacco for wrappers, the leaves of the plant are commonly speckled in this way. Several patents have been obtained for methods of spotting tobacco leaves artificially. A St. Louis firm uses a solution composed of:

Sodium carbonate.....	3 parts
Calx chlorinata.......	1 part
Hot water...........	8 parts

Dissolve the washing soda in the hot water, add the chlorinated lime, and heat the mixture to a boiling temperature for 3 minutes. When cool, decant into earthenware or stoneware jugs, cork tightly, and keep in a cool place. The corks of jugs not intended for immediate use should be covered with a piece of bladder or strong parchment paper, and tightly tied down to prevent the escape of gas, and consequent weakening of the bleaching power of the fluid. The prepared liquor is sprinkled on the tobacco, the latter being then exposed to light and air, when, it is said, the disagreeable odor produced soon disappears.

CINCHONA:
See Wines and Liquors.

CINNAMON ESSENCE:
See Essences and Extracts.

CINNAMON OIL AS AN ANTISEPTIC:
See Antiseptics.

CITRATE OF MAGNESIUM:
See Magnesium Citrate.

CLARET LEMONADE AND CLARET PUNCH:
See Beverages, under Lemonades.

CLARIFICATION OF GELATIN AND GLUE:

See Gelatin.

CLARIFYING.

Clarification is the process by which any solid particles suspended in a liquid are either caused to coalesce together or to adhere to the medium used for clarifying, that they may be removed by filtration (which would previously have been impossible), so as to render the liquid clear.

One of the best agents for this purpose is albumen. When clarifying vegetable extracts, the albumen which is naturally present in most plants accomplishes this purpose easily, provided the vegetable matter is extracted in the cold, so as to get as much albumen as possible in solution.

Egg albumen may also be used. The effect of albumen may be increased by the addition of cellulose, in the form of a fine magma of filtering paper. This has the further advantage that the subsequent filtration is much facilitated.

Suspended particles of gum or pectin may be removed by cautious precipitation with tannin, of which only an exceedingly small amount is usually necessary. It combines with the gelatinous substances better with the aid of heat than in the cold. There must be no excess of tannin used.

Another method of clarifying liquids turbid from particles of gum, albumen, pectin, etc., is to add to them a definite quantity of alcohol. This causes the former substances to separate in more or less large flakes. The quantity of alcohol required varies greatly according to the nature of the liquid. It should be determined in each case by an experiment on a small scale.

Resinous or waxy substances, such as are occasionally met with in honey, etc., may be removed by the addition of bole, pulped filtering paper, and heating to boiling.

In each case the clarifying process may be hastened by making the separating particles specifically heavier; that is, by incorporating some heavier substance, such as talcum, etc., which may cause the flocculi to sink more rapidly, and to form a compact sediment.

Clarifying powder for alcoholic liquids:

Egg albumen, dry....	40 parts
Sugar of milk........	40 parts
Starch..............	20 parts

Reduce them to very fine powder, and mix thoroughly.

For clarifying liquors, wines, essences, etc., take for every quart of liquid 75 grains of the above mixture, shake repeatedly in the course of a few days, the mixture being kept in a warm room, then filter.

Powdered talcum renders the same service, and has the additional advantage of being entirely insoluble. However, the above mixture acts more energetically.

CLAY:

Claying Mixture for Forges.—Twenty parts fire clay; 20 parts cast-iron turnings; 1 part common salt; ½ part sal ammoniac; all by measure.

The materials should be thoroughly mixed dry and then wet down to the consistency of common mortar, constantly stirring the mass as the wetting proceeds. A rough mold shaped to fit the tuyère opening, a trowel, and a few minutes' time are all that are needed to complete the successful claying of the forge. This mixture dries hard and when glazed by the fire will last.

Plastic Modeling Clay.—A permanently plastic clay can be obtained by first mixing it with glycerine, turpentine, or similar bodies, and then adding vaseline or petroleum residues rich in vaseline. The proportion of clay to the vaseline varies according to the desired consistency of the product, the admixture of vaseline varying from 10 to 50 per cent. It is obvious that the hardness of the material decreases with the amount of vaseline added, so that the one richest in vaseline will be the softest. By the use of various varieties of clay and the suitable choice of admixtures, the plasticity, as well as the color of the mass, may be varied.

Cleaning Preparations and Methods

(See also Soaps, Polishes, and Household Formulas).

TO REMOVE STAINS FROM THE HANDS:

Removal of Aniline-Dye Stains from the Skin.—Rub the stained skin with a pinch of slightly moistened red crystals of chromic trioxide until a distinct sensation of warmth announces the destruction of the dye stuff by oxidation and an incipient irritation of the skin. Then rinse with soap and water. A single application usually suffices to remove the stain. It is hardly necessary to call attention to the poisonousness and strong caustic action of chromic trioxide; but only moderate caution is required to avoid evil effects.

Pyrogallic-Acid Stains on the Fingers (see also Photography).—Pyro stains may be prevented fairly well by rubbing in a little wool fat before beginning work. A very effective way of eliminating developer stains is to dip the finger tips occasionally during development into the clearing bath. It is best to use the clearing bath, with ample friction, before resorting to soap, as the latter seems to have a fixing effect upon the stain. Lemon peel is useful for removing pyro stains, and so are the ammonium persulphate reducer and the thiocarbamide clearer.

To Clean Very Soiled Hands.—In the morning wash in warm water, using a stiff brush, and apply glycerine. Repeat the application two or three times during the day, washing and brushing an hour or so afterwards, or apply a warm solution of soda or potash, and wash in warm water, using a stiff brush as before. Finally, rub the hands with pumice or infusorial earth. There are soaps made especially for this purpose, similar to those for use on woodwork, etc., in which infusorial earth or similar matter is incorporated.

To Remove Nitric-Acid Stains.—One plan to avoid stains is to use rubber finger stalls, or rubber gloves. Nitric-acid stains can be removed from the hands by painting the stains with a solution of permanganate of potash, and washing off the permanganate with a 5 per cent solution of hydrochloric (muriatic) acid. After this wash the hands with pure castile soap. Any soap that roughens the skin should be avoided at all times. Castile soap is the best to keep the skin in good condition.

CLEANING GILDED ARTICLES:

To Clean Gilt Frames and Gilded Surfaces Generally.—Dip a soft brush in alcohol to which a few drops of ammonia water has been added, and with it go over the surface. Do not rub—at least, not roughly, or harshly. In the course of five minutes the dirt will have become soft, and easy of removal. Then go over the surface again gently with the same or a similar brush dipped in rain water. Now lay the damp article in the sunlight to dry. If there is no sunlight, place it near a warm (but not *hot*) stove, and let dry completely. In order to avoid streaks, take care that the position of the article, during the drying, is not exactly vertical.

To Clean Fire-Gilt Articles.—Fire-gilt articles are cleaned, according to their condition, with water, diluted hydrochloric acid, ammonia, or potash solution. If hydrochloric acid is employed thorough dilution with water is especially necessary. The acidity should hardly be noticeable on the tongue.

To clean gilt articles, such as gold moldings, etc., when they have become tarnished or covered with flyspecks, etc., rub them slowly with an onion cut in half and dipped in rectified alcohol, and wash off lightly with a moist soft sponge after about 2 hours.

Cleaning Gilded and Polychromed Work on Altars.—To clean bright gold a fine little sponge is used which is moistened but lightly with tartaric acid and passed over the gilding. Next go over the gilt work with a small sponge saturated with alcohol to remove all dirt. For matt gilding, use only a white flannel dipped in lye, and carefully wipe off the dead gold with this, drying next with a fine linen rag. To clean polychromed work sponge with a lye of rain water, 1,000 parts, and calcined potash, 68 parts, and immediately wash off with a clean sponge and water, so that the lye does not attack the paint too much.

SPOT AND STAIN REMOVERS:

To Remove Aniline Stains.—

I.—Sodium nitrate...... 7 grains
Diluted sulphuric acid 15 grains
Water............. 1 ounce

Let the mixture stand a day or two before using. Apply to the spot with a sponge, and rinse the goods with plenty of water.

II.—An excellent medium for the removal of aniline stains, which are often very stubborn, has been found to be liquid opodeldoc. After its use the stains are said to disappear at once and entirely.

Cleansing Fluids.—A spot remover is made as follows:

I.—Saponine......... 7 parts
Water............. 130 parts
Alcohol........... 70 parts
Benzine.......... 1,788 parts
Oil mirbane...... 5 parts

II.—Benzene (benzol).. 89 parts
Ascetic ether...... 10 parts
Pear oil......... 1 part

This yields an effective grease eradicator, of an agreeable odor.

III.—To Remove Stains of Sulphate of copper, or of salts of mercury, silver, or gold from the hands, etc., wash them first with a dilute solution either of ammonia, iodide, bromide, or cyanide of potassium, and then with plenty of water; if the stains are old ones they should first be rubbed with the strongest acetic acid and then treated as above.

Removal of Picric-Acid Stains.—I.— Recent stains of picric acid may be removed readily if the stain is covered with a layer of magnesium carbonate, the carbonate moistened with a little water to form a paste, and the paste then rubbed over the spot.

II.—Apply a solution of

Boric acid.......... 4 parts
Sodium benzoate.... 1 part
Water............. 100 parts

III.—Dr. Prieur, of Besançon, recommends lithium carbonate for the removal of picric-acid stains from the skin or from linen. The method of using it is simply to lay a small pinch on the stain, and moisten the latter with water. Fresh stains disappear almost instantly, and old ones in a minute or two.

To Remove Finger Marks from Books, etc.—I.—Pour benzol (not benzine or gasoline, but Merck's "c. p." crystallizable) on calcined magnesia until it becomes a crumbling mass, and apply this to the spot, rubbing it in lightly, with the tip of the finger. When the benzol evaporates, brush off. Any dirt that remains can be removed by using a piece of soft rubber.

II.—If the foregoing fails (which it sometimes, though rarely, does), try the following: Make a hot solution of sodium hydrate in distilled water, of strength of from 3 per cent to 5 per cent, according to the age, etc., of the stain. Have prepared some bits of heavy blotting paper somewhat larger than the spot to be removed; also, a blotting pad, or several pieces of heavy blotting paper. Lay the soiled page face downward on the blotting pad, then, saturating one of the bits of blotter with the hot sodium hydrate solution, put it on the stain and go over it with a hot smoothing iron. If one application does not remove all the grease or stain, repeat the operation. Then saturate another bit of blotting paper with a 4 per cent or 5 per cent solution of hydrochloric acid in distilled water, apply it to the place, and pass the iron over it to neutralize the strong alkali. This process will instantly restore any faded writing or printing, and make the paper bright and fresh again.

Glycerine as a Detergent.—For certain kinds of obstinate spots (such as coffee and chocolate, for instance) there is no better detergent than glycerine, especially for fabrics with delicate colors. Apply the glycerine to the spot, with a sponge or otherwise, let stand a minute or so, then wash off with water or alcohol. Hot glycerine is even more efficient than cold.

CLEANING SKINS AND LEATHER:
See also Leather.

To Clean Colored Leather.—Pour carbon bisulphide on non-vulcanized gutta-percha, and allow it to stand about 24 hours. After shaking actively add more gutta-percha gradually until the solution becomes of gelatinous consistency. This mixture is applied in suitable quantity to oil-stained, colored leather and allowed to dry two or three hours. The subsequent operation consists merely in removing the coat of gutta-percha from the surface of the leather—that is, rubbing it with the fingers, and rolling it off the surface.

The color is not injured in the least by the sulphuret of carbon; only those leathers on which a dressing containing starch has been used look a little lighter in color, but the better class of leathers are not so dressed. The dry gutta-percha can be redissolved in sulphuret of carbon and used over again.

To Clean Skins Used for Polishing Purposes.—First beat them thoroughly to get rid of dust, then go over the surface on both sides with a piece of good white soap and lay them in warm water in which has been put a little soda. Let them lie here for 2 hours, then wash them in plenty of tepid water, rubbing them vigorously until perfectly clean. This bath should also be made alkaline with soda. The skins are finally rinsed in warm water, and dried quickly. Cold water must be avoided at all stages of the cleansing process, as it has a tendency to shrink and harden the skins.

The best way to clean a chamois skin is to wash and rinse it out in clean water immediately after use, but this practice is apt to be neglected so that the skin becomes saturated with dirt and grime. To clean it, first thoroughly soak in clean, soft water. Then, after soaping it and rolling it into a compact wad, beat with a small round stick—a buggy spoke, say—turning the wad over repeatedly, and keeping it well wet and soaped. This should suffice to loosen the dirt. Then rinse in clean water until the skin is clean. As wringing by hand is apt to injure the chamois skin, it is advisable to use a small clothes wringer. Before using the skin again rinse it in clear water to which a little pulverized alum has been added.

STRAW-HAT RENOVATION:

To Renovate Straw Hats.—I.—Hats made of natural (uncolored) straw, which have become soiled by wear, may be cleaned by thoroughly sponging with a weak solution of tartaric acid in water, followed by water alone. The hat after being so treated should be fastened by the rim to a board by means of pins, so that it will keep its shape in drying.

II.—Sponge the straw with a solution of

	By weight
Sodium hyposulphite.	10 parts
Glycerine	5 parts
Alcohol	10 parts
Water	75 parts

Lay aside in a damp place for 24 hours and then apply

	By weight
Citric acid	2 parts
Alcohol	10 parts
Water	90 parts

Press with a moderately hot iron, after stiffening with weak gum water, if necessary.

III.—If the hat has become much darkened in tint by wear the fumes of burning sulphur may be employed. The material should be first cleaned by thoroughly sponging with an aqueous solution of potassium carbonate, followed by a similar application of water, and it is then suspended over the sulphur fumes. These are generated by placing in a metal or earthen dish, so mounted as to keep the heat from setting fire to anything beneath, some brimstone (roll sulphur), and sprinkling over it some live coals to start combustion. The operation is conducted in a deep box or barrel, the dish of burning sulphur being placed at the bottom, and the article to be bleached being suspended from a string stretched across the top. A cover not fitting so tightly as to exclude all air is placed over it, and the apparatus allowed to stand for a few hours.

Hats so treated will require to be stiffened by the application of a little gum water, and pressed on a block with a hot iron to bring them back into shape.

Waterproof Stiffening for Straw Hats. —If a waterproof stiffening is required use one of the varnishes for which formulas follow:

I.—Copal	450 parts
Sandarac	75 parts
Venice turpentine	40 parts
Castor oil	5 parts
Alcohol	800 parts

II.—Shellac	500 parts
Sandarac	175 parts
Venice turpentine	50 parts
Castor oil	15 parts
Alcohol	2,000 parts

III.—Shellac	750 parts
Rosin	150 parts
Venice turpentine	150 parts
Castor oil	20 parts
Alcohol	2,500 parts

How to Clean a Panama Hat.—Scrub with castile soap and warm water, a nail brush being used as an aid to get the dirt away. The hat is then placed in the hot sun to dry and in the course of two or three hours is ready for use. It will not only be as clean as when new, but it will retain its shape admirably. The cleaned hat will be a trifle stiff at first, but will soon grow supple under wear.

A little glycerine added to the rinsing water entirely prevents the stiffness and brittleness acquired by some hats in drying, while a little ammonia in the washing water materially assists in the scrubbing process. Ivory, or, in fact, any good white soap, will answer as well as castile for the purpose. It is well to rinse a second time, adding the glycerine to the water used the second time. Immerse the hat completely in the rinse water, moving it about to get rid of traces of the dirty water. When the hat has been thoroughly rinsed, press out the surplus water, using a Turkish bath towel for the purpose, and let it rest on the towel when drying.

PAINT, VARNISH, AND ENAMEL REMOVERS:

To Remove Old Oil, Paint, or Varnish Coats.—I.—Apply a mixture of about 5 parts of potassium silicate (water glass, 36 per cent), about 1 part of soda lye (40 per cent), and 1 part of ammonia. The composition dissolves the old varnish coat, as well as the paint, down to the bottom. The varnish coatings which are to be removed may be brushed off or left for days in a hardened state. Upon being thoroughly moistened with water the old varnish may be readily washed off, the lacquer as well as the oil paint coming off completely. The ammonia otherwise employed dissolves the varnish, but not the paint.

II.—Apply a mixture of 1 part oil of turpentine and 2 parts of ammonia. This is effective, even if the coatings withstand the strongest lye. The two liquids are shaken in a bottle until they mix like milk. The mixture is applied to the coating with a little oakum; after a few minutes the old paint can be wiped off.

To Clean Brushes and Vessels of Dry Paint (see also Brushes and Paints).— The cleaning of the brushes and vessels in which the varnish or oil paint had dried is usually done by boiling with soda solution. This frequently spoils the brushes or cracks the vessels if of glass; besides, the process is rather slow and dirty. A much more suitable remedy is amyl acetate, which is a liquid with a pleasant odor of fruit drops, used mainly for dissolving and cementing celluloid. If amyl acetate is poured over a paint brush the varnish or hardened paint dissolves almost immediately and the brush is again rendered serviceable at once. If necessary, the process is repeated. For cleaning vessels shake the liquid about in them, which softens the paint so that it can be readily removed with paper. In this manner much labor can be saved. The amyl acetate can be easily removed from the brushes, etc., by alcohol or oil of turpentine.

Varnish and Paint Remover.—Dissolve 20 parts of caustic soda (98 per cent) in 100 parts of water, mix the solution with 20 parts of mineral oil, and stir in a kettle provided with a mechanical stirrer, until the emulsion is complete. Now add, with stirring, 20 parts of sawdust and pass the whole through a paint mill to obtain a uniform intermixture. Apply the paste moist.

To Remove Varnish from Metal.—To remove old varnish from metals, it suffices to dip the articles in equal parts of ammonia and alcohol (95 per cent).

To Remove Water Stains from Varnished Furniture.—Pour olive oil into a dish and scrape a little white wax into it. This mixture should be heated until the wax melts and rubbed sparingly on the stains. Finally, rub the surface with a linen rag until it is restored to brilliancy.

To Remove Paint, Varnish, etc., from Wood.—Varnish, paint, etc., no matter how old and hard, may be softened in a few minutes so that they can be easily scraped off, by applying the following mixture;

Water glass.......... 5 parts
Soda lye, 40° B. (27 per cent).......... 1 part
Ammonia water...... 1 part

Mix.

Removing Varnish, etc.—A patent has been taken out in England for a liquid for removing varnish, lacquer, tar, and paint. The composition is made by mixing 4 ounces of benzol, 3 ounces of fusel oil, and 1 ounce of alcohol. It is stated by the inventor that this mixture, if applied to a painted or varnished surface, will make the surface quite clean in less than 10 minutes, and that a paint-soaked brush "as hard as iron" can be made as soft and pliable as new by simply soaking for an hour or so in the mixture.

To Remove Enamel and Tin Solder.—Pour enough of oil of vitriol (concentrated sulphuric acid) over powdered fluorspar in an earthen or lead vessel, so as just to cover the parts whereby hydrofluoric acid is generated. For use, dip the article suspended on a wire into the liquid until the enamel or the tin is eaten away or dissolved, which does not injure the articles in any way. If heated, the liquid acts more rapidly. The work should always be conducted in the open air, and care should be taken not to inhale the fumes, which are highly injurious to the health, and not to get any liquid on the skin, as hydrofluoric acid is one of the most dangerous poisons. Hydrofluoric acid must be kept in earthen or leaden vessels, as it destroys glass.

Removing Paint and Varnish from Wood.—The following compound is given as one which will clean paint or varnish from wood or stone without injuring the material:

Flour or wood pulp.. 385 parts
Hydrochloric acid... 450 parts
Bleaching powder... 160 parts
Turpentine......... 5 parts

This mixture is applied to the surface and left on for some time. It is then brushed off, and brings the paint away with it. It keeps moist quite long enough to be easily removed after it has acted.

Paste for Removing Old Paint or Varnish Coats.—

I.—Sodium hydrate..... 5 parts
Soluble soda glass ... 3 parts
Flour paste......... 6 parts
Water............. 4 parts

II.—Soap.............. 10 parts
Potassium hydrate... 7 parts
Potassium silicate.... 2 parts

To Remove Old Enamel.—Lay the articles horizontally in a vessel containing a concentrated solution of alum and boil them. The solution should be just sufficient to cover the pieces. In 20 or 25 minutes the old enamel will fall into dust, and the article can be polished with emery. If narrow and deep vessels are used the operation will require more time.

INK ERADICATORS:

Two-Solution Ink Remover.—

I.—(a) Citric acid . . . .   1 part
    Concentrated solu-
      tion of borax . . .   2 parts
    Distilled water. . . .   16 parts

Dissolve the acid in the water, add the borax solution, and mix by agitation.

(b) Chloride of lime. . .   3 parts
    Water . . . .   16 parts
    Concentrated bor-
      ax solution . . . . .   2 parts

Add the chloride of lime to the water, shake well and set aside for a week, then decant the clear liquid and to it add the borax solution.

For use, saturate the spot with solution (a), apply a blotter to take off the excess of liquid, then apply solution (b). When the stain has disappeared, apply the blotter and wet the spot with clean water; finally dry between two sheets of blotting paper.

II.—(a) Mix, in equal parts, potassium chloride, potassium hypochlorite, and oil of peppermint. (b) Sodium chloride, hydrochloric acid and water, in equal parts.

Wet the spot with (a), let dry, then brush it over lightly with (b), and rinse in clear water.

A good single mixture which will answer for most inks is made by mixing citric acid and alum in equal parts. If desired to vend in a liquid form add an equal part of water. In use, the powder is spread well over the spot and (if on cloth or woven fabrics) well rubbed in with the fingers. A few drops of water are then added, and also rubbed in. A final rinsing with water completes the process.

Ink Erasers.—I.—Inks made with nutgalls and copperas can be removed by using a moderately concentrated solution of oxalic acid, followed by use of pure water and frequent drying with clean blotting paper. Most other black inks are erased by use of a weak solution of chlorinated lime, followed by dilute acetic acid and water, with frequent drying with blotters. Malachite green ink is bleached by ammonia water; silver inks by potassium cyanide or sodium hyposulphite. Some aniline colors are easily removed by alcohol, and nearly all by chlorinated lime, followed by diluted acetic acid or vinegar. In all cases apply the substances with camel's-hair brushes or feathers, and allow them to remain no longer than necessary, after which rinse well with water and dry with blotting paper.

II.—Citric acid . . . .   1 part
    Water, distilled . . . . . .   10 parts
    Concentrated solution
      of borax . . . .   2 parts

Dissolve the citric acid in the water and add the borax. Apply to the paper with a delicate camel's-hair pencil, removing any excess of water with a blotter. A mixture of oxalic, citric, and tartaric acids, in equal parts, dissolved in just enough water to give a clean solution, acts energetically on most inks.

Erasing Powder or Pounce.—Alum, 1 part; amber, 1 part; sulphur, 1 part; saltpeter, 1 part. Mix well together and keep in a glass bottle. If a little of this powder is placed on an ink spot or fresh writing, rubbing very lightly with a clean linen rag, the spot or the writing will disappear at once.

Removing Ink Stains.—I.—The material requiring treatment should first be soaked in clean, warm water, the superfluous moisture removed, and the fabric spread over a clean cloth. Now allow a few minims of liquor ammoniæ fortis, specific gravity 0.891, to drop on the ink spot, then saturate a tiny tuft of absorbent cotton-wool with acidum phosphoricum dilutum, B. P., and apply repeatedly and with firm pressure over the stain; repeat the procedure two or three times, and finally rinse well in warm water, afterwards drying in the sun, when every trace of ink will have vanished. This method is equally reliable for old and fresh ink stains, is rapid in action, and will not injure the most delicate fabric.

II.—To remove ink spots the fabric is soaked in warm water, then it is squeezed out and spread upon a clean piece of linen. Now apply a few drops of liquid ammonia of a specific gravity of 0.891 to the spot, and dab it next with a wad of cotton which has been saturated with dilute phosphoric acid. After repeating the process several times and drying the piece in the sun, the ink spot will have disappeared without leaving the slightest trace.

III.—Ink spots may be removed by the following mixture:

Oxalic acid	10 parts
Stannic chloride	2 parts
Acetic acid	5 parts
Water to make	500 parts

Mix.

IV.—The customary method of cleansing ink spots is to use oxalic acid. Thick blotting paper is soaked in a concentrated solution and dried. It is then laid immediately on the blot, and in many instances will take the latter out without leaving a trace behind. In more stubborn cases the cloth is dipped in boiling water and rubbed with crystals of oxalic acid, after which it is soaked in a weak solution of chloride of lime—say 1 ounce to a quart of water. Under such circumstances the linen should be thoroughly rinsed in several waters afterwards. Oxalic acid is undesirable for certain fabrics because it removes the color.

V.—Here is a more harmless method: Equal parts of cream of tartar and citric acid, powdered fine, and mixed together. This forms the "salts of lemon" sold by druggists. Procure a hot dinner plate, lay the part stained in the plate, and moisten with hot water; next rub in the above powder with the bowl of a spoon until the stains disappear; then rinse in clean water and dry.

To Remove Red (Aniline) Ink.— Stains of red anilines, except eosine, are at once removed by moistening with alcohol of 94 per cent, acidulated with acetic acid. Eosine does not disappear so easily. The amount of acetic acid to be used is ascertained by adding it, drop by drop, to the alcohol, testing the mixture from time to time, until when dropped on the stain, the latter at once disappears.

CLEANING OF WALLS, CEILINGS, AND WALL PAPER:

See also Household Formulas.

To Renovate Brick Walls.—Dissolve glue in water in the proportion of 1 ounce of glue to every gallon of water; add, while hot, a piece of alum the size of a hen's egg, ¼ pound Venetian red, and 1 pound Spanish brown. Add more water if too dark; more red and brown if too light.

Cleaning Painted Doors, Walls, etc.— The following recipe is designed for painted objects that are much soiled. Simmer gently on the fire, stirring constantly, 30 parts, by weight, of pulverized borax, and 450 parts of brown soap of good quality, cut in small pieces, in 3,000 parts of water. The liquid is applied by means of flannel and rinsed off at once with pure water.

To Remove Aniline Stains from Ceilings, etc.—In renewing ceilings, the old aniline color stains are often very annoying, as they penetrate the new coating. Painting over with shellac or oil paint will bring relief, but other drawbacks appear. A very practical remedy is to place a tin vessel on the floor of the room, and to burn a quantity of sulphur in it after the doors and windows of the room have been closed. The sulphur vapors destroy the aniline stains, which disappear entirely.

Old Ceilings.—In dealing with old ceilings the distemper must be washed off down to the plaster face, all cracks raked out and stopped with putty (plaster of Paris and distemper mixed), and the whole rubbed smooth with pumice stone and water; stained parts should be painted with oil color, and the whole distempered. If old ceilings are in bad condition it is desirable that they should be lined with paper, which should have a coat of weak size before being distempered.

Oil Stains on Wall Paper.—Make a medium thick paste of pipe clay and water, applying it carefully flat upon the oil stain, but avoiding all friction. The paste is allowed to remain 10 to 12 hours, after which time it is very carefully removed with a soft rag. In many cases a repeated action will be necessary until the purpose desired is fully reached. Finally, however, this will be obtained without blurring or destroying the design of the wall paper, unless it be of the cheapest variety. In the case of a light, delicate paper, the paste should be composed of magnesia and benzine.

To Clean Painted Walls. A simple method is to put a little aqua ammonia in moderately warm water, dampen a flannel with it, and gently wipe over the painted surface. No scrubbing is necessary.

Treatment of Whitewashed Walls.— It is suggested that whitewashed walls which it is desired to paper, with a view to preventing peeling, should be treated with water, after which the scraper should be vigorously used. If the whitewash has been thoroughly soaked it can easily be removed with the scraper. Care should be taken that every part of the wall is well scraped.

Cleaning Wall Paper.—I.—To clean wall paper the dust should first be removed by lightly brushing, preferably with a feather duster, and the surface then gently rubbed with slices of moderately stale bread, the discolored surface of the bread being removed from time to time, so as to expose a fresh portion for use. Care should be taken to avoid scratching the paper with the crust of the bread, and the rubbing should be in one direction, the surface being systematically gone over, as in painting, to avoid the production of streaks.

II.—Mix 4 ounces of powdered pumice with 1 quart of flour, and with the aid of water make a stiff dough. Form the dough into rolls 2 inches in diameter and 6 inches long; sew each roll separately in a cotton cloth, then boil for 40 or 50 minutes, so as to render the mass firm. Allow to stand for several hours, remove the crust, and they are ready for use.

III.—Bread will clean paper; but unless it is properly used the job will be a very tedious one. Select a "tin" loaf at least two days old. Cut off the crust at one end, and rub down the paper, commencing at the top. Do not rub the bread backwards and forwards, but in single strokes. When the end gets dirty take a very sharp knife and pare off a thin layer; then proceed as before.

It is well to make sure that the walls are quite dry before using the bread, or it may smear the pattern. If the room is furnished it will, of course, be necessary to place cloths around the room to catch the crumbs.

IV.—A preparation for cleansing wall paper that often proves much more effectual than ordinary bread, especially when the paper is very dirty, is made by mixing ⅔ dough and ⅓ plaster of Paris. This should be made a day before it is needed for use, and should be very gently baked.

If there are any grease spots they should be removed by holding a hot flatiron against a piece of blotting paper placed over them. If this fails, a little fuller's earth or pipe clay should be made into a paste with water, and this should then be carefully plastered over the grease spots and allowed to remain till quite dry, when it will be found to have absorbed the grease.

V.—Mix together 1 pound each of rye flour and white flour into a dough, which is partially cooked and the crust removed. To this 1 ounce common salt and ½ ounce of powdered naphthaline are added, and finally 1 ounce of corn meal, and ¼ ounce of burnt umber. The composition is formed into a mass, of the proper size to be grasped in the hand, and in use it should be drawn in one direction over the surface to be cleaned.

VI.—Procure a soft, flat sponge, being careful that there are no hard or gritty places in it, then get a bucket of new, clean, dry, wheat bran. Hold the sponge flat side up, and put a handful of bran on it, then quickly turn against the wall, and rub the wall gently and carefully with it; then repeat the operation. Hold a large pan or spread down a drip cloth to catch the bran as it falls, but never use the same bran twice. Still another way is to use Canton flannel in strips a foot wide and about 3 yards long. Roll a strip around a stick 1 inch thick and 10 inches long, so as to have the ends of the stick covered, with the nap of the cloth outside. As the cloth gets soiled, unroll the soiled part and roll it up with the soiled face inside.

In this way one can change places on the cloth when soiled and use the whole face of the cloth. To take out a grease spot requires care. First, take several thicknesses of brown wrapping paper and make a pad, place it against the grease spot, and hold a hot flatiron against it to draw out the grease, which will soak into the brown paper. Be careful to have enough layers of brown paper to keep the iron from scorching or discoloring the wall paper. If the first application does not take out nearly all the grease, repeat with clean brown paper or a blotting pad Then take an ounce vial of washed sulphuric ether and a soft, fine, clean sponge and sponge the spot carefully until all the grease disappears. Do not wipe the place with the sponge and ether, but dab the sponge carefully against the place. A small quantity of ether is advised, as it is very inflammable.

CLOTHES AND FABRIC CLEANERS:

Soaps for Clothing and Fabrics.—When the fabric is washable and the color fast, ordinary soap and water are sufficient for removing grease and the ordinarily attendant dirt; but special soaps are made which may possibly be more effectual.

I.—Powdered borax....	30 parts
Extract of soap bark	30 parts
Ox gall (fresh).....	120 parts
Castile soap.......	450 parts

First make the soap-bark extract by boiling the crushed bark in water until it has assumed a dark color, then strain the liquid into an evaporating dish, and by the aid of heat evaporate it to a solid extract; then powder and mix it with the borax and the ox gall. Melt the castile soap by adding a small quantity of water and warming, then add the other ingredients and mix well.

About 100 parts of soap bark make 20 parts of extract.

II.—Castile soap.......... 2 pounds
Potassium carbonate.. ½ pound
Camphor............. ½ ounce
Alcohol............. ½ ounce
Ammonia water...... ½ ounce
Hot water, ½ pint, or sufficient.

Dissolve the potassium carbonate in the water, add the soap previously reduced to thin shavings, keep warm over a water bath, stirring occasionally, until dissolved, adding more water if necessary, and finally, when of a consistence to become semisolid on cooling, remove from the fire. When nearly ready to set, stir in the camphor, previously dissolved in the alcohol and the ammonia.

The soap will apparently be quite as efficacious without the camphor and ammonia.

If a paste is desired, a potash soap should be used instead of the castile in the foregoing formula, and a portion or all of the water omitted. Soaps made from potash remain soft, while soda soaps harden on the evaporation of the water which they contain when first made.

A liquid preparation may be obtained, of course, by the addition of sufficient water, and some more alcohol would probably improve it.

Clothes-Cleaning Fluids :
See also Household Formulas.

I.—Borax.............. 1 ounce
Castile soap........ 1 ounce
Sodium carbonate... 3 drachms
Ammonia water..... 5 ounces
Alcohol............ 4 ounces
Acetone............ 4 ounces
Hot water to make... 4 pints

Dissolve the borax, sodium bicarbonate, and soap in the hot water, mix the acetone and alcohol together, unite the two solutions, and then add the ammonia water. The addition of a couple of ounces of rose water will render it somewhat fragrant.

II.—A strong decoction of soap bark, preserved by the addition of alcohol, forms a good liquid cleanser for fabrics of the more delicate sort.

III.—Chloroform........ 15 parts
Ether............. 15 parts

Alcohol...........120 parts
Decoction of quillaia
bark of 30°.... 4,500 parts

IV.—Acetic ether....... 10 parts
Amyl acetate....... 10 parts
Liquid ammonia.... 10 parts
Dilute alcohol...... 70 parts

V.—Another good non-inflammable spot remover consists of equal parts of acetone, ammonia, and diluted alcohol. For use in large quantities carbon tetrachloride is suggested.

VI.—Castile soap...... 4 av. ounces
Water, boiling.... 32 fluidounces

Dissolve and add:

Water........... 1 gallon
Ammonia........ 8 fluidounces
Ether........... 2 fluidounces
Alcohol........ 4 fluidounces

To Remove Spots from Tracing Cloth. —It is best to use benzine, which is applied by means of a cotton rag. The benzine also takes off lead-pencil marks, but does not attack India and other inks. The places treated with benzine should subsequently be rubbed with a little talcum, otherwise it would not be possible to use the pen on them.

Removal of Paint from Clothing.— Before paint becomes "dry" it can be removed from cloth by the liberal application of turpentine or benzine. If the spot is not large, it may be immersed in the liquid; otherwise, a thick, folded, absorbent cloth should be placed under the fabric which has been spotted, and the liquid sponged on freely enough that it may soak through, carrying the greasy matter with it. Some skill in manipulation is requisite to avoid simply spreading the stain and leaving a "ring" to show how far it has extended.

When benzine is used the operator must be careful to apply it only in the absence of light or fire, on account of the extremely inflammable character of the vapor.

Varnish stains, when fresh, are treated in the same way, but the action of the solvent may possibly not be so complete on account of the gum rosins present.

When either paint or varnish has dried, its removal becomes more difficult. In such case soaking in strong ammonia water may answer. An emulsion, formed by shaking together 2 parts of ammonia water and 1 of spirits of turpentine, has been recommended.

To Remove Vaseline Stains from Clothing.—Moisten the spots with a mixture of 1 part of aniline oil, 1 of powdered soap, and 10 of water. After allowing the cloth to lie for 5 or 10 minutes, wash with water.

To Remove Grease Spots from Plush. —Place fresh bread rolls in the oven, break them apart as soon as they have become very hot, and rub the spots with the crumbs, continuing the work by using new rolls until all traces of fat have disappeared from the fabric. Purified benzine, which does not alter even the most delicate colors, is also useful for this purpose.

To Remove Iron Rust from Muslin and Linen.—Wet with lemon juice and salt and expose to the sun. If one application does not remove the spots, a second rarely fails to do so.

Keroclean. — This non-inflammable cleanser removes grease spots from delicate fabrics without injury, cleans all kinds of jewelry and tableware by removing fats and tarnish, kills moths, insects, and household pests by suffocation and extermination, and cleans ironware by removing rust, brassware by removing grease, copperware by removing verdigris. It is as clear as water and will stand any fire test.

Kerosene............ 1 ounce
Carbon tetrachloride
   (commercial)....... 3 ounces
Oil of citronella....... 2 drachms

Mix, and filter if necessary. If a strong odor of carbon bisulphide is detected in the carbon tetrachloride first shake with powdered charcoal and filter.

To Clean Gold and Silver Lace.— I.—Alkaline liquids sometimes used for cleaning gold lace are unsuitable, for they generally corrode or change the color of the silk. A solution of soap also interferes with certain colors, and should therefore not be employed. Alcohol is an effectual remedy for restoring the luster of gold, and it may be used without any danger to the silk, but where the gold is worn off, and the base metal exposed, it is not so successful in accomplishing its purpose, as by removing the tarnish the base metal becomes more distinguishable from the fine gold.

II.—To clean silver lace take alabaster in very fine powder, lay the lace upon a cloth, and with a soft brush take up some of the powder, and rub both sides with it till it becomes bright and clean, afterwards polish with another brush until all remnants of the powder are removed, and it exhibits a lustrous surface.

III.—Silver laces are put in curdled milk for 24 hours. A piece of Venetian soap, or any other good soap, is scraped and stirred into 2 quarts of rain water. To this a quantity of honey and fresh ox gall is added, and the whole is stirred for some time. If it becomes too thick, more water is added. This mass is allowed to stand for half a day, and the wet laces are painted with it. Wrap a wet cloth around the roller of a mangle, wind the laces over this, put another wet cloth on top, and press, wetting and repeating the application several times. Next, dip the laces in a clear solution of equal parts of sugar and gum arabic, pass them again through the mangle, between two clean pieces of cloth, and hang them up to dry thoroughly, attaching a weight to the lower end.

IV.—Soak gold laces over night in cheap white wine and then proceed as with silver laces. If the gold is worn off, put 771 grains of shellac, 31 grains of dragon's blood, 31 grains of turmeric in strong alcohol and pour off the ruby-colored fluid. Dip a fine hair pencil in this, paint the pieces to be renewed, and hold a hot flatiron a few inches above them, so that only the laces receive the heat.

V.—Silver embroideries may also be cleaned by dusting them with Vienna lime, and brushing off with a velvet brush.

For gildings the stuff is dipped in a solution of gold chloride, and this is reduced by means of hydrogen in another vessel.

For silvering, one of the following two processes may be employed: (a) Painting with a solution of 1 part of phosphorus in 15 parts bisulphide of carbon and dipping in a solution of nitrate of silver; (b) dipping for 2 hours in a solution of nitrate of silver, mixed with ammonia, then exposing to a current of pure hydrogen.

To Remove Silver Stains from White Fabrics.—Moisten the fabric for two or three minutes with a solution of 5 parts of bromine and 500 parts of water. Then rinse in clear water. If a yellowish stain remains, immerse in a solution of 150 parts of sodium hyposulphite in 500 parts of water, and again rinse in clear water.

Rust-Spot Remover.—Dissolve potassium bioxalate, 200 parts, in distilled water, 8,800 parts; add glycerine, 1,000 parts, and filter. Moisten the rust or ink spots with this solution; let the linen, etc., lie for 3 hours, rubbing the moistened spots frequently, and then wash well with water.

To Clean Quilts.—Quilts are cleaned by first washing them in lukewarm soapsuds, then laying them in cold, soft (rain) water over night. The next day they are pressed as dry as possible and hung up; the ends, in which the moisture remains for a long time, must be wrung out from time to time.

It is very essential to beat the drying quilts frequently with a smooth stick or board. This will have the effect of swelling up the wadding, and preventing it from felting. Furthermore, the quilts should be repeatedly turned during the drying from right to left and also from top to bottom. In this manner streaks are avoided.

Removal of Peruvian-Balsam Stains.—The fabric is spread out, a piece of filter paper being placed beneath the stain, and the latter is then copiously moistened with chloroform, applied by means of a tuft of cotton wool. Rubbing is to be avoided.

Solution for Removing Nitrate of Silver Spots.—

Bichloride of mercury	5 parts
Ammonium chloride.	5 parts
Distilled water......	40 parts

Apply the mixture to the spots with a cloth, then rub. This removes, almost instantaneously, even old stains on linen, cotton, or wool. Stains on the skin thus treated become whitish yellow and soon disappear.

Cleaning Tracings.—Tracing cloth can be very quickly and easily cleaned, and pencil marks removed by the use of benzine, which is applied with a cotton swab. It may be rubbed freely over the tracing without injury to lines drawn in ink, or even in water color, but the pencil marks and dirt will quickly disappear. The benzine evaporates almost immediately, leaving the tracing unharmed. The surface, however, has been softened and must be rubbed down with talc, or some similar substance, before drawing any more ink lines.

The glaze may be restored to tracing cloth after using the eraser by rubbing the roughened surface with a piece of hard wax from an old phonograph cylinder. The surface thus produced is superior to that of the original glaze, as it is absolutely oil- and water-proof.

Rags for Cleaning and Polishing.—Immerse flannel rags in a solution of 20 parts of dextrine and 30 parts of oxalic acid in 20 parts of logwood decoction; gently wring them out, and sift over them a mixture of finely powdered tripoli and pumice stone. Pile the moist rags one upon another, placing a layer of the powder between each two. Then press, separate, and dry.

Cleaning Powder.—

Bole.............	500 parts
Magnesium carbonate.............	50 parts

Mix and make into a paste with a small quantity of benzine or water; apply to stains made by fats or oils on the clothing and when dry remove with a brush.

CLEANING PAINTED AND VARNISHED SURFACES:

Cleaning and Preserving Polished Woodwork.—Rub down all the polished work with a very weak alcoholic solution of shellac (1 to 20 or even 1 to 30) and linseed oil, spread on a linen cloth. The rubbing should be firm and hard. Spots on the polished surface, made by alcohol, tinctures, water, etc., should be removed as far as possible and as soon as possible after they are made, by the use of boiled linseed oil. Afterwards they should be rubbed with the shellac and linseed oil solution on a soft linen rag. If the spots are due to acids go over them with a little dilute ammonia water. Ink spots may be removed with dilute or (if necessary) concentrated hydrochloric acid, following its use with dilute ammonia water. In extreme cases it may be necessary to use the scraper or sandpaper, or both.

Oak as a general thing is not polished, but has a matt surface which can be washed with water and soap. First all stains and spots should be gone over with a sponge or a soft brush and very weak ammonia water. The carved work should be freed of dust, etc., by the use of a stiff brush, and finally washed with dilute ammonia water. When dry it should be gone over very thinly and evenly with brunoline applied with a soft pencil. If it is desired to give an especially handsome finish, after the surface is entirely dry, give it a preliminary coat of brunoline and follow this on the day after with a second. Brunoline may be purchased of any dealer in paints. To make it, put 70 parts of linseed oil in a very capacious vessel (on account of the foam that ensues) and add to it 20 parts of powdered litharge, 20 parts of powdered minium, and 10 parts of lead acetate, also powdered. Boil until the oil is completely oxidized, stirring constantly. When completely oxidized the oil is no longer red, but is of a dark brown color. When it acquires this color, remove from the fire, and add 160 parts of turpentine oil, and stir well. This brunoline serves splendidly for polishing furniture or other polished wood.

To Clean Lacquered Goods.—Papier-maché and lacquered goods may be cleaned perfectly by rubbing thoroughly with a paste made of wheat flour and olive oil. Apply with a bit of soft flannel or old linen, rubbing hard; wipe off and polish by rubbing with an old silk handkerchief.

Polish for Varnished Work.—To renovate varnished work make a polish of 1 quart good vinegar, 2 ounces butter of antimony, 2 ounces alcohol, and 1 quart oil. Shake well before using.

To Clean Paintings.—To clean an oil painting, take it out of its frame, lay a piece of cloth moistened with rain water on it, and leave it for a while to take up the dirt from the picture. Several applications may be required to secure a perfect result. Then wipe the picture very gently with a tuft of cotton wool damped with absolutely pure linseed oil. Gold frames may be cleaned with a freshly cut onion; they should be wiped with a soft sponge wet with rain water a few hours after the application of the onion, and finally wiped with a soft rag.

Removing and Preventing Match Marks.—The unsightly marks made on a painted surface by striking matches on it can sometimes be removed by scrubbing with soapsuds and a stiff brush. To prevent match marks dip a bit of flannel in alboline (liquid vaseline), and with it go over the surface, rubbing it hard. A second rubbing with a dry bit of flannel completes the job. A man may "strike" a match there all day, and neither get a light nor make a mark.

GLOVE CLEANERS:

Powder for Cleaning Gloves.—

I.—White bole or pipe
   clay............ 60.0 parts
Orris root (pow-
   dered)......... 30.0 parts
Powdered grain
   soap.......... 7.5 parts
Powdered borax... 15.0 parts
Ammonium chlor-
   ide........... 2.5 parts

Mix the above ingredients. Moisten the gloves with a damp cloth, rub on the powder, and brush off after drying.

II.—Four pounds powdered pipe clay, 2 pounds powdered white soap, 1 ounce lemon oil, thoroughly rubbed together. To use, make powder into a thin cream with water and rub on the gloves while on the hands. This is a cheaply produced compound, and does its work effectually.

Soaps and Pastes for Cleaning Gloves.—

I.—Soft soap.......... 1 ounce
Water............. 4 ounces
Oil of lemon....... ½ drachm
Precipitated chalk, a
   sufficient quantity.

Dissolve the soap in the water, add the oil, and make into a stiff paste with a sufficient quantity of chalk.

II.—White hard soap.... 1 part
Talcum.......... 1 part
Water............ 4 parts

Shave the soap into ribbons, dissolve in the water by the aid of heat, and incorporate the talcum.

III.—Curd soap....... 1 av. ounce
Water.......... 4 fluidounces
Oil of lemon..... ½ fluidrachm
French chalk, a sufficient quantity.

Shred the soap and melt it in the water by heat, add the oil of lemon, and make into a stiff paste with French chalk.

IV.—White castile soap,
   old and dry...... 15 parts
Water............ 15 parts
Solution of chlorin-
   ated soda........ 16 parts
Ammonia water.... 1 part

Cut or shave up the soap, add the water, and heat on the water bath to a smooth paste. Remove, let cool, and add the other ingredients and mix thoroughly.

V.—Castile soap, white,
   old, and dry..... 100 parts
Water............ 75 parts
Tincture of quillaia 10 parts
Ether, sulphuric... 10 parts
Ammonia water,
   FF............ 5 parts
Benzine, deodorized 75 parts

Melt the soap, previously finely shaved, in the water, bring to a boil and remove from the fire. Let cool down, then add the other ingredients, incorporating them thoroughly. This should be put up in collapsible tubes or tightly closed metallic boxes. This is also useful for clothing.

Liquid Cloth and Glove Cleaner.—

Gasoline............ 1 gallon
Chloroform......... 1 ounce
Carbon disulphide... 1 ounce

Essential oil almond..   5 drops
Oil bergamot........   1 drachm
Oil cloves..........   5 drops

Mix. To be applied with a sponge or soft cloth.

STONE CLEANING:

Cleaning and Polishing Marble.—
I.—Marble that has become dirty by ordinary use or exposure may be cleaned by a simple bath of soap and water.

If this does not remove stains, a weak solution of oxalic acid should be applied with a sponge or rag, washing quickly and thoroughly with water to minimize injury to the surface.

Rubbing well after this with chalk moistened with water will, in a measure, restore the luster. Another method of finishing is to apply a solution of white wax in turpentine (about 1 in 10), rubbing thoroughly with a piece of flannel or soft leather.

If the marble has been much exposed, so that its luster has been seriously impaired, it may be necessary to repolish it in a more thorough manner. This may be accomplished by rubbing it first with sand, beginning with a moderately coarse-grained article and changing this twice for finer kinds, after which tripoli or pumice is used. The final polish is given by the so-called putty powder. A plate of iron is generally used in applying the coarse sand; with the fine sand a leaden plate is used; and the pumice is employed in the form of a smooth-surfaced piece of convenient size. For the final polishing coarse linen or bagging is used, wedged tightly into an iron planing tool. During all these applications water is allowed to trickle over the face of the stone.

The putty powder referred to is binoxide of tin, obtained by treating metallic tin with nitric acid, which converts the metal into hydrated metastannic acid. This, when heated, becomes anhydrous. In this condition it is known as putty powder. In practice putty powder is mixed with alum, sulphur, and other substances, the mixture used being dependent upon the nature of the stone to be polished.

According to Warwick, colored marble should not be treated with soap and water, but only with the solution of beeswax above mentioned.

II.—Take 2 parts of sodium bicarbonate, 1 part of powdered pumice stone, and 1 part of finely pulverized chalk. Pass through a fine sieve to screen out all particles capable of scratching the marble, and add sufficient water to form a pasty mass. Rub the marble with it vigorously, and end the cleaning with soap and water.

III.—Ox gall..........   1 part
Saturated solution
of sodium carbo-
nate..........   4 parts
Oil of turpentine..   1 part
Pipe clay enough to form a paste.

IV.—Sodium carbonate.   2 ounces
Chlorinated lime..   1 ounce
Water..........   14 ounces

Mix well and apply the magma to the marble with a cloth, rubbing well in, and finally rubbing dry. It may be necessary to repeat this operation.

V.—Wash the surface with a mixture of finely powdered pumice stone and vinegar, and leave it for several hours; then brush it hard and wash it clean. When dry, rub with whiting and wash leather.

VI.—Soft soap..........   4 parts
Whiting ....   4 parts
Sodium bicarbonate 1 part
Copper sulphate...   2 parts

Mix thoroughly and rub over the marble with a piece of flannel, and leave it on for 24 hours, then wash it off with clean water, and polish the marble with a piece of flannel or an old piece of felt.

VII.—A strong solution of oxalic acid effectually takes out ink stains. In handling it the poisonous nature of this acid should not be forgotten.

VIII.—Iron mold or ink spots may be taken out in the following manner: Take $\frac{1}{2}$ ounce of butter of antimony and 1 ounce of oxalic acid and dissolve them in 1 pint of rain water; add enough flour to bring the mixture to a proper consistency. Lay it evenly on the stained part with a brush, and, after it has remained for a few days, wash it off and repeat the process if the stain is not wholly removed.

IX.—To remove oil stains apply common clay saturated with benzine. If the grease has remained in long the polish will be injured, but the stain will be removed.

X.—The following method for removing rust from iron depends upon the solubility of the sulphide of iron in a solution of cyanide of potassium. Clay is made into a thin paste with ammonium sulphide, and the rust spot smeared with the mixture, care being taken that the spot is only just covered. After ten minutes this paste is washed off and replaced by one consisting of white bole mixed with a solution of potassium cyanide (1 to 4), which is in its turn washed off after about 2½ hours. Should a reddish spot remain after washing off the first paste, a second layer may be applied for about 5 minutes.

XI.—Soft soap......... 4 ounces
Whiting......... 4 ounces
Sodium carbonate. 1 ounce
Water, a sufficient quantity.

Make into a thin paste, apply on the soiled surface, and wash off after 24 hours.

XII.—In a spacious tub place a tall vessel upside down. On this set the article to be cleaned so that it will not stand in the water, which would loosen the cemented parts. Into this tub pour a few inches of cold water—hot water renders marble dull—take a soft brush and a piece of Venetian soap, dip the former in the water and rub on the latter carefully, brushing off the article from top to bottom. When in this manner dust and dirt have been dissolved, wash off all soap particles by means of a watering pot and cold water, dab the object with a clean sponge, which absorbs the moisture, place it upon a cloth and carefully dry with a very clean, soft cloth, rubbing gently. This treatment will restore the former gloss to the marble.

XIII.—Mix and shake thoroughly in a bottle equal quantities of sulphuric acid and lemon juice. Moisten the spots and rub them lightly with a linen cloth and they will disappear.

XIV.—Ink spots are treated with acid oxalate of potassium; blood stains by brushing with alabaster dust and distilled water, then bleaching with chlorine solution. Alizarine ink and aniline ink spots can be moderated by laying on rags saturated with Javelle water, chlorine water, or chloride of lime paste. Old oil stains can only be effaced by placing the whole piece of marble for hours in benzine. Fresh oil or grease spots are obliterated by repeated applications of a little damp, white clay and subsequent brushing with soap water or weak soda solution. For many other spots an application of benzine and magnesia is useful.

XV.—Marble slabs keep well and do not lose their fresh color if they are cleaned with hot water only, without the addition of soap, which is injurious to the color. Care must be taken that no liquid dries on the marble. If spots of wine, coffee, beer, etc., have already appeared, they are cleaned with diluted spirit of sal ammoniac, highly diluted oxalic acid, Javelle water, ox gall, or, take a quantity of newly slaked lime, mix it with water into a paste-like consistency,

apply the paste uniformly on the spot with a brush, and leave the coating alone for two to three days before it is washed off. If the spots are not removed by a single application, repeat the latter. In using Javelle water 1 or 2 drops should be carefully poured on each spot, rinsing off with water.

To Remove Grease Spots from Marble. —If the spots are fresh, rub them over with a piece of cloth that has been dipped into pulverized china clay, repeating the operation several times, and then brush with soap and water. When the spots are old brush with distilled water and finest French plaster energetically, then bleach with chloride of lime that is put on a piece of white cloth. If the piece of marble is small enough to permit it, soak it for a few hours in refined benzine.

Preparation for Cleaning Marble, Furniture, and Metals, Especially Copper.—This preparation is claimed to give very quickly perfect brilliancy, persisting without soiling either the hand or the articles, and without leaving any odor of copper. The following is the composition for 100 parts of the product: Wax, 2.4 parts; oil of turpentine, 9.4 parts; acetic acid, 42 parts; citric acid, 42 parts; white soap, 42 parts.

Removing Oil Stains from Marble.— Saturate fuller's earth with a solution of equal parts of soap liniment, ammonia, and water; apply to the greasy part of the marble; keep there for some hours, pressed down with a smoothing iron sufficiently hot to warm the mass, and as it evaporates occasionally renew the solution. When wiped off dry the stain will have nearly disappeared. Some days later, when more oil works toward the surface repeat the operation. A few such treatments should suffice.

Cleaning Terra Cotta.—After having carefully removed all dust, paint the terra cotta, by means of a brush, with a mixture of slightly gummed water and finely powdered terra cotta.

Renovation of Polished and Varnished Surfaces of Wood, Stone, etc.—This is composed of the following ingredients, though the proportions may be varied: Cereal flour or wood pulp, 38½ parts; hydrochloric acid, 45 parts; chloride of lime, 16 parts; turpentine, ½ part. After mixing the ingredients thoroughly in order to form a homogeneous paste, the object to be treated is smeared with it and allowed to stand for some time. The paste on the surface is then removed by passing over it quickly a piece of soft leather or a brush, which will remove dirt, grease, and other deleterious substances. By rubbing gently with a cloth or piece of leather a polished surface will be imparted to wood, and objects of metal will be rendered lustrous.

The addition of chloride of lime tends to keep the paste moist, thus allowing the ready removal of the paste without damaging the varnish or polish, while the turpentine serves as a disinfectant and renders the odor less disagreeable during the operation.

The preparation is rapid in its action, and does not affect the varnished or polished surfaces of wood or marble. While energetic in its cleansing action on brass and other metallic objects, it is attended with no corrosive effect.

Nitrate of Silver Spots.—To remove these spots from white marble, they should be painted with Javelle water, and after having been washed, passed over a concentrated solution of thiosulphate of soda (hyposulphite).

To Remove Oil-Paint Spots from Sandstones.—This may be done by washing the spots with pure turpentine oil, then covering the place with white argillaceous earth (pipe clay), leaving it to dry, and finally rubbing with sharp soda lye, using a brush. Caustic ammonia also removes oil-paint spots from sandstones.

RUST REMOVERS:

To Remove Rust from Iron or Steel Utensils.—

I.—Apply the following solution by means of a brush, after having removed any grease by rubbing with a clean, dry cloth: 100 parts of stannic chloride are dissolved in 1,000 parts of water; this solution is added to one containing 2 parts tartaric acid dissolved in 1,000 parts of water, and finally 20 cubic centimeters indigo solution, diluted with 2,000 parts of water, are added. Afte allowing the solution to act upon the stain for a few seconds, it is rubbed clean, first with a moist cloth, then with a dry cloth; to restore the polish use is made of silver sand and jewelers' rouge.

II.—When the rust is recent it is removed by rubbing the metal with a cork charged with oil. In this manner a perfect polish is obtained. To take off old rust, mix equal parts of fine tripoli and flowers of sulphur, mingling this mixture with olive oil, so as to orm a paste. Rub the iron with this preparation by means of a skin.

III.—The rusty piece is connected with a piece of zinc and placed in water containing a little sulphuric acid. After the articles have been in the liquid for several days or a week, the rust will have completely disappeared. The length of time will depend upon the depth to which the rust has penetrated. A little sulphuric acid may be added from time to time, but the chief point is that the zinc always has good electric contact with the iron. To insure this an iron wire may be firmly wound around the iron object and connected with the zinc. The iron is not attacked in the least, as long as the zinc is kept in good electric contact with it. When the articles are taken from the liquid they assume a dark gray or black color and are then washed and oiled.

IV.—The rust on iron and steel objects, especially large pieces, is readily removed by rubbing the pieces with oil of tartar, or with very fine emery and a little oil, or by putting powdered alum in strong vinegar and rubbing with this alumed vinegar.

V.—Take cyanide of calcium, 25 parts; white soap, powdered, 25 parts; Spanish white, 50 parts; and water, 200 parts. Triturate all well and rub the piece with this paste. The effect will be quicker if before using this paste the rusty object has been soaked for 5 to 10 minutes in a solution of cyanide of potassium in the ratio of 1 part of cyanide to 2 parts of water.

VI.—To remove rust from polished steel cyanide of potassium is excellent. If possible, soak the instrument to be cleaned in a solution of cyanide of potassium in the proportion of 1 ounce of cyanide to 4 ounces of water. Allow this to act till all loose rust is removed, and then polish with cyanide soap. The latter is made as follows: Potassium cyanide, precipitated chalk, white castile soap. Make a saturated solution of the cyanide and add chalk sufficient to make a creamy paste. Add the soap cut in fine shavings and thoroughly incorporate in a mortar. When the mixture is stiff cease to add the soap. It should be remembered that potassium cyanide is a virulent poison.

VII.—Apply turpentine or kerosene oil, and after letting it stand over night, clean with finest emery cloth.

VIII.—To free articles of iron and steel from rust and imbedded grains of sand the articles are treated with fluorhydric acid (about 2 per cent) 1 to 2 hours, whereby the impurities but not the metal are dissolved. This is followed by a washing with lime milk, to neutralize any fluorhydric acid remaining.

To Remove Rust from Nickel.—First grease the articles well; then, after a few days, rub them with a rag charged with ammonia. If the rust spots persist, add a few drops of hydrochloric acid to the ammonia, rub and wipe off at once. Next rinse with water, dry, and polish with tripoli.

Removal of Rust.—To take off the rust from small articles which glass or emery paper would bite too deeply, the ink-erasing rubber used in business offices may be employed. By beveling it, or cutting it to a point as needful, it can be introduced into the smallest cavities and windings, and a perfect cleaning be effected.

To Remove Rust from Instruments.—
I.—Lay the instruments over night in a saturated solution of chloride of tin. The rust spots will disappear through reduction. Upon withdrawal from the solution the instruments are rinsed with water, placed in a hot soda-soap solution, and dried. Cleaning with absolute alcohol and polishing chalk may also follow.

II.—Make a solution of 1 part of kerosene in 200 parts of benzine or carbon tetrachloride, and dip the instruments, which have been dried by leaving them in heated air, in this, moving their parts, if movable, as in forceps and scissors, about under the liquid, so that it may enter all the crevices. Next lay the instruments on a plate in a dry room, so that the benzine can evaporate. Needles are simply thrown in the paraffine solution, and taken out with tongs or tweezers, after which they are allowed to dry on a plate.

III.—Pour olive oil on the rust spots and leave for several days; then rub with emery or tripoli, without wiping off the oil as far as possible, or always bringing it back on the spot. Afterwards remove the emery and the oil with a rag, rub again with emery soaked with vinegar, and finally with fine plumbago on a piece of chamois skin.

To Preserve Steel from Rust.—To preserve steel from rust dissolve 1 part caoutchouc and 16 parts turpentine with a gentle heat, then add 8 parts boiled oil, and mix by bringing them to the heat of boiling water. Apply to the steel with a brush, the same as varnish. It can be removed again with a cloth soaked in turpentine.

METAL CLEANING:

Cleaning and Preserving Medals, Coins, and Small Iron Articles.—The coating of silver chloride may be reduced with molten potassium cyanide. Then boil the article in water, displace the water with alcohol, and dry in a drying closet. When dry brush with a soft brush and cover with "zaponlack" (any good transparent lacquer or varnish will answer).

Instead of potassium cyanide alone, a mixture of that and potassium carbonate may be used. After treatment in this way, delicate objects of silver become less brittle. Another way is to put the article in molten sodium carbonate and remove the silver carbonate thus formed, by acetic acid of 50 per cent strength. This process produces the finest possible polish.

The potassium-cyanide process may be used with all small iron objects. For larger ones molten potassium rhodanide is recommended. This converts the iron oxide into iron sulphide that is easily washed off and leaves the surface of a fine black color.

Old coins may be cleansed by first immersing them in strong nitric acid and then washing them in clean water. Wipe them dry before putting away.

To Clean Old Medals.—Immerse in lemon juice until the coating of oxide has completely disappeared; 24 hours is generally sufficient, but a longer time is not harmful.

Steel Cleaner.—Smear the object with oil, preferably petroleum, and allow some days for penetration of the surface of the metal. Then rub vigorously with a piece of flannel or willow wood. Or, with a paste composed of olive oil, sulphur flowers, and tripoli, or of rotten stone and oil. Finally, a coating may be employed, made of 10 parts of potassium cyanide and 1 part of cream of tartar; or 25 parts of potassium cyanide, with the addition of 55 parts of carbonate of lime and 20 parts of white soap.

Restoring Tarnished Gold.—

Sodium bicarbonate.	20 ounces
Chlorinated lime....	1 ounce
Common salt.......	1 ounce
Water.............	16 ounces

Mix well and apply with a soft brush. A very small quantity of the solution is sufficient, and it may be used either cold or lukewarm. Plain articles may be brightened by putting a drop or two of the liquid upon them and lightly brushing the surface with fine tissue paper.

Cleaning Copper.—

I.—Use Armenian bole mixed into a paste with oleic acid.

II.—Rotten stone....... 1 part
Iron subcarbonate.. 3 parts
Lard oil, a sufficient quantity.

III.—Iron oxide......... 10 parts
Pumice stone...... 32 parts
Oleic acid, a sufficient quantity.

IV.—Soap, cut fine....... 16 parts
Precipitated chalk.. 2 parts
Jewelers' rouge..... 1 part
Cream of tartar..... 1 part
Magnesium carbonate 1 part
Water, a sufficient quantity.

Dissolve the soap in the smallest quantity of water that will effect solution over a water bath. Add the other ingredients to the solution while still hot, stirring constantly.

To Remove Hard Grease, Paint, etc., from Machinery.—To remove grease, paint, etc., from machinery add half a pound of caustic soda to 2 gallons of water and boil the parts to be cleaned in the fluid. It is possible to use it several times before its strength is exhausted.

Solutions for Cleaning Metals.—

I.—Water............. 20 parts
Alum .... 2 parts
Tripoli............ 2 parts
Nitric acid......... 1 part

II.—Water............. 40 parts
Oxalic acid......... 2 parts
Tripoli .... 7 parts

To Cleanse Nickel.—I.—Fifty parts of rectified alcohol; 1 part of sulphuric acid; 1 part of nitric acid. Plunge the piece in the bath for 10 to 15 seconds, rinse it off in cold water, and dip it next into rectified alcohol. Dry with a fine linen rag or with sawdust.

II.—Stearine oil........ 1 part
Ammonia water..... 25 parts
Benzine........... 50 parts
Alcohol........... 75 parts

Rub up the stearine with the ammonia, add the benzine and then the alcohol, and agitate until homogeneous. Put in wide-mouthed vessels and close carefully.

To Clean Petroleum Lamp Burners.—Dissolve in a quart of soft water an ounce or an ounce and a half of washing soda, using an old half-gallon tomato can. Into this put the burner after removing the wick, set it on the stove, and let it boil strongly for 5 or 6 minutes, then take out, rinse under the tap, and dry.

Every particle of carbonaceous matter will thus be got rid of, and the burner be as clean and serviceable as new. This ought to be done at least every month, but the light would be better if it were done every 2 weeks.

Gold-Ware Cleaner.—

Acetic acid........ 2 parts
Sulphuric acid..... 2 parts
Oxalic acid........ 1 part
Jewelers' rouge...... 2 parts
Distilled water...... 200 parts

Mix the acids and water and stir in the rouge, after first rubbing it up with a portion of the liquid. With a clean cloth, wet with this mixture, go well over the article. Rinse off with hot water and dry.

Silverware Cleaner.—Make a thin paste of levigated (not precipitated) chalk and sodium hyposulphite, in equal parts, rubbed up in distilled water. Apply this paste to the surface, rubbing well with a soft brush. Rinse in clear water and dry in sawdust. Some authorities advise the cleaner to let the paste dry on the ware, and then to rub off and rinse with hot water.

Silver-Coin Cleaner.—Make a bath of 10 parts of sulphuric acid and 90 parts of water, and let the coin lie in this until the crust of silver sulphide is dissolved. From 5 to 10 minutes usually suffice. Rinse in running water, then rub with a soft brush and castile soap, rinse again, dry with a soft cloth, and then carefully rub with chamois.

Cleaning Silver-Plated Ware.—Into a wide-mouthed bottle provided with a good cork put the following mixture:

Cream of tartar...... 2 parts
Levigated chalk..... 2 parts
Alum............. 1 part

Powder the alum and rub up with the other ingredients, and cork tightly. When required for use wet sufficient of the powder and with soft linen rags rub the article, being careful not to use much pressure, as otherwise the thin layer of plating may be cut through. Rinse in hot suds, and afterwards in clear water, and dry in sawdust. When badly blackened with silver sulphide, if small, the article may be dipped for an instant in hydrochloric acid and immediately rinsed in running water. Larger articles may be treated as coins are—immersed for 2 or 3 minutes in a 10 per cent aqueous solution of sulphuric acid, or the surface may be rapidly wiped with a swab carrying nitric acid and instantly rinsed in running water.

Cleaning Gilt Bronze Ware.—If greasy, wash carefully in suds, or, better, dip into a hot solution of caustic potash, and then wash in suds with a soft rag, and rinse in running water. If not then clean and bright, dip into the following mixture:

Nitric acid	10 parts
Aluminum sulphate	1 part
Water	40 parts

Mix. Rinse in running water.

Britannia Metal Cleaner.—Rub first with jewelers' rouge made into a paste with oil; wash in suds, rinse, dry, and finish with chamois or wash leather.

To Remove Ink Stains on Silver.—Silver articles in domestic use, and especially silver or plated inkstands, frequently become badly stained with ink. These stains cannot be removed by ordinary processes, but readily yield to a paste of chloride of lime and water. Javelle water may be also used.

Removing Egg Stains.—A pinch of table salt taken between the thumb and finger and rubbed on the spot with the end of the finger will usually remove the darkest egg stain from silver.

To Clean Silver Ornaments.—Make a strong solution of soft soap and water, and in this boil the articles for a few minutes—five will usually be enough. Take out, pour the soap solution into a basin, and as soon as the liquid has cooled down sufficiently to be borne by the hand, with a soft brush scrub the articles with it. Rinse in boiling water and place on a porous substance (a bit of tiling, a brick, or unglazed earthenware) to dry. Finally give a light rubbing with a chamois. Articles thus treated look as bright as new.

Solvent for Iron Rust.—Articles attacked by rust may be conveniently cleaned by dipping them into a well-saturated solution of stannic chloride. The length of time of the action must be regulated according to the thickness of the rust. As a rule 12 to 24 hours will suffice, but it is essential to prevent an excess of acid in the bath, as this is liable to attack the iron itself. After the objects have been removed from the bath they must be rinsed with water, and subsequently with ammonia, and then quickly dried. Greasing with vaseline seems to prevent new formation of rust. Objects treated in this manner are said to resemble dead silver.

Professor Weber proposed a diluted alkali, and it has been found that after employing this remedy the dirt layer is loosened and the green platina reappears. Potash has been found to be an efficacious remedy, even in the case of statues that had apparently turned completely black.

To Clean Polished Parts of Machines.—Put in a flask 1,000 parts of petroleum; add 20 parts of paraffine, shaved fine; cork the bottle and stand aside for a couple of days, giving it an occasional shake. The mixture is now ready for use. To use, shake the bottle, pour a little of the liquid upon a woolen rag and rub evenly over the part to be cleaned; or apply with a brush. Set the article aside and, next day, rub it well with a dry, woolen rag. Every particle of rust, resinified grease, etc., will disappear provided the article has not been neglected too long. In this case a further application of the oil will be necessary. If too great pressure has not been made, or the rubbing continued too long, the residual oil finally leaves the surface protected by a delicate layer of paraffine that will prevent rusting for a long time.

To Clean Articles of Nickel.—Lay them for a few seconds in alcohol containing 2 per cent of sulphuric acid; remove, wash in running water, rinse in alcohol, and rub dry with a linen cloth. This process gives a brilliant polish and is especially useful with plated articles on the plating of which the usual polishing materials act very destructively. The yellowest and brownest nickeled articles are restored to pristine brilliancy by leaving them in the alcohol and acid for 15 seconds. Five seconds suffice ordinarily.

How to Renovate Bronzes.—For gilt work, first remove all grease, dirt, wax, etc., with a solution in water of potassium or sodium hydrate, then dry, and with a soft rag apply the following:

Sodium carbonate	7 parts
Spanish whiting	15 parts
Alcohol, 85 per cent	50 parts
Water	125 parts

Go over every part carefully, using a brush to get into the minute crevices. When this dries on, brush off with a fine linen cloth or a supple chamois skin.

Or the following plan may be used: Remove grease, etc., as directed above, dry and go over the spots where the gilt surface is discolored with a brush dipped in a solution of two parts of alum in 250 parts of water and 65 parts of nitric acid. As soon as the gilding reappears or the surface becomes bright, wash off, and dry in the direct sunlight.

Still another cleaner is made of nitric acid, 30 parts; aluminum sulphate, 4 parts; distilled or rain water, 125 parts. Clean of grease, etc., as above, and apply the solution with a camel's-hair pencil. Rinse off and dry in sawdust. Finally, some articles are best cleaned by immersing in hot soap suds and rubbing with a soft brush. Rinse in clear, hot water, using a soft brush to get the residual suds out of crevices. Let dry, then finish by rubbing the gilt spots or places with a soft, linen rag, or a bit of chamois.

There are some bronzes gilt with imitation gold and varnished. Where the work is well done and the gilding has not been on too long, they will deceive even the practiced eye. The deception, however, may easily be detected by touching a spot on the gilt surface with a glass rod dipped in a solution of corrosive sublimate. If the gilding is true no discoloration will occur, but if false a brown spot will be produced.

To Clean a Gas Stove. — An easy method of removing grease spots consists in immersing the separable parts for several hours in a warm lye, heated to about 70° C. (158° F.), said lye to be made of 9 parts of caustic soda and 180 parts of water. These pieces, together with the fixed parts of the stove, may be well brushed with this lye and afterwards rinsed in clean, warm water. The grease will be dissolved, and the stove restored almost to its original state.

Cleaning Copper Sinks. — Make rotten stone into a stiff paste with soft soap and water. Rub on with a woolen rag, and polish with dry whiting and rotten stone. Finish with a leather and dry whiting. Many of the substances and mixtures used to clean brass will effectively clean copper. Oxalic acid is said to be the best medium for cleaning copper, but after using it the surface of the copper must be well washed, dried, and then rubbed with sweet oil and tripoli, or some other polishing agent. Otherwise the metal will soon tarnish again.

Treatment of Cast-Iron Grave Crosses. — The rust must first be thoroughly removed with a steel-wire brush. When this is done apply one or two coats of red lead or graphite paint. After this priming has become hard, paint with double-burnt lampblack and equal parts of oil of turpentine and varnish. This coating is followed by one of lampblack ground with coach varnish. Now paint the single portions with "mixtion" (gilding oil) and gild as usual. Such crosses look better when they are not altogether black. Ornaments may be very well treated in colors with oil paint and then varnished. The crosses treated in this manner are preserved for many years, but it is essential to use good exterior or coach varnish for varnishing, and not the so-called black varnish, which is mostly composed of asphalt or tar.

Cleaning Inferior Gold Articles. — The brown film which forms on low-quality gold articles is removed by coating with fuming hydrochloric acid, whereupon they are brushed off with Vienna lime and petroleum. Finally, clean the objects with benzine, rinse again in pure benzine, and dry in sawdust.

To Clean Bronze. — Clean the bronze with soft soap; next wash it in plenty of water; wipe, let dry, and apply light encaustic mixture composed of spirit of turpentine in which a small quantity of yellow wax has been dissolved. The encaustic is spread by means of a linen or woolen wad. For gilt bronze, add 1 spoonful of alkali to 3 spoonfuls of water and rub the article with this by means of a ball of wadding. Next wipe with a clean chamois, similar to that employed in silvering.

How to Clean Brass and Steel. — To clean brasses quickly and economically, rub them with vinegar and salt or with oxalic acid. Wash immediately after the rubbing, and polish with tripoli and sweet oil. Unless the acid is washed off the article will tarnish quickly. Copper kettles and saucepans, brass andirons, fenders, and candlesticks and trays are best cleaned with vinegar and salt. Cooking vessels in constant use need only to be well washed afterwards. Things for show — even pots and pans — need the oil polishing, which gives a deep, rich, yellow luster, good for six months. Oxalic acid and salt should be employed for furniture brasses — if it touches the wood it only improves the tone. Wipe the brasses well with a wet cloth, and polish thoroughly with oil and tripoli. Sometimes powdered rotten stone does better than the tripoli. Rub, after using, either with a dry cloth or leather, until there is no trace of oil. The brass to be cleaned must be freed completely from grease, caked dirt, and grime. Wash with strong ammonia suds and rinse dry before beginning with the acid and salt.

The best treatment for wrought steel is to wash it very clean with a stiff brush and ammonia soapsuds, rinse well, dry by heat, oil plentifully with sweet oil, and dust thickly with powdered quicklime. Let the lime stay on 2 days, then brush it off with a clean, very stiff brush. Polish with a softer brush, and rub with cloths until the luster comes out. By leaving the lime on, iron and steel may be kept from rust almost indefinitely.

Before wetting any sort of bric-a-brac, and especially bronzes, remove all the dust possible. After dusting, wash well in strong white soapsuds and ammonia, rinse clean, polish with just a suspicion of oil and rotten stone, and rub off afterwards every trace of the oil. Never let acid touch a bronze surface, unless to eat and pit it for antique effects.

Composition for Cleaning Copper, Nickel, and other Metals.—Wool grease, 46 parts, by weight; fire clay, 30 parts, by weight; paraffine, 5 parts, by weight; Canova wax, 5 parts, by weight; cocoanut oil, 10 parts, by weight; oil of mirbane, 1 part, by weight. After mixing these different ingredients, which constitute a paste, this is molded in order to give a cylindrical form, and introduced into a case so that it can be used like a stick of cosmetic.

Putz Pomade.—I.—Oxalic acid, 1 part; caput mortuum, 15 parts (or, if white pomade is desired, tripoli, 12 parts); powdered pumice stone, best grade, 20 parts; palm oil, 60 parts; petroleum or oleine, 4 parts. Perfume with mirbane oil.

II.—Oxalic acid........   1 part
    Peroxide of iron
        (jewelers' rouge).. 15 parts
    Rotten stone.......   20 parts
    Palm oil...........   60 parts
    Petrolatum........   5 parts

Pulverize the acid and the rotten stone and mix thoroughly with the rouge. Sift to remove all grit, then make into a paste with the oil and petrolatum. A little nitro-benzol may be added to scent the mixture.

III.—Oleine............   40 parts
    Ceresine...........   5 parts
    Tripoli............   40 parts
    Light mineral oil
        (0.870).......... 20 parts

Melt the oleine, ceresine, and mineral oil together, and stir in the tripoli; next, grind evenly in a paint mill.

To Clean Gummed Parts of Machinery.—Boil about 10 to 15 parts of caustic soda or 100 parts of soda in 1,000 parts of water, immerse the parts to be cleaned in this for some time, or, better, boil them with it. Then rinse and dry. For small shops this mode of cleaning is doubtless the best.

To Remove Silver Plating.—I.—Put sulphuric acid 100 parts and potassium nitrate (saltpeter) 10 parts in a vessel of stoneware or porcelain, heated on the water bath. When the silver has left the copper, rinse the objects several times. This silver stripping bath may be used several times, if it is kept in a well-closed bottle. When it is saturated with silver, decant the liquid, boil it to dryness, then add the residue to the deposit, and melt in the crucible to obtain the metal.

II.—Stripping silvered articles of the silvering may be accomplished by the following mixture: Sulphuric acid, 60° B., 3 parts; nitric acid, 40° B., 1 part; heat the mixture to about 166° F., and immerse the articles by means of a copper wire. In a few seconds the acid mixture will have done the work. A thorough rinsing off is, of course, necessary.

To Clean Zinc Articles.—In order to clean articles of zinc, stir rye bran into a paste with boiling water, and add a handful of silver sand and a little vitriol. Rub the article with this paste, rinse with water, dry, and polish with a cloth.

To Remove Rust from Nickel.—Smear the rusted parts well with grease (ordinary animal fat will do), and allow the article to stand several days. If the rust is not thick the grease and rust may be rubbed off with a cloth dipped in ammonia. If the rust is very deep, apply a diluted solution of hydrochloric acid, taking care that the acid does not touch the metal, and the rust may be easily rubbed off. Then wash the article and polish in the usual way.

Compound for Cleaning Brass.—To make a brass cleaning compound use oxalic acid, 1 ounce; rotten stone, 6 ounces; enough whale oil and spirits of turpentine of equal parts, to mix, and make a paste.

To Clean Gilt Objects.—I.—Into an ordinary drinking glass pour about 20 drops of ammonia, immerse the piece to be cleaned repeatedly in this, and brush with a soft brush. Treat the article with pure water, then with alcohol, and wipe with a soft rag.

II.—Boil common alum in soft, pure water, and immerse the article in the solution, or rub the spot with it, and dry with sawdust.

III.—For cleaning picture frames,

moldings, and, in fact, all kinds of gilded work, the best medium is liquor potassæ, diluted with about 5 volumes of water. Dilute alcohol is also excellent. Methylated wood spirit, if the odor is not objectionable, answers admirably.

To Scale Cast Iron.—To remove the scale from cast iron use a solution of 1 part vitriol and 2 parts water; after mixing, apply to the scale with a cloth rolled in the form of a brush, using enough to wet the surface well. After 8 or 10 hours wash off with water, when the hard, scaly surface will be completely removed.

Cleaning Funnels and Measures.— Funnels and measures used for measuring varnishes, oils, etc., may be cleaned by soaking them in a strong solution of lye or pearlash. Another mixture for the same purpose consists of pearlash with quicklime in aqueous solution. The measures are allowed to soak in the solution for a short time, when the resinous matter of the paint or varnish is easily removed. A thin coating of petroleum lubricating oils may be removed, it is said, by the use of naphtha or petroleum benzine.

To Clean Aluminum.—I.—Aluminum articles are very hard to clean so they will have a bright, new appearance. This is especially the case with the matted or frosted pieces. To restore the pieces to brilliancy place them for some time in water that has been slightly acidulated with sulphuric acid.

II.—Wash the aluminum with coal-oil, gasoline or benzine, then put it in a concentrated solution of caustic potash, and after washing it with plenty of water, dip it in the bath composed of $\frac{2}{3}$ nitric acid and $\frac{1}{3}$ water. Next, subject it to a bath of concentrated nitric acid, and finally to a mixture of rum and olive oil. To render aluminum capable of being worked like pure copper, $\frac{2}{3}$ of oil of turpentine and $\frac{1}{3}$ stearic acid are used. For polishing by hand, take a solution of 30 parts of borax and 1,000 parts of water, to which a few drops of spirits of ammonia have been added.

How to Clean Tarnished Silver.—I.— If the articles are only slightly tarnished, mix 3 parts of best washed and purified chalk and 1 part of white soap, adding water, till a thin paste is formed, which should be rubbed on the silver with a dry brush, till the articles are quite bright. As a substitute, whiting, mixed with caustic ammonia to form a paste, may be used. This mixture is very effective, but it irritates the eyes and nose.

II.—An efficacious preparation is obtained by mixing beech-wood ashes, 2 parts; Venetian soap, $\frac{4}{100}$ part; cooking salt, 2 parts; rain water, 8 parts. Brush the silver with this lye, using a somewhat stiff brush.

III.—A solution of crystallized potassium permanganate has been recommended.

IV.—A grayish violet film which silverware acquires from perspiration, can be readily removed by means of ammonia.

V.—To remove spots from silver lay it for 4 hours in soapmakers' lye, then throw on fine powdered gypsum, moisten the latter with vinegar to cause it to adhere, dry near the fire, and wipe off. Next rub the spot with dry bran. This not only causes it to disappear, but gives extraordinary gloss to the silver.

VI.—Cleaning with the usual fine powders is attended with some difficulty and inconvenience. An excellent result is obtained without injury to the silver by employing a saturated solution of hyposulphite of soda, which is put on with a brush or rag. The article is then washed with plenty of water.

VII.—Never use soap on silverware, as it dulls the luster, giving the article more the appearance of pewter than silver. When it wants cleaning, rub it with a piece of soft leather and prepared chalk, made into a paste with pure water, entirely free from grit.

To Clean Dull Gold.—I.—Take 80 parts, by weight, of chloride of lime, and rub it up with gradual addition of water in a porcelain mortar into a thin, even paste, which is put into a solution of 80 parts, by weight, of bicarbonate of soda, and 20 parts, by weight, of salt, in 3,000 parts, by weight, of water. Shake it, and let stand a few days before using. If the preparation is to be kept for any length of time the bottle should be placed, well corked, in the cellar. For use, lay the tarnished articles in a dish, pour the liquid, which has previously been well shaken, over them so as just to cover them, and leave them therein for a few days.

II.—

Bicarbonate of soda.	31	parts
Chloride of lime....	15.5	parts
Cooking salt.......	15	parts
Water.............	240	parts

Grind the chloride of lime with a little water to a thin paste, in a porcelain vessel, and add the remaining chemicals. Wash the objects with the aid of a soft brush with the solution, rinse several times in water, and dry in fine sawdust.

Cleaning Bronze Objects.—Employ powdered chicory mixed with water, so as to obtain a paste, which is applied with a brush. After the brushing, rinse off and dry in the sun or near a stove.

Cleaning Gilded Bronzes.—I.—Commence by removing the spots of grease and wax with a little potash or soda dissolved in water. Let dry, and apply the following mixture with a rag: Carbonate of soda, 7 parts; whiting, 15 parts; alcohol (85°), 50 parts; water, 125 parts. When this coating is dry pass a fine linen cloth or a piece of supple skin over it. The hollow parts are cleaned with a brush.

II.—After removing the grease spots, let dry and pass over all the damaged parts a pencil dipped in the following mixture: Alum, 2 parts; nitric acid, 65; water, 250 parts. When the gilding becomes bright, wipe, and dry in the sun or near a fire.

III.—Wash in hot water containing a little soda, dry, and pass over the gilding a pencil soaked in a liquid made of 30 parts nitric acid, 4 parts of aluminum phosphate, and 125 parts of pure water. Dry in sawdust.

IV.—Immerse the objects in boiling soap water, and facilitate the action of the soap by rubbing with a soft brush; put the objects in hot water, brush them carefully, and let them dry in the air; when they are quite dry rub the shining parts only with an old linen cloth or a soft leather, without touching the others.

Stripping Gilt Articles.—Degilding or stripping gilt articles may be done by attaching the object to the positive pole of a battery and immersing it in a solution composed of 1 pound of cyanide dissolved in about 1 gallon of water. Desilvering may be effected in the same manner.

To Clean Tarnished Zinc.—Apply with a rag a mixture of 1 part sulphuric acid with 12 parts of water. Rinse the zinc with clear water.

Cleaning Pewter Articles.—Pour hot lye of wood ashes upon the tin, throw on sand, and rub with a hard, woolen rag, hat felt, or whisk until all particles of dirt have been dissolved. To polish pewter plates it is well to have the turner make similar wooden forms fitting the plates, and to rub them clean this way. Next they are rinsed with clean water and placed on a table with a clean linen cover on which they are left to dry without being touched, otherwise spots will appear. This scouring is not necessary so often if the pewter is rubbed with wheat bran after use and cleaned perfectly. New pewter is polished with a paste of whiting and brandy, rubbing the dishes with it until the mass becomes dry.

To Clean Files.—Files which have become clogged with tin or lead are cleaned by dipping for a few seconds into concentrated nitric acid. To remove iron filings from the file cuts, a bath of blue vitriol is employed. After the files have been rinsed in water they are likewise dipped in nitric acid. File-ridges closed up by zinc are cleaned by immersing the files in diluted sulphuric acid. Such as have become filled with copper or brass are also treated with nitric acid, but here the process has to be repeated several times. The files should always be rinsed in water after the treatment, brushed with a stiff brush, and dried in sawdust or by pouring alcohol over them, and letting it burn off on the file.

Scale Pan Cleaner.—About the quickest cleaner for brass scale pans is a solution of potassium bichromate in dilute sulphuric acid, using about 1 part of chromate, in powder, to 3 parts of acid and 6 parts of water. In this imbibe a cloth wrapped around a stick (to protect the hands), and with it rub the pans. Do this at tap or hydrant, so that no time is lost in placing the pan in running water after having rubbed it with the acid solution. For pans not very badly soiled rubbing with ammonia water and rinsing is sufficient.

Tarnish on Electro-Plate Goods.—This tarnish can be removed by dipping the article for from 1 to 15 minutes—that is, until the tarnish shall have been removed—in a pickle of the following composition: Rain water 2 gallons and potassium cyanide ½ pound. Dissolve together, and fill into a stone jug or jar, and close tightly. The article, after having been immersed, must be taken out and thoroughly rinsed in several waters, then dried with fine, clean sawdust. Tarnish on jewelry can be speedily removed by this process; but if the cyanide is not completely removed it will corrode the goods.

OIL-, GREASE-, PAINT-SPOT ERADICATORS:

Grease- and Paint-Spot Eradicators.—

I.—Benzol............ 500 parts
  Benzine.......... 500 parts
  Soap, best white,
    shaved......... 5 parts
  Water, warm, sufficient.

Dissolve the soap in the warm water, using from 50 to 60 parts. Mix the benzol and benzine, and add the soap solution, a little at a time, shaking up well after each addition. If the mixture is slow in emulsifying, add at one time from 50 to 100 parts of warm water, and shake violently. Set the emulsion aside for a few days, or until it separates, then decant the superfluous water, and pour the residual pasty mass, after stirring it up well, into suitable boxes.

II.—Soap spirit........ 100 parts
    Ammonia solution,
      10 per cent...... 25 parts
    Acetic ether....... 15 parts

III.—Extract of quillaia . 1 part
    Borax........... 1 part
    Ox gall, fresh..... 6 parts
    Tallow soap...... 15 parts

Triturate the quillaia and borax together, incorporate the ox gall, and, finally, add the tallow soap and mix thoroughly by kneading. The product is a plastic mass, which may be rolled into sticks or put up into boxes.

Removing Oil Spots from Leather.—To remove oil stains from leather, dab the spot carefully with spirits of sal ammoniac, and after allowing it to act for a while, wash with clean water. This treatment may have to be repeated a few times, taking care, however, not to injure the color of the leather. Sometimes the spot may be removed very simply by spreading the place rather thickly with butter and letting this act for a few hours. Next scrape off the butter with the point of a knife, and rinse the stain with soap and lukewarm water.

To Clean Linoleum.—Rust spots and other stains can be removed from linoleum by rubbing with steel chips.

To Remove Putty, Grease, etc., from Plate Glass.—To remove all kinds of greasy materials from glass, and to leave the latter bright and clean, use a paste made of benzine and burnt magnesia of such consistence that when the mass is pressed between the fingers a drop of benzine will exude. With this mixture and a wad of cotton, go over the entire surface of the glass, rubbing it well. One rubbing is usually sufficient. After drying, any of the substance left in the corners, etc., is easily removed by brushing with a suitable brush. The same preparation is very useful for cleaning mirrors and removing grease stains from books, papers, etc.

Removing Spots from Furniture.—White spots on polished tables are removed in the following manner: Coat the spot with oil and pour on a rag a few drops of "mixtura balsamica oleosa," which can be bought in every drug store, and rub on the spot, which will disappear immediately.

To Remove Spots from Drawings, etc.—Place soapstone, fine meerschaum shavings, amianthus, or powdered magnesia on the spot, and, if necessary, lay on white filtering paper, saturating it with peroxide of hydrogen. Allow this to act for a few hours, and remove the application with a brush. If necessary, repeat the operation. In this manner black coffee spots were removed from a valuable diagram without erasure by knife or rubber.

WATCHMAKERS' AND JEWELERS' CLEANING PREPARATIONS:

To Clean the Tops of Clocks in Repairing.—Sprinkle whiting on the top. Pour good vinegar over this and rub vigorously. Rinse in clean water and dry slowly in the sun or at the fire. A good polish will be obtained.

To Clean Watch Chains.—Gold or silver watch chains can be cleaned with a very excellent result, no matter whether they be matt or polished, by laying them for a few seconds in pure aqua ammonia; they are then rinsed in alcohol, and finally shaken in clean sawdust, free from sand. Imitation gold and plated chains are first cleaned in benzine, then rinsed in alcohol, and afterwards shaken in dry sawdust. Ordinary chains are first dipped in the following pickle: Pure nitric acid is mixed with concentrated sulphuric acid in the proportion of 10 parts of the former to 2 parts of the latter; a little table salt is added. The chains are boiled in this mixture, then rinsed several times in water, afterwards in alcohol, and finally dried in sawdust.

Cleaning Brass Mountings on Clock Cases, etc.—The brass mountings are first cleaned of dirt by dipping them for a short time into boiling soda lye, and next are pickled, still warm, if possible, in a mixture consisting of nitric acid, 60 parts; sulphuric acid, 40 parts; cooking salt, 1 part; and shining soot (lampblack), ½ part, whereby they acquire a handsome golden-yellow coloring. The pickling mixture, however, must not be employed immediately after pouring together the acids, which causes a strong generation of heat, but should settle for at least

1 day. This makes the articles handsomer and more uniform. After the dipping the objects are rinsed in plenty of clean water and dried on a hot, iron plate, and at the same time warmed for lacquering. Since the pieces would be lacquered too thick and unevenly in pure gold varnish, this is diluted with alcohol, 1 part of gold varnish sufficing for 10 parts of alcohol. Into this liquid dip the mountings previously warmed and dry them again on the hot plate.

Gilt Zinc Clocks.—It frequently happens that clocks of gilt zinc become covered with green spots. To remove such spots the following process is used: Soak a small wad of cotton in alkali and rub it on the spot. The green color will disappear at once, but the gilding being gone, a black spot will remain. Wipe off well to remove all traces of the alkali. To replace the gilding, put on, by means of liquid gum arabic, a little bronze powder of the color of the gilding. The powdered bronze is applied dry with the aid of a brush or cotton wad. When the gilding of the clock has become black or dull from age, it may be revived by immersion in a bath of cyanide of potassium, but frequently it suffices to wash it with a soft brush in soap and water, in which a little carbonate of soda has been dissolved. Brush the piece in the lather, rinse in clean water, and dry in rather hot sawdust. The piece should be dried well inside and outside, as moisture will cause it to turn black.

To Clean Gummed Up Springs.—Dissolve caustic soda in warm water, place the spring in the solution and leave it there for about one half hour. Any oil still adhering may now easily be taken off with a hard brush; next, dry the spring with a clean cloth. In this manner gummed up parts of tower clocks, locks, etc., may be quickly and thoroughly cleaned, and oil paint may be removed from metal or wood. The lye is sharp, but free from danger, nor are the steel parts attacked by it.

To Clean Soldered Watch Cases.—Gold, silver, and other metallic watch cases which in soldering have been exposed to heat, are laid in diluted sulphuric acid (1 part acid to 10 to 15 parts water), to free them from oxide. Heating the acid accelerates the cleaning process. The articles are then well rinsed in water and dried. Gold cases are next brushed with powdered tripoli moistened with oil, to remove the pale spots caused by the heat and boiling, and to restore the original color. After that they are cleaned with soap water and finally polished with rouge. Silver cases are polished after boiling, with a scratch brush dipped in beer.

A Simple Way to Clean a Clock.—Take a bit of cotton the size of a hen's egg, dip it in kerosene and place it on the floor of the clock, in the corner; shut the door of the clock, and wait 3 or 4 days. The clock will be like a new one—and if you look inside you will find the cotton batting black with dust. The fumes of the oil loosen the particles of dust, and they fall, thus cleaning the clock.

To Restore the Color of a Gold or Gilt Dial.—Dip the dial for a few seconds in the following mixture: Half an ounce of cyanide of potassium is dissolved in a quart of hot water, and 2 ounces of strong ammonia, mixed with $\frac{1}{2}$ an ounce of alcohol, are added to the solution. On removal from this bath, the dial should immediately be immersed in warm water, then brushed with soap, rinsed, and dried in hot boxwood dust. Or it may simply be immersed in dilute nitric acid; but in this case any painted figures will be destroyed.

A Bath for Cleaning Clocks.—In an enameled iron or terra-cotta vessel pour 2,000 parts of water, add 50 parts of scraped Marseilles soap, 80 to 100 parts of whiting, and a small cup of spirits of ammonia. To hasten the process of solution, warm, but do not allow to boil.

If the clock is very dirty or much oxidized, immerse the pieces in the bath while warm, and as long as necessary. Take them out with a skimmer or strainer, and pour over them some benzine, letting the liquid fall into an empty vessel. This being decanted and bottled can be used indefinitely for rinsing.

If the bath has too much alkali or is used when too hot, it may affect the polish and render it dull. This may be obviated by trying different strengths of the alkali. Pieces of blued steel are not injured by the alkali, even when pure.

To Remove a Figure or Name from a Dial.—Oil of spike lavender may be employed for erasing a letter or number. Enamel powder made into a paste with water, oil, or turpentine is also used for this purpose. It should be previously levigated so as to obtain several degrees of fineness. The powder used for repolishing the surface, where an impression has been removed, must be extremely fine. It is applied with a piece of pegwood or ivory. The best method is to employ diamond powder. Take a little of the powder, make into a paste with fine oil, on the end of a copper polisher the surface of which has been freshly filed and slightly rounded. The marks will rapidly disappear when rubbed with this. The surface is left a little dull; it may be rendered bright by rubbing with the same powder mixed with a greater quantity of oil, and applied with a stick of pegwood. Watchmakers will do well to try on disused dials several degrees of fineness of the diamond powder.

Cleaning Pearls.—Pearls turn yellow in the course of time by absorbing perspiration on account of being worn in the hair, at the throat, and on the arms. There are several ways of rendering them white again.

I.—The best process is said to be to put the pearls into a bag with wheat bran and to heat the bag over a coal fire, with constant motion.

II.—Another method is to bring 8 parts each of well-calcined, finely powdered lime and wood charcoal, which has been strained through a gauze sieve, to a boil with 500 parts of pure rain water, suspend the pearls over the steam of the boiling water until they are warmed through, and then boil them in the liquid for 5 minutes, turning frequently. Let them cool in the liquid, take them out, and wash off well with clean water.

III.—Place the pearls in a piece of fine linen, throw salt on them, and tie them up. Next rinse the tied-up pearls in lukewarm water until all the salt has been extracted, and dry them at an ordinary temperature.

IV.—The pearls may also be boiled about ¼ hour in cow's milk into which a little cheese or soap has been scraped; take them out, rinse off in fresh water, and dry them with a clean, white cloth.

V.—Another method is to have the pearls, strung on a silk thread or wrapped up in thin gauze, mixed in a loaf of bread of barley flour and to have the loaf baked well in an oven, but not too brown. When cool remove the pearls.

VI.—Hang the pearls for a couple of minutes in hot, strong, wine vinegar or highly diluted sulphuric acid, remove, and rinse them in water. Do not leave them too long in the acid, otherwise they will be injured by it.

GLASS CLEANING:

Cleaning Preparation for Glass with Metal Decorations.—Mix 1,000 parts of denaturized spirit (96 per cent) with 150 parts, by weight, of ammonia; 20 parts of acetic ether; 15 parts of ethylic ether; 200 parts of Vienna lime; 950 parts of bolus; and 550 parts of oleine. With this mixture both glass and metal can be quickly and thoroughly cleaned. It is particularly recommended for show windows ornamented with metal.

Paste for Cleaning Glass.—

Prepared chalk......	6 pounds
Powdered French chalk............	1½ pounds
Phosphate calcium...	2¼ pounds
Quillaia bark.......	2¼ pounds
Carbonate ammonia..	18 ounces
Rose pink...........	6 ounces

Mix the ingredients, in fine powder, and sift through muslin. Then mix with soft water to the consistency of cream, and apply to the glass by means of a soft rag or sponge; allow it to dry on, wipe off with a cloth, and polish with chamois.

Cleaning Optical Lenses.—For this purpose a German contemporary recommends vegetable pith. The medulla of rushes, elders, or sunflowers is cut out, the pieces are dried and pasted singly alongside of one another upon a piece of cork, whereby a brush-like apparatus is obtained, which is passed over the surface of the lens. For very small lenses pointed pieces of elder pith are employed. To dip dirty and greasy lenses into oil of turpentine or ether and rub them with a linen rag, as has been proposed, seems hazardous, because the Canada balsam with which the lenses are cemented might dissolve.

To Remove Glue from Glass.—If glue has simply dried upon the glass hot water ought to remove it. If, however, the spots are due to size (the gelatinous wash used by painters) when dried they become very refractory and recourse must be had to chemical means for their removal. The commonest size being a solution of gelatin, alum, and rosin dissolved in a solution of soda and combined with starch, hot solutions of caustic soda or of potash may be used. If that fails to remove them, try diluted hydrochloric, sulphuric, or any of the stronger acids. If the spots still remain some abrasive powder (flour of emery) must be used and the glass repolished with jewelers' rouge applied by means of a chamois skin. Owing to the varied nature of sizes used the above are only suggestions.

Cleaning Window Panes.—Take diluted nitric acid about as strong as strong vinegar and pass it over the glass pane, leave it to act a minute and throw on pulverized whiting, but just enough to give off a hissing sound.   Now rub both with the hand over the whole pane and polish with a dry rag.   Rinse off with clean water and a little alcohol and polish dry and clear.   Repeat the process on the other side.   The nitric acid removes all impurities which have remained on the glass at the factory, and even with inferior panes a good appearance is obtained.

To Clean Store Windows.—For cleaning the large panes of glass of store windows, and also ordinary show cases, a semiliquid paste may be employed, made of calcined magnesia and purified benzine.   The glass should be rubbed with a cotton rag until it is brilliant.

Cleaning Lamp Globes.—Pour 2 spoonfuls of a slightly heated solution of potash into the globe, moisten the whole surface with it, and rub the stains with a fine linen rag; rinse the globe with clean water and carefully dry it with a fine, soft cloth.

To Clean Mirrors.—Rub the mirror with a ball of soft paper slightly dampened with methylated spirits, then with a duster on which a little whiting has been sprinkled, and finally polish with clean paper or a wash leather.   This treatment will make the glass beautifully bright.

To Clean Milk Glass.—To remove oil spots from milk glass panes and lamp globes, knead burnt magnesia with benzine to a plastic mass, which must be kept in a tight-closing bottle.   A little of this substance rubbed on the spot with a linen rag will make it disappear.

To Remove Oil-Paint Spots from Glass.—If the window panes have been bespattered with oil paint in painting walls, the spots are, of course, easily removed while wet.   When they have become dry the operation is more difficult and alcohol and turpentine in equal parts, or spirit of sal ammoniac should be used to soften the paint.   After that go over it with chalk.   Polishing with salt will also remove paint spots.   The salt grates somewhat, but it is not hard enough to cause scratches in the glass; a subsequent polishing with chalk is also advisable, as the drying of the salt might injure the glass.   For scratching off soft paint spots sheet zinc must be used, as it cannot damage the glass on account of its softness.   In the case of silicate paints (the so-called weather-proof coatings) the panes must be especially protected, because these paints destroy the polish of the glass.   Rubbing the spots with brown soap is also a good way of removing the spots, but care must be taken in rinsing off that the window frames are not acted upon.

Removing Silver Stains.—The following solution will remove silver stains from the hands, and also from woolen, linen, or cotton goods:

Mercuric chloride....	1 part
Ammonia muriate....	1 part
Water..............	8 parts

The compound is poisonous.

MISCELLANEOUS CLEANING METHODS AND PROCESSES:

Universal Cleaner.—

Green soap.......	20 to 25 parts
Boiling water......	750 parts
Liquid ammonia, caustic.........	30 to 40 parts
Acetic ether........	20 to 30 parts

Mix.

To Clean Playing Cards.—Slightly soiled playing cards may be made clean by rubbing them with a soft rag dipped in a solution of camphor.   Very little of the latter is necessary.

To Remove Vegetable Growth from Buildings.—To remove moss and lichen from stone and masonry, apply water in which 1 per cent of carbolic acid has been dissolved.   After a few hours the plants can be washed off with water.

Solid Cleansing Compound.—The basis of most of the solid grease eradicators is benzine and the simplest form is a benzine jelly made by shaking 3 ounces of tincture of quillaia (soap bark) with enough benzine to make 16 fluidounces. Benzine may also be solidified by the use of a soap with addition of an excess of alkali.   Formulas in which soaps are used in this way follow:

I.—Cocoanut-oil soap.	2 av. ounces
Ammonia water...	3 fluidounces
Solution of potassium...........	1½ fluidounces
Water enough to make..........	12 fluidounces

Dissolve the soap with the aid of heat in 4 fluidounces of water, add the ammonia and potassa and the remainder of the water.

If the benzine is added in small portions, and thoroughly agitated, 2½ fluidounces of the above will be found sufficient to solidify 32 fluidounces of benzine.

II.—Castile soap, white.   3½ av. ounces
    Water, boiling.....   3½ fluidounces
    Water of ammonia   5 fluidrachms
    Benzine enough to
        make..........   16 fluidounces

Dissolve the soap in the water, and when cold, add the other ingredients.

To Clean Oily Bottles.—Use 2 heaped tablespoonfuls (for every quart of capacity) of fine sawdust or wheat bran, and shake well to cover the interior surface thoroughly; let stand a few minutes and then add about a gill of cold water. If the bottle be then rotated in a horizontal position, it will usually be found clean after a single treatment. In the case of drying oils, especially when old, the bottles should be moistened inside with a little ether, and left standing a few hours before the introduction of sawdust. This method is claimed to be more rapid and convenient than the customary one of using strips of paper, soap solution, etc.

Cork Cleaner.—Wash in 10 per cent solution of hydrochloric acid, then immerse in a solution of sodium hyposulphite and hydrochloric acid. Finally the corks are washed with a solution of soda and pure water. Corks containing oil or fat cannot be cleaned by this method.

To Clean Sponges.—Rinse well first in very weak, warm, caustic-soda lye, then with clean water, and finally leave the sponges in a solution of bromine in water until clean. They will whiten sooner if exposed to the sun in the bromine water. Then repeat the rinsings in weak lye and clean water, using the latter till all smell of bromine has disappeared. Dry quickly and in the sun if possible.

CLEARING BATHS:
See Photography.

CLICHÉ METALS:
See Alloys.

CLOCK-DIAL LETTERING:
See Watchmakers' Formulas.

CLOCK-HAND COLORING:
See Metals.

CLOCK OIL:
See Oil.

CLOCK REPAIRING:
See Watchmaking.

CLOCKMAKERS' CLEANING PROCESSES.
See Cleaning Preparations and Methods.

CLOTH TO IRON, GLUEING:
See Adhesives.

CLOTHES CLEANERS:
See Cleaning Preparations and Methods; also, Household Formulas.

CLOTHS FOR POLISHING:
See Polishes.

CLOTH, WATERPROOFING:
See Waterproofing.

CLOTHING, CARE OF:
See Household Formulas.

COACH VARNISH:
See Varnishes.

COALS, TO EAT BURNING:
See Pyrotechnics.

COAL OIL:
See Oil.

COBALTIZING:
See Plating.

COCOAS:
See Beverages.

COCOA CORDIAL:
See Wines and Liquors.

COCOANUT CAKE:
See Household Formulas and Recipes.

COCHINEAL INSECT REMEDY:
See Insecticides.

COD-LIVER OIL AND ITS EMULSION:
See Oil, Cod-Liver.

COFFEE, SUBSTITUTES FOR.

I.—Acorn.—From acorns deprived of their shells, husked, dried, and roasted.

II.—Bean.—Horse beans roasted along with a little honey or sugar.

III.—Beet Root.—From the yellow beet root, sliced, dried in a kiln or oven, and ground with a little coffee.

IV.—Dandelion.—From dandelion roots, sliced, dried, roasted, and ground with a little caramel.

All the above are roasted, before grinding them, with a little fat or lard. Those which are larger than coffee berries are cut into small slices before being roasted. They possess none of the exhilarating properties or medicinal virtues of the genuine coffee.

V.—Chicory.—This is a common adulterant. The roasted root is prepared by cutting the full-grown root into slices, and exposing it to heat in iron cylinders, along with about 1½ per cent or 2 per cent of lard, in a similar way to that adopted for coffee. When ground to powder in a mill it constitutes the chicory coffee so generally employed both as a substitute for coffee and as an adulterant. The addition of 1 part of good, fresh, roasted chicory to 10 or 12 parts of coffee forms a mixture which yields a beverage of a fuller flavor, and of a deeper color than that furnished by an equal quantity of pure or unmixed coffee. In this way a less quantity of coffee may be used, but it should be remembered that the article substituted for it does not possess in any degree the peculiar exciting, soothing, and hunger-staying properties of that valuable product. The use, however, of a larger proportion of chicory than that just named imparts to the beverage an insipid flavor, intermediate between that of treacle and licorice; while the continual use of roasted chicory, or highly chicorized coffee, seldom fails to weaken the powers of digestion and derange the bowels.

COFFEE CORDIAL:
See Wines and Liquors.

COFFEE EXTRACTS:
See Essences and Extracts.

COFFEE SYRUPS:
See Syrups.

COFFEE FOR THE SODA FOUNTAIN:
See Beverages.

COIL SPRING:
See Steel.

COIN CLEANING:
See Cleaning Preparations and Methods.

COINS, IMPRESSIONS OF:
See Matrix Mass.

COIN METAL:
See Alloys.

COLAS:
See Veterinary Formulas.

Cold and Cough Mixtures

Cough Syrup.—The simplest form of cough syrup of good keeping quality is syrup of wild cherry containing ammonium chloride in the dose of $2\frac{1}{2}$ grains to each teaspoonful. Most of the other compounds contain ingredients that are prone to undergo fermentation.

I.—Ipecacuanha wine   1 fluidounce
Spirit of anise....   1 fluidrachm
Syrup...........   16 fluidounces
Syrup of squill....   8 fluidounces
Tincture of Tolu.   4 fluidrachms
Distilled water
enough to make   30 fluidounces

II.—Heroin..........   6 grains
Aromatic sulphuric acid....... $1\frac{1}{2}$ fluidounces
Concentrated acid infusion of roses   4 fluidounces
Distilled water....   5 fluidounces
Glycerine........   5 fluidounces
Oxymel of squill..   10 fluidounces

III.—Glycerine........   2 fluidounces
Fluid extract of wild cherry....   4 fluidounces
Oxymel.........   10 fluidounces
Syrup...........   10 fluidounces
Cochineal, a sufficient quantity.

Benzoic-Acid Pastilles.—

Benzoic acid.......   105 parts
Rhatany extract....   525 parts
Tragacanth........   35 parts
Sugar.............   140 parts

The materials, in the shape of powders, are mixed well and sufficient fruit paste added to bring the mass up to 4,500 parts. Roll out and divide into lozenges weighing 20 grains each.

Cough Balsam with Iceland Moss.—

Solution of morphine acetate..........   12 parts
Sulphuric acid, dilute   12 parts
Cherry-laurel water.   12 parts
Orange-flower water, triple............   24 parts
Syrup, simple......   128 parts
Glycerine..........   48 parts
Tincture of saffron..   8 parts
Decoction of Iceland moss............   112 parts

Mix. Dose: One teaspoonful.

Balsamic Cough Syrup.—

Balsam of Peru......   2 drachms
Tincture of Tolu....   4 drachms
Camphorated tincture of opium ....   4 ounces
Powdered extract licorice.............   1 ounce
Syrup squill.........   4 ounces
Syrup dextrine (glucose) sufficient to make............   16 ounces

Add the balsam of Peru to the tinctures, and in a mortar rub up the extract of licorice with the syrups. Mix together and direct to be taken in teaspoonful doses.

Whooping-Cough Remedies.—The following mixture is a spray to be used in the sick room in cases of whooping cough:

Thymol	1.0
Tincture of eucalyptus.	30.0
Tincture of benzoin	30.0
Alcohol	100.0
Water enough to make	1000.0

Mix. Pour some of the mixture on a cloth and hold to mouth so that the mixture is inhaled, thereby giving relief.

Expectorant Mixtures.—

I.—

Ammon. chloride.	1 drachm
Potass. chlorate	30 grains
Paregoric	2 fluidrachms
Syrup of ipecac	2 fluidrachms
Syrup wild cherry enough to make	2 fluidounces

Dose: One teaspoonful.

II.—

Potass. chlorate	1 drachm
Tincture guaiac	3½ drachms
Tincture rhubarb.	1½ drachms
Syrup wild cherry enough to make	3 fluidounces

Dose: One teaspoonful.

Eucalyptus Bonbons for Coughs.—

Eucalyptus oil	5 parts
Tartaric acid	15 parts
Extract of malt	24 parts
Cacao	100 parts
Peppermint oil	1.4 parts
Bonbon mass	2,203 parts

Mix and make into bonbons weighing 30 grains each.

COLD CREAM:
See Cosmetics.

COLIC IN CATTLE:
See Veterinary Formulas.

COLLODION.

Turpentine	5 parts
Ether and alcohol	10 parts
Collodion	94 parts
Castor oil	1 part

Dissolve the turpentine in the ether and alcohol mixture (in equal parts) and filter, then add to the mixture of collodion and castor oil. This makes a good elastic collodion.
See also Court Plaster, Liquid.

COLOGNE:
See Perfumes.

COLOGNE FOR HEADACHES:
See Headaches.

COLORS:
See Dyes and Pigments.

COLORS, FUSIBLE ENAMEL:
See Enameling.

COLORS FOR PAINTS:
See Paint.

COLOR PHOTOGRAPHY:
See Photography.

COLORS FOR SYRUPS:
See Syrups.

CONCRETE:
See Stone, Artificial.

Condiments

Chowchow.—

Curry powder	4 ounces
Mustard powder	6 ounces
Ginger	3 ounces
Turmeric	2 ounces
Cayenne	2 drachms
Black pepper powder.	2 drachms
Coriander	1 drachm
Allspice	1 drachm
Mace	30 grains
Thyme	30 grains
Savory	30 grains
Celery seed	2 drachms
Cider vinegar	2 gallons

Mix all the powders with the vinegar, and steep the mixture over a very gentle fire for 3 hours. The pickles are to be parboiled with salt, and drained, and the spiced vinegar, prepared as above, is to be poured over them while it is still warm. The chowchow keeps best in small jars, tightly covered.

Essence of Extract of Soup Herbs.—Thyme, 4 ounces; winter savory, 4 ounces; sweet marjoram, 4 ounces; sweet basil, 4 ounces; grated lemon peel, 1 ounce; eschalots, 2 ounces; bruised celery seed, 1 ounce; alcohol (50 per cent), 64 ounces. Mix the vegetables, properly bruised, add the alcohol, close the container and set aside in a moderately warm place to digest for 15 days. Filter and press out. Preserve in 4-ounce bottles, well corked.

Tomato Bouillon Extract.—Tomatoes, 1 quart; arrowroot, 2 ounces; extract of beef, 1 ounce; bay leaves, 1 ounce; cloves, 2 ounces; red pepper, 4 drachms; Worcestershire sauce, quantity sufficient to flavor. Mix.

Mock Turtle Extract.—Extract of beef, 2 ounces; concentrated chicken, 2 ounces; clam juice, 8 ounces; tincture of black pepper, 1 ounce; extract of celery, 3 drachms; extract of orange peel, soluble, 1 drachm; hot water enough to make 2 quarts,

RELISHES:

Digestive Relish.—

I.—Two ounces Jamaica ginger; 2 ounces black peppercorns; 1 ounce mustard seed; 1 ounce coriander fruit (seed); 1 ounce pimento (allspice); ½ ounce mace; ½ ounce cloves; ½ ounce nutmegs; ½ ounce chili pods; 3 drachms cardamom seeds; 4 ounces garlic; 4 ounces eschalots; 4 pints malt vinegar.

Bruise spices, garlic, etc., and boil in vinegar for 15 minutes and strain. To this add 2½ pints mushroom ketchup; 1½ pints India soy.

Again simmer for 15 minutes and strain through muslin.

II.—One pound soy; 50 ounces best vinegar; 4 ounces ketchup; 4 ounces garlic; 4 ounces eschalots; 4 ounces capsicum; ½ ounce cloves; ½ ounce mace; ¼ ounce cinnamon; 1 drachm cardamom seeds. Boil well and strain.

Lincolnshire Relish.—

Two ounces garlic; 2 ounces Jamaica ginger; 3 ounces black peppercorns; ¾ ounce cayenne pepper; ¼ ounce ossein; ¾ ounce nutmeg; 2 ounces salt; 1½ pints India soy. Enough malt vinegar to make 1 gallon. Bruise spices, garlic, etc., and simmer in ½ a gallon of vinegar for 20 minutes, strain and add soy and sufficient vinegar to make 1 gallon, then boil for 5 minutes. Keep in bulk as long as possible.

Curry Powder.—

I.—

Coriander seed	6	drachms
Turmeric	5	scruples
Fresh ginger	4½	drachms
Cumin seed	18	grains
Black pepper	54	grains
Poppy seed	94	grains
Garlic	2	heads
Cinnamon	1	scruple
Cardamom	5	seeds
Cloves	8	only
Chillies	1 or 2	pods
Grated cocoanut	½	nut

II.—

Coriander seed	¼	pound
Turmeric	¼	pound
Cinnamon seed	2	ounces
Cayenne	½	ounce
Mustard	1	ounce
Ground ginger	1	ounce
Allspice	½	ounce
Fenugreek seed	2	ounces

TABLE SAUCES:

Worcestershire Sauce.—

Pimento	2	drachms
Clove	1	drachm
Black pepper	1	drachm
Ginger	1	drachm
Curry powder	1	ounce
Capsicum	1	drachm
Mustard	2	ounces
Shallots, bruised	2	ounces
Salt	2	ounces
Brown sugar	8	ounces
Tamarinds	4	ounces
Sherry wine	1	pint
Wine vinegar	2	pints

The spices must be freshly bruised. The ingredients are to simmer together with the vinegar for an hour, adding more of the vinegar as it is lost by evaporation; then add the wine, and if desired some caramel coloring. Set aside for a week, strain, and bottle.

Table Sauce.—

Brown sugar, 16 parts; tamarinds, 16 parts; onions, 4 parts; powdered ginger, 4 parts; salt, 4 parts; garlic, 2 parts; cayenne, 2 parts; soy, 2 parts; ripe apples, 64 parts; mustard powder, 2 parts; curry powder, 1 part; vinegar, quantity sufficient. Pare and core the apples, boil them in sufficient vinegar with the tamarinds and raisins until soft, then pulp through a fine sieve. Pound the onions and garlic in a mortar and add the pulp to that of the apples. Then add the other ingredients and vinegar, 60 parts; heat to boiling, cool, and add sherry wine, 10 parts, and enough vinegar to make the sauce just pourable. If a sweet sauce is desired add sufficient treacle before the final boiling.

Epicure's Sauce.—

Eight ounces tamarinds; 12 ounces sultana raisins; 2 ounces garlic; 4 ounces eschalots; 4 ounces horse-radish root; 2 ounces black pepper; ½ ounce chili pods; 3 ounces raw Jamaica ginger; 1½ pounds golden syrup; 1 pound burnt sugar (caramel); 1 ounce powdered cloves; 1 pint India soy; 1 gallon malt vinegar. Bruise roots, spices, etc., and boil in vinegar for 15 minutes, then strain. To the strained liquor add golden syrup, soy, and burnt sugar, then simmer for 10 minutes.

Piccalilli Sauce.—

One drachm chili pods; 1½ ounces black peppercorns; ½ ounce pimento; ¾ ounce garlic; ½ gallon malt vinegar. Bruise spices and garlic, boil in the vinegar for 10 minutes, and strain.

One ounce ground Jamaica ginger; 1 ounce turmeric; 2 ounces flower of mustard; 2 ounces powdered natal arrowroot; 8 ounces strong acetic acid. Rub powders in a mortar with acetic acid and add to above, then boil for 5 minutes, or until it thickens.

FLAVORING SPICES.

I.—Five ounces powdered cinnamon bark; 2½ ounces powdered cloves; 2½ ounces powdered nutmegs; 1¼ ounces powdered caraway seeds; 1¼ ounces powdered coriander seeds; 1 ounce powdered Jamaica ginger; ½ ounce powdered allspice. Let all be dry and in fine powder. Mix and pass through a sieve.

II.—Pickling Spice.—Ten pounds small Jamaica ginger; 2½ pounds black peppercorns; 1¼ pounds white peppercorns; 1½ pounds allspice; ¾ pound long pepper; 1¼ pounds mustard seed; ½ pound chili pods. Cut up ginger and long pepper into small pieces, and mix all the other ingredients intimately.

One ounce to each pint of boiling vinegar is sufficient, but it may be made stronger if desired hot.

Essence of Savory Spices.—Two and one-half ounces black peppercorns; 1 ounce pimento; ¾ ounce nutmeg; ½ ounce mace; ½ ounce cloves; ¼ ounce cinnamon bark; ¼ ounce caraway seeds; 20 grains cayenne pepper; 15 ounces spirit of wine; 5 ounces distilled water. Bruise all the spices and having mixed spirit and water, digest in mixture 14 days, shaking frequently, then filter.

MUSTARD:

The Prepared Mustards of Commerce.—The mustard, i. e., the flower or powdered seed, used in preparing the different condiments, is derived from three varieties of Brassica (*Cruciferæ*)—*Brassica alba* L., *Brassica nigra*, and *Brassica juncea*. The first yields the "white" seed of commerce, which produces a mild mustard; the second the "black" seed, yielding the more pungent powder; and the latter a very pungent and oily mustard, much employed by Russians. The pungency of the condiment is also affected by the method of preparing the paste, excessive heat destroying the sharpness completely. The pungency is further controlled and tempered, in the cold processes, by the addition of wheat or rye flour, which also has the advantage of serving as a binder of the mustard. The mustard flour is prepared by first decorticating the seed, then grinding to a fine powder, the expression of the fixed oil from which completes the process. This oil, unlike the volatile, is of a mild, pleasant taste, and of a greenish color, which, it is said, makes it valuable in the sophistication and imitation of "olive" oils, refined, cottonseed, or peanut oil being thus converted into *huile vierge de* Lucca, Florence, or some other noted brand of olive oil. It is also extensively used for illuminating purposes, especially in southern Russia.

The flavors, other than that of the mustard itself, of the various preparations are imparted by the judicious use of spices—cinnamon, nutmeg, cloves, pimento, etc.—aromatic herbs, such as thyme, sage, chervil, parsley, mint, marjoram, tarragon, etc., and finally chives, onions, shallots, leeks, garlic, etc.

In preparing the mustards on a large scale, the mustard flower and wheat or rye flour are mixed and ground to a smooth paste with vinegar, must (unfermented grape juice), wine, or whatever is used in the preparation, a mill similar to a drug or paint mill being used for the purpose. This dough immediately becomes spongy, and in this condition, technically called "cake," is used as the basis of the various mustards of commerce.

Mustard Cakes.—In the mixture, the amount of flour used depends on the pungency of the mustard flower, and the flavor desired to be imparted to the finished product. The cakes are broadly divided into the yellow and the brown. A general formula for the yellow cake is:

Yellow mustard, from 20 to 30 per cent; salt, from 1 to 3 per cent; spices, from ¼ to ½ of 1 per cent; wheat flour, from 8 to 12 per cent.

Vinegar, must, or wine, complete the mixture.

The brown cake is made with black mustard, and contains about the following proportions:

Black mustard, from 20 to 30 per cent; salt, from 1 to 3 per cent; spices, from ¼ to ½ of 1 per cent; wheat or rye flour, from 10 to 15 per cent.

The variations are so wide, however, that it is impossible to give exact proportions. In the manufacture of table mustards, in fact, as in every other kind of manufacture, excellence is attained only by practice and the exercise of sound judgment and taste by the manufacturer.

Moutarde des Jesuittes.—Twelve sardels and 280 capers are crushed into a paste and stirred into 3 pints of boiling wine vinegar. Add 4 ounces of brown cake and 8 ounces of yellow cake and mix well.

Kirschner Wine Mustard.—Reduce 30 quarts of freshly expressed grape juice to half that quantity, by boiling over a moderate fire, on a water bath. Dissolve in the boiling liquid 5 pounds of sugar, and pour the syrup through a colander containing 2 or 3 large horse-radishes cut into very thin slices and laid on a coarse towel spread over the bottom and sides of the colander. To the colate add the following, all in a state of fine powder:

Cardamom seeds	2½	drachms
Nutmeg...........	2½	drachms
Cloves.............	4½	drachms
Cinnamon.........	1	ounce
Ginger............	1	ounce
Brown mustard cake.	6	pounds
Yellow mustard cake.	9	pounds

Grind all together to a perfectly smooth paste, and strain several times through muslin.

Duesseldorff Mustard.— •

Brown mustard cake.	10	ounces
Yellow mustard cake.	48	ounces
Boiling water........	96	ounces
Wine vinegar.......	64	ounces
Cinnamon..........	5	drachms
Cloves.............	15	drachms
Sugar.............	64	ounces
Wine, good white....	64	ounces

Mix after the general directions given above.

German Table Mustard.—

Laurel leaves........	8	ounces
Cinnamon..........	5	drachms
Cardamom seeds....	2	drachms
Sugar.............	64	ounces
Wine vinegar.......	96	ounces
Brown cake........	10	ounces
Yellow cake........	48	ounces

Mix after general directions as given above.

Krems Mustard, Sweet.—

Yellow cake........	10	pounds
Brown cake........	20	pounds
Fresh grape juice	6	pints

Mix and boil down to the proper consistency.

Krems Mustard, Sour.—

Brown mustard flour.	30	parts
Yellow mustard flour.	10	parts
Grape juice, fresh....	8	parts

Mix and boil down to a paste and then stir in 8 parts of wine vinegar.

Tarragon Mustard.—

Brown mustard flour.	40	parts
Yellow mustard flour.	20	parts
Vinegar...........	6	parts
Tarragon vinegar....	6	parts

Boil the mustard in the vinegar and add the tarragon vinegar.

Tarragon Mustard, Sharp.—This is prepared by adding to every 100 pounds of the above 21 ounces of white pepper, 5 ounces of pimento, and 2½ ounces of cloves, mixing thoroughly by grinding together in a mill, then put in a warm spot and let stand for 10 days or 2 weeks. Finally strain.

Moutarde aux Epices.—

Mustard flour, yellow.	10	pounds
Mustard flour, brown.	40	pounds
Tarragon...........	1	pound
Basil, herb.........	5	ounces
Laurel leaves.......	12	drachms
White pepper........	3	ounces
Cloves.............	12	drachms
Mace..............	2	drachms
Vinegar...........	1	gallon

Mix the herbs and macerate them in the vinegar to exhaustion, then add to the mustards, and grind together. Set aside for a week or ten days, then strain through muslin.

In all the foregoing formulas where the amount of salt is not specified, it is to be added according to the taste or discretion of the manufacturer.

Mustard Vinegar.—

Celery, chopped fine.	32 parts
Tarragon, the fresh herb...........	6 parts
Cloves, coarsely powdered...........	6 parts
Onions, chopped fine	6 parts
Lemon peel, fresh, chopped fine.....	3 parts
White-wine vinegar..	575 parts
White wine........	515 parts
Mustard seed, crushed..........	100 parts

Mix and macerate together for a week or 10 days in a warm place, then strain off.

Ravigotte Mustard.—

Parsley............	2 parts
Chervil............	2 parts
Chives............	2 parts
Cloves.............	1 part
Garlic.............	1 part
Thyme............	1 part
Tarragon..........	1 part
Salt..............	8 parts
Olive oil..........	4 parts
White-wine vinegar..	128 parts
Mustard flower, sufficient.	

Cut or bruise the plants and spices, and macerate them in the vinegar for 15 or 20 days. Strain the liquid through a cloth and add the salt. Rub up mustard with the olive oil in a vessel set in ice, adding a little of the spiced vinegar from time to time, until the whole is incorporated and the complete mixture makes 384 parts.

CONDIMENTS, TESTS FOR ADULTERATED:
See Foods.

CONDITION POWDERS FOR CATTLE:
See Veterinary Formulas.

CONDUCTIVITY OF ALUMINUM ALLOYS:
See Alloys.

Confectionery

Cream Bonbons for Hoarseness.—Stir into 500 parts of cream 500 parts of white sugar. Put in a pan and cook, with continuous stirring, until it becomes brown and viscid. Now put in a baking tin and smooth out, as neatly as possible, to the thickness of, say, twice that of the back of a table knife and let it harden. Before it gets completely hard draw lines with a knife across the surface in such manner that when it is quite hard it will break along them, easily, into bits the size of a lozenge.

Nut Candy Sticks.—Cook to 320° F. 8 pounds best sugar in 2 pints water, with 4 pounds glucose added. Pour out on an oiled slab and add 5 pounds almonds, previously blanched, cut in small pieces, and dried in the drying room. Mix up well together to incorporate the nuts thoroughly with the sugar. When it has cooled enough to be handled, form into a round mass on the slab and spin out in long, thin sticks.

Fig Squares.—Place 5 pounds of sugar and 5 pounds of glucose in a copper pan, with water enough to dissolve the sugar. Set on the fire, and when it starts to boil add 5 pounds of ground figs. Stir and cook to 210° on the thermometer. Set off the fire, and then add 5 pounds of fine cocoanuts; mix well and pour out on greased marble, roll smooth, and cut like caramels.

Caramels.—Heat 10 pounds sugar and 8 pounds glucose in a copper kettle until dissolved. Add cream to the mixture, at intervals, until 2½ quarts are used. Add 2¼ pounds caramel butter and 12 ounces paraffine wax to the mixture. Cook to a rather stiff ball, add nuts, pour out between iron bars and, when cool enough, cut into strips. For the white ones flavor with vanilla, and add 2 pounds melted chocolate liquor for the chocolate caramel when nearly cooked.

Candy Orange Drops.—It is comparatively easy to make a hard candy, but to put the material into "drop" form apparently requires experience and a machine. To make the candy itself, put, say, a pint of water into a suitable pan or kettle, heat to boiling, and add gradually to it 2 pounds or more of sugar, stirring well so as to avoid the risk of burning the sugar. Continue boiling the syrup so formed until a little of it poured on a cold slab forms a mass of the required hardness. If the candy is to be of orange flavor, a little fresh oil of orange is added just before the mass is ready to set and the taste is improved according to the general view at least by adding, also, say, 2 drachms of citric acid dissolved in a very little water. As a coloring an infusion of safflower or tincture of turmeric is used.

To make such a mass into tablets, it is necessary only to pour out on a well-greased slab, turning the edges back if inclined to run, until the candy is firm, and then scoring with a knife so that it can easily be broken into pieces when cold. To make "drops" a suitable mold is necessary.

Experiment as to the sufficiency of the boiling in making candy may be saved and greater certainty of a good result secured by the use of a chemical thermometer. As the syrup is boiled and the water evaporates the temperature of the liquid rises. When it reaches 220° F., the sugar is then in a condition to yield the "thread" form; at 240° "soft ball" is formed; at 245°, "hard ball"; at 252°, "crack"; and at 290°, "hard crack." By simply suspending the thermometer in the liquid and observing it from time to time, one may know exactly when to end the boiling.

Gum Drops. Grind 25 pounds of Arabian or Senegal gum, place it in a copper pan or in a steam jacket kettle, and pour 3 gallons of boiling water over it; stir it up well. Now set the pan with the gum into another pan containing boiling water and stir the gum slowly until dissolved, then strain it through a No. 40 sieve. Cook 19 pounds of sugar with sufficient water, 2 pounds of glucose, and a teaspoonful of cream of tartar to a stiff ball, pour it over the gum, mix well, set the pan on the kettle with the hot water, and let it steam for 1½ hours, taking care that the water in the kettle does not run dry; then open the door of the stove and cover the fire with ashes, and let the gum settle for nearly an hour, then remove the scum which has settled on top, flavor and run out with the funnel dropper into the starch impressions, and place the trays in the drying room for 2 days, or until dry; then take the drops out of the starch, clean them off well and place them in crystal pans, one or two layers. Cook sugar and water to 34½° on the syrup gauge and pour over the drops lukewarm. Let stand in a moderately warm place over night, then drain the syrup off, and about an hour afterwards knock the gum drops out on a clean table, pick them apart, and place on trays until dry, when they are ready for sale.

A Good Summer Taffy.—Place in a kettle 4 pounds of sugar, 3 pounds of glucose, and 1½ pints of water; when it boils drop in a piece of butter half the size of an egg and about 2 ounces of paraffine wax. Cook to 262°, pour on a slab, and when cool enough, pull, flavor, and color if you wish. Pull until light, then spin out on the table in strips about 3 inches wide and cut into 4- or 4½-inch lengths. Then wrap in wax paper for the counter. This taffy keeps long without being grained by the heat.

Chewing Candy.—Place 20 pounds of sugar in a copper pan, add 20 pounds of glucose, and enough water to easily dissolve the sugar. Set on the fire or cook in the steam pan in 2 quarts of water. Have a pound of egg albumen soaked in 2 quarts of water. Beat this like eggs into a very stiff froth, add gradually the sugar and glucose; when well beaten up, add 5 pounds of powdered sugar, and beat at very little heat either in the steam beater or on a pan of boiling water until light, and does not stick to the back of the hand, flavor with vanilla, and put in trays dusted with fine sugar. When cold it may be cut, or else it may be stretched out on a sugar-dusted table, cut, and wrapped in wax paper. This chewing candy has to be kept in a very dry place, or else it will run and get sticky.

Montpelier Cough Drops.—

Brown sugar	10	pounds
Tartaric acid	2	ounces
Cream of tartar	½	ounce
Water	1½	quarts
Anise-seed flavoring, quantity sufficient.		

Melt the sugar in the water, and when at a sharp boil add the cream of tartar. Cover the pan for 5 minutes. Remove the lid and let the sugar boil up to crack degree. Turn out the batch on an oiled slab, and when cool enough to handle mold in the acid and flavoring. Pass it through the acid drop rollers, and when

the drops are chipped up, and before sifting, rub some icing with them.

Medicated Cough Drops.—

Light-brown sugar	14	pounds
Tartaric acid	1½	ounces
Cream of tartar	½	ounce
Water	2	quarts
Anise-seed, cayenne, clove, and peppermint flavoring, a few drops of each.		

Proceed as before prescribed, but when sufficiently cool pass the batch through the acid tablet rollers and dust with sugar.

Horehound Candy.—

Dutch crushed sugar	10	pounds
Dried horehound leaves	2	ounces
Cream of tartar	¾	ounce
Water	2	quarts
Anise-seed flavoring, quantity sufficient.		

Pour the water on the leaves and let it gently simmer till reduced to 3 pints; then strain the infusion through muslin, and add the liquid to the sugar. Put the pan containing the syrup on the fire, and when at a sharp boil add the cream of tartar. Put the lid on the pan for 5 minutes; then remove it, and let the sugar boil to stiff boil degree. Take the pan off the fire and rub portions of the sugar against the side until it produces a creamy appearance; then add the flavoring. Stir all well, and pour into square tin frames, previously well oiled.

Menthol Cough Drops.—

Gelatin	1	ounce
Glycerine (by weight)	2½	ounces
Orange-flower water	2½	ounces
Menthol	5	grains
Rectified spirits	1	drachm

Soak the gelatin in the water for 2 hours, then heat on a water bath until dissolved, and add 1½ ounces of glycerine. Dissolve the menthol in the spirit, mix with the remainder of the glycerine, add to the glyco-gelatin mass, and pour into an oiled tin tray (such as the lid of a biscuit box). When the mass is cold divide into 10 dozen pastilles.

Menthol pastilles are said to be an excellent remedy for tickling cough as well as laryngitis. They should be freshly prepared, and cut oblong, so that the patient may take half of one, or less, as may be necessary.

Violet Flavor for Candy.—Violet flavors, like violet perfumes, are very complex mixtures, and their imitation is a correspondingly difficult undertaking. The basis is vanilla (or vanillin), rose, and orris, with a very little of some pungent oil to bring up the flavor. The following will give a basis upon which a satisfactory flavor may be built:

Oil of orris	1 drachm
Oil of rose	1 drachm
Vanillin	2 drachms
Cumarin	30 grains
Oil of clove	30 minims
Alcohol	11 ounces
Water	5 ounces

Make a solution, adding the water last.

CONFECTIONERY COLORS. — The following are excellent and entirely harmless coloring agents for the purposes named:

Red.—Cochineal syrup prepared as follows:

Cochineal, in coarse powder	6 parts
Potassium carbonate	2 parts
Distilled water	15 parts
Alcohol	12 parts
Simple syrup enough to make	500 parts

Rub up the potassium carbonate and the cochineal together, adding the water and alcohol, little by little, under constant trituration. Set aside over night, then add the syrup and filter.

Pink.—

Carmine	1 part
Liquor potassæ	6 parts
Rose water, enough to make	48 parts

Mix. Should the color be too high, dilute with water until the requisite tint is acquired.

Orange.—Tincture of red sandalwood, 1 part; ethereal tincture of orlean, quantity sufficient. Add the tincture of orlean to the sandalwood tincture until the desired shade of orange is obtained.

A red added to any of the yellows gives an orange color.

The aniline colors made by the "Aktiengesellschaft für Anilin - Fabrikation," of Berlin, are absolutely non-toxic, and can be used for the purposes recommended, i. e., the coloration of syrups, cakes, candies, etc., with perfect confidence in their innocuity.

Pastille Yellow.—

Citron yellow II	7 parts
Grape sugar, first quality	1 part
White dextrine	2 parts

Sap-Blue Paste.—

Dark blue	3 parts
Grape sugar	1 part
Water	6 parts

Sugar-Black Paste.—

Carbon black	3 parts
Grape sugar	1 part
Water	6 parts

Cinnabar Red.*—

Scarlet	65 parts
White dextrine	30 parts
Potato flour	5 parts

Bluish Rose.*—

Grenadine	65 parts
White dextrine	30 parts
Potato flour	5 parts

Yellowish Rose.—

Rosa II	60 parts
Citron yellow	5 parts
White dextrine	30 parts
Potato flour	5 parts

Violet.—

Red violet	65 parts
White dextrine	30 parts
Potato flour	5 parts

Carmine Green.—

Woodruff (Waldmeister) green	55 parts
Rosa II	5 parts
Dextrine	35 parts
Potato flour	5 parts

To the colors marked with an asterisk (*) add, for every 4 pounds, 4½ ounces, a grain and a half each of potassium iodide and sodium nitrate. Colors given in form of powders should be dissolved in hot water for use.

Yellow.—Various shades of yellow may be obtained by the maceration of Besiello saffron, or turmeric, or grains d'Avignon in alcohol until a strong tincture is obtained. Dilute with water until the desired shade is obtained. An aqueous solution of quercitrine also gives an excellent yellow.

Blue.—

Indigo carmine	1 part
Water	2 parts

Mix.

Indigo carmine is a beautiful, powerful, and harmless agent. It may usually be bought commercially, but if it cannot be readily obtained, proceed as follows:

Into a capsule put 30 grains of indigo in powder, place on a water bath, and heat to dryness. When entirely dry put into a large porcelain mortar (the substance swells enormously under subsequent treatment—hence the necessity for a large, or comparatively large, mortar) and cautiously add, drop by drop, 120 grains, by weight, of sulphuric acid, C. P., stirring continuously during the addition. Cover the swollen mass closely, and set aside for 24 hours. Now add 3 fluidounces of distilled water, a few drops at a time, rubbing or stirring continuously. Transfer the liquid thus obtained to a tall, narrow, glass cylinder or beaker, cover and let stand for 4 days, giving the liquid an occasional stirring. Make a strong solution of sodium carbonate or bicarbonate, and at the end of the time named cautiously neutralize the liquid, adding the carbonate a little at a time, stirring the indigo solution and testing it after each addition, as the least excess of alkali will cause the indigo to separate out, and fall in a doughy mass. Stop when the test shows the near approach of neutrality, as the slight remaining acidity will not affect the taste or the properties of the liquid. Filter, and evaporate in the water bath to dryness. The resultant matter is sulphindigotate of potassium, or the "indigo carmine" of commerce.

Tincture of indigo may also be used as a harmless blue.

Green.—The addition of the solution indigo carmine to an infusion of any of the matters given under "yellow" will produce a green color. Tincture of crocus and glycerine in equal parts, with the addition of indigo-carmine solution, also gives a fine green. A solution of commercial chlorophyll gives grass-green, in shades varying according to the concentration of the solution.

Voice and Throat Lozenges.—

Catechu	191 grains
Tannic acid	273 grains
Tartaric acid	273 grains
Capsicin	30 minims
Black-currant paste	7 ounces
Refined sugar,	
Mucilage of acacia,	
of each a sufficient	
quantity.	

Mix to produce 7 pounds of lozenges.

CONSTIPATION IN BIRDS:
See Veterinary Formulas.

COOKING TABLE:
See Tables.

COOLING SCREEN:
See Refrigeration.

Copper

Annealing Copper.—

Copper is almost universally annealed in muffles, in which it is raised to the desired temperature, and subsequently allowed to cool either in the air or in water. A muffle is nothing more or less than a reverberatory furnace. It is necessary to watch the copper carefully, so that when it has reached the right temperature it may be drawn from the muffle and allowed to cool. This is important, for if the copper is heated too high, or is left in the muffle at the ordinary temperature of annealing too long, it is burnt, as the workmen say. Copper that has been burnt is yellow, coarsely granular, and exceedingly brittle—even more brittle at a red heat than when cold.

In the case of coarse wire it is found that only the surface is burnt, while the interior is damaged less. This causes the exterior to split loose from the interior when bent or rolled, thus giving the appearance of a brittle copper tube with a copper wire snugly fitted into it. Cracks a half inch in depth have been observed on the surface of an ingot on its first pass through the rolls, all due to this exterior burning. It is apparent that copper that has been thus overheated in the muffle is entirely unfit for rolling. It is found that the purer forms of copper are less liable to be harmed by overheating than samples containing even a small amount of impurities. Even the ordinary heating in a muffle will often suffice to burn in this manner the surface of some specimens of copper, rendering them unfit for further working. Copper that has been thus ruined is of use only to be refined again.

As may be inferred only the highest grades of refined copper are used for drawing or for rolling. This is not because the lower grades, when refined, cannot stand sufficiently high tests, but because methods of working are not adequate to prevent these grades of copper from experiencing the deterioration due to overheating.

The process of refining copper consists in an oxidizing action followed by a reducing action which, since it is performed by the aid of gases generated by stirring the melted copper with a pole, is called poling. The object of the oxidation is to oxidize and either volatilize or turn to slag all the impurities contained in the copper. This procedure is materially aided by the fact that the suboxide of copper is freely soluble in metallic copper and thus penetrates to all parts of the copper, and parting with its oxygen, oxidizes the impurities. The object of the reducing part of the refining process is to change the excess of the suboxide of copper to metallic copper. Copper containing even less than 1 per cent of the suboxide of copper shows decreased malleability and ductility, and is both cold-short and red-short. If the copper to be refined contains any impurities, such as arsenic or antimony, it is well not to remove too much of the oxygen in the refining process. If this is done, overpoled copper is produced. In this condition it is brittle, granular, of a shining yellow color, and more red-short than cold-short. When the refining has been properly done, and neither too much nor too little oxygen is present, the copper is in the condition of "tough pitch," and is in a fit state to be worked.

Copper is said to be "tough pitch" when it requires frequent bending to break it, and when, after it is broken, the color is pale red, the fracture has a silky luster, and is fibrous like a tuft of silk. On hammering a piece to a thin plate it should show no cracks at the edge. At tough pitch copper offers the highest degree of malleability and ductility of which a given specimen is capable. This is the condition in which refined copper is (or should be) placed on the market, and if it could be worked without changing this tough pitch, any specimen of copper that could be brought to this condition would be suitable for rolling or drawing. But tough pitch is changed if oxygen is either added or taken from refined copper.

By far the more important of these is the removal of oxygen, especially from those specimens that contain more than a mere trace of impurities. This is shown by the absolutely worthless condition of overpoled copper. The addition of carbon also plays a very important part in the production of overpoled copper.

That the addition of oxygen to refined copper is not so damaging is shown by the fact that at present nearly all the copper that is worked is considerably oxidized at some stage of the process, and not especially to its detriment.

Burnt copper is nothing more or less than copper in the overpoled condition. This is brought about by the action of reducing gases in the muffle. By this means the small amount of oxygen necessary to give the copper its tough pitch is removed. This oxygen is combined with impurities in the copper, and thus renders them inert. For example, the oxide of arsenic or antimony is incapable of combining more than mechanically with the copper, but when its oxygen is removed the arsenic or antimony is left free to combine with the copper. This forms a brittle alloy, and one that corresponds almost exactly in its properties with overpoled copper. To be sure overpoled copper is supposed to contain carbon, but that this is not the essential ruling principle in case of annealing is shown by the fact that pure copper does not undergo this change under conditions that ruin impure copper, and also by the fact that the same state may be produced by annealing in pure hydrogen and thus removing the oxygen that renders the arsenic or antimony inert. No attempt is made to deny the well-known fact that carbon does combine with copper to the extent of 0.2 per cent and cause it to become exceedingly brittle. It is simply claimed that this is probably not what occurs in the production of so-called burnt copper during annealing. The amount of impurities capable of rendering copper easily burnt is exceedingly small. This may be better appreciated when it is considered that from 0.01 to 0.2 per cent expresses the amount of oxygen necessary to render the impurities inert. The removal of this very small amount of oxygen, which is often so small as to be almost within the limits of the errors of analysis, will suffice to render copper overpoled and ruin it for any use.

There are methods of avoiding the numerous accidents that may occur in the annealing of copper, due to a change of pitch. As already pointed out, the quality of refined copper is lowered if oxygen be either added to or taken from it. It is quite apparent, therefore, that a really good method of annealing copper will prevent any change in the state of oxidation. It is necessary to prevent access to the heated copper both of atmospheric air, which would oxidize it, and of the reducing gases used in heating the muffle, which would take oxygen away from it. Obviously the only way of accomplishing this is to inclose the copper when heated and till cool in an atmosphere that can neither oxidize nor deoxidize copper. By so doing copper may be heated to the melting point and allowed to cool again without suffering as regards its pitch. There are comparatively few gases that can be used for this purpose, but, fortunately, one which is exceedingly cheap and universally prevalent fulfills all requirements, viz., steam. In order to apply the principles enunciated it is necessary only to anneal copper in the ordinary annealing pots such as are used for iron, care being taken to inclose the copper while heating and while cooling in an atmosphere of steam. This will effectually exclude air and prevent the ingress of gases used in heating the annealer. Twenty-four hours may be used in the process, as in the annealing of iron wire, with no detriment to the wire. This may seem incredible to those manufacturers who have tried to anneal copper wire after the manner of annealing iron wire. By this method perfectly bright annealed wire may be produced. Such a process of annealing copper offers many advantages. It allows the use of a grade of copper that has hitherto been worked only at a great disadvantage, owing to its tendency to get out of pitch. It allows the use of annealers such as are ordinarily employed for annealing iron, and thus cheapens the annealing considerably as compared with the present use of muffles. There is no chance of producing the overpoled condition from the action of reducing gases used in heating the muffles. There is no chance of producing the underpoled condition due to the absorption of suboxide of copper. None of the metal is lost as scale, and the saving that is thus effected amounts to a considerable percentage of the total value of the copper. The expense and time of cleaning are wholly saved. Incidentally bright annealed copper is produced by a process which is applicable to copper of any shape, size, or condition—a product that has hitherto been obtained only by processes (mostly secret) which are too cumbersome and too expensive for extensive use; and, as is the case with at least one process, with the danger of producing the overpoled condition, often in only a small section of the wire, but thus ruining the whole piece.

COPPER COLORING:

Blacking Copper.—To give a copper article a black covering, clean it with emery paper, heat gently in a Bunsen or a spirit flame, immerse for 10 seconds in solution of copper filings in dilute nitric acid, and heat again.

Red Coloring of Copper.—A fine red color may be given to copper by gradually heating it in an air bath. Prolonged heating at a comparatively low temperature, or rapid heating at a high temperature, produces the same result. As soon as the desired color is attained the metal should be rapidly cooled by quenching in water. The metal thus colored may be varnished.

To Dye Copper Parts Violet and Orange.—Polished copper acquires an orange-like color leaning to gold, when dipped for a few seconds into a solution of crystallized copper acetate. A handsome violet is obtained by placing the metal for a few minutes in a solution of antimony chloride and rubbing it afterwards with a piece of wood covered with cotton. During this operation the copper must be heated to a degree bearable to the hand. A crystalline appearance is produced by boiling the article in copper sulphate.

Pickle for Copper.—Take nitric acid, 100 parts; kitchen salt, 2 parts; calcined soot, 2 parts; or nitric acid, 10 parts; sulphuric acid, 10 parts; hydrochloric acid, 1 part. As these bleaching baths attack the copper quickly, the objects must be left in only for a few seconds, washing them afterwards in plenty of water, and drying in sawdust, bran, or spent tan.

Preparations of Copper Water.—I.—Water, 1,000 parts; oxalic acid, 30 parts; spirit of wine, 100 parts; essence of turpentine, 50 parts; fine tripoli, 100 parts.

II.—Water, 1,000 parts; oxalic acid, 30 parts; alcohol, 50 parts; essence of turpentine, 40 parts; fine tripoli, 50 parts.

III.—Sulphuric acid, 300 parts; sulphate of alumina, 80 parts; water, 520 parts.

Tempered Copper.—Objects made of copper may be satisfactorily tempered by subjecting them to a certain degree of heat for a determined period of time and bestrewing them with powdered sulphur during the heating. While hot the objects are plunged into a bath of blue vitriol; after the bath they may be heated again.

COPPER ALLOYS:
See Alloys.

COPPER CLEANING:
See Cleaning Preparations and Methods.

COPPER ETCHING:
See Etching.

COPPER IN FOOD:
See Food.

COPPER LACQUERS:
See Lacquers.

COPPER PAPER:
See Paper, Metallic.

COPPER PATINIZING AND PLATING:
See Plating.

COPPER POLISHES:
See Polishes.

COPPER, SEPARATION OF GOLD FROM:
See Gold.

COPPER SOLDER:
See Solders.

COPPER VARNISHES:
See Varnishes.

COPYING PRINTED PICTURES.

The so-called "metallic" paper used for steam-engine indicator cards has a smooth surface, chemically prepared so that black lines can be drawn upon it with pencils made of brass, copper, silver, aluminum, or any of the softer metals. When used on the indicator it receives the faint line drawn by a brass point at one end of the pencil arm, and its special advantage over ordinary paper is that the metallic pencil slides over its surface with very little friction, and keeps its point much longer than a graphite pencil.

This paper can be used as a transfer paper for copying engravings or sketches, or anything printed or written in ink or drawn in pencil.

The best copies can be obtained by following the directions below: Lay the metallic transfer paper, face up, upon at least a dozen sheets of blank paper, and lay the print face down upon it. On the back of the print place a sheet of heavy paper, or thin cardboard, and run the rubbing tool over this protecting sheet. In this manner it is comparatively easy to prevent slipping, and prints 8 or 10 inches on a side may be copied satisfactorily.

Line drawings printed from relief plates, or pictures with sharp contrast of black and white, without any half-tones, give the best copies. Very few half-tones can be transferred satisfactorily; almost all give streaked, indistinct copies, and many of the results are worthless.

The transfer taken off as described is a reverse of the original print. If the question of right and left is not important this reversal will seldom be objectionable, for it is easy to read backward what few letters generally occur. However, if desired, the paper may be held up to the light and examined from the back, or placed before a mirror and viewed by means of its reflected image, when the true relations of right and left will be seen. Moreover, if sufficiently important, an exact counterpart of the original may be taken from the reversed copy by laying another sheet face downward upon it, and rubbing on the back of the fresh sheet just as was done in making the reversed copy. The impression thus produced will be fainter than the first, but almost always it can be made dark enough to show a distinct outline which may afterwards be retouched with a lead pencil.

For indicator cards the paper is prepared by coating one surface with a suitable compound, usually zinc oxide mixed with a little starch and enough glue to make it adhere. After drying it is passed between calendar rolls under great pressure. The various brands manufactured for the trade, though perhaps equally good for indicator diagrams, are not equally well suited for copying. If paper of firmer texture could be prepared with the same surface finish, probably much larger copies could be produced.

Other kinds of paper, notably the heavy plate papers used for some of the best trade catalogues, possess this transfer property to a slight degree, though they will not receive marks from a metallic pencil. The latter feature would seem to recommend them for transfer purposes, making them less likely to become soiled by contact with metallic objects, but so far no kind has been found which will remove enough ink to give copies anywhere near as dark as the indicator paper.

Fairly good transfers can be made from almost any common printers' ink, but some inks copy much better than others, and some yield only the faintest impressions. The length of time since a picture was printed does not seem to determine its copying quality. Some very old prints can be copied better than new ones; in fact, it was by accidental transfer to an indicator card from a book nearly a hundred years old that the peculiar property of this "metallic" paper was discovered.

Copying Process on Wood.—If wood surfaces are exposed to direct sunlight the wood will exhibit, after 2 weeks action, a browning of dark tone in the exposed places. Certain parts of the surface being covered up during the entire exposure to the sun, they retain their original shade and are set off clearly and sharply against the parts browned by the sunlight. Based on this property of the wood is a sun-copying process on wood. The method is used for producing tarsia in imitation on wood. A pierced stencil of tin, wood, or paper is laid on a freshly planed plate of wood, pasting it on in places to avoid shifting, and put into a common copying frame. To prevent the wood from warping a stretcher is employed, whereupon expose to the sun for from 8 to 14 days. After the brown shade has appeared the design obtained is partly fixed by polishing or by a coating of varnish, lacquer, or wax. Best suited for such works are the pine woods, especially the 5-year fir and the cembra pine, which, after the exposure, show a yellowish brown tone of handsome golden gloss, that stands out boldly, especially after subsequent polishing, and cannot be replaced by any stain or by pyrography. The design is sharper and clearer than that produced by painting. In short, the total effect is pleasing.

How to Reproduce Old Prints.—Prepare a bath as follows: Sulphuric acid, 3 to 5 parts (according to the antiquity of print, thickness of paper, etc.); alcohol, 3 to 5 parts; water, 100 parts. In this soak the print from 5 to 15 minutes (the time depending on age, etc., as above), remove, spread face downward on a glass or ebonite plate, and wash thoroughly in a gentle stream of running water. If the paper is heavy, reverse the sides, and let the water flow over the face of the print. Remove carefully and place on a heavy sheet of blotting paper, cover with another, and press out every drop of water possible. Where a wringing machine is convenient and sufficiently wide, passing the blotters and print through the rollers is better than mere pressing with the hands. The print, still moist, is then laid face upward on a heavy glass plate (a marble slab or a lithographers' stone answers equally well), and smoothed out. With a very soft sponge go over the surface with a thin coating of gum-arabic water. The print is now ready for inking, which is done exactly as in lithographing, with a roller and printers' or lithographers' ink, cut with oil of turpentine. Suitable paper is then laid on and rolled with a dry roller. This gives a reverse image of the print, which is then applied to a zinc plate or a lithographers' stone, and as many prints as desired pulled off in the usual lithographing method. When carefully done and the right kind of paper used, it is said that the imitation of the original is perfect in every detail.

To Copy Old Letters, Manuscripts, etc. —If written in the commercial ink of the period from 1860 to 1864, which was almost universally an iron and tannin or gallic-acid ink, the following process may succeed: Make a thin solution of glucose, or honey, in water, and with this wet the paper in the usually observed way in copying recent documents in the letter book, put in the press, and screw down tightly. Let it remain in the press somewhat longer than in copying recent documents. When removed, before attempting to separate the papers, expose to the fumes of strong water of ammonia, copy side downward.

CORDAGE:

See also Ropes.

Strong Twine.—An extraordinarily strong pack thread or cord, stronger even than the so-called "Zuckerschnur," may be obtained by laying the thread of fibers in a strong solution of alum, and then carefully drying them.

Preservation of Fishing Nets.—The following recipe for the preservation of fishing nets is also applicable to ropes, etc., in contact with water. Some have been subjected to long test.

For 40 parts of cord, hemp, or cotton, 3 parts of kutch, 1 part of blue vitriol, $\frac{1}{2}$ part of potassium chromate, and $2\frac{1}{2}$ parts of wood tar are required. The kutch is boiled with 150 parts of water until dissolved, and then the blue vitriol is added. Next, the net is entered and the tar added. The whole should be stirred well, and the cordage must boil 5 to 8 minutes. Now take out the netting, lay it in another vessel, cover up well, and leave alone for 12 hours. After that it is dried well, spread out in a clean place, and coated with linseed oil. Not before 6 hours have elapsed should it be folded together and put into the water. The treatment with linseed oil may be omitted.

CORDAGE LUBRICANT:

See Lubricants.

CORDAGE WATERPROOFING:

See Waterproofing.

CORDIALS:

See Wines and Liquors.

CORKS:

Impervious Corks.—Corks which have been steeped in petrolatum are said to be an excellent substitute for glass stoppers. Acid in no way affects them and chemical fumes do not cause decay in them, neither do they become fixed by a blow or long disuse.

Non-Porous Corks.—For benzine, turpentine, and varnish cans, immerse the corks in hot melted paraffine. Keep them under about 5 minutes; hold them down with a piece of wire screen cut to fit the dish in which you melt the paraffine. When taken out lay them on a screen till cool. Cheap corks can in this way be made gas- and air-tight, and can be cut and bored with ease.

Substitute for Cork.—Wood pulp or other ligneous material may be treated to imitate cork. For the success of the composition it is necessary that the constituents be mingled and treated under special conditions. The volumetric proportions in which these constituents combine with the best results are the following: Wood pulp, 3 parts; cornstalk pith, 1 part; gelatin, 1 part; glycerine, 1 part; water, 4 parts; 20 per cent formicaldehyde solution, 1 part; but the proportions may be varied. After disintegrating the ligneous substances, and while these are in a moist and hot condition they are mingled with the solution of gelatin, glycerine, and water. The mass is stirred thoroughly so as to obtain a homogeneous mixture. The excess of moisture is removed. As a last operation the formic aldehyde is introduced, and the mass is left to coagulate in this solution. The formic aldehyde renders the product insoluble in nearly all liquids. So it is in this last operation that it is necessary to be careful in producing the composition properly. When the operation is terminated the substance is submitted to pressure during its coagulation, either by molding it at once into a desired form, or into a mass which is afterwards converted into the finished product.

CORKS, TO CLEAN:
See Cleaning Preparations and Methods, under Miscellaneous Methods.

CORK TO METAL, FASTENING:
See Adhesives, under Pastes.

CORK AS A PRESERVATIVE:
See Preserving.

CORKS, WATERPROOFING:
See Waterproofing.

CORN CURES:

I.—**Salicylic-Acid Corn Cure.**—Extract cannabis indica, 1 part, by measure; salicylic acid, 10 parts, by measure; oil of turpentine, 5 parts, by measure; acetic acid, glacial, 2 parts, by measure; cocaine, alkaloidal, 2 parts, by measure; collodion, elastic, sufficient to make 100 parts. Apply a thin coating every night, putting each layer directly on the preceding one. After a few applications, the mass drops off, bringing the indurated portion, and frequently the whole of the corn, off with it.

II.—**Compound Salicylated Collodion Corn Cure.**—Salicylic acid, 11 parts, by weight; extract of Indian hemp, 2 parts, by weight; alcohol, 10 parts, by weight; flexible collodion, U. S. P., a sufficient quantity to make 100 parts, by weight. The extract is dissolved in the alcohol and the acid in about 50 parts, by weight, of collodion, the solutions mixed, and the liquid made up to the required amount. The Indian hemp is presumably intended to prevent pain; whether it serves this or any other useful purpose seems a matter of doubt. The acid is frequently used without this addition.

III.—Extract of cannabis indica, 90 grains; salicylic acid, 1 ounce; alcohol, 1 ounce; collodion enough to make 10 ounces. Soften the extract with the alcohol, then add the collodion, and lastly the acid.

IV.—Resorcin, 1 part, by weight; salicylic acid, 1 part, by weight; lactic acid, 1 part, by weight; collodion elasticum, 10 parts, by weight. Paint the corn daily for 5 or 6 days with the above solution and take a foot bath in very hot water. The corn will readily come off.

Corn Plaster.—Yellow wax, 24 parts, by weight; Venice turpentine, 3 parts, by weight; rosin, 2 parts, by weight; salicylic acid, 2 parts, by weight; balsam of Peru, 2 parts, by weight; lanolin, 4 parts, by weight.

Corn Cure.—Melt soap plaster, 85 parts, by weight, and yellow wax, 5 parts by weight, in a vapor bath, and stir finely ground salicylic acid, 10 parts, by weight, into it.

Removal of Corns.—The liquid used by chiropodists with pumice stone for the removal of corns and callosities is usually nothing more than a solution of potassa or concentrated lye, the pumice stone being dipped into the solution by the operator just before using.

Treatment of Bunions.—Wear right and left stockings and shoes, the inner edges of the sole of which are perfectly straight. The bunion is bathed night and morning in a 4 per cent solution of carbolic acid for a few minutes, followed by plain water. If, after several weeks, the bursa is still distended with fluid, it is aspirated. If the bunion is due to flatfoot, the arch of the foot must be restored by a plate. When the joints are enlarged because of gout or rheumatism, the constitutional conditions must be treated. In other cases, osteotomy and tenotomy are required.

The Treatment of Corns.—Any corn may be speedily and permanently cured. The treatment is of three kinds—preventive, palliative, and curative.

I.—The preventive treatment lies in adopting such measures as will secure freedom from pressure and friction for the parts most liable to corns. To this end a well-fitting shoe is essential. The shoes should be of well-seasoned leather, soft and elastic, and should be cut to a proper model.

II.—The palliative treatment is generally carried out with chemical substances. The best method, is, briefly, as follows: A ring of glycerine jelly is painted around the circumference of the corn, to form a raised rampart. A piece of salicylic plaster mull is then cut to the size and shape of the central depression, and applied to the surface of the corn. This is then covered with a layer of glycerine jelly, and before it sets a pad of cotton wool is applied to the surface. This process is repeated as often as is necessary, until the horny layer separates and is cast off.

If the point of a sharp, thin-bladed knife be introduced at the groove which runs around the margin of the corn, and be made to penetrate toward its central axis, by the exercise of a little manual dexterity the horny part of the corn can be easily made to separate from the parts beneath.

III.—Any method of treatment to be curative must secure the removal of the entire corn, together with the underlying bursa. It is mainly in connection with the latter structure that complications, which alone make a corn a matter of serious import, are likely to arise. Freeland confidently advises the full and complete excision of corns, on the basis of his experience in upward of 60 cases.

Every precaution having been taken to render the operation aseptic, a spot is selected for the injection of the anæsthetic solution. The skin is rendered insensitive with ethyl chloride, and 5 minims of a 4 per cent solution of cocaine is injected into the subcutaneous tissue beneath the corn. After a wait of a few minutes the superficial parts of the site of the incision are rendered insensitive with ethyl chloride. Anæsthesia is now complete.

Two semielliptical incisions meeting at their extremities are made through the skin around the circumference of the growth, care being taken that they penetrate well into the subcutaneous tissue. Seizing the parts included in the incision with a pair of dissecting forceps, a wedge-shaped piece of tissue—including the corn, a layer of skin and subcutaneous tissue, and the bursa if present—is dissected out. The oozing is pretty free, and it is sometimes necessary to torsion a small vessel; but the hemorrhage is never severe. The edges of the wound are brought together by one or two fine sutures; an antiseptic dressing is applied, and the wound is left to heal—primary union in a few days being the rule. The rapidity of the healing is often phenomenal. There is produced a scar tissue at the site of the corn, but this leads to no untoward results.

Cosmetics

COLD CREAM.

I.—Oil of almonds	425	parts
Lanolin	185	parts
White wax	62	parts
Spermaceti	62	parts
Borax	4.5	parts
Rose water	300	parts

Melt together the first four ingredients, then incorporate the solution of borax in the rose water.

II.—Tragacanth	125	parts
Boric acid	100	parts
Glycerine	140	parts
Expressed oil of almonds	50	parts
Glyconine	50	parts
Oil of lavender	0.5	parts
Water enough to make	1,000	parts

Mix the tragacanth and the boric acid with the glycerine; add the almond oil, lavender oil, and egg glycerite, which have been previously well incorporated, and, lastly, add the water in divided portions until a clear jelly of the desired consistency is obtained.

III.—Oil of almonds	26 ounces
Castor oil (odorless)	6 ounces
Lard (benzoated)	8 ounces
White wax	8 ounces
Rose water (in winter less, in summer more, than quantity named)	12 ounces
Orange-flower water	8 ounces
Oil of rose	15 minims
Extract of jasmine	6 drachms
Extract of cassia	4 drachms
Borax	2 ounces
Glycerine	4 ounces

Melt the oil of sweet almonds, wax, and lard together, and stir in the castor oil; make a solution of the borax in the glycerine and rose and orange-flower waters; add this solution, a little at a time, to the melted fat, stirring constantly to insure thorough incorporation; finally add the oil of rose dissolved in the extracts, and beat the ointment until cold.

IV.—Spermaceti (pure), ¼ ounce; white wax (pure), ¼ ounce; almond oil, ¼ pound; butter of cocoa, ¼ pound; lanolin, 2 ounces.

Melt and stir in 1 drachm of balsam of Peru. After settling, pour off the clear portion and add 2 fluidrachms of orange-flower water and stir briskly until it concretes.

Camphorated Cold Cream.—

Oil of sweet almonds	8	fluidounces
White wax	1	ounce
Spermaceti	1	ounce
Camphor	1	ounce
Rose water	5	fluidounces
Borax (in fine powder)	4	drachms
Oil of rose	10	drops

Melt the wax and spermaceti, add the oil of sweet almonds, in which the camphor has been dissolved with very gentle heat; then gradually add the rose water, in which the borax has previously been dissolved, beating or agitating constantly with a wooden spatula until cold. Lastly add the oil of rose.

Petrolatum Cold Cream.—

Petrolatum (white)	7	ounces
Paraffine	½	ounce
Lanolin	2	ounces
Water	3	ounces
Oil of rose	3	drops
Alcohol	1	drachm

A small quantity of borax may be added, if desirable, and the perfume may be varied to suit the taste.

LIP SALVES:

Pomades for the Lips.—Lip pomatum which is said always to retain a handsome red color and never to grow rancid is prepared as follows:

I.—Paraffine	80.0	parts
Vaseline	80.0	parts
Anchusine	0.5	parts
Bergamot oil	1.0	part
Lemon peel	1.0	part

II.—Vaseline Pomade.—

Vaseline oil, white	1,000	parts
Wax, white	300	parts

Geranium oil, African	40	parts
Lemon oil	20	parts

III.—Rose Pomade.—

Almond oil	1,000	parts
Wax, white	300	parts
Alkannin	3	parts
Geranium oil	20	parts

IV.—Yellow Pomade.—

Vaseline oil, white	1,000	parts
Wax, white	200	parts
Spermaceti	200	parts
Saffron surrogate	10	parts
Clove oil	20	parts

V.—White Pomade.—

Vaseline oil, white	1,000	parts
Wax, white	300	parts
Bitter almond oil, genuine	10	parts
Lemon oil	2	parts

VI.—Paraffine	49.0	parts
Vaseline	49.0	parts
Oil of lemon	0.75	parts
Oil of violet	0.75	parts
Carmine, quantity sufficient.		

Lipol.—For treating sore, rough, or inflamed lips, apply the following night and morning, rubbing in well with the finger tips: Camphor, ½ ounce; menthol, ½ ounce; eucalyptol, 1 drachm; petrolatum (white), 1 pound; paraffine, ¼ pound; alkanet root, ½ ounce; oil of bitter almonds, 15 drops; oil of cloves, 10 drops; oil of cassia, 5 drops. Digest the root in the melted paraffine and petrolatum, strain, add the other ingredients and pour into lip jars, hot.

MANICURE PREPARATIONS:

Powdered Nail Polishes.—

I.—Tin oxide	8	drachms
Carmine	¼	drachm
Rose oil	6	drops
Neroli oil	5	drops

II.—Cinnabar	1	drachm
Infusorial earth	8	drachms

III.—Putty powder (fine)	4	drachms
Carmine	2	grains
Oil of rose	1	drop

IV.—White castile soap	1	part
Hot water	16	parts
Zinc chloride solution, 10 per cent, quantity sufficient.		

Dissolve the soap in the water and to the solution add the zinc-chloride solution until no further precipitation occurs. Let stand over night; pour off the supernatant fluid, wash the precipitate well with water, and dry at the ordinary temperature. Carmine may be added if desired.

Polishing Pastes for the Nails.—

I.—Talcum........... 5 drachms
Stannous oxide..... 3 drachms
Powdered tragacanth 5 grains
Glycerine.......... 1 drachm
Rose water, quantity sufficient.
Solution of carmine sufficient to tint.

Make paste.

For softening the nails, curing hang-nails, etc., an ointment is sometimes used consisting of white petrolatum, 8 parts; powdered castile soap, 1 part; and perfume to suit.

II.—Eosine............. 10 grains
White wax........ ½ drachm
Spermaceti........ ½ drachm
Soft paraffine....... 1 ounce
Alcohol, a sufficient quantity.

Dissolve the eosine in as little alcohol as will suffice, melt the other ingredients together, add the solution, and stir until cool.

Nail-Cleaning Washes.—

I.—Tartaric acid....... 1 drachm
Tincture of myrrh.. 1 drachm
Cologne water...... 2 drachms
Water............. 3 ounces

Dissolve the acid in the water; mix the tincture of myrrh and cologne, and add to the acid solution.

Dip the nails in this solution, wipe, and polish with chamois skin.

II.—Oxalic acid........ 30 grains
Rose water........ 1 ounce

Nail Varnish.—

Paraffine wax...... 60 grains
Chloroform....... 2 ounces
Oil of rose........ 3 drops

POMADES:

I.—Beef-Marrow Pomade.—
Vaseline oil, yellow .... 20,000 parts
Ceresine, yellow 3,000 parts
Beef marrow .. 2,000 parts
Saffron substitute .... 15 parts
Lemon oil..... 50 parts
Bergamot oil... 20 parts
Clove oil ...... 5 parts
Lavender oil... 10 parts

II.—China Pomade.—
Vaseline oil, yellow..... 20,000 parts
Ceresine, yellow........ 5,000 parts

Brilliant, brown....... 12 parts
Peru balsam... 50 parts
Lemon oil..... 5 parts
Bergamot oil .. 5 parts
Clove oil...... 5 parts
Lavender oil.. 5 parts

III.—Crystalline Honey Pomade.— Nut oil, 125 drachms; spermaceti, 15 drachms; gamboge, 2 drachms; vervain oil, 10 drops; cinnamon oil, 20 drops; bergamot oil, 30 drops; rose oil, 3 drops. The spermaceti is melted in the nut oil on a water bath and digested with the gamboge for 20 minutes; it is next strained, scented, and poured into cans which are standing in water. The cooling must take place very slowly. Instead of gamboge, butter color may be used. Any desired scent mixture may be employed.

IV.—Herb Pomade.—
Vaseline oil, yellow .... 20,000 parts
Ceresine, yellow 5,000 parts
Chlorophyll... 20 parts
Lemon oil..... 50 parts
Clove oil...... 20 parts
Geranium oil, African..... 12 parts
Curled mint oil. 4 parts

V.—Rose Pomade.—
Vaseline oil, white....... 20,000 parts
Ceresine, white 5,000 parts
Alkannin..... 15 parts
Geranium oil, African..... 50 parts
Palmarosa oil. 30 parts
Lemon oil.... 20 parts

VI.—Strawberry Pomade.—When the strawberry season is on, and berries are plenty and cheap, the following is timely:
Strawberries, ripe and fresh...... 4 parts
Lard, sweet and fresh.......... 25 parts
Tallow, fresh..... 5 parts
Alkanet tincture, quantity sufficient.
Essential oil, quantity sufficient to perfume.

Melt lard and tallow together on the water bath at the temperature of boiling water. Have the strawberries arranged on a straining cloth. Add the alkanet tincture to the melted grease, stir in, and then pour the mixture over the berries. Stir the strained fats until the mass begins to set, then add the perfume and stir in. A little artificial essence of strawberries may be added. The odor usually employed is rose, about 1 drop to every 2 pounds.

VII.—Stick Pomade.—

Tallow........	500 parts
Ceresine.......	150 parts
Wax, yellow.....	50 parts
Rosin, light......	200 parts
Paraffine oil (thick)........	300 parts
Oil of cassia.....	5 parts
Oil of bergamot..	5 parts
Oil of clove......	2 parts

VIII.—Vaseline Pomade.—

Melt 250 parts of freshly rendered lard and 25 parts of white wax at moderate heat and mix well with 200 parts of vaseline. Add 15 parts of bergamot oil, 3 parts of lavender oil, 2 parts of geranium oil, and 2 parts of lemon oil, mixing well.

IX.—Witch-Hazel Jelly.—

Oil of sweet almonds........	256 parts
Extract of witch-hazel fluid.....	10 parts
Glycerine.......	32 parts
Soft soap........	20 parts
Tincture of musk, quantity sufficient to perfume.	

Mix in a large mortar the glycerine and soft soap and stir until incorporated. Add and rub in the witch-hazel, and then add the oil, slowly, letting it fall in a very thin, small stream, under constant agitation; add the perfume, keeping up the agitation until complete incorporation is attained. Ten drops of musk to a quart of jelly is sufficient. Any other perfume may be used.

Colors for Pomade.—Pomade may be colored red by infusing alkanet in the grease; yellow may be obtained by using annotto in the same way; an oil-soluble chlorophyll will give a green color by admixture.

In coloring grease by means of alkanet or annotto it is best to tie the drug up in a piece of coarse cloth, place in a small portion of the grease, heat gently, squeezing well with a rod from time to time; and then adding this strongly colored grease to the remainder. This procedure obviates exposing the entire mass to heat, and neither decantation nor straining is needed.

Brocq's Pomade for Itching.—

Acid phenic.........	1 part
Acid salicylic........	2 parts
Acid tartaric...	3 parts
Glycerole of starch........	60 to 100 parts

Mix and make a pomade.

White Cosmetique.—

Jasmine pomade.....	2 ounces
Tuberose pomade....	2 ounces
White wax..........	2 ounces
Refined suet.........	4 ounces
Rose oil............	15 minims

Melt the wax and suet over a water bath, then add the pomades, and finally the otto.

Glycerine and Cucumber Jelly.—

Gelatin.....160 to 240 grains	
Boric acid........	240 grains
Glycerine........	6 fluidounces
Water..........	10 fluidounces

Perfume to suit. The perfume must be one that mixes without opalescence, otherwise it mars the beauty of the preparation. Orange-flower water or rose water could be substituted for the water if desired, or another perfume consisting of

Spirit of vanillin (15 grains per ounce).	2 fluidrachms
Spirit of coumarin (15 grains per ounce).........	2 fluidrachms
Spirit of bitter almonds (⅛)......	8 minims

to the quantities given above would prove agreeable.

Cucumber Pomade.—

Cucumber pomade...	2 ounces
Powdered white soap.	½ ounce
Powdered borax.....	2 drachms
Cherry-laurel water..	3 ounces
Rectified spirit......	3 ounces
Distilled water to make 48 ounces	

Rub the pomade with the soap and borax until intimately mixed, then add the distilled water (which may be warmed to blood heat), ounce by ounce, to form a smooth and uniform cream. When 40 ounces of water have been so incorporated, dissolve any essential oils desired as perfume in the spirit, and add the cherry-laurel water, making up to 48 ounces with plain water.

ROUGES AND PAINTS:

Grease Paints.—Theatrical face paints are sold in sticks, and there are many varieties of color. Yellows are obtained with ocher; browns with burnt umber; and blue is made with ultramarine. These colors should in each case be levigated finely along with their own weight of equal parts of precipitated chalk and oxide of zinc and diluted with the same to the tint required, then made into sticks with mutton suet (or vaseline or paraffine, equal parts) well perfumed. By blending these colors, other tints may thus be obtained.

White Grease Paints.—

I.—Prepared chalk ..   4 av. ounces
Zinc oxide . . . .   4 av. ounces
Bismuth subnitrate . . . .   4 av. ounces
Asbestos powder .   4 av. ounces
Sweet almond oil, about . . . .   2½ fluidounces
Camphor . . . .   40 grains
Oil peppermint . . .   3 fluidrachms
Esobouquet extract . . . .   3 fluidrachms

Sufficient almond oil should be used to form a mass of proper consistence.

II.—Zinc oxide . . . .   8 parts
Bismuth subnitrate . .   8 parts
Aluminum oxychloride . . . .   8 parts
Almond oil, quantity sufficient, or 5–6 parts.
Perfume, quantity sufficient.

Mix the zinc, bismuth, and aluminum oxychloride thoroughly; make into a paste with the oil. Any perfume may be added, but that generally used is composed of 1 drachm of essence of bouquet, 12 grains of camphor, and 12 minims of oil of peppermint for every 3½ ounces of paste.

Bright Red.—

Zinc oxide . . . .   10 parts
Bismuth subnitrate . . .   10 parts
Aluminum oxychloride . . . .   10 parts
Almond oil, quantity sufficient.

Mix the zinc, bismuth, and aluminum salts, and to every 4 ounces of the mixture add 2¼ grains of eosine dissolved in a drachm of essence of bouquet, 12 minims oil of peppermint, and 12 grains of camphor. Make the whole into a paste with almond oil.

Red.—

Cacao butter . . . . .   4 av. ounces
White wax . . . .   4 av. ounces
Olive oil . . . .   2 fluidounces
Oil of rose . . . .   8 drops
Oil of bergamot . .   3 drops
Oil of neroli . . . . . .   2 drops
Tincture musk . . .   2 drops
Carmine . . . .   90 grains
Ammonia water . .   3 fluidrachms

Deep, or Bordeaux, Red.—

Zinc oxide . . . .   30 parts
Bismuth subnitrate . . .   30 parts
Aluminum oxychloride . . . .   30 parts
Carmine . . . .   1 part
Ammonia water . . . . .   5 parts
Essence bouquet . . . . .   3 parts
Peppermint, camphor, etc., quantity sufficient.

Mix the zinc, bismuth, and aluminum salts. Dissolve the carmine in the ammonia and add solution to the mixture. Add 24 grains of camphor, and 24 minims of oil of peppermint dissolved in the essence bouquet, and make the whole into a paste with oil of sweet almonds.

Vermilion.—

Vermilion . . . .   18 parts
Tincture of saffron . .   12 parts
Orris root, powdered   30 parts
Chalk, precipitated . .   120 parts
Zinc oxide . . . .   120 parts
Camphor . . . .   2 parts
Essence bouquet . . . .   9 parts
Oil of peppermint . . .   2 parts
Almond oil, quantity sufficient.

Mix as before.

Pink.—

Zinc carbonate . . . . .   250 parts
Bismuth subnitrate . .   250 parts
Asbestos . . . .   250 parts
Expressed oil of almonds . . . .   100 parts
Camphor . . . .   55 parts
Oil of peppermint . .   55 parts
Perfume . . . .   25 parts
Eosine . . . .   1 part

Dark Red.—Like the preceding, but colored with a solution of carmine.

Rouge.—

Zinc oxide . . . .   2½ ounces
Bismuth subnitrate . . .   2½ ounces
Aluminum plumbate .   2½ ounces
Eosine . . . .   1 drachm
Essence bouquet . . . . .   2 drachms
Camphor . . . .   6 drachms
Oil of peppermint . . . .   20 minims
Almond oil, quantity sufficient.

Dissolve the eosine in the essence bouquet, and mix with the camphor and peppermint; add the powder and make into a paste with almond oil.

Black Grease Paints.—

I.—Soot . . . .   2 av. ounces
Sweet almond oil .   2 fluidounces
Cacao butter . . . . .   6 av. ounces
Perfume, sufficient.

The soot should be derived from burning camphor and repeatedly washed with alcohol. It should be triturated to a smooth mixture with the oil; then add to the melted cacao butter; add the perfume, and form into sticks.

Brown or other colors may be obtained by adding appropriate pigments, such as finely levigated burned umber, sienna, ocher, jeweler's rouge, etc., to the foregoing base instead of lampblack.

II.—Best lampblack..... 1 drachm
    Cacao butter....... 3 drachms
    Olive oil.......... 3 drachms
    Oil of neroli....... 2 drops

Melt the cacao butter and oil, add the lampblack, and stir constantly as the mixture cools, adding the perfume toward the end.

III.—Lampblack........ 1 part
    Cacao butter ...... 6 parts
    Oil neroli, sufficient.

Melt the cacao butter and the lampblack, and while cooling make an intimate mixture, adding the perfume toward the last.

IV.—Lampblack........ 1 part
    Expressed oil of almonds........... 1 part
    Oil cocoanut....... 1 part
    Perfume, sufficient.

Beat the lampblack into a stiff paste with glycerine. Apply with a sponge; if necessary, mix a little water with it when using.

V.—Beat the finest lampblack into a stiff paste with glycerine and apply with a sponge; if necessary, add a little water to the mixture when using. Or you can make a grease paint as follows: Drop black, 2 drachms; almond oil, 2 drachms; cocoanut oil, 6 drachms; oil of lemon, 5 minims; oil of neroli, 1 minim. Mix.

Fatty Face Powders.—These have a small percentage of fat mixed with them in order to make the powder adhere to the skin.

Dissolve 1 drachm anhydrous lanolin in 2 drachms of ether in a mortar. Add 3 drachms of light magnesia. Mix well, dry, and then add the following: French chalk, 2 ounces; powdered starch, 1½ ounces; boric acid, 1 drachm; perfume, a sufficient quantity. A good perfume is coumarin, 2 grains, and attar of rose, 2 minims.

Nose Putty.—I.—Mix 1 ounce wheat flour with 2 drachms of powdered tragacanth and tint with carmine. Take as much of the powder as necessary, knead into a stiff paste with a little water and apply to the nose, having previously painted it with spirit gum.

II.—White wax, 8 parts; rosin, white, 8 parts; mutton suet, 4 parts; color to suit. Melt together.

Rose Powder.—As a base take 200 parts of powdered iris root, add 600 parts of rose petals, 100 parts of sandalwood, 100 parts of patchouli, 3 parts of oil of geranium, and 2 parts of true rose oil.

Rouge Tablets.—There are two distinct classes of these tablets: those in which the coloring matter is carmine, and those in which the aniline colors are used. The best are those prepared with carmine, or ammonium carminate, to speak more correctly. The following is an excellent formula:

    Ammonium carminate... 10 parts
    Talc, in powder .... 25 parts
    Dextrin. .... 8 parts
    Simple syrup, sufficient.
    Perfume, to taste, sufficient.

Mix the talc and dextrin and add the perfume, preferably in the shape of an essential oil (attar of rose, synthetic oil of jasmine, or violet, etc.), using 6 to 8 drops to every 4 ounces of other ingredients. Incorporate the ammonium carminate and add just enough simple syrup to make a mass easily rolled out. Cut into tablets of the desired size. The ammonium carminate is made by adding 1 part of carmine to 2½ parts of strong ammonia water. Mix in a vial, cork tightly, and set aside until a solution is formed, shaking occasionally. The ammonium carminate is made by dissolving carmine in ammonia water to saturation.

Rouge Palettes.—To prepare rouge palettes rub up together:

    Carmine........... 9 parts
    French chalk....... 50 parts
    Almond oil........ 12 parts

Add enough tragacanth mucilage to make the mass adhere and spread the whole evenly on the porcelain palette.

Liquid Rouge.—

I.—Carmine........... 4 parts
    Stronger ammonia water........... 4 parts
    Essence of rose .... 16 parts
    Rose water to make. 500 parts

Mix. A very delightful violet odor, if this is preferred, is obtained by using ionone in place of rose essence. A cheaper preparation may be made as follows:

II.—Eosine............ 1 part
Distilled water..... 20 parts
Glycerine.......... 5 parts
Cologne water...... 75 parts
Alcohol...........100 parts

Mix.

Rub together with 10 parts of almond oil and add sufficient mucilage of tragacanth to make the mass adhere to the porcelain palette.

III.—Carmine.......... 1 part
Stronger ammonia
water.......... 1 part
Attar of rose...... 4 parts
Rose water....... 125 parts

Mix. Any other color may be used in place of rose, violet (ionone), for instance, or heliotrope. A cheaper preparation may be made by substituting eosine for the carmine, as follows:

IV.—Eosine........... 1 part
Distilled water.... 20 parts
Glycerine......... 5 parts
Cologne water..... 75 parts
Alcohol.......... 100 parts

Mix.

Peach Tint.—

a.—Buffalo eosine .... 4 drachms
Distilled water...... 16 fluidounces

Mix.

b.—Pure hydrochloric
acid............ 2½ drachms
Distilled water..... 64 fluidounces

Mix.

Pour a into b, shake, and set aside for a few hours; then pour off the clear portion and collect the precipitate on a filter. Wash with the same amount of b and immediately throw the precipitate into a glass measure, stirring in with a glass rod sufficient of b to measure 16 ounces in all. Pass through a hair sieve to get out any filtering paper. To every 16 ounces add 8 ounces of glycerine.

Theater Rouge.—Base:

Cornstarch.......... 4 drachms
Powdered white tal-
cum.............. 6 drachms

Mix.

a.—Carminoline .... 10 grains
Base.............. 6 drachms
Water.............. 4 drachms

Dissolve the carminoline in the water, mix with the base and dry.

b.—Geranium red .... 10 grains
Base.............. 6 drachms
Water.............. 4 drachms

Mix as above and dry.

SKIN FOODS.

Wrinkles on the face yield to a wash consisting of 50 parts milk of almonds (made with rose water) and 4 parts aluminum sulphate. Use morning and night.

Rough skin is to be washed constantly in Vichy water. Besides this, rough places are to have the following application twice daily—either a few drops of:

I.—Rose water........ 100 parts
Glycerine.......... 25 parts
Tannin........... ¾ part

Mix. Or use:

II.—Orange-flower water 100 parts
Glycerine.......... 10 parts
Borax............. 2 parts

Mix. Sig.: Apply twice daily.

"Beauty Cream."—This formula gives the skin a beautiful, smooth, and fresh appearance, and, at the same time, serves to protect and preserve it:

Alum, powdered..... 10 grams
Whites of.......... 2 eggs
Boric acid.......... 3 grams
Tincture of benzoin.. 40 drops
Olive oil........... 40 drops
Mucilage of acacia... 5 drops
Rice flour, quantity sufficient.
Perfume, quantity sufficient.

Mix the alum and the white of eggs, without any addition of water whatever, in an earthen vessel, and dissolve the alum by the aid of very gentle heat (derived from a lamp, or gaslight, regulated to a very small flame), and constant, even, stirring. This must continue until the aqueous content of the albumen is completely driven off. Care must be taken to avoid coagulation of the albumen (which occurs very easily, as all know). Let the mass obtained in this manner get completely cold, then throw into a Wedgwood mortar, add the boric acid, tincture of benzoin, oil, mucilage (instead of which a solution of fine gelatin may be used), etc., and rub up together, thickening it with the addition of sufficient rice flour to give the desired consistence, and perfuming at will. Instead of olive oil any pure fat, or fatty oil, may be used, even vaseline or glycerine.

Face Bleach or Beautifier.—

Syrupy lactic acid.... 40 ounces
Glycerine.......... 80 ounces
Distilled water....... 5 gallons

Mix. Gradually add

Tincture of benzoin.. 3 ounces

Color by adding

Carmine No. 40...... 40 grains
Glycerine.......... 1 ounce
Ammonia solution... ½ ounce
Water to........... 3 ounces

Heat this to drive off the ammonia, and mix all. Shake, set aside; then filter, and add

Solution of ionone.... 1 drachm

Add a few drachms of kaolin and filter until bright.

BLACKHEAD REMEDIES.

I.—Lactic acid........ 1 drachm
Boric acid......... 1 drachm
Ceresine........... 1 drachm
Paraffine oil....... 6 drachms
Hydrous wool fat... 1½ ounces
Castor oil .... 6 drachms

II.—Unna advises hydrogen dioxide in the treatment of blackheads, his prescription being:

Hydrogen dioxide 20 to 40 parts
Hydrous wool fat.. 10 parts
Petrolatum....... 30 parts

III.—Thymol.......... 1 part
Boric acid........ 2 parts
Tincture of witch-
hazel.......... 18 parts
Rose water suffi-
cient to make... 200 parts

Mix. Apply to the face night and morning with a sponge, first washing the face with hot water and castile soap, and drying it with a coarse towel, using force enough to start the dried secretions. An excellent plan is to steam the face by holding it over a basin of hot water, keeping the head covered with a cloth.

IV.—Ichthyol........... 1 drachm
Zinc oxide......... 2 drachms
Starch............ 2 drachms
Petrolatum....... 3 drachms

This paste should be applied at night. The face should first be thoroughly steamed or washed in water as hot as can be comfortably borne. All pustules should then be opened and blackheads emptied with as little violence as possible. After careful drying the paste should be thoroughly rubbed into the affected areas. In the morning, after removing the paste with a bland soap, bathe with cool water and dry with little friction.

HAND CREAMS AND LOTIONS:

Chapped Skin.—

I.—Glycerine .... 8 parts
Bay rum........... 4 parts
Ammonia water ..... 4 parts
Rose water .... 4 parts

Mix the bay rum and glycerine, add the ammonia water, and finally the rose water. It is especially efficacious after shaving.

II.—As glycerine is bad for the skin of many people, here is a recipe which will be found more generally satisfactory as it contains less glycerine: Bay rum, 3 ounces; glycerine, 1 ounce; carbolic acid, ½ drachm (30 drops). Wash the hands well and apply while hands are soft, preferably just before going to bed. Rub in thoroughly. This rarely fails to cure the worst "chaps" in two nights.

III.—A sure remedy for chapped hands consists in keeping them carefully dry and greasing them now and then with an anhydrous fat (not cold cream). The best substances for the purpose are unguentum cereum or oleum olivarum.

If the skin of the hands is already cracked the following preparation will heal it:

Finely ground zinc oxide, 5.0 parts; bismuth oxychloride, 2.0 parts; with fat oil, 12.0 parts; next add glycerine, 5.0 parts; lanolin, 30.0 parts; and scent with rose water, 10.0 parts.

IV.—Wax salve (olive oil 7 parts, and yellow wax 3 parts), or pure olive oil.

Hand-Cleaning Paste.—Cleaning pastes are composed of soap and grit, either with or without some free alkali. Any soap may be used, but a white soap is preferred. Castile soap does not make as firm a paste as soap made from animal fats, and the latter also lather better. For grit, anything may be used, from powdered pumice to fine sand.

A good paste may be made by dissolving soap in the least possible quantity of hot water, and as it cools and sets stirring in the grit. A good formula is:

White soap.......... 2½ pounds
Fine sand.......... 1 pound
Water............. 5½ pints

Lotion for the Hands.—

Boric acid.......... 1 drachm
Glycerine.......... 6 drachms

Dissolve by heat and mix with

Lanolin............ 6 drachms
Vaseline.......... 1 ounce

Add any perfume desired. The borated glycerine should be cooled before mixing it with the lanolin.

Cosmetic Jelly.—

Tragacanth (white rib-
bon)............ 60 grains
Rose water......... 14 ounces

Macerate for two days and strain forcibly through coarse muslin or cheese cloth. Add glycerine and alcohol, of each 1 ounce. Perfume to suit. Use immediately after bathing, rubbing in well until dry.

Perspiring Hands.—I.—Take rectified eau de cologne, 50 parts (by weight); belladonna dye, 8 parts; glycerine, 3 parts; rub gently twice or three times a day with half a tablespoonful of this mixture. One may also employ chalk, carbonate of magnesia, rice starch, hot and cold baths of the hands (as hot and as cold as can be borne), during 6 minutes, followed by a solution of 4 parts of tannin in 32 of glycerine.

II.—Rub the hands several times per day with the following mixture:

	By weight
Rose water	125 parts
Borax	10 parts
Glycerine	8 parts

Hand Bleach.—Lanolin, 30 parts; glycerine, 20 parts; borax, 10 parts; eucalyptol, 2 parts; essential oil of almonds, 1 part. After rubbing the hands with this mixture, cover them with gloves during the night.

For the removal of developing stains, see Photography.

MASSAGE CREAMS:

Massage Application.—

White potash soap, shaved	20 parts
Glycerine	30 parts
Water	30 parts
Alcohol (90 per cent)	10 parts

Dissolve the soap by heating it with the glycerine and water, mixed. Add the alcohol, and for every 30 ounces of the solution add 5 or 6 drops of the mistura oleoso balsamica, German Pharmacopœia. Filter while hot.

Medicated Massage Balls.—They are the balls of paraffine wax molded with a smooth or rough surface with menthol, camphor, oil of wintergreen, oil of peppermint, etc., added before shaping. Specially useful in headaches, neuralgias, and rheumatic affections, and many other afflictions of the skin and bones. The method of using them is to roll the ball over the affected part by the aid of the palm of the hand with pressure. Continue until relief is obtained or a sensation of warmth. The only external method for the treatment of all kinds of headaches is the menthol medicated massage ball. This may be made with smooth or corrugated surfaces. Keep wrapped in foil in cool places.

Casein Massage Cream.—The basis of the modern massage cream is casein. Casein is now produced very cheaply in the powdered form, and by treatment with glycerine and perfumes it is possible to turn out a satisfactory cream. The following formula is suggested:

Skimmed milk	1 gallon
Water of ammonia	1 ounce
Acetic acid	1 ounce
Oil of rose geranium	1 drachm
Oil of bitter almond	1 drachm
Oil of anise	2 drachms
Cold cream (see below), enough.	
Carmine enough to color.	

Add the water of ammonia to the milk and let it stand 24 hours. Then add the acetic acid and let it stand another 24 hours. Then strain through cheese cloth and add the oils. Work this thoroughly in a Wedgwood mortar, adding enough carmine to color it a delicate pink. To the product thus obtained add an equal amount of cold cream made by the formula herewith given:

White wax	4 ounces
Spermaceti	4 ounces
White petrolatum	12 ounces
Rose water	14 ounces
Borax	80 grains

Melt the wax, spermaceti, and petrolatum together over a water bath; dissolve the borax in the rose water and add to the melted mass at one time. Agitate violently. Presumably the borax solution should be of the same temperature as the melted mass.

Massage Skin Foods.—

This preparation is used in massage for removing wrinkles:

I.—White wax	$\frac{1}{2}$ ounce
Spermaceti	$\frac{1}{2}$ ounce
Cocoanut oil	1 ounce
Lanolin	1 ounce
Oil of sweet almonds	2 ounces

Melt in a porcelain dish, remove from the fire, and add

Orange-flower water	1 ounce
Tincture of benzoin	3 drops

Beat briskly until creamy.

II.—Snow-white cold cream	4 ounces
Lanolin	4 ounces
Oil of theobroma	4 ounces
White petrolatum oil	4 ounces
Distilled water	4 ounces

In hot weather add

Spermaceti	$1\frac{1}{2}$ drachms
White wax	$2\frac{1}{2}$ drachms

In winter the two latter are left out and the proportion of cocoa butter is modified. Prepared and perfumed in proportion same as cold cream.

III.—White petrolatum 7 av. ounces
    Paraffine wax.... ½ ounce
    Lanolin......... 2 av. ounces
    Water........... 3 fluidounces
    Oil of rose....... 3 drops
    Vanillin......... 2 grains
    Alcohol......... 1 fluidrachm

Melt the paraffine, add the lanolin and petrolatum, and when these have melted pour the mixture into a warm mortar, and, with constant stirring, incorporate the water. When nearly cold add the oil and vanillin, dissolved in the alcohol.

Preparations of this kind should be rubbed into the skin vigorously, as friction assists the absorbed fat in developing the muscles, and also imparts softness and fullness to the skin.

SKIN BLEACHES, BALMS, LOTIONS, ETC. :

See also Cleaning Methods and Photography for removal of stains caused by photographic developers.

Astringent Wash for Flabby Skin.— This is used to correct coarse pores, and to remedy an oily or flabby skin. Apply with sponge night and morning:

Cucumber juice ...... 1½ ounces
Tincture of benzoin .. ½ ounce
Cologne .... 1 ounce
Elder-flower water... 5 ounces

Put the tincture of benzoin in an 8-ounce bottle, add the other ingredients, previously mixed, and shake slightly. There will be some precipitation of benzoin in this mixture, but it will settle out, or it may be strained out through cheese cloth.

Bleaching Skin Salves.—A skin-bleaching action, due to the presence of hydrogen peroxide, is possessed by the following mixtures:

I.—Lanolin............. 30 parts
    Bitter almond oil.... 10 parts

Mix and stir with this salve base a solution of

Borax.............. 1 part
Glycerine........... 15 parts
Hydrogen peroxide.. 15 parts

For impure skin the following composition is recommended:

II.—White mercurial ointment............. 5 grams
    Zinc ointment....... 5 grams
    Lanolin............. 30 grams
    Bitter almond oil.... 10 grams

And gradually stir into this a solution of

Borax.............. 2 grams
Glycerine........... 30 grams
Rose water......... 10 grams
Concentrated nitric acid .... 5 drops

III.—Lanolin........... 30 grams
    Oil sweet almond... 10 grams
    Borax.............. 1 gram
    Glycerine........... 15 grams
    Solution hydrogen peroxide......... 15 grams

Mix the lanolin and oil, then incorporate the borax previously dissolved in the mixture of glycerine and peroxide solution.

IV.—Ointment ammoniac mercury......... 5 grams
    Ointment zinc oxide. 5 grams
    Lanolin........... 30 grams
    Oil sweet almond... 10 grams
    Borax .... 2 grams
    Glycerine........... 30 grams
    Rose water......... 10 grams
    Nitric acid, C. P.... 5 drops

Prepare in a similar manner as the foregoing. Rose oil in either ointment makes a good perfume. Both ointments may, of course, be employed as a general skin bleach, which, in fact, is their real office—cosmetic creams.

Emollient Skin Balm.—

Quince seed......... ½ ounce
Water............. 7 ounces
Glycerine........... 1½ ounces
Alcohol............. 4½ ounces
Salicylic acid........ 6 grains
Carbolic acid........ 10 grains
Oil of bay........... 10 drops
Oil of cloves........ 5 drops
Oil of orange peel.... 10 drops
Oil of wintergreen.... 8 drops
Oil of rose........ 2 drops

Digest the quince seed in the water for 24 hours, and then press through a cloth; dissolve the salicylic acid in the alcohol; add the carbolic acid to the glycerine; put all together, shake well, and bottle.

Skin Lotion.—

Zinc sulphocarbolate............ 30 grains
Alcohol (90 per cent) 4 fluidrachms
Glycerine........ 2 fluidrachms
Tincture of cochineal............. 1 fluidrachm
Orange-flower water .... 1½ fluidounces
Rose water (triple) to make........ 6 fluidounces

Skin Discoloration.—Discoloration of the neck may be removed by the use of acids, the simplest of which is that in buttermilk, but if the action of this is too slow try 4 ounces of lactic acid, 2 of glycerine, and 1 of rose water. These will mix without heating. Apply several times daily with a soft linen rag; pour a small quantity into a saucer and dip the cloth into this. If the skin becomes sore use less of the remedy and allay the redness and smarting with a good cold cream. It is always an acid that removes freckles and discolorations, by burning them off. It is well to be slow in its use until you find how severe its action is. It is not wise to try for home making any of the prescriptions which include corrosive sublimate or any other deadly poison. Peroxide of hydrogen diluted with 5 times as much water, also will bleach discolorations. Do not try any of these bleaches on a skin freshly sunburned. For that, wash in hot water, or add to the hot water application enough witch-hazel to scent the water, and after that has dried into the skin it will be soon enough to try other applications.

Detergent for Skin Stains.—Moritz Weiss has introduced a detergent paste which will remove stains from the skin without attacking it, is non-poisonous, and can be used without hot water. Moisten the hands with a little cold water, apply a small quantity of the paste to the stained skin, rub the hands together for a few minutes, and rinse with cold water. The preparation is a mixture of soft soap and hard tallow, melted together over the fire and incorporated with a little emery powder, flint, glass, sand, quartz, pumice stone, etc., with a little essential oil to mask the smell of the soap. The mixture sets to a mass like putty, but does not dry hard. The approximate proportions of the ingredients are: Soft soap, 30 per cent; tallow, 15 per cent; emery powder, 55 per cent, and a few drops of essential oil.

If an extra detergent quality is desired, 4 ounces of sodium carbonate may be added, and the quantity of soap may be reduced. Paste thus made will attack grease, etc., more readily, but it is harder on the skin.

Removing Inground Dirt.—

Egg albumen..........	8 parts
Boric acid............	1 part
Glycerine	32 parts
Perfume to suit.	
Distilled water to make.	50 parts

Dissolve the boric acid in a sufficient quantity of water; mix the albumen and glycerine and pass through a silk strainer. Finally, mix the two fluids and add the residue of water.

Every time the hands are washed, dry on a towel, and then moisten them lightly but thoroughly with the liquid, and dry on a soft towel without rubbing. At night, on retiring, apply the mixture and wipe slightly or just enough to take up superfluous liquid; or, better still, sleep in a pair of cotton gloves.

TOILET CREAMS:

Almond Cold Creams.—A liquid almond cream may be made by the appended formula. It has been known as milk of almond:

I.—Sweet almonds....	5 ounces
White castile soap.	2 drachms
White wax.......	2 drachms
Spermaceti.......	2 drachms
Oil of bitter almonds.........	10 minims
Oil of bergamot....	20 minims
Alcohol	6 fluidounces
Water, a sufficient quantity.	

Beat the almonds in a smooth mortar until as much divided as their nature will admit; then gradually add water in very small quantities, continuing the beating until a smooth paste is obtained; add to this, gradually, one pint of water, stirring well all the time. Strain the resulting emulsion without pressure through a cotton cloth previously well washed to remove all foreign matter. If new, the cloth will contain starch, etc., which must be removed. Add, through the strainer, enough water to bring the measure of the strained liquid to 1 pint. While this operation is going on let the soap be shaved into thin ribbons, and melted, with enough water to cover it, over a very gentle fire or on a water bath. When fluid add the wax and spermaceti in large pieces, so as to allow them to melt slowly, and thereby better effect union with the soap. Stir occasionally. When all is melted place the soapy mixture in a mortar, run into it slowly the emulsion, blending the two all the while with the pestle. Care must be taken not to add the emulsion faster than it can be incorporated with the soap. Lastly add the alcohol in which the perfumes have been previously dissolved, in the same manner, using great care.

This preparation is troublesome to make and rather expensive, and it is perhaps no better for the purpose than glycerine. The mistake is often made of applying the latter too freely, its "stickiness" being unpleasant, and it is best to dilute it largely with water. Such a lotion may be made by mixing

Glycerine............ 1 part
Rose water .... 9 parts

Plain water may, of course, be used as the diluent, but a slightly perfumed preparation is generally considered more desirable. The perfume may easily be obtained by dissolving a very small proportion of handkerchief "extract" or some essential oil in the glycerine, and then mixing with plain water.

II.—White wax....... ½ ounce
Spermaceti....... 2½ ounces
Oil of sweet al-
monds......... 2½ ounces

Melt, remove from the fire, and add

Rose water....... 1½ ounces

Beat until creamy: not until cold. When the cream begins to thicken add a few drops of oil of rose. Only the finest almond oil should be used. Be careful in weighing the wax and spermaceti. These precautions will insure a good product.

III.—White wax..... 4 ounces
Spermaceti..... 3 ounces
Sweet almond
oil.......... 6 fluidounces
Glycerine...... 4 fluidounces
Oil of rose gera-
nium........ 1 fluidrachm
Tincture of ben-
zoin......... 4 fluidrachms

Melt the wax and spermaceti, add the oil of sweet almonds, then beat in the glycerine, tincture of benzoin, and oil of rose geranium. When all are incorporated to a smooth, creamy mass, pour into molds.

IV.—Sweet almonds,
blanched..... 5 ounces
Castile soap,
white........ 120 grains
White wax..... 120 grains
Spermaceti .... 120 grains
Oil of bitter al-
monds....... 10 drops
Oil of bergamot 20 drops
Alcohol .... 6 fluidounces
Water, sufficient.

Make an emulsion of the almonds with water so as to obtain 16 fluidounces of product, straining through cotton which has previously been washed to remove starch. Dissolve the soap with the aid of heat in the necessary amount of water to form a liquid, add the wax and spermaceti, continue the heat until the latter is melted, transfer to a mortar, and incorporate the almond emulsion

slowly with constant stirring until all has been added and a smooth cream has been formed. Finally, add the two volatile oils.

V.—Melt, at moderate heat,

	By weight.
White wax	100 parts
Spermaceti.....	1,000 parts

Then stir in

	By weight.
Almond oil......	500 parts
Rose water......	260 parts

And scent with

	By weight.
Bergamot oil....	10 parts
Geranium oil....	5 parts
Lemon oil.......	4 parts

	By weight.
VI.—Castor oil.......	500 parts
White wax......	100 parts
Almond oil......	150 parts

Melt at moderate heat and scent with

	By weight.
Geranium oil....	6 parts
Lemon oil.......	5 parts
Bergamot oil....	10 parts

	By weight.
VII.—Almond oil......	400 parts
Lanoline........	200 parts
White wax.......	60 parts
Spermaceti......	60 parts
Rose water......	300 parts

	By weight.
VIII.—White wax......	6 parts
Tallow, freshly
tried out......	4 parts
Spermaceti......	2 parts
Oil of sweet al-
monds........	6 parts

Melt together and while still hot add, with constant stirring, 1 part of sodium carbonate dissolved in 79 parts of hot water. Stir until cold. Perfume to the taste.

IX.—Ointment of
rose water... 1 ounce
Oil of sweet
almonds .... 1 fluidounce
Glycerine ..... 1 fluidounce
Boric acid..... 100 grains
Solution of
soda........ 2¼ fluidounces
Mucilage of
quince seed . 4 fluidounces
Water enough to
make....... 40 fluidounces
Oil of rose, oil of bitter almonds,
of each sufficient to perfume.

Heat the ointment, oil, and solution of soda together, stirring constantly until an emulsion or saponaceous mixture is formed. Then warm together the glycerine, acid, and mucilage and about 30 fluidounces of water; mix with the emulsion, stir until cold, and add the remainder of the water. Lastly, add the volatile oils.

The rose-water ointment used should be the "cold cream" of the United States Pharmacopœia.

X. —Spermaceti....	2	ounces
White wax....	2	ounces
Sweet almond oil.........	14	fluidounces
Water, distilled	7	fluidounces
Borax, powder	60	grains
Coumarin.....	½	grain
Oil of bergamot	24	drops
Oil of rose.....	6	drops
Oil of bitter almonds....	8	drops
Tincture of ambergris.....	5	drops

Melt the spermaceti and wax, add the sweet almond oil, incorporate the water in which the borax has previously been dissolved, and finally add the oils of bergamot, rose, and bitter almond.

XI. — Honey........	2	av. ounces
Castile soap, white powder	1	av. ounce
Oil sweet almonds......	26	fluidounces
Oil bitter almonds......	1	fluidrachm
Oil bergamot..	½	fluidrachm
Oil cloves	15	drops
Peru balsam...	1	fluidrachm
Liquor potassa.	
Solution carmine, of each sufficient.	

Mix the honey with the soap in a mortar, and add enough liquor potassa (about 1 fluidrachm) to produce a nice cream. Mix the volatile oils and balsam with the sweet almond oil, mix this with the cream, and continue the trituration until thoroughly mixed. Finally add, if desired, enough carmine solution to impart a rose tint.

XII. — White wax.....	800	parts
Spermaceti.....	800	parts
Sweet almond oil..........	5,600	parts
Distilled water..	2,800	parts
Borax.........	50	parts
Bergamot oil...	20	parts
Attar of rose....	5	parts
Coumarin......	0.1	part

Add for each pound of the cream 5 drops of etheric oil of bitter almonds, and 3 drops tincture of ambra. Proceed as in making cold cream.

The following also makes a fine cream:

XIII.—Spermaceti.......	3	parts
White wax.......	2	parts
Oil of almonds, fresh..........	12	parts
Rose water, double	1	part
Glycerine, pure...	1	part

Melt on a water bath the spermaceti and wax, add the oil (which should be fresh), and pour the whole into a slightly warmed mortar, under constant and lively stirring, to prevent granulation. Continue the trituration until the mass has a white, creamy appearance, and is about the consistence of butter at ordinary temperature. Add, little by little, under constant stirring, the orange-flower water and glycerine mixed, and finally the perfume as before. Continue the stirring for 15 or 20 minutes, then immediately put into containers.

Chappine Cream.—

Quince seed.........	2	drachms
Glycerine...........	1½	ounces
Water..............	1½	ounces
Lead acetate........	10	grains
Flavoring, sufficient.	

Macerate the quince seed in water, strain, add the glycerine and lead acetate, previously dissolved in sufficient water; flavor with jockey club or orange essence.

Cucumber Creams.—

I.—White wax..........	3	ounces
Spermaceti..........	3	ounces
Benzoinated lard....	8	ounces
Cucumbers.........	3	ounces

Melt together the wax, spermaceti, and lard, and infuse in the liquid the cucumbers previously grated. Allow to cool, stirring well; let stand a day, remelt, strain and again stir the "cream" until cold.

II.—Benzoinated lard....	5	ounces
Suet...............	3	ounces
Cucumber juice.....	10	ounces
Proceed as in making cold cream.	

Glycerine Creams.—

I.—Oil of sweet almonds..........	100	parts
White wax.........	13	parts
Glycerine, pure.....	25	parts
Add a sufficient quantity of any suitable perfume.	

Melt, on the water bath, the oil, wax, and glycerine together, remove and as the mass cools down add the perfume in sufficient quantity to make a creamy mass.

II.—Quince seed...... 1 ounce
    Boric acid........ 16 grains
    Starch........... 1 ounce
    Glycerine......... 16 ounces
    Carbolic acid..... 30 minims
    Alcohol.......... 12 ounces
    Oil of lavender..... 30 minims
    Oil of rose......... 10 drops
    Extract of white rose 1 ounce
    Water enough to make 64 ounces

Dissolve the boric acid in a quart of water and in this solution macerate the quince seed for 3 hours; then strain. Heat together the starch and the glycerine until the starch granules are broken, and mix with this the carbolic acid. Dissolve the oils and the extract of rose in the alcohol, and add to the quince-seed mucilage; then mix all together, strain, and add water enough to make the product weigh 64 ounces.

III.—Glycerine.......... 1 ounce
    Borax............. 2 drachms
    Boracic acid....... 1 drachm
    Oil rose geranium.. 30 drops
    Oil bitter almond... 15 drops
    Milk.............. 1 gallon

Heat the milk until it curdles and allow it to stand 12 hours. Strain it through cheese cloth and allow it to stand again for 12 hours. Mix in the salts and glycerine and triturate in a mortar, finally adding the odors and coloring if wanted. The curdled milk must be entirely free from water to avoid separation. If the milk will not curdle fast enough the addition of 1 ounce of water ammonia to a gallon will hasten it. Take a gallon of milk, add 1 ounce ammonia water, heat (not boil), allow to stand 24 hours, and no trouble will be found in forming a good base for the cream.

IV.—This is offered as a substitute for cucumber cream for toilet uses. Melt 15 parts, by weight, of gelatin in hot water containing 15 parts, by weight, of boracic acid as well as 150 parts, by weight, of glycerine; the total amount of water used should not exceed 300 parts, by weight. It may be perfumed or not.

Lanolin Creams.—

I.—Anhydrous lanolin. 650 parts
    Peach-kernel oil... 200 parts
    Water............ 150 parts

Perfume with about 15 drops of ionone or 20 drops of synthetic ylang-ylang.

II.—Lanolin............ 40 parts
    Olive oil.......... 15 parts
    Paraffine ointment.. 10 parts

    Aqua naphæ........ 10 parts
    Distilled water..... 15 parts
    Glycerine......... 5 parts
    Boric acid........ 4 parts
    Borax............. 4 parts
    Geranium oil, sufficient.
    Extract, triple, of ylang-ylang,
      quantity sufficient.

III.—Anhydrous lanolin. 650 drachms
    Almond oil........ 200 drachms
    Water............ 150 drachms
    Oil of ylang-ylang. 5 drops

Preparations which have been introduced years ago for the care of the skin and complexion are the glycerine gelées, which have the advantage over lanolin that they go further, but present the drawback of not being so quickly absorbed by the skin. These products are filled either into glasses or into tubes. The latter way is preferable and is more and more adopted, owing to the convenience of handling.

A good recipe for such a gelée is the following:

Moisten white tragacanth powder, 50 parts, with glycerine, 200 parts, and spirit of wine, 100 parts, and shake with a suitable amount of perfume; then quickly mix and shake with warm distilled water, 650 parts.

A transparent slime will form immediately which can be drawn off at once.

Mucilage Creams.—

I.—Starch............ 30 parts
    Carrageen mucilage. 480 parts
    Boric acid......... 15 parts
    Glycerine......... 240 parts
    Cologne water...... 240 parts

Boil the starch in the carrageen mucilage, add the boric acid and the glycerine. Let cool, and add the cologne water.

II.—Linseed mucilage... 240 parts
    Boric acid......... 2 parts
    Salicylic acid....... 1.3 parts
    Glycerine......... 60 parts
    Cologne water...... 120 parts
    Rose water......... 120 parts

Instead of the cologne water any extracts may be used. Lilac and ylang-ylang are recommended.

Witch-Hazel Creams.—

I.—Quince seed...... 90 grains
    Boric acid........ 8 grains
    Glycerine......... 4 fluidounces
    Alcohol.......... 6 fluidounces
    Carbolic acid..... 6 drachms
    Cologne water.... 4 fluidounces
    Oil lavender flow-
      ers............. 40 drops

Glycerite starch... 4 av. ounces
Distilled witch-hazel extract enough
to make 32 fluidounces

Dissolve the boric acid in 16 ounces of
the witch-hazel extract, macerate the
quince seed in the solution for 3 hours,
strain, add the glycerine, carbolic acid,
and glycerite, and mix well. Mix the
alcohol, cologne water, lavender oil, and
mucilages, incorporate with the previous
mixture, and add enough witch-hazel
extract to bring to the measure of 32
fluidounces.

II.—Quince seed .... 4 ounces
Hot water......... 16 ounces
Glycerine.......... 32 ounces
Witch-hazel water.. 128 ounces
Boric acid......... 6 ounces
Rose extract....... 2 ounces
Violet extract...... 1 ounce

Macerate the quince seed in the hot
water; add the glycerine and witch-hazel,
in which the boric acid has been previously dissolved; let the mixture stand
for 2 days, stirring occasionally; strain
and add the perfume.

Skin Cream for Collapsible Tubes.—

I.—White vaseline..... 6 ounces
White wax......... 1 ounce
Spermaceti........ 5 drachms
Subchloride bismuth 6 drachms
Attar of rose........ 6 minims
Oil of bitter almonds 1 minim
Rectified spirit...... ½ ounce

Melt the vaseline, wax, and spermaceti together, and while cooling incorporate the subchloride of bismuth (in
warm mortar). Dissolve the oils in the
alcohol, and add to the fatty mixture,
stirring all until uniform and cold. In
cold weather the quantities of wax and
spermaceti may be reduced.

II.—Lanolin........... 1 ounce
Almond oil........ 1 ounce
Oleate of zinc (pow-
der) .... 3 drachms
Extract of white rose 1½ drachms
Glycerine......... 2 drachms
Rose water........ 2 drachms

Face Cream Without Grease.—

Quince seed........ 10 parts
Boiling water.......1,000 parts
Borax............. 5 parts
Boric acid......... 5 parts
Glycerine .... 100 parts
Alcohol, 94 per cent. 125 parts
Attar of rose, quantity sufficient to
perfume.

Macerate the quince seed in half of
the boiling water, with frequent agitations, for 2 hours and 30 minutes, then
strain off. In the residue of the boiling
water dissolve the borax and boric acid,
add the glycerine and the perfume, the
latter dissolved in the alcohol. Now
add, little by little, the colate of quince
seed, under constant agitation, which
should be kept up for 5 minutes after
the last portion of the colate is added.

TOILET MILKS:

Cucumber Milk.—

Simple cerate........ 2 pounds
Powdered borax..... 11½ ounces
Powdered castile soap 10 ounces
Glycerine.......... 26 ounces
Alcohol............. 24 ounces
Cucumber juice...... 32 ounces
Water to............ 5 gallons
Ionone............. 1 drachm
Jasmine............. ½ drachm
Neroli .... ½ drachm
Rhodinol........... 15 minims

To the melted cerate in a hot water
bath add the soap and stir well, keeping
up the heat until perfectly mixed. Add
8 ounces of borax to 1 gallon of boiling
water, and pour gradually into the hot
melted soap and cerate; add the remainder of the borax and hot water, then
the heated juice and glycerine, and
lastly the alcohol. Shake well while
cooling, set aside for 48 hours, and siphon
off any water that may separate. Shake
well, and repeat after standing again if
necessary; then perfume.

Cucumber Juice.—It is well to make
a large quantity, as it keeps indefinitely.
Washed unpeeled cucumbers are grated
and pressed; the juice is heated, skimmed
and boiled for 5 minutes, then cooled
and filtered. Add 1 part of alcohol to
2 parts of juice, let stand for 12 hours or
more, and filter until clear.

Glycerine Milk.—

Glycerine......... 1,150 parts
Starch, powdered.. 160 parts
Distilled water..... 400 parts
Tincture of benzoin 20 parts

Rub up 80 parts of the starch with the
glycerine, then put the mixture on the
steam bath and heat, under continuous
stirring, until it forms a jellylike mass.
Remove from the bath and stir in the
remainder of the starch. Finally, add
the water and tincture and stir till homogeneous.

Lanolin Toilet Milk.—

White castile soap,
powdered......... 22 grains
Lanolin............. 1 ounce
Tincture benzoin.... 12 drachms
Water, enough.

Dissolve the soap in 2 fluidounces of warm water, also mix the lanolin with 2 fluidounces of warm water; then incorporate the two with each other, finally adding the tincture. The latter may be replaced by 90 grains of powdered borax.

Jasmine Milk.—To 25 parts of water add gradually, with constant stirring, 1 part of zinc white, 2 quarts of grain spirit, and 0.15 to 0.25 part of glycerine; finally stir in 0.07 to 0.10 part of jasmine essence. Filter the mixture and fill into glass bottles. For use as a cosmetic, rub on the raspberry paste on retiring at night, and in the morning use the jasmine milk to remove the paste from the skin. The two work together in their effect.

SUNBURN AND FRECKLE REMEDIES.

I.—Apply over the affected skin a solution of corrosive sublimate, 1 in 500, or, if the patient can stand it, 1 in 300, morning and evening, and for the night apply emplastrum hydrargyri compositum to the spots. In the morning remove the plaster and all remnants of it by rubbing fresh butter or cold cream over the spots.

For redness of the skin apply each other day zinc oxide ointment or ointment of bismuth subnitrate.

II.—Besnier recommends removal of the mercurial ointment with green soap, and the use, at night, of an ointment composed of vaseline and Vigo's plaster (emplastrum hydrargyri compositum), in equal parts. In the morning wash off with soap and warm water, and apply the following:

Vaseline, white......	20 parts
Bismuth carbonate...	5 parts
Kaolin.............	5 parts

Mix, and make an ointment.

III.—Leloir has found the following of service. Clean the affected part with green soap or with alcohol, and then apply several coats of the following:

Acid chrysophanic..	15 parts
Chloroform........	100 parts

Mix. Apply with a camel's-hair pencil.

When the application dries thoroughly, go over it with a layer of traumaticine. This application will loosen itself in several days, when the process should be repeated.

IV.—When the skin is only slightly discolored use a pomade of salicylic acid, or apply the following:

Acid chrysophanic, from..........	1 to 4 parts
Acid salicylic......	1 to 2 parts
Collodion.........	40 parts

V.—When there is need for a more complicated treatment, the following is used:

(a) Corrosive sublimate	1 part
Orange - flower water..........	7,500 parts
Acid, hydrochloric, dilute.........	500 parts
(b) Bitter almonds....	4,500 parts
Glycerine..........	2,500 parts
Orange-flower water..........	25,000 parts

Rub up to an emulsion in a porcelain capsule. Filter and add, drop by drop, and under constant stirring, 5 grams of tincture of benzoin. Finally mix the two solutions, adding the second to the first.

This preparation is applied with a sponge, on retiring, to the affected places, and allowed to dry on.

VI.—According to Brocq the following should be penciled over the affected spots:

Fresh pure milk......	50 parts
Glycerine..........	30 parts
Acid, hydrochloric, concentrated......	5 parts
Ammonium chlorate.	3 parts

VII.—Other external remedies that may be used are lactic acid diluted with 3 volumes of water, applied with a glass rod: dilute nitric acid, and, finally, peroxide of hydrogen, which last is a very powerful agent. Should it cause too much inflammation, the latter may be assuaged by using an ointment of zinc oxide or bismuth subnitrate—or one may use the following:

Kaolin.............	4 parts
Vaseline............	10 parts
Glycerine.........:.	4 parts
Magnesium carbonate	2 parts
Zinc oxide..........	2 parts

Freckle Remedies.—

I.—Poppy oil...........	1 part
Lead acetate........	2 parts
Tincture benzoin....	1 part
Tincture quillaia....	5 parts
Spirit nitrous ether...	1 part
Rose water........	95 parts

Saponify the oil with the lead acetate; add the rose water, and follow with the tinctures.

II.—Chloral hydrate.....	2 drachms
Carbolic acid.......	1 drachm

Tincture iodine...... 60 drops
Glycerine........... 1 ounce

Mix and dissolve. Apply with a camel's-hair pencil at night.

III.—Distilled vinegar... 660 parts
Lemons, cut in
small pieces..... 135 parts
Alcohol, 85 per
cent........... 88 parts
Lavender oil...... 23 parts
Water........... 88 parts
Citron oil........ 6 parts

This mixture is allowed to stand for 3 or 4 days in the sun and filtered. Coat, by means of a sponge before retiring, the places of the skin where the freckles are and allow to dry.

Freckles and Liver Spots.—Modern dermatological methods of treating freckles and liver spots are based partly on remedies that cause desquamation and those that depigmentate (or destroy or neutralize pigmentation). Both methods may be distinguished in respect to their effects and mode of using into the following: The active ingredients of the desquamative pastes are reductives which promote the formation of epithelium and hence expedite desquamation.

There are many such methods, and especially to be mentioned is that of Unna, who uses resorcin for the purpose. Lassar makes use of a paste of naphthol and sulphur.

Sunburn Remedies.—

I.—Zinc sulphocarbo-
late........... 1 part
Glycerine........ 20 parts
Rose water....... 70 parts
Alcohol, 90 per
cent........... 8 parts
Cologne water.... 1 part
Spirit of camphor. 1 part

II.—Borax........... 4 parts
Potassium chlorate 2 parts
Glycerine........ 10 parts
Alcohol.......... 4 parts
Rose water to make 90 parts

III.—Citric acid....... 2 drachms
Ferrous sulphate
(cryst.)........ 18 grains
Camphor......... 2 grains
Elder-flower water 3 fluidounces

IV.—Potassium carbon-
ate........... 3 parts
Sodium chloride.. 2 parts
Orange-flower
water.......... 15 parts
Rose water....... 65 parts

V.—Boroglycerine, 50
per cent......... 1 part
Ointment of rose
water........... 9 parts

VI.—Sodium bicarbon-
ate............. 1 part
Ointment of rose
water........... 7 parts

VII.—Bicarbonate of soda 2 drachms
Powdered borax... 1 drachm
Compound tincture
of lavender...... 1½ drachms
Glycerine........ 1 ounce
Rose water........ 4 ounces

Dissolve the soda and borax in the glycerine and rose water, and add the tincture. Apply with a small piece of sponge 2 or 3 times a day. Then gently dry by dabbing with a soft towel.

VIII.—Quince seeds..... 2 drachms
Distilled water.... 10 ounces
Glycerine........ 2 ounces
Alcohol, 94 per
cent........... 1 ounce
Rose water...... 2 ounces

Boil the seeds in the water for 10 minutes, then strain off the liquid, and when cold add to it the glycerine, alcohol, and rose water.

IX.—White soft soap... 2½ drachms
Glycerine........ 1½ drachms
Almond oil....... 11 drachms

Well mix the glycerine and soap in a mortar, and very gradually add the oil, stirring constantly until perfectly mixed.

X.—Subnitrate of bis-
muth.......... 1½ drachms
Powdered French
chalk.......... 30 grains
Glycerine........ 2 drachms
Rose water...... 1½ ounces

Mix the powders, and rub down carefully with the glycerine; then add the rose water. Shake the bottle before use.

XI.—Glycerine cream.. 2 drachms
Jordan almonds.. 4 drachms
Rose water...... 5 ounces
Essential oil of al-
monds......... 3 drops

Blanch the almonds, and then dry and beat them up into a perfectly smooth paste; then mix in the glycerine cream and essential oil. Gradually add the rose water, stirring well after each addition; then strain through muslin.

Tan and Freckle Lotion.—

Solution A:
Potassium iodide, iodine, glycerine, and infusion rose.
Dissolve the potassium iodide in a

small quantity of the infusion and a drachm of the glycerine; with this fluid moisten the iodine in a glass of water and rub it down, gradually adding more liquid, until complete solution has been obtained; then stir in the remainder of the ingredients, and bottle the mixture.

Solution B:

Sodium thiosulphate and rose water. With a small camel's-hair pencil or piece of fine sponge apply a little of solution A to the tanned or freckled surface, until a slight or tolerably uniform brownish yellow skin has been produced. At the expiration of 15 or 20 minutes moisten a piece of cambric, lint, or soft rag with B and lay it upon the affected part, removing, squeezing away the liquid, soaking it afresh, and again applying until the iodine stain has disappeared. Repeat the process thrice daily, but diminish the frequency of application if tenderness be produced.

A Cure for Tan.—Bichloride of mercury, in coarse powder, 10 grains; distilled water, 1 pint. Agitate the two together until a complete solution is obtained. Add ½ ounce of glycerine. Apply with a small sponge as often as agreeable. This is not strong enough to blister and skin the face in average cases. It may be increased or reduced in strength by adding to or taking from the amount of bichloride of mercury. Do not forget that this last ingredient is a powerful poison and should be kept out of the reach of children and ignorant persons.

Improved Carron Oil.—Superior to the old and more suitable. A desirable preparation for burns, tan, freckle, sunburn, scalds, abrasions, or lung affections. Does not oxidize so quickly or dry up so rapidly and less liable to rancidity.

Linseed oil	2 ounces
Limewater	2 ounces
Paraffine, liquid	1 ounce

Mix the linseed oil and water, and add the paraffine. Shake well before using.

LIVER SPOTS.

I.—
Corrosive sublimate	1 part
White sugar	190 parts
White of egg	34 parts
Lemon juice	275 parts
Water to make	2,500 parts

Mix the sublimate, sugar, and albumen intimately, then add the lemon juice and water. Dissolve, shake well, and after standing an hour, filter. Apply in the morning after the usual ablutions, and let dry on the face.

II.—Bichloride of mercury, in coarse powder, 8 grains; witch-hazel, 2 ounces; rose water, 2 ounces.

Agitate until a solution is obtained. Mop over the affected parts. Keep out of the way of ignorant persons and children.

TOILET POWDERS:

Almond Powders for the Toilet.—

I.—
Almond meal	6,000 parts
Bran meal	3,000 parts
Soap powder	600 parts
Bergamot oil	50 parts
Lemon oil	15 parts
Clove oil	15 parts
Neroli oil	6 parts

II.—
Almond meal	7,000 parts
Bran meal	2,000 parts
Violet root	900 parts
Borax	350 parts
Bitter almond oil	18 parts
Palmarosa oil	36 parts
Bergamot oil	10 parts

III.—
Almond meal	3,000 parts
Bran meal	3,000 parts
Wheat flour	3,000 parts
Sand	100 parts
Lemon oil	40 parts
Bitter almond oil	10 parts

Bath Powder.—

Borax	4 ounces
Salicylic acid	1 drachm
Extract of cassia	1 drachm
Extract of jasmine	1 drachm
Oil of lavender	20 minims

Rub the oil and extracts with the borax and salicylic acid until the alcohol has evaporated. Use a heaping teaspoonful to the body bath.

Brunette or Rachelle.—

Base	9 pounds
Powdered Florentine orris	1 pound
Perfume the same.	
Powdered yellow ocher	(av.) 3 ounces 120 grains
Carmine No. 40	60 grains

Rub down the carmine and ocher with alcohol in a mortar, and spread on glass to dry; then mix and sift.

Violet Poudre de Riz.—

I.—
Cornstarch	7 pounds
Rice flour	1 pound
Powdered talc	1 pound
Powdered orris root	1 pound
Extract of cassia	3 ounces
Extract of jasmine	1 ounce

II.—Cheaper.

Potato starch	8	pounds
Powdered talc	1	pound
Powdered orris	1	pound
Extract of cassia	3	ounces

Barber's Powder.—

Cornstarch	5	pounds
Precipitated chalk	3	pounds
Powdered talc	2	pounds
Oil of neroli	1	drachm
Oil of cedrat	1	drachm
Oil of orange	2	drachms
Extract of jasmine	1	ounce

Rose Poudre de Riz.—

I.—

Cornstarch	9	pounds
Powdered talc	1	pound
Oil of rose	1¼	drachms
Extract of jasmine	6	drachms

II.—

Potato starch	9	pounds
Powdered talc	1	pound
Oil of rose	½	drachm
Extract of jasmine	½	ounce

Ideal Cosmetic Powder.—The following combines the best qualities that a powder for the skin should have:

Zinc, white	50	parts
Calcium carbonate, precipitated	300	parts
Steatite, best white	50	parts
Starch, wheat, or rice	100	parts
Extract white rose, triple	3	parts
Extract jasmine, triple	3	parts
Extract orange flower, triple	3	parts
Extract of cassia, triple	3	parts
Tincture of myrrh	1	part

Powder the solids and mix thoroughly by repeated siftings.

Flesh Face Powder.—

Base	9	pounds
Powdered Florentine orris	1	pound
Carmine No. 40	250	grains
Extract of jasmine	100	minims
Oil of neroli	20	minims
Vanillin	5	grains
Artificial musk	30	grains
White heliotropin	30	grains
Coumarin	1	grain

Rub the carmine with a portion of the base and alcohol in a mortar, mixing the perfume the same way in another large mortar, and adding the orris. Mix and sift all until specks of carmine disappear on rubbing.

White Face Powder.—

Base	9	pounds
Powdered Florentine orris	1	pound

Perfume the same. Mix and sift.

Talcum Powders.—Talc, when used as a toilet powder should be in a state of very fine division. Antiseptics are sometimes added in small proportion, but these are presumably of little or no value in the quantity allowable, and may prove irritating. For general use, at all events, the talcum alone is the best and the safest. As a perfume, rose oil may be employed, but on account of its cost, rose geranium oil is probably more frequently used. A satisfactory proportion is ½ drachm of the oil to a pound of the powder. In order that the perfume may be thoroughly disseminated throughout the powder, the oil should be triturated first with a small portion of it; this should then be further triturated with a larger portion, and, if the quantity operated on be large, the final mixing may be effected by sifting. Many odors besides that of rose would be suitable for a toilet powder. Ylang-ylang would doubtless prove very attractive, but expensive.

The following formulas for other varieties of the powder may prove useful:

Violet Talc.—

I.—

Powdered talc	14	ounces
Powdered orris root	2	ounces
Extract of cassia	½	ounce
Extract of jasmine	¼	ounce

Rose Talc.—

II.—

Powdered talc	5	pounds
Oil of rose	½	drachm
Extract of jasmine	4	ounces

Tea-Rose Talc.—

III.—

Powdered talc	5	pounds
Oil of rose	50	drops
Oil of wintergreen	4	drops
Extract of jasmine	2	ounces

Borated Apple Blossom.—

IV.—

Powdered talc	22	pounds
Magnesium carbonate	2¾	pounds
Powdered boric acid	1	pound

Mix.

Carnation pink blossom (Schimmel's)	2	ounces
Extract of trefle	2	drachms

To 12 drachms of this mixture add:

Neroli	1	drachm
Vanillin	½	drachm
Alcohol to	3	ounces

Sufficient for 25 pounds.

V.—Talcum............ 8   ounces
    Starch 8   ounces
    Oil of neroli....... 10 drops
    Oil of ylang-ylang. 5   drops

VI.—Talcum............ 12 ounces
    Starch............ 4   ounces
    Orris root......... 2   ounces
    Oil of bergamot.... 12 drops

VII.—Talcum............ 14 ounces
    Starch............ 2   ounces
    Lanolin........... ½   ounce
    Oil of rose........ 10 drops
    Oil of neroli....... 5   drops

TOILET VINEGARS:

Pumillo Toilet Vinegar.—

    Alcohol, 80 per cent 1,600 parts
    Vinegar, 10 per
      cent............. 840 parts
    Oil of pinu spumillo 44 parts
    Oil of lavender..... 4 parts
    Oil of lemon....... 2 parts
    Oil of bergamot.... 2 parts

Dissolve the oils in the alcohol, add the vinegar, let stand for a week and filter.

Vinaigre Rouge.—

    Acetic acid........ 24 parts
    Alum............. 3 parts
    Peru balsam...... 1 part
    Carmine, No. 40... 12 parts
    Ammonia water... 6 parts
    Rose water, dis-
      tilled........... 575 parts
    Alcohol.......... 1,250 parts

Dissolve the balsam of Peru in the alcohol, and the alum in the rose water. Mix the two solutions, add the acetic acid, and let stand overnight. Dissolve the carmine in the ammonia water and add to mixture. Shake thoroughly, let stand for a few minutes, then decant.

TOILET WATERS:

" Beauty Water."—

    Fresh egg albumen.. 500 parts
    Alcohol............ 125 parts
    Lemon oil.......... 2 parts
    Lavender oil....... 2 parts
    Oil of thyme....... 2 parts

Mix the ingredients well together. When first mixed the liquid becomes flocculent, but after standing for 2 or 3 days clears up—sometimes becomes perfectly clear, and may be decanted. It forms a light, amber-colored liquid that remains clear for months.

At night, before retiring, pour about a teaspoonful of the water in the palm of the hand, and rub it over the face and neck, letting it dry on. In the morning, about an hour before the bath, repeat the oper-ation, also letting the liquid dry on the skin. The regular use of this preparation for 4 weeks will give the skin an extraordinary fineness, clearness, and freshness.

Rottmanner's Beauty Water.—Koller says that this preparation consists of 1 part of camphor, 5 parts of milk of sulphur, and 50 parts of rose water.

Birch Waters. — Birch water, which has many cosmetic applications, especially as a hair wash, or an ingredient in hair washes, may be prepared as follows:

I.—Alcohol, 96 per cent 3,500 parts
    Water............ 700 parts
    Potash soap....... 200 parts
    Glycerine......... 150 parts
    Oil of birch buds... 50 parts
    Essence of spring
      flowers.......... 100 parts
    Chlorophyll, quantity sufficient to color.

Mix the water with 700 parts of the alcohol, and in the mixture dissolve the soap. Add the essence of spring flowers and birch oil to the remainder of the alcohol, mix well, and to the mixture add, little by little, and with constant agitation, the soap mixture. Finally, add the glycerine, mix thoroughly, and set aside for 8 days, filter and color the filtrate with chlorophyll, to which is added a little tincture of saffron. To use, add an equal volume of water to produce a lather.

II.—Alcohol, 96 per
      cent.......... 2,000 parts
    Water........... 500 parts
    Tincture of can-
      tharides........ 25 parts
    Salicylic acid..... 25 parts
    Glycerine........ 100 parts
    Oil of birch buds. 40 parts
    Bergamot oil..... 30 parts
    Geranium oil..... 5 parts

Dissolve the oils in the alcohol, add the acid and tincture of cantharides; mix the water and glycerine and add, and, finally, color as before.

III.—Alcohol........ 30,000 parts
    Birch juice...... 3,000 parts
    Glycerine....... 1,000 parts
    Bergamot oil.... 90 parts
    Vanillin........ 10 parts
    Geranium oil.... 50 parts
    Water.......... 14,000 parts

IV.—Alcohol......... 40,000 parts
    Oil of birch..... 150 parts
    Bergamot oil.... 100 parts
    Lemon oil...... 50 parts

Palmarosa oil...	100 parts
Glycerine......	2,000 parts
Borax..........	150 parts
Water..........	20,000 parts

Violet Ammonia Water.—Most preparations of this character consist of either coarsely powdered ammonium carbonate, with or without the addition of ammonia water, or of a coarsely powdered mixture, which slowly evolves the odor of ammonia, the whole being perfumed by the addition of volatile oil, pomade essences, or handkerchief extract. The following are typical formulas:

I.—Moisten coarsely powdered ammonium carbonate, contained in a suitable bottle, with a mixture of concentrated tincture of orris root, 2½ ounces; aromatic spirit of ammonia, 1 drachm; violet extract, 3 drachms.

II.—Fill suitable bottles with coarsely powdered ammonium carbonate and add to the salt as much of the following solution as it will absorb: Oil of orris, 5 minims; oil of lavender flowers, 10 minims; violet extract, 30 minims; stronger water of ammonia, 2 fluidounces.

III.—The following is a formula for a liquid preparation: Extract violet, 8 fluidrachms; extract cassia, 8 fluidrachms; spirit of rose, 4 fluidrachms; tincture of orris, 4 fluidrachms; cologne spirit, 1 pint; spirit of ammonia, 1 ounce. Spirit of ionone may be used instead of extract of violet.

Violet Witch-Hazel.—

Spirit of ionone......	½ drachm
Rose water..........	6 ounces
Distilled extract of witch-hazel enough to make..........	16 ounces

Cotton

BLEACHING OF COTTON:

I.—**Bleaching by Steaming.**—The singed and washed cotton goods are passed through hydrochloric acid of 2° Bé. Leave them in heaps during 1 hour, wash, pass through sodium hypochlorite of 10° Bé. diluted with 10 times the volume of water. Let the pieces lie in heaps for 1 hour, wash, pass through caustic soda lye of 38° Bé. diluted with 8 times its volume of water, steam, put again through sodium chloride, wash, acidulate slightly with hydrochloric acid, wash and dry. Should the whiteness not be sufficient, repeat the operations.

II.—**Bleaching with Calcium Sulphite.**—The cotton goods are impregnated with 1 part, by weight, of water, 1 part of caustic lime, and ½ part of bisulphite of 40° Bé.; next steamed during 1–2 hours at a pressure of ½ atmosphere, washed, acidulated, washed and dried. The result is as white a fabric as by the old method with caustic lime, soda, and calcium chloride. The bisulphite may also be replaced by calcium hydrosulphite, and, instead of steaming, the fabric may be boiled for several hours with calcium sulphite.

III.—**Bleaching of Vegetable Fibers with Hydrogen Peroxide.**—Pass the pieces through a solution containing caustic soda, soap, hydrogen peroxide, and burnt magnesia. The pieces are piled in heaps on carriages; the latter are shoved into the well-known apparatus of Mather & Platt (kier), and the liquid is pumped on for 6 hours, at a pressure of ⅔ atmosphere. Next wash, acidulate, wash and dry. The bleaching may also be done on an ordinary reeling vat. For 5 pieces are needed about 1,000 parts, by weight, of water; 10 parts, by weight, of solid caustic soda; 1 part of burnt magnesia; 30 parts, by weight, of hydrogen peroxide. After 3–4 hours' boiling, wash, acidulate, wash and dry. The bleaching may also be performed by passing through barium peroxide, then through sulphuric acid or hydrochloric acid, and next through soda lye. It is practicable also to commence with the latter and finally give a treatment with hydrogen peroxide.

The whiteness obtained by the above process is handsomer than that produced by the old method with hypochlorites, and the fabric is weakened to a less extent.

TESTS FOR COTTON.

I.—Cotton, when freed from extraneous matter by boiling with potash, and afterwards with hydrochloric acid, yields pure cellulose or absorbent cotton, which, according to the U. S. P., is soluble in copper ammonium sulphate solution. The B. P. is more specific and states that cotton is soluble in a concentrated solution of copper ammonium sulphate. The standard test solution (B. P.) is made by dissolving 10 parts of copper sulphate in 160 parts of distilled water, and cautiously adding solution of ammonia to the liquid until the precipitate first formed is nearly dissolved. The product is then filtered and the filtrate made up to 200 parts with distilled water. The concentrated solution is prepared by using a smaller quantity of distilled water.

II.—Schweitzer's reagent for textile fibers and cellulose is made by dissolving 10 parts of copper sulphate in 100 parts of water and adding a solution of 5 parts of potassium hydrate in 50 parts of water; then wash the precipitate and dissolve in 20 per cent ammonia until saturated. This solution dissolves cotton, linen, and silk, but not wool. The reagent is said to be especially useful in microscopy, as it rapidly dissolves cellulose, but has no action on lignin.

III.—Jandrier's Test for Cotton in Woolen Fabrics.—Wash the sample of fabric and treat with sulphuric acid (20 Bé.) for half an hour on the water bath. To 100 to 200 parts of this solution add 1 part resorcin, and overlay on concentrated sulphuric acid free from nitrous products. The heat developed is sufficient to give a color at the contact point of the liquids, but intensity of color may be increased by slightly heating. If the product resulting from treating the cotton is made up 1 in 1,000, resorcin will give an orange color; alphanaphtol a purple; gallic acid a green gradually becoming violet down in the acid; hydroquinone or pyrogallol a brown; morphine or codeine, a lavender; thymol or menthol a pink. Cotton may be detected in colored goods, using boneblack to decolorize the solution, if necessary.

IV.—Overbeck's test for cotton in woolen consists in soaking the fabric in an aqueous solution of alloxantine (1 in 10), and after drying expose to ammonia vapor and rinse in water. Woolen material is colored crimson, cotton remains blue.

V.—Liebermann's Test.—Dye the fabric for half an hour in fuchsine solution rendered light yellow by caustic soda solution and then washed with water—silk is colored dark red; wool, light red; flax, pink; and cotton remains colorless.

To Distinguish Cotton from Linen.—Take a sample about an inch and a half square of the cloth to be tested and plunge it into a tepid alcoholic solution of cyanine. After the coloring matter has been absorbed by the fiber, rinse it in water and then plunge into dilute sulphuric acid. If it is of cotton the sample will be almost completely bleached, while linen preserves the blue color almost unchanged. If the sample be then plunged in ammonia, the blue will be strongly reinforced.

Aromatic Cotton.—Aromatic cotton is produced as follows: Mix camphor, 5 parts; pine-leaf oil, 5 parts; clove oil, 5 parts; spirit of wine (90 per cent), 80 parts; and distribute evenly on cotton, 500 parts, by means of an atomizer. The cotton is left pressed together in a tightly closed tin vessel for a few days.

Cotton Degreasing.—Cotton waste, in a greasy condition, is placed in an acid-proof apparatus, where it is simultaneously freed from grease, etc., and prepared for bleaching by the following process, which is performed without the waste being removed from the apparatus: (1) treatment with a solvent, such as benzine; (2) steaming, for the purpose of vaporizing and expelling from the cotton waste the solvent still remaining in it after as much as possible of this has been recovered by draining; (3) treatment with a mineral acid; (4) boiling with an alkali lye; (5) washing with water.

COTTONSEED HULLS AS STOCK FOOD.

Cottonseed hulls or other material containing fiber difficult of digestion are thoroughly mixed with about 5 per cent of their weight of hydrochloric acid (specific gravity, 1.16), and heated in a closed vessel, provided with a stirrer, to a temperature of 212° to 300° F. The amount of acid to be added depends on the material employed and on the duration of the heating. By heating for 30 minutes the above percentage of acid is required, but the quantity may be reduced if the heating is prolonged. After heating, the substance is ground and at the same time mixed with some basic substances such as sodium carbonate, chalk, cottonseed kernel meal, etc., to neutralize the acid. During the heating, the acid vapors coming from the mixture may be led into a second quantity of material contained in a separate vessel, air being drawn through both vessels to facilitate the removal of the acid vapors.

COTTONSEED OIL:
See Oil.

COTTONSEED OIL IN FOOD, TESTS FOR:
See Foods.

COTTONSEED OIL IN LARD, DETECTION OF:
See Foods and Lard.

COUGH CANDY:
See Confectionery.

COUGH MIXTURES FOR CATTLE:
See Veterinary Formulas.

COUGH MIXTURES AND REMEDIES:
See Cold and Cough Mixtures.

Court Plasters

(See also Plasters.)

Liquid Court Plaster.—I.—If soluble guncotton is dissolved in acetone in the proportion of about 1 part, by weight, of the former to 35 or 40 parts, by volume, of the latter, and half a part each of castor oil and glycerine be added, a colorless, elastic, and flexible film will form on the skin wherever it is applied. Unlike ordinary collodion it will not be likely to dry and peel off. If tinted very slightly with alkanet and saffron it can be made to assume the color of the skin so that when applied it is scarcely observable. A mixture of warm solution of sodium silicate and casein, about 9 parts of the former to 1 part of the latter, gelatinizes and forms a sort of liquid court plaster.

II.—In order to make liquid court plaster flexible, collodion, U. S. P., is the best liquid that can possibly be recommended. It may be made by weighing successively into a tarred bottle:

Collodion	4 av. ounces
Canada turpentine..	95 grains
Castor oil..........	57 grains

Before applying, the skin should be perfectly dry; each application or layer should be permitted to harden. Three or four coats are usually sufficient.

III.—Procure an ounce bottle and fill it three-fourths full of flexible collodion, and fill up with ether. Apply to cuts, bruises, etc., and it protects them and will not wash off. If the ether evaporates, leaving it too thick for use, have more ether put in to liquefy it. It is a good thing to have in the house and in the tool chest.

COW DISEASES AND THEIR REMEDIES:
See Veterinary Formulas.

CRAYONS:
See Pencils.

CRAYONS FOR GRAINING AND MARBLING.

Heat 4 parts of water and 1 part of white wax over a fire until the wax has completely dissolved. Stir in 1 part of purified potash. When an intimate combination has taken place, allow to cool and add a proportionate quantity of gum arabic. With this mixture the desired colors are ground thick enough so that they can be conveniently rolled into a pencil with chalk. The desired shades must be composed on the grinding slab as they are wanted, and must not be simply left in their natural tone. Use, for instance, umber, Vandyke brown, and white lead for oak; umber alone would be too dark for walnut use. All the earth colors can be conveniently worked up. It is best to prepare 2 or 3 crayons of each set, mixing the first a little lighter by the addition of white lead and leaving the others a little darker. The pencils should be kept in a dry place and are more suitable for graining and marbling than brushes, since they can be used with either oil or water.

CRAYONS FOR WRITING ON GLASS:
See Etching, and Glass.

Cream

(See also Milk.)

Whipped Cream.—There are many ways to whip cream. The following is very highly indorsed: Keep the cream on ice until ready to whip. Take 2 earthen vessels about 6 inches in diameter. Into 1 bowl put 1 pint of rich sweet cream, 2 teaspoonfuls powdered sugar, and 5 drops of best vanilla extract. Add the white of 1 egg and beat with large egg beater or use whipping apparatus until 2 inches of froth has formed; skim off the froth into the other vessel and so proceed whipping and skimming until all the cream in the first vessel has been exhausted. The whipped cream will stand up all day and should be let stand in the vessel on ice.

Special machines have been constructed for whipping cream, but most dispensers prepare it with an ordinary egg beater. Genuine whipped cream is nothing other than pure cream into which air has been forced by the action of the different apparatus manufactured for the purpose; care must, however, be exercised in order that butter is not produced instead of whipped cream. To avoid this the temperature of the cream must be kept at a low degree and the whipping must not be too violent or prolonged; hence the following rules must be observed in order to produce the desired result:

1. Secure pure cream and as fresh as possible.

2. Surround the bowl in which the cream is being whipped with cracked ice, and perform the operation in a cool place.

3. As rapidly as the whipped cream arises, skim it off and place it in another bowl, likewise surrounded with ice.

4. Do not whip the cream too long or too violently.

5. The downward motion of the beater should be more forcible than the upward, as the first has a tendency to force the air into the cream, while the second, on the contrary, tends to expel it.

6. A little powdered sugar should be added to the cream after it is whipped, in order to sweeten it.

7. Make whipped cream in small quantities and keep it on ice.

I.—Cummins's Whipped Cream.—Place 12 ounces of rich cream on the ice for about 1 hour; then with a whipper beat to a consistency that will withstand its own weight.

II.—Eberle's Whipped Cream.—Take a pint of fresh, sweet cream, which has been chilled by being placed on the ice, add to it a heaping tablespoonful of powdered sugar and 2 ounces of a solution of gelatin (a spoonful dissolved in 2 ounces of water), whip slowly for a minute or two until a heavy froth gathers on top. Skim off the dense froth, and put in container for counter use; continue this until you have frothed all that is possible.

III.—Foy's Whipped Cream.—Use only pure cream; have it ice cold, and in a convenient dish for whipping with a wire whipper. A clear, easy, quick, and convenient way is to use a beater. Fill about one-half full of cream, and beat vigorously for 2 or 3 minutes; a little powdered sugar may be added before beating. The cream may be left in the beater, and placed on ice.

IV.—American Soda Fountain Company's Whipped Cream.—Take 2 earthen bowls and 2 tin pans, each 6 or 8 inches greater in diameter than the bowls; place a bowl in each pan, surround it with broken ice, put the cream to be whipped in 1 bowl, and whip it with a whipped cream churn. The cream should be pure and rich, and neither sugar nor gelatin should be added to it. As the whipped cream rises and fills the bowl, remove the churn, and skim off the whipped cream into the other bowl. The philosophy of the process is that the churn drives air into the cream, and blows an infinity of tiny bubbles, which forms the whipped cream; therefore, in churning, raise the dasher gently and slowly, and bring it down quickly and forcibly. When the second bowl is full of whipped cream, pour off the liquid cream, which has settled to the bottom, into the first bowl, and whip it again. Keep the whipped cream on ice.

The addition of an even teaspoonful of salt to 1 quart of sweet cream, before whipping, will make it whip up very readily and stiff, and stand up much longer and better.

CRESOL EMULSION.

One of the best starting points for the preparation is the "creosote" obtained from blast furnaces, which is rich in cresols and contains comparatively little phenols. The proportions used are: Creosote, 30 parts; soft soap, 10 parts; and solution of soda (10 per cent), 30 parts. Boil the ingredients together for an hour, then place aside to settle. The dark fluid is afterwards drained from any oily portion floating upon the top.

CREAM, COLD:
See Cosmetics.

CREAMS FOR THE FACE AND SKIN:
See Cosmetics.

CREOSOTE SOAP:
See Soap.

CROCKERY:
See Ceramics.

CROCKERY CEMENTS:
See Adhesives.

CROCUS.

The substance known as "crocus," which is so exceedingly useful as a polishing medium for steel, etc., may be very generally obtained in the cinders produced from coal containing iron. It will be easily recognized by its rusty color, and should be collected and reduced to a powder for future use. Steel burnishers may be brought to a high state of polish with this substance by rubbing them upon a buff made of soldiers' belt or hard wood. After this operation, the burnisher should be rubbed on a second buff charged with jewelers' rouge.

CRYSTAL CEMENTS FOR REUNITING BROKEN PIECES:
See Adhesives, under Cements.

CRYSTALLIZATION, ORNAMENTAL:
See Gardens, Chemical.

CUCUMBER ESSENCE:
See Essences and Extracts.

CUCUMBER JELLY, JUICE, AND MILK:
See Cosmetics.

CURAÇOA CORDIAL:
See Wines and Liquors.

CURTAINS, COLORING OF:
See Laundry Preparations.

CURRY POWDER:
See Condiments.

CUSTARD POWDER:

Corn flour	7 pounds
Arrowroot	8 pounds
Oil of almond	20 drops
Oil of nutmegs	10 drops
Tincture of saffron to color.	

Mix the tincture with a little of the mixed flours; then add the essential oils and make into a paste; dry this until it can be reduced to a powder, and then mix all the ingredients by sifting several times through a fine hair sieve.

CUTLERY CEMENTS:
See Adhesives.

CYLINDER OIL:
See Lubricants.

CYMBAL METAL:
See Alloys.

Damaskeening

Damaskeening, practiced from most ancient times, consists in ornamentally inlaying one metal with another, followed usually by polishing. Generally gold or silver is employed for inlaying. The article to be decorated by damaskeening is usually of iron (steel) or copper; in Oriental (especially Japanese) work, also frequently of bronze, which has been blackened, or, at least, darkened, so that the damaskeening is effectively set off from the ground. If the design consists of lines, the grooves are dug out with the graver in such a manner that they are wider at the bottom, so as to hold the metal forced in. Next, the gold or silver pieces suitably formed are laid on top and hammered in so as to fill up the opening. Finally the surface is gone over again, so that the surface of the inlay is perfectly even with the rest. If the inlays, however, are not in the form of lines, but are composed of larger pieces of certain outlines, they are sometimes allowed to project beyond the surface of the metal decorated. At times there are inlays again in the raised portions of another metal; thus, Japanese bronze articles often contain figures of raised gold inlaid with silver.

Owing to the high value which damaskeening imparts to articles artistically decorated, many attempts have been made to obtain similar effects in a cheaper manner. One is electro-etching, described further on. Another process for the wholesale manufacture of objects closely resembling damaskeened work is the following: By means of a steel punch, on which the decorations to be produced project in relief, the designs are stamped by means of a drop hammer or a stamping press into gold plated or silver plated sheet metal on the side which is to show the damaskeening, finally grinding off the surface, so that the sunken portions are again level. Naturally, the stamped portion, as long as the depth of the stamping is at least equal to the thickness of the precious metal on top, will appear inlaid.

It is believed that much of the early damaskeening was done by welding together iron and either a steel or an impure or alloyed iron, and treating the surface with a corroding acid that affected the steel or alloy without changing the iron.

The variety or damaskeening known as koftgari or kuft-work, practiced in India, was produced by rough-etching a metallic surface and laying on gold-leaf, which was imbedded so that it adhered only to the etched parts of the design.

Damaskeening by Electrolysis.—Damaskeening of metallic plates may be done by electrolysis. A copper plate is covered with an isolating layer of feeble thickness, such as wax, and the desired design is scratched in it by the use of a pointed tool. The plate is suspended in a bath of sulphate of copper, connecting it with the positive pole of a battery, while a second copper plate is connected with the negative pole. The current etches grooves wherever the wax has been removed. When enough has been eaten away, remove the plate from the bath, cleanse it with a little hydrochloric acid to remove any traces of oxide of copper which might appear on the lines of the design; then wash it in plenty of water and place it in a bath of silver or nickel, connecting it now with the negative pole, the positive pole being represented by a leaf of platinum. After a certain time the hollows are completely filled with a deposit of silver or nickel, and it only remains to polish the plate, which has the appearance of a piece damaskeened by hand.

Damaskeening on Enamel Dials.— Dip the dial into molten yellow wax, trace on the dial the designs desired, penetrating down to the enamel. Dip the dial in a fluorhydric acid a sufficient length of time that it may eat to the desired depth. Next, wash in several waters, remove the wax by means of turpentine, i. e., leave the piece covered with wax immersed in essence of turpentine. By filling up the hollows thus obtained with enamel very pretty effects are produced.

DANDRUFF CURE:

See Hair Preparations.

DECALCOMANIA PROCESSES:

See also Chromos, Copying Processes, and Transfer Processes.

The decalcomania process of transferring pictures requires that the print (usually in colors) be made on a specially prepared paper. Prints made on decalcomania paper may be transferred in the reverse to chinaware, wood, celluloid, metal, or any hard smooth surface, and being varnished after transfer (or burnt in, in the case of pottery) acquire a fair degree of permanence. The original print is destroyed by the transfer.

Applying Decalcomania Pictures on Ceramic Products under a Glaze.—A biscuit-baked object is first coated with a mixture of alcohol, shellac, varnish, and liquid glue. Then the prepared picture print is transferred on to this adhesive layer in the customary manner. The glaze, however, does not adhere to this coating and would, therefore, not cover the picture when fused on. To attain this, the layer bearing the transfer picture, as well as the latter, are simultaneously coated with a dextrin solution of about 10 per cent. When this dextrin coating is dry, the picture is glazed.

The mixing proportions of the two solutions employed, as well as of the adhesive and the dextrin solutions, vary somewhat according to the physical conditions of the porcelain, its porosity, etc. The following may serve for an example: Dissolve 5 parts of shellac or equivalent gum in 25 parts of spirit and emulsify this liquid with 20 parts of varnish and 8 parts of liquid glue. After drying, the glaze is put on and the ware thus prepared is placed in the grate fire.

The process described is especially adapted for film pictures, i. e., for such as bear the picture on a cohering layer, usually consisting of collodion. It cannot be employed outright for gum pictures, i. e., for such pictures as are composed of different pressed surfaces, consisting mainly of gum or similar material. If this process is to be adapted to these pictures as well, the ware, which has been given the biscuit baking, is first provided with a crude glaze coating, whereupon the details of the process are carried out as described above with the exception that there is another glaze coating between the adhesive coat and the biscuit-baked ware. In this case the article is also immediately placed in the grate fire. It is immaterial which of the two kinds of metachromatypes (transfer pictures) is used, in every case the baking in the muffle, etc., is dropped. The transfer pictures may also be produced in all colors for the grate fire.

Decalcomania Paper.—Smooth unsized paper, not too thick, is coated with the following solutions:

I.—Gelatin, 10 parts, dissolved in 300 parts warm water. This solution is applied with a sponge. The paper should be dried flat.

II.—Starch, 50 parts; gum tragacanth, dissolved in 600 parts of water. (The gum tragacanth is soaked in 300 parts of water; in the other 300 parts the starch is boiled to a paste; the two are then poured together and boiled.) The dried paper is brushed with this paste uniformly, a fairly thick coat being applied. The paper is then allowed to dry again.

III.—One part blood albumen is soaked in 3 parts water for 24 hours. A small quantity of sal ammoniac is added.

The paper, after having been coated with these three solutions and dried, is run through the printing press, the pictures, however, being printed reversed so that it may appear in its true position when transferred. Any colored inks may be used.

IV.—A transfer paper, known as "décalque rapide," invented by J. B. Duramy, consists of a paper of the kind generally used for making pottery transfers, but coated with a mixture of gum and arrowroot solutions in the proportion of $2\frac{1}{2}$ parts of the latter to 100 of the former. The coating is applied in the ordinary manner, but the paper is only semi-glazed. Furthermore, to decorate pottery ware by means of this new transfer paper, there is no need to immerse the ware in a bath in order to get the paper to draw off, as it will come away when moistened with a damp sponge, after having been in position for less than 5 minutes, whereas the ordinary papers require a much longer time.

Picture Transferrer.—A very weak solution of soft soap and pearlashes is used to transfer recent prints, such as illustrations from papers, magazines, etc., to unglazed paper, on the decalcomania principle. Such a solution is:

I.—Soft soap......... $\frac{1}{2}$ ounce
　　Pearlash......... 2 drachms
　　Distilled water.... 16 fluidounces

The print is laid upon a flat surface, such as a drawing board, and moistened with the liquid. The paper on which the reproduction is required is laid over this, and then a sheet of thicker paper placed on the top, and the whole rubbed evenly and hard with a blunt instrument, such as the bowl of a spoon, until the desired depth of color in the transferrer is obtained. Another and more artistic process is to cover the print with a transparent sheet of material coated with wax, to trace out the pictures with a point and to take rubbings of the same after powdering with plumbago.

II.—Hard soap....... 1 drachm
　　Glycerine........ 30 grains
　　Alcohol......... 4 fluidrachms
　　Water.....,..... 1 fluidounce

Dampen the printed matter with the solution by sponging, and proceed as with I.

DEHORNERS:
　　See Horn.

DELTA METAL:
　　See Alloys.

DEMON BOWLS OF FIRE:
　　See Pyrotechnics.

DENTAL CEMENTS:
　　See Cements.

Dentifrices

TOOTH POWDERS:

A perfect tooth powder that will clean the teeth and mouth with thoroughness need contain but few ingredients and is easily made. For the base there is nothing better than precipitated chalk; it possesses all the detergent and polishing properties necessary for the thorough cleansing of the teeth, and it is too soft to do any injury to soft or to defective or thinly enameled teeth. This cannot be said of pumice, cuttlebone, charcoal, kieselguhr, and similar abradants that are used in tooth powders. Their use is reprehensible in a tooth powder. The use of pumice or other active abradant is well enough occasionally, by persons afflicted with a growth of tartar on the teeth, but even then it is best applied by a competent dentist. Abrading powders have much to answer for in hastening the day of the toothless race.

Next in value comes soap. Powdered white castile soap is usually an ingredient of tooth powders. There is nothing so effective for removing sordes or thickened mucus from the gums or mouth. But used alone or in too large proportions, the taste is unpleasant. Orris possesses no cleansing properties, but is used for its flavor and because it is most effective for masking the taste of the soap. Sugar or saccharine may be used for sweetening, and for flavoring almost anything can be used. Flavors should, in the main, be used singly, though mixed flavors lack the clean taste of simple flavors.

The most popular tooth powder sold is the white, saponaceous, wintergreen-flavored powder, and here is a formula for this type:

I.—Precipitated chalk... 1 pound
　　White castile soap... 1 ounce
　　Florentine orris...... 2 ounces
　　Sugar (or saccharine,
　　　2 grains)......... 1 ounce
　　Oil of wintergreen... $\frac{1}{4}$ ounce

The first four ingredients should be in the finest possible powder and well dried. Triturate the oil of wintergreen with part of the chalk, and mix this with the balance of the chalk. Sift each ingredient separately through a sieve (No. 80 or finer), and mix well together, afterwards sifting the mixture 5 or 6 times. The finer the sieve and the more the mixture is sifted, the finer and lighter the powder will be.

This powder will cost about 15 cents a pound.

Pink, rose-flavored powder of the Caswell and Hazard, Hudnut, or McMahan type, once so popular in New York. It was made in two styles, with and without soap.

II.—
Precipitated chalk...	1	pound
Florentine orris......	2	ounces
Sugar...............	1½	ounces
White castile soap...	1	ounce
No. 40 carmine......	15	grains
Oil of rose.........	12	drops
Oil of cloves........	4	drops

Dissolve the carmine in an ounce of water of ammonia and triturate this with part of the chalk until the chalk is uniformly dyed. Then spread it in a thin layer on a sheet of paper and allow the ammonia to evaporate. When there is no ammoniacal odor left, mix this dyed chalk with the rest of the chalk and sift the whole several times until thoroughly mixed. Then proceed to make up the powder as in the previous formula, first sifting each ingredient separately and then together, being careful thoroughly to triturate the oils of rose and cloves with the orris after it is sifted and before it is added to the other powders. The oil of cloves is used to back up the oil of rose. It strengthens and accentuates the rose odor. Be careful not to get a drop too much, or it will predominate over the rose.

Violet Tooth Powder.—
Precipitated chalk....	1	pound
Florentine orris......	4	ounces
Castile soap.........	1	ounce
Sugar...............	1½	ounces
Extract of violet.....	¼	ounce
Evergreen coloring, R. & F., quantity sufficient.		

Proceed as in the second formula, dyeing the chalk with the evergreen coloring to the desired shade before mixing.

III.—
Precipitated chalk.	16	pounds
Powdered orris....	4	pounds
Powdered cuttlefish bone..........	2	pounds
Ultramarine......	9½	ounces
Geranium lake....	340	grains
Jasmine..........	110	minims
Oil of neroli......	110	minims
Oil of bitter almonds.........	35	minims
Vanillin..........	50	grains
Artificial musk (Lautier's)......	60	grains
Saccharine........	140	grains

Rub up the perfumes with 2 ounces of alcohol, dissolve the saccharine in warm water, add all to the orris, and set aside to dry. Rub the colors up with water and some chalk, and when dry pass all through a mixer and sifter twice to bring out the color.

Camphorated and Carbolated Powders. —A camphorated tooth powder may be made by leaving out the oil of wintergreen in the first formula and adding 1½ ounces of powdered camphor.

Carbolated tooth powder may likewise be made with the first formula by substituting 2 drachms of liquefied carbolic acid for the oil of wintergreen. But the tooth powder gradually loses the odor and taste of the acid. It is not of much utility anyway, as the castile soap in the powder is of far greater antiseptic power than the small amount of carbolic acid that can safely be combined in a tooth powder. Soap is one of the best antiseptics.

Alkaline salts, borax, sodium bicarbonate, etc., are superfluous in a powder already containing soap. The only useful purpose they might serve is to correct acidity of the mouth, and that end can be reached much better by rinsing the mouth with a solution of sodium bicarbonate. Acids have no place in tooth powders, the French Codex to the contrary notwithstanding.

Peppermint as a Flavor.—In France and all over Europe peppermint is the popular flavor, as wintergreen is in this country.

English apothecaries use sugar of milk and heavy calcined magnesia in many of their tooth powders. Neither has any particular virtue as a tooth cleanser, but both are harmless. Cane sugar is preferable to milk sugar as a sweetener, and saccharine is more efficient, though objected to by some; it should be used in the proportion of 2 to 5 grains to the pound of powder, and great care taken to have it thoroughly distributed throughout.

An antiseptic tooth powder, containing the antiseptic ingredients of listerine, is popular in some localities.

IV.—
Precipitated chalk..	1	pound
Castile soap........	5	drachms
Borax..............	3	drachms
Thymol............	20	grains
Menthol...........	20	grains
Eucalyptol.........	20	grains
Oil of wintergreen..	20	grains
Alcohol............	½	ounce

Dissolve the thymol and oils in the alcohol, and triturate with the chalk, and proceed as in the first formula.

One fault with this powder is the disagreeable taste of the thymol. This may be omitted and the oil of wintergreen increased to the improvement of the taste, but with some loss of antiseptic power.

Antiseptic Powder.—

V.—Boric acid	50 parts
Salicylic acid	50 parts
Dragon's blood	20 parts
Calcium carbonate	1,000 parts
Essence spearmint	12 parts

Reduce the dragon's blood and calcium carbonate to the finest powder, and mix the ingredients thoroughly. The powder should be used twice a day, or even oftener, in bad cases. It is especially recommended in cases where the enamel has become eroded from the effects of iron.

Menthol Tooth Powder. — Menthol leaves a cool and pleasant sensation in the mouth, and is excellent for fetid breath. It may be added to most formulas by taking an equal quantity of oil of wintergreen and dissolving in alcohol.

Menthol	1 part
Salol	8 parts
Soap, grated fine	20 parts
Calcium carbonate	20 parts
Magnesia carbonate	60 parts
Essential oil of mint	2 parts

Powder finely and mix. If there is much tartar on the teeth it will be well to add to this formula from 10 to 20 parts of pumice, powdered very finely.

Tooth Powders and Pastes.—Although the direct object of these is to keep the teeth clean and white, they also prevent decay, if it is only by force of mere cleanliness, and in this way (and also by removing decomposing particles of food) tend to keep the breath sweet and wholesome. The necessary properties of a tooth powder are cleansing power unaccompanied by any abrading or chemical action on the teeth themselves, a certain amount of antiseptic power to enable it to deal with particles of stale food, and a complete absence of any disagreeable taste or smell. These conditions are easy to realize in practice, and there is a very large number of efficient and good powders, as well as not a few which are apt to injure the teeth if care is not taken to rinse out the mouth very thoroughly after using. These powders include some of the best cleansers, and have hence been admitted in the following recipes, mostly taken from English collections.

I.—Charcoal and sugar, equal weights. Mix and flavor with clove oil.

II.—Charcoal	156 parts
Red kino	156 parts
Sugar	6 parts

Flavor with peppermint oil.

III.—Charcoal	270 parts
Sulphate of quinine	1 part
Magnesia	1 part

Scent to liking.

IV.—Charcoal	30 parts
Cream of tartar	8 parts
Yellow cinchona bark	4 parts
Sugar	15 parts

Scent with oil of cloves.

V.—Sugar	120 parts
Alum	10 parts
Cream of tartar	20 parts
Cochineal	3 parts

VI.—Cream of tartar	1,000 parts
Alum	190 parts
Carbonate of magnesia	375 parts
Sugar	375 parts
Cochineal	75 parts
Essence Ceylon cinnamon	90 parts
Essence cloves	75 parts
Essence English peppermint	45 parts

VII.—Sugar	200 parts
Cream of tartar	400 parts
Magnesia	400 parts
Starch	400 parts
Cinnamon	32 parts
Mace	11 parts
Sulphate of quinine	16 parts
Carmine	17 parts

Scent with oil of peppermint and oil of rose.

| VIII.—Bleaching powder | 11 parts |
| Red coral | 12 parts |

IX.—Red cinchona bark	12 parts
Magnesia	50 parts
Cochineal	9 parts
Alum	6 parts
Cream of tartar	100 parts

English pep-
permint oil.   4 parts
Cinnamon oil   2 parts

Grind the first five ingredients separately, then mix the alum with the cochineal, and then add to it the cream of tartar and the bark. In the meantime the
magnesia is mixed with the essential oils,
and finally the whole mass is mixed
through a very fine silk sieve.

X.—Whitewood
   charcoal...   250 parts
   Cinchona
   bark......   125 parts
   Sugar......   250 parts
   Peppermint
   oil........   12 parts
   Cinnamon oil   8 parts

XI.—Precipitated
   chalk......   750 parts
   Cream of tar-
   tar........   250 parts
   Florence or-
   ris root....   250 parts
   Sal ammoniac   60 parts
   Ambergris...   4 parts
   Cinnamon...   4 parts
   Coriander....   4 parts
   Cloves......   4 parts
   Rosewood ...   4 parts

XII.—Dragon's
   blood......   250 parts
   Cream of tar-
   tar........   30 parts
   Florence or-
   ris root....   30 parts
   Cinnamon...   16 parts
   Cloves......   8 parts

XIII.—Precipitated
   chalk......   500 parts
   Dragon's
   blood......   250 parts
   Red sandal-
   wood......   125 parts
   Alum........   125 parts
   Orris root....   250 parts
   Cloves......   15 parts
   Cinnamon...   15 parts
   Vanilla......   8 parts
   Rosewood...   15 parts
   Carmine lake   250 parts
   Carmine.....   8 parts

XIV.—Cream of tar-
   tar........   150 parts
   Alum........   25 parts
   Cochineal....   12 parts
   Cloves......   25 parts
   Cinnamon...   25 parts
   Rosewood...   6 parts

Scent with essence of rose.

XV.—Coral........   20 parts
   Sugar.......   20 parts
   Wood char-
   coal......   6 parts
   Essence of ver-
   vain.......   1 part

XVI.—Precipitated
   chalk......   500 parts
   Orris root....   500 parts
   Carmine.....   1 part
   Sugar.......   1 part
   Essence of
   rose.......   4 parts
   Essence of ne-
   roli.......   4 parts

XVII.—Cinchona
   bark......   50 parts
   Chalk.......   100 parts
   Myrrh.......   50 parts
   Orris root....   100 parts
   Cinnamon...   50 parts
   Carbonate of
   ammonia..   100 parts
   Oil of cloves.   2 parts

XVIII.—Gum arabic..   30 parts
   Cutch.......   80 parts
   Licorice juice.   550 parts
   Cascarilla....   20 parts
   Mastic......   20 parts
   Orris root...   20 parts
   Oil of cloves..   5 parts
   Oil of pepper-
   mint......   15 parts
   Extract of
   amber.....   5 parts
   Extract of
   musk......   5 parts

XIX.—Chalk........   200 parts
   Cuttlebone...   100 parts
   Orris root....   100 parts
   Bergamot oil..   2 parts
   Lemon oil....   4 parts
   Neroli oil....   1 part
   Portugal oil..   2 parts

XX.—Borax ....   50 parts
   Chalk.......   100 parts
   Myrrh.......   25 parts
   Orris root....   22 parts
   Cinnamon...   25 parts

XXI.—Wood char-
   coal......   30 parts
   White honey.   30 parts
   Vanilla sugar   30 parts
   Cinchona
   bark......   16 parts

Flavor with oil of peppermint.

XXII.—Syrup of 33°B.   38 parts
   Cuttlebone...   200 parts
   Carmine lake   30 parts
   English oil of
   peppermint   5 parts

XXIII.—

Red coral....	50	parts
Cinnamon...	12	parts
Cochineal....	6	parts
Alum........	2⅕	parts
Honey.......	125	parts
Water.......	6	parts

Triturate the cochineal and the alum with the water. Then, after allowing them to stand for 24 hours, put in the honey, the coral, and the cinnamon. When the effervescence has ceased, which happens in about 48 hours, flavor with essential oils to taste.

XXIV.—

Well-skimmed honey.....	50 parts
Syrup of peppermint...	50 parts
Orris root....	12 parts
Sal ammoniac	12 parts
Cream of tartar........	12 parts
Tincture of cinnamon..	3 parts
Tincture of cloves.....	3 parts
Tincture of vanilla	3 parts
Oil of cloves.	1 part

XXV.—

Cream of tartar........	120 parts
Pumice......	120 parts
Alum........	30 parts
Cochineal....	30 parts
Bergamot oil.	3 parts
Clove.......	3 parts

Make to a thick paste with honey or sugar.

XXVI.—

Honey.......	250 parts
Precipitated chalk......	250 parts
Orris root....	250 parts
Tincture of opium.....	7 parts
Tincture of myrrh.....	7 parts
Oil of rose...	2 parts
Oil of cloves..	2 parts
Oil of nutmeg	2 parts

XXVII.—

Florentine orris........	6 parts
Magnesium carbonate..	2 parts
Almond soap	12 parts
Calcium carbonate....	60 parts
Thymol.....	1 part
Alcohol, quantity sufficient.	

Powder the solids and mix. Dissolve the thymol in as little alcohol as possible, and add perfume in a mixture in equal parts of oil of peppermint, oil of clove, oil of lemon, and oil of eucalyptus. About 1 minim of each to every ounce of powder will be sufficient.

XXVIII.—Myrrh, 10 parts; sodium chloride, 10 parts; soot, 5 parts; soap, 5 parts; lime carbonate, 500 parts.

XXIX.—Camphor, 5 parts; soap, 10 parts; saccharine, 0.25 parts; thymol, 0.5 parts; lime carbonate, 500 parts. Scent, as desired, with rose oil, sassafras oil, wintergreen oil, or peppermint oil.

XXX.—Powdered camphor, 6 parts; myrrh, 15 parts; powdered Peruvian bark, 6 parts; distilled water, 12 parts; alcohol of 80° F., 50 parts. Macerate the powders in the alcohol for a week and then filter.

XXXI.—Soap, 1; saccharine, 0.025; thymol, 0.05; lime carbonate, 50; sassafras essence, enough to perfume.

XXXII.—Camphor, 0.5; soap, 1; saccharine, 0.025; calcium carbonate, 50; oil of sassafras, or cassia, or of gaultheria, enough to perfume.

XXXIII.—Myrrh, 1; sodium chloride, 1; soap, 50; lime carbonate, 50; rose oil as required.

XXXIV.—Precipitated calcium carbonate, 60 parts; quinine sulphate, 2 parts; saponine, 0.1 part; saccharine, 0.1 part; carmine as required; oil of peppermint, sufficient.

XXXV.—Boracic acid, 100 parts; powdered starch, 50 parts; quinine hydrochlorate, 10 parts; saccharine, 1 part; vanillin (dissolved in alcohol), 1.5 parts.

Neutral Tooth Powder.—Potassium chlorate, 200 parts; starch, 200 parts; carmine lake, 40 parts; saccharine (in alcoholic solution), 1 part; vanillin (dissolved in alcohol), 1 part.

Tooth Powder for Children.—

Magnesia carbonate..	10 parts
Medicinal soap......	10 parts
Sepia powder........	80 parts
Peppermint oil, quantity sufficient to flavor.	

Flavorings for Dentifrice.—

I.—

Sassafras oil, true....	1 drachm
Pinus pumilio oil	20 minims
Bitter orange oil.....	20 minims
Wintergreen oil......	2 minims
Anise oil...........	4 minims
Rose geranium oil...	1 minim
Alcohol............	1 ounce

Use according to taste.

II.—

Oil of peppermint, English.........	4 parts
Oil of aniseed........	6 parts

Oil of clove	1 part
Oil of cinnamon	1 part
Saffron	1 part
Deodorized alcohol	350 parts
Water	300 parts

Or, cassia, 4 parts, and vanilla, ½ part, may be substituted for the saffron.

LIQUID DENTIFRICES AND TOOTH WASHES:

A French Dentifrice.—I.—A preparation which has a reputation in France as a liquid dentifrice is composed of alcohol, 96 per cent, 1,000 parts; Mitcham peppermint oil, 30 parts; aniseed oil, 5 parts; oil of Acorus calamus, 0.5 parts. Finely powdered cochineal and cream of tartar, 5 parts each, are used to tint the solution. The mixed ingredients are set aside for 14 days before filtering.

Sozodont.—

II.—The liquid tooth preparation "Sozodont" is said to contain: Soap powder, 60 parts; glycerine, 60 parts; alcohol, 360 parts; water, 220 parts; oils of peppermint, of aniseed, of clover, and of cinnamon, 1 part each; oil of wintergreen, 1-200 part.

III.—Thymol	2 grains
Benzoic acid	24 grains
Tincture eucalyptus	2 drachms
Alcohol quantity sufficient to make 2 ounces.	

Mix. Sig.: A teaspoonful diluted with half a wineglassful of water.

IV.—Carbolic acid, pure	2	ounces
Glycerine, 1,260°	1	ounce
Oil wintergreen	6	drachms
Oil cinnamon	3	drachms
Powdered cochineal		½ drachm
S. V. R.	40	ounces
Distilled water	40	ounces

Dissolve the acid in the glycerine with the aid of a gentle heat and the essential oils in the spirit; mix together, and add the water and cochineal; then let the preparation stand for a week and filter.

A mixture of caramel and cochineal coloring, N. F., gives an agreeable red color for saponaceous tooth washes. It is not permanent, however.

Variations of this formula follow:

V.—White castile soap	1	ounce
Tincture of asarum	2	drachms
Oil of peppermint	½	drachm
Oil of wintergreen	½	drachm
Oil of cloves	5	drops
Oil of cassia	5	drops
Glycerine	4	ounces
Alcohol	14	ounces
Water	14	ounces

VI.—White castile soap	1½	ounces
Oil of orange	10	minims
Oil of cassia	5	minims
Oil of wintergreen	15	minims
Glycerine	3	ounces
Alcohol	8	ounces
Water enough to make 1 quart.		

VII.—White castile soap	3	ounces
Glycerine	5	ounces
Water	20	ounces
Alcohol	30	ounces
Oil of peppermint	1	drachm
Oil of wintergreen	1	drachm
Oil of orange peel	1	drachm
Oil of anise	1	drachm
Oil of cassia	1	drachm

Beat up the soap with the glycerine; dissolve the oils in the alcohol and add to the soap and glycerine. Stir well until the soap is completely dissolved.

VIII.—White castile soap	1	ounce
Orris root	4	ounces
Rose leaves	4	ounces
Oil of rose	½	drachm
Oil of neroli	½	drachm
Cochineal	½	ounce
Diluted alcohol	2	quarts

If the wash is intended simply as an elixir for sweetening the breath, the following preparation, resembling the celebrated *eau de botot*, will be found very desirable:

IX.—Oil of peppermint	30 minims
Oil of spearmint	15 minims
Oil of cloves	5 minims
Oil of red cedar wood	60 minims
Tincture of myrrh	1 ounce
Alcohol	1 pint

Care must be taken not to confound the oil of cedar tops with the oil of cedar wood. The former has an odor like turpentine; the latter has the fragrance of the red cedar wood.

For a cleansing wash, a solution of soap is to be recommended. It may be made after the following formula:

X.—White castile soap	1	ounce
Alcohol	6	ounces
Glycerine	4	ounces
Hot water	6	ounces
Oil of peppermint	15	minims
Oil of wintergreen	20	minims
Oil of cloves	5	minims
Extract of vanilla	½	ounce

Dissolve the soap in the hot water and add the glycerine and extract of vanilla. Dissolve the oils in the alcohol, mix the solutions, and after 24 hours filter through paper.

It is customary to color such preparations. An agreeable brown-yellow tint may be given by the addition of a small quantity of caramel. A red color may be given by cochineal. The color will fade, but will be found reasonably permanent when kept from strong light.

TOOTH SOAPS AND PASTES:

Tooth Soaps.—

I.—White castile soap.. 225 parts
Precipitated chalk .. 225 parts
Orris root......... 225 parts
Oil of peppermint.. 7 parts
Oil of cloves ...... 4 parts
Water, a sufficient quantity.

II.—Castile soap........ 100 drachms
Precipitated chalk.. 100 drachms
Powdered orris root. 100 drachms
White sugar....... 50 drachms
Rose water .... 50 drachms
Oil of cloves .... 100 drops
Oil of peppermint... 3 drachms

Dissolve the soap in water, add the rose water, then rub up with the sugar with which the oils have been previously triturated, the orris root and the precipitated chalk.

III.—Potassium chlorate, 20 drachms; powdered white soap, 10 drachms; precipitated chalk, 20 drachms; peppermint oil, 15 drops; clove oil, 5 drops; glycerine, sufficient to mass. Use with a soft brush.

Saponaceous Tooth Pastes.—

I.—Precipitated carbonate of lime.. 90 parts
Soap powder..... 30 parts
Ossa sepia, powdered.......... 15 parts
Tincture of cocaine 45 parts
Oil of peppermint. 6 parts
Oil of ylang-ylang. 0.3 parts
Glycerine........ 30 parts
Rose water to cause liquefaction. Carmine solution to color.

II.—Precipitated carbonate of lime.. 150 parts
Soap powder..... 45 parts
Arrowroot....... 45 parts
Oil of eucalyptus . 2 parts
Oil of peppermint. 1 part
Oil of geranium .. 1 part
Oil of cloves..... 0.25 parts
Oil of aniseed.... 0.25 parts
Glycerine .... 45 parts
Chloroform water to cause liquefaction. Carmine solution to color.

Cherry Tooth Paste.—

III.—Clarified honey .. 100 drachms
Precipitated chalk 100 drachms
Powdered orris root.......... 100 drachms
Powdered rose leaves........ 60 drops
Oil of cloves..... 55 drops
Oil of mace..... 55 drops
Oil of geranium.. 55 drops

Chinese Tooth Paste.—

IV.—Powdered pumice 100 drachms
Starch......... 20 drachms
Oil of peppermint 40 drops
Carmine........ ¼ drachm

Eucalyptus Paste. — Forty drachms precipitated chalk, 11 drachms soap powder, 11 drachms wheaten starch, ¼ drachm carmine, 30 drops oil of peppermint, 30 drops oil of geranium, 60 drops eucalyptus oil, 2 drops oil of cloves, 12 drops oil of anise mixed together and incorporated to a paste, with a mixture of equal parts of glycerine and spirit.

Myrrh Tooth Paste.—

Precipitated chalk 8 ounces
Orris........... 8 ounces
White castile soap. 2 ounces
Borax........... 2 ounces
Myrrh.......... 1 ounce
Glycerine, quantity sufficient.

Color and perfume to suit.

A thousand grams of levigated powdered oyster shells are rubbed up with 12 drachms of cochineal to a homogeneous powder. To this is added 1 drachm of potassium permanganate and 1 drachm boric acid and rubbed well up. Foam up 200 drachms castile soap and 5 drachms chemically pure glycerine and mix it with the foregoing mass, adding by teaspoonful 150 grams of boiling strained honey. The whole mass is again thoroughly rubbed up, adding while doing so 200 drops honey. Finally the mass should be put into a mortar and pounded for an hour and then kneaded with the hands for 2 hours.

Tooth Paste to be put in Collapsible Tubes.—

Calcium carbonate, levigated..........100 parts
Cuttlefish bone, in fine powder.......... 25 parts
Castile soap, old white, powdered......... 25 parts
Tincture of carmine, ammoniated....... 4 parts
Simple syrup........ 25 parts

Menthol............... 2 parts
Alcohol............... 5 parts
Attar of rose or other perfume, quantity sufficient.
Rose water sufficient to make a paste.

Beat the soap with a little rose water, then warm until softened, add syrup and tincture of carmine. Dissolve the perfume and menthol in the alcohol and add to soap mixture. Add the solids and incorporate thoroughly. Finally, work to a proper consistency for filling into collapsible tubes, adding water, if necessary.

MOUTH WASHES.

I.—Quillaia bark.... 125 parts
    Glycerine....... 95 parts
    Alcohol......... 155 parts

Macerate for 4 days and add:

Acid. carbol.
   cryst .... 4 parts
Ol. geranii .... 0.6 parts
Ol. caryophyll .. 0.6 parts
Ol. rosæ........ 0.6 parts
Ol. cinnam..... 0.6 parts
Tinct. ratanbæ.. 45 parts
Aqua rosæ...... 900 parts

Macerate again for 4 days and filter.

Thymol .... 20 parts
Peppermint oil.. 10 parts
Clove oil .... 5 parts
Sage oil........ 5 parts
Marjoram oil... 3 parts
Sassafras oil .... 3 parts
Wintergreen oil. 0.5 parts
Coumarin ...... 0.5 parts
Alcohol, dil.....1,000 parts

A teaspoonful in a glass of water.

II.—Tincture orris (1
    in 4).......... 1½ parts
Lavender water... ½ part
Tinct. cinnamon
   (1 in 8)........ 1 part
Tinct. yellow cinch
   bark.......... 1 part
Eau de cologne.... 2 parts

Orris and Rose.—

III.—Orris root....... 30 drachms
    Rose leaves...... 8 drachms
    Soap bark....... 8 drachms
    Cochineal....... 3½ drachms
    Diluted alcohol.. 475 drachms
    Oil rose........ 30 drops
    Oil neroli....... 40 drops

Myrrh Astringent.—

IV.—Tincture myrrh.. 125 drachms
    Tincture benzoin. 50 drachms
    Tincture cinchona 8 drachms
    Alcohol......... 225 drachms
    Oil of rose....... 30 drops

Borotonic.—

V.—Acid boric..... 20 parts
    Oil wintergreen. 10 parts
    Glycerine...... 110 parts
    Alcohol........ 150 parts
    Distilled water
    enough to make 600 parts

Sweet Salicyl.—

VI.—Acid salicylic... 4 parts
    Saccharine..... 1 part
    Sodium bicarbonate ...... 1 part
    Alcohol........ 200 parts

Foaming Orange.—

VII.—Castile soap.... 29 drachms
    Oil orange ..... 10 drops
    Oil cinnamon... 5 drops
    Distilled water.. 30 drachms
    Alcohol........ 90 drachms

Australian Mint.—

VIII.—Thymol.. .... 0.25 parts
    Acid benzoic... 3 parts
    Tincture eucalyptus.......... 15 parts
    Alcohol........ 100 parts
    Oil peppermint. 0.75 parts

Fragrant Dentine.—

IX.—Soap bark...... 125 parts
    Glycerine...... 95 parts
    Alcohol........ 155 parts
    Rose water..... 450 parts

Macerate for 4 days and add:

Carbolic acid,
   cryst .... 4 parts
Oil geranium... 0.6 parts
Oil cloves...... 0.6 parts
Oil rose........ 0.6 parts
Oil cinnamon... 0.6 parts
Tincture rhatany 45 parts
Rose water..... 450 parts

Allow to stand 4 days; then filter.

Aromantiseptic.—

X.—Thymol........ 20 parts
    Oil peppermint. 10 parts
    Oil cloves...... 5 parts
    Oil sage....... 5 parts
    Oil marjoram... 3 parts
    Oil sassafras.... 3 parts
    Oil wintergreen. 0.5 parts
    Coumarin...... 0.5 parts
    Diluted alcohol. 1,000 parts

The products of the foregoing formulas are used in the proportion of 1 teaspoonful in a half glassful of water.

Foaming.—

XI.—Soap bark, powder 2 ounces
    Cochineal powder. 60 grains
    Glycerine........ 3 ounces

Alcohol.......... 10 ounces
Water sufficient
    to make....... 32 ounces

Mix the soap, cochineal, glycerine, alcohol, and water together; let macerate for several days; filter and flavor; if same produces turbidity, shake up the mixture with magnesium carbonate, and filter through paper.

Odonter.—

XII.—Soap bark, powder   2 ounces
    Cudbear, powder.   4 drachms
    Glycerine.......   4 ounces
    Alcohol.........   14 ounces
    Water sufficient
      to make......   32 ounces

Mix, and let macerate with frequent agitation, for several days; filter; add flavor; if necessary filter again through magnesium carbonate or paper pulp.

Sweet Anise.—

XIII.—Soap bark........   2 ounces
    Aniseed ....   4 drachms
    Cloves...........   4 drachms
    Cinnamon.......   4 drachms
    Cochineal........   60 grains
    Vanilla..........   60 grains
    Oil of peppermint.   1 drachm
    Alcohol..........   16 ounces
    Water sufficient to
      make ....   32 ounces

Reduce the drugs to coarse powder, dissolve the oil of peppermint in the alcohol, add equal parts of water, and macerate therein the powders for 5 to 6 days, with frequent agitation; place in percolator and percolate until 32 fluidounces have been obtained. Let stand for a week and filter through paper; if necessary to make it perfectly bright and clear, shake up with some magnesia, and again filter.

Saponaceous.—

XIV.—White castile soap   2 ounces
    Glycerine........   2 ounces
    Alcohol..........   8 ounces
    Water...........   4 ounces
    Oil peppermint...   20 drops
    Oil wintergreen...   30 drops
    Solution of carmine N. F. sufficient to color.

Dissolve the soap in the alcohol and water, add the other ingredients, and filter.

XV.—Crystallized carbolic acid......   4   parts
    Eucalyptol.......   1   part
    Salol............   2   parts
    Menthol.........   0.25 parts
    Thymol..........   0.1   part
    Alcohol..........100   parts
Dye with cochineal (1½ per cent).

Jackson's Mouth Wash.—Fresh lemon peel, 10 parts; fresh sweet orange peel, 10 parts; angelica root, 10 parts; guaiacum wood, 30 parts; balsam of Tolu, 12 parts; benzoin, 12 parts; Peruvian balsam, 4 parts; myrrh, 3 parts; alcohol (90 per cent), 500 parts.

Tablets for Antiseptic Mouth Wash.—Heliotropine, 0.01 part; saccharine, 0.01 part; salicylic acid, 0.01 part; menthol, 1 part; milk sugar, 5 parts. These tablets may be dyed green, red, or blue, with chlorophyll, eosine, and indigo carmine, respectively.

Depilatories

Depilatory Cream.—The depilatory cream largely used in New York hospitals for the removal of hair from the skin previous to operations:

I.—Barium sulphide....   3 parts
    Starch............   1 part
    Water, sufficient quantity.

The mixed powders are to be made into a paste with water, and applied in a moderately thick layer to the parts to be denuded of hair, the excess of the latter having been previously trimmed off with a pair of scissors. From time to time a small part of the surface should be examined, and when it is seen that the hair can be removed, the mass should be washed off. The barium sulphide should be quite fresh. It can be prepared by making barium sulphate and its own weight of charcoal into a paste with linseed oil, rolling the paste into the shape of a sausage, and placing it upon a bright fire to incinerate. When it has ceased to burn, and is a white hot mass, remove from the fire, cool, and powder.

The formula is given with some reserve, for preparations of this kind are usually unsafe unless used with great care. It should be removed promptly when the skin begins to burn.

II.—Barium sulphide....   25 parts
    Soap.............   5 parts
    Talc.............   35 parts
    Starch...........   35 parts
    Benzaldehyde sufficient to make...   120 parts

Powder the solids and mix. To use, to a part of this mixture add 3 parts of water, at the time of its application, and with a camel's-hair pencil paint the mixture evenly over the spot to be freed of hair. Let remain in contact with the skin for 5 minutes, then wash off with a sponge, and in the course of 5 minutes longer the hair will come off on slight friction with the sponge.

Strontium sulphide is an efficient depilatory. A convenient form of applying it is as follows:

III.—Strontium sulphide .   2 parts
     Zinc oxide.........   3 parts
     Powdered starch ...   3 parts

Mix well and keep in the dry state until wanted for use, taking then a sufficient quantity, forming into a paste with warm water and applying to the surface to be deprived of hair. Allow to remain from 1 to 5 minutes, according to the nature of the hair and skin; it is not advisable to continue the application longer than the last named period. Remove in all cases at once when any caustic action is felt. After the removal of the paste, scrape the skin gently but firmly with a blunt-edged blade (a paper knife, for instance) until the loosened hair is removed. Then immediately wash the denuded surface well with warm water, and apply cold cream or some similar emollient as a dressing.

	By weight	
IV.—Alcohol............	12	parts
Collodion	35	parts
Iodine	0.75	parts
Essence of turpentine	1.5	parts
Castor oil.........	2	parts

Apply with a brush on the affected parts for 3 or 4 days in thick coats. When the collodion plaster thus formed is pulled off, the hairs adhere to its inner surface.

V.—Rosin sticks are intended for the removal of hairs and are made from colophony with an admixture of 10 per cent of yellow wax. The sticks are heated like a stick of sealing wax until soft or semi-liquid (142° F.), and lightly applied on the place from which the hair is to be removed, and the mass is allowed to cool. These rosin sticks are said to give good satisfaction.

DEPTHINGS, VERIFICATION OF:
See Watchmakers' Formulas.

DESILVERING:
See Plating.

DETERGENTS:
See Cleaning Preparations and Methods.

DEVELOPERS FOR PHOTOGRAPHIC PURPOSES:
See Photography.

DEXTRIN PASTES AND MUCILAGES:
See Adhesives.

DIAL CEMENTS:
See Adhesives, under Jewelers' Cements.

DIAL CLEANERS:
See Cleaning Preparations and Methods.

DIAL REPAIRING:
See Watchmakers' Formulas.

DIAMALT:
See Milk.

DIAMOND TESTS:
See also Gems and Jewelers' Formulas.

To Distinguish Genuine Diamonds.— If characters or marks of any kind are drawn with an aluminum pencil on glass, porcelain, or any substance containing silex, the marks cannot be erased by rubbing, however energetic the friction, and even acids will not cause them to disappear entirely, unless the surface is entirely freed from greasy matter, which can be accomplished by rubbing with whiting and passing a moistened cloth over the surface at the time of writing. So, in order to distinguish the true diamond from the false, it is necessary only to wipe the stone carefully and trace a line on it with an aluminum pencil, and then rub it briskly with a moistened cloth. If the line continues visible, the stone is surely false. If, on the contrary, the stone is a true diamond, the line will disappear without leaving a trace, and without injury to the stone.

The common test for recognizing the diamond is the file, which does not cut it, though it readily attacks imitations. There are other stones not affected by the file, but they have characteristics of color and other effects by which they are readily distinguished.

This test should be confirmed by others. From the following the reader can select the most convenient:

A piece of glass on which the edge of a diamond is drawn, will be cut without much pressure; a slight blow is sufficient to separate the glass. An imitation may scratch the glass, but this will not be cut as with the diamond.

If a small drop of water is placed upon the face of a diamond and moved about by means of the point of a pin, it will preserve its globular form, provided the stone is clean and dry. If the attempt is made on glass, the drop will spread.

A diamond immersed in a glass of water will be distinctly visible, and will shine clearly through the liquid. The imitation stone will be confounded with the water and will be nearly invisible.

By looking through a diamond with a glass at a black point on a sheet of white paper, a single distinct point will be seen. Several points, or a foggy point will appear if the stone is spurious.

Hydrofluoric acid dissolves all imitations, but has no effect on true diamonds. This acid is kept in gutta-percha bottles.

For an eye practiced in comparisons it is not difficult to discern that the facets in the cut of a true diamond are not as regular as are those of the imitation; for in cutting and polishing the real stone an effort is made to preserve the original as much as possible, preferring some slight irregularities in the planes and edges to the loss in the weight, for we all know that diamonds are sold by weight. In an imitation, however, whether of paste or another less valuable stone, there is always an abundance of cheap material which may be cut away and thereby form a perfect-appearing stone.

Take a piece of a fabric, striped red and white, and draw the stone to be tested over the colors. If it is an imitation, the colors will be seen through it, while a diamond will not allow them to be seen.

A genuine diamond, rubbed on wood or metal, after having been previously exposed to the light of the electric arc, becomes phosphorescent in darkness, which does not occur with imitations.

Heat the stone to be tested, after giving it a coating of borax, and let it fall into cold water. A diamond will undergo the test without the slightest damage; the glass will be broken in pieces.

Finally, try with the fingers to crush an imitation and a genuine diamond between two coins, and you will soon see the difference.

DIAMOND CEMENT:

See Adhesives, under Jewelers' Cements.

DIARRHEA IN BIRDS:

See Veterinary Formulas.

DIARRHEA REMEDIES:

See Cholera Remedies.

Die Venting.—Many pressmen have spent hours and days in the endeavor to produce sharp and full impressions on figured patterns. If all the deep recesses in deep-figured dies are vented to allow the air to escape when the blow is struck, it will do much to obtain perfect impressions, and requires only half the force that is necessary in unvented dies. This is not known in many shops and consequently this little air costs much in power and worry.

DIGESTIVE POWDERS AND TABLETS.

I.—Sodium bicarbonate. 93 parts
Sodium chlorate.... 4 parts
Calcium carbonate.. 3 parts
Pepsin............. 5 parts
Ammonium carbonate............. 1 part

II.—Sodium bicarbonate. 120 parts
Sodium chlorate.... 5 parts
Sal physiologic (see below).......... 4 parts
Magnesium carbonate............. 10 parts

III.—Pepsin, saccharated (U. S. P.)........ 10 drachms
Pancreatin......... 10 drachms
Diastase........... 50 drachms
Acid, lactic........ 40 drops
Sugar of milk...... 40 drachms

IV.—Pancreatin......... 3 parts
Sodium bicarbonate. 15 parts
Milk sugar......... 2 parts

Sal Physiologicum.—The formula for this ingredient, the so-called nutritive salt (*Nahrsalz*), is as follows:

Calcium phosphate.	40	parts
Potassium sulphate.	2	parts
Sodium phosphate..	20	parts
Sulphuric, precipitated.	5	parts
Sodium chlorate....	60	parts
Magnesium phosphate...........	5	parts
Carlsbad salts, artificial.............	60	parts
Silicic acid.........	10	parts
Calcium fluoride....	$2\frac{1}{2}$	parts

Digestive Tablets.—

Powdered double refined sugar......	300 parts
Subnitrate bismuth	60 parts
Saccharated pepsin	45 parts
Pancreatin........	45 parts
Mucilage.........	35 parts
Ginger..........	30 parts

Mix and divide into suitable sizes,

DIOGEN DEVELOPER:
See Photography.

DIP FOR BRASS:
See Plating and Brass.

DIPS:
See Metals.

DIPS FOR CATTLE:
See Disinfectants and Veterinary Formulas.

DISH WASHING:
See Household Formulas

Disinfectants

Disinfecting Fluids.—

I.—Creosote	40 gallons
Rosin, powdered	56 pounds
Caustic soda lye, 38° Tw	9 gallons
Boiling water	12 gallons
Methylated spirit	1 gallon
Black treacle	14 pounds

Melt the rosin and add the creosote; run in the lyes; then add the matter and methylated spirit mixed together, and add the treacle; boil all till dissolved and mix well together.

II.—Hot water	120 pounds
Caustic soda lye, 38° B	120 pounds
Rosin	300 pounds
Creosote	450 pounds

Boil together the water, lye, and rosin, till dissolved; turn off steam and stir in the creosote; keep on steam to nearly boiling all the time, but so as not to boil over, until thoroughly incorporated.

III.—Fresh-made soap (hard yellow)	7 pounds
Gas tar	21 pounds
Water, with 2 pounds soda	21 pounds

Dissolve soap (cut in fine shavings) in the gas tar; then add slowly the soda and water which has been dissolved.

IV.—Rosin	1 cwt.
Caustic soda lye, 18° B	16 gallons
Black tar oil	½ gallon
Nitro-naphthalene dissolved in boiling water (about ½ gallon)	2 pounds

Melt the rosin, add the caustic lye; then stir in the tar oil and add the nitro-naphthalene.

V.—Camphor	1 ounce
Carbolic acid (75 per cent)	12 ounces
Aqua ammonia	10 drachms
Soft salt water	8 drachms

To be diluted when required for use.

VI.—Heavy tar oil	10 gallons
Caustic soda dissolved in 5 gallons water 600° F	30 pounds

Mix the soda lyes with the oil, and heat the mixture gently with constant stirring; add, when just on the boil, 20 pounds of refuse fat or tallow and 20 pounds of soft soap; continue the heat until thoroughly saponified, and add water gradually to make up 40 gallons. Let it settle; then decant the clear liquid.

Disinfecting Fluids or Weed-Killers.—
I.—Cold water, 20 gallons; powdered rosin, 56 pounds; creosote oil, 40 gallons; sulphuric acid, ½ gallon; caustic soda lye, 30° B., 9 gallons.

Heat water and dissolve the rosin; then add creosote and boil to a brown mass and shut off steam; next run in sulphuric acid and then the lyes.

II.—Water	40 gallons
Powdered black rosin	56 pounds
Sulphuric acid	2½ gallons
Creosote	10 gallons
Melted pitch	24 pounds
Pearlash boiled in 10 gallons water	56 pounds

Boil water and dissolve rosin and acid; then add creosote and boil well again; add pitch and run in pearlash solution (boiling); then shut off steam.

III. (White).—Water, 40 gallons; turpentine, 2 gallons; ammonia, ½ gallon; carbolic crystals, 14 pounds; caustic lyes, 2 gallons; white sugar, 60 pounds, dissolved in 40 pounds water.

Heat water to boiling, and add first turpentine, next ammonia, and then carbolic crystals. Stir well until thoroughly dissolved, and add lyes and sugar solution.

DISINFECTING POWDERS.

I.—Sulphate of iron	100 parts
Sulphate of zinc	50 parts
Oak bark, powder	40 parts
Tar	5 parts
Oil	5 parts

II.—Mix together chloride of lime and burnt umber, add water, and set on plates.

Blue Sanitary Powder.—

Powdered alum.....	2 pounds
Oil of eucalyptus...	12 ounces
Rectified spirits of tar..............	6 ounces
Rectified spirit of turpentine.......	2 ounces
Ultramarine blue (common).......	¾ ounces
Common salt.......	14 pounds

Mix alum with about 3 pounds of salt in a large mortar, gradually add oil of eucalyptus and spirits, then put in the ultramarine blue, and lastly remaining salt, mixing all well, and passing through a sieve.

Carbolic Powder. (Strong).—

Slaked lime in fine powder, 1 cwt.; carbolic acid, 75 per cent, 2 gallons.

Color with aniline dye and then pass through a moderately fine sieve and put into tins or casks and keep air-tight.

Pink Carbolized Sanitary Powder.—

Powdered alum.....	6 ounces
Powdered green copperas...........	5 pounds
Powdered red lead..	5 pounds
Calvert's No. 5 carbolic acid........	12½ pounds
Spirit of turpentine.	1½ pounds
Calais sand........	10 pounds
Slaked lime........	60 pounds

Mix carbolic acid with turpentine and sand, then add the other ingredients, lastly the slaked lime and, after mixing, pass through a sieve. It is advisable to use lime that has been slaked some time.

Cuspidor Powder.—

Peat rubble is ground to a powder, and 100 parts put into a mixing machine, which can be hermetically sealed. Then 15 parts of blue vitriol are added either very finely pulverized or in a saturated aqueous solution. Next are added 2 parts of formalin, and lastly 1 part of ground cloves, orange peel, or a sufficient quantity of some volatile oil, to give the desired perfume. The mixing machine is then closed, and kept at work until the constituents are perfectly mixed; the powder is then ready to be put up for the market. Its purpose is to effect a rapid absorption of the sputum, with simultaneous destruction of any microbes present, and to prevent decomposition and consequent unpleasant odors.

Deodorants for Water-Closets.—

I.—Ferric chloride.....	4 parts
Zinc chloride.......	5 parts
Aluminum chloride.	5 parts
Calcium chloride....	4 parts
Magnesium chloride.	3 parts
Water sufficient to make.............	90 parts

Dissolve, and add to each gallon 10 grains thymol and ¼ ounce oil of rosemary, previously dissolved in about 6 quarts of alcohol, and filter.

II.—Sulphuric acid, fuming	90 parts
Potassium permanganate.........	45 parts
Water...........	4,200 parts

Dissolve the permanganate in the water, and add under the acid. This is said to be a most powerful disinfectant, deodorizer, and germicide. It should not be used where there are metal trimmings.

Formaldehyde for Disinfecting Books, Papers, etc.—

The property of formaldehyde of penetrating all kinds of paper, even when folded together in several layers, may be utilized for a perfect disinfection of books and letters, especially at a temperature of 86° to 122° F. in a closed room. The degree of penetration as well as the disinfecting power of the formaldehyde depend upon the method of generating the gas. Letters, paper in closed envelopes, are completely disinfected only in 12 hours, books in 24 hours at a temperature of 122° F. when 70 cubic centimeters of formochloral—17.5 g. of gas—per cubic meter of space are used. Books must be stood up in such a manner that the gas can enter from the sides. Bacilli of typhoid preserve their vitality longer upon unsized paper and on filtering paper than on other varieties.

There is much difference of opinion as to the disinfecting and deodorizing power of formaldehyde when used to disinfect wooden tierces. While some have found it to answer well, others have got variable results, or failed of success. The explanation seems to be that those who have obtained poor results have not allowed time for the disinfectant to penetrate the pores of the wood, the method of application being wrong. The solution is thrown into the tierce, which is then steamed out at once, whereby the aldehyde is volatilized before it has had time to do its work. If the formal and the steam, instead of being used in succession, were used together, the steam would carry the disinfectant into the pores of the wood. But a still better plan is to give the aldehyde more time.

Another point to be remembered in all cases of disinfection by formaldehyde is that a mechanical cleansing must precede the action of the antiseptic. If there are thick deposits of organic matter which can be easily dislodged with a scrubbing brush, they can only be disinfected by the use of large quantities of formaldehyde used during a long period of time.

General Disinfectants.—

I.—Alum............. 10 ounces
Sodium carbonate.. 10 ounces
Ammonium chloride 2 ounces
Zinc chloride....... 1 ounce
Sodium chloride.... 2 ounces
Hydrochloric acid, quantity sufficient.
Water to make 1 gallon.

Dissolve the alum in one half gallon of boiling water, and add the sodium carbonate; then add hydrochloric acid until the precipitate formed is dissolved. Dissolve the other salt in water and add to the previous solution. Finally add enough water to make the whole measure 1 gallon, and filter.

In use, this is diluted with 7 parts of water.

II.—For the Sick Room.—In using this ventilate frequently: Guaiac, 10 parts; eucalyptol, 8 parts; phenol, 6 parts; menthol, 4 parts; thymol, 2 parts; oil of cloves, 1 part; alcohol of 90 per cent, 170 parts.

Atomizer Liquid for Sick Rooms.—

III.—Eucalyptol......... 10 ⎫
Thyme oil......... 5 ⎪ Parts
Lemon oil......... 5 ⎬ by
Lavender oil....... 5 ⎪ weight.
Spirit, 90 per cent...110 ⎭

To a pint of water a teaspoonful for evaporation.

Non-Poisonous Sheep Dips.—Paste.—

I.—Creosote (containing
15 per cent to 20
per cent of carbolic acid)....... 2 parts
Stearine or Yorkshire
grease.......... 1 part
Caustic soda lyes,
specific gravity,
1340............ 1 part
Black rosin, 5 per cent to 10 per cent.

Melt the rosin and add grease and soda lyes, and then add creosote cold.

II.—Creosote.......... 1 part
Crude hard rosin oil 1 part

Put rosin oil in copper and heat to about 220° F., and add as much caustic soda powder, 98 per cent strength, as the oil will take up. The quantity depends upon the amount of acetic acid in the oil. If too much soda is added it will remain at the bottom. When the rosin oil has taken up the soda add creosote, and let it stand.

Odorless Disinfectants.—

I.—Ferric chloride..... 4 parts
Zinc chloride....... 5 parts
Aluminum chloride. 5 parts
Calcium chloride... 4 parts
Manganese chloride 3 parts
Water............ 69 parts

If desired, 10 grains thymol and 2 fluidrachms oil of rosemary, previously dissolved in about 12 fluidrachms of alcohol, may be added to each gallon.

II.—Alum............. 10 parts
Sodium carbonate.. 10 parts
Ammonium chloride 2 parts
Sodium chloride.... 2 parts
Zinc chloride....... 1 part
Hydrochloric acid, sufficient.
Water............100 parts

Dissolve the alum in about 50 parts boiling water and add the sodium carbonate. The resulting precipitate of aluminum hydrate dissolve with the aid of just sufficient hydrochloric acid, and add the other ingredients previously dissolved in the remainder of the water.

III.—Mercuric chloride... 1 part
Cupric sulphate.... 10 parts
Zinc sulphate....... 50 parts
Sodium chloride.... 65 parts
Water to make 1,000 parts.

Paris Salts.—The disinfectant known by this name is a mixture made from the following recipe:

Zinc sulphate...... 49 parts
Ammonia alum..... 49 parts
Potash permanganate............ 1 part
Lime............. 1 part

The ingredients are fused together, mixed with a little calcium chloride, and perfumed with thymol.

Platt's Chlorides.—

I.—Aluminum sulphate. 6 ounces
Zinc chloride....... 1½ ounces
Sodium chloride.... 2 ounces
Calcium chloride... 3 ounces
Water enough to make 2 pints.

II.—A more elaborate formula for a preparation said to resemble the proprietary article is as follows:

Zinc, in strips	4	ounces
Lead carbonate	2	ounces
Chlorinated lime	1	ounce
Magnesium carbonate	$\frac{1}{2}$	ounce
Aluminum hydrate	$1\frac{1}{2}$	ounces
Potassium hydrate	$\frac{1}{2}$	ounce
Hydrochloric acid	16	ounces
Water	16	ounces
Whiting, enough.		

Dissolve the zinc in the acid; then add the other salts singly in the order named, letting each dissolve before the next is added. When all are dissolved add the water to the solution, and after a couple of hours add a little whiting to neutralize any excess of acid; then filter.

Zinc chloride ranks very low among disinfectants, and the use of such solutions as these, by giving a false sense of security from disease germs, may be the means of spreading rather than of checking the spread of sickness.

Disinfecting Coating.—Carbolic acid, 2 parts; manganese, 3 parts; calcium chloride, 2 parts; china clay, 10 parts; infusorial earth, 4 parts; dextrin, 2 parts; and water, 10 parts.

DISTEMPER IN CATTLE:
See Veterinary Formulas.

DIURETIC BALL:
See Veterinary Formulas.

DOG APPLICATIONS:
See Insecticides.

DOG BISCUIT.

The waste portions of meat and tallow, including the skin and fiber, have for years been imported from South American tallow factories in the form of blocks. Most of the dog bread consists principally of these remnants, chopped and mixed with flour. They contain a good deal of firm fibrous tissue, and a large percentage of fat, but are lacking in nutritive salts, which must be added to make good dog bread, just as in the case of the meat flour made from the waste of meat extract factories. The flesh of dead animals is not used by any reputable manufacturers, for the reason that it gives a dark color to the dough, has an unpleasant odor, and if not properly sterilized would be injurious to dogs as a steady diet.

Wheat flour, containing as little bran as possible, is generally used, oats, rye, or Indian meal being only mixed in to make special varieties, or, as in the case of Indian meal, for cheapness. Rye flour would give a good flavor, but it dries slowly, and the biscuits would have to go through a special process of drying after baking, else they would mold and spoil. Dog bread must be made from good wheat flour, of a medium sort, mixed with 15 or 16 per cent of sweet, dry chopped meat, well baked and dried like pilot bread or crackers. This is the rule for all the standard dog bread on the market. There are admixtures which affect more or less its nutritive value, such as salt, vegetables, chopped bones, or bone meal, phosphate of lime, and other nutritive salts. In preparing the dough and in baking, care must be taken to keep it light and porous.

DOG DISEASES AND THEIR REMEDIES:
See Veterinary Formulas.

DOG SOAP:
See Soap.

DONARITE:
See Explosives.

DOORS, TO CLEAN:
See Cleaning Preparations and Methods.

DOSES FOR ADULTS AND CHILDREN.

The usual method pursued by medical men in calculating the doses of medicine for children is to average the dose in proportion to their approximate weight or to figure out a dose upon the assumption that at 12 years of age half of an adult dose will be about right. Calculated on this basis the doses for those under 12 will be in direct proportion to the age in years plus 12, divided into the age. By this rule a child 1 year old should get 1 plus 12, or 13, dividing 1, or $\frac{1}{13}$ of an adult dose. If the child is 2 years old it should get 2 plus 12, or 14, dividing 2, or $\frac{1}{7}$ of an adult dose. A child of 3 years should get 3 plus 12, or 15, dividing 3, or $\frac{1}{5}$ of an adult dose. A child of 4 should get 4 plus 12, or 16, dividing 4, or $\frac{1}{4}$ of an adult dose.

As both children and adults vary materially in size when of the same age the calculation by approximate weights is the more accurate way. Taking the weight of the average adult as 150 pounds, then a boy, man, or woman, whatever the age, weighing only 75 pounds should receive only one-half of an adult dose, and a man of 300 pounds, provided his weight is the result of a properly proportioned body, and not due to mere adipose tissue, should be double that of the average adult. If the weight is due to mere fat or to some diseased condition of the body, such a calculation would be entirely wrong. The object of the calculation is to get as nearly as possible to the amount of dilution the dose undergoes in the blood or in the intestinal contents of the patient. Each volume of blood should receive exactly the same dose in order to give the same results, other conditions being equal.

DOSE TABLE FOR VETERINARY PURPOSES:

See Veterinary Formulas.

DRAWINGS, PRESERVATION OF.

Working designs and sketches are easily soiled and rendered unsuitable for further use. This can be easily avoided by coating them with collodion, to which 24 per cent of stearine from a good stearine candle has been added. Lay the drawing on a glass plate or a board, and pour on the collodion, as the photographer treats his plates. After 10 or 20 minutes the design will be dry and perfectly white, possessing a dull luster, and being so well protected that it may be washed off with water without fear of spoiling it.

DRAWINGS, TO CLEAN:

See Cleaning Preparations and Methods.

DRIERS:

See Siccatives.

DRILLING, LUBRICANT FOR:

See Lubricants.

DRINKS FOR SUMMER AND WINTER:

See Beverages.

DROPS, TABLE OF:

See Tables.

DRYING OILS:

See Oil.

DRY ROT:

See Rot.

DUBBING FOR LEATHER:

See Lubricants.

DUST-LAYING:

See Oil.

DUST PREVENTERS AND DUST CLOTHS:

See Household Formulas.

Dyes

In accordance with the requirements of dyers, many of the following recipes describe dyes for large quantities of goods, but to make them equally adapted for the use of private families they are usually given in even quantities, so that it is an easy matter to ascertain the quantity of materials required for dyeing, when once the weight of the goods is known, the quantity of materials used being reduced in proportion to the smaller quantity of goods.

Employ soft water for all dyeing purposes, if it can be procured, using 4 gallons water to 1 pound of goods; for larger quantities a little less water will do. Let all the implements used in dyeing be kept perfectly clean. Prepare the goods by scouring well with soap and water, washing out the soap well, and dipping in warm water, before immersion in the dye or mordant. Goods should be well aired, rinsed, and properly hung up after dyeing. Silks and fine goods should be tenderly handled, otherwise injury to the fabric will result.

Aniline Black.—Water, 20 to 30 parts; chlorate of potassa, 1 part; sal ammoniac, 1 part; chloride of copper, 1 part; aniline and hydrochloric acid, each 1 part, previously mixed together. It is essential that the preparation should be acid, and the more acid it is the more rapid will be the production of the blacks; if too much so, it may injure the fabric. The fabric or yarn is dried in ageing rooms at a low temperature for 24 hours, and washed afterwards.

Black on Cotton.—For 40 pounds goods, use sumac, 30 pounds; boil ¾ of an hour; let the goods steep overnight, and immerse them in limewater, 40 minutes, remove, and allow them to drip ¾ of an hour; add copperas, 4 pounds, to the sumac liquor, and dip 1 hour more; next work them through limewater for 20 minutes; then make a new dye of logwood, 20 pounds, boil 2½ hours, and enter the goods 3 hours; then add bichromate of potash, 1 pound, to the new dye, and dip 1 hour more. Work in clean cold water and dry out of the sun.

Black Straw Hat Varnish.—Best alcohol, 4 ounces; pulverized black sealing wax, 1 ounce. Place in a phial, and put the phial into a warm place, stirring or shaking occasionally until the wax is dissolved. Apply it when warm before the fire or in the sun. This makes a beautiful gloss.

Chrome Black for Wool. — For 40 pounds of goods, use blue vitriol, 3 pounds; boil a short time, then dip the wool or fabric ¾ of an hour, airing frequently. Take out the goods, and make a dye with logwood, 24 pounds; boil ½ hour, dip ¾ of an hour, air the goods, and dip ¼ of an hour longer; then wash in strong soapsuds. A good fast color.

Black Dye on Wool, for Mixtures. — For 50 pounds of wool, take bichromate of potash, 1 pound, 4 ounces; ground argal, 15 ounces; boil together and put in the fabric, stirring well, and let it remain in the dye 5 hours. Take it out, rinse slightly in clean water, then make a new dye, into which put logwood, 1½ pounds. Boil 1¼ hours, adding chamber lye, 5 pints. Let the fabric remain in all night, and wash out in clean water.

Bismarck Brown. — Mix together 1 pound Bismarck, 5 gallons water, and ¼ pound sulphuric acid. This paste dissolves easily in hot water and may be used directly for dyeing. A liquid dye may be prepared by making the bulk of the above mixture to 2 gallons with alcohol. To dye, sour with sulphuric acid; add a quantity of sulphate of soda, immerse the wool, and add the color by small portions, keeping the temperature under 212° F. Very interesting shades may be developed by combining the color with indigo paste or picric acid.

Chestnut Brown for Straw Bonnets. — For 25 hats, use ground sanders, 1½ pounds; ground curcuma, 2 pounds; powdered gallnuts or sumac, ¾ pound; rasped logwood, ⅒ pound. Boil together with the hats in a large kettle (so as not to crowd), for 2 hours, then withdraw the hats, rinse, and let them remain overnight in a bath of nitrate of 4° Bé., when they are washed. A darker brown may be obtained by increasing the quantity of sanders. To give the hats the desired luster, they are brushed with a brush of couchgrass, when dry.

Cinnamon or Brown for Cotton and Silk. — Give the goods as much color, from a solution of blue vitriol, 2 ounces, to water, 1 gallon, as they will take up in dipping 15 minutes; then turn them through limewater. This will make a beautiful sky blue of much durability. The fabric should next be run through a solution of prussiate of potash, 1 ounce, to water, 1 gallon.

Brown Dye for Cotton or Linen. — Give the pieces a mixed mordant of acetate of alumina and acetate of iron, and then dye them in a bath of madder, or madder and fustic. When the acetate of alumina predominates, the dye has an amaranth tint. A cinnamon tint is obtained by first giving a mordant of alum, next a madder bath, then a bath of fustic, to which a little green copperas has been added.

Brown for Silk. — Dissolve annatto, 1 pound; pearlash, 4 pounds, in boiling water, and pass the silk through it for 2 hours; then take it out, squeeze well, and dry. Next give it a mordant of alum, and pass through a bath of brazil wood, and afterwards through a bath of logwood, to which a little green copperas has been added; wring it out and dry; afterwards rinse well.

Brown Dye for Wool. — This may be induced by a decoction of oak bark, with variety of shade according to the quantity employed. If the goods be first passed through a mordant of alum the color will be brightened.

Brown for Cotton. — Catechu or terra japonica gives cotton a brown color; blue vitriol turns it to the bronze; green copperas darkens it, when applied as a mordant and the stuff is boiled in the bath. Acetate of alumina as a mordant brightens it. The French color Carmelite is given with catechu, 1 pound; verdigris, 4 ounces; and sal ammoniac, 5 ounces.

Dark Snuff Brown for Wool. — For 50 pounds of goods, take camwood, 10 pounds, boil for 20 minutes, then dip the goods for ¾ of an hour; take them out, and add to the dye, fustic, 25 pounds, boil 12 minutes, and dip the goods ¾ of an hour; then add blue vitriol, 10 ounces, copperas, 2 pounds, 8 ounces; dip again 40 minutes. Add more copperas if the shade is required darker.

Brown for Wool and Silk. — Infusion or decoction of walnut peels dyes wool and silk a brown color, which is brightened by alum. Horse-chestnut peels also impart a brown color; a mordant of muriate of tin turns it on the bronze, and sugar of lead the reddish brown.

Alkali Blue and Nicholson's Blue. — Dissolve 1 pound of the dye in 10 gallons boiling water, and add this by small portions to the dye bath, which should be rendered alkaline by borax. The fabric should be well worked about between each addition of the color. The temperature must be kept under 212° F. To develop the color, wash with water and pass through a bath containing sulphuric acid.

Aniline Blue.—To 100 pounds of fabric, dissolve 1¼ pounds aniline blue in 3 quarts hot alcohol, strain through a filter, and add it to a bath of 130° F.; also 10 pounds Glauber's salts, and 5 pounds acetic acid. Immerse the goods and handle them well for 20 minutes. Next heat slowly to 200° F.; then add 5 pounds sulphuric acid diluted with water. Let the whole boil 20 minutes longer; then rinse and dry. If the aniline be added in 2 or 3 proportions during the process of coloring, it will facilitate the evenness of the color.

Blue on Cotton.—For 40 pounds of goods, use copperas, 2 pounds; boil and dip 20 minutes; dip in soapsuds, and return to the dye 3 or 4 times; then make a new bath with prussiate of potash, ½ pound; oil of vitriol, 1¼ pints; boil ½ hour, rinse out and dry.

Sky Blue on Cotton.—For 60 pounds of goods, blue vitriol, 5 pounds. Boil a short time, then enter the goods, dip 3 hours, and transfer to a bath of strong limewater. A fine brown color will be imparted to the goods if they are then put through a solution of prussiate of potash.

Blue Dye for Hosiery.—One hundred pounds of wool are colored with 4 pounds Guatemala or 3 pounds Bengal indigo, in the soda or wood vat. Then boil in a kettle a few minutes, 5 pounds of cudbear or 8 pounds of archil paste; add 1 pound of soda, or, better, 1 pail of urine; then cool the dye to about 170° F. and enter the wool. Handle well for about 20 minutes, then take it out, cool, rinse, and dry. It makes no difference whether the cudbear is put in before or after the indigo. Three ounces of aniline purple dissolved in alcohol, ½ pint, can be used instead of the cudbear. Wood spirit is cheaper than alcohol, and is much used by dyers for the purpose of dissolving aniline colors. It produces a very pretty shade, but should never be used on mixed goods which have to be bleached.

Dark-Blue Dye.—This dye is suitable for thibets and lastings. Boil 100 pounds of the fabric for 1½ hours in a solution of alum, 25 pounds; tartar, 4 pounds; mordant, 6 pounds; extract of indigo, 6 pounds; cool as usual. Boil in fresh water from 8 to 10 pounds of logwood, in a bag or otherwise, then cool the dye to 170° F. Reel the fabric quickly at first, then let it boil strongly for 1 hour. This is a very good imitation of indigo blue.

Saxon Blue.—For 100 pounds thibet or comb yarn, use alum, 20 pounds; cream of tartar, 3 pounds; mordant, 2 pounds; extract of indigo, 3 pounds; or carmine, 1 pound, makes a better color. When all is dissolved, cool the kettle to 180° F.; enter and handle quickly at first, then let the fabric boil ½ hour, or until even. Long boiling dims the color. Zephyr worsted yarn ought to be prepared, first, by boiling it in a solution of alum and sulphuric acid; the indigo is added afterwards.

Logwood and Indigo Blue.—For 100 pounds of cloth. Color the cloth first by one or two dips in the vat of indigo blue, and rinse it well, and then boil it in a solution of 20 pounds of alum, 2 pounds of half-refined tartar, and 5 pounds of mordant, for 2 hours; finally take it out and cool. In fresh water boil 10 pounds of good logwood for half an hour in a bag or otherwise; cool off to 170° F. before entering. Handle well over a reel, let it boil for half an hour; then take it out, cool and rinse. This is a very firm blue.

Blue Purple for Silk.—For 40 pounds of goods, take bichromate of potash, 8 ounces; alum, 1 pound; dissolve all and bring the water to a boil, and put in the goods; boil 1 hour. Then empty the dye, and make a new dye with logwood, 8 pounds, or extract of logwood, 1 pound 4 ounces, and boil in this 1 hour longer. Grade the color by using more or less logwood, as dark or light color is wanted.

Blue Purple for Wool.—One hundred pounds of wool are first dipped in the blue vat to a light shade, then boiled in a solution of 15 pounds of alum and 3 pounds of half-refined tartar, for 1½ hours, the wool taken out, cooled, and let stand 24 hours. Then boil in fresh water 8 pounds of powdered cochineal for a few minutes, cool the kettle to 170° F. Handle the prepared wool in this for 1 hour, when it is ready to cool, rinse and dry. By coloring first with cochineal, as aforesaid, and finishing in the blue vat, the fast purple or dahlia, so much admired in German broadcloths, will be produced. Tin acids must not be used in this color.

To Make Extract of Indigo Blue.—Take of vitriol, 2 pounds, and stir into it finely pulverized indigo, 8 ounces, stirring briskly for the first half hour; then cover up, and stir 4 or 5 times daily for a few days. Add a little pulverized chalk, stirring it up, and keep adding it as long as it foams; it will neutralize the acid. Keep it closely corked.

Light Silver Drab.—For 50 pounds of goods, use logwood, $\frac{1}{2}$ pound; alum, about the same quantity; boil well, enter the goods, and dip them for 1 hour. Grade the color to any desired shade by using equal parts of logwood and alum.

GRAY DYES:

Slate Dye for Silk.—For a small quantity, take a pan of warm water and about a teacupful of logwood liquor, pretty strong, and a piece of pearlash the size of a nut; take gray-colored goods and handle a little in this liquid, and it is finished. If too much logwood is used, the color will be too dark.

Slate for Straw Hats.—First, soak in rather strong warm suds for 15 minutes to remove sizing or stiffening; then rinse in warm water to get out the soap. Scald cudbear, 1 ounce, in sufficient water to cover the hat; work it in this dye at 180° F., until a light purple is obtained. Have a vessel of cold water, blued with the extract of indigo, $\frac{1}{2}$ ounce, and work or stir the bonnet in this until the tint pleases. Dry, then rinse out with cold water, and dry again in the shade. If the purple is too deep in shade the final slate will be too dark.

Silver Gray for Straw.—For 25 hats, select the whitest hats and soften them in a bath of crystallized soda to which some clean limewater has been added. Boil for 2 hours in a large vessel, using for a bath a decoction of the following: Alum, 4 pounds; tartaric acid, $\frac{3}{4}$ pound; some ammoniacal cochineal, and carmine of indigo. A little sulphuric acid may be necessary in order to neutralize the alkali of the cochineal dye. If the last-mentioned ingredients are used, let the hats remain for an hour longer in the boiling bath, then rinse in slightly acidulated water.

Dark Steel.—Mix black and white wool together in the proportion of 50 pounds of black wool to 7$\frac{1}{2}$ pounds of white. For large or small quantities, keep the same proportion, mixing carefully and thoroughly.

GREEN DYES:

Aniline Green for Silk.—Iodine green or night green dissolves easily in warm water. For a liquid dye 1 pound may be dissolved in 1 gallon alcobol, and mixed with 2 gallons water, containing 1 ounce sulphuric acid.

Aniline Green for Wool.—Prepare two baths, one containing the dissolved dye and a quantity of carbonate of soda or borax. In this the wool is placed, and the temperature raised to 212° F. A grayish green is produced, which must be brightened and fixed in a second bath of water 100° F., to which some acetic acid has been added. Cotton requires preparation by sumac.

Green for Cotton.—For 40 pounds of goods, use fustic, 10 pounds; blue vitriol, 10 ounces; soft soap, 2$\frac{1}{2}$ quarts; and logwood chips, 1 pound 4 ounces. Soak the logwood overnight in a brass vessel, and put it on the fire in the morning, adding the other ingredients. When quite hot it is ready for dyeing; enter the goods at once, and handle well. Different shades may be obtained by letting part of the goods remain longer in the dye.

Green for Silk.—Boil green ebony in water, and let it settle. Take the clear liquor as hot as the hands can bear, and handle the goods in it until of a bright yellow. Take water and put in a little sulphate of indigo; handle goods in this till of the shade desired. The ebony may previously be boiled in a bag to prevent it from sticking to the silk.

Green for Wool and Silk.—Take equal quantities of yellow oak and hickory bark, make a strong yellow bath by boiling, and shade to the desired tint by adding a small quantity of extract of indigo.

Green Fustic Dye.—For 50 pounds of goods, use 50 pounds of fustic with alum, 11 pounds. Soak in water until the strength is extracted, put in the goods until of a good yellow color, remove the chips, and add extract of indigo in small quantities at a time, until the color is satisfactory.

PURPLE AND VIOLET DYES:

Aniline Violet and Purple.—Acidulate the bath by sulphuric acid, or use sulphate of soda; both these substances render the shade bluish. Dye at 212° F. To give a fair middle shade to 10 pounds of wool, a quantity of solution equal to $\frac{1}{2}$ to $\frac{3}{4}$ ounces of the solid dye will be required. The color of the dyed fabric is improved by washing in soap and water, and then passing through a bath soured by sulphuric acid.

Purple.—For 40 pounds of goods, use

alum, 3 pounds; muriate of tin, 4 tea-cups; pulverized cochineal, 1 pound; cream of tartar, 2 pounds. Boil the alum, tin, and cream of tartar, for 20 minutes, add the cochineal and boil 5 minutes; immerse the goods 2 hours; remove and enter them in a new dye composed of brazil wood, 3 pounds; logwood, 7 pounds; alum, 4 pounds, and muriate of tin, 8 cupfuls, adding a little extract of indigo.

Purple for Cotton.—Get up a tub of hot logwood liquor, enter 3 pieces, give them 5 ends, and hedge out. Enter them in a clean alum tub, give them 5 ends, and hedge out. Get up another tub of logwood liquor, enter, give them 5 ends, and hedge out; renew the alum tub, give 5 ends in that, and finish.

Purple for Silk.—For 10 pounds of goods, enter the goods in a blue dye bath, and secure a light-blue color, dry, and dip in a warm solution containing alum, 2½ pounds. Should a deeper color be required, add a little extract of indigo.

Solferino and Magenta for Woolen, Silk, or Cotton.—For 1 pound of woolen goods, magenta shade, 96 grains, apothecaries' weight, of aniline red, will be required. Dissolve in a little warm alcohol, using, say, 6 fluidounces, or about 6 gills alcohol per ounce of aniline. Many dyers use wood spirits because of its cheapness. For a solferino shade, use 64 grains aniline red, and dissolve in 4 ounces alcohol, to each 1 pound of goods. Cold water, 1 quart, will dissolve these small quantities of aniline red, but the cleanest and quickest way will be found by using the alcohol, or wood spirits. Clean the cloth and goods by steeping at a gentle heat in weak soapsuds, rinse in several masses of clean water and lay aside moist. The alcoholic solution of aniline is to be added from time to time to the warm or hot dye bath, till the color on the goods is of the desired shade. The goods are to be removed from the dye bath before each addition of the alcoholic solution, and the bath is to be well stirred before the goods are returned. The alcoholic solution should be first dropped into a little water, and well mixed, and the mixture should then be strained into the dye bath. If the color is not dark enough after working from 20 to 30 minutes, repeat the removal of the goods from the bath, and the addition of the solution, and the re-immersion of the goods from 15 to 30 minutes more, or until suited, then remove from the bath and rinse in several masses of clean water, and dry in the shade. Use about 4 gallons water for dye bath for 1 pound of goods; less water for larger quantities.

Violet for Silk or Wool.—A good violet dye may be given by passing the goods first through a solution of verdigris, then through a decoction of logwood, and lastly through alum water. A fast violet may be given by dyeing the goods crimson with cochineal, without alum or tartar, and after rinsing passing them through the indigo vat. Linens or cottons are first galled with 18 per cent of gallnuts, next passed through a mordant of alum, iron liquor, and sulphate of copper, working them well, then worked in a madder bath made with an equal weight of root, and lastly brightened with soap or soda.

Violet for Straw Bonnets.—Take alum, 4 pounds; tartaric acid, 1 pound; chloride of tin, 1 pound. Dissolve and boil, allowing the hats to remain in the boiling solution 2 hours; then add enough decoction of logwood, carmine, and indigo to induce the desired shade, and rinse finally in water in which some alum has been dissolved.

Wine Color.—For 50 pounds of goods, use camwood, 10 pounds, and boil 20 minutes; dip the goods ½ hour, boil again, and dip 40 minutes; then darken with blue vitriol, 15 ounces, and 5 pounds of copperas.

Lilac for Silk.—For 5 pounds of silk, use archil, 7½ pounds, and mix well with the liquor. Make it boil ¼ hour, and dip the silk quickly; then let it cool, and wash in river water. A fine half violet, or lilac, more or less full, will be obtained.

RED, CRIMSON, AND PINK DYES:

Aniline Red.—Inclose the aniline in a small muslin bag. Have a kettle (tin or brass) filled with moderately hot water and rub the substance out. Then immerse the goods to be colored, and in a short time they are done. It improves the color to wring the goods out of strong soapsuds before putting them in the dye. This is a permanent color on wool or silk.

Red Madder.—To 100 pounds of fabric, use 20 pounds of alum, 5 pounds of tartar, and 5 pounds of muriate of tin. When these are dissolved, enter the goods and let them boil for 2 hours, then take out, let cool, and lay overnight. Into fresh water, stir 75 pounds of good madder, and enter the fabric at 120° F. and bring it up to 200° F. in the course of an hour. Handle well to secure evenness, then rinse and dry.

Red for Wool.—For 40 pounds of goods, make a tolerably thick paste of lac dye and sulphuric acid, and allow it to stand for a day. Then take tartar, 4 pounds, tin liquor, 2 pounds 8 ounces, and 3 pounds of the paste; make a hot bath with sufficient water, and enter the goods for ¾ hour; afterwards carefully rinse and dry.

Crimson for Silk.—For 1 pound of goods, use alum, 3 ounces; dip at hand heat 1 hour; take out and drain, while making a new dye, by boiling for 10 minutes, cochineal, 3 ounces; bruised nutgalls, 2 ounces; and cream of tartar, ¼ ounce, in 1 pail of water. When a little cool begin to dip, raising the heat to a boil, continuing to dip 1 hour. Wash and dry.

Aniline Scarlet.—For every 40 pounds of goods, dissolve 5 pounds white vitriol (sulphate of zinc) at 180° F., place the goods in this bath for 10 minutes, then add the color, prepared by boiling for a few minutes, 1 pound aniline scarlet in 3 gallons water, stirring the same continually. This solution has to be filtered before being added to the bath. The goods remain in the latter for 15 minutes, when they have become browned and must be boiled for another half hour in the same bath after the solution of sal ammoniac. The more of this is added the deeper will be the shade.

Scarlet with Cochineal.—For 50 pounds of wool, yarn, or cloth, use cream of tartar, 1 pound 9 ounces; cochineal, pulverized, 12½ ounces; muriate of tin or scarlet spirit, 8 pounds. After boiling the dye, enter the goods, work them well for 15 minutes, then boil them 1½ hours, slowly agitating the goods while boiling, wash in clean water, and dry out of the sun.

Scarlet with Lac Dye.—For 100 pounds of flannel or yarn, take 25 pounds of ground lac dye, 15 pounds of scarlet spirit (made as per directions below), 5 pounds of tartar, 1 pound of flavine, or according to shade, 1 pound of tin crystals, 5 pounds of muriatic acid. Boil all for 15 minutes, then cool the dye to 170° F. Enter the goods, and handle them quickly at first. Let boil 1 hour, and rinse while yet hot, before the gum and impurities harden. This color stands scouring with soap better than

cochineal scarlet. A small quantity of sulphuric acid may be added to dissolve the gum.

Muriate of Tin or Scarlet Spirit.—Take 16 pounds muriatic acid, 22° Bé.; 1 pound feathered tin, and water, 2 pounds. The acid should be put in a stoneware pot, and the tin added, and allowed to dissolve. The mixture should be kept a few days before using. The tin is feathered or granulated by melting in a suitable vessel, and pouring it from a height of about 5 feet into a pailful of water. This is a most powerful agent in certain colors, such as scarlets, oranges, pinks, etc.

Pink for Cotton.—For 40 pounds of goods, use redwood, 20 pounds; muriate of tin, 2½ pounds. Boil the redwood 1 hour, turn off into a large vessel, add the muriate of tin, and put in the goods. Let it stand 5 or 10 minutes, and a good fast pink will be produced.

Pink for Wool.—For 60 pounds of goods, take alum, 5 pounds 12 ounces; boil and immerse the goods 50 minutes; then add to the dye cochineal well pulverized, 1 pound, 4 ounces; cream of tartar, 5 pounds; boil and enter the goods while boiling, until the color is satisfactory.

YELLOW, ORANGE, AND BRONZE DYES:

Aniline Yellow.—This color is slightly soluble in water, and for dyers' use may be used directly for the preparation of the bath dye, but is best used by dissolving 1 pound of dye in 2 gallons alcohol. Temperature of bath should be under 200° F. The color is much improved and brightened by a trace of sulphuric acid.

Yellow for Cotton.—For 40 pounds goods, use sugar of lead, 3 pounds 8 ounces; dip the goods 2 hours. Make a new dye with bichromate of potash, 2 pounds; dip until the color suits, wring out and dry. If not yellow enough repeat the operation.

Yellow for Silk.—For 10 pounds of goods, use sugar of lead, 7½ ounces; alum, 2 pounds. Enter the goods, and let them remain 12 hours; remove them, drain, and make a new dye with fustic, 10 pounds. Immerse until the color suits.

Orange.—I.—For 50 pounds of goods, use argal, 3 pounds; muriate of tin, 1 quart; boil and dip 1 hour; then add to the dye, fustic, 25 pounds; madder, 2½ quarts; and dip again 40 minutes. If preferred, cochineal, 1 pound 4 ounces, may be used instead of the madder, as a better color is induced by it.

II.—For 40 pounds of goods, use sugar of lead, 2 pounds, and boil 15 minutes. When a little cool, enter the goods, and dip for 2 hours, wring them out, make a fresh dye with bichromate of potash, 4 pounds; madder, 1 pound, and immerse until the desired color is secured. The shade may be varied by dipping in limewater.

Bronze.—Sulphate or muriate of manganese dissolved in water with a little tartaric acid imparts a beautiful bronze tint. The stuff after being put through the solution must be turned through a weak lye of potash, and afterwards through another of chloride of lime, to brighten and fix it.

Prussiate of copper gives a bronze or yellowish-brown color to silk. The piece well mordanted with blue vitriol may be passed through a solution of prussiate of potash.

Mulberry for Silk.—For 5 pounds of silk, use alum, 1 pound 4 ounces; dip 50 minutes, wash out, and make a dye with brazil wood, 5 ounces, and logwood, $1\frac{1}{4}$ ounces, by boiling together. Dip in this $\frac{1}{2}$ hour; then add more brazil wood and logwood, equal parts, until the color suits.

FEATHER DYES.

I.—Cut some white curd soap in small pieces, pour boiling water on them, and add a little pearlash. When the soap is quite dissolved, and the mixture cool enough for the hand to bear, plunge the feathers into it, and draw them through the hand till the dirt appears squeezed out of them; pass them through a clean lather with some blue in it; then rinse them in cold water with blue to give them a good color. Beat them against the hand to shake off the water, and dry by shaking them near a fire. When perfectly dry, coil each fiber separately with a blunt knife or ivory folder.

II.—Black.—Immerse for 2 or 3 days in a bath, at first hot, of logwood, 8 parts, and copperas or acetate of iron, 1 part.

III.—Blue.—Same as II, but with the indigo vat.

IV.—Brown.—By using any of the brown dyes for silk or woolen.

V.—Crimson.—A mordant of alum, followed by a hot bath of brazil wood, afterwards by a weak dye of cudbear.

VI.—Pink or Rose.—With safflower or lemon juice.

VII.—Plum.—With the red dye, followed by an alkaline bath.

VIII.—Red.—A mordant of alum, followed by a bath of brazil wood.

IX.—Yellow.—A mordant of alum, followed by a bath of turmeric or weld.

X.—Green.—Take of verdigris and verditer, of each 1 ounce; gum water, 1 pint; mix them well and dip the feathers, they having been first soaked in hot water, into the said mixture.

XI.—Purple.—Use lake and indigo.

XII.—Carnation.—Vermilion and smalt.

DYES FOR ARTIFICIAL FLOWERS.

The French employ velvet, fine cambric, and kid for the petals, and taffeta for the leaves. Very recently thin plates of bleached whalebone have been used for some portions of the artificial flowers.

Colors and Stains.—I.—Blue.—Indigo dissolved in oil of vitriol, and the acid partly neutralized with salt of tartar or whiting.

II.—Green.—A solution of distilled verdigris.

III.—Lilac.—Liquid archil.

IV.—Red.—Carmine dissolved in a solution of salt of tartar, or in spirts of hartshorn.

V.—Violet.—Liquid archil mixed with a little salt of tartar.

VI.—Yellow.—Tincture of turmeric. The colors are generally applied with the fingers.

DYES FOR FURS:

I.—Brown.—Use tincture of logwood.

II.—Red.—Use ground brazil wood, $\frac{1}{2}$ pound; water, $1\frac{1}{2}$ quarts; cochineal, $\frac{1}{2}$ ounce; boil the brazil wood in the water 1 hour; strain and add the cochineal; boil 15 minutes.

III.—Scarlet.—Boil $\frac{1}{2}$ ounce saffron in $\frac{1}{2}$ pint of water, and pass over the work before applying the red.

IV.—Blue.—Use logwood, 7 ounces; blue vitriol, 1 ounce; water, 22 ounces; boil.

V.—Purple.—Use logwood, 11 ounces; alum, 6 ounces; water, 29 ounces.

VI.—Green.—Use strong vinegar, $1\frac{1}{2}$ pints; best verdigris, 2 ounces, ground fine; sap green, $\frac{1}{4}$ ounce; mix all together and boil.

DYES FOR HATS.

The hats should be at first strongly galled by boiling a long time in a decoction of galls with a little logwood so that the dye may penetrate into their substance; after which a proper quantity of vitriol and decoction of logwood, with a little verdigris, are added, and the hats kept in this mixture for a considerable time. They are afterwards put into a fresh liquor of logwood, galls, vitriol, and verdigris, and, when the hats are costly, or of a hair which with difficulty takes the dye, the same process is repeated a third time. For obtaining the most perfect color, the hair or wool is dyed blue before it is formed into hats.

The ordinary bath for dyeing hats, employed by London manufacturers, consists, for 12 dozen, of 144 pounds of logwood; 12 pounds of green sulphate of iron or copperas; 7½ pounds verdigris. The logwood having been introduced into the copper and digested for some time, the copperas and verdigris are added in successive quantities, and in the above proportions, along with every successive 2 or 3 dozen of hats suspended upon the dripping machine. Each set of hats, after being exposed to the bath with occasional airings during 40 minutes, is taken off the pegs, and laid out upon the ground to be more completely blackened by the peroxydizement of the iron with the atmospheric oxygen. In 3 or 4 hours the dyeing is completed. When fully dyed, the hats are well washed in running water.

Straw hats or bonnets may be dyed black by boiling them 3 or 4 hours in a strong liquor of logwood, adding a little copperas occasionally. Let the bonnets remain in the liquor all night; then take out to dry in the air. If the black is not satisfactory, dye again after drying. Rub inside and out with a sponge moistened in fine oil; then block.

I.—Red Dye.—Boil ground brazil wood in a lye of potash, and boil your straw hats in it.

II.—Blue Dye.—Take a sufficient quantity of potash lye, 1 pound of litmus or lacmus, ground; make a decoction and then put in the straw, and boil it.

TO DYE, STIFFEN, AND BLEACH FELT HATS.

Felt hats are dyed by repeated immersion, drawing and dipping in a hot watery solution of logwood, 38 parts; green vitriol, 3 parts; verdigris, 2 parts; repeat the immersions and drawing with exposure to the air 13 or 14 times, or until the color suits, each step in the process lasting from 10 to 15 minutes. Aniline colors may be advantageously used instead of the above. For a stiffening, dissolve borax, 10 parts; carbonate of potash, 3 parts, in hot water; then add shellac, 50 parts, and boil until all is dissolved; apply with a sponge or a brush, or by immersing the hat when it is cold, and dip at once in very dilute sulphuric or acetic acid to neutralize the alkali and fix the shellac. Felt hats can be bleached by the use of sulphuric acid gas.

LIQUID DYE COLORS.

These colors, thickened with a little gum, may be used as inks in writing, or as colors to tint maps, foils, artificial flowers, etc., or to paint on velvet:

I.—Blue.—Dilute Saxon blue or sulphate of indigo with water. If required for delicate work, neutralize with chalk.

II.—Purple.—Add a little alum to a strained decoction of logwood.

III.—Green.—Dissolve sap green in water and add a little alum.

IV.—Yellow.—Dissolve annatto in a weak lye of subcarbonate of soda or potash.

V.—Golden Color. — Steep French berries in hot water, strain, and add a little gum and alum.

VI.—Red.—Dissolve carmine in ammonia, or in weak carbonate of potash water, or infuse powdered cochineal in water, strain, and add a little gum in water.

UNCLASSIFIED DYERS' RECIPES:

To Cleanse Wool.—Make a hot bath composed of water, 4 parts; and urine, 1 part; enter the wool, teasing and opening it out to admit the full action of the liquid. After 20 minutes' immersion, remove from the liquid and allow it to drain; then rinse in clean running water, and spread out to dry. The liquid is good for subsequent operations, only keep up the proportions, and use no soap.

To Extract Oil Spots from Finished Goods.—Saturate the spot with benzine; then place two pieces of very soft blotting paper under and two upon it, press well with a hot iron, and the grease will be absorbed.

New Mordant for Aniline Colors.—Immerse the goods for some hours in a bath of cold water in which chloride or acetate of zinc has been dissolved until the solution shows 2° Bé. For the wool the mordanting bath should be at a boiling heat, and the goods should also be placed in a warm bath of tannin, 90° F., for half an hour. In dyeing, a hot solution of the color must be used to which should be added, in the case of the cotton, some chloride of zinc, and, in the case of the wool, a certain amount of tannin solution.

To Render Aniline Colors Soluble in Water.—A solution of gelatin in acetic acid of almost the consistence of syrups is first made, and the aniline in fine is gradually added, stirring all the time so as to make a homogeneous paste. The mixture is then to be heated over a water bath to the temperature of boiling water and kept at that heat for some time.

Limewater for Dyers' Use.—Put some lime, 1 pound, and strong limewater, 1½ pounds, into a pail of water; rummage well for 7 or 8 minutes. Then let it rest until the lime is precipitated and the water clear; add this quantity to a tubful of clear water.

To Renew Old Silks.—Unravel and put them in a tub, cover with cold water, and let them remain 1 hour. Dip them up and down, but do not wring; hang up to drain, and iron while very damp.

Fuller's Purifier for Cloths.—Dry, pulverize, and sift the following ingredients: Fuller's earth, 6 pounds; French chalk, 4 ounces; pipe clay, 1 pound. Make into a paste with rectified oil of turpentine, 1 ounce; alcohol, 2 ounces; melted oil soap, 1½ pounds. Compound the mixture into cakes of any desired size, keeping them in water, or small wooden boxes.

To Fix Dyes.—Dissolve 20 ounces of gelatin in water, and add 3 ounces of bichromate of potash. This is done in a dark room. The coloring matter is then added and the goods submitted thereto, after which they are exposed to the action of light. The pigment thus becomes insoluble in water and the color is fast.

DYES AND DYESTUFFS.

Prominent among natural dyestuffs is the coloring matter obtained from logwood and known as "hæmatein." The color-forming substance (or chromogen), hæmatoxylin, exists in the logwood partly free and partly as a glucoside. When pure, hæmatoxylin forms nearly colorless crystals, but on oxidation, especially in the presence of an alkali, it is converted into the coloring matter hæmatein, which forms colored lakes with metallic bases, yielding violets,

blues, and blacks with various mordants. Logwood comes into commerce in the form of logs, chips, and extracts. The chips are moistened with water and exposed in heaps so as to induce fermentation, alkalies and oxidizing agents being added to promote the "curing" or oxidation. When complete and the chips have assumed a deep reddish-brown color, the decoction is made which is employed in dyeing. The extract offers convenience in transportation, storage, and use. It is now usually made from logwood chips that have not been cured. The chips are treated in an extractor, pressure often being used. The extract is sometimes adulterated with chestnut, hemlock, and quercitron extracts, and with glucose or molasses.

Fustic is the heart-wood of certain species of trees indigenous to the West Indies and tropical South America. It is sold as chips and extract, yields a coloring principle which forms lemon-yellow lakes with alumina and is chiefly used in dyeing wool. Young fustic is the heart-wood of a sumac native to the shores of the Mediterranean, which yields an orange-colored lake with alumina and tin salts.

Cutch, or catechu, is obtained from the wood and pods of the *Acacia catechu*, and from the betel nut, both native in India. Cutch appears in commerce in dark-brown lumps, which form a dark-brown solution with water. It contains catechu-tannic acid, as tannin and catechin, and is extensively used in weighting black silks, as a mordant for certain basic coal-tar dyes, as a brown dye on cotton, and for calico printing.

Indigo, which is obtained from the glucoside indican existing in the indigo plant and in woad, is one of the oldest dyestuffs. It is obtained from the plant by a process of fermentation and oxidation. Indigo appears in commerce in dark-blue cubical cakes, varying very much in composition as they often contain indigo red and indigo brown, besides moisture, mineral matters, and glutinous substances. Consequently the color varies. Powdered indigo dissolves in concentrated fuming sulphuric acid, forming monosulphonic and disulphonic acids. On neutralizing these solutions with sodium carbonate and precipitating the indigo carmine with common salt there is obtained the indigo extract, soluble indigo, and indigo carmine of commerce. True indigo carmine is the sodium salt of the disulphonic acid, and when sold dry it is called "indigotine." One of the most important of the recent achievements of chemistry is the synthetic production of indigo on a commercial scale.

Artificial dyestuffs assumed preponderating importance with the discovery of the lilac color mauve by Perkin in 1856, and fuchsine or magenta by Verguin in 1895, for with each succeeding year other colors have been discovered, until at the present time there are several thousand artificial organic dyes or colors on the market. Since the first of these were prepared from aniline or its derivatives the colors were known as "aniline dyes," but as a large number are now prepared from other constituents of coal tar than aniline they are better called "coal-tar dyestuffs." There are many schemes of classification. Benedikt-Knecht divides them into I, aniline or amine dyes; II, phenol dyes; III, azo dyes; IV, quinoline and acridine derivatives; V, anthracene dyes; and VI, artificial indigo.

Of the anthracene dyes, the alizarine is the most important, since this is the coloring principle of the madder. The synthesis of alizarine from anthracene was effected by Gräbe and Liebermann in 1868. This discovery produced a complete revolution in calico printing, turkey-red dyeing, and in the manufacture of madder preparations. Madder finds to-day only a very limited application in the dyeing of wool.

In textile dyeing and printing, substances called mordants are largely used, either to fix or to develop the color on the fiber. Substances of mineral origin, such as salts of aluminum, chromium, iron, copper, antimony, and tin, principally, and many others to a less extent and of organic origin, like acetic, oxalic, citric, tartaric, and lactic acid, sulphonated oils, and tannins are employed as mordants.

Iron liquor, known as black liquor or pyrolignite of iron, is made by dissolving scrap iron in pyroligneous acid. It is used as a mordant in dyeing silks and cotton and in calico printing.

Red liquor is a solution of aluminum acetate in acetic acid, and is produced by acting on calcium or lead acetate solutions with aluminum sulphate or the double alums, the supernatant liquid forming the red liquor. The red liquor of the trade is often the sulpho-acetate of alumina resulting when the quantity of calcium or lead acetate is insufficient to completely decompose the aluminum salt. Ordinarily the solutions have a dark-brown color and a strong pyroligneous odor. It is called red liquor because it was first used in dyeing reds.

It is employed as a mordant by the cotton dyer and largely by the printer.

Non-Poisonous Textile and Egg Dyes for Household Use.—The preparation of non-poisonous colors for dyeing fabrics and eggs at home constitutes a separate department in the manufacture of dyestuffs.

Certain classes of aniline dyes may be properly said to form the materials. The essence of this color preparation consists chiefly in diluting or weakening the coal-tar dyes, made in the aniline factories, and bringing them down to a certain desired shade by the addition of certain chemicals suited to their varying characteristics, which, though weakening the color, act at the same time as the so-called mordants.

The anilines are divided with reference to their characteristic reactions into groups of basic, acid, moderately acid, as well as dyes that are insoluble in water.

In cases where combinations of one or more colors are needed, only dyes of similar reaction can be combined, that is, basic with basic, and acid with acid.

For the purpose of reducing the original intensity of the colors, and also as mordants, dextrin, Glauber's salt, alum, or aluminum sulphate is pressed into service. Where Glauber's salt is used, the neutral salt is exclusively employed, which can be had cheaply and in immense quantities in the chemical industry. Since it is customary to pack the color mixtures in two paper boxes, one stuck into the other, and moreover since certain coal-tar dyes are only used in large crystals, it is only reasonable that the mordants should be calcined and not put up in the shape of crystallized salts, particularly since these latter are prone to absorb the moisture from the air, and when thus wet likely to form a compact mass very difficult to dissolve. This inconvenience often occurs with the large crystals of fuchsine and methyl violet. Because these two colors are mostly used in combination with dextrin to color eggs, and since dextrin is also very hygroscopic, it is better in these individual cases to employ calcined Glauber's salt. In the manufacture of egg colors the alkaline coloring coal-tar dyes are mostly used, and they are to be found in a great variety of shades.

Of the non-poisonous egg dyes, there are some ten or a dozen numbers, new red, carmine, scarlet, pink, violet, blue, yellow, orange, green, brown, black, heliotrope, etc., which when mixed will enable the operator to form shades almost without number.

The manufacture of the egg dyes as carried on in the factory consists in a mechanical mixing of basic coal-tar dyestuffs, also some direct coloring benzidine dyestuffs, with dextrin in the ratio of about 1 part of aniline dye to 8 parts of dextrin; under certain circumstances, according to the concentrated state of the dyes, the reducing quantity of the dextrin may be greatly increased. As reducing agents for these colors insoluble substances may also be employed. A part also of the egg dyes are treated with the neutral sulphate; for instance, light brilliant green, because of its rubbing off, is made with dextrin and Glauber's salt in the proportion of 1:3:3.

For the dyeing of eggs such color mixtures are preferably employed as contain along with the dye proper a fixing agent (dextrin) as well as a medium for the superficial mordanting of the eggshell. The colors will then be very brilliant.

Here are some recipes:

Color	Dyestuff	Parts by Weight	Cit. Acid	Dextrin
Blue	Marine blue B. N...	3.5	35.0	60.0
Brown	Vesuvin S.........	30.0	37.5	30.0
Green	Brilliant green O...	13.5	18.0	67.5
Orange	Orange II........	9.0	18.0	75.0
Red	Diamond fuchsine I.	3.5	18.0	75.0
Pink	Eosin A...........	4.5	—	90.0
Violet	Methyl violet 6 B..	3.6	18.0	75.0
Yellow	Naphthol yellow S.	13.5	36.0	67.5

Very little of these mixtures suffices for dyeing five eggs. The coloring matter is dissolved in 600 parts by weight of boiling water, while the eggs to be dyed are boiled hard, whereupon they are placed in the dye solution until they seem sufficiently colored. The dyes should be put up in waxed paper.

Fast Stamping Color.—Rub up separately, 20 parts of cupric sulphate and 20 parts of anilic hydrochlorate, then mix carefully together, after adding 10 parts of dextrin. The mixture is next ground with 5 parts of glycerine and sufficient water until a thick, uniform, paste-like mass results, adapted for use by means of stencil and bristle-brush. Aniline black is formed thereby in and upon the fiber, which is not destroyed by boiling.

New Mordanting Process.—The ordinary method of mordanting wool with a bichromate and a reducing agent always makes the fiber more or less tender, and Amend proposed to substitute the use of a solution of chromic acid containing 1 to 2 per cent of the weight of the wool, at a temperature not exceeding 148° F., and to treat it afterwards with a solution of sodium bisulphite. According to a recent French patent, better results are obtained with neutral or slightly basic chromium sulphocyanide. This salt, if neutral or only slightly basic, will mordant wool at 148° F. The double sulphocyanide of chromium and ammonium, got by dissolving chromic oxide in ammonium sulphocyanide, can also be used. Nevertheless, in order to precipitate chromium chromate on the fiber, it is advisable to have a soluble chromate and a nitrate present, as well as a soluble copper salt and a free acid. One example of the process is as follows: Make the bath with 2 to 3 per cent of ammonio-chromium sulphocyanide, one-half of 1 per cent sodium bichromate, one-third of 1 per cent sodium nitrite, one-third of 1 per cent sulphate of copper, and 1.5 per cent sulphuric acid—percentages based on the weight of the wool. Enter cold and slowly heat to about 140° to 150° F. Then work for half an hour, lift and rinse. The bath does not exhaust and can be reinforced and used again.

Process for Dyeing in Khaki Colors.— Bichromate of potash or of soda, chloride of manganese, and a solution of acetate of soda or formiate of soda (15° Bé.) are dissolved successively in equal quantities.

The solution thus composed of these three salts is afterwards diluted at will, according to the color desired, constituting a range from a dark brown to a light olive green shade. The proportions of the three salts may be increased or diminished, in order to obtain shades more or less bister.

Cotton freed from its impurities by the usual methods, then fulled as ordinarily, is immersed in the bath. After a period, varying according to the results desired, the cotton, threads, or fabrics of cotton, are washed thoroughly and plunged, still wet, into an alkaline solution, of which the concentration ought never to be less than 14° Bé. This degree of concentration is necessary to take hold of the fiber when the cotton comes in contact with the alkaline bath, and by the contraction which takes place the oxides of chrome and of manganese remain fixed in the fibers.

This second operation is followed by washing in plenty of water, and then the cotton is dried in the open air. If the color is judged to be too pale, the threads or fabrics are immersed again in the initial bath, left the necessary time for obtaining the desired shade, and then washed, but without passing them through an alkaline bath. This process furnishes a series of khaki colors, solid to light, to fulling and to chlorine.

LAKES:

Scarlet Lake.—In a vat holding 120 gallons provided with good agitating apparatus, dissolve 8 pounds potash alum in 10 gallons hot water and add 50 gallons cold water. Prepare a solution of 2 pounds ammonia soda and add slowly to the alum solution, stirring all the time. In a second vessel dissolve 5 pounds of brilliant scarlet aniline, by first making it into a paste with cold water and afterwards pouring boiling water over it; now let out steam into the vat until a temperature of 150° to 165° F. is obtained. Next dissolve 10 pounds barium chloride in 10 gallons hot water in a separate vessel, add this very slowly, stir at least 3 hours, keeping up temperature to the same figures. Fill up vat with cold water and leave the preparation for the night. Next morning the liquor (which should be of a bright red color) is drawn off, and cold water again added. Wash by decantation 3 times, filter, press gently, and make into pulp.

It is very important to precipitate the aluminum cold, and heat up before adding the dyestuff. The chemicals used for precipitating must be added very slowly and while constantly stirring. The quantity used for the three washings is required each time to be double the quantity originally used.

I.—Madder Lakes. — Prepare from the root 1 pound best madder, alum water (1 pound alum with 1½ gallons of water), saturated solution of carbonate of potash (¾ pound carbonate of potash to ½ gallon of water).

The madder root is inclosed in a linen bag of fine texture, and bruised with a pestle in a large mortar with 2 gallons of water (free from lime) added in small quantities at a time, until all the coloring matter is extracted. Make this liquor boil, and gradually pour into the boiling water solution. Add the carbonate of potash solution gradually, stirring all the time. Let the mixture stand for 12 hours and drop and dry as required.

II.—Garancine Process.—This is the method usually employed in preference to that from the root. Garancine is prepared by steeping madder root in sulphate of soda and washing.

Garancine	2	pounds
Alum (dissolved in a little water)	2	pounds

Chloride of tin	½	ounce

Sufficient carbonate of potash or soda to precipitate the alum.

Boil the garancine in 4 gallons of pure water; add the alum, and continue boiling from 1 to 2 hours. Allow the product to partially settle and filter through flannel before cooling. Add to the filtrate the chloride of tin, and sufficient of the potash or soda solution to precipitate the alum; filter through flannel and wash well. The first filtrate may be used for lake of an inferior quality, and the garancine originally employed may also be treated as above, when a lake slightly inferior to the first may be obtained.

Maroon Lake.—Take of a mixture made of:

⅔ Sapan wood } ⅓ Lima wood }	....	56 parts
Soda crystals		42 parts
Alum		56 parts

Extract the color from the woods as for rose pink, and next boil the soda and alum together and add to the woods solution cold. This must be washed clean before adding to the wood liquor.

Carnation Lake.—

Water	42	gallons
Cochineal	12	pounds
Salts of tartar	1½	pounds
Potash alum	¾	pound
Nitrous acid, nitromuriate of tin	44	pounds
Muriatic acid, nitromuriate of tin	60	pounds
Pure block tin, nitromuriate of tin	22	pounds

Should give specific gravity 1.310.

Boil the water with close steam, taking care that *no iron* touches it; add the cochineal and boil for not more than five minutes; then turn off the steam and add salts of tartar and afterwards carefully add the alum. If it should not rise, put on steam until it does, pass through a 120-mesh sieve into a settling vat, and let it stand for 48 hours (not for precipitation). Add gradually nitromuriate of tin until the test on blotting paper (given below) shows that the separation is complete. Draw off clear water after it has settled, and filter. To test, rub a little of the paste on blotting paper, then dry on steam chest or on the hand, and if on bending it cracks, too much tin has been used.

To Test the Color to See if it is Precipitating.—Put a drop of color on white blotting paper, and if the color spreads, it is not precipitating. If there is a colorless ring around the spot of color it shows that precipitation is taking place; if the white ring is too strong, too much has been used.

BLACK LAKES FOR WALL-PAPER MANUFACTURE:

Bluish-Black Lake.—Boil well 220 parts of Domingo logwood in 1,000 parts of water to which 2 parts of ammonia soda have been added; to the boiling logwood add next 25 parts of green vitriol and then 3.5 parts of sodium bichromate. The precipitated logwood lake is washed out well twice and then filtered.

Black Lake A1.—Logwood extract, Sanford, 120 parts; green vitriol, 30 parts; acetic acid, 7° Bé., 10 parts; sodium bichromate, 16 parts; powdered alum, 20 parts. The logwood extract is first dissolved in boiling water and brought to 25° Bé. by the addition of cold water. Then the remaining ingredients are added in rotation, the salts in substance, finely powdered, with constant stirring. After the precipitation, wash twice and filter.

Aniline Black Lake.—In the precipitating vat filled with 200 parts of cold water enter with constant stirring in the order mentioned the following solutions kept in readiness: Forty parts of alum dissolved in 800 parts of water; 10 parts of calcined soda dissolved in 100 parts of water; 30 parts of azo black dissolved in 1,500 parts of water; 0.6 parts of "brilliant green" dissolved in 100 parts of water; 0.24 parts of new fuchsine dissolved in 60 parts of water; 65 parts of barium chloride dissolved in 1,250 parts of water. Allow to settle for 24 hours, wash the lake three times and filter it.

Carmine Lake for Wall Paper and Colored Papers.—Ammonia soda (98 per cent), 57.5 parts by weight; spirits (96 per cent), 40 parts by weight; corallin (dark), 10 parts by weight; corallin (pale), 5 parts by weight; spirit of sal ammoniac (16° Bé.), 8 parts by weight; sodium phosphate, 30 parts by weight; stannic chloride, 5 parts by weight; barium chloride, 75 parts by weight. Dissolve the corallin in the spirit, and filter the solution carefully into eight bottles, each containing 1 part of the above quantity of spirit of sal ammoniac, and let stand. The soda should meanwhile be dissolved in hot water and the solution run into the stirring vat, in which there is cold water to the height of 17 inches. Add the sodium phosphate, which has been dissolved in a copper vessel, then the corallin solution, and next the stannic chloride diluted with 3 pailfuls of cold water. Lastly the barium chloride solution is added. The day previous barium chloride is dissolved in a cask in as little boiling water as possible, and the receptacle is filled entirely with cold water. On the day following, allow the same to run in slowly during a period of three-fourths of an hour, stir till evening, allow to settle for 2 days, draw off and filter.

English Pink.—

Quercitron bark....	200 parts
Lime...............	10 parts
Alum..............	10 parts
Terra alba.........	300 parts
Whiting............	200 parts
Sugar of lead.......	7 parts

Put the bark into a tub, slake lime in another tub, and add the clear limewater to wash the bark; repeat this 3 times, letting the bark stand in each water 24 hours. Run liquor into the tub below and add the terra alba and whiting; wash well in the top tub and run into liquor below through a hair sieve, stirring well.

Dissolve the sugar of lead in warm water and pour gently into the tub, stirring all the time; then dissolve the alum and run in while stirring; press slightly, drop, and dry as required.

Dutch Pink.—

I.—Quercitron bark...	200 parts
Lime.............	20 parts
Alum.............	20 parts
Whiting..........	100 parts
Terra alba........	200 parts
White sugar of lead	10 parts
II.—Quercitron bark...	300 parts
Lime.............	10 parts
Alum.............	10 parts
Terra alba........	400 parts
Whiting..........	100 parts
Sugar of lead......	7 parts

Put the bark into a tub with cold water, slake 28 pounds of lime, and add the limewater to the bark. (This draws all the color out of the wood.) Dissolve alum in water and run it into bark liquor. The alum solution must be just warm. Dissolve sugar of lead and add it to above, and afterwards add the terra alba and whiting. The product should now be in a pulp, and must be dropped and dried as required.

Rose Pink.—I.—Light.

Sapan wood.......	100 parts
Lima.............	100 parts
Paris white........	200 parts
Alum	210 parts

II.—Deep.

Sapan wood	300 parts
Lima	300 parts
Terra alba	400 parts
Paris white	120 parts
Lime	12 parts
Alum	200 parts

III.—

Sapan wood	200 parts
Alum	104 parts
Whiting	124 parts

Boil the woods together in 4 waters and let the products stand until cold; wash in the whiting and terra alba through a hair sieve, and afterwards run in the alum. If a deep color is required slake 12 pounds lime and run it in at the last through a hair sieve. Let the alum be just warm or it will show in the pink.

DYES, COLORS, ETC., FOR TEXTILE GOODS:

Aniline Black.—This black is produced by carefully oxidizing aniline hydrochloride. The exact stage of oxidation must be carefully regulated or the product will be a different body (quinone). There are several suitable oxidizing agents, such as chromic acid, potassic bichromate, ferrocyanide of potassium, etc., but one of the easiest to manipulate is potassic chlorate, which by reacting on copper sulphate produces potassic sulphate and copper chlorate. This is easily decomposed, its solution giving off gases at 60° F. which consist essentially of chloride anhydrate. But one of the most useful agents for the production of aniline black is vanadate of ammonia, 1 part of which will do the work of 4,000 parts of copper. Many other salts besides copper may be used for producing aniline black, but the following method is one of the best to follow in making this dye:

Aniline hydrochloride	40 parts
Potassic chlorate	20 parts
Copper sulphate	40 parts
Chloride of ammonia (sal ammoniac)	16 parts
Warm water at 60° F.	500 parts

After warming a few minutes the mass froths up. The vapor should not be inhaled. Then set aside, and if the mass is not totally black in a few hours, again heat to 60° F., and expose to the air for a few days, and finally wash away all the soluble salts and the black is fit for use.

Aniline Black Substitutes.—I.—Make a solution of

Aniline (fluid measure)	30 parts
Toluidine (by weight)	10 parts
Pure hydrochloric acid, B. P. (fluid measure)	60 parts
Soluble gum arabic (fluid measure)	60 parts

Dissolve the toluidine in the aniline and add the acid, and finally the mucilage.

II.—Mix together at gentle heat:

Starch paste	13 quarts
Potassic chlorate	350 scruples
Sulphate of copper	300 scruples
Sal ammoniac	300 scruples
Aniline hydrochloride	800 scruples

Add 5 per cent of alizarine oil, and then steep it for 2 hours in the dye bath of red liquor of $2\frac{1}{2}°$ Tw. Dye in a bath made up of $\frac{1}{2}$ ounce of rose bengal and $1\frac{1}{2}$ ounces of red liquor to every 70 ounces of cotton fabric dyed, first entering the fabric at 112° F., and raising it to 140° F., working for 1 hour, or until the desirable shade is obtained; then rinse and dry.

Blush Pink on Cotton Textile.—Rose bengal or fast pink will give this shade. The mordant to use is a 5 per cent solution of stannate of soda and another 5 per cent solution of alum.

Dissolve in a vessel (a) $8\frac{1}{2}$ parts of chloride of copper in 30 parts of water, and then add 10 parts chloride of sodium and $9\frac{1}{2}$ parts liquid ammonia.

In a second vessel dissolve (b) 30 parts aniline hydrochlorate in 20 parts of water, and add 20 parts of a solution of gum arabic prepared by dissolving 1 part of gum in 2 parts of water.

Finally mix 1 part of a with 4 parts of b; expose the mixture to the air for a few days to develop from a greenish to a black color. Dilute for use, or else dry the thick compound to a powder.

If new liquor is used as the mordant, mix 1 part of this with 4 parts of water, and after working the fabric for 1 to 2 hours in the cold liquor, wring or squeeze it out and dry; before working it in the dye liquor, thoroughly wet the fabric by rinsing it in hot water at a spring boil; then cool by washing in the dye bath until the shade desired is attained, and again rinse and dry.

The red liquor or acetate of aluminum may be made by dissolving 13 ounces of alum in 69 ounces of water and mixing this with a solution made by dissolving $7\frac{1}{2}$ ounces of acetate of lime, also dissolved in 69 ounces of water. Stir well, allow it to settle, and filter or decanter off the clear fluid for use, and use this mixture $2\frac{1}{2}°$ Tw.

The fabric is first put into the stannate of soda mordant for a few minutes, then wrung out and put into the alum mordant for about the same time; then it is again wrung out and entered in the dye bath at 120° F. and dyed to shade desired, and afterwards rinsed in cold water and dried.

The dye bath is made of $\frac{1}{4}$ ounce of rose bengal per gallon of water. If fast pink is the dye used, the mordant used would be Turkey red oil and red liquor. Use 8 ounces of Turkey red oil per gallon of water. Put the fabric into this, then wring out the textile and work in red liquor of 7° Tw. for about 2 hours, then wring out and dye in a separate bath made up of cosine, or fast pink, in water in which a little alum has been dissolved.

To Dye Woolen Yarns, etc., Various Shades of Magenta.—To prepare the dye bath dissolve 1 pound of roscine in 15 gallons of water. For a concentrated solution use only 10 gallons of water, while if a very much concentrated color is needed, dissolve the dye in methylated spirit of wine, and dilute this spirituous tincture with an equal quantity of water.

No mordant is required in using this color in dyeing woolen goods. The dyeing operation consists simply in putting the goods into the dye bath at 190° F. and working them therein until the desired shade is obtained, then rinsing in cold water and drying.

If the water used in preparing the dye is at all alkaline, make use of the acid roseine dissolved in water in which a little sulphuric acid has been mixed, and work, gradually raising to the boiling point, and keep up the temperature for 30 minutes, or according to the shade desired. Put about 20 per cent sulphate of soda into the dye bath.

Maroon Dye for Woolens.—To prepare the dye bath, dissolve about 1 pound of maroon dye in boiling water, with or without the addition of methylated spirit of wine. For dark shades dissolve in boiling water, only slightly acidulated with hydrochloric acid, and filter before use. No mordant is required with this dye when dyeing wool, but for the bright shade a little curd soap may be dissolved in the dye bath before proceeding to dye the wool, while for the dark shade it is best to put in a little acetate of soda. To use the dye, first dye in a weak bath and gradually strengthen it until the desired shade is obtained, at the same time gradually increasing the temperature until just below the boiling point.

To Dye Woolens with Blue de Lyons.—Dissolve 8 ounces of blue dye in 1 gallon of methylated spirit, which has been slightly soured with sulphuric acid, and boil the solution over a water bath until it is perfectly clear. To prepare the dye bath, add more or less of the spirituous tincture to a 10- or 15-gallon dye bath of water, which has been slightly soured with sulphuric acid.

Rich Orange on Woolen.—Dissolve 1 pound of phosphine in 15 gallons of boiling water, and stir the fluid until the acid has dissolved. No mordant is required to dye wool. First work the goods about in a weak solution, and finally in one of full strength, to which a little acetate of soda has been added. Keep up the temperature to just below the boiling point while working the goods in the dye bath.

DYEING SILK OR COTTON FABRICS WITH ANILINE DYES:

Aniline Blue on Cotton.—Prepare a dye bath by dissolving 1 pound of aniline blue (soluble in spirit) in 10 gallons of water, and set it aside to settle. Meanwhile prepare a mordant while boiling 35 ounces of sumac (or $5\frac{1}{2}$ ounces tannic acid in 30 gallons of water) and then dissolve therein 17 ounces of curd soap. Boil up and filter. Put the cotton goods in the hot liquid and let them remain therein for 12 hours. Then wring them out and make up a dye bath of $2\frac{1}{2}°$ Tw. with red liquor. Add dye color according to the shade desired. Put in the goods and work them until the color is correct, keeping the temperature at the boiling point.

To Dye Silk a Delicate Greenish Yellow.—Dissolve 2 ounces of citronine in 1 gallon of methylated spirit and keep the solution hot over a water bath until perfectly clear.

To prepare silk fabrics, wash them in a weak soap liquor that has been just sweetened (i. e., its alkalinity turned to a slight sourness) with a little sulphuric acid. Work the goods until dyed to shade, and then rinse them in cold water that has been slightly acidulated with acetic, tartaric, or citric acid.

To Dye Cotton Dark Brown.—Prepare a mordant bath of 10 pounds of catechu, 2 pounds of logwood extract, and $\frac{1}{4}$ pound magenta (roseine), and bring to a boil; work the goods therein for 3 hours at that temperature; then put into a fresh dye bath made up of 3 pounds of bichromate of potash and 2 pounds of sal soda, and dye to shade. These proportions are for a dye bath to dye 100 pounds of cotton goods at a time.

To Dye Silk Peacock Blue.—Make up a dye bath by putting 1 pint of sulphuric acid at 170° Tw., and 10 ounces of methylin blue crystal dye liquor of 120° to 160° Tw., with a dye bath that will hold 80 pounds of goods. Put in the silk at 130° F., and raise to 140° F., and work up to shade required.

To Dye Felt Goods.—Owing to this material being composed of animal and vegetable fiber it is not an easy matter always to produce evenness of shade. The best process to insure success is to steep well the felt in an acid bath of from 6° to 12° Bé., and then wash away all traces of acid. Some dyers make the fulling stork the medium of conveying the dye, while others partially dye before fulling, or else dye after that process.

The fulling stock for 72 ounces of beaver consists of a mixture of

Black lead or plum-
bago.............. 16 ounces
Venetian red........ 48 ounces
Indigo extract (fluid). 5 ounces

Ordinary Drab.—

Common plumbago.. 12 ounces
Best plumbago...... 12 ounces
Archil extract (fluid).. 15 ounces
Indigo extract........ 10 ounces

Mix into fluid paste with water and add sulphuric acid at 30° Tw. For the dye liquor make a boiling-hot solution of the aniline dye and allow it to cool; then put into an earthenware vessel holding water and heat to 83° F., and add sufficient dye liquor to give the quantity of felt the desired shade. First moisten well the felted matter (or the hair, if dyed before felting) with water, and then work it about in the above dye bath at 140° F. To deepen the shade, add more dye liquor, lifting out the material to be dyed before adding the fresh dye liquor, so that it can be well stirred up and thoroughly mixed with the exhausted bath.

Brown Shades.—Bismarck brown will give good results, particularly if the dyed goods are afterwards steeped or passed through a weak solution (pale straw color) of bichromate of potash. This will give a substantial look to the color. Any of the aniline colors suitable for cotton or wool, or those suited for mixed cotton and wool goods may be used.

Blue.—Use either China blue, dense ferry blue, or serge blue, first making the material acid before dyeing.

Green.—Use brilliant green and have the material neutral, i. e., neither acid nor alkali; or else steep in a bath of sumac before dyeing.

Plum Color.—Use maroon (neutral or acid) and work in an acid bath or else sumac.

Black.—Use negrosin in an acid bath, or else mordant in two salts and dye slightly acid.

Soluble Blue, Ball Blue, etc.—A soluble blue has for many years been readily obtainable in commerce which is similar in appearance to Prussian blue, but, unlike the latter, is freely soluble in water. This blue is said to be potassium ferriferrocyanide.

To prepare instead of buying it ready made, gradually add to a boiling solution of potassium ferricyanide (red prussiate of potash) an equivalent quantity of hot solution of ferrous sulphate, boiling for 2 hours and washing the precipitate on a filter until the washings assume a dark-blue color. The moist precipitate can at once be dissolved by the further addition of a sufficient quantity of water. About 64 parts of the iron salt is necessary to convert 100 parts of the potassium salt into the blue compound.

If the blue is to be sent out in the liquid form, it is desirable that the solution should be a perfect one. To attain that end the water employed should be free from mineral substances, and it is best to filter the solution through several thicknesses of fine cotton cloth before bottling; or if made in large quantities this method may be modified by allowing it to stand some days to settle, when the top portion can be siphoned off for use, the bottom only requiring filtration.

The ball blue sold for laundry use consists of ultramarine. Balls or tablets of this substance are formed by mixing it with glucose or glucose and dextrin, and pressing into shape. When glucose alone is used, the product has a tendency to become soft on keeping, which tendency may be counteracted by a proper proportion of dextrin. Bicarbonate of sodium is added as a filler to cheapen the product, the quantity used and the quality of the ultramarine employed being both regulated by the price at which the product is to sell.

New Production of Indigo.—Forty parts of a freshly prepared ammonium sulphide solution containing 10 per cent of hydrogen sulphide are made to flow quickly and with constant stirring into a heated solution of 20 parts of isatine anilide in 60 parts of alcohol. With spontaneous heating and temporary green and blue coloration, an immediate separation of indigo in small crystalline needles of a faint copper luster takes place. Boil for a short time, whereupon the indigo is filtered off, rewashed with alcohol, and dried.

To Dye Feathers.—A prerequisite to the dyeing of feathers appears to be softening them, which is sometimes accomplished by soaking them in warm water, and sometimes an alkali, such as ammonium or sodium carbonate, is added. This latter method would apparently be preferable on account of the removal of any greasy matter that may be present.

When so prepared the feathers may be dyed by immersion in any dye liquor. An old-time recipe for black is immersion in a bath of ferric nitrate suitably diluted with water, and then in an infusion of equal parts of logwood and quercitron. Doubtless an aniline dye would prove equally efficient and would be less troublesome to use.

After dyeing, feathers are dipped in an emulsion formed by agitating any bland fixed oil with water containing a little potassium carbonate, and are then dried by gently swinging them in warm air. This operation gives the gloss.

Curling where required is effected by slightly warming the feathers before a fire, and then stroking with a blunt metallic edge, as the back of a knife. A certain amount of manual dexterity is necessary to carry the whole process to a successful ending.

DYES FOR FOOD:
See Foods.

DYES FOR LEATHER:
See Leather.

DYE STAINS, THEIR REMOVAL FROM THE SKIN:
See Cleaning Preparations and Methods.

DYNAMITE:
See Explosives.

EARTHENWARE:
See Ceramics.

EAU DE QUININE:
See Hair Preparations.

EBONY:
See Wood.

EBONY LACQUER:
See Lacquers.

ECZEMA DUSTING POWDER FOR CHILDREN.
Starch, French chalk, lycopodium, of each, 40 parts; bismuth subnitrate, 2 parts; salicylic acid, 2 parts; menthol, 1 part. Apply freely to the affected parts.

Eggs

The age of eggs may be approximately judged by taking advantage of the fact that as they grow old their density decreases through evaporation of moisture. According to Siebel, a new-laid egg placed in a vessel of brine made in the proportion of 2 ounces of salt to 1 pint of water, will at once sink to the bottom. An egg 1 day old will sink below the surface, but not to the bottom, while one 3 days old will swim just immersed in the liquid. If more than 3 days old the egg will float on the surface, the amount of shell exposed increasing with age; and if 2 weeks old, only a little of the shell will dip in the liquid.

The New York State Experiment Station studied the changes in the specific gravity of the eggs on keeping and found that on an average fresh eggs had a specific gravity of 1.090; after they were 10 days old, of 1.072; after 20 days, of 1.053; and after 30 days, of 1.035. The test was not continued further. The changes in specific gravity correspond to the changes in water content. When eggs are kept they continually lose water by evaporation through the pores in the shell. After 10 days the average loss was found to be 1.60 per cent of the total water present in the egg when perfectly fresh; after 20 days, 3.16 per cent; and after 30 days, 5 per cent. The average temperature of the room where the eggs were kept was 63.8° F. The evaporation was found to increase somewhat with increased temperature. None of the eggs used in the 30-day test spoiled.

Fresh eggs are preserved in a number of ways which may, for convenience, be grouped under two general classes: (1) Use of low temperature, i. e., cold storage; and (2) excluding the air by coating, covering, or immersing the eggs, some material or solution being used which may or may not be a germicide. The two methods are often combined. The first method owes its value to the fact that microörganisms, like larger forms of plant life, will not grow below a certain temperature, the necessary degree of cold varying with the species. So far as experiment shows, it is impossible to kill these minute plants, popularly called "bacteria" or "germs," by any degree of cold; and so, very low temperature is unnecessary for preserving eggs, even if it were not undesirable for other reasons, such as injury by freezing and increased cost. According to a report of the Canadian commission of agriculture and dairying:

Eggs are sometimes removed from the shells and stored in bulk, usually on a commercial scale, in cans containing about 50 pounds each. The temperature recommended is about 30° F., or a little below freezing, and it is said they will keep any desired length of time. They must be used soon after they have been removed from storage and have been thawed.

Water glass or soluble glass is the popular name for potassium silicate, or sodium silicate, the commercial article often being a mixture of the two. The commercial water glass is used for preserving eggs, as it is much cheaper than the chemically pure article which is required for many scientific purposes. Water glass is commonly sold in two forms, a syrup-thick liquid of about the consistency of molasses, and a powder. The thick syrup, the form perhaps most usually seen, is sometimes sold wholesale as low as 1¾ cents per pound in carboy lots. The retail price varies, though 10 cents per pound, according to the North Dakota Experiment Station, seems to be the price commonly asked. According to the results obtained at this station a solution of the desired strength for preserving eggs may be made by dissolving 1 part of the syrup-thick water glass in 10 parts, by measure, of water. If the water-glass powder is used, less is required for a given quantity of water. Much of the water glass offered for sale is very alkaline. Such material should not be used, as the eggs preserved in it will not keep well. Only pure water should be used in making the solution, and it is best to boil it and cool it before mixing with the water glass.

The solution should be carefully poured over the eggs packed in a suitable vessel, which must be clean and sweet, and if wooden kegs or barrels are used they should be thoroughly scalded before packing the eggs in them. The packed eggs should be stored in a cool place. If they are placed where it is too warm, silicate deposits on the shell and the eggs do not keep well. The North Dakota Experiment Station found it best not to wash the eggs before packing, as this removes the natural mucilaginous coating on the outside of the shell. The station states that 1 gallon of the solution is sufficient for 50 dozen eggs if they are properly packed.

It is, perhaps, too much to expect that eggs packed in any way will be just as satisfactory for table use as the fresh article. The opinion seems to be, however, that those preserved with water glass are superior to most of those preserved otherwise. The shells of eggs preserved in water glass are apt to crack in boiling. It is stated that this may be prevented by puncturing the blunt end of the egg with a pin before putting it into the water.

To Discover the Age of Eggs.—The most reliable method of arriving at the age of hens' eggs is that by specific gravity. Make a solution of cooking salt (sodium chloride) in rain or distilled water, of about one part of salt to two parts of water, and in this place the eggs to be tested. A perfectly fresh egg (of from 1 to 36 hours old) will sink completely, lying horizontally on the bottom of the vessel; when from two to three days old, the egg also sinks, but not to the bottom, remaining just below the surface of the water, with a slight tendency of the large end to rise. In eggs of four or five days old this tendency of the large end to rise becomes more marked, and it increases from day to day, until at the end of the fifth day the long axis of the egg (an imaginary line drawn through the center lengthwise) will stand at an angle of 20° from the perpendicular. This angle is increased daily, until at the end of the eighth day it is at about 45°; on the fourteenth day it is 60°; on the twenty-first day it is 75°, while at the end of 4 weeks the egg stands perfectly upright in the liquid, the point or small end downward.

This action is based on the fact that the air cavity in the big end of the egg increases in size and capacity, from day to day, as the egg grows older. An apparatus (originally devised by a German poultry fancier) based on this principle, and by means of which the age of an egg maintained at ordinary temperature may be told approximately to within a day, is made by placing a scale of degrees, drawn from 0° to 90° (the latter representing the perpendicular) behind the vessel containing the solution, and observing the angle made by the axis of the egg with the perpendicular line. This gives the age of the egg with great accuracy.

Weights of Eggs.—The following table shows the variation in weight between eggs of the same family of chickens and of the comparative value of the product of different kinds of fowls:

	Weight of Whole Eggs, Grains.	Shell, Grains.	Net.
Common hen, small..	635.60	84.86	550.54
Common hen, mean..	738.35	92.58	645.77
Common hen, large..	802.36	93.25	709.11
Italian hen........	840.00	92.50	747.50
Houdan............	956.60	93.50	853.10
La Flesche.........	926.50	94.25	835.25
Brahma............	1,025.50	114.86	910.64

From this it will be seen that the Houdans and Brahmas are the most profitable producers, as far as food value of the product is concerned—provided, of course, they are equally prolific with the ordinary fowl.

Another calculation is the number of eggs to the pound, of the various weights. This is as follows:

Small ordinary eggs
  (635 grains).....12.20 to pound
Large ordinary eggs
  (802 grains)..... 9.25 to pound
Houdan eggs...... 8.0 to pound
Brahma, mean.... 7.4 to pound
Brahma, large..... 7.1 to pound

Dried Yolk of Egg.—To prepare this, the yolks of eggs, separated from the whites, are thoroughly mixed with ⅓ their weight of water. The resulting emulsion is strained and evaporated under reduced pressure at a temperature of 87° to 122° F., to a paste. The latter is further dried over quicklime or a similar absorbent of moisture, at a temperature of 77° to 86° F., and ground to a fine powder.

Egg Oil.—

Yolks of eggs (about
  250)............. 5.0 parts
Distilled water...... 0.3 parts

Beat this together and heat the mass with constant stirring in a dish on the water bath until it thickens and a sample exhibits oil upon pressing between the fingers. Squeeze out between hot plates, mix the turbid oil obtained with 0.05 parts of dehydrated Glauber's salt, shake repeatedly, and finally allow to settle. The oil, which must be decanted clear from the sediment, gives a yield of at least 0.5 parts of egg oil.

Artificial Egg Oil.—

Yellow beeswax..... 0.2 parts
Cacao oil.......... 0.5 parts

Melt on the water bath and gradually add 9 parts of olive oil.

Egg Powder.—

Sodium bicarbonate.. 8 ounces
Tartaric acid........ 3 ounces
Cream tartar........ 5 ounces
Turmeric, powdered. 3 drachms
Ground rice......... 16 ounces

Mix and pass through a fine sieve. One teaspoonful to a dessertspoonful (according to article to be made), to be mixed with each half pound of flour.

The Preservation of Eggs.—The spoiling of eggs is due to the entrance of air carrying germs through the shells. Normally the shell has a surface coating of mucilaginous matter, which prevents for a time the entrance of these harmful organisms into the egg. But if this coating is removed or softened by washing or otherwise the keeping quality of the egg is much reduced. These facts explain why many methods of preservation have not been entirely successful, and suggest that the methods employed should be based upon the idea of protecting and rendering more effective the natural coating of the shell, so that air bearing the germs that cause decomposition may be completely excluded.

Eggs are often packed in lime, salt, or other products, or are put in cold storage for winter use, but such eggs are very far from being perfect when they come upon the market. German authorities declare that water glass more closely conforms to the requirements of a good preservative than any of the substances commonly employed. A 10 per cent solution of water glass is said to preserve eggs so effectually that at the end of three and one-half months eggs still appeared to be perfectly fresh. In most packed eggs the yolk settles to one side, and the egg is then inferior in quality. In eggs preserved in water glass the yolk retained its normal position in the egg, and in taste they were not to be distinguished from fresh, unpacked store eggs.

Of twenty methods tested in Germany, the three which proved most effective were coating the eggs with vaseline, preserving them in limewater, and preserving them in water glass. The conclusion was reached that the last is preferable, because varnishing the eggs with vaseline takes considerable time, and treating them with limewater is likely to give the eggs a limy flavor.

Other methods follow:

I.—Eggs can be preserved for winter use by coating them, when perfectly fresh, with paraffine. As the spores of fungi get into eggs almost as soon as they are laid, it is necessary to rub every egg with chloroform or wrap it a few minutes in a chloroform soaked rag before dipping it into the melted paraffine. If only a trace of the chloroform enters the shell the development of such germs as may have gained access to freshly laid eggs is prevented. The paraffine coating excludes all future contamination from germ-laden air, and with no fungi growing within, they retain their freshness and natural taste.

II.—**Preserving with Lime.**—Dissolve in each gallon of water 12 ounces of quicklime, 6 ounces of common salt, 1 drachm of soda, ½ drachm saltpeter, ½ drachm tartar, and 1½ drachms of borax. The fluid is brought into a barrel and sufficient quicklime to cover the bottom is then poured in. Upon this is placed a layer of eggs, quicklime is again thrown in and so on until the barrel is filled so that the liquor stands about 10 inches deep over the last layer of eggs. The barrel is then covered with a cloth, upon which is scattered some lime.

III.—Melt 4 ounces of clear beeswax in a porcelain dish over a gentle fire, and stir in 8 ounces of olive oil. Let the solution of wax in oil cool somewhat, then dip the fresh eggs one by one into it so as to coat every part of the shell. A momentary dip is sufficient, all excess of the mixture being wiped off with a cotton cloth. The oil is absorbed in the shell, the wax hermetically closing all the pores.

IV.—The Reinhard method is said to cause such chemical changes in the surface of the eggshell that it is closed up perfectly air-tight and an admittance of air is entirely excluded, even in case of long-continued storing. The eggs are for a short time exposed to the direct action of sulphuric acid, whereby the surface of the eggshell, which consists chiefly of lime carbonate, is transformed into lime sulphate. The dense texture of the surface thus produced forms a complete protection against the access of the outside air, which admits of storing the egg for a very long time, without the contents of the egg suffering any disadvantageous changes regarding taste and odor. The egg does not require any special treatment to prevent cracking on boiling, etc.

Some object to this on the ground that sulphuric acid is a dangerous poison, that might, on occasion, penetrate the shell.

V.—Take about half a dozen eggs and place them in a netting (not so many as would chill the water below the boiling point, even for an instant), into a boiling solution of boric acid, withdraw immediately, and pack. Or put up, in oil, carrying 2 per cent or 3 per cent of salicylic acid. Eggs treated in this way are said to taste, after six months, absolutely as fresh as they were when first put up. The eggs should be as fresh as possible, and should be thoroughly clean before dipping. The philosophy of the process is that the dipping in boiling boric acid solution not only kills all bacteria existing on, or in, the shell and membrane, but reinforces these latter by a very thin layer of coagulated albumen; while the packing in salicylated oil prevents the admission of fresh germs from the atmosphere. Salicylic acid is objected to on the same grounds as sulphuric acid.

VI.—Dissolve sodium silicate in boiling water, to about the consistency of a syrup (or about 1 part of the silicate to 3 parts water). The eggs should be as fresh as possible, and must be thoroughly clean. They should be immersed in the solution in such manner that every part of each egg is covered with the liquid, then removed and let dry. If the solution is kept at or near the boiling temperature, the preservative effect is said to be much more certain and to last longer.

EGG CHOCOLATE:
  See Beverages.

EGG DYES:
  See Dyes.

EGG LEMONADE:
  See Beverages, under Lemonade.

EGG PHOSPHATE:
  See Beverages.

EGG-STAIN REMOVER:
  See Cleaning Preparations and Methods.

EGGS, TESTS FOR:
  See Foods.

EIKONOGEN DEVELOPER:
  See Photography.

EKTOGAN:
  See Antiseptics.

ELAINE SUBSTITUTE.

A substitute for elaine for woolen yarns is obtained by boiling 4 pounds carrageen moss in 25 gallons water for 3 hours. The soda is then put in and the boiling continued for another half hour; 2 pounds fleabane seeds are gradually added, and a little water to make up for the evaporation. After a further 1½ hours boiling, the extract is passed through a fine sieve and well mixed with 25 pounds cottonseed oil, 12½ pounds sweet oil, and 12½ pounds ammonia solution of 0.96 specific gravity. Next day stir in 25 pounds saponified elaine and 13 pounds of odorless petroleum of 0.885 specific gravity. The resulting emulsion keeps well, dissolves perfectly in lukewarm water, and answers its purpose excellently.

ELECTRODEPOSITION PROCESSES:
See Plating.

ELECTROLYSIS IN BOILERS:
See Boiler Compounds.

Electroplating and Electro-typing

(See also Plating.)

PROCESS OF ELECTROPLATING.

First, clean the articles to be plated. To remove grease, warm the pieces before a slow fire of charcoal or coke, or in a dull red stove. Delicate or soldered articles should be boiled in a solution of caustic potash, the latter being dissolved in 10 times its weight of water.

The scouring bath is composed of 100 parts of water to from 5 to 20 parts of sulphuric acid. The articles may be put in hot and should be left in the bath till the surface turns to an ocher red tint.

The articles, after having been cleansed of grease by the potash solution, must be washed in water and rinsed before being scoured. Copper or glass tongs must then be used for moving the articles, as they must not afterwards be handled. For small pieces, suitable earthenware or porcelain strainers may be used.

The next stage is the spent nitric acid bath. This consists of nitric acid weakened by previous use. The articles are left in until the red color disappears, so that after rinsing they show a uniform metallic tint. The rinsing should be thoroughly carried out.

Having been well shaken and drained, the articles are next subjected to the strong nitric acid bath, which is made up as follows:

Nitric acid of 36° Bé..100 volumes
Chloride of sodium
(common salt)..... 1 volume
Calcined soot (lamp-
black)........... 1 volume

The articles must be immersed in this bath for only a few seconds. Avoid overheating or using too cold a bath. They are next rinsed thoroughly with cold water and are again subjected to a strong nitric acid bath to give them a bright or dull appearance as required.

To produce a bright finish, plunge them for a few seconds (moving them about rapidly at the same time) in a cold bath of the following composition:

Nitric acid...........100 volumes
Sulphuric acid.......100 volumes
Chloride of sodium... 1 volume

Again rinse thoroughly in cold water. The corresponding bath giving a dull or matt appearance is composed of:

Nitric acid......... 200 volumes
Sulphuric acid..... 100 volumes
Sea salt........... 1 volume
Sulphate of zinc...1 to 5 volumes

The duration of immersion in this bath varies from 5 to 20 minutes, according to the dullness required. Wash with plenty of water. The articles will then have an unpleasant appearance, which will disappear on plunging them for a moment into the brightening bath and rinsing quickly.

The pieces are next treated with the nitrate of mercury bath for a few seconds.

Plain water....... 10,000 parts
Nitrate of mercury 10 parts
Sulphuric acid..... 20 parts

It is necessary to stir this bath before using it. For large articles the proportion of mercury should be greater. An article badly cleaned will come out in various shades and lacking its metallic brightness. It is better to throw a spent bath away than attempt to strengthen it.

The various pieces, after having passed through these several processes, are then ready for the plating bath.

A few words on the subject of gilding may not be amiss. Small articles are gilded hot, large ones cold. The cold cyanide of gold and potassium bath is composed as follows:

Distilled water..... 10,000 parts
Pure cyanide of po-
tassium......... 200 parts
Pure gold........ 100 parts

The gold, transformed into chloride, is dissolved in 2,000 parts of water and the cyanide in 8,000 parts. The two solutions are then mixed and boiled for half an hour.

The anode must be entirely submerged in the bath, suspended from platinum wires and withdrawn immediately the bath is out of action.

Hot Gold Bath.—Zinc, tin, lead, antimony and the alloys of these metals are better if previously covered with copper.

The following are the formulas for the other metals per 10,000 parts of distilled water:

Crystallized phosphate of soda, 600 parts; alloys rich in copper castings, 500 parts.

Bisulphide of soda, 100 parts; alloys rich in copper, 125 parts.

Pure cyanide of potassium, 10 parts; alloys rich in copper, 5 parts. Pure gold transformed into chloride, 10 parts; alloys rich in copper, 10 parts.

Dissolve the phosphate of soda hot in 8,000 parts water, let the chloride of gold cool in 1,000 parts water; mix little by little the second solution with the first; dissolve the cyanide and bisulphide in 1,000 parts water and mix this last solution with the other two. The temperature of the bath may vary between 122° and 175° F.

Silvering.—For amateurs a bath of 10 parts silver per 1,000 is sufficient. Dissolve 150 parts nitrate of silver, equivalent to 100 parts pure silver, in 10,000 parts of water and add 250 parts pure cyanide of potassium. Stir it up until completely dissolved, and then filter the solution. Silvering is generally effected cold, except in the case of small articles. Iron, steel, zinc, lead, and tin are better if previously copper-plated and then silvered hot. The cleaned articles are first treated in a nitrate of mercury bath, being kept continually in motion.

With excess of current the pieces become gray, and blacken. In the cold bath anodes of platinum or silver should be employed. Old baths are, in this case, preferable to new. They may, if required, be artificially aged by the addition of 1 or 2 parts in 1,000 of liquid ammonia.

If the anode blackens, the bath is too weak. If it becomes white, there is too much current, and the deposit, being too rapid, does not adhere. The deposit may be taken as normal and regular when the anode becomes gray during the passage of the current and white again when it ceases to flow.

The nickel vat should be of glass, porcelain, or earthenware, or a case lined with impermeable gum. The best nickel bath is prepared by dissolving to saturation, in hot distilled water, nickel sulphate and ammonium, free from oxides or alkalies and alkaline earthy metals. The proportion of salt to dissolve is 1 part, by weight, to 10 of water. Filter after cooling and the bath is then ready for use.

When the bath is ready and the battery-set up, the wires from the latter are joined by binding screws to two metal bars resting on the edge of the vat. The bar joined to the positive pole of the battery supports, through the intervention of a nickel-plated copper hook, a plate of nickel, constituting the soluble anode, which restores to the bath the metal deposited on the cathode by the electrolytic action. From the other bar are suspended the articles to be plated. These latter should be well polished before being put into the bath. To remove all grease, scrub them with brushes soaked in a hot solution of whiting, boiled in water and carbonate of soda.

Copper and its alloys are cleaned well in a few seconds by immersion in a bath composed of 10 parts, by weight, of water, and 1 part of nitric acid. For rough articles, 2 parts water, 1 nitric acid, and 1 sulphuric acid. For steel and polished castings, 100 parts water to 1 sulphuric acid. The articles should remain in the bath until the whole surface is of a uniform gray tint. They are then rubbed with powdered pumice stone till the solid metal appears. Iron and steel castings are left in the bath for three or four hours and then scrubbed with well-sifted sand.

If the current be too strong, the nickel is deposited gray or even black. An hour or so is time enough to render the coat sufficiently thick and in a condition to stand polishing. When the articles are removed from the bath they are washed in water and dried in hot sawdust.

To polish the articles they should be taken in one hand and rubbed rapidly backward and forward on a strip of cloth soaked in polishing powder boiled in water, the cloth being firmly fixed at one end and held in the other hand. The hollow parts are polished by means of cloth pads of various sizes fixed on sticks. These pads must be dipped in the polishing paste when using them. The articles, when well brightened, are washed in water to get rid of the paste and the wool threads, and finally dried in sawdust.

SOME NOTES ON ELECTROTYPING, PLATING, AND GILDING.

The first step in the process is the preparation of the mold. The substance originally used for the construction of this was plaster of Paris. This substance is, however, porous and must be rendered impermeable. The materials most commonly used of later years are stearine, wax, marine glue, gelatin, india rubber, and fusible alloys. With hollow molds it is a good plan to arrange an internal skeleton of platinum, for ultimate connection with the anodes, in order to secure a good electrical contact with all parts of the mold. When covering several pieces at once, it is as well to connect each of them with the negative pole by an iron or lead wire of suitable dimensions.

Having prepared the molds in the usual way—by obtaining an impression in the material when soft, and allowing it to set—they should be given a metallic coating on their active surfaces of pure powdered plumbago applied with a polishing brush.

For delicate and intricate objects, the wet process is most suitable. It consists in painting the object with two or more coats of nitrate of silver and ultimately reducing it by a solution of phosphorus in bisulphide of carbon.

The plating baths are prepared as follows:

A quantity of water is put in a jar and to it is added from 8 to 10 parts in 100 of sulphuric acid, in small quantities, stirring continually in order to dissipate the heat generated by the admixture of acid and water. Sulphate of copper (bluestone) is then dissolved in the acidulated water at the normal temperature until it will take up no more. The solution is always used cold and must be maintained in a saturated condition by the addition of copper sulphate crystals or suitable anodes.

For use it should be poured into vessels of clay, porcelain, glass, hard brown earthenware, or india rubber. For large baths wood may be used, lined on the interior with an impervious coating of acid-proof cement, india rubber, marine glue, or even varnished lead sheets.

If the solution be too weak and the current on the other hand be too strong, the resulting deposit will be of a black color. If too concentrated a solution and too weak a current be employed, a crystalline deposit is obtained. To insure a perfect result, a happy medium in all things is necessary.

During the process of deposition, the pieces should be moved about in the bath as much as possible in order to preserve the homogeneity of the liquid. If this be not attended to, stratification and circulation of the liquid is produced by the decomposition of the anode, and is rendered visible by the appearance of long, vertical lines on the cathode.

For amateurs and others performing small and occasional experiments, the following simple apparatus will be serviceable. Place the solution of sulphate of copper in an earthenware or porcelain jar, in the center of which is a porous pot containing amalgamated zinc and a solution of sulphuric acid and water, about 2 or 3 parts in 100. At the top of the zinc a brass rod is fixed, supporting a circle of the same metal, the diameter of which is between that of the containing vessel and the porous pot. From this metallic circle the pieces are suspended in such a manner that the parts to be covered are turned toward the porous pot. Two small horsehair bags filled with copper sulphate crystals are suspended in the solution to maintain its saturation.

ELM TEA.

Powdered slippery
elm bark . . . . 2 teaspoonfuls
(or the equivalent in whole bar)
Boiling water. . . . . . 1 cup
Sugar, enough.
Lemon juice, enough.

Pour the water upon the bark. When cool, strain and flavor with lemon juice and add sugar. This is soothing in case of inflammation of the mucous membrane.

EMBALMING FLUIDS.

Success in the use of any embalming fluid depends largely on manipulation, an important part of the process being the thorough removal of fluid from the circulatory system before undertaking the injection of the embalming liquid.

I.—Solution zinc
chloride (U. S.
P.). . . . . 1 gallon
Solution sodium
chloride 6
ounces to pint. 6 pints
Solution mercury
bichloride, 1
ounce to pint. . 4 pints
Alcohol. . . . . 4 pints
Carbolic acid
(pure). . . . . 8 ounces
Glycerine. . . . . 24 fluidounces

Mix the glycerine and carbolic acid, then all the other ingredients, when a clear solution of 3 gallons results, which is the proper amount for a body weighing 150 pounds.

II.—Arsenious acid...100 parts
Sodium hydrate . 50 parts
Carbolic acid and water, of each a sufficient quantity.

Dissolve the arsenious acid and the soda in 140 parts of water by the aid of heat. When the solution is cold, drop carbolic acid into it until it becomes opalescent, and finally add water until the finished product measures 700 parts.

III.—Salicylic acid....   4 drachms
Boric acid.......   5 drachms
Potassium c a r -
bonate........   1 drachm
Oil of cinnamon.   3 drachms
Oil of cloves.....   3 drachms
Glycerine.......   5 ounces
Alcohol.........   12 ounces
Hot water........   12 ounces

Dissolve the first 3 ingredients in the water and glycerine, the oils in the alcohol, and mix the solutions.

IV.—Thymol.........   15 grains
Alcohol.........   ½ ounce
Glycerine.......   10 ounces
Water..........   5 ounces

V.—Cooking salt.....   500 parts
Alum...........   750 parts
Arsenious acid...   350 parts
Zinc chloride....   120 parts
Mercury chloride   90 parts
Formaldehyde
solution, 40 per
cent.......... 6,000 parts
Water, up to..... 24,000 parts

VI.—Arsenious acid.... 360 grains
Mercuric chloride.   1¼ ounces
Alcohol...........   9 ounces
Sol. ac. carbolic, 5
per cent........ 120 ounces

From 10 to 12 pints are injected into the carotid artery—at first slowly and afterwards at intervals of from 15 to 30 minutes.

EMERALD (IMITATION):

See Gems, Artificial.

EMERY:

Emery Grinder.—Shellac, melted together with emery and fixed to a short metal rod, forms the grinder used for opening the holes in enameled watch dials and similar work. The grinder is generally rotated with the thumb and forefinger, and water is used to lubricate its cutting part, which soon wears away. The grinder is reshaped by heating the shellac and molding the mass while it is in a plastic condition.

Preparing Emery for Lapping.—To prepare emery for lapping screw-gages, plugs, etc., fill a half-pint bottle with machine oil and flour emery, 7 parts oil to 1 part emery, by bulk. Mix thoroughly and let stand for 20 minutes to settle. Take the bottle and pour off one-half the contents without disturbing the settlings. The portion poured off contains only the finest emery and will never scratch the work.

For surface lapping put some flour emery in a linen bag and tie up closely with a string. Dust out the emery by striking the bag against the surface plate; use turpentine for rough lapping and the dry surface plate for finishing.

Removing Glaze from Emery Wheels. —If the wheel is not altogether too hard, it can sometimes be remedied by reducing the face of the wheel to about ⅛ inch, or by reducing the speed, or by both. Emery wheels should be turned off so that they will run true before using. A wheel that glazes immediately after it has been turned off, can sometimes be corrected by loosening the nut, and allowing the wheel to assume a slightly different position, when it is again tightened.

Emery Substitute.—For making artificial emery, 1,634 parts of the following substances may be employed: Seven hundred and fifty-nine parts of bauxite, 700 parts of coke, and 96 parts of a flux, which may be a carbonate of lime, of potash, or of soda, preferably carbonate of lime on account of its low price. These materials are arranged in alternate layers and fused in an oven having a good draught. They are said to yield an artificial emery similar to the natural emery of Smyrna and Naxos, and at low cost.

EMULSIFIERS:

Rosin Soap as an Emulsifier.—The soap should be made by boiling gently for 2 hours, in an evaporating dish, a mixture of 1,800 grains rosin and 300 caustic soda with 20 fluidounces water. Upon cooling, the soap separates as a yellow mass, which is drained from the liquid, squeezed, then heated on a water bath until it is dry and friable. Fixed oils may be emulsified by adding 1 ounce to a solution of 10 grains soap in 1 ounce water. Volatile oils require 10 grains rosin soap, $2\frac{1}{4}$ ounces water, and 2 drachms oil. Creosote requires double this amount of soap. Thymol may be rendered miscible with water by dissolving 18 grains together with 20 grains soap in 3 fluidounces alcohol, then adding enough water to make 6 fluidounces. Of course many other substances may be emulsified with the same emulsifier.

Yolk of Egg as an Emulsifier.—The domestic ointment of Unona, consisting of a mixture of oil and yolk of egg, is miscible in all proportions with water. It is proposed to utilize this fact by substituting a diluted ointment for the gum emulsions in general use, the following being given as a general formula:

Yolk of egg......	10 parts
Balsam Peru.....1 to	2 parts
Zinc oxide.......5 to	10 parts
Distilled water....	100 parts

If desired, 33 parts of vinegar may be substituted for the same amount of water, while oil of cade, oil of birch, lianthral or storax may be substituted for the balsam Peru, and an equal quantity of talc, magnesium carbonate, sulphur of bismuth subcarbonate, may be introduced in place of the oxide of zinc. A further variation in the character of the liquid may be introduced by the use of medicated or perfumed waters instead of the plain distilled water. Where so diluted, as in the above formula, the yolk of egg separates out after long standing, but the mixture quickly reëmulsifies upon shaking. Tar and balsams can be emulsified by mixing with double their quantity of yolk of egg, then diluting by the addition of small quantities of water or milk.

Emulgen.—This emulsifying agent has the following composition: Gluten, 5; gum acacia, 5; gum tragacanth, 20; glycerine, 20; water, 50; alcohol, 10. This mixture forms a clear grayish jelly.

EMULSIONS OF PETROLEUM:
See Petroleum.

Enameling

(See also Ceramics, Glazes, Paints, Waterproofing, and Varnishes.)

COMMERCIAL ENAMELING.

Commercial enameling includes: (1) Hollow ware enameling for domestic use; (2) hollow ware enameling for chemical use; (3) enameling locomotive and other tubes; (4) enameling drain and water pipes; (5) signboard enameling.

There is one defect to which all enamel ware is subject, and that is chipping. This may be caused by (1) imperfect mixing of the enamels; (2) imperfect fusing; (3) imperfect pickling of the iron; (4) rough usage. With ordinary care a well-enameled article has been known to last in daily use for 10 or 12 years, whereas defective enameling, say, on a sign tablet—which is exempt from rough usage—may not have a life exceeding a few months. All enameled articles, such as hollow ware and sign tablets, first receive a coating of a composition chiefly composed of glass called "gray," and this is followed by a deposit of "white," any additional color required being laid above the white. In the mixing and depositing of these mixtures lie the secrets of successful enameling. The "gray" has to be fused not only on but also into the metal at a bright red—almost white—heat, and it is obvious that its constituents must be arranged and proportioned to expand and contract in a somewhat uniform manner with the iron itself. The "white" has to be fused on the surface of the gray, but the gray being much harder is not affected by the second firing. If it were liquid it would become mixed with the white and destroy its purity. Frequently, owing to inferior chemicals, imperfect mixing or fusing, a second coating of white is necessary, in order to produce a surface of the necessary purity and luster. The difficulties of enameling are thus easily understood. Unless the metals and chemicals are so arranged and manipulated that their capacities of expansion and contraction are approximately the same, inferior work will be produced. Oxide of iron on the surface of the plates, inferior chemicals, incorrect mixings, insufficient or overheating in the process of fusing, prevent that chemical combination which is essential to successful enameling. The coatings will be laid on and not combined, with the result that there will be inequalities in expansion and contraction which will cause the enamel to chip off immediately if submitted to anything approaching rough usage, and in a very short time if submitted to chemical or ordinary atmospheric conditions.

The manufacture of sign tablets is the simplest form to which this important art is adapted. Sign-tablet enameling is, however, kept as great a secret as any other type. This branch of the industry is divided up as follows: (1) Setting the plates; (2) scaling and pickling the plates; (3) mixing the enamel constituents; (4) melting the enamel constituents; (5) grinding the enamel constituents; (6) applying the enamel; (7) drying the enamel coatings; (8) fusing the enamel on the articles; (9) lettering—including alphabetical and other drawing, spacing, and artistic art in arrangement; (10) stencil cutting on paper and stencil metal; (11) brushing; (12) refusing. Distinctive branches of this work have distinctive experts, the arrangement being generally as follows: Nos. 1 and 2 may or may not be combined; Nos. 3 and 5 may or may not be combined; Nos. 4, 7, 8, and 12 generally combined; No. 6 generally the work of girls; Nos. 9 and 10 generally combined; No. 11 generally the work of girls and boys. The twelve processes, therefore, require six classes of trained workpeople, and incompetence or carelessness at any section can only result in imperfect plates or "wasters."

A brief description of these processes will enable the reader to understand the more detailed and technical description to follow, and is, therefore, not out of place. Ordinary iron sheets will do for the manufacture of sign tablets; but a specially prepared charcoal plate can be had at a slightly increased price. The latter type is the best, for in many cases the scaling and pickling may, to a certain extent, be dispensed with. To make this article, however, as complete as possible, we shall begin from the lowest rung of the manufacturing ladder—i. e., from the first steps in the working of suitable iron.

I.—Setting.—The plates may be received in sheets, and cut to the required size at the enameling factory, or, what is more general, received in sizes according to specification. The former are more liable to have buckled slightly or become dented, and have to be restored to a smooth and uniform surface by hammering on a flat plate. The operation seems simple, but an inexperienced operator may entirely fail to produce the desired result, and, if he does succeed, it is with the expenditure of a great amount of time. An expert setter with comparatively few and well-directed strokes brings an imperfect plate into truth and in readiness for the next operation.

II.—Scaling and Pickling.—The annealing of the sheets in special furnaces loosens the scale, which can then be easily removed, after which immersion for some time in diluted sulphuric or muriatic acid thoroughly cleans the plate.

Firing to a red heat follows, and then a generous course of scrubbing, and the last traces of acid are removed by dipping in boiling soda solution. Scouring with sand and washing in clean water may follow, and the metal has then a perfect and chemically clean surface.

III.—Mixing the Enamel Constituents.—Ground, foundation, or gray.—All articles, whether hollow ware or plates, are operated upon in a very similar manner. Both require the foundation coating generally called "gray." The gray constituents vary considerably in different manufactures; but as regards the use of lead, it is universally conceded that while it may in many instances be used with advantage in the enameling of sign tablets, etc., it should under no circumstances be introduced into the coating of articles for culinary purposes, or in which acids are to be used. The first successful commercial composition of this covering was: Cullet (broken glass), carbonate of soda, and boracic acid. This composition remained constant for many years, but ultimately gave place to the following: Cullet, red lead, borax, niter. The borax and red lead form the fluxes, while the niter is to "purify" the mass. Some of the later mixings consist of the following: Silica powder, crystallized or calcium borax, white lead, fused together. This would be called a frit, and with it should be pulverized powdered silica, clay, magnesia. This recipe is one requiring a very high temperature for fusing: Silica powder, borax, fused and ground with silica, clay, magnesia. This requires a slightly lower temperature: Frit of silica powder, borax, feldspar, fused together, and then ground with clay, feldspar, and magnesia.

The approximate quantities of each constituent will be given later, but it must always be remembered that no hard-and-fast line can be laid down. Chemicals vary in purity, the furnaces vary in temperature, the pounding, grinding, and mixing are not always done alike, and each of these exerts a certain influence on the character of the "melt." These compositions may be applied to the metal either in the form of a powder or of a liquid. Some few years ago the powder coating was in general use, but at the present time the liquid form is in favor, as it is considered easier of application, capable of giving a coating more uniform in thickness and less costly. In using the powder coating the plate is rubbed with a cloth dipped in a gum solution, and the powder then carefully dusted through a sieve over the surface. In this condition the plate is submitted to the fusing process. In using the liquid material the plate surface is dipped into or has the liquid mixing carefully poured over it, any surplus being drained off, and any parts which are not to be coated being wiped clean by a cloth. The coating is then dried in suitable stoves, after which it is ready for fusing on to the iron. The gray coating should be fairly uniform and smooth, free from holes or blisters, and thoroughly covering every part of the iron which is to be subjected to any outside influence. Cooling slowly is important. Rapid cooling frequently causes chipping of the coating, and in any case it will greatly reduce the tenacity of the connection existing between the glaze and the metal.

Generally the next surface is a white one, and it depends upon the class of article, the character of the enamels, and the efficiency of application, whether one coat or two will be required. Roughly speaking, the coating is composed of a glass to which is added oxide of tin, oxide of lead, or some other suitable opaque white chemical. The mixture must be so constituted as to fuse at a lower temperature than the foundation covering. If its temperature of fusion were the same the result would be that the gray would melt on the iron and become incorporated with the white, thus loosening the attachment of the mass to the iron and also destroying the purity of the white itself. Bone ash is sometimes used, as it becomes uniformly distributed throughout the melt, and remains in suspension instead of settling. Bone ash and oxide of lead are, however, in much less demand than oxide of tin. The lead is especially falling into disfavor, for the following reasons: Firstly, it requires special and laborious treatment; secondly, it gives a yellowish-white color; thirdly, it cannot resist the action of acids. The following is a recipe which was in very general use for some years: Glass (cullet), powdered flint, lead, soda (crystals), niter, arsenic. Another consists of the following: Borax, glass, silica powder, oxide of tin, niter, soda, magnesia, clay. These are fused together, and when being ground a mixture of Nos. 1, 3, 7, and boracic acid is added.

Enamel mixings containing glass or china are now generally in use, although for several years the experience of manufacturers using glass was not satisfactory Improved compositions and work-

ing now make this constituent a most useful, and, in fact, an almost essential element. The glass should be white broken glass, and as uniform in character as possible, as colored glass wou. impart a tinge of its own color to the mixing.

The following are two distinct glazes which do not contain glass or porcelain: Feldspar, oxide of tin, niter, soda. This is free from any poisonous body and requires no additions: Silica powder, oxide of tin, borax, soda, niter, carbonate of ammonia, or magnesia.

Alkalies.—Of the alkalies which are necessary to produce complete fusion of and combination with the quartz, soda is chiefly applied in enamel manufactures, as the fusing temperature is then lower.

Bone Ash.—This material will not add opacity, but only semi-transparency to the enamel, and is therefore not much used.

Boracic Acid.—Boracic acid is sometimes substituted for silicic acid, but generally about 15 per cent of the former to 85 per cent of the latter is added. Borax as a flux is, however, much more easily used and is therefore largely employed in enamel factories.

Borax.—Calcined borax, that is, borax from which a large proportion of the natural moisture has been eliminated, is best for enamel purposes. It is a flux that melts at medium heat, and enters into the formation of the vitreous basis. Borax has also the property of thoroughly distributing oxide colors in the enamels.

Clay.—Only a fairly pure clay can be used in enamel mixings, and the varieties of clay available are therefore limited. The two best are pipe—or white—clay and china clay—kaolin. The latter is purer than the former, and in addition to acting as a flux, it is used to increase the viscosity of mixings and therefore the opacity. It is used in much the same way as oxide of tin.

Cryolite.—Ground cryolite is a white mineral, easily fusible, and sometimes used in enamel mixings. It is closely associated with aluminum.

Cullet.—This is the general material used as a basis. Clear glass only should be introduced; and as the compositions of glass vary greatly, small experimental frits should always be made to arrive at the correct quantity to be added.

Feldspar.—The introduction of feldspar into an enamel frit increases consistency. The common white variety is generally used, and its preliminary treatment by pounding is similar to that adopted with quartz.

Fluor-Spar.—In this mineral we have another flux, which fuses at a red heat.

Fluxes.—These are for the purpose of regulating the temperature of fusion of a mixing—frit—some being better adapted for this purpose than others. This, however, is not the only consideration, for the character of the flux depends upon the composition or chemical changes to which the ingredients are to be subjected. The fluxes are borax, clays, cullet, porcelain, feldspar, gypsum, and fluor-spar.

Glass.—Glass is composed of lime, silicic acid, and soda or potash. The use of the glass is to form the hard, crystal-like foundation.

Gypsum.—This mineral is sometimes used in conjunction with baryta and fluor-spar.

Lead.—Crystallized carbonate of lead, or "lead white," is frequently used in enamels when a low temperature for fusion is required. It should never be used on articles to be submitted to chemical action, or for culinary use. Minium is a specially prepared oxide of lead, and suitable for enameling purposes, but is expensive.

Lime.—Lime is in the form of carbonate of calcium when used.

Magnesium Carbonate is used only in small quantities in enamel mixings. It necessitates a higher temperature for fusion, but does not affect the color to the slightest extent if pure.

Manganese.—As a decolorant, this mineral is very powerful, and therefore only small quantities must be used. Purity of the mineral is essential—i. e., it should contain from 95 to 98 per cent of binoxide of manganese.

Niter.—At a certain temperature niter shows a chemical change, which, when affected by some of the other constituents, assists in the formation of the vitreous base.

Porcelain.—Broken uncolored porcelain is sometimes used in enamel manufacture. Its composition: Quartz, china clay, and feldspar. It increases viscosity.

Red Lead.—This decolorant is sometimes called purifier. It will, however, interfere with certain coloring media, and when this is the case its use should at once be discontinued.

Silicic Acid.—Quartz, sand, rock crystal, and flint stone are all forms of this acid in crystallized form. By itself it is practically infusible, but it can be incorporated with other materials to form mixings requiring varying temperatures for fusion.

Soda.—The soda in general use is carbonate of soda—58 per cent—or enameling soda. The latter is specially prepared, so as to free it almost entirely from iron, and admit of the production of a pure white enamel when such is required.

Tin Oxide.—All enamels must contain white ingredients to produce opacity, and the most generally used is oxide of tin. By itself it cannot be fused, but with proper manipulation it becomes diffused throughout the enamel mass. On the quantity added depends the denseness or degree of opacity imparted to the enamel.

It will be understood that the enamel constituents are divided into four distinct groups : I. Fundamental media. II. Flux media. III. Decolorant media. IV. Coloring media. We have briefly considered the three first named, and we will now proceed to No. IV. The coloring material used is in every case a metallic oxide, so that, so far as this goes, the coloring of an enamel frit is easy enough. Great care is, however, necessary, and at times many difficulties present themselves, which can only be overcome by experience. Coloring oxides are very frequently adulterated, and certain kinds of the adulterants are injurious to the frit and to the finish of the color.

Comparison of Hollow Ware and Sign-Tablet Enameling.—The enameling for sign tablets is much the same as for hollow ware; the mixings are practically alike, but, as a general rule, the mixing is applied in a much more liquid form on the latter. It is easy to understand that hollow ware in everyday use receives rougher usage than tablets. By handling, it is submitted to compression, expansion, and more or less violence due to falls, knocks, etc., and unless, therefore, the enamel coating follows the changes of the metal due to these causes, the connection between the two will become loosened and chipping will take place.

The enamel, therefore, though much alike for both purposes, should be so prepared for hollow ware that it will be capable of withstanding the changes to which we have referred. In all cases it must be remembered that the thinner the coat of the enamel the better it will be distributed over the iron, and the greater will be its adherence to the iron. Any article heavily enameled is always liable to chip, especially if submitted to the slightest bending action, and therefore any excess of material added to a plate means that it will always be readily liable to separate from the plate. In hollow-ware enameling the preparation of each frit generally receives somewhat more attention than for plate enameling. The grinding is more effectively carried out, in order to remove almost every possibility of roughness on any part of the surface, especially the inside surface.

The iron used in tablet and hollow-ware manufacture is rolled sheet iron. It is supplied in a variety of qualities. Charcoal iron is purer than ordinary plate iron, more ductile, and therefore capable of being driven out to various forms and depths by stamping presses. The surface of the charcoal iron is not so liable to become oxidized, and therefore can be more readily made chemically clean for the reception of the enamels. Some manufacturers use charcoal plates for tablet work, but these are expensive; the ordinary plates, carefully pickled and cleaned, adapt themselves to the work satisfactorily.

The sheet irons generally used for the enameling purposes referred to vary in gauge. The finer the iron the greater must be the care used in coating it with enamel. Thin iron will rapidly become hot or cool, the temperatures changing much more quickly than that of the mixing. Unless care, therefore, is used, the result of fusing will be that the enamel mass will not have become thoroughly liquid, and its adherence to the iron will be imperfect.

If, however, the temperature is gradually raised to the maximum, and sympathetic combination takes place, the dangers of rapid cooling are avoided. Again, the iron, in losing its temperature more rapidly than the enamel, will contract, thus loosening its contact with the glaze, and the latter will either then, or after a short period of usage, chip off. We then arrive at the following hard-and-fast rules: (1) In all classes of enameling, but particularly where thin iron sheets are used, the temperature of the plate and its covering must be raised very gradually and very uniformly. (2) In all cases a plate which has had a glaze fused on its surface must be cooled very gradually and very uniformly. The importance of these rules cannot be over-estimated, and will, therefore, be referred to in a more practical way later.

In enameling factories no causes are more prolific in the production of waste than these, and in many cases the defects produced are erroneously attributed to something else. Cast iron is much easier to enamel than wrought iron. This is due to the granular character of its composition. It retains the enamels in its small microscopic recesses, and greater uniformity can be arrived at with greater ease. Cast-iron enameled sign tablets and hollow ware were at one time made, but their great weight made it impossible for them ever to come into general use.

Wrought-iron plates, if examined microscopically, will show that they are of a fibrous structure, the fibers running in the direction in which they have been rolled. The enamels, therefore, will be more liable to flow longitudinally than transversely, and this tendency will be more accentuated at some places than at others. This, however, is prevented by giving the iron sheets what might be described as a cast-iron finish. The sheets to be enameled should be thoroughly scoured in all directions by quartz or flint sand, no part of the surface being neglected. This thorough scrubbing will roughen the surface sufficiently to make it uniformly retentive of enamel mixture, and in no cases should it be omitted or carelessly carried out.

Copper Enameling.—On a clean copper surface the enameling process is easy. The foundation glaze is not essential, and when required the most beautiful results of blended colors can be obtained by very little additional experience to ordinary enameling.

When the vase or other article has been hammered out to the required shape in copper, it is passed on to another class of artisans, who prepare it for the hands of the enameler. The design or designs are sketched carefully. The working appliances consist only of a pointed tool, two or three small punches of varying sizes, and a hammer. With this small equipment the operator sets to work. The spaces between each dividing line are gradually lowered by hammering, and when this has been uniformly completed, each little recess is ready to receive its allotment of enamel. More accurate work even than this can be obtained by the introduction of flat wire. This wire is soldered or fixed on the vase, and forms the outline for the entire design. It may be of brass, copper, or gold, but is fixed and built round every item of the whole design with the most laborious care. It stands above the surface of the design on the copper articles, but the little recesses formed by it are then gradually filled up by enamel in successive fusings. The whole surface of the article is now ground perfectly smooth and polished until its luster is raised to the highest point possible, and when this stage has been reached the article is ready for the market.

From the Sheet to the Sign Tablet.— The plates are generally in lengths of 6 feet by 2 feet, 6 feet by 3 feet, etc., the gauge generally being from 14 to 22, according to the size and class of plates to be enameled. These must be cut, but some enamelers prefer to order their plates in specified sizes, which does away with the necessity of cutting at the enameling factory. In order, however, to make this article complete, we will assume that a stock of large plates is kept on hand, the sizes being 6 feet by 3 feet and 6 feet by 2 feet. An order for sign tablets is given; particulars, say as follows: Length, 2 feet by 12 inches, white letters on blue ground; lettering, The Engineer, 33 Norfolk Street; block letters, no border line, 2 holes. For ordinary purposes these particulars would be sufficient for the enameler.

Stage I.—Cutting the plate is the first operation. The plates 6 feet by 2 feet would first be cut down the center in a circular cutting machine, thus forming two strips, 6 feet by 12 inches. Each strip would then be cut into three lengths of 2 feet each. If a guillotine had to be used instead of a circular cutter, the plate would be first cut transversely at distances of 2 feet, thus forming three square pieces of 2 feet by 2 feet. These would then be subdivided longitudinally into two lengths each, the pieces being then 2 feet by 12 inches. Each sheet would thus be cut into six plates.

Stage II.—The cut plates should next have any roughness removed from the edges, then punched with two holes— one at each end, followed by leveling or setting. This is done by hammering carefully on a true flat surface.

Stage III.—The plates should then be taken and dipped into a hydrochloric acid bath made up of equal quantities of the acid and water. The plates are then raised to a red heat in the stoves, and on removal it will be found that the scale— iron oxide—has become loosened, and will readily fall off, leaving a clean metallic surface. A second course of cleaning then follows in diluted sulphuric acid—1 part acid to 20 parts water. In this bath the iron may be kept for about 12 hours. In some cases a much stronger bath is used, and the plates are left in only a very short time. The bath is constructed of hard wood coated inside with suitable varnish.

In mixing the sulphuric acid bath it must be remembered that the acid should be slowly poured into the water under continuous stirring. Following the bath, the metal is rinsed in water, after which it is thoroughly scoured with fine flinty sand. Rinsing again follows, but in boiling water, and then the metal is allowed to dry. The enameling process should immediately follow the drying, for if kept for any length of time the surface of the metal again becomes oxidized. In hollow-ware enameling the hydrochloric acid bath may be omitted.

Stage IV.—The plates are now ready for the reception of the foundation or gray coating. If powder is used the plate is wiped over with a gum solution, and then the powder is carefully and uniformly dusted through a fine sieve over the surface. The plate is then reversed and the operation repeated on the other side. If a liquid "gray" is to be used it should have a consistency of cream, and be poured or brushed with equal care over the two surfaces in succession, after the plate has been heated to be only just bearable to the touch. The plates are then put on rests, or petits, in a drying stove heated to about 160° F., and when thoroughly dry they are ready for the fusing operation. The petits, with the plates, are placed on a long fork fixed on a wagon, which can be moved backward and forward on rails; the door of the fusing oven is then raised and the wagon moved forward. The fork enters the oven just above fire clay brick supports arranged to receive the petits. The fork is then withdrawn and the door closed. The stove has a cherry-red, almost white, heat and in a few minutes the enamel coating has been uniformly melted, and the plates are ready to be removed on the petits and fork in the same manner as they were inserted. Rapid cooling must now be carefully avoided, otherwise the enamel and the iron will be liable to separate, and chipping will result. The temperature of fusion should be about 2,192° F.* When all the plates have been thus prepared they are carefully examined and defective ones laid aside, the others being now ready for the next operation.

*Melting a piece of copper will approximately represent this temperature.

Stage V.—The coating of the plate with white is the next stage. The temperature of fusion of the white glaze is lower than that of the gray, so that the plate will remain a shorter time in the stove, or be submitted to a somewhat lower temperature. The latter system is to be strongly recommended in order to prevent any possibility of fusion of the ground mass. The white should be made as liquid as possible consistent with good results. The advantages of thin coatings have already been explained, but if the mixing is too thin the ground coating will not only be irregularly covered, but, in fusion, bubbles will be produced, owing to the steam escaping, and these are fatal to the sale of any kind of enameled ware. When the plate has been thoroughly dried and fusion has taken place, slow and steady cooling is absolutely essential. Special muffles are frequently built for this purpose, and their use is the means of preventing a large number of wasters. Before putting on the glaze, care must be taken to remove the gray from any part which is not to be coated. The temperature of fusion should be about 1,890° F.,* and the time taken is about 5 minutes.

Stage VI.—The stencil must be cut with perfect exactitude. The letters should be as clear as possible, proportioned, and spaced to obtain the best effects as regards boldness and appearance. Stencils may be cut either from paper or from specially prepared soft metal, called stencil metal. The former are satisfactory enough when only a few plates are required from one stencil, but when large quantities are required, say, 60 upward, metal stencils should be used. The paper should be thick, tough, and strong, and is prepared in the following manner: Shellac is dissolved in methylated spirits to the ordinary liquid gum form, and this is spread over both sides of the paper with a brush. When thoroughly dry a second protective coating is added, and the paper is then ready for stencil work. The stencil cutter's outfit consists of suitable knives, steel rule, scales of various fractions to an inch, a large sheet of glass on which the cutting is done, and alphabets and numerals of various characters and types. For ordinary lettering one stencil is enough, but for more intricate designs 2, 3, and even 4 stencils may be required. In the preparation of the plates referred to in the paragraph preceding Stage I, only 1 stencil would be necessary. The paper before preparation would be measured out to the exact size of the plate, and the letters would be drawn in. The cutting would then be done, and the result shown at Fig. 1 would be obtained, the

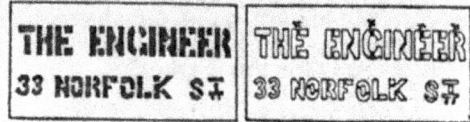

Fig. 1   Fig. 2

black parts being cut out. The lines or corners of each letter or figure should be perfectly clear and clean, for any flaw in the stencil will be reproduced on the plate.

Stage VII.—The next stage is the application of the blue enamel. The operation is almost identical with that of the white, but when the coating has been applied and dried, the lettering must be brushed out before it is fused. The coating is generally applied by a badger brush after a little gum water has been added; the effect of this is to make the blue more compact.

Stage VIII.—The next operation is brushing; the stencil is carefully placed over the plate, and held in position, and with a small hand brush with hard bristles the stencil is brushed over. This brushing removes all the blue coating, which shows the lettering and leaves the rest of the white intact. When this has been done, the stencil is removed and the connecting ribs of the lettering—some of which are marked X in Fig. 2—are then removed by hand, the instrument generally being a pointed stick of box or other similar wood.

Stage IX.—Fusing follows as in the case of the white glaze, and the plate is complete. One coat of blue should be sufficient, but if any defects are apparent a second layer is necessary.

The white and blue glazes are applied only on the front side of the plate, the back side being left coated with gray only.

From the Sheet to the Hollow Ware.— In hollow-ware enameling, the iron is received in squares, circles, or oblongs, of the size required for the ware to be turned out. It is soft and ductile, and by means of suitable punches and dies it is driven in a stamping press to the necessary shape. For shallow articles only one operation is necessary, but for deeper articles from 2 to 6 operations may be

* Melting a piece of brass will represent this temperature.

required, annealing in a specially constructed furnace taking place between each. Following the "drawing" operations comes that of trimming; this may be done in a press or spinning lathe, the object being to trim the edges and remove all roughness. The articles are now ready for enameling. For explanation, let us suppose they are tumblers, to be white inside, and blue outside. The gray is first laid on, then the white, and lastly the blue—that is, after the pickling and cleaning operations have been performed. The line of demarcation between the blue and white must be clear, otherwise the appearance of the article will not be satisfactory. The process of enameling is exactly the same as for sign-plate enameling, but more care must be exercised in order to obtain a smoother surface. While the liquid enamels are being applied, circular articles should be steadily rotated in order to let the coating flow uniformly and prevent thick and thin places. The enameling of "whole drawn" ironware presents no difficulty to the ordinary enameler, but with articles which are seamed or riveted, special care and experience is necessary.

Seamed or riveted parts are, of course, thicker than the ordinary plate, will expand and contract differently, will take longer to heat and longer to cool, and the conclusion, therefore, that must be arrived at is that the thickness should be reduced as much as possible, and the joints be made as smooth as possible. Unless special precautions are taken, cracks will be seen on articles of this kind running in straight lines from the rivets or seams. To avoid these, the enamel liquid must be reduced to the greatest stage of liquidity, the heat must be raised slowly, and in cooling the articles should pass through, say, 2 or 3 muffles, each one having a lower temperature than the preceding one. It is now generally conceded that the slower and more uniform the cooling process, the greater will be the durability of the enamel. Feldspar is an almost absolutely necessary addition to the gray in successful hollow-ware enameling, and the compositions of both gray and white should be such as to demand a high temperature for fusion. The utensils with the gray coating should first be raised to almost a red heat in a muffle, and then placed in a furnace raised to a white heat. The white should be treated similarly, and in this way the time taken for complete fusion at the last stage will be about 4 minutes.

The outside enamel on utensils is less viscous than the inside enamel, and should also be applied as thinly as possible.

Stoves and Furnaces.—Fritting and Fusing.—The best results are obtained in enameling when the thoroughly ground and mixed constituents are fused together, reground, and then applied to the metal surface. In cheap enamels the gray is sometimes applied without being previously melted, but it lacks the durability which is obtained by thorough fusion and regrinding. In smelting enamel one of two kinds of furnaces may be used, viz., tank or crucible. The former is better adapted to the melting of considerable quantities of ordinary enamel, while the latter is more suitable for smaller quantities or for finer enamels as the mixture is protected from the direct action of the flames by covers on the crucibles. The number of tanks and crucibles in connection with each furnace depends upon the heating capacity of the furnace and upon the out-turn required. They are so arranged that all or any of them can be used or put out of use readily by means of valves and dampers. Generally, they are arranged in groups of from 6 to 12, placed in a straight or circular line, but the object aimed at is complete combustion of the fuel, and the utilization of the heat to the fullest extent. One arrangement is to have the flame pass along the bottom and sides of the tank and then over the top to the chimney.

The general system in use is, however, the crucible system. The crucibles are made from the best fire clay, and the most satisfactory are sold under the name of "Hessian crucibles." The chief objection to the use of the crucibles is that of cost. They are expensive, and in many factories the life of the crucible is very short, in some cases not extending beyond one period of fusion. When this, however, is the rule rather than the exception, the results are due to carelessness. Sudden heating or cooling of the crucible will cause it to crack or fall to pieces, but for this there is no excuse. Running the molten material quickly out of the crucible and replacing it hurriedly with a fresh cold mixing is liable—in fact, almost certain—to produce fracture, not only causing the destruction of the crucible, but also the loss of the mixing. New crucibles should be thoroughly dried in a gentle heat for some days and then gradually raised to the requisite temperature which they must sustain for the purposes of fusion. Sometimes unglazed porcelain crucibles specially prepared with a large proportion of china clay are used. These are, however, expensive and require special attention during the first melt. The life of all crucibles can be lengthened by: (1) Gradually heating them before putting them into the fire; (2) never replacing a frit with a cold mass for the succeeding one; it should first be heated in a stove and then introduced into the crucible; (3) carefully protecting the hot crucibles from cold draughts or rapid cooling.

Melting and Melting Furnaces.—The arrangement of the melting furnace must be such as to protect the whole of the crucible from chills. The usual pit furnaces, with slight modifications, are suitable for this purpose. The crucible shown at b in Fig. 3 is of the type already

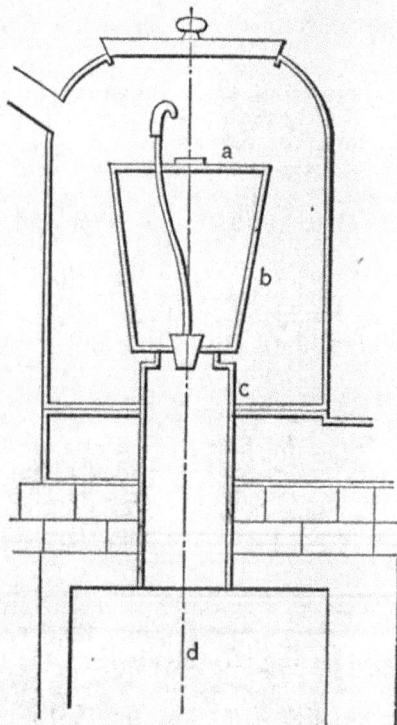

Fig. 3

described; at the top it is fitted with a lid, a, hinged at the middle, and at the bottom it is pierced by a 2-inch conical hole.* The hole, while melting is going on, is plugged up with a specially prepared stopper. The crucible stands on

* Two inches for gray, one inch for glaze; the hole should be wider at the top

a tubular fireproof support, c, which allows the molten mass to be easily run off into a tub of water, which is placed in the chamber, d. The fuel is thrown in from the top, and the supply must be kept uniform. From 4 to 6 of these furnaces are connected with the same chimney; but before passing to the chimney the hot gases are in some cases used for heating purposes in connection with the drying stove. The plug used may be either a permanent iron one coated with a very hard enamel or made from a composition of quartz powder and water. An uncovered iron plug would be unsuitable owing to the action of the iron on the ingredients of the mixing.

In some cases only a very small hole is made in the crucible and no stopper used, the fusion of the mixing automatically closing up the hole. In some other factories no hole is made in the crucible, and when fusion is complete the crucible is removed and the mixing poured out. The two latter systems are bad; in the first there is always some waste of material through leakage, and in the latter the operation of removing the crucible is clumsy and difficult, while the exposure to the colder atmosphere frequently causes rupture.

The plug used should be connected with a rod, as shown in Fig. 3, which passes through a slot in one-half of the hinged lid, a. . When fusion is complete this half is turned over, and the plug pulled up, thus allowing the molten mass to fall through into the vat of water placed underneath. The mixing in the crucibles, as it becomes molten, settles down, and more material can then be added until the crucible is nearly full. If the mixing is correctly composed, and has been thoroughly fused, it should flow freely from the crucible when the plug is withdrawn. Fusing generally requires only to be done once, but for fine enamels the operation may be repeated. The running off into the water is necessary in order to make the mass brittle and easy to grind. If this was not done it would again form into hard flinty lumps and require much time and labor to reduce to a powder.

A careful record should be kept of the loss in weight of the dried material at each operation. The weighings should be made at the following points: (1) Before and after melting; (2) after crushing.

The time required for melting varies greatly, but from 6 to 9 hours may be considered as the extreme limits. Gas is much used for raising the necessary heat for melting. The generator may be placed in any convenient position, but a very good system is to have it in the center of a battery of muffles, any or all of which can be brought into use. When quartz stoppers are used there is considerable trouble in their preparation, and as each new batch of material requires a fresh stopper, wrought-iron stoppers have been introduced in many factories. These are coated with an enamel requiring a much higher temperature of fusion than the fundamental substance, and this coating prevents the iron having any injurious action on the frit.

Fusing.—For fusing the enamel muffle furnaces are used; these furnaces are simple in construction, being designed specially for: (1) Minimum consumption of fuel; (2) maximum heat in the muffle; (3) protection of the inside of the muffle from dust, draughts, etc.

The muffle furnaces may be of any size, but in order to economize fuel, it is obvious that they should be no larger than is necessary for the class and quantity of work being turned out. For sign-plate enameling the interior of the muffle may be as much as 10 feet by 5 feet wide by 3 feet in height, but a furnace of this kind would be absolutely ruinous for a concern where only about a dozen small hollow-ware articles were enameled at a time. The best system is to have 2 or 3 muffle furnaces of different dimensions, as in this way all or any one of them can be brought into use as the character and number of the articles may require. The temperature throughout the muffle is not uniform, the end next to the furnace being hotter than that next to the door. In plate enameling it is therefore necessary that the plates should be turned so that uniform fusion of the enamel may take place. In the working of hollow ware the articles should be first placed at the front of the muffle and then moved toward the back. The front of the furnace is closed in by a vertically sliding door or lid, and in this an aperture is cut, through which the process of fusion can be inspected. All openings to the muffle should be used as little as possible; otherwise cold air is admitted, and the inside temperature rapidly lowered.

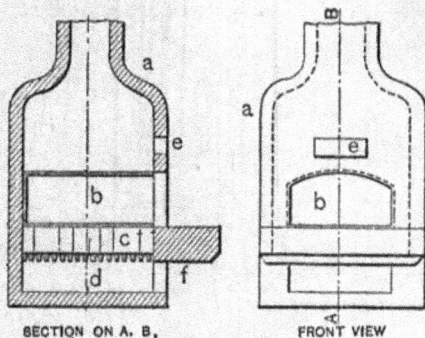

SECTION ON A. B.   FRONT VIEW

Fig. 4

Fig. 4 shows a simple arrangement of a muffle furnace; a is the furnace itself, with an opening, e, through which the fuel is fed; b is the muffle; c shows the firebars, and d the cinder box; f is a rest or plate on which is placed the articles to be enameled. The plate or petits on which the articles rest while being put into the muffle should be almost red hot, as the whole heat of the muffle in this way begins to act immediately on the enamel coating. The articles inside the muffles can be moved about when necessary, either by a hook or a pair of tongs, but care must be taken that every part of the vessel or plate is submitted to the same amount of heat.

In Figs. 5, 6, and 7 are given drawings of an arrangement of furnaces, etc., connected with an enameling factory at

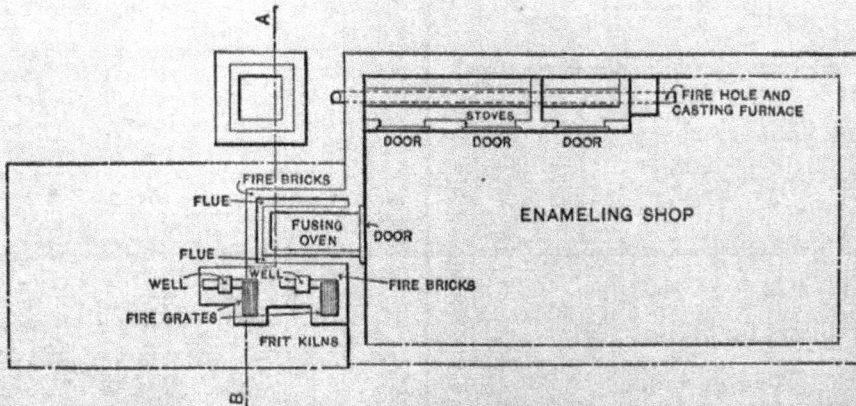

Fig. 5

present working. The stoves shown in Fig. 5 are drying stoves fired from the end by charcoal, and having a temperature of about 160° F. Fig. 6 shows the arrangement of the flues for the passage of the gases round the fusing oven. The section through the line *A B*, Fig. 5, as shown in Fig. 7, and the section through

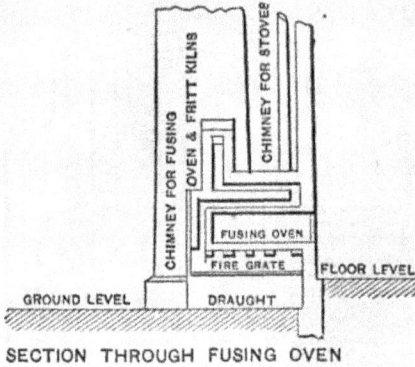

SECTION THROUGH FUSING OVEN

Fig. 6

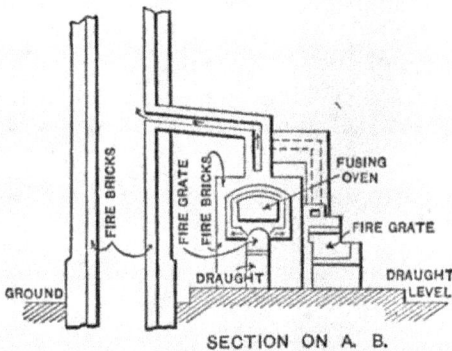

SECTION ON A. B.

Fig. 7

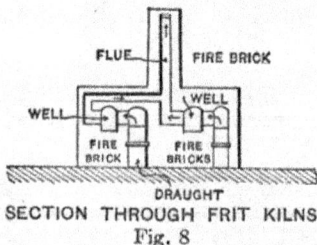

SECTION THROUGH FRIT KILNS

Fig. 8

the frit kilns, as shown in Fig. 8, are sufficiently explanatory. The frit kilns and the fusing oven flues both lead to the brick chimney, but the stoves are connected to a wrought-iron chimney shown in Fig. 6. Another arrangement would have been to so arrange the stoves that the gases from the frit kilns could have been utilized for heating purposes.

Fuel.—The consumption of fuel in an enameling factory is the most serious

item of the expenditure. Ill-constructed or badly proportioned stoves may represent any loss of coal from a quarter to one ton per day, and as great and uniform temperatures must be maintained, fuel of low quality and price is not desirable. In the melting stoves either arranged as tank or crucible furnaces, the character of the coal must not be neglected, as light dust, iron oxide, or injurious gases will enter into the crucibles through any opening, especially if the draught is not very great. Almost any of the various kinds of fuel may be used, provided that the system of combustion is specially arranged for in the construction of the furnaces. Charcoal is one of the best fuels available, its calorific value being so great; but its cost is in some places almost prohibitive. Wood burns too quickly, and is therefore expensive, and necessitates incessant firing.

For practical purposes we are thus often left to a selection of some type of coal. A coal with comparatively little heating power at a cheap price will be found more expensive in the end than one costing more, but capable of more rapid combustion and possessing more heat yielding gases. Cheap and hard coals give the fireman an amount of labor which is excessive. The proper maintenance of the temperature of the stove is almost impossible. Anthracite is excellent in every way, as it consists of nearly pure carbon, giving off a high degree of heat without smoke. Its use, of course, necessitates the use of a blower, but to this there can be no objection. Any coal which will burn freely and clean, giving off no excessive smoke, and capable of almost complete combustion, will give satisfaction in enameling; but it must not be forgotten that the consumption of fuel is so large that both price and quality must be carefully considered. Experimental tests must be made from time to time. A cheap, common coal will never give good results, and a good expensive coal will make the cost of manufacture so great that the prices of the enameled articles will render them unsalable. Any ordinary small factory will use from 2 to 4 tons per day of coal, and it will thus be seen that the financial success of a concern lies to a very great extent at the mouth of the furnace. Coke is a good medium for obtaining the necessary heat required in enameling if it can be got at a reasonable price. With a good draught a uniform temperature can be easily kept up, and the use of this by-product is, therefore, to be recommended.

With good coal and a furnace constructed to utilize the heat given off to the fullest extent, there may still be unnecessary waste. The arrangement of the bars should only be made by those who fully understand the character of the coal and the objects in view. The fireman in charge should be thoroughly experienced and reliable, as much waste is frequently traced to imperfect feeding of the fuel.

Each charge of articles should be as large as possible, as fusing will take place equally as well on many articles as on few. The charges should follow one another as rapidly as can be conveniently carried out; and where this is not done there is a lack of organization which should be immediately remedied.

Mills.—Any hard substances must first be broken up and pounded in a pounding or stamping mill, or in any other suitable manner, thus reducing the lumps to a granular condition. When this has been done, the coarse is separated from the fine parts and the former again operated on. The next process is roller grinding for reducing the hard fritted granular particles to a fine powder. These mills vary in construction, but a satisfactory type is shown in Fig. 9. Motion is con-

GRINDING MILL
Fig. 9

veyed by a belt to the driving pulley, and this is transmitted from the pinion to the large bevel, which is connected by a shaft to the ground plate. As this revolves the material causes the mill wheels to revolve, and in this way the material is reduced to a powder. The rollers are of reduced diameter on the inner side to prevent slippage, and when all the parts are made of iron, the metal must be close grained and of very hard structure, so as to reduce the amount removed by wear to a minimum. When the materials are ground wet, the powder should be carefully protected from dust and

thoroughly dried before passing to the next operation.

The glazing or enamel mills are shown in Fig. 10. These mills consist of a

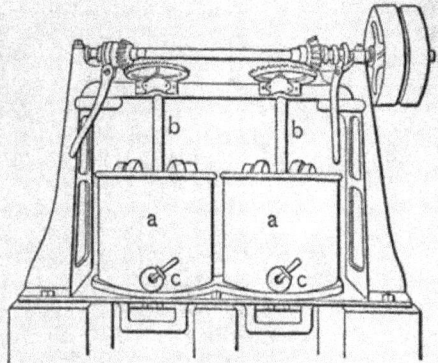

GLAZING MILL
Fig. 10

strong iron frame securely bolted to a stone foundation. In the sketch shown the framing carries 2 mills, but 3 or 4 can be arranged for. A common arrangement for small factories consists of 2 large mills, and 1 smaller mill, driven from the same shaft. One of the mills is used for foundation or gray mixings, the second for white, and the smallest one for colored mixings. In these mills it is essential that the construction is such as to prevent any iron fitting coming into contact with the mixing, for, as has already been explained, the iron will cause discoloration. The ground plate is composed of quartz and is immovable. It is surrounded by a wooden casing—as shown at a—and bound together by iron hoops. The millstones are heavy, rectangular blocks of quartz, called "French burr stone," and into the center the spindle, b, is led. The powdered material mixed with about three times its bulk of water is poured into the vats, a, and the grinding stones are then set in motion. When a condition ready for enameling has been reached the mixture is run off through the valves, c. Each mill can be thrown out of gear when required, by means of a clutch box, without interfering with the working of the others. The grinding stones wear rapidly and require to be refaced from time to time. To avoid stoppage of the work, therefore, it is advisable to always have a spare set in readiness to replace those removed for refacing. The composition of the stones should not be neglected, for, in many cases, faults in the enamel have been traced to the wearing away of stones containing earthy or metallic matter.

Enamel Mixing.—All constituents of which an enamel glaze is composed must be intimately mixed together. This can only be done by reducing each to a fine powder and thoroughly stirring them up together. This part of the work is often carried out in a very superficial manner, one material showing much larger lumps than another. Under circumstances such as these it is absurd to imagine that in fusion equal distribution will take place. What really happens is that some parts of the mass are insufficiently supplied with certain properties while others have too much. A mixture of this class can produce only unsatisfactory results in every respect, for the variations referred to will produce variations in the completeness of fusion in the viscous character of the mass, and in the color.

The mixing can be done by thoroughly stirring the various ingredients together, and a much better and cheaper system is mixing in rotating barrels or churns. These are mounted on axles which rest in bearings, one axle being long enough to carry a pulley. From the driving shaft a belt is led to the cask, which then rotates at a speed of from 40 to 60 revolutions per minute, and in about a quarter of an hour the operation is complete. The cask should not exceed the 5-gallon size, and should at no time be more than two-thirds full. Two casks of this kind give better results than one twice the size. The materials are shot into the cask in their correct proportions through a large bung hole, which is then closed over by a close-fitting lid.

Mixings.—For gray or fundamental coatings:

I.—Almost any kind of		
glass	49	per cent
Oxide of lead	47	per cent
Fused borax	4	per cent

II.—Glass (any kind)	61	per cent
Red lead	22	per cent
Borax	16	per cent
Niter	1	per cent

III.—Quartz	67.5	per cent
Borax	29.5	per cent
Soda (enameling)	3	per cent

The above is specially adapted for iron pipes.

IV.—Frit of silica pow-		
der	60	per cent
Borax	33	per cent
White lead	7	per cent

Fused and then ground with—
Three-tenths weight of silica frit.
Clay, three-tenths weight of silica frit.
Magnesia, one-sixth weight of white lead.

V.—Silica	65	per cent
Borax	14	per cent
Oxide of lead	4	per cent
Clay	15	per cent
Magnesia	2	per cent

No. V gives a fair average of several mixings which are in use, but it can be varied slightly to suit different conditions of work.

Defects in the Gray or Ground Coating.—Chipping is the most disastrous. This may be prevented by the addition of some bitter salt, say from 3 to 4 per cent of the weight of the frit.

The addition of magnesia when it has been omitted from the frit may also act as a preventive, but it should only be added in very small quantities, not exceeding 2.5 per cent, otherwise the temperature required for fusion will be very great.

Coating and Fusion.—Difficulties of either may generally be done away with by reducing the magnesia used in the frit to a minimum.

A soft surface is always the outcome of a mixing which can be fused at a low temperature. It is due to too much lead or an insufficiency of clay or silica powder.

A hard surface is due to the quantity of lead in the mixing being too small. Increase the quantity and introduce potash, say about 2.5 per cent.

The gray or fundamental mixing should be kept together in a condition only just sufficiently liquid to allow of being poured out. When required to be applied to the plate, the water necessary to lower it to the consistency of thick cream can then be added gradually, energetic stirring of the mass taking place simultaneously in order to obtain uniform distribution.

The time required for fusion may vary from 15 minutes to 25 minutes, but should never exceed the latter. If it does, it shows that the mixing is too viscous, and the remedy would be the addition and thorough intermixture of calcined borax or boracic acid. Should this fail, then remelting or a new frit is necessary.

A highly glazed surface on leaving the muffle shows that the composition is too fluid and requires the addition of clay, glass, silica powder or other substance to increase the viscosity.

As has been already explained, the glaze is much more important than the fundamental coating. Discoloration or slight flaws which could be tolerated in the latter would be fatal to the former.

In glazes, oxide of lead need not be used. It should never be used in a coating for vessels which are to contain acids or be used as cooking utensils. It may be used in sign-tablet production.

For pipes the following glaze gives good results:

I.—Feldspar	33	per cent
Borax	22.5	per cent
Quartz	16.5	per cent
Oxide of tin	15	per cent
Soda	8	per cent
Fluorspar	3.75	per cent
Saltpeter	2.25	per cent

For sign tablets the following gives fair results, although some of the succeeding ones are in more general use:

II.—Cullet	20	per cent
Powdered flint	15	per cent
Lead	52	per cent
Soda	4.5	per cent
Arsenic	4.5	per cent
Niter	4	per cent

III.—Frit of silica powder	30	per cent
Oxide of tin	18	per cent
Borax	17	per cent
Soda	8.6	per cent
Niter	7.5	per cent
White lead	5.5	per cent
Carbonate of ammonia	5.5	per cent
Magnesia	4	per cent
Silica powder	4	per cent

The following are useful for culinary utensils, as they do not contain lead:

IV.—Frit of silica powder	26	per cent
Oxide of tin	21	per cent
Borax	20	per cent
Soda	10.25	per cent
Niter	7	per cent
Carbonate of ammonia	5	per cent
Magnesia	3.25	per cent

This should be ground up with the following:

Silica powder	4.25	per cent
Oxide of tin	2.25	per cent
Soda	0.5	per cent
Magnesia	0.5	per cent

V.—Feldspar	41	per cent
Borax	35	per cent
Oxide of tin	17	per cent
Niter	7	per cent

VI.—Borax	30	per cent
Feldspar	22	per cent
Silicate powder	17.5	per cent
Oxide of tin	15	per cent
Soda	13.5	per cent
Niter	2	per cent

Borax will assist fusion. Quartz mixings require more soda than feldspar mixings.

VII.—Borax	28	per cent
Oxide of tin	19.5	per cent
Cullet (powdered white glass)	18	per cent
Silica powder	17.5	per cent
Niter	9.5	per cent
Magnesia	5	per cent
Clay	2.5	per cent

VIII.—Borax	26.75	per cent
Cullet	19	per cent
Silica powder	18.5	per cent
Oxide of tin	19	per cent
Niter	9.25	per cent
Magnesia	4.5	per cent
Soda	3	per cent

To No. VII must be added—while being ground—the following percentages of the weight of the frit:

Silica powder	18	per cent
Borax	9	per cent
Magnesia	5.25	per cent
Boracic acid	1.5	per cent

To No. VIII should be similarly added the following percentages of the frit:

Silica powder	1.75	per cent
Magnesia	1.75	per cent
Soda	1	per cent

This mixing is one which is used in the production of some of the best types of hollow ware for culinary purposes. The glaze should be kept in tubs mixed with water until used, and it should be carefully protected from dust.

Defects in the Glaze or White.—A bad white may be due to its being insufficiently opaque. More oxide of tin is required. Cracks may be prevented by the addition of carbonate of ammonia. Insufficient luster can be avoided by adding to the quantity of soda and reducing the borax. If the gray shows through the white it proves that the temperature of fusion is too high or the viscosity of the mixing is too great. If the coating is not uniformly spread it may be due to the glaze being too thin; add magnesia. If the glaze separates from the gray add some bitter salt. Viscosity will be increased by reducing the quantity of borax. Immunity against chemical reaction is procured by increasing the quantity of borax. An improved luster will be obtained by adding native carbonate of soda. The greater the quantity of silicic acid the greater must be the temperature for fusion. To reduce the temperature add borax. Clay will increase the difficulty of fusion. Oxide of lead will make a frit more easily fusible. A purer white can be obtained by adding a small quantity of smalt.

Water.—The character of the water used in the mixing of enamels is too frequently taken for granted, for unsuitable water may render a mixing almost entirely useless. Clean water, and with little or no sulphur present, is essential. For very fine enamels it is advisable to use carefully filtered water which has shown, after analysis, that it is free from any matter which is injurious to any of the enamel constituents.

How to Tell the Character of Enamel. —In the case of sign tablets the characteristics looked to are appearance and the adherence of the coatings to the iron. For the latter the tests are simple. The plate if slightly bent should not crack the coating. An enamel plate placed in boiling water for some time and then plunged into very cold water should not show any cracks, however small, even after repeated treatment of this kind.

Culinary utensils, and those to hold chemicals, should not only look well, but should be capable of resisting the action of acids. Lead should never enter into the composition of enamels of this class, as they then become easily acted upon, and in the case of chipping present a menace to health. The presence of lead is easily detected. Destroy the outside coating of the enamel at some spot by the application of strong nitric acid. Wash the part and apply a drop of ammonium sulphide. If lead is present, the part will become almost black, but remains unchanged in color if it is absent.

Another simple test is to switch up an egg in a vessel and allow it to stand for about 24 hours. When poured out and rinsed with water a dark stain will remain if lead is present in the enamel. To test the power of chemical resistance is equally simple. Boil diluted vinegar in the vessel for several minutes, and if a sediment is formed and the luster and smoothness of the glaze destroyed or partially destroyed, it follows that it is incapable of resisting the attacks of acids for any length of time. There are several other tests adopted, but those given present little difficulty in carrying out, and give reliable results.

Wasters and Seconds: Repairing Old Articles.—In all enameling there must be certain articles turned out which are defective, but the percentage should never be very great. The causes which most frequently tend to the production of wasters are new mixings and a temperature of fusion which is either too high or too low. There are two ways of disposing of defective articles, viz.: (1) Chipping off the bad spots, patching them up and selling them as "seconds"; (2) throwing the articles into the waste heap. The best firms adopt the latter course, because the recoating and firing of defective parts practically means a repetition of the whole process, thus adding greatly to the cost, while the selling price is reduced. Overheating in fusion is generally shown by blisters or by the enamel being too thin in various places. Chipping may be also due to this cause, the excessive heat having practically fused the fundamental coating.

At this stage the defects may be remedied by breaking off the faulty parts, patching them up, and then recoating the whole. With sign tablets there is no objection to doing so, but with hollow ware the fact remains that the article is faulty, no matter how carefully defects may be hidden. As white is the most general coating used, and shows up the defects more than the colored coatings, the greatest care is necessary at every stage of the manufacture. While glowing on the article, it should appear uniformly yellow, but on cooling it should revert to a pure white shade. On examining different makes of white coated articles, it will be found that some are more opaque than others. The former are less durable than the latter, because they contain a large percentage of oxide of tin, which reduces the elasticity. To ensure hardness the mixing must be very liquid, and this cannot be arrived at when a large quantity of oxide of tin is introduced.

Old utensils which have become broken or chipped can be repaired, although, except in the case of large articles, this is rarely done. The operations necessary are: (1) The defective parts chipped off; (2) submitted to a red heat for a few moments; (3) coated with gray on the exposed iron; (4) fused; (5) coated with the glaze on the gray; (6) fused.

To Repair Enameled Signs.—

Copal	5 parts
Damar	5 parts
Venice turpentine	4 parts

Powder the rosins, mix with the turpentine and add enough alcohol to form a thick liquid. To this add finely powdered zinc white in sufficient quantity to yield a plastic mass. Coloring matter may, of course, be added if desired.

The mass after application is polished when it has become sufficiently hard.

Enamel for Copper Cooking Vessels.— White fluorspar is ground to a fine powder and strongly calcined with an equal volume of unburnt gypsum, at a light glowing heat, stirring diligently. Grind the mixture to a paste with water, paint the vessel with it, using a brush, or pour in the paste like a glaze and dry the same. Increase the heat gradually and bring the vessels with the glass substance quickly into strong heat, under a suitable covering or a mantle of burnt clay. The substance soon forms a white opaque enamel, which ahderes firmly to the copper. It can stand pretty hard knocks without cracking, is adapted for cooking purposes and not attacked by acid matters. If the glassy substance is desired to cling well and firmly to the copper, a sudden and severe heat must be observed.

To Pickle Black Iron-Plate Scrap Before Enameling.—The black iron-plate scraps are first dipped clean in a mixture of about 1 part of sulphuric acid and 20 to 22 parts of water heated to 30° to 40° C. (86° to 104° F.), and sharp quartz sand is then used for scouring. They are then plunged for a few seconds in boiling water, taken out, and allowed to dry. Rinsing with cold water and allowing to dry thus may cause rust. The grains of quartz cut grooves in the fibers of the iron; this helps the grounding to adhere well. With many kinds of plate it is advisable to anneal after pickling, shutting off the air; by this means the plates will be thoroughly clean and free from oxidation. Much practice is required.—*The Engineer.*

ENAMELED IRON RECIPES.

The first thing is to produce a flux to fuse at a moderate heat, which, by flowing upon the plate, forms a uniform surface for the white or colored enamels to work upon.

Flux for Enameled Iron.—

White lead	10	parts
Ball clay	1	part
Flint glass	10	parts
Whiting	1	part

The plates may then be coated with any of the following mixtures, which may either be spread on as a powder with a little gum, as in the case of the flux, or the colors may be mixed with oil and the plates dipped therein when coated; the plate requires heating sufficiently to run the enamels bright.

Soft Enamels for Iron, White.—

Flint glass	16	parts
Oxide of tin	1½	parts
Niter	1½	parts
Red lead	4	parts
Flint or china clay	1	part

Black.—

Red oxide of iron	1¼	parts
Carbonate of cobalt	1¼	parts
Red lead	6	parts
Borax	2	parts
Lynn sand	2	parts

Yellow Coral.—

Chromate of lead	1	part
Red lead	2¾	parts
Flint	1	part
Borax	¼	part

Canary.—

Oxide of uranium	1	part
Red lead	4½	parts
Flint	1½	parts
Flint glass	1	part

Turquoise.—

Red lead	40	parts
Flint glass	12	parts
Borax	16	parts
Flint	12	parts
Enamel white	14	parts
Oxide of copper	7	parts
Oxide of cobalt	¼	part

Red Brown.—

Calcined sulphate of iron	1	part
Flux No. 8 (see page 307)	3	parts

Mazarine Blue.—

Oxide of cobalt	10	parts
Paris white	9	parts
Sulphate barytes	1	part

Fire the above at an intense heat and for use take

Above stain	1	part
Flux No. 8 (see page 307)	3	parts

Sky Blue.—

Flint glass	30	parts
White lead	10	parts
Pearlash	2	parts
Common salt	2	parts
Oxide of cobalt	4	parts
Enamel, white	4	parts

Chrome Green.—

Borax	10	parts
Oxide of chrome	4½	parts
White lead	9	parts
Flint glass	9	parts
Oxide of cobalt	2	parts
Oxide of tin	1	part

Coral Red.—

Bichromate potash ..	1	part
Red lead............	4½	parts
Sugar of lead........	1½	parts
Flint................	1½	parts
Flint glass..........	1	part

Enamel White.—Soft:

Red lead............	80	parts
Opal glass..........	50	parts
Flint................	50	parts
Borax..............	24	parts
Arsenic.............	8	parts
Niter...............	6	parts

Enamel White.—

Red lead............	10	parts
Flint................	6	parts
Boracic acid........	4	parts
Niter...............	1	part
Soda crystals........	1	part

Where the enameled work is intended to be exposed to the weather do not use flux No. 8, but substitute the following:

White lead..........	1	part
Ground flint glass....	1	part

All the enamels should, after being mixed, be melted in crucibles, poured out when in liquid, and powdered or ground for use.

FUSIBLE ENAMEL COLORS.

The following colors are fusible by heat, and are all suitable for the decoration of china and glass. In the following collection of recipes certain terms are employed which may not be quite understood by persons who are not connected with either the glass or porcelain industries, such as "glost fire" and "run down," and in such cases reference must be made to the following definitions:

"Run down." Sufficient heat to melt into liquid.

"Glost fire." Ordinary glaze heat.

"Grind only." No calcination required.

"Hard fire." Highest heat attainable.

"Frit." The ingredients partly composing a glaze, which require calcination.

"Stone." Always best Cornwall stone.

"Paris white." Superior quality of whiting.

"Parts." Always so many parts *by weight*, unless otherwise stated.

"D. L. Zinc." Particular brand not essential. Any good quality oxide of zinc will do.

Ruby and Maroon.—Preparation of silver:

Nitric acid..........	1	ounce
Water..............	1	ounce

Dissolve the silver till saturated, then put a plate of copper in the solution to precipitate the silver in a metallic state. Wash well with water to remove the acetate of copper.

Flux for Above.—Six dwts. white lead to 1 ounce prepared silver.

Tin Solution.—Put the acid (aqua regia) in a bottle, add tin in small quantities until it becomes a dark-red color; let it stand about 4 days before use. When the acid becomes saturated it will turn red at the bottom of the bottle, then shake it up and add more tin; let it stand and it will become clear.

Aqua Regia.—

Nitric acid..........	2	parts
Muriatic acid........	1	part

Dissolve grain gold in the aqua regia so as to make a saturated solution. Take a basin and fill it 3 parts full of water; drop the solution of gold into it till it becomes an amber color. Into this solution of gold gradually drop the solution of tin, until the precipitate is complete. Wash the precipitate until the water becomes tasteless, then dry slowly and flux as follows:

Flux No. 1.—

Borax..............	3	parts
Red lead............	3	parts
Flint................	2	parts

Run down.

Rose Mixture.—

Purple of Cassius....	1	ounce
Flux No. 1..........	6	ounces
Prepared silver......	3	dwts.
Flint glass..........	2	ounces

Grind.

Purple Mixture.—

Purple of Cassius	1	ounce
Flux No. 8 (see page 307)	2½	ounces
Flint glass..........	2	ounces

Grind.

Ruby.—

Purple mixture......	2½	parts
Rose mixture........	1½	parts

Grind.

Maroon.—

Rose mixture........	1	part
Purple mixture........	2	parts

Grind.

Black—Extra quality.—

Red oxide of iron	12 parts
Carbonate of cobalt ..	12 parts
Oxide of cobalt......	1 part
Black flux A (see next formula).........	80 parts

Glost fire.

Black Flux A.—

Red lead...........	3 parts
Calcined borax......	½ part
Lynn sand..........	1 part

Run down.

Black No. 2.—

Oxide of copper.....	1 part
Carbonate of cobalt..	½ part
Flux No. 8 (see next column)...........	4 parts

Grind only.

Enamel White.—

Arsenic.............	2½ parts
Niter...............	1½ parts
Borax..............	4 parts
Flint...............	16 parts
Glass..............	16 parts
Red lead...........	32 parts

Glost fire.

Turquoise.—China:

Calcined copper.....	5 parts
Whiting............	5 parts
Phosphate of soda....	8 parts
Oxide of zinc........	16 parts
Soda crystals........	4 parts
Magnesia...........	2 parts
Red lead...........	8 parts
Flux T (see next formula)	52 parts

Glost fire.

Flux T.—

Borax..............	2 parts
Sand...............	1 part

Run down.

Orange.—

Orange U. G.......	1 part
Flux No. 8 (see next column)...........	3 parts

Grind only.

Blue Green.—

Flint glass..........	8 parts
Enamel white.......	25 parts
Borax.............	8 parts
Red lead...........	24 parts
Flint..............	6 parts
Oxide of copper.....	2½ parts

Glost heat.

Coral Red.—

Chromate of potash..	1 part
Sugar of lead.......	1¼ parts

Dissolve in hot water, then dry. Take 1 part of above, 3 parts flux for coral. Grind.

Flux for Coral.—

Red lead............	4½ parts
Flint...............	1½ parts
Flint glass.........	1½ parts

Run down.

Turquoise.—

Oxide of copper.....	5 parts
Borax.............	10 parts
Flint..............	12 parts
Enamel white.......	14 parts
Red lead...........	40 parts

Glost fire.

Flux No. 8.—

Red lead...........	6 parts
Borax.............	4 parts
Flint..............	2 parts

Run down.

Russian Green.—

Malachite green.....	10 parts
Enamel yellow.......	5 parts
Majolica white......	5 parts
Flux No. 8 (see previous formula)......	2 parts

Grind only.

Amber.—

Oxide of uranium ...	1 part
Coral flux..........	8 parts

Grind only.

Gordon Green.—

Yellow U. G........	5 parts
Flux No. 8 (see above)	15 parts
Malachite green.....	10 parts

Grind only.

Celadon.—

Enamel light blue ...	1 part
Malachite green.....	1 part
Flux No. 8 (see above)	15 parts

Grind only.

Red Brown.—

Sulphate of iron, fired	1 part
Flux No. 8 (see above)	3 parts

Grind only.

Matt Blue.—

Flux No. 8 (see above)	10½ parts
Oxide of zinc........	5 parts
Oxide of cobalt......	4 parts

Glost fire, then take

Of above base.......	1 part
Flux No. 8 (see above)	1⅛ parts

Grind only.

PREPARATION OF ENAMELS.

The base of enamel is glass, colored different shades by the addition of metallic oxides mixed and melted with it.

The oxide of cobalt produces blue; red is obtained by the Cassius process. The purple of Cassius, which is one of the most brilliant of colors, is used almost exclusively in enameling and miniature painting; it is produced by adding to a solution of gold chloride a solution of tin chloride mixed with ferric chloride until a green color appears. The oxide of iron and of copper also produces red, but of a less rich tone; chrome produces green, and manganese violet; black is produced by the mixture of these oxides. Antimony and arsenic also enter into the composition of enamels.

Enamels are of two classes—opaque and transparent. The opacity is caused by the presence of tin.

When the mingled glass and oxides have been put in the crucible, this is placed in the furnace, heated to a temperature of 1,832° or 2,200° F. When the mixture becomes fused, it is stirred with a metal rod. Two or three hours are necessary for the operation. The enamel is then poured into water, which divides it into grains, or formed into cakes or masses, which are left to cool.

For applying enamels to metals, gold, silver, or copper, it is necessary to reduce them to powder, which is effected in an agate mortar with the aid of a pestle of the same material. During the operation the enamel ought to be soaked in water.

For dissolving the impurities which may have been formed during the work, a few drops of nitric acid are poured in immediately afterwards, well mixed, and then got rid of by repeated washing with filtered water. This should be carefully done, stirring the enamel powder with a glass rod, in order to keep the particles in suspension.

The powder is allowed to repose at the bottom of the vessel, after making sure by the taste of the water that it does not contain any trace of acid; only then is the enamel ready for use.

For enameling a jewel or other object it is necessary, first to heat it strongly, in order to burn off any fatty matter, and afterwards to cleanse it in a solution of nitric acid diluted with boiling water. After rinsing with pure water and wiping with a very clean cloth, it is heated slightly and is then ready to receive the enamel.

Enamels are applied with a steel tool in the form of a spatula; water is the vehicle. When the layers of enamel have been applied, the contained water is removed by means of a fine linen rag, pressing slightly on the parts that have received the enamel. The tissue absorbs the water, and nothing remains on the object except the enamel powder. It is placed before the fire to remove every trace of moisture. Thus prepared and put on a fire-clap slab, it is ready for its passage to the heat which fixes the enamel. This operation is conducted in a furnace, with a current of air whose temperature is about 1,832° F. In this operation the fire-chamber ought not to contain any gas.

Enamels are fused at a temperature of 1,292° to 1,472° F. Great attention is needed, for experience alone is the guide, and the duration of the process is quite short. On coming from the fire, the molecules composing the enamel powder have been fused together and present to the eye a vitreous surface covering the metal and adhering to it perfectly. Under the action of the heat the metallic oxides contained in the enamel have met the oxide of the metal and formed one body with it, thus adhering completely.

JEWELERS' ENAMELS.

Melt together:

Transparent Red.—Cassius gold purple, 65 parts, by weight; crystal glass, 30 parts, by weight; borax, 4 parts, by weight.

Transparent Blue.—Crystal glass, 34 parts, by weight; borax, 6 parts, by weight; cobalt oxide, 4 parts, by weight.

Dark Blue.—Crystal glass, 30 parts, by weight; borax, 6 parts, by weight; cobalt oxide, 4 parts, by weight; bone black, 4 parts, by weight; arsenic acid, 2 parts, by weight.

Transparent Green.—Crystal glass, 80 parts, by weight; cupric oxide, 4 parts, by weight; borax, 2 parts, by weight.

Dark Green.—Crystal glass, 30 parts, by weight; borax, 8 parts, by weight; cupric oxide, 4 parts, by weight; bone black, 4 parts by weight; arsenic acid, 2 parts, by weight.

Black.—Crystal glass, 30 parts, by weight; borax, 8 parts, by weight; cupric oxide, 4 parts, by weight; ferric oxide, 3 parts, by weight; cobalt oxide, 4 parts, by weight; manganic oxide, 4 parts, by weight.

White.—I.—Crystal glass, 30 parts, by weight; stannic oxide, 6 parts, by weight; borax, 6 parts, by weight; arsenic acid, 2 parts, by weight.

II.—Crystal glass, 30 parts, by weight; sodium antimonate, 10 parts, by weight.

The finely pulverized colored enamel is applied with a brush and lavender oil on the white enamel already fused in and then only heated until it melts. For certain purposes, the color compositions may also be fused in without a white ground. The glass used for white, No. 2, must be free from lead, otherwise the enamel will be unsightly.

Various Enamels for Precious Metals:
White.—Crystal glass, 30 parts, by weight; oxide of tin, 6 parts, by weight; borax, 6 parts, by weight; dioxide of arsenic, 2 parts, by weight, or silicious sand, 50 parts, by weight; powder, consisting of 15 of tin per 100 of lead, 100 parts, by weight; carbonate of potassium, 40 parts, by weight. Fuse the whole with a quantity of manganese. To take away the accidental coloring, pour it into water, and after having pulverized it, melt again 3 or 4 times.

Opaque Blue.—Crystal glass, 30 parts, by weight; borax, 6 parts, by weight; cobalt oxide, 4 parts, by weight; calcined bone, 4 parts, by weight; dioxide of arsenic, 2 parts, by weight.

Transparent Green.—Crystal glass, 30 parts, by weight; blue verditer, 4 parts, by weight; borax, 2 parts, by weight.

Opaque Green.—Crystal glass, 30 parts, by weight; borax, 8 parts, by weight; blue verditer, 4 parts, by weight; calcined bone, 4 parts, by weight; dioxide of arsenic, 2 parts, by weight.

Black.—I.—Crystal glass, 30 parts, by weight; borax, 8 parts, by weight; oxide of copper, 4 parts, by weight; oxide of iron, 3 parts, by weight; oxide of cobalt, 4 parts, by weight; oxide of manganese, 4 parts, by weight.

II.—Take $\frac{1}{2}$ part, by weight, of silver; $2\frac{1}{2}$ parts of copper; $3\frac{1}{2}$ parts of lead, and $2\frac{1}{2}$ parts of muriate of ammonia. Melt together and pour into a crucible with twice as much pulverized sulphur; the crucible is then to be immediately covered that the sulphur may not take fire, and the mixture is to be calcined over a smelting fire until the superfluous sulphur is burned away. The compound is then to be coarsely pounded, and, with a solution of muriate of ammonia, to be formed into a paste which is to be placed upon the article it is designed to enamel. The article must then be held over a spirit lamp till the compound upon it melts and flows. After this it may be smoothed and polished up in safety.

See also Varnishes and Ceramics for other enamel formulas.

ENAMEL COLORS, QUICK DRYING:
See Varnishes.

ENAMEL REMOVERS:
See Cleaning Preparations and Methods.

ENAMELING ALLOYS:
See Alloys.

ENGINES (GASOLINE), ANTI-FREEZING SOLUTION FOR:
See Freezing Preventives.

ENGRAVING SPOON HANDLES.

After the first monogram has been engraved, rub it with a mixture of 3 parts of beeswax, 3 of tallow, 1 of Canada balsam, and 1 of olive oil. Remove any superfluous quantity, then moisten a piece of paper with the tongue, and press it evenly upon the engraving. Lay a dry piece of paper over it, hold both firmly with thumb and forefinger of left hand, and rub over the surface with a polishing tool of steel or bone. The wet paper is thereby pressed into the engraving, and, with care, a clear impression is made. Remove the paper carefully, place it in the same position on another handle, and a clear impression will be left. The same paper can be used 2 dozen times or more.

ENGRAVING ON STEEL:
See Steel.

Engravings: Their Preservation

(See also Pictures, Prints, and Lithographs.)

Cleaning of Copperplate Engravings.—Wash the sheet on both sides by means of a soft sponge or brush with water to which 40 parts of ammonium carbonate has been added per 1,000 parts of water, and rinse the paper each time with clear water. Next moisten with water in which a little wine vinegar has been admixed, rinse the sheet again with water containing a little chloride of lime, and dry in the air, preferably in the sun. The paper will become perfectly clear without the print being injured.

Restoration of Old Prints.—Old engravings, woodcuts, or printed sheets that have turned yellow may be rendered white by first washing carefully in water containing a little hyposulphite of soda, and then dipping for a minute in javelle water. To prepare the latter, put 4 pounds of bicarbonate of soda in a pan, pour over it 1 gallon of boiling water; boil for 15 minutes, then stir in 1 pound of chloride of lime. When cold, pour off the clear liquid, and keep in a jug ready for use.

Surprising results are obtained from the use of hydrogen peroxide in the restoration of old copper or steel engravings or lithographs which have become soiled or yellow, and this without the least injury to the picture. The cellulose which makes the substance of the paper resists the action of ozone, and the black carbon color of these prints is indestructible.

To remove grease or other spots of dirt before bleaching, the engravings are treated with benzine. This is done by laying each one out flat in a shallow vessel and pouring the benzine over it. As benzine evaporates very rapidly, the vessel must be kept well covered, and since its vapors are also exceedingly inflammable, no fire or smoking should be allowed in the room. The picture is left for several hours, then lifted out and dried in the air, and finally brushed several times with a soft brush. The dust which was kept upon the paper by the grease now lies more loosely upon it and can easily be removed by brushing.

In many cases the above treatment is sufficient to improve the appearance of the picture. In the case of very old or badly soiled engravings, it is followed by a second, consisting in the immersion of the picture in a solution of sodium carbonate or a very dilute solution of caustic soda, it being left as before for several hours. After the liquid has been poured off, the picture must be repeatedly rinsed in clear water, to remove any remnant of the soda.

By these means the paper is so far cleansed that only spots of mold or other discolorations remain. These may be removed by hydrogen peroxide, in a fairly strong solution. The commercial peroxide may be diluted with 2 parts water.

The picture is laid in a shallow vessel, the peroxide poured over it, and the vessel placed in a strong light. Very soon the discolorations will pale.

To Reduce Engravings.—Plaster casts, as we know, can be perceptibly reduced in size by treatment with water or alcohol, and if this is properly done, the reduction is so even that the cast loses nothing of its clear outline, but sometimes even gains in this respect by contraction. If it is desired to reduce an engraved plate, make a plaster cast of it, treat this with water or alcohol, and fill the new cast with some easily fusible metal. This model, which will be considerably smaller than the original, is to be made again in plaster, and again treated, until the desired size is reached. In this way anything of the kind, even medallions, can be reproduced on a smaller scale.

ENLARGEMENTS:
See Photography.

ENVELOPE GUM:
See Adhesives, under Mucilages.

EPIZOOTY:
See Veterinary Formulas.

Essences and Extracts of Fruits

Preservation of Fruit Juices.—The juices of pulpy fruits, when fresh, contain an active principle known as pectin, which is the coagulating substance that forms the basis of fruit jellies. This it is which prevents the juice of berries and similar fruits from passing through filtering media. Pectin may be precipitated by the addition of alcohol, or by fermentation. The latter is the best, as the addition of alcohol to the fresh juices destroys their aroma and injures the taste. The induction of a light fermentation is far the better method, not only preserving, when carefully conducted, the taste and aroma of the fruit, but yielding far more juice. The fruit is crushed and the juice subsequently carefully but strongly pressed out. Sometimes the crushed fruit is allowed to stand awhile, and to proceed to a light fermentation before pressure is applied; but while a greater amount of juice is thus obtained, the aroma and flavor of the product are very sensibly injured by the procedure.

To the juice thus obtained, add from 1 to 2 per cent of sugar, and put away in a cool place (where the temperature will not rise over 70° or 75° F.). Fermentation soon begins, and will proceed for a few days. As soon as the development of carbonic acid gas ceases, the juice begins to clear itself, from the surface downward, and in a short time all solid matter will lie in a mass at the bottom, leaving the liquid bright and clear. Draw off the latter with a siphon, very carefully, so as not to disturb the sedimentary matter. Fermentation should be induced in closed vessels only, as when conducted in open containers a fungoid growth is apt to form on the surface, sometimes causing putrefactive, and at others, an acetic, fermentation, in either event spoiling the juice for subsequent use, except as a vinegar. The vessels, to effect the end desired, should be filled only two-thirds or three-fourths full, and then carefully closed with a tight-fitting cork, through which is passed a tube of glass, bent at the upper end, the short end of which passes below the surface of a vessel filled with water. As soon as fermentation commences the carbonic acid developed thereby escapes through the tube into the water, whence it passes off into the atmosphere. When bubbles no longer pass off from the tube the operation should be interrupted, and decantation or siphoning, with subsequent filtration, commenced.

By proceeding in this manner all the aroma and flavor of the juices are retained. If it is intended for preservation for any length of time the juice should be heated on a water bath to about 176° F. and poured, while hot, into bottles which have been asepticized by filling with cold water, and placing in a vessel similarly filled, bringing to a boiling temperature, and maintaining at this temperature until the juice, while still hot, is poured into them. If now closed with corks similarly asepticized, or by dipping into hot melted paraffine, the juice may be kept unaltered for years. It is better, however, to make the juice at once into syrup, using the best refined sugar, and boiling in a copper kettle (iron or tin spoil the color), following the usual precautions as to skimming, etc. The syrup should be poured hot into the bottles previously heated as before described.

Ripe fruit may be kept in suitable quantities for a considerable time if covered with a solution of saccharine and left undisturbed, this, too, without deteriorating the taste, color, or aroma of the fruit if packed with care.

Whole fruit may be stored in bulk, by carefully and without fracture filling into convenient-sized jars or bottles, and pouring thereon a solution containing a quarter of an ounce of refined saccharine to the gallon of water, so filling each vessel that the solution is within an inch of the cork when pressed into position. The corks should first of all be immersed in melted paraffine wax, then drained, and allowed to cool. When fruit juices alone are required for storage purposes they are prepared by subjecting the juicy fruits to considerable pressure, by which process the juices are liberated.

The sound ripe fruits are crushed and packed into felt or flannel bags. The fruit should be carefully selected, rotten or impaired portions being carefully removed; this is important, or the whole stock would be spoiled. Several methods are adopted for preserving and clarifying fruit juices.

A common way in which they are kept from fermenting is by the use of salicylic acid or other antiseptic substance, which destroys the fermentative germ, or otherwise retards its action for a considerable time. The use of this acid is seriously objected to by some as injurious to the consumer. About 2 ounces of salicylic acid, previously dissolved in alcohol, to 25 gallons of juice, or 40 grains to the gallon, is generally considered the proper proportion.

Another method adopted is to fill the freshly prepared cold juice into bottles until it reaches the necks, and on the top of this fruit juice a little glycerine is placed.

Juices thus preserved will keep in an unchanged condition in any season. Probably one of the best methods of preserving fruit juices is to add 15 per cent of 95 per cent alcohol. On such an addition, albumen and mucilaginous matter will be deposited. The juice may then be stored in large bottles, jars, or barrels, if securely closed, and when clear, so that further clarification is unnecessary, the juice should finally be decanted or siphoned off.

A method applicable to most berries is as follows:

Take fresh, ripe berries, stem them, and rub through a No. 8 sieve, rejecting all soft and green fruit. Add to each gallon of pulp thus obtained 8 pounds of granulated sugar. Put on the fire and bring just to a boil, stirring constantly. Just before removing from the fire, add to each gallon 1 ounce of a saturated alcoholic solution of salicylic acid, stirring well. Remove the scum, and, while still hot, put into jars and hermetically seal. Put the jars in cold water, and raise them to the boiling point, to prevent them from bursting by sudden expansion on pouring hot fruit into them. Fill the jars entirely full, so as to leave no air space when fruit cools and contracts.

Prevention of Foaming and Partial Caramelization of Fruit Juices.—Fresh fruit juices carry a notable amount of free carbonic acid, which must make its escape on heating the liquid. This will do easily enough if the juice be heated in its natural state, but the addition of the sugar so increases the density of the fluid that the acid finds escape difficult, and often the result is foaming. As to the burning or partial caramelization of the syrup, that is easily accounted for in the greater density of the syrup at the bottom of the kettle—the lighter portion, or that still carrying imprisoned gases, remaining on top until it is freed from them. Constant stirring can prevent this only partially, since it cannot entirely overcome the results of the natural forces in action. The consequence is more or less caramelization. The remedy is very simple. Boil the juices first, adding distilled water to make up for the loss by evaporation, and add the sugar afterwards.

ESSENCES AND EXTRACTS:

Almond Extracts.—

I.—Oil of bitter almonds 90 minims
    Alcohol, 94 per cent, quantity sufficient to make 8 ounces.

II.—Oil of bitter almonds 80 minims
    Alcohol. . . . . 7 ounces
    Distilled water, quantity sufficient to make 8 ounces.

III.—Oil of bitter almonds, deprived of its hydrocyanic acid . . . . 1 ounce
    Alcohol. . . . . 15 ounces

In order to remove the hydrocyanic acid in oil of bitter almonds, dissolve 2 parts of ferrous sulphate in 16 parts of distilled water; in another vessel slake 1 part freshly burned quicklime in a similar quantity of distilled water, and to this add the solution of iron sulphate, after the same has cooled. In the mixture put 4 parts of almond oil, and thoroughly agitate the liquids together. Repeat the agitation at an interval of 5 minutes, then filter. Put the filtrate into a glass retort and distil until all the oil has passed over. Remove any water that may be with the distillate by decantation, or otherwise.

Apricot Extract.—

Linalyl formate. . . . . . 90 minims
Glycerine. . . . . 1 ounce
Amyl valerianate. . . . . 4 drachms
Alcohol. . . . . 11 ounces
Fluid extract orris. . . . 1 ounce
Water, quantity sufficient to make 1 pint.

Apple Extract.—

Glycerine. . . . . 1 ounce
Amyl valerianate. . . . . 4 drachms
Linalyl formate. . . . . . 45 minims
Fluid extract orris. . . . 1 ounce
Alcohol. . . . . 11 ounces
Water, quantity sufficient to make 1 pint.

Apple Syrup.—

I.—Peel and remove the cores of, say, 5 parts of apples and cut them into little bits. Put in a suitable vessel and pour over them a mixture of 5 parts each of common white wine and water, and let macerate together for 5 days at from 125° to 135° F., the vessel being closed during the time. Then strain the liquid through a linen cloth, using gentle pressure on the solid matter, forcing as much as possible of it through the cloth. Boil 30 parts of sugar and 20 parts of water together, and when boiling add to the resulting syrup the apple juice; let it boil up for a minute or so, and strain through flannel.

II.—Good ripe apples are cut into small pieces and pounded to a pulp in a mortar of any metal with the exception of iron. To 1 part of this pulp add 11 parts of water. Allow this to stand for 12 hours. Colate. To 11 parts of the colature add 1 part of sugar. Boil for 5 minutes. Skim carefully. Bottle slightly warm. A small quantity of tartaric acid may be added to heighten the flavor.

Banana Syrup.—

Cut the fruit in slices and place in a jar; sprinkle with sugar and cover the jar, which is then enveloped in straw and placed in cold water and the latter is heated to the boiling point. The jar is then removed, allowed to cool, and the juice poured into bottles.

Cinnamon Essence.—

Oil of cinnamon. . . . . 2 drachms
Cinnamon, powdered 4 ounces
Alcohol, deodorized. . 16 ounces
Distilled water. . . . . . . 16 ounces

Dissolve the oil in the alcohol, and add the water, an ounce at a time, with agitation after each addition. Moisten the cinnamon with a little of the water, add, and agitate. Cork tightly, and put in a warm place, to macerate, 2 weeks, giving the flask a vigorous agitation several times a day. Finally, filter through paper, and keep in small vials, tightly stoppered.

Chocolate Extract.—

Probably the best form of chocolate extract is made as follows:

Curaçao cocoa. . . . . 400 parts
Vanilla, chopped fine. . . . . 1 part
Alcohol of 55 per cent. . . . . 2,000 parts

Mix and macerate together for 15 days, express and set aside. Pack the residue in a percolator, and pour on boiling water (soft) and percolate until 575 parts pass through. Put the percolate in a flask, cork, and let cool, then mix with the alcoholic extract. If it be desired to make a syrup, before mixing the extract, add 1,000 parts of sugar to the percolate, and with gentle heat dissolve the sugar. Mix the syrup thus formed, after cooling, with the alcoholic extract.

Coffee Extracts.—In making coffee extract, care must be used to avoid extracting the bitter properties of the coffee, as this is where most manufacturers fail; in trying to get a strong extract they succeed only in getting a bitter one.

I.—The coffee should be a mixture of Mocha, 3 parts; Old Government Java, 5 parts; or, as some prefer, Mocha, 3 parts; Java, 3 parts; best old Rio, 2 parts.

```
Coffee, freshly roasted
    and pulverized..... 100 parts
Boiling water........ 600 parts
```

Pack the coffee, moistened with boiling water, in a strainer, or dipper, placed in a vessel standing in the water bath at boiling point, and let 400 parts of the water, in active ebullition, pass slowly through it. Draw off the liquid as quickly as possible (best into a vessel previously heated by boiling water to nearly the boiling point), add 200 parts of boiling water, and pass the whole again through the strainer (the container remaining in the water bath). Remove from the bath; add 540 parts of sugar, and dissolve by agitation while still hot.

II.—The following is based upon Liebig's method of making coffee for table use: Moisten 50 parts of coffee, freshly roasted and powdered as before, with cold water, and add to it a little egg albumen and stir in. Pour over the whole 400 parts of boiling water, set on the fire, and let come to a boil. As the liquid foams, stir down with a spoon, but let it come to a boil for a moment; add a little cold water, cover tightly, and set aside in a warm place. Exhaust the residual coffee with 300 parts of boiling water, as detailed in the first process, and to the filtrate add carefully the now clarified extract, up to 600 parts, by adding boiling water. Proceed to make the syrup by the method detailed above.

III.—To make a more permanent extract of coffee saturate 600 parts of freshly roasted coffee, ground moderately fine, with any desired quantity of a 1 in 3 mixture of alcohol of 94 per cent and distilled water, and pack in a percolator. Close the faucet and let stand, closely stoppered, for 24 hours; then pour on the residue of the alcohol and water, and let run through, adding sufficient water, at the last, so as to compensate for what boils away. Set this aside, and continue the percolation, with boiling water, until the powder is exhausted. Evaporate the resultant percolate down to the consistency of the alcoholic extract, and mix the two. If desired, the result may be evaporated down to condition of an extract. To dissolve, add boiling water.

IV.—This essence is expressly adapted to boiling purposes. Take 3 pounds of good coffee, 4 ounces of granulated sugar, 4 pints of pure alcohol, 6 pints of hot water. Have coffee fresh roasted and of a medium grinding. Pack in a glass percolator, and percolate it with a menstruum, consisting of the water and the alcohol. Repeat the percolation until the desired strength is obtained, or the coffee exhausted; then add the sugar and filter.

```
V.—Mocha coffee....... 1 pound
    Java coffee......... 1 pound
    Glycerine, quantity sufficient.
    Water, quantity sufficient.
```

Grind the two coffees fine, and mix, then moisten with a mixture of 1 part of glycerine and 3 parts of water, and pack in a glass percolator, and percolate slowly until 30 ounces of the percolate is obtained. It is a more complete extraction if the menstruum be poured on in the condition of boiling, and it be allowed to macerate for 20 minutes before percolation commences. Coffee extract should, by preference, be made in a glass percolator. A glycerine menstruum is preferable to one of dilute alcohol, giving a finer product.

```
VI.—Coffee, Java, roast-
    ed, No. 20 pow-
    der............ 4 ounces
    Glycerine, pure.... 4 fluidounces
    Water, quantity sufficient.
    Boiling, quantity sufficient.
```

Moisten the coffee slightly with water, and pack firmly in a tin percolator; pour on water, gradually, until 4 fluidounces are obtained, then set aside. Place the coffee in a clean tin vessel, with 8 fluidounces of water, and boil for 5 minutes. Again place the coffee in the percolator with the water (infusion), and when the liquid has passed, or drained off, pack the grounds firmly, and pour on boiling water until 8 fluidounces are obtained. When cold, mix the first product, and add the glycerine, bottle, and cork well.

The excellence of this extract of coffee, from the manner of its preparation, will be found by experience to be incomparably superior to that made by the formulas usually recommended, the reason being apparent in the first step in the process.

Coffee Essence.—

Best ground Mocha coffee.............	4 pounds
Best ground chicory..	2 pounds

Boil with 2 gallons of water in a closed vessel and when cold, strain, press, and make up to 2 gallons, and to this add

Rectified spirit of wine	8 ounces
Pure glycerine (fluid)	16 ounces

Add syrup enough to make 4 gallons, and mix intimately.

Cucumber Essence.—Press the juice from cucumbers, mix with an equal volume of alcohol and distil. If the distillate is not sufficiently perfumed, more juice may be added and the mixture distilled. It is said that the essence thus prepared will not spoil when mixed with fats in the preparation of cosmetics.

Fruit Jelly Extract.—Fill into separate paper bags:

Medium finely powdered gelatin......	18 parts
Medium finely powdered citric acid....	3 parts

Likewise into a glass bottle a mixture of any desired

Fruit essence........	1 part
Spirit of wine........	1 part

and dissolve in the mixture for obtaining the desired color, raspberry red or lemon yellow, $\frac{1}{10}$ part.

For use, dissolve the gelatin and the citric acid in boiling water, adding

Sugar.............	125 parts

and mixing before cooling with the fruit essence mixture.

Ginger Extracts.—The following is an excellent method of preparing a soluble essence or extract of ginger:

I.—Jamaica ginger.....	24 ounces
Rectified spirits, 60 per cent........	45 ounces
Water.............	15 ounces

Mix and let macerate together with frequent agitations for 10 days, then percolate, press off, and filter. The yield should be 45 ounces. Of this take 40 ounces and mix with an equal amount of distilled water. Dissolve 6 drachms of sodium phosphate in 5 ounces of boiling water; let cool and add the solution to the filtrate and water, mixing well. Add 2 drachms of calcium chloride dissolved in 5 ounces of water, nearly cold, and again

thoroughly shake the whole. Let stand for 12 hours; then filter.

Put the filtrate in a still, and distil off, at as slow a temperature as possible, 30 ounces. Set this distillate to one side, and continue the distillation till another 40 ounces have passed, then let the still cool. The residue in the still, some 18 ounces, is the desired essence. Pour out all that is possible and wash the still with the 30 ounces of distillate first set aside. This takes up all that is essential. Finally, filter once more, through double filter paper and preserve the filtrate—about 40 ounces, of an amber-colored liquid containing all of the essentials of Jamaica ginger.

Soluble Essence of Ginger.—II.—The following is Harrop's method of proceeding:

Fluid extract of ginger (U. S.)........	4 ounces
Pumice, in moderately fine powder..	1 ounce
Water enough to make	12 ounces

Pour the fluid extract into a bottle, add the pumice and shake the mixture and repeat the shaking in the course of several hours. Now add the water in proportion of about 2 ounces, shaking well and frequently after each addition. When all is added repeat the agitation occasionally during 24 hours, then filter, returning the last portion of the filtrate until it comes through clear, and if necessary add sufficient water to make 12 ounces.

III.—Jamaica ginger,	
ground..........	2 pounds
Pumice stone, ground	2 ounces
Lime, slaked........	2 ounces
Alcohol, dilute.....	4 pints

Rub the ginger with the pumice stone and lime until thoroughly mixed. Moisten with the dilute alcohol until saturated and place in a narrow percolator, being careful not to use force in packing, but simply putting it in to obtain the position of a powder to be percolated, so that the menstruum will go through uniformly. Finally, add the dilute alcohol and proceed until 4 pints of percolate are obtained. Allow the liquid to stand for 24 hours; then filter if necessary.

IV.—Tincture ginger.....	480 parts
Tincture capsicum..	12 parts
Oleoresin ginger....	8 parts
Magnesium carbonate.............	16 parts

Rub the oleoresin with the magnesia, and add the tinctures; add about 400 parts of water, in divided portions, stirring vigorously the while. Transfer the mixture to a bottle, and allow to stand 1 week, shaking frequently; then filter, and make up 960 parts with water.

V.—Fluid extract of ginger
(U. S. P.)........... 4 ounces
Pumice, powdered and
washed............ 1 ounce
Water enough to make 12 ounces

Pour the fluid extract of ginger into a bottle, and add the pumice, shake thoroughly, set aside, and repeat the operation in the course of several hours. Add the water, in the proportion of about 2 ounces at a time, agitating vigorously after each addition. When all is added, repeat the agitation occasionally during 24 hours, then filter, returning the first portion of the filtrate until it comes through bright and clear. If necessary, pass water through the filter, enough to make 12 fluidounces of filtrate.

VI.—Strongest tincture
of ginger....... 1 pint
Fresh slaked lime. 1½ ounces
Salt of tartar...... ¼ ounce

VII.· Jamaica ginger,
ground........ 32 parts
Pumice stone, powdered.......... 32 parts
Lime, slaked..... 2 parts
Alcohol, dilute,
sufficient to make 32 parts

Rub the ginger with the pumice stone and lime, then moisten with alcohol until it is saturated with it. Put in a narrow percolator, using no force in packing. Allow the mass to stand for 24 hours, then let run through. Filter if necessary.

VIII.—The following is insoluble:

Cochin ginger,
cut fine...... 1,000 parts
Alcohol, 95 per
cent.......... 2,500 parts
Water......... 1,250 parts
Glycerine...... 250 parts

Digest together for 8 days in a very warm, not to say hot, place. Decant, press off the roots, and add to the colature, then filter through paper. This makes a strong, natural tasting essence.

IX.—Green Ginger Extract.—The green ginger root is freed from the epidermis and surface dried by exposure to the air for a few hours. It is then cut into thin slices and macerated for some days with an equal weight of rectified spirit, which when filtered will yield an essence possessing a very fine aroma and forming an almost perfectly clear solution in water. If the ginger is allowed to dry more than the few hours mentioned it will not produce a soluble essence. It is used in some of the imported ginger ales as a flavoring only, and makes a lovely ginger flavor.

Hop Syrup.—A palatable preparation not inferior to many of the so-called hop bitters:

Hops.............. 2 parts
Dandelion.......... 2 parts
Gentian............ 2 parts
Chamomile......... 2 parts
Stillingia............ 2 parts
Orange peel......... 2 parts
Alcohol............. 75 parts
Water.............. 75 parts
Syrup, simple....... 50 parts

Coarsely powder the drugs and exhaust with the water and alcohol mixed. Decant, press out and filter, and finally add the syrup. The dose is a wineglassful 2 or 3 times daily.

Lemon Essences.—I.—Macerate the cut-up fresh peelings of 40 lemons and 30 China oranges in 8 quarts of alcohol and 2 quarts of water, for 2 or 3 days, then distil off 8 quarts. Every 100 parts of this distillate is mixed with 75 parts of citric acid dissolved in 200 parts of water, colored with a trace of orange and filtered through talc. Each 200 parts of the filtrate is then mixed with 2 quarts of syrup.

II.—Twenty-five middle-sized lemons are thinly peeled, the peelings finely cut, and the whole, lemons and peels, put to macerate in a mixture of 3 pints 90 per cent alcohol and 5 quarts water. Let macerate for 24 hours. Add 10 drops lemon and 10 drops orange oil; then slowly distil off 4 quarts. The distillate will be turbid, but if left to stand in a cool, dark place for a week it will filter off clear, and should make a clear mixture with equal parts of water and simple syrup. If it does not, add with a pipette, drop by drop, sufficient alcohol to make it do so. Finally, dissolve in the mixture 4 drachms of vanillin, and color with a few drops of tincture of turmeric and a little caramel.

III.—Peel thinly and lightly, 25 medium-sized fresh lemons and 1 orange, and cut the peelings into very small pieces. Macerate in 55 drachms 96 per cent alcohol, for 6 hours. Filter off the macerate without pressing. Dilute the filtrate with 3 pints water and set aside for eight days, shaking frequently. At the end of this time filter. The filtrate is usually clear, and if so, add 4 drachms of vanillin. If not, proceed as in the second formula above.

IV.—Oil of lemon, select, 8 fluidounces; oil of lemon grass (fresh), 1 fluidrachm; peel, freshly grated, of 12 lemons; alcohol, 7 pints; boiled water, 1 pint.

Mix and macerate for 7 days. If in a hurry for the product, percolate through the lemon peel and filter. The addition of any other substance than the oil and rind of the lemon is not recommended.

V.—Fresh oil of lemon   64 parts
    Lemon peel (outer
        rind) freshly
        grated .... 32 parts
    Oil of lemon grass   1 part
    Alcohol......... 500 parts

Mix, let macerate for 14 days, and filter.

VI.—Essence of lemon $1\frac{3}{4}$ ounces
    Rectified spirit of
        wine......... 6 ounces
    Pure glycerine... 3 ounces
    Pure phosphate
        calcium....... 4 ounces
    Distilled water to make 1 pint.

Mix essence of lemon, spirit of wine, glycerine, and 8 ounces of distilled water, agitate briskly in a quart bottle for 10 minutes, and introduce phosphate of calcium and again shake. Put in a filter and let it pass through twice. Digest in filtrate for 2 or 3 days, add $1\frac{1}{2}$ ounces fresh lemon peel, and again filter.

VII.—Oil of lemon...... 6 parts
    Lemon peel (fresh-
        ly grated)...... 4 parts
    Alcohol, sufficient.

Dissolve the oil of lemon in 90 parts of alcohol, add the lemon peel, and macerate for 24 hours. Filter through paper, adding through the filter enough alcohol to make the filtrate weigh 100 parts.

VIII.—Exterior rind of
        lemon .... 2 ounces
    Alcohol, 95 per
        cent, deodorized 32 ounces
    Oil of lemon, re-
        cent......... 3 fluidounces

Expose the lemon rind to the air until perfectly dry, then bruise in a wedgwood mortar, and add it to the alcohol, agitating until the color is extracted; then add the lemon oil.

Natural Lemon Juice.—I.—Take 4.20 parts of crystallized citric acid; 2 parts

essence of lemons; 3 parts of alcohol of 96 per cent; $\frac{1}{2}$ part calcium carbonate; $50\frac{1}{20}$ parts sodium phosphate, and $\frac{1}{210}$ part calcium citrate, and dissolve the whole in sufficient water to make 60 parts.

II.—Squeeze out the lemon juice, strain it to get rid of the seeds and larger particles of pulp, etc., heat it to the boiling point, let it cool down, add talc, shake well together and filter. If it is to be kept a long time (as on a sea voyage) a little alcohol is added.

Limejuice.—This may be clarified by heating it either alone or mixed with a small quantity of egg albumen, in a suitable vessel, without stirring, to near the boiling point of water, until the impurities have coagulated and either risen to the top or sunk to the bottom. It is then filtered into clean bottles, which should be completely filled and closed (with pointed corks), so that each cork has to displace a portion of the liquid to be inserted. The bottles are sealed and kept at an even temperature (in a cellar). In this way the juice may be satisfactorily preserved.

Nutmeg Essence.—Oil of nutmeg, 2 drachms; mace, in powder, 1 ounce; alcohol, 95 per cent, deodorized, 32 ounces.

Dissolve the oil in the alcohol by agitation, add the mace, agitate, then stopper tightly, and macerate 12 hours. Filter through paper.

Orange Extract.—Grated peel of 24 oranges; alcohol, 1 quart; water, 1 quart; oil of orange, 4 drachms. Macerate the orange peel and oil of orange with alcohol for 2 weeks. Add distilled water and filter.

Orange Extract, Soluble.—I.—Pure oil of orange, $1\frac{1}{4}$ fluidounces; carbonate of magnesium, 2 ounces; alcohol, 12 fluidounces; water, quantity sufficient to make 2 pints.

II.—Dissolve oil of orange in the alcohol, and rub it with the carbonate of magnesium, in a mortar. Pour the mixture into a quart bottle, and fill the bottle with water. Allow to macerate for a week or more, shaking every day. Then filter through paper, adding enough water through the paper to make filtrate measure 2 pints.

Orange Peel, Soluble Extract.—
    Freshly grated orange
        rind............. 1 part
    Deodorized alcohol... 1 part

Macerate for 4 days and express. Add the expressed liquid to 10 per cent of its weight of powdered magnesium carbonate in a mortar, and rub thoroughly until a smooth, creamy mixture results; then gradually add the water, constantly stirring. Let stand for 48 hours, then filter through paper. Keep in an amber bottle and cool place. To make syrup of orange, add 1 part of this extract to 7 parts of heavy simple syrup.

Peach Extract.—

Linalyl formate.....	120 minims
Amyl valerianate....	8 drachms
Fluid extract orris...	2 ounces
Oenanthic ether....	2 drachms
Oil rue (pure German)............	30 minims
Chloroform........	2 drachms
Glycerine..........	2 ounces
Alcohol, 70 per cent, to 3 pints.	

Pineapple Essence.—

A ripe, but not too soft, pineapple, weighing about, say, 1 pound, is mashed up in a mortar with Tokay wine, 6 ounces. The mass is then brought into a flask with 1 pint of water, and allowed to stand 2 hours. Alcohol, 90 per cent, ¾ pint, is then added and the mixture distilled until 7 quarts of distillate have been collected. Cognac, 9 ounces, is then added to the distillation.

Pistachio Essence.—

I.—Essence of almond	2	fluidounces
Tincture of vanilla	4	fluidounces
Oil of neroli......	1	drop

II.—Oil of orange peel .	4	fluidrachms
Oil of cassia......	1	fluidrachm
Oil of bitter almond	15	minims
Oil of calamus....	15	minims
Oil of nutmeg.....	1½	fluidrachms
Oil of clove......	30	minims
Alcohol..........	12	fluidounces
Water............	4	fluidounces
Magnesium carbonate........	2	drachms

Shake together, allow to stand 24 hours, and filter.

Pomegranate Essence.—

Oil of sweet orange	3 parts
Oil of cloves......	3 parts
Tincture of vanilla.	15 parts
Tincture of ginger.	10 parts
Maraschino liqueur	150 parts
Tincture of coccionella............	165 parts
Distilled water.....	150 parts
Phosphoric acid, dilute	45 parts
Alcohol, 95 per cent, quantity sufficient to make 1,000 parts.	

Mix and dissolve.

Quince Extract.—

Fluid extract orris....	2 ounces
Oenanthic ether.....	1½ ounces
Linalyl formate......	90 minims
Glycerine...........	2 ounces
Alcohol, 70 per cent, to 3 pints.	

Raspberry Syrup, without Alcohol or Antiseptics.—

The majority of producers of fruit juices are firmly convinced that the preservation of these juices without the addition of alcohol, salicylic acid, etc., is impossible. Herr Steiner's process to the contrary is here reproduced:

The fruit is crushed and pressed; the juice, with 2 per cent of sugar added, is poured into containers to about three-quarters of their capacity, and there allowed to ferment. The containers are stoppered with a cork through which runs a tube, whose open end is protected by a bit of gum tubing, the extremity of which is immersed in a glass filled with water. It should not go deeper than $\frac{4}{10}$ of an inch high. The evolution of carbonic gas begins in about 4 hours and is so sharp that the point of the tube must not be immersed any deeper.

Ordinarily fermentation ceases on the tenth day, a fact that may be ascertained by shaking the container sharply, when, if it has ceased, no bubbles of gas will appear on the surface of the water.

The fermented juice is then filtered to get rid of the pectinic matters, yeast, etc., and the filtrate should be poured back on the filter several times. The juice filters quickly and comes off very clear. The necessary amount of sugar to make a syrup is now added to the liquid and allowed to dissolve gradually for 12 hours. At the end of this time the liquid is put on the fire and allowed to boil up at once, by which operation the solution of the sugar is made complete. Straining through a tin strainer and filling into heated bottles completes the process.

The addition of sugar to the freshly pressed juice has the advantage of causing the fermentation to progress to the full limit, and also to preserve, by the alcohol produced by fermentation, the beautiful red color of the juice.

Any fermentation that may be permitted prior to the pressing out of the juices is at the expense of aroma and flavor; but whether fermentation occurs before or after pressure of the berry, the ordinary alcohol test cannot determine whether the juice has been completely fermented (and consequently whether the pectins have been completely separated) or not. Since, in spite of the fact that the liquid remains limpid after 4 days' fermentation, the production of alcohol is progressing all the time—a demonstration that fermentation cannot then be completed, and that at least 10 days will be required for this purpose.

An abortive raspberry syrup is always due to an incomplete or faulty fermentation, for too often does it occur that incompletely fermented juices after a little time lose color and become turbid.

The habit of clarifying juices by shaking up with a bit of paper, talc, etc., or boiling with albumen is a useless waste of time and labor. By the process indicated the entire process of clarification occurs automatically, so to speak.

Deep Red Raspberry Syrup.—A much deeper and richer color than that ordinarily attained may be secured by adding to crushed raspberries, before fermentation, small quantities of sugar, sifted over the surface in layers. The ethylic alcohol produced by fermentation in this manner aids in the extraction of the red coloring matter of the fruit. Moreover, the fermented juice should never be cooked over a fire, but by superheated steam. Only in this way can caramelization be completely avoided. Only sugar free from ultramarine and chalk should be used in making the syrup, as these impurities also have a bad influence on the color.

Raspberry Essences.—

I.—Raspberries, fresh..16 ounces
   Angelica (California) 6 fluidounces
   Brandy (California) 6 ounces
   Alcohol .... 6 ounces
   Water, quantity sufficient.

Mash the berries to a pulp in a mortar or bowl, and transfer to a flask, along with the Angelica, brandy, alcohol, and about 8 ounces of water. Let macerate overnight, then distil off until 32 ounces have passed over. Color red. The addition of a trifle of essence of vanilla improves this essence.

II.—Fresh raspberries... 200 grams
    Water, distilled..... 100 grams
    Vanilla essence..... 2 grams

Pulp the raspberries, let stand at a temperature of about 70° F. for 48 hours, and then add 100 grams of water. Fifty grams are then distilled off, and alcohol, 90 per cent, 25 grams, in which 0.01 vanillin has been previously dissolved, is added to the distillate.

Sarsaparilla, Soluble Extract.—
   Pure oil of winter-
      green.......... 5 fluidrachms
   Pure oil of sassa-
      fras.......... 5 fluidrachms
   Pure oil of anise.. 5 fluidrachms
   Carbonate of mag-
      nesium........ 2½ ounces
   Alcohol.......... 1 pint
   Water, quantity sufficient to make
      2 pints.

Dissolve the various oils in the alcohol, and rub with carbonate of magnesium in a mortar. Pour the mixture into a quart bottle, and fill the bottle with water. Allow to macerate for a week or more, shaking every day. Then filter through the paper, adding enough water through the paper to make the finished product measure 2 pints.

Strawberry Juice.—Put into the water bath 1,000 parts of distilled water and 600 parts of sugar and boil, with constant skimming, until no more scum arises. Add 5 parts of citric acid and continue the boiling until about 1,250 parts are left. Stir in, little by little, 500 parts of fresh strawberries, properly stemmed, and be particularly careful not to crush the fruit. When all the berries are added, cover the vessel, remove from the fire, put into a warm place and let stand, closely covered, for 3 hours, or until the mass has cooled down to the surrounding temperature, then strain off through flannel, being careful not to crush the berries. Prepare a sufficient number of pint bottles by filling them with warm water, putting them into a kettle of the same and heating them to boiling, then rapidly emptying and draining as quickly as possible. Into these pour the hot juice, cork and seal the bottles as rapidly as possible. Juice thus prepared retains all the aroma and flavor of the fresh berry, and if carefully corked and sealed up will retain its properties a year.

Strawberry Essence.—
   Strawberries, fresh.. 16 ounces
   Angelica (California) 6 fluidounces
   Brandy (California). 6 ounces
   Alcohol .... 8 ounces
   Water, quantity sufficient.

Mash the berries to a pulp in a mortar or bowl, and transfer to a flask, along with the Angelica, brandy, alcohol, and about 8 ounces of water. Let macerate overnight, then distil off until 32 ounces have passed over. Color strawberry red. The addition of a little essence of vanilla and a hint of lemon improves this essence.

Tea Extract.—

I.—

Best Souchong tea	175 parts
Cinnamon	3 parts
Cloves	3 parts
Vanilla	1 part
Arrack	800 parts
Rum	200 parts

Coarsely powder the cinnamon, clove, etc., mix the ingredients, and let macerate for 3 days, then filter, press off, and make up to 1,000 parts, if necessary, by adding rum. The Souchong may be replaced by any other brand of tea, and the place of the arrack may be occupied by Santa Cruz, or New England rum. The addition of fluid extract of kola nut not only improves the taste, but gives the drink a remarkably stimulating property. The preparation makes a clear solution with either hot or cold water and keeps well.

II.—Tea, any desirable variety, 16 ounces; glycerine, 4 ounces; hot water, 4 pints; water, sufficient to make 1 pint.

Reduce the tea to a powder, moisten with sufficient of the glycerine and alcohol mixed, with 4 ounces of water added, pack in percolator, and pour on the alcohol (diluted with glycerine and water) until 12 ounces of percolate have been obtained. Set this aside, and complete the percolation with the hot water. When this has passed through, evaporate to 4 ounces, and add it to the percolate first obtained.

Tonka Extract.—

Tonka beans	1 ounce
Magnesium carbonate, quantity sufficient.	
Balsam of Peru	2 drachms
Sugar	4 ounces
Alcohol	8 ounces

Water sufficient to make 16 ounces.

Mix the tonka, balsam of Peru, and magnesia, and rub together, gradually adding the sugar until a homogeneous powder is obtained. Pack in a percolator; mix the alcohol with an equal amount of water, and pour over the powder, close the exit of the percolator, and let macerate for 24 to 36 hours, then open the percolator, and let pass through, gradually adding water until 16 ounces pass through.

Vanilla Extracts.—I.—Vanilla, in fine bits, 250 parts, is put into 1,350 parts of mixture, of 2,500 parts 95 per cent alcohol, and 1,500 parts distilled water. Cover tightly, put on the water bath, and digest for 1 hour, at 140° F. Pour off the liquid and set aside. To the residue in the bath, add half the remaining water, and treat in the same manner. Pack the vanilla in an extraction apparatus, and treat with 250 parts of alcohol and water, mixed in the same proportions as before. Mix the results of the three infusions first made, filter, and wash the filter paper with the results of the percolation, allowing the filtered percolate to mingle with the filtrate of the mixed infusions.

II.—Take 60 parts of the best vanilla beans, cut into little pieces, and put into a deep vessel, wrapped with a cloth to retain the heat as long as possible. Shake over the vanilla 1 part of potassium carbonate in powder, and immediately add 240 parts distilled water, in an active state of ebullition. Cover the vessel closely, set aside until it is completely cold, and then add 720 parts alcohol. Cover closely, and set aside in a moderately warm place for 15 days, when the liquid is strained off, the residue pressed, and the whole colate filtered. The addition of 1 part musk to the vanilla before pouring on the hot water improves this essence.

To prepare vanilla fountain syrup with extracts I or II, mix 25 minims of the extract with 1 pint simple syrup. Color with caramel.

III.—

Vanilla beans, cut fine	1 ounce
Sugar	3 ounces
Alcohol, 50 per cent.	1 pint

Beat sugar and vanilla together to a fine powder. Pour on the dilute alcohol, cork the vessel, and let stand for 2 weeks, shaking it up 2 or 3 times a day.

IV.—

Vanilla beans, chopped fine	30 parts
Potassium carbonate	1 part
Boiling water	1,450 parts
Alcohol	450 parts
Essence of musk	1 part

Dissolve the potassium carbonate in the boiling water, add the vanilla, cover the vessel, and let stand in a moderately warm place until cold. Transfer to a wide-mouthed jar, add the alcohol, cork, and let macerate for 15 days; then decant the clear essence and filter the remainder. Mix the two liquids and add the essence of musk.

V.—Cut 60 parts of best vanilla beans into small bits; put into a deep vessel, which should be well wrapped in a woolen cloth to retain heat as long as possible. Shake over the beans 1 part of potassium carbonate, in powder, then pour over the mass 240 parts distilled water, in an active state of ebullition, cover the vessel closely, and set aside in a moderately warm place. When quite cold add 720 parts alcohol, close the vessel tightly, and set aside in a moderately warm place, to macerate for 15 days, then strain off, press out, and set aside for a day or two. The liquid may then be filtered and bottled. The addition of a little musk to the beans before pouring on the hot water, is thought by many to greatly improve the product. One part of this extract added to 300 parts simple syrup is excellent for fountain purposes.

VI.—Vanilla beans....... 8 ounces
    Glycerine.......... 6 ounces
    Granulated sugar... 1 pound
    Water............. 4 pints
    Alcohol of cologne
      spirits.......... 4 pints

Cut or grind the beans very fine; rub with the glycerine and put in a wooden keg; dissolve the sugar in the water, first heating the water, if convenient; mix the water and spirits, and add to the vanilla; pour in keg. Keep in a warm place from 3 to 6 months before using. Shake often. To clear, percolate through the dregs. If a dark, rich color is desired add a little sugar coloring.

VII.—Vanilla beans,
    good quality.. 16 ounces
    Alcohol........ 64 fluidounces
    Glycerine...... 24 fluidounces
    Water......... 10 fluidounces
    Dilute alcohol, quantity sufficient.

Mix and macerate, with frequent agitation, for 3 weeks, filter, and add dilute alcohol to make 1 gallon.

VIII.—Vanilla beans,
    good quality... 8 ounces
    Pumice stone,
      lump......... 1 ounce
    Rock candy..... 8 ounces
    Alcohol and water, of each a suffi-
      ciency.

Cut the beans to fine shreds and triturate well with the pumice stone and rock candy. Place the whole in a percolator and percolate with a menstruum composed of 9 parts alcohol and 7 parts water until the percolate passes through clear. Bring the bulk up to 1 gallon with the same menstruum and set aside to ripen.

IX.—Cut up, as finely as possible, 20 parts of vanilla bean and with 40 parts of milk sugar (rendered as dry as possible by being kept in a drying closet until it no longer loses weight) rub to a coarse powder. Moisten with 10 parts of dilute alcohol, pack somewhat loosely in a closed percolator and let stand for 2 hours. Add 40 parts of dilute alcohol, close the percolator, and let stand 8 days. At the end of this time add 110 parts of dilute alcohol, and let pass through. The residue will repay working over. Dry it well, add 5 parts of vanillin, and 110 parts of milk sugar and pass through a sieve, then treat as before.

The following are cheap extracts:

X.—Vanilla beans,
    chopped fine.. 5 parts
    Tonka beans,
      powdered..... 10 parts
    Sugar, powdered. 14 parts
    Alcohol, 95 per
      cent.......... 25 parts
    Water, quantity sufficient to
      make 100 parts.

Rub the sugar and vanilla to a fine powder, add the tonka beans, and incorporate. Pack into a filter, and pour on 10 parts of alcohol, cut with 15 parts of water; close the faucet, and let macerate overnight. In the morning percolate with the remaining alcohol, added to 80 parts of water, until 100 parts of percolate pass through.

XI.—Vanilla beans..... 4 ounces
    Tonka beans..... 8 ounces
    Deodorized alcohol 8 pints
    Simple syrup..... 2 pints

Cut and bruise the vanilla beans, afterwards bruising the tonka beans. Macerate for 14 days in one-half of the spirit, with occasional agitation. Pour off the clear liquor and set aside; pour the remaining spirits in the magma, and heat by means of the water bath to about 170° F. in a loosely covered vessel. Keep at this temperature 2 or 3 hours, and strain through flannel, with slight pressure. Mix the two portions of liquid, and filter through felt. Add the syrup.

White Pine and Tar Syrup.—

White pine bark....	75 parts
Wild cherry bark....	75 parts
Spikenard root....	10 parts
Balm of Gilead buds	10 parts
Sanguinaria root....	8 parts
Sassafras bark......	7 parts
Sugar.............	750 parts
Chloroform........	6 parts
Syrup of tar........	75 parts
Alcohol, enough.	
Water, enough.	
Syrup enough to make 1,000 parts.	

Reduce the first six ingredients to a coarse powder and by using a menstruum composed of 1 in 3 alcohol, obtain 500 parts of a tincture from them. In this dissolve the sugar, add the syrup of tar and the chloroform, and, finally, enough syrup to bring the measure of the finished product up to 1,000 parts.

Wild Cherry Extract.—

Oenanthic ether..	2 fluidrachms
Amyl acetate.....	2 fluidrachms
Oil of bitter almonds (free from hydrocyanic acid)	1 fluidrachm
Fluid extract of wild cherry..........	3 fluidounces
Glycerine........	2 fluidounces
Deodorized alcohol enough to make 16 fluidounces.	

HARMLESS COLORS FOR USE IN SYRUPS, ETC.:

Red.—Cochineal syrup, prepared as follows:

I.—Cochineal in coarse powder..........	6 parts
Potassium carbonate..............	3 parts
Distilled water.....	15 parts
Alcohol, 95 per cent	12 parts
Simple syrup to make 500 parts.	

Rub the cochineal and potassium together, adding the water and alcohol little by little, under constant trituration. Let stand overnight, add the syrup, and filter.

II.—Carmine, in fine powder..........	1 part
Stronger ammonia water...........	4 parts
Distilled water to make 24 parts.	

Rub up the carmine and ammonia and to the solution add the water, little by little, under constant trituration. If in standing this shows a tendency to separate, a drop or two of ammonia will correct the trouble.

Besides these there is caramel, which, of course, you know.

Pink.—

III.—Carmine..........	1 part
Liquor potassæ....	6 parts
Distilled water.....	40 parts

Mix. If the color is too high, dilute with distilled water until the requisite color is obtained.

To Test Fruit Juices and Syrups for Aniline Colors.—Add to a sample of the syrup or juice, in a test tube, its own volume of distilled water, and agitate to get a thorough mixture, then add a few drops of the standard solution of lead diacetate, shake, and filter. If the syrup is free from aniline coloring matter the filtrate will be clear as crystal, since the lead salt precipitates natural coloring matters, but has no effect upon the aniline colors.

To Test Fruit Juices for Salicylic Acid.—Put a portion of the juice to be tested in a large test tube, add the same volume of ether, close the mouth of the tube and shake gently for 30 seconds. Set aside until the liquid separates into two layers. Draw off the supernatant ethereal portion and evaporate to dryness in a capsule. Dissolve the residue in alcohol, dilute with 3 volumes of water, and add 1 drop of tincture of iron chloride. If salicylic acid be present the characteristic purple color will instantly disappear.

Syrups Selected from the Formulary of the Pharmaceutical Society of Antwerp.—

Dionine Syrup.—Dionine, 1 part; distilled water, 19 parts; simple syrup, 1,980 parts. Mix.

Jaborandi Syrup.—Tincture of jaborandi, 1 part; simple syrup, 19 parts. Mix.

Convallaria Syrup. — Extract of convallaria, 1 part; distilled water, 4 parts; simple syrup, 95 parts. Dissolve the extract in the water and mix.

Codeine Phosphate Syrup.—Codeine phosphate, 3 parts; distilled water, 17 parts; simple syrup, 980 parts. Dissolve the codeine in the water and mix with the syrup.

Licorice Syrup.—Incised licorice root, 4 parts; dilute solution of ammonia, 1 part; water, 20 parts. Mix and macerate for 12 hours at 58° to 66° F. with frequent agitation; press, heat the liquid to boiling, then evaporate to two parts on the water bath; add alcohol, 2 parts; allow to stand for 12 hours; then filter. Add to the filtrate enough simple syrup to bring the final weight to 20 parts.

Maize Stigma Syrup.—Extract of maize stigmas, 1 part; distilled water, 4 parts; simple syrup, 95 parts. Dissolve the extract in the water, filter, and add the syrup.

Ammonium Valerianate Solution.—Ammonium valerianate, 2 parts; alcoholic extract of valerian, 1 part; distilled water, 47 parts.

Kola Tincture.—Powdered kola nuts, 1 part; alcohol, 60 per cent, 5 parts. Macerate for 6 days, press, and filter.

Bidet's Liquid Vesicant.—Tincture of cantharides, tincture of rosemary, chloroform, equal parts.

Peptone Wine.—Dried peptone, 1 part; Malaga wine, 19 parts. Dissolve without heat and filter after standing for several days.

Etching

General Instructions for Etching.—
In etching, two factors come into consideration, (1) that which covers that part of the metal not exposed to the etching fluid (the resist), and (2) the etching fluid itself.

In the process, a distinction is to be made between etching in relief and etching in intaglio. In relief etching, the design is drawn or painted upon the surface with the liquid etching-ground, so that after etching and removal of the etching-ground, it appears raised. In intaglio etching, the whole surface is covered with the etching-ground, and the design put on with a needle; the ground being thus removed at the points touched by the drawing, the latter, after etching and removal of the etching-ground, is sunken.

Covering Agents or Resists.—The plate is enclosed by a border made of grafting wax (yellow beeswax, 8 parts; pine rosin, 10 parts; beef tallow, 2 parts; turpentine, 10 parts); or a mixture of yellow wax, 8 parts; lard, 3 parts; Burgundy pitch, ½ part. This mixture is also used to cover the sides of vessels to be etched. Another compound consists of wax, 5 parts; cobbler's wax, 2½ parts; turpentine, 1 part.

Etching-Ground. — I. — Soft: Wax, 2 parts; asphalt, 1 part; mastic, 1 part. II.—Wax, 3 parts; asphalt, 4 parts. III.—Mastic, 16 parts; Burgundy pitch, 50 parts; melted wax, 125 parts; and melted asphalt, 200 parts added successively, and, after cooling, turpentine oil, 500 parts. If the ground should be deep black, lampblack is added.

Hard: Burgundy pitch, 125 parts; rosin, 125 parts, melted; and walnut oil, 100 parts, added, the whole to be boiled until it can be drawn out into long threads.

Etching-Ground for Copper Engraving.—White wax, 120 parts; mastic, 15 parts; Burgundy pitch, 60 parts; Syrian asphalt, 120 parts, melted together; and 5 parts concentrated solution of rubber in rubber oil added.

Ground for Relief Etching.—I.—Syrian asphalt, 500 parts, dissolved in turpentine oil, 1,000 parts. II.—Asphalt, rosin, and wax, 200 parts of each, are melted, and dissolved in turpentine oil, 1,200 parts. The under side of the metal plate is protected by a coating of a spirituous shellac solution, or by a solution of asphalt, 300 parts, in benzol, 600 parts.

For Strongly Acid Solutions.—I.—Black pitch, 1 part; Japanese wax, 2 parts; rosin, 1½ parts; Damar rosin, 1 part, melted together and mixed with turpentine oil, 1 part. II.—Heavy black printers' ink, 3 parts; rosin, 1 part; wax, 1 part.

For electro-etching, the following ground is recommended: Wax, 4 parts; asphalt, 4 parts; pitch, 1 part.

If absolute surety is required respecting the resistance of the etching-ground to the action of the etching fluids, several etching-grounds are put on, one over the other; first (for instance), a solution of rubber in benzol, then a spirituous shellac solution, and a third stratum of asphalt dissolved in turpentine oil.

If the etching is to be of different degrees of depth, the places where it is to be faint are stopped out with varnish, after they are deep enough, and the object is put back into the bath for further etching.

For putting on a design before the etching, the following method may be used: Cover the metal plate, tin plate for example, with a colored or colorless spirit varnish; after drying, cover this, in a dark room, with a solution of gelatin, 5 parts, and red potassium chromate, 1 part, in water, 100 parts; or with a solution of albumen, 2 parts; ammonium bichromate, 2 parts, in water, 200 parts. After drying, put the plate, covered with a stencil, in a copying or printing frame, and expose to light. The sensitive gelatin stratum will become insoluble at the places exposed. Place in water, and the gelatin will be dissolved at the places covered by the stencil; dry, and remove the spirit varnish from the places with spirit, then put into the etching fluid.

Etching Fluids.—The etching fluid is usually poured over the metallic surface, which is enclosed in a border, as described before. If the whole object is to be put into the fluid, it must be entirely covered with the etching-ground. After etching it is washed with pure water, dried with a linen cloth, and the etching-ground is then washed off with turpentine oil or a light volatile camphor oil. The latter is very good for the purpose.

Etching Fluids for Iron and Steel.—
I.—Pure nitric acid, diluted for light etching with 4 to 8 parts of water, for deep etching with an equal weight of water.

II.—Tartaric acid, 1 part, by weight; mercuric chloride, 15 parts, by weight; water, 420 parts; nitric acid, 16 to 20 drops, if 1 part equals 28½ grains.

III.—Spirit, 80 per cent, 120 parts, by weight; pure nitric acid, 8 parts; silver nitrate, 1 part.

IV.—Pure acetic acid, 30 per cent, 40 parts, by weight; absolute alcohol, 10 parts; pure nitric acid, 10 parts.

V.—Fuming nitric acid, 10 parts, by weight; pure acetic acid, 30 per cent, 50 parts, diluted with water if necessary or desired.

VI.—A chromic acid solution.

VII.—Bromine, 1 part; water, 100 parts. Or—mercuric chloride, 1 part; water, 30 parts.

VIII.—Antimonic chloride, 1 part; water, 6 parts; hydrochloric acid, 6 parts.

For Delicate Etchings on Steel.—I.—Iodine, 2 parts; potassium iodide, 4 parts; water, 40 parts.

II.—Silver acetate, 8 parts, by weight; alcohol, 250 parts; water, 250 parts; pure nitric acid, 260 parts; ether, 64 parts; oxalic acid, 4 parts.

III.—A copper chloride solution.

Etching Powder for Iron and Steel.—Blue vitriol, 50 parts; common salt, 50 parts; mixed and moistened with water.

For lustrous figures on a dull ground, as on sword blades, the whole surface is polished, the portions which are to remain bright covered with stencils and the object exposed to the fumes of nitric acid. This is best done by pouring sulphuric acid, 20 parts, over common salt, 10 parts.

Relief Etching of Copper, Steel, and Brass.—Instead of nitric acid, which has a tendency to lift up the etching-ground, by evolution of gases, it is better to use a mixture of potassium bichromate, 150 parts; water, 800 parts; and concentrated sulphuric acid, 200 parts. The etching is slow, but even, and there is no odor.

For Etching Copper, Brass, and Tombac.—Pure nitric acid diluted with water to 18° Bé. The bubbles of gas given out should immediately be removed with a feather that the etching may be even.

Another compound consists of a boiling solution of potassium chlorate, 2 parts, in water, 20 parts, poured into a mixture of nitric acid, 10 parts, and water, 70 parts. For delicate etchings dilute still more with 100 to 200 parts of water.

Etching Fluid for Copper.—Weak: A boiling solution of potassium chlorate, 20 parts, in water, 200 parts, poured into a mixture of pure hydrochloric acid, 20 parts; water, 500 parts.

Stronger: A boiling solution of potassium chlorate, 25 parts, in water, 250 parts, poured into a mixture of pure hydrochloric acid, 250 parts; water, 400 parts.

Very strong: A boiling solution of potassium chlorate, 30 parts, in water, 300 parts, poured into a mixture of pure hydrochloric acid, 300 parts; water, 300 parts.

For etching on copper a saturated solution of bromine in dilute hydrochloric acid may also be used; or a mixture of potassium bichromate, $\frac{1}{2}$ part; water, 1 part; crude nitric acid, 3 parts.

The following are also much used for copper and copper alloys:

I.—A copper chloride solution acidified with hydrochloric acid.

II.—Copper nitrate dissolved in water.

III.—A ferric chloride solution of 30° to 45° Bé. If chrome gelatin or chrome albumen is used for the etching-ground, a spirituous ferric chloride solution is employed. The etching process can be made slower by adding common salt to the ferric chloride solution.

Matt Etching of Copper.—White vitriol, 1 to 5 parts; common salt, 1 part; concentrated sulphuric acid, 100 parts; nitric acid (36° Bé.), 200 parts, mixed together. The sulphuric acid is to be poured carefully into the nitric acid, not the reverse.

Etching Fluid for Brass.—Nitric acid, 8 parts; mixed with water, 80 parts; into this mixture pour a hot solution of potassium chlorate, 3 parts, in water, 50 parts.

Etching Fluid for Brass to Make Stencils.—Mix nitric acid, of 1.3 specific weight, with enough fuming nitric acid to give a deep yellow color. This mixture acts violently, and will eat through the strongest sheet brass.

Etching Fluid for Zinc.—Boil pounded gallnuts, 40 parts, with water, 560 parts, until the whole amounts to 200 parts; filter, and add nitric acid, 2 parts, and a few drops of hydrochloric acid. Ferric chloride and antimonic chloride solutions may also be used to etch zinc.

Relief Etching of Zinc.—The design is to be drawn with a solution of platinum chloride, 1 part, and rubber, 1 part, in water, 12 parts. The zinc plate is placed in dilute sulphuric acid (1 in 16). The black drawing will remain as it is.

Another compound for the drawing is made of blue vitriol, 2 parts; copper chloride, 3 parts; water, 64 parts; pure hydrochloric acid, 1.1 specific weight. After the drawing is made, lay the plate in dilute nitric acid (1 in 8).

Etching Fluid for Aluminum.—Dilute hydrochloric acid serves this purpose. Aluminum containing iron can be matted with soda lye, followed by treatment with nitric acid. The lye dissolves the aluminum, and the nitric acid dissolves the iron. Aluminum bronze is etched with nitric acid.

Etching Fluid for Tin or Pewter.—Ferric chloride, or highly diluted nitric acid.

Etching Fluids for Silver.—I.—Dilute pure nitric acid.

II.—Nitric acid (specific weight, 1.185), 172 parts; water, 320 parts; potassium bichromate, 30 parts.

Etching Fluid for Gold.—Dilute aqua regia (= nitric and sulphuric acids, in the proportion of 1 in 3).

Etching Fluid for Copper, Zinc, and Steel.—A mixture of 4 parts of acetic acid (30 per cent), and alcohol, 1 part; to this is added gradually, nitric acid, 1 part.

Etching Fluid for Lead, Antimony, and Britannia Metal.—Dilute nitric acid.

Etching Powder for Metals (Tin, Silver, Iron, German Silver, Copper, and Zinc).—Blue vitriol, 1 part; ferric oxide, 4 parts. The powder, moistened, is applied to the places to be etched, as, for instance, knife blades. Calcined green vitriol can also be used.

Electro-Etching.—This differs from ordinary etching in the use of a bath, which does not of itself affect the metal, but is made capable of doing so by the galvanic current.

Ordinary etching, seen under the microscope, consists of a succession of uneven depressions, which widen out considerably at a certain depth. In electro-etching, the line under the microscope appears as a perfectly even furrow, not eaten out beneath, however deeply cut. The work is, accordingly, finer and sharper; the fumes from the acids are also avoided, and the etching can be modified by regulation of the current. The preparation of the surface, by covering, stopping-out, etc., is the same as in ordinary etching. At some uncovered place a conducting wire is soldered on with soft solder, and covered with a coat of varnish. The plate is then suspended in the bath, and acts as the anode, with another similar plate for the cathode. If gradations in etching are desired, the plates are taken out after a time, rinsed, and covered, and returned to the bath. For the bath dilute acids are used,

or saline solutions. Thus, for copper, dilute sulphuric acid, 1 in 20. For copper and brass, a blue vitriol solution. For zinc, white vitriol or a zinc chloride solution. For steel and iron, green vitriol, or an ammonium chloride solution. For tin, a tin-salt solution. For silver, a silver nitrate or potassium cyanide solution. For gold and platinum, gold chloride and platinum chloride solutions, or a potassium cyanide solution. For electro-etching a Leclauché or Bunsen battery is to be recommended. In the former, the negative zinc pole is connected with a plate of the same metal as that to be etched, and the positive iron pole with the plate to be etched. In the Bunsen battery, the carbon pole is connected with the object to be etched, the zinc pole with the metal plate.

Etching Bath for Brass.—1.—Mix nitric acid (specified gravity, 1.4), 8 parts, with water, 80 parts. 2.—Chlorate of potash, 3 parts, dissolved in 50 parts of water. Mix 1 and 2. For protecting those portions which are not to be etched, any suitable acid-proof composition can be used.

Etching on Copper.—I.—In order to do regular and quick etching on copper take a copper plate silvered on the etching side. Trace on this plate, either with varnish or lithographic ink, the design. When the tracing is dry, place the plate in an iron bath, using a battery. The designs traced with the varnish or ink are not attacked by the etching fluid. When the plate is taken from the bath and has been washed and dried, remove the varnish or ink with essence of turpentine; next pour mercury on the places reserved by the varnish or ink; the mercury will attack the silvered portions and the etching is quickly made. When the mercury has done its duty gather up the excess and return to the bottle with a paper funnel. Wash the plate in strong alum water, and heat.

II.—The plate must be first polished either with emery or fine pumice stone, and after it has been dried with care, spread thereon a varnish composed of equal parts of yellow wax and essence of turpentine. The solution of the wax in the essence is accomplished in the cold; next a little oil of turpentine and some lampblack are added. This varnish is allowed to dry on, away from dust and humidity. When dry, trace the design with a very fine point. Make a border with modeling wax, so as to prevent the acid from running off. Pour on nitric acid if the plate is of copper, or hydrochloric acid diluted with water if the plate is of zinc, allow the acid to act according to the desired depth of the engraving; wash several times and remove the varnish by heating the plate lightly. Wash with essence of turpentine and dry well in sawdust or in the stove. For relief engraving the designs are traced before the engraving on the plate with the resist varnish instead of covering the plate entirely. These designs must be delicately executed and without laps, as the acid eats away all the parts not protected by the varnish.

Etching Fluids for Copper.—I.—A new etching fluid for copper plate is hydrogen peroxide, to which a little dilute ammonia water is added. It is said to bite in very rapidly and with great regularity and uniformity.

II.—Another fluid is fuming hydrochloric acid (specific gravity, 1.19), 10 parts; water, 70 parts. To this add a solution of potassium chlorate, 2 parts, dissolved in 20 parts of hot water. If the articles to be etched are very delicate and fine this should be diluted with from 100 to 200 parts of water.

ETCHING ON GLASS.

Names, designs, etc., can be etched on glass in three ways: First, by means of an engraving wheel, a method which requires some manual skill. Second, by means of a sand blast, making a stencil of the name, fixing this on the glass, and then, by means of a blast of air, blowing sand on the glass. Third, by the use of hydrofluoric acid. The glass is covered with beeswax, paraffine wax, or some acid resisting ink or varnish; the name or device is then etched out of the wax by means of a knife, and the glass dipped in hydrofluoric acid, which eats away the glass at those parts where the wax has been cut away.

Fancy work, ornamental figures, lettering, and monograms are most easily and neatly cut into glass by the sandblast process. Lines and figures on tubes, jars, etc., may be deeply etched by smearing the surface of the glass with beeswax, drawing the lines with a steel point, and exposing the glass to the fumes of hydrofluoric acid. This acid is obtained by putting powdered fluorspar into a tray made of sheet lead and pouring sulphuric acid on it, after which the tray is slightly warmed. The proportions will vary with the purity of the materials used, fluorspar (except when in crystals) being generally mixed with a large quantity of other matter. Enough acid to make a thin paste with the powdered spar will be about right. Where a lead tray is not at hand, the powdered spar may be poured on the glass and the acid poured on it and left for some time. As a general rule, the marks are opaque, but sometimes they are transparent. In this case cut them deeply and fill up with black varnish, if they are required to be very plain, as in the case of graduated vessels. Liquid hydrofluoric acid has been recommended for etching, but is not always suitable, as it leaves the surface on which it acts transparent.

There are two methods of marking bottles—dry etching, or by stamping with etching inks. The first process is usually followed in glass factories. A rubber stamp is necessary for this process, and the letters should be made as large and clean cut as possible without crowding them too much. Besides this, an etching powder is required.

A small quantity of the powder is poured into a porcelain dish, and this is placed on a sand bath or over a gentle fire, and heated until it is absolutely dry, so that it can be rubbed down to an impalpable powder.

The bottle or other glass to be marked must be perfectly clean and dry. The etching powder takes better when the vessel is somewhat warm. The stamp should be provided with a roller which is kept constantly supplied with a viscid oil which it distributes on the stamp and which the stamp transfers to the glass surface. The powder is dusted on the imprint thus made, by means of a camel's-hair brush. Any surplus falling on the unoiled surface may be removed with a fine long-haired pencil. The printed bottle is transferred to a damp place and kept for several minutes, the dampness aiding the etching powder in its work on the glass surface. The bottle is then well washed in plain water.

Glass cylinders, large flasks, carboys, etc., may be treated in a somewhat different manner. The stamp here is inserted, face upward, between two horizontal boards, in such a manner that its face projects about a quarter of a millimeter (say 0.01 inch) above the surface. Oil is applied to the surface, after which the cylinder, carboy, or what not, is rolled along the board and over the stamp. The design is thus neatly transferred to the glass surface, and the rest of the operation is as in the previous case.

For an etching ink for glassware the following is recommended:

Ammonium fluoride..	2 drachms
Barium sulphate.....	2 drachms

Reduce to a fine powder in a mortar,

then transfer to a lead dish and make into a thin writing-cream with hydrofluoric acid or fuming sulphuric acid. Use a piece of lead to stir the mixture. The ink may be put up in bottles coated with paraffine, which can be done by heating the bottle, pouring in some melted paraffine, and letting it flow all around. The writing is done with a quill, and in about half a minute the ink is washed off.

Extreme caution must be observed in handling the acid, since when brought in contact with the skin it produces dangerous sores very difficult to heal. The vapor is also dangerously poisonous when inhaled.

Hydrofluoric Formulas.—I.—Dissolve about 0.72 ounces fluoride of soda with 0.14 ounces sulphate of potash in $\frac{1}{2}$ pint of water. Make another solution of 0.28 ounces chloride of zinc and 1.30 ounces hydrochloric acid in an equal quantity of water. Mix the solutions and apply to the glass vessel with a pin or brush. At the end of half an hour the design should be sufficiently etched.

II.—A mixture consisting of ammonium fluoride, common salt, and carbonate of soda is prepared, and then placed in a gutta-percha bottle containing fuming hydrofluoric acid and concentrated sulphuric acid. In a separate vessel which is made of lead, potassium fluoride is mixed with hydrochloric acid, and a little of this solution is added to the former, along with a small quantity of sodium silicate and ammonia. Some of the solution is dropped upon a rubber pad, and by means of a suitable rubber stamp, bearing the design which is to be reproduced, is transferred to the glass vessel that is to be etched.

Etching with Wax.—Spread wax or a preservative varnish on the glass, and trace on this wax or varnish the letters or designs. If letters are desired, trace them by hand or by the use of letters cut out in tin, which apply on the wax, the inside contours being taken with a fine point. When this is done, remove the excess of wax from the glass, leaving only the full wax letters undisturbed. Make an edge of wax all along the glass plate so as to prevent the acid from running over when you pour it on to attack the glass. At the end of 3 to 4 hours remove the acid, wash the glass well with hot water, next pour on essence of turpentine or alcohol to take off the wax or the preservative varnish. Pass again through clean water; the glass plate will have become dead wherever the acid has eaten in, only the letters remaining polished. For fancy designs it suffices to put on the back of the plate a black or colored varnish, or tin foil, etc., to obtain a brilliant effect.

Etching Glass by Means of Glue.—It is necessary only to cover a piece of ordinary or flint glass with a coat of glue dissolved in water in order to see that the layer of glue, upon contracting through the effect of drying, becomes detached from the glass and removes therefrom numerous scales of varying thickness. The glass thus etched presents a sort of regular and decorative design similar to the flowers of frost deposited on window-panes in winter. When salts that are readily crystallizable and that exert no chemical action upon the gelatin are dissolved in the latter the figures etched upon the glass exhibit a crystalline appearance that recalls fern fronds.

Hyposulphite of soda and chlorate and nitrate of potash produce nearly the same effects. A large number of mineral substances are attacked by gelatin. Toughened glass is easily etched, and the same is the case with fluorspar and polished marble. A piece of rock crystal, cut at right angles with the axis and coated with isinglass, the action of which seems to be particularly energetic, is likewise attacked at different points, and the parts detached present a conchoidal appearance. The contraction of the gelatin may be rendered visible by applying a coating of glue to sheets of cardboard or lead, which bend backward in drying and assume the form of an irregular cylinder.

Such etching of glass and different mineral substances by the action of gelatin may be employed for the decoration of numerous objects.

Dissolve some common glue in ordinary water, heated by a water bath, and add 6 per cent of its weight of potash alum. After the glue has become perfectly melted, homogeneous, and of the consistency of syrup, apply a layer, while it is still hot, to a glass object by means of a brush. If the object is of ground glass the action of the glue will be still more energetic. After half an hour apply a second coat in such a way as to obtain a smooth, transparent surface destitute of air bubbles. After the glue has become so hard that it no longer yields to the pressure of the finger nail (say, in about 24 hours), put the article in a warmer place, in which the temperature must not exceed 105° F. When the object is removed from the oven, after a few hours, the glue will detach itself with a noise and removes with it numerous flakes of glass. All that the piece then requires is to be carefully washed and dried.

The designs thus obtained are not always the same, the thickness of the coat of glue, the time of drying, and various other conditions seeming to act to modify the form and number of the flakes detached.

It is indispensable to employ glass objects of adequate thickness, since, in covering mousseline glass with a layer of glue, the mechanical action that it has to support during desiccation is so powerful that it will break with an explosion. Glue, therefore, must not be allowed to dry in glass vessels, since they would be corroded and broken in a short time.

Indelible Labels on Bottles.—To affix indelible labels on bottles an etching liquid is employed which is produced as follows:

Liquid I, in one bottle.—Dissolve 36 parts of sodium fluoride in 500 parts of distilled water and add 7 parts of potassium sulphate.

Liquid II, in another bottle.—Dissolve zinc chloride, 14 parts, in 500 parts of distilled water, and add 65 parts of concentrated hydrochloric acid.

For use mix equal parts together and add a little dissolved India ink to render the writing more visible.

The mixing cannot, however, be conducted in a vessel. It is best to use a cube of paraffine which has been hollowed out.

Etching on Marble or Ivory (see also Ivory).—Cover the objects with a coat of wax dissolved in 90 per cent alcohol, then trace the desired designs by removing the wax with a sharp tool and distribute on the tracing the following mixture: Hydrochloric acid, 1 part; acetic acid, 1 part. Repeat this operation several times, until the desired depth is attained. Then take off the varnish with alcohol. The etching may be embellished, filling up the hollows with any colored varnish, by wiping the surface with a piece of linen fixed on a stick, to rub the varnish into the cavities after it has been applied with a brush. The hollows may be gilded or silvered by substituting "mixtion" for the varnish and applying on this mixtion a leaf of gold or silver, cut in pieces a little larger than the design to be covered; press down the gold by means of a soft brush so as to cause it to penetrate to the bottom; let dry and remove the protruding edges.

Etching on Steel.—The print should be heavily inked and powdered with dragon's blood several times. After each powdering heat slightly and additional powder will stick, forming a heavy coating in 2 or 3 operations. Before proceeding to heat up, the plate should receive a light etching in a weak solution of the acid described later on. The purpose of this preliminary etching is to clean up the print, so that the lines will not tend to thicken, as would be the case otherwise. Next a good strong heating should be given. On top the dragon's blood plumbago may be used in addition. For etching use nitric acid mixed with an even amount of acetic acid. Some operators use vinegar, based on the same theory. When commencing the etching, start with a weak solution and increase as soon as the plate is deep enough to allow another powdering. If the operator is familiar with lithography, and understands rolling up the print with a litho-roller, the etching of steel is not harder than etching on zinc.

Liquids for Etching Steel.—

I.—Iodine	2 parts
Potassium iodide	5 parts
Water	40 parts
II.—Nitric acid	60 parts
Water	120 parts
Alcohol	200 parts
Copper nitrate	8 parts
III.—Glacial acetic acid	4 parts
Nitric acid	1 part
Alcohol	1 part

IV.—Mix 1 ounce sulphate of copper, ½ ounce alum, ½ teaspoonful of salt (reduced to powder), with 1 gill of vinegar and 20 drops of nitric acid. This fluid can be used either for etching deeply or for frosting, according to the time it is allowed to act. The parts of the work which are not to be etched should be protected with beeswax or some similar substance.

V.—Nitric acid, 60 parts; water, 120 parts; alcohol, 200 parts; and copper nitrate, 8 parts. Keep in a glass-stoppered bottle. To use the fluid, cover the surface to be marked with a thin even coat of wax and mark the lines with a machinist's scriber. Wrap clean cotton waste around the end of the scriber or a stick, and dip in the fluid, applying it to the marked surface. In a few minutes the wax may be scraped off, when fine lines will appear where the scriber marked the wax. The drippings from a lighted wax candle can be used for the coating, and this may be evenly spread with a knife heated in the candle flame.

VI.—For Hardened Steel.—Heat an iron or an old pillar-file with a smooth side, and with it spread a thin, even coat of beeswax over the brightened surface to be etched. With a sharp lead pencil (which is preferable to a scriber) write or mark as wanted through the wax so as to be sure to strike the steel surface. Then daub on with a stick etching acid made as follows: Nitric acid, 3 parts; muriatic acid, 1 part. If a lead pencil has been used the acid will begin to bubble immediately. Two or three minutes of the bubbling or foaming will be sufficient for marking; then soak up the acid with a small piece of blotting paper and remove the beeswax with a piece of cotton waste wet with benzine, and if the piece be small enough dip it into a saturated solution of sal soda, or if the piece be large swab over it with a piece of waste. This neutralizes the remaining acid and prevents rusting, which oil will not do.

If it is desired to coat the piece with beeswax without heating it, dissolve pure beeswax in benzine until of the consistency of thick cream and pour on to the steel, and even spread it by rocking or blowing, and lay aside for it to harden; then use the lead pencil, etc., as before. This method will take longer. Keep work from near the fire or an open flame.

EUCALYPTUS BONBONS FOR COLDS AND COUGHS:
See Cold and Cough Mixtures.

EXPECTORANTS:
See Cold and Cough Mixtures.

Explosives

Explosives may be divided into two great classes—mechanical mixtures and chemical compounds. In the former the combustible substances are intimately mixed with some oxygen supplying material, as in the case of gunpowder, where carbon and sulphur are intimately mixed with potassium nitrate; while gun cotton and nitro-glycerine are examples of the latter class, where each molecule of the substance contains the necessary oxygen for the oxidation of the carbon and hydrogen present, the oxygen being in feeble combination with nitrogen. Many explosives are, however, mechanical mixtures of compounds which are themselves explosive, e. g., cordite, which is mainly composed of gun cotton and nitro-glycerine.

The most common and familiar of explosives is undoubtedly gunpowder. The mixture first adopted appears to have consisted of equal parts of the three ingredients—sulphur, charcoal, and niter; but some time later the proportions, even now taken for all ordinary purposes, were introduced, namely:

Potassium nitrate....	75 parts
Charcoal...........	15 parts
Sulphur.............	10 parts
	100 parts

Since gunpowder is a mechanical mixture, it is clear that the first aim of the maker must be to obtain perfect incorporation, and, necessarily, in order to obtain this, the materials must be in a very finely divided state. Moreover, in order that uniformity of effect may be obtained, purity of the original substances, the percentage of moisture present, and the density of the finished powder are of importance.

The weighed quantities of the ingredients are first mixed in gun metal or copper drums, having blades in the interior capable of working in the opposite direction to that in which the drum itself is traveling. After passing through a sieve, the mixture (green charge) is passed on to the incorporating mills, where it is thoroughly ground under heavy metal rollers, a small quantity of water being added to prevent dust and facilitating incorporation, and during this process the risk of explosion is greater possibly than at any other stage in the manufacture. There are usually 6 mills working in the same building, with partitions between. Over the bed of each mill is a horizontal board, the "flash board," which is connected with a tank of water overhead, the arrangement being such that the upsetting of one tank discharges the contents of the other tanks onto the corresponding mill beds below, so that in the event of an accident the charge is drowned in each case. The "mill cake" is now broken down between rollers, the "meal" produced being placed in strong oak boxes and subjected to hydraulic pressure, thus increasing its density and hardness, at the same time bringing the ingredients into more intimate contact. After once more breaking down the material (press cake), the powder only requires special treatment to adapt it for the various purposes for which it is intended.

The products of the combustion of powder and its manner of burning are largely influenced by the pressure, a property well illustrated by the failure of a red-hot platinum wire to ignite a mass of powder in a vacuum, only a few grains actually in contact with the platinum undergoing combustion.

Nitro-glycerine is a substance of a similar chemical nature to gun cotton, the principles of its formation and purification being very similar, only in this case the materials and product are liquids, thereby rendering the operations of manufacture and washing much less difficult. The glycerine is sprayed into the acid mixture by compressed-air injectors, care being taken that the temperature during nitration does not rise above 86° F. The nitro-glycerine formed readily separates from the mixed acids, and being insoluble in cold water, the washing is comparatively simple.

Nitro-glycerine is an oily liquid readily soluble in most organic solvents, but becomes solid at 3° or 4° above the freezing point of water, and in this condition is less sensitive. It detonates when heated to 500° F., or by a sudden blow, yielding carbon dioxide, oxygen, nitrogen, and water. Being a fluid under ordinary conditions, its uses as an explosive were limited, and Alfred Nobel conceived the idea of mixing it with other substances which would act as absorbents, first using charcoal and afterwards an infusorial earth, "kieselguhr," and obtaining what he termed "dynamite." Nobel found that "collodion cotton"—soluble gun cotton—could be converted by treatment with nitro-glycerine into a jellylike mass which was more trustworthy in action than the components alone, and from its nature the substance was christened "blasting gelatin."

Nobel took out a patent for a smokeless powder for use in guns, in which these ingredients were adopted with or without the use of retarding agents. The powders of this class are ballistite and filite, the former being in sheets, the latter in threads. Originally camphor was introduced, but its use has been abandoned, a small quantity of aniline taking its place.

Sir Frederick Abel and Prof. Dewar patented in 1889 the use of trinitrocellulose and nitro-glycerine, for although, as is well known, this form of nitro-cellulose is not soluble in nitro-glycerine, yet by dissolving the bodies in a mutual solvent, perfect incorporation can be attained. Acetone is the solvent used in the preparation of "cordite," and for all ammunition except blank charges a certain proportion of vaseline is also added. The combustion of the powder without vaseline gives products so free from solid or liquid substances that excessive friction of the projectile in the gun causes rapid wearing of the rifling, and it is chiefly to overcome this that the vaseline is introduced, for on explosion a thin film of solid matter is deposited in the gun, and acts as a lubricant.

The proportion of the ingredients are:

Nitro-glycerine ...... 58 parts
Gun-cotton .... 37 parts
Vaseline .... 5 parts

Gun cotton to be used for cordite is prepared as previously described, but the alkali is omitted, and the mass is not submitted to great pressure, to avoid making it so dense that ready absorption of nitro-glycerine would not take place. The nitro-glycerine is poured over the dried gun cotton and first well mixed by hand, afterwards in a kneading machine with the requisite quantity of acetone for $3\frac{1}{2}$ hours. A water jacket is provided, since, on mixing, the temperature rises. The vaseline is now added, and the kneading continued for a similar period. The cordite paste is first subjected to a preliminary pressing, and is finally forced through a hole of the proper size in a plate either by hand or by hydraulic pressure. The smaller sizes are wound on drums, while the larger cordite is cut off in suitable lengths, the drums and cut material being dried at 100° F., thus driving off the remainder of the acetone.

Cordite varies from yellow to dark brown in color, according to its thickness. When ignited it burns with a strong flame, which may be extinguished by a vigorous puff of air. Macnab and Ristori give the yield of permanent gases from English cordite as 647 cubic centimeters, containing a much higher per cent of carbon monoxide than the gases evolved from the old form of powder. Sir Andrew Noble failed in attempts to detonate the substance, and a rifle bullet fired into the mass only caused it to burn quietly.

Dynamite.—Dynamite is ordinarily made up of 75 per cent nitro-glycerine, 25 per cent infusorial earth; dualine contains 80 per cent nitro-glycerine, 20 per cent nitro-cellulose; rend-rock has 40 per cent nitro glycerine, 40 per cent nitrate of potash, 13 per cent cellulose, 7 per cent paraffine; giant powder, 36 per cent nitro-glycerine, 48 per cent nitrate of potash, 8 per cent sulphur, 8 per cent rosin or charcoal.

Smokeless Powder. — The base of smokeless powders is nitrated cellulose, which has been treated in one of various ways to make it burn slower than gun cotton, and also to render it less sensitive to heat and shocks. As a rule, these powders are not only less inflammable than gun cotton, but require stronger detonators. As metallic salts cause smoke, they are not used in these powders. The smokeless powders now in use may be divided into three groups: (1) Those consisting of mixtures of nitro-glycerine and nitrated cellulose, which have been converted into a hard, hornlike mass, either with or without the aid of a solvent. To this group belongs ballistite, containing 50 per cent of nitro-glycerine, 49 per cent of nitrated cellulose, and 1 per cent of diphenylamin; also cordite (see further on), Lenord's powder, and amberite. This last contains 40 parts of nitro-glycerine and 56 parts of nitrated cellulose. (2) Those consisting mainly of nitrated cellulose of any kind, which has been rendered hard and horny by treatment with some solvent which is afterwards evaporated. These are prepared by treating nitrated cellulose with ether or benzine, which dissolves the collodion, and when evaporated leaves a hard film of collodion on the surface of each grain. Sometimes a little camphor is added to the solvent, and, remaining in the powder, greatly retards its combustion. (3) Those consisting of nitro-derivatives of the aromatic hydrocarbons, either with or without the admixture of nitrated cellulose; to this group belong Dupont's powder, consisting of nitrated cellulose dissolved in nitro-benzine; indurite, consisting of cellulose hexanitrate (freed from collodion by extraction with methyl alcohol), made into a paste with nitrobenzine, and hardened by treatment with steam until the excess of nitro-benzine is removed; and plastomeite, consisting of dinitrotoluene and nitrated wood pulp.

Cordite is the specific name of a smokeless powder which has been adopted by the English government as a military explosive. It contains nitro-glycerine, 58 parts; gun cotton, 37 parts; and petrolatum, 5 parts. The nitro-glycerine and gun cotton are first mixed, 19.2 parts of acetone added, and the pasty mass kneaded for several hours. The petrolatum is then added and the mixture again kneaded. The paste is then forced through fine openings to form threads, which are dried at about 105° F. until the acetone evaporates. The threads, which resemble brown twine, are then cut into short lengths for use.

Another process for the manufacture of smokeless powder is as follows: Straw, preferably oat-straw, is treated in the usual way with a mixture of nitric acid and concentrated sulphuric acid, and then washed in water to free it from these, then boiled with water, and again with a solution of potassium carbonate. It is next subjected, for 2 to 6 hours, to the action of a solution composed of 1,000 parts of water, 12.5 parts of potassium nitrate, 3.5 parts of potassium chlorate, 12.5 parts of zinc sulphate, and 12.5 parts of potassium permanganate. The excess of solution is pressed out, and the mass is then pulverized, granulated, and finally dried.

The warning as to the danger of experimenting with the manufacture of ordinary gunpowder applies with renewed force when nitro-glycerine is the subject of the experiment.

Berge's Blasting Powder. — This is composed of chlorate of potash, 1 part; chromate of potash, 0.1 part; sugar, 0.45 parts; yellow wax, 0.09 parts. The proportions indicated may vary within certain limits, according to the force desired. For the preparation, the chlorate and the chromate of potash, as well as the sugar, are ground separately and very finely, and sifted so that the grains of the different substances may have the same size. At first any two of the substances are mixed as thoroughly as possible, then the third is added. The yellow wax, cut in small pieces, is finally added, and all the substances are worked together to produce a homogeneous product. The sugar may be replaced with charcoal or any other combustible body. For commercial needs, the compound may be colored with any inert matter, also pulverized.

Safety in Explosives. — Ammoniacal salts have been used in the manufacture of explosives to render them proof against firedamp, but not with the full success desired. Ammonium chloride has been utilized, but inconveniences are met with, and the vapor is quite disagreeable. In coöperation with equivalent quantities of soda and potash, its action is regarded as favorable. Tests employing benzine vapor and coal dust were made, and the comparative security calculated to be as given below.

I. — Donarite, composed as follows: 80 per cent of nitrate of ammonia, 12 of trinitrotoluol, 4 of flour, 3.8 of nitro-glycerine, and 0.2 per cent of cotton collodion. Security: Donarite alone, 87 parts; 95 per cent of donarite and 5 per cent of ammonium chloride, 125 parts; 90 per cent of donarite and 10 per cent of ammonium chloride, 250 parts; 86 per cent of donarite and 5.5 per cent of ammonium chloride, with 8.5 per cent of nitrate of soda, 425 parts. The force of the explosion is decreased about 8 per cent, while the security is quintupled.

II.—Roburite, with the following composition: 72.5 per cent nitrate of ammonia; 12 binitro-benzol; 10 nitrate of potash; 5 sulphate of ammonia; 0.5 per cent permanganate of potash. Security: Roburite only, 325 parts; ammonium chloride, taking the place of sulphate of ammonia, 400 parts. Here an intensification of the explosive force is simultaneously produced.

III.—Ammon carbonite I, composed thus: 4 per cent nitro-glycerine; 75.5 nitrate of ammonia; 9.5 nitrate of potash; 9.5 coal dust; 10.5 flour. Security: Ammon carbonite I only, 250 parts; 95 per cent A. C. I. and 5 per cent ammonium chloride, 400 parts; 92 per cent A. C. I. and 8 per cent ammonium chloride, 500 parts. The addition of 5 per cent ammonium chloride diminishes the explosive force only 3 per cent.

IV.—An explosive of nitro-glycerine base composed thus: 30 per cent nitroglycerine; 1 per cent cotton collodion; 52.6 nitrate of ammonia; 13 nitrate of potash; 3 to 4 per cent starch. Security of this mixture, 150 parts.

V.—Thirty per cent nitro-glycerine; 1 per cent cotton collodion; 47.3 nitrate of ammonia; 11.6 nitrate of potash; 3.1 starch; 7 per cent ammonium chloride. This mixture has a security of 350 parts.

Inflammable Explosive with Chlorate of Potash.—Take as an agent promoting combustion, potassium chlorate; as a combustible agent, an oxidized, nitrated, or natural rosin. If, to such a mixture, another body is added in order to render it soft and plastic, such as oil, nitro-benzine, glucose, glycerine, the benefit of the discovery is lost, for the mixture is rendered combustible with nitro-benzine, fecula, sulphur, etc., and inexplosive with glycerine, glucose, and the oil.

Of all the chlorates and perchlorates, potassium chlorate ($KClO_3$) responds the best to what is desired. As to the rosins, they may be varied, or even mixed. To obtain the oxidation or nitration of the rosins, they are heated with nitric acid, more or less concentrated, and with or without the addition of sulphuric acid. An oxidation, sufficient and without danger, can be secured by a simple and practical means. This is boiling them for several hours in water containing nitric acid, which is renewed from time to time in correspondence with its decomposition. The rosins recommended by M. Turpin are of the terebinthine group, having for average formula $C_{20}H_{30}O_2$. Colophony is the type.

The products, thus nitrated, are washed with boiling water, and, on occasion, by a solution slightly alkaline, with a final washing with pure water, and dried at a temperature of 230° F. or in the open air.

The mixing of the constituents of this explosive is preferably cold. For this purpose they are used in the state of fine powder, and when mixed in the tub, $2\frac{1}{2}$ to 5 per cent of a volatile dissolvent is added, as alcohol, carbon sulphide, ether, or benzine. As soon as thoroughly mingled, the mass is put either in an ordinary grainer, or in a cylinder of wire cloth revolving horizontally on its axis, with glass gobilles forming a screen, by the aid of which the graining is rapidly accomplished. Thus a powder more or less finely granulated is produced free from dust.

The proportions preferably employed are:

1. Potassium chlorate... 85 parts
   Natural rosin........ 15 parts

2. Potassium chlorate... 80 parts
   Nitrated rosin....... 20 parts

For employment in firedamp mines, there is added to these compounds from 20 to 40 per cent of one of the following substances: Ammonium oxalate, ammonium carbonate, oxalic acid, sodium bicarbonate, calcium fluoride, or other substance of the nature to lower sufficiently the temperature of the explosive flame.

Gun Cotton.—For the production of a high-grade gun cotton, it is important that the cotton used should approach as near as possible pure cellulose. The waste from cotton mills, thoroughly purified, is usually employed. After careful chemical examination has been made to ascertain its freedom from grease and other impurities, the cotton waste is picked over by hand to remove such impurities as wood, cardboard, string, etc. The cotton is then passed through the "teasing machine," which opens out all knots and lumps, thereby reducing it to a state more suitable for the acid treatment and exposing to view any foreign substances which may have escaped notice in the previous picking. The cotton is then dried. When perfectly dry, it is removed to air-tight iron cases, in which it is allowed to cool. The iron cases are taken to the dipping houses, and the cotton waste weighed into small portions, which are then transferred as rapidly as possible to the mixed acids, allowed to remain a few minutes, then removed to the grating and the excess of acid squeezed out. The cotton now containing about ten times its weight of acid is placed in an earthenware pot and transferred to the steeping pits, where it is allowed to remain for 24 hours, a low temperature being maintained by a stream of cold water.

The cotton is now wholly converted into nitro-cellulose. The superfluous acid is next removed by a centrifugal extractor, after which the gun cotton is taken out of the machine and immediately immersed in a large volume of water, and thoroughly washed until it shows no acid reaction. The moisture is then run out and the gun cotton is conveyed by tramway to the boiling vats, where it undergoes several boilings by means of steam. When the " heat test " shows that a sufficient degree of stability has been obtained, the gun cotton is removed to a beating engine, and reduced to a very fine state of division. When this process is completed the pulp is run by gravity along wooden shoots, provided with "grit traps" and electro-magnets, which catch any traces of sand, iron, etc., into large "poachers," in which the gun cotton is continuously agitated, together with a large quantity of water. In this way it is thoroughly washed and a blend made of a large quantity of gun cotton.

Soluble Gun Cotton.—Soluble gun cotton is made on the same lines, except that greater attention has to be paid to the physical condition of the cotton used, and to the temperature and strength of acid mixture, etc.

The term "soluble" usually implies that the gun cotton is dissolved by a mixture of ethyl-ether and ethyl-alcohol, 2 parts of the former to 1 of the latter being the proportions which yield the best solvent action. The classification of nitro-celluloses according to their solubility in ether-alcohol is misleading, except when the nitrogen contents are also quoted.

The number of solvents for gun cotton which have at various times been proposed is very large. Among the more important may be mentioned the following: Alcohols (used chiefly in conjunc-

tion with other solvents), methyl, ethyl, propyl, and amyl, methyl-amyl ether, acetic ether, di-ethyl-ketone, methyl-ethyl ketone, amyl nitrate and acetate, nitro-benzole, nitro-toluol, nitrated oils, glacial acetic acid, camphor dissolved in alcohol, etc.

Some of the above may be called selective solvents, i. e., they dissolve one particular variety of gun cotton better than others, so that solubility in any given solvent must not be used to indicate solubility in another. No nitro-cotton is entirely soluble in any solvent. The solution, after standing some time, always deposits a small amount of insoluble matter. Therefore, in making collodion solutions, care should be taken to place the containing bottles in a place free from vibration and shock. After standing a few weeks the clear supernatant liquid may be decanted off. On a larger scale collodion solutions are filtered under pressure through layers of tightly packed cotton wool. The state of division is important. When the end in view is the production of a strong film or thread, it is advisable to use unpulped or only slightly pulped nitro-cellulose. In this condition it also dissolves more easily than the finely pulped material.

FULMINATES:

Fulminating Antimony.—Tartar emetic (dried), 100 parts; lampblack or charcoal powder, 3 parts. Triturate together, put into a crucible that it will three-fourths fill (previously rubbed inside with charcoal powder). Cover it with a layer of dry charcoal powder, and lute on the cover. After 3 hours' exposure to a strong heat in a reverberatory furnace, and 6 or 7 hours' cooling, cautiously transfer the solid contents of the crucible, as quickly as possible, without breaking, to a wide-mouthed stoppered phial, where, after some time, it will spontaneously crumble to a powder. When the above process is properly conducted, the resulting powder contains potassium, and fulminates violently on contact with water. A piece the size of a pea introduced into a mass of gunpowder explodes it on being thrown into water, or on its being moistened in any other manner.

Fulminating Bismuth.—Take bismuth, 120 parts; carbureted cream of tartar, 60 parts; niter, 1 part.

Fulminating Copper.—Digest copper (in powder of filings) with fulminate of mercury or of silver, and a little water.

It forms soluble green crystals that explode with a green flame.

Fulminating Mercury.—Take mercury, 100 parts; nitric acid (specific gravity, 1.4), 1,000 parts (or 740 parts, by measure). Dissolve by a gentle heat, and when the solution has acquired the temperature of 130° F., slowly pour it through a glass funnel tube into alcohol (specific gravity, .830), 830 parts (or 1,000 parts, by measure). As soon as the effervescence is over, and white fumes cease to be evolved, filter through double paper, wash with cold water, and dry by steam (not hotter than 212° F.) or hot water. The fulminate is then to be packed in 100-grain paper parcels, and these stored in a tight box or corked bottle. Product 130 per cent of the weight of mercury employed.

Fulminating Powder.—I.—Niter, 3 parts; carbonate of potash (dry), 2 parts; flowers of sulphur, 1 part; reduce them separately to fine powder, before mixing them. A little of this compound (20 to 30 grains), slowly heated on a shovel over the fire, first fuses and becomes brown, and then explodes with a deafening report.

II.—Sulphur, 1 part; chlorate of potassa, 3 parts. When triturated, with strong pressure, in a marble or wedgwood-ware mortar, it produces a series of loud reports. It also fulminates by percussion.

III.—Chlorate of potassa, 6 parts; pure lampblack, 4 parts; sulphur, 1 part. A little placed on an anvil detonates with a loud report when struck with a hammer.

EXPOSURES IN PHOTOGRAPHING:
See Photography.

EXTRACTS:
See Essences and Extracts.

EXTRACTS, TESTS FOR:
See Foods.

EYE LOTIONS:
" Black Eye " Lotion.—"Black eyes" or other temporary discolorations of the skin may be disguised by the application of pink grease paint, or collodion colored by means of a little carmine. As lotions the following have been recommended:

I.—Ammonium chloride............ 1 part
Alcohol........... 1 part
Water............ 10 parts

Diluted acetic acid may be substituted for half of the water, and the alcohol may be replaced by tincture of arnica, with advantage.

II.—Potassium nitrate... 15 grains
Ammonium chloride 30 grains
Aromatic vinegar... 4 drachms
Water to make 8 ounces.

III.—The following is to be applied with camel's-hair pencil every 1, 2, or 3 hours. Be careful not to get it in the eyes, as it smarts. It will remove the black discoloration overnight:

Oxalic acid......... 15 grains
Distilled water....... 1 ounce

Foreign Matter in the Eye.—If a piece of iron or other foreign matter in the eye irritates it, and there is no way of removing it until morning, take a raw Irish potato, grate it, and use as a poultice on the eye. It will ease the eye so one can sleep, and sometimes draws the piece out.

Drops of Lime in the Eye.—If lime has dropped in the eye, the pouring-in of or the wiping-out with a few drops of oil is the best remedy, as the causticity of the lime is arrested thereby. Poppy-seed oil or olive oil is prescribed, but pure linseed oil ought to render the same service, as it is also used in the household. Subsequently, the eye may be rinsed out with syrup, as the saccharine substance will harden any remaining particles of lime and destroy all causticity entirely.

FABRIC CLEANERS:
See Cleaning Preparations and Methods and also Household Formulas.

FABRICS, WATERPROOFING OF:
See Waterproofing.

FACE BLACK AND FACE POWDER:
See Cosmetics.

Fats

Bear Fat.—Fresh bears' fat is white and very similar to lard in appearance. The flank fat is softer and more transparent than the kidney fat, and its odor recalls that of fresh bacon. Bears' fat differs from the fats of the dog, fox, and cat in having a lower specific gravity, a very low melting point, and a fairly high iodine value.

Bleaching Bone Fat.—Bone fat, which is principally obtained from horse bones, is very dark colored in the crude state, and of an extremely disagreeable smell. To remedy these defects it may be bleached by the air or chemicals, the former method only giving good results when the fat has been recovered by means of steam. It consists in cutting up the fat into small fragments and exposing it to the air for several days, the mass being turned over at intervals with a shovel. When sufficiently bleached in this manner, the fat is boiled with half its own weight of water, which done, about 3 or 4 per cent of salt is added, and the whole is boiled over again. This treatment, which takes 2 or 3 weeks, sweetens the fat, makes it of the consistency of butter, and reduces the color to a pale yellow. Light seems to play no part in the operation, the change being effected solely by the oxygen of the air. The chemical treatment has the advantage of being more rapid, sufficient decoloration being produced in a few hours. The fat, which should be free from gelatin, phosphate of lime, and water, is placed in an iron pan along with an equal weight of brine of 14° to 15° Bé. strength, with which it is boiled for 3 hours and left to rest overnight. Next day the fat is drawn off into a wooden vessel, where it is treated by degrees with a mixture of 2 parts of potassium bichromate, dissolved in 6 of boiling water, and 8 parts of hydrochloric acid (density 22° Bé.), this quantity being sufficient for 400 parts of fat. Decoloration proceeds gradually, and when complete the fat is washed with hot water.

Bleaching Tallows and Fats.—Instead of exposing to the sun, which is always attended with danger of rendering fats rancid, it is better to liquefy these at a gentle heat, and then add ⅛ in weight of a mixture of equal parts of kaolin and water. The fatty matter should be worked up for a time and then left to separate. Kaolin has the advantage of cheapness in price and of being readily procured.

Freshly burned animal charcoal would perhaps be a more satisfactory decolorizer than kaolin, but it is more expensive to start with, and not so easy to regenerate.

Exposure of tallow to the action of steam under high pressure (a temperature of 250° or 260° F.) is also said to render it whiter and harder.

Coloring Matter in Fats.—A simple method for the detection of the addition of coloring matter to fats is here described. Ten parts, by measure, of the melted fat are put into a small separating funnel and dissolved in 10 parts, by measure, of petroleum ether. The solution is then treated with 15 parts, by measure,

of glacial acetic acid and the whole shaken thoroughly. The addition of coloring matter is known by the red or yellow coloration which appears in the lower layer of acetic acid after the contents of the funnel have been allowed to settle. If only a slight addition of coloring matter is suspected, the acetic acid solution is run off into a porcelain basin and the latter heated on a water bath, when the coloration will be seen more readily. This test is intended for butter and margarine, but is also suitable for tallow, lard, etc.

Fatty Acid Fermentation Process.— The production of fatty acids from fats and oils by fermentation is growing in importance. These particulars, which are the actual results from recent experiments on a somewhat extended scale, are given: Seven hundred and fifty pounds of cottonseed oil are mixed with 45 gallons of water and 3½ pounds of acetic acid; this mixture is heated to a temperature of 85° F. Castor-oil seeds, 53 pounds, decorticated and ground, are mixed thoroughly with 3 gallons of water and 4½ gallons of the oil, and this mixture is stirred into the oil and water; the whole mass is then kept mixed for 12 hours by blowing air through, after which it is allowed to stand for another 12 hours, being given a gentle stir by hand at the end of every hour. After 24 hours the mass is heated to a temperature of 180° F., which stops the fermentation and at the same time allows the fatty acids to separate more freely. To assist in this effect there is added 1 gallon of sulphuric acid (1 in 3) solution.

After 2 hours' standing, the mass will have separated into three layers—fatty acids on the top, glycerine water below, and a middle, undefined layer. The glycerine water is run away, and the whole mass left to stand for 2 hours. The middle portion is run off from the separated fatty acids into another vessel, where it is mixed with 10 gallons of hot water, thoroughly stirred, and allowed to stand for 16 hours or more. The watery layer at the bottom, which contains some glycerine, is then run off, while the residue is mixed with a further quantity of 10 gallons of water, and again allowed to stand. The water which separates out, also the layer of fatty acids that forms on the top, are run off and mixed with the portions previously obtained. The various glycerine waters are treated to recover the glycerine, while the fatty acids are made marketable in any convenient way.

Preservation of Fats.—To produce fats and oils containing both iodine and sulphur, whereby they are preserved from going rancid, and consequently can be utilized to more advantage for the usual purposes, such as the manufacture of soaps, candles, etc., following is the Loebell method:

The essential feature of the process is that the iodine is not merely held in solution by the oil or fat, but enters into chemical combination with the same; the sulphur also combines chemically with the oil or fat, and from their reactions the preserving properties are derived.

The process consists of heating, for example, 6 parts of oil with 1 part of sulphur to a temperature varying between 300° and 400° F., then, when at about 195° F., a solution of iodine and oil is added to the mixture, which is constantly agitated until cool to prevent lumps forming. A product is thus obtained which acquires the consistency of butter, and contains both iodine and sulphur in combination.

Purifying Oils and Fats.—In purifying fatty oils and fats for edible purposes the chief thing is to remove the free fatty acids, which is done by the aid of solutions of alkalies and alkaline earths. The subsequent precipitation of the resulting soapy emulsions, especially when lime is used, entails prolonged heating to temperatures sometimes as high as the boiling point of water. Furthermore, the amount of alkalies taken is always greater than is chemically necessary, the consequence being that some of the organic substances present are attacked, and malodorous products are formed, a condition necessitating the employment of animal charcoal, etc., as deodorizer.

To prevent the formation of these untoward products, which must injuriously affect the quality of edible oils, C. Fresenius proposes to accelerate the dispersion of the said emulsions by subjecting the mixtures to an excess pressure of 1 to 1½ atmospheres and a corresponding temperature of about 220° F., for a short time, the formation of decomposition products, and any injurious influence on the taste and smell of the substance being prevented by the addition of fresh charcoal, etc., beforehand. Charcoal may, and must in certain cases, be replaced for this purpose by infusorial earth or fuller's earth. When this process is applied to cottonseed oil, 100 parts of the oil are mixed with $\frac{1}{10}$ part of fresh, pure charcoal, and ½ part of pure fuller's earth.

The mixture is next neutralized with lime-water, and placed in an autoclave, where it is kept for an hour under pressure, and at a temperature of 220° F. Under these conditions the emulsion soon separates, and when this is accomplished the whole is left to cool down in a closed vessel.

FATS, DECOMPOSITION OF:
See Oil.

FEATHER BLEACHING AND COLORING:
See also Dyes.

Bleaching and Coloring Feathers.—Feathers, in their natural state, are not adapted to undergo the processes of dyeing and bleaching; they must be prepared by removing their oil and dirt. This is usually done by washing them in moderately warm soap and water, and rinsing in warm and cold water; or the oil may be chemically removed by the use of benzine. To remove it entirely, the feathers must be left in the cleansing fluid from a half hour to an hour, when they may be subjected to the process of bleaching.

Bleaching Plumes.—Plumes may be almost entirely bleached by the use of hydrogen peroxide, without injuring their texture.

In specially constructed glass troughs, made the length of an average ostrich feather, 15 or 20 of these feathers can be treated at a time. The bleaching fluid is made from a 30 per cent solution of hydrogen peroxide, with enough ammonia added to make it neutral; in other words when neutral, blue litmus paper will not turn red, and red will take a pale violet tinge. The previously cleansed feathers are entirely immersed in this bleaching bath, which may be diluted if desired. The trough is covered with a glass plate and put in a dark place. From time to time the feathers are stirred and turned, adding more hydrogen peroxide. This process requires 10 to 12 hours and if necessary should be repeated. After bleaching they are rinsed in distilled water or rain water, dried in the air, and kept in motion while drying.

To insure success in coloring feathers in delicate tints, they must be free from all impurities, and evenly white. It has been found of advantage to rub the quill of heavy ostrich plumes while still moist with carbonate of ammonia before the dyeing is begun.

Methods of Dyeing Feathers.—I.—A boiling hot neutral solution, the feathers to be dried in a rotating apparatus. Suitable dyes for this method are chrysoidin,

A, C; crystal vesuvin, 4 B C; phosphin extra, leather yellow, O H; leather red, O, G B; leather brown, O; morocco red, O; azophocphine, G O, B R O; fuchsine, cerise, G R; grenadine, O; safranine, O; methylene violet, malachite green, crystal brilliant green, methylene green, methylene gray, coal black II.

II.—A boiling hot sulphuric solution. Dyes, acid fuchsine, orseilline, R B; acid cerise, O; acid maroon, O; opal blue, blue de lyon, R B; cotton blue, No. 2, China blue No. 2, naphthalene green, O; patent blue, V A; fast blue, O R; fast blue black, O; deep black, G; azo yellow, victorine yellow, orange No. 2, fast brown O, ponceau G R R R, fast red O, Bordeaux, G B R.

III.—An acetic solution. Dyes, Bengal pink G B, phloxine G O, rosolan O B O F, rhodamine O 4 G, eosine A G, erythrosine.

By appropriate mixtures of the dyes of any one class, plumes can be dyed every possible color. After dyeing they are rinsed, and dried in a rotating apparatus. The final process is that of curling, which is done by turning them round and round over a gentle heat. For white feathers a little sulphur may be burned in the fire; for black or colored ones a little sugar.

IV.—The spray method. The solution of the dye to be used is put into an atomizer, and the spray directed to that part of the feather which it is desired to color. By using different colors the most marvelous effects and most delicate transitions from one color to another are obtained. Any kind of an atomizer can be used, the rubber bulb, pump, or bellows; the result is the same.

FELT WATERPROOFING:
See Waterproofing.

FERMENTATION PROCESS, FATTY ACID:
See Fats.

FERMENTATION, PREVENTION OF:
See Anti-Ferments and Wines and Liquors.

FERROUS OXALATE DEVELOPER:
See Photography.

Fertilizers

(See also Phosphate, Artificial.)

Plant Fertilizers.—Plants are as sensitive to excessively minute quantities of nutrient substances, such as salts of potassium, in the soil, as they are to minute quantities of poisonous substances. Poisons are said to be infinitely more sensitive reagents for the presence of certain metallic salts than the most delicate chemical, the statement having been made that a trace of copper which might be obtained by distilling in a copper retort is fatal to the white and yellow lupin, the castor-oil plant, and spirogyra. Coupin has found salts of silver, mercury, copper, and cadmium especially fatal to plants. With copper sulphate the limit of sensitiveness is placed at 1 in 700,000,000. Devaux asserts that both phanerogams and cryptogams are poisoned by solutions of salts of lead or copper diluted to the extent of 1 in 10,000,000, or less.

As a result of a series of experiments, Schloesing stated that the nitrification of ammonium salts is not for all plants a necessary preliminary to the absorption of nitrogen by the plant. While for some plants, as for example buckwheat, the preferable form of the food material is that of a nitrate, others, for instance, tropeolum, thrive even better when the nitrogen is presented to them in an ammoniacal form.

Artificial Fertilizers for Pot Plants.—Experiments on vegetation have shown that a plant will thrive when the lacking substances are supplied in a suitable form, e. g., in the following combinations:

I.—Calcium nitrate, potassium nitrate, potassium phosphate, magnesium phosphate, ferric phosphate (sodium chloride).

II.—Calcium nitrate, ammonium nitrate, potassium sulphate, magnesium phosphate, iron chloride (or sulphate) (sodium silicate).

It is well known that in nature nitrates are formed wherever decomposition of organic nitrogenous substances takes place in the air, the ammonia formed by the decomposition being oxidized to nitric acid. These conditions for the formation of nitrates are present in nearly every cornfield, and they are also the cause of the presence of nitrates in water that has its source near stables, etc. In Peruvian guano nitrogen is present partly in the form of potassium nitrate, partly as ammonium phosphate and sulphate. As a nitrate it acts more rapidly than in the form of ammonia, but in the latter case the effect is more lasting. Phosphoric acid occurs in guano combined with ammonia, potash, and chiefly with lime, the last being slower and more lasting in action than the others.

Nearly all artificial fertilizers conform, more or less, to one of the following general formulas:

I.—Artificial Flower Fertilizer.—

	1	2	3
Ammonium nitrate	0.40	1.60	40.0 parts
Ammonium phosphate	0.20	0.80	20.0 parts
Potassium nitrate	0.25	1.00	25.0 parts
Ammonium chloride	0.05	0.20	5.0 parts
Calcium sulphate	0.06	0.24	6.0 parts
Ferrous sulphate	0.04	0.16	4.0 parts
	1.00	4.00	100.0 parts

Dissolve 1 part in 1,000 parts water, and water the flowers with it 2 or 3 times weekly. Dissolve 4 parts in 1,000 parts water, and water with this quantity 10 or 12 pots of medium size.

II.—Compost for Indoor Plants.—

	1	2	3
Ammonium sulphate	0.30	1.20	30.0 parts
Sodium chloride	0.30	1.20	30.0 parts
Potassium nitrate	0.15	0.60	15.0 parts
Magnesium sulphate	0.15	0.60	15.0 parts
Magnesium phosphate	0.04	0.20	4.0 parts
Sodium phosphate	0.06	0.24	6.0 parts
	1.00	4.00	100.0 parts

One part to be dissolved in 1,000 parts water and the flowers watered up to 3 times daily. Dissolve 4 parts in 1,000 parts water, and water with this solution daily:

III.—Plant Food Solution.—

	1		2
Potassium chloride	0.16	or	12.5 parts
Calcium nitrate	0.71	or	58.0 parts
Magnesium sulphate	0.125	or	12.0 parts
Potassium phosphate	0.133	or	15.0 parts
Iron phosphate, recently precipitated	0.032	or	2.5 parts
	1.160	or	100.0 parts

This turbid mixture (1 part in 1,000 parts) is used alternately with water for watering a pot of about 1 quart capacity; for smaller or larger pots in proportion. After using the amount indicated, the watering is continued with water alone.

IV.—Fertilizer with Organic Matter, for Pot Flowers.—

Potassium nitrate	100.0 parts
Ammonium phosphate	100.0 parts
Phosphoric acid	2.5 parts
Simple syrup	1,000 parts

Add not more than 10 parts to 1,000 parts water, and water alternately with this and with water alone. For cactaceæ, crassulaceæ, and similar plants, which do not assimilate organic matter directly, use distilled water instead of syrup.

Chlorotic plants are painted with a dilute iron solution or iron is added to the soil, which causes them to assume their natural green color. The iron is used in form of ferric chloride or ferrous sulphate.

V.—

Sodium phosphate	4 ounces
Sodium nitrate	4 ounces
Ammonium sulphate	2 ounces
Sugar	1 ounce

Use 2 teaspoonfuls to a gallon of water.

VI.—

Ammonium phosphate	30 parts
Sodium nitrate	25 parts
Potassium nitrate	25 parts
Ammonium sulphate	20 parts
Water	100,000 parts

One application of this a week is enough for the slower growing plants, and 2 for the more rapid growing herbaceous ones.

VII.—

Calcium phosphate	4 ounces
Potassium nitrate	1 ounce
Potassium phosphate	1 ounce
Magnesium sulphate	1 ounce
Iron (ferric) phosphate	100 grains

VIII.—Pot plants, especially flowering plants kept around the house, should be treated to an occasional dose of the following:

Ammonium chloride	2 parts
Sodium phosphate	4 parts
Sodium nitrate	3 parts
Water	80 parts

Mix and dissolve. To use, add 25 drops to the quart of water, and use as in ordinary watering.

IX.—

Sugar	1 part
Potassium nitrate	2 parts
Ammonium sulphate	4 parts

X.—

Ferric phosphate	1 part
Magnesium sulphate	2 parts
Potassium phosphate	2 parts
Potassium nitrate	2 parts
Calcium acid phosphate	8 parts

About a teaspoonful of either of these mixtures is added to a gallon of water, and the plants sprinkled with the liquid.

For hastening the growth of flowers, the following fertilizer is recommended:

XI.—Potassium nitrate. 30 parts
Potassium phosphate.......... 25 parts
Ammonium sulphate.......... 10 parts
Ammonium nitrate 35 parts

The following five are especially recommended for indoor use:

XII.—Sodium chloride.. 10 parts
Potassium nitrate. 5 parts
Magnesium sulphate.......... 5 parts
Magnesia........ 1 part
Sodium phosphate 2 parts

Mixed and bottled. Dissolve a teaspoonful daily in a quart of water and water the plants with the solution.

XIII.—Ammonium nitrate 40 parts
Potassium nitrate. 90 parts
Ammonium phosphate.......... 50 parts

Two grams is sufficient for a medium-sized flower pot.

XIV.—Ammonium sulphate.......... 10 parts
Sodium chloride.. 10 parts
Potassium nitrate. 5 parts
Magnesium sulphate.......... 5 parts
Magnesium carbonate .... 1 part
Sodium phosphate 20 parts

One teaspoonful to 1 quart of water.

XV.—Ammonium nitrate 40 parts
Ammonium phosphate.......... 20 parts
Potassium nitrate.0.25 parts
Ammonium chloride.......... 5 parts
Calcium sulphate. 6 parts
Ferrous sulphate.. 4 parts

Dissolve 2 parts in 1,000 of water, and water the plants with the solution.

XVI.—Potassium nitrate. 20 parts
Potassium phosphate.......... 25 parts
Ammonium sulphate.......... 10 parts
Ammonium nitrate 35 parts

This mixture produces a luxuriant foliage. If blooms are desired, dispense with the ammonium nitrate.

XVII.—Saltpeter, 5 parts; cooking salt, 10 parts; bitter salt, 5 parts; magnesia, 1 part; sodium phosphate, 2 parts. Mix and fill in bottles. Dissolve a teaspoonful in 1¼ pints of hot water, and water the flower pots with it each day.

XVIII.—Ammonium sulphate, 30 parts; sodium chloride, 30 parts; potash niter, 15 parts; magnesium sulphate, 15 parts; magnesium phosphate, 4 parts; sodium phosphate, 6 parts. Dissolve 1 part in 1,000 parts water, and apply 3 times per day.

XIX.—Calcium nitrate, 71 parts; potassium chlorate, 15 parts; magnesium sulphate, 12.5 parts; potassium phosphate, 13.3 parts; freshly precipitated ferric phosphate, 3.2 parts. A solution of 1 in 1,000 of this mixture is applied, alternating with water, to the plants. After using a certain quantity, pour on only water.

XX.—Ammonium phosphate, 300 parts; sodium nitrate, 250 parts; potassium nitrate, 250 parts; and ammonium sulphate, 200 parts, are mixed together. To every 1,000 parts of water dissolve 2 parts of the mixture, and water the potted plants once a week with this solution.

XXI.—Potash niter, 20 parts; calcium carbonate, 20 parts; sodium chlorate, 20 parts; calcium phosphate, 20 parts; sodium silicate, 14 parts; ferrous sulphate, 1.5 parts. Dissolve 1 part of the mixture in 1,000 parts water.

Preparing Bone for Fertilizer.—Bone, in its various forms, is the only one of the insoluble phosphates that is now used directly upon the soil, or without other change than is accomplished by mechanical action or grinding. The terms used to indicate the character of the bone have reference rather to their mechanical form than to the relative availability of the phosphoric acid contained in them. The terms raw bone, fine bone, boiled and steamed bone, etc., are used to indicate methods of preparation, and inasmuch as bone is a material which is useful largely in proportion to its rate of decay, its fineness has an important bearing upon availability, since the finer the bone the more surface is exposed to the action of those forces which cause decay or solution, and the quicker will the constituents become available. In the process of boiling or steaming, not only is the bone made finer but its physical character in other respects is also changed, the particles, whether fine or coarse, being made soft and crumbly rather than dense or hard; hence it is more likely to act quickly than if the same degree of fineness be obtained by simple grinding. The phosphoric acid in fine steamed bone may all become available in 1 or 2 years, while the coarser fatty raw bone sometimes resists final decay for 3 or 4 years or even longer.

Bone contains considerable nitrogen, a fact which should be remembered in its use, particularly if used in comparison with other phosphatic materials which do not contain this element. Pure raw bone contains on an average 22 per cent of phosphoric acid and 4 per cent of nitrogen. By steaming or boiling, a portion of the organic substance containing nitrogen is extracted, which has the effect of proportionately increasing the phosphoric acid in the product; hence a steamed bone may contain as high as 28 per cent of phosphoric acid and as low as 1 per cent of nitrogen. Steamed bone is usually, therefore, much richer in phosphoric acid and has less nitrogen than the raw bone.

Brewers' Yeast and Fertilizers.—A mixture is made of about 2 parts of yeast with 1 part of sodium chloride and 5 parts of calcium sulphate, by weight, for use as a manure. Pure or impure yeast, or yeast previously treated for the extraction of a portion of its constituents, may be used, and the gypsum may be replaced by other earthy substances of a similar non-corrosive nature.

Authorities seem to agree that lime is necessary to the plant, and if it be wholly lacking in the soil, even though an abundance of all the other essential elements is present, it cannot develop normally. Many soils are well provided with lime by nature and it is seldom or never necessary for those who cultivate them to resort to liming. It would be just as irrational to apply lime where it is not needed as to omit it where it is required, and hence arises the necessity of ascertaining the needs of particular soils in this respect.

The method usually resorted to for ascertaining the amount of lime in soils is to treat them with some strong mineral acid, such as hydrochloric acid, and determine the amount of lime which is thus dissolved. The fact that beets of all kinds make a ready response to liming on soils which are deficient in lime may be utilized as the basis of testing.

FEVER IN CATTLE:
See Veterinary Formulas.

FIG SQUARES:
See Confectionery.

Files

Composition Files.—These files, which are frequently used by watchmakers and other metal workers for grinding and polishing, and the color of which resembles silver, are composed of 8 parts copper, 2 parts tin, 1 part zinc, 1 part lead. They are cast in forms and treated upon the grindstone; the metal is very hard, and therefore worked with difficulty with the file.

To Keep Files Clean (see also Cleaning Preparations and Methods).—The uneven working of a file is usually due to the fact that filings clog the teeth of the file. To obviate this evil, scratch brush the files before use, and then grease them with olive oil. A file prepared in this manner lasts for a longer time, does not become so quickly filled with filings and can be conveniently cleaned with an ordinary rough brush.

Recutting Old Files.—Old files may be rendered useful again by the following process: Boil them in a potash bath, brush them with a hard brush and wipe off. Plunge for half a minute into nitric acid, and pass over a cloth stretched tightly on a flat piece of wood. The effect will be that the acid remains in the grooves, and will take away the steel without attacking the top, which has been wiped dry. The operation may be repeated according to the depth to be obtained. Before using the files thus treated they should be rinsed in water and dried.

FILE METAL:
See Alloys.

FILLERS FOR LETTERS:
See Lettering.

FILLERS FOR WOOD:
See Wood.

FILTERS FOR WATER.

A filter which possesses the advantages of being easily and cheaply cleaned when dirty, and which frees water from mechanical impurities with rapidity, may be formed by placing a stratum of sponge between two perforated metallic plates, united by a central screw, and arranged in such a manner as to permit of the sponge being compressed as required. Water, under gentle pressure, flows with such rapidity through the pores of compressed sponge, that it is said that a few square feet of this substance will perfectly filter several millions of gallons of water daily.

The sponges are cleaned thoroughly, rolled together as much as possible, and placed in the escape pipe of a percolator in such a manner that the larger portion of the sponge is in the pipe while the smaller portion, spreading by itself, protrudes over the pipe toward the interior of the percolator, thus forming a flat filter covering it. After a thorough moistening of the sponge it is said to admit of a very quick and clear filtration of large quantities of tinctures, juices, etc.

For filtering water on a small scale, and for domestic use, "alcarrazas," diaphragms of porous earthenware and filtering-stone and layers of sand and charcoal, etc., are commonly employed as filtering.

A cheap, useful form of portable filter is the following, given in the proceedings of the British Association: "Take any common vessel, perforated below, such as a flower pot, fill the lower portion with coarse pebbles, over which place a layer of finer ones, and on these a layer of clean coarse sand. On the top of this a piece of burnt clay perforated with small holes should be put, and on this again a stratum of 3 or 4 inches thick of well-burnt, pounded animal charcoal. A filter thus formed will last a considerable time, and will be found particularly useful in removing noxious and putrescent substances held in solution by water."

The "portable filters," in stoneware, that are commonly sold in the shops, contain a stratum of sand, or coarsely powdered charcoal; before, however, having access to this, the water has to pass through a sponge, to remove the coarser portion of the impurities.

Alum Process of Water Purification.— Water may be filtered and purified by precipitation, by means of alum, by adding a 4 per cent solution to the water to be clarified until a precipitate is no longer produced. After allowing the turbid mixture to stand for 8 hours, the clear portion may be decanted or be siphoned off. About 2 grains of alum is ordinarily required to purify a gallon of water. Potassa alum only should be used, as ammonia alum cannot be used for this purpose. The amount of alum required varies with the water, so that an initial experiment is required whenever water from a new source is being purified. If the purification is properly done, the water will not contain any alum, but only a trace of potassium sulphate, for the aluminum of the double sulphate unites with the various impurities to form an insoluble compound which gradually settles out, mechanically carrying with it suspended matter, while the sulphuric acid radical unites with the calcium in the water to form insoluble calcium sulphate.

FILTER PAPER:
See Paper.

FILM-STRIPPING:
See Photography.

FINGER-TIPS, SPARKS FROM:
See Pyrotechnics.

FIRES, COLORED:
See Pyrotechnics.

FIREARM LUBRICANTS:
See Lubricants.

FIRE EXTINGUISHERS:

I.—Calcium chloride. 184 parts
Magnesium chloride........... 57 parts
Sodium chloride.. 13 parts
Potassium bromide .... 22 parts
Barium chloride.. 3 parts
Water to make... 1,000 parts

Dissolve and fill into hand grenades.

II.—Iron sulphate ..... 4 parts
Ammonium sulphate.......... 16 parts
Water............ 100 parts

Mix, dissolve, and fill into flasks.

III.—Sodium chloride... 430 parts
Alum............ 195 parts
Glauber salts...... 50 parts
Sodium carbonate, impure......... 35 parts
Water glass....... 266 parts
Water............ 233 parts

Mix, etc.

IV.—Sodium chloride... 90 parts
Ammonium chloride............ 45 parts
Water............ 300 parts

Mix, dissolve, and put into quart flasks of very thin glass, which are to be kept conveniently disposed in the dwelling rooms, etc., of all public institutions.

V.—Make 6 solutions as follows:
a.—Ammonium chloride...... 20 parts
Water .... 2,000 parts
b.—Alum, calcined and powdered 35 parts
Water......... 1,000 parts
c.—Ammonium sulphate, powdered........ 30 parts
Water......... 500 parts
d.—Sodium chloride 20 parts
Water......... 4,000 parts
e.—Sodium carbonate.......... 35 parts
Water......... 500 parts
f.—Liquid water glass........ 450 parts

Mix the solutions in the order named and to the mixture, while still yellow and turbid, add 2,000 parts of water, and let stand. When the precipitate has subsided fill off the clear liquid into thin glass (preferably blue, to deter decomposition) containers each of 3 pints to a half gallon capacity.

VI.—Calcium chloride.. 30 parts
    Magnesium chloride........... 10 parts
    Water........... 60 parts

VII.—Sodium chloride.. 20 parts
    Ammonium chloride........... 9 parts
    Water........... 71 parts

VIII.—Sodium carbonate 16 parts
    Sodium chloride. 64 parts
    Water..........920 parts

The most effective of all extinguishers is ammonia water. It is almost instantaneous in its effect, and a small quantity only is required to extinguish any fire. Next in value is carbonic acid gas. This may be thrown from siphons or soda-water tanks. The vessel containing it should be thrown into the fire in such a way as to insure its breaking.

Dry Powder Fire Extinguishers.—The efficacy of these is doubted by good authorities. They should be tested before adoption.

I.—Alum............. 24 parts
    Ammonium sulphate 52 parts
    Ferrous sulphate.... 4 parts

II.—Sodium chloride.... 8 parts
    Sodium bicarbonate 6 parts
    Sodium sulphate.... 2 parts
    Calcium chloride.... 2 parts
    Sodium silicate..... 2 parts

III.—Sodium chloride.... 6 parts
    Ammonium chloride 6 parts
    Sodium bicarbonate.. 8 parts

IV.—Ammonium chloride 10 parts
    Sodium sulphate.... 6 parts
    Sodium bicarbonate 4 parts

Oil Extinguisher.—To extinguish oils which have taken fire, a fine-meshed wire net of the size of a boiling pan should be kept on hand in every varnish factory, etc. In the same moment when the netting is laid upon the burning surface, the flame is extinguished because it is a glowing mass of gas, which the iron wire quickly cools off so that it cannot glow any more. The use of water is excluded, and that of earth and sand undesirable, because both dirty the oil.

Substitute for Fire Grenades.—A common quart bottle filled with a saturated solution of common salt makes a cheap and efficient substitute for the ordinary hand grenade. The salt forms a coating on all that the water touches and makes it nearly incombustible.

Fireproofing

For Textiles.—I.—Up to the present this has generally been accomplished by the use of a combination of water glass or soluble glass and tungstate of soda. The following is cheaper and more suitable for the purpose:

Equal parts, by weight, of commercial white copperas, Epsom salt, and sal ammoniac are mingled together and mixed with three times their weight of ammonia alum. This mixture soon changes into a moist pulp or paste, that must be dried by a low heat. When dressing the material, add $\frac{1}{2}$ part of this combination to every 1 part of starch.

II.—Good results are also obtained from the following formula: Supersaturate a quantity of superphosphate of lime with ammonia, filter, and decolorize it with animal charcoal. Concentrate the solution and mix with it 5 per cent of gelatinous silica, evaporate the water, dry, and pulverize. For use mix 30 parts of this powder with 35 parts of gum and 35 parts of starch in sufficient water to make of suitable consistency.

III.—As a sample of the Melunay process, introduced in France, the following has been published: Apply to a cotton fabric like flannellet, or other cotton goods, a solution of stannate of soda (or a salt chemically equivalent) of the strength of 5 to 10° Bé., then dry the fabric and saturate it again, this time with a solution of a titanium salt; any soluble titanium salt is suitable. This salt should be so concentrated that each 1,000 parts may contain about 62 parts of titanium oxide. The fabrics are again dried, and the titanium is ultimately fixed by means of a suitable alkaline bath. It is advantageous to employ for this purpose a solution of silicate of soda of about 14° Bé., but a mixed bath, composed of tungstate of soda and ammonium chloride, may be employed. The objects are afterwards washed, dried, and finished as necessary for trade. A variation consists in treating the objects in a mixed bath containing titanium, tungsten, and a suitable solvent.

IV.—Boil together, with constant

stirring, the following ingredients until a homogeneous mass results:

Linseed oil	77	parts
Litharge	10	parts
Sugar of lead	2	parts
Lampblack	4	parts
Oil turpentine	2	parts
Umber	0.4	parts
Japanese wax	0.3	parts
Soap powder	1.2	parts
Manila copal	0.7	parts
Caoutchouc varnish	2	parts

V.—For Light Woven Fabrics.— Ammonium sulphate, 8 parts, by weight; ammonium carbonate, 2.5 parts; borax, 2; boracic acid, 3; starch, 2; or dextrin, 0.4, or gelatin, 0.4; water, 100. The fabric is to be saturated with the mixture, previously heated to 86° F., and dried; it can then be calendered in the ordinary way. The cost is only 2 or 3 cents for 16 yards or more of material.

VI.—For Rope and Straw Matting.- Ammonium chloride (sal ammoniac), 15 parts, by weight; boracic acid, 6 parts; borax, 3; water, 100. The articles are to be left in the solution, heated to 212° F. for about 3 hours, then squeezed out and dried. The mixture costs about 5 cents a quart.

VII.—For Clothing.— The following starch is recommended: Sodium tungstate, perfectly neutral, 30 parts; borax, 20; wheat or rice starch, 60. The constituents are to be finely pulverized, sharply dried, and mixed, and the starch used like any other. Articles stiffened with it, if set on fire, will not burst into flame, but only smolder.

VIII.—For Tents.—

Water	100	
Ammonium sulphate, chemically pure	14	Parts by weight.
Boracic acid	1	
Hartshorn salt	1	
Borax	3	
Glue water	2	

Boil the water, put ammonium sulphate into a vat, pour a part of the boiling water on and then add the remaining materials in rotation. Next follow the rest of the hot water. The vat should be kept covered until the solution is complete.

IX.—For Stage Decorations.— Much recommended and used as a fireproofing composition is a cheap mixture of borax, bitter salt, and water; likewise for canvas a mixture of ammonium sulphate, gypsum, and water. Ammonium sulphate and sodium tungstate are also named for impregnating the canvas before painting.

X.—For Mosquito Netting.— Immerse in a 20 per cent solution of ammonium sulphate. One pound of netting will require from 20 to 24 ounces of the solution to thoroughly saturate. After withdrawing from the bath, do not wring it out, but spread it over a pole or some such object, and let it get about half dry, then iron it out with a hot iron. The material (ammonium sulphate) is inoffensive.

Fireproofing of Wood.— Strictly speaking, it is impossible to render wood completely incombustible, but an almost absolute immunity against the attacks of fire can be imparted.

Gay-Lussac was one of the first to lay down the principal conditions indispensable for rendering organic matters in general, and wood in particular, uninflammable.

During the whole duration of the action of the heat the fibers must be kept from contact with the air, which would cause combustion. The presence of borates, silicates, etc., imparts this property to organic bodies.

Combustible gases, disengaged by the action of the heat, must be mingled in sufficient proportion with other gases difficult of combustion in such a way that the disorganization of bodies by heat will be reduced to a simple calcination without production of flame. Salts volatile or decomposable by heat and not combustible, like certain ammoniacal salts, afford excellent results.

Numerous processes have been recommended for combating the inflammability of organic tissues, some consisting in external applications, others in injection, under a certain pressure, of saline solutions.

By simple superficial applications only illusory protection is attained, for these coverings, instead of fireproofing the objects on which they are applied, preserve them only for the moment from a slight flame. Resistance to the fire being of only short duration, these coatings scale off or are rapidly reduced to ashes and the parts covered are again exposed. It often happens, too, that such coatings have disappeared before the occurrence of a fire, so that the so-called remedy becomes injurious from the false security occasioned.

Some formulas recommended are as follows:

I.—For immersion or imbibition the following solution is advised: Ammonium phosphate, 100 parts; boracic acid, 10 parts per 1,000; or ammonium sulphate, 135 parts; sodium borate, 15 parts; boracic acid, 5 parts per 1,000. For each of these formulas two coats are necessary.

II.—For application with the brush the following compositions are the best:

a. Apply hot, sodium silicate, 100 parts; Spanish white, 50 parts; glue, 100 parts.

b. Apply successively and hot; for first application, water, 100 parts; aluminum sulphate, 20 parts; second application, water, 100 parts; liquid sodium silicate, 50 parts.

c. First application, 2 coats, hot; water, 100 parts; sodium silicate, 50 parts; second application, 2 coatings; boiling water, 75 parts; gelatin, white, 200 parts; work up with asbestos, 50 parts; borax, 30 parts; and boracic acid, 10 parts.

Oil paints rendered uninflammable by the addition of phosphate of ammonia and borax in the form of impalpable powders incorporated in the mass, mortar of plaster and asbestos and asbestos paint, are still employed for preserving temporarily from limited exposure to a fire.

III.—Sodium silicate,
  solid........... 350 parts
  Asbestos, pow-
  dered.......... 350 parts
  Water, boiling....1,000 parts

Mix. Give several coatings, letting each dry before applying the next.

IV.—Asbestos, powdered 35 parts
  Sodium borate.... 20 parts
  Water............ 100 parts
  Gum lac........10 to 15 parts

Dissolve the borax in the water by the aid of heat, and in the hot solution dissolve the lac. When solution is complete incorporate the asbestos. These last solutions give a superficial protection, the efficiency of which depends upon the number of coatings given.

V.—Prepare a syrupy solution of sodium silicate, 1 part, and water, 3 parts, and coat the wood 2 to 3 times, thus imparting to it great hardness. After drying, it is given a coating of lime of the consistency of milk, and when this is almost dry, is fixed by a strong solution of soluble glass, 2 parts of the syrupy mass to 3 parts of water. If the lime is applied thick, repeat the treatment with the soluble glass.

VI.—Subject the wood or wooden objects for 6 to 8 hours to the boiling heat of a solution of 33 parts of manganese chloride, 20 parts of orthophosphoric acid, 12 parts of magnesium carbonate, 10 parts of boracic acid, and 25 parts of ammonium chloride in 1,000 parts of water. The wood thus treated is said to be perfectly incombustible even at great heat, and, besides, to be also protected by this method against decay, injury by insects, and putrefaction.

VII.—One of the simplest methods is to saturate the timber with a solution of tungstate of soda; if this is done in a vacuum chamber, by means of which the wood is partly deprived of the air contained in its cells, a very satisfactory result will be obtained. Payne's process consists in treating wood under these conditions first with solution of sulphate of iron, and then with chloride of calcium; calcium sulphate is thus precipitated in the tissues of the timber, which is rendered incombustible and much more durable. There are several other methods besides these, phosphate of ammonia and tungstate being most useful. A coat of common whitewash is an excellent means of lessening the combustibility of soft wood.

Fireproofing Wood Pulp.—The pulp is introduced into a boiler containing a hot solution of sulphate and phosphate of ammonia and provided with a stirring and mixing apparatus, as well as with an arrangement for regulating the temperature. After treatment, the pulp is taken out and compressed in order to free it from its humidity. When dry, it may be used for the manufacture of paper or for analogous purposes. Sawdust treated in the same manner may be used for packing goods, for deadening walls, and as a jacketing for steam pipes.

Fireproofing for Wood, Straw, Textiles, etc.—The material to be made fireproof is treated with a solution of 10 to 20 parts of potassium carbonate and 4 to 8 parts of ammonium borate in 100 parts of water. Wherever excessive heat occurs, this compound, which covers the substance, is formed into a glassy mass, thus protecting the stuff from burning; at the same time a considerable amount of carbonic acid is given off, which smothers the flames.

MISCELLANEOUS FORMULAS FOR FIREPROOFING.

I.—In coating steel or other furnaces, first brush over the brickwork to be covered a solution made by boiling 1 pound each of silicate of soda and alum in 4 gallons of water, and follow immediately with composition:

Silica	50 parts
Plastic fire clay	10 parts
Ball clay	3 parts

Mix well.

Fireproof Compositions.—II.—For furnaces, etc.:

Pure silica (in grain)	60 parts
Ground flint	8 parts
Plaster of Paris	3 parts
Ball clay	3 parts

Mix well together by passing once or more through a fine sieve, and use in the same way as cement.

Fireproof Paper.—Paper is rendered fireproof by saturating it with a solution of

Ammonium sulphate	8 parts
Boracic acid	3 parts
Borax	2 parts
Water	100 parts

For the same purpose sodium tungstate may also be employed.

Fireproof Coating.—A fireproof coating (so-called) consists of water, 100 parts; strong glue, 20 parts; silicate of soda, 38° Bé., 50 parts; carbonate of soda, 35 parts; cork in pieces of the size of a pea, 100 parts.

Colored Fireproofing. — I. — Ammonium sulphate, 70 parts; borax, 50 parts; glue, 1 part; and water up to 1,000 parts.

II.—Solution of glue, 5 parts, zinc chloride, 2 parts; sal ammoniac, 80 parts; borax, 57 parts; and water up to 700 parts.

If the coating is to be made visible by coloration, an addition of 10 parts of Cassel brown and 6 parts of soda per 1,000 parts is recommended, which may be dissolved separately in a portion of the water used.

FIREPROOFING CELLULOID:
See Celluloid.

FIREPROOFING OF PAPER:
See Paper.

FIREWORKS:
See Pyrotechnics.

FILIGREE GILDING:
See Plating.

FISH BAIT.

Oil of rhodium	3 parts
Oil of cumin	2 parts
Tincture of musk	1 part

Mix. Put a drop or two on the bait, or rub trigger of trap with the solution.

FIXATIVES FOR CRAYON DRAWINGS, ETC.

I.—

Shellac	40	Parts
Sandarac	20	by
Spirit of wine	940	weight.

II.—During the Civil War, when both alcohol and shellac often were not purchasable, and where, in the field especially, ink was almost unknown, and sized paper, of any description, a rarity, men in the field were compelled to use the pencil for correspondence of all sorts. Where the communication was of a nature to make its permanency desirable, the paper was simply dipped in skim milk, which effected the purpose admirably. Such documents written with a pencil on unsized paper have stood the wear and rubbing of upward of 40 years.

To Fix Pounced Designs.—Take beer or milk or alcohol, in which a little bleached shellac has been dissolved, and blow one of these liquids upon the freshly pounced design by means of an atomizer. After drying, the drawing will have the desired fixedness.

FIXING BATHS FOR PAPER AND NEGATIVES:
See Photography.

FLANNELS, WHITENING OF:
See Laundry Preparations.

FLASH-LIGHT APPARATUS AND POWDERS:
See Photography.

FLAVORINGS:
See Condiments.

FLEA DESTROYERS:
See Insecticides.

FLIES IN THE HOUSE:
See Household Formulas.

FLIES AND PAINT:
See Paint.

Floor Dressings

(See also Paint, Polishes, Waxes, and Wood.)

Oil Stains for Hard Floors.—I.—Burnt sienna, slate brown, or wine black, is ground with strong oil varnish in the paint mill. The glazing color obtained is thinned with a mixture of oil of turpentine and applied with a brush on the respective object. The superfluous stain is at once wiped away with a rag, so that only the absorbed stain remains in the wood. If this is uneven, go over the light places again with dark stain. In a similar manner all otherwise tinted and colored oil stains are produced by merely grinding the respective color with the corresponding addition of oil. Thus, green, red, and even blue and violet shades on wood can be obtained, it being necessary only to make a previous experiment with the stains on a piece of suitable wood. In the case of soft wood, however, it is advisable to stain the whole previously with ordinary nut stain (not too dark), and only after drying to coat with oil stain, because the autumn rings of the wood take no color, and would appear too light, and, therefore, disturb the effect.

II.—Boil 25 parts, by weight, of fustic and 12 parts of Brazil wood with 2,400 parts of soapmakers' lye and 12 parts of potash, until the liquid measures about 12 quarts. Dissolve in it, while warm, 30 parts of annatto and 75 of wax, and stir until cold. There will be a sufficient quantity of the brownish-red stain to keep the floor of a large room in good order for a year. The floor should be swept with a brush broom daily, and wiped up twice a week with a damp cloth, applying the stain, when necessary, to places where there is much wear, and rubbing it in with a hard brush. Every 6 weeks put the stain all over the floor, and brush it in well.

III.—Neatsfoot oil . . . . 1 part
    Cottonseed oil . . . . 1 part
    Petroleum oil . . . . 1 part

IV.—Beeswax . . . . 8 parts
    Water . . . . 56 parts
    Potassium carbonate 4 parts

Dissolve the potash in 12 parts of water; heat together the wax and the remaining water till the wax is liquefied; then mix the two and boil together until a perfect emulsion is effected. Color, if desired, with a solution of annatto.

V.—Paraffine oil . . . . 8 parts
    Kerosene . . . . 1 part
    Limewater . . . . 1 part

Mix thoroughly. A coat of the mixture is applied to the floor with a mop.

Paraffining of Floors.—The cracks and joints of the parquet floor are filled with a putty consisting of Spanish white, 540 parts; glue, 180 parts; sienna, 150 parts;

umber, 110 parts; and calcareous earth, 20 parts. After 48 hours apply the paraffine, which is previously dissolved in petroleum, or preferably employed in a boiling condition, in which case it will enter slightly into the floor. When solidification sets in, the superfluous paraffine is scratched off and an even, smooth surface of glossy color results, which withstands acids and alkalies.

Ball-Room Floor Powder.—

Hard paraffine . . . .   1 pound
Powdered boric acid . .   7 pounds
Oil lavender . . . .   1 drachm
Oil neroli . . . .   20 minims

Melt the paraffine and add the boric acid and the perfumes. Mix well, and sift through a $\frac{1}{15}$ mesh sieve.

Renovating Old Parquet Floors.—Caustic soda lye, prepared by boiling for 45 minutes with 1 part calcined soda, and 1 part slaked lime with 15 parts water, in a cast-iron pot, is applied to the parquet to be renovated by means of a cloth attached to a stick. After a while rub off the floor with a stiff brush, fine sand, and a sufficient quantity of water, to remove the dirt and old wax. Spread a mixture of concentrated sulphuric acid and water in the proportion of 1 to 8 on the floor. The sulphuric acid will remove the particles of dirt and wax which have entered the floor and enliven the color of the wood. Finally, wax the parquet after it has been washed off with water and dried completely.

FLOOR OIL:
See Oils.

FLOOR PAPER:
See Paper.

FLOOR POLISH:
See Polishes.

FLOOR VARNISHES:
See Varnishes.

FLOOR WATERPROOFING:
See Waterproofing.

FLOOR WAX:
See Waxes.

FLORICIN OIL:
See Oil.

FLOWER PRESERVATIVES.

I.—To preserve flowers they should be dipped in melted paraffine, which should be just hot enough to maintain its fluidity. The flowers should be dipped one at a time, held by the stalks and moved about for an instant to get rid of air bubbles. Fresh cut flowers, free from moisture, are said to make excellent specimens when treated in this way. A solution in which cut flowers may be kept immersed is made as follows:

Salicylic acid....... 20 grains
Formaldehyde...... 10 minims
Alcohol........... 2 fluidounces
Distilled water..... 1 quart

II.—The English method of preserving flowers so as to retain their form and color is to imbed the plants in a mixture of equal quantities of plaster of Paris and lime, and gradually heat them to a temperature of 100° F. After this the flower looks dusty, but if it is laid aside for an hour so as to absorb sufficient moisture to destroy its brittleness, it can be dusted without injury. To remove the hoary appearance which is often left, even after dusting, a varnish composed of 5 ounces of dammar and 16 ounces of oil of turpentine should be used and a second coat given if necessary. When the gum has been dissolved in the turpentine, 16 ounces of benzoline should be added, and the whole should be strained through fine muslin.

III.—Five hundred parts ether, 20 parts transparent copal, and 20 parts sand. The flowers should be immersed in the varnish for 2 minutes, then allowed to dry for 10 minutes, and this treatment should be repeated 5 or 6 times.

IV.—Place the flowers in a solution of 30 grains of salicylic acid in 1 quart of water.

V.—Moisten 1,000 parts of fine white sand that has been previously well washed and thoroughly dried and sifted, with a solution consisting of 3 parts of stearine, 3 parts of paraffine, 3 parts of salicylic acid, and 100 parts of alcohol. Work the sand up thoroughly so that every grain of it is impregnated with the mixture, and then spread it out and let it become perfectly dry. To use, place the flowers in a suitable box, the bottom of which has been covered with a portion of the prepared sand, and then dust the latter over them until all the interstices have been completely filled with it. Close the box lightly and put it in a place where it can be maintained at a temperature of from 86° to 104° F. for 2 or 3 days. At the expiration of this time remove the box and let the sand escape. The flowers can then be put into suitable receptacles or glass cases without fear of deterioration. Wilted or withered flowers should be freshened up by dipping into a suitable aniline solution, which will restore their color.

VI.—Stand the flowers upright in a box of proper size and pour over and around them fine dry sand, until the flowers are completely surrounded in every direction. Leave them in this way for 8 or 10 days, then carefully pour off the sand. The flowers retain their color and shape perfectly, but in very fleshy, juicy specimens the sand must be renewed. To be effective the sand must be as nearly dry as possible.

VII.—A method of preserving cut flowers in a condition of freshness is to dissolve small amounts of ammonium chloride, potassium nitrate, sodium carbonate or camphor in the water into which the stems are inserted. The presence of one or more of these drugs keeps the flowers from losing their turgidity by stimulating the cells to action and by opposing germ growth. Flowers that have already wilted are said to revive quickly if the stems are inserted in a weak camphor water.

Stuccoed Gypsum Flowers.—Take natural flowers, and coat the lower sides of their petals and stamens with paraffine or with a mixture of glue, gypsum, and lime, which is applied lightly. Very fine parts of the flowers, such as stamens, etc., may be previously supported by special attachments of textures, wire, etc. After the drying of the coating the whole is covered with shellac solution or with a mixture of glue, gypsum, lime with lead acetate, oil, mucilage, glycerine, colophony, etc. If desired, the surface may be painted with bronzes in various shades. Such flowers are much employed in the shape of festoons for decorating walls, etc.

Artificial Coloring of Flowers.—A method employed by florists to impart a green color to the white petals of "carnation pinks" consists in allowing longstemmed flowers to stand in water containing a green aniline dye. When the flowers are fresh they absorb the fluid readily, and the dye is carried to the petals.

Where the original color of the flower is white, colored stripes can be produced upon the petals by putting the cut ends into water impregnated with a suitable aniline dye. Some dyes can thus be taken up by the capillary action of the stem and deposited in the tissue of the petal. If flowers are placed over a basin of water containing a very small amount of ammonia in a bell glass, the colors of the petals will generally show some marked change. Many violet-colored flowers when so treated will become green, and if the petals contain several tints they will show greens where reds were, yellows where they were white, and deep carmine will become black. When such flowers are put into water they will retain their changed colors for hours.

If violet asters are moistened with very dilute nitric acid, the ray florets become red and acquire an agreeable odor.

FLUID MEASURES:
See Weights and Measures.

FLUORESCENT LIQUIDS.
Æsculin gives pale blue by (1) reflected light, straw color by (2) transmitted light.

Amido-phthalic acid, pale violet (1), pale yellow (2). Amido-terephthalic acid, bright green (1), pale green (2).

Eosine, yellow green (1), orange (2).

Fluorescein, intense green (1), orange yellow (2).

Fraxin, blue green (1), pale green (2).

Magdala red, opaque scarlet (1), brilliant carmine (2).

Quinine, pale blue (1), no color (2).

Safranine, yellow red (1), crimson (2).

FLUXES USED IN ENAMELING:
See Enameling.

FLUXES FOR SOLDERING:
See Soldering.

Fly-Papers and Fly-Poisons
(See also Insecticides.)

Sticky Fly-Papers.—The sticky material applied to the paper is the following:

I.—Boiled linseed oil. 5 to 7 parts
    Gum thus....... 2 to 3 parts
    Non-drying oil... 3 to 7 parts

For the non-drying oil, cottonseed, castor, or neatsfoot will answer—in fact, any of the cheaper oils that do not readily dry or harden will answer. The proper amount of each ingredient depends upon the condition of the boiled oil. If it is boiled down very stiff, more of the other ingredients will be necessary, while if thin, less will be required.

II.—Rosin............. 8 parts
    Rapeseed oil....... 4 parts
    Honey........... 1 part

Melt the rosin and oil together, and incorporate the honey. Two parts of raw linseed oil and 2 parts of honey may be used along with 8 parts of rosin instead of the foregoing. Use paper already sized, as it comes from the mills, on which to spread the mixture.

III.—Castor oil......... 12 ounces
     Rosin........... 27 ounces

Melt together and spread on paper sized with glue, using 12 ounces glue to 4 pints water.

IV.—Rosin............. 8 ounces
    Venice turpentine... 2 ounces
    Castor oil......... 2 ounces

Spread on paper sized with glue.

Poisonous Fly-Papers.—

I.—Quassia chips..... 150 parts
   Chloride of cobalt.. 10 parts
   Tartar emetic..... 2 parts
   Tincture of long
     pepper (1 to 4).. 80 parts
   Water........... 400 parts

Boil the quassia in the water until the liquid is reduced one-half, strain, add the other ingredients, saturate common absorbent paper with the solution, and dry. The paper is used in the ordinary way.

II.—Potassium bichromate 10 ounces
    Sugar .... 3 drachms
    Oil of black pepper.. 2 drachms
    Alcohol .... 2 ounces
    Water........... 14 ounces

Mix and let stand for several days, then soak unsized paper with the solution.

III.—Cobalt chloride..... 4 drachms
     Hot water........ 16 ounces
     Brown sugar....... 1 ounce

Dissolve the cobalt in the water and add the sugar, saturate unsized paper in the solution, and hang up to dry.

IV.—Quassia chips..... 150 parts
    Cobalt chloride.... 10 parts
    Tartrate antimony. 2 parts
    Tincture of pepper. 80 parts
    Water........... 400 parts

Boil chips in the water until the volume of the latter is reduced one-half, add other ingredients and saturate paper and dry.

Fly-Poison.—

    Pepper........... 4 ounces
    Quassia.......... 4 ounces
    Sugar........... 8 ounces
    Diluted alcohol..... 4 ounces

Mix dry and sprinkle around where the flies can get it.

Non-Poisonous Fly-Papers.—I.—Mix 25 parts of quassia decoction (1:10) with 6 parts of brown sugar and 3 parts of ground pepper, and place on flat dishes.

II.—Mix 1 part of ground pepper and 1 part of brown sugar with 16 parts milk or cream, and put the mixture on flat plates.

III.—Macerate 20 parts of quassia wood with 100 parts of water for 24 hours, boil one-half hour, and squeeze off 24 hours. The liquid is mixed with 3 parts of molasses, and evaporated to 10 parts. Next add 1 part of alcohol. Soak blotting paper with this mixture, and put on plates.

IV.—Dissolve 5 parts of potassium bichromate, 15 parts of sugar, and 1 part of essential pepper oil in 60 parts of water, and add 10 parts of alcohol. Saturate unsized paper with this solution and dry well.

V.—Boil together for half an hour

Ground quassia wood	18 pounds
Broken colocynth	3 pounds
Ground long pepper	5 pounds
Water	80 pounds

Then percolate and make up to 60 pounds if necessary with more water. Then add 4 pounds of syrup. Unsized paper is soaked in this, and dried as quickly as possible to prevent it from getting sour.

VI.—Mix together

Ordinary syrup	100 ounces
Honey	30 ounces
Extract of quassia wood	4 ounces
Oil of aniseed, a few drops.	

Removing the Gum of Sticky Fly-Paper.—The "gum" of sticky fly-paper that has "leaked" over furniture and shelfware can be removed without causing injury to either furniture or bottles. The "gum" of sticky fly-paper, while being quite adhesive, is easily dissolved with alcohol (grain or wood) or oil of turpentine. Alcohol will not injure the shelfware, but it should not be used on varnished furniture; in the latter case turpentine should be used.

FLY PROTECTIVES FOR ANIMALS:
See Insecticides.

FOAM PREPARATIONS.
A harmless gum cream is the following:

I.—Digest 100 parts of Panama wood for 8 days with 400 parts of water and 100 parts of spirits of wine (90 per cent). Pour off without strong pressure and filter.

For every 5 parts of lemonade syrup take 5 parts of this extract, whereby a magnificent, always uniform foam is obtained on the lemonade.

II.—Heat 200 parts of quillaia bark with distilled water during an hour in a vapor bath, with frequent stirring, and squeeze out. Thin with water if necessary and filter.

FOOD ADULTERANTS, SIMPLE TESTS FOR THEIR DETECTION.
Abstract of a monograph by W. D. Bigelow and Burton J. Howard, published by the Department of Agriculture.

Generally speaking, the methods of chemical analysis employed in food laboratories can be manipulated only by one who has had at least the usual college course in chemistry, and some special training in the examination of foods is almost as necessary. Again, most of the apparatus and chemicals necessary are entirely beyond the reach of the home, and the time consumed by the ordinary examination of a food is in itself prohibitive.

Yet there are some simple tests which serve to point out certain forms of adulteration and can be employed by the careful housewife with the reagents in her medicine closet and the apparatus in her kitchen. The number may be greatly extended by the purchase of a very few articles that may be procured for a few cents at any drug store. In applying these tests, one general rule must always be kept carefully in mind. Every one, whether layman or chemist, must familiarize himself with a reaction before drawing any conclusions from it. For instance, before testing a sample of supposed coffee for starch, the method should be applied to a sample of pure coffee (which can always be procured unground) and to a mixture of pure coffee and starch prepared by the operator.

Many manufacturers and dealers in foods have the ordinary senses so highly developed that by their aid alone they can form an intelligent opinion of the nature of a product, or of the character, and sometimes even of the proportion of adulterants present. This is especially true of such articles as coffee, wine, salad oils, flavoring extracts, butter, and milk. The housewife finds herself constantly submitting her purchases to this test. Her broad experience develops her senses of taste and smell to a high degree, and her discrimination is often sharper and more accurate than she herself realizes. The manufacturer who has developed his natural senses most highly appreciates best the assistance or collaboration of the chemist, who can often come to his relief when his own powers do not avail. So the housewife, by a few simple chemical tests, can broaden her field of vision and detect many impurities that are not evident to the senses.

There are here given methods adapted to this purpose, which may be applied to milk, butter, coffee, spices, olive oil, vinegar, jams and jellies, and flavoring extracts. In addition to this some general methods for the detection of coloring matter and preservatives will be given. All of the tests here described may be performed with utensils found in any well-appointed kitchen. It will be convenient, however, to secure a small glass funnel, about 3 inches in diameter, since filtration is directed in a number of methods prescribed. Filter paper can best be prepared for the funnel by cutting a circular piece about the proper size and folding it once through the middle, and then again at right angles to the first fold. The paper may then be opened without unfolding in such a way that three thicknesses lie together on one side and only one thickness on the other. In this way the paper may be made to fit nicely into the funnel.

Some additional apparatus, such as test tubes, racks for supporting them, and glass rods, will be found more convenient for one who desires to do considerable work on this subject, but can be dispensed with. The most convenient size for test tubes is a diameter of from $\frac{1}{2}$ to $\frac{5}{8}$ inch, and a length of from 5 to 6 inches. A graduated cylinder will also be found very convenient. If this is graduated according to the metric system, a cylinder containing about 100 cubic centimeters will be found to be convenient; if the English liquid measure is used it may be graduated to from 3 to 8 ounces.

Chemical Reagents.—The word "reagent" is applied to "any substance used to effect chemical change in another substance for the purpose of identifying its component parts or determining its percentage composition." The following reagents are required in the methods here given:

Turmeric paper.
Iron alum (crystal or powdered form).
Hydrochloric acid (muriatic acid), concentrated.

Caution.—All tests in which hydrochloric acid is used should be conducted in glass or earthenware, for this acid attacks and will injure metal vessels. Care must also be taken not to bring it into contact with the flesh or clothes. If, by accident, a drop of it falls upon the clothes, ammonia, or in its absence a solution of saleratus or sal soda (washing soda), in water, should be applied promptly.

Iodine tincture.
Potassium permanganate, 1 per cent solution.
Alcohol (grain alcohol).
Chloroform.
Boric acid or borax.
Ammonia water.
Halphen's reagent.

With the exception of the last reagent mentioned, these substances may be obtained in any pharmacy. The Halphen reagent should be prepared by a druggist, certainly not by an inexperienced person.

It is prepared as follows: An approximately 1 per cent solution of sulphur is made by dissolving about $\frac{1}{8}$ of a teaspoonful of precipitated sulphur in 3 or 4 ounces of carbon bisulphide. This solution mixed with an equal volume of amyl alcohol forms the reagent required by the method. A smaller quantity than that indicated by these directions may, of course, be prepared.

If turmeric paper be not available it may be made as follows: Place a bit of turmeric powder (obtainable at any drug store) in alcohol, allow it to stand for a few minutes, stir, allow it to stand again until it settles, dip a strip of filter paper into the solution, and dry it.

Determination of Preservatives.—The following methods cover all of the more important commercial preservatives with the exception of sulphites and fluorides. These are quite frequently used for preserving foods—the former with meat products and the latter with fruit products—but, unfortunately, the methods for their detection are not suitable for household use.

Detection of Salicylic Acid.—The determination of salicylic acid can best be made with liquids. Solid and semisolid foods, such as jelly, should be dissolved, when soluble, in sufficient water to make them thinly liquid. Foods containing insoluble matter, such as jam, marmalade, and sausage, may be macerated with water and strained through a piece of white cotton cloth. The maceration may be performed by rubbing in a teacup or other convenient vessel with a heavy spoon.

Salicylic acid is used for preserving

fruit products of all kinds, including beverages. It is frequently sold by drug stores as fruit acid. Preserving powders consisting entirely of salicylic acid are often carried from house to house by agents. It may be detected as follows:

Between 2 and 3 ounces of the liquid obtained from the fruit products, as described above, are placed in a narrow bottle holding 5 ounces, about a quarter of a teaspoonful of cream of tartar (or, better, a few drops of sulphuric acid) is added, the mixture shaken for 2 or 3 minutes, and filtered into a second small bottle. Three or 4 tablespoonfuls of chloroform are added to the clear liquid in the second bottle and the liquids mixed by a somewhat vigorous rotary motion, poured into an ordinary glass tumbler, and allowed to stand till the chloroform settles out in the bottom. Shaking is avoided, as it causes an emulsion which is difficult to break up. As much as possible of the chloroform layer (which now contains the salicylic acid) is removed (without any admixture of the aqueous liquid) by means of a medicine dropper and placed in a test tube or small bottle with about an equal amount of water and a small fragment—a little larger than a pinhead—of iron alum. The mixture is thoroughly shaken and allowed to stand till the chloroform again settles to the bottom. The presence of salicylic acid is then indicated by the purple color of the upper layer of liquid.

Detection of Benzoic Acid.

Benzoic acid is also used for preserving fruit products. Extract the sample with chloroform as in the case of salicylic acid; remove the chloroform layer and place it in a white saucer, or, better, in a plain glass sauce dish. Set a basin of water—as warm as the hand can bear—on the outside window ledge and place the dish containing the chloroform extract in it, closing the window until the chloroform has completely evaporated. In this manner the operation may be conducted with safety even by one who is not accustomed to handling chloroform. In warm weather the vessel of warm water may, of course, be omitted. Benzoic acid, if present in considerable amount, will now appear in the dish in characteristic flat crystals. On warming the dish the unmistakable irritating odor of benzoic acid may be obtained. This method will detect benzoic acid in tomato catsup or other articles in which it is used in large quantities. It is not sufficiently delicate, however, for the smaller amount used with some articles, such as wine. It is often convenient to extract a larger quantity of the sample and divide the chloroform layer into two portions, testing one for salicylic acid and the other for benzoic acid.

Detection of Boric Acid and Borax.

Boric acid (also called boracic acid) and its compound with sodium (borax) are often used to preserve animal products, such as sausage, butter, and sometimes milk. For the detection of boric acid and borax, solids should be macerated with a small amount of water and strained through a white cotton cloth. The liquid obtained by treating solids in this manner is clarified somewhat by thoroughly chilling and filtering through filter paper.

In testing butter place a heaping teaspoonful of the sample in a teacup, add a couple of teaspoonfuls of hot water, and stand the cup in a vessel containing a little hot water until the butter is thoroughly melted. Mix the contents of the cup well by stirring with a teaspoon and set the cup with the spoon in it in a cold place until the butter is solid. The spoon with the butter (which adheres to it) is now removed from the cup and the turbid liquid remaining strained through a white cotton cloth, or, better, through filter paper. The liquid will not all pass through the cloth or filter paper, but a sufficient amount for the test may be secured readily.

In testing milk for boric acid 2 or 3 tablespoonfuls of milk are placed in a bottle with twice that amount of a solution of a teaspoonful of alum in a pint of water, shaken vigorously, and filtered through filter paper. Here again a clear or only slightly turbid liquid passes through the paper.

About a teaspoonful of the liquid obtained by any one of the methods mentioned above is placed in any dish, not metal, and 5 drops of hydrochloric (muriatic) acid added. A strip of turmeric paper is dipped into the liquid and then held in a warm place—near a stove or lamp—till dry. If boric acid or borax was present in the sample the turmeric paper becomes bright cherry red when dry. A drop of household ammonia changes the red color to dark green or greenish black. If too much hydrochloric acid is used the turmeric paper may take on a brownish-red color even in the absence of boric acid. In this case, however, ammonia changes the color to brown just as it does turmeric paper which has not been dipped into the acid solution.

Detection of Formaldehyde.—Formaldehyde is rarely used with other foods than milk. The method for its detection in milk is given later. For its detection in other foods it is usually necessary first to separate it by distillation, a process which is scarcely available for the average person without laboratory training and special apparatus. For this reason no method is suggested here for the detection of formaldehyde in other foods than milk.

Detection of Saccharine.—Saccharine has a certain preservative power, but it is used not so much for this effect as because of the very sweet taste which it imparts. It is extracted by means of chloroform, as described under the detection of salicylic acid. In the case of solid and semi-solid foods, the sample must, of course, be prepared by extraction with water, as described under salicylic acid. The residue left after the evaporation of the chloroform, if a considerable amount of saccharine is present, has a distinctly sweet taste.

The only other substance having a sweet taste which may be present in foods, i. e., sugar, is not soluble in chloroform, and therefore does not interfere with this reaction. Certain other bodies (tannins) which have an astringent taste are present, and as they are soluble in chloroform may sometimes mask the test for saccharine, but with practice this difficulty is obviated.

Determination of Artificial Colors: Detection of Coal-Tar Dyes.—Coloring matters used with foods are usually soluble in water. If the food under examination be a liquid, it may therefore be treated directly by the method given below. If it be a solid or a pasty substance, soluble in water either in the cold or after heating, it may be dissolved in sufficient water to form a thin liquid. If it contains some insoluble material, it may be treated with sufficient water to dissolve the soluble portion with the formation of a thin liquid and filtered, and then strained through a clean white cotton cloth to separate the insoluble portion. About a half teacupful of the liquid thus described is heated to boiling, after adding a few drops of hydrochloric acid and a small piece of white woolen cloth or a few strands of white woolen yarn. (Before using, the wool should be boiled with water containing a little soda, to remove any fat it may contain, and then washed with water.) The wool is again washed, first with hot and then with cold water, the water pressed out as completely as possible, and the color of the fabric noted. If no marked color is produced, the test may be discontinued and the product considered free from artificial colors. If the fabric is colored, it may have taken up coal-tar colors, some foreign vegetable colors, and if a fruit product is being examined, some of the natural coloring matter of the fruit. Rinse the fabric in hot water, and then boil for 2 or 3 minutes in about one-third of a teacupful of water and 2 or 3 teaspoonfuls of household ammonia. Remove and free from as much of the liquid as possible by squeezing or wringing. Usually the fabric will retain the greater part of the natural fruit color, while the coal-tar color dissolves in dilute ammonia. The liquid is then stirred with a splinter of wood and hydrochloric acid added, a drop or two at a time, until there is no longer any odor of ammonia. (The atmosphere of the vessel is sometimes charged with the ammonia for several minutes after it has all been driven out of the liquid; therefore one should blow into the dish to remove this air before deciding whether the ammonia odor has been removed or not.) When enough acid has been added the liquid has a sour taste, as may be determined by touching the splinter, used in stirring, to the tongue.

A fresh piece of white woolen cloth is boiled in this liquid and thoroughly washed. If this piece of cloth has a distinct color the food under examination is artificially colored. The color used may have been a coal-tar derivative, commonly called an aniline dye, or an artificial color chemically prepared from some vegetable color. If of the first class the dyed fabric is usually turned purple or blue by ammonia. In either case, if the second fabric has a distinct color, it is evident that the product under examination is artificially colored. Of course a dull, faint tint must be disregarded.

Detection of Copper.—The presence of copper, often used to deepen the green tint of imported canned peas, beans, spinach, etc., may be detected as follows:

Mash some of the sample in a dish with a stiff kitchen spoon. Place a teaspoonful of the pulp in a teacup with 3 teaspoonfuls of water and add 30 drops of strong hydrochloric acid with a medicine dropper. Set the cup on the stove in a saucepan containing boiling water. Drop a bright iron brad or nail (wire nails are the best and tin carpet tacks will not answer the purpose) into the cup and keep the water in the saucepan boiling for 20 minutes, stirring the contents of the cup frequently with a splinter of wood. Pour out the contents of the cup and examine the nail. If present in an appreciable amount the nail will be heavily plated with copper.

Caution.—Be careful not to allow the hydrochloric acid to come in contact with metals or with the flesh or clothing.

Detection of Turmeric.—In yellow spices, especially mustard and mace, turmeric is often employed. This is especially true of prepared mustard to which a sufficient amount of starch adulterant has been added to reduce the natural color materially. If turmeric be employed to restore the normal shade an indication of that fact may sometimes be obtained by mixing a half teaspoonful of the sample in a white china dish and mixing with it an equal amount of water, and a few drops (4 to 10) of household ammonia, when a marked brown color, which does not appear in the absence of turmeric, is formed. At the present time turmeric or a solution of curcuma (the coloring matter of turmeric) is sometimes added to adulterated mustard in sufficient amount to increase its color, but not to a sufficient extent to give the brown appearance with ammonia described above. In such cases a teaspoonful of the suspected sample may be thoroughly stirred with a couple of tablespoonfuls of alcohol, the mixture allowed to settle for 15 minutes or more, and the upper liquid poured off into a clean glass or bottle. To about 1 tablespoonful of the liquid thus prepared and placed in a small, clear dish (a glass salt cellar serves excellently) add 4 or 5 drops of a concentrated solution of boric acid or borax and about 10 drops of hydrochloric acid, and mix the solution by stirring with a splinter of wood. A wedge-shaped strip of filter paper, about 2 or 3 inches long, 1 inch wide at the upper end, and ¼ inch at the lower end, is then suspended by pinning, so that its narrow end is immersed in the solution, and is allowed to stand for a couple of hours. The best results are obtained if the paper is so suspended that air can circulate freely around it, i. e., not allowing it to touch anything except the pin and the liquid in the dish. If turmeric be present a cherry-red color forms on the filter paper a short distance below the upper limit to which the liquid is absorbed by the paper, frequently from ¾ of an inch to an inch above the surface

of the liquid itself. A drop of household ammonia changes this red color to a dark green, almost black. If too much hydrochloric acid is used a dirty brownish color is produced.

Detection of Caramel.—A solution of caramel is used to color many substances, such as vinegar and some distilled liquors. To detect it two test tubes or small bottles of about equal size and shape should be employed and an equal amount (2 or 3 tablespoonfuls or more) of the suspected sample placed in each. To one of these bottles is added a teaspoonful of fuller's earth, the sample shaken vigorously for 2 or 3 minutes, and then filtered through filter paper, the first portion of the filtered liquid being returned to the filter paper and the sample finally collected into the test tube or bottle in which it was originally placed, or a similar one. The filtered liquid is now compared with the untreated sample. If it is markedly lighter in color it may be taken for granted that the color of the liquid is due to caramel, which is largely removed by fuller's earth. In applying this test, however, it must be borne in mind that caramel occurs naturally in malt vinegar, being formed in the preparation of the malt. It is evident that the tests require practice and experience before they can be successfully performed. The housewife can use them, but must repeat them frequently in order to become proficient in their use.

EXAMINATION OF CERTAIN CLASSES OF FOODS:

Canned Vegetables.—These are relatively free from adulteration by means of foreign substances. The different grades of products may with care be readily detected by the general appearance of the sample. The purchaser is, of course, at the disadvantage of not being able to see the product until the can is opened. By a study of the different brands available in the vicinity, however, he can readily select those which are preferable. As stated in an earlier part of this article, canned tomatoes sometimes contain an artificial coloring matter, which may be detected as described.

Canned sweet corn is sometimes sweetened with saccharine, which may be detected as described.

It is believed that, as a rule, canned vegetables are free from preservatives, although some instances of chemical preservation have recently been reported in North Dakota, and some imported tomatoes have been found to be artificially preserved. The presence of copper, often used for the artificial greening of imported canned peas, beans, spinach, etc., may be detected as described.

Coffee.—There are a number of simple tests for the presence of the adulterants of ground coffee. These are called simple because they can be performed without the facilities of the chemical laboratory, and by one who has not had the experience and training of a chemist. It must be understood that they require careful observation and study, and that one must perform them repeatedly in order to obtain reliable results. Before applying them to the examination of an unknown sample, samples of known character should be secured and studied. Unground coffee may be ground in the home and mixed with various kinds of adulterants, which can also be secured separately. Thus the articles themselves in known mixtures may be studied, and when the same results are obtained with unknown samples they can be correctly interpreted. These tests are well known in the laboratory and may be used in the home of the careful housewife who has the time and perseverance to master them.

Physical Tests.—The difference between the genuine ground coffee and the adulterated article can often be detected by simple inspection with the naked eye. This is particularly true if the product be coarsely crushed rather than finely ground. In such condition pure coffee has a quite uniform appearance, whereas the mixtures of peas, beans, cereals, chicory, etc., often disclose their heterogeneous nature to the careful observer. This is particularly true if a magnifying glass be employed. The different articles composing the mixture may then be separated by the point of a pen-knife. The dark, gummy-looking chicory particles stand out in strong contrast to the other substances used, and their nature can be determined by one who is familiar with them by their astringent taste.

The appearance of the coffee particles is also quite distinct from that of many of the coffee substitutes employed. The coffee has a dull surface, whereas some of its substitutes, especially leguminous products, often present the appearance of having a polished surface.

After a careful inspection of the sample with the naked eye, or, better, with a magnifying glass, a portion of it may be placed in a small bottle half full of water and shaken. The bottle is then placed on the table for a moment. Pure coffee contains a large amount of oil, by reason of which the greater portion of the sample will float. All coffee substitutes and some particles of coffee sink to the bottom of the liquid. A fair idea of the purity of the sample can often be determined by the proportion of the sample which floats or sinks.

Chicory contains a substance which dissolves in water, imparting a brownish-red color. When the suspected sample is dropped into a glass of water, the grains of chicory which it contains may be seen slowly sinking to the bottom, leaving a train of a dark-brown colored liquid behind them. This test appears to lead to more errors in the hands of inexperienced operators than any other test here given. Wrong conclusions may be avoided by working first with known samples of coffee and chicory as suggested above.

Many coffee substitutes are now sold as such and are advertised as more wholesome than coffee. Notwithstanding the claims that are made for them, a few of them contain a considerable percentage of coffee. This may be determined by shaking a teaspoonful in a bottle half full of water, as described above. The bottle must be thoroughly shaken so as to wet every particle of the sample. Few particles of coffee substitutes will float.

Chemical Tests.—Coffee contains no starch, while all of the substances, except chicory, used for its adulteration and in the preparation of coffee substitutes contain a considerable amount of starch. The presence of such substitutes may, therefore, be detected by applying the test for starch. In making this test less than a quarter of a teaspoonful of ground coffee should be used, or a portion of the ordinary infusion prepared for the table may be employed after dilution. The amount of water that should be added can only be determined by experience.

Condimental Sauces.—Tomato catsup and other condimental sauces are frequently preserved and colored artificially. The preservatives employed are usually salicylic acid and benzoic acid or their sodium salts. These products may be detected by the methods given.

Coal-tar colors are frequently employed with this class of goods, especially with those of a reddish tint, like tomato catsup. They may be detected by the methods given.

DAIRY PRODUCTS:

Butter.—Methods are available which, with a little practice, may be employed to distinguish between fresh butter, renovated or process butter, and oleomargarine.

These methods are commonly used in food and dairy laboratories. They give reliable results. At the same time considerable practice is necessary before we can interpret correctly the results obtained. Some process butters are on the market which can be distinguished from fresh butter only with extreme difficulty. During the last few years considerable progress has been made in the attempt to renovate butter in such a way that it will appear like fresh butter in all respects. A study must be made of these methods if we would obtain reliable results.

The "spoon" test has been suggested as a household test, and is commonly used by analytical chemists for distinguishing fresh butter from renovated butter and oleomargarine. A lump of butter, 2 or 3 times the size of a pea, is placed in a large spoon and heated over an alcohol or Bunsen burner. If more convenient the spoon may be held above the chimney of an ordinary kerosene lamp, or it may even be held over an ordinary illuminating gas burner. If the sample in question be fresh butter it will boil quietly, with the evolution of many small bubbles throughout the mass which produce a large amount of foam. Oleomargarine and process butter, on the other hand, sputter and crackle, making a noise similar to that heard when a green stick is placed in a fire. Another point of distinction is noted if a small portion of the sample be placed in a small bottle and set in a vessel of water sufficiently warm to melt the butter. The sample is kept melted from half an hour to an hour, when it is examined. If renovated butter or oleomargarine, the fat will be turbid, while if genuine fresh butter the fat will almost certainly be entirely clear.

To manipulate what is known as the "Waterhouse" or "milk" test, about 2 ounces of sweet milk are placed in a wide-mouthed bottle, which is set in a vessel of boiling water. When the milk is thoroughly heated, a teaspoonful of butter is added, and the mixture stirred with a splinter of wood until the fat is melted. The bottle is then placed in a dish of ice water and the stirring continued until the fat solidifies. If the sample be butter, either fresh or renovated, it will be solidified in a granular condition and distributed through the milk in small particles. If, on the other hand, the sample consist of oleomargarine it solidifies practically in one piece and may be lifted by the stirrer from the milk.

By these two tests, the first of which distinguishes fresh butter from process or renovated butter and oleomargarine, and the second of which distinguishes oleomargarine from either fresh butter or renovated butter, the nature of the sample under examination may be determined.

Milk.—The oldest and simplest method of adulterating milk is by dilution with water. This destroys the natural yellowish-white color and produces a bluish tint, which is sometimes corrected by the addition of a small amount of coloring matter.

Another form of adulteration is the removal of the cream and the sale as whole milk of skimmed or partially skimmed milk. Again, the difficulty experienced in the preservation of milk in warm weather has led to the widespread use of chemical preservatives.

Detection of Water.—If a lactometer or hydrometer, which can be obtained of dealers in chemical apparatus, be available, the specific gravity of milk will afford some clew as to whether the sample has been adulterated by dilution with water. Whole milk has a specific gravity between 1.027 and 1.033. The specific gravity of skimmed milk is higher, and milk very rich in cream is sometimes lower than these figures. It is understood, of course, that by specific gravity is meant the weight of a substance with reference to the weight of an equal volume of water. The specific gravity of water is 1. It is obvious that if water be added to a milk with the specific gravity of 1.030, the specific gravity of the mixture will be somewhat below those figures.

An indication by means of a hydrometer or lactometer below the figure 1.027 therefore indicates either that the sample in question is a very rich milk or that it is a milk (perhaps normal, perhaps skimmed) that has been watered. The difference in appearance and nature of these two extremes is sufficiently obvious to make use of the lactometer or hydrometer of value as a preliminary test of the purity of milk.

Detection of Color.—As previously stated, when milk is diluted by means of water the natural yellowish-white color is changed to a bluish tint, which is sometimes corrected by the addition of coloring matter. Coal-tar colors are usually employed for this purpose. A reaction for these colors is often obtained in the method given below for the detection of formaldehyde. When strong hydrochloric acid is added to the milk in approximately equal proportions before the mixture is heated a pink tinge sometimes is evident if a coal-tar color has been added.

Detection of Formaldehyde.—Formaldehyde is the substance most commonly used for preserving milk and is rarely, if ever, added to any other food. Its use is inexcusable and especially objectionable in milk served to infants and invalids.

To detect formaldehyde in milk 3 or 4 tablespoonfuls of the sample are placed in a teacup with at least an equal amount of strong hydrochloric acid and a piece of ferric alum about as large as a pinhead, the liquids being mixed by a gentle rotary motion. The cup is then placed in a vessel of boiling water, no further heat being applied, and left for 5 minutes. At the end of this time, if formaldehyde be present, the mixture will be distinctly purple. If too much heat is applied, a muddy appearance is imparted to the contents of the cup.

Caution.—Great care must be exercised in working with hydrochloric acid, as it is strongly corrosive.

Edible Oils.—With the exception of cottonseed oil, the adulterants ordinarily used with edible oils are of such a nature that the experience of a chemist and the facilities of a chemical laboratory are essential to their detection. There is, however, a simple test for the detection of cottonseed oil, known as the Halphen test, which may be readily applied.

Great care must be taken in the manipulation of this test, as one of the reagents employed—carbon bisulphide—is very inflammable. The chemicals employed in the preparation of the reagent used for this test are not household articles. They may, however, be obtained in any pharmacy. The mixture should be prepared by a druggist rather than by an inexperienced person who desires to use it.

In order to perform the test 2 or 3 tablespoonfuls of this reagent are mixed in a bottle with an equal volume of the suspected sample of oil and heated in a vessel of boiling salt solution (prepared by dissolving 1 tablespoonful of salt in a pint of water) for 10 or 15 minutes. At the end of that time, if even a small percentage of cottonseed oil be present, the mixture will be of a distinct reddish color, and if the sample consists largely or entirely of cottonseed oil, the color will be deep red.

Eggs.—There is no better method for the testing of the freshness of an egg than the familiar one of "candling," which has long been practiced by dealers. The room is darkened and the egg held between the eye and a light; the presence of dark spots indicates that the egg is not perfectly fresh, one that is fresh presenting a homogeneous, translucent appearance. Moreover, there is found in the larger end of a fresh egg, between the shell and the lining membrane, a small air cell which, of course, is distinctly transparent. In an egg which is not perfectly fresh this space is filled and hence presents the same appearance as the rest of the egg.

It is now a matter of considerable importance to be able to distinguish between fresh eggs and those that have been packed for a considerable time. Until recently that was not a difficult matter. All of the solutions that were formerly extensively used for that purpose gave the shell a smooth, glistening appearance which is not found in the fresh egg. This characteristic, however, is of less value now than formerly, owing to the fact that packed eggs are usually preserved in cold storage. There is now no means by which a fresh egg can be distinguished from a packed egg without breaking it. Usually in eggs that have been packed for a considerable time the white and yolk slightly intermingle along the point of contact, and it is a difficult matter to separate them. Packed eggs also have a tendency to adhere to the shell on one side and when opened frequently have a musty odor.

FLAVORING EXTRACTS.

Although a large number of flavoring extracts are on the market, vanilla and lemon extracts are used so much more commonly than other flavors that a knowledge of their purity is of the greatest importance. Only methods for the examination of those two products will be considered.

Vanilla Extract.—Vanilla extract is made by extracting vanilla beans with alcohol. It consists of an alcoholic solution of vanillin (the characteristic flavoring matter of the vanilla bean) and several other products, chiefly rosins, which, though present in but small amount and having only a slight flavor in themselves, yet affect very materially the flavor of the product. Vanilla extract is sometimes adulterated with the extract of the Tonka bean. This extract, to a certain extent, resembles vanilla extract. The extract of the Tonka bean, however, is far inferior to that of the vanilla bean. It has a relatively penetrating, almost pungent odor, standing in sharp contrast to the flavor of the vanilla extract. This odor is so different that one who has given the matter some attention may readily distinguish the two, and the quality of the vanilla extract may often be judged with a fair degree of accuracy by means of the odor alone.

Another form of adulteration, and one that is now quite prevalent, is the use of artificial vanillin in place of the extract of either vanilla or Tonka beans. Artificial vanillin has, of course, the same composition and characteristics as the natural vanillin of the vanilla bean. Extracts made from it, however, are deficient in the rosins and other products which are just as essential to the true vanilla, as is vanillin itself. Since vanillin is thus obtained from another source so readily, methods for the determination of the purity of vanilla extract must depend upon the presence of other substances than vanillin.

Detection of Caramel.—The coloring matter of vanilla extract is due to substances naturally present in the vanilla bean and extracted therefrom by alcohol. Artificial extracts made by dissolving artificial vanillin in alcohol contain no color of themselves, and to supply it caramel is commonly employed. Caramel may be detected in artificial extracts by shaking and observing the color of the resulting foam after a moment's standing. The foam of pure extracts is colorless. If caramel is present a color persists at the points of contact between the bubbles until the last bubble has disappeared. The test with fuller's earth given for caramel in vinegar is also very satisfactory, but of course requires the loss of the sample used for the test.

Examination of the Rosin.—If pure vanilla extract be evaporated to about one-third its volume the rosins become insoluble and settle to the bottom of the dish. Artificial extracts remain clear under the same conditions. In examining vanilla extract the character of these rosins is studied. For this purpose a dish containing about an ounce of the extract is placed on a teakettle or other vessel of boiling water until the liquid

evaporates to about one-third or less of its volume. Owing to the evaporation of the alcohol the rosins will then be insoluble. Water may be added to restore the liquid to approximately its original volume. The rosin will then separate out as a brown flocculent precipitate. A few drops of hydrochloric acid may be added and the liquid stirred and the insoluble matter allowed to settle. It is then filtered and the rosin on the filter paper washed with water. The rosin is then dissolved in a little alcohol, and to 1 portion of this solution is added a small particle of ferric alum, and to another portion a few drops of hydrochloric acid. If the rosin be that of the vanilla bean, neither ferric alum nor hydrochloric acid will produce more than a slight change of color. With rosins from most other sources, however, one or both of these substances yield a distinct color change.

For filtering, a piece of filter paper should be folded once through the middle and again at right angles to the first fold. It may now be opened with one fold on one side and three on the other and fitted into a glass funnel. When the paper is folded in this manner the precipitated rosins may be readily washed with water. When the washing is completed the rosins may be dissolved by pouring alcohol through the filter. This work with the rosins will require some practice before it can be successfully performed. It is of considerable value, however, in judging of the purity of vanilla extract.

Lemon Extract.—By lemon extract is understood a solution of lemon oil in strong alcohol. In order to contain as much lemon oil as is supposed to be found in high-grade extracts the alcohol should constitute about 80 per cent of the sample. The alcohol is therefore the most valuable constituent of lemon extract, and manufacturers who turn out a low-grade product usually do so because of their economy of alcohol rather than of lemon oil. Owing to the fact that lemon extract is practically a saturated solution of oil of lemon in strong alcohol the sample may be examined by simple dilution with water. A teaspoonful of the oil in question may be placed in the bottom of an ordinary glass tumbler and 2 or 3 teaspoonfuls of water added. If the sample in question be real lemon extract the lemon oil should be thrown out of solution by reason of its insolubility in the alcohol after its dilution with water. The result is at first a marked turbidity and later the separation of the oil of lemon on the top of the aqueous liquid. If the sample remains perfectly clear after the addition of water, or if a marked turbidity is not produced, it is a low-grade product and contains very little, if any, oil of lemon.

Fruit Products.—Adulteration of fruit products is practically confined to jellies and jams. Contrary to the general belief, gelatin is never used in making fruit jelly. In the manufacture of the very cheapest grade of jellies starch is sometimes employed. Jellies containing starch, however, are so crude in their appearance that the most superficial inspection is sufficient to demonstrate that they are not pure fruit jellies. From their appearance no one would think it worth while to examine them to determine their purity.

Natural fruit jellies become liquid on being warmed. A spoonful dissolves readily in warm water, although considerable time is required with those that are especially firm. The small fruits contain practically no starch, as apples do, and the presence of starch in a jelly indicates that some apple juice has probably been used in its preparation.

Detection of Starch.—Dissolve a teaspoonful of jelly in a half teacupful of hot water, heat to boiling and add, drop by drop, while stirring with a teaspoon, a solution of potassium permanganate until the solution is almost colorless. Then allow the solution to cool and test for starch with tincture of iodine, as directed later. Artificially colored jellies are sometimes not decolorized by potassium permanganate. Even without decolorizing, however, the blue color can usually be seen.

Detection of Glucose.—For the detection of glucose, a teaspoonful of the jelly may be dissolved in a glass tumbler or bottle in 2 or 3 tablespoonfuls of water. The vessel in which the jelly is dissolved may be placed in hot water if necessary to hasten the solution. In case a jam or marmalade is being examined, the mixture is filtered to separate the insoluble matter. The solution is allowed to cool, and an equal volume or a little more of strong alcohol is added. If the sample is a pure fruit product the addition of alcohol causes no precipitation, except that a very slight amount of proteid bodies is thrown down. If glucose has been employed in its manufacture, however, a dense white precipitate separates and, after a time, settles to the bottom of the liquid.

Detection of Foreign Seeds.—In addition to the forms of adulteration to which jellies are subject, jams are sometimes manufactured from the exhausted fruit pulp left after removing the juice for making jelly. When this is done residues from different fruits are sometimes mixed. Exhausted raspberry or blackberry pulp may be used in making "strawberry" jam and *vice versa*. Some instances are reported of various small seeds, such as timothy, clover, and alfalfa seed, having been used with jams made from seedless pulp.

With the aid of a small magnifying glass such forms of adulteration may be detected, the observer familiarizing himself with the seeds of the ordinary fruits.

Detection of Preservatives and Colors.—With jellies and jams salicylic and benzoic acids are sometimes employed. They may be detected by the methods given.

Artificial colors, usually coal-tar derivatives, are sometimes used and may be detected as described.

Meat Products.—As in many other classes of foods, certain questions important in the judgment of meats require practical experience and close observation rather than chemical training. This is especially true of meat products. The general appearance of the meat must largely guide the purchaser. If, however, the meat has been treated with preservatives and coloring matter its appearance is so changed as to deceive him. The preservatives employed with meat products are boric acid, borax, and sulphites. The methods for the detection of sulphites are not suitable for household use.

Detection of Boric Acid and Borax.—To detect boric acid (if borax has been used the same reaction will be obtained), about a tablespoonful of the chopped meat is thoroughly macerated with a little hot water, pressed through a bag, and 2 or 3 tablespoonfuls of the liquid placed in a sauce dish with 15 or 20 drops of strong hydrochloric acid for each tablespoonful. The liquid is then filtered through filter paper, and a piece of turmeric paper dipped into it and dried near a lamp or stove. If boric acid or borax were used for preserving the sample, the turmeric paper should be changed to a bright cherry-red color. If too much hydrochloric acid has been employed a dirty brownish-red color is obtained, which interferes with the color due to the presence of boric acid. When a drop of household ammonia is added to the colored turmeric paper, it is turned a dark green, almost black color, if boric acid is present. If the reddish color, however, was caused by the use of too much hydrochloric acid this green color does not form.

Caution.—The corrosive nature of hydrochloric acid must not be forgotten. It must not be allowed to touch the flesh, clothes, or any metal.

Detection of Colors.—The detection of coloring matter in sausage is often a difficult matter without the use of a compound microscope. It may sometimes be separated, however, by macerating the meat with a mixture of equal parts of glycerine and water to which a few drops of acetic or hydrochloric acid have been added. After macerating for some time the mixture is filtered and the coloring matter detected by means of dyeing wool in the liquid thus obtained.

Spices.—Although ground spices are very frequently adulterated, there are few methods that may be used by one who has not had chemical training, and who is not skilled in the use of a compound microscope, for the detection of the adulterants employed. The majority of the substances used for the adulteration of spices are of a starchy character. Unfortunately for our purposes, most of the common spices also contain a considerable amount of starch. Cloves, mustard, and cayenne, however, are practically free from starch, and the presence of starch in the ground article is proof of adulteration.

Detection of Starch in Cloves, Mustard, and Cayenne.—A half teaspoonful of the spice in question is stirred into half a cupful of boiling water, and the boiling continued for 2 or 3 minutes. The mixture is then cooled. If of a dark color, it is diluted with a sufficient amount of water to reduce the color to such an extent that the reaction formed by starch and iodine may be clearly apparent if starch be present. The amount of dilution can only be determined by practice, but usually the liquid must be diluted with an equal volume of water, or only $\frac{1}{4}$ of a teaspoonful of the sample may be employed originally. A single drop of tincture of iodine is now added. If starch is present, a deep blue color, which in the presence of a large amount of starch appears black, is formed. If no blue color appears, the addition of the iodine tincture should be continued, drop by drop, until the liquid shows by its color the presence of iodine in solution.

Detection of Colors.—Spice substitutes are sometimes colored with coal-tar colors. These products may be detected by the methods given.

Vinegar.—A person thoroughly familiar with vinegar can tell much regarding the source of the article from its appearance, color, odor, and taste.

If a glass be rinsed out with the sample of vinegar and allowed to stand for a number of hours or overnight, the odor of the residue remaining in the glass is quite different with different kinds of vinegar. Thus, wine vinegar has the odor characteristic of wine, and cider vinegar has a peculiar fruity odor. A small amount of practice with this test enables one to distinguish with a high degree of accuracy between wine and cider vinegars and the ordinary substitutes.

If a sample of vinegar be placed in a shallow dish on a warm stove or boiling teakettle and heated to a temperature sufficient for evaporation and not sufficient to burn the residue, the odor of the warm residue is also characteristic of the different kinds of vinegar. Thus, the residue from cider vinegar has the odor of baked apples and the flavor is acid and somewhat astringent in taste, and that from wine vinegar is equally characteristic. The residue obtained by evaporating vinegar made from sugarhouse products and from spirit and wood vinegar colored by means of caramel has the peculiar bitter taste characteristic of caramel.

If the residue be heated until it begins to burn, the odor of the burning product also varies with different kinds of vinegar. Thus, the residue from cider vinegar has the odor of scorched apples, while that of vinegars made from sugarhouse wastes and of distilled and wood vinegars colored with a large amount of caramel has the odor of burnt sugar. In noting these characteristics, however, it must be borne in mind that, in order to make them conform to these tests, distilled and wood vinegars often receive the addition of apple jelly.

The cheaper forms of vinegar, especially distilled and wood vinegar, are commonly colored with caramel, which can be detected by the method given.

FOOD COLORANTS.

(Most, if not all, of these colorants are injurious and should therefore be used with extreme caution.)

Sausage Color.—To dye sausage red, certain tar dyestuffs are employed, especially the azo dyes, preference being given to the so-called genuine red. For this purpose about 100 parts of dyestuff are dissolved in 1,000 to 2,000 parts of hot water; when the solution is complete, add a likewise hot solution of 45 to 50 parts of boracic acid, whereupon the mixture should be stirred well for some time; then filter, allow to cool, and preserve in tightly closing bottles. It is absolutely necessary in using aniline colors to add a disinfectant to the dyestuff solution, the object of which is, in case the sausage should commence to decompose, to prevent the decomposition azo dyestuff by the disengaged hydrogen. Instead of boracic acid, formalin may be used as a disinfectant. Of this formalin, 38 per cent, add about 25 to 30 parts to the cooled and filtered dyestuff solution. This sausage color is used by adding about 1½ to 2 tablespoonfuls of it to the preserving salt measured out for 100 kilos of sausage mass, stirring well. The sausage turns neither gray nor yellow on storing.

Cheese Color.—I.—To produce a suitable, pretty yellow color, boil 100 parts of orlean or annatto with 75 parts of potassium carbonate in 1½ to 2 liters of water, allow to cool, and filter after settling, whereupon 15 to 18 parts of boracic acid are added to give keeping qualities to the solution. According to another method, digest about 200 parts of orlean, 200 parts of potassium carbonate, and 100 parts of turmeric for 10 to 12 days in 1,500 to 2,000 parts of 60 per cent alcohol, filter, and keep in bottles. To 100,000 parts of milk to be made into cheese add 1½ to 2 small spoonfuls of this dye, which imparts to the cheese a permanent and natural yellow appearance.

II.—To obtain a handsome yellow color for cheese, such as is demanded for certain sorts, boil together 100 parts of annatto and 75 parts of potassium carbonate in from 1,500 to 2,000 parts of pure water; let it cool, stand it aside for a time, and filter, adding finally from 12 to 15 parts of boracic acid as a preservative. For coloring butter, there is in the trade a mixture of bicarbonate of soda with 12 per cent to 15 per cent of sodium chloride, to which is added from 1½ per cent to 2 per cent of powdered turmeric.

Butter Color.—For the coloring of butter there is in the market under the name of butter powder a mixture of sodium bicarbonate with 12 to 15 per cent of sodium chloride and 1½ to 2 per cent of powdered turmeric; also a mix-ture of sodium bicarbonate, 1,500 parts; saffron surrogate, 8 parts; and salicylic acid, 2 parts. For the preparation of liquid butter color use a uniform solution of olive oil, 1,500 parts; powdered turmeric, 300 parts; orlean, 200 parts. The orlean is applied on a plate of glass or tin in a thin layer and allowed to dry perfectly, whereupon it is ground very fine and intimately mixed with the powdered turmeric. This mixture is stirred into the oil with digestion for several hours in the water bath. When a uniform, liquid mass has resulted, it is filtered hot through a linen filter with wide meshes. After cooling, the filtrate is filled into bottles. Fifty to 60 drops of this liquid color to 1½ kilos of butter impart to the latter a handsome golden yellow shade.

INFANT FOODS:

Infants' (Malted) Food.—

I.—Powdered malt .... 1 ounce
Oatmeal (finest ground)......... 2 ounces
Sugar of milk...... 4 ounces
Baked flour........ 1 pound

Mix thoroughly.

II.—Infantine is a German infant food which is stated to contain egg albumen, 5.5 per cent; fat, 0.08 per cent; water, 4.22 per cent; carbohydrates, 86.58 per cent (of which 54.08 per cent is soluble in water); and ash, 2.81 per cent (consisting of calcium, 10.11 per cent; potassium, 2.64 per cent; sodium, 25.27 per cent; chlorine, 36.65 per cent; sulphuric acid, 3.13 per cent; and phosphoric acid, 18.51 per cent).

MEAT PRESERVATIVES.

(Most of these are considered injurious by the United States Department of Agriculture and should therefore be used with extreme caution.)

The Preservation of Meats.—Decomposition of the meat sets in as soon as the blood ceases to pulse in the veins, and it is therefore necessary to properly preserve it until the time of its consumption.

The nature of preservation must be governed by circumstances such as the kind and quality of the article to be preserved, length of time and climatic condition, etc. While salt, vinegar, and alcohol merit recognition on the strength of a long-continued usage as preservatives, modern usage favors boric acid and borax, and solutions containing salicylic acid and sulphuric acid are common, and have been the subject of severe criticism.

Many other methods of preservation have been tried with variable degrees of success; and of the more thoroughly tested ones the following probably include all of those deserving more than passing mention or consideration.

1. The exclusion of external, atmospheric electricity, which has been observed to materially reduce the decaying of meat, milk, butter, beer, etc.

2. The retention of occluded electric currents. Meats from various animals packed into the same packages, and surrounded by a conducting medium, such as salt and water, liberate electricity.

3. The removal of the nerve centers. Carcasses with the brains and spinal cord left therein will be found more prone to decomposition than those wherefrom these organs have been removed.

4. Desiccation. Dried beef is an excellent example of this method of preservation. Other methods coming under this heading are the application of spices with ethereal oils, various herbs, coriander seed extracted with vinegar, etc.

5. Reduction of temperature, i. e., cold storage.

6. Expulsion of air from the meat and the containers. Appert's, Willaumez's, Redwood's, and Prof. A. Vogel's methods are representative for this category of preservation. Phenyl paper, Dr. Busch's, Georges's, and Medlock and Baily's processes are equally well known.

7. The application of gases. Here may be mentioned Dr. Gamgee's and Bert and Reynoso's processes, applying carbon dioxide and other compressed gases, respectively.

Air-drying, powdering of meat, smoking, pickling, sugar or vinegar curing are too well known to receive any further attention here. Whatever process may be employed, preference should be given to that which will secure the principal objects sought for, the most satisfactory being at the same time not deleterious to health, and of an easily applicable and inexpensive nature.

To Preserve Beef, etc., in Hot Weather.—Put the meat into a hot oven and let it remain until the surface is browned all over, thus coagulating the albumen of the surface and inclosing the body of the meat in an impermeable envelope of cooked flesh. Pour some melted lard or suet into a jar of sufficient size, and roll the latter around until the sides are evenly coated to the depth of half

an inch with the material. Put in the meat, taking care that it does not touch the sides of the jar (thus scraping away the envelope of grease), and fill up with more suet or lard, being careful to completely cover and envelop the meat. Thus prepared, the meat will remain absolutely fresh for a long time, even in the hottest weather. When required for use the outer portion may be left on or removed. The same fat may be used over and over again by melting and retaining in the melted state a few moments each time, by which means not only all solid portions of the meat which have been retained fall to the bottom, but all septic microbes are destroyed.

Meat Preservatives. — I. — *Barmenite Corning Agent:* For every 100 parts, by weight, take 25.2 parts, by weight, of saltpeter; 46.8 parts, by weight, sodium chloride; 25.7 parts, by weight, cane sugar; 0.8 parts, by weight, plaster of Paris or gypsum; 0.1 part, by weight, of some moistening material, and a trace of magnesia.

II.—*Carniform, A:* For every 100 parts, by weight, take 3.5 parts, by weight, sodium diphosphate; 3.1 parts, by weight, water of crystallization; 68.4 parts, by weight, sodium chloride; 24.9 parts, by weight, saltpeter; together with traces of calcium phosphate, magnesia, and sulphuric acid.

III.—*Carniform, B:* For every 100 parts, by weight, take 22.6 parts, by weight, sodium diphosphate; 17.3 parts, by weight, water of crystallization; 59.7 parts, by weight, saltpeter; 0.6 parts, by weight, calcium phosphate; with traces of sulphuric acid and magnesia.

IV.—*"Cervelatwurst"* (*spice powder*): For 100 parts, by weight, take 0.7 parts, by weight, of moistening; 3.5 parts, by weight, spices—mostly pepper; 89 parts, by weight, sodium chloride; 5 parts, by weight, saltpeter; 0.7 parts, by weight, gypsum; and traces of magnesia.

V.—*Cervelatwurst Salt* (*spice powder*): For 100 parts, by weight, take 7.5 parts, by weight, spices—mostly pepper; 1.6 parts, by weight, moistener; 81.6 parts, by weight, sodium chloride; 2.5 parts, by weight, saltpeter; 6.2 parts, by weight, cane sugar; and traces of magnesia.

VI.—*Rubrolin Sausage* (*spice powder*): For 100 parts by weight, take 53.5 parts, by weight, sal ammoniac, and 45.2 parts, by weight, of saltpeter.

VII.—*Servator Special Milk and Butter Preserving Salt:* 80.3 per cent of crystallized boracic acid; 10.7 per cent sodium chloride; and 9.5 per cent of benzoic acid. (Its use is, however, prohibited in Germany.)

VIII.—*Wittenberg Pickling Salt:* For 100 parts, by weight, take 58.6 parts, by weight, sodium chloride; 40.5 parts, by weight, saltpeter; 0.5 parts, by weight, gypsum; traces of moisture and magnesia.

IX.—*Securo:* For a quart take 3.8 parts, by weight, aluminum oxide, and 8 parts, by weight, acetic acid; basic acetate of alumina, 62 parts, by weight; sulphuric acid, 0.8 parts, by weight; sodium oxide, with substantially traces of lime and magnesia.

X.—*Michels Cassala Salt:* This is partially disintegrated. 30.74 per cent sodium chloride; 15.4 per cent sodium phosphate; 23.3 per cent potassio-sodic tartrate; 16.9 per cent water of crystallization; 1.2 per cent aluminum oxide; and 2.1 per cent acetic acid as basic acetate of alumina; 8.4 per cent sugar; 0.98 per cent benzoic acid; 0.5 per cent sulphuric acid; and traces of lime.

XI.—*Corning Salt:* Sodium nitrate, 50 parts; powdered boracic acid, 45 parts; salicylic acid, 5 parts.

XII.—*Preservative Salt:* Potassium nitrate, 70 parts; sodium bicarbonate, 15 parts; sodium chloride, 15 parts.

XIII.—*Another Corning Salt:* Potassium nitrate, 50 parts; sodium chloride, 20 parts; powdered boracic acid, 20 parts; sugar, 10 parts.

XIV.—*Maciline (offered as condiment and binding agent for sausages):* A mixture of wheat flour and potato flour dyed intensely yellow with an azo dyestuff and impregnated with oil of mace.

XV.—Borax.......... 80 parts
Boric acid....... 17 parts
Sodium chloride. 3 parts

Reduce the ingredients to a powder and mix thoroughly.

XVI.—Sodium sulphite, powdered .... 80 parts
Sodium sulphate, powdered.... 20 parts

XVII.—Sodium chloride. 80 parts
Borax.......... 8 parts
Potassium nitrate 12 parts

Reduce to a powder and mix.

XVIII.—Sodium nitrate.. 50 parts
Salicylic acid.... 5 parts
Boric acid...... 45 parts

XIX.—Potassium nitrate........ 70 parts

Sodium bicarbonate ...... 15 parts
Sodium chloride. 15 parts

XX.—Potassium nitrate......... 50 parts
Sodium chloride. 20 parts
Boric acid...... 20 parts
Sugar.......... 10 parts

A German Method of Preserving Meat. —Entire unboweled cattle or large, suitably severed pieces are sprinkled with acetic acid and then packed and transported in sawdust impregnated with cooking salt and sterilized.

Extract of Meat Containing Albumen. —In the ordinary production of meat extract, the albumen is more or less lost, partly through precipitation by the acids or the acid salts of the meat extract, partly through salting out by the salts of the extract, and partly by coagulation at a higher temperature. A subsequent addition of albumen is impracticable because the albumen is likewise precipitated, insolubly, by the acids and salts contained in the extract. This precipitation can be prevented, according to a French patent, by neutralizing the extract before mixing with albumen, by the aid of sodium bicarbonate. The drying of the mixture is accomplished in a carbonic acid atmosphere. The preparation dissolves in cold or hot water into a white, milky liquid and exhibits the smell and taste of meat extract, if the albumen added was tasteless. The taste which the extract loses by the neutralization returns in its original strength after the mixture with albumen. In this manner a meat preparation is obtained which contains larger quantities of albumen and is more nutritious and palatable than other preparations.

Foot-Powders and Solutions

The following foot-powders have been recommended as dusting powders:

I.—Boric acid......... 2 ounces
Zinc oleate......... 1 ounce
Talcum........... 3 ounces

II.—Oleate of zinc (powdered).......... ½ ounce
Boric acid......... 1 ounce
French chalk....... 5 ounces
Starch........... 1½ ounces

III.—Dried alum........ 1 drachm
Salicylic acid....... ½ drachm
Wheat starch....... 4 drachms
Powdered talc...... 1½ ounces

IV.—Formaldehyde solution............. 1 part
Thymol.......... $\frac{1}{10}$ part
Zinc oxide........ 35 parts
Powdered starch.... 65 parts

V.—Salicylic acid....... 7 drachms
Boric acid. 2 ounces, 440 grains
Talcum.......... 38 ounces
Slippery elm bark... 1 ounce
Orris root.......... 1 ounce

VI.—Talc.............. 12 ounces
Boric acid......... 10 ounces
Zinc oleate........ 1 ounce
Salicylic acid....... 1 ounce
Oil of eucalyptus... 2 drachms

VII.—Salicylic acid...... 7 drachms
Boric acid........ 3 ounces
Talcum .... 38 ounces
Slippery elm, powdered .... 1 ounce
Orris, powdered.... 1 ounce

Salicylated Talcum.—

I.—Salicylic acid...... 1 drachm
Talcum .... 6 ounces
Lycopodium....... 6 drachms
Starch .... 3 ounces
Zinc oxide........ 1 ounce
Perfume, quantity sufficient.

II.—Tannoform .... 1 drachm
Talcum........... 2 drachms
Lycopodium....... 30 grains

Use as a dusting powder.

Solutions for Perspiring Feet.—

I.—Balsam Peru....... 15 minims
Formic acid........ 1 drachm
Chloral hydrate.... 1 drachm
Alcohol to make 3 ounces.

Apply by means of absorbent cotton.

II.—Boric acid......... 15 grains
Sodium borate..... 6 drachms
Salicylic acid....... 6 drachms
Glycerine.......... 1½ ounces
Alcohol to make 3 ounces.

For local application.

FOOTSORES ON CATTLE:
See Veterinary Formulas.

FORMALDEHYDE:
See also Disinfectants, Foods, and Milk.
Commercial Formaldehyde.—This extremely poisonous preservative is obtained by passing the vapors of wood spirit, in the presence of air, over copper heated to redness. The essential parts of the apparatus employed are a metal chamber into which a feed-tube enters, and from which 4 parallel copper tubes or oxidizers discharge by a common exit tube. This chamber is fitted with inspection apertures, through which the course of the process may be watched and controlled. The wood spirit, stored in a reservoir, falls into a mixer where it is volatilized and intimately mixed with air from a chamber which is connected with a force pump. The gases after traversing the oxidizer are led into a condensing coil, and the crude formaldehyde is discharged into the receiver beneath.

The small amount of uncondensed gas is then led through a series of two washers. The "formol" thus obtained is a mixture of water, methyl alcohol, and 30 to 40 per cent of formaldehyde. It is rectified in a still, by which the free methyl alcohol is removed and pure formol obtained, containing 40 per cent of formaldehyde, chiefly in the form of the acetal. Rectification must not be pushed too far, otherwise the formaldehyde may become polymerized into trioxmethylene. When once oxidation starts, the heat generated is sufficient to keep the oxidizers red hot, so that the process works practically automatically.

Determination of the Presence of Formaldehyde in Solutions.—Lemme makes use, for this purpose, of the fact that formaldehyde, in neutral solutions of sodium sulphite, forms normal bisulphite salts, setting free a corresponding quantity of sodium hydrate, that may be titrated with sulphuric acid and phenolphthalein. The sodium sulphite solution has an alkaline reaction toward phenolphthalein, and must be exactly neutralized with sodium bisulphite. Then to 100 cubic centimeters of this solution of 250 grams of sodium sulphite ($Na_2SO_3 + 7H_2O$) in 750 grams water, add 5 cubic centimeters of the suspected formaldehyde solution. A strong red color is instantly produced. Titrate with normal sulphuric acid until the color disappears. As the exact disappearance of the color is not easily determined, a margin of from 0.1 to 0.2 cubic centimeters may be allowed without the exactness of the reaction being injured, since 1 cubic centimeter of normal acid answers to only 0.03 grams of formaldehyde.

FORMALIN FOR GRAIN SMUT:
See Grain.

FRAMES: THEIR PROTECTION FROM FLIES.

Since there is great risk of damaging the gilt when trying to remove fly-specks with spirits of wine, it has been found serviceable to cover gilding with a copal varnish. This hardens and will stand rough treatment, and may be renewed wherever removed.

FRAME CLEANING:

See Cleaning Preparations and Methods.

FRAME POLISHES:

See Polishes.

FRAMING, PASSE-PARTOUT:

See Passe-Partout.

FRECKLE LOTIONS:

See Cosmetics.

FREEZING MIXTURES:

See also Refrigeration and Refrigerants.

Freezing Preventives

Liquid for Cooling Automobile Engines.—In order to prevent freezing of the jacket water, when the engine is not in operation in cold weather, solutions are used, notably of glycerine and of calcium chloride ($CaCl_2$). The proportions for the former solution are equal parts of water and glycerine, by weight; for the latter, approximately $\frac{1}{2}$ gallon of water to 8 pounds of $CaCl_2$, or a saturated solution at 60° F. This solution ($CaCl_2 + 6H_2O$) is then mixed with equal parts of water, gallon for gallon. Many persons complain that $CaCl_2$ corrodes the metal parts, but this warning need do no more than urge the automobilist to use only the chemically pure salt, carefully avoiding the "chloride of lime" ($CaOCl_2$).

A practical manufacturing chemist of wide experience gives this:

A saturated solution of common salt is one of the best things to use. It does not affect the metal of the engine, as many other salts would, and is easily renewed. It will remain fluid down to 0° F., or a little below.

Equal parts of glycerine and water is also good, and has the advantage that it will not crystallize in the chambers, or evaporate readily. It is the most convenient solution to use on this account, and may repay the increased cost over brine, in the comfort of its use. It needs only the occasional addition of a little water to make it last all winter and leave the machinery clean when it is drawn off. With brine an incrustation of salt as the water evaporates is bound to occur which reduces the efficiency of the solution until it is removed. Water frequently must be added to keep the original volume, and to hold the salt in solution. A solution of calcium chloride is less troublesome so far as crystallizing is concerned, but is said to have a tendency to corrode the metals.

Anti-Freezing Solution for Automobilists.—Mix and filter $4\frac{1}{2}$ pounds pure calcium chloride and a gallon of warm water and put the solution in the radiator or tank. Replace evaporation with clean water, and leakage with solution. Pure calcium chloride retails at about 8 cents per pound, or can be procured from any wholesale drug store at 5 cents.

Anti-Freezing, Non-Corrosive Solution.—A solution for water-jackets on gas engines that will not freeze at any temperature above 20° below zero (F.) may be made by combining 100 parts of water, by weight, with 75 parts of carbonate potash and 50 parts of glycerine. This solution is non-corrosive and will remain perfectly liquid at all temperatures above its congealing point.

Anti-Frost Solution.—As an excellent remedy against the freezing of shop windows, apply a mixture consisting of 55 parts of glycerine dissolved in 1,000 parts of 62 per cent alcohol, containing, to improve the odor, some oil of amber. As soon as the mixture clarifies, it is rubbed over the inner surface of the glass. This treatment, it is claimed, not only prevents the formation of frost, but also stops sweating.

Protection of Acetylene Apparatus from Frost.—Alcohol, glycerine, and calcium chloride have been recommended for the protection of acetylene generators from frost. The employment of calcium chloride, which must not be confounded with chloride of lime, appears preferable in all points of view. A solution of 20 parts of calcium chloride in 80 parts of water congeals only at 5° F. above zero. But as this temperature does not generally penetrate the generators, it will answer to use 10 or 15 parts of the chloride for 100 parts of water, which will almost always be sufficient to avoid congelation. Care must be taken not to use sea salt or other alkaline or metallic salts, which deteriorate the metal of the apparatus.

FROST BITE.

When the skin is as yet unbroken, Hugo Kuhl advises the following:

I.—Carbolized water...   4 drachms
Nitric acid.........   1 drop
Oil of geranium....   1 drop

Mix. Pencil over the skin and then hold the penciled place near the fire until the skin is quite dry.

If the skin is already broken, use the following ointment:

II.—Hebra's ointment..   500 parts
Glycerine.........   100 parts
Liquefied carbolic
acid...........   15 parts

Mix. Apply to the broken skin occasionally.

III.—Camphor.........   25 parts
Iodine, pure......   50 parts
Olive oil.........   500 parts
Paraffine, solid....   450 parts
Alcohol, enough.

Dissolve the camphor in the oil and the iodine in the least possible amount of alcohol. Melt the paraffine and add the mixed solutions. When homogeneous pour into suitable molds. Wrap the pencils in paraffine paper or tin foil, and pack in wooden boxes. By using more or less olive oil the pencils may be made of any desired consistency.

IV.—Dissolve 5 parts of camphor in a mixture consisting of 5 parts of ether and 5 parts of alcohol; then add collodion sufficient to make 100 parts.

V.—Dissolve 1 part of thymol in 5 parts of a mixture of ether and alcohol, then add collodion sufficient to make 100 parts.

VI.—Carbolic acid.....   2   parts
Lead ointment....   40   parts
Lanolin.........   40   parts
Olive oil.........   20   parts
Lavender oil......   1½   parts

VII.—Tannic acid......   15   parts
Lycopodium.....   15   parts
Lard...........   30   parts

VIII.—Zinc oxide........   15   parts
Glycerine........   45   parts
Lanolin.........   40   parts

IX.—Ichthyol.........   10   parts
Resorcin.........   10   parts
Tannic acid......   10   parts
Distilled water....   50   parts

Any of these is to be applied about twice a day.

FROSTED GLASS:
See Glass.

FROST PREVENTIVE:
See Freezing Preventives.

FROST REMOVERS:
See Glass.

FRUIT ESSENCES AND EXTRACTS:
See Essences and Extracts.

Fruit Preserving

(See also Essences, Extracts, and Preserves.)

How to Keep Fruit.—According to experiments of Max de Nansouty, fruit carefully wrapped in silk paper and then buried in dry sand will preserve a fresh appearance with a fresh odor or flavor, almost indefinitely. It may also be preserved in dry excelsior, but not nearly so well. In stubble or straw fruit rots very quickly, while in shavings it mildews quickly. In short, wheat-straw fruit often takes on a musty taste and odor, even when perfectly dry. Finally, when placed on wooden tablets and exposed to the air, most fruit decays rapidly.

I.—Crushed Strawberry.—Put up by the following process, the fruit retains its natural color and taste, and may be exposed to the air for months, without fermenting:

Take fresh, ripe berries, stem them, and rub through a No. 8 sieve, rejecting all soft and green fruit. Add to each gallon of pulp thus obtained, 8 pounds of granulated sugar. Put on the fire and bring just to a boil, stirring constantly. Just before removing from the fire, add to each gallon 1 ounce of a saturated alcoholic solution of salicylic acid, stirring well. Remove the scum, and, while still hot, put into jars, and hermetically seal. Put the jars in cold water, and raise them to the boiling point, to prevent them from bursting by sudden expansion on pouring hot fruit into them. Fill the jars entirely full, so as to leave no air space when fruit cools and contracts.

II.—Crushed Raspberry.—Prepare in the same manner as for crushed strawberry, using ½ red raspberries and ½ black, to give a nice color, and using 7 pounds of sugar to each gallon of pulp.

III.—Crushed Pineapple.—Secure a good brand of canned grated pineapple, and drain off about one-half of the liquor, by placing on a strainer. Add to each pound of pineapple 1 pound of granulated sugar. Place on the fire, and bring to boiling point, stirring constantly. Just before removing from the fire, add to each gallon of pulp 1 ounce saturated alcoholic solution of salicylic acid.

Put into air-tight jars until wanted for use.

IV.—Crushed Peach.—Take a good brand of canned yellow peaches, drain off liquor, and rub through a No. 8 sieve. Add sugar, bring to the boiling point, and when ready to remove from fire add to each gallon 1 ounce saturated alcoholic solution of salicylic acid. Put into jars and seal hermetically.

V.—Crushed Apricot.—Prepared in similar manner to crushed peach, using canned apricots.

VI. — Crushed Orange. — Secure oranges with a thin peel, and containing plenty of juice. Remove the outer, or yellow peel, first, taking care not to include any of the bitter peel. (The outer peel may be used in making orange phosphate, or tincture of sweet orange peel.) Next remove the inner, bitter peel, quarter, and remove the seeds. Extract part of the juice, and grind the pulp through an ordinary meat grinder. Add sugar, place on the fire, and bring to the boiling point. When ready to remove, add to each gallon 1 ounce of saturated alcoholic solution of salicylic acid and 1 ounce of glycerine. Put into air-tight jars.

VII.—Crushed Cherries.—Stone the cherries and grind them to a pulp. Add sugar, and place on the fire, stirring constantly. Before removing, add to each gallon 1 ounce of the saturated solution of salicylic acid. Put into jars and seal.

VIII.—Fresh Crushed Fruits in Season.—In their various seasons berries and fruits may be prepared in fresh lots for the soda fountain each morning, by reducing the fruit to a pulp, and mixing this pulp with an equal quantity of heavy simple syrup.

Berries should be rubbed through a sieve. In selecting berries, it is better to use the medium-sized berries for the pulp, reserving the extra large specimens for garnishing and decorative effects.

Mash the berries with a wooden masher, never using iron or copper utensils, which may discolor the fruit.

Pineapple may be prepared by removing the rough outer skin and grating the pulp upon an ordinary tin kitchen grater. The grater should be scrupulously clean, and care should be taken not to grate off any of the coarse, fibrous matter comprising the fruit's core.

All crushed fruits are served as follows: Mix equal quantities of pulp and simple syrup in the counter bowl; use 1½ to 2 ounces to each glass, adding the usual quantity of cream, or ice cream. Draw soda, using a fine stream freely.

IX.—Glacés.—Crushed fruits, served in the following manner, make a delicious and refreshing drink:

> Crushed fruit........ 12 drachms
> Juice of half a lemon.
> Shaved ice.

Put the ice into a small glass, add the fruit and lemon juice, stir well, and serve with a spoon and straws.

FRUIT PRODUCTS, TESTS FOR:
See Foods.

FRUIT SYRUPS:
See Syrups.

FRUIT VINEGAR:
See Vinegar.

Fumigants

(See also Disinfectants.)

Fumigating Candles.—I.—Lime wood charcoal, 6,000 parts, by weight, saturated with water (containing saltpeter, 150 parts, by weight, in solution), and dried again, is mixed with benzoin, 750 parts, by weight; styrax, 700 parts, by weight; mastic, 100 parts, by weight; cascarilla, 450 parts, by weight; Peruvian balsam, 40 parts, by weight; Mitcham oil, lavender oil, lemon oil, and bergamot oil, 15 parts, by weight, each; and neroli oil, 3 parts, by weight.

II.—Charcoal, 7,500 parts, by weight; saltpeter, 150 parts, by weight; Tolu balsam, 500 parts, by weight; musk, 2 parts, by weight; rose oil, 1 part. The mixtures are crushed with thick tragacanth to a solid mass.

III.—Sandal wood, 48 parts, by weight; clove, 6 parts, by weight; benzoin, 6 parts, by weight; licorice juice, 4 parts, by weight; potash saltpeter, 2 parts, by weight; cascarilla bark, 1.5 parts, by weight; cinnamon bark, 1.5 parts, by weight; musk, 0.05 parts, by weight. All these substances are powdered and mixed, whereupon the following are added: Styrax (liquid), 5 parts, by weight; cinnamon oil, 0.05 parts, by weight; clove oil, 0.05 parts, by weight; geranium oil, 0.5 parts, by weight; lavender oil, 0.2 parts, by weight; Peruvian balsam, 0.2 parts, by weight. The solid ingredients are each powdered separately, then placed in the respective proportion in a spacious porcelain dish and intimately mixed by means of a flat spatula. The dish must be covered up with a cloth in this operation. After the mixture has been accomplished, add the essential oils and just enough solution of gum arabic so that by subsequent kneading with the pestle a moldable dough results which possesses sufficient solidity after drying. The mass is pressed into metallic molds in the shape of cones not more than ¾ of an inch in height.

IV.—Red Fumigating Candles.—Sandal wood, 1 part; gum benzoin, 1.5 parts; Tolu balsam, 0.250 parts; sandal oil, .025 parts; cassia oil, .025 parts; clove oil, 25 parts; saltpeter, .090 parts. The powder is mixed intimately, saturated with spirit of wine, in which the oils are dissolved, and shaped into cones.

V.—Wintergreen oil.	1 part
Tragacanth	20 parts
Saltpeter	50 parts
Phenol, crystallized	100 parts
Charcoal, powdered	830 parts
Water.	

Dissolve the saltpeter in the water, stir the solution together with the powdered charcoal and dry. Then add the tragacanth powder, also the wintergreen oil and the phenol, and prepare from the mixture, by means of a tragacanth solution containing 2 per cent of saltpeter, a mass which can be shaped into candles.

Fumigating Perfumes.—These are used for quickly putting down bad odors in the sick room, etc. They are decidedly antiseptic, and fulfil their purpose admirably.

I.—Select good white blotting paper, and cut each large sheet lengthwise into 3 equal pieces. Make a solution of 1 ounce of potassium nitrate in 12 ounces of boiling water; place this solution in a large plate, and draw each strip of paper over the solution so as to saturate it. Then dry by hanging up. The dried paper is to be saturated in a similar manner with either of the following solutions:

(1) Siam benzoin	1 ounce
Storax	3 drachms
Olibanum	2 scruples
Mastic	2 scruples
Cascarilla	2 drachms
Vanilla	1 drachm
Rectified spirit	8 ounces

Bruise the solids and macerate in the spirit 5 days, filter, and add

Oil of cinnamon	8 parts
Oil of cloves	8 parts
Oil of bergamot	5 parts
Oil of neroli	5 parts

Mix.

(2) Benzoin	1½ ounces
Sandal wood	1 ounce
Spirit	8 ounces

Macerate as No. 1, and add

| Essence of vetiver | 3 ounces |
| Oil of lemon grass | 40 drops |

Mix.

After the paper is dry, cut up into suitable sized pieces to go into commercial envelopes.

II.—Benzoin	1 av. ounce
Storax	1 av. ounce
Fumigating essence	2 fluidounces
Ether	1 fluidounce
Acetic acid, glacial	20 drops
Alcohol	2 fluidounces

Dissolve the benzoin and storax in a mixture of the alcohol and ether, filter and add the fumigating and the acetic acid. Spread the mixture upon filtering or bibulous paper and allow it to dry. To prevent sticking, dust the surface with talcum and preserve in wax paper. When used the paper is simply warmed, or held over a lamp.

III.—Musk	0.2 parts
Oil of rose	1 part
Benzoin	100 parts
Myrrh	12 parts
Orris root	250 parts
Alcohol (90 per cent)	500 parts

IV.—Benzoin	80 parts
Balsam Tolu	20 parts
Storax	20 parts
Sandal wood	20 parts
Myrrh	10 parts
Cascarilla bark	20 parts
Musk	0.2 parts
Alcohol	250 parts

Fumigating Ribbon.—I.—Take ½-inch cotton tape and saturate it with niter; when dry, saturate with the following tincture:

Benzoin	1 ounce
Orris root	1 ounce
Myrrh	2 drachms
Tolu balsam	2 drachms
Musk	10 grains
Rectified spirit	10 ounces

Macerate for a week, filter, and add 10 minims of attar of rose.

II.—Another good formula which may also be used for fumigating paper, is:

Olibanum......... 2 ounces
Storax............ 1 ounce
Benzoin........... 6 drachms
Peruvian balsam... ½ ounce
Tolu balsam....... 3 drachms
Rectified spirit..... 10 ounces

Macerate 10 days, and filter.

Perfumed Fumigating Pastilles.—

I.—Vegetable charcoal.. 6 ounces
Benzoin........... 1 ounce
Nitrate of potash... ½ ounce
Tolu balsam....... 2 drachms
Sandal wood....... 2 drachms
Mucilage of tragacanth, a sufficiency.

Reduce the solids to fine powder, mix, and make into a stiff paste with the mucilage. Divide this into cones 25 grains in weight, and dry with a gentle heat.

II.—Powdered willow
charcoal......... 8 ounces
Benzoic acid....... 6 ounces
Nitrate of potash... 6 drachms
Oil of thyme....... ½ drachm
Oil of sandal wood.. ½ drachm
Oil of caraway ...... ½ drachm
Oil of cloves .... ½ drachm
Oil of lavender..... ½ drachm
Oil of rose......... ½ drachm
Rose water........ 10 ounces

Proceed as in I, but this recipe is better for the addition of 20 grains of powdered tragacanth.

III.—Benzoin......... 10 av. ounces
Charcoal......... 24 av. ounces
Potassium nitrate. 1 av. ounce
Sassafras........ 2 av. ounces
Mucilage of acacia, sufficient.

Mix the first four in fine powder, add the mucilage, form a mass, and make into conical pastilles.

IV.—Potassium nitrate 375 grains
Water.......... 25 fluidounces
Charcoal wood,
powder....... 30 av. ounces
Tragacanth, pow-
der.......... 375 grains
Storax.......... 300 grains
Benzoin........ 300 grains
Vanillin........ 8 grains
Coumarin...... 3 grains
Musk.......... 3 grains
Civet.......... 1½ grains
Oil of rose...... 20 drops
Oil of bergamot. 15 drops
Oil of ylang-ylang 10 drops
Oil of rhodium.. 10 drops
Oil of sandal
wood........ 5 drops
Oil of cinnamon. 5 drops
Oil of orris...... 1 drop
Oil of cascarilla. · 1 drop

Saturate the charcoal with the potassium nitrate dissolved in the water, dry the mass, powder, add the other ingredients, and mix thoroughly. Beat the mixture to a plastic mass with the addition of sufficient mucilage of tragacanth containing 2 per cent of saltpeter in solution, and form into cone-shaped pastilles. In order to evenly distribute the storax throughout the mass, it may be previously dissolved in a small amount of acetic ether.

V.—Benzoin........ 2 av. ounces
Cascarilla...... 1 av. ounce
Myrrh......... 1 av. ounce
Potassium ni-
trate........ ½ av. ounce
Potassium chlo-
rate......... 60 grains
Charcoal, wood. 4 av. ounces
Oil of cloves.... 1 fluidrachm
Oil of cinnamon 1 fluidrachm
Oil of lavender. 1 fluidrachm
Mucilage of tragacanth, sufficient.

Mix the first six ingredients previously reduced to fine powder, add the oils, and then incorporate enough mucilage to form a mass. Divide this into pastilles weighing about 60 grains and dry.

VI.—Charcoal, pow-
der......... 30 av. ounces
Potassium ni-
trate....... ½ av. ounce
Water....... 33 fluidounces
Tragacanth,
powder..... 300 grains
Tincture of
benzoin..... 1½ fluidounces
Peru balsam.. 300 grains
Storax, crude.. 300 grains
Tolu balsam.. 300 grains
Oleo-balsamic
mixture..... 2½ fluidrachms
Coumarin .... 8 grains

Saturate the charcoal with the potassium nitrate dissolved in the water, then dry, reduce to powder, and incorporate the tragacanth and then the remaining ingredients. Form a mass by the addition of sufficient mucilage of tragacanth containing 2 per cent of potassium nitrate in solution and divide into pastilles.

VII.—Powdered nitrate of
potassium....... ½ ounce
Powdered gum ara-
bic............ ½ ounce
Powdered cascarilla
bark (fresh)..... ½ ounce
Powdered benzoin
(fresh)......... 4 ounces

Powdered charcoal. 7 ounces
Oil of eucalyptus... 25 drops
Oil of cloves....... 25 drops
Water, a sufficiency.

Make a smooth paste, press into molds and dry.

FURS:

To Clean Furs.—For dark furs, warm a quantity of new bran in a pan, taking care that it does not burn, to prevent which it must be briskly stirred. When well warmed rub it thoroughly into the fur with the hand. Repeat this 2 or 3 times, then shake the fur, and give it another sharp rubbing until free from dust. For white furs: Lay them on a table, and rub well with bran made moist with warm water; rub until quite dry, and afterwards with dry bran. The wet bran should be put on with flannel, then dry with book muslin. Light furs, in addition to the above, should be well rubbed with magnesia or a piece of book muslin, after the bran process, against the way of the fur.

To Preserve Furs.—I.—Furs may be preserved from moths and other insects by placing a little colocynth pulp (bitter apple), or spice (cloves, pimento, etc.), wrapped in muslin, among them; or they may be washed in a very weak solution of corrosive sublimate in warm water (10 to 15 grains to the pint), and afterwards carefully dried. As well as every other species of clothing, they should be kept in a clean, dry place, from which they should be taken out occasionally, well beaten, exposed to the air, and returned.

II.—Sprinkle the furs or woolen stuffs, as well as the drawers or boxes in which they are kept, with spirits of turpentine, the unpleasant scent of which will speedily evaporate on exposure of the stuffs to the air. Some persons place sheets of paper moistened with spirits of turpentine, over, under, or between pieces of cloth, etc., and find it a very effectual method. Many woolen drapers put bits of camphor, the size of a nutmeg, in papers, on different parts of the shelves in their shops, and as they brush their cloths every 2, 3, or 4 months, this keeps them free from moths; and this should be done in boxes where the furs, etc., are put. A tallow candle is frequently put within each muff when laid by. Snuff or pepper is also good.

FURNACE JACKET.

A piece of asbestos millboard—10 inches by 4 inches by ⅜ inch—is perforated in about a dozen or more places with glycerined cork borers, then nicked about an inch from each short end and immersed in water until saturated; next the board is bent from the nicks at right angles and the perforated portion shaped by bending it over a bottle with as little force as possible. The result should be a perforated arched tunnel, resting on narrow horizontal ledges at each side. Dry this cover in the furnace, after setting it in position, and pressing it well to the supports. Three such covers, weighing 1 pound, replaced 24 fire clay tiles, weighing 13 pounds, and a higher temperature was obtained than with the latter.

FURNITURE CLEANERS:
See Cleaning Preparations and Methods.

FURNITURE, ITS DECORATION:
See Wood.

FURNITURE ENAMEL:
See Varnishes.

FURNITURE POLISHES:
See Polishes.

FURNITURE WAX:
See Waxes.

FUSES:
See Pyrotechnics.

FUSES FOR ELECTRICAL CIRCUITS:
See Alloys.

FUNNELS, TO CLEAN:
See Cleaning Preparations and Methods.

GALVANIZED PAPER:
See Paper, Metallic.

GAMBOGE STAIN:
See Lacquers.

GAPES IN POULTRY:
See Veterinary Formulas.

GARANCINE PROCESS:
See Dyes.

GARDENS, CHEMICAL:
See also Sponges.

I.—Put some sand into a fish-globe or other suitable glass vessel to the depth of 2 or 3 inches; in this place a few pieces of sulphate of copper, aluminum, and iron; pour over the whole a solution of sodium silicate (water glass), 1 part, and water, 3 parts, care being taken not to disarrange the chemicals. Let this stand a week or so, when a dense growth of the silicates of the various bases used will be seen in various colors. Now displace the solution of the sodium silicate with clear water, by conveying a stream of water through a very small rubber tube into the vessel. The water will gradually displace the sodium silicate solution. Care must be taken not to disarrange or break down the growth with the stream of water. A little experimenting, experience and expertness will enable the operator to produce a very pretty garden.

II.—This is a permanent chemical garden, which may be suspended by brass chains with a lamp behind.

Prepare a small beaker or jar full of cold saturated solution of Glauber's salt, and into the solution suspend by means of threads a kidney bean and a non-porous body, such as a marble, stone, glass, etc. Cover the jar, and in a short time there will be seen radiating from the bean small crystals of sulphate of sodium which will increase and give the bean the aspect of a sea urchin, while the non-porous body remains untouched. The bean appears to have a special partiality for the crystals, which is due to the absorption of water by the bean, but not of the salt. In this way a supersaturated solution is formed in the immediate neighborhood of the bean, and the crystals, in forming, attach themselves to its surface.

III.—A popular form of ornamental crystallization is that obtained by immersing a zinc rod in a solution of a lead salt, thus obtaining the "lead tree." To prepare this, dissolve lead acetate in water, add a few drops of nitric acid, and then suspend the zinc rod in the solution. The lead is precipitated in large and beautiful plates until the solution is exhausted or the zinc dissolved. In this case the action is electro-chemical, the first portions of the lead precipitated forming with the zinc a voltaic arrangement of sufficient power to decompose the salt.

It is said that by substituting chloride of tin for the lead salt a "tin tree" may be produced, while nitrate of silver under the same conditions would produce a "silver tree." In the latter case distilled water should be used to prevent precipitation of the silver by possible impurities contained in ordinary water.

GAS FIXTURES:
See Brass.

GAS FIXTURES, BRONZING OF:
See Plating.

GAS SOLDERING:
See Soldering.

GAS-STOVES, TO CLEAN:
See Cleaning Preparations and Methods.

GAS TRICK:
See Pyrotechnics.

GEAR LUBRICANT:
See Lubricants.

GELATIN:

French Gelatin.—Gelatin is derived from two sources, the parings of skins, hides, etc., and from bones. The latter are submitted to the action of dilute hydrochloric acid for several days, which attacks the inorganic matters—carbonates, phosphates, etc., and leaves the ossein, which is, so to say, an isomer of the skin substance. The skin, parings of hide, etc., gathered from the shambles, butcher shops, etc., are brought into the factory, and if not ready for immediate use are thrown into quicklime, which preserves them for the time being. From the lime, after washing, they pass into dilute acid, which removes the last traces of lime, and are now ready for the treatment that is to furnish the pure gelatin. The ossein from bones goes through the same stages of treatment, into lime, washed and laid in dilute acid again. From the acid bath the material goes into baths of water maintained at a temperature not higher than from 175° to 195° F.

The gelatin manufacturer buys from the button-makers and manufacturers of knife handles and bone articles generally, those parts of the bone that they cannot use, some of which are pieces 8 inches long by a half inch thick.

Bones gathered by the ragpickers furnish the strongest glue. The parings of skin, hide, etc., are from those portions of bullock hides, calf skins, etc., that cannot be made use of by the tanner, the heads, legs, etc.

The gelatin made by Coignet for the Pharmacie Centrale de France is made from skins procured from the tawers of Paris, who get it directly from the abattoirs, which is as much as to say that the material is guaranteed fresh and healthy, since these institutions are under rigid inspection and surveillance of government inspectors and medical men.

There is a gelatin or glue, used exclusively for joiners, inside carpenters, and ceiling makers (*plafonneurs*), called *rabbit vermicelli*, and derived from rabbit skins. As the first treatment of these skins is to saturate them with mercury bichloride, it is needless to say the product is not employed in pharmacy.

To Clarify Solutions of Gelatin, Glues, etc.—If 1 per cent of ammonium fluoride be added to turbid solutions of gelatin or common glue, or, in fact, of any gums, it quickly clarifies them. It causes a deposition of ligneous matter, and also very materially increases the adhesive power of such solutions.

Air Bubbles in Gelatin.—The presence of minute air bubbles in cakes of commercial gelatin often imparts to them an unpleasant cloudy appearance. These minute air bubbles are the result of the rapid, continuous process of drying the sheets of gelatin by a counter-current of hot air. Owing to the rapid drying a hard skin is formed on the outside of the cake, leaving a central layer from which the moisture escapes only with difficulty, and in which the air bubbles remain behind. Since the best qualities of gelatin dry most rapidly, the presence of these minute bubbles is, to a certain extent, an indication of superiority, and they rarely occur in the poorer qualities of gelatin. If dried slowly in the old way gelatin is liable to be damaged by fermentation; in such cases large bubbles of gas are formed in the sheets, and are a sign of bad quality.

GEMS, ARTIFICIAL:
See also Diamonds.

The raw materials for the production of artificial gems are the finest silica and, as a rule, finely ground rock crystals; white sand and quartz, which remain pure white even at a higher temperature, may also be used.

Artificial borax is given the preference, since the native variety frequently contains substances which color the glass. Lead carbonate or red lead must be perfectly pure and not contain any protoxide, since the latter gives the glass a dull, greenish hue. White lead and red lead have to dissolve completely in dilute nitric acid or without leaving a residue; the solution, neutralized as much as possible, must not be reddened by prussiate of potash. In the former case tin is present, in the latter copper. Arsenious acid and saltpeter must be perfectly pure; they serve for the destruction of the organic substances. The materials, without the coloring oxide, furnish the starting quantity for the production of artificial gems; such glass pastes are named "strass."

The emerald, a precious stone of green color, is imitated by melting 1,000 parts of strass and 8 parts of chromic oxide. Artificial emeralds are also obtained with cupric acid and ferric oxides, consisting of 43.84 parts of rock crystal; 21.92 parts of dry sodium carbonate; 7.2 parts of calcined and powdered borax; 7.2 parts of red lead; 3.65 parts of saltpeter; 1.21 parts of red ferric oxide, and 0.6 parts of green copper carbonate.

Agates are imitated by allowing fragments of variously colored pastes to flow together, and stirring during the deliquation.

The amethyst is imitated by mixing 300 parts of a glass frit with 0.6 parts of gray manganese ore, or from 300 parts of frit containing 0.8 per cent of manganic oxide, 36.5 parts of saltpeter, 15 parts of borax, and 15 parts of minium (red lead). A handsome amethyst is obtained by melting together 1,000 parts of strass, 8 parts of manganese oxide, 5 parts of cobalt oxide, and 2 parts of gold purple.

Latterly, attempts have also been made to produce very hard glasses for imitation stones from alumina and borax with the requisite coloring agents.

Besides imitation stones there are also produced opaque glass pastes bearing the name of the stones they resemble, e. g., aventurine, azure-stone (lapis lazuli), chrysoprase, turquoise, obsidian, etc. For these, especially pure materials, as belonging to the most important ingredients of glassy bodies, are used, and certain quantities of red lead and borax are also added.

GEM CEMENTS:
See Adhesives, under Jewelers' Cements.

GERMAN SILVER:
See Alloys.

GERMAN SILVER SOLDERS:
See Solders.

GILDING:
See Paints, Plating, and Varnishes.

GILDING GLASS:
See Glass.

GILDING, TO CLEAN:
See Cleaning Preparations and Methods.

GILDING, RENOVATION OF:
See Cleaning Compounds.

GILDING SUBSTITUTE:
See Plating.

GILT, TEST FOR:
See Gold.

GILT WORK, TO BURNISH:
See Gold.

GINGERADE:
See Beverages.

GINGER ALE AND GINGER BEER:
See Beverages.

GINGER CORDIAL:
See Wines and Liquors.

GINGER EXTRACTS:
See Essences and Extracts.

Glass

Bent Glass.—This was formerly used for show cases; its use in store fronts is becoming more and more familiar, large plates being bent for this purpose. It is much used in the construction of dwellings, in windows, or rounded corners, and in towers; in coach fronts and in rounded front china closets. Either plain glass or beveled glass may be bent, and to any curve.

The number of molds required in a glass-bending establishment is large.

The bending is done in a kiln. Glass melts at 2,300° F.; the heat employed in bending is 1,800° F. No pyrometer would stand long in that heat, so the heat of the kiln is judged from the color of the flame and other indications. Smaller pieces of glass are put into the molds in the kilns with forks made for the purpose. The great molds used for bending large sheets of glass are mounted on cars, that may be rolled in and out of kilns. The glass is laid upon the top of the mold or cavity, and is bent by its own weight. As it is softened by the heat it sinks into the mold and so is bent. It may take an hour or two to bend the glass, which is then left in the kiln from 24 to 36 hours to anneal and cool. Glass of any kind or size is put into the kilns in its finished state; the great heat to which it is subjected does not disturb the polished surface. Despite every precaution more or less glass is broken in bending. Bent glass costs about 50 per cent more than the flat.

The use of bent glass is increasing, and there are 4 or 5 glass-bending establishments in the United States, of which one is in the East.

Colored Glass.—R. Zsigmondy has made some interesting experiments in coloring glass with metallic sulphides, such as molybdenite, and sulphides of antimony, copper, bismuth, and nickel. Tests made with batches of 20 to 40 pounds and with a heat not too great, give good results as follows:

Sand, 65 parts; potash, 15 parts; soda, 5 parts; lime, 9 parts; molybdenite, 3 parts; sulphide of sodium, 2 parts, gave a dark reddish-brown glass. In thinner layers this glass appeared light brownish yellow. Flashed with opal, it became a smutty black brown.

Sand, 50 parts; potash, 15 parts; soda, 5 parts; lime, 9 parts; molybdenite, 1 part; sulphide of sodium, 2 parts, gave a yellow glass.

Sand, 10 parts; potash, 3.3 parts; soda, 0.27 parts; lime, 1.64 parts; molybdenite, 0.03 parts, gave a reddish-yellow glass with a fine tinge of red.

Sand, 100 parts; potash, 26 parts; soda, 108 parts; lime, 12 parts; sulphide of copper, 1.7 parts; sulphide of sodium, 2.3 parts, gave a dark-brown color, varying from sepia to sienna. In thick layers it was no longer transparent, but still clear and unclouded. When heated this glass became smutty black brown and clouded.

A fine copper red was obtained from sand, 10 parts; potash, 3 parts; lime, 1.2 parts; soda, 0.25 parts; sulphide of copper, 7.5 parts; sulphide of sodium, 10.5 parts; borax, 9.5 parts.

Attempts to color with sulphides of antimony and bismuth failed. But the addition of 7 per cent of sulphide of nickel to an ordinary batch gave a glass of fine amethyst color.

Coloring Electric-Light Bulbs and Globes.—Two substances suggest themselves as excellent vehicles of color, and both water soluble—water glass (potassium or sodium silicate) and gelatin. For tinting, water-soluble aniline colors should be tried. The thickness of the solution must be a matter of experimentation. Prior to dipping the globes they should be made as free as possible from all grease, dirt, etc. The gelatin solution should not be so thick that any appreciable layer of it will form on the surface of the glass, and to prevent cracking, some non-drying material should be added to it, say glycerine.

Rose-Tint Glass.—Selenium is now used for coloring glass. Rose-tinted glass is made by adding selenium directly to the ingredients in the melting pot. By mixing first with cadmium sulphide, orange red is produced. This process is stated not to require the reheating of the glass and its immersion in the coloring mixture, as in the ordinary process of making red glass.

CUTTING, DRILLING, GRINDING, AND SHAPING GLASS:

To Cut Glass.—I.—Glass may be cut without a diamond. Dip a piece of common string in alcohol and squeeze it reasonably dry. Then tie the string tightly around the glass on the line of cutting. Touch a match to the string and let it burn off. The heat of the burning string will weaken the glass in this particular place. While it is hot plunge the glass under water, letting the arm go well under to the elbow, so there will be no vibration when the glass is struck. With the free hand strike the glass outside the line of cutting, giving a quick, sharp stroke with a stick of wood, a long-bladed knife, or the like, and the cut will be as clean and straight as if made by a regular glass cutter.

The same principle may be employed to cut bottles into vases, and to form all sorts of pretty things, such as jewelry boxes, picture panes, trays, small tablets, windows for a doll house, etc.

II.—Scratch the glass around the shape you desire with the corner of a file or graver; then, having bent a piece of wire into the same shape, heat it red hot and lay it upon the scratch and sink the glass into cold water just deep enough for the water to come almost on a level with its upper surface. It will rarely fail to break perfectly true.

To Cut Glass Under Water.—It is possible to cut a sheet of glass roughly to any desired shape with an ordinary pair of scissors, if the operation be performed under water. Of course, a smooth edge cannot be obtained by such means, but it will be found satisfactory.

Drilling, Shaping, and Filing Glass.—Take any good piece of steel wire, file to the shape of a drill, and then hold it in a flame till it is at a dull red heat; then quench in metallic mercury. A piece of good steel, thus treated, will bore through glass almost as easily as through soft brass. In use, lubricate with oil of turpentine in which camphor has been dissolved. When the point of the drill has touched the other side put the glass in water, and proceed with the drilling very slowly. If not possible to do this, reverse the work—turn the glass over and drill, very carefully, from the opposite side. By proceeding with care you can easily drill three holes through glass $\frac{3}{16}$ inch thick $\frac{1}{4}$ of an inch apart. In making the drill be careful not to make the point and the cutting edges too acute. The drill cuts more slowly, but more safely, when the point and cutting edges are at a low angle.

To Make Holes in Thin Glass.—To produce holes in panes of thin or weak glass, provide the places to be perforated with a ring of moist loam, whose center leaves free a portion of glass exactly the size of the desired hole. Pour molten lead into the ring, and the glass and lead will fall through at once. This process is based upon the rapid heating of the glass.

To Grind Glass.—For the grinding of glass, iron, or steel laps and fine sand are first used; after that, the sand is replaced by emery. Then the polishing is started with pure lead or pure tin laps, and finished with willow wood laps. The polishing powder is tin putty, but peroxide of iron or dioxide of tin is a good polishing medium.

Pohl asserts that if glass is polished with crocus (Paris red) it appears of a dark or a yellowish-brown tint. He contends that the crocus enters the pores of the glass, and, to prevent this, he uses zinc white with the most satisfactory results.

A Home-Made Outfit for Grinding Glass.—Provide two pieces of cork, one concave and one convex (which may be cut to shape after fitting to the lathe). Take a copper cent or other suitable article and soft-solder a screw to fit the lathe, and then wax it to the cork; get a cheap emery wheel, such as is used on sewing machines. Polish the edge on the zinc collar of the emery wheel (or use a piece of zinc). The other cork should be waxed to a penny and centered. Spectacle lenses may be cut on the same emery wheel if the wheel is attached to the lathe so as to revolve. Another method is to take a common piece of window glass (green glass is the best) and make a grindstone of that, using the flat surface for grinding. Cement it on a large chuck, the glass being from 2 to $2\frac{1}{2}$ inches in diameter.

To Drill Optical Glass.—A graver sharpened to a long point is twisted between the fingers, and pressed against the glass, the point being moistened from time to time with turpentine. When the hole is finished half way, the drilling should be commenced from the other side. The starting should be begun with care, as otherwise the graver is likely to slide out and scratch the lens. It is advisable to mark the point of drilling with a diamond, and not to apply too great a pressure when twisting the graver.

Lubricants for Glass Drilling.—I.—Put garlic, chopped in small pieces, into spirit of turpentine and agitate the mixture from time to time. Filter at the end of a fortnight, and when you desire to pierce the glass dip your bit or drill into this liquid, taking care to moisten it constantly to prevent the drill, etc., from becoming heated.

II.—Place a little alum in acetic acid, dip your drill into this and put a drop of it on the spot where the glass is to be pierced.

GILDING GLASS.

When it is desired to gild glass for decorative purposes use a solution of gelatin in hot water, to which an equal quantity of alcohol has been added. The glass to be gilded is covered with this solution and the gold leaf put on while wet. A sheet of soft cotton must be pressed and smoothed over the leaf until the gelatin below is evenly distributed. This prevents spots in gilding. Careful apportionment of the gelatin is necessary. If too much be used, the gold may become spotted; if too little, the binding may be too weak to allow the gold to be polished. The glass should be cleaned thoroughly before gilding. After the gold leaf is put on the whole is allowed to dry for 10 or 20 minutes, when the luster of the gold can be raised by a cautious rubbing with cotton. Then another layer of gelatin is spread on with one stroke of a soft brush, and, if especially good work be required, a second layer of gold is put on and covered as before. In this case, however, the gelatin is used hot. After the gilding has become perfectly dry the letters or ornamentation are drawn and the surplus gold around the edges is taken off. The gilding does not become thoroughly fixed until after several months, and until then rough handling, washing, etc., should be avoided.

The best backing for glass gilding is asphaltum, with a little lampblack, this to be mixed up with elastic varnish; outside finishing varnish is the best, as the addition of this material gives durability.

GLASS MANUFACTURING:

See also Ceramics.

The blue tint of the common poison bottle is got by the addition of black oxide of cobalt to the molten glass; the green tint of the actinic glass bottle is obtained in the same way by the addition of potassium bichromate, which is reduced to the basylous condition, and the amber tint is produced by the addition of impure manganese dioxide, a superior tint being produced by suphur

in one form or another. The formulas for various kinds of bottle glass, which indicate the general composition of almost all glasses, are:

White Glass for Ordinary Molded Bottles.—

Sand.....................	64	Parts
Lime.....................	6	by
Carbonate of sodium...	23	weight.
Nitrate of sodium.......	5	

White Flint Glass Containing Lead.—

Sand...................	63	
Lime...................	5	Parts
Carbonate of sodium....	21	by
Nitrate of sodium.......	3	weight.
Red lead...............	8	

Ordinary Green Glass for Dispensing Bottles.—

Sand...................	63	Parts
Carbonate of sodium....	26	by
Lime...................	11	weight.

A mixture for producing a good green flint glass is much the same as that for the ordinary white flint glass, except that the lime, instead of being the purest, is ordinary slaked lime, and the sodium nitrate is omitted. Sand, lime, and sodium carbonate are the ordinary bases of glass, while the sodium nitrate is the decolorizing agent.

Glass Refractory to Heat.—Fine sand, 70 parts; potash, 30 parts; kaolin, 25 parts.

Transparent Ground Glass.—Take hold of the glass by one corner with an ordinary pair of fire tongs. Hold it in front of a clear fire, and heat to about 98° F., or just hot enough to be held comfortably in the hand. Then hold the glass horizontally, ground side uppermost, and pour in the center a little photographer's dry-plate negative varnish. Tilt the glass so that the varnish spreads over it evenly, then drain back the surplus varnish into the bottle from one corner of the glass. Hold the glass in front of the fire again for a few minutes and the varnish will crystallize on its surface, making it transparent. The glass should not be made too hot before the varnish is put on, or the varnish will not run evenly. This method answers very well for self-made magic-lantern slides. Ground glass may be made temporarily transparent by wiping with a sponge dipped in paraffine or glycerine.

WATER-TIGHT GLASS:

Water-Tight Glass Roofs.—Glass roofs, the skeletons of which are constructed of iron, are extremely difficult to keep water-tight, as the iron expands and contracts with atmospheric changes. To meet this evil, it is necessary to use an elastic putty, which follows the variations of the iron. A good formula is: Two parts rosin and one part tallow, melted together and stirred together thoroughly with a little minium. This putty is applied hot upon strips of linen or cotton cloth, on top and below, and these are pasted while the putty is still warm, with one edge on the iron ribs and the other, about one-fourth inch broad, over the glass.

Tightening Agent for Acid Receptacles. —Cracked vessels of glass or porcelain, for use in keeping acids, can be made tight by applying a cement prepared in the following manner: Take finely sifted sand, some asbestos with short fiber, a little magnesia and add enough concentrated water glass to obtain a readily kneadable mass. The acid renders the putty firm and waterproof.

PENCILS FOR MARKING GLASS:

See also Etching and Frosted Glass.

Crayons for Writing on Glass.—I.— The following is a good formula:

Spermaceti..........	4 parts
Tallow.............	3 parts
Wax...............	2 parts
Red lead...........	6 parts
Potassium carbonate.	1 part

Melt the spermaceti, tallow, and wax together over a slow fire, and when melted stir in, a little at a time, the potassium carbonate and red lead, previously well mixed. Continue the heat for 20 or 30 minutes, stirring constantly. Withdraw from the source of heat, and let cool down somewhat, under constant stirring, at the temperature of about 180° F.; before the mixture commences to set, pour off into molds and let cool. The latter may be made of bits of glass tubing of convenient diameter and length. After the mixture cools, drive the crayons out by means of a rod that closely fits the diameter of the tubes.

II.—Take sulphate of copper, 1 part, and whiting, 1 part. Reduce these to a fine powder and mix with water; next roll this paste into the shape of crayons and let dry. When it is desired to write on the glass use one of these crayons and wipe the traced designs. To make them reappear breathe on the glass.

III.—Melt together, spermaceti, 3 parts; talc, 3 parts, and wax, 2 parts. When melted stir in 6 parts of minium

and 1 part of caustic potash. Continue heating for 30 minutes, then cast in suitable molds. When formed and ready to be put away dust them with talc powder, or roll each pencil in paraffine powder.

PREVENTION OF FOGGING, DIMMING, AND CLOUDING.

I.—Place a few flat glass or porcelain dishes with calcium chloride in each window. This substance eagerly absorbs all moisture from the air. The contents of the dishes have to be renewed every 2 or 3 days, and the moist calcium chloride rigorously dried, whereupon it may be used over again.

II.—Apply to the inside face of the glass a thin layer of glycerine, which does not permit the vapor to deposit in fine drops and thus obstruct the light. Double glass may also be used. In this way the heat of the inside is not in direct contact with the cold outside.

III.—By means of the finger slightly moistened, apply a film of soap of any brand or kind to the mirror; then rub this off with a clean, dry cloth; the mirror will be as bright and clear as ever; breathing on it will not affect its clearness.

IV.—Window glass becomes dull during storage by reason of the presence of much alkali. This can be avoided by taking sand, 160 parts; calcined sodium sulphate, 75; powdered marble, 50; and coke, 4 to 5 parts. About 3 parts of the sodium sulphate may be replaced by an equal quantity of potash.

FROSTED GLASS.

I.—A frosted appearance may be given to glass by covering it with a mixture of

Magnesium sulphate.	6 ounces
Dextrin.............	2 ounces
Water..............	20 ounces

When this solution dries, the magnesium sulphate crystallizes in fine needles.

II.—Another formula directs a strong solution of sodium or magnesium sulphate, applied warm, and afterwards coated with a thin solution of acacia.

III.—A more permanent "frost" may be put on the glass by painting with white lead and oil, either smooth or in stipple effect. The use of lead acetate with oil gives a more pleasing effect, perhaps, than the plain white lead.

IV.—If still greater permanency is desired, the glass may be ground by rubbing with some gritty substance.

V.—For a temporary frosting, dip a piece of flat marble into glass cutter's sharp sand, moistened with water; rub over the glass, dipping frequently in sand and water. If the frosting is required very fine, finish off with emery and water. Mix together a strong, hot solution of Epsom salt and a clear solution of gum arabic; apply warm. Or use a strong solution of sodium sulphate, warm, and when cool, wash with gum water. Or daub the glass with a lump of glazier's putty, carefully and uniformly, until the surface is equally covered. This is an excellent imitation of ground glass, and is not disturbed by rain or damp.

VI.—This imitates ground glass:

Sandarac.........	2½ ounces
Mastic...........	½ ounce
Ether...........	24 ounces
Benzine.......16 to 18 ounces

VII.—Take white lead ground in a mixture of ¾ varnish and ¼ oil of turpentine, to which burnt white vitriol and white sugar of lead are added for drier. The paint must be prepared exceedingly thin and applied to the glass evenly, using a broad brush. If the windows require a new coat, the old one is first removed by the use of a strong lye, or else apply a mixture of hydrochloric acid, 2 parts; vitriol, 2 parts; copper sulphate, 1 part; and gum arabic 1 part, by means of a brush. The production of this imitation frosting entails little expense and is of special advantage when a temporary use of the glass is desired.

VIII.—A little Epsom salt (sulphate of magnesia) stirred in beer with a small dose of dextrin and applied on the panes by means of a sponge or a brush permits of obtaining mat panes.

Hoarfrost Glass.—The feathery foams traced by frost on the inside of the windows in cold weather may be imitated as follows:

The surface is first ground either by sand-blast or the ordinary method, and is then covered with a sort of varnish. On being dried either in the sun or by artificial heat, the varnish contracts strongly, taking with it the particles of glass to which it adheres; and as the contraction takes places along definite lines, the pattern given by the removal of the particles of glass resembles very closely the branching crystals of frostwork. A single coat gives a small, delicate effect, while a thick film, formed by putting on 2, 3 or more coats, contracts so strongly as to produce a large and bold design.

By using colored glass, a pattern in half-tint may be made on the colored ground, and after decorating white glass, the back may be silvered or gilded.

Engraving, Matting, and Frosting.—Cover the glass with a layer of wax or of varnish on which the designs are traced with a graver or pen-point; next, hydrofluoric acid is poured on the tracings. This acid is very dangerous to handle, while the following process, though furnishing the same results, does not present this drawback: Take powdered fluoride of lime, 1 part, and sulphuric acid, 2 parts. Make a homogeneous paste, which is spread on the parts reserved for the engraving or frosting. At the end of 3 or 4 hours wash with water to remove the acid, next with alcohol to take off the varnish, or with essence of turpentine if wax has been employed for stopping off.

To Render Window Panes Opaque.—I.—Panes may be rendered mat and non-transparent by painting them on one side with a liquid prepared by grinding whiting with potash water-glass solution. After one or two applications, the panes are perfectly opaque, while admitting the light.

II.—Paint the panes with a solution of

Dextrin.............	200	Parts by weight.
Zinc vitriol..........	800	
Bitter salt...........	300	
In water..	2,000	

III.—For deadening panes already set in frames the following is suitable: Dissolve 1 part of wax in 10 parts of oil of turpentine, adding 1 part of varnish and 1 part of siccative. With this mixture coat the panes on the outside and dab, while still wet, with a pad of cotton wadding. If desired small quantities of Paris blue, madder lake, etc., may be added to the wax solution.

IV.—For deadening window panes in factories and workshops: To beeswax dissolved in oil of turpentine, add some dryer and varnish to obtain a quicker drying and hardening. After the window pane has been coated with this mixture on the outside, it is dabbed uniformly with a pad of wadding. The wax may be tinted with glazing colors.

Frosted Mirrors.—I.—Cover with a solution of Epsom salts in stale beer; apply with a sponge to the mirror, first wiping it clean and dry. On drying, the Epsom salt crystallizes, giving very handsome frosted effects, but the solution must not be applied on humid days when the glass is liable to be damp, for in that case the effect will be a blurred one. When it is desirable to remove the coating, lukewarm water will serve the purpose without damage to the luster of the mirror.

II.—The following mixture, when applied to a mirror and left to dry, will form in many shapes, all radiating from a focus, this focus forming anywhere on the glass, and when all dry tends to form a most pleasing object to the eye.

Sour ale............ 4 ounces
Magnesium sulphate. 1 ounce

Put on the mirror with a small, clean sponge and let dry. It is now ready for the artist, and he may choose his own colors and subject.

Crystalline Coatings or Frostwork on Glass or Paper.—Dissolve a small quantity of dextrin (gum arabic and tragacanth are not so suitable) in aqueous salt solution as concentrated as possible, for instance, in sulphate of magnesia (bitter salt), sulphate of zinc or any other readily crystallizing salt; filter the solution through white blotting paper and coat glass panes uniformly thin with the clear filtrate, using a fine, broad badger brush; leave them lying at an ordinary medium temperature about one-quarter hour in a horizontal position.

As the water slowly evaporates during this short time, handsome crystalline patterns, closely resembling frostwork, will develop gradually on the glass panes, which adhere so firmly to the glass or the paper (if well-sized glazed paper had been used) that they will not rub off easily. They can be permanently fixed by a subsequent coat of alcoholic shellac solution.

Especially handsome effects are produced with colored glass panes thus treated, and in the case of reflected light by colored paper.

For testing crystals as regards their optical behavior, among others their behavior to polarized light, it is sufficient to pour a solution of collodion wool (soluble peroxide lime for the preparation of collodion) over the surface of glass with the crystalline designs, and to pull off the dry collodion film carefully. If this is done cautiously it is not difficult to lift the whole crystalline group from the glass plate and to incorporate it with the glass-like, thin collodion film.

REMOVING WINDOW FROST.

Here are fourteen methods of preventing frost on windows, arranged in the order of their efficacy: 1, Flame of an alcohol lamp; 2, sulphuric acid; 3, aqua ammonia; 4, glycerine; 5, aqua regia; 6, hydrochloric acid; 7, benzine; 8, hydriodic acid; 9, boric acid; 10, alcohol; 11, nitric acid; 12, cobalt nitrate; 13, infusion of nutgalls; 14, tincture of ferrous sulphate. By the use of an alcohol lamp (which, of course, has to be handled with great care) the results are immediate, and the effect more nearly permanent than by any other methods. The sulphuric acid application is made with a cotton cloth swab, care being taken not to allow any dripping, and so with all other acids. The effect of the aqua ammonia is almost instantaneous, but the window is frosted again in a short time. With the glycerine there are very good results—but slight stains on the window which may be easily removed.

The instructions for glycerine are: Dissolve 2 ounces of glycerine in 1 quart of 62 per cent alcohol containing, to improve the odor, some oil of amber. When the mixture clarifies it is rubbed over the inner surface of the glass. This, it is claimed, not only prevents the formation of frost, but also prevents sweating.

To Prevent Dimming of Eyeglasses, etc.—Mix olein-potash soap with about 3 per cent of glycerine and a little oil turpentine. Similar mixtures have also been recommended for polishing physicians' reflectors, show-windows, etc., to prevent dimming.

WRITING ON GLASS:

See also Etching and Inks.

Composition for Writing on Glass.—To obtain mat designs on glass, take sodium fluoride, 35 parts; potassium sulphate, 7 parts; zinc chloride, 15 parts; hydrochloric acid, 65 parts; distilled water, 1,000 parts. Dissolve the sodium fluoride and the potassium sulphate in half the water; dissolve the zinc chloride in the remaining water and add the hydrochloric acid. Preserve these two solutions separately. For use, mix a little of each solution and write on the glass with a pen or brush.

Ink for Writing on Glass.—

Shellac............	20 parts
Alcohol...........	150 parts
Borax.............	35 parts
Water.............	250 parts
Water-soluble dye sufficient to color.	

Dissolve the shellac in the alcohol, the borax in the water, and pour the shellac solution slowly into that of the borax. Then add the coloring matter previously dissolved in a little water.

GLASS AND GLASSWARE CEMENT:
See Adhesives and Amalgams.

GLASS CLEANERS:
See Cleaning Preparations and Methods.

GLASS, COPPERING, GILDING, AND PLATING:
See Plating.

GLASS ETCHING:
See Etching.

GLASS, HOW TO AFFIX SIGN-LETTERS ON:
See Adhesives under Sign-Letter Cements.

GLASS, FASTENING METALS ON:
See Adhesives.

GLASS LETTERING:
See Lettering.

GLASS LUBRICANTS:
See Lubricants.

GLASS, PERCENTAGE OF LIGHT ABSORBED BY:
See Light.

GLASS POLISHES:
See Polishes.

GLASS, SILVERING OF:
See Mirrors.

GLASS SOLDERS:
See Solders.

GLASS, SOLUBLE, AS A CEMENT:
See Adhesives.

GLASS, TO AFFIX PAPER ON:
See Adhesives, under Water-Glass Cements.

GLASS, TO SILVER:
See Silver.

Glazes

(See also Ceramics, Enamels, Paints, and Varnishes.)

Glazes for Cooking Vessels.—Melt a frit of red lead, 22.9 parts (by weight); crystallized boracic acid, 31 parts; enamel soda, 42.4 parts; cooking salt, 10 parts; gravel, 12 parts; feldspar, 8 parts. According to the character of the clay, this frit is mixed with varying quantities of sand, feldspar and kaolin, in the following manner:

Frit	84	84	84	84
Red lead	1.5	1.5	1.5	1.5
Gravel	8	6	3	—
Feldspar	—	2	5	8
Kaolin, burnt	6.5	6.5	6.5	6.5

Glazes which are produced without addition of red lead to the frit, are prepared as follows. Melt a frit of the following composition: Red lead, 22.9 parts (by weight); boracic acid in crystals, 24.8 parts; enamel soda, 37.1 parts; calcined potash, 6.9 parts; cooking salt, 10 parts; chalk, 10 parts; gravel, 12 parts; feldspar, 8 parts.

From the frit the following glazes are prepared:

Frit	86.5	86.5	86.5	86.5
Gravel	7	4.5	3	—
Feldspar	—	2.5	4	7
Kaolin, burnt	6.5	6.5	6.5	6.5

Glazing on Size Colors.—The essential condition for this work is a well-sized foundation. For the glazing paint, size is likewise used as a binder, but a little dissolved soap is added, of about the strength employed for coating ceilings. Good veining can be done with this, and a better effect can be produced in executing pieces which are to appear in relief, such as car-touches, masks, knobs, etc., than with the ordinary means. A skillful grainer may also impart to the work the pleasant luster of natural wood. The same glazing method is applicable to colored paintings. If the glazing colors are prepared with wax, dissolved in French turpentine, one may likewise glaze with them on a size-paint ground. Glazing tube-oil colors thinned with turpentine and siccative, are also useful for this purpose. For the shadows, asphalt and Van Dyke brown are recommended, while the contour may be painted with size-paint.

Coating Metallic Surfaces with Glass.—Metallic surfaces may be coated with glass by melting together 125 parts (by weight) of flint-glass fragments, 20 parts of sodium carbonate, and 12 parts of boracic acid. The molten mass is next poured on a hard and cold surface, stone or metal. After it has cooled, it is powdered. Make a mixture of 50° Bé. of this powder and sodium silicate (water glass). The metal to be glazed is coated with this and heated in a muffle or any other oven until the mixture melts and can be evenly distributed. This glass coating adheres firmly to iron and steel.

Glaze for Bricks.—A glazing color for bricks patented in Germany is a composition of 12 parts (by weight) lead; 4 parts litharge; 3 parts quartzose sand; 4 parts white argillaceous earth; 2 parts kitchen salt; 2 parts finely crushed glass, and 1 part saltpeter. These ingredients are all reduced to a powder and then mixed with a suitable quantity of water. The color prepared in this manner is said to possess great durability, and to impart a fine luster to the bricks.

GLAZES FOR LAUNDRY:
See Laundry Preparations.

GLOBES, HOW TO COLOR:
See Glass-Coloring.

GLOBES, PERCENTAGE OF LIGHT ABSORBED BY:
See Light.

GLOBES, SILVERING OF:
See Mirrors.

GLOSS FOR PAPER:
See Paper.

GLOVE-CLEANERS:
See Cleaning Compounds.

GLOVES, SUBSTITUTE FOR RUBBER:
See Antiseptics.

GLOVES, TESTING:
See Rubber.

GLUCOSE IN JELLY:
See Foods.

Glue

(Formulas for Glues and methods of manufacturing Glue will be found under Adhesives.)

Rendering Glue Insoluble in Water.— Stuebling finds that the usual mixture of bichromate and glue when used in the ordinary way does not possess the waterproof properties with which it is generally credited. If mixed in the daylight, it sets hard before it can be applied to the surfaces to be glued, and if mixed and applied in the dark room it remains just as soluble as ordinary glue, the light being unable to penetrate the interior of the joints. Neither is a mixture of linseed oil and glue of any use for this purpose. Happening to upset a strong solution of alum—prepared for wood staining—into an adjacent glue pot, he stirred up the two together out of curiosity and left them. Wishing to use the glue a few days later, he tried to thin it down with water, but unsuccessfully, the glue having set to a waterproof mass. Fresh glue was then mixed with alum solution and used to join two pieces of wood, these resisting the action of the water completely.

To Bleach Glue.—Dissolve the glue in water, by heat, and while hot, add a mixture in equal parts of oxalic acid and zinc oxide, to an amount equal to about 1 per cent of the glue. After the color has been removed, strain through muslin.

Method of Purifying Glue.—The glue is soaked in cold water and dissolved in a hot 25 per cent solution of magnesium sulphate. The hot solution is filtered, and to the filtrate is added a 25 per cent solution of magnesium sulphate containing 0.5 per cent of hydrochloric acid (or, if necessary, sulphuric acid). A white flocculent precipitate is obtained which is difficult to filter. The remainder of the glue in the saline solution is extracted by treatment with magnesium sulphate.

The viscous matter is washed, then dissolved in hot water, and allowed to cool, a quantity of weak alcohol acidulated by 1 per cent of hydrochloric acid being added just before the mass solidifies. From 2 to 3 parts, by volume, of strong alcohol (methyl or ethyl) are then added and the solution filtered, charcoal being used if necessary. The glue is finally precipitated from this solution by neutralizing with ammonia and washing with alcohol or water.

To Distinguish Glue and Other Adhesive Agents.—The product to be examined is heated with hydrofluoric acid (50 per cent). If bone glue is present in any reasonable quantity, an intense odor of butyric acid arises at once, similar to that of Limburger cheese. But if dextrin or gum arabic is present, only an odor of dextrine or fluorhydric acid will be perceptible. Conduct the reaction with small quantities; otherwise the smell will be so strong that it is hard to remove from the room.

GLUE CLARIFIER:
See Gelatin.

Glycerine

Recovering Glycerine from Soap Boiler's Lye.—I.—Glycerine is obtained as a by-product in making soap. For many years the lyes were thrown away as waste, but now considerable quantities of glycerine are recovered, which are much used in making explosive compounds.

When a metallic salt or one of the alkalies, as caustic soda, is added to tallow, a stearite of the metal (common soap is stearite of sodium) is formed, whereby the glycerine is eliminated.

This valuable by-product is contained in the waste lye, and has formed the subject of several patents.

Draw the lye off from the soap-pans; this contains a large quantity of water, some salt and soap and a small quantity of glycerine, and the great trouble is to concentrate the lye so that the large quantity of water is eliminated, sometimes 10 to 12 days being occupied in doing this. The soap and salt are easily removed.

To remove the soap, run the lye into a series of tanks alternating in size step-like, so that as the first, which should be the largest, becomes full, the liquor will flow into the second, from that into the third, and so on; by this arrangement the rosinous and albuminous matters will settle, and the soap still contained in the lyes will float on the surface, from which it is removed by skimming.

After thus freeing the lye of the solid impurities, convey the purified lye to the glycerine recovering department (wooden troughs or pipes may be used to do this), and after concentrating by heating it in a steam-jacketed boiler, and allowing it to cool somewhat, ladle out the solid salt that separates, and afterwards concentrate the lye by allowing it to flow into a tank, but before doing so let the fluid come in contact with a hot blast of air or superheated steam, whereby the crude discolored glycerine is obtained. This is further purified by heating with animal charcoal to decolorize it, then distilling several times in copper stills with superheated steam. The chief points to attend to are: (1) The neutralizing and concentrating the lye as much as possible and then separating the salts and solid matters; (2) concentrating the purified lye, and mixing this fluid with oleic acid, oil, tallow, or lard, and heating the mixture to 338° F., in a still, by steam, and gradually raise the heat to 372° F.; (3) stirring the liquor while being heated, and allowing the aqueous vapor to escape, and when thus concentrated, saponifying the liquid with lime to eliminate the glycerine; water is at the same time expelled, but this is removed from the glycerine by evaporating the mixture.

II.—In W. E. Garrigues's patent for the recovering of glycerine from spent soap lyes, the liquid is neutralized with a mineral acid, and after separation of the insoluble fatty acids it is concentrated and then freed from mineral salts and volatile fatty acids, and the concentrated glycerine solution treated with an alkaline substance and distilled. Thus the soap lye may be neutralized with sulphuric acid, and aluminum sulphate added to precipitate the insoluble fatty acids. The filtrate from these is concentrated and the separated mineral salts removed, after which barium chloride is added and then sufficient sulphuric acid to liberate the volatile fatty acids combined with the alkali. These acids are partially enveloped in the barium sulphate, with which they can be separated from the liquid by filtration, while the remaining portion can be expelled by evaporating the liquid in a vacuum evaporator. Finally, the solution is treated with sodium carbonate, and the glycerine distilled.

Glycerine Lotion.—

Glycerine	4	ounces
Essence bouquet	$\frac{1}{4}$	ounce
Water	4	ounces
Cochineal coloring,	a	sufficient
quantity.		

(See also Cosmetics for Glycerine Lotions.)

GLYCERINE APPLICATIONS:
See Cosmetics.

GLYCERINE AS A DETERGENT:
See Cleaning Preparations and Methods.

GLYCERINE PROCESS:
See Photography.

GLYCERINE SOAP:
See Soap.

GLYCERINE DEVELOPER:
See Photography.

Gold

(See also Jewelers' Formulas.)

Gold Printing on Oilcloth and Imitation Leather.—Oilcloth can very easily be gilt if the right degree of heat is observed. After the engraving has been put in the press, the latter is heated slightly, so that it is still possible to lay the palm of the hand on the heated plate without any unpleasant sensation. Go over the oilcloth with a rag in which a drop of olive oil has been rubbed up, which gives a greasy film. No priming with white of egg or any other priming agent should be done, since the gold leaf would stick. Avoid sprinkling on gilding powder. The gold leaf is applied directly on the oilcloth; then place in the lukewarm press, squeezing it down with a quick jerky motion and opening it at once. If the warm plate remains too long on the oilcloth, the gold leaf will stick. When the impression is done, the gold leaf is not swept off at once, but the oilcloth is first allowed to cool completely for several minutes, since there is a possibility that it has become slightly softened under the influence of the heat, especially at the borders of the pressed figures, and the gold would stick there if swept off immediately. The printing should be sharp and neat and the gold glossy. For bronze printing on oilcloth, a preliminary treatment of printing with varnish ground should be given. The bronze is dusted on this varnish.

Imitation leather is generally treated in the same manner. The tough paper substance is made to imitate leather perfectly as regards color and pressing, especially the various sorts of calf, but the treatment in press gilding differs entirely from that of genuine leather. The stuff does not possess the porous, spongy nature of leather, but on the contrary is very hard, and in the course of manufacture in stained-paper factories is given an almost waterproof coating of color and varnish. Hence the applied ground of white of egg penetrates but slightly into this substance, and a thin layer of white of egg remains on the surface. The consequence is that in gilding the gold leaf is prone to become attached, the ground of albumen being quickly dissolved under the action of the heat and put in a soft sticky state even in places where there is no engraving. In order to avoid this the ground is either printed only lukewarm, or this imitation leather is not primed at all, but the gold is applied immediately upon going over the surface with the oily rag. Print with a rather hot press, with about the same amount of heat as is employed for printing shagreen and title paper. A quick jerky printing, avoiding a long pressure of the plate, is necessary.

Liquid Gold.—Take an evaporating dish, put into it 880 parts, by weight, of pure gold; then 4,400 parts, by weight, of muriatic acid, and 3,520 parts, by weight, nitric acid; place over a gas flame until the gold is dissolved, and then add to it 22 parts, by weight, of pure tin; when the tin is dissolved add 42 parts, by weight, of butter of antimony. Let all remain over the gas until the mixture begins to thicken. Now put into a glass and test with the hydrometer, which should give about 1,800 specific gravity.

Pour into a large glass and fill up with water until the hydrometer shows 1090; pour all the solution into a chemical pot and add to it 1,760 parts, by weight, balsam of sulphur, stirring well all the while, and put it over the gas again; in an hour it should give, on testing, 125° F.; gradually increase the heat up to 185° F., when it should be well stirred and then left to cool about 12 hours. Pour the watery fluid into a large vessel and wash the dark-looking mass 5 or 6 times with hot water; save each lot of water as it contains some portion of gold. Remove all moisture from the dark mass by rolling on a slab and warming before the fire occasionally so as to keep it soft. When quite dry add 2¼ times its weight of turpentine and put it over a small flame for about 2 hours; then slightly increase the heat for another hour and a half. Allow this to stand about 24 hours, and then take a glazed bowl and spread over the bottom of it 1,760 parts, by weight, of finely powdered bismuth; pour the prepared gold over it in several places. Now take a vessel containing water and place inside the other vessel containing the gold, and heat it so as to cause the water to boil for 3 hours; allow it to remain until settled and pour off the gold from the settlings of the bismuth, and try it; if not quite right continue the last process with bismuth until good; the bismuth causes the gold to adhere.

Preparation of Balsam of Sulphur.—Take 16 parts oil of turpentine; 2½ parts spirits of turpentine; 8 parts flour of sulphur.

Place all in a chemical pot and heat until it boils; continue the boiling until no sulphur can be seen in it; now remove from the heat and thin it with turpentine until about the thickness of treacle, then warm it again, stirring well; allow it to cool until it reaches 45° F., then test it with the hydrometer, and if specific gravity is not 995 continue the addition of turpentine and warming until correct, let it thoroughly cool, then bottle, keeping it air-tight.

To Purify Bismuth.—Take 6 parts bismuth metal, ¾ part saltpeter. Melt together in a biscuit cup, pour out on to a slab, and take away all dirt, then grind into a fine powder.

To Recover the Gold from the Remains of the Foregoing Process.—Put all the "watery" solutions into a large vessel and mix with a filtered saturated solution of copperas; this will cause a precipitate of pure metallic gold to gradually subside; wash it with cold water and dry in an evaporating dish.

All rags and settlings that are thick should be burnt in a crucible until a yellow mass is seen; then take this and dissolve it in 2 parts muriatic acid and 1 part nitric acid. Let it remain in a porcelain dish until it begins to thicken, and crystals form on the sides. Add a little nitric acid, and heat until crystals again form. Now take this and mix with cold water, add a solution of copperas to it and allow it to settle; pour off the water, and with fresh water wash till quite free from acid. The gold may then be used again, and if great care is exercised almost one-half the original quantity may be recovered.

The quantities given in the recipe should produce about 13 to 15 parts of the liquid gold. It does not in use require any burnishing, and should be fired at rose-color heat. If desired it can be fluxed with Venice turpentine, oil of lavender, or almonds.

Treatment of Brittle Gold.—I.—Add to every 100 parts, by weight, 5 to 8 parts, by weight, of cupric chloride and melt until the oily layer which forms has disappeared. Then pour out, and in most cases a perfectly pliable gold will have been obtained. If this should not be the case after the first fusion, repeat the operation with the same quantity of cupric chloride. The cupric chloride must be kept in a well-closed bottle, made tight with paraffine, and in a dry place.

II.—Pass chlorine gas through the molten gold, by which treatment most of the gold which has otherwise been set aside as unfit for certain kinds of work may be redeemed.

Assaying of Gold.—To determine the presence of gold in ores, etc., mix a small quantity of the finely powdered ore in a flask with an equal volume of tincture of iodine, shake repeatedly and well, and leave in contact about 1 hour, with repeated shaking. Next allow the mixture to deposit and dip a narrow strip of filtering paper into the solution. Allow the paper to absorb, next to dry; then dip it again into the solution, repeating this 5 to 6 times, so that the filtering paper is well saturated and impregnated. The strip is now calcined, as it were, and the ashes, if gold is present, show a purple color. The coloring disappears immediately if the ashes are moistened with bromine water. The same test may also be modified as follows: Cover the finely pulverized ore with bromine water, shake well and repeatedly during about 1 hour of the contact, and filter. Now add to the solution stannic protochloride in solution, whereby, in case gold is present, a purple color (gold purple of Cassius) will at once appear. In case the ore to be assayed contains sulphides, it is well to roast the ore previously, and should it contain lime carbonate, it is advisable to calcine the ore before in the presence of ammonium carbonate.

Gold Welding.—Gold may be welded together with any metal, if the right methods are employed, but best with copper. Some recipes for welding agents are here given.

I.—Two parts by weight (16 ounces equal 1 pound) of green vitriol; 1 part by weight (16 ounces equal 1 pound) of saltpeter; 6 parts by weight (16 ounces equal 1 pound) of common salt; 1 part by weight (16 ounces equal 1 pound) of black manganic oxide or pulverized, and mixed with 48 parts by weight (16 ounces equal 1 pound) of good welding sand.

II.—Filings of the metal to be used in welding are mixed with melted borax in the usual proportion. To be applied in the thickness desired.

III.—A mixture of 338 parts of sodium phosphate and 124 parts of boracic acid is used when the metal is at dark-red heat. The metal is then to be brought to a bright-red heat, and hammered at the same time. The metal easily softens at a high temperature, and a wooden mallet is best. All substances containing carbon should be removed from the surface, as success depends upon the formation of a fusible copper phosphate, which dissolves a thin layer of oxide on the surface, and keeps the latter in good condition for welding.

To Recover Gold-Leaf Waste.—To recover the gold from color waste, gold brushes, rags, etc., they are burned up to ashes. The ashes are leached with boiling water containing hydrochloric acid. The auriferous residuum is then boiled with aqua regia (1 part nitric acid and 3 parts hydrochloric acid), whereby the gold is dissolved and gold chloride results. After filtration and evaporation to dryness the product is dissolved in water and precipitated with sulphate of protoxide of iron. The precipitated gold powder is purified with hydrochloric acid.

Gold from Acid Coloring Baths.—I.—Different lots are to be poured together and the gold in them recovered. The following method is recommended: Dissolve a handful of phosphate of iron in boiling water, to which liquor add the coloring baths, whereby small particles of gold are precipitated. Then draw off the water, being careful not to dissolve the auriferous sediment at the bottom. Free this from all traces of acid by washing with plenty of boiling water; it will require 3 or 4 separate washings, with sufficient time between each to allow the water to cool and the sediment to settle before pouring off the water. Then dry in an iron vessel by the fire and fuse in a covered skittlepot with a flux.

II.—The collected old coloring baths are poured into a sufficiently large pot, an optional quantity of nitro-muriatic acid is added, and the pot is placed over the fire, during which time the fluid is stirred with a wooden stick. It is taken from the fire after a while, diluted largely with rain water and filtered through coarse paper. The gold is recovered from the filtered solution with a solution of green vitriol which is stored in air-tight bottles, then freshened with hot water, and finally smelted with borax and a little saltpeter.

Parting with Concentrated Sulphuric Acid.—It is not necessary scrupulously to observe the exact proportion of the gold to the silver. After having prepared the auriferous silver, place it in a quantity of concentrated sulphuric acid contained in a porcelain vessel, and let it come to a violent boil. When the acid has either become saturated and will dissolve no more, or when solution is complete, remove the dissolving vessel from the fire, let it cool, and, for the purpose of clarifying, pour dilute sulphuric acid into the solution. The dissolved silver is next carefully decanted from the gold sediment upon the bottom, another portion of concentrated acid is poured in, and the gold is well boiled again, as it will still contain traces of silver; this operation may be repeated as often as is deemed necessary. The solution, poured into the glass jars, is well diluted with water, and the silver is then precipitated by placing a sheet of copper in the solution. The precipitate is then freshened with hot water, which may also be done by washing upon the filter; the granulated silver (sulphate of silver) is pressed out in linen, dried and smelted. The freshened gold, after drying, is first smelted with bisulphate of soda, in order to convert the last traces of silver into sulphate, and then smelted with borax and a little saltpeter.

To Remove Gold from Silver.—I.—Gold is taken from the surface of silver by spreading over it a paste, made of powdered sal ammoniac with aqua fortis and heating it till the matter smokes and is nearly dry, when the gold may be separated by rubbing it with the scratch brush.

II.—The alloy is to be melted and poured from a height into a vessel of cold water, to which a rotary motion is imparted, or else it is to be poured through a broom. By this means the metal is reduced to a fine granular condition. The metallic substance is then treated with nitric acid, and gently heated. Nitrate of silver is produced, which can be reduced by any of the ordinary methods; while metallic gold remains as a black sediment, which must be washed and melted.

Simple Specific Gravity Test.—A certain quantity of the metal is taken and drawn out into a wire, which is to be exactly of the same length as one from fine silver; of course, both must have been drawn through the same hole, silver being nearly $\frac{1}{2}$ lighter than gold, it is natural that the one of fine silver must be lighter, and the increased weight of the wire under test corresponds to the percentage of gold contained in it.

To Make Fat Oil Gold Size.—First thin up the fat oil with turpentine to workable condition; then mix a little very finely ground pigment with the gold size, about as much as in a thin priming coat. Make the size as nearly gold color as is convenient; chrome yellow tinted with vermilion is as good as anything for this purpose. Then thin ready for the brush with turpentine, and it will next be in order to run the size through a very fine strainer. Add japan, as experience or experiment may teach, to make it dry tacky about the time the leaf is to be laid. Dry slowly, because the slower the size dries, the longer it will hold its proper tackiness when it is once in that condition.

To Dissolve Copper from Gold Articles.—Take 2 ounces of proto-sulphate of iron and dissolve it in $\frac{1}{2}$ a pint of water, then add to it in powder 2 ounces of nitrate of potash; boil the mixture for some time, and afterwards pour it into a shallow vessel to cool and crystallize; then to every part of the crystallized salt add 8 ounces of muriatic acid, and preserve in a bottle for use. Equal parts of the above preparation and of boiling water is a good proportion to use in dissolving copper, or 1 part by weight of nitric acid may be used to 4 parts by weight of boiling water as a substitute.

GOLD PURPLE.

I.—The solution of stannous chloride necessary for the preparation of gold purple is produced by dissolving pure tin in pure hydrochloric acid (free from iron), in such a manner that some of the tin remains undissolved, and evaporating the solution, into which a piece of tin is laid, to crystallization.

II.—**Recipe for Pale Purple.**—Dissolve 2 parts by weight of tin in boiling aqua regia, evaporate the solution at a moderate heat until it becomes solid, dissolve in distilled water and add 2 parts by weight of a solution of stannous chloride (specific gravity 1.7) dilute with 9,856 parts by weight of water, stir into the liquid a solution of gold chloride prepared from 0.5 parts by weight of gold and containing no excess of acid (the latter being brought about by evaporating the solution of gold chloride to dryness and heating for some time to about 320° F.). This liquid is dimmed by the admixture of 50 parts by weight of liquid ammonia which eliminates the purple. The latter is quickly filtered off, washed out and while still moist rubbed up with the glass paste. This consists of enamel of lead 20 parts by weight; quartzose sand, 1 part by weight; red lead, 2 parts by weight; and calcined borax, 1 part by weight, with silver carbonate, 3 parts by weight.

III.—**Recipe for Dark Gold Purple.**—Gold solution of 0.5 parts by weight of gold, solution of stannous chloride (specific gravity 1.7) 7.5 parts by weight; thin with 9,856 parts by weight of water, separate the purple by a few drops of sulphuric acid, wash out the purple and mix same with enamel of lead 10 parts by weight and silver carbonate, 0.5 parts by weight.

IV.—**Recipe for Pink Purple.**—Gold solution of 1 part by weight of gold; solution of 50 parts by weight of alum in 19,712 parts by weight of water; add 1.5 parts by weight of stannous chloride solution (specific gravity 1.7) and enough ammonia until no more precipitate is formed; mix the washed out precipitate, while still moist, with 70 parts by weight of enamel of lead and 2.5 parts by weight of silver carbonate. According to the composition of the purple various reds are obtained in fusing it on; the latter may still be brightened up by a suitable increase of the flux.

To Render Pale Gold Darker.—Take verdigris, 50 parts by weight and very strong vinegar, 100 parts by weight. Dissolve the verdigris in the vinegar, rub the pieces with it well, heat them and dip them in liquid ammonia diluted with water. Repeat the operation if the desired shade does not appear the first time. Rinse with clean water and dry.

To Color Gold.—Gilt objects are improved by boiling in the following solution: Saltpeter, 2 parts by weight; cooking salt, 1 part by weight; alum, 1 part by weight; water, 24 parts by weight; hydrochloric acid, 1 part by weight (1.12 specific gravity). In order to impart a rich appearance to gilt articles, the following paste is applied: Alum, 3 parts by weight; saltpeter, 2 parts by weight; zinc vitriol, 1 part by weight; cooking salt, 1 part by weight; made into a paste with water. Next, heat until black, on a hot iron plate, wash with water, scratch with vinegar and dry after washing.

Gold-Leaf Striping.—To secure a good job of gilding depends largely for its beauty upon the sizing. Take tube chrome yellow ground in oil, thin with wearing body varnish, and temper it ready for use with turpentine. Apply in the evening with an ox-tail striper, and let it stand until the next morning, when, under ordinary circumstances, it will be ready for the gold leaf, etc. After the gilding is done, let the job stand 24 hours before varnishing.

Composition of Aqua Fortis for the Touch-Stone.—Following are the three compositions mostly in use: I.—Nitric acid, 30 parts; hydrochloric acid, 3 parts; distilled water, 20 parts.

II.—Nitric acid, 980 parts by weight; hydrochloric acid, 20 parts by weight.

III.—Nitric acid, 123 parts by weight; hydrochloric acid, 2 parts by weight.

To Remove Soft Solder from Gold.—Place the work in spirits of salts (hydrochloric acid) or remove as much as possible with the scraper, using a gentle heat to remove the solder more easily.

Tipping Gold Pens.—Gold pens are usually tipped with iridium. This is done by soldering very small pieces to the points and filing to the proper shape.

To Recognize Whether an Article is Gilt.—Simply touch the object with a glass rod previously dipped into a solution of bichloride of copper. If the article has been gilt the spot touched should remain intact, while it presents a brown stain if no gold has been deposited on its surface.

To Burnish Gilt Work.—Ale has proved a very good substitute for soap and water in burnishing gilt as it increases the ease and smoothness with which it is accomplished. Vinegar is a somewhat poorer substitute for ale.

White-Gold Plates Without Solder.—The gold serving as a background for white-gold is rolled in the desired dimensions and then made perfectly even under a powerful press. It is then carefully treated with a file until a perfectly smooth surface is obtained. After a white-gold plate of the required thickness has been produced in the same manner, the surfaces of the two plates to be united are coated with borax and then pressed together by machine, which causes the harder metal to be squeezed slightly into the surface of the other, furnishing a more solid and compact mass. The metals, now partially united, are firmly fastened together by means of strong iron wire and a little more borax solution is put on the edges. Then heat to the temperature necessary for a complete adhesion, but the heat must not be so great as to cause an alloyage by fusing. The whole is finally rolled out into the required thickness.

To Fuse Gold Dust.—Use such a crucible as is generally used for melting brass; heat very hot; then add the gold dust mixed with powdered borax; after some time a scum or slag will be on top, which may be thickened by the addition of a little lime or bone ash. If the dust contains any of the more oxidizable metals, add a little niter, and skim off the slag or scum very carefully; when melted, grasp the crucible with strong iron tongs, and pour off immediately into molds, slightly greased. The slag and crucibles may be afterwards pulverized, and the auriferous matter recovered from the mass through cupellation by means of lead.

GOLD ALLOYS:
See Alloys.

GOLD, EXTRACTION OF, BY AMALGAMATION:
See Amalgams.

GOLD LETTERS ON GLASS, CEMENTS FOR AFFIXING:
See Adhesives, under Sign-Letter Cements.

GOLD, REDUCTION OF OLD PHOTOGRAPHIC:
See Photography.

GOLD FOIL SUBSTITUTES AND GOLD LEAF:
See Metal Foil.

GOLD-LEAF ALLOYS:
See Alloys.

GOLD LEAF AND ITS APPLICATION:
See Paints.

GOLD PLATING:
See Plating.

GOLD, RECOVERY OF WASTE:
See Jewelers' Formulas.

GOLD RENOVATOR:
See Cleaning Preparations and Methods.

GOLD, SEPARATION OF PLATINUM FROM:
See Platinum.

GOLD SOLDERS:
See Solders.

GOLD TESTING:
See Jewelers' Formulas.

GOLD VARNISH:
See Varnishes.

GOLDWASSER:
See Wines and Liquors.

GONG METAL:
See Alloys.

GRAIN.

Formalin Treatment of Seed Grain for Smut.—Smut is a parasitic fungus, and springs from a spore (which corresponds to a seed in higher plants). This germinates when the grain is seeded and, penetrating the little grain plant when but a few days old, grows up within the grain stem. After entering the stem there is no evidence of its presence until the grain begins to head. At this time the smut plant robs the developing kernels of their nourishment and ripens a mass of smut spores.

These spores usually ripen before the grain, and are blown about the field, many spores becoming lodged on the ripening grain kernels. The wholesale agent of infection is the threshing machine. For this reason the safest plan is to treat all seed wheat and oats each year.

Secure a 40 per cent solution of formalin (the commercial name for formaldehyde gas held in a water solution). About 1 ounce is required for every 5 bushels of grain to be treated.

Clean off a space on the barn floor or sweep a clean space on the hard level ground and lay a good-sized canvas down, on which to spread out the wheat. See that the place where the grain is to be treated is swept clean and thoroughly sprinkled with the formalin solution before placing the seed grain there.

Prepare the formalin solution immediately before use, as it is volatile, and if kept may disappear by evaporation.

Use 4 ounces of formalin for 10 gallons of water. This is sufficient for 600 pounds of grain. Put the solution in a barrel or tub, thoroughly mixing.

The solution can be applied with the garden sprinkler. Care must be taken to moisten the grain thoroughly. Sprinkle, stir the grain up thoroughly and sprinkle again, until every kernel is wet.

After sprinkling, place the grain in a conical pile and cover with horse-blankets, gunny sacks, etc. The smut that does the damage lies just under the glume of the oats or on the basal hairs of the wheat. Covering the treated grain holds the gas from the formalin *within* the pile, where it comes in contact with the kernels, killing such smut spores as may have survived the previous treatment. After the grain has remained in a covered pile 2 to 4 hours, spread it out again where the wind can blow over it, to air and dry.

As soon as the grain can be taken in the hand without the kernels sticking together, it can be sown in the field. The grain may be treated in the forenoon and seeded in the afternoon.

Since this treatment swells the kernels it hastens germination and should be done in the spring just before seeding time.

While the copper sulphate or blue-stone treatment is valuable in killing smut, the formalin treatment can be given in less time, is applied so easily and is so effectual that it is recommended as a sure and ready means of killing smut in wheat and oats.

GRAINING CRAYONS:
See Crayons.

GRAINING COLORS:
See Pigments.

GRAINING WITH PAINT:
See Paint.

GRAINING, PALISANDER:
See Palisander.

GRAPE JUICE, PRESERVATION OF:
See Wines and Liquors.

GRAPHITE AS A LUBRICANT:
See Lubricants.

GRAVEL WALKS.

For cleaning gravel walks any of the following may be used: I.—Gas-tar liquor.

II.—Rock salt (cattle salt).

III.—Hydrochloric acid.

IV.—Sulphuric acid.

V.—Fresh limewater. The gas-tar liquor must be poured out a few times in succession, and must not touch the tree roots and borders of the paths. This medium is cheap. Cattle salt must likewise be thrown out repeatedly. The use of hydrochloric and sulphuric acids is somewhat expensive. Mix 60 parts of water with 10 parts of unslaked lime and 1 part of sulphuric acid in a kettle, and sprinkle the hot or cold mixture on the walks by means of a watering pot. If limewater is used alone it must be fresh —1 part of unslaked lime in 10 parts of water.

GRAVERS:

To Prepare Gravers for Bright-Cutting. —Set the gravers after the sharpening on the oilstone on high-grade emery (tripoli) paper. Next, hone them further on the rouge leather, but without tearing threads from it. In this manner the silver and aluminum engravers grind their gravers. A subsequent whetting of the graver on the touchstone is not advisable, since it is too easily injured thereby. A graver prepared as described gives excellent bright engraving and never fails.

In all bright-cutting the graver must be highly polished; but when bright-cutting aluminum a lubricant like coal-oil or vaseline is generally employed with the polished tool; a mixture of vaseline and benzine is also used for this purpose. Another formula which may be recommended for bright-cutting aluminum is composed of the following ingredients: Mix 4 parts of oil of turpentine and 1 part of rum with 1 ounce of stearine. Immerse the graver in any of the mixtures before making the bright-cut.

GREASES:
See Lubricants.

GREASE ERADICATORS:
See Cleaning Preparations and Methods.

GREASE PAINTS:
See Cosmetics.

GREEN, TO DISTINGUISH BLUE FROM, AT NIGHT:
See Blue.

GREEN GILDING:
See Plating.

GRENADES:
See Fire Extinguishers.

GRINDING:
See Tool Setting.

GRINDER DISK CEMENT, SUBSTITUTE FOR:
See Adhesives.

GRINDSTONES:

To Mend Grindstones.—The mending of defective places in grindstones is best done with a mass consisting of earth-wax (so-called stone-pitch), 5 parts, by weight; tar, 1 part; and powdered sandstone or cement, 3 parts, which is heated to the boiling point and well stirred together. Before pouring in the mass the places to be mended must be heated by laying red-hot pieces of iron on them. The substance is, in a tough state, poured into the hollows of the stone, and the pouring must be continued, when it commences to solidify, until even with the surface.

Treatment of the Grindstone.—The stone should not be left with the lower part in the water. This will render it brittle at this spot, causing it to wear off more quickly and thus lose its circularity. It is best to moisten the stone only when in use, drop by drop from a vessel fixed above it and to keep it quite dry otherwise. If the stone is no longer round, it should be made so again by turning by means of a piece of gas pipe or careful trimming, otherwise it will commence to jump, thus becoming useless. It is important to clean all tools and articles before grinding, carefully removing all grease, fat, etc., as the pores of the stone become clogged with these impurities, which destroy its grain and diminish its strength. Should one side of the grindstone be lighter, this irregularity can be equalized by affixing pieces of lead, so as to obtain a uniform motion of the stone. It is essential that the stone should be firm on the axis and not move to and fro in the bearings.

Grindstone Oil.—Complaints are often heard that grindstones are occasionally harder on one side than the other, the softer parts wearing away in hollows, which render grinding difficult, and soon make the stone useless. This defect can be remedied completely by means of boiled linseed oil. When the stone is thoroughly dry, the soft side is turned uppermost, and brushed over with boiled oil, which sinks into the stone, until the latter is saturated. The operation takes about 3 to 4 hours in summer. As soon as the oil has dried, the stone may be damped, and used without any further delay. Unlike other similar remedies, this one does not prevent the stone from biting properly in the oiled parts, and the life of the stone is considerably lengthened, since it does not have to be dressed so often.

GROUNDS FOR GRAINING COLORS:
See Pigments.

GUMS:
(See also Adhesives, under Mucilages.)

Gums, their Solubility in Alcohol.—The following table shows the great range of solubility of the various gums, and of various specimens of the same gum, in 60 per cent alcohol:

Acajou	6.94 to 42.92
Aden	0.60 to 26.90
Egyptian	46.34
Yellow Amrad	26.90 to 32.16
White Amrad	0.54 to 1.50
Kordofan	1.40 to 6.06
Australian	10.67 to 20.85
Bombay	22.06 to 46.14
Cape	1.67 to 1.88
Embavi	25.92
Gedda	1.24 to 1.30
Ghatti	31.60 to 70.32
Gheziereh	1.50 to 12.16
Halebi	3.70 to 22.60
La Plata	9.65
Mogadore	27.66
East Indian	3.24 to 74.84
Persian	1.74 to 17.34
Senegal	0.56 to 14.30

Substitute for Gum Arabic.—Dissolve 250 parts of glue in 1,000 parts of boiling water and heat this glue solution on the water bath with a mixture of about 10 parts of barium peroxide of 75 per cent BaO_2 and 5 parts of sulphuric acid (66°) mixed with 115 parts of water, for about 24 hours. After the time has elapsed, pour off from the barium sulphate, whereby a little sulphurous acid results owing to reduction of the sulphuric acid, which has a bleaching action and makes the glue somewhat paler. If this solution is mixed, with stirring, and dried upon glass plates in the drying-room, a product which can hardly be distinguished from gum arabic is obtained. An envelope sealed with this mucilage cannot be opened by moistening the envelope. The traces of free acid which it contains prevent the invasion of bacteria, hence all putrefaction.

The adhesive power of the artificial gum is so enormous that the use of cork stoppers is quite excluded, since they crumble off every time the bottle is opened, so that finally a perfect wreath around the inner neck of the bottle is formed. Only metallic or porcelain stoppers should be used.

GUM ARABIC, INCREASING ADHESION OF:
See Adhesives, under Mucilages.

GUM BICHROMATE PROCESS:
See Photography.

GUM DROPS:
See Confectionery.

GUM-LAC:
See Oil.

GUMS USED IN MAKING VARNISH:
See Varnishes.

GUN BARRELS, TO BLUE:
See Steel.

GUN BRONZE:
See Alloys, under Phosphor Bronze.

GUN COTTON:
See Explosives.

GUN LUBRICANTS:
See Lubricants.

GUNPOWDER:
See Explosives.

GUNPOWDER STAINS.

A stain produced by the embedding of grains of gunpowder in the skin is practically the same thing as a tattoo mark. The charcoal of the gunpowder remains unaffected by the fluids of the tissues, and no way is known of bringing it into solution there. The only method of obliterating such marks is to take away with them the skin in which they are embedded. This has been accomplished by the application of an electric current, and by the use of caustics. When the destruction of the true skin has been accomplished, it becomes a foreign body, and if the destruction has extended to a sufficient depth, the other foreign body, the coloring matter which has been tattooed in, may be expected to be cast off with it.

Recently pepsin and papain have been proposed as applications to remove the cuticle. A glycerole of either is tattooed into the skin over the disfigured part; and it is said that the operation has proved successful.

It is scarcely necessary to say that suppuration is likely to follow such treatment, and that there is risk of scarring. In view of this it becomes apparent that any such operation should be undertaken only by a surgeon skilled in dermatological practice. An amateur might not only cause the patient suffering without success in removal, but add another disfigurement to the tattooing.

Carbolic acid has been applied to small portions of the affected area at a time, with the result that the powder and skin were removed simultaneously and, according to the physician reporting the case, with little discomfort to the patient.

Rubbing the affected part with moistened ammonium chloride once or twice a day has been reported as a slow but sure cure.

GUTTA-PERCHA.

Gutta-Percha Substitute.—I.—A decoction of birch bark is first prepared, the external bark by preference, being evaporated. The thick, black residue hardens on exposure to the air, and is said to possess the properties of gutta-percha without developing any cracks. It can be mixed with 50 per cent of India rubber or gutta-percha. The compound is said to be cheap, and a good non-conductor of electricity. Whether it possesses all the good qualities of gutta-percha is not known.

II.—A new method of making gutta-percha consists of caoutchouc and a rosin soap, the latter compounded of 100 parts of rosin, 100 parts of Carnauba wax, and 40 parts of gas-tar, melted together and passed through a sieve. They are heated to about 355° to 340° F., and slowly saponified by stirring with 75 parts of limewater of specific gravity 1.06. The product is next put into a kneading machine along with an equal quantity of caoutchouc cuttings, and worked in this machine at a temperature of 195° F. or over. When sufficiently kneaded, the mass can be rolled to render it more uniform.

GUTTER CEMENT:
See Cement and Putty.

GYPSUM:
See also Plaster.

Method of Hardening Gypsum and Rendering it Weather-Proof.—Gypsum possesses only a moderate degree of strength even after complete hardening, and pieces are very liable to be broken off. Various methods have been tried, with a view to removing this defect and increasing the hardness of gypsum. Of these methods, that of Wachsmuth, for hardening articles made of gypsum and rendering them weather-proof, deserves special notice. All methods of hardening articles made of gypsum have this in common: the gypsum is first deprived of its moisture, and then immersed in a solution of certain salts, such as alum, green vitriol, etc. Articles treated by the methods hitherto in vogue certainly acquire considerable hardness, but are no more capable of resistance to the effects of water than crude gypsum. The object of Wachsmuth's process is not merely to harden the gypsum, but to transform it on the surface into insoluble combinations. The process is as follows: The article is first put into the required shape by mechanical means, and then deprived of its moisture by heating to 212° to 302° F. It is then plunged into a heated solution of barium hydrate, in which it is allowed to remain for a longer or shorter time, according to its strength. When this part of the process is complete, the article is smoothed by grinding, etc., and then placed in a solution of about 10 per cent of oxalic acid in water. In a few hours it is taken out, dried, and polished. It then possesses a hardness surpassing that of marble, and is impervious to the action of water. Nor does the polish sustain any injury from contact with water, whereas gypsum articles hardened by the usual methods lose their polish after a few minutes' immersion in water. Articles treated by the method described have the natural color of gypsum, but it is possible to add a color to the gypsum during the hardening process. This is done by plunging the gypsum, after it has been deprived of its moisture, and before the treatment with the barium solution, into a solution of a colored metallic sulphate, such as iron, copper, or chrome sulphate, or into a solution of some coloring matter. Pigments soluble in the barium or oxalic-acid solutions may also be added to the latter.

Gypsum may be hardened and rendered insoluble by ammonium borate as follows: Dissolve boric acid in hot water and add sufficient ammonia water to the solution that the borate at first separated is redissolved. The gypsum to be cast is stirred in with this liquid, and the mass treated in the ordinary way. Articles already cast are simply washed with the liquid, which is quickly absorbed. The articles withstand the weather as well as though they were of stone.

GYPSUM FLOWERS:
See Flowers.

GYPSUM, PAINT FOR:
See Paint.

HAIR FOR MOUNTING.

The microscopist or amateur, who shaves himself, need never resort to the trouble of embedding and cutting hairs in the microtome in order to secure very thin sections of the hair of the face. If he will first shave himself closely "with the hair," as the barbers say (i. e., in the direction of the natural growth of the hair), and afterwards lightly "against the hair" (in the opposite direction to above), he will find in the "scrapings" a multitude of exceedingly thin sections. The technique is very simple. The lather and "scrapings" are put into a saucer or large watch-glass and carefully washed with clean water. This breaks down and dissolves the lather, leaving the hair sections lying on the bottom of the glass. The after-treatment is that usually employed in mounting similar objects.

Hair Preparations

DANDRUFF CURES.

The treatment of that condition of the scalp which is productive of dandruff properly falls to the physician, but unfortunately the subject has not been much studied. One cure is said to be a sulphur lotion made by placing a little sublimed sulphur in water, shaking well, then allowing to settle, and washing the head every morning with the clear liquid.

Sulphur is said to be insoluble in water; yet a sulphur water made as above indicated has long been in use as a hair wash. A little glycerine improves the preparation, preventing the hair from becoming harsh by repeated washings.

The exfoliated particles of skin or "scales" should be removed only when entirely detached from the cuticle. They result from an irritation which is increased by forcible removal, and hence endeavors to clean the hair from them by combing or brushing it in such a way as to scrape the scalp are liable to be worse than useless. It follows that gentle handling of the hair is important when dandruff is present.

I.—Chloral hydrate..... 2 ounces
  Resorcin.......... 1 ounce
  Tannin............ 1 ounce
  Alcohol............ 8 ounces
  Glycerine.......... 4 ounces
  Rose water to make . 4 pints

II.—White wax......... 3½ drachms
  Liquid petrolatum .. 2½ ounces
  Rose water........ 1 ounce
  Borax..............15 grains
  Precipitated sulphur. 3½ drachms

Pine-Tar Dandruff Shampoo.—

  Pine tar............ 4 parts
  Linseed oil......... 40 parts

Heat these to 140° F.; make solution of potassa, U. S. P., 10 parts, and water, 45 parts; add alcohol, 5 parts, and gradually add to the heated oils, stirring constantly. Continue the heat until saponified thoroughly; and make up with water to 128 parts. When almost cool, add ol. lavender, ol. orange, and ol. bergamot, of each 2 parts.

HAIR-CURLING LIQUIDS.

It is impossible to render straight hair curly without the aid of the iron or paper and other curlers. But it is possible, on the other hand, to make artificial curls more durable and proof against outside influences, such as especially dampness of the air. Below are trustworthy recipes:

	I	II
Water...............	70	80
Spirit of wine.......	30	20
Borax..............	2	—
Tincture of benzoin..		3
Perfume...........ad. lib.		ad. lib.

HAIR DRESSINGS AND WASHES:

Dressings for the Hair.—

I.—Oil of wintergreen . 20 drops
  Oil of almond, essential.......... 35 drops
  Oil of rose, ethereal 1 drop
  Oil of violets...... 30 drops
  Tincture of cantharides .... 50 drops
  Almond oil....... 2,000 drops
Mix.

Hair Embrocation.—

II.—Almond oil, sweet . 280 parts
  Spirit of sal ammoniac........ 280 parts
  Spirit of rosemary.. 840 parts
  Honey water...... 840 parts

Mix. Rub the scalp with it every morning by means of a sponge.

Hair Restorer.—

III.—Tincture of cantharides........ 7 parts
  Gall tincture...... 7 parts
  Musk essence...... 1 part
  Carmine.......... 0.5 part
  Rectified spirit of wine........... 28 parts
  Rose water........ 140 parts

To be used at night.

Rosemary Water.—

IV.—Rosemary oil...... 1½ parts
  Rectified spirit of wine........... 7 parts
  Magnesia......... 7 parts
  Distilled water.....1,000 parts

Mix the oil with the spirit of wine and rub up with the magnesia in a mortar; gradually add the water and finally filter.

Foamy Scalp Wash.—Mix 2 parts of soap spirit, 1 part of borax-glycerine (1+2), 6 parts of barium, and 7 parts of orange-flower water.

Lanolin Hair Wash.—Extract 4 parts quillaia bark with 36 parts water for several days, mix the percolate with 4 parts alcohol, and filter after having settled. Agitate 40 parts of the filtrate at a temperature at which wool grease becomes liquid, with 12 parts anhydrous lanolin, and fill up with water to which 15 per cent spirit of wine has been added, to 300 parts. Admixture, such as cinchona extract, Peru balsam, quinine, tincture of cantharides, bay-oil, ammonium carbonate, menthol, etc., may be made. The result is a yellowish-white, milky liquid, with a cream-like fat layer floating on the top, which is finely distributed by agitating.

Birch Water.—Birch water, which has many cosmetic applications, especially as a hair wash or an ingredient in hair washes, may be prepared as follows:

  Alcohol, 96 per cent ..3,500 parts
  Water............... 700 parts
  Potash soap......... 200 parts
  Glycerine........... 150 parts
  Oil of birch buds..... 50 parts
  Essence of spring flowers........... 100 parts
  Chlorophyll, q. s. to color.

Mix the water with 700 parts of the alcohol, and in the mixture dissolve the soap. Add the essence of spring flowers and birch oil to the remainder of the alcohol, mix well, and to the mixture add, little by little, and with constant agitation, the soap mixture. Finally add the glycerine, mix thoroughly, and set aside for 8 days, filter and color the filtrate with chlorophyll, to which add a little tincture of saffron. To use, add an equal volume of water to produce a lather.

Petroleum Hair Washes.—I.—Deodorized pale petroleum, 10 parts; citronella oil, 10 parts; castor oil, 5 parts; spirit of wine, 90 per cent, 50 parts; water, 75 parts.

II.—Quinine sulphate, 10 parts; acetic acid, 4 parts; tincture of cantharides, 30 parts; tincture of quinine, 3 parts; spirit of rosemary, 60 parts; balm water, 90 parts; barium, 120 parts; spirit of wine, 150 parts; water, 1,000 parts.

III.—Very pure petroleum, 1 part; almond oil, 2 parts.

Brilliantine.—I.—Olive oil, 4 parts: glycerine, 3 parts; alcohol, 3 parts; scent as desired. Shake before use.

II.—Castor oil, 1 part; alcohol, 2 parts; saffron to dye yellow. Scent as desired.

III.—Lard, 7 parts; spermaceti, 7 parts; almond oil, 7 parts; white wax, 1 part.

A Cheap Hair Oil.—I.—Sesame oil or sunflower oil, 1,000 parts; lavender oil, 15 parts; bergamot oil, 10 parts; and geranium oil, 5 parts.

II.—Sesame oil or sunflower oil, 1,000 parts; lavender oil, 12 parts; lemon oil, 20 parts; rosemary oil, 5 parts; and geranium oil, 2 parts.

HAIR DYES.

There is no hair dye which produces a durable coloration; the color becomes gradually weaker in the course of time. Here are some typical formulas in which a mordant is employed:

I.—Nitrate of silver..... ½ ounce
 Distilled water..... 3 ounces
Mordant:
 Sulphuret of potassium............ ½ ounce
 Distilled water..... 3 ounces

II.—
(a) Nitrate of silver (crystal) .... 1½ ounces
 Distilled water ..... 12 ounces
 Ammonia water sufficient to make a clear solution.

Dissolve the nitrate of silver in the water and add the ammonia water until the precipitate is redissolved.

(b) Pyrogallic acid..... 2 drachms
 Gallic acid......... 2 drachms
 Cologne water...... 2 ounces
 Distilled water..... 4 ounces
III.—Nitrate of silver..... 20 grains
 Sulphate of copper.. 2 grains
 Ammonia, quantity sufficient.

Dissolve the salts in ½ ounce of water and add ammonia until the precipitate which is formed is redissolved. Then make up to 1 ounce with water. Apply to the hair with a brush. This solution slowly gives a brown shade. For darker shades, apply a second solution, composed of:

IV.—Yellow sulphide ammonium......... 2 drachms
 Solution of ammonia 1 drachm
 Distilled water..... 1 ounce

Black Hair Dye without Silver.—

V.—Pyrogallic acid .... 3.5 parts
 Citric acid........ 0.3 parts
 Boro-glycerine.... 11 parts
 Water........... 100 parts

If the dye does not impart the desired intensity of color, the amount of pyrogallic acid may be increased. The wash is applied evenings, followed in the morning by a weak ammoniacal wash.

One Bottle Preparation.—

VI.—Nitrate of copper.. 360 grains
 Nitrate of silver... 7 ounces
 Distilled water.... 60 ounces
 Water of ammonia, a sufficiency.

Dissolve the salts in the water and add the water of ammonia carefully until the precipitate is all redissolved. This solution, properly applied, is said to produce a very black color; a lighter shade is secured by diluting the solution. Copper sulphate may be used instead of the nitrate.

Brown Hair Dyes.—A large excess of ammonia tends to produce a brownish dye. Various shades of brown may be produced by increasing the amount of water in the silver solution. It should be remembered that the hair must, previously to treatment, be washed with warm water containing sodium carbonate, well rinsed with clear water, and dried.

I.—Silver nitrate..... 480 grains
 Copper nitrate... 90 grains
 Distilled water... 8 fluidounces
 Ammonia water, sufficient.

Dissolve the two salts in the distilled water and add the ammonia water until the liquid becomes a clear fluid.

In using apply to the hair carefully

with a tooth-brush, after thoroughly cleansing the hair, and expose the latter to the rays of the sun.

II.—
Silver nitrate	30 parts
Copper sulphate, crystals	20 parts
Citric acid	20 parts
Distilled water	950 parts
Ammonia water, quantity sufficient to dissolve the precipitate first formed.	

Various shades of brown may be produced by properly diluting the solution before it be applied.

Bismuth subnitrate	200 grains
Water	2 fluidounces
Nitric acid, sufficient to dissolve, or about	420 grains

Use heat to effect solution. Also:

Tartaric acid	150 grains
Sodium bicarbonate	168 grains
Water	32 fluidounces

When effervescence of the latter has ceased, mix the cold liquids by pouring the latter into the former with constant stirring. Allow the precipitate to subside; transfer it to a filter or strainer, and wash with water until free from the sodium nitrate formed.

Chestnut Hair Dye.—

Bismuth nitrate	230 grains
Tartaric acid	75 grains
Water	100 minims

Dissolve the acid in the water, and to the solution add the bismuth nitrate and stir until dissolved. Pour the resulting solution into 1 pint of water and collect the magma on a filter. Remove all traces of acid from the magma by repeated washings with water; then dissolve it in:

| Ammonia water | 2 fluidrachms |

And add:

Glycerine	20 minims
Sodium hyposulphite	75 grains
Water, enough to make	4 fluidounces.

HAIR RESTORERS AND TONICS:

Falling of the Hair.—After the scalp has been thoroughly cleansed by the shampoo, the following formula is to be used:

Salicylic acid	1 part
Precipitate of sulphur	2½ parts
Rose water	25 parts

The patient is directed to part the hair, and then to rub in a small portion of the ointment along the part, working it well into the scalp. Then another part is made parallel to the first, and more ointment rubbed in. Thus a series of first, longitudinal, and then transverse parts are made, until the whole scalp has been well anointed. Done in this way, it is not necessary to smear up the whole shaft of the hair, but only to reach the hair roots and the sebaceous glands, where the trouble is located. This process is thoroughly performed for six successive nights, and the seventh night another shampoo is taken. The eighth night the inunctions are commenced again, and this is continued for six weeks. In almost every case the production of dandruff is checked completely after six weeks' treatment, and the hair, which may have been falling out rapidly before, begins to take firmer root. To be sure, many hairs which are on the point of falling when treatment is begun will fall anyway, and it may even seem for a time as if the treatment were increasing the hair-fall, on account of the mechanical dislodgment of such hairs, but this need never alarm one.

After six weeks of such treatment the shampoo may be taken less frequently.

Next to dandruff, perhaps, the most common cause of early loss of hair is heredity. In some families all of the male members, or all who resemble one particular ancestor, lose their hair early. Dark-haired families and races, as a rule, become bald earlier than those with light hair. At first thought it would seem as though nothing could be done to prevent premature baldness when heredity is the cause, but this is a mistake. Careful hygiene of the scalp will often counterbalance hereditary predisposition for a number of years, and even after the hair has actually begun to fall proper stimulation will, to a certain extent, and for a limited time, often restore to the hair its pristine thickness and strength. Any of the rubefacients may be prescribed for this purpose for daily use, such as croton oil, 1½ per cent; tincture of cantharides, 15 per cent; oil of cinnamon, 40 per cent; tincture of capsicum, 15 per cent; oil of mustard, 1 per cent; or any one of a dozen others. Tincture of capsicum is one of the best, and for a routine prescription the following has served well:

Resorcin	5 parts
Tincture capsicum	15 parts
Castor oil	10 parts
Alcohol	100 parts
Oil of roses, sufficient.	

It is to be recommended that the stimulant be changed from time to time, so as not to rely on any one to the exclusion of others. Jaborandi, oxygen gas, quinine, and other agents have enjoyed a great reputation as hair-producers for a time, and have then taken their proper position as aids, but not specifics, in restoring the hair.

It is well known that after many fevers, especially those accompanied by great depression, such as pneumonia, typhoid, puerperal, or scarlet fever, the hair is liable to fall out. This is brought about in a variety of ways: In scarlatina, the hair papilla shares in the general desquamation; in typhoid and the other fevers the baldness may be the result either of the excessive seborrhea, which often accompanies these diseases, or may be caused by the general lowering of nutrition of the body. Unless the hairfall be accompanied by considerable dandruff (in which case the above-mentioned treatment should be vigorously employed), the ordinary hygiene of the scalp will result in a restoration of the hair in most cases, but the employment of moderate local stimulation, with the use of good general tonics, will hasten this end. It seems unwise to cut the hair of women short in these cases, because the baldness is practically never complete, and a certain proportion of the hairs will retain firm root. These may be augmented by a switch made of the hair which has fallen out, until the new hair shall have grown long enough to do up well. In this way all of that oftentimes most annoying short-hair period is avoided.

For Falling Hair.—

I.—Hydrochloric acid   75 parts
     Alcohol......... 2,250 parts

The lotion is to be applied to the scalp every evening at bedtime.

II.—Tincture of cinchona   1 part
     Tincture of rosemary............ 1 part
     Tincture of jaborandi............ 1 part
     Castor oil.......... 2 parts
     Rum.............. 10 parts

Mix.

Jaborandi Scalp Waters for Increasing the Growth of Hair.—First prepare a jaborandi tincture from jaborandi leaves, 200 parts; spirit, 95 per cent, 700 parts; and water, 300 parts. After digesting for a week, squeeze out the leaves and filter the liquid. The hair wash is now prepared as follows:

I.—Jaborandi tincture, 1,000 parts: spirit, 95 per cent, 700 parts; water, 300 parts; glycerine, 150 parts; scent essence, 100 parts; color with sugar color.

II.—Jaborandi tincture, 1,000 parts; spirit, 95 per cent, 1,500 parts; quinine tannate, 4 parts; Peru balsam, 20 parts; essence heliotrope, 50 parts. Dissolve the quinine and the Peru balsam in the spirit and then add the jaborandi tincture and the heliotrope essence. Filter after a week. Rub into the scalp twice a week before retiring.

POMADES:

I.—Cinchona Pomade.—

Ox marrow........	100 drachms
Lard..............	70 drachms
Sweet almond oil....	17 drachms
Peru balsam........	1 drachm
Quinine sulphate...	1 drachm
Clover oil..........	2 drachms
Rose essence.......	25 drops

II.—Cantharides Pomade.—

Ox marrow........	300 drachms
White wax.........	30 drachms
Mace oil...........	1 drachm
Clove oil..........	1 drachm
Rose essence or geranium oil.......	25 drops
Tincture of cantharides.............	8 drachms

Pinaud Eau de Quinine.—The composition of this nostrum is not known. Dr. Tsheppe failed to find in it any constituent of cinchona bark. The absence of quinine from the mixture probably would not hurt it, as the "tonic" effect of quinine on the hair is generally regarded as a myth.

On the other hand, it has been stated that this preparation contains:

Quinine sulphate...	2 parts
Tincture of krameria	4 parts
Tincture of cantharides.............	2 parts
Spirit of lavender...	10 parts
Glycerine..........	15 parts
Alcohol............	100 parts

SHAMPOOS:

A Hair Shampoo is usually a tincture of odorless soft soap. It is mostly perfumed with lavender and colored with green aniline. Prepared the same as tr. sapon. virid. (U. S. P.), using an inexpensive soft soap, that is a good foam producer. Directions: Wet the hair well in warm water and rub in a few teaspoonfuls of the following formulas. No. I is considered the best:

	I	II	III	IV
		Parts used		
Cottonseed oil	—	24	26	14
Linseed oil	20	—	—	—
Malaga olive oil	20	—	—	—
Caustic potash	9½	8	6	3
Alcohol	5	4½	5	2
Water	30	26	34	16½

Warm the mixed oils on a large water bath, then the potash and water in another vessel, heating both to 158° F., and adding the latter hot solution to the hot oil while stirring briskly. Now add and thoroughly mix the alcohol. Stop stirring, keeping the heat at 158° F., until the mass becomes clear and a small quantity dissolves in boiling water without globules of oil separating. If stirred after the alcohol has been mixed the soap will be opaque. Set aside for a few days in a warm place before using to make liquid shampoo.

Liquid Shampoos.—

I.—Fluid extract of
    soap-bark ...... 10 parts
    Glycerine .... 5 parts
    Cologne water .... 10 parts
    Alcohol .... 20 parts
    Rose water .... 30 parts

II.—Soft soap .... 24 parts
    Potassium carbon-
        ate .... 5 parts
    Alcohol .... 48 parts
    Water enough to
        make .... 400 parts

Shampoo Pastes.—

I.—White castile soap,
    in shavings .... 2 ounces
    Ammonia water .. 2 fluidounces
    Bay rum, or co-
        logne water.... 1 fluidounce
    Glycerine .... 1 fluidounce
    Water .... 12 fluidounces

Dissolve the soap in the water by means of heat; when nearly cold stir in the other ingredients.

II.—Castile soap, white. 4 ounces
    Potassium carbon-
        ate .... 1 ounce
    Water .... 6 fluidounces
    Glycerine .... 2 fluidounces
    Oil of lavender
        flowers .... 5 drops
    Oil of bergamot ... 10 drops

To the water add the soap, in shavings, and the potassium carbonate, and heat on a water bath until thoroughly softened; add the glycerine and oils. If necessary to reduce to proper consistency, more water may be added.

Egg Shampoo.—

    Whites of .... 2 eggs
    Water .... 5 fluidounces
    Water of ammonia. 3 fluidounces
    Cologne water ..... ⅓ fluidounce
    Alcohol .... 4 fluidounces

Beat the egg whites to a froth, and add the other ingredients in the order in which they are named, with a thorough mixing after each addition.

Imitation Egg Shampoos.—Many of the egg shampoos are so called from their appearance. They usually contain no egg and are merely preparations of perfumed soft soap. Here are some formulas:

I.—White castile soap.... 4 ounces
    Powdered curd soap.. 2 ounces
    Potassium carbonate. 1 ounce
    Honey .... 1 ounce

Make a homogeneous paste by heating with water.

II.—Melt 3½ pounds of lard over a salt-water bath and run into it a lye formed by dissolving 8 ounces of caustic potassa in 1½ pints of water. Stir well until saponification is effected and perfume as desired.

HAIR REMOVERS:
    See Depilatories.

HAMBURG BITTERS:
    See Wines and Liquors.

HAMMER HARDENING:
    See Steel.

HAND CREAMS:
    See Cosmetics.

HANDS, TO REMOVE STAINS FROM THE:
    See Cleaning Preparations.

HARE-LIP OPERATION, ANTISEPTIC PASTE FOR:
    See Antiseptics.

HARNESS DRESSINGS AND PREPARATIONS:
    See Leather Dressings.

HARNESS WAX:
    See Waxes.

HAT-CLEANING COMPOUNDS:
    See Cleaning Compounds.

HAT WATERPROOFING:
    See Waterproofing.

HATS:

Dyeing Straw Hats.—The plan generally followed is that of coating the hats with a solution of varnish in which a suitable aniline dye has dissolved. The following preparations are in use:

I.—For dark varnishes prepare a basis consisting of orange shellac, 900 parts; sandarac, 225 parts; Manila copal, 225 parts; castor oil, 55 parts; and wood-spirit, 9,000 parts. To color, add to the foregoing amount alcohol-soluble, coal-tar dyes as follows: Black, 55 parts of soluble ivory-black (modified by blue or green). Olive-brown, 15 parts of brilliant-green, 55 parts of Bismarck brown R, 8 parts of spirit blue. Olive-green, 28 parts of brilliant-green, 28 parts of Bismarck-brown R. Walnut, 55 parts of Bismarck-brown R, 15 parts of nigrosin. Mahogany, 28 parts of Bismarck-brown R, which may be deepened by a little nigrosin.

II.—For light colors prepare a varnish as follows: Sandarac, 1,350 parts; elemi, 450 parts; rosin, 450 parts; castor oil, 110 parts; wood-spirit, 9,000 parts. For this amount use dyes as follows: Gold, 55 parts of chrysoidin, 55 parts of aniline-yellow. Light green, 55 parts of brilliant-green, 7 parts of aniline-yellow. Blue, 55 parts of spirit blue. Deep blue, 55 parts of spirit blue, 55 parts of indulin. Violet, 28 parts of methyl-violet, 3 B. Crimson, 55 parts of safranin. Chestnut, 55 parts of safranin, 15 parts of indulin.

III.—
Shellac	4 ounces
Sandarac	1 ounce
Gum thus	1 ounce
Methyl spirit	1 pint

In this dissolve aniline dyes of the requisite color, and apply. For white straw, white shellac must be used.

To Extract Shellac from Fur Hats.—Use the common solvents, as carbon bisulphide, benzine, wood alcohol, turpentine, and so forth, reclaiming the spirit and shellac by a suitable still.

HEADACHE REMEDIES:
See also Pain Killers.

Headache Cologne.—As a mitigant of headache, cologne water of the farina type is refreshing.

Oil of neroli	6 drachms
Oil of rosemary	3 drachms
Oil of bergamot	3 drachms
Oil of cedrat	7 drachms
Oil of orange peel	7 drachms
Deodorized alcohol	1 gallon

To secure a satisfactory product from the foregoing formula it is necessary to look carefully to the quality of the oils. Oil of cedrat is prone to change, and oil of orange peel, if exposed to the atmosphere for a short time, becomes worthless, and will spoil the other materials.

A delightful combination of the acetic odor with that of cologne water may be had by adding to a pint of the foregoing, 2 drachms of glacial acetic acid. The odor so produced may be more grateful to some invalids than the neroli and lemon bouquet.

Still another striking variation of the cologne odor, suitable for the use indicated, may be made by adding to a pint of cologne water an ounce of ammoniated alcohol.

Liquid Headache Remedies.—

Acetanilid	60 grains
Alcohol	4 fluidrachms
Ammonium carbonate	30 grains
Water	2 fluidrachms
Simple elixir to make	2 fluidounces

Dissolve the acetanilid in the alcohol, the ammonium carbonate in the water, mix each solution with a portion of the simple elixir, and mix the whole together.

HEAT-INDICATING PAINT:
See Paint.

HEAT INSULATION:
See Insulation.

HEAT, PRICKLY:
See Household Formulas.

HEAT-RESISTANT LACQUERS:
See Lacquers.

HEAVES:
See Veterinary Formulas.

HEDGE MUSTARD.

Hedge mustard (erysimum) was at one time a popular remedy in France for hoarseness, and is still used in country districts, but is not often prescribed.

Liquid ammonia	10 drops
Syrup of erysimum	1½ ounces
Infusion of lime flowers	3 ounces

To be taken at one dose.

HERBARIUM SPECIMENS, MOUNTING.

A matter of first importance, after drying the herbarium specimens, is to poison them, to prevent the attacks of insects. This is done by brushing them over on both sides, using a camel's-hair pencil, with a solution of 2 grains of corrosive sublimate to an ounce of methylated spirit. In tropical climates the solution is generally used of twice this strength. There are several methods of mounting them. Leaves with a waxy surface and coriaceous texture are best stitched through the middle after they have been fastened on with an adhesive mixture. Twigs of leguminous trees will often throw off their leaflets in drying. This may, in some measure, be prevented by dipping them in boiling water before drying, or if the leaves are not very rigid, by using strong pressure at first, without the use of hot water. If the specimens have to be frequently handled, the most satisfactory preparation is Lepage's fish glue, but a mixture of glue and paste, with carbolic acid added, is used in some large herbaria. The disadvantage of using glue, gum, or paste is that it is necessary to have some of the leaves turned over so as to show the under surface of the leaf, and some of the flowers and seeds placed loose in envelopes on the same sheet for purposes of comparison or microscopic examination. Another plan is to use narrow slips of gummed stiff but thin paper, such as very thin parchment paper. These strips are either gummed over the stems, etc., and pinched in round the stem with forceps, or passed through slits made in the sheet and fastened at the back. If the specimens are mounted on cards and protected in glass frames, stitching in the principal parts with gray thread produces a very satisfactory appearance.

Hectograph Pads and Inks

The hectograph is a gelatin pad used for duplicating letters, etc., by transfer. The pad should have a tough elastic consistency, similar to that of a printer's roller. The letter or sketch to be duplicated is written or traced on a sheet of heavy paper with an aniline ink (which has great tinctorial qualities). When dry this is laid, inked side down, on the pad and subjected to moderate and uniform pressure for a few minutes. It may then be removed, when a copy of the original will be found on the pad which has absorbed a large quantity of the ink. The blank sheets are laid one by one on the pad, subjected to moderate pressure over the whole surface with a wooden or rubber roller, or with the hand, and lifted off by taking hold of the corners and stripping them gently with an even movement. If this is done too quickly the composition may be torn. Each succeeding copy thus made will

be a little fainter than its predecessor. From 40 to 60 legible copies may be made. When the operation is finished the surface of the pad should be gone over gently with a wet sponge and the remaining ink soaked out. The superfluous moisture is then carefully wiped off, when the pad will be ready for another operation.

The pad or hectograph is essentially a mixture of glue (gelatin) and glycerine. This mixture has the property of remaining soft yet firm for a long time and of absorbing and holding certain coloring matters in such a way as to give them up slowly or in layers, so to speak, on pressure.

Such a pad may be made by melting together 1 part of glue, 2 parts of water and 4 parts of glycerine (all by weight, of course), evaporating some of the water and tempering the mixture with more glue or glycerine if the season or climate require. The mass when of proper consistency, which can be ascertained by cooling a small portion, is poured into a shallow pan and allowed to set. Clean glue must be used or the mixture strained; and air bubbles should be removed by skimming the surface with a piece of card-board or similar appliance.

Variations of this formula have been proposed, some of which are appended:

I.—Glycerine	12	ounces
Gelatin	2	ounces
Water	7½	ounces
Sugar	2	ounces
II.—Water	10	ounces
Dextrin	1½	ounces
Sugar	2	ounces
Gelatin	15	ounces
Glycerine	15	ounces
Zinc oxide	1½	ounces
III.—Gelatin	10	ounces
Water	40	ounces
Glycerine	120	ounces
Barium sulphate	8	ounces

The Tokacs patent composition, besides the usual ingredients, such as gelatin, glycerine, sugar, and gum, contains soap, and can therefore be washed off much easier for new use. The smoothness of the surface is also increased, without showing more sticking capacity with the first impressions.

Hectograph Inks (see also Inks).—The writing to be copied by means of the hectograph is done on good paper with an aniline ink. Formulas for suitable ones are appended. It is said that more copies can be obtained from writing with the purple ink than with other kinds:

Purple.—

I.—Methyl violet........ 2 parts
Alcohol............. 2 parts
Sugar.............. 1 part
Glycerine........... 4 parts
Water.............. 24 parts

Dissolve the violet in the alcohol mixed with the glycerine; dissolve the sugar in the water; mix both solutions.

II.—A good purple hectograph ink is made as follows: Dissolve 1 part methyl violet in 8 parts of water and add 1 part of glycerine. Gently warm the solution for an hour, and add, when cool, ¼ part alcohol. Or take methyl violet, 1 part; water, 7 parts; and glycerine, 2 parts.

Black.—

Methyl violet........ 10 parts
Nigrosin............ 20 parts
Glycerine........... 30 parts
Gum arabic......... 5 parts
Alcohol............. 60 parts

Blue.—

Resorcin blue M..... 10 parts
Dilute acetic acid.... 1 part
Water.............. 85 parts
Glycerine........... 4 parts
Alcohol............. 10 parts

Dissolve by heat.

Red.—

Fuchsin............ 10 parts
Alcohol............ 10 parts
Glycerine.......... 10 parts
Water.............. 50 parts

Green.—

Aniline green, water
soluble........... 15 parts
Glycerine.......... 10 parts
Water.............. 50 parts
Alcohol............ 10 parts

Repairing Hectographs.—Instead of remelting the hectograph composition, which is not always successful, it is recommended to pour alcohol over the surface of the cleaned mass and to light it. After solidifying, the surface will be again ready for use.

HEMORRHOIDS:
See Piles.

HERB VINEGAR:
See Vinegar.

HIDES:
See Leather.

HIDE BOUND:
See Veterinary Formulas.

HIDE-CLEANING PROCESSES:
See Cleaning Preparations and Methods.

HOARHOUND CANDY:
See Confectionery.

HOARSENESS, CREAM BON-BONS FOR:
See Confectionery.

HOARSENESS, REMEDY FOR:
See Cough and Cold Mixtures and Turpentine.

HONEY:

Honey Clarifier.—For 3,000 parts of fresh honey, take 875 parts of water, 150 parts of washed, dried, and pulverized charcoal, 70 parts of powdered chalk, and the whites of 3 eggs beaten in 90 parts of water. Put the honey and the chalk in a vessel capable of containing ⅓ more than the mixture and boil for 3 minutes; then introduce the charcoal and stir up the whole. Add the whites of the eggs while continuing to stir, and boil again for 3 minutes. Take from the fire, and after allowing the liquid to cool for a quarter of an hour, filter, and to secure a perfectly clear liquid refilter on flannel.

Detecting Dyed Honey.—For the detection of artificial yellow dyestuff in honey, treat the aqueous yellow solution with hydrochloric acid, as well as with ammonia; also extract the dyestuff from the acid or ammoniacal solution by solvents, such as alcohol or ether, or conduct the Arata wool test in the following manner: Dissolve 10 parts of honey in 50 parts of water, mix with 10 parts of a 10 per cent potassium-bisulphate solution and boil the woolen thread in this liquid for 10 minutes.

HONEY WINE:
See Mead.

HONING:
See Whetstones.

HOOF SORES:
See Veterinary Formulas.

HOP BITTER BEER:
See Beverages.

HOP SYRUP:
See Essences and Extracts.

HORN:

Artificial Horn.—To prepare artificial horn from compounds of nitro-cellulose and casein, by hardening them and removing their odor of camphor, the compounds are steeped in formaldehyde from several hours to as many days, according to the thickness of the object treated. When the formaldehyde has penetrated through the mass and dissolved the camphor, the object is taken out of the liquid and dried. Both the camphor extracted and the formaldehyde used can be recovered by distillation, and used over again, thus cheapening the operation.

Dehorners or Horn Destroyers.—The following are recommended by the Board of Agriculture of Great Britain:

Clip the hair from the top of the horn when the calf is from 2 to 5 days old. Slightly moisten the end of a stick of caustic potash with water or saliva (or moisten the top of the horn bud) and rub the tip of each horn firmly with the potash for about a quarter of a minute, or until a slight impression has been made on the center of the horn. The horns should be treated in this way from 2 to 4 times at intervals of 5 minutes. If, during the interval of 5 minutes after one or more applications, a little blood appears in the center of the horn, it will then only be necessary to give another very slight rubbing with the potash.

The following directions should be carefully observed: The operation is best performed when the calf is under 5 days old, and should not be attempted after the ninth day. When not in use the caustic potash should be kept in a stoppered glass bottle in a dry place, as it rapidly deteriorates when exposed to the air. One man should hold the calf while an assistant uses the caustic. Roll a piece of tin foil or brown paper round the end of the stick of caustic potash, which is held by the fingers, so as not to injure the hand of the operator. Do not moisten the stick too much, or the caustic may spread to the skin around the horn and destroy the flesh. For the same reason keep the calf from getting wet for some days after the operation. Be careful to rub on the center of the horn and not around the side of it.

Staining Horns.—A brown stain is given to horns by covering them first with an aqueous solution of potassium ferrocyanide, drying them, and then treating with a hot dilute solution of copper sulphate. A black stain can be produced in the following manner:

After having finely sandpapered the horns, dissolve 50 to 60 grains of nitrate of silver in 1 ounce of distilled water. It will be colorless. Dip a small brush in, and paint the horns where they are to be black. When dry, put them where the sun can shine on them, and you will find that they will turn jet black, and may then be polished.

To Soften Horn.—Lay the horn for 10 days in a solution of water, 1 part; nitric acid, 3 parts; wood vinegar, 2 parts; tannin, 5 parts; tartar, 2 parts; and zinc vitriol, 2.5 parts.

HORN BLEACHES:
See Bone and Ivory.

HORN, UNITING GLASS WITH:
See Adhesives.

HORSES, THE TREATMENT OF THEIR DISEASES:
See Veterinary Formulas.

Household Formulas

How to Lay Galvanized Iron Roofing.—The use of galvanized iron for general roofing work has increased greatly during the past few years. It has many features which commend it as a roofing material, but difficulties have been experienced by beginners as to the proper method of applying it to the roof. The weight of material used is rather heavy to permit of double seaming, but a method has been evolved that is satisfactory. Galvanized iron roofing can be put on at low cost, so as to be water-tight and free from buckling at the joints. The method does away with double seaming, and is considered more suitable than the latter for roofing purposes wherever it can be laid on a roof steeper than 1 to 12.

Galvanized iron of No. 28 and heavier gauges is used, the sheets being lap-seamed and soldered together in strips in the shop the proper length to apply to the roof. After the sheets are fastened together a 1¼-inch edge is turned up the entire length of one side of the sheet, as indicated in Fig. 1. This operation is

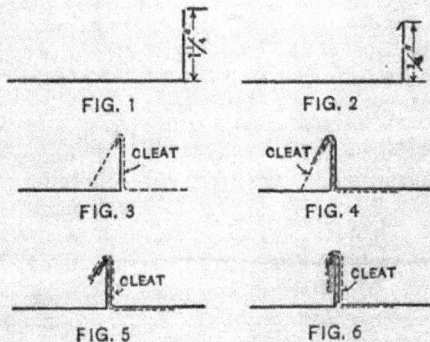

FIG. 1   FIG. 2

FIG. 3   FIG. 4

FIG. 5   FIG. 6

done with tongs having gauge pins set at the proper point. The second operation consists in turning a strip $\frac{1}{4}$ inch wide toward the sheet, as shown in Fig. 2. This sheet is then laid on the roof, and a cleat about 8 inches long and 1 inch wide, made of galvanized iron, is nailed to the roof close to the sheet and bent over it, as shown in Fig. 3.

A second sheet having $1\frac{1}{2}$ inches turned up is now brought against the first sheet and bent over both sheet and cleat, as shown in Fig. 4. The cleat is then bent backward over the second sheet and cut off close to the roof, as in Fig. 5, after which the seams are drawn together by double seaming tools, as the occasion demands, and slightly hammered with a wooden mallet. The finished seam is shown in Fig. 6. It will be seen that the second sheet of galvanized iron, cut $\frac{1}{4}$ inch longer than the first, laps over the former, making a sort of bead which prevents water from driving in. Cleats hold both sheets firmly to the roof and are nailed about 12 inches apart. Roofs of this character, when laid with No. 28 gauge iron, cost very little more than the cheaper grades of tin, and do not have to be painted.

Applications for Prickly Heat.—Many applications for this extremely annoying form of urticaria have been suggested and their efficacy strongly urged by the various correspondents of the medical press who propose them, but none of them seem to be generally efficacious. Thus, sodium bicarbonate in strong, aqueous solution, has long been a domestic remedy in general use, but it fails probably as often as it succeeds. A weak solution of copper sulphate has also been highly extolled, only to disappoint a very large proportion of those who resort to it. And so we might go on citing remedies which may sometimes give relief, but fail in the large proportion of cases. In this trouble, as in almost every other, the idiosyncrasies of the patient play a great part in the effects produced by any remedy. It is caused, primarily by congestion of the capillary vessels of the skin, and anything that tends to relieve this congestion will give relief, at least temporarily. Among the newer suggestions are the following:

Alcohol.............. 333 parts
Ether................ 333 parts
Chloroform......... 333 parts
Menthol .... 1 part

Mix. Directions: Apply occasionally with a sponge.

Among those things which at least assist one in bearing the affliction is frequent change of underwear. The undergarments worn during the day should never be worn at night. Scratching or rubbing should be avoided where possible. Avoid stimulating food and drinks, especially alcohol, and by all means keep the bowels in a soluble condition.

Cleaning and Polishing Linoleum.—Wash the linoleum with a mixture of equal parts of milk and water, wipe dry, and rub in the following mixture by means of a cloth rag: Yellow wax, 5 parts; turpentine oil, 11 parts; varnish, 5 parts. As a glazing agent, a solution of a little yellow wax in turpentine oil is also recommended. Other polishing agents are:

I.—Palm oil, 1 part; paraffine, 18; kerosene, 4.

II.—Yellow wax, 1 part; carnauba wax, 2; turpentine oil, 10; benzine, 5.

Lavatory Deodorant.—

Sodium bicarbonate.. 5 ounces
Alum.............. $5\frac{1}{2}$ ounces
Potassium bromide... 4 ounces
Hydrochloric acid enough.
Water enough to make 4 pints.

To 3 parts of boiling water add the alum and then the bicarbonate. Introduce enough hydrochloric acid to dissolve the precipitate of aluminum hydrate which forms and then add the potassium bromide. Add enough water to bring the measure of the finished product up to 4 pints.

Removal of Odors from Wooden Boxes, Chests, Drawers, etc.—This is done by varnishing them with a solution of shellac, after the following manner: Make a solution of shellac, 1,000 parts; alcohol, 90 per cent to 95 per cent, 1,000 parts; boric acid, 50 parts; castor oil, 50 parts. The shellac is first dissolved in the alcohol and the acid and oil added afterwards. For the first coating use 1 part of the solution cut with from 1 to 2 parts of alcohol, according to the porosity of the wood—the more porous the less necessity for cutting. When the first coat is absorbed and dried in, repeat the application, if the wood is very porous, with the diluted shellac, but if of hard, dense wood, the final coating may be now put on, using the solution without addition of alcohol. If desired, the solution may be colored with any of the alcohol soluble aniline colors. The shellac solution, by the way, may be applied to the outside of chests, etc., and finished off after the fashion of "French polish."

When used this way, a prior application of 2 coats of linseed oil is advisable.

Stencil Marking Ink that will Wash Out.—Triturate together 1 part of fine soot and 2 parts of Prussian blue, with a little glycerine; then add 3 parts of gum arabic and enough glycerine to form a thin paste.

Washing Fluid.—Take 1 pound sal soda, ½ pound good stone lime, and 5 quarts of water; boil a short time, let it settle, and pour off the clear fluid into a stone jug, and cork for use; soak the white clothes overnight in simple water, wring out and soap wristbands, collars, and dirty or stained places. Have the boiler half filled with water just beginning to boil, then put in 1 common teacupful of fluid, stir and put in your clothes, and boil for half an hour, then rub lightly through one suds only, and all is complete.

Starch Luster.—A portion of stearine, the size of an old-fashioned cent, added to starch, ½ pound, and boiled with it for 2 or 3 minutes, will add greatly to the beauty of linen, to which it may be applied.

To Make Loose Nails in Walls Rigid.—As soon as a nail driven in the wall becomes loose and the plastering begins to break, it can be made solid and firm by the following process: Saturate a bit of wadding with thick dextrin or glue; wrap as much of it around the nail as possible and reinsert the latter in the hole, pressing it home as strongly as possible. Remove the excess of glue or dextrin, wiping it cleanly off with a rag dipped in clean water; then let dry. The nail will then be firmly fastened in place. If the loose plastering be touched with the glue and replaced, it will adhere and remain firm.

How to Keep Lamp Burners in Order.—In the combustion of coal oil a carbonaceous residue is left, which attaches itself very firmly to the metal along the edge of the burner next the flame. This is especially true of round burners, where the heat of the flame is more intense than in flat ones, and the deposit of carbon, where not frequently removed, soon gets sufficiently heavy to interfere seriously with the movement of the wick up or down. The deposit may be scraped off with a knife blade, but a much more satisfactory process of getting rid of it is as follows: Dissolve sodium carbonate, 1 part, in 5 or 6 parts of water, and in this boil the burner for 5 minutes or so. When taken out the burner will look like a new one, and acts like one, provided that the apparatus for raising and lowering the wick has not previously been bent and twisted by attempting to force the wick past rough deposits.

To Remove the Odor from Pasteboard.—Draw the pasteboard through a 3 per cent solution of viscose in water. The pasteboard must be calendered after drying.

To Remove Woody Odor—To get rid of that frequently disagreeable smell in old chests, drawers, etc., paint the surface over with the following mixture:

Acetic ether	100 parts
Formaldehyde	6 parts
Acid, carbolic	4 parts
Tincture of eucalyptus leaves	60 parts

Mix. After applying the mixture expose the article to the open air in the sunlight.

To Keep Flies Out of a House.—Never allow a speck of food to remain uncovered in dining room or pantry any length of time after meals. Never leave remnants of food exposed that you intend for cat or hens. Feed at once or cover their food up a distance from the house. Let nothing decay near the house. Keep your dining room and pantry windows open a few inches most of the time. Darken your room and pantry when not in use. If there should be any flies they will go to the window when the room is darkened, where they are easily caught, killed, or brushed out.

An Easy Way to Wash a Heavy Comfortable.—Examine the comfortable, and if you find soiled spots soap them and scrub with a small brush. Hang the comfortable on a strong line and turn the hose on. When one side is washed turn and wash the other. The water forces its way through cotton and covering, making the comfortable as light and fluffy as when new. Squeeze the corners and ends as dry as possible.

Preservation of Carpets.—Lay sheets of brown paper under the carpet. This gives a soft feeling to the foot, and by diminishing the wear adds longer life to the carpet; at the same time it tends to keep away the air and renders the apartments warm.

To Do Away with Wiping Dishes.—Make a rack by putting a shelf over the kitchen sink, slanting it so that the water will drain off into the sink. Put a lattice railing about 6 inches high at the front and ends of the shelf so that dishes can be set against it on their edges without falling out. Have 2 pans of hot water. Wash the dishes in one and rinse them in the other. Set them on edge in the rack and leave until dry.

A Convenient Table.—

Ten common-sized eggs weigh 1 pound.

Soft butter, the size of an egg, weighs 1 ounce.

One pint of coffee and of sugar weighs 12 ounces.

One quart of sifted flour (well heaped) weighs 1 pound.

One pint of best brown sugar weighs 12 ounces.

How to Make a Cellar Waterproof.— The old wall surface should be roughened and perfectly cleaned before plastering is commenced. It may be advisable to put the first coat on not thicker than ¼ inch, and after this has set it may be cut and roughened by a pointing trowel. Then apply a second ¼-inch coat and finish this to an even and smooth surface. Proportion of plaster: One-half part slaked lime, 1 part Portland cement, part fine, sharp sand, to be mixed well and applied instantly.

Removing Old Wall Paper.—Some paper hangers remove old paper from walls by first dampening it with water in which a little baking soda has been dissolved, the surface being then gone over with a "scraper" or other tool. However, the principle object of any method is to soften the old paste. This may be readily accomplished by first wetting a section of the old paper with cold or tepid water, using a brush, repeating the wetting until the paper and paste are soaked through, when the paper may easily be pulled off, or, if too tender, may be scraped with any instrument of a chisel form shoved between the paper and the wall. The wall should then be washed with clean water, this operation being materially assisted by wetting the wall ahead of the washing.

Stained Ceilings.—Take unslaked white lime, dilute with alcohol, and paint the spots with it. When the spots are dry— which will be soon, as the alcohol evaporates and the lime forms a sort of insulating layer—one can proceed painting with size color, and the spots will not show through again.

To Overcome Odors in Freshly Papered Rooms.—After the windows and doors of such rooms have been closed, bring in red-hot coal and strew on this several handfuls of juniper berries. About 12 hours later open all windows and doors, so as to admit fresh air, and it will be found that the bad smell has entirely disappeared.

Treatment of Damp Walls.—I.—A good and simple remedy to obviate this evil is caoutchouc glue, which is prepared from rubber hose. The walls to be laid dry are first to be thoroughly cleaned by brushing and rubbing off; then the caoutchouc size, which has been previously made liquid by heating, is applied with a broad brush in a uniform layer—about 8 to 12 inches higher than the wall appears damp — and finally paper is pasted over the glue when the latter is still sticky. The paper will at once adhere very firmly. Or else, apply the liquefied glue in a uniform layer upon paper (wall paper, caoutchouc paper, etc.). Upon this, size paint may be applied, or it may be covered with wall paper or plaster.

If the caoutchouc size is put on with the necessary care—i. e., if all damp spots are covered with it—the wall is laid dry for the future, and no peeling off of the paint or the wall paper needs to be apprehended. In cellars, protection from dampness can be had in a like manner, as the caoutchouc glue adheres equally well to all surfaces, whether stone, glass, metal, or wood.

II.—The walls must be well cleaned before painting. If the plaster should be worn and permeated with saltpeter in places it should be renewed and smoothed. These clean surfaces are coated twice with a water-glass solution, 1.1, using a brush and allowed to dry well. Then they are painted 3 times with the following mixture: Dissolve 100 parts, by weight, of mastic in 10 parts of absolute alcohol; pour 1,000 parts of water over 200 parts of isinglass; allow to soak for 6 hours; heat to solution and add 100 parts of alcohol (50 per cent). Into this mixture pour a hot solution of 50 parts of ammonia in 250 parts of alcohol (50 per cent), stir well, and subsequently add the mastic solution and stand aside warm, stirring diligently. After 5 minutes take away from the fire and painting may be commenced. Before a fresh application, however, the solution should be removed.

When this coating has dried completely it is covered with oil or varnish paint, preferably the latter. In the same manner the exudation of so-called saltpeter in fresh masonry or on the exterior of façades, etc., may be prevented, size paint or lime paint being employed instead of the oil-varnish paint. New walls which are to be painted will give off no more saltpeter after 2 or 3 applications of the isinglass solution, so that the colors of the wall paper will not be injured either. Stains caused by smoke, soot, etc., on ceilings of rooms, kitchens, or corridors which are difficult to cover up with size paint, may also be completely isolated by applying the warm isinglass solution 2 or 3 times. The size paint is, of course, put on only after complete drying of the ceilings.

To Protect Papered Walls from Vermin.—It is not infrequent that when the wall paper becomes defective or loose in papered rooms, vermin, bed bugs, ants, etc., will breed behind it. In order to prevent this evil a little colocynth powder should be added to the paste used for hanging the paper, in the proportion of 50 or 60 parts for 3,000 parts.

Care of Refrigerators.—See that the sides or walls of all refrigerators are occasionally scoured with soap, or soap and slaked lime.

Dust Preventers.—Against the beneficial effects to be observed in the use of most preparations we must place the following bad effects: The great smoothness and slipperiness of the boards during the first few days after every application of the dressing, which forbids the use of the latter on steps, floors of gymnasia, dancing floors, etc. The fact that the oil or grease penetrates the soles of the boots or shoes, the hems of ladies' dresses, and things accidentally falling to the floor are soiled and spotted. Besides these there is, especially during the first few days after application, the dirty dark coloration which the boards take on after protracted use of the oils. Finally, there is the considerable cost of any process, especially for smaller rooms and apartments. In schoolrooms and railroad waiting rooms and other places much frequented by children and others wearing shoes set with iron, the boards soon become smooth from wear, and for such places the process is not suited.

According to other sources of information, these evil tendencies of the application vanish altogether, or are reduced to a minimum, if (1) entirely fresh, or at least, not rancid oils be used; (2) if, after each oiling, a few days be allowed to elapse before using the chamber or hall, and finally (3), if resort is not had to

costly foreign special preparations, but German goods, procurable at wholesale in any quantity, and at very low figures.

The last advice (to use low-priced preparations) seems sensible since according to recent experiments, none of the oils experimented upon possess any especial advantages over the others.

An overwhelming majority of the laboratories for examination have given a verdict in favor of oil as a dust-suppressing application for floors, and have expressed a desire to see it in universal use. The following is a suggestion put forth for the use of various preparations:

This dust-absorbing agent has for its object to take up the dust in sweeping floors, etc., and to prevent its development. The production is as follows: Mix in an intimate manner 12 parts, by weight, of mineral sperm oil with 88 parts, by weight, of Roman or Portland cement, adding a few drops of mirbane oil. Upon stirring a uniform paste forms at first, which then passes into a greasy, sandy mass. This mass is sprinkled upon the surface to be swept and cleaned of dust, next going over it with a broom or similar object in the customary manner, at which operation the dust will mix with the mass. The preparation can be used repeatedly.

HUNYADI WATER:
See Water.

HYDROCHINON DEVELOPER:
See Photography.

HYDROGEN, AMALGAMS AS A SOURCE OF NASCENT:
See Amalgams.

HYDROGEN PEROXIDE AS A PRESERVATIVE:
See Preserving.

HYDROMETER AND ITS USE.

Fill the tall cylinder or test glass with the spirit to be tested and see that it is of the proper temperature (60° F.). Should the thermometer indicate a higher temperature wrap the cylinder in cloths which have been dipped in cold water until the temperature falls to the required degree. If too low a temperature is indicated, reverse the process, using warm instead of cold applications. When 60° is reached note the specific gravity on the floating hydrometer. Have the cylinder filled to the top and look across the top of the liquid at the mark on the hydrometer. This is to preclude an incorrect reading by possible refraction in the glass cylinder.

HYGROMETERS AND HYGROSCOPES:

Paper Hygrometers.—Paper hygrometers are made by saturating white blotting paper with the following liquid and then hanging up to dry:

Cobalt chloride......	1	ounce
Sodium chloride.....	½	ounce
Calcium chloride.....	75	grains
Acacia..............	¼	ounce
Water..............	3	ounces

The amount of moisture in the atmosphere is roughly indicated by the changing color of the papers, as follows:

Rose red........	rain
Pale red........	very moist
Bluish red.......	moist
Lavender blue....	nearly dry
Blue............	very dry

Colored Hygroscopes.—These instruments are often composed of a flower or a figure, of light muslin or paper, immersed in one of the following solutions:

I.—Cobalt chloride....	1	part
Gelatin..........	10	parts
Water...........	100	parts

The normal coloring is pink; this color changes into violet in medium humid weather and into blue in very dry weather.

II.—Cupric chloride...	1	part
Gelatin..........	10	parts
Water...........	100	parts

The color is yellow in dry weather.

III.—Cobalt chloride....	1	part
Gelatin..........	20	parts
Nickel oxide......	75	parts
Cupric chloride....	25	parts
Water...........	200	parts

The color is green in dry weather.

HYOSCYAMUS, ANTIDOTE TO:
See Atropine.

ICE:
See also Refrigeration.

Measuring the Weight of Ice. — A close estimate of the weight of ice can be reached by multiplying together the length, breadth, and thickness of the block in inches, and dividing the product by 30. This will be very closely the weight in pounds. Thus, if a block is 10 x 10 x 9, the product is 900, and this divided by 30 gives 30 pounds as correct weight. A block 10 x 10 x 6 weighs 20 pounds. This simple method can be easily applied, and it may serve to remove unjust suspicions, or to detect short weight.

To Keep Ice in Small Quantities.—To keep ice from melting, attention is called to an old preserving method. The ice is cracked with a hammer between 2 layers of a strong cloth. Tie over a common unglazed flower-pot, holding about 2 to 4 quarts and placed upon a porcelain dish, a piece of white flannel in such a manner that it is turned down funnel-like into the interior of the pot without touching the bottom. Placed in this flannel funnel the cracked ice keeps for days.

ICE FLOWERS.

Make a 2 per cent solution of the best clear gelatin in distilled water, filter, and flood the filtrate over any surface which it is desired to ornament. Drain off slightly, and if the weather is sufficiently cold, put the plate, as nearly level as possible, out into the cold air to freeze. In freezing, water is abstracted from the colloidal portion, which latter then assumes an efflorescent form, little flowers, with exuberant, graceful curves of crystals, showing up as foliage, from all over the surface. To preserve in permanent form all that is necessary is to flood them with absolute alcohol. This treatment removes the ice, thus leaving a lasting framework of gelatin which may be preserved indefinitely. In order to do this, as soon as the gelatin has become quite dry it should be either varnished, flowed with an alcoholic solution of clear shellac, or the gelatin may be rendered insoluble by contact, for a few moments, with a solution of potassium bichromate, and subsequent exposure to sunlight.

IMOGEN DEVELOPER:
See Photography.

INCENSE:
See Fumigants.

INCRUSTATION, PREVENTION OF:
See Boiler Compounds.

INDIGO:
See Dyes.

INFANT FOODS:
See Foods.

INFLUENZA IN CATTLE:
See Veterinary Formulas.

INK ERADICATORS:

See Cleaning Preparations and Methods.

IGNITING COMPOSITION.

Eight parts of powdered manganese, 10 parts of amorphous phosphorus, and 5 parts of glue. The glue is soaked in water, dissolved in the heat, and the manganese and the phosphorus stirred in, so that a thinly liquid paste results, which is applied by means of a brush. Allow to dry well. This, being free from sulphur, can be applied on match-boxes.

Inks

BLUEPRINT INKS.

I.—For red-writing fluids for blueprints, take a piece of common washing soda the size of an ordinary bean, and dissolve it in 4 tablespoonfuls of ordinary red-writing ink, to make a red fluid. To keep it from spreading too much, use a fine pen to apply it with, and write fast so as not to allow too much of the fluid to get on the paper, for it will continue eating until it is dry.

II.—For red and white solutions for writing on blueprints, dissolve a crystal of oxalate of potash about the size of a pea in an ink-bottle full of water. This will give white lines on blueprints; other potash solutions are yellowish. If this shows a tendency to run, owing to too great strength, add more water and thicken slightly with mucilage. Mix this with red or any other colored ink about half and half, and writing may be done on the blueprints in colors corresponding to the inks used.

III.—Add to a small bottle of water enough washing soda to make a clear white line, then add enough gum arabic to it to prevent spreading and making ragged lines. To make red lines dip the pen in red ink and then add a little of the solution by means of the quill.

IV.—For white ink, grind zinc oxide fine on marble and incorporate with it a mucilage made with gum tragacanth. Thin a little for use. Add a little oil of cloves to prevent mold, and shake from time to time.

V.—A fluid which is as good as any for writing white on blueprints is made of equal parts of sal soda and water.

VI.—Mix equal parts of borax and water.

Both these fluids, V and VI, must be used with a fine-pointed pen; a pen with a blunt point will not work well.

DRAWING INKS:

Blue Ruling Ink.—Good vitriol, 4 ounces; indigo, 1 ounce. Pulverize the indigo, add it to the vitriol, and let it stand exposed to the air for 6 days, or until dissolved; then fill the pots with chalk, add fresh gall, $\frac{1}{2}$ gill, boiling it before use.

Black Ruling Ink.—Take good black ink, and add gall as for blue. Do not cork it, as this prevents it from turning black.

Carbon Ink.—Dissolve real India ink in common black ink, or add a small quantity of lampblack previously heated to redness, and ground perfectly smooth, with a small portion of the ink.

Carmine.—The ordinary solution of carmine in ammonia water, after a short time in contact with steel, becomes blackish red, but an ink may be made that will retain its brilliant carmine color to the last by the following process, given by Dingler: Triturate 1 part of pure carmine with 15 parts of acetate of ammonia solution, with an equal quantity of distilled water in a porcelain mortar, and allow the whole to stand for some time. In this way, a portion of the alumina, which is combined with the carmine dye, is taken up by the acetic acid of the ammonia salt, and separates as a precipitate, while the pure pigment of the cochineal remains dissolved in the half-saturated ammonia. It is now filtered and a few drops of pure white sugar syrup added to thicken it. A solution of gum arabic cannot be used to thicken it, since the ink still contains some acetic acid, which would coagulate the bassorine, one of the constituents of the gum.

Liquid Indelible Drawing Ink.—Dissolve, by boiling, 2 parts of blond (golden yellow) shellac in 1.6 parts, by weight, of sal ammoniac, 16°, with 10 parts, by weight, of distilled water, and filter the solution through a woolen cloth. Now dissolve or grind 0.5 parts, by weight, of shellac solution with 0.01 part, by weight, of carbon black. Also dissolve .03 parts of nigrosin in 0.4 parts of distilled water and pour both solutions together. The mixture is allowed to settle for 2 days and the ready ink is drawn off from the sediment.

GLASS, CELLULOID, AND METAL INKS:

See also Etching.

Most inks for glass will also write on celluloid and the metals. The following

I and II are the most widely known recipes:

I.—In 500 parts of water dissolve 36 parts of sodium fluoride and 7 parts of sodium sulphate. In another vessel dissolve in the same amount of water 14 parts of zinc chloride and to the solution add 56 parts of concentrated hydrochloric acid. To use, mix equal volumes of the two solutions and add a little India ink; or, in the absence of this, rub up a little lampblack with it. It is scarcely necessary to say that the mixture should not be put in glass containers, unless they are well coated internally with paraffine, wax, gutta-percha, or some similar material. To avoid the inconvenience of keeping the solutions in separate bottles, mix them and preserve in a rubber bottle. A quill pen is best to use in writing with this preparation, but metallic pens may be used, if quite clean and new.

II.—In 150 parts of alcohol dissolve 20 parts of rosin, and add to this, drop by drop, stirring continuously, a solution of 35 parts of borax in 250 parts of water. This being accomplished, dissolve in the solution sufficient methylene blue to give it the desired tint.

Ink for Writing on Glazed Cardboard. —The following are especially recommended for use on celluloid:

I.—Dissolve 4 drachms of brown shellac in 4 ounces of alcohol. Dissolve 7 drachms of borax in 6 ounces of distilled water. Pour the first solution slowly into the second and carefully mix them, after which add 12 grains of aniline dye of the desired color. Violet, blue, green, red, yellow, orange, or black aniline dyes can be used.

Such inks may be used for writing on bottles, and the glass may be cleaned with water without the inscription being impaired.

II.—Ferric chloride.... 10 parts
    Tannin.......... 15 parts
    Acetone......... 100 parts

Dissolve the ferric chloride in a portion of the acetone and the tannin in the residue, and mix the solutions.

III.—Dissolve a tar dyestuff of the desired color in anhydrous acetic acid.

Indelible Inks for Glass or Metal.— Schobel recommends the following inks for marking articles of glass, glass slips for microscopy, reagent flasks, etc., in black:

I.—Sodium silicate.....1 to 2 parts
    Liquid India ink.... 1 part

For white:

II.—Sodium water glass 3 to 4 parts
    Chinese white..... 1 part

Instead of Chinese white, a sufficient amount of the so-called permanent white (barium sulphate) may be used. The containers for these inks should be kept air-tight. The writing in either case is not attacked by any reagent used in microscopical technique but may be readily scraped away with a knife. The slips or other articles should be as near chemically clean as possible, before attempting to write on them.

According to Schuh, a mixture of a shellac solution and whiting or precipitated chalk answers very well for marking glass. Any color may be mixed with the chalk. If the glass is thoroughly cleaned with alcohol or ether, either a quill pen or a camel's-hair pencil (or a fresh, clean steel pen) may be used.

Ink on Marble.—Ink marks on marble may be removed with a paste made by dissolving an ounce of oxalic acid and half an ounce of butter of antimony in a pint of rain water, and adding sufficient flour to form a thin paste. Apply this to the stains with a brush; allow it to remain on 3 or 4 days and then wash it off. Make a second application, if necessary.

Perpetual Ink.—I.—Pitch, 3 pounds; melt over the fire, and add of lampblack, ¾ pound; mix well.

II.—Trinidad asphaltum and oil of turpentine, equal parts. Used in a melted state to fill in the letters on tombstones, marbles, etc. Without actual violence, it will endure as long as the stone itself.

Ink for Steel Tools.—Have a rubber stamp made with white letters on a black ground. Make up an ink to use with this stamp, as follows:

Ordinary rosin, ½ pound; lard oil, 1 tablespoonful; lampblack, 2 tablespoonfuls; turpentine, 2 tablespoonfuls. Melt the rosin, and stir in the other ingredients in the order given. When the ink is cold it should look like ordinary printers' ink. Spread a little of this ink over the pad and ink the rubber stamp as usual, and press it on the clean steel—saw blade, for instance. Have a rope of soft putty, and make a border of putty around the stamped design as close up to the lettering as possible, so that no portion of the steel inside the ring of putty is exposed but the lettering. Then pour into the putty ring the etching mixture, composed of 1 ounce of nitric acid, 1 ounce of muriatic acid, and 12 ounces of water. Allow it to rest for only a minute, draw off the acid with a glass or rubber syringe, and soak up the last trace of acid with a moist sponge. Take off the putty, and wipe off the design with potash solution first, and then with turpentine, and the job is done.

Writing on Ivory, Glass, etc.—Nitrate of silver, 3 parts; gum arabic, 20 parts; distilled water, 30 parts. Dissolve the gum arabic in two-thirds of the water, and the nitrate of silver in the other third. Mix and add the desired color.

Writing on Zinc (see also Horticultural Inks).—Take 1 part sulphate of copper (copper vitriol), 1 part chloride of potassium, both dissolved in 35 parts water. With this blue liquid, writing or drawing may be done with a common steel pen upon zinc which has been polished bright with emery paper. After the writing is done the plates are put in water and left in it for some time, then taken out and dried. The writing will remain intact as long as the zinc. If the writing or drawing should be brown, 1 part sulphate of iron (green vitriol) is added to the above solution. The chemicals are dissolved in warm water and the latter must be cold before it can be used.

GOLD INK.

I.—The best gold ink is made by rubbing up gold leaf as thoroughly as possible with a little honey. The honey is then washed away with water, and the finely powdered gold leaf left is mixed to the consistency of a writing ink with weak gum water. Everything depends upon the fineness of the gold powder, i. e., upon the diligence with which it has been worked with the honey. Precipitated gold is finer than can be got by any rubbing, but its color is wrong, being dark brown. The above gold ink should be used with a quill pen.

II.—An imitation gold or bronze ink is composed by grinding 1,000 parts of powdered bronze of handsome color with a varnish prepared by boiling together 500 parts of nut oil, 200 parts of garlic, 500 parts of cocoanut oil, 100 parts of Naples yellow, and as much of sienna.

HORTICULTURAL INK.

I.—Chlorate of platinum, ¼ ounce; soft water, 1 pint. Dissolve and preserve it in glass. Used with a clean quill to write on zinc labels. It almost immediately turns black, and cannot be removed by washing. The addition of gum and lampblack, as recommended in certain books, is unnecessary, and even prejudicial to the quality of the ink.

II.—Verdigris and sal ammoniac, of each ½ ounce; levigated lampblack, ½ ounce; common vinegar, ¼ pint; mix thoroughly. Used as the last, for either zinc, iron, or steel.

III.—Blue vitriol, 1 ounce; sal ammoniac, ½ ounce (both in powder); vinegar, ¼ pint; dissolve. A little lampblack or vermilion may be added, but it is not necessary. Use No. I, for iron, tin, or steel plate.

INDELIBLE INKS.

These are also frequently called waterproof, incorrodible, or indestructible inks. They are employed for writing labels on bottles containing strong acids and alkaline solutions. They may be employed with stamps, types or stencil plates, by which greater neatness will be secured than can be obtained with either a brush or pen.

The following is a superior preparation for laundry use:

Aniline oil	85 parts
Potassium chlorate	5 parts
Distilled water	44 parts
Hydrochloric acid, pure (specific gravity, 1.124)	68 parts
Copper chloride, pure	6 parts

Mix the aniline oil, potassium chlorate, and 26 parts of the water and heat in a capacious vessel, on the water bath, at a temperature of from 175° to 195° F., until the chlorate is entirely dissolved, then add one-half of the hydrochloric and continue the heat until the mixture begins to take on a darker color. Dissolve the copper chloride in the residue of the water, add the remaining hydrochloric acid to the solution, and add the whole to the liquid on the water bath, and heat the mixture until it acquires a fine red-violet color. Pour into a flask with a well-fitting ground-glass stopper, close tightly and set aside for several days, or until it ceases to throw down a precipitate. When this is the case, pour off the clear liquid into smaller (one drachm or a drachm and a half) containers.

This ink must be used with a quill pen, and is especially good for linen or cotton fabrics, but does not answer so well for silk or woolen goods. When first used, it appears as a pale red, but on washing with soap or alkalies, or on exposure to the air, becomes a deep, dead black. The following is a modification of the foregoing:

Blue Indelible Ink.—This ink has the reputation of resisting not only water and oil, but alcohol, oxalic acid, alkalies, the chlorides, etc. It is prepared as follows: Dissolve 4 parts of gum lac in 36 parts of boiling water carrying 2 parts of borax. Filter and set aside. Now dissolve 2 parts of gum arabic in 4 parts of water and add the solution to the filtrate. Finally, after the solution is quite cold, add 2 parts of powdered indigo and dissolve by agitation. Let stand for several hours, then decant, and put in small bottles.

Red Indelible Inks.—By proceeding according to the following formula, an intense purple-red color may be produced on fabrics, which is indelible in the customary sense of the word:

1.—Sodium carbonate..	3 drachms
Gum arabic........	3 drachms
Water............	12 drachms
2.—Platinic chloride....	1 drachm
Distilled water.....	2 ounces
3.—Stannous chloride...	1 drachm
Distilled water.....	4 drachms

Moisten the place to be written upon with No. 1 and rub a warm iron over it until dry; then write with No. 2, and, when dry, moisten with No. 3. An intense and beautiful purple-red color is porduced in this way. A very rich purple color—the purple of Cassius—may be produced by substituting a solution of gold chloride for the platinic chloride in the above formula.

Crimson Indelible Ink.—

The following formula makes an indelible crimson ink:

Silver nitrate.......	50 parts
Sodium carbonate, crystal..........	75 parts
Tartaric acid.......	16 parts
Carmine..........	1 part
Ammonia water, strongest.........	288 parts
Sugar, white, crystallized............	36 parts
Gum arabic, powdered...........	60 parts
Distilled water, quantity sufficient to make..........	400 parts

Dissolve the silver nitrate and the sodium carbonate separately, each in a portion of the distilled water, mix the solutions, collect the precipitate on a filter, wash, and put the washed precipitate, still moist, into a mortar. To this add the tartaric acid, and rub together until effervescence ceases. Now, dissolve the carmine in the ammonia water (which latter should be of specific gravity .882, or contain 34 per cent of ammonia), filter, and add the filtrate to the silver tartrate magma in the mortar. Add the sugar and gum arabic, rub up together, and add gradually, with constant agitation, sufficient distilled water to make 400 parts.

Gold Indelible Ink.—Make two solutions as follows:

1.—Chloride of gold and sodium.........	1	part
Water............	10	parts
Gum.............	2	parts
2.—Oxalic acid	1	part
Water............	5	parts
Gum.............	2	parts

The cloth or stuff to be written on should be moistened with liquid No. 2. Let dry, and then write upon the prepared place with liquid No. 1, using preferably a quill pen. Pass a hot iron over the mark, pressing heavily.

INDIA, CHINA, OR JAPAN INK.

Ink by these names is based on lampblack, and prepared in various ways. Many makes flow less easily from the pen than other inks, and are less durable than ink that writes paler and afterwards turns black. The ink is usually unfitted for steel pens, but applies well with a brush.

I.—Lampblack (finest) is ground to a paste with very weak liquor of potassa, and this paste is then diffused through water slightly alkalized with potassa, after which it is collected, washed with clean water, and dried; the dry powder is next levigated to a smooth, stiff paste, with a strong filtered decoction of carrageen or Irish moss, or of quince seed, a few drops of essence of musk, and about half as much essence of ambergris being added, by way of perfume, toward the end of the process; the mass is, lastly, molded into cakes, which are ornamented with Chinese characters and devices, as soon as they are dry and hard.

II.—A weak solution of fine gelatin is boiled at a high temperature in a digester for 2 hours, and then in an open vessel for 1 hour more. The liquid is next filtered and evaporated to a proper consistency, either in a steam- or salt-water bath. It is, lastly, made into a paste, as before, with lampblack which has been previously heated to dull redness in a well-closed crucible. Neither of the above gelatinizes in cold weather, like the ordinary imitations.

To Keep India Ink Liquid.—If one has to work with the ink for some time, a small piece should be dissolved in warm water and the tenth part of glycerine added, which mixes intimately with the ink after shaking for a short time. India ink thus prepared will keep very well in a corked bottle, and if a black jelly should form in the cold, it is quickly dissolved by heating. The ink flows well from the pen and does not wipe.

INK POWDERS AND LOZENGES.

Any of these powders may, by the addition of mucilage of gum arabic, be made into lozenges or buttons—the "ink buttons" or "ink stones" in use abroad and much affected by travelers.

The following makes a good serviceable black ink, on macerating the powder in 100 times its weight of rain or distilled water for a few days:

I.—Powdered gallnuts .. 16 parts
Gum arabic......... 8 parts
Cloves............. 1 part
Iron sulphate....... 10 parts

Put into an earthenware or glass vessel, cover with 100 parts of rain or distilled water, and set aside for 10 days or 2 weeks, giving an occasional shake the first 3 or 4 days. Decant and bottle for use.

The following is ready for use instantly on being dissolved in water:

II.—Aleppo gallnuts ..... 84 parts
Dutch Madder...... 6 parts

Powder, mix, moisten, and pack into the percolator. Extract with hot water, filter, and press out. To the filtrate add 4 parts of iron acetate (or pyroacetate) and 2½ parts of tincture of indigo. Put into the water bath and evaporate to dryness and powder the dry residue.

LITHOGRAPHIC INKS.

These are for writing on lithographic stones or plates:

I.—Mastic (in tears), 8 ounces; shellac, 12 ounces; Venice turpentine, 1 ounce. Melt together, add wax, 1 pound; tallow, 6 ounces. When dissolved, add hard tallow soap (in shavings), 6 ounces; and when the whole is perfectly combined, add lampblack, 4 ounces. Mix well, cool a little, and then pour it into molds, or upon a slab, and when cold cut it into square pieces.

II. (Lasteyrie).—Dry tallow soap, mastic (in tears), and common soda (in fine powder), of each, 30 parts; shellac, 150 parts; lampblack, 12 parts. Mix as indicated in Formula I.

MARKING OR LABELING INKS:

Black Marking Inks.—

I.—Borax........... 60 parts
Shellac........... 180 parts
Boiling water..... 1,000 parts
Lampblack, a sufficient quantity.

Dissolve the borax in the water, add the shellac to the solution and stir until dissolved. Rub up a little lampblack with sufficient of the liquid to form a paste, and add the rest of the solution a little at a time and with constant rubbing. Test, and if not black enough, repeat the operation. To get the best effect—a pure jet-black—the lampblack should be purified and freed from the calcium phosphate always present in the commercial article to the extent, frequently, of 85 to 87 per cent, by treating with hydrochloric acid and washing with water.

II.—An ink that nothing will bleach is made by mixing pyrogallic acid and sulphate of iron in equal parts. Particularly useful for marking labels on bottles containing acids. Varnish the label after the ink is dry so that moisture will not affect it.

COLORED MARKING INKS:

Eosine Red.—

Eosine B............ 1 drachm
Solution of mercuric
chloride.......... 2 drachms
Mucilage of acacia... 2 drachms
Rectified spirit....... 4 ounces
Oil of lavender...... 1 drop
Distilled water...... 8 ounces

Dissolve the eosine in the solution and 2 ounces of water, add the mucilage, and mix, then the oil dissolved in the spirit, and finally make up.

Orange.—

Aniline orange....... 1 drachm
Sugar............. 2 drachms
Distilled water to.... 4 ounces

Blue.—

I.—Resorcin blue....... 1 drachm
Distilled water....... 6 drachms

Mix and agitate occasionally for 2 hours, then add:

Hot distilled water...	24	ounces
Oxalic acid.........	10	grains
Sugar.............	½	ounce

Shake well. This and other aniline inks can be perfumed by rubbing up a drop of attar of rose with the sugar before dissolving it in the hot water.

II.—A solid blue ink, or marking paste, to be used with a brush for stenciling, is made as follows: Shellac, 2 ounces; borax, 2 ounces; water, 25 ounces; gum arabic, 2 ounces; and ultramarine, sufficient. Boil the borax and shellac in some of the water till they are dissolved, and withdraw from the fire. When the solution has become cold, add the rest of the 25 ounces of water, and the ultramarine. When it is to be used with the stencil, it must be made thicker than when it is to be applied with a marking brush.

III.—In a suitable kettle mix well, stirring constantly, 50 parts of liquid logwood extract (80 per cent) with 3 parts of spirit previously mingled with 1 part of hydrochloric acid, maintaining a temperature of 68° F. Dissolve 5 parts of potassium chromate in 15 parts of boiling water; to this add 10 parts of hydrochloric acid, and pour this mixture, after raising the temperature to about 86° F., very slowly and with constant stirring into the kettle. Then heat the whole to 185° F. This mass, which has now assumed the nature of an extract, is stirred a little longer, and next 15 parts of dextrin mixed with 10 parts of fine white earth (white bole) are added. The whole is well stirred throughout. Transfer the mass from the kettle into a crusher, where it is thoroughly worked through.

PRINTING INKS.

Black printing inks owe their color to finely divided carbon made from lampblack, pine-wood, rosin oil, etc., according to the quality of the ink desired. The finest inks are made from flamelampblack. There are, however, certain requirements made of all printing inks alike, and these are as follows: The ink must be a thick and homogeneous liquid, it must contain no solid matter but finely divided carbon, and every drop when examined microscopically must appear as a clear liquid containing black grains uniformly distributed.

The consistency of a printing ink must be such that it passes on to the printing rollers at the proper rate. It will be obvious that various consistencies are demanded according to the nature of the machine used by the printer. For a rotary machine which prints many thousands of copies an hour a much thinner ink will be necessary than that required for art printing or for slow presses. As regards color, ordinary printing ink should be a pure black. For economy's sake, however, newspaper printers often use an ink so diluted that it does not look deep black, but a grayish black, especially in large type.

The question of the time that the ink takes to dry on the paper is a very important one, especially with ink used for printing newspapers which are folded and piled at one operation. If then the ink does not dry very quickly, the whole impression smudges and "sets off" so much that it becomes illegible in places. Although it is essential to have a quick drying ink for this purpose, it is dangerous to go too far, for a too quickly drying ink would make the paper stick to the forms and tear it. A last condition which must be fulfilled by a good printing ink is that it must be easy of removal from the type, which has to be used again.

No one composition will answer every purpose and a number of different inks are required. Makers of printing inks are obliged, therefore, to work from definite recipes so as to be able to turn out exactly the same ink again and again. They make newspaper ink for rotary presses, book-printing inks, half-tone inks, art inks, etc. As the recipes have been attained only by long, laborious, and costly experiments, it is obvious that the makers are not disposed to communicate them, and the recipes that are offered and published must be looked upon with caution, as many of them are of little or no value. In the recipes given below for printing inks, the only intention is to give hints of the general composition, and the practical man will easily discover what, if any, alterations have to be made in the recipe for his special purpose.

Many different materials for this manufacture are given in recipes, so many, in fact, that it is impossible to discover what use they are in the ink. The following is a list of the articles commonly in use for the manufacture of printing ink:

Boiled linseed oil, boiled without driers.

Rosin oil from the dry distillation of rosin.

Rosin itself, especially American pine rosin.

Soap, usually rosin-soap, but occasionally ordinary soap.

Lampblack and various other pigments.

By the most time-honored method, linseed oil was very slowly heated over an open fire until it ignited. It was allowed to burn for a time and then extinguished by putting a lid on the pot. In this way a liquid was obtained of a dark brown or black color with particles of carbon, and with a consistency varying with the period of heating, being thicker, the longer the heating was continued. If necessary, the liquid was then thinned with unboiled, or only very slightly boiled, linseed oil. Lampblack in the proper quantity was added and the mixture was finally rubbed up on a stone in small quantities at a time to make it uniform.

Boiling the Linseed Oil.—This process, although it goes by the name of boiling, is not so in the proper sense of the word, but a heating having for its object an initial oxidation of the oil, so that it will dry better. Linseed oil is a type of the drying oils, those which when exposed in thin coats to the air absorb large quantities of oxygen and are thereby converted into tough, solid sheets having properties very similar to those of soft India rubber. The process goes on much faster with the aid of heat than at the ordinary temperature, and the rate at which the boiled oil will dry in the ink can be exactly regulated by heating it for a longer or shorter time. Prolonged heating gives an oil which will dry very quickly on exposure in thin coats to the air, the shorter the heating the more slowly will the ink afterwards made with the oil dry.

Linseed oil must always be boiled in vessels where it has plenty of room, as the oil soon swells up and it begins to decompose so energetically at a particular temperature that there is considerable risk of its boiling over and catching fire. Various contrivances have been thought out for boiling large quantities of the oil with safety, such as pans with an outlet pipe in the side, through which the oil escapes when it rises too high instead of over the edge of the pan, and fires built on a trolley running on rails, so that they can at once be moved from under the pan if there is any probability of the latter boiling over. The best apparatus for preparing thickened linseed oil is undoubtedly one in which the oil offers a very large surface to the air, and on that account requires to be moderately heated only. The oil soon becomes very thick under these conditions and if necessary can be diluted to any required consistency with unboiled oil.

In boiling linseed oil down to the proper thickness by the old method there are two points demanding special attention. One is the liability of the oil to boil over, and the other consists in the development of large quantities of vapor, mostly of acroleine, which have a most powerful and disagreeable smell, and an intense action upon the eyes. The attendant must be protected from these fumes, and the boiling must therefore be done where there is a strong draught to take the fumes as fast as they are produced. There are various contrivances to cope with boiling over.

Savage's Printing Ink.—Pure balsam of copaiba, 9 ounces; lampblack, 3 ounces; indigo and Prussian blue, each 5 drachms; drachms; Indian red, ¾ ounce; yellow soap, 3 ounces. Mix, and grind to the utmost smoothness.

Toning Black Inks.—Printers' inks consisting solely of purified lampblack and vehicle give, of course, impressions which are pure black. It is, however, well known that a black which has to a practiced eye a tinge of blue in it looks much better than a pure black. To make such an ink many makers mix the lampblack with a blue pigment, which is added in very fine powder before the first grinding. Prussian blue is the pigment usually chosen and gives very attractive results. Prussian blue is, however, not a remarkable stable substance, and is very apt to turn brown from the formation of ferric oxide. Hence an ink made with Prussian blue, although it may look very fine at first, often assumes a dull brown hue in the course of time. Excellent substitutes for Prussian blue are to be found in the Induline blues. These are very fast dyes, and inks tinted with them do not change color. As pure indigo is now made artificially and sold at a reasonable price, this extremely fast dye can also be used for tinting inks made with purified lampblack.

To Give Dark Inks a Bronze or Changeable Hue.—Dissolve 1½ pounds gum shellac in 1 gallon 65 per cent alcohol or cologne spirits for 24 hours. Then add 14 ounces aniline red. Let it stand a few hours longer, when it will be ready for use. Add this to good blue, black, or other dark ink, as needed in quantities to suit, when if carefully done they will be found to have a rich bronze or changeable hue.

Quick Dryer for Inks Used on Bookbinders' Cases.—Beeswax, 1 ounce; gum arabic (dissolved in sufficient acetic acid to make a thin mucilage), ¼ ounce; brown japan, ¼ ounce. Incorporate with 1 pound of good cut ink.

INKS FOR STAMP PADS.

The ink used on vulcanized rubber stamps should be such that when applied to a suitable pad it remains sufficiently fluid to adhere to the stamp. At the same time the fluidity should cease by the time the stamp is pressed upon an absorbing surface such as paper. Formerly these inks were made by rubbing up pigments in fat to a paste. Such inks can hardly be prevented, however, from making impressions surrounded by a greasy mark caused by the fat spreading in the pores of the paper. Now, most stamping inks are made without grease and a properly prepared stamping ink contains nothing but glycerine and coaltar dye. As nearly all these dyes dissolve in hot glycerine the process of manufacture is simple enough. The dye, fuchsine, methyl violet, water blue, emerald green, etc., is put into a thin porcelain dish over which concentrated glycerine is poured, and the whole is heated to nearly 212° F. with constant stirring. It is important to use no more glycerine than is necessary to keep the dye dissolved when the ink is cold. If the mass turns gritty on cooling it must be heated up with more glycerine till solution is perfect.

In dealing with coal-tar dyes insoluble in glycerine, or nearly so, dissolve them first in the least possible quantity of strong, hot alcohol. Then add the glycerine and heat till the spirit is evaporated.

To see whether the ink is properly made spread some of it on a strip of cloth and try it with a rubber stamp. On paper, the separate letters must be quite sharp and distinct. If they run at the edges there is too much glycerine in the ink and more dye must be added to it. If, on the contrary, the impression is indistinct and weak, the ink is too thick and must be diluted by carefully adding glycerine.

Aniline colors are usually employed as the tinting agents. The following is a typical formula, the product being a black ink:

I.—Nigrosin............ 3 parts
  Water.............. 15 parts
  Alcohol............ 15 parts
  Glycerine.......... 70 parts

Dissolve the nigrosin in the alcohol, add the glycerine previously mixed with the water, and rub well together.

Nigrosin is a term applied to several compounds of the same series which differ in solubility. In the place of these compounds it is probable that a mixture would answer to produce black as suggested by Hans Wilder for making writing ink. His formula for the mixture is:

II.—Methyl violet........ 3 parts
  Bengal green........ 5 parts
  Bismarck green..... 4 parts

A quantity of this mixture should be taken equivalent to the amount of nigrosin directed. These colors are freely soluble in water, and yield a deep greenish-black solution.

The aniline compound known as brilliant green answers in place of Bengal green. As to the permanency of color of this or any aniline ink, no guarantee is offered. There are comparatively few coloring substances that can be considered permanent even in a qualified sense. Among these, charcoal takes a foremost place. Lampblack remains indefinitely unaltered. This, ground very finely with glycerine, would yield an ink which would perhaps prove serviceable in stamping; but it would be liable to rub off to a greater extent than soluble colors which penetrate the paper more or less. Perhaps castor oil would prove a better vehicle for insoluble coloring matters. Almost any aniline color may be substituted for nigrosin in the foregoing formula, and blue, green, red, purple, and other inks obtained. Insoluble pigments might also be made to answer as suggested for lampblack.

The following is said to be a cushion that will give color permanently. It consists of a box filled with an elastic composition, saturated with a suitable color. The cushion fulfils its purpose for years without being renewed, always contains sufficient moisture, which is drawn from the atmosphere, and continues to act as a color stamp cushion so long as a remnant of the mass or composition remains in the box or receptacle. This cushion or pad is too soft to be self-supporting, but should be held in a low, flat pan, and have a permanent cloth cover.

III.—The composition consists preferably of 1 part gelatin, 1 part water, 6 parts glycerine, and 6 parts coloring matter. A suitable black color can be made from the following materials: One part gelatin glue, 3 parts lampblack, aniline black, or a suitable quantity of logwood extract, 10 parts of glycerine, 1 part absolute alcohol, 2 parts water, 1 part Venetian soap, ⅙ part salicylic acid. For red, blue, or violet: One part gelatin glue, 2 parts aniline of desired color, 1 part absolute alcohol, 10 parts glycerine, 1 part Venetian soap, and ⅙ part salicylic acid.

The following are additional recipes used for this purpose:

IV.—Mix and dissolve 2 to 4 drachms aniline violet, 15 ounces alcohol, 15 ounces glycerine. The solution is poured on the cushion and rubbed in with a brush. The general method of preparing the pad is to swell the gelatin with cold water, then boil and add the glycerine, etc.

V.—Mix well 16 pounds of hot linseed oil, 3 ounces of powdered indigo, or a like quantity of Berlin blue, and 8 pounds of lampblack. For ordinary sign-stamping an ink without the indigo might be used. By substituting ultramarine or Prussian blue for the lampblack, a blue "ink" or paint would result.

Inks for Hand Stamps.—As an excipient for oily inks, a mixture of castor oil and crude oleic acid, in parts varying according to the coloring material used, is admirable. The following are examples:

Black.—Oil soluble nigrosin and crude oleic acid in equal parts. Add 7 to 8 parts of castor oil.

Red.—Oil soluble aniline red, 2 parts; crude oleic acid, 3 parts; castor oil, from 30 to 60 parts, according to the intensity of color desired.

Red.—Dissolve ¼ ounce of carmine in 2 ounces strong water of ammonia, and add 1 drachm of glycerine and ¾ ounce dextrin.

Blue.—Rub 1 ounce Prussian blue with enough water to make a perfectly smooth paste; then add 1 ounce dextrin, incorporate it well, and finally add sufficient water to bring it to the proper consistency.

Blue.—Oil soluble aniline blue, 1 part; crude oleic acid, 2 parts; castor oil, 30 to 32 parts.

Violet.—Alcohol, 15 ounces: glycerine, 15 ounces; aniline violet, 2 to 4 drachms. Mix, dissolve, pour the solution on the cushion, and dab on with a brush.

Color Stamps for Rough Paper.—It has hitherto been impossible to get a satisfactory application for printing with rubber stamps on rough paper. Fatty vehicles are necessary for such paper, and they injure the India rubber. It is said, however, that if the rubber is first soaked in a solution of glue, and then in one of tannin, or bichromate of potash, it becomes impervious to the oils or fats. Gum arabic can be substituted for the glue.

Indelible Hand-Stamp Ink.—

I.—Copper sulphate.... 20 parts
Aniline chlorate.... 20 parts

Rub up separately to a fine powder, then carefully mix, and add 10 parts of dextrin and incorporate. Add 5 parts of glycerine and rub up, adding water, a little at a time, until a homogeneous viscid mass is obtained. An aniline color is produced in the material, which boiling does not destroy.

II.—Sodium carbonate.. 22 parts
Glycerine........... 85 parts
Gum arabic, in powder............. 20 parts
Silver nitrate....... 11 parts
Ammonia water.... 20 parts
Venetian turpentine 10 parts

Triturate the carbonate of sodium, gum arabic, and glycerine together. In a separate flask dissolve the silver nitrate in the ammonia water, mix the solution with the triturate, and heat to boiling, when the turpentine is to be added, with constant stirring. After stamping, expose to the sunlight or use a hot iron. The quantity of glycerine may be varied to suit circumstances.

White Stamping Ink for Embroidery.—
Zinc white.......... 2 drachms
Mucilage........... 1 drachm
Water.............. 6 drachms

Triturate the zinc white with a small quantity of water till quite smooth, then add the mucilage and the remainder of the water.

STENCIL INKS.

I.—Dissolve 1 ounce of gum arabic in 6 ounces water, and strain. This is the mucilage. For *Black Color* use drop black, powdered, and ground with the mucilage to extreme fineness; for *Blue,* ultramarine is used in the same manner; for *Green,* emerald green: for *White,* flake white; for *Red,* vermilion, lake, or carmine; for *Yellow,* chrome yellow. When ground too thick they are thinned with a little water. Apply with a small brush.

II.—Triturate together 1 pint pine soot and 2 pints Prussian blue with a little glycerine, then add 3 pints gum arabic and sufficient glycerine to form a thin paste.

Blue Stencil Inks.—The basis of the stencil inks commonly used varies to some extent, some preferring a mixture of pigments with oils, and others a watery shellac basis. The basis:

I.—Shellac............ 2 ounces
Borax............. 1½ ounces
Water............ 10 ounces

Boil together until 10 ounces of solution is obtained. The coloring:

Prussian blue...... 1 ounce
China clay......... ½ ounce
Powdered acacia... ½ ounce

Mix thoroughly and gradually incorporate the shellac solution.

II.—Prussian blue...... 2 ounces
Lampblack........ 1 ounce
Gum arabic........ 3 ounces
Glycerine, sufficient.

Triturate together the dry powders and then make into a suitable paste with glycerine.

Indelible Stencil Inks.—I.—Varnish such as is used for ordinary printing ink, 1 pound; black sulphuret of mercury, 1 pound; nitrate of silver, 1 ounce; sulphate of iron, 1 ounce; lampblack, 2 tablespoonfuls. Grind all well together; thin with spirits turpentine as desired.

II.—Sulphate of manganese, 2 parts; lampblack, 1 part; sugar, 4 parts; all in fine powder and triturated to a paste in a little water.

III.—Nitrate of silver, ¼ ounce; water, ¾ ounce. Dissolve, add as much of the strongest liquor of ammonia as will dissolve the precipitate formed on its first addition. Then add of mucilage, 1½ drachms, and a little sap green, syrup of buckthorn, or finely powdered indigo, to color. This turns black on being held near the fire, or touched with a hot iron.

SYMPATHETIC INKS:

Table of Substances Used in Making Sympathetic Inks.—

For writing and for bringing out the writing:

Cobalt chloride, heat.

Cobalt acetate and a little saltpeter, heat.

Cobalt chloride and nickel chloride mixed, heat.

Nitric acid, heat.

Sulphuric acid, heat.

Sodium chloride, heat.

Saltpeter, heat.

Copper sulphate and ammonium chloride, heat.

Silver nitrate, sunlight.

Gold trichloride, sunlight.

Ferric sulphate, infusion of gallnuts or ferrocyanide of potassium.

Copper sulphate, ferrocyanide of potassium.

Lead vinegar, hydrogen sulphide.

Mercuric nitrate, hydrogen sulphide.

Starch water, tincture of iodine or iodine vapors.

Cobalt nitrate, oxalic acid.

Fowler's solution, copper nitrate.

Soda lye or sodium carbonate, phenolphthaleine.

A sympathetic ink is one that is invisible when written, but which can be made visible by some treatment. Common milk can be used for writing, and exposure to strong heat will scorch and render the dried milk characters visible.

The following inks are developed by exposure to the action of reagents:

I.—Upon writing with a very clear solution of starch on paper that contains but little sizing, and submitting the dry characters to the vapor of iodine (or passing over them a weak solution of potassium iodide), the writing becomes blue, and disappears under the action of a solution of hyposulphite of soda (1 in 1,000).

II.—Characters written with a weak solution of the soluble chloride of platinum or iridium become black when the paper is submitted to mercurial vapor. This ink may be used for marking linen, as it is indelible.

III.—Sulphate of copper in very dilute solution will produce an invisible writing, which may be turned light blue by vapors of ammonia.

IV.—Soluble compounds of antimony will become red by hydrogen sulphide vapor.

V.—Soluble compounds of arsenic and of peroxide of tin will become yellow by the same vapor.

VI.—An acid solution of iron chloride is diluted until the writing is invisible when dry. This writing has the property of becoming red by sulphocyanide vapors (arising from the action of sulphuric acid on potassium sulphocyanide in a long-necked flask), and it disappears by ammonia, and may alternately be made to appear and disappear by these two vapors.

VII.—Write with a solution of paraffine in benzol. When the solvent has evaporated, the paraffine is invisible, but becomes visible on being dusted with lampblack or powdered graphite or smoking over a candle flame.

VIII.—Dissolve 1 part of a lead salt, 0.1 part of uranium acetate, and the same quantity of bismuth citrate in 100 parts of water. Then add, drop by drop, a solution of sal ammoniac until the whole becomes transparent. Afterwards, mix with a few drops of gum arabic. To reveal the characters traced with this ink, expose them to the fumes of sulphuric acid, which turns them immediately to a dark brown. The characters fade away in a few minutes, but can be renewed by a slight washing with very dilute nitric acid.

TYPEWRITER RIBBON INKS.

I.—Take vaseline (petrolatum) of high boiling point, melt it on a water bath or slow fire, and incorporate by constant stirring as much lamp or powdered drop black as it will take up without becoming granular. If the vaseline remains in excess, the print is liable to have a greasy outline; if the color is in excess, the print will not be clear. Remove the mixture from the fire, and while it is cooling mix equal parts of petroleum, benzine, and rectified oil of turpentine, in which dissolve the fatty ink, introduced in small portions, by constant agitation. The volatile solvents should be in such quantity that the fluid ink is of the consistence of fresh oil paint. One secret of success lies in the proper application of the ink to the ribbon. Wind the ribbon on a piece of cardboard, spread on a table several layers of newspaper, then unwind the ribbon in such lengths as may be most convenient, and lay it flat on the paper. Apply the ink, after agitation, by means of a soft brush, and rub it well into the interstices of the ribbon with a toothbrush. Hardly any ink should remain visible on the surface. For colored inks use Prussian blue, red lead, etc., and especially the aniline colors.

II.—Aniline black . . . . . .   ½ ounce
Pure alcohol . . . . 15 ounces
Concentrated glycer-
ine . . . . 15 ounces

Dissolve the aniline black in the alcohol, and add the glycerine. Ink as before. The aniline inks containing glycerine are copying inks.

III.—Alcohol . . . . 2 ounces
Aniline color . . . . ¼ ounce
Water . . . . 2 ounces
Glycerine . . . . 4 ounces

Dissolve the aniline in the alcohol and add the water and glycerine.

IV.—Castor oil . . . . 2 ounces
Cassia oil . . . . ½ ounce
Carbolic acid . . . . ½ ounce

Warm them together and add 1 ounce of aniline color. Indelible typewriter inks may be made by using lampblack in place of the aniline, mixing it with soft petrolatum and dissolving the cooled mass in a mixture of equal parts of benzine and turpentine.

COLORING AGENTS:

Red.—

I.—Bordeaux red, O. S. 15 parts
Aniline red, O. S. . . . . 15 parts
Crude oleic acid . . . . 45 parts
Castor oil enough to make 1,000 parts

Rub the colors up with the oleic acid, add the oil, warming the whole to 100° to 110° F. (not higher), under constant stirring. If the color is not sufficiently intense for your purposes, rub up a trifle more of it with oleic acid, and add it to the ink. By a little experimentation you can get an ink exactly to your desire in the matter.

Blue-Black.—

II.—Aniline black, O. S. . . 5 parts
Oleic acid, crude . . . . 5 parts
Castor oil, quantity sufficient to 100 parts.

Violet.—

III.—Aniline violet, O. S. . . 3 parts
Crude oleic acid . . . . 5 parts
Castor oil, quantity sufficient to 100 parts.

The penetration of the ink may be increased *ad libitum* by the addition of a few drops of absolute alcohol, or, better, of benzol.

Reinking.—For reinking ribbons use the following recipe for black: One ounce aniline black; 15 ounces pure grain alcohol; 15 ounces concentrated glycerine. Dissolve the aniline black in the alcohol and then add the glycerine. For blue use Prussian blue, and for red use red lead instead of the aniline black. This ink is also good for rubber stamp pads.

WRITING INKS.

The common writing fluids depend mostly upon galls, logwood, or aniline for coloring. There are literally thousands of formulas. A few of the most reliable have been gathered together here:

I.—Aleppo galls (well bruised), 4 ounces; clean soft water, 1 quart; macerate in a clean corked bottle for 10 days or a fortnight or longer, with frequent agitation; then add of gum arabic (dissolved in a wineglassful of water), 1½ ounces; lump sugar, ½ ounce. Mix well, and afterwards further add of sulphate of iron (green copperas crushed small), 1½ ounces. Agitate occasionally for 2 or 3 days, when the ink may be decanted for use, but is better if the whole is left to digest together for 2 or 3 weeks. When time is an object, the whole of the ingredients may at once be put into a bottle, and the latter agitated daily until the ink is made; and boiling water instead of cold water may be employed. Product, 1 quart of excellent ink, writing pale at first, but soon turning intensely black.

II.—Aleppo galls (bruised), 12 pounds; soft water, 6 gallons. Boil in a copper vessel for 1 hour, adding more water to make up for the portion lost by evaporation; strain, and again boil the galls with water, 4 gallons, for ½ hour; strain off the liquor, and boil a third time with water, 2½ gallons, and strain. Mix the several liquors, and while still hot add of green copperas (coarsely powdered), 4½ pounds; gum arabic (bruised small), 4 pounds. Agitate until dissolved, and after defecation strain through a hair sieve, and keep in a bunged cask for use. Product, 12 gallons.

III.—Aleppo galls (bruised), 14 pounds; gum, 5 pounds. Put them in a small cask, and add boiling soft water, 15 gallons. Allow the whole to macerate, with frequent agitation, for a fortnight, then further add of green copperas, 5 pounds, dissolved in water, 7 pints. Again mix well, and agitate the whole once daily for 2 or 3 weeks. Product, 15 gallons.

Brown Ink.—I.—To make brown ink, use for coloring a strong decoction of catechu; the shade may be varied by the cautious addition of a little weak solution of bichromate of potash.

II.—A strong decoction of logwood, with a very little bichromate of potash.

Blue Ink.—To make blue ink, substitute for the black coloring sulphate of indigo and dilute it with water till it produces the required color.

Anticorrosive or Asiatic Ink.—I.—Galls, 4 pounds; logwood, 2 pounds; pomegranate peel, 2 pounds; soft water, 5 gallons. Boil as usual; then add to the strained, decanted cold liquor, 1 pound of gum arabic, lump sugar or sugar candy, ¼ pound; dissolved in water, 3 pints. Product, 4½ gallons. Writes pale, but flows well from the pen, and soon darkens.

II.—Bruised galls, 14 pounds; gum, 5 pounds. Put them in a small cask, and add of boiling water, 15 gallons, Allow the whole to macerate, with frequent agitation, for 2 weeks, then further add green copperas, 5 pounds, dissolved in 7 pints water. Again mix well, and agitate the whole daily for 2 or 3 weeks.

Blue-Black Ink.—Blue Aleppo galls (free from insect perforations), 4½ ounces; bruised cloves, 1 drachm; cold water, 40 ounces; purified sulphate of iron, 1½ ounces; pure sulphuric acid (by measure), 35 minims; sulphate of indigo (in the form of a paste), which should be neutral, or nearly so, 1 ounce. The weights used are avoirdupois, and the measures apothecaries'. Place the galls, then bruised with the cloves, in a 50-ounce bottle, pour upon them the water, and digest, often daily shaking for a fortnight. Then filter through paper in another 50-ounce bottle. Get out also the refuse galls, and wring out of it the remaining liquid through a strong, clean linen or cotton cloth, into the filter, in order that as little as possible may be lost. Next put in the iron, dissolve completely, and filter through paper. Then the acid, and agitate briskly. Lastly, the indigo, and thoroughly mix by shaking. Pass the whole through paper; just filter out of one bottle into another until the operation is finished.

NOTE.—No gum or sugar is proper and on no account must the acid be omitted. When intended for copying, 5½ ounces of galls is the quantity. On the large scale this fine ink is made by percolation.

Colored Inks.—Inks of various colors may be made from a strong decoction of the ingredients used in dyeing, mixed with a little alum or other substance used as a mordant, and gum arabic. Any of the ordinary water-color cakes employed in drawing diffused through water may also be used for colored ink.

COPYING INK.

This is usually prepared by adding a little sugar to ordinary black ink, which for this purpose should be very rich in color, and preferably made galls prepared by heat. Writing executed with this ink may be copied within the space of 5 or 6 hours, by passing it through a copying press in contact with thin, unsized paper, slightly damped, enclosed between 2 sheets of thick oiled or waxed paper, when a reversed transcript will be obtained, which will read in proper order when the back of the copy is turned upwards. In the absence of a press a copy may be taken, when the ink is good and the writing very recent, by rolling the sheets, duly arranged on a ruler, over the surface of a flat, smooth table, employing as much force as possible, and avoiding any slipping or crumbling of the paper. Another method is to pass a warm flatiron over the paper laid upon the writing. The following proportions are employed:

I.—Sugar candy or lump sugar, 1 ounce; or molasses or moist sugar, $1\frac{1}{4}$ ounces; rich black ink, $1\frac{1}{2}$ pints; dissolve.

II.—Malt wort, 1 pint; evaporate it to the consistence of a syrup, and then dissolve it in good black ink, $1\frac{1}{4}$ pints.

III.—Solazza juice, 2 ounces; mild ale, $\frac{1}{2}$ pint; dissolve, strain, and triturate with lampblack (previously heated to dull redness in a covered vessel), $\frac{1}{4}$ ounce; when the mixture is complete, add of strong black, $1\frac{1}{2}$ pints; mix well, and in 2 or 3 hours decant the clear.

After making the above mixtures, they must be tried with a common steel pen, and if they do not flow freely, some more unprepared ink should be added until they are found to do so.

Alizarine Blue.—In 20 parts of fuming sulphuric acid dissolve 5 parts of indigo, and to the solution add 100 parts of extract of aqueous myrobalous and 10.5 parts iron filings or turning shavings. Finally add:

Gum arabic.......	1.5 parts
Sugar...........	7.5 parts
Sulphuric acid, 66°	
B.............	10.5 parts
Aniline blue.......	1.5 parts
Carbolic acid......	0.5 parts
Mirobalan extract to make 1,000 parts.	

This ink when first used has a bluish tint, afterwards becoming black.

Alizarine Green.—In 100 parts of aqueous extract of gall apples dissolve:

Iron sulphate........	30	parts
Copper sulphate.....	0.5	parts
Sulphuric acid.......	2	parts
Sugar.............	8	parts
Wood vinegar, rectified............	50	parts
Indigo carmine......	30	parts

Copying Ink for Copying Without a Press.—An ordinary thin-paper copying book may be used, and the copying done by transferrence. It is only necessary to place the page of writing in the letter book, just as one would use a leaf of blotting paper. The superfluous ink that would go into the blotting paper goes on to the leaf of the letter book, and showing through the thin paper gives on the other side of the leaf a perfect transcript of the letter. Any excess of ink on the page, either of the letter or of the copying paper, is removed by placing a sheet of blotting paper between them, and running one's hand firmly over the whole in the ordinary manner. This ready transcription is accomplished by using ink which dries slowly. Obviously the ink must dry sufficiently slowly for the characters at the top of a page of writing to remain wet when the last line is being written, while it must dry sufficiently to preclude any chance of the copied page being smeared while subsequent pages are being covered. The drying must also be sufficiently rapid to prevent the characters "setting off," as printers term it, from one page on to another after folding. The formula for the requisite ink is very simple:

Reduce by evaporation 10 volumes of any good ink to 6, then add 4 volumes of glycerine. Or manufacture some ink of nearly double strength, and add to any quantity of it nearly an equal volume of glycerine.

Gold Ink.—Mosaic gold, 2 parts; gum arabic, 1 part; rubbed up to a proper condition.

Green Ink.—A good, bright green, aniline ink may be made as follows:

Aniline green (soluble).............	2 parts
Glycerine..........	16 parts
Alcohol............	112 parts
Mucilage of gum arabic.............	4 parts

Dissolve the aniline in the alcohol, and add the other ingredients. Most of the gum arabic precipitates, but according to the author of the formula (Nelson) it has the effect of rendering the ink slow-flowing enough to write with. Filter.

Hectograph Inks (see also Hectograph).
—I.—Black.—Methyl violet, 10 parts; nigrosin, 20 parts; glycerine, 30 parts; gum arabic, 5 parts; alcohol, 60 parts.

II.—Blue.—Resorcin blue M, 10 parts. Dissolve by means of heat in a mixture of:

Dilute acetic acid.... 1 part
Distilled water....... 85 parts
Glycerine.......... 4 parts
Alcohol, 90 per cent.. 10 parts

III.—Green.—Aniline green, water solution, 15 parts; glycerine, 10 parts; Water, 50 parts; alcohol, 10 parts.

Paste Ink to Write with Water.—I.—Black.—Take 4 parts of bichromate of potash, pulverized, and mixed with 25 parts of acetic acid; 50 parts of liquid extract of logwood; ¼ part of picric acid; 10 parts of pulverized sal sorrel; 10 parts of mucilage; and ½ part of citrate of iron, and mix well. The liquid extract of logwood is prepared by mixing 3 parts of an extract of common commercial quality with 2 parts of water.

II.—Red.—Take 1 part of red aniline mixed with 10 parts of acetic acid; 5 parts of citric acid, and 25 parts of mucilage, all well mixed. For use, mix 1 part of the paste with 16 parts of water.

III.—Blue.—Take 2 parts of aniline blue mixed with 10 parts of acetic acid; 5 parts of citric acid, and 40 parts of mucilage, all well mixed. For use, mix 1 part of the paste with 8 parts of water.

IV.—Violet.—Use the same ingredients in the same proportions as blue, with the difference that violet aniline is used instead of blue aniline.

V.—Green.—Take 1 part of aniline blue; 3 parts of picric acid, mixed with 10 parts of acetic acid; 3 parts of citric acid, and 80 parts of mucilage. For use, 1 part of this paste is mixed with 8 parts of water.

VI.—Copying.—Take 6 parts of pulverized bichromate of potash, mixed with 10 parts of acetic acid and 240 parts of liquid extract of logwood, and add a pulverized mixture of 35 parts of alum, 20 parts of sal sorrel, and 20 parts mucilage. Mix well. For use, 1 part of this paste is mixed with 4 parts of hot water.

Purple Ink.—I.—A strong decoction of logwood, to which a little alum or chloride of tin has been added.

II. (Normandy).—To 12 pounds of Campeachy wood add as many gallons of boiling water. Pour the solution through a funnel with a strainer made of coarse flannel, or 1 pound of hydrate, or acetate of deutoxide of copper finely powdered (having at the bottom of the funnel a piece of sponge); then add immediately 14 pounds of alum, and for every 340 gallons of liquid add 80 pounds of gum arabic or gum senegal. Let these remain for 3 or 4 days, and a beautiful purple color will be produced.

Red Ink.—Brazil wood, ground, 4 ounces; white wine vinegar, hot, 1¼ pints. Digest in a glass or a well-tinned copper or enamel saucepan, until the next day; then gently simmer for half an hour, adding toward the end gum arabic and alum, of each, ½ ounce.

Inks for Shading Pen.—The essential feature in the ink for use with a shading pen is simply the addition of a sufficient quantity of acacia or other mucilaginous substance to impart a proper degree of consistency to the ink. A mixture of 2 parts of mucilage of acacia with 8 of ink gives about the required consistency. The following formulas will probably be found useful:

I.—Water-soluble nigro-
sin .... 1 part
Water............. 9 parts
Mucilage acacia.... 1 part

II.—Paris violet........ 2 parts
Water............. 6 parts
Mucilage acacia.... 2 parts

III.—Methyl violet....... 1 part
Distilled water..... 7 parts
Mucilage acacia.... 2 parts

IV.—Bordeaux red...... 3 parts
Alcohol........... 2 parts
Water............ 20 parts
Mucilage acacia.... 2 parts

V.—Rosaniline acetate.. 2 parts
Alcohol........... 1 part
Water............ 10 parts
Mucilage acacia.... 2 parts

Silver Ink.—I.—Triturate in a mortar equal parts of silver foil and sulphate of potassa, until reduced to a fine powder; then wash the salt out, and mix the residue with a mucilage of equal parts of gum arabic water.

II.—Make as gold ink, but use silver leaf or silver bronze powder.

III.—Oxide of zinc ...... 30 grains
Mucilage .... 1 ounce
Spirit of wine ...... 40 drops
Silver bronze ...... 3 drachms

Rub together, until perfectly smooth,

the zinc and mucilage, then add the spirit of wine and silver bronze and make up the quantity to 2 ounces with water.

Violet Ink.—I.—For 2 gallons, heat 2 gills of alcohol on a water bath. Add to the alcohol 2 ounces of violet aniline, and stir till dissolved; then add the mixture to 2 gallons of boiling water; mix well, and it is ready for use. Smaller quantities in proportion.

II.—Another good violet ink is made by dissolving some violet aniline in water to which some alcohol has been added. It takes very little aniline to make a large quantity of the ink.

White Ink (for other White Inks see Blueprint Inks).—So-called white inks are, properly speaking, white paints, as a white solution cannot be made. A paint suitable for use as an "ink" may be made by grinding zinc oxide very fine on a slab with a little tragacanth mucilage, and then thinning to the required consistency to flow from the pen. The mixture requires shaking or stirring from time to time to keep the pigment from separating. The "ink" may be preserved by adding a little oil of cloves or other antiseptic to prevent decomposition of the mucilage.

White marks may sometimes be made on colored papers by the application of acids or alkalies. The result, of course, depends on the nature of the coloring matter in each instance, and any "ink" of this kind would be efficacious or otherwise, according to the coloring present in the paper.

Yellow Ink.—I.—Gamboge (in coarse powder), 1 ounce; hot water, 5 ounces. Dissolve, and when cold, add of spirit, ¾ ounce.

II.—Boil French berries, ½ pound, and alum, 1 ounce, in rain water, 1 quart, for ¼ an hour, or longer, then strain and dissolve in the hot liquor gum arabic, 1 ounce.

Waterproof Ink (see also Indelible Inks).—Any ordinary ink may be made waterproof by mixing with it a little ordinary glue. After waterproofing ink in this way it is possible to wash drawings with soap and water, if necessary, without the ink running at all.

White Stamping Ink.—

Zinc white	2	drachms
White precipitate	5	grains
Mucilage	1	drachm
Water	6	drachms

Triturate the zinc white with a small quantity of water till quite smooth, then add the mucilage and the remainder of the water.

INK FOR THE LAUNDRY:
See Laundry Preparations.

INK FOR LEATHER FINISHERS:
See Leather.

INKS FOR TYPEWRITERS:
See Typewriter Ribbons.

INK FOR WRITING ON GLASS:
See Etching and Glass.

INLAYING BY ELECTROLYSIS.
See also Electro-etching, under Etching.

The process consists in engraving the design by means of the sand-blast and stencils on the surface of the article. The design or pattern is rendered conductive and upon this conductive surface a precipitate of gold, silver, platinum, etc., is applied, and fills up the hollows. Subsequently the surface is ground smooth.

Insect Bites

REMEDIES FOR INSECT BITES.

I.—	Carbolic acid	15	grains
	Glycerine	2	drachms
	Rose water	4	ounces
II.—	Salicylic acid	15	grains
	Collodion	2½	drachms
	Spirit of ammonia	5½	drachms
III.—	Fluid extract rhus toxicodendron	1	drachm
	Water	8	ounces
IV.—	Ipecac, in powder	1	drachm
	Alcohol	1	ounce
	Ether	1	ounce
V.—	Betanaphthol	30	grains
	Camphor	30	grains
	Lanolin cold cream	1	ounce

VI.—Spirit of sal ammoniac, whose favorable action upon fresh insect bites is universally known, is often unavailable. A simple means to alleviate the pain and swelling due to such bites, when still fresh, is cigar ashes. Place a little ashes upon the part stung, add a drop of water—in case of need beer, wine, or coffee may be used instead—and rub the resulting paste thoroughly into the skin. It is preferable to use fresh ashes of tobacco, because the recent heat offers sufficient guarantee for absolute freedom from impurities. The action of the tobacco ashes is due to the presence of potassium carbonate, which, like spirit of sal ammoniac, deadens the effect of the small quantities of acid (formic acid, etc.) which have been introduced into the small wound by the biting insect.

Insecticides

(See also Petroleum.)

The Use of Hydrocyanic Acid Gas for Exterminating Household Insects.—Recent successful applications of hydrocyanic acid gas for the extermination of insects infecting greenhouse plants have suggested the use of the same remedy for household pests. It is now an established fact that 1½ grains of 98 per cent pure cyanide of potassium volatilized in a cubic foot of space, will, if allowed to remain for a period of not less than 3 hours, kill all roaches and similar insects.

It may be stated that a dwelling, office, warehouse, or any building may be economically cleared of all pests, provided that the local conditions will permit the use of this gas. It probably would be dangerous to fumigate a building where groceries, dried fruits, meats, or prepared food materials of any kind are stored. Air containing more than 25 per cent of the gas is inflammable; therefore it would be well to put out all fire in an inclosure before fumigating. Hydrocyanic acid, in all its forms, is one of the most violent poisons known, and no neglect should attend its use. There is probably no sure remedy for its effects after it has once entered the blood of any of the higher animals. When cyanide of potassium is being used it should never be allowed to come in contact with the skin, and even a slight odor of the gas should be avoided. Should the operator have any cut or break in the skin of the hands or face it should be carefully covered with court-plaster to prevent the gas coming in contact with the flesh, or a small particle of the solid compound getting into the cut might cause death by poisoning in a few minutes' time.

Hydrocyanic acid gas should not be used in closely built apartments with single walls between, as more or less of the gas will penetrate a brick wall. An inexperienced person should never use cyanide of potassium for any purpose, and if it be found practicable to treat buildings in general for the extermination of insects, the work should be done only under the direction of competent officials. Experiments have shown that a smaller dose and a shorter period of exposure are required to kill mice than for roaches and household insects generally, and it readily follows that the larger animals and human beings would be more quickly overcome than mice, since a smaller supply of pure air would be required to sustain life in mice, and small openings are more numerous than large ones.

The materials employed and the method of procedure are as follows: After ascertaining the cubic content of the inclosure, provide a glass or stoneware (not metal) vessel of 2 to 4 gallons capacity for each 5,000 cubic feet of space to be fumigated. Distribute the jars according to the space, and run a smooth cord from each jar to a common point near an outside door where they may all be fastened; support the cord above the jar by means of the back of a chair or other convenient object in such a position that when the load of cyanide of potassium is attached it will hang directly over the center of the jar. Next weigh out upon a piece of soft paper about 17 ounces of 98 per cent pure cyanide of potassium, using a large pair of forceps for handling the lumps; wrap up and place in a paper bag and tie to the end of the cord over the jar. After the load for each jar has been similarly provided, it is well to test the working of the cords to see that they do not catch or bind. Then remove the jar a short distance from under the load of cyanide and place in it a little more than a quart of water, to which slowly add 1½ pints of commercial sulphuric acid, stirring freely. The action of the acid will bring the temperature of the combination almost to the boiling point. Replace the jars beneath the bags of cyanide, spreading a large sheet of heavy paper on the floor to catch any acid that may possibly fly over the edge of the jar when the cyanide is dropped, or as a result of the violent chemical action which follows. Close all outside openings and open up the interior of the apartment as much as possible, in order that the full strength of the gas may reach the hiding places of the insects. See that all entrances are locked or guarded on the outside to prevent persons entering; then leave the building, releasing the cords as you go. The gas will all be given off in a few minutes, and should remain in the building at least 3 hours.

When the sulphuric acid comes in contact with the cyanide of potassium the result is the formation of sulphate of potash, which remains in the jar, and the hydrocyanic acid is liberated and escapes into the air. The chemical action is so violent as to cause a sputtering, and frequently particles of the acid are thrown over the sides of the jar; this may be prevented by supporting a sheet of stiff paper over the jar by means of a hole in the center, through which the cord supporting the cyanide of potassium is passed, so that when the cord is released the paper will descend with the cyanide and remain at rest on the top of the jar, but will not prevent the easy descent of the cyanide into the acid. The weight of this paper will in no way interfere with the escape of the gas.

At the end of the time required for fumigation, the windows and doors should be opened from the outside and the gas allowed to escape before anyone enters the building. A general cleaning should follow, as the insects leave their hiding places and, dying on the floors, are easily swept up and burned. The sulphate of potash remaining in the jars is poisonous and should be immediately buried and the jars themselves filled with earth or ashes. No food that has remained during fumigation should be used, and thorough ventilation should be maintained for several hours. After one of these experiments it was noted that ice water which had remained in a closed cooler had taken up the gas, and had both the odor and taste of cyanide.

For dwellings one fumigation each year would be sufficient, but for storage houses it may be necessary to make an application every 3 or 4 months to keep them entirely free from insect pests. The cost of materials for one application is about 50 cents for each 5,000 cubic feet of space to be treated. The cyanide of potassium can be purchased at about 35 cents per pound, and the commercial sulphuric acid at about 4 cents per pound. The strength of the dose may be increased and the time of exposure somewhat shortened, but this increases the cost and does not do the work so thoroughly. In no case, however, should the dose remain less than 1 hour.

The application of this method of controlling household insects and pests generally is to be found in checking the advance of great numbers of some particular insect, or in eradicating them where they have become thoroughly established. This method will be found very advantageous in clearing old buildings and ships of cockroaches.

APPLICATIONS FOR CATTLE, POULTRY, ETC.:

See also Veterinary Formulas.

Fly Protectives for Animals.—

I.—

Oil of cloves	3 parts
Bay oil	5 parts
Eucalyptus tincture	5 parts
Alcohol	150 parts
Water	200 parts

II.—Tar well diluted with grease of any kind is as effective an agent as any for keeping flies from cattle. The mixture indicated has the advantage of being cheap. Applying to the legs, neck, and ears will usually be sufficient.

Cattle Dip for Ticks.—Dr. Noorgard of the Bureau of Animal Industry finds the following dip useful, immersion lasting one minute:

Sulphur	86 pounds
Extra dynamo oil	1,000 gallons

Insecticides for Animals.—

I.—Bay oil	500	
Naphthalene	100	
Camphor	60	Parts by weight.
Animal oil	25	
II.—Bay oil, pressed	400	
Naphthalene	100	
Crude carbolic acid	10	

For Dogs, Cats, etc.—The following is an excellent powder for the removal of fleas from cats or dogs:

Naphthalene	4 av. ounces
Starch	12 av. ounces

Reduce to fine powder. A few grains of lampblack added will impart a light gray color, and if desirable a few drops of oil of pennyroyal or eucalyptus will disguise the naphthalene odor.

Rub into the skin of the animal and let the powder remain for a day or two, when the same can be removed by combing or giving a bath, to which some infusion of quassia or quassia chips has been added. This treatment is equally efficient for lice and ticks.

Poultry Lice Destroyer.—I.—Twenty pounds sublimed sulphur; 8 pounds fuller's earth; 2 pounds powdered naphthalene; ½ ounce liquid carbolic acid. Mix thoroughly and put up in half-pound tins or boxes. Sprinkle about the nest for use.

II.—Oil of eucalyptus smeared about the coop will cause the parasites to leave. To drive them out of the nests of sitting hens, place in the nest an egg that has been emptied, and into which has been inserted a bit of sponge imbibed in essence of eucalyptus. There may be used also a concentrated solution of extract of tobacco, to which phenol has been added.

III.—Cover the floor or soil of the house with ground or powdered plaster, taken from old walls, etc.

ANT DESTROYERS:

A most efficacious means of getting rid of ants is spraying their resorts with petroleum. The common oil is worth more for this purpose than the refined. Two thorough sprayings usually suffice.

In armoires, dressing cases, etc., oil of turpentine should be employed. Pour it in a large plate, and let it evaporate freely. Tobacco juice is another effective agent, but both substances have the drawback of a very penetrating and disagreeable odor.

Boiling water is deadly to ants wherever it can be used (as in the garden, or yard around the house). So is carbon disulphide injected into the nests by aid of a good, big syringe. An emulsion of petroleum and water (oil, 1 part; water, 3 parts) poured on the earth has proven very efficacious, when plentifully used (say from 1 ounce to 3 ounces to the square yard). A similar mixture of calcium sulphide and water (calcium sulphide, 100 parts; water, 1,000 parts; and the white of 1 egg to every quart of water) poured into their holes is also effective.

A weak solution of corrosive sublimate is very deadly to ants. Not only does it kill them eventually, but it seems to craze them before death, so that ants of the same nest, after coming into contact with the poison, will attack each other with the greatest ferocity.

Where ants select a particular point for their incursions it is a good plan to surround it with a "fortification" of obnoxious substance. Sulphur has been used successfully in this way, and so has coal oil. The latter, however, is not a desirable agent, leaving a persistent stain and odor.

The use of carbon disulphide is recommended to destroy ants' nests on lawns. A little of the disulphide is poured into the openings of the hills, stepping on each as it is treated to close it up. The volatile vapors of the disulphide will penetrate the chambers of the nest in every direction, and if sufficient has been used will kill not only the adult insects but the larvæ as well. A single treatment is generally sufficient.

Formulas to Drive Ants Away.—

I.—Water.............. 1 quart
Cape aloes......... 4 ounces

Boil together and add:
Camphor in small
pieces............ 1½ ounces

II.—Powdered cloves.... 1 ounce
Insect powder...... 1 ounce

Scatter around where ants infest.

III.—Cape aloes......... ½ pound
Water............. 4 pints

Boil together and add camphor gum, 3 ounces. Sprinkle around where the ants infest.

BEDBUG DESTROYERS.

A good bug killer is benzine, pure and simple, or mixed with a little oil of mirbane. It evaporates quickly and leaves no stain. The only trouble is the inflammability of its vapor.

The following is a popular preparation: To half a gallon of kerosene oil add a quart of spirit of turpentine and an ounce of oil of pennyroyal. This mixture is far less dangerous than benzine. The pennyroyal as well as the turpentine are not only poisonous but exceedingly distasteful to insects of all kinds. The kerosene while less quickly fatal to bugs than benzine is cheaper and safer, and when combined with the other ingredients becomes as efficient.

Where the wall paper and wood work of a room have become invaded, the usual remedy is burning sulphur. To be efficient the room must have every door, window, crevice, and crack closed. The floor should be wet in advance so as to moisten the air. A rubber tube should lead from the burning sulphur to a key-hole or auger-hole and through it, and by aid of a pair of bellows air should be blown to facilitate the combustion of the sulphur.

Pastes.—Some housewives are partial to corrosive sublimate for bedbugs; but it is effective only if the bug eats the poison. The corrosive sublimate cannot penetrate the waxy coat of the insect. But inasmuch as people insist on having this a few formulas are given.

I.—Common soap...... 1 av. ounce
Ammonium chlo-
ride .... 3 av. ounces
Corrosive sublimate 3 av. ounces
Water enough to make 32 fluid-
ounces.

Dissolve the salts in the water and add the soap.

This will make a paste that can be painted with a brush around in the cracks and crevices. Besides, it will make an excellent filling to keep the cracks of the wall and wainscoting free from bugs of all kinds. The formula could be modified so as to permit the use of Paris green or London purple, if desired. A decoction of quassia could be used to dissolve the soap. The latter paste would, of course, not be poisonous, and in many instances it would be preferred. It is possible to make a cold infusion of white hellebore of 25 per cent strength, and in 1 quart of infusion dissolve 1 ounce of common soap. The advantage of the soap paste is simply to keep the poisonous substance thoroughly distributed throughout the mass at all times. The density of the paste can be varied to suit. Kerosene oil or turpentine could replace 6 ounces or 8 ounces of the water in making the paste, and either of these would make a valuable addition.

Another paste preparation which will meet with hearty recommendation is blue ointment. This ointment, mixed with turpentine or kerosene oil, can be used to good advantage; especially so as the turpentine is so penetrating that both it and the mercury have a chance to act more effectually. It can be said that turpentine will kill the bedbug if the two come in contact; and kerosene is not far behindhand in its deadly work.

II.—Blue ointment...... 1 ounce
    Turpentine......... 3 ounces

Stir well together.

Liquid Bedbug Preparations.—There is no doubt that the liquid form is the best to use; unlike a powder, or even a paste, it will follow down a crack into remote places where bugs hide, and will prevent their escape, and it will also kill the eggs and nits. The following substances are the most employed, and are probably the best: Kerosene, turpentine, benzine, carbolic acid, corrosive sublimate solution, oil pennyroyal, and strong solution of soap. Here are several good formulas that can be depended upon:

I.—Oil of pennyroyal... 1 drachm
    Turpentine........ 8 ounces
    Kerosene oil, enough to make 1 gallon.

Put up in 8-ounce bottles as a bedbug exterminator.

II.—Oil of eucalyptus... 1 drachm
    Eucalyptus leaves... 1 ounce
    Benzine.......... 2 ounces
    Turpentine........ 2 ounces
    Kerosene enough to make 16 ounces.

Mix the turpentine, benzine, and kerosene oil, and macerate the eucalyptus leaves in it for 24 hours; then strain and make up the measure to 1 pint, having first added the oil of eucalyptus.

FLY-KILLERS.

A fly poison that is harmless to man may be made from quassia wood as follows:

Quassia......... 1,000 parts
Molasses .... 150 parts
Alcohol......... 50 parts
Water.......... 5,750 parts

Macerate the quassia in 500 parts of water for 24 hours, boil for half an hour, set aside for 24 hours, then press out the liquid. Mix this with the molasses and evaporate to 200 parts. Add the alcohol and the remaining 750 parts of water, and without filtering, saturate absorbent paper with it.

This being set out on a plate with a little water attracts the flies, which are killed by partaking of the liquid.

Sticky Preparations.—

I.—Rosin........... 150 parts
    Linseed oil....... 50 parts
    Honey.......... 18 parts

Melt the rosin and oil together and stir in the honey.

II.—Rapeseed oil..... 70 parts
    Rosin........... 30 parts

Mix and melt together.

III.—Rosin........... 60 parts
    Linseed oil....... 38 parts
    Yellow wax...... 2 parts

IV.—Rosin........... 10 parts
    Turpentine...... 5 parts
    Rapeseed oil..... 5 parts
    Honey.......... 1 part

Sprinkling Powders for Flies.—

I.—Long peppers, powdered............ 5 parts
    Quassia wood, powdered............ 5 parts
    Sugar, powdered.... 10 parts

Mix, moisten the mixture with 4 parts of alcohol, dry, and again powder. Keep the powder in closely stoppered jars, taking out a sufficient quantity as desired.

II.—Orris root, powdered 4 parts
    Starch, powdered.... 15 parts
    Eucalyptol.......... 1 part

Mix. Keep in a closely stoppered jar or box. Strew in places affected by flies.

Fly Essences.—

I.—Eucalyptol......... 10 parts
    Bergamot oil....... 3 parts
    Acetic ether........ 10 parts
    Cologne water...... 50 parts
    Alcohol, 90 per cent. 100 parts

Mix. One part of this "essence" is

to be added to 10 parts of water and sprayed around the rooms frequently.

II.—Eucalyptol........ 10 parts
    Acetic ether....... 5 parts
    Cologne water..... 40 parts
    Tincture of insect
        powder (1:5).... 50 parts

REMEDIES AGAINST HUMAN PARASITES:

By weight
I.—Yellow wax....... 85 parts
    Spermaceti....... 60 parts
    Sweet oil........ 500 parts

Melt and add:
    Boiling distilled
        water.......... 150 parts

After cooling add:
    Clove oil.......... 2 parts
    Thyme oil........ 3 parts
    Eucalyptus oil.... 4 parts

II.—Bay oil, pressed... 100 parts
    Acetic ether....... 12 parts
    Clove oil.......... 4 parts
    Eucalyptus oil..... 3 parts

For Head Lice in Children.—One of the best remedies is a vinegar of sabadilla. This is prepared as follows: Sabadilla seed, 5 parts; alcohol, 5 parts; acetic acid, 9 parts; and water, 36 parts. Macerate for 3 days, express and filter. The directions are: Moisten the scalp and hair thoroughly at bedtime, binding a cloth around the head, and let remain overnight. If there are any sore spots on the scalp, these should be well greased before applying the vinegar.

To Exterminate Mites.—Mix together 10 parts of naphthalene, 10 parts of phenic acid, 5 parts of camphor, 5 parts of lemon oil, 2 parts of thyme oil, 2 parts of oil of lavender, and 2 parts of the oil of juniper, in 500 parts of pure alcohol.

Vermin Killer.—
    Sabadilla, powder.. 2 av. ounces
    Acetic acid........ ½ fluidounce
    Wood alcohol..... 2 fluidounces
    Water sufficient to make 16 fluid
        ounces.

Mix the acetic acid with 14 fluidounces of water and boil the sabadilla in this mixture for 5 to 10 minutes, and when nearly cold add the alcohol, let stand, and decant the clear solution and bottle.

Directions: Shake the bottle and apply to the affected parts night and morning.

INSECTICIDES FOR PLANTS.

Two formulas for insecticides with especial reference to vermin which attack plants:

I.—Kerosene.......... 2 gallons
    Common soap...... ½ pound
    Water............. 1 gallon

Heat the solution of soap, add it boiling hot to the kerosene and churn until it forms a perfect emulsion. For use upon scale insects it is diluted with 9 parts of water; upon other ordinary insects with 15 parts of water, and upon soft insects, like plant lice, with from 20 to 25 parts of water.

For lice, etc., which attack the roots of vines and trees the following is recommended:

II.—Caustic soda .... 5 pounds
    Rosin.............. 40 pounds
    Water, a sufficient quantity.

Dissolve the soda in 4 gallons of water, by the aid of heat, add the rosin and after it is dissolved and while boiling add, slowly, enough water to make 50 gallons. For use, 1 part of this mixture is diluted with 10 parts of water and about 5 gallons of the product poured into a depression near the root of the vine or tree.

For Cochineal Insects.—An emulsion for fumagine (malady of orange trees caused by the cochineal insect) and other diseases caused by insects is as follows:

Dissolve, hot, 4 parts of black soap in 15 parts of hot water. Let cool to 104° F., and pour in 10 parts of ordinary petroleum, shaking vigorously. Thus an emulsion of *café au lait* color is obtained, which may be preserved indefinitely. For employment, each part of the emulsion is diluted, according to circumstances, with from 10 to 20 parts of water.

For Locusts.—Much trouble is experienced in the Transvaal and Natal with locust pests, the remedies used being either a soap spray, containing 1 pound ordinary household soap in 5 gallons of water, or arsenite of soda, the latter being issued by the government for the purpose, and also used for the destruction of prickly pear, and as a basis of tick dips. A solution of 1 pound in 10 gallons of water is employed for full-grown insects, and of 1 pound in 20 gallons of water for newly hatched ones, 1 pound of sugar being added to each pound of arsenite dissolved. The solution sometimes causes sores on the skin, and the natives employed in its use are given grease to rub over themselves as a measure of protection. An advantage of the arsenite solution over soap is that much less liquid need be used.

A composition for the destruction of pear blight, which has been patented in the United States, is as follows: Peppermint oil, 16 parts; ammonia water, 60 parts; calomel, 30 parts; and linseed oil, 1,000 parts.

For Moths and Caterpillars.—

I.—		
Venice turpentine	200	parts
Rosin	1,000	parts
Turpentine	140	parts
Tar	80	parts
Lard	500	parts
Rape oil	240	parts
Tallow	200	parts

II.—		
Rosin	50	parts
Lard	40	parts
Stearine oil	40	parts

For Non-Masticating Insects.—For protection against all non-masticating and many mandibulate insects, kerosene oil is much used. It is exhibited in the form of emulsion, which may be made as follows:

Kerosene	2 gallons
Common soap	8 ounces
Water	1 gallon

Dissolve the soap in the water by the aid of heat, bring to the boiling point, and add the kerosene in portions, agitating well after each addition. This is conveniently done by means of the pump to be used for spraying the mixture.

For Scale Insects.—For destroying scale insects dilute the cochineal emulsion (see above) with 9 times its volume of water; in the case of most others, except lice, dilute with 14 volumes, and for the latter with 20 to 25 volumes.

For the extermination of scale insects, resinous preparations are also employed, which kill by covering them with an impervious coating. Such a wash may be made as follows:

Rosin	3½	pounds
Caustic soda	1	pound
Fish oil	8	ounces
Water	20	gallons

Boil the rosin, soda, and oil with a small portion of the water, adding the remainder as solution is effected.

For the San José scale a stronger preparation is required, the proportion of water being decreased by half, but such a solution is applied only when the tree is dormant.

Scale Insects on Orange Trees.—Scale insect enemies of orange trees are directly controlled in two ways: (1) By spraying the infested trees with some liquid insecticide, and (2) by subjecting them to the fumes of hydrocyanic acid gas, commonly designated as "gassing." The latter method is claimed to be the most effective means known of destroying scale insects. In practice the method consists in closing a tree at night with a tent and filling the latter with the poisonous fumes generated by treating refined potassium cyanide (98 per cent) with commercial sulphuric acid (66 per cent) and water. The treatment should continue from 30 to 40 minutes, the longer time being preferable. The work is done at night to avoid the scalding which follows day applications, at least in bright sunshine.

The oily washes are said to be the best for the use by the spraying method. "Kerosene emulsion" is a type of these washes. A formula published by the United States Department of Agriculture follows: Kerosene, 2 gallons; whale-oil soap, ½ pound; water, 1 gallon. The soap is dissolved in hot water, the kerosene added, and the whole thoroughly emulsified by means of a power pump until a rather heavy, creamy emulsion is produced. The quantity of soap may be increased if desired. The insecticide is applied by spraying the infected tree with an ordinary force pump with spraying nozzle.

Coating Against the Plant Louse.—(a)—Mix 75 parts of green soap, 50 parts of linseed oil, and 25 parts of carbolic acid. Afterwards mix the mass with 15,000 parts of water.

(b) Mix 4 parts of carbolic acid with 100 parts water glass.

Louse Washes.—

Unslaked lime	18	parts
Sulphur	9	parts
Salt	6.75	parts

Mix as follows: A fourth part of the lime is slaked and boiled for ⅔ of an hour with the sulphur in 22.6 parts of water. The remainder of the lime is then slaked and added with the salt to the hot mixture. The whole is burned for another half hour or an hour, and then diluted to 353 parts. The fluid is applied lukewarm when the plants are not in active growth.

For Slugs on Roses.—

Powdered pyrethrum	8 ounces
Powdered colocynth	4 ounces
Powdered hellebore	16 ounces

Flea Powder.—

Naphthalene	4 ounces
Talcum	10 ounces
Tobacco dust	2 ounces

To Keep Flaxseed Free from Bugs.— As a container use a tin can with a close-fitting top. At the bottom of the can place a small vial of chloroform with a loose-fitting cork stopper. Then pour the flaxseed, whole or ground, into the can, covering the vial. Enough of the chloroform will escape from the vial to kill such insects as infest the flaxseed.

INSECT POWDERS.

Pyrethrum, whale oil (in the form of soap), fish oil (in the form of soap), soft soap, paraffine, Prussic acid, Paris green, white lead, sulphur, carbon bisulphide, acorus calamus, camphor, Cayenne pepper, tobacco, snuff, asafetida, white hellebore, eucalyptol, quassia, borax, acetic ether are most important substances used as insecticides, alone, or in combination of two or more of them. The Prussic acid and Paris green are dangerous poisons and require to be used with extreme care:

Insect powder is used for all small insects and as a destroyer of roaches. The observations of some experimenters seem to show that the poisonous principle of these flowers is non-volatile, but the most favorable conditions under which to use them are in a room tightly closed and well warmed. There may be two poisonous principles, one of which is volatile. Disappointment sometimes arises in their use from getting powder either adulterated, or which has been exposed to the air and consequently lost some of its efficiency.

The dust resulting from the use of insect powder sometimes proves irritating to the mucous membranes of the one applying the powder. This is best avoided by the use of a spray atomizer.

Persistence in the use of any means is an important element in the work of destroying insects. A given poison may be employed and no visible result follow at first, when in reality many may have been destroyed, enough being left to deceive the observer as to numbers. They multiply very rapidly, too, it must be remembered, and vigorous work is required to combat this increase. Where they can easily migrate from one householder's premises to those of another, as in city "flats," it requires constant vigilance to keep them down, and entire extermination is scarcely to be expected.

The ordinary insect powder on the market is made from pyrethrum carneum, pyrethrum roseum, and pyrethrum cinerariæ-folium. The first two are generally ground together and are commercially called Persian insect powder; while the third is commonly called Dalmatian insect powder. These powders are sold in the stores under many names and in combination with other powders under proprietary names.

The powder is obtained by crushing the dried flowers of the pellitory (pyrethrum). The leaves, too, are often used. They are cultivated in the Caucasus, whence the specific name Caucasicum sometimes used. Pyrethrum belongs to the natural order compositæ, and is closely allied to the chrysanthemum. The active principle is not a volatile oil, as stated by some writers, but a rosin, which can be dissolved out from the dry flowers by means of ether. The leaves also contain this rosin but in smaller proportions than the flowers. Tincture of pyrethrum is made by infusing the dried flowers in five times their weight of rectified spirit of wine. Diluted with water it is used as a lotion.

Borax powder also makes a very good insectifuge. It appears to be particularly effective against the common or kitchen cockroach. Camphor is sometimes used, and the powdered dried root of acorus calamus, the sweet flag. A mixture of white lead with four times its weight of chalk is also highly recommended. The fish-oil soaps used in a powdered form are made from various recipes, of which the following is a typical example:

Powdered rosin......	2 pounds
Caustic soda........	8 ounces
Fish or whale oil.....	4 ounces

Boil together in a gallon of water for at least an hour, replacing some of the water if required.

The following insect-powder formulas are perfectly safe to use. In each instance insect powder relates to either one of the pyrethrum plants powdered, or to a mixture:

I.—Insect powder.... 8 ounces av.
    Powdered borax.. 8 ounces av.
    Oil of pennyroyal. 2 fluidrachms

II.—Insect powder.... 8 ounces av.
    Borax.......... 8 ounces av.
    Sulphur........ 4 ounces av.
    Oil of eucalyptus. 2 fluidrachms

This formula is especially good for cockroaches:

III.—Insect powder....14 ounces av.
    Quassia in fine
      powder........ 6 ounces av.
    White hellebore,
      powdered...... 2 ounces av.

Beetle Powder.—

Cocoa powder.......	4 ounces
Starch.............	8 ounces
Borax.............	37 ounces

Mix thoroughly.

Remedies Against Mosquitoes.—A remedy to keep off mosquitoes, etc., is composed as follows: Cinnamon oil, 1 part; patchouli oil, 1 part; sandal oil, 4 parts; alcohol, 400 parts. This has a pleasant odor.

Oil of pennyroyal is commonly used to keep mosquitoes away. Some form of petroleum rubbed on the skin is even more efficient, but unpleasant to use, and if left on long enough will burn the skin.

A 40 per cent solution of formaldehyde for mosquito bites gives remarkably quick and good results. It should be applied to the bites as soon as possible with the cork of the bottle, and allowed to dry on. Diluted ammonia is also used to rub on the bites.

Roach Exterminators.— Borax, starch, and cocoa are said to be the principal ingredients of some of the roach foods on the market. A formula for a poison of this class is as follows:

Borax..............	37 ounces
Starch.............	9 ounces
Cocoa.............	4 ounces

Moth Exterminators.—Cold storage is the most effective means of avoiding the ravages of moths. Where this is impracticable, as in bureau drawers, camphor balls may be scattered about with satisfactory result. The following is also effective:

Spanish pepper.....	100 parts
Turpentine oil......	50 parts
Camphor..........	25 parts
Clove oil..........	10 parts
Alcohol, 96 per cent.	900 parts

Cut the Spanish pepper into little bits, and pour over them the alcohol and oil of turpentine. Let stand 2 or 3 days, then decant, and press out. To the liquid thus obtained add the camphor and clove oil, let stand a few days, then filter and fill into suitable bottles. To use, imbibe bits of bibulous paper in the liquid and put them in the folds of clothing to be protected.

Protecting Stuffed Furniture from Moths.—The stuffing, no matter whether consisting of tow, hair, or fiber, as well as the covering, should be coated with a 10 per cent solution of sulphur in carbon sulphide. The carbon sulphide dissolves the sulphur so as to cause a very fine division and to penetrate the fibers completely.

Powder to Keep Moths Away.—

Cloves..............	2 ounces
Cinnamon..........	2 ounces
Mace..............	2 ounces
Black pepper........	2 ounces
Orris root..........	2 ounces

Powder coarsely and mix well together.

Book-Worms.—When these insects infest books they are most difficult to deal with, as the ordinary destructive agents injuriously affect the paper of the book. The books should be well beaten and exposed to the sun, and a rag moistened with formalin passed through the binding and the covers where possible. In other cases the bottom edge of the binding should be moistened with formalin before putting on the shelves, so that formaldehyde vapor can be diffused.

INSECT POWDERS:
See Insecticides.

INSECT TRAP.

Into a china wash-basin, half filled with water, pour a glass of beer; cover the basin with a newspaper, in the center of which a small round hole is cut. Place it so that the edges of the paper lie on the floor and the hole is over the center of the basin. At night beetles and other insects, attracted by the smell of beer, climb the paper and fall through the hole into the liquid.

INSTRUMENT ALLOYS:
See Alloys.

INSTRUMENT CLEANING:
See Cleaning Preparations and Methods.

INSTRUMENT LACQUER:
See Lacquers.

Insulation

ELECTRIC INSULATION:

Insulating Varnishes. — For earth cables and exposed strong current wires:

I.—Melt 2 parts of asphalt together with 0.4 parts of sulphur, add 5 parts of linseed-oil varnish, linseed oil or cottonseed oil, keep at 320° F. for 6 hours; next pour in oil of turpentine as required.

II.—Maintain 3 parts of elaterite with 2 parts of linseed-oil varnish at 392° F. for 5 to 6 hours; next melt 3 parts of asphalt, pour both substances together, and again maintain the temperature of

392° F. for 3 to 4 hours, and then add 1 part of linseed-oil varnish and oil of turpentine as required.

III.—Insulating Varnish for Dynamos and Conduits with Low Tension.—Shellac, 4 parts; sandarac, 2 parts; linoleic acid, 2 parts; alcohol, 15 parts.

IV.—An insulating material which contains no caoutchouc is made by dissolving natural or coal-tar asphalt in wood oil, adding sulphur and vulcanizing at 572° F. The mixture of asphalt and wood oil may also be vulcanized with chloride of sulphur by the ordinary process used for caoutchouc. Before vulcanizing, a solution of rubber scraps in naphthalene is sometimes added and the naphthalene expelled by a current of steam. Substitutes for hard rubber are made of natural or artificial asphalt combined with heavy oil of tar and talc or infusorial earth.

Most of the insulating materials advertised under alluring names consist of asphalt combined with rosin, tar, and an inert powder such as clay or asbestos. Some contain graphite, which is a good conductor and therefore a very undesirable ingredient in an insulator.

INSULATION AGAINST HEAT.

An asbestos jacket is the usual insulator for boilers, steampipes, etc. The thicker the covering around the steampipe, the more heat is retained. A chief requirement for such protective mass is that it contains air in fine channels, so that there is no connection with the closed-in air. Most substances suitable for insulating are such that they can only with difficulty be used for a protective mass. The most ordinary way is to mix infusorial earth, kieselguhr, slag-wool, hair, ground cork, etc., with loam or clay, so that this plastic mass may be applied moist on the pipes. In using such substances care should be taken carefully to clean and heat the surfaces to be covered. The mass for the first coating is made into a paste by gradual addition of water and put on thick with a brush. After drying each time a further coating is applied. This is repeated until the desired thickness is reached. The last layer put on is rubbed smooth with the flat hand. Finally, strips of linen are wound around, which is coated with tar or oil paint as a protection against outside injuries. Cork stones consist of crushed cork with a mineral binding agent, and are sold pressed into various shapes.

Leather Waste Insulation.—Portions of leather, such as the fibers of sole leather of any size and form, are first rendered soft. The surface is then carded or the surface fibers scratched or raised in such a manner that when several pieces are pressed together their surface fibers adhere, and a compact, durable piece of leather is produced. The carding can be done by an ordinary batting machine, the action of which is so regulated that not only are the pieces of leather softened, but the fibers on their surfaces raised. The structure of the separate pieces of leather remains essentially unaltered. The raised fibers give the appearance of a furry substance to the leather. The batted pieces of leather are well mixed with paste or some suitable gum, either in or outside of the machine, and are then put into specially shaped troughs, where they are pressed together into layers of the required size and thickness. The separate pieces of leather adhere and are matted together. An agglutinant, if accessible, will contribute materially to the strength and durability of the product. The layers are dried, rolled, and are then ready for use. The pieces need not be packed together promiscuously. If larger portions of waste can be secured, the separate pieces can be arranged one upon another in rows. The larger pieces can also be used for the top and bottom of a leather pad, the middle portion of which consists of smaller pieces.

INSULATION AGAINST MOISTURE, WEATHER, ETC.

Experiments have shown that with the aid of red lead a very serviceable, resistive, and weatherproof insulation material may be produced from inferior fibers, to take the place, in many cases, of guttapercha and other substances employed for insulating purposes, and particularly to effect the permanent insulation of aerial conductors exposed to the action of the weather. Hackethal used for the purpose any vegetable fiber which is wrapped around the conductors to be insulated. The fiber is then saturated with liquid red lead. The latter is accomplished in the proportion of 4 to 5 parts of red lead, by weight, to 1 part, by weight, of linseed oil, by the hot or cold process, by mere immersion or under pressure. All the three substances, fiber, oil, and red lead, possess in themselves a certain insulating capacity, but none of them is alone of utility for such purposes. Even the red lead mixed with linseed oil does not possess in the liquid state a high degree of insulating power.

Only when both substances, the ingredients of the linseed oil capable of absorbing oxygen and the lead oxide rich in oxygen, oxidize in the air, a new gummy product of great insulating capacity results.

INTENSIFIERS:
See Photography.

IODINE SOLVENT.

Iodine is quickly dissolved in oils by first rubbing up the iodine with one-fourth of its weight of potassium iodide and a few drops of glycerine, then adding a little oil and rubbing up again. The addition of the resultant liquid to the rest of the oil and a sharp agitation finishes the process.

IODINE SOAP:
See Soap.

IODOFORM DEODORIZER.

Rub the part with about a teaspoonful of wine vinegar, after a previous thorough washing with soap.

Iron

(See also Metals and Steel.)

To Color Iron Blue.—One hundred and forty parts of hyposulphite of soda are dissolved in 1,000 parts of water; 35 parts of acetate of lead are dissolved in 1,000 parts of water; the two solutions are mixed, boiled, and the iron is immersed therein. The metal takes a blue color, such as is obtained by heating.

To Distinguish Iron from Steel.—The piece of metal to be tested is washed and then plunged into a solution of bichromate of potash, with the addition of considerable sulphuric acid. In half a minute or a minute the metal can be taken out, washed, and wiped. Soft steels and cast iron assume under this treatment an ash-gray tint. Tempered steels become almost black, without any metallic reflection. Puddled and refined irons remain nearly white and always have metallic reflections on the part of their surface previously filed, the remainder of the surface presenting irregular blackish spots.

Another method is to apply a magnet. Steel responds much more quickly and actively to the magnetic influence than does iron.

Powder for Hardening Iron and Steel.—For wrought iron place in the charge 20 parts, by weight, of common salt; 2 parts, by weight, of potassium cyanide; 0.3 parts, by weight, of potassium bichromate; 0.15 parts, by weight, of broken glass; and 0.1 part, by weight, of potassium nitrate for case-hardening. For cooling and hardening cast iron: To 60 parts, by weight, of water add 2.5 parts, by weight, of vinegar; 3 parts, by weight, of common salt; and 0.25 parts, by weight, of hydrochloric acid.

Preventing the Peeling of Coatings for Iron.—To obviate the scaling of coatings on iron, if exposed to the attacks of the weather, it is advisable to wash the iron thoroughly and to paint it next with a layer of boiling linseed oil. If thus treated, the paint never cracks off. If the iron objects are small and can be heated, it is advantageous to heat them previously and to dip them into linseed oil. The boiling oil enters all the pores of the metal and drives out the moisture. The coating adheres so firmly that frost, rain, nor wind can injure it.

To Soften Iron Castings.—To soften hard iron castings, heat the object to a high temperature, cover it over with fine coal dust or some similar substance, and allow it to cool gradually. When the articles are of small size, a number of them are packed in a crucible with substances yielding carbon to iron at a glowing heat. The crucible is then tightly closed, and placed in a stove or on an open fire. It is gradually heated and kept at a red heat for several hours, and then allowed to cool slowly. Cast-iron turnings, carbonate of soda, and unrefined sugar are recommended as substances suitable for packing in the crucible with the castings. If unrefined sugar alone is added, the quantity must not be too small. By this process the iron may be rendered extremely soft.

To Whiten Iron.—Mix ammoniacal salt in powder with an equal volume of mercury. This is dissolved in cold water and mixed thoroughly. Immerse the metal, heated to redness, in this bath and it will come out possessing the whiteness and beauty of silver. Care should be taken not to overheat the article and thus burn it.

IRON, BITING OFF RED HOT:
See Pyrotechnics.

IRON, CEMENTS FOR:
See Adhesives.

IRON, TO CLEAN:
See Cleaning Preparations and Methods.

IRON TO CLOTH, GLUING:
See Adhesives.

IRON, HOW TO ATTACH RUBBER TO:
See Adhesives, under Rubber Cements.

IRON OXALATE DEVELOPER:
See Photography.

IRON SOLDERS:
See Solders.

IRONING WAX:
See Laundry Preparations.

IRON VARNISHES:
See Varnishes.

ITCH, BARBERS':
See Ointments.

Ivory

(See also Bones, Shell, and Horn.)

TO COLOR IVORY:

Red.—The article is placed for 24 hours in water, 1,000 parts of which carry 100 parts of vinegar (acetic acid, 6 per cent), and from 1 to 5 parts of aniline red. As soon as it acquires the desired color pour off the liquid, let the ivory dry, and polish with Vienna lime.

Black.—Wash the article first in potash or soda lye and then put into a neutral solution of silver nitrate. Drain off the liquid and lay in the direct sunshine.

Red-Purple.—Put the article in a weak solution of triple gold chloride and then into direct sunshine.

Red.—For a different shade of red (from the first given), place the article for a short time in water weakly acidified with nitric acid and then in a solution of cochineal in ammonia.

Yellow.—Leave for several hours in a solution of lead acetate, rinse and dry. When quite dry place in a solution of potassium chromate.

To Color Billiard Balls Red.—

Fiery Red.—Wash the article first in a solution of carbonate of soda, then plunge for a few seconds in a bath of equal parts of water and nitric acid. Remove, rinse in running water; then put in an alcoholic solution of fuchsine and let it remain until it is the required color.

Cherry Red.—Clean by washing in the sodium carbonate solution, rinse and lay in a 2 per cent solution of tin chloride, for a few moments, then boil in a solution of logwood. Finally lay in a solution of potassium carbonate until it assumes the desired color.

Pale Red.—Wash in soda solution, rinse and lay for 25 minutes in a 5 per cent solution of nitric acid, rinse, then lay for several minutes in a weak solution of tin chloride. Finally boil in the following solution: Carmine, 2 parts; sodium carbonate, 12 parts; water, 200 parts; acetic acid enough to saturate.

Brown.—Apply several coats of an ammoniacal solution of potassium permanganate. Similar results are obtained if the solution is diluted with vinegar, and the ivory article allowed to remain in the liquid for some time.

Etching on Ivory (see also Etching).—Although decorations on ivory articles, such as umbrella handles, cuff-buttons, fans, book-covers, boxes, etc., are generally engraved, the work is frequently done by etching. The patterns must be very delicate, and are executed in lines only. The simplest way is to cover the surface with a thin rosin varnish. Then transfer the pattern and scratch it out accurately with a pointed needle. Otherwise proceed same as in etching on metal and stone, making an edge of modeling wax around the surface to be etched and pouring on the acid, which consists, in this case, of sulphuric acid, 1 part, to which 5 to 6 parts of water are added. It acts very quickly. The lines turn a deep black. If brown lines are desired, dissolve 1 part of silver nitrate in 5 parts of water, etch for a short time, and expose the article for a few hours to the light, until the design turns brown. Very often etchings in ivory are gilded. For this purpose, fill the etched patterns accurately with siccatives, using a writing pen, dry, and dab on gold leaf. After a few hours remove the superfluous gold with wadding, and the design will be nicely gilded. Etched ivory articles present a very handsome appearance if they are first covered with a silvery gloss, the design being gilded afterwards. For the former purpose the etched object is laid in the above described solution of silver nitrate until it has acquired a dark yellow color. Then rinse it off in clean water and, while still moist, expose to direct sunlight. After 3 to 4 hours the surface becomes entirely black, but will take on a fine silvery luster if rubbed with soft leather.

Flexible Ivory.—To soften ivory and render it flexible put pure phosphoric acid (specific gravity, 1.13) into a wide-mouthed bottle or jar that can be covered, and steep the ivory in this until it partially loses its opacity; then wash the ivory in cold, soft water and dry, when the ivory will be found soft and flexible.

It regains its hardness in course of time when freely exposed to air, although its flexibility can be restored by immersing the ivory in hot water.

Another softening fluid is prepared by mixing 1 ounce of spirit of niter with 5 ounces of water and steeping the ivory in the fluid for 4 or 5 days.

Hardened Ivory.—To restore the hardness to ivory that has been softened by the above methods, wrap it in a sheet of white writing paper, cover it with dry decrepitated salt, and let it remain thus covered for 24 hours. The decrepitated salt is prepared by strewing common kitchen salt on a plate or dish and standing same before a fierce fire, when the salt loses its crystalline appearance and assumes a dense opaque whiteness.

IMITATION IVORY:
See also Casein and Plaster.

Manufacture of Compounds Imitating Ivory, Shell, etc.—Casein, as known, may act the part of an acid and combine with bases to form caseinates or caseates; among these compounds, caseinates of potash, of soda, and of ammonia are the only ones soluble in water; all the others are insoluble and may be readily prepared by double decomposition. Thus, for example, to obtain caseinate of alumina, it is sufficient to add to a solution of casein in caustic soda a solution of sulphate of alumina; an insoluble precipitate of casein, or caseinate of alumina, is instantly formed. This precipitate ought to be freed from the sulphate of soda (formed by double decomposition) by means of prolonged washing.

When pure, ordinary cellulose may be incorporated with it by this process, producing a new compound, cheaper than pure cellulose, although possessing the same properties, and capable of replacing it in all its applications. According to the results desired, in transparency, color, hardness, etc., the most suitable cascinate should be selected. Thus, if a translucent compound is to be obtained, the caseinate of alumina yields the best. If a white compound is desired, the caseinate of zinc or of magnesia should be chosen; and for colored products the caseinates of iron, copper, and nickel will give varied tints.

The process employed for the new products, with a base of celluloid and caseinate, is as follows: On one hand casein is dissolved in a solution of caustic soda (100 of water for 10 to 25 of soda), and this liquid is filtered, to sepa-

rate the matters not dissolved and the impurities.

On the other hand, a salt (of the base of which the cascinate is desired) is dissolved, and the solution filtered. It is well not to operate on too concentrated a solution. The two solutions are mixed in a reservoir furnished with a mechanical stirrer, in order to obtain the insoluble caseinate precipitate in as finely divided a state as possible. This precipitate should be washed thoroughly so as to free it from the soda salt formed by double decomposition, but on account of its gummy or pasty state, this washing presents certain difficulties, and should be done carefully. After the washing it should be freed from the greater part of water contained by draining, followed by drying, or energetic pressing; then it is washed in alcohol, dried or pressed again, and is ready to be incorporated in the mass of the celluloid.

For the latter immersion and washing, it has been found that an addition of 1 to 5 per cent of borax is advantageous, for it renders the mass more plastic, and facilitates the operation of mixing. This may be conducted in a mixing apparatus; but, in practice, it is found preferable to effect it with a rolling mill, operated as follows:

The nitro-cellulose is introduced in the plastic state, and moistened with a solution of camphor in alcohol (40 to 50 parts of camphor in 50 to 70 parts of alcohol for 100 parts of nitro-cellulose) as it is practiced in celluloid factories.

This plastic mass of nitro-cellulose is placed in a rolling mill, the cylinders of which are slightly heated at the same time as the caseinate, prepared as above; then the whole mass is worked by the cylinders until the mixture of the two is perfectly homogeneous, and the final mass is sufficiently hard to be drawn out in leaves in the same way as practiced for pure celluloid. These leaves are placed in hydraulic presses, where they are compressed, first hot, then cold, and the block thus formed is afterwards cut into leaves of the thickness desired. These leaves are dried in an apparatus in the same way as ordinary celluloid. The product resembles celluloid, and has all its properties. At 195° to 215° F. it becomes quite plastic, and is easily molded. It may be sawed, filed, turned, and carved without difficulty, and takes on a superb polish. It burns less readily than celluloid, and its combustibility diminishes in proportion as the percentage of caseinate increases; finally, the cost price is less than that of celluloid, and by using a large proportion of caseinate, products may be manufactured at an extremely low cost.

IVORY AND BONE BLEACHES.

If simply dirty, scrub with soap and tepid water, using an old tooth or nail brush for the purpose. Grease stains may be sometimes removed by applying a paste of chalk or whiting and benzol, covering the article so that the benzol may not dry too rapidly. Carbon disulphide (the purified article) may be used in place of benzol. When dry, rub off with a stiff brush. If not removed with the first application, repeat the process. Delicately carved articles that show a tendency to brittleness should be soaked for a short time in dilute phosphoric acid before any attempt to clean them is made. This renders the minuter portions almost ductile, and prevents their breaking under cleaning.

The large scratched brush should be treated as follows: If the scratches are deep, the surface may be carefully rubbed down to the depth of the scratch, using the finest emery cloth, until the depth is nearly reached, then substituting crocus cloth.

To restore the polish nothing is superior to the genuine German putz pomade, following by rubbing first with chamois and finishing off with soft old silk. The more "elbow grease" put into the rubbing the easier the task, as the heat generated by friction seems to lend a sort of ductility to the surface. To remove the yellow hue due to age, proceed as follows: Make a little tripod with wire, to hold the object a few inches above a little vessel containing lime chloride moistened with hydrochloric acid; put the object on the stand, cover the whole with a bell glass, and expose to direct sunlight. When bleached, remove and wash in a solution of sodium bicarbonate, rinse in clear water and dry.

Like mother-of-pearl, ivory is readily cleaned by dipping in a bath of oxygenized water or immersing for 15 minutes in spirits of turpentine, and subsequently exposing to the sun for 3 or 4 days. For a simple cleaning of smooth articles, wash them in hot water, in which there has been previously dissolved 100 parts (by weight) of bicarbonate of soda per 1,000 parts of water. To clean carved ivory make a paste of very fine, damp sawdust, and put on this the juice of 1 or 2 lemons, according to the article to be treated. Now apply a layer of this sawdust on the ivory, and when dry brush it off and rub the object with a chamois.

IVORY TESTS.

Many years ago an article was introduced in the industrial world which in contradistinction to the genuine animal ivory, has its origin in the vegetable kingdom, being derived from the nut of a palm-like shrub called phytelephasmacrocarpa, whose fruit reaches the size of an apple. This fruit has a very white, exceedingly hard kernel which can be worked like ivory. A hundred of these fruits only costing about $1, their use offers great advantages. Worked on the lathe this ivory can be passed off as the genuine article, it being so much like it that it is often sold at the same price. It can also be colored just like genuine ivory.

To distinguish the two varieties of ivory, the following method may be employed: Concentrated sulphuric acid applied to vegetable ivory will cause a pink coloring in about 10 or 12 minutes, which can be removed again by washing with water. Applied on genuine ivory, this acid does not affect it in any manner.

IVORY BLACK:
See Bone Black.

IVORY CEMENT:
See Adhesives.

IVORY GILDING:
See Plating.

IVORY POLISHES:
See Polishes.

JAPAN BLACK:
See Paints.

JAPANNING AND JAPAN TINNING:
See Varnishes.

JASMINE MILK:
See Cosmetics.

JELLY (FRUIT) EXTRACT:
See Essences and Extracts.

JEWELERS' CEMENTS:
See Adhesives.

JEWELERS' CLEANING PROCESSES:
See Cleaning Preparations and Methods.

Jewelers' Formulas

(See also Gems, Gold, and Watchmakers' Recipes.)

Coloring Gold Jewelry.—Following are several recipes for coloring: Saltpeter, 40 parts; alum, 30 parts; sea salt, 30 parts; or, liquid ammonia, 100 parts; sea salt, 3 parts; water, 100 parts. Heat without allowing to boil and plunge the objects into it for 2 or 3 minutes, stirring constantly; rinse in alum water and then in clean water. Another recipe: Calcium bromide, 100 parts; bromine, 5 parts. Place the articles in this solution, with stirring, for 2 to 3 minutes; next wash in a solution of hyposulphite of sodium and rinse in clean water. Another: Verdigris, 30 parts; sea salt, 30 parts; blood stone, 30 parts; sal ammoniac, 30 parts; alum, 5 parts. Grind all and stir with strong vinegar; or, verdigris, 100 parts; hydrochlorate of ammonia, 100 parts; saltpeter, 65 parts; copper filings, 40 parts. Bray all and mix with strong vinegar.

To Widen a Jewel Hole.—Chuck the hole in a lathe with cement. Place a spirit lamp underneath to prevent the cement from hardening. Hold the pointed bit against the hole, while the lathe is running, until the hole is true, when the lamp should be removed. The broach to widen the hole should be made of copper, of the required size and shape, and the point, after being oiled, should be rolled in diamond dust until it is entirely covered. The diamond dust should then be beaten in with a burnisher, using very light blows so as not to bruise the broach. After the hole is widened as desired, it requires polishing with a broach made of ivory and used with oil and the finest diamond dust, loose, not driven into the broach.

To Clean Jet Jewelry.—Reduce bread crumbs into small particles, and introduce into all the curves and hollows of the jewelry, while rubbing with a flannel.

Coloring Common Gold.—In coloring gold below 18 carat, the following mixture may be used with success, and if carefully employed, even 12 carat gold may be colored by it: Take nitrate of potassa (saltpeter), 4 parts, by weight; alum, 2 parts; and common salt, 2 parts. Add sufficient warm water to mix the ingredients into a thin paste; place the mixture in a small pipkin or crucible and allow to boil. The article to be colored should be suspended by a wire and dipped into the mixture, where it should remain from 10 to 20 minutes. The article should then be removed and well rinsed in hot water, when it must be scratch brushed, again rinsed and returned to the coloring salts for a few minutes; it is then to be again rinsed in hot water, scratch brushed, and finally brushed with soap and hot water, rinsed in hot water, and placed in boxwood sawdust. The object being merely to remove the alloy, as soon as the article has acquired the proper color of fine gold it may be considered sufficiently acted upon by the above mixture. The coloring salts should not be used for gold of a lower standard than 12 carat, and, even for this quality of gold, some care must be taken when the articles are of a very slight make.

Shades of Red, etc., on Matt Gold Bijouterie.—For the production of the red and other shades on matt gold articles, the so-called gold varnishes are employed, which consist of shellac dissolved in alcohol and are colored with gum rosins. Thus a handsome golden yellow is obtained from shellac, 35 parts; seed-lac, 35 parts; dragon's blood, 50 parts; gamboge, 50 parts; dissolved in 400 parts of alcohol; the clear solution is decanted and mixed with 75 parts of Venice turpentine. By changing the amounts of the coloring rosins, shades from bright gold yellow to copper color are obtained. The varnish is applied evenly and after drying is wiped off from the raised portions of the article by means of a pad of wadding dipped into alcohol, whereby a handsome patination effect is produced, since the lacquer remains in the cavities. Chased articles are simply rubbed with earth colors ground into a paste with turpentine oil, for which purpose burnt sienna, fine ochers of a golden color, golden yellow, and various shades of green are employed.

I.—Yellow wax	32 parts
Red bole	3 parts
Crystallized verdigris	2 parts
Alum	2 parts

II.—Yellow wax	95 parts
Red bole	64 parts
Colcothar	2 parts
Crystallized verdigris	32 parts
Copper ashes	20 parts
Zinc vitriol	32 parts
Green vitriol	16 parts
Borax	1 part

The wax is melted and the finely powdered chemicals are stirred in, in rotation. If the gilt bronze goods are to obtain a lustrous orange shade, apply a mixture of ferric oxide, alum, cooking salt, and vinegar in the heated articles by means of a brush, heating to about 266° F. until the shade commences to turn black and water sprinkled on will evaporate with a hissing sound, then cool in water, dip in a mixture of 1 part of nitric acid with 40 parts of water, rinse thoroughly, dry, and polish. For the production of a pale-gold shade use a wax preparation consisting of:

III.—Yellow wax........ 19 parts
    Zinc vitriol......... 10 parts
    Burnt borax........ 3 parts

Green-gold color is produced by a mixture of:

IV.—Saltpeter .... 6 parts
    Green vitriol....... 2 parts
    Zinc vitriol......... 1 part
    Alum............. 1 part

To Matt Gilt Articles.—If it is desired to matt gilt articles partly or entirely, the portions which are to remain burnished are covered with a mixture of chalk, sugar, and mucilage, heating until this "stopping-off" covering shows a black color. On the places not covered apply a matting powder consisting of:

    Saltpeter.......... 40 parts
    Alum............. 25 parts
    Cooking salt....... 35 parts

Heat the objects to about 608° F., whereby the powder is melted and acquires the consistency of a thin paste. In case of too high a temperature decomposition will set in.

To Find the Number of Carats.—To find the number of carats of gold in an object, first weigh the gold and mix with seven times its weight in silver. This alloy is beaten into thin leaves, and nitric acid is added; this dissolves the silver and copper. The remainder (gold) is then fused and weighed; by comparing the first and last weights the number of carats of pure gold is found. To check repeat several times.

Acid Test for Gold.—The ordinary ready method of ascertaining whether a piece of jewelry is made of gold consists in touching it with a glass stopper wetted with nitric acid, which leaves gold untouched, but colors base alloys blue from the formation of nitrate of copper.

Imitation Diamonds. — 1. — Minium, 75 parts (by weight); washed white sand, 50 parts; calcined potash, 18 parts; calcined borax, 6 parts; bioxide of arsenic, 1 part. The sand must be washed in hydrochloric acid and then several times in clean water. The specific gravity of this crystal glass is almost the same as that of the diamond.

II. — Washed white sand, 100 parts (by weight); minium, 35 parts; calcined potash, 25 parts; calcined borax, 20 parts; nitrate of potash (crystals), 10 parts; peroxide of manganese, 5 parts. The sand must be washed as above stated.

Diamantine.—This substance consists of crystallized boron, the basis of borax. By melting 100 parts of boracic acid and 80 parts of aluminum crystals is obtained the so-called bort, which even attacks diamond. The diamantine of commerce is not so hard.

To Refine Board Sweepings.—The residue resulting from a jobbing jeweler's business, such as board sweepings and other residuum, which is continually accumulating and which invariably consists of all mixed qualities of standard, may have the precious metals recovered therefrom in a very simple manner, as follows: Collect the residue and burn it in an iron ladle or pan, until all grease or other organic matter is destroyed. When cool mix with ½ part soda-ash, and melt in a clay crucible. When the metal is thoroughly melted it will leave the flux and sink to the bottom of the crucible; at this stage the flux assumes the appearance of a thin fluid, and then is the time to withdraw the pot from the fire. The metal in the crucible—but not the flux—may now be poured into a vessel of water, stirring the water in a circular direction while the metal is being poured in, which causes it to form into small grains, and so prepares it for the next process. Dissolve the grains in a mixture of nitric acid and water in equal quantities. It takes about four times the quantity of liquid as metal to dissolve. The gold remains undissolved in this mixture, and may be recovered by filtering or decanting the liquid above it in the dissolving vessel; it is then dried, mixed with a little flux, and melted in the usual manner, whereupon pure gold will be obtained. To recover the silver, dilute the solution which has been withdrawn from the gold with six times its bulk of water, and add by degrees small quantities of finely powdered common salt, and this will throw down the silver into a white, curdy powder of chloride of silver. Continue to add salt until no cloudiness is observed in the solution, when the water above the sediment may be poured off; the sediment is next well washed with warm water several times, then dried and melted in the same manner as the gold, and you will have a lump of pure silver.

Restoration of the Color of Turquoises.—After a certain time turquoises lose a part of their fine color. It is easy to restore the color by immersing them in a solution of carbonate of soda. But it seems that the blue cannot be restored anew after this operation, if it again becomes dull. The above applies to common turquoises, and not to those of the Orient, of which the color does not change.

Colorings for Jewelers' Work.—I.—Take 40 parts of saltpeter; 30 parts of alum; 30 parts of sea salt; or 100 grams of liquid ammonia; 3 parts sea salt; and 100 parts water. This is heated without bringing it to a boil, and the articles dipped into it for from 2 to 3 minutes, stirring the liquid constantly; after this bath they are dipped in alum water and then thoroughly rinsed in clean water.

II.—One hundred parts of calcium bromide and 2 parts of bromium. The objects are allowed to remain in this solution (which must be also constantly stirred) for from 2 to 3 minutes, then washed in a solution of sodium hyposulphite, after which they must be rinsed in clean water.

III.—Thirty parts of verdigris; 30 parts of sea salt; 30 parts of hematite; 30 parts of sal ammoniac, and 5 parts of alum. This must be all ground up together and mixed with strong vinegar; or we may also use 100 parts of verdigris; 100 parts of hydrochlorate of ammonia; 65 parts of saltpeter, and 40 parts of copper filings, all of which are to be well mixed with strong vinegar.

22-Carat Solder.—Soldering is a process which, by means of a more fusible compound, the connecting surfaces of metals are firmly secured to each other, but, for many practical purposes, it is advisable to have the fusing point of the metal and solder as near each other as possible, which, in the majority of cases, preserves a union more lasting, and the joint less distinguishable, in consequence of the similarity of the metal and solder in color, which age does not destroy, and this is not the case with solders the fusible points of which are very low. The metal to be soldered together must have an affinity for the solder, otherwise the union will be imperfect; and the solder should likewise act upon the metal, partly by this affinity or chemical attraction, and partly by cohesive force, to unite the connections soundly and firmly together. Solders should therefore be prepared suitable to the work in hand, if a good and lasting job is to be made. It should always be borne in mind that the higher the fusing point of the gold alloy —and this can be made to vary considerably, even with any specified quality— the harder solder must be used, for, in the case of a more fusible mixture of gold, the latter would melt before the solder

and cause the work to be destroyed. A very good formula for the first, or ordinary, 22-carat alloy is this:

	dwts.	grs.
Fine gold	1	0
Fine silver	0	3
Fine copper	0	2
	1	5

This mixture will answer all the many purposes of the jobber; for soldering high quality gold wares that come for repairs, particularly wedding rings, it will be found admirably suited. If an easier solder is wanted, and such is very often the case with jobbing jewelers, especially where several solderings have to be accomplished, it is as well to have at hand a solder which will not disturb the previous soldering places, for if this is not prevented a very simple job is made very difficult, and a lot of time and patience wholly wasted. To guard against a thing of this kind the following solder may be employed on the top of the previous one:

	dwts.	grs.
Fine gold	1	0
Fine silver	0	3
Yellow brass	0	2
	1	5

This solder is of the same value as the previous one, but its melting point is lower, and it will be found useful for many purposes that can be turned to good account in a jobbing jeweler's business.

JEWELERS' ALLOYS:

See also Alloys and Solders.

18-Carat Gold for Rings.—Gold coin, 19½ grains; pure copper, 3 grains; pure silver, 1½ grains.

Cheap Gold, 12 Carat.—Gold coin, 25 grains; pure copper, 13½ grains; pure silver, 7⅓ grains.

Very Cheap 4-Carat Gold.—Copper, 18 parts; gold, 4 parts; silver, 2 parts.

Imitations of Gold.—I.—Platina, 4 pennyweights; pure copper, 2¼ pennyweights; sheet zinc, 1 pennyweight; block tin, 1¾ pennyweights; pure lead, 1½ pennyweight. If this should be found too hard or brittle for practical use, remelting the composition with a little sal ammoniac will generally render it malleable as desired.

II.—Platina, 2 parts; silver, 1 part; copper, 3 parts. These compositions, when properly prepared, so nearly resemble pure gold that it is very difficult to distinguish them therefrom. A little powdered charcoal, mixed with metals while melting, will be found of service.

Best Oreide of Gold.—Pure copper, 4 ounces; sheet zinc, 1¾ ounces; magnesia, ⅝ ounce; sal ammoniac, 1½ ounce; quicklime, ¾ ounce; cream tartar, ⅞ ounce. First melt the copper at as low a temperature as it will melt; then add the zinc, and afterwards the other articles in powder, in the order named. Use a charcoal fire to melt these metals.

Bushing Alloy for Pivot Holes, etc.—Gold coin, 3 pennyweights; silver, 1 pennyweight, 20 grains; copper, 3 pennyweights, 20 grains; palladium, 1 pennyweight. The best composition known for the purpose named.

Gold Solder for 14- to 16-Carat Work.—Gold coin, 1 pennyweight; pure silver, 9 grains; pure copper, 6 grains; brass, 3 grains.

Darker Solder.—Gold coin, 1 pennyweight; pure copper, 8 grains; pure silver, 5 grains; brass, 2 grains. Melt together in charcoal fire.

Solder for Gold.—Gold, 6 pennyweights; silver, 1 pennyweight; copper, 2 pennyweights.

Soft Gold Solder.—Gold, 4 parts; silver, 1 part; copper, 1 part.

Solders for Silver (for the use of jewelers).—Fine silver, 19 pennyweights; copper, 1 pennyweight; sheet brass, 10 pennyweights.

White Solder for Silver.—Silver, 1 ounce; tin, 1 ounce.

Silver Solder for Plated Metal.—Fine silver, 1 ounce; brass, 10 pennyweights.

Solders for Gold.—I.—Silver, 7 parts; copper, 1 part; with borax.

II.—Gold, 2 parts; silver, 1 part; copper, 1 part.

III.—Gold, 3 parts; silver, 3 parts; copper, 1 part; zinc, ½ part.

For Silver.—Silver, 2 parts; brass, 1 part; with borax; or, silver, 4 parts; brass, 3 parts; zinc, ⅟₁₆ part; with borax.

Gold Solders (see also Solders).—I.—Copper, 24.24 parts; silver, 27.57 parts; gold, 48.19 parts.

II.—**Enamel Solder.**—Copper, 25 parts; silver, 7.07 parts; gold, 67.93 parts.

III.—Copper, 26.55 parts; zinc, 6.25 parts; silver, 31.25 parts; gold, 36 parts.

IV.—**Enamel Solder.**—Silver, 19.57 parts; gold, 80.43 parts.

Solder for 22-Carat Gold.—Gold of 22 carats, 1 pennyweight; silver, 2 grains; copper, 1 grain.

For 18-Carat Gold.—Gold of 18 carats, 1 pennyweight; silver, 2 grains; copper, 1 grain.

For Cheaper Gold.—I.—Gold, 1 pennyweight; silver, 10 grains; copper, 8 grains.

II.—Fine gold, 1 pennyweight; silver, 1 pennyweight; copper, 1 pennyweight.

Silver Solders (see also Solders).—I. (Hard.)—Copper, 30 parts; zinc, 12.85 parts; silver, 57.15 parts.

II.—Copper, 23.33 parts; zinc, 10 parts; silver, 66.67 parts.

III.—Copper, 26.66 parts; zinc, 10 parts; silver, 63.34 parts.

IV. (Soft.)—Copper, 14.75 parts; zinc, 8.50 parts; silver, 77.05 parts.

V.—Copper, 22.34 parts; zinc, 10.48 parts; silver, 67.18 parts.

VI.—Tin, 63 parts; lead, 37 parts.

FOR SILVERSMITHS:

I.—**Sterling Silver.**—Fine silver, 11 ounces, 2 pennyweights; fine copper, 18 pennyweights.

II.—**Equal to Sterling.**—Fine silver, 1 ounce; fine copper, 1 pennyweight, 12 grains.

III.—Fine silver, 1 ounce; fine copper, 5 pennyweights.

IV.—**Common Silver for Chains.**—Fine silver, 6 pennyweights; fine copper, 4 pennyweights.

V.—**Solder.**—Fine silver, 16 pennyweights; fine copper, 12 grains; pin brass, 3 pennyweights, 12 grains.

VI.—**Alloy for Plating.**—Fine silver, 1 ounce; fine copper, 10 pennyweights.

VII.—**Silver Solder.**—Fine silver, 1 ounce; pin brass, 10 pennyweights; pure spelter, 2 pennyweights.

VIII.—**Copper Solder for Plating.**—Fine silver, 10 pennyweights; fine copper, 10 pennyweights.

IX.—**Common Silver Solder.**—Fine silver, 10 ounces; pin brass, 6 ounces, 12 pennyweights; spelter, 12 pennyweights.

X.—**Silver Solder for Enameling.**—Fine silver, 14 pennyweights; fine copper, 8 pennyweights.

XI.—**For Filling Signet Rings.**—Fine silver, 10 ounces; fine copper, 1 ounce, 16 pennyweights; fine pin brass, 6 ounces, 12 pennyweights; spelter, 12 pennyweights.

XII.—Silver Solder for Gold Plating. —Fine silver, 1 ounce; fine copper, 5 pennyweights; pin brass, 5 pennyweights.

XIII.—Mercury Solder.—Fine silver, 1 ounce; pin brass, 10 pennyweights; bar tin, 2 pennyweights.

XIV.—Imitation Silver.—Fine silver, 1 ounce; nickel, 1 ounce, 11 grains; fine copper, 2 ounces, 9 grains.

XV.—Fine silver, 3 ounces; nickel, 1 ounce, 11 pennyweights; fine copper, 2 ounces, 9 grains; spelter, 10 pennyweights.

XVI.—Fine Silver Solder for Filigree Work.—Fine silver, 4 pennyweights, 6 grains; pin brass, 1 pennyweight.

Bismuth Solder.—Bismuth, 3 ounces; lead, 3 ounces, 18 pennyweights; tin, 5 ounces, 6 pennyweights.

BRASS:

I.—Yellow Brass for Turning.—(Common article.)—Copper, 20 pounds; zinc, 10 pounds; lead, 4 ounces.

II.—Copper, 32 pounds; zinc, 10 pounds; lead, 1 pound.

III.—Red Brass Free, for Turning.— Copper, 100 pounds; zinc, 50 pounds; lead, 10 pounds; antimony, 44 ounces.

IV.—Best Red Brass for Fine Castings.—Copper, 24 pounds; zinc, 5 pounds; bismuth, 1 ounce.

V.—Red Tombac.—Copper, 10 pounds; zinc, 1 pound.

VI.—Tombac.—Copper, 16 pounds; tin, 1 pound; zinc, 1 pound.

VII.—Brass for Heavy Castings.— Copper, 6 to 7 parts; tin, 1 part; zinc, 1 part.

VIII.—Malleable Brass.—Copper, 70.10 parts; zinc, 29.90 parts.

IX.—Superior Malleable Brass.—Copper, 60 parts; zinc, 40 parts.

X.—Brass.—Copper, 73 parts; zinc, 27 parts.

XI.—Copper, 65 parts; zinc, 35 parts.

XII.—Copper, 70 parts; zinc, 30 parts.

XIII.—German Brass.—Copper, 1 pound; zinc, 1 pound.

XIV.—Watchmakers' Brass.—Copper, 1 part; zinc, 2 parts.

XV.—Brass for Wire.—Copper, 34 parts; calamine, 56 parts.

XVI.—Brass for Tubes.—Copper, 2 parts; zinc, 1 part.

XVII.—Brass for Heavy Work.— Copper, 100 parts; tin, 15 parts; zinc, 15 parts.

XVIII.—Copper, 112 parts; tin, 13 parts; zinc, 1 part.

XIX.—Tombac or Red Brass.—Copper, 8 parts; zinc, 1 part.

XX.—Brass.—Copper, 3 parts; melt, then add zinc, 1 part.

XXI.—Buttonmakers' Fine Brass.— Brass, 8 parts; zinc, 5 parts.

XXII. — Buttonmakers' Common Brass.—Button brass, 6 parts; tin, 1 part; lead, 1 part. Mix.

XXIII.—Mallet's Brass.—Copper, 25.4 parts; zinc, 74.6 parts. Used to preserve iron from oxidizing.

XXIV.—Best Brass for Clocks.— Rose copper, 85 parts; zinc, 14 parts; lead, 1 part.

GOLD ALLOYS:

See also Gold Alloys, under Alloys.

Gold of 22 carats fine being so little used is intentionally omitted.

I.—Gold of 18 Carats, Yellow Tint. —Gold, 15 pennyweights; silver, 2 pennyweights, 18 grains; copper, 2 pennyweights, 6 grains.

II.—Gold of 18 Carats, Red Tint.— Gold, 15 pennyweights; silver, 1 pennyweight, 18 grains; copper, 3 pennyweights, 6 grains.

III.—Spring Gold of 16 Carats.— Gold, 1 ounce, 16 pennyweights; silver, 6 pennyweights; copper, 12 pennyweights. This when drawn or rolled very hard makes springs little inferior to steel.

IV.—Jewelers' Fine Gold, Yellow Tint, 16 Carats Nearly.—Gold, 1 ounce; silver, 7 pennyweights; copper, 5 pennyweights.

V.—Gold of Red Tint, 16 Carats.— Gold, 1 ounce; silver, 2 pennyweights; copper, 8 pennyweights.

Sterling Gold Alloys.—**I.**—Fine gold, 18 pennyweights, 12 grains; fine silver, 1 pennyweight; fine copper, 12 grains.

II.—Dry Colored Gold Alloys, 17 Carat.—Fine gold, 15 pennyweights; fine silver, 1 pennyweight, 10 grains; fine copper, 4 pennyweights, 17 grains.

III.—18 Carat.—Fine gold, 1 ounce; fine silver, 4 pennyweights, 10 grains; fine copper, 2 pennyweights, 5 grains.

IV.—18 Carat.—Fine gold, 15 pennyweights; fine silver, 2 pennyweights, 4 grains; fine copper, 2 pennyweights, 19 grains.

V.—18 Carat.—Fine gold, 18 pennyweights; fine silver, 2 pennyweights, 18 grains; fine copper, 3 pennyweights, 18 grains.

VI.—19 **Carat.**—Fine gold, 1 ounce; fine silver, 2 pennyweights, 6 grains; fine copper, 3 pennyweights, 12 grains.

VII.—20 **Carat.**—Fine gold, 1 ounce; fine silver, 2 pennyweights; fine copper, 2 pennyweights, 4 grains.

VIII.—22 **Carat.**—Fine gold, 18 pennyweights; fine silver, 12 grains; fine copper, 1 pennyweight, 3 grains.

IX.—**Gold Solder for the Foregoing Alloys.**—Take of the alloyed gold you are using, 1 pennyweight; fine silver, 6 grains.

X.—**Alloy for Dry Colored Rings.**—Fine gold, 1 ounce; fine silver, 4 pennyweights, 6 grains; fine copper, 4 pennyweights, 6 grains.

XI.—**Solder.**—Scrap gold, 2 ounces; fine silver, 3 pennyweights; fine copper, 3 pennyweights.

XII.—**Dry Colored Scrap Reduced to 35s. Gold.**—Colored scrap, 1 ounce, 9 pennyweights, 12 grains; fine silver, 2 pennyweights; fine copper, 17 pennyweights, 12 grains; spelter, 4 pennyweights.

To Quickly Remove a Ring from a Swollen Finger.—If the ring is of gold, pull the folds of the swollen muscles apart, so that it can be seen, then drop on it a little absolute alcohol and place the finger in a bowl of metallic mercury. In a very few minutes the ring will snap apart. If the ring is of brass, scrape the surface slightly, or put on a few drops of a solution of oxalic acid, or even strong vinegar, let remain in contact for a moment or two, then put into the mercury, and the result will be as before.

Soldering a Jeweled Ring.—In order to prevent the bursting of the jewels of a ring while the latter is being soldered, cut a juicy potato into halves and make a hollow in both portions in which the part of the ring having jewels may fit exactly. Wrap the jeweled portion in soft paper, place it in the hollow, and bind up the closed potato with binding wire. Now solder with easy-flowing gold solder, the potato being held in the hand. Another method is to fill a small crucible with wet sand, bury the jeweled portion in the sand, and solder in the usual way.

JEWELRY, TO CLEAN:
See Cleaning Preparations and Methods.

Kalsomine

Sodium carbonate...	8 parts
Linseed oil.........	32 parts
Hot water..........	8 parts
White glue.........	12 parts
Whiting............	160 parts

Dissolve the sodium carbonate in the hot water, add the oil and saponify by heating and agitation. Cover the glue, broken into small pieces, with cold water and let soak overnight. In the morning pour the whole on a stout piece of stuff and let the residual water drain off, getting rid of as much as possible by slightly twisting the cloth. Throw the swelled glue into a capsule, put on the water bath, and heat gently until it is melted. Add the saponified oil and mix well; remove from the bath, and stir in the whiting, a little at a time, adding hot water as it becomes necessary. When the whiting is all stirred in, continue adding hot water, until a liquid is obtained that flows freely from the kalsomining brush.

The addition of a little soluble blue to the mixture increases the intensity of the white.

Sizing Walls for Kalsomine.—A size to coat over "hot walls" for the reception of the kalsomine is made by using shellac, 1 part; sal soda, ½ part. Put these ingredients in ½ gallon of water and dissolve by steady heat. Another size is made of glue size prepared in the usual way, and alum. To ½ pound of white glue add ¾ pound of alum, dissolving the alum in hot water before adding it to the glue size.

KARATS, TO FIND NUMBER OF:
See Jewelers' Formulas.

KERAMICS:
See Ceramics.

KERIT:
See Rubber.

KEROCLEAN:
See Cleaning Preparations and Methods.

KEROSENE DEODORIZER:
See also Benzine, Oils, and Petroleum.

Various processes have been recommended for masking the odor of kerosene such as the addition of various essential oils, artificial oil of mirbane, etc., but none of them seems entirely satisfactory. The addition of amyl acetate in the proportion of 10 grams to the liter (1 per cent) has also been suggested, several experimenters reporting very successful results therefrom. Some years ago Beringer proposed a process for removing sulphur compounds from benzine, which would presumably be equally applicable to kerosene. This process is as follows:

Potassium permanganate..............	1 ounce
Sulphuric acid.......	½ pint
Water..............	3½ pints

Mix the acid and water, and when the mixture has become cold pour it into a 2-gallon bottle. Add the permanganate and agitate until it is dissolved. Then add benzine, 1 gallon, and thoroughly agitate. Allow the liquids to remain in contact for 24 hours, frequently agitating the mixture. Separate the benzine and wash in a similar bottle with a mixture of

Potassium permanganate..............	¼ ounce
Caustic soda........	½ ounce
Water..............	2 pints

Agitate the mixture frequently during several hours; then separate the benzine and wash it thoroughly with water. On agitating the benzine with the acid permanganate solution an emulsion-like mixture is produced, which separates in a few seconds, the permanganate slowly subsiding and showing considerable reduction. In the above process it is quite probable that the time specified (24 hours) is greatly in excess of what is necessary, as the reduction takes place almost entirely in a very short time. It has also been suggested that if the process were adopted on a manufacturing scale, with mechanical agitation, the time could be reduced to an hour or two.

KEROSENE-CLEANING COMPOUNDS:
See Cleaning Preparations, under Miscellaneous Methods.

KEROSENE EMULSIONS:
See Petroleum.

KETCHUP (ADULTERATED), TESTS FOR:
See Foods.

KHAKI COLORS:
See Dyes.

KID:
See Leather.

KISSINGEN SALTS:
See Salts (Effervescent).

KISSINGEN WATER:
See Waters.

KNIFE-SHARPENING PASTES:
See Razor Pastes.

KNOCKENPLOMBE:
See Adhesives.

KNOTS:
See Paint.

KOLA CORDIAL:
See Wines and Liquors.

KOUMISS SUBSTITUTE:
See also Beverages.

To prepare a substitute for koumiss from cow's milk: Dissolve ½ ounce grape sugar in 3 fluid ounces water. Mix 18 grains well washed and pressed beer yeast with 2 fluid ounces of cow's milk. Mix the two liquids in a champagne bottle, fill with milk, stopper securely, and keep for 3 to 4 days at a temperature not exceeding 50° F., shaking frequently. The preparation does not keep longer than 4 to 5 days.

KÜMMEL:
See Wines and Liquors.

KWASS:
See Beverages.

LABEL PASTES, GLUES, AND MUCILAGES:
See Adhesives.

LABEL VARNISHES:
See Varnishes.

LACE LEATHER:
See Leather.

LACE, TO CLEAN GOLD AND SILVER:
See Cleaning Preparations and Methods.

LACES, WASHING AND COLORING OF:
See Laundry Preparations.

Lacquers

(See also Enamels, Glazes, Paints, Varnishes, and Waterproofing.)

LAC AND THE ART OF LACQUERING.

The art of lacquering includes various steps, which are divulged as little as possible. Without them nothing but a varnish of good quality would be realized Thus in Tonkin, where the abundant production is the object of an important trade with the Chinese, it is so used only for varnishing, while in China the same product from the same sources contributes to most artistic applications.

When the Annamites propose to lacquer an object, a box, for example, they first stop up the holes and crevices, covering all the imperfections with a coating of diluted lac, by means of a flat, close, short brush. Then they cover the whole with a thick coating of lac and white clay. This clay, oily to the touch, is found at the bottom of certain lakes in Tonkin; it is dried, pulverized, and sifted with a piece of fine silk before being embodied with the lac. This operation is designed to conceal the inequalities of the wood and produce a uniform surface which, when completely dry, is rendered smooth with pumice stone.

If the object has portions cut or sunk the clayey mixture is not applied, for it would make the details clammy, but in its place a single, uniform layer of pure lac.

In any case, after the pumicing, a third coating, now pure lac, is passed over the piece, which at this time has a mouse-gray color. This layer, known under the name of *sou lot*, colors the piece a brilliant black. As the lac possesses the remarkable property of not drying in dry air, the object is left in a damp place. When perfectly dried the piece is varnished, and the desired color imparted by a single operation. If the metallic applications are excepted, the lac is colored only black, brown, or red.

The following formulas are in use:

Black.—One part of turpentine is warmed for 20 minutes beyond the fusing point; then poured into 3 parts of lac; at the same time *pheu deu* (copperas) is added. The mixture is stirred for at least a day, sometimes more, by means of a large paddle.

Maroon.—This is prepared by a process similar to the preceding, replacing half of the copperas by an equal quantity of China vermilion.

Red.—The lac, previously stirred for 6 hours, is mixed with hot oil of *trau*, and the whole is stirred for a day, after which vermilion is added. The latter should be of good quality, so as to have it brilliant and unchangeable.

The operation of lacquering is then ended, but there are parts to be gilded. These are again covered with a mixture of lac and oil of *trau*. When this layer is dry the metallic leaves are applied, which are themselves protected by a coating, composed also of lac and oil of *trau*. All these lac and oil of *trau* mixtures are carefully filtered, which the natives effect by pressing the liquid on a double filtering surface formed of wadding and of a tissue on which it rests. It can only be applied after several months when the metallic leaf is of gold. In the case of silver or tin the protecting coat can be laid on in a few days. It favorably modifies the white tints of these two metals by communicating a golden color. The hue, at first reddish, gradually improves and acquires its full brilliancy in a few months.

Little information is procurable concerning the processes employed by the Chinese. The wood to be lacquered should be absolutely dry. It receives successive applications, of which the number is not less than 33 for perfect work. When the lac coating attains the thickness of $\frac{1}{4}$ of an inch it is ready for the engravers. The Chinese, like the inhabitants of Tonkin, make use of oil of *trau* to mix with the lac, or oil of *aleurites*, and the greatest care is exercised in the drying of the different layers. The operation is conducted in dim-lighted rooms specially fitted up for the purpose; the moisture is maintained to a suitable extent by systematically watering the earth which covers the walls of this " cold stove."

Lacquer for Aluminum.—Dissolve 100 parts of gum lac in 300 parts of ammonia, and heat the solution for about 1 hour moderately on the water bath. After cooling, the mixture is ready for use. The aluminum to be coated is cleaned in the customary manner. After it has been painted with the varnish, it is heated in the oven to about 572° F. The coating and heating may be repeated.

Lacquer for Brass.—

Annatto................	$\frac{1}{4}$	ounce
Saffron................	$\frac{1}{4}$	ounce
Turmeric..............	1	ounce
Seed lac in coarse powder................	3	ounces
Alcohol................	1	pint

Digest the annatto, saffron, and turmeric in the alcohol for several days, then strain into a bottle containing the seed lac; cork and shake until dissolved.

Lacquer for Bronze.—I.—The following process yields a protective varnish for bronze articles and other metallic objects in various shadings, the lacquer produced excelling in high luster and permanency: Fill 40 parts of best pale shellac; 12 parts of pulverized Florentine lake; 30 parts gamboge; and 6 parts of dragon's blood, likewise powdered, into a bottle and add 400 parts of spirit. Allow this mixture to form a solution preferably by heating the flask on the water bath, to nearly the boiling point of the water, and shaking now and then until all has dissolved. After the cooling pour off the liquid from the sediment, if any is present; this liquid constitutes a lacquer of dark-red color. In a second bottle dissolve in the same manner 24 parts of gamboge in 400 parts of spirit, which affords a lacquer of golden yellow color. According to the desired shade, the red lacquer is now mixed with the yellow one, thus producing any hue required from the deepest red to a golden tone. If necessary, thin with spirit of wine. The varnish is applied, as usual, on the somewhat warmed article, a certain temperature having to be adhered to, which can be ascertained by trials and is easily regulated by feeling.

II.—The following is equally suitable for boots and leather goods as for application on iron, stone, glass, paper, cloth, and other surfaces. The inexperienced should note before making this liquid that it does not give a yellowish bronze like gold paint, but a darkish iridescent one, and as it is a pleasing variation in aids to home decoration, it would doubtless sell well. Some pretty effects are obtained by using a little phloxine instead of part of the violet aniline, or phloxine alone will produce a rich reddish bronze, and a lustrous peacock green is obtained with brilliant aniline green crystals.

Quantities: Flexile methylated collodion, 1 gallon; pure violet aniline, 1 pound. Mix, stand away for a few days to allow the aniline to dissolve and stir frequently, taking care to bung down securely, as the collodion is a volatile liquid, then strain and bottle off. It is applied with a brush, dries rapidly, and does not rub off or peal.

Celluloid Lacquer.—Dissolve uncolored celluloid in a mixture of strong alcohol and ether. The celluloid first swells up in the solvent, and after vigorous shaking, the bottle is allowed to stand quietly for the undissolved portion to settle, when the clear, supernatant fluid is poured off. The latter may be immediately used; it yields a colorless glossy lacquer, or may be colored, as desired, with aniline colors.

Colored Lacquer.—Make a strong solution of any coloring matter which is soluble in methylated spirit, such as cochineal, saffron, the aniline dyes, etc. Filter through fine cambric, and to this filtered solution add brown shellac in flakes in the proportion of 4 to 5 ounces of shellac to each pint of methylated spirit. Shake once a day for about 8 days. If too thick it may be thinned by adding more colored spirit or plain spirit as required, and any lighter shade can be obtained by mixing with plain lacquer mixed in the above proportions. Lacquer works best in a warm, dry place, and the process is improved by slightly warming the articles, which must be absolutely free from grease, dirt, or moisture. The best results are obtained by applying many coats of thin, light-colored lacquer, each coat to be thoroughly dry before applying the next. Apply with a soft camel's-hair brush; it is better to use too small a brush than too large. When complete, warm the articles for a few seconds before a clear fire; the hotter the better; if too hot, however, the colors will fade. This makes the lacquer adhere firmly, especially to metallic surfaces. Aniline green works very well.

Lacquer for Copper.—A lacquer which to a certain degree resists heat and acid liquids, but not alkaline ones, is obtained by heating fine, thickly liquid amber varnish, whereby it is rendered sufficiently liquid to be applied with the brush. The copper article is coated with this and left to stand until the lacquer has dried perfectly. Next, the object is heated until the lacquer commences to smoke and turns brown. If the operation is repeated twice, a coating is finally obtained, which, as regards resisting qualities to acid bodies, excels even enamel, but which is strongly attacked even by weakly alkaline liquids.

Ebony Lacquer.—The ebony lacquer recommended by the well-known English authority, Mr. H. C. Standage, consists of $\frac{1}{3}$ ounce aniline hydrochloride, $\frac{1}{3}$ ounce alcohol, 1 part sulphate of copper, 100 parts of water. The aniline dye is dissolved in the alcohol and the copper sulphate in the water. The wood is first coated with the copper sulphate solution, and after this coating has been given plenty of time to dry the aniline salt tincture is applied. Shortly the copper salt absorbed by the wood will react on the aniline hydrochloride, developing a deep, rich black which acids or alkalies are powerless to destroy. Coat with shellac and give a French polish, thus bringing the ebony finish up to a durable and unsurpassed luster.

GOLD LACQUERS:

I.—For Brassware.—A gold lacquer to improve the natural color of brassware is prepared from 16 parts gum lac, 4 parts dragon's blood, and 1 part curcuma powder dissolved in 320 parts spirits of wine in the warmth and filtered well. The articles must be thoroughly cleaned by burning, grinding, or turning either dull or burnished, and then coated with a thin layer of the above mixture, applied with a soft hair brush or a pad of wadding. If the objects are colored the lacquer must be laid on by stippling. Should the color be too dark, it may be lightened by reduction with a little spirit until the correct shade is produced. The most suitable temperature for the metal during the work is about the warmth of the hand; if too hot or too cold, the lacquer may smear, and will then have to be taken off again with spirit or hot potash lye, the goods being dried in sawdust or recleaned as at first, before applying the lacquer again. Round articles may be fixed in the lathe and the lacquer laid on with a pad of wadding. In order to color brassware, a solution of 30 parts caustic soda; 10 parts cupric carbonate; 200 parts water (or 200 parts ammonia neutralized by acetic acid); 100 parts verdigris, and 60 parts sal ammoniac is employed, into which the warmed articles are dipped. After having dried they are coated with colorless shellac varnish.

II.—For Tin.—Transparent gold lacquer for tin (all colors) may be made as follows: Take ½ pint of alcohol, add 1 ounce gum shellac; ¼ ounce turmeric; 1¼ ounce red sanders. Set the vessel in a warm place and shake frequently for half a day. Then strain off the liquor, rinse the bottle and return it, corking tightly for use. When this is used, it must be applied to the work freely and flowed on full, or if the work admits it, it may be dipped. One or more coats may be given as the color is required light or dark. For rose color substitute ¼ ounce of finely ground lake in place of the turmeric. For blue, substitute Prussian blue. For purple, add a little of the blue to the turmeric.

For Bottle Caps, etc.—

I.—

Gum gutta	10 parts
Shellac	100 parts
Turpentine	10 parts
Alcohol	450 parts

I.—

Gum gutta	40 parts
Dragon's blood	5 parts
Alcoholic extract of sandalwood	5 parts
Sandarac	75 parts
Venice turpentine	25 parts
Alcohol, 95 per cent.	900 parts

Mix and dissolve by the aid of a gentle heat.

Liquid Bottle Lac.—Into a half-gallon bottle put 8 ounces of shellac, and pour over it 1½ pints of alcohol of 94 per cent, and 2½ ounces of sulphuric ether. Let stand, with occasional shaking, until the shellac is melted, and then add 4 ounces of thick turpentine and ½ ounce of boric acid. Shake until dissolved. To color, use the aniline colors soluble in alcohol—for red, eosine; blue, phenol blue; black, negrosin; green, aniline green; violet, methyl violet, etc. If it is desired to have the lac opaque, add 8 ounces of pulverized steatite, but remember to keep the lac constantly stirred while using, as otherwise the steatite falls to the bottom.

Lithographic Lacquer.—Dissolve 15 parts, by weight, of red lithol R or G in paste of 17 per cent, in 150 parts, by weight, of hot water. Boil for 2 minutes, shaking with 2.5 parts, by weight, of barium chloride. Dissolve in 25 parts, by weight, of water. Add to the mixture 100 parts, by weight, of aluminum hydrate of about 4 per cent. Cool, filter, and dry.

Lacquer for Microscopes, Mathematical Instruments, etc.—Pulverize 160 parts, by weight, turmeric root, cover it with 1,700 parts alcohol, digest in a warm place for 24 hours, and then filter. Dissolve 80 parts dragon's blood, 80 parts sandarac, 80 parts gum elemi, 50 parts gum gutta, and 70 parts seed lac, put in a retort with 250 parts powdered glass, pour over them the colored alcohol first made, and hasten solution by warming in the sand or water bath. When completely dissolved, filter.

To Fix Alcoholic Lacquers on Metallic Surfaces.—Dissolve 0.5 parts of crystallized boracic acid in 100 parts of the respective spirit varnish whereby the latter after being applied forms so hard a coating upon a smooth tin surface that it cannot be scratched off even with the finger-nails. The aforementioned percentage of boracic acid should not be exceeded in preparing the solution; otherwise the varnish will lose in intensity of color.

Lacquer for Oil Paintings.—Dilute 100 parts of sulphate of baryta with 600 parts of water containing in solution 60 parts of red lithol R or G in paste of 17 per cent. Boil the mixture for several minutes in a solution of 10 parts of barium chloride in 100 parts of water. After cooling, filter and dry.

Lacquers for Papers.—I.—With base of baryta: Dissolve 30 parts of red lithol R or G in paste of 17 per cent, in 300 parts of hot water. Add an emulsion obtained by mixing 10 parts of sulphate of alumina in 100 parts of water and 5 parts of calcined soda dissolved in 50 parts of water. Precipitate with a solution of 17.5 parts of barium chloride in 125 parts of water. Cool and filter.

II.—With base of lime: Dissolve 30 parts red lithol R or G in paste of 17 per cent, in 300 parts of hot water. Boil for a few minutes with an emulsion prepared by mixing 10 parts sulphate of alumina with 100 parts of water and 2.5 parts of slaked lime in 100 parts of water. Filter after cooling.

Lacquer for Stoves and other Articles to Withstand Heat.—This is not altered by heat, and does not give off disagreeable odors on heating: Thin 1 part of sodium water glass with 2 parts of water in order to make the vehicle. This is to be thickened with the following materials in order to get the desired color: White, barium sulphate or white lead; yellow, baryta chromate, ocher, or uranium yellow; green, chromium oxide or ultramarine green; brown, cadmium oxide, manganese oxide, or sienna brown; red, either iron or chrome red. The coloring materials must be free from lumps, and well ground in with the vehicle. Bronze powders may also be used either alone or mixed with other coloring stuffs, but care must be taken, in either instance, to secure a sufficient quantity. The colors should be made up as wanted, and no more than can conveniently be applied at the time should be prepared. An excellent way to use the bronze powders is to lay on the coloring matter, and then to dust on the powder before the glass sets. Lines or ornamentation of any sort may be put on by allowing the coating of enamel to dry, and then drawing the lines or any desired design with a fresh solution of the water glass colored to suit the taste, or dusted over with bronze.

MISCELLANEOUS RECIPES:

Russian Polishing Lac.—

I.—Sticklac .... 925 parts
    Sandarac .... 875 parts
    Larch turpentine... 270 parts
    Alcohol, 96 per cent 3,500 parts

The sticklac is broken up and mixed with the sandarac, put into a suitable container with a wide mouth, the spirit poured over it and set aside. After standing for a week in a warm place, frequently stirring in the meantime (best with a glass rod) and fully dissolving, stir in the turpentine. Let stand 2 or 3 days longer, then filter through glass wool. The sandarac dissolves completely in the spirit, but the stick leaves a slight residue which may be added to the next lot of lac made up and thus be treated to a fresh portion of spirit. The larch turpentine should be of the best quality. This lac is used by woodcarvers and turners and is very much prized by them.

Mastic Lac.—

II.—Mastic, select...... 150 parts
    Sandarac......... 400 parts
    Camphor........ 15 parts
    Alcohol, 96 per cent 1,000 parts

Prepare as directed in the first recipe.

Leather Polish Lac.—

III.—Shellac........... 16 parts
    Venice turpentine.. 8 parts
    Sandarac......... 4 parts
    Lampblack, Swedish............. 2 parts
    Turpentine oil.... 4 parts
    Alcohol, 96 per cent 960 parts

The alcohol and turpentine oil are mixed and warmed under constant stirring in the sand or water bath. The shellac and sandarac are now stirred in, the stirring being maintained until both are dissolved. Finally add the turpentine and dissolve. Stir the lampblack with a little vinegar and then add and stir in. Instead of lampblack 125 to 150 parts of nigrosin may be used. This lac should be well shaken before application.

LACQUERED WARE, TO CLEAN:

See Cleaning Preparations and Methods.

LAKES:

See Dyes.

LAMPBLACK:

Production of Lampblack.—The last oil obtained in the distillation of coal tar, and freed from naphthalene as far as possible, viz., soot oil, is burned in a special furnace for the production of various grades of lampblack. In this furnace is an iron plate, which must always be kept glowing; upon this plate the soot oil trickles through a small tube fixed above it. It is decomposed and the smoke (soot) rises into four chambers through small apertures. When the quantity of oil destined for decomposition has been used up, the furnace is allowed to stand undisturbed for a few days, and only after this time has elapsed are the chambers opened by windows provided for that purpose. In the fourth chamber is the very finest lampblack, which the lithographers use, and in the third the fine grade employed by manufacturers of printers' ink, while the first and second contain the coarser soot, which, well sifted, is sold as flame lampblack.

From grade No. 1 the calcined lampblack for paper makers is also produced. For preparing this black capsules of iron plate with closing lid are filled, the stuff is stamped firmly into them and the cover smeared up with fine loam. The capsules are next placed in a well drawing stove and calcined, whereby the empyreumatic oils evaporate and the remaining lampblack becomes odorless. Allow the capsules to cool for a few days before opening them, as the soot dries very slowly, and easily ignites again as soon as air is admitted if the capsules are opened before. This is semi-calcined lampblack.

For the purpose of preparing completely calcined lampblack, the semi-calcined article is again jammed into fresh capsules, closing them up well and calcining thoroughly once more. After 2 days the capsules are opened containing the all-calcined lampblack in compact pieces.

For the manufacture of coal soot another furnace is employed. Asphalt or pitch is burned in it with exclusion of air as far as practicable. It is thrown in through the doors, and the smoke escapes through the chimney to the soot chambers, 1, 2, 3, 4, and 5, assorting itself there.

When the amount of asphalt pitch destined for combustion has burned up completely, the furnace is left alone for several days without opening it. After this time has elapsed the outside doors are slowly opened and some air is admitted. Later on they can be opened altogether after one is satisfied that the soot has cooled completely. Chamber 4 contains the finest soot black, destined for the manufacture of leather cloth and oil cloth. In the other chambers is fine and ordinary flame black, which is sifted and packed in suitable barrels. Calcined lampblack may also be produced from it, the operation being the same as for oil black.

LAMP BURNERS AND THEIR CARE:
See Household Formulas.

LAMPS:

Coloring Incandescent Lamps.—Incandescent light globes are colored by dipping the bulbs into a thin solution of collodion previously colored to suit with anilines soluble in collodion. Dip and rotate quickly, bulb down, till dry.

For office desks, room lights, and in churches, it appears often desirable to modify the glaring yellowish rays of the incandescent light. A slight collodion film of a delicate bluish, greenish, or pink shade will do that.

For advertising purposes the bulbs are often colored in two or more colors. It is also easy with a little practice to paint words or pictures, etc., on the bulbs with colored collodion with a brush.

Another use of colored collodion in pharmacy is to color the show globes on their inside, thus avoiding freezing and the additional weight of the now used colored liquids. Pour a quantity of colored collodion into the clean, dry globe, close the mouth and quickly let the collodion cover all parts of the inside. Remove the balance of the collodion at once, and keep it to color electric bulbs for your trade.

LANOLINE CREAMS:
See Cosmetics.

LANOLINE SOAP:
See Soap.

LANTERN SLIDES:
See Photography.

LARD:

Detection of Cottonseed Oil in Lard.—Make a 2 per cent solution of silver nitrate in distilled water, and acidify it by adding 1 per cent of nitrate acid, C. P. Into a test tube put a sample of the suspected lard and heat gently until it liquefies. Now add an equal quantity of the silver nitrate solution, agitate a little, and bring to a boil. Continue the boiling vigorously for about 8 minutes. If the lard remain clear and colorless, it may be accepted as pure. The presence of cottonseed oil or fat will make itself known by a coloration, varying from yellow, grayish green to brown, according to the amount present.

LATHE LUBRICANT:
See Lubricants.

LAUNDRY INKS:
See Household Formulas.

Laundry Preparations

BLUING COMPOUNDS:

Laundry Blue.—The soluble blue of commerce, when properly made, dissolves freely in water, and solutions so made are put up as liquid laundry blue. The water employed in making the solution should be free from mineral substances, especially lime, or precipitation may occur. If rain water or distilled water and a good article of blue be used, a staple preparation ought apparently to result; but whether time alone affects the matter of solubility it is impossible to state. As it is essential that the solution should be a perfect one, it is best to filter it through several thicknesses of fine cotton cloth before bottling; or if made in large quantities this method may be modified by allowing it to stand some days to settle, when the top portion can be siphoned off for use, the bottom only requiring filtration.

This soluble blue is said to be potassium ferri-ferrocyanide, and is prepared by gradually adding to a boiling solution of potassium ferricyanide (red prussiate of potash) an equivalent quantity of hot solution of ferrous sulphate, boiling for 2 hours and washing the precipitate on a filter until the washings assume a dark-blue color; the moist precipitate can then at once be dissolved by the further addition of a sufficient quantity of water. About 64 parts of the iron salt are necessary to convert 100 parts of the potassium salt into the blue compound.

Leaf bluing for laundry use may be prepared by coating thick sized paper with soluble blue formed into a paste with a mixture of dextrin mucilage and glycerine. Dissolve a given quantity of dextrine in water enough to make a solution about as dense as ordinary syrup, add about as much glycerine as there was dextrine, rub the blue smooth with a sufficient quantity of this vehicle and coat the sheets with the paint. The amount of blue to be used will depend of course on the intended cost of the product, and the amount of glycerine will require adjustment so as to give a mixture which will not "smear" after the water has dried out and yet remain readily soluble.

Ultramarine is now very generally used as a laundry blue where the insoluble or "bag blue" is desired. It is mixed with glucose, or glucose and dextrine, and pressed into balls or cakes. When glucose alone is used, the product has a tendency, it is said, to become soft on keeping, which tendency may be counteracted by a proper proportion of dextrin. Bicarbonate of sodium is added as a "filler" to cheapen the product, the quantity used and the quality of the ultramarine employed being both regulated by the price at which the product is to sell.

The coal-tar or aniline blues are not offered to the general public as laundry blues, but laundry proprietors have them frequently brought under their notice, chiefly in the form of solutions, usually 1 to 1½ per cent strong. These dyes are strong bluing materials, and, being in the form of solution, are not liable to speck the clothes. Naturally their properties depend upon the particular dye used; some are fast to acids and alkalies, others are fast to one but not to another; some will not stand ironing, while others again are not affected by the operation; generally they are not fast to light, but this is only of minor importance. The soluble, or cotton, blues are those most favored; these are made in a great variety of tints, varying from a reddish blue to a pure blue in hue, distinguished by such brands as 3R, 6B, etc. Occasionally the methyl violets are used, especially the blue tints. Blackley blue is very largely used for this purpose, being rather faster than the soluble blues. It may be mentioned that a 1 per cent solution of this dye is usually strong enough. Unless care is taken in dissolving these dyes they are apt to produce specks. The heat to which the pure blues are exposed in ironing the clothes causes some kinds to assume a purple tinge.

The cheapest aniline blue costs about three times as much as soluble blue, yet the tinctorial power of the aniline colors is so great that possibly they might be cheapened.

Soluble Blue.—I.—Dissolve 217 parts of prussiate of potash in 800 parts of hot water and bring the whole to 1,000 parts. Likewise dissolve 100 parts of ferric chloride in water and bring the solution also to 1,000 parts. To each of these solutions add 2,000 parts of cooking salt or Glauber's salt solution saturated in the cold and mix well. The solutions thus prepared of prussiate of potash and ferric chloride are now mixed together with stirring. Allow to settle and remove by suction the clear liquid containing undecomposed ferrocyanide of potassium and Glauber's salt; this is kept and used for the next manufacture by boiling it down and allowing the salts to crystallize out. The percentage of ferrocyanide of potassium is estimated by analysis, and for the next production proportionally less is used, employing that obtained by concentration.

After siphoning off the solution the precipitate is washed with warm water, placed on a filter and washed out on the latter by pouring on cold water until the water running off commences to assume a strong blue color. The precipitate is then squeezed out and dried at a moderate heat (104° F.). The Paris blue thus obtained dissolves readily in water and can be extensively employed in a similar manner as indigo carmine.

II.—Make ordinary Prussian blue (that which has been purified by acids, chlorine, or the hypochlorites) into a thick paste with distilled or rain water, and add a saturated solution of oxalic acid sufficient to dissolve. If time be of no consequence, by leaving this solution exposed to the atmosphere, in the course of 60 days the blue will be entirely precipitated in soluble form. Wash with weak alcohol and dry at about 100° F. The resultant mass dissolves in pure water and remains in solution indefinitely. It gives a deep, brilliant blue, and is not injurious to the clothing or the hands of the washwoman.

The same result may be obtained by precipitating the soluble blue from its oxide solution by the addition of alcohol of 95 per cent, or with a concentrated solution of sodium sulphate. Pour off the mother liquid and wash with very dilute alcohol; or throw on a filter and wash with water until the latter begins to come off colored a deep blue.

Liquid Laundry Blue.—This may be prepared either with liquid Prussian blue or indigo carmine. Make a solution of gum dragon (gum tragacanth) by dissolving 1 to 2 ounces of the powdered gum in 1 gallon of cold water in which $\frac{1}{2}$ ounce oxalic acid has been dissolved. The gum will take several days to dissolve, and will require frequent stirring and straining before use. To the strained portion add as much Prussian blue in fine powder as the liquid will dissolve without precipitating, and the compound is ready for use.

Instead of powdered Prussian blue, soluble Prussian blue may be used. This is made by dissolving solid Prussian blue in a solution of oxalic acid, but as the use of oxalic acid is to be deprecated for the use of laundresses, as it would set up blood poisoning should it get into any cuts in the flesh, it is best to prepare liquid blue by making a solution of yellow prussiate of potash (ferrocyanide of potassium) with water, and then by adding a sufficient quantity of chloride of iron to produce a blue, but not enough to be precipitated.

Ball Blue.—The ball sold for laundry use consists usually, if not always, of ultramarine. The balls are formed by compression, starch or some other excipient of like character being added to render the mass cohesive. Blocks of blue can, of course, be made by the same process. The manufacturers of ultramarine prepare balls and cubes of the pigment on a large scale, and it does not seem likely that there would be a sufficient margin of profit to justify the making of them in a small way from the powdered pigment. Careful experiments, however, would be necessary to determine this positively. Ultramarine is of many qualities, and it may be expected that the balls will vary also in the amount of "filling" according to the price at which they are to be sold.

Below is a "filled" formula:

Ultramarine......... 6 ounces
Sodium carbonate.... 4 ounces
Glucose............. 1 ounce
Water, a sufficient quantity.

Make a thick paste, roll into sheets, and cut into tablets. The balls in bulk can be obtained only in large packages of the manufacturers, say barrels of 200 pounds; but put up in 1-pound boxes they can be bought in cases as small as 28 pounds.

Laundry Blue Tablets.—

Ultramarine......... 6 ounces
Sodium carbonate.... 4 ounces
Glucose............. 1 ounce
Water, a sufficient quantity.

Make a thick paste, roll into sheets, and cut into tablets.

Polishes or Glazes for Laundry Work.—I.—To a mixture of 200 parts each of Japan wax and paraffine, add 100 parts of stearic acid, melt together, and cast in molds. If the heated smoothing iron be rubbed with this wax the iron will not merely get over the surface much more rapidly, but will leave a handsome polish.

Laundry Gloss Dressing.—

II.—Dissolve white wax, 5.0 parts, in ether, 20.5 parts, and add spirit, 75.0 parts. Shake before use.

Heat until melted, in a pot, 1,000 parts

of wax and 1,000 parts of stearine, as well as a few drops of an essential oil. To the hot liquid add with careful stirring 250 parts of ammonia lye of 10 per cent, whereby a thick, soft mass results immediately. Upon further heating same turns thin again, whereupon it is diluted with 20,000 parts of boiling water, mixed with 100 parts of starch and poured into molds.

STARCHES.

Most laundry starches now contain some polishing mixture for giving a high luster.

I.—Dissolve in a vessel of sufficient capacity, 42 parts of crystallized magnesium chloride in 30 parts of water. In another vessel stir 12 parts of starch in 20 parts of water to a smooth paste. Mix the two and heat under pressure until the starch is fluidified.

II.—Pour 250 parts, by weight, of water, over 5 parts, by weight, of powdered gum tragacanth until the powder swells uniformly; then add 750 parts, by weight, of boiling water, dissolve 50 parts, by weight, of borax in it, and stir 50 parts, by weight, of stearine and 50 parts, by weight, of talcum into the whole. Of this fluid add 250 parts to 1,000 parts of boiled starch, or else the ironing oil is applied by means of a sponge on the starched wash, which is then ironed.

	By weight
III.—Starch	1,044 parts
Borax	9 parts
Common salt	1 part
Gum arabic	8 parts
Stearine	20 parts

WASHING FLUIDS, BRICKS AND POWDERS:

Washing Fluids.—Rub up 75 parts of milk of sulphur with 125 parts of glycerine in a mortar, next add 50 parts of camphorated spirit and 1 part of lavender oil, and finally stir in 250 parts of rose water and 1,000 parts of distilled water. The liquid must be stirred constantly when filling it into bottles, since the sulphur settles rapidly and would thus be unevenly distributed.

Grosser's Washing Brick.—

Water	54 parts
Sodium hydrate	38.21 parts
Sodium biborate	6.61 parts
Sodium silicate	1.70 parts

Haenkel's Bleaching Solution.—

Water	36.15 parts
Sodium hydrate	40.22 parts
Sodium silicate	23.14 parts

Luhn's Washing Extract.—

Water	34.50 parts
Sodium hydrate	25.33 parts
Soap	39.40 parts

Washing Powders.—

I.—Sodium carbonate, partly effloresced	2 parts	
Soda ash	1 part	
II.—Sodium carbonate, partly effloresced	6 parts	
Soda ash	3 parts	
Yellow soap	1 part	
III.—Sodium carbonate, partly effloresced	3 parts	
Soap bark	1 part	
IV.—Sodium carbonate, partly effloresced		
Borax		Equal parts.
Yellow soap		

V.—A good powder can be made from 100 parts of crystal soda, 25 parts of dark-yellow rosin-cured soap, and 5 parts of soft soap. The two latter are placed in a pan, along with one-half the soda (the curd soap being cut into small lumps), and slowly heated, with continual crutching, until they are thoroughly melted—without, however, beginning to boil. The fire is then drawn and the remaining soda crutched in until it, too, is melted, this being effected by the residual heat of the mass and the pan. The mass will be fairly thick by the time the soda is all absorbed. After leaving a little longer, with occasional stirring, the contents are spread out on several thin sheets of iron in a cool room, to be then turned over by the shovel at short intervals, in order to further cool and break down the mixture. The soap will then be in a friable condition, and can be rubbed through the sieve, the best results being obtained by passing through a coarse sieve first, and one of finer mesh afterwards. With these ingredients a fine yellow-colored powder will be obtained. White stock soap may also be used, and, if desired, colored with palm oil and the same colorings as are used for toilet soaps. The object of adding soft soap is to increase the solubility and softness of the powder, but the proportion used should not exceed one-third of the hard soap, or the powder will be smeary and handle moist. The quality of the foregoing product is good, the powder being stable and not liable to ball, even after prolonged storage; neither does it wet the paper in which it is packed, nor swell up, and therefore the packets retain their appearance.

In making ammonia-turpentine soap powder the ammonia and oil of turpentine are crutched into the mass shortly before removing it from the pan, and if the powder is scented—for which purpose oil of mirbane is mostly used—the perfume is added at the same stage.

To Whiten Flannels.—Dissolve, by the aid of heat, 40 parts of white castile soap, shaved fine, in 1,200 parts of soft water, and to the solution, when cold, gradually add, under constant stirring, 1 part of the strongest water of ammonia. Soak the goods in this solution for 2 hours, then let them be washed as usual for fine flannels. A better process, in the hands of experts, is to soak the goods for an hour or so in a dilute solution of sodium hyposulphite, remove, add to the solution sufficient dilute hydrochloric acid to decompose the hyposulphite. Replace the goods, cover the tub closely, and let remain for 15 minutes longer. Then remove the running water, if convenient, and if not, wring out quickly, and rinse in clear water. One not an expert at such work must be very careful in the rinsing, as care must be taken to get out every trace of chemical. This is best done by a second rinsing.

Ink for the Laundry.—The following is said to make a fine, jet-black laundry ink:

a. Copper chloride, crystals .... 85 parts
Sodium chlorate.... 106 parts
Ammonium chloride 53 parts
Water, distilled..... 600 parts

b. Glycerine.......... 100 parts
Mucilage gum arabic (gum, 1 part; water, 2 parts).... 200 parts
Aniline hydrochlorate.............. 200 parts
Distilled water...... 300 parts

Make solutions a and b and preserve in separate bottles. When wanted for use, mix 1 part of solution a with 4 parts of solution b.

Laces, Curtains, etc.—I.—To give lace curtains, etc., a cream color, take 1 part of chrysoidin and mix with 2 parts of dextrin and dissolve in 250 parts of water. The articles to be washed clean are plunged in this solution. About an ounce of chrysoidin is sufficient for 5 curtains.

II.—Washing curtains in coffee will give them an ecru color, but the simplest way to color curtains is with "Philadel-

phia yellow" (G. or R. of the Berlin *Aktiengesellschaft's* scale).

LAUNDRY SOAP:
See Soap.

LAVATORY DEODORANT:
See Household Formulas.

LAXATIVES FOR CATTLE AND HORSES:
See Veterinary Formulas.

LEAD:
See also Metals.

Simple Test for Red Lead and Orange Lead.—Take a little of the sample in a test tube, add pure, strong nitric acid and heat by a Bunsen burner until a white, solid residue is obtained. Then add water, when a clear, colorless solution will be obtained. A white residue would indicate adulteration with barytes, a red residue or a yellow solution with oxide of iron. The presence of iron may be ascertained by adding a few drops of a solution of potassium ferrocyanide (yellow prussiate of potash) to the solution, when a blue precipitate will be obtained if there be the least trace of iron present.

LEAD, TO TAKE BOILING, IN THE MOUTH:
See Pyrotechnics.

LEAD ALLOYS:
See Alloys.

LEAD PAPER:
See Paper.

LEAD PLATE, TINNED:
See Plating.

LEAKS, IN BOILERS, STOPPING:
See Putties.

LEAKS:

To Stop Leakage in Iron Hot-Water Pipes.—Take some fine iron borings or filings and mix with them sufficient vinegar to form a sort of paste, though the mixture is not adhesive. With this mixture fill up the cracks where the leakage is found, having previously dried the pipe. It must be kept dry until the paste has become quite hard. If an iron pipe should burst, or there should be a hole broken into it by accident, a piece of iron may be securely fastened over it, by bedding it on in paste made of the borings and vinegar as above, but the pipe should not be disturbed until it has become perfectly dry.

To Prevent Wooden Vessels from Leaking. (See also Casks.)—Wooden vessels, such as pails, barrels, etc., often become so dry that the joints do not meet, thus causing leakage. In order to obviate this evil stir together 60 parts hog's lard, 40 parts salt, and 33 parts wax, and allow the mixture to dissolve slowly over a fire. Then add 40 parts charcoal to the liquid mass. The leaks in the vessels are dried off well and filled up with putty while still warm. When the latter has become dry, the barrels, etc., will be perfectly tight. If any putty is left, keep in a dry place and heat it to be used again.

Leather

(See also Shoes.)

Artificial Leather.—Pure Italian hemp is cut up fine; 1 part of this and $\frac{1}{2}$ part of coarse, cleaned wool are carded together and formed into wadding. This wadding is packed in linen and felted by treatment with hot acid vapors. The resulting felt is washed out, dried, and impregnated with a substance whose composition varies according to the leather to be produced. Thus, good sole leather, for instance, is produced according to a Danish patent, in the following manner: Mix together 50 parts of boiled linseed oil; 20 parts of colophony; 25 parts of French turpentine; 10 parts of glycerine, and 10 parts of vegetable wax, and heat over a water bath with some ammonia water. When the mass has become homogeneous, add 25 parts of glue, soaked in water, as well as a casein solution, which latter is produced by dissolving 50 parts, by weight, of moist, freshly precipitated casein in a saturated solution of 16 parts of borax and adding 10 parts of potassium bichromate, the last two also by weight. Finally, mineral dyestuffs as well as antiseptic substances may be added to the mass. The whole mixture is now boiled until it becomes sticky and the felt is impregnated with it by immersion. The impregnated felt is dried for 24 hours at an ordinary temperature; next laid into a solution of aluminum acetate and finally dried completely, dyed, and pressed between hot rollers.

Black Dye for Tanned Leather.—This recipe takes the place of the ill-smelling iron blacking, and is not injurious to the leather. Gallnuts, pulverized, 150 parts; vitriol, green or black, 10 parts; rock candy, 60 parts; alum, 15 parts; vinegar, 250 parts; cooking salt, 20 parts. Dissolve with 4,000 parts of distilled water. Boil this solution slowly and the

blacking is done. When it has cooled and settled, pour through linen, thus obtaining a pure, good leather blacking.

Bronze Leather.—All sorts of skins—sheepskins, goatskins, coltskins, and light calfskins—are adapted for the preparation of bronze leather. In this preparation the advantage lies not only in the use of the faultless skins, but scarified skins and those of inferior quality may also be employed. The dressing of the previously tanned skin must be carried out with the greatest care, to prevent the appearance of spots and other faults. After tanning, the pelts are well washed, scraped, and dried. Then they are bleached. For coloring, it is customary to employ methyl violet which has previously been dissolved in hot water, taking 100 parts, by weight, of the aniline color to 8,000 parts, by weight, of water. If in the leather-dressing establishment a line of steam piping be convenient, it is advisable to boil up all the coloring dyes, rather than simply to dissolve them; for in this way complete solution is effected. Where steam is used no special appliance is required for boiling up the dyes, for this may take place without inconvenience in the separate dye vats. A length of steam hose and a brass nozzle with a valve is all that is needed. It may be as well to add here that the violet color for dyeing may be made cheaper than as above described. To 3,000 parts, by weight, of pretty strong logwood decoction add 50 parts, by weight, of alum and 100 parts, by weight, of methyl violet. This compound is almost as strong as the pure violet solution, and instead of 8,000 parts, by weight, we now have 30,000 parts, by weight, of color.

The color is applied and well worked in with a stiff brush, and the skins allowed to stand for a short time, sufficient to allow the dye to penetrate the pores, when it is fulled. As for the shade of the bronze, it may be made reddish, bluish, or brownish, according to taste.

For a reddish or brownish ground the skins are simply fulled in warm water, planished, fulled again, and then dyed. According to the color desired, the skins are treated with cotton blue and methyl violet R, whereupon the application of the bronze follows.

The bronze is dissolved in alcohol, and it is usual to take 200 parts, by weight, of bronze to 1,000 of alcohol. By means of this mixture the peculiar component parts of the bronze are dissolved. For a fundamental or thorough solution a fortnight is required. All bronze mixtures are to be well shaken or agitated before using. Skins may be bronzed, however, without the use of the bronze colors, for it is well known that all the aniline dyes present a bronze appearance when highly concentrated, and this is particularly the case with the violet and red dyes. If, therefore, the violet be applied in very strong solutions, the effect will be much the same as when the regular bronze color is employed.

Bronze color on a brown ground is the most beautiful of all, and is used to the greatest advantage when it is desirable to cover up defects. Instead of warm clear water in such a case, use a decoction of logwood to which a small quantity of alum has been added, and thus, during the fulling, impart to the skins a proper basic tint, which may, by the application of a little violet or bronze color, be converted into a most brilliant bronze. By no means is it to be forgotten that too much coloring matter will never produce the desired results, for here, as with the other colors, too much will bring out a greenish tint, nor will the gloss turn out so beautiful and clear. Next rinse the skins well in clean water, and air them, after which they may be dried with artificial heat. Ordinary as well as damaged skins which are not suitable for chevreaux (kid) and which it is desirable to provide with a very high polish, in order the more readily to conceal the defects in the grain, and other imperfections, are, after the drying, coated with a mixture, compounded according to the following simple formula: Stir well 1 pint of ox blood and 1 pint of unboiled milk in 10 quarts of water, and with a soft sponge apply this to the surface of the skin. The blood has no damaging effect upon the color. Skins thus moistened must not be laid one upon another, but must be placed separately in a thoroughly well-warmed chamber to dry. When dry they are glossed, and may then be pressed into shagreen or pebbled. The thin light goatskins are worked into kid or chevreaux. Properly speaking, they are only imitation chevreaux (kid), for although they are truly goatskins, under the term chevreaux one understands only such skins as have been cured in alum and treated with albumen and flour.

After drying, these skins are drawn over the perching stick with the round knife, then glossed, stretched, glossed again, and finally vigorously brushed upon the flesh side with a stiff brush. The brushing should be done preferably by hand, for the brushing machines commonly pull the skins out of all shape. Brushing is intended only to give the flesh side more of a flaky appearance.

During the second glossing care must be taken that the pressure is light, for the object is merely to bring the skin back into its proper shape, lost in the stretching; the glossing proper should have been accomplished during the first operation.

Cracked Leather.—The badly cracked and fissured carriage surface greets the painter on every hand. The following is the recipe for filling up and facing over such a surface: Finest pumice stone, 6 parts; lampblack (in bulk), 1 part; common roughstuff filler, 3 parts. Mix to stiff paste in good coach japan, 5 parts; hard drying rubbing varnish, 1 part. Thin to a brushing consistency with turpentine, and apply 1 coat per day. Put on 2 coats of this filler and then 2 coats of ordinary roughstuff. Rub with lump pumice stone and water. This process does not equal burning off in getting permanently rid of the cracks, but when the price of painting forbids burning off, it serves as an effective substitute. Upon a job that is well cared for, and not subjected to too exacting service, this filler will secrete the cracks and fissures for from 3 to 5 months.

DRESSINGS FOR LEATHER:

For Carriage Tops.—I.—Here is an inexpensive and quickly prepared dressing for carriage tops or the like: Take 2 parts of common glue; soak and liquefy it over a fire. Three parts of castile soap are then dissolved over a moderate heat. Of water, 120 parts are added to dissolve the soap and glue, after which an intimate mixture of the ingredients is effected. Then 4 parts of spirit varnish are added; next, 2 parts of wheat starch, previously mixed in water, are thrown in. Lampblack in a sufficient quantity to give the mixture a good coloring power, without killing the gloss, is now added. This preparation may be used as above prepared, or it may be placed over a gentle fire and the liquid ingredients slowly evaporated. The evaporated mass is then liquefied with beer as shop needs demand.

II.—Shabby dark leather will look like new if rubbed over with either linseed oil or the well-beaten white of an egg mixed with a little black ink. Polish with soft dusters until quite dry and glossy.

Polishes.—I.—Dissolve sticklac, 25

parts; shellac, 20 parts; and gum benzoin, 4 parts, all finely powdered, in a rolling cask containing 100 parts of 96 per cent alcohol; perfume with 1 part of oil of rosemary. Upon letting stand for several days, filter the solution, whereupon a good glossy polish for leather, etc., will be obtained.

II.—Dissolve 2 pounds of borax in 4 gallons of water and add 5 pounds of shellac to the boiling liquid in portions, till all is dissolved. Then boil half an hour, and finally stir in 5 pounds of sugar, $2\frac{1}{2}$ pounds of glycerine, and $1\frac{1}{2}$ pounds of soluble nigrosin. When cold add 4 pounds of 95 per cent methylated spirit.

III.—Ox blood, fresh,
clean.......... 1,000 parts
Commercial glyc-
erine .... 200 parts
Oil of turpentine. 300 parts
Pine oil (rosin
oil)........... 5,000 parts
Ox gall.......... 200 parts
Formalin........ 15 parts

Mix in the order named, stirring in each ingredient. When mixed strain through linen.

Kid Leather Dressings.—Creams for greasing fine varieties of leather, such as kid, patent leather, etc., are produced as follows, according to tried recipes:

White Cream.—
Lard .... 75 parts
Glycerine, technical . 25 parts
Mirbane oil, ad libitum.

Black Cream.—
Lard .... 100 parts
Yellow vaseline..... 20 parts
Glycerine, technical. 10 parts
Castor oil, technical . 10 parts

Dye black with lampblack and perfume with oil of mirbane.

Colored Cream.—
Lard............... 100 parts
Castor oil.......... 20 parts
Yellow wax......... 25 parts
White vaseline...... 30 parts

Dye with any desired dyestuff, e. g., red with anchusine, green with chlorophyl. In summer it is well to add some wax to the first and second prescriptions. These are for either Morocco or kid:

I.—Shellac.......... 2 parts
Benzoin.......... 2 parts
Yellow wax........ 5 parts
Soap liniment..... 7 parts
Alcohol.......... 600 parts

Digest until solution is effected, then allow the liquid to stand in a cool place for 12 hours and strain. Apply with a bit of sponge or soft rag; spread thinly and evenly over the surface, without rubbing much. If dirty, the leather should first be washed with a little soft soap and warm water, wiped well, and allowed to dry thoroughly before the dressing is put on.

II.—Oil of turpentine.... 8 ounces
Suet.............. 2 pounds
Soft soap .... 8 ounces
Water............. 16 ounces
Lampblack........ 4 ounces

Patent Leather Dressings.—

I.—Wax............. 22 parts
Olive oil........... 60 parts
Oil turpentine, best. 20 parts
Lavender oil....... 10 parts

With gentle heat, melt the wax in the oil, and as soon as melted remove from the fire. Add the turpentine oil, incorporate, and when nearly cold, add and incorporate the lavender oil.

II.—Wax............. 22 parts
Olive oil........... 60 parts
Oil of turpentine.... 30 parts

With gentle heat, melt the wax in the olive oil, and as soon as melted remove from the fire. When nearly cold stir in the turpentine.

Red Russia Leather Varnish.—
Shellac............. 1.20 parts
Dammar rosin, pow-
dered .... 0.15 parts
Turpentine, Venice.. 0.60 parts

Dissolve with frequent shaking in 12 parts of alcohol (95 per cent), add 1.8 parts of powdered red sanders wood, let stand for 3 days and filter. The object of this varnish is to restore the original color to worn Russia leather boots, previously cleaned with benzine.

Russet Leather Dressing.—The following formulas are said to yield efficient preparations that are at once detersive and polishing, thus rendering the use of an extra cleaning liquid unnecessary.

I.—Soft soap.......... 2 parts
Linseed oil......... 3 parts
Annatto solution (in
oil).............. 8 parts
Beeswax........... 3 parts
Turpentine........ 8 parts
Water............. 8 parts

Dissolve the soap in the water, and add the annatto; melt the wax in the oil and turpentine; and gradually stir in the soap solution, stirring until cold.

II.—
Palm oil	16 parts
Common soap	48 parts
Oleic acid	32 parts
Glycerine	10 parts
Tannic acid	1 part

Melt the soap and palm oil together at a gentle heat, and add the oleic acid; dissolve the tannic acid in the glycerine, add to the hot soap and oil mixture, and stir until perfectly cold.

Shoe Leather Dressing.—Over a water bath melt 50 parts, by weight, of oil of turpentine; 100 parts, by weight, of olive oil; 100 parts, by weight, of train oil; 40 parts, by weight, of carnauba wax; 15 parts, by weight, of asphaltum; and 2 parts, by weight, of oil of bitter almonds.

DYEING LEATHER.

In dyeing leather, aniline or coal-tar colors are generally used. These dyes, owing to their extremely rapid action on organic substances, such as leather, do not readily adapt themselves to the staining process, because a full brushful of dye liquor would give a much deeper coloration than a half-exhausted brush would give. Consequently, to alter and to color leather by the staining process results in a patchy coloration of the skin. In the dyeing operation a zinc shallow trough, 4 to 6 inches deep, is used, into which the dye liquor is put, and to produce the best results the contents of the trough are kept at a uniform temperature by means of a heating apparatus beneath the trough, such as a gas jet or two, which readily allows of a heat being regulated. The skins to be dyed are spread out flat in the dye trough, one at a time, each skin remaining in the dye liquor the time prescribed by the recipe. The best coloration of the skin is produced by using 3 dye troughs of the same dye liquor, each of different strength, the skin being put in the weakest liquor first, then passed into the second, and from there into the third dye liquor, where it is allowed to remain until its full depth of color is obtained. Very great skill is required in the employment of aniline dyes, as if the heat be too great, or the skins remain too long in the final bath, "bronzing" of the color occurs. The only remedy for this (and that not always effectual) is to sponge the skin with plenty of cold, clean water, directly it is taken out of the final dye bath. The dyed skins are dried and finished as before.

Leather Brown.—

Extract of fustic	5 ounces
Extract of hypernic	1 ounce
Extract of logwood	½ ounce
Water	2 gallons

Boil all these ingredients for 15 minutes, and then dilute with water to make 10 gallons of dye liquor. Use the dye liquor at a temperature of 110° F.

Mordant.—Dissolve 3 ounces of white tartar and 4 ounces of alum in 10 gallons of water.

Fast Brown.—Prepare a dye liquor by dissolving 1½ ounces fast brown in 1 gallon of water, and make a 10-gallon bulk of this. Use at a temperature of 110° F., and employ the same mordanting liquor as in last recipe.

Bismarck Brown.—

Extract of fustic	4 ounces
Extract of hypernic	1 ounce
Extract of logwood	½ ounce
Water	2 gallons

Preparation.—Boil all together for 15 minutes.

Method of Dyeing.—First mordant the skins with a mordanting fluid made by dissolving 3 ounces tartar and ½ ounce borax in 10 gallons of water. Then put the skins into the above foundation bath at a temperature of 100° F. Take them out, and then put in 1 ounce of Bismarck brown, dissolved in boiling water. Put the skins in again until colored deep enough, then lift out, drip and dry.

HARNESS PREPARATIONS:

Blacking for Harness.—I.—In a water bath dissolve 90 parts of yellow wax in 900 parts of oil of turpentine; aside from this mix well together, all the ingredients being finely powdered, 10 parts of Prussian blue, 5 parts of indigo, 50 parts of bone black, and work this into a portion of the above-mentioned waxy solution. Now throw this into the original solution, which still remains in the water bath, and stir it vigorously until the mass becomes homogeneous, after which pour it into any convenient earthenware receptacle.

II.—Best glue, 4 ounces; good vinegar, 1½ pints; best gum arabic, 2 ounces; good black ink, ½ pint; best isinglass, 2 drachms. Dissolve the gum in the ink, and melt the isinglass in another vessel in as much hot water as will cover it. Having first steeped the glue in the vinegar until soft, dissolve it completely by the aid of heat, stirring to prevent burning. The heat should not exceed 180° F. Add the gum and ink, and allow the mixture to rise again to the same temperature. Lastly mix the solution in isinglass, and remove from fire. When used, a small portion must be heated until fluid, and then applied with a sponge and allowed to dry on.

Dressings for Harness.—

I.—Ox blood, fresh and
well purified.......100 parts
Glycerine, technical. 20 parts
Turpentine oil...... 30 parts
Pine oil............ 50 parts
Ox gall............. 20 parts
Formalin.......... 1½ parts

The raw materials are stirred together cold in the order named. Pour the mixture through thin linen. It imparts a wonderful mild, permanent gloss.

II.—A French harness dressing of good quality consists of oil of turpentine, 900 parts; yellow wax, 90 parts; Berlin blue, 10 parts; indigo, 5 parts; and bone black, 50 parts. Dissolve the yellow wax in the oil of turpentine with the aid of moderate heat in a water bath, mix the remaining substances, which should previously be well pulverized, and work them with a small portion of the wax solution. Finally, add the rest of the wax solution, and mix the whole well in the water bath. When a homogeneous liquid has resulted, pour it into earthen receptacles.

Harness Oils.—

I.—Neatsfoot oil...... 10 ounces
Oil of turpentine.... 2 ounces
Petrolatum........ 4 ounces
Lampblack........ ½ ounce

Mix the lampblack with the turpentine and the neatsfoot oil, melt the petrolatum and mix by shaking together.

II.—Black aniline.... 35 grains
Muriatic acid... 50 minims
Bone black..... 175 grains
Lampblack..... 18 grains
Yellow wax..... 2½ av. ounces
Oil of turpentine 22 fluidounces

III.—Oil of turpentine 8 fluidounces
Yellow wax..... 2 av. ounces
Prussian blue... ½ av. ounce
Lampblack..... ¼ av. ounce

Melt the wax, add the turpentine, a portion first to the finely powdered Prussian blue and lampblack, and thin with neatsfoot oil.

Harness Pastes.—

I.—Ceresine, natural
yellow.......... 1.5 parts
Yellow beeswax.... 1.5 parts
Japan wax........ 1.5 parts

Melt on the water bath, and when half cooled stir in 8 parts of turpentine oil.

Harness Grease.—

	By weight
II.—Ceresine, natural yellow...........	2.5 parts
Beeswax, yellow....	0.8 parts
French colophony, pale............	0.4 parts

	By weight
III.—French oil turpentine.............	2.0 parts
Intimately mixed in the cold with American lampblack...........	1.5 parts

Put mixture I in a kettle and melt over a fire. Remove from the fire and stir in mixture II in small portions. Then pour through a fine sieve into a second vessel, and continue pouring from one kettle into the other until the mass is rather thickish. Next fill in cans.

Should the mixture have become too cold during the filling of the cans, the vessel containing the grease need only be placed in hot water, whereby the contents are rendered liquid again, so that pouring out is practicable. For perfuming, use cinnamon oil as required.

This harness grease is applied by means of a rag and brushed.

Waterproof Harness Composition.—
See also Waterproofing.

	By weight
Rosin spirit......	27¼ parts
Dark mineral oil..	13½ parts
Paraffine scales...	16.380 parts
Lampblack	7.940 parts
Dark rosin.......	5.450 parts
Dark syrup......	5.450 parts
Naphthalene black	2.500 parts
Berlin blue.......	0.680 parts
Mirbane oil......	0.170 parts

Melt the paraffine and the rosin, add the mineral oil and the rosin spirit, stir the syrup and the pigments into this, and lastly add the mirbane oil.

PATENT AND ENAMELED LEATHER.

Patent leather for boots and shoes is prepared from sealskins, enameled leather for harness from heavy bullock's hides. The process of tanning is what is called "union tannage" (a mixture of oak and hemlock barks). These tanned skins are subjected to the process of soaking, unhairing, liming, etc., and are then subjected to the tanning process. When about one-third tanned a buffing is taken off (if the hides are heavy), and the hide is split into three layers. The top or grain side is reserved for enameling in fancy colors for use on tops of carriages; the middle layer is finished for splatter boards and carriage trimmings, and some parts of harness; the underneath layer, or flesh side is used for shoe uppers and other purposes. The tanning of the splits is completed by subjecting them to a gambier liquor instead of a bark liquor.

When the splits are fully tanned they are laid on a table and scored, and then stretched in frames and dried, after which each one is covered on one side with the following compound, so as to close the pores of the leather that it may present a suitable surface for receiving the varnish: Into 14 parts of raw linseed oil put 1 part dry white lead and 1 part silver litharge, and boil, stirring constantly until the compound is thick enough to dry in 15 or 20 minutes (when spread on a sheet of iron or china) into a tough, elastic mass, like caoutchouc. This compound is laid on one side of the leather while it is still stretched in the frame. If for enameled leather (i. e., not the best patent), chalk or yellow ocher may be mixed in the above compound while boiling, or afterwards, but before spreading it on the leather.

The frames are then put into a rack in a drying closet, and the coated leather dried by steam heat at 80° to 160° F., the heat being raised gradually. After removal from the drying closet, the grounding coat previously laid on is pumiced, to smooth out the surface, and then given 2 or 3 coats of the enameling varnish, which consists of Prussian blue and lampblack boiled with linseed oil and diluted with turpentine, so as to enable it to flow evenly over the surface of the coated leather. When spread on with a brush, each coating of the enamel is dried before applying the next, and pumiced or rubbed with tripoli powder on a piece of flannel (the coat last laid on is not subjected to this rubbing), when the leather is ready for market.

To prepare the enameling composition, boil 1 part asphaltum with 20 parts raw linseed oil until thoroughly combined; then add 10 parts thick copal varnish, and when this mixture is homogeneous dilute with 20 parts spirit of turpentine.

Instead of the foregoing enameling varnish the following is used for superior articles:

Prussian blue...... 18 ounces
Vegetable black... 4 ounces
Raw linseed oil.... 160 fluidounces

Boil together as previously directed, and dilute with turpentine as occasion requires. These enameling varnishes should be made and kept several weeks in the same room as the varnishing is carried on, so that they are always subjected to the same temperature.

STAINS FOR PATENT LEATHER:

Black Stain.—

Vinegar............ 1 gallon
Ivory black......... 14 ounces
Ground iron scales... 6 pounds

Mix well and allow to stand a few days.

Red Stain.—Water, 1 quart; spirit of hartshorn, 1 quart; cochineal, $\frac{1}{4}$ pound. Heat the water to near the boiling point, and then dissolve in it the cochineal, afterwards adding the spirit of hartshorn. Stir well to incorporate.

Liquid Cochineal Stain.—

Good French carmine 2$\frac{1}{2}$ drachms
Solution of potash..... $\frac{1}{2}$ ounce
Rectified spirit of wine 2 ounces
Pure glycerine........ 4 ounces
Distilled water to make 1 pint.

To the carmine in a 20-ounce bottle add 14 ounces of distilled water. Then gradually introduce solution of potash, shaking now and again until dissolved. Add glycerine and spirit of wine, making up to 20 ounces with distilled water, and filter.

Blue Black.—Ale droppings, 2 gallons; bruised galls, $\frac{1}{2}$ pound; logwood extract, $\frac{1}{4}$ pound; indigo extract, 2 ounces; sulphate of iron, 3$\frac{1}{2}$ ounces. Heat together and strain.

Finishers' Ink.—Soft water, 1 gallon; logwood extract, 1$\frac{1}{4}$ ounces; green vitriol, 2$\frac{1}{2}$ ounces; potassium bichromate, $\frac{1}{2}$ ounce; gum arabic, $\frac{1}{2}$ ounce.

Grind the gum and potassium bichromate to powder and then add all the coloring ingredients to the water and boil.

To Restore Patent Leather Dash.— Take raw linseed oil, 1 part; cider vinegar, 4 ounces; alcohol, 2 ounces; butter of antimony, 1 ounce; aqua ammonia, $\frac{1}{2}$ ounce; spirits of camphor, $\frac{1}{2}$ ounce; lavender, $\frac{1}{2}$ ounce. Shake well together; apply with a soft brush.

PRESERVATIVES FOR LEATHER.

I.—Mutton suet....... 50 parts
Sweet oil.......... 50 parts
Turpentine......... 1 part
Melt together.

The application should be made on the dry leather warmed to the point where it will liquefy and absorb the fat.

II.—Equal parts of mutton fat and linseed oil, mixed with one-tenth their weight of Venice turpentine, and melted together in an earthen pipkin, will produce a "dubbin" which is very efficacious in preserving leather when exposed to wet or snow, etc. The mixture should be applied when the leather is quite dry and warm.

III.—A solution of 1 ounce of solid paraffine in 1 pint light naphtha, to which 6 drops of sweet oil have been added, is put cold on the soles, until they will absorb no more. One dressing will do for the uppers. This process is claimed to vastly increase the tensile strength.

Patent Leather Preserver.—

Carnauba wax	1.0	part
Turpentine oil	9.5	parts
Aniline black, soluble in fat	0.06	parts

Melt the wax, stir in the turpentine oil and the dye and scent with a little mirbane oil or lavender oil. The paste is rubbed out on the patent leather by means of a soft rag, and when dry should be polished with a soft brush.

REVIVERS AND REGENERATORS.

	By weight.	
I.—Methylic alcohol	22½	parts
Ground ruby shellac	2.250	parts
Dark rosin	0.910	parts
Gum rosin	0.115	parts
Sandarac	0.115	parts
Lampblack	0.115	parts
Aniline black, spirit-soluble	0.115	parts

The gums are dissolved in spirit and next the aniline black soluble in spirit is added; the lampblack is ground with a little liquid to a paste, which is added to the whole, and filtering follows.

Kid Reviver.—

	By weight.	
II.—Clear chloride of lime solution	3.5	parts
Spirit of sal ammoniac	0.5	parts
Scraped Marseilles soap	4.5	parts
Water	6.0	parts

Mix chloride of lime solution and spirit of sal ammoniac and stir in the soap dissolved in water. Revive the gloves with the pulpy mass obtained, by means of a flannel rag.

TANNING LEATHER.

Pickling Process.—Eitner and Stiazny have made a systematic series of experiments with mixtures of salt and various acids for pickling skins preparatory to tanning. Experiments with hydrochloric acid, acetic and lactic acids showed that these offered no advantages over sulphuric acid for use in pickling, the pickled pelts and the leather produced from them being similar in appearance and quality. By varying the concentration of the pickle liquors, it was found that the amount of salt absorbed by the pelt from the pickle liquor was controlled by the concentration of the solution, 23 to 25 per cent of the total amount used being taken up by the pelt, and that the absorption capacity of the pelt for acid was limited.

The goods pickled with the largest amount of acid possessed a more leathery feel and after drying were fuller and stretched much better than those in which smaller amounts of acids were employed. Dried, pickled pieces, containing as much as 3 per cent of sulphuric acid, showed no deterioration or tendering of fiber. The pickled skins after chrome tanning still retained these characteristics. An analysis of the leather produced by tanning with sumac showed that no free acid was retained in the finished leather. An Australian pickled pelt was found to contain 19.2 per cent of salt and 2.8 per cent of sulphuric acid.

From a very large number of experiments the following conclusions were drawn: 1. That sulphuric acid is quite equal in efficiency to other acids for the purpose. 2. To a certain limit increasing softness is produced by increasing the quantity of acid used. 3. For naturally soft skins and when a leather not very soft is required the best results are obtained by using 22 pounds of salt, 2.2 pounds of sulphuric acid, and 25 gallons of water for 110 pounds of pelt in the drum. 4. For material which is naturally hard and when a soft leather is required, the amount of acid should be increased to 4.4 pounds, using similar amounts as those given above of pelt, salt, and water.

French Hide Tanning Process.—I.—The prepared pelts are submitted to a 3 to 4 hours' immersion in a solution of rosin soap, containing 5 to 10 per cent of caustic soda. The goods are afterwards placed in a 6 to 12 per cent solution of a salt of chromium, iron, copper, or aluminum (preferably aluminum sulphate) for 3 to 4 hours.

II.—The hides are soaked in a solution of sodium carbonate of 10° Bé. for 3 to 6 hours. After washing with water they are allowed to remain for 5 hours in a bath of caustic soda, the strength of which may vary from 2° to 30° Bé. From this they are transferred to a bath of hydrochloric acid (1° to 5° Bé.) in which they remain for 2 hours. Finally the hides are washed and the beam-work finished in the usual way. The tannage consists of a special bath of sodium or ammonium sulphoricinoleate (2 to 30 per cent) and sumac extract, or similar tanning material (2 to 50 per cent). The strength of this bath is gradually raised from 4° to 30° or 40° Bé.

Tanning Hides for Robes.—The hides should be very thoroughly soaked in order to soften them completely. For dry hides this will require a longer time than for salted. A heavy hide requires longer soaking than a skin. Thus it is impossible to fix a certain length of time. After soaking, the hide is fleshed clean, and is now ready to go into the tan liquor, which is made up as follows: One part alum; 1 part salt; $\frac{1}{4}$ to $\frac{1}{2}$ part japonica. These are dissolved in hot water in sufficient quantity to make a 35° liquor. The hide, according to the thickness, is left in the tan from 5 to 10 days. Skins are finished in about 2 or 3 days. The hide should be run in a drum for about 2 hours before going into tan, and again after that process. In tanning hides for robes, shaving them down is a main requisite for success, as it is impossible to get soft leather otherwise. After shaving put back into the tan liquor again for a day or two and hang up to dry. When good and hard, shave again and lay away in moist sawdust and give a heavy coat of oil. When dry, apply a solution of soft soap; roll up and lay away in moist sawdust again. Run the hides on a drum or wheel until thoroughly soft. The composition of the tan liquor may be changed considerably. If the brownish tinge of the japonica be objectionable, that article may be left out entirely. The japonica has the effect of making the robe more able to resist water, as the alum and salt alone are readily soaked out by rain.

Lace Leather.—Take cow hides averaging from 25 to 30 pounds each; 35 hides will make a convenient soak for a vat containing 1,000 gallons of water, or 25 hides to a soak of 700 gallons. Soak 2 days or more, as required. Change water every 24 hours. Split and flesh; resoak if necessary. When thoroughly soft put in limes. Handle and strengthen once a day, for 5 or 6 days. Unhair and wash. Bathe in hen manure, 90° F. Work out of drench, wash well, drain 4

of 5 hours. Then process, using 45 pounds vitriol and 600 pounds of soft water to 700 gallons of water. In renewing process for second or consecutive packs, use 15 pounds vitriol and 200 pounds salt, always keeping stock constantly in motion during time of processing. After processing, drain over night, then put in tan in agitated liquors, keeping the stock in motion during the whole time of tanning. Pack down overnight. Use 200 pounds dry leather to each mill in stuffing.

For stuffing, use 3 gallons curriers' hard grease and 3 gallons American cod oil. Strike out from mill, on flesh. Set out on grain. Dry slowly. Trim and board, length and cross. The stock is then ready to cut. The time for soaking the hides may be reduced one-half by putting the stock into a rapidly revolving reel pit, with a good inflow of water, so that the dirty water washes over and runs off. After 10 hours in the soak, put the stock into a drum, and keep it tumbling 5 hours. This produces soft stock.

In liming, where the saving of the hair is no object, softer leather is obtainable by using 35 pounds sulphide of sodium with 60 pounds lime. Then, when the stock comes from the limes, the hair is dissolved and immediately washes off, and saves the labor of unhairing and caring for the hair, which in some cases does not pay.

MISCELLANEOUS RECIPES:

Russian Leather.—This leather owes its name to the country of its origin. The skins used for its production are goat, large sheep, calfskin, and cow or steer hide. The preliminary operations of soaking, unhairing, and fleshing are done in the usual manner, and then the hides are permitted to swell in a mixture of rye flour, oat flour, yeast, and salt. This compound is made into a paste with water, and is then thinned with sufficient water to steep a hundred hides in the mixture. The proportions of ingredients used for this mixture are 22 pounds rye flour, 10 pounds oat flour, a little salt, and sufficient yeast to set up fermentation.

The hides are steeped in this compound for 2 days, until swelled up, and then put into a solution of willow and poplar barks, in which they are allowed to remain 8 days, being frequently turned about. The tanning process is then completed by putting them into a tanning liquor composed of pine and willow barks, equal parts. They are steeped 8 days in this liquor, and then a fresh liquor of the same ingredients and proportions is made up. The hides are hardened and split, and then steeped in the freshly made liquor for another 8 days, when they are sufficiently tanned.

The hides are then cut down the middle (from head to tail) into sides, and scoured, rinsed, and dried by dripping, and then passed on to the currier, who slightly dampens the dry sides and puts them in a heap or folds them together for a couple of days to temper, and then impregnates them with a compound consisting of $\frac{2}{3}$ parts birch oil and $\frac{1}{3}$ parts seal oil. This is applied on the flesh side for light leather, and on the grain side also for heavy leather. The leather is then "set out," "whitened," and well boarded and dried before dyeing.

A decoction of sandalwood, alone or mixed with cochineal, is used for producing the Russian red color, and this dye liquor is applied several times, allowing each application to dry before applying the following one. A brush is used, and the dye liquor is spread on the grain side. A solution of tin chloride is used in Russia as a mordant for the leather before laying on the dye. The dye liquor is prepared by boiling 18 ounces of sandalwood in 13 pints of water for 1 hour, and then filtering the liquid and dissolving in the filtering fluid 1 ounce of prepared tartar and soda, which is then given an hour's boiling and set aside for a few days before use.

After dyeing, the leather is again impregnated with the mixture of birch and seal oils (applied to the grain side on a piece of flannel) and when the dyed leather has dried, a thin smear of gum-dragon mucilage is given to the dyed side to protect the color from fading, while the flesh side is smeared with bark-tan juice and the dyed leather then grained for market.

Toughening Leather. — Leather is toughened and also rendered impervious by impregnating with a solution of 1 part of caoutchouc or gutta-percha in 16 parts of benzene or other solvent, to which is added 10 parts of linseed oil. Wax and rosin may be added to thicken the solution.

Painting on Leather. — When the leather is finished in the tanneries it is at the same time provided with the necessary greasy particles to give it the required pliancy and prevent it from cracking. It is claimed that some tanners strive to obtain a greater weight thereby, thus increasing their profit, since a pound of fat is only one-eighth as dear as a pound of leather.

If such leather, so called kips, which are much used for carriage covers and knee caps, is to be prepared for painting purposes, it is above all necessary to close up the pores of the leather, so that the said fat particles cannot strike through. They would combine with the applied paint and prevent the latter from drying, as the grease consists mainly of fish oil. For this reason an elastic spirit leather varnish is employed, which protects the succeeding paint coat sufficiently from the fat.

For further treatment take a good coach varnish to which $\frac{1}{4}$ of stand oil (linseed oil which has thickened by standing) has been added and allow the mixture to stand for a few days. With this varnish grind the desired colors, thinning them only with turpentine oil. Put on 2 coats. In this manner the most delicate colors may be applied to the leather, only it is needful to put on pale and delicate shades several times. In some countries the legs or tops of boots are painted yellow, red, green, or blue in this manner. Inferior leather, such as sheepskin and goat leather, which is treated with alum by the tanner, may likewise be provided with color in the manner stated. Subsequently it can be painted, gilded, or bronzed.

Stains for Oak Leather. — I. — Apply an intimate mixture of 4 ounces of umber (burnt or raw); $\frac{1}{2}$ ounce of lampblack, and 17 fluidounces ox gall.

II. — The moistened leather is primed with a solution of 1 part, by weight, of copper acetate in 50 parts of water, slicked out and then painted with solution of yellow prussiate potash in feebly acid water.

LEATHER AS AN INSULATOR:
See Insulation.

LEATHER CEMENTS:
See Adhesives, under Cements.

LEATHER-CLEANING PROCESSES:
See Cleaning Preparations and Methods.

LEATHER, GLUES FOR:
See Adhesives.

LEATHER LAC:
See Lacquers.

LEATHER LUBRICANTS:
See Lubricants.

LEATHER VARNISH:
See Varnish.

LEATHER WATERPROOFING:
See Waterproofing.

LEMONS:
See also Essences, Extracts, and Fruits.

Preservation of Fresh Lemon Juice.—The fresh juice is cleared by gently heating it with a little egg albumen, without stirring the mixture. This causes all solid matter to sink with the coagulated white, or to make its way to the surface. The juice is then filtered through a woolen cloth and put into bottles, filled as full as possible, and closed with a cork stopper, in such a way that the cork may be directly in contact with the liquid. Seal at once and keep in a cool place. The bottles should be asepticized with boiling water just before using.

LEMON EXTRACT (ADULTERATED), TESTS FOR:
See Foods.

LEMON SHERBET POWDER:
See Salts, Effervescent.

LEMONADES, LEMONADE POWDERS, AND LEMONADE DROPS:
See Beverages.

LEMONADE POWDER:
See Salts, Effervescent.

LENSES AND THEIR CARE:

Unclean Lenses (see also Cleaning Preparations and Methods).—If in either objective or eyepiece the lenses are not clean, the definition may be seriously impaired or destroyed. Uncleanliness may be due to finger marks upon the front lens of the objective, or upon the eyepiece lenses; dust which in time may settle upon the rear lens of the objective or on the eye lens; a film which forms upon one or the other lens, due occasionally to the fact that glass is hygroscopic, but generally to the exhalation from the interior finish of the mountings, and, in immersion objectives, because the front lens is not properly cleaned; or oil that has leaked on to its rear surface, or air bubbles that have formed in the oil between the cover glass and front lens.

Remedy.—Keep all lenses scrupulously clean. For cleaning, use well-washed linen (an old handkerchief) or Japanese lens paper.

Eyepieces.—To find impurities, revolve the eyepieces during the observation; breathe upon the lenses, and wipe gently with a circular motion and blow off any particles which may adhere.

Dry Objectives.—Clean the front lens as described. To examine the rear and interior lenses use a 2-inch magnifier, looking through the rear. Remove the dust from the rear lens with a camel's-hair brush.

Oil Immersion Objectives.—Invariably clean the front lens after use with moistened linen or paper, and wipe dry.

In applying oil examine the front of the objective with a magnifier, and if there are any air bubbles, remove them with a pointed quill, or remove the oil entirely and apply a fresh quantity.

LETTERS, TO REMOVE FROM CHINA:
See Cleaning Preparations and Methods, under Miscellaneous Methods.

LETTER-HEAD SENSITIZERS:
See Photography, under Paper-Sensitizing Processes.

Lettering

CEMENTS FOR ATTACHING LETTERS ON GLASS:
See Adhesives, under Sign-Letter Cements.

Gold Lettering.—This is usually done by first drawing the lettering, then covering with an adhesive mixture, such as size, and finally applying gold bronze powder or real gold leaf. A good method for amateurs to follow in marking letters on glass is to apply first a coat of whiting, mixed simply with water, and then to mark out the letters on this surface, using a pointed stick or the like. After this has been done the letters may easily be painted or gilded on the reverse side of the glass. When done, wash off the whiting from the other side, and the work is complete.

Bronze Lettering.—The following is the best method for card work: Write with asphaltum thinned with turpentine until it flows easily, and, when nearly dry, dust bronze powder over the letters. When the letters are perfectly dry tap the card to take off the extra bronze, and it will leave the letters clean and sharp. The letters should be made with a camel's-hair brush and not with the automatic pen, as oil paints do not work satisfactorily with these pens.

For bronzed letters made with the pen, use black letterine or any water color.

If a water color is used add considerable gum arabic. Each letter should be bronzed as it is made, as the water color dries much more quickly than the asphaltum.

Another method is to mix the bronze powder with bronze sizing to about the consistency of the asphaltum. Make the letter with a camel's-hair brush, using the bronze paint as one would any oil paint. This method requires much skill, as the gold paint spreads quickly and is apt to flood over the edge of the letter. For use on oilcloth this is the most practical method.

Bronzes may be purchased at any hardware store. They are made in copper, red, green, silver, gold, and copper shades.

Lettering on Glass.—White lettering on glass and mirrors produces a rich effect. Dry zinc, chemically pure, should be used. It can be obtained in any first-class paint store and is inexpensive. To every teaspoonful of zinc, 10 drops of mucilage should be added. The two should be worked up into a thick paste, water being gradually added until the mixture is about the consistency of thick cream. The paint should then be applied with a camel's-hair brush.

Another useful paint for this purpose is Chemnitz white. If this distemper color is obtained in a jar, care should be exercised to keep water standing above the color to prevent drying. By using mucilage as a sizing these colors will adhere to the glass until it is washed off. Both mixtures are equally desirable for lettering on block card-board.

Any distemper color may be employed on glass without in any way injuring it. An attractive combination is—first to letter the sign with Turkey red, and then to outline the letters with a very narrow white stripe. The letter can be rendered still more attractive by shading one side in black.

Signs on Show Cases.—Most show cases have mirrors at the back, either in the form of sliding panels or spring doors. Lettering in distemper colors on these mirrors can easily be read through the fronts or tops of cases. If the mirror is on a sliding panel, it will be necessary to detach it from the case in order to letter it. When the mirror is on a spring door the sign can be lettered with less trouble.

By tracing letters in chalk on the outside of the glass, and then painting them on the inside, attractive signs can be produced on all show cases; but painting letters on the inside of a show case glass is more or less difficult, and it is not advisable to attempt it in very shallow cases.

"Spatter" Work.—Some lettering which appears very difficult to the uninitiated is, in fact, easily produced. The beautiful effect of lettering and ornamentation in the form of foliage or conventional scrolls in a speckled ground is simple and can be produced with little effort. Pressed leaves and letters or designs, cut from newspapers or magazines may be tacked or pasted on cardboard or a mat with flour paste. As little paste as possible should be used—only enough to hold the design in place. When all the designs are in the positions desired, a toothbrush should be dipped in the ink or paint to be employed. A toothpick or other small piece of wood is drawn to and fro over the bristles, which are held toward the sign, the entire surface of which should be spattered or sprinkled with the color. When the color is dry the designs pasted on should be carefully removed and the paste which held them in place should be scraped off. This leaves the letters and other designs clean cut and white against the "spatter" background. The beginner should experiment first with a few simple designs. After he is able to produce attractive work with a few figures or letters he may confidently undertake more elaborate combinations.

Lettering on Mirrors.—From a bar of fresh common brown soap cut off a one-inch-wide strip across its end. Cut this into 2 or 3 strips. Take one strip and with a table-knife cut from two opposite sides a wedge-shaped point resembling that of a shading pen, but allow the edge to be fully $\frac{1}{8}$ inch thick. Clean the mirror thoroughly and proceed to letter in exactly the same manner as with a shading pen.

To Fill Engraved Letters on Metal Signs.—Letters engraved on metal may be filled in with a mixture of asphaltum, brown japan, and lampblack, the mixture being so made as to be a putty-like mass. It should be well pressed down with a spatula. Any of the mass adhering to the plate about the edges of the letters is removed with turpentine, and when the cement is thoroughly dried the plate may be polished.

If white letters are desired, make a putty of dry white lead, with equal parts of coach japan and rubbing varnish. Fill the letters nearly level with the surface, and when hard, apply a stout coat of flake white in japan thinned with turpentine. This will give a clean white finish that may be polished.

The white cement may be tinted to any desired shade, using coach colors ground in japan.

Tinseled Letters, or Chinese Painting on Glass.—This is done by painting the groundwork with any color, leaving the letter or figure naked. When dry, place tin foil or any of the various colored copper foils over the letters on the back of the glass, after crumpling them in the hand, and then partially straightening them out.

LICE KILLERS:
See Insecticides.

LICHEN REMOVERS:
See Cleaning Preparations and Methods, under Miscellaneous Methods and Household Formulas.

LICORICE:
Stable Solutions of Licorice Juice.— A percolator, with alternate layers of broken glass, which have been well washed, first with hydrochloric acid and plentifully rinsed with distilled water, is the first requisite. This is charged with pieces of crude licorice juice, from the size of a hazel nut to that of a walnut, which are weighted down with well-washed pebbles. The percolate is kept for 3 days in well corked flasks which have been rinsed out with alcohol beforehand. Decant and filter and evaporate down rapidly, under constant stirring, or *in vacuo*. The extract should be kept in vessels first washed with alcohol and closed with parchment paper, in a dry place—never in the cellar.

To dissolve this extract, use water, first boiled for 15 minutes. The solution should be kept in small flasks, first rinsed with alcohol and well corked. If to be kept for a long time, the flasks should be subjected for 3 consecutive days, a half hour each day, to a stream of steam, and the corks paraffined.

There is frequently met with in commerce a purified juice that remains clear in the *mixtura solvens*. It is usually obtained by supersaturation with pure ammonia, allowing to stand for 3 days, decanting, filtering the decanted liquor, and quick evaporation. Since solutions with water alone rapidly spoil, it is well to observe with them the precautions common for narcotic extracts.

To Test Extract of Licorice.—Mere solubility is no test for the purity of extract of licorice. It is, therefore, proposed to make the glycyrrhizin content and the nature of the ash the determining test. To determine the glycyrrhizin quantitatively proceed as follows: Macerate $\frac{1}{10}$ ounce of the extract, in coarse powder, in 10 fluidounces distilled water for several hours, with more or less frequent agitation. When solution is complete, add 10 fluidounces alcohol of 90 per cent, filter and wash the filter with alcohol of 40 per cent until the latter comes off colorless. Drive off the alcohol, which was added merely to facilitate filtration, by evaporation in the water bath; let the residue cool down and precipitate the glycyrrhizin by addition of sulphuric acid. Filter the liquid and wash the precipitate on the filter with distilled water until the wash water comes off neutral. Dissolve the glycyrrhizin from the filter by the addition of ammonia water, drop by drop, collecting the filtered solution in a tared capsule. Evaporate in the water bath, dry the residual glycyrrhizin at 212° F., and weigh. Repeated examinations of known pure extracts have yielded a range of percentage of glycyrrhizin running from 8.06 per cent to 11.90 per cent. The ash should be acid in reaction and a total percentage of from 5.64 to 8.64 of the extract.

LIGHT, INACTINIC:
See Photography.

LIGNALOE SOAP:
See Soap.

LIMEADE:
See Beverages, under Lemonades.

LIME AS A FERTILIZER:
See Fertilizers.

LIME, BIRD.
Bird lime is a thick, soft, tough, and sticky mass of a greenish color, has an unpleasant smell and bitter taste, melts easily on heating, and hardens when exposed in thin layers to the air. It is difficult to dissolve in alcohol, but easily soluble in hot alcohol, oil of turpentine, fat oils, and also somewhat in vinegar. The best quality is prepared from the inner green bark of the holly (*Ilex aquifolium*), which is boiled, then put in barrels, and submitted for 14 days to slight fermentation until it becomes sticky. Another process of preparing it is to mix the boiled bark with juice of mistletoe berries and burying it in the ground until

fermented. The bark is then pulverized, boiled, and washed. Artificial bird lime is prepared by boiling and then igniting linseed oil, or boiling printing varnish until it is very tough and sticky. It is also prepared by dissolving cabinet-makers' glue in water and adding a concentrated solution of chloride of zinc. The mixture is very sticky, does not dry on exposure to the air, and has the advantage that it can be easily washed off the feathers of the birds.

LIME JUICE:
See Essences and Extracts

LIME-JUICE CORDIAL:
See Wines and Liquors.

LIME WAFERS:
See Confectionery.

LINEN, TO DISTINGUISH COTTON FROM:
See Cotton.

LINEN DRESSING:
See Laundry Preparations.

LINIMENTS:
See also Ointments.

For external use only.—I.—The following penetrating oily liniment reduces all kinds of inflammatory processes:

Paraffine oil......... 4 ounces
Capsicum powder.... ½ ounce

Digest on a sand bath and filter. To this may be added directly the following: Oil of wintergreen or peppermint, phenol, thymol, camphor or eucalyptol, etc.

II.—Camphor.......... 2 ounces
Menthol........... 1 ounce
Oil of thyme....... 1 ounce
Oil of sassafras.... 1 ounce
Tincture of myrrh.. 1 ounce
Tincture of capsicum 1 ounce
Chloroform........ 1 ounce
Alcohol........... 2 pints

LINIMENTS FOR HORSES:
See Veterinary Formulas.

LINOLEUM:
See also Oilcloth.

Composition for Linoleum, Oilcloth, etc.—This is composed of whiting, dried linseed oil, and any ordinary dryer, such as litharge, to which ingredients a proportion of gum tragacanth is to be added, replacing a part of the oil and serving to impart flexibility to the fabric, and to the composition in a pasty mass the property of drying more rapidly. In the production of linoleum, the whiting is replaced in whole or in part by pulverized cork. The proportions are approximate-ly the following by weight: Whiting or powdered cork, 13 parts; gum tragacanth, 5 parts; dried linseed oil, 5½ parts; siccative, ½ part.

Dressings for Linoleum.—A weak solution of beeswax in spirits of turpentine has been recommended for brightening the appearance of linoleum. Here are some other formulas:

I.—Palm oil........... 1 ounce
    Paraffine.......... 18 ounces
    Kerosene.......... 4 ounces

Melt the paraffine and oil, remove from the fire and incorporate the kerosene.

II.—Yellow wax........ 5 ounces
     Oil turpentine...... 11 ounces
     Amber varnish..... 5 ounces

Melt the wax, add the oil, and then the varnish. Apply with a rag.

Treatment of Newly Laid Linoleum.—The proper way to cleanse a linoleum flooring is first to sweep off the dust and then wipe up with a damp cloth. Several times a year the surface should be well rubbed with floor wax. Care must be had that the mass is well pulverized and free from grit. Granite linoleum and figured coverings are cleansed without the application of water. A floor covering which has been treated from the beginning with floor wax need only be wiped off daily with a dry cloth, either woolen or felt, and afterwards rubbed well with a cloth filled with the mass. It will improve its appearance, too, if it be washed several times a year with warm water and a neutral soap.

LINOLEUM, CLEANING AND POLISHING:
See Household Formulas.

LINOLEUM ON IRON STAIRS OR CEMENT FLOORS, TO GLUE:
See Adhesives, under Glues.

LINSEED OIL:
See also Oils.

Bleaching of Linseed Oil and Poppyseed Oil.—In order to bleach linseed oil and poppyseed oil for painting purposes, thoroughly shake 2.5 parts of it in a glass vessel with a solution of potassium permanganate, 50 parts, in 1,250 parts of water; let stand for 24 hours in a warm temperature, and then mix with 75 parts of pulverized sodium sulphite. Now shake until the latter has dissolved and add 100 parts of crude hydrochloric acid, 20°. Agitate frequently and wash, after the previously brown mass has become light colored, with water, in which a little chalk has been finely distributed, until the water is neutral. Finally filter over calcined Glauber's salt.

Adulteration of Linseed Oil.—This is common, and a simple and cheap method of testing is by nitric acid. Pour equal parts of the linseed oil and nitric acid into a flask, shake vigorously, and let it stand for 20 minutes. If the oil is pure, the upper stratum is of straw yellow color and the lower one colorless. If impure, the former is dark brown or black, the latter pale orange or dark yellow, according to the admixtures to the oil.

The addition of rosin oil to linseed oil or other paint oils can be readily detected by the increase in specific gravity, the low flash point, and the odor of rosin on heating; while the amount may be approximately ascertained from the amount of unsaponifiable oil left after boiling with caustic soda.

LIP SALVES AND LIPOL:
See Cosmetics.

LIPOWITZ METAL:
See Alloys.

LIQUEURS:
See Wines and Liquors.

LIQUOR AMMONII ANISATUS:
See Ammonia.

LIQUORS:
See Wines and Liquors.

LITHOGRAPHERS' LACQUER:
See Lacquers.

LITHOGRAPHS:
See Pictures and Engravings.

LIVER-SPOT REMEDIES:
See Cosmetics.

LOCKSMITH'S VARNISH:
See Varnishes.

LOCOMOTIVE LUBRICANTS:
See Lubricants.

LOCUST KILLER:
See Insecticides.

LOUSE WASH:
See Insecticides.

Lubricants

Oil for Firearms.—Either pure vaseline oil, white, 0.870, or else pure white-bone oil, proof to cold, is employed for this purpose, since these two oils are not only free from acid, but do not oxidize or resinify.

Leather Lubricants.—Russian tallow, 1 pound; beeswax, 6 ounces; black pitch, 4 ounces; common castor oil, 3 pounds; soft paraffine, ½ pound; oil of citronella, ½ ounce. Melt all together in a saucepan, except the citronella, which add on cooling. Stir occasionally.

Machinery Oils.—I.—The solid fat, called bakourine, a heavy lubricant which possesses extraordinary lubricating qualities, has a neutral reaction and melts only at about 176° to 188° F. It is prepared as follows:

A mixture is made of 100 parts of Bienne petroleum or crude naphtha, with 25 parts of castor oil or some mineral oil, and subjected to the action of 60 or 70 parts of sulphuric acid of 66° Bé. The acid is poured in a small stream into the oil, while carefully stirring. The agitation is continued until a thick and blackish-brown mass is obtained free from non-incorporated petroleum. Very cold water of 2 or 3 times the weight of the mass is then added, and the whole is stirred until the mass turns white and becomes homogeneous. It is left at rest for 24 hours, after which the watery liquid, on the surface of which the fat is floating, must be poured off. After resting again from 3 to 4 days, the product is drawn off, carefully neutralized with caustic potash, and placed in barrels ready for shipping.

II.—Melt in a kettle holding 2 to 4 times as much as the volume of the mass which is to be boiled therein, 10 parts, by weight, of tallow in 20 parts of rape oil on a moderate fire; add 10 parts of freshly and well burnt lime, slaked in 30 or 40 parts of water; increase the fire somewhat, and boil with constant stirring until a thick froth forms and the mass sticks to the bottom of the kettle. Burning should be prevented by diligent stirring. Then add in portions of 10 parts each, gradually, 70 parts of rape oil and boil with a moderate fire, until the little lumps gradually forming have united to a whole uniform mass. With this operation it is of importance to be able to regulate the fire quickly. Samples are now continually taken, which are allowed to cool quickly on glass plates. The boiling down must not be carried so far that the samples harden on cooling; they must spin long, fine threads, when touched with the finger. When this point is reached add, with constant stirring, when the heat has abated sufficiently (which may be tested by pouring in a few drops of water), 25 to 30 parts of water. Now raise the fire, without ceasing to stir, until the mass comes to a feeble, uniform boil. In order to be able to act quickly in case of a sudden boiling over, the fire must be such that it can be removed quickly, and a little cold water must always be kept on hand. Next, gradually add in small portions, so as not to disturb the boiling of the mass, 500 parts of paraffine oil (if very thick, 800 to 900 parts may be added), remove from the fire, allow the contents of the kettle to clarify, and skim off the warm grease from the sediment into a stirring apparatus. Agitate until the mass begins to thicken and cool; if the grease should still be too solid, stir in a little paraffine oil the second time. The odor of the paraffine oil may be disguised by the admixture of a little mirbane oil.

For Cutting Tools.—The proportion of ingredients of a lubricating mixture for cutting tools is 6 gallons of water, $3\frac{1}{2}$ pounds of soft soap, and $\frac{1}{2}$ gallon of clean refuse oil. Heat the water and mix with the soap, preferably in a mechanical mixer; afterwards add the oil. A cast-iron circular tank to hold 12 gallons, fitted with a tap at the bottom and having three revolving arms fitted to a vertical shaft driven by bevels and a fast and loose pulley, answers all requirements for a mixer. This should be kept running all through the working day.

For Highspeed Bearings.—To prevent heating and sticking of bearings on heavy machine tools due to running continuously at high speeds, take about $\frac{1}{8}$ of flake graphite, and the remainder kerosene oil. As soon as the bearing shows the slightest indication of heating or sticking, this mixture should be forcibly squirted through the oil hole until it flows out between the shaft and bearing, when a small quantity of thin machine oil may be applied.

For Heavy Bearings.—An excellent lubricant for heavy bearings can be made from either of the following recipes:

I.—Paraffine........... 6 pounds
    Palm oil........... 12 pounds
    Oleonaphtha....... 8 pounds

II.—Paraffine........... 8 pounds
    Palm oil........... 20 pounds
    Oleonaphtha....... 12 pounds

The oleonaphtha should have a density of 0.9. First dissolve the paraffine in the oleonaphtha at a temperature of about 158° F. Then gradually stir in the palm oil a little at a time. The proportions will show that No. II gives a less liquid product than No. I. Quicklime may be added if desired.

For Lathe Centers.—An excellent lubricant for lathe centers is made by using 1 part graphite and 4 parts tallow thoroughly mixed.

Sewing Machine Oil.—I.—Petroleum oils are better adapted for the lubrication of sewing machines than any of the animal oils. Sperm oil has for a long time been considered the standard oil for this purpose, but it is really not well adapted to the conditions to which a sewing machine is subjected. If the machine were operated constantly or regularly every day, probably sperm oil could not be improved on. The difficulty is, however, that a family sewing machine will frequently be allowed to stand untouched for weeks at a time and will then be expected to run as smoothly as though just oiled. Under this kind of treatment almost any oil other than petroleum oil will become gummy. What is known in the trade as a "neutral" oil, of high viscosity, would probably answer better for this purpose than anything else. A mixture of 1 part of petrolatum and 7 parts of paraffine oil has also been recommended.

II.—Pale oil of almonds. 9 ounces
    Rectified benzoline.. 3 ounces
    Foreign oil of laven-
      der............. 1 ounce

PETROLEUM JELLIES AND SOLIDIFIED LUBRICANTS.

Petroleum jelly, vaseline, and petrolatum are different names for the same thing.

The pure qualities are made from American stock thickened with hot air until the desired melting point is attained. Three colors are made: white, yellow, and black of various qualities. Cheaper qualities are made by using ceresine wax in conjunction with the genuine article and pale mineral oil. This is the German method and is approved of by their pharmacopœia. Machinery qualities are made with cylinder oils, pale mineral oils, and ceresine wax.

I.—Yellow ceresine wax 11 parts
    White ceresine wax. 6 parts
    American mineral
      oil, $\frac{343}{367}$ .... 151 parts

Melt the waxes and stir in the oil. To make white, use all white ceresine wax. To color, use aniline dyes soluble in oil to any shade required.

II.—Ceresine wax....... 1 pound
    Bloomless mineral
      oil, Sq. 910....... 1 gallon

Melt the wax and add the oil, varying according to the consistency required. To color black, add 28 pounds lampblack to 20 gallons oil. Any wax will do, according to quality of product desired.

White Petroleum Jelly.—

White tasteless oil ..	4 parts
White ceresine wax.	1 part

Solidified Lubricants.—

I.—Refined cotton oil...   2 parts
    American mineral
    oil, ⁸⁸¾ ....   2 parts
Oleate of alumina ..   1 part

Gently heat together.

II.—Petroleum jelly....   120 parts
    Ceresine wax......   5 parts
    Slaked lime.......   ½ part
    Water...........   4½ parts

Heat the wax and the petroleum jelly gently until liquid; then mix together the water and lime. Decant the former into packing receptacles, and add lime and water, stirring until it sets. For cheaper qualities use cream cylinder oil instead of petroleum jelly.

WAGON AND AXLE GREASES:

For Axles of Heavy Vehicles.—I.— Tallow (free from acid), 19½ parts; palm oil, 14 parts; sal soda, 5¼ parts; water, 3 parts, by weight. Dissolve the soda in the water and separately melt the tallow, then stir in the palm oil. This may be gently warmed before adding, as it greatly facilitates its incorporation with the tallow, unless the latter be made boiling hot, when it readily melts the semi-solid palm oil. When these two greases are thoroughly incorporated, pour the mixture slowly into the cold lye (or soda solution), and stir well until the mass is homogeneous. This lubricant can be made less solid by decreasing the tallow or increasing the palm oil.

II.—Slaked lime (in powder), 8 parts, is slowly sifted into rosin oil, 10 parts. Stir it continuously to incorporate it thoroughly, and gently heat the mixture until of a syrupy consistency. Color with lampblack, or a solution of turmeric in a strong solution of sal soda. For blue grease, 275 parts of rosin oil are heated with 1 part of slaked lime and then allowed to cool. The supernatant oil is removed from the precipitated matter, and 5 or 6 parts of the foregoing rosin-oil soap are stirred in until all is a soft, unctuous mass.

For Axles of Ordinary Vehicles.—I.— Mix 80 parts of fat and 20 parts of very fine black lead; melt the fat in a varnished earthen vessel; add the black lead while constantly stirring until it is cold, for otherwise the black lead, on account of its density, would not remain in suspension in the melted fat. Axles lubricated with this mixture can make 80 miles without the necessity of renewing the grease.

II.—Mix equal parts of red American rosin, melted tallow, linseed oil, and caustic soda lye (of 1.5 density).

III.—Melt 20 parts of rosin oil in 50 parts of yellow palm oil, saponify this with 25 parts of caustic soda lye of 15° Bé., and add 25 parts of mineral oil or paraffine.

IV.—Mix residue of the distillation of petroleum, 60 to 80 parts; tallow, 10 parts; colophony, 10 parts; and caustic soda solution of 40° Bé., 15 parts.

A Grease for Locomotive Axles.— Saponify a mixture of 50 parts tallow, 28 parts palm oil, 2 parts sperm oil. Mix in soda lye made by dissolving 12 parts of soda in 137 parts of water.

MISCELLANEOUS LUBRICANTS:

For Cotton Belts.— Carefully melt over a slow fire in a closed iron or self-regulating boiler 250 parts of caoutchouc or gum elastic, cut up in small pieces; then add 200 parts of colophony; when the whole is well melted and mixed, incorporate, while carefully stirring, 200 parts of yellow wax. Then heat 850 parts of train oil, mixing with it 250 parts of talc, and unite the two preparations, constantly stirring, until completely cold.

Chloriding Mineral Lubricating Oils.— A process has been introduced for producing industrial vaselines and mineral oils for lubrication, based on the treatment of naphthas, petroleums, and similar hydrocarbides, by means of chlorine or mixtures of chlorides and hypochlorides, known under the name of decoloring chlorides. Mix and stir thoroughly 1,000 parts of naphtha of about 908 density; 55 parts of chloride of lime, and 500 parts of water. Decant and wash.

Glass Stop Cock Lubricant.—(See also Stoppers).

Pure rubber........	14 parts
Spermaceti.........	5 parts
Petroleum.........	1 part

Melt the rubber in a covered vessel and then stir in the other ingredients. A little more petroleum will be required when the compound is for winter use.

Hard Metal Drilling Lubricant.—For drilling in hard metal it is recommended to use carbolic acid instead of another fatty substance as a lubricant, since the latter, by decreasing the friction, diminishes the "biting" of the drill, whereas the carbolic acid has an etching action.

Plaster Model Lubricant.—Take linseed oil, 1,000 parts; calcined lead, 50 parts; litharge, 60 parts; umber, 30 parts; talc, 25 parts. Boil for 2 hours on a moderate fire; skim frequently and keep in well-closed flasks.

Graphite Lubricating Compound.—Graphite mixed with tallow gives a good lubricating compound that is free from any oxidizing if the tallow be rendered free from rancidity. The proportions are: Plumbago, 1 part; tallow, 4 parts. The plumbago being stirred into the melted tallow and incorporated by passing it through a mixing mill, add a few pounds per hundredweight of camphor in powder to the hot compound.

Lubricants for Redrawing Shells.—Zinc shells should be clean and free from all grit and should be immersed in boiling hot soap water. They must be redrawn while *hot* to get the best results. On some shells hot oil is used in preference to soap water.

For redrawing aluminum shells use a cheap grade of vaseline. It may not be amiss to add that the draw part of the redrawing die should not be made too long, so as to prevent too much friction, which causes the shells to split and shrivel up.

For redrawing copper shells use good thick soap water as a lubricant. The soap used should be of a kind that will produce plenty of "slip." If none such is to be had, mix a quantity of lard oil with the soap water on hand and boil the two together. Sprinkling graphite over the shells just before redrawing sometimes helps out on a mean job.

Rope Grease.—For hemp ropes, fuse together 20 pounds of tallow and 30 pounds of linseed oil. Then add 20 pounds of paraffine, 30 pounds of vaseline, and 60 pounds of rosin. Finally mix with 10 pounds of graphite, first rubbed up with 50 pounds of boiled oil. For wire ropes fuse 100 pounds of suint with 20 pounds of dark colophony (rosin). Then stir in 30 pounds of rosin oil and 10 pounds of dark petroleum.

Sheet Metal Lubricant.—Mix 1 quart of whale oil, 1 pound of white lead, 1 pint of water, and 3 ounces of the finest graphite. This is applied to the metal with a brush before it enters the dies.

Steam Cylinder Lubricant.—To obtain a very viscous oil that does not decompose in the presence of steam even at a high temperature, it is necessary to expose neutral wool fats, that have been freed from wool-fatty acids, such as crude lanolin or wool wax, either quite alone or in combination with mineral oils, to a high heat. This is best accomplished in the presence of ordinary steam or superheated steam at a heat of 572° F., and a pressure of 50 atmospheres, corresponding with the conditions in the cylinder in which it is to be used. Instead of separating any slight quantities of acid that may arise, they may be dissolved out as neutral salts.

Wooden Gears.—An excellent lubricating agent for wooden gears consists of tallow, 30 parts (by weight); palm oil, 20 parts; fish oil, 10 parts; and graphite, 20 parts. The fats are melted at moderate heat, and the finely powdered and washed graphite mixed with them intimately by long-continued stirring. The teeth of wooden combs are kept in a perfectly serviceable condition for a much longer time if to the ordinary tallow or graphite grease one-tenth part of their weight of powdered glass is added.

TESTS FOR LUBRICANTS.

In testing lubricants in general, a great deal depends upon the class of work in which they are to be employed. In dealing with lubricating greases the specific gravity should always be determined. The viscosity is, of course, also a matter of the utmost importance. If possible the viscosity should be taken at the temperature at which the grease is to be subjected when used, but this cannot always be done; 300° F. will be found to be a very suitable temperature for the determination of the viscosity of heavy lubricants. Although one of the standard viscosimeters is the most satisfactory instrument with which to carry out the test, yet it is not a necessity. Provided the test be always conducted in exactly the same manner, and at a fixed temperature, using a standard sample for comparison, the form of apparatus used is not of great importance. Most dealers in scientific apparatus will provide a simple and cheap instrument, the results obtained with which will be found reliable. With the exercise of a little ingenuity any one can fit up a viscosimeter for himself at a very small outlay.

Acidity is another important point to

note in dealing with lubricating greases. Calculated as sulphuric acid, the free acid should not exceed .01 per cent, and free fatty acids should not be present to any extent. Cylinder oil should dissolve completely in petroleum benzine (specific gravity, .700), giving a clear solution. In dealing with machine oils the conditions are somewhat different. Fatty oils in mixture with mineral oils are very useful, as they give better lubrication and driving power, especially for heavy axles, for which these mixtures should always be used. The specific gravity should be from .900 to .915 and the freezing point should not be above 58° F. The flash point of heavy machine oils is not a matter of great importance. The viscosity of dynamo oils, taken in Engler's apparatus, should be 15–16 at 68° F. and 3½–4 at 122° F. In dealing with wagon oils and greases it should be remembered that the best kinds are those which are free from rosin and rosin products, and their flash point should be above 212° F.

To Test Grease.—To be assured of the purity of grease, its density is examined as compared with water; a piece of fat of the size of a pea is placed in a glass of water. If it remains on the surface or sinks very slowly the fat is pure; if it sinks rapidly to the bottom the fat is mixed with heavy matters and coom is the result.

LUBRICANTS FOR WATCHMAKERS:
See Watchmakers' Formulas.

LUPULINE BITTERS:
See Wines and Liquors.

LUSTER PASTE.
This is used for plate glass, picture frames, and metal. Five parts of very finely washed and pulverized chalk; 5 parts of Vienna lime, powdered; 5 parts of bolus, powdered; 5 parts of wood ashes, powdered; 5 parts of English red, powdered; 5 parts of soap powder. Work all together in a kneading machine, to make a smooth, even paste, adding spirit. The consistency of the paste can be varied, by varying the amount of spirit, from a solid to a soft mass.

LUTES:
See Adhesives.

MACHINE OIL:
See Lubricants.

MACHINERY, TO CLEAN:
See Cleaning Preparations and Methods.

MAGIC:
See Pyrotechnics.

MAGNESIUM CITRATE.

Magnesium carbonate	10	ounces
Citric acid	20	ounces
Sugar	21	ounces
Oil of lemon	½	drachm
Water enough to make	240	ounces

Introduce the magnesium carbonate into a wide-mouthed 2-gallon bottle, drop the oil of lemon on it, stir with a wooden stick: then add the citric acid, the sugar, and water enough to come up to a mark on the bottle indicating 240 ounces. For this purpose use cold water, adding about half of the quantity first, and the remainder when the substances are mostly dissolved. By allowing the solution to stand for a half to a whole day, it will filter better and more quickly than when hot water is used.

MAGNESIUM ORGEAT POWDER:
See Salts, Effervescent.

MAGNESIUM FLASH-LIGHT POWDERS:
See Photography.

MAGNETIC CURVES OF IRON FILINGS, THEIR FIXATION.

One of the experiments made in every physical laboratory in teaching the elements of magnetism and electricity is the production of the magnetic curves by sprinkling iron filings over a glass plate, after the well-known method.

For fixing these curves so that they may be preserved indefinitely, a plate of glass is warmed on the smooth upper surface of a shallow iron chest containing water raised to a suitable temperature by means of a spirit-lamp. A piece of paraffine is placed on the glass, and in the course of 3 or 4 minutes spreads itself evenly in a thin layer over the surface. The glass plate is removed, the surplus paraffine running off. The image is formed with iron filings on the cooled paraffine, which does not adhere to the iron, so that if the image is unsatisfactory the filings may be removed and a new figure taken. To fix the curves, the plate of glass is again placed on the warming stove. Finally, the surface of the paraffine is covered with white paint, so that the curves appear black on a white ground. Very well-defined figures may thus be obtained. A similar though much simpler process consists in covering one surface of stiff white paper with a layer of paraffine, by warming over an iron plate, spreading the filings over the cooled surface, and fixing them with a hot iron or a gas flame.

MAGNOLIA METAL:
See Alloys.

MAHOGANY:
See Wood.

MALTED FOOD:
See Foods.

MALTED MILK:
See Milk.

MALT, HOT:
See Beverages.

MANGANESE ALLOYS:
See Alloys.

MANGANESE STEEL:
See Steel.

MANGE CURES:
See Veterinary Formulas.

MANICURE PREPARATIONS:
See Cosmeties.

MANTLES.

These are prepared after processes differing slightly from one another, but all based on the original formula of Welsbach—the impregnation of vegetable fibers with certain mineral oxides in solution, drying out, and arranging on platinum wire.

Lanthanum oxide...	30 parts
Yttrium oxide......	20 parts
Burnt magnesia.....	50 parts
Acetic acid.........	50 parts
Water, distilled.....	100 parts

The salts are dissolved in the water, and to the solution another 150 parts of distilled water are added and the whole filtered. The vegetable fiber (in its knitted or woven form) is impregnated with this solution dried, and arranged on platinum wire. In the formula the acetic acid may be replaced with dilute nitric acid. The latter seems to have some advantages over the former, among which is the fact that the residual ash where acetic acid is used has a tendency to ball up and make a vitreous residue, while that of the nitric acid remains in powdery form.

Self-Igniting Mantles.—A fabric of platinum wire and cotton thread is sewed or woven into the tissue of the incandescent body; next it is impregnated with a solution of thorium salts and dried. The thorium nitrate in glowing gives a very loose but nevertheless fireproof residue. A mixture of thorium nitrate with platinic chloride leaves after incandescence a fire-resisting sponge possessing to a great extent the property of igniting gas mixtures containing oxygen. Employ a mixture of 1 part of thorium nitrate to $2\frac{1}{2}$ parts of platinic chloride.

MANURES:
See Fertilizers.

MANUSCRIPT COPYING:
See Copying.

MAPLE:
See Wood.

MARASCHINO:
See Wines and Liquors

MARBLE CEMENTS:
See Adhesives.

MARBLE CLEANING:
See Cleaning Preparations and Methods.

MARBLE COLORS:
See Stone.

MARBLE ETCHING:
See Etching.

MARBLE, IMITATION:
See Plaster.

MARBLE, PAINTING ON:
See Painting.

MARBLE POLISHING:
See Polishes.

MARBLING CRAYONS:
See Crayons.

MARGERINE:
See Butter.

MARKING FLUID:
See also Inks and Etching.

For laying out work on structural iron or castings a better way than chalking the surface is to mix whiting with benzine or gasoline to the consistency of paint, and then apply it with a brush; in a few minutes the benzine or gasoline will evaporate, leaving a white surface ready for scribing lines.

MASSAGE APPLICATIONS:
See Cosmetics.

MASSAGE SOAPS:
See Soaps.

Matches

(See also Phosphorus.)

Manufacture of Matches.—Each factory uses its own methods and chemical mixtures, though, in a general way the latter do not vary greatly. It is impossible here to give a full account of the different steps of manufacture, and of all the precautions necessary to turn out good, marketable matches. In the manufacture of the ordinary safety match, the wood is first comminuted and reduced to the final shape and then steeped in a solution of ammonium phosphate (2 per cent of this salt with 1 or 1½ per cent of phosphoric acid), or in a solution of ammonium sulphate (2½ per cent), then drained and dried. The object of this application is to prevent the match from continuing to glow after it has been burned out. Next the matches are dipped into a paraffine or stearine bath, and after that into the match bath proper, which is best done by machines constructed for the purpose. Here are a few formulas:

I.—Potassium chlorate......... 2,000 parts
Lead binoxide.... 1,150 parts
Red lead........ 2,500 parts
Antimony trisulphide......... 1,250 parts
Gum arabic...... 670 parts
Paraffine........ 250 parts
Potassium bichromate...... 1,318 parts

Directions: See No. II.

II.—Potassium chlorate......... 2,000 parts
Lead binoxide.... 2,150 parts
Red lead........ 2,500 parts
Antimony trisulphide......... 1,250 parts
Gum arabic...... 670 parts
Paraffine........ 250 parts

Rub the paraffine and antimony trisulphide together, and then add the other ingredients. Enough water is added to bring the mass to a proper consistency when heated. Conduct heating operations on a water bath. The sticks are first dipped in a solution of paraffine in benzine and then are dried. For striking surfaces, mix red phosphorus, 9 parts; pulverized iron pyrites, 7 parts; pulverized glass, 3 parts; and gum arabic or glue, 1 part, with water, quantity sufficient. To make the matches water or damp proof, employ glue instead of gum arabic in the above formula, and conduct the operations in a darkened room. For parlor matches dry the splints and immerse the ends in melted stearine. Then dip in the following mixture and dry:

Red phosphorus..... 3.0 parts
Gum arabic or tragacanth............ 0.5 parts
Water............. 3.0 parts
Sand (finely ground). 2.0 parts
Lead binoxide....... 2.0 parts

Perfume by dipping in a solution of benzoic acid.

III.—M. O. Lindner, of Paris, has patented a match which may be lighted by friction upon any surface whatever, and which possesses the advantages of being free from danger and of emitting no unpleasant odor. The mixture into which the splints are first dipped consists of

Chlorate of potash... 6 parts
Sulphide of antimony. 2 parts
Gum................ 1½ parts
Powdered clay....... 1½ parts

The inflammable compound consists of

Chlorate of potash. 2 to 3 parts
Amorphous phosphorus.......... 6 parts
Gum............. 1½ parts
Aniline.......... 1½ parts

Red or amorphous is substituted for yellow phosphorus in the match heads. The composition of the igniting paste is given as follows:

By weight
Soaked glue (1 to 5 of water)............. 37.0 parts
Powdered glass...... 7.5 parts
Whiting............ 7.5 parts
Amorphous phosphorus (pure)......... 10.0 parts
Paraffine wax........ 4.0 parts
Chlorate of potash... 27.0 parts
Sugar or lampblack .. 7.0 parts

Silicate of soda may be substituted for the glue, bichromate of potash added for damp climates, and sulphur for large matches.

The different compositions for tipping the matches in use in different countries and factories all consist essentially of emulsions of phosphorus in a solution of glue or gum, with or without other matters for increasing the combustibility, for coloring, etc.

I.—English.—Fine glue, 2 parts, broken into small pieces, and soaked in water till quite soft, is added to water, 4 parts, and heated by means of a water bath until it is quite fluid, and at a temperature of 200° to 212° F. The vessel is then removed from the fire, and phosphorus, 1½ to 2 parts, is gradually added, the mixture being agitated briskly and continually with a stirrer having wooden pegs or bristles projecting at its lower end. When a uniform emulsion is obtained, chlorate of potassa, 4 to 5 parts; powdered glass, 3 to 4 parts; and red lead, smalt, or other coloring matter, a sufficient quantity (all in a state of very fine powder), are added, one at a time, to prevent accidents, and the stirring continued until the mixture is comparatively cool. The above proportions are those of the best quality of English composition. The matches tipped with it deflagrate with a snapping noise.

II. — German (Böttger). — Dissolve gum arabic, 16 parts, in the least possible quantity of water; add of phosphorus (in powder), 9 parts, and mix by trituration. Then add niter, 14 parts; vermilion or binoxide of manganese, 16 parts, and form the whole into a paste as directed above. Into this the matches are to be dipped, and then exposed to dry. As soon as they are quite dry they are to be dipped into very dilute copal varnish or lac varnish, and again exposed to dry, by which means they are rendered waterproof, or at least less likely to suffer from exposure in damp weather.

III. (Böttger.) — Glue, 6 parts, is soaked in a little cold water for 24 hours, after which it is liquefied by trituration in a heated mortar; phosphorus, 4 parts, is added, and rubbed down at a heat not exceeding 150° F.; niter (in fine powder), 10 parts, is next mixed in, and afterwards red ocher, 5 parts, and smalt, 2 parts, are further added, and the whole formed into a uniform paste, into which the matches are dipped, as before. This is cheaper than the previous one.

IV. (Diesel.)—Phosphorus, 17 parts; glue, 21 parts; red lead, 24 parts; niter, 38 parts. Proceed as above.

Matches tipped with II, III, or IV, inflame without fulmination when rubbed against a rough surface, and are hence termed noiseless matches by the makers.

Safety Paste for Matches.—The danger of explosion during the preparation of match composition may be minimized by addition to the paste of the following mixture: Finely powdered cork, 3 parts, by weight; oxide of iron, 15 parts; flour, 23 parts; and water, about 40 parts. In practice, 30 parts of gum arabic are dissolved in water, 40 parts, and to the solution are added powdered potassium chlorate, 57 parts, and when this is well distributed, amorphous phosphorus, 7 parts, and powdered glass, 15 parts, are stirred in. The above mixture is then immediately introduced, and when mixing is complete, the composition can be applied to wooden sticks which need not have been previously dried or paraffined. The head of the match is finally coated with tallow, which prevents atmospheric action and also spontaneous ignition.

Most chemists agree that the greatest improvement of note in the manufacture of matches is that of Landstrom, of Jonkoping, in Sweden. It consists in dividing the ingredient of the match mixture into two separate compositions, one being placed on the ends of the splints, as usual, and the other, which contains the phosphorus, being spread in a thin layer upon the end or lid of the box. The following are the compositions used: (a) For the splints: Chlorate of potassa, 6 parts; sulphuret of antimony, 2 to 3 parts; glue, 1 part. (b) For the friction surface: Amorphous phosphorus, 10 parts; sulphuret of antimony or peroxide of manganese, 8 parts; glue, 3 to 6 parts; spread thinly upon the surface, which has been previously made rough by a coating of glue and sand. By thus dividing the composition the danger of fire arising from ignition of the matches by accidental friction is avoided, as neither the portion on the splint nor that on the box can be ignited by rubbing against an unprepared surface. Again, by using the innocuous red or amorphous phosphorus, the danger of poisoning is entirely prevented.

MATCH MARKS ON PAINT, TO REMOVE:
See Cleaning Preparations and Methods.

MATCH PHOSPHORUS, SUBSTITUTE FOR:
See Phosphorus Substitute.

Matrix Masses

Matrix for Medals, Coins, etc.—I.—Sharp impressions of coins, medals, etc., are obtained, according to Böttger, with the following: Mix molten, thinly liquid sulphur with an equal quantity of infusorial earth, adding some graphite. If a sufficient quantity of this mass, made liquid over a flame, is quickly applied with a spatula or spoon on the coin, etc., an impression of great sharpness is obtained after cooling, which usually takes place promptly. Owing to the addition of graphite the articles do not become dull or unsightly.

II.—Bronze and silver medals should always be coated with a separating grease layer. The whole coin is greased slightly and then carefully wiped off again with a little wadding, but in such a manner that a thin film of grease remains on the surface. Next, a ring of strong cardboard or thin pasteboard is placed around the edge, and the ends are sealed together. Now stir up a little gypsum in a small dish and put a teaspoonful of it on the surface of which the mold is to be taken, distributing it carefully with a badger's-hair brush, entering the finest cavities, which operation will be assisted by blowing on it. When the object is covered with a thin layer of plaster of Paris, the plaster, which has meanwhile become somewhat stiffer, is poured on, so that the thickness of the mold will be about $\frac{1}{20}$ of an inch. The removal of the cast can be effected only after a time, when the plaster has become warm, has cooled again, and has thoroughly hardened. If it be attempted to remove the cast from the metal too early and by the use of force, fine pieces are liable to break off and remain adhering to the model. In order to obtain a positive mold from the concave one, it is laid in water for a short time, so that it becomes saturated with the water it absorbs. The dripping, wet mold is again provided with an edge, and plaster of Paris is poured on. The latter readily flows out on the wet surface, and only in rare cases blisters will form. Naturally this casting method will furnish a surface of pure gypsum, which is not the case if the plaster is poured into a greased mold. In this case the surface of the cast contains a soapy layer, for the liquid plaster forms with oil a subsequently rather hard lime soap. The freshly cast plaster must likewise be taken off only when a quarter of an hour has elapsed, after it has become heated and has cooled again.

MATS FOR METALS:
See Metals.

MATZOON.

Add 2 tablespoonfuls of bakers' yeast to 1 pint of rich milk, which has been slightly warmed, stirring well together and setting aside in a warm room in a pitcher covered with a wet cloth for a time varying from 6 to 12 hours, according to the season or temperature of the room. Take from this, when curdled, 6 tablespoonfuls, add to another pint of milk, and again ferment as before, and continue for five successive fermentations in all, when the product will have become free from the taste of the yeast. As soon as the milk thickens, which is finally to be kept for use, it should be stirred again and then put into a refrigerator to prevent further fermentation. It should be smooth, of the consistence of thick cream, and of a slightly acid taste.

The milk should be prepared fresh every day, and the new supply is made by adding 6 tablespoonfuls of the previous day's lot to a pint of milk and proceeding as before.

The curd is to be eaten with a spoon, not drunk, and preferably with some bread broken into it. It is also sometimes eaten with sugar, which is said not to impair its digestibility.

MAY WINE:
See Wines and Liquors.

MEAD.

In its best form Mead is made as follows: 12 gallons of pure, soft water (clean rain water is, next to distilled water, best) are mixed with 30 gallons of expressed honey in a big caldron, 4 ounces of hops added, and the whole brought to a boil. The boiling is continued with diligent skimming, for at least an hour and a half. The fire is then drawn, and the liquid allowed to cool down slowly. When cold, it is drawn off into a clean barrel, which it should fill to the bung, with a little over. A pint of fresh wine yeast or ferment is added, and the barrel put in a moderately warm place, with the bung left out, to ferment for from 8 to 14 days, according to the weather (the warmer it is the shorter the period occupied in the primary or chief fermentation). Every day the foam escaping from the bung should be carefully skimmed off, and every 2 or 3 days there should be added a little honey and water to keep the barrel quite full, and in the meantime a pan or cup should be inverted over the hole, to keep out dust, insects, etc. When fermentation ceases, the procedure varies. Some merely drive in the bung securely and let the liquor stand for a few weeks, then bottle; but the best German makers proceed as follows, this being a far superior process: The liquor is removed from the barrel in which it fermented to another, clean, barrel, being strained through a haircloth sieve to prevent the admission of the old yeast. A second portion of yeast is added, and the liquid allowed to pass through the secondary fermentation, lasting usually as long as the first. The bung is driven into the barrel, the liquid allowed to stand a few days to settle thoroughly and then drawn off into bottles and stored in the usual way. Some add nutmeg, cinnamon, etc., prior to the last fermentation.

MEASURES:
See Weights and Measures.

MEASURES, TO CLEAN:
See Cleaning Preparations and Methods.

MEAT EXTRACT CONTAINING ALBUMEN:
See Foods.

MEAT PEPTONOIDS:
See Peptonoids.

MEAT PRESERVATIVES:
See Foods.

MEAT PRODUCTS (ADULTERATED):
See Foods.

MEDAL IMPRESSIONS:
See Matrix Mass.

MEDALS, CLEANING AND PRESERVING:
See Cleaning Compounds.

MEDALLION METAL:
See Alloys.

MEDICINE DOSES:
See Doses.

MEERSCHAUM:

To Color a Meerschaum Pipe.—I.—Fill the pipe and smoke down about one-third, or to the height to which you wish to color. Leave the remainder of the tobacco in the pipe, and do not empty or disturb it for several weeks, or until the desired color is obtained. When smoking put fresh tobacco on the top and smoke to the same level. A new pipe should never be smoked outdoors in extremely cold weather.

II.—The pipe is boiled in a preparation of wax, 8 parts; olive oil, 2 parts; and nicotine, 1 part, for 10 or 15 minutes. The pipe absorbs this, and a thin coating of wax is held on the surface of the pipe, and made to take a high polish. Under the wax is retained the oil of tobacco, which is absorbed by the pipe; and its hue grows darker in proportion to the tobacco used. A meerschaum pipe at first should be smoked very slowly, and before a second bowlful is lighted the pipe should cool off. This is to keep the wax as far up on the bowl as possible; rapid smoking will overheat, driving the wax off and leaving the pipe dry and raw.

To Repair Meerschaum Pipes.—To cement meerschaum pipes, make a glue of finely powdered and sifted chalk and white of egg. Put a little of this glue on the parts to be repaired and hold them pressed together for a moment.
See also Adhesives under Cements.

To Tell Genuine Meerschaum.—For the purpose of distinguishing imitation meerschaum from the true article, rub with silver. If the silver leaves lead pencil-like marks on the mass, it is not genuine but artificial meerschaum. If no such lines are produced, the article is genuine.

MENTHOL COUGH DROPS:
See Confectionery.

MENTHOL TOOTH POWDER:
See Dentifrices.

MERCURY SALVES:
See Ointments.

MERCURY STAINS, TO REMOVE:
See Cleaning Preparations and Methods.

METACARBOL DEVELOPER:
See Photography.

Metals and Their Treatment

METAL CEMENTS:
See Adhesives and Lutes.

METAL CLEANING:
See Cleaning Preparations and Methods.

METAL INLAYING:
See Damaskeening.

METAL POLISHES:
See Polishes.

METAL PROTECTIVES:
See Rust Preventives.

METAL VARNISHES:
See Varnishes.

METALS, HOW TO ATTACH TO RUBBER:
See Adhesives, under Rubber Cements.

METALS, SECURING WOOD TO:
See Adhesives.

METALS, BRIGHTENING AND DEADENING, BY DIPPING:

Brightening Pickle.—To brighten articles by dipping, the dipping liquid must not be too hot, otherwise the pickled surface turns dull; neither must it be prepared too thin, nor must wet articles be entered, else only tarnished surfaces will be obtained.

For a burnish-dip any aqua fortis over 33° Bé., i. e., possessing a specific gravity of 1.30, may be employed. It is advisable not to use highly concentrated aqua fortis, to reduce the danger of obtaining matt work. It is important that the quantity of oil of vitriol, which is added, is correct. It is added because the action of the aqua fortis is very uncertain. Within a short time it becomes so heated in acting on the metals that it turns out only dull work, and pores or even holes are apt to be the result of the violent chemical action. If the aqua fortis is diluted with water the articles do not become bright, but tarnish. For this reason sulphuric acid should be used. This does not attack the metals; it only dilutes the aqua fortis and distributes the heat generated in pickling over a larger space. It is also much cheaper, and it absorbs water from the aqua fortis and, therefore, keeps it in a concentrated state and yet distributed over the space.

In the case of too much oil of vitriol the dilution becomes too great and the goods are tarnished; if too little is added, the mixture soon ceases to turn out bright articles, because of overheating. On this experience are based the formulas given below.

Dip the articles, which must be free from grease, into the pickle, after they have been either annealed and quenched in diluted sulphuric acid or washed out with benzine. Leave them in the dipping mixture until they become covered with a greenish froth. Then quickly immerse them in a vessel containing plenty of water, and wash them out well with running water. Before entering the dipped articles in the baths it is well to remove all traces of acid, by passing them through a weak soda or potassium cyanide solution and washing them out again. If the brightly dipped goods are to remain bright they must be coated with a thin spirit or zapon acquer.

Following are two formulas for the pickle:

I.—Aqua fortis, 36° Bé.,
  by weight....... 100 parts
Oil of vitriol (sulphuric acid), 66°
  Bé., by weight .. 70 parts
Cooking salt, by
  volume......... 1½ parts
Shining soot (lampblack), by volume.......... 1½ parts

II.—Aqua fortis, 40° Bé.,
  by weight....... 100 parts
Oil of vitriol, 66°
  Bé., by weight ... 100 parts
Cooking salt, by
  volume......... 2 parts
Shining soot, by
  volume......... 2 parts

Matting or Deadening Pickle.—When, instead of brilliancy, a matted appearance is desired for metals, the article is corroded either mechanically or chemically. In the first case it is pierced with fine holes near together, rubbed with emery powder or pumice stone and tamponned. In the other case the corrosion is effected in acid baths thus composed:

Nitric acid of 36° Bé., 200 parts, by volume; sulphuric acid of 56° Bé., 200 parts, by volume; sea salt, 1 part, by volume; zinc sulphate, 1 to 5 parts, by volume.

With this proportion of acids the articles can remain from 5 to 20 minutes in the mixture cold; the prominence of the matt depends on the length of time of the immersion. The pieces on being taken from the bath have an earthy appearance which is lightened by dipping them quickly in a brightening acid. If left too long the matted appearance is destroyed.

Cotton Matt.—This matt, thus called on account of its soft shade, is rarely employed except for articles of stamped brass, statuettes, or small objects. As much zinc is dissolved in the bath as it will take. The pieces are left in it from 15 to 30 minutes. On coming from the bath they are dull, and to brighten them somewhat they are generally dipped into acids as before described.

Silver Matt.—Articles of value for which gilding is desired are matted by covering them with a light coating of silver by the battery. It is known that this deposit is always matt, unless the bath contains too large a quantity of potassium cyanide. A brilliant silvering can be regularly obtained with electric baths only by adding carbon sulphide. Four drachms are put in an emery flask containing a quart of the bath fluid and allowed to rest for 24 hours, at the end of which a blackish precipitate is formed. After decanting, a quart is poured into the electric bath for each quart before every operation of silvering.

Dangers of Dipping.—The operation of dipping should be carried out only in a place where the escaping fumes of hyponitric acid and chlorine can pass off without molesting the workmen, e. g., under a well-drawing chimney, preferably in a vapor chamber. If such an arrangement is not present the operator should choose a draughty place and protect himself from the fumes by tying a wet sponge under his nose. The vapors are liable to produce very violent and dangerous inflammations of the respiratory organs, coming on in a surprisingly quick manner after one has felt no previous injurious effect at all.

COLORING METALS:

See also Plating.

Processes by Oxidation.—By heat:—Coloration of Steel.—The steel, heated uniformly, is covered in the air with a pellicle of oxide and has successively the following colors: Straw yellow, blue (480° to 570° F.), violet, purple, water-green, disappearance of the color; lastly the steel reddens. For producing the blue readily, plunge the object into a bath of 25 parts of lead and 1 part of tin; its temperature is sufficient for bluing small pieces.

Bronzing of Steel.—I.—The piece to be bronzed is wet by the use of a sponge with a solution formed of iron perchloride, cupric sulphate, and a nitric acid. It is dried in a stove at 86° F., then kept for 20 minutes over boiling water. It is dried again at 86° F., and rubbed with a scratch brush.

This operation is repeated several times.

Bronzing of Steel.—II.—Rust and grease are removed from the objects with a paste of whiting and soda. They are immersed in a bath of dilute sulphuric acid, and rubbed with very fine pumice-stone powder. They are then exposed from 2 to 3 minutes to the vapor of a mixture of equal parts of concentrated chlorhydric and nitric acids.

The object is heated to 570° to 660° F. until the bronze color appears. When cooled, it is covered with paraffine or vaseline while rubbing, and heated a second time until the vaseline or paraffine commences to decompose. The operation is repeated. The shades obtained are beautiful, and the bronzing is not changeable. By subjecting the object to the vapors of the mixture of chlorhydric and nitric acids, shades of a light reddish brown are obtained. By adding to these two acids acetic acid, beautiful yellow bronze tints are procured. By varying the proportion of these three acids, all the colors from light reddish brown to deep brown, or from light yellow bronze to deep yellow bronze, are produced at will.

Bronzing.—III.—Under the name of Tuker bronze, a colored metal is found in trade which imitates ornamental bronze perfectly. It is obtained by deoxidizing or, if preferred, by burnishing cast iron. A thin layer of linseed oil or of linseed-oil varnish is spread on. It is heated at a temperature sufficient for producing in the open air the oxidation of the metal. The temperature is raised more or less, according as a simple yellow coloration or a deep brown is desired.

Lustrous Black.—In a quantity of oil of turpentine, sulphuric acid is poured drop by drop, stirring continually until a precipitate is no longer formed. Then the whole is poured into water, shaken, decanted, and the washing of the precipitate commenced again until blue litmus paper immersed in the water is no longer reddened. The precipitate will thus be completely freed from acid. After having drained it on a cloth, it is ready for use. It is spread on the iron and burned at the fire.

If the precipitate spreads with difficulty over the metal, a little turpentine can be added. It is afterwards rubbed with a linen rag, soaked with linseed oil, until the surface assumes a beautiful lustrous black. This covering is not liable to be detached.

Bluish Black.—Make a solution composed of nitric acid, 15 parts; cupric sulphate, 8 parts; alcohol, 20 parts; and water, 125 parts. Spread over the metal when well cleaned and grease removed. Dry and rub with linen rag.

Black.—Make a solution composed of cupric sulphate, 80 parts; alcohol, 40 parts; ferric chloride, 30 parts; nitric acid, 20 parts; ether, 20 parts; water, 400 to 500 parts, and pass over the object to be blackened.

Magnetic Oxide.—I.—A coating of magnetic oxide preserves from rust. To obtain it, heat the object in a furnace to a temperature sufficient to decompose steam. Then inject from 4 to 6 hours superheated steam at 1,100° F. The thickness of the layer of oxide formed varies with the duration of the operation. This process may replace zincking, enameling, or tinning.

II.—A deposit of magnetic oxide may be obtained by electrolysis. The iron object is placed at the anode in a bath of distilled water heated to 176° F. The cathode is a plate of copper, or the vessel itself if it is of iron or copper. By electrolysis a layer of magnetic oxide is formed.

In the same way other peroxides may be deposited. With an alkaline solution of litharge a brilliant black deposit of lead peroxide, very adherent, is obtained.

The employment of too strong a current must be avoided. It will produce a pulverulent deposit. To obtain a good coating, it is necessary after leaving the objects for a moment at the opposite pole, to place them at the other pole until the outside is completely reduced, then bring them back to the first place.

Processes by Sulphuration.—Oxidized Brown Color.—The object is plunged into some melted sulphur mingled with lampblack, or into a liquid containing the flowers of sulphur mingled with lampblack. It is drained and dried. The bronzing obtained resists acids, and may acquire a beautiful polish which has the appearance of oxidized bronze, due perhaps to the formation of ferric sulphide, a sort of pyrites remarkable for its beautiful metallic reflections and its resistance to chemical agents.

Brilliant Black.—Boil 1 part of sulphur and 10 parts turpentine oil. A sulphurous oil is obtained of disagreeable odor. Spread this oil with the brush as lightly as possible, and heat the object in the flame of an alcohol lamp until the patina takes the tint desired. This process produces on iron and steel a brilliant black patina, which is extremely solid.

Blue.—Dissolve 500 drachms of hyposulphite of soda in 1 quart of water, and 35 grains of lead acetate in 1 quart of water. The two solutions mingled are heated to the boiling point. The iron is immersed, and assumes a blue coloration similar to that obtained by annealing.

Deposit of a Metal or of a Non-Oxidizable Compound.—Bronze Color.—Rub the iron smartly with chloride of antimony. A single operation is not sufficient. It is necessary to repeat it, heating the object slightly.

Black.—I.—Make a paste composed of equal parts of chloride of antimony and linseed oil. Spread on the object, previously heated, with a brush or rag; then pass over it a coating of wax and brush it. Finally varnish with gum lac.

II.—Prepare a solution of bismuth chloride, 10 parts; mercury chloride, 20 parts; cupric chloride, 10 parts; hydrochloric acid, 60 parts; alcohol, 50 parts; water, 500 parts. Add fuchsine in sufficient quantity to mask the color.

The mercury chloride is poured into the hydrochloric acid, and the bismuth chloride and cupric chloride added; then the alcohol. Employ this mixture with a brush or a rag for smearing the object. The object may also be immersed in the liquid if it is well cleaned and free from grease. It is dried and afterwards submitted to boiling water for half an hour. The operation is repeated until the wished-for tint is obtained; then the object is passed into the oil bath and

taken to the fire without wiping. The object may also be placed for 10 minutes in boiling linseed oil.

Brown Tint.—A solution is made of chloride of mercury, 20 parts; cupric chloride, 10 parts; hydrochloric acid, 60 parts; alcohol, 50 parts; water, 500 parts. The object is plunged into this solution after being well cleaned. The solution may also be applied with a brush, giving two coats. It is afterwards put into hot water. The surface of the object is covered with a uniform layer of vegetable oil. It is placed in a furnace at a high temperature, but not sufficient for carbonizing the oil. The iron is covered with a thin layer of brown oxide, which adheres strongly to the metal, and which can be beautifully burnished, producing the appearance of bronze.

Brilliant Black.—The process begins by depositing on the object, perfectly clean and free from grease, a layer of metallic copper. For this purpose the following solutions are prepared: (a) Cupric sulphate, 1 part; water, 16 parts. Add ammonia until complete dissolution. (b) Chloride of tin, 1 part; water, 2 parts; and chlorhydric acid, 2 parts. The object is immersed in solution b, and afterwards in solution a. In this way there is deposited on the iron a very adherent coating of copper. The object, washed with water, is afterwards rubbed with sulphur, or immersed in a solution of ammonium sulphhydrate. A dull black coating of cupric sulphide is produced, which becomes a brilliant black by burnishing.

Blue Black.—The iron object is first heated according to the previous recipe, but the copper is converted into cupric sulphide, not by a sulphhydrate, but by a hyposulphite. It is sufficient to dip the coppered object into a solution of sodium hyposulphite, acidulated with chlorhydric acid, and raised to the temperature of 175° to 195° F.

Thus a blue-black coating is obtained, unchangeable in air and in water. After polishing, it has the color of blue steel. It adheres strongly enough to resist the action of the scratch brush.

Deposition of Molybdenum.—Iron is preserved from rust by covering it with a coating of molybdenum, as follows: Water, 1,000 parts; ammonium molybdate, 1 part; ammonium nitrate, 15 to 20 parts. Suspend the object at the negative pole of a battery. The current ought to have a strength of 2 to 5 amperes per cubic decimeter.

Deposit of Manganese Peroxide.—The

iron or steel is first covered with a coating of manganese peroxide by immersing as an anode in a bath containing about 0.05 per cent of chloride or sulphate of manganese and from 5 to 25 per cent of ammonium nitrate. The bath is electrolyzed cold, making use of a cathode of charcoal. Feeble currents (1 or 2 amperes) produce an adherent and unchangeable deposit.

Bronzing of Cannon.—Prepare a solution of ferric chloride of density 1.281, 14 parts; mercury chloride, 3 parts; fuming nitric acid, 3 parts; cupric sulphate, 3 parts; water, 80 parts. Give to the piece of ordnance 2 or 3 coatings of the solution, taking care always to scratch the preceding layer with a steel brush before spreading the second. Afterwards, the object is plunged in a solution of potassium sulphide in 900 parts of water. It is left in this for 10 days. It is removed by washing with soap and hot water. The object is rinsed, dried, and finally brushed with linseed-oil varnish.

Green Bronzing. — Dissolve 1 part of acetate of silver in 20 parts of essence of lavender; coat the surface of iron with this liquid by means of a brush and raise the temperature to 292° F. A brilliant green color is developed on the surface.

Coating on Steel Imitating Gilding.— The object is first covered by the galvanic method by means of a solution of cyanide of copper and potassium, then covered electrolytically with a thin deposit of zinc. It is dried and cleaned with a little washed chalk and finally immersed in boiling linseed oil. The surface of the piece after a few seconds, at a temperature of 310° F., appears as if there had been a real penetration of copper and zinc; that is to say, as though there were a formation of tombac.

Bronzing of Cast Iron.—The piece, when scraped, is coppered with the following bath: Cupric chloride, 10 parts; hydrochloric acid, 80 parts; nitric acid, 10 parts. It is rubbed with a rag and washed with pure water, and then rubbed with the following solution: Ammonium chlorhydrate, 4 parts; oxalic acid, 1 part; water, 30 parts.

Gilding of Iron and Steel.—Chloride of gold is dissolved either in oil of turpentine or in ether, and this solution is applied with the brush on the metallic surface, after being perfectly scraped. It is allowed to dry, and then heated more or less strongly for obtaining the necessary adherence. When it is dry the gilding is burnished.

Process by Deposit of a Color or Varnish.—Beautiful colorations, resistive to light, may be given to metals by the following method:

The metallic objects are immersed in a colorless varnish with pyroxyline, and dried in a current of hot air at 176° F. When the varnish is sufficiently dry, the objects are bathed for a few minutes in a 2 per cent alcoholic solution of alizarine or of a color of the same group. By washing with water the yellowish color covering the object on coming from the coloring bath passes to the golden red.

Coloring Copper.—To redden copper hang it from a few minutes to an hour, according to the shade wanted, in a 5 to 10 per cent solution of ferrocyanide of potassium in water. By adding a little hydrochloric acid to the solution the color given to the copper may be made to assume a purple shade. On removing the copper, dry it in the air or in fine sawdust, rinse, and polish with a brush or chamois leather, after drying it again.

Coloring Brass.—To redden brass, dip in solution of 5 ounces of sulphate of copper and 6 to 7 ounces of permanganate of potash in 500 ounces of water.

To blue copper or brass any one of the following recipes may be used:

I.—Dip the article in a solution of 2 ounces of liver of sulphur and 2 ounces of chlorate soda in 1,000 ounces of water.

II.—Dip the article in a solution of ferrocyanide of potassium very strongly acidulated with hydrochloric acid.

III.—Stir the article about constantly in a solution of liver of sulphur in 50 times its weight of water.

Fusion Point of Metals.—The point of fusion of common metals is as follows: Antimony, 808° F.; aluminum, 1,160° F.; bismuth, 517° F.; copper, 1,931° F.; gold, 1,913° F.; iron, 2,912° F.; lead, 850° F.; nickel, 2,642° F.; platinum, 3,225° F.; silver, 1,750° F.; tin, 551° F.; zinc, 812° F. Mercury, which is normally fluid, congeals at 38° below zero, F., this being its point of fusion.

To Produce Fine Leaves of Metal.— The metal plate is laid between parchment leaves and beaten out with hammers. Although films obtained in this manner reach a high degree of fineness, yet the mechanical production has its limit. If very fine films are desired the galvano-plastic precipitation is employed in the following manner:

A thin sheet of polished copper is entered in the bath and connected with the electric conduit. The current precipitates gold on it. In order to loosen it, the gilt copper plate is placed in a solution of ferric chloride, which dissolves the copper and leaves the gold behind. In this manner gold leaf can be hammered out to almost incredible thinness.

METAL FOIL.

Tin foil is the most common foil used, being a combination of tin, lead, and copper, sometimes with properties of other metals.

	I	II	III
	Per cent	Per cent	Per cent
Tin	97.60	98.47	96.21
Copper	2.11	0.38	0.95
Lead	0.04	0.84	2.41
Iron	0.11	0.12	0.09
Nickel	....	....	0.30

I is a mirror foil; III is a tin foil.

Tin Foils for Capsules.—

	I	II
	Per cent	Per cent
Tin	20	22
Lead	80	77
Copper	..	1

Tin Foils for Wrapping Cheese, etc.—

	I	II	III
	Per cent	Per cent	Per cent
Tin	97	90	92
Lead	2.5	7.8	7
Copper	0.5	0.2	1

Tin Foils, for Fine Wrapping, I and II; for Tea Boxes, III.—

	I	II	III
	Per cent	Per cent	Per cent
Tin	60	65	40
Lead	40	35	58.5
Copper	..	..	1.5

Imitation Gold Foils.—

	Deep gold	Pure gold	Pale gold
	Per cent	Per cent	Per cent
Copper	84.5	78	76
Zinc	15.5	22	14

	Deep gold	Deep gold	Gold
	Per cent	Per cent	Per cent
Copper	91	86	83
Zinc	9	14	17
		dark reddish yellow	pale yellow

Imitation Silver Foil.—Alloy of tin and zinc; harder than tin and softer than zinc: Zinc, 1 part; tin, 11 parts.

To Attach Gold Leaf Permanently.— Dissolve finely cut isinglass in a little water, with moderate heat, which must not be increased to a boil, and add as much nitric acid as has been used of the isinglass. The adhesive will not penetrate the cardboard or paper.

METH:
See Mead.

METHEGLIN:
See Mead.

METHYL SALICYLATE, TO DISTINGUISH FROM OIL OF WINTERGREEN:
See Wintergreen.

METOL DEVELOPER:
See Photography.

METRIC WEIGHTS:
See Weights and Measures.

MICE POISON:
See Rat Poison.

MICROPHOTOGRAPHS
See Photography.

MILK:
See also Foods.

Determining Cream.—An apparatus for determining cream in milk consists of a glass cylinder having a mark about half its height, and a second mark a little above the first. The milk is added up to the lower mark, and water up to the second. The amount of water thus added is about one-fourth the volume of the milk, and causes the cream to rise more quickly. The tube is graduated between the two marks in percentages of cream on the undiluted milk. A vertical blue strip in the side of the cylinder aids the reading of the meniscus.

Formaldehyde in Milk, Detection of.— To 10 parts of milk add 1 part of fuchsine sulphurous acid. Allow to stand 5 minutes, then add 2 parts of pure hydrochloric acid and shake. If formaldehyde is not present, the mixture remains yellowish white, while if present a blue-violet color is produced. This test will detect 1 grain of anhydrous formaldehyde in 1 quart of milk.

Malted Milk.—To malt milk, add the following:

Powdered malt	1 ounce
Powdered oat meal	2 ounces
Sugar of milk	4 ounces
Roasted flour	1 pound

Milk Extracts.—These are made from skimmed milk freed from casein, sugar and albumen, and resemble meat extracts. The milk is slightly acidulated with phosphoric or hydrochloric acid, and evaporated *in vacuo* to the consistency of thick syrup. During the crystallization of the sugar, the liquid is sterilized.

Modification of Milk for Infants.—For an ill child note the percentages of milk taken; decide, if indigestion is present, which ingredient of the milk, fat or proteid, or both, is at fault, and make formula accordingly.

After allowing the milk to stand 8 hours, remove the top 8 ounces from a quart jar of 4 per cent fat milk by means of a dipper, and count this as 12 per cent fat cream. Count the lowest 8 ounces of the quart fat-free milk. From these the following formula may be obtained, covering fairly well the different percentages required for the different periods of life.

First Week.

12 per cent cream. Fat-free milk.

Fat......	2.00	Cream....	3¼ oz.
Sugar....	5.00	Milk......	1½ oz.
Proteids..	0.75	Milk sugar	2 meas.

Second Week.

Fat......	2.50	Cream....	4¼ oz.
Sugar....	6.00	Milk......	1¼ oz.
Proteids..	1.00	Milk sugar	2½ meas.

Third Week.

Fat......	3.00	Cream....	5 oz.
Sugar....	6.00	Milk......	1 oz.
Proteids..	1.00	Milk sugar.	2½ meas.

Four to Six Weeks.

Fat......	3.50	Cream....	5¾ oz.
Sugar....	6.50	Milk......	1¾ oz.
Proteids..	1.00	Milk sugar	2½ meas.

Six to Eight Weeks.

Fat......	3.50	Cream....	5¾ oz.
Sugar....	6.50	Milk......	3¼ oz.
Proteids..	1.50	Milk sugar	2¼ meas.

Two to Four Months.

Fat......	4.00	Cream....	6¾ oz.
Sugar....	7.00	Milk......	2¼ oz.
Proteids..	1.50	Milk sugar	2½ meas.

Four to Eight Months.

Fat......	4.00	Cream....	6¾ oz.
Sugar....	7.00	Milk......	4¾ oz.
Proteids..	2.00	Milk sugar	2¼ meas.

Eight to Nine Months.

Fat......	4.00	Cream....	6¾ oz.
Sugar....	7.00	Milk......	7½ oz.
Proteids..	2.50	Milk sugar	2 meas.

Nine to Ten Months.

Fat......	4.00	Cream....	6¾ oz.
Sugar....	7.00	Milk....	10½ oz.
Proteids..	3.00	Milk sugar	1½ meas.

Ten to Twelve Months.

Fat......	4.00	Cream....	6¾ oz.
Sugar....	5.00	Milk......	11¾ oz.
Proteids..	3.50	Milk sugar	½ meas.

After Twelve Months.

Unmodified cow's milk.

Preservation of Milk (see also Foods).—I.—Shortly after the milk is strained add to it from 1 per cent to 2 per cent of a 12-volume solution of hydrogen peroxide, and set it aside for 10 to 12 hours. It thus acquires the property of keeping perfectly sweet and fresh for 3 or 4 days, and is far preferable to milk sterilized by heat. Two points are worthy of notice in the process. The addition of oxygenated water should be made as soon after it is taken from the cow, strained, etc., as possible; the peroxide appears to destroy instantly all anaërobic microbes (such as the bacillus of green diarrhea of childhood), but has no effect upon the bacillus of tuberculosis. This process is to be especially recommended in the heat of summer, and at all times in the milk of cattle known to be free of tuberculosis.

II.—Fresh milk in bottles has been treated with oxygen and carbonic acid under pressure of some atmospheres. By this method it is said to be possible to preserve milk fresh 50 to 60 days. The construction of the bottle is siphon-like.

Milk Substitute.—Diamalt is a thick syrupy mass of pleasant, strong, somewhat sourish odor and sweetish taste, which is offered as a substitute for milk. The preparation has been analyzed. Its specific gravity is 1.4826; the percentage of water fluctuates between 24 and 28 per cent; the amount of ash is 1.3 per cent. There are present: Lactic acid, 0.718 to 1.51; nitrogenous matter, 4.68 to 5.06 per cent; and constituents rich in nitrogen, about 68 per cent. The latter consist principally of maltose. Dissolved in water it forms a greenish-yellow mixture. Turbidness is caused by starch grains, yeast cells, bacteria, and a shapeless coagulum.

MILK AS A SUBSTITUTE FOR CELLULOID, BONE, AND IVORY:

See Casein.

MILK, CUCUMBER:

See Cosmetics.

MILK OF SOAP:

See Cleaning Preparations and Methods, under Miscellaneous Methods.

MINARGENT:

See Alloys.

MINERAL WATERS:
See Waters.

MINOFOR METAL:
See Alloys.

MINT CORDIAL:
See Wines and Liquors.

Mirrors

(See also Glass.)

Mirror Silvering.—Mirror silvering is sometimes a misnomer, inasmuch as the coating applied to glass in the manufacture of mirrors does not always contain silver. In formula I it is an amalgam of mercury and tin.

I.—A sheet of pure tin foil, slightly larger than the glass plate to be silvered, is spread evenly on a perfectly plane stone table having a raised edge, and is well cleaned from all dust and impurity. The foil must be free from the slightest flaw or crack. The tin is next covered uniformly to a depth of $\frac{1}{8}$ of an inch with mercury, preference being given by some to that containing a small proportion of tin from a previous operation. The glass plate, freed from all dust or grease, and repolished if necessary, is then carefully slid over the mercury. This part of the work requires skill and experience to exclude all air bubbles, and even the best workmen are not successful every time. If there is a single bubble or scratch the operation must be repeated and the tin foil is lost; not a small expense for large sizes. When this step has been satisfactorily accomplished the remainder is easy. The glass plate is loaded with heavy weights to press out the excess of mercury which is collected and is used again. After 24 hours the mirror is lifted from the table and placed on edge against a wall, where it is left to drain well.

II.—Solution No. 1 is composed as follows: To 8 ounces of distilled water, brought to a boil, add 12 grains of silver nitrate and 12 grains of Rochelle salts. Let it come to a boil for 6 to 7 minutes; then cool and filter.

Solution No. 2 is made as follows: Take 8 ounces of distilled water, and into a small quantity poured into a tumbler put 19 grains of silver nitrate. Stir well until dissolved. Then add several drops of 26° ammonia until the solution becomes clear. Add 16 grains more of nitrate of silver, stirring well until dissolved. Add balance of distilled water and filter. The filtering must be done through a glass funnel, in which the filter paper is placed. The solution must be stirred with a glass rod. Keep the solutions in separate bottles marked No. 1 and No. 2.

Directions for Silvering: Clean the glass with ammonia and wipe with a wet chamois. Then take half and half of the two solutions in a graduating glass, stirring well with a glass rod. Pour the contents on the middle of the glass to be silvered. It will spread over the surface of itself if the glass is laid flat. Leave it until the solution precipitates.

Silvering Globes.—The insides of globes may be silvered, it is said, by the following methods:

I.—Take $\frac{1}{3}$ ounce of clean lead, and melt it with an equal weight of pure tin; then immediately add $\frac{1}{2}$ ounce of bismuth, and carefully skim off the dross; remove the alloy from the fire, and before it grows cold add 5 ounces of mercury, and stir the whole well together; then put the fluid amalgam into a clean glass, and it is fit for use. When this amalgam is used for silvering, it should be first strained through a linen rag; then gently pour some ounces of it into the globe intended to be silvered; the alloy should be poured into the globe by means of a paper or glass funnel reaching almost to the bottom of the globe, to prevent it splashing the sides; the globe should be turned every way very slowly, to fasten the silvering.

II.—Make an alloy of 3 ounces of lead, 2 ounces of tin, and 5 ounces of bismuth. Put a portion of this alloy into the globe and expose it to a gentle heat until the compound is melted; it melts at 197° F.; then by turning the globe slowly round, an equal coating may be laid on, which, when cold, hardens and firmly adheres.

Resilvering Mirrors—If mirrors coated with amalgam become damaged they may sometimes be successfully repaired by one of the following processes:

I.—Place the old mirror in a weak solution of nitric acid—say 5 per cent—which immediately removes the silver. Rinse it a little, and then clean very thoroughly with a pledget of cotton-wool and a mixture of whiting and ammonia. Rouge will answer in place of whiting, or, as a last extreme, finest levigated pumice, first applied to a waste glass to crush down any possible grit. This cleaning is of the utmost importance, as upon its thoroughness depends eventual success. Front, back, and edges must alike be left in a state above suspicion. The plate is then again flowed with weak acid, rinsed under the tap, then flowed back and front with distilled water, and kept immersed in a glass-covered dish of distilled water until the solutions are ready.

The depositing vessel is the next consideration, and it should be realized that unless most of the silver in the solution finds its way on to the face of the mirror it were cheaper that the glass should be sent to the professional mirror-maker. The best plan is to use a glass dish allowing a $\frac{1}{16}$ inch margin all round the mirror, inside. But such a glass dish is expensive, having to be made specially, there being no regular sizes near enough to 4 x 7 or 8 x 5 (usual mirror sizes). If too large, a dish must perforce be used, the sides or ends of which should be filled up with sealing wax. Four strips of glass are temporarily bound together with 2 or 3 turns of string, so as to form a hollow square. The side pieces are $\frac{1}{4}$ inch longer outside, and the end pieces $\frac{1}{2}$ inch wider than the mirror glass. This frame is placed in about the center of the dish, moistened with glycerine, and the molten wax flowed outside of it to a depth of about $\frac{3}{4}$ of an inch or more. For economy's sake, good "parcel wax" may be used, but best red sealing wax is safer. This wax frame may be used repeatedly, being cleaned prior to each silvering operation. It is the only special appliance necessary, and half an hour is a liberal time allowance for making it.

Use a stock solution of silver nitrate of the strength of 25 grains to 1 ounce of distilled water: Take 2 drachms of silver nitrate stock solution and convert it to ammonia nitrate, by adding ammonia drop by drop until the precipitate is redissolved. Add $3\frac{1}{2}$ ounces of distilled water.

In another measure take 80 drops (approximately 74 minims) of 40 per cent formalin. Pour the solution of ammonio nitrate of silver into the measure containing the formalin, then back into the original measure, and finally into the dish containing the glass to be silvered. This should be done rapidly, and the dish containing the mirror well rocked until the silvering is complete, which may be ascertained by the precipitation of a black, flocculent deposit, and the clearing of the solution. The actual process of silvering takes about 2 minutes.

Cleanliness throughout is of the greatest importance. The vessels in which the solutions are mixed should be well rinsed with a solution of bichromate of potash and sulphuric acid, then washed out three or four times under the tap, and finally with distilled water. For cleansing, dip the glass for a short time in a solution of bichromate of potash, to which a little sulphuric acid is added. The glass is afterwards well rinsed for a minute or two under the tap, flooded with distilled water, and dried with a clean linen cloth. A little absolute alcohol is then rubbed on with a soft linen handkerchief, which is immediately rolled into a pad and used for well polishing the surface. The cleaning with alcohol is repeated to avoid risk of failure.

After the mirror has been silvered, hold it under the tap and allow water to flow over it for about 3 minutes. Rinse it with distilled water, and stand it up on edge on blotting paper. When it is quite dry take a pad of very soft washleather, spread a small quantity of finest opticians' rouge on a sheet of clean glass, and well coat the pad with rouge by polishing the sheet of glass. A minute quantity of rouge is sufficient. Afterwards polish the mirror by gently rubbing the surface with the pad, using a circular stroke.

It will be seen that with this process it is unnecessary to suspend the mirror in the silvering solution, as usually recommended. The mirror is laid in the dish, which is a distinct advantage, as the progress of the silvering may be watched until complete. The film also is much more robust than that obtained by the older methods.

II.—Clean the bare portion of the glass by rubbing it gently with fine cotton, taking care to remove any trace of dust and grease. If this cleaning be not done very carefully, defects will appear around the place repaired. With the point of a penknife cut upon the back of another looking glass around a portion of the silvering of the required form, but a little larger. Upon it place a small drop of mercury; a drop the size of a pin's head will be sufficient for a surface equal to the size of the nail. The mercury spreads immediately, penetrates the amalgam to where it was cut off with the knife, and the required piece may be now lifted and removed to the place to be repaired. This is the most difficult part of the operation. Then press lightly the renewed portion with cotton; it hardens almost immediately, and the glass presents the same appearance.

Clouding of Mouth Mirrors.—By means of the finger, slightly moistened, apply a film of soap of any brand or kind to the mirror; then rub this off with a clean, dry cloth; the mirror will be as bright and clear as ever. Breathing on it will not affect its clearness and the mirror does not suffer from the operation.

Magic Mirrors. — Among the many amusing and curious articles which the amateur mechanic can turn out, metallic mirrors having concealed designs on them, and which can be brought into view by breathing on the polished surface, are both funny and easy to produce. To produce steel mirrors either tough bronze or good cast mottled iron discs should be used, and the design should be on the bottom of the cast disc, as this is the soundest and densest part of the metal. The method of working is different with bronze and iron, and bronze will be dealt with first.

The cast disk of bronze should be turned up level on both sides, and the edges should be turned or shaped up, the metal being about half an inch thick. On the side which was at the bottom in casting, a line should be drawn to allow for working up the border or frame of the mirror, and on the rest of the smooth surface the design should be drawn, not having too much detail. It is best to mark the lines with a sharp scriber, to prevent their effacement during working. When the disk is marked out, it should be laid on a smoothly planed iron block, and the lines punched to a depth of about ¼ inch, a punch with round edges being used. Then the disk should be turned down to just below the surface of the punched-in metal, and the border or edge formed, finishing smoothly, but without burnishing. The back can be turned down and, with the outer edge, burnished; but the inside of the edge and the face of the mirror should be polished with fine abrasive powder, and finished with fine rouge. When dry, the mirror will appear equally bright all over; but when breathed on the design will show, again disappearing as the moisture is removed. The metal punched in will be more dense than the rest of the surface, and will also be very slightly raised, this being imperceptible unless the polishing has been too long continued.

With iron mirrors a good mottled iron must be used, selecting hematite for preference; but in any case it must be chillable metal. Preferably it should be melted in a crucible, as this causes the least change in the metallic content, and as the metal can be made hot and fluid, it works well. The design must be worked out in iron of about ¼ inch in thickness, and must be level, as it has to touch the molten metal in the bottom of the mold. If preferred, the design may be cast and ground flat, but this depends largely on the design. The chill pattern should be coated with plumbago, and in molding the disk pattern of about ¼ inch in thickness should be laid on a board, and on this the design—chill—should be placed, and the mold should be rammed up from the back in the ordinary manner. The casting should be allowed to get cold in the mold, and should then be removed and dressed in the usual way. It should then be ground bright all over on emery wheels of successively finer grades, and the mirror surface should be buffed and polished until a steely mirror surface is produced. With a good mottled iron the chilled design will not show until the surface is breathed on or rubbed with a greasy rag, but will then show clearly.

MIRROR ALLOYS:
See Alloys.

MIRRORS, FROSTED:
See Glass.

MIRROR-LETTERING:
See Lettering.

MIRROR POLISHES:
See Polishes.

MIRRORS, TO CLEAN:
See Cleaning Preparations and Methods.

MIRRORS, TO PREVENT DIMMING OF:
See Glass.

MIRROR VARNISH:
See Varnishes.

MITE KILLER:
See Insecticides.

MIXING STICKS FOR PAINT:
See Paint.

MODELING WAX:
See Wax, Modeling.

MOISTURE:
See Insulation.

MOLDS:
See also Casting and Matrix.

Molding Sand.—A high grade of molding sand should be fat, i. e., strongly mixed with clay. Naturally the molds of this sand should be employed only in a perfectly dry state. The fat molding sand is prepared artificially from quartz sand (fine sprinkling sand), fat clay, free from lime and ferric oxide (red ocher). The molding sand is fixed by breaking up the loose pieces in which it is partly dug; next it is passed through a fine sieve and mixed up to one-third of its volume with charcoal dust, or, better still, with lampblack, which, owing to its looseness and fatness, does not detract so much from the binding qualities of the sand. The utility of the sand may be tested by pressing the finger into it, whereupon the fine lines of the skin should appear sharply defined; its binding power is ascertained by dropping a lump pressed together with the hand from a height, which is increased until it breaks.

MOLDS OF PLASTER:
See Plaster.

MOLES:
See also Warts.

Lunar caustic is frequently used to remove warts and moles. It should be wrapped in tin foil or placed in a quill so that it will not touch the bare flesh. Moisten the raised surface and touch with the caustic night and morning. Successive layers of skin will dry up and peel off. When on a level with the surrounding flesh apply a healing ointment. Let the last crust formed drop without touching it. Unless carefully done this process may leave a white scar.

A simple remedy for warts consists in wetting and rubbing them several times a day in a strong solution of common washing soda. The electric treatment, however, is now the most popular.

MORDANTS:
See also Dyes.

Mordant for Cement Surfaces.—Take green vitriol and dissolve it in hot water. If the cement is rather fresh add 1 part of vinegar for each part of green vitriol. Best suited, however, is triple vinegar (vinegar containing $\frac{1}{3}$ per cent of acetic acid), which is alone sufficient for well-dried places. For such surfaces that have been smoothed with a steel tool and have hardly any pores, take alcohol, 1 part, and green vitriol, 10 parts, and apply this twice until the iron has acquired a yellowish color. This mordant forms a neutral layer between cement and paint, and causes the latter to dry well.

Mordant for Gold Size.—A mordant for gold size gilding that has been thoroughly tested and found to be often preferable to the shellac-mixed article, is prepared from yolk of egg and glycerine. The yolk of an egg is twirled in a cup and up to 30 drops of glycerine are added to it. The more glycerine added, the longer the mordant will take to dry. Or else an equal portion of ordinary syrup is mixed with the yolk of egg. Same must be thinly liquid. If the mass becomes too tough it is warmed a little or thinned with a few drops of warm water. A single application is sufficient. Naturally, this style of gilding is only practicable indoors; it cannot withstand the influence of moisture.

MORTAR, ASBESTOS.
Asbestos mortar consists of a mixture of asbestos with 10 per cent of white lime. Canadian asbestos is generally used, which is composed of 80 per cent of asbestos and 20 per cent of serpentine. The asbestos is ground and the coarse powder used for the first rough cast, while the finer material is employed for the second top-plastering. This mortar is highly fire-resisting and waterproof, is only half as heavy as cement mortar, and tough enough to admit of nails being driven in without breaking it.

MOSQUITO REMEDIES:
See Insecticides.

MOSS REMOVERS:
See Cleaning Preparations and Methods, under Miscellaneous Methods.

MOTHS:
See Turpentine.

MOTH PAPER:
See Paper.

MOTH TRAPS AND MOTH KILLERS:
See Household Formulas.

MOTHER-OF-PEARL:
See Pearl.

MOTORS, ANTI-FREEZING SOLUTION FOR:
See Freezing Preventives.

MOUNTANTS:
See also Adhesives and Photography.

Mounting Drawings, Photos, etc., upon Fine Pasteboard.—It frequently happens that the pasteboard will warp toward the face of the picture, even if left in a press till the gluing medium is perfectly dry. This fault can be obviated by moistening the back of the pasteboard moderately with a sponge, and, while this is still wet, pasting the picture on with good, thin glue. If moistening the pasteboard is impracticable (with sensitive drawings, paintings, etc.), paste which has been pressed through a fine cloth is rubbed on, always in the same direction, and the picture is carefully and evenly pressed on. Then bend the pasteboard backward in a wide semicircle, and place it between two heavy objects on the table. After a few hours, when the paste is completely dry, put the picture down flat and load proportionately. Papers of large size, which cannot conveniently be placed between two objects, are wrapped up, and twine is stretched around, thus keeping them bent.

Mounting Prints on Glass.—Take 4 ounces of gelatin; soak ½ hour in cold water; then place in a glass jar, adding 16 ounces of water; put the jar in a large dish of warm water and dissolve the gelatin. When dissolved pour in a shallow tray; have the prints rolled on a roller, albumen side up; take the print by the corners and pass rapidly through the gelatin, using great care to avoid air bubbles. Squeeze carefully onto the glass. The better the quality of glass, the finer the effect.

MOUTH ANTISEPTICS:
See Antiseptics.

MOUTH WASHES:
See Dentifrices.

MOVING OBJECTS AND HOW TO PHOTOGRAPH THEM:
See Photography.

MUCILAGE:
See Adhesives.

MUSIC BOXES.
Care must be exercised in taking apart, for if the box is wound up and the fly is removed, the cylinder is ruined. The spring relaxes at a bound, causing the cylinder to turn with such rapidity that the pins cannot resist the teeth, whose force is intensified by the velocity of the cylinder. The pins originally bent forward are broken, or pressed backwards; as they are hardened, they cannot be bent forward again without breaking. This accident involves the cost of a new cylinder, the most expensive part of the apparatus. Besides, the comb almost always loses some teeth and the wheelwork also suffers in its turn.
To avoid such mishaps the careful operator will take the parts asunder in the following order:

1. Remove the comb.
2. Take the apparatus from the box and completely disarm the spring.
3. Remove the barrel.
4. Remove the escapement.
5. Remove the cylinder.

The barrel and the wheels are cleaned like those of a watch.
The cylinder should be handled carefully. The holes should be well cleaned. Oil should be put only on the pivots, especially none on the part of the arbor to which the cylinder is attached. It is the first piece to be replaced, care being taken to see that the arbor turns freely, but without play, between the bridges. When it is in position, put in the escapement, then the barrel, and finally the comb.
The comb, representing the musical part of a simple box, cannot receive too much care. Before replacing it examine the springs closely, and in supplying the ones that are lacking, take for the model of size and form those resembling them the most. If the parts have been put together properly, then, as soon as the comb is screwed in its place, these should be found in good working order: the *levée* (lift)—that is, that the pins do not lift the teeth too much or too little; the *tombée* (fall)—that is, that the chords, the bass, the medium, and the treble, fall together; and the *visée* (pointing)—that the pins catch at the center of the ends of the teeth.

MUSLIN, PAINTING ON:
See Painting.

MUSTACHE FIXING FLUID.

Balsam of Tolu	1 part
Rectified spirit	3 parts
Jockey club	1 part

Dissolve the balsam in the liquids. Apply a few drops to the mustache with a brush, then twist into the desired shape.

MUSTARD PAPER.

I.—India rubber	1 part
Benzol	49 parts
Black mustard in powder, a sufficiency.	

Dissolve the India rubber in the benzol, then stir in the mustard until the mixture is of a suitable consistence for spreading. It was further recommended to remove the fixed oil from the mustard by percolation with benzol. Mustard paper thus made is of good quality, very active, and keeps well.

II.—Black and white mustard, in No. 60 powder, deprived of fixed oil. ... 1 part
Benzol solution of India rubber (1 in 40) .... 4 parts

Mix to a smooth mass, and spread the same over one side of a suitable paper by means of a plaster-spreading machine, or passing the paper over the mass contained in a suitable shallow vessel. Expose to warm air for a short time to dry. Preserve the dry paper in well-closed boxes. It may be useful to know that mustard paper, after spreading, should not be long exposed to light and air. By so doing not only does the mustard bleach but the rubber soon perishes. Moreover, mustard paper is hygroscopic, so that in a moist atmosphere it soon loses its virtue. It is, therefore, highly important that mustard paper should be rapidly dried in a warm atmosphere with free ventilation, then at once stored in well-closed packets. Thus prepared they keep well and remain active for many years.

MUSTARDS:
See Condiments.

MYRRH ASTRINGENT:
See Dentifrices.

NAIL, INGROWING.

Copious applications of dried powdered alum are sufficient to cure every case of ingrowing nail in about 5 days. The applications are not painful in the least, and the destruction of the pathologic tissue results in the formation of a hard, resistant, and non-sensitive bed for the nail, a perfect cure for the ingrowing tendency. Apply a fomentation of soap and water for 24 hours beforehand and then pour the alum into the space between the nail and its bed, tamponing with cotton to keep the alum in place, and repeating the application daily. The suppuration rapidly dries up, and pain and discomfort are relieved almost at once.

NAIL POLISHES:
See Cosmetics.

NAPOLEON CORDIAL:
See Wines and Liquors.

NAPHTHOL SOAP:
See Soap.

NEATSFOOT OIL.

Crude neatsfoot oil	5,000 parts
Alcohol, 90 per cent	2,500 parts
Tannin..........	5 parts

Place in a clearing flask, agitate vigorously and allow to stand for 8 days in a warm room with daily repetition of the shaking. Then draw off the spirit of wine on top, rinse again with 1,000 parts of spirit of wine (90 per cent) and place the oil in a temperature of about $53\frac{1}{2}°$ F. Allow to stand in this temperature for at least 6 weeks, protected from the light, and then filter.

NEEDLES, ANTI-RUST PAPER FOR:
See Rust Preventives.

NEGATIVES, HOW TO USE SPOILED:
See Photography.

NERVE PASTE:
See also Dental Cements, under Cements.

Arsenious acid.......	4 parts
Morphine sulphate...	2 parts
Clove oil............	1 part
Creosote, quantity sufficient to make a paste.	

After the nerve is destroyed the following paste is to be put in the cavity:

Alum..............	1 part
Thymol............	1 part
Zinc oxide..........	1 part
Glycerine..........	1 part

NERVINE OINTMENT:
See Ointments.

NESSELRODE PUDDING:
See Ice Creams.

NETS:
See Cordage.

NICKEL-TESTING.

Pure nickel will remain nearly white, while "patent nickel," or nickel-copper will not retain its primitive brilliancy, but soon becomes slightly oxidized and grayish in color. The magnet furnishes a good means of testing. The unadulterated nickel is distinctly sensitive to magnetism, while that much alloyed is destitute of this property.

NICKEL ALLOYS:
See Alloys.

NICKEL, TO REMOVE RUST FROM:
See Cleaning Preparations and Methods.

NICKEL-PLATING:
See Plating.

NICKEL STEEL:
See Steel.

NICKELING, TEST FOR:
See Plating.

NIELLO:
See Steel.

NITROGLYCERINE:
See Explosives.

NOYAUX LIQUEUR:
See Wines and Liquors.

NUT CANDY STICKS:
See Confectionery.

NUTMEG CORDIAL:
See Wines and Liquors.

NUTMEG ESSENCE:
See Essences and Extracts.

OAK:
See Wood.

ODONTER:
See Dentifrices.

Oils

Clock Oil.—Put 2,000 parts, by weight, of virgin oil in a decanting vessel, add a solution of 40 parts of ether tannin in 400 parts of water and shake until completely emulsified. Let stand for 8 days, with frequent shaking; next, add 100 parts of talcum and, when this has also been well shaken, 1,600 parts of water. Allow to settle for 24 hours, and then run off the lower water layer, repeating the washing as long as the wash water still shows a coloration with ferric chloride. Pour the contents of the decanting vessel into an evaporating dish; then add 200 parts of thoroughly dried and finely ground cooking salt; let stand for 24 hours and filter through paper. The clock oil is now ready, and should be filled in brown glass bottles, holding 20 to 25 parts (about 1 ounce), which must be corked up well and kept at a cool temperature.

COD-LIVER OIL:

Aromatic Cod-Liver Oil.—

Coumarin	0.01	parts
Saccharine	0.50	parts
Vanillin	0.10	parts
Alcohol, absolute.	5.40	parts
Oil of lemon	5.00	parts
Oil of peppermint.	1.00	part
Oil of neroli	1.00	part
Cod-liver oil to make	1,000	parts

Deodorized Cod-Liver Oil.—Mix 400 parts of cod-liver oil with 20 parts of ground coffee and 10 parts of bone black, warm the mixture in an open vessel to 140° F., let it stand 5 days, shaking occa-

sionally, and strain through linen. The oil acquires the taste of coffee.

Cod-Liver Oil Emulsions.—

I.—

Calcium hypophosphite	80 grains
Sodium hypophosphite	120 grains
Sodium chloride	60 grains
Gum acacia, in powder	2 ounces
Elixir of glucoside	20 minims
Essential oil of almonds	15 minims
Glycerine	2 fluidounces
Cod-liver oil	8 fluidounces

Distilled water, a sufficient quantity to produce 16 fluidounces.

II.—Mix 190 parts of powdered sugar with 5 parts of acacia and 500 parts of tragacanth in a mortar. Mix in a large bottle and shake thoroughly together 500 parts of cod-liver oil and 200 parts of a cold infusion of coffee. Gradually add a part of this mixture to the powder in the mortar and triturate until emulsified. To the remaining liquid mixture add 100 parts of rum, then gradually incorporate with the contents of the mortar by trituration.

Extracting Oil from Cottonseed.—Claim is made for a process of extraction, in an English patent, in which the seeds are placed in a rotable vessel mounted on a hollow shaft divided into compartments by means of a partition. The solvent is introduced at one end of this shaft and passes into the vessel, which is then made to rotate. After the extraction the bulk of the solvent and the extracted oil pass away through an exit pipe, and steam is then introduced through the same opening as the solvent, in order to cook the seeds and expel the residual solvent. The steam and the vapors pass through perforations in a scraper fixed to the shaft and thence through connected pipes into the other compartment of the shaft, the end of which is attached to a condenser.

Silver Nitrate Test for Cottonseed Oil.—Investigations of Charabout and March throw some light on the value of this test in presence of olive oil. The free-fat acids obtained from cottonseed oil by saponification were treated in accordance with the method of Milliau on a water bath with a 3 per cent solution of silver nitrate, and the brown precipitate thus formed subjected to a chemical examination. It was found to consist chiefly of a brown silver salt composed of a fat acid melting at 52° F., and congealing at 120° to 122° F., and of sulphide of silver. Olive oil, which contains a sulphur compound of an analogous composition, is also capable of forming a more or less distinct precipitate of a dark colored silver sulphide with nitrate of silver. It is important to bear this fact in mind when examining olive oil for cottonseed oil.

Floral Hair Oil.—

White vaseline.....	5,000 parts
Floricin, pure......	800 parts
Linalool rosé......	60 parts
Terpineol.........	50 parts
Aubepine (haw-thorne), liquid...	12 parts

Floral Hair Pomade.—

White ceresine.....	250 parts
Floricin, pure......	1,600 parts
Vanillin..........	3 parts
Geranium oil......	5 parts
Isoeugenol........	4 parts

Floricin Brilliantine.—

Floricin oil........	2,100 parts
White ceresine.....	250 parts
Ylang-ylang oil....	2 parts
Kananga oil.......	5 parts
Oil of rose, artificial	1 part
Cheirantia........	5 parts

Solid Linseed Oil.—Cements for the manufacture of linoleum and other similar substances are composed to a large extent of linseed oil, oxidized or polymerized until it has become solid. The old process of preparing this solid oil is tedious, costly, and invites danger from fire. It consists in running linseed oil over sheets of thin cloth hung from the top of a high building. The thin layer of oil upon the cloth dries, and then a second layer is obtained in the same way. This is continued until a thick skin of solid oil is formed on either side of the cloth. A new method of solidifying linseed oil is by means of alkalies. The drying oils, when heated with basic substances such as the alkalies, polymerize and become solid. Hertkorn makes use of the oxides of the alkaline earths, or their salts with weak acids, such as their soaps. When chalk or lime is added to the oil during the process of oxidation, either during the liquid or the plastic stage, it forms a calcium soap, and causes polymerization to set in in the partially oxidized oil. Similarly, if caustic soda or caustic potash be added, the action is not caused by them in the free state, but by the soaps which they form. Oxidized oil is more readily saponified than raw oil, and the greater the oxidation, the more readily does saponification take place. Lime soaps are not soluble in water, whereas soda and potash soaps are. Consequently a cement made with the latter, if exposed to the weather, will be acted upon by rain and moisture, owing to the soluble soap contained in it, while a cement made with lime will not be acted upon. It is suggested that the action of the bases on linseed oil is simply due to their neutralization of the free acid. The acidity of linseed oil increases as it becomes oxidized. When the basic matter is added part of the free acid is neutralized, and polymerization sets in. The presence of a large amount of free acid must therefore hinder polymerization. From 5 to 10 per cent of chalk or lime is considered to be the amount which gives the best result in practice.

Decolorizing or Bleaching Linseed Oil.—Linseed oil may be bleached by the aid of chemical bodies, the process of oxidizing or bleaching being best performed by means of peroxide of hydrogen. For this purpose, the linseed oil to be bleached is mixed with 5 per cent peroxide of hydrogen in a tin or glass bottle, and the mixture is shaken repeatedly. After a few days have elapsed the linseed oil is entirely bleached and clarified, so that it can be poured off from the peroxide of hydrogen, which has been reduced to oxide of hydrogen, i. e., water, by the process of oxidation. The use of another oxidizing medium, such as chloride of lime and hydrochloric acid or bichromate of calcium and sulphuric acid, etc., cannot be recommended to the layman, as the operation requires more care and is not without danger. If there is no hurry about the preparation of bleached linseed oil, sun bleaching seems to be the most recommendable method. For this only a glass bottle is required, or, better still, a flat glass dish, of any shape, which can be covered with a protruding piece of glass. For the admission of air, lay some sticks of wood over the dish and the glass on top. The thinner the layer of linseed oil, the quicker will be the oxidation process. It is, of course, necessary to place the vessel in such a manner that it is exposed to the rays of the sun for many hours daily.

Linseed Oil for Varnish-Making.—Heat in a copper vessel 50 gallons Baltic oil to 280° F., add 2¼ pounds calcined white vitriol, and stir well together. Keep the oil at the above temperature for half an hour, then draw the fire, and in 24 hours decant the clear oil. It should stand for at least 4 weeks.

Refining Linseed Oil.—Put 236 gallons of oil into a copper boiler, pour in 6 pounds of oil of vitriol, and stir them together for 3 hours, then add 6 pounds fuller's earth well mixed with 14 pounds hot lime, and stir for 3 hours. The oil must be put in a copper vessel with an equal quantity of water. Now boil for 3 hours, then extinguish the fire. When cold draw off the water. Let the mixture settle for a few weeks.

MINERAL OIL:

See also Petroleum.

Production of Consistent Mineral Oils.—

	By weight
I.—Mineral oil	100 parts
Linseed oil	25 parts
Ground nut oil	25 parts
Lime	10 parts
II.—Mineral oil	100 parts
Rosin oil	100 parts
Rape seed oil	50 parts
Linseed oil	75 parts
Lime	25 parts

Mixing Castor Oil with Mineral Oils.—Castor oil is heated for 6 hours in an autoclave at a temperature of 500° to 575° F., and under a pressure of 4 to 6 atmospheres. When cold the resulting product mixes in all proportions with mineral oils.

BLEACHING OILS:

Linseed Oil or Poppy Oil.—Agitate in a glass balloon 25,000 parts, by weight, of oil with a solution of 50 parts, by weight, potassium permanganate in 1,250 parts, by volume, of water. Let stand for 24 hours at a gentle warmth and add 75 parts, by weight, of powdered sodium sulphite. Agitate strongly and add 100 parts, by weight, of hydrochloric acid and again agitate. Let stand until decolorization takes place, then wash the oil with a sufficiency of water, carrying in suspension chalk, finely powdered, until the liquid no longer has an acid reaction. Finally filter off over anhydrous sodium sulphate.

Boiled Oil.—The following is especially adapted for zinc painting, but will also answer for any paint: Mix 1 part binoxide of manganese, in coarse powder, but not dusty, with 10 parts nut or linseed oil. Keep it gently heated and frequently stirred for about 30 hours, or until the oil begins to turn reddish.

British Oil.—

I.—Oil of turpentine	40 parts
Barbadoes pitch	26 parts
Oil of rosemary	1 part
Oil of origanum	1 part
II.—Oil of turpentine	2 parts
Rape oil	20 parts
Spirit of tar	2 parts
Alkanet root, quantity sufficient.	

Macerate the alkanet root in the rape oil until the latter is colored deep red; then strain off and add the other ingredients.

Decolorizing and Deodorizing Oils.—I.—One may partially or completely deodorize and decolorize rank fish and other oils by sending a current of hot air or of steam through them, after having heated them from 175° to 200° F. To decolorize palm oil pass through it a current of steam under pressure corresponding to a temperature of 230° F., agitating the oil constantly. The vapor is then passed through leaden tuyeres of about 2 inches diameter, 10 hours being sufficient for deodorizing 4 tons of oil.

II.—Another method that may be applied to almost all kinds of fats and oils with excellent results is the following: Melt say 112 parts, by weight, of palm oil in a boiler. When the mass is entirely liquefied add to it a solution of calcium chloride, made by dissolving 7 parts, by weight, of lime chloride for every 84 parts, by weight, of oil in water, and mix intimately. After cooling, the mass hardens and is cut into small bits and exposed to the air for a few weeks. After this exposure the material is reassembled in a boiler of iron, jacketed on the inside with lead; a quantity of sulphuric acid diluted to 5 per cent, equal in amount to the lime chloride previously used, is added, and heat is applied until the oil melts and separates from the other substances. It is then left to cool off and solidify.

Decomposition of Oils, Fats, etc.—In many of the processes at present in use, whereby oils and fats are decomposed by steam at a high pressure, the time during which the oil or fat has to be exposed to high pressure and temperature has the effect of considerably darkening the resulting product. Hannig's process claims to shorten the time required, by bringing the steam and oil into more intimate contact. The oil to be treated is projected in fine streams into the chamber containing steam at 8 to 10 atmospheres pressure. The streams of oil are projected with sufficient force to cause them to strike against the walls of the chamber, and they are thus broken up into minute globules which mix intimately with the steam. In this way the most satisfactory conditions for the decomposition of the oil are obtained.

Driffield Oils.—

Barbadoes tar	1	ounce
Linseed oil	16	ounces
Oil turpentine	3	ounces
Oil vitriol	$\frac{1}{2}$	ounce

Add the oil of vitriol to the other ingredients very gradually, with constant stirring.

Drying Oils.—To dry oils for varnishes, paintings, etc., the most economical means is to boil them with shot, to leave them for some time in contact with shot, or else to boil them with litharge. Another method consists in boiling the oils with equal parts of lead, tin, and sulphate of zinc in the ratio of $\frac{1}{10}$ part (weight) of the united metals to 1 part of oil to be treated. These metals must be granulated, which is easily accomplished by melting them separately and putting them in cold water. They will be found at the bottom of the water in the shape of small balls. It is in this manner, by the way, that shot is produced.

Dust-Laying Oil.—A process has been patented for rendering mineral oils miscible in all proportions of water. The method consists of forming an intimate mixture of the oil with a soap which is soluble in water. The most simple method is as follows: The oil is placed in a tank provided with an agitator. The latter is set in motion and the fatty oil or free fatty acid from which the soap is to be formed is added, and mixed intimately with the mineral oil. When the mixture is seen to be thoroughly homogeneous, the alkali, in solution in water, is added little by little and the stirring continued until a thorough emulsion is obtained, of which the constituents do not separate, even after prolonged standing at ordinary temperatures. The agitation may be produced either by a mechanical apparatus or by forcing air in under pressure. As a rule, the operation can be carried out in the cold, but in certain cases the solution of the fatty body and its saponification requires the application of moderate heat. This may be obtained by using either a steam-jacketed pan, or by having the steam coil within the pan, or live steam may be blown through the mixture, serving at the same time both as a heating and stirring agent. Any fatty matter or fatty acid suitable for soap-making may be used, and the base may be any one capable of forming a soluble soap, most commonly the alkaline hydroxides, caustic soda, and caustic potash, as also ammonia. The raw materials are chosen according to the use to which the finished product is to be applied. A good formula, suitable for preparing an oily liquid for watering dusty roads, is as follows:

	By weight	
Heavy mineral oil	75	parts
Commercial olein	2	parts
Commercial ammonia	1.5	parts
Water	21.5	parts

Floor Oils.—

I.—Neatsfoot oil	1	part
Cottonseed oil	1	part
Petroleum oil	1	part

II.—Beeswax	8	parts
Water	56	parts
Potassium carbonate	4	parts

Dissolve the potash in 12 parts of water; heat together the wax and the remaining water till the wax is liquefied; then mix the two and boil together until a perfect emulsion is effected. Color, if desired, with a solution of annatto.

Ground-Laying Oil for Ceramics.—Boil together until thoroughly incorporated 1 pint of linseed oil, 1 pint of dissolved gum mastic, $\frac{1}{2}$ ounce of red lead, $\frac{1}{2}$ ounce of rosin. In using mix with Venice turpentine.

Oil Suitable for Use with Gold.—Heat and incorporate linseed oil, 1 quart; rape oil, 1 pint; Canadian balsam, 3 pints; rectified spirits of tar, 1 quart.

Wool Oil.—These are usually produced by the distillation in retorts of Yorkshire grease and other greases. The distilled oil is tested for quality, and is brought down to 70 per cent or 50 per cent grades by the addition of a suitable quantity of mineral oil. The lower the quality of the grease used the lower is the grade of the resulting wool oil.

OIL, CASTOR:
See Castor Oil.

OIL FOR FORMING A BEAD ON LIQUORS:
See Wines and Liquors.

OILS FOR HARNESS:
See Leather.

OILS (EDIBLE), TESTS FOR:
See Foods.

OIL, HOW TO POUR OUT:
See Castor Oil.

OIL, LUBRICATING:
See Lubricants.

OILS, PURIFICATION OF:
See Fats.

OILCLOTH:
See Linoleum.

OILCLOTH ADHESIVES:
See Adhesives.

OILCLOTH VARNISHES:
See Varnishes.

OILING FIBERS AND FABRICS:
See Waterproofing.

OILSKINS:
See Waterproofing.

OIL REMOVERS:
See Cleaning Preparations and Methods.

OIL, SOLIDIFIED:
See Lubricants.

Ointments

Arnica Salve.—

Solid extract of arnica	2 parts
Rosin ointment	16 parts
Petrolatum	4 parts
Sultanas	16 parts
Fine cut tobacco	1 part

Boil the raisins and the tobacco in 40 ounces of water until exhausted, express the liquid, and evaporate down to 8 ounces. Soften the arnica extract in a little hot water and mix in the liquid. Melt the rosin ointment and petrolatum together, and add the liquid to the melted mass and incorporate thoroughly.

Barbers' Itch.—

Ichthyol	30 grains
Salicylic acid	12 grains
Mercury oleate (10 per cent)	3 drachms
Lanolin	1 ounce

Mix. To be kept constantly applied to the affected parts.

Brown Ointment.—

Rosin	1 ounce
Lead plaster	4 ounces
Soap cerate	8 ounces
Yellow beeswax	1 ounce
Olive oil	7½ fluidounces

Chilblains.—The following are for unbroken chilblains:

I.—		
	Sulphurous acid	3 parts
	Glycerine	1 part
	Water	1 part

II.—		
	Balsam Peru	1 part
	Alcohol	24 parts
	Hydrochloric acid	1 part
	Tincture benzoin compound	8 parts

Dissolve the balsam in the alcohol, and add the acid and tincture. Apply morning and evening.

Domestic Ointments.—

I.—		
	Vaseline	80 parts
	Diachylon ointment	30 parts
	Carbolic acid	4 parts
	Camphor	5 parts

II.—		
	Butter, fresh (unsalted)	750 parts
	Wax, yellow	125 parts
	Rosin, white	100 parts
	Nutmeg oil	15 parts
	Peru balsam	1 part

III.—		
	Lead plaster, simple	6,090 parts
	Vaseline, yellow	1,000 parts
	Camphor	65 parts
	Carbolic acid	50 parts

Mix.

Green Salve.—

White pine turpentine	8 ounces
Lard, fresh	8 ounces
Honey	4 ounces
Beeswax, yellow	4 ounces

Melt, stir well, and add

Verdigris, powdered	4 drachms

Apply locally.
This cannot be surpassed when used for deep wounds, as it prevents the formation of proud flesh and keeps up a healthy discharge.

Salve for all Wounds.—

Lard, fresh	16 ounces
White lead, dry	3 ounces
Red lead, dry	1 ounce
Beeswax, yellow	3 ounces
Black rosin	2 ounces

Mix, melt, and boil for 45 minutes, then add

Common turpentine	4 ounces

Boil for 3 minutes and cool.
Apply locally to cuts, burns, sores, ulcers, etc. It first draws, then heals.

Irritating Plaster.—

Tar, purified	16 ounces
Burgundy pitch	1 ounce
White pine turpentine	1 ounce
Rosin, common	2 ounces

Melt and add

Mandrake root, powdered	1 drachm
Bloodroot, powdered	1 ounce
Poke root, powdered	1 ounce
Indian turnip root, powdered	1 ounce

Apply to the skin in the form of a

plaster (spread on muslin) and renew it daily.

This salve will raise a sore which is to be wiped with a dry cloth to remove matter, etc. The sore must not be wetted. This is a powerful counter-irritant for removing internal pains, and in other cases where an irritating plaster is necessary.

Mercury Salves.—I.—Red Salve.—Red mercury oxide, 1 part; melted lard, 9 parts.

II.—White Salve.—Mercury precipitate, 1 part; melted lard, 9 parts.

Pink salve.

Ammoniated mercury	1 ounce
Mercuric oxide, precipitated	2½ ounces
Red mercuric sulphide (vermilion)	60 grains
Perfume	½ fluidounce
Lard	1½ pounds
Prepared suet	½ pound

Antiseptic Nervine Ointment.—

Iodoform	2 parts
Salol	4 parts
Boric acid	5 parts
Antipyrine	5 parts
Vaseline	80 parts

Photographers' Ointment.—The following protects the hands from photographic chemicals:

Best castile soap, in fine shavings	1 ounce
Water	1 ounce
Wax	1 ounce
Ammonia	45 minims
Lanolin	1 ounce

The soap is dissolved in the water heated for that purpose, the wax mixed in with much stirring, and, when all is in solution, the ammonia is added. When clear, the lanolin is put in, and then, if the mixture is very thick, water is added until the whole has the consistency of honey. Keep in a covered stoneware jar. The hands should be first washed with ordinary soap, and then, while the lather is still on them, a bit of the mixture about the size of a hazel nut is rubbed in until all is absorbed, and the hands are dry. At the close of the work, the film of wax is washed off in warm water and a little lanolin rubbed into the hands.

Pain-Subduing Ointment.—The following is an excellent formula:

Tincture of capsicum	5 parts
Tincture of camphor	1 part
Ammonia water	2 parts
Alcohol	2 parts
Soap liniment	2 parts

Skin Ointment.—I.—Add about 2 per cent of phenol to petrolatum, perfuming it with oil of bergamot and color a dull green. It has been suggested that a mixture of Prussian blue and yellow ocher would answer as the coloring agent.

II.—Phenol	40 grains
Boric acid	2 drachms
Oil of bergamot	90 minims
Petrolatum	1 pound
Color with chlorophyll.	

OINTMENTS FOR VETERINARY PURPOSES:
See Veterinary Formulas.

OLEIN SOAP:
See Soap.

OLEOMARGARINE:
See Butter.

OLIVE-OIL PASTE:
See Butter Substitutes.

ONYX CEMENTS:
See Adhesives.

ORANGEADE:
See Beverages, under Lemonades.

ORANGE BITTERS AND CORDIAL:
See Wines and Liquors.

ORANGE DROPS:
See Confectionery.

ORANGE EXTRACT:
See Essences and Extracts.

ORANGE FRAPPÉ:
See Beverages, under Lemonades.

ORANGE PHOSPHATE:
See Beverages.

ORGEAT PUNCH:
See Beverages, under Lemonades.

ORTOL DEVELOPER:
See Photography.

OXIDIZING:
See Bronzing, Plating, Painting.

OXIDE, MAGNETIC:
See Rust Preventives.

OXOLIN:
See Rubber.

OZONATINE:
See Air Purifying.

PACKAGE POP:
See Beverages, under Ginger Ale.

PACKAGE WAX:
See Waxes.

PACKINGS:

Packing for Stuffing Boxes.—

Tallow	10 parts
Barrel soap, non-filled	30 parts
Cylinder oil	10 parts
Talcum Venetian, finely powdered	20 parts
Graphite, finely washed	6 parts
Powdered asbestos	6 parts

Melt the tallow and barrel soap together, add the other materials in rotation, mix intimately in a mixing machine, and fill in 4-pound cans.

Packing for Gasoline Pumps.—For packing pumps on gasoline engines use asbestos wick-packing rubbed full of regular laundry soap; it will work without undue friction and will pack tightly. Common rubber packing is not as good, as the gasoline cuts it out.

PADS OF PAPER:
See Paper Pads.

PAIN-SUBDUING OINTMENT:
See Ointments.

PAINTING PROCESSES:

Painting Ornaments or Letters on Cloth and Paper.—Dissolve gum shellac in 95 per cent alcohol at the rate of 1 pound of shellac to 3 pints of alcohol, and mix with it any dry color desired. If it becomes too thick, thin with more alcohol. This works free, does not bleed out, imparts brilliancy to the color, and wears well. The preparation can be used also on paper.

Painting on Marble.—To paint marble in water colors, it must be first thoroughly cleaned and all grease completely removed. The slab is washed well, and then rubbed off with benzine by means of a rag or sponge. In order to be quite sure, add a little ox gall or aguoline to the colors. After marble has been painted with water colors it cannot be polished any more.

Painting on Muslin.—To paint on muslin requires considerable skill. Select a smooth wall or partition, upon which tack the muslin, drawing the fabric taut and firm. Then make a solution of starch and water, adding one-fourth starch to three-fourths water, and apply a glaze of this to the muslin. To guard against the striking in of the paint, and to hold it more securely in place and texture, mix the pigment with rubbing varnish to the consistency of a stiff paste, and then thin with turpentine to a free working condition. A double thick camel's-hair brush, of a width to correspond properly with the size of the surface to be coated, is the best tool with which to coat fine muslin. A fitch-hair tool is probably best suited to the coarser muslin. Many painters, when about to letter on muslin, wet the material with water; but this method is not so reliable as sizing with starch and water. Wetting canvas or duck operates very successfully in holding the paint or color in check, but these materials should not be confounded with muslin, which is of an entirely different texture.

PAINTING ON LEATHER:
See Leather.

PAINTINGS:

Protection for Oil Paintings.—Oil paintings should under no circumstances be varnished over before the colors are surely and unmistakably dry, otherwise the fissuring and early decay of the surface may be anticipated. The contention of some people that oil paintings need the protection of a coat of varnish is based upon the claim that the picture, unvarnished, looks dead and lusterless in parts and glossy in still others, the value and real beauty of the color being thus unequally manifested. It is not to be inferred, however, that a heavy coating of varnish is required. When it is deemed advisable to varnish over an oil painting the varnish should be mastic, with perhaps 3 or 4 drops of refined linseed oil added to insure against cracking. A heavy body of varnish used over paintings must be strictly prohibited, inasmuch as the varnish, as it grows in age, naturally darkens in color, and in so doing carries with it a decided clouding and discoloration of the delicate pigments. A thinly applied coat of mastic varnish affords the required protection from all sorts and conditions of atmospheric impurities, besides fulfilling its mission in other directions.

Oil paintings, aquarelles, etc., may be also coated with a thin layer of Canada balsam, and placed smoothly on a pane of glass likewise coated with Canada balsam, so that both layers of balsam come together. Then the pictures are pressed down from the back, to remove all air bubbles.

To Renovate Old Oil Paintings.—When old oil paintings have become dark and cracked, proceed as follows: Pour alcohol in a dish and put the picture over it, face downward. The fumes of the alcohol dissolve the paint of the picture, the fissures close up again, and the color assumes a freshness which is surprising. Great caution is absolutely necessary, and one must look at the painting very often, otherwise it may happen that the colors will run together or even run off in drops.

PAINTINGS, TO CLEAN:

See Cleaning Preparations and Methods.

Paints

(See also Acid-Proofing, Ceramics, Enamels, Fireproofing, Glazing, Painting Processes, Pigments, Rust Preventives, Varnishes, and Waterproofing.)

PAINT BASES:

Dry Bases for Paints.—The following colors and minerals, mixed in the proportions given and then ground to fine powder, make excellent dry paints, and may be thinned with turpentine oil, and a small percentage of cheap varnish to consistency required.

Buff.—

Yellow ocher	44	pounds
Whiting	6	pounds
Oxide of zinc	5	pounds
Plaster of Paris	½	pound

Brick Brown.—

Yellow ocher	26	pounds
Calcined copperas	4	pounds
Red hematite	1¼	pounds
Best silica	7	pounds
Whiting	18	pounds

Gray.—

Oxide of zinc	30	pounds
White lead	6	pounds
Whiting	12	pounds
Bone black	¼	pound
Yellow ocher	2	pounds

Crimson.—

Indian red	25	pounds
Crocus martis	7	pounds
Oxide of zinc	6	pounds
Whiting	6	pounds

Vandyke Brown.—

Yellow ocher	25	pounds
Whiting	18	pounds
Umber	4	pounds
Oxide of zinc	7	pounds
Purple oxide of iron	1	pound

Blood Red.—

Crocus martis	30	pounds
Whiting	20	pounds
Hematite	3	pounds
Silica	6	pounds
Venetian red	2	pounds

Drab.—

Yellow ocher	40	pounds
Whiting	10	pounds
Oxide of zinc	8½	pounds
Sulphate of barytes	1	pound

Paint for Blackboards.—

Shellac	1	pound
Alcohol	1	gallon
Lampblack (fine quality)	4	ounces
Powdered emery	4	ounces
Ultramarine blue	4	ounces

Dissolve the shellac in the alcohol. Place the lampblack, emery, and ultramarine blue on a cheese-cloth strainer, pour on part of the shellac solution, stirring constantly and gradually adding the solution until all of the powders have passed through the strainer.

Dark-Green Paint for Blackboards.— Mix 1 part Prussian blue and 1 part chrome green with equal parts of gilders' size and alcohol to a thin cream consistency. Apply with a large, stiff brush and after an hour a second coat is given. After 24 to 48 hours smooth the surface with a felt cloth. This renders it rich and velvety. The shade must be a deep black green and the quantities of the colors have to be modified accordingly if necessary. Old blackboards should be previously thoroughly cleaned with soda.

BRONZING SOLUTIONS FOR PAINTS.

I.—The so-called "banana solution" (the name being derived from its odor) which is used in applying bronzes of various kinds, is usually a mixture of equal parts of amyl acetate, acetone, and benzine, with just enough pyroxyline dissolved therein to give it body. Powdered bronze is put into a bottle containing this mixture and the paint so formed applied with a brush. The thin covering of pyroxyline that is left after the evaporation of the liquid protects the bronze from the air and keeps it from being wiped off by the cleanly housemaid. Tarnished picture frames and tarnished chandeliers to which a gold bronze has been applied from such a solution will look fresh and new for a long time. Copper bronze as well as gold bronze and the various colored bronze powders can be used in the "banana solution" for making very pretty advertising signs for use in the drug store. Lettering and bordering work upon the signs can be done with it. Several very small, stiff painters' brushes are needed for such work and they must be either kept in the solution when not in use, or, better still, washed in benzine or acetone immediately after use and put away for future service. As the "banana solution" is volatile, it must be kept well corked.

II.—A good bronzing solution for paint tins, applied by dipping, is made by dissolving Syrian asphaltum in spirits of turpentine, etc., and thinning it down with these solvents to the proper bronze color and consistency. A little good boiled oil will increase the adherence.

Paint Brushes.—To soften a hard paint brush, stand the brush overnight in a pot of soft soap and clean in warm water. Afterwards clean in benzine. If the brush is wrapped with a string do not let the string touch the soap.

Paint brushes which have dried up as hard as stone can be cleaned in the following manner: Dissolve 1 part soda in 3 parts water; pour the solution in a cylinder glass, and suspend in it the brushes to be cleaned, so that they are about 2 inches from the bottom of the vessel. Let it remain undisturbed at a temperature of 140° to 158° F., 12 to 24 hours, after which the most indurated brushes will have become soft, so that they can be readily cleaned with soap. It is essential, however, to observe the temperature, as bristle brushes will be injured and spoiled if the heat is greater.

Black.—A Permanent Black of Rich Luster for Metal Boxes. — Dissolve chlorate of potassium and blue vitriol, equal parts, in 36 times as much water, and allow the solution to cool. The parts to be blacked may be either dipped in the solution, or the solution may be flowed on and allowed to remain until the metal becomes black, after which the fixtures should be rinsed in clean water and allowed to dry. Those parts of the surface which show imperfections in the black should be recoated.

Dead White on Silver Work, etc.— Bruise charcoal very finely and mix it with calcined borax in the proportion of 4 parts of charcoal to 1 of borax. Of this make a paste with water; apply this paste on the parts to be deadened; next expose the piece to the fire of well-lit coal until it acquires a cherry-red shade; allow to cool and then place it in water slightly acidulated with sulphuric acid. The bath must not be more than 5° Bé. Leave the piece in the bath about 2 hours, then rinse off several times.

White Coating for Signs, etc.—A white color for signs and articles exposed to the air is prepared as follows for the last coat: Thin so-called Dutch "stand" oil with oil of turpentine to working consistency, and grind in it equal parts of zinc white and white lead, not adding much siccative, as the white lead assists the drying considerably. If the paint is smoothed well with a badger brush, a very durable white color of great gloss is obtained. Linseed oil, or varnish which has thickened like "stand" oil by long open storing, will answer equally well.

To Prevent Crawling of Paints.— Probably the best method to pursue will be to take an ordinary flannel rag and carefully rub it over the work previous to varnishing, striping, or painting. This simple operation will obviate the possibility of crawling.

In some instances, however, crawling may be traced to a defective varnish. The latter, after drying evenly on a well-prepared paint surface will at times crawl, leaving small pitmarks. For this, the simple remedy consists in purchasing varnish from a reputable manufacturer.

FIREPROOF PAINTS:

See also Fireproofing.

Fireproofing paints of effective quality are prepared in different ways. Naturally no oily or greasy substances enter into their composition, the blending agent being simply water.

I.—One of the standing paints consists of 40 pounds of powdered asbestos, 10 pounds of aluminate of soda, 10 pounds of lime, and 30 pounds of silicate of soda, with the addition of any non-rosinous coloring matter desired. The whole is thoroughly mixed with enough water to produce a perfect blend and render an easy application. Two or more coats of this is the rule in applying it to any wood surface, inside or outside of building.

II.—Another formula involves the use of 40 pounds of finely ground glass, a like amount of ground porcelain, and similarly of China clay or the same quantity of powdered asbestos, and 20 pounds of quicklime. These materials are ground very fine and then mixed in 60 pounds of liquid silicate of soda with water, as in the preceding formula. Two or more coats, if necessary, are given.

Each of these paints is applied with a brush in the ordinary way, the drying being accomplished in a few hours, and, if coloring matter is desired, the above proportions are varied accordingly.

III.—A surface coated with 3 coats of water glass, these 3 coats being subsequently coated with water glass containing enough whiting or ground chalk to make it a trifle thicker than ordinary paint, is practically non-inflammable, only yielding to fierce consuming flames after a somewhat protracted exposure.

IV.—Zinc white, 70 pounds; air-slaked lime, 39 pounds; white lead, 50 pounds; sulphate of zinc, 10 pounds; silicate of soda, 7 gallons. The zinc white and lime are mixed together, then ground in elastic oil, after which the silicate of soda is added, this addition being followed by the white lead and sulphate of zinc. This white paint can be colored to meet any desired shade and it may be classed as a good working paint and probably fireproof to the same extent that most of the pretentiously sounded pigments on the markets are.

Fireproof and Waterproof Paints.—The following recipes are claimed to resist both fire and water: A preparation for protecting wood against the action of fire and of moisture, and also for producing on the surface of wood and metal a coat, insulating with reference to electricity and preservative from corrosion, has been introduced in France by Louis Bethisy and Myrthil Rose. The bases or fundamental raw materials quite distinct from those hitherto employed for the same purpose, are 100 parts, by weight, of nitro-cellulose and 30 parts, by weight, of chloride of lime, dissolved in 50 per cent alcohol.

Preparation of the Bases.—The cellulose (of wood, paper, cotton, linen, ramie, or hemp) is put in contact with two-thirds part of sulphuric acid of 66° Bé. and one-third part of nitric acid of 42° Bé. for some 20 or 30 minutes, washed with plenty of water, and kept for 24 hours in a tank of water supplied with an energetic current.

The nitro-cellulose thus obtained is bleached for this purpose; a double hypochlorite of aluminum and magnesium is employed. This is obtained by grinding together 100 parts of chloride of lime, 60 parts of aluminum sulphate, 23 parts of magnesium sulphate, with 200 parts of water.

When the nitro-cellulose is bleached and rewashed, it is reduced to powder and dried as thoroughly as possible. It is then placed in a vat hermetically closed and put in contact with the indicated proportion of calcium chloride dissolved in alcohol. This solution of calcium chloride should be prepared at least 24 hours in advance and filtered.

Composition of the Coating.—This has the following constituents: Bases (nitro-cellulose and solution of calcium chloride), 1 part; amyl acetate (solvent of the bases), 5 parts, by weight; sulphuric ether of 65°, 1.650 parts, by weight; alcohol, 0.850 parts, by weight; one of these powders, alum, talc, asbestos, or mica, 0.100 parts. Other solvents may be employed instead of amyl acetate; for example, acetone, acetic acid, ether alcohol, or methylic alcohol. The ether alcohol furnishes a product drying very quickly. If a very pliant coating is desired, the amyl acetate is employed preferably, with addition of vaseline oil, 0.20 parts, and lavender oil, 0.010 parts.

Method of Operating.—The sulphuric acid is mixed with the alcohol, and left for an hour in contact, shaking from time to time. Afterwards the amyl acetate is added, and left in contact for another hour under similar agitation. In case of the employment of vaseline oil and lavender oil, these two are mingled in ether alcohol. The base is introduced and left in contact for 24 hours, with frequent agitation. The fluidity of the product is augmented by increasing the quantity of the solvent.

Properties.—Wood covered with this coating is fireproof, non-hygrometric, and refractory to the electric current. It also resists the action of acids and alkalies. Metals covered with it are sheltered from oxidation, and effectually insulated on their surface from the electric current. The coating is liquid in form, and applied like collodions, either by the brush or by immersion or other suitable method.

Paint Deadening.—In order to obtain an even dullness of large walls, proceed as follows: After all the dirt has been carefully swept off, oil with 2 parts linseed oil and 1 part turpentine and rub down the smooth places in the wet oil with pumice stone. When the oil coating is dry, mix the ground paint, consisting of whiting, 2 parts; and white lead, 1 part; both finely ground and diluted as above. Do not apply the grounding too thin, because the chalk in itself possesses little covering power. It is not the mission of the chalk, however, to adulterate the material, but to afford a hard foundation for the subsequent coats. For the third coating take white lead, 1 part; and zinc white, 1 part; thin as above and blend with a soft hair pencil. For the final application use only zinc white, ground stiff in oil with any desired mixing color and thinned with turpentine and rain water. Mix the water and the turpentine with the color at the same time, and this coat may be dabbed instead of blended. By the addition of water the paint becomes dull more slowly and is a little more difficult to lay on; but it does not show a trace of gloss after a few days and never turns yellow, even in places less exposed to the air, and besides excels by great permanency.

Another way is to add white wax instead of water to the last coating. This wax paint also gives a handsome dullness but is more difficult of treatment. A nice matt coating is also obtained by addition of Venetian soap, dissolved in water instead of the wax. This is very desirable for church decorations where exceptionally large surfaces are to be deadened.

PAINT DRYERS:

I.—Ordinary barytes...	25	pounds
Whiting...........	4	pounds
Litharge...........	2	pounds
Sulphate of zinc....	2	pounds
Sugar of lead.......	2	pounds
Boiled linseed oil....	5	pounds
Plaster of Paris.....	½	pound

II.—Whiting...........	16	pounds
Barytes...........	16	pounds
White lead.........	3	pounds
Boiled linseed oil...	¾	gallon

PAINTS FOR GOLD AND GILDING:

Gold Paints.—The formulas of the various gold paints on the market are carefully guarded trade secrets. Essentially they consist of a bronze powder mixed with a varnish. The best bronze powder for the purpose is what is known in the trade as "French flake," a deep gold bronze. This bronze, as seen under the microscope, consists of tiny flakes or spangles of the bronze metal. As each minute flake forms a facet for the reflection of color, the paint made with it is much more brilliant than that prepared from finely powdered bronze.

For making gold paint like the so-called "washable gold enamel" that is sold by the manufacturers at the present time, it is necessary to mix a celluloid varnish with the French flake bronze powder. This varnish is made by dissolving transparent celluloid in amyl acetate in the proportion of about 5 per cent of celluloid.

Transparent celluloid, finely shredded.............	1 ounce
Acetone, sufficient quantity.	
Amyl acetate to make 20 ounces.	

Digest the celluloid in the acetone until dissolved and add the amyl acetate. From 1 to 4 ounces of flake bronze is to be mixed with this quantity of varnish. For silver paint or "aluminum enamel," flake aluminum bronze powder should be used in place of the gold. The celluloid varnish incloses the bronze particles in an impervious coating, air-tight and water-tight. As it contains nothing that will act upon the bronze, the latter retains its luster for a long period, until the varnished surface becomes worn or abraded and the bronze thus exposed to atmospheric action.

All of the "gold" or, more properly, gilt furniture that is sold so cheaply by the furniture and department stores is gilded with a paint of this kind, and for that reason such furniture can be offered at a moderate price. The finish is surprisingly durable, and in color and luster is a very close imitation of real gold-leaf work. This paint is also used on picture frames of cheap and medium grades, taking the place of gold leaf or the lacquered silver leaf formerly used on articles of the better grades; it is also substituted for "Dutch metal," or imitation gold leaf, on the cheapest class of work.

A cheaper gold paint is made by using an inexpensive varnish composed of gutta percha, gum dammar, or some other varnish gum, dissolved in benzole, or in a mixture of benzole and benzine. The paints made with a celluloid-amyl-acetate varnish give off a strong banana-like odor when applied, and may be readily recognized by this characteristic.

The impalpably powdered bronzes are called "lining" bronzes. They are chiefly used for striping or lining by carriage painters; in bronzing gas fixtures and metal work; in fresco and other interior decoration, and in printing; the use of a very fine powder in inks or paints admits of the drawing or printing of very delicate lines.

Lining bronze is also used on picture frames or other plastic ornamental work. Mixed with a thin weak glue sizing it is applied over "burnishing clay," and when dry is polished with agate burnishers. The object thus treated, after receiving a finishing coat of a thin transparent varnish, imitates very closely in appearance a piece of finely cast antique bronze. To add still more to this effect the burnishing clay is colored the greenish black that is seen in the deep parts of real antique bronzes, and the bronze powder, mixed with size, is applied only to the most prominent parts or "high lights" of the ornament.

Since the discovery of the celluloid-amyl-acetate varnish, or bronze liquid, and its preservative properties on bronze powders, manufacturers have discontinued the use of liquids containing oils, turpentine, or gums, since their constituents corrode the bronze metal, causing the paint finally to turn black.

Gilding in Size.—The old painters and gilders used to prepare the gold size themselves, but nowadays it is usually bought ready made, barring the white of egg additional. The best and most reliable, and especially suited for fine work, is undoubtedly the red French gold size. It is cleaned, as far as possible, of all impurities, and powdered. For 246 grains take 1 white of egg; put it into a glass, taking care to exclude the yolk entirely—otherwise the burnish will show black spots. Beat the white of egg to a froth with a long, well-cleaned bristle brush; add the froth to the size and grind finely together, which is soon done. When grinding, a little water and red size, if necessary, may be added (use only water for thinning). After being ground, the size is forced through a very fine hair sieve into a perfectly clean vessel, and covered up well, for immediate or subsequent use.

The raw stuff of the red size is bolus, which is dug in France and Armenia in excellent quality. Besides the red size there are yellow, white (pipe clay), blue, and gray (alumina), which are used for certain purposes, to enumerate which here would lead too far.

For burnish gold, always take yellow size for ground work. Dip a finely ground bristle brush in the gold size prepared for use; fill a well-cleaned glass (holding 1 pint) half full of water, and add the size contained in the brush, also about 4 to 5 spoonfuls of pure alcohol. It is advisable not to take too much size; the liquid, when applied, must hardly have a yellow tint. When this is dry soon after, commence applying the size, for which a hair pencil is used. The essentials are to paint evenly and not too thickly, so that the tone remains uniform. Apply three coats of size.

When the size is laid on correctly and has become dry, brush the whole with a special brush, or rub with a flannel rag, so as to obtain the highest possible luster. The size must not stand too long; otherwise no gloss can be developed. After brushing, coat the work with weak glue water and wrap it up in tissue paper if the gilding is not to be done at once. The strictest cleanliness is essential, as the red gold size is very sensitive. The parts where the size has been applied must not be touched with the hand, else grease spots will ensue, which will make a flawless gloss in gilding impossible. The least relaxation of the necessary attention may spoil the whole job, so that everything has to be ground off again.

The necessary tools for the application of gold leaf are: Hair pencils of various sizes, tip, cushion, and gilding knife, as with oil-gilding. Take pure alcohol or grain brandy, and dilute with two-thirds water. When ready to apply the gold leaf, dip a hair pencil of suitable size into the fluid, but do not have it full enough that the alcohol will run on the size ground. Moisten a portion of the ground surface as large as the gold leaf, which is laid on immediately after. Proceed in the same manner, first moistening, then applying the ready-cut gold leaf. The latter must not be pressed on, but merely laid down lightly, one leaf a little over the edge of the previous one, without using up too much gold. Technical practice in gold-leaf gilding is presupposed; through this alone can any skill be acquired, reading being of no avail.

The leaf of gold being applied, all dust must be swept off by means of a light, fine hair pencil (but never against the overlapping edges), and the burnishing is commenced. For this purpose there are special agate tools of the shape of a horn. Flint stone, blood stone, and wolf's teeth are sometimes, but gradually more seldom, employed. Burnish till a full, fine luster appears; but very carefully avoid dents and lines, not to speak of scratches, which would be very hard to mend.

Gold Enamel Paints.—

I.—Pure turps......... 6 pints
    Copal varnish...... 1 pint
    Good gold bronze... 6½ pounds
    Calcis hydrate (dry-
      slaked lime)...... ½ ounce

Mix the varnish and turps at a gentle heat, then slake well with the lime, and settle for a few days, then pour off the clean portion and mix with the powder.

II.—White hard varnish. 1 gallon
    Methylated spirit... ¼ gallon
    Gold bronze....... 12 pounds
    Finely powdered
      mica............ 3 ounces

Mix the varnish and the spirit, reduce the mica to an impalpable powder, mix with the gold, then add to the liquid. Many bronze powders contain a goodly proportion of mica, as it imparts brilliancy. Powdered mother-of-pearl is used also.

GRAINING WITH PAINT:

See also Wood.

Oak Graining.—Prepare a paint of two-thirds of white lead and one-third of golden ocher with the requisite amount of boiled linseed oil and a little drier, and cover the floor twice with this mixture, which possesses great covering power. When the last coating is dry, paint the floor with a thinly liquid paint consisting of varnish and sienna, applying the same in the longitudinal direction of the boards. Treat a strip about 20 inches wide at a time, and draw at once a broad paint brush or, in the absence of such, an ordinary brush or goose feather along the planks through the wet paint, whereupon the floor will acquire a nicely grained appearance. The paint requires several days to dry. A subsequent coating of varnish will cause the graining to stand out still more prominently.

Birch.—Imitations of birch are usefully employed for furniture. The ground should be a light, clean buff, made from white lead, stained with either yellow ocher or raw sienna in oil. In graining, brush over the surface with a thin wash of warm brown, making the panel of 2 or 3 broad color shades. Then take a large mottler and mottle the darker parts into the light, working slantwise, as for maple, but leaving a broad and stiff mark. While this is still wet soften the panel and then slightly mottle across the previous work to break it up. When thoroughly dry, carefully wet the work over with clean water and clean mottler, and put in darker overgrain with a thin oak overgrainer or overgrainer in tubes.

Maple.—Sixty pounds white lead; 1 ounce deep vermilion; 1 ounce lemon chrome.

Ash.—Sixty pounds white lead; 1 ounce deep vermilion; 1 ounce lemon chrome.

Medium Oak.—Sixty pounds white lead; 2 pounds French ocher; 1 ounce burnt umber.

Light Oak.—Sixty pounds white lead; 1 ounce lemon chrome; ½ pound French ocher.

Dark Oak.—Sixty pounds white lead; 10 pounds burnt umber; 1½ pounds medium Venetian red.

Satin Wood.—Sixty pounds white lead; 1 ounce deep vermilion; 1½ pounds lemon chrome.

Pollard Oak.—Seventy-five pounds white lead; 20 pounds French ocher; 3 pounds burnt umber; 2½ pounds medium Venetian red.

Pitch Pine.—Sixty pounds white lead; ¼ pound French ocher; ½ pound medium Venetian red.

Knotted Oak.—Sixty pounds white lead; 9 pounds French ocher; 3½ pounds burnt umber.

Italian Walnut.—Sixty pounds white lead; 6 pounds French ocher; 1½ pounds burnt umber; 1¼ pounds medium Venetian red.

Rosewood.—Nine and one-half pounds burnt umber; 40 pounds medium Venetian red; 10 pounds orange chrome.

Dark Mahogany.—Nine and one-half pounds burnt umber; 40 pounds medium Venetian red; 10 pounds orange chrome.

Light Mahogany.—Sixty pounds white lead; 3 pounds burnt umber; 10 pounds medium Venetian red.

American Walnut.—Thirty pounds white lead; 9 pounds French ocher; 4 pounds burnt umber; 1 pound medium Venetian red.

LUMINOUS PAINTS.

The illuminating power of the phosphorescent masses obtained by heating strontium thiosulphate or barium thiosulphate is considerably increased by the addition, before heating, of small quantities of the nitrates of uranium, bismuth, or thorium. Added to calcium thiosulphate, these nitrates do not heighten the luminosity or phosphorescence. The product from strontium thiosulphate is more luminous than that of the barium compound. Among the best luminous paints are the following:

I.—**Lennord's.**—One hundred parts, by weight, of strontium carbonate; 100 parts, by weight, of sulphur; 0.5 parts, by weight, of potassium chloride; 0.5 parts, by weight, of sodium chloride; 0.4 parts, by weight, of manganese chloride. The materials are heated for three-quarters of an hour to one hour, to about 2,372° F. The product gives a violet light.

II.—**Mourel's.**—One hundred parts, by weight, of strontium carbonate; 30 parts, by weight, of sulphur; 2 parts, by weight, of sodium carbonate; 0.5 parts, by weight, of sodium chloride; 0.2 parts, by weight, of manganese sulphate. The method of treatment is the same as in the first, the phosphorescence deep yellow.

III.—Vanino's.—Sixty parts, by weight, of strontium thiosulphate; 12 parts, by weight, of a 0.5 per cent acidified alcoholic solution of bismuth nitrate; 6 parts, by weight, of a 0.5 per cent alcoholic solution of uranium nitrate. The materials are mixed, dried, brought gradually to a temperature of 2,372° F., and heated for about an hour. The phosphorescence is emerald green.

IV.—Balmain's.—Twenty parts, by weight, of calcium oxide (burnt lime), free from iron; 6 parts, by weight, of sulphur; 2 parts, by weight, of starch; 1 part, by weight, of a 0.5 per cent solution of bismuth nitrate; 0.15 parts, by weight, of potassium chloride; 0.15 parts, by weight, of sodium chloride. The materials are mixed, dried, and heated to 1,300° C. (2,372° F.). The product gives a violet light.

To make these phosphorescent substances effective, they are exposed for a time to direct sunlight; or a mercury lamp may be used. Powerful incandescent gas light also does well, but requires more time.

PAINTS FOR METAL SURFACES:

Blackening Ornaments of Iron.—I.—To give iron ornaments a black-brown to black color, proceed in the following manner: The articles are treated with corrosives, cleaned of all adhering grease, and placed in a 10 per cent solution of potassium bichromate, dried in the air, and finally held over an open, well-glowing, non-sooting fire for 2 minutes. The first coloring is usually black brown, but if this process is repeated several times, a pure black shade is obtained. Special attention has to be paid to removing all grease, otherwise the greasy spots will not be touched by the liquid, and the coloring produced will become irregular. Benzine is employed for that purpose and the articles must not be touched with the fingers afterwards.

II.—This process protects the iron from rust for a long time. The treatment consists in coating the objects very uniformly with a thin layer of linseed-oil varnish, and burning it off over a charcoal fire. During the deflagration the draught must be stopped. The varnish will first go up in smoke with a strong formation of soot, and finally burn up entirely. The process is repeated, i. e., after one coating is burned off a new one is applied. until the parts exhibit a uniformly handsome. deep-black color. Next. wipe off the covering with a dry rag. and heat again, but only moderately. Finally, the articles are taken from the fire and rubbed with a rag well saturated with linseed-oil varnish. The black turns completely dull, and forms a real durable covering for the objects.

Black for Polished Iron Pieces.—Apply successive layers of a very concentrated solution of nitrate of manganese dissolved in alcohol over a gentle fire and the water bath. The surfaces to be blackened should be previously heated. By repeating the layers all the tints between brownish black and bluish black may be obtained.

Glossy Black for Bicycles, etc.—

Amber	8	ounces
Linseed oil	4	ounces
Asphaltum	1½	ounces
Rosin	1½	ounces
Oil turpentine	8	ounces

Heat the linseed oil to boiling point, add the amber, asphaltum, and rosin, and when all melted remove from the fire and gradually add the turpentine.

Japan Black.—The following is a good japan black for metal surfaces: Take 12 ounces of amber and 2 ounces of asphaltum. Fuse by heat, and add ½ pint boiled oil and 2 ounces of rosin. When cooling add 16 ounces of oil of turpentine.

Brass and Bronze Protective Paint.—As a protective covering, especially for brass and bronze objects, a colorless celluloid solution is recommended, such as is found in trade under the name of "Zapon" (q. v.).

Paint for Copper.—Dissolve 1 ounce of alum in 1 quart of warm soft water. When cold add flour to make it about the consistency of cream, then add ½ thimble of rosin and ½ ounce of sugar of lead.

Priming Iron.—The following, if carefully carried out, gives the best satisfaction: The first step consists in thoroughly cleaning the surface of the iron, removing all adhesions in the way of dirt, rust, etc., before the question of priming is considered. As paint in this instance is applied more with a view of protecting the iron from atmospheric influences, rather than for a decorative effect, careful attention should be devoted for securing a base or surface which is calculated to produce a thorough and permanent application. A great deal depends upon the nature of the metal to be painted. Common cast iron, for instance, possessing a rough exterior, with ordinary precautions can be more readily painted with the prospect of a permanent adhesion of the paint, than a planed steel or wrought-iron surface. With the latter it has been demonstrated that a hard and elastic paint is needed, while with regard to cast iron, other paints containing iron oxides are more suitable. For good drying and covering properties, as well as elasticity, a good boiled oil to which has been added an adequate proportion of red lead will be found to form an excellent paint for smooth metal surfaces. The primary object is to protect the surface of the iron from moisture for the purpose of avoiding rust. The priming must therefore be carried out so that it will stick, after which subsequent coats may be added if desired.

It is advisable that articles made of iron should first be coated with linseed-oil varnish. It dries slowly, hardens, and enables the operator afterwards to exercise an effective control over the condition of his material. Iron must be absolutely dry and free from rust when it is to be painted. It is best to apply next a coating of hot linseed oil; when dry this should be followed by a priming of pure red lead in good linseed oil, and the iron should then be painted as desired, using ground oil paints and leaving an interval of a week between each coating. Cementing should be done after the red lead priming, but the last coat must not be given until the whole is thoroughly dry. Bright oil paints and an upper coating with plenty of oil resist the effects of heat better than thin coatings; moreover, rust can be detected in its early stages with the former. Coatings of tar and asphalt (asphalt dissolved in turpentine) are practicable for underground pipes, but are not adapted for pipes exposed to the air, as they are quickly spoiled. Asphalt varnish, used for coating coal scuttles, fire screens, etc., consists of asphalt dissolved in linseed-oil varnish. Iron stoves and stovepipes are best coated with graphite.

Galvanized Iron.—For galvanized iron there has been recommended a wash consisting simply of dilute hydrochloric acid, which produces chloride of zinc, that in combination with the oxygen of the air is said to produce a film upon which oil color takes as good a hold as it would upon ordinary sheet iron.

Another method which has been tested and found effective is to make a solution as follows: One ounce of chloride of copper; 1 ounce nitrate of copper; 1 ounce sal ammoniac, dissolved in 2 quarts of soft water, to which is added 1 ounce of crude or commercial hydrochloric acid. This solution should be made in an earthenware dish or pot, or in glass or stoneware, as tin will precipitate the copper salts and make the solution imperfect. To large surfaces this solution is applied with a broad brush, when the surface assumes a deep black color, which in drying out in from 12 to 24 hours becomes a gray white, upon which the properly prepared primer will take a permanent grip. On the film so produced a much thinner paint will cover very much better than a stouter paint would on the untreated galvanized or ordinary iron surface. A single trial will convince the craftsman that this treatment is a method that will give lasting results, provided he tries the same priming paint on the treated and untreated surface.

To Paint Wrought Iron with Graphite.—In order to make wrought iron look like new mix fine graphite with equal parts of varnish and turpentine oil, adding a little siccative. Paint the iron parts with this twice, allowing to dry each time. Especially the second coating must be perfectly dry before further treatment. The latter consists in preparing graphite with spirit and applying it very thinly over the first coat. After the drying or evaporation of the spirit the graphite last applied is brushed vigorously, whereby a handsome, durable gloss is produced.

Paint for Iron Bodies Exposed to Heat.—Dilute 1 part soda water glass with 2 parts water and mix intimately with the following pigments:

White.—White lead or sulphate of barium.

Yellow.—Chromate of barium, ocher, or uranium yellow.

Green.—Chromic oxide or ultramarine green.

Blue.—Ultramarine.

Brown.—Oxide of cadmium, oxide of manganese or terra di sienna.

Red.—English red or chrome red.

Bronze powder in a suitable quantity may be added to the mixture, but not more paint should be prepared than can be used up in a few hours. The bronze powder may also be strewn on the fresh paint, or applied with a dry brush, to enhance the gloss. This paint is not affected by heat, and is inodorous.

Protective Coating for Bright Iron Articles.—Zinc white, 30 parts; lampblack, 2 parts; tallow, 7 parts; vaseline, 1 part; olive oil, 3 parts; varnish, 1 part. Boil together ¼ hour and add ½ part of benzine and ¼ part of turpentine, stirring the mass carefully and boiling for some time. The finished paste-like substance can be readily removed with a rag without the use of solvents.

Rust Paints.—I.—A new rust paint is produced by the following process: Mix 100 parts dry iron sulphate and 87 parts sodium chlorate and heat to 1,500° to 1,800° F. The chlorine set free seems to have a very favorable action on the color of the simultaneously forming iron oxide. In order to avoid, however, too far-reaching an effect of the chlorine gas, about 18 pounds of a substance which absorbs the same mechanically, such as kaolin, ground pumice stone, ocher, etc., are added to the mixture.

II.—A material known under the names of lardite, steatite, agalmatolite, pagodite, is excellently adapted as a substitute for the ordinary metallic protective agent of the pigments and has the property of protecting iron from rust in an effective manner. In China, lardite is used for protecting edifices of sandstone, which crumbles under the action of the atmosphere. Likewise a thin layer of powdered steatite, applied in the form of paint, has been found valuable there as a protector against the decay of obelisks, statues, etc. Lardite, besides, possesses the quality of being exceedingly fine-grained, which renders this material valuable for use in ship painting. Ground steatite is one of the finest materials which can be produced, and no other so quickly and firmly adheres to the fibers of iron and steel. Furthermore, steatite is lighter than metallic covering agents, and covers, mixed in paint, a larger surface than zinc white, red lead, or iron oxide. Steatite as it occurs in Switzerland is used there and in the Tyrol for stoves, since it is fireproof.

Steel.—An excellent coating for steel, imitating the blue color of natural steel, is composed of white shellac, 5 parts; borax, 1 part; alcohol, 5 parts; water, 4 parts; and a sufficient quantity of methylene blue. The borax is dissolved in water, the shellac in alcohol. The aqueous solution of the borax is heated to a boil and the alcoholic solution of the shellac is added with constant stirring. Next add the blue color, continuing to stir. Before this coating is applied to the steel, e. g., the spokes of a bicycle, the latter are first rubbed off with fine emery paper. The coat is put on with a soft rag. The quantity of pigment to be added is very small. By varying the quantity a paler or darker coloring of the steel can be produced.

PAINTS FOR ROOFS AND ROOF PAPER:

Carbolineum.—This German preparation is made in three colors.

I.—Pale.—Melt together in an iron kettle, over a naked fire, 30 parts of American rosin F and 150 parts of pale paraffine oil and stir in 10 parts of single rectified rosin oil.

II.—Dark.—Melt 100 parts of anthracene oil and 20 parts of American rosin F on a slow fire. Next stir in 2 parts of Para rubber solution (or solution of caoutchouc waste) and keep on boiling until all is dissolved. When this is done there should be still added 5 parts of crude concentrated carbolic acid and 5 parts of zinc chloride lye, 50° Bé., stirring until cool. The last-named admixture is not absolutely necessary, but highly advisable, owing to its extraordinary preservative and bactericidal properties.

III.—Colored.—For red, melt 100 parts of coal-tar oil, then stir in 50 parts of pale paraffine oil, and finally 75 parts of bole or iron minium, and pass through the paint mill. Although the addition of iron minium is very desirable, it is considerably more expensive. For gray, proceed as above, with the exception that metallic gray is used in place of the bole. For green, metallic green is employed. The colors are identical with those used in the manufacture of roof varnish. To increase the antiseptic properties of the colored carbolineum, any desired additions of phenol or zinc chloride solutions may be made, but the chief requirement in the case of colored carbolineum is good covering power of he coating.

Paints for Roofs Covered with Tar Paper, for Roofing Paper, etc.—

I.—Distilled coal tar....	70 parts
Heavy mineral oil (lubricating oil)..	10 parts
American rosin.....	20 parts

II.—Distilled coal tar....	50 parts
Trinidad asphalt...	15 parts
Mineral oil, containing paraffine.....	10 parts
Dry clay, finely ground.........	25 parts

Imitation Oil Paint.—Schulz's German patent paint is cheap, and claimed to be durable, weatherproof, and glossy, like oil paint. The application consists of a ground coat, upon which the surface coat proper is applied after the former is dry. For the preparation of the grounding dissolve 1,000 parts, by weight, of Marseilles soap in 10,000 parts of boiling water and stir. In a separate vessel dissolve 2,000 parts of glue in 10,000 parts of boiling water, adding 17,500 parts of spirit of sal ammoniac. These two solutions are poured together and well stirred. Then dissolve 400 parts of chrome alum in 5,000 parts of water, and pour into the above mixture. To this mixture add 10,000 parts of pipe clay, stirring the whole well and tinting with earth colors, ocher, Vandyke brown, etc. The solid ingredients must be dissolved in boiling hot water, and sifted so as to obtain a finely divided ground color. This priming is applied in a warm state. The coating proper is put on the ground coat after it is dry, in about one-half to one hour. For this coat dissolve 2,000 parts of crystallized alum in 10,000 parts of boiling water and add to this liquid a solution of 2,000 parts of glue in 10,000 parts of water; in a special vessel prepare soapsuds of 1,000 parts of Marseilles soap in 12,000 parts of boiling water; dissolve 120 parts of chrome alum in 1,500 parts of boiling water, and mix the three solutions together with diligent stirring. This paint or liquid should also be put on hot, and assures a durable exterior paint.

PAINTS, STAINS, ETC., FOR SHIPS.

Anti-Fouling Composition.—Make an agglutinant by heating together

	By weight
White lead, ground in oil	2 parts
Red lead, dry	1 part
Raw linseed oil	14 parts

While hot stir in yellow ocher, kaolin, baked clay in powder, or any inert body, such as silica, barytes, gypsum, etc., to form a stiff dough, and, without allowing this compound to become cold (the vessel should not be removed from the source of heat), dilute with more or less manganese linoleate to the required consistency.

Marine Paint to Resist Sea Water.—First prepare the water-resisting agglutinant by heating together

Dry white lead, carbonate only	1 part
Litharge	1 part
Linseed oil (fluid measure)	14 parts

Heat these and stir until of the consistency of thick glue, and for every 36 parts, by weight, of this compound add 3 parts, by weight, of turpentine, and 1 part, by weight, of mastic varnish (mastic rosin dissolved in turpentine); reheat the whole, and for every 32 parts, by weight, stir in and mix the following:

Baked and powdered clay	4 parts
Portland cement	16 parts
Zinc white	1 part
Red lead	1 part

After well mixing, dilute with more or less turpentine (not exceeding 25 per cent of the whole), or linoleate of manganese, the latter being preferable, as it has greater binding power. For colored paints use red oxide of iron or green oxide of chrome, but do not use chrome green or lead, as they will not stand the action of the sea water.

Compositions for Ships' Bottoms.—

Green.

Pale rosin	25	pounds
Prepared mineral green	8	pounds
D. L. zinc	13	pounds
Boiled oil	2	pounds
Mineral naphtha	1	gallon
Petroleum spirit	1½	gallons

Prepared Mineral Green.

Dry levigated mineral green	28 pounds
Turpentine	7 pounds
Turpentine varnish	7 pounds
Refined linseed oil	7 pounds

Copper Color.

Pale rosin	25	pounds
Light Italian ocher	15	pounds
D. L. zinc	5	pounds
Turkey red paint	½	pound
Petroleum spirit	1½	pounds
Mineral naphtha	1	pound

Pink.

Pale rosin	25	pounds
D. L. zinc	16	pounds
Deep vermilion	7	pounds
Mineral naphtha	1	gallon
Petroleum spirit	1½	gallons

PAINTS FOR WALLS OF CEMENT, PLASTER, HARD FINISH, ETC.

Coating for Bathrooms.—As a rule cement plastering, as well as oil paint, suffices for the protection of walls and ceilings in bathrooms, but attention must be called to the destructive action of medicinal admixtures. For such rooms as well as for laboratories, an application of Swedish wood tar, made into a flowing consistency with a little oil of turpentine and put on hot, has been found very excellent. It is of advantage previously to warm the wall slightly. To the second coat add some wax. A very durable coating is obtained, which looks so pleasing that it is only necessary to draw some stripes with a darker paint so as to divide the surface into fields.

Cement, to Paint Over Fresh.—The wall should be washed with dilute sulphuric acid several days before painting. This will change the surplus caustic lime to sulphate of lime or gypsum. The acid should be about one-half chamber acid and one-half water. This should be repeated before painting, and a coat of raw linseed oil flowed on freely should be given for the first coat. While this cannot be always guaranteed as effectual for making the paint hold, it is the best method our correspondent has heard of for the purpose, and is worth trying when it is absolutely necessary to paint over fresh cement.

Damp Walls, Coating for.—Thirty parts of tin are dissolved in 40 parts of hydrochloric acid, and 30 parts of sal ammoniac are added. A powder composed of freestone, 50 parts; zinc oxide, 20 parts; pounded glass, 15 parts; powdered marble, 10 parts; and calcined magnesia, 5 parts, is prepared, and made into a paste with the liquid above mentioned. Coloring matter may be added. The composition may be used as a damp-proof coating for walls, or for repairing stonework, or for molding statues or ornaments.

Façade Paint.—For this zinc oxide is especially adapted, prepared with size or casein. Any desired earth colors may also be added. The surfaces are coated 3 times with this mass. After the third application is dry, put on a single coating of zinc chloride solution of 30° Bé. to which 3 per cent borax is added.

This coating is very solid, can be washed, and is not injured by hydrogen sulphide.

Hard-Finished Walls.—The treatment for hard-finished walls which are to be painted in flat colors is to prime with a thin coat of lead and oil well brushed into the wall. Next put on a thin coat of glue size; next a coat mixed with $\frac{1}{2}$ oil and $\frac{3}{4}$ turpentine; next a coat of flat paint mixed with turpentine. If you use any dry pigment mix it stiff in oil and thin with turps. If in either case the paint dries too fast, and is liable to show laps, put a little glycerine in, to retard the drying.

PAINTS, WATERPROOF AND WEATHERPROOF:
See also Fireproof Paint.

The following are claimed to be both waterproof and weatherproof:

I.—In 50 parts, by weight, of spirit of 96 per cent, dissolve 16 parts, by weight, of shellac, orange, finely powdered; 3 parts, by weight, of silver lake, finely powdered; and 0.6 parts, by weight, of gamboge, finely powdered. This paint may be employed without admixture of any siccative, and is excellently adapted for painting objects which are exposed to the inclemencies of the weather, as it is perfectly weatherproof.

II.—Mix glue water with zinc oxide (zinc white) and paint the respective object with this mixture. When this is dry (after about 2 hours) it is followed up with a coating of glue water and zinc chloride in a highly diluted state. Zinc oxide enters into a chemical combination with zinc chloride, which acquires the hardness of glass and a mirror-like bright surface. Any desired colors can be prepared with the glue water (size) and are practically imperishable. This zinc coating is very durable, dries quickly, and is 50 per cent cheaper than oil paint.

Water- and Acid-Resisting Paint.—Caoutchouc is melted with colophony at a low temperature, after the caoutchouc has been dried in a drying closet (stove) at 158° to 176° F., until no more considerable increase in weight is perceptible, while the colophony has completely lost its moisture by repeated melting. The raw products thus prepared will readily melt upon slight heating. To the melted colophony and caoutchouc add in a hot liquid state zinc white or any similar pigment. Thin with a varnish consisting of 50 parts of perfectly anhydrous colophony, 40 parts of absolute alcohol, and 40 parts of benzine. The whole syrupy mass is worked through in a paint mill to obtain a uniform product, at which operation more or less colophony varnish is added according to the desired consistency.

Water- and Air-Proof Paint.—An airproof and waterproof paint, the subject of a recent French patent, is a compound of 30 parts, by weight, acetone; 100 parts acetic ether; 50 parts sulphuric ether; 100 parts camphor; 50 parts gum lac; 200 parts cotton; 100 parts paper (dissolved in sulphuric acid); 100 parts mastic in drops. These proportions may fluctuate according to need. The paper is reduced well and dissolved without heat with sufficient sulphuric ether; the cotton is dissolved in the acetone and the whole is mixed together with the other ingredients and stirred well. The application is performed as with any other varnish. The coating is said not to crack or shrink and to be particularly useful as a protection against moisture for all stuffs.

PAINTS FOR WOOD:
See also Wood.

Floor Coating.—A new paint for floors, especially those of soft wood: Mix together 2.2 pounds joiners' glue; a little over 1 ounce powdered bichromate of potash; $3\frac{1}{2}$ ounces aniline brown; and $10\frac{1}{2}$ quarts water in a tin vessel. After 6 hours have elapsed (when the glue is completely soaked), heat gradually to the boiling point. The coating becomes perfectly water-tight after 2 or 3 days; it is not opaque, as the earthy body is lacking. The glue causes the wood fibers to be firmly united. It becomes insoluble by the addition of bichromate of potash, under the influence of light. Without this admixture a simple glue coat has formerly not been found satisfactory, as it dissolves if cleaned with water.

Durable House Paint. — I. — New houses should be primed once with pure linseed oil, then painted with a thin paint from white lead and chalk, and thus gradually covered. The last coat is prepared of well-boiled varnish, white lead, and chalk. The chalk has the mission to moderate the saponification of the linseed oil by the white lead. Mixing colors such as ocher and black, which take up plenty of oil, materially assist in producing a durable covering.

II.—Prime with zinc white and let this be succeeded by a coating with zinc chloride in glue water (size). The zinc oxide forms with the zinc chloride an oxy-chloride of great hardness and glossy surface. By admixture of pigments any desired shade may be produced. The zinc coating is indestructible, dries quickly, does not peel, is free from the smell of fresh oil paint, and more than 5 per cent cheaper.

Ivory Coating for Smooth, Light Wood. —In order to cover the articles, which may be flat or round, with this coating, they must first be polished quite smooth and clean; then they are coated with thin, hot, white glue. When the coat is thoroughly dry, the glue is rubbed off again with fine glass paper. The mass is prepared as follows: Take 3 pounds (more or less, according to the number of articles) of the purest and best collodion; grind upon a clean grinding stone twice the quantity that can be taken up with the point of a knife of Krems white, with enough good pale linseed oil as is necessary to grind the white smooth and fine. Take a clean bottle, into which one-half of the collodion is poured; to this add the ground white, which can be removed clean from the stone by means of a good spatula and put in the bottle. Add about 100 drops of linseed oil, and shake the mass till it looks like milk.

Now painting with this milky substance may be commenced, using a fine hair pencil of excellent quality. The pencil is not dipped in the large bottle; but a glass is kept at hand with an opening of about 1 inch, so as to be able to immerse the pencil quickly. The substance is not flowing like the alcohol lacquers, for which reason it may be put on thick, for the ether, chiefly constituting the mass, evaporates at once and leaves but a very thin film which becomes noticeable only after about 10 such applications have been made. Shake the bottle well each time before filling the small glass, as the heavy Krems white is very apt to sink to the bottom of the bottle. If it is observed that the substance becomes too thick, which may easily occur on account of the evaporation, a part of the remaining ether is added, to which in turn 30 to 40 drops of oil are added, shaking it till the oil appears to be completely dissolved.

The operator must put on the mass in quick succession and rather thick. After about 10 coats have been applied the work is allowed to rest several hours; then 3 or 4 coats of pure collodion, to which likewise several drops of oil have been added, are given. Another pause of several hours having been allowed to intervene, application of the mass is once more begun.

When it is noticed that a layer of the thickness of paper has formed, the articles, after drying thoroughly, should be softly rubbed off with very fine glass paper, after which they require to be wiped off well with a clean linen rag, so that no dust remains. Then coating is continued till the work seems serviceable.

A few applications of pure collodion should be made, and when this has become perfectly hard, after a few hours, it can be rubbed down with a rag, tripoli, and oil, and polished by hand, like horn or ivory. This work can be done only in a room which is entirely free from dust. The greatest cleanliness must be observed.

MISCELLANEOUS RECIPES, PAINTS, ETC.:

Bathtub Paint.—Take white keg lead, tint to any desired color and then add, say, $\frac{1}{8}$ boiled oil (pure linseed) to $\frac{7}{8}$ hard drying durable body varnish. Clean the surface of the tub thoroughly before applying the paint. Benzine or lime wash are good cleaning agents. Coat up until a satisfactorily strong, pure color is reached. This will give good gloss and will also wear durably.

Coating for Name Plates.—A durable coating for name plates in nurseries is produced as follows: Take a woolen rag, saturate it with joiners' polish, lay it into a linen one, and rub the wooden surface with this for some time. Rub down with sandpaper and it can be written on almost like paper. When all is dry, coat with dammar lacquer for better protection. If the wood is to receive a color it is placed in the woolen rag before rubbing down, in this case chrome yellow.

To Keep Flies from Fresh Paint.—For the purpose of keeping flies and other insects away from freshly painted surfaces mix a little bay oil (laurel oil) with the oil paint, or place a receptacle containing same in the vicinity of the painted objects. The pungent odor keeps off the flies.

Heat-Indicating Paint.—A heat-indicating paint composed of a double iodide of copper and mercury was first discovered years ago by a German physicist. At ordinary temperatures the paint is red, but when heated to 206° F. it turns black. Paper painted with this composition and warmed at a stove exhibits the change in a few seconds. A yellow double iodide of silver and mercury is even more sensitive to heat, changing from yellow to dark red.

To Keep Liquid Paint in Workable Condition.—To prevent liquid paint which, for convenience sake, is kept in small quantities and flat receptacles, from evaporating and drying, give the vessels such a shape that they can be placed one on top of the other without danger of falling over, and provide the under side with a porous mass—felt or very porous clay, etc.—which, if mois-

tened, will retain the water for a long time. Thus, in placing the dishes one on top of the other, a moist atmosphere is created around them, which will inhibit evaporation and drying of the paint. A similar idea consists in producing covers with a tight outside and porous inside, for the purpose of covering up, during intermission in the work, clay models and like objects which it is desired to keep soft. In order to avoid the formation of fungous growth on the constantly wet bottom, it may be saturated with non-volatile disinfectants, or with volatile ones if their vapors are calculated to act upon the objects kept underneath the cover. If the cover is used to cover up oil paints, it is moistened on the inside with volatile oil, such as oil of turpentine, oil of lavender, or with alcohol.

Peeling of Paints.—For the prevention of peeling of new coatings on old oil paintings or lakes, the latter should be rubbed with roughly ground pumice stone, wet by means of felt rags, and to the first new coat there should be added fine spirit in the proportion of about $\frac{1}{10}$ of the thinning necessary for stirring (turpentine, oil, etc.). This paint dries well and has given good results, even in the most difficult cases. The subsequent coatings are put on with the customary paint. Fat oil glazes for graining are likewise mixed with spirit, whereby the cracking of the varnish coating is usually entirely obviated.

Polychroming of Figures.—This paint consists of white wax, 1 part, and powdered mastic, 1 part, melted together upon the water bath and mixed with rectified turpentine. The colors to be used are first ground stiffly in turpentine on the grinding slab, and worked into consistency with the above solution.

Priming Coat for Water Spots.—A very simple way to remove rain spots, or such caused by water soaking through ceilings, has been employed with good results. Take unslaked white lime, dilute with alcohol, and paint the spots with it. When the spots are dry—which ensues quickly, as the alcohol evaporates and the lime forms a sort of insulating layer—one can proceed painting with size color, and the spots will not show through again.

PAINT FOR PROTECTING CEMENT AGAINST ACID:
See Acid-Proofing.

PAINT, GREASE:
See Cosmetics.

PAINT REMOVERS:
See Cleaning Compounds.

PALLADIUM ALLOYS:
See Alloys.

PALLADIUMIZING:
See Plating.

PALMS, THEIR CARE.

Instead of washing the leaves of palms with water, many florists employ a mixture of milk and water, the object being to prevent the formation of disfiguring brown stains.

Paper

Paper Pads (see also Adhesives, under Glue).

I.—Glue.............. 3½ ounces
Glycerine.......... 8 ounces
Water, a sufficient quantity.

Pour upon the glue more than enough water to cover it and let stand for several hours, then decant the greater portion of the water; apply heat until the glue is dissolved, and add the glycerine. If the mixture is too thick, add more water.

II.—Glue.............. 6 ounces
Alum.............. 30 grains
Acetic acid......... ½ ounce
Alcohol........... 1½ ounces
Water............. 6½ ounces

Mix all but the alcohol, digest on a water bath till the glue is dissolved, allow to cool, and add the alcohol.

Papier Maché.—The following are the ingredients necessary to make a lump of papier maché a little larger than an ordinary baseball and weighing 17 ounces:

Wet paper pulp, dry paper, 1 ounce; water, 3 ounces; 4 ounces (avoirdupois); dry plaster Paris, 8 ounces (avoirdupois); hot glue, ½ gill, or 4½ tablespoonfuls.

While the paper pulp is being prepared, melt some best Irish glue in the glue pot and make it of the same thickness and general consistency as that used by cabinet makers. On taking the paper pulp from the water squeeze it gently, but do not try to dry it. Put in a bowl, add about 3 tablespoonfuls of the hot glue, and stir the mass up into a soft and very sticky paste. Add the plaster of Paris and mix thoroughly. By the time about 3 ounces of the plaster have been used, the mass is so dry and thick that it can hardly be worked. Add the remainder of the glue, work it up again until it becomes sticky once more, and then add the remainder of the plaster. Squeeze it vigorously through the fingers to thoroughly mix the mass, and work it until free from lumps, finely kneaded and sticky enough to adhere to the surface of a planed board. If it is too dry to stick fast add a few drops of either glue or water, and work it up again. When the paper pulp is poor and the maché is inclined to be lumpy, lay the mass upon a smooth board, take a hammer and pound it hard to grind it up fine.

If the papier maché is not sticky enough to adhere firmly to whatever it is rubbed upon, it is a failure, and requires more glue. In using it the mass should be kept in a lump and used as soon as possible after making. Keep the surface of the lump moist by means of a wet cloth laid over it, for if you do not, the surface will dry rapidly. If it is to be kept overnight, or longer, wrap it up in several thicknesses of wet cotton cloth, and put under an inverted bowl. If it is desired to keep a lump for a week, to use daily, add a few drops of glycerine when making, so that it will dry more slowly.

The papier maché made according to this formula has the following qualities: When tested by rubbing between the thumb and finger, it was sticky and covered the thumb with a fine coating. (Had it left the thumb clean, it would have been because it contained too much water.) When rubbed upon a pane of glass it sticks tightly and dries hard in 3 hours without cracking, and can only be removed with a knife. When spread in a layer as thin as writing paper it dries in half an hour. A mass actually used dried hard enough to coat with wax in 18 hours, and, without cracking, became as hard as wood; yet a similar quantity wrapped in a wet cloth and placed under an inverted bowl kept soft and fit for use for an entire week.

Parchment Paper.—I.—Dip white unsized paper for half a minute in strong sulphuric acid, specific gravity, 1.842, and afterwards in water containing a little ammonia.

II.—Plunge unsized paper for a few seconds into sulphuric acid diluted with half to a quarter its bulk of water (this solution being of the same temperature as the air), and afterwards wash with weak ammonia.

Razor Paper.—I.—Smooth unsized paper, one of the surfaces of which, while in a slightly damp state, has been rubbed over with a mixture of calcined peroxide of iron and emery, both in impalpable powder. It is cut up into pieces (about 5 x 3 inches), and sold in packets. Used to wipe the razor on, which thus does not require stropping.

II.—From emery and quartz (both in impalpable powder), and paper pulp (estimated in the dry state), equal parts, made into sheets of the thickness of drawing paper, by the ordinary process. For use, a piece is pasted on the strop and moistened with a little oil.

Safety Paper.—White paper pulp mixed with an equal quantity of pulp tinged with any stain easily affected by chlorine, acids, alkalies, etc., and made into sheets as usual, serves as a safety paper on which to write checks or the like. Any attempt to wash out the writing affects the whole surface, showing plainly that it has been tampered with.

Tracing Paper.—Open a quire of smooth, unsized white paper, and place it flat upon a table. Apply, with a clean sash tool to the upper surface of the first sheet, a coat of varnish made of equal parts of Canada balsam and oil of turpentine, and hang the prepared sheet across the line to dry; repeat the operation on fresh sheets until the proper quantity is finished. If not sufficiently transparent, a second coat of varnish may be applied as soon as the first has become quite dry.

Strengthened Filter Paper.—When ordinary filter paper is dipped into nitric acid (specific gravity, 1.42), thoroughly washed and dried, it becomes a tissue of remarkable properties, and one that deserves to be better known by chemists and pharmacists. It shrinks somewhat in size and in weight, and gives, on burning, a diminished ash. It yields no nitrogen, nor does it in the slightest manner affect liquids. It remains perfectly pervious to liquids, its filtering properties being in no wise affected, which, it is needless to say, is very different from the behavior of the same paper "parchmented" by sulphuric acid. It is as supple as a rag, yet may be very roughly handled, even when wet, without tearing or giving way. These qualities make it very valuable for use in filtration under pressure or exhaust. It fits closely to the funnel, upon which it may be used direct, without any supports, and it thus prevents undue access of air. As to strength, it is increased upward of 10 times. A strip of ordinary white Swedish paper, ⅛ of an inch wide, will sustain a load of from ½ to ¾ of a pound avoirdupois, according to the quality of the paper. A similar strip of the toughened paper

broke, in 3 trials, with 5 pounds, 7 ounces, and 3 drachms; 5 pounds, 4 ounces, and 36 grains; and 5 pounds, 10 ounces respectively. These are facts that deserve to be better known than they seem to be to the profession at large.

Blotting Paper.—A new blotting paper which will completely remove wet as well as dry ink spots, after moistening the paper with water, is produced as follows: Dissolve 100 parts of oxalic acid in 400 parts of alcohol, and immerse porous white paper in this solution until it is completely saturated. Next hang the sheets up separately to dry over threads. Such paper affords great advantages, but in its characteristic application is serviceable for ferric inks only, while aniline ink spots cannot be removed with it, after drying.

Carbon Paper.—Many copying papers act by virtue of a detachable pigment, which, when the pigmented paper is placed between two sheets of white paper, and when the uppermost paper is written on, transfers its pigment to the lower white paper sheet along lines which correspond to those traced on the upper paper, and therefore gives an exact copy of them on the lower paper.

The pigments used are fine soot or ivory black, indigo carmine, ultramarine, and Paris blue, or mixtures of them. The pigment is intimately mixed with grain soap, and then rubbed on to thin but strong paper with a stiff brush. Fatty oils, such as linseed or castor oil, may be used, but the grain soap is preferable. Graphite is frequently used for black copying paper. It is rubbed into the paper with a cotton pad until a uniform light-gray color results. All superfluous graphite is then carefully brushed off.

It is sometimes desired to make a copying paper which will produce at the same time a positive copy, which is not required to be reproduced, and a negative or reversed copy from which a number of direct copies can be taken. Such paper is covered on one side with a manifolding composition, and on the other with a simple copying composition, and is used between 2 sheets of paper with the manifolding side undermost.

The manifolding composition is made by mixing 5 ounces of printers' ink with 40 of spirits of turpentine, and then mixing it with a fused mixture of 40 ounces of tallow and 5 ounces of stearine. When the mass is homogeneous, 30 ounces of the finest powdered protoxide of iron, first mixed with 15 ounces of pyrogallic acid and 5 ounces of gallic acid, are stirred in till a perfect mixture is obtained. This mass will give at least 50 copies on damp paper in the ordinary way. The copying composition for the other side of the prepared paper consists of the following ingredients:

Printers' ink	5 parts
Spirits of turpentine	40 parts
Fused tallow	30 parts
Fused wax	3 parts
Fused rosin	2 parts
Soot	20 parts

It goes without saying that rollers or stones or other hard materials may be used for the purpose under consideration as well as paper. The manifolding mass may be made blue with indigotin, red with magenta, or violet with methyl violet, adding 30 ounces of the chosen dye to the above quantities of pigment. If, however, they are used, the oxide of iron and gallic acids must be replaced by 20 ounces of carbonate of magnesia.

Celloidin Paper.—Ordinary polished celluloid and celloidin paper are difficult to write upon with pen and ink. If, however, the face is rubbed over with a chalk crayon, and the dust wiped off with a clean rag, writing becomes easy.

Cloth Paper.—This is prepared by covering gauze, calico, canvas, etc., with a surface of paper pulp in a Foudrinier machine, and then finishing the compound sheet in a nearly similar manner to that adopted for ordinary paper.

Drawing Paper.—The blue drawing paper of commerce, which is frequently employed for technical drawings, is not very durable. For the production of a serviceable and strong drawing paper, the following process is recommended. Mix a solution of

Gum arabic	2 parts
Ammonia iron citrate	3 parts
Tartaric acid	2 parts
Distilled water	20 parts

After still adding 4 parts of solution of ammonia with a solution of

Potassium ferricyanide	2.5 parts
Distilled water	10.0 parts

allow the mixture to stand in the dark half an hour. Apply the preparation on the paper by means of a soft brush, in artificial light, and dry in the dark. Next, expose the paper to light until it appears dark violet, place in water for 10 seconds, air a short time, wash with water, and finally dip in a solution of

Eau de javelle	50 parts
Distilled water	1,000 parts

until it turns dark blue.

Filter Paper.—This process consists in dipping the paper in nitric acid of 1.433 specific gravity, subsequently washing it well and drying it. The paper thereby acquires advantageous qualities. It shrinks a little and loses in weight, while on burning only a small quantity of ash remains. It possesses no traces of nitrogen and does not in any way attack the liquid to be filtered. Withal, this paper remains perfectly pervious for the most varying liquids, and its filtering capacity is in no wise impaired. It is difficult to tear, and still elastic and flexible like linen. It clings completely to the funnel. In general it may be said that the strength of the filtering paper thus treated increases 100 per cent.

Fireproof Papers.—I.—Ammonium sulphate, 8 parts, by weight; boracic acid, 3 parts; borax, 2 parts; water, 100 parts. The temperature should be about 122° F.

II.—For paper, either printed or unprinted, bills of exchange, deeds, books, etc., the following solution is recommended: Ammonium sulphate, 8 parts; boracic acid, 3 parts; sodium borate, 1.7 parts; water, 10,000 parts. The solution is heated to 122° F., and may be used when the paper is manufactured. As soon as the paper leaves the machine it is passed through this solution, then rolled over a warm cylinder and dried. If printed or in sheets, it is simply immersed in the solution, at a temperature of 122° F., and spread out to dry, finally pressed to restore the luster.

Hydrographic Paper.—This is paper which may be written on with simple water or with some colorless liquid having the appearance of water.

I.—A mixture of nut galls, 4 parts, and calcined sulphate of iron, 1 part (both perfectly dry and reduced to very fine powder), is rubbed over the surface of the paper, and is then forced into its pores by powerful pressure, after which the loose portion is brushed off. The writing shows black when a pen dipped in water is used.

II.—A mixture of persulphate of iron and ferrocyanide of potassium may be employed as in formula I. This writes blue.

Iridescent Paper.—Sal ammoniac and sulphate of indigo, of each 1 part; sulphate of iron, 5 parts; nut galls, 8 parts; gum arabic, ½ part. Boil them in water, and expose the paper washed with the liquid to (the fumes of) ammonia.

Lithographic Paper.—I.—Starch, 6 ounces; gum arabic, 2 ounces; alum, 1 ounce. Make a strong solution of each separately, in hot water, mix, strain through gauze, and apply it while still warm to one side of leaves of paper, with a clean painting brush or sponge; a second and a third coat must be given as the preceding one becomes dry. The paper must be, lastly, pressed, to make it smooth.

II.—Give the paper 3 coats of thin size, 1 coat of good white starch, and 1 coat of a solution of gamboge in water, the whole to be applied cold, with a sponge, and each coat to be allowed to dry before the other is applied. The solutions should be freshly made.

Lithographic paper is written on with lithographic ink. The writing is transferred simply by moistening the back of the paper, placing it evenly on the stone, and then applying pressure. A reversed copy is obtained, which, when printed from, yields corrected copies resembling the original writing or drawing. In this way the necessity of executing the writing or drawing in a reversed direction is obviated.

MARBLING PAPER FOR BOOKS.

Provide a wooden trough 2 inches deep and the length and width of any desired sheet; boil in a brass or copper pan a quantity of linseed and water until a thick mucilage is formed; strain it into a trough, and let cool; then grind on a marble slab any of the following colors in small beer:

For Blue.—Prussian blue or indigo.

Red.—Rose pink, vermilion, or drop lake.

Yellow.—King's yellow, yellow ocher, etc.

White.—Flake white.

Black.—Burnt ivory or lampblack.

Brown. — Umber, burnt; terra di sienna, burnt.

Black mixed with yellow or red also makes brown.

Green.—Blue and yellow mixed.

Orange.—Red and yellow mixed.

Purple.—Red and blue mixed.

For each color have two cups, one for the color after grinding, the other to mix it with ox gall, which must be used to thin the colors at discretion. If too much gall is used, the colors will spread. When they keep their place on the surface of the trough, when moved with a quill, they are fit for use. All things in readiness, the colors are successively sprinkled on the surface of the mucilage in the trough with a brush, and are waved or drawn about with a quill or a stick, according to taste. When the design is just formed, the book, tied tightly between cutting boards of the same size, is lightly pressed with its edge on the surface of the liquid pattern, and then withdrawn and dried. The covers may be marbled in the same way, only letting the liquid colors run over them. In marbling paper the sides of the paper are gently applied to the colors in the trough. The film of color in the trough may be as thin as possible, and if any remains after the marbling it may be taken off by applying paper to it before you prepare for marbling again. To diversify the effects, colors are often mixed with a little sweet oil before sprinkling them on, by which means a light halo or circle appears around each spot.

WATERPROOF PAPERS.

I.—Wall papers may be easily rendered washable, either before or after they are hung, by preparing them in the following manner: Dissolve 2 parts of borax and 2 parts of shellac in 24 parts of water, and strain through a fine cloth. With a brush or a sponge apply this to the surface of the paper, and when it is dry, polish it to a high gloss with a soft brush. Thus treated the paper may be washed without fear of removing the colors or even smearing or blurring them.

II.—This is recommended for drawing paper. Any kind of paper is lightly primed with glue or a suitable binder, to which a finely powdered inorganic body, such as zinc white, chalk, lime, or heavy spar, as well as the desired coloring matter for the paper, are added. Next the paper thus treated is coated with soluble glass—silicate of potash or of soda—to which small amounts oi magnesia have been admixed, or else it is dipped into this mixture, and dried for about 10 days in a temperature of 77° F. Paper thus prepared can be written or drawn upon with lead pencil, chalk, colored crayons, charcoal, India ink, and lithographic crayon, and the writing or drawing may be washed off 20 or more times, entirely or partly, without changing the paper materially. It offers the convenience that anything may be readily and quickly removed with a moist sponge and immediately corrected, since the washed places can be worked on again at once.

Wax Paper.—I.—Place cartridge paper or strong writing paper, on a hot iron plate, and rub it well with a lump of beeswax. Used to form extemporaneous steam or gas pipes, to cover the joints of vessels, and to tie over pots, etc.

II.—For the production of waxed or ceresine paper, saturate ordinary paper with equal parts of stearine and tallow or ceresine. If it is desired to apply a business stamp on the paper before saturation and after stamping, it should be dried well for 24 hours, so as to prevent the aniline color from spreading.

Wrapping Paper for Silverware.—Make a solution of 6 parts of sodium hydrate in sufficient water to make it show about 20° B. (specific gravity, 1.60). To it add 4 parts zinc oxide, and boil together until the latter is dissolved. Now add sufficient water to reduce the specific gravity of the solution to 1.075 (10° B.). The bath is now ready for use. Dip each sheet separately, and hang on threads stretched across the room, to dry. Be on your guard against dust, as particles of sand adhering to the paper will scratch the ware wrapped in it. Ware, either plated or silver, wrapped in this paper, will not blacken.

Varnished Paper.—Before proceeding to varnish paper, card-work, pasteboard, etc., it is necessary to give it 2 or 3 coats of size, to prevent the absorption of the varnish, and any injury to the color or design. The size may be made by dissolving a little isinglass in boiling water, or by boiling some clean parchment cuttings until they form a clear solution. This, after being strained through a piece of clean muslin, or, for very nice purposes, clarified with a little white of egg, is applied by means of a small clean brush called by painters a sash tool. A light, delicate touch must be adopted, especially for the first coat, lest the ink or colors be started or smothered. When the prepared surface is quite dry it may be varnished.

Impregnation of Papers with Zapon Varnish.—For the protection of important papers against the destructive influences of the atmosphere, of water fungi, and light, but especially against the consequences of the process of molding, a process has been introduced under the name of zapon impregnation.

The zaponizing may be carried out by dipping the papers in zapon or by coating them with it by means of a brush or pencil. Sometimes the purpose may also be reached by dripping or sprinkling it on, but in the majority of cases a painting of the sheets will be the simplest method.

Zapon in a liquid state is highly inflammable, for which reason during the application until the evaporation of the solvent, open flames and fires should be kept away from the vicinity. When the drying is finished, which usually takes a few hours where both sides are coated, the zaponized paper does not so easily ignite at an open flame any more or at least not more readily than non-impregnated paper. For coating with and especially for dipping in zapon, a contrivance which effects a convenient suspension and dripping off with collection of the excess is of advantage.

The zapon should be thinned according to the material to be treated. Feebly sized papers are coated with ordinary, i. e., undiluted zapon. For dipping purposes, the zapon should be mixed with a diluent, if the paper is hard and well sized. The weaker the sizing, the more careful should be the selection of the zapon.

Zapon to be used for coating purposes should be particularly thick, so that it can be thinned as desired. Unsized papers require an undiluted coating.

The thick variety also furnishes an excellent adhesive agent as cement for wood, glass, porcelain, and metals which is insoluble in cold and hot water, and binds very firmly. Metallic surfaces coated with zapon do not oxidize or alter their appearance, since the coating is like glass and only forms a very thin but firmly adhering film, which, if applied on pliable sheet metal, does not crack on bending.

For the preparation of zapon the following directions are given: Pour 20 parts of acetone over 2 parts of colorless celluloid waste—obtainable at the celluloid factories—and let stand several days in a closed vessel, shaking frequently, until the whole has dissolved into a clear, thick mass. Next admix 78 parts of amyl acetate and completely clarify the zapon varnish by allowing to settle for weeks.

Slate Parchment.—Soak good paper with linseed-oil varnish (boiled oil) and apply the following mass, mentioned below, several times in succession: Copal varnish, 1 part, by weight; turpentine oil, 2 parts; finest sprinkling sand, 1 part; powdered glass, 1 part; ground slate as used for slates, 2 parts; and lampblack, 1 part, intimately mixed together, and repeatedly ground very fine. After drying and hardening, the plates can be written upon with lead or slate pencils.

Paper Floor Covering.—The floor is carefully cleaned, and all holes and cracks are filled up with a mass which is prepared by saturating newspapers with a paste that is made by mixing thoroughly 17⅝ ounces wheat flour, 3.17 quarts water, and 1 spoonful of pulverized alum. The floor is coated with this paste throughout, and covered with a layer of manilla paper, or other strong hemp paper. If something very durable is desired, paint the paper layer with the same paste and put on another layer of paper, leaving it to dry thoroughly. Then apply another coat of paste, and upon this place wall paper of any desired kind. In order to protect the wall paper from wear, give it 2 or more coats of a solution of 8½ ounces white glue in 2.11 quarts hot water, allow them to dry, and finish the job with a coating of hard oil varnish.

METALLIC PAPER.

This paper, made by transferring, pasting, or painting a coating of metal on ordinary paper, retains a comparatively dull and dead appearance even after glazing or polishing with the burnisher or agate. Galvanized or electroplated metal paper, on the other hand, in which the metal has penetrated into the most minute pores of the paper, possesses an extraordinarily brilliant polish, fully equal to that of a piece of compact polished metal. It is much more extensively used than the kind first mentioned.

The following solutions are recommended for making "galvanized" metal paper:

I.—For silver paper: Twenty parts argento-cyanide of potassium; 13 parts cyanide of potassium; 980 parts water.

II.—For gold paper: Four parts auro-cyanide of potassium; 9 parts cyanide of potassium; 900 parts water.

Moth Paper.—

Naphthalene	4 ounces
Paraffine wax	8 ounces

Melt together and while warm paint unsized paper and pack away with the goods.

Lead Paper.—Lay rough drawing paper (such as contains starch) on an 8 per cent potassium iodide solution. After a moment take it out and dry. Next, in a dark room, float the paper face downward on an 8 per cent lead nitrate solution. This sensitizes the paper. Dry again. The paper is now ready for printing. This process should be carried on till all the detail is out in a grayish color. Then develop in a 10 per cent ammonium chloride solution. The tones obtained are of a fine blue black.

Aluminum Paper.—Aluminum paper is not leaf aluminum, but real paper glazed with aluminum powder. It is said to keep food materials fresh. The basic material is artificial parchment, coated with a solution of rosin in alcohol or ether. After drying, the paper is warmed until the rosin has again softened to a slight degree. The aluminum powder is dusted on and the paper then placed under heavy pressure to force the powder firmly into it. The metallic coating thus formed is not affected by air or greasy substances.

PAPER (ANTI-RUST) FOR NEEDLES:
See Rust Preventives.

PAPER CEMENTS:
See Adhesives.

PAPER DISINFECTANT:
See Disinfectants.

PAPER, FIREPROOF:
See Fireproofing.

PAPER, FROSTED:
See Glass (Frosted).

PAPER ON GLASS, TO AFFIX:
See Adhesives, under Water-Glass Cements.

PAPERS, IGNITING:
See Pyrotechnics.

PAPER ON METALLIC SURFACES, PASTING:
See Adhesives.

PAPER AS PROTECTION FOR IRON AND STEEL:
See Rust Preventives.

PAPERHANGERS' PASTES:
See Adhesives.

PAPER, PHOTOGRAPHIC:
See Photography.

PAPER VARNISHES:
See Varnishes.

PAPER WATERPROOFING:
See Waterproofing.

PAPIER MACHÉ:
See Paper.

PARAFFINE:

Rendering Paraffine Transparent.—A process for rendering paraffine and its mixtures with other bodies (ceresine, etc.) used in the manufacture of transparent candles consists essentially in adding a naphthol, particularly beta-naphthol, to the material which is used for the manufacture of the candles, tapers, etc. The quantity added varies according to the material and the desired effect. One suitable mixture is made by heating 100 parts of paraffine and 2 parts of beta-naphthol at 175° to 195° F. The material can be colored in the ordinary way.

Removal of Dirt from Paraffine.—Filtration through felt will usually remove particles of foreign matter from paraffine. It may be necessary to use a layer of fine sand or of infusorial earth. If discolored by any soluble matter, try freshly heated animal charcoal. To keep the paraffine fluid, if a large quantity is to be handled, a jacketed funnel will be required, either steam or hot water being kept in circulation in the jacket.

Paraffine Scented Cakes.

Paraffine, 1 ounce; white petrolatum, 2 ounces; heliotropin, 10 grains; oil of bergamot, 5 drops; oil of lavender, 5 drops; oil of cloves, 2 drops. Melt the first two substances, then add the next, the oils last, and stir all until cool. After settling cut into blocks and wrap in tin foil. This is a disseminator of perfume. It perfumes where it is rubbed. It kills moths and perfumes the wardrobe. It is used by rubbing on cloth, clothes, and the handkerchief.

PARCHMENT AND PARCHMENT PAPER:
See Paper.

PARCHMENT CEMENT:
See Adhesives.

PARCHMENT PASTE:
See Adhesives.

PARFAITS:
See Ice Creams.

PARFAIT D'AMOUR CORDIAL:
See Wines and Liquors.

PARIS GREEN:
See Pigments.

PARIS RED:
See Polishes.

PARIS SALTS:
See Disinfectants.

PARISIAN CEMENT:
See Adhesives.

PASSE-PARTOUT FRAMING.

It is hardly correct to call the passe-partout a frame, as it is merely a binding together of the print, the glass, and the backing with a narrow edge of paper. This simple arrangement lends to the picture when complete a much greater finish and a more important appearance than might be anticipated.

In regard to the making of a passe-partout frame, the first thing is to decide as to the width of the mount or matt to be used. In some cases, of course, the print is framed with no mount being visible; but, unless the picture is of large size, it will usually be found more becoming to have one, especially should the wall paper be of an obtrusive design. When the print and mount are both neatly trimmed to the desired size, procure a piece of clear white picture glass— most amateur framers will have discovered that there is a variance in the quality of this—and a piece of stout cardboard, both of exactly the same dimensions as the picture. Next prepare or buy the paper to be used for binding the edges together. This may now be bought at most all stationery stores in a great variety of colors. If it is prepared at home a greater choice of colors is available, and it is by no means a difficult task with care and sharp scissors. The tint should be chosen to harmonize with the print and the mount, taking also into consideration the probable surroundings—brown for photographs of brown tone, dark gray for black, pale gray for lighter tones; dark green is also a good color. All stationers keep colored papers suitable for the purpose, while plain wall papers or thin brown paper answers equally well.

Cut the paper, ruling it carefully, into even strips an inch wide, and then into four pieces, two of them the exact length of the top and bottom of the frame, and the other two half an inch longer than the two sides. Make sure that the print is evenly sandwiched between the glass and the back. Cut some tiny strips of thin court-plaster, and with these bind the corners tightly together. Brush over the two larger pieces of paper with mountant, and with them bind tightly together the three thicknesses—print, glass, and cardboard—allowing the paper to project over about a third of an inch on the face side, and the ends which were left a little longer must be neatly turned over and stuck at the back. Then, in the same manner, bind the top and bottom edges together, mitering the corners neatly.

It should not be forgotten, before binding the edges together, to make two slits in the cardboard back for the purpose of inserting little brass hangers, having flat ends like paper fasteners, which may be bought for the purpose; or, where these are not available, two narrow loops of tape may be used instead, sticking the ends firmly on the inside of the cardboard by means of a little strong glue.

These are the few manipulations necessary for the making of a simple passe-partout frame, but there are numberless variations of the idea, and a great deal of variety may be obtained by means of using different mounts. Brown paper answers admirably as a mount for some subjects, using strips of paper of a darker shade as binding. A not too obtrusive design in pen and ink is occasionally drawn on the mount, while a more ambitious scheme is to use paint and brushes in the same way. An ingenious idea which suits some subjects is to use a piece of hand-blocked wall paper as a mount.

PARQUET POLISH:
See Polishes.

PASTES:
See Adhesives for Adhesive Purposes.

Pastes, Razor.—I.—From jewelers' rouge, plumbago, and suet, equal parts, melted together and stirred until cold.

II.—From prepared putty powder (levigated oxide of tin), 3 parts; lard, 2 parts; crocus martis, 1 part; triturated together.

III.—Prepared putty powder, 1 ounce; powdered oxalic acid, $\frac{1}{4}$ ounce; powdered gum, 20 grains; make a stiff paste with water, quantity sufficient, and evenly and thinly spread it over the strop, the other side of which should be covered with any of the common greasy mixtures. With very little friction this paste gives a fine edge to the razor, and its action is still further increased by slightly moistening it, or even breathing on it. Immediately after its use, the razor should receive a few turns on the other side of the strop.

PASTE FOR PAPER:
See Paper.

PASTES FOR POLISHING METALS:
See Soaps.

PASTEBOARD CEMENT:
See Adhesives.

PASTEBOARD DEODORIZERS:
See Household Formulas.

PASTILLES, FUMIGATING:
See Fumigants.

PATINAS:
See Bronzing and Plating.

PATENT LEATHER:
See Leather.

PEACH EXTRACT:
See Essences and Extracts.

PEARLS, TO CLEAN:
See Cleaning Preparations and Methods.

PEGAMOID.
Camphor, 100 parts; mastic, 100 parts; bleached shellac, 50 parts; gun cotton, 200 parts; acetone, 200 parts; acetic ether, 100 parts; ethylic ether, 50 parts.

PEN METAL:
See Alloys.

PENCILS, ANTISEPTIC:
See Antiseptics.

PENCILS FOR MARKING GLASS:
See Etching, Frosted Glass, and Glass.

PENS, GOLD:
See Gold.

PEONY ROOTS, THEIR PRESERVATION:
See Roots.

PERCENTAGE SOLUTION.
Multiply the percentage by 5; the product is the number of grains to be added to an ounce of water to make a solution of the desired percentage. This is correct for anything less than 15 per cent.

Perfumes

DRY PERFUMES:
Sachet Powders.—

I.—Orris root	6 ounces
Lavender flowers	2 ounces
Talcum	4 drachms
Musk	20 grains
Terpinol	60 grains

II.—Orange peel	2 ounces
Orris root	1 ounce
Sandalwood	4 drachms
Tonka	2 drachms
Musk	6 grains

Lavender Sachets.—

I.—Lavender flowers... 16 ounces
Gum benzoin...... 4 ounces
Oil lavender....... 2 drachms

II.—Lavender flowers, 150 parts; orris root, 150 parts; benzoin, 150 parts; Tonka beans, 150 parts; cloves, 100 parts; "Neugenwerz," 50 parts; sandalwood, 50 parts; cinnamon, 50 parts; vanilla, 50 parts; and musk, ½ part. All is bruised finely and mixed.

Violet Sachet.—

Powdered orris root 500 parts
Rice flour......... 250 parts
Essence bouquet... 10 parts
Spring flowers extract.......... 10 parts
Violet extract...... 20 parts
Oil of bergamot... 4 parts
Oil of rose....... 2 parts

Borated Talcum.—

I.—Purified talcum, N. F........... 2 pounds
Powdered boric acid 1 ounce

To perfume add the following:

Powered orris root.. 1½ ounces
Extract jasmine.... 2 drachms
Extract musk...... 1 drachm

II.—A powder sometimes dispensed under this name is the salicylated powder of talcum of the National Formulary, which contains in every 1,000 parts 30 parts of salicylic acid and 100 parts of boric acid.

Rose.—

I.—Cornstarch......... 9 pounds
Powdered talc...... 1 pound
Oil of rose......... 80 drops
Extract musk...... 2 drachms
Extract jasmine.... 6 drachms

II.—Potato starch....... 9 pounds
Powdered talc...... 1 pound
Oil rose........... 45 drops
Extract jasmine.... ½ ounce

Rose Talc.—

I.—Powdered talc...... 5 pounds
Oil rose........... 50 drops
Oil wintergreen.... 4 drops
Extract jasmine.... 2 ounces

II.—Powdered talc...... 5 pounds
Oil rose........... 32 drops
Oil jasmine........ 4 ounces
Extract musk...... 1 ounce

Violet Talc.—

I.—Powdered talc...... 14 ounces
Powdered orris root. 2 ounces
Extract cassie...... ½ ounce
Extract jasmine..... ¼ ounce
Extract musk...... 1 drachm

II.—Starch.......... 5,000 parts
Orris root........ 1,000 parts
Oil of lemon..... 14 parts
Oil of bergamot.. 14 parts
Oil of clove..... 4 parts

Smelling Salts.—I.—Fill small glasses having ground stopper with pieces of sponge free from sand and saturate with a mixture of spirit of sal ammoniac (0.910), 9 parts, and oil of lavender, 1 part. Or else fill the bottles with small dice of ammonium sesquicarbonate and pour the above mixture over them.

II.—Essential oil of lavender.......... 18 parts
Attar of rose...... 2 parts
Ammonium carbonate.......... 480 parts

Violet Smelling Salts.—I.—Moisten coarsely powdered ammonia carbonate, contained in a suitable bottle, with a mixture of concentrated tincture of orris root, 2½ ounces; aromatic spirit of ammonia, 1 drachm; violet extract, 3 drachms.

II.—Moisten the carbonate, and add as much of the following solution as it will absorb: Oil of orris, 5 minims; oil of lavender flowers, 10 minims; violet extract, 30 minims; stronger water of ammonia, 2 fluidounces.

To Scent Advertising Matter, etc.— The simplest way of perfuming printed matter, such as calendars, cards, etc., is to stick them in strongly odorous sachet powder. Although the effect of a strong perfume is obtained thereby, there is a large loss of powder, which clings to the printed matter. Again, there are often little spots which are due to the essential oils added to the powder.

Another way of perfuming, which is used especially in France for scenting cards and other articles, is to dip them in very strong "extraits d'odeur," leaving them therein for a few days. Then the cards are taken out and laid between filtering paper, whereupon they are pressed vigorously, which causes them not only to dry, but also to remain straight. They remain under strong pressure until completely dry.

Not all cardboard, however, can be subjected to this process, and in its choice one should consider the perfuming operation to be conducted. Nor can the cards be glazed, since spirit dissolves the glaze. It is also preferable to have lithographed text on them, since in the case of ordinary printing the letters often partly disappear or the colors are changed.

For pocket calendars, price lists, and voluminous matter containing more leaves than one, another process is recommended. In a tight closet, which should be lined with tin, so that little air can enter, tables composed of laths are placed on which nets stretched on frames are laid. Cover these nets with tissue paper, and proceed as follows: On the bottom of the closet sprinkle a strongly odorous and reperfumed powder; then cover one net with the printed matter to be perfumed and shove it to the closet on the lath. The next net again receives powder, the following one printed matter, and so on until the closet is filled. After tightly closing the doors, the whole arrangement is left to itself. This process presents another advantage in that all sorts of residues may be employed for scenting, such as the filters of the odors and infusions, residues of musk, etc. These are simply laid on the nets, and will thus impart their perfume to the printed matter.

Such a scenting powder is produced as follows:

By weight

Iris powder, finely
ground......... 5,000 parts
Residues of musk.. 1,000 parts
Ylang-ylang oil.... 10 parts
Bergamot oil...... 50 parts
Artificial musk.... 2 parts
Ionone...........2 to 5 parts
Tincture of benzoin 100 parts

The powder may subsequently be employed for filling cheap sachets, etc.

LIQUID PERFUMES:

Coloring Perfumes.—Chlorophyll is a suitable agent for coloring liquid perfumes green. Care must be taken to procure an article freely soluble in the menstruum. As found in the market it is prepared (in form of solutions) for use in liquids strongly alcoholic; in water or weak alcohol; and in oils. Aniline greens of various kinds will answer the same purpose, but in a trial of any one of these it must be noted that very small quantities should be used, as their tinctorial power is so great that liquids in which they are incautiously used may stain the handkerchief.

Color imparted by chlorophyll will be found fairly permanent; this term is a relative one, and not too much must be expected. Colors which may suffer but little change by long exposure to diffused light may fade perceptibly by short exposure to the direct light of the sun.

Chlorophyll may be purchased or it may be prepared as follows: Digest leaves of grass, nettles, spinach, or other green herb in warm water until soft; pour off the water and crush the herb to a pulp. Boil the pulp for a short time with a half per cent solution of caustic soda, and afterwards precipitate the chlorophyll by means of dilute hydrochloric acid; wash the precipitate thoroughly with water, press and dry it, and use as much for the solution as may be necessary. Or a tincture made from grass as follows may be employed:

Lawn grass, cut fine.. 2 ounces
Alcohol............. 16 ounces

Put the grass in a wide-mouthed bottle, and pour the alcohol upon it. After standing a few days, agitating occasionally, pour off the liquid. The tincture may be used with both alcoholic and aqueous preparations.

Among the anilines, spirit soluble malachite green has been recommended.

A purple or violet tint may be produced by using tincture of litmus or ammoniated cochineal coloring. The former is made as follows:

Litmus............. 2½ ounces
Boiling water........ 16 ounces
Alcohol............. 3 ounces

Pour the water upon the litmus, stir well, allow to stand for about an hour, stirring occasionally, filter, and to the filtrate add the alcohol.

The aniline colors "Paris violet" or methyl violet B may be similarly employed. The amount necessary to produce a desired tint must be worked out by experiment. Yellow tints may best be imparted by the use of tincture of turmeric or saffron, fustic, quercitron, etc.

If a perfumed spirit, as, for instance, a mouth wash, is poured into a wineglassful of water, the oils will separate at once and spread over the surface of the water. This liquid being allowed to stand uncovered, one oil after another will evaporate, according to the degree of its volatility, until at last the least volatile remains behind.

This process sometimes requires weeks, and in order to be able to watch the separate phases of this evaporation correctly, it is necessary to use several glasses and to conduct the mixtures at certain intervals. The glasses must be numbered according to the day when set up, so that they may be readily identified.

If we assume, for example, that a mouth wash is to be examined, we may probably prepare every day for one week a mixture of about 100 grams of water and 10 drops of the respective liquid. Hence, after a lapse of 7 days we will have before us 7 bouquets, of different odor, according to the volatility of the oils contained in them. From these different bouquets the qualitative composition of the liquid may be readily recognized, provided that one is familiar enough with the character of the different oils to be able to tell them by their odors.

The predominance of peppermint oil—to continue with the above example—will soon be lost and other oils will rise one after the other, to disappear again after a short time, so that the 7 glasses afford an entire scale of characteristic odors, until at last only the most lasting are perceptible. Thus it is possible with some practice to tell a bouquet pretty accurately in its separate odors.

In this manner interesting results are often reached, and with some perseverance even complicated mixtures can be analyzed and recognized in their distinctiveness. Naturally the difficulty in recognizing each oil is increased in the case of oils whose volatility is approximately the same. But even in this case changes, though not quite so marked, can be determined in the bouquet.

In a quantitative respect this method also furnishes a certain result as far as the comparison of perfumed liquids is concerned.

According to the quantity of the oils present the dim zone on the water is broader or narrower, and although the size of this layer may be changed by the admixture of other substances, one gains an idea regarding the quantity of the oils by mere smelling. It is necessary, of course, to choose glasses with equally large openings and to count out the drops of the essence carefully by means of a dropper.

When it is thought that all the odors have been placed, a test is made by preparing a mixture according to the recipe resulting from the trial.

Not pure oils, always alcoholic dilutions in a certain ratio should be used, in order not to disturb the task by a surplus of the different varieties, since it is easy to add more, but impossible to take away.

It is true this method requires patience, perseverance, and a fine sense of smell. One smelling test should not be considered sufficient, but the glasses should be carried to the nose as often as possible.

Fixing Agents in Perfumes.—The secret of making perfumery lies mainly in the choice of the fixing agents—i. e., those bodies which intensify and hold the floral odors. The agents formerly employed were musk, civet, and ambergris, all having a heavy and dull animal odor, which is the direct antithesis of a floral fragrance. A free use of these bodies must inevitably mean a perfume which requires a label to tell what it is intended for, to say nothing of what it is. To-day there is no evidence that the last of these (ambergris) is being used at all in the newer perfumes, and the other two are employed very sparingly, if at all. The result is that the newer perfumes possess a fragrance and a fidelity to the flowers that they imitate which is far superior to the older perfumes. Yet the newer perfume is quite as prominent and lasting as the old, while it is more pleasing. It contains the synthetic odors, with balsams or rosinous bodies as fixatives, and employs musk and civet only in the most sparing manner in some of the more sensitive odors. As a fixing agent benzoin is to be recommended. Only the best variety should be used, the Siamese, which costs 5 or 6 times as much as that from Sumatra. The latter has a coarse pungent odor.

Musk is depressing, and its use in cologne in even the minutest quantity will spoil the cologne. The musk lingers after the lighter odors have disappeared, and a sick person is pretty sure to feel its effects. Persons in vigorous health will not notice the depressing effects of musk, but when lassitude prevails these are very unpleasant. Moreover, it is not a necessity in these toilet accessories, either as a blending or as a fixing agent. Its place is better supplied by benzoin for both purposes.

As to alcohol, a lot of nonsense has been written about the necessity of extreme care in selecting it, such as certain kinds requiring alcohol made from grapes and others demanding extreme purification, etc. A reasonable attention to a good quality of alcohol, even at a slight increase in cost, will always pay, but, other things being equal, a good quality of oils in a poor quality of alcohol will give far better satisfaction than the opposite combination. The public is not composed of exacting connoisseurs, and it does not appreciate extreme care or expense in either particular. A good grade of alcohol, reasonably free from heavy and lingering foreign odors, will answer practically all the requirements.

General Directions for Making Perfumes.—It is absolutely essential for obtaining the best results to see that all vessels are perfectly clean. Always employ alcohol, 90 per cent, deodorized by means of charcoal. When grain musk is used as an ingredient in liquid perfumes, first rub down with pumice stone, then digest in a little *hot* water for 2 or 3 hours; finally add to alcohol. The addition of 2 or 3 minims of acetic acid will improve the odor and also prevent accumulation of NH$_3$. Civet and ambergris should also be thoroughly rubbed down with some coarse powder, and transferred directly to alcohol.

Seeds, pods, bark rhizomes, etc., should be cut up in small pieces or powdered.

Perfumes improve by storing. It is a good plan to tie over the mouth of the containing vessel some fairly thick porous material, and to allow the vessel to stand for a week or two in a cool place, instead of corking at once.

It is perhaps unnecessary to add that as large a quantity as possible should be decanted, and then the residue filtered. This obviously prevents loss by evaporation. Talc or kieselguhr (amorphous SiO$_2$) are perhaps the best substances to add to the filter in order to render liquid perfumes bright and clear, and more especially necessary in the case of aromatic vinegars.

The operations involved in making perfumes are simple; the chief thing to be learned, perhaps, is to judge of the quality of materials.

The term "extract," when used in most formulas, means an alcoholic solution of the odorous principles of certain flowers obtained by enfluerage; that is, the flowers are placed in contact with prepared grease which absorbs the odorous matter, and this grease is in turn macerated with alcohol which dissolves out the odor. A small portion of the grease is taken up also at ordinary temperatures; this is removed by filtering the "extract" while "chilled" by a freezing mixture. The extracts can be either purchased or made directly from the pomade (as the grease is called). To employ the latter method successfully some experience may be necessary.

The tinctures are made with 95 per cent deodorized alcohol, enough menstruum being added through the marc when filtering to bring the finished preparation to the measure of the menstruum originally taken.

The glycerine is intended to act as a "fixing" agent—that is, to lessen the volatility of the perfumes.

Tinctures for Perfumes.—

a. Ambergris, 1 part; alcohol, 96 per cent, 15 parts.

b. Benzoin, Sumatra, 1 part; alcohol, 96 per cent, 6 parts.

c. Musk, 1 part; distilled water, 25 parts; spirit, 96 per cent, 25 parts.

d. Musk, 1 part; spirit, 96 per cent, 50 parts; for very oleiferous compositions.

e. Peru balsam, 1 part in spirit, 96 per cent, 7 parts; shake vigorously.

f. Storax, 1 part in spirit, 96 per cent, 15 parts.

g. Powdered Tolu balsam, 1 part; spirit, 96 per cent, 6 parts.

h. Chopped Tonka beans, 1 part; spirit, 60 per cent, 6 parts; for compositions containing little oil.

i. Chopped Tonka beans, 1 part; spirit, 96 per cent, 6 parts; for compositions containing much oil.

j. Vanilla, 1 part; spirit, 60 per cent, 6 parts; for compositions containing little oil.

k. Vanilla, 1 part; spirit, 96 per cent, 6 parts; for compositions containing much oil.

l. Vanillin, 20 parts; spirit, 96 per cent, 4,500 parts.

m. Powdered orris root, 1 part; spirit, 96 per cent, 5 parts.

n. Grated civet, 1 part in spirit, 96 per cent, 10 parts.

Bay Rum.—Bay rum, or more properly bay spirit, may be made from the oil with weak alcohol as here directed:

I.—		
Oil of bay leaves....	3	drachms
Oil of orange peel...	½	drachm
Tincture of orange peel.............	2	ounces
Magnesium carbonate.	½	ounce
Alcohol............	4	pints
Water.............	4	pints

Triturate the oils with the magnesium carbonate, gradually adding the other ingredients previously mixed, and filter.

The tincture of orange peel is used chiefly as a coloring for the mixture.

Oil of bay leaves as found in the market varies in quality. The most costly will presumably be found the best, and its use will not make the product expensive. It can be made from the best oil and deodorized alcohol and still sold at a moderate price with a good profit.

Especial care should be taken to use only perfectly fresh oil of orange peel. As is well known, this oil deteriorates rapidly on exposure to the air, acquiring an odor similar to that of turpentine. The oil should be kept in bottles of such size that when opened the contents can be all used in a short time.

II.—Bay oil, 15 parts; sweet orange oil, 1 part; pimento oil, 1 part; spirit of wine, 1,000 parts; water, 750 parts; soap spirit or quillaia bark, ad libitum.

III.—Bay oil, 12.5 parts; sweet orange oil, 0.5 part; pimento oil, 0.5 part; spirit of wine, 200 parts; water, 2,800 parts; Jamaica rum essence, 75 parts; soap powder, 20 parts; quillaia extract, 5 parts; borax, 10 parts; use sugar color.

Colognes.—In making cologne water, the alcohol used should be that obtained from the distillation of wine, provided a first-class article is desired. It is possible, of course, to make a good cologne with very highly rectified and deodorized corn or potato spirits, but the product never equals that made from wine spirits. Possibly the reason for this lies in the fact that the latter always contains a varying amount of oenanthic ether.

I.—
Oil of bergamot..	10 parts
Oil of neroli......	15 parts
Oil of citron......	5 parts
Oil of cedrat.....	5 parts
Oil of rosemary...	1 part
Tincture of ambergris........:	5 parts
Tincture of benzoin..........	5 parts
Alcohol.........	1,000 parts

II.—The following is stated to be the "original" formula:

Oil of bergamot.	96 parts
Oil of citron	96 parts
Oil of cedrat....	96 parts
Oil of rosemary.	48 parts
Oil of neroli.....	48 parts
Oil of lavender..	48 parts
Oil of cavella....	24 parts
Absolute alcohol.	1,000 parts
Spirit of rosemary........	25,000 parts

III.—
Alcohol, 90 per cent..........	5,000 parts
Bergamot oil.....	220 parts
Lemon oil.......	75 parts
Neroli oil........	20 parts
Rosemary oil.....	5 parts
Lavender oil, French........	5 parts

The oils are well dissolved in spirit and left alone for a few days with frequent shaking. Next add about 40 parts of acetic acid and filter after a while.

IV.—
Alcohol, 90 per cent..........	5,000 parts
Lavender oil, French........	35 parts
Lemon oil.......	30 parts

Portugallo oil....	30 parts
Neroli oil........	15 parts
Bergamot oil.....	15 parts
Petit grain oil....	4 parts
Rosemary oil.....	4 parts
Orange water....	700 parts

Cologne Spirits or Deodorized Alcohol.—This is used in all toilet preparations and perfumes. It is made thus:

Alcohol, 95 per cent..	1 gallon
Powdered unslaked lime..............	4 drachms
Powdered alum	2 drachms
Spirit of nitrous ether	1 drachm

Mix the lime and alum, and add them to the alcohol, shaking the mixture well together; then add the sweet spirit of niter and set aside for 7 days, shaking occasionally; finally filter.

Florida Waters.—

Oil of bergamot...	3 fluidounces
Oil of lavender ...	1 fluidounce
Oil of cloves......	1¼ fluidrachms
Oil of cinnamon ..	2½ fluidrachms
Oil of neroli......	½ fluidrachm
Oil of lemon	1 fluidounce
Essence of jasmine	6 fluidounces
Essence of musk..	2 fluidounces
Rose water......	1 pint
Alcohol..........	8 pints

Mix, and if cloudy, filter through magnesium carbonate.

Lavender Water.—This, the most famous of all the perfumed waters, was originally a distillate from a mixture of spirit and lavender flowers. This was the perfume. Then came a compound water, or "palsy water," which was intended strictly for use as a medicine, but sometimes containing ambergris and musk, as well as red sanders wood. Only the odor of the old compound remains to us as a perfume, and this is the odor which all perfume compounders endeavor to hit. The most important precaution in making lavender water is to use well-matured oil of lavender. Some who take pride in this perfume use no oil which is less than 5 years old, and which has had 1 ounce of rectified spirit added to each pound of oil before being set aside to mature. After mixing, the perfume should stand for at least a month before filtering through gray filtering paper. This may be taken as a general instruction:

I.—
Oil of lavender.....	1½ ounces
Oil of bergamot	4 drachms
Essence ambergris..	4 drachms
Proof spirit........	3 pints

II.—English oil of laven-
der............... 1 ounce
Oil of bergamot.... 1½ drachms
Essence of musk
(No. 2)........... ½ ounce
Essence of amber-
gris.............. ½ ounce
Proof spirit........ 2 pints

III.—English oil of laven-
der.............. ½ ounce
Oil of bergamot.... 2 drachms
Essence of amber-
gris.............. 1 drachm
Essence of musk
(No. 1).......... 3 drachms
Oil of angelica...... 2 minims
Attar of rose....... 6 minims
Proof spirit........ 1 pint

IV.—Oil of lavender..... 4 ounces
Grain musk........ 15 grains
Oil of bergamot.... 2½ ounces
Attar of rose....... 1½ drachms
Oil of neroli........ ½ drachm
Spirit of nitrous
ether.......... 2½ ounces
Triple rose water... 12 ounces
Proof spirit........ 5 pints
Allow to stand 5 weeks before filtering.

LIQUID PERFUMES FOR THE HAND-KERCHIEF, PERSON, ETC.:

Acacia Extract.—

French acacia...... 400 parts
Tincture of amber
(1 in 10)......... 3 parts
Eucalyptus oil...... 0.5 parts
Lavender oil....... 1 part
Bergamot oil...... 1 part
Tincture of musk... 2 parts
Tincture of orris root 150 parts
Spirit of wine, 80 per
cent............. 500 parts

Bishop Essence.—

Fresh green peel of
unripe oranges.. 60.0 grams
Curaçao orange peel 180.0 grams
Malaga orange peel 90.0 grams
Ceylon cinnamon.. 2.0 grams
Cloves........... 7.5 grams
Vanilla........... 11.0 grams
Orange flower oil.. 4 drops
Spirit of wine...... 1,500.0 grams
Hungarian wine... 720.0 grams
A dark-brown tincture of pleasant taste and smell.

Caroline Bouquet.—

Oil of lemon......... 15 minims
Oil of bergamot...... 1 drachm
Essence of rose...... 4 ounces
Essence of tuberose.. 4 ounces
Essence of violet..... 4 ounces
Tincture of orris..... 2 ounces

Alexandra Bouquet.—

Oil of bergamot...... 3½ drachms
Oil of rose geranium ½ drachm
Oil of rose......... ½ drachm
Oil of cassia....... 15 minims
Deodorized alcohol... 1 pint

Navy Bouquet.—

Spirit of sandalwood.. 10 ounces
Extract of patchouli.. 10 ounces
Spirit of rose........ 10 ounces
Spirit of vetivert..... 10 ounces
Extract of verbena... 12 ounces

Bridal Bouquet.—

Sandal oil, 30 minims; rose extract, 4 fluidounces; jasmine extract, 4 fluidounces; orange flower extract, 16 fluidounces; essence of vanilla, 1 fluidounce; essence of musk, 2 fluidounces; tincture of storax, 2 fluidounces. (The tincture of storax is prepared with liquid storax and alcohol [90 per cent], 1:20, by macerating for 7 days.)

Irish Bouquet.—

White rose essence. 5,000 parts
Vanilla essence .... 450 parts
Rose oil.......... 5 parts
Spirit............. 100 parts

Essence Bouquet.—

I.—Spirit............ 8,000 parts
Distilled water.... 2,000 parts
Iris tincture....... 250 parts
Vanilla herb tinc-
ture............ 100 parts
Benzoin tincture... 40 parts
Bergamot oil...... 50 parts
Storax tincture.... 50 parts
Clove oil......... 15 parts
Palmarosa oil..... 12 parts
Lemon-grass oil... 15 parts

II.—Extract of rose (2d).. 64 ounces
Extract of jasmine
(2d).............. 12 ounces
Extract of cassie (2d). 8 ounces
Tincture of orris (1
to 4).............. 64 ounces
Oil of bergamot...... ½ ounce
Oil of cloves........ 1 drachm
Oil of ylang-ylang.... ½ drachm
Tincture of benzoin
(1 to 8).......... 2 ounces
Glycerine.......... 4 ounces

Bouquet Canang.—

Ylang-ylang oil... 45 minims
Grain musk....... 3 grains
Rose oil......... 15 minims
Tonka beans..... 3
Cassie oil........ 5 minims
Tincture orris rhi-
zome.......... 1 fluidounce

Civet............... 1   grain
Almond oil........   ½   minim
Storax tincture... 3   fluidrachms
Alcohol,90 per cent 9   fluidounces

Mix, and digest 1 month. The above is a very delicious perfume.

Cassie oil or otto is derived from the flowers of *Acacia farnesiana Mimosa farnesiana*, L. (N. O. Leguminosæ, sub-order Mimoseæ). It must not be confounded with cassia otto, the essential oil obtained from *Cinnamomum cassia*.

Cashmere Nosegay.—

I.—Essence of violet,
     from pomade..... 1 pint
   Essence of rose,
     from pomade..... 1½ pints
   Tincture of benzoin,
     (1 to 4).......... ½ pint
   Tincture of civet (1
     to 64) .... ¼ pint
   Tincture of Tonka (1
     to 4)............ ¼ pint
   Benzoic acid....... ½ ounce
   Oil of patchouli.... ¼ ounce
   Oil of sandal...... ½ ounce
   Rose water........ ½ pint

II.—Essence violet..... 120 ounces
   Essence rose...... 180 ounces
   Tincture benjamin
     (1 in 4).......... 60 ounces
   Tincture civet (1 in
     62)............. 30 ounces
   Tincture Tonka (1 in
     4).............. 30 ounces
   Oil patchouli...... 3 ounces
   Oil sandalwood..... 6 ounces
   Rose water........ 60 ounces

Clove Pink.—

I.—Essence of rose..... 2 ounces
   Essence of orange
     flower.......... 6 ounces
   Tincture of vanilla.. 3½ ounces
   Oil of cloves....... 20 minims

II.—Essence of cassie.... 5 ounces
   Essence of orange
     flower.......... 5 ounces
   Essence of rose..... 10 ounces
   Spirit of rose....... 7 ounces
   Tincture of vanilla.. 3 ounces
   Oil of cloves....... 12 minims

Frangipanni.—

I.—Grain musk..... 10 grains
   Sandal otto..... 25 minims
   Rose otto....... 25 minims
   Orange flower
     otto (neroli) 30 minims
   Vetivert otto.... 5 minims
   Powdered orris
     rhizome ...... ½ ounce

Vanilla .... 30 grains
Alcohol (90 per
  cent) .... 10 fluidounces

Mix and digest for 1 month. This is a lasting and favorite perfume.

II.—Oil of rose......... 2 drachms
   Oil of neroli....... 2 drachms
   Oil of sandalwood .. 2 drachms
   Oil of geranium
     (French)........ 2 drachms
   Tincture of vetivert
     (1¼ to 8)........ 96 ounces
   Tincture of Tonka (1
     to 8)............ 16 ounces
   Tincture of orris (1
     to 4)............ 64 ounces
   Glycerine.......... 6 ounces
   Alcohol............ 64 ounces

Handkerchief Perfumes.—

I.—Lavender oil..... 10 parts
   Neroli oil......... 10 parts
   Bitter almond oil.. 2 parts
   Orris root........ 200 parts
   Rose oil.......... 5 parts
   Clove oil .... 5 parts
   Lemon oil........ 1 part
   Cinnamon oil..... 2 parts

Mix with 2,500 parts of best alcohol, and after a rest of 3 days heat moderately on the water bath, and filter.

II.—Bergamot oil....... 10 parts
   Orange peel oil..... 10 parts
   Cinnamon oil...... 2 parts
   Rose geranium oil.. 1 part
   Lemon oil......... 4 parts
   Lavender oil....... 4 parts
   Rose oil........... 1 part
   Vanilla essence..... 5 parts

Mix with 2,000 parts of best spirit, and after leaving undisturbed for 3 days, heat moderately on the water bath, and filter.

Honeysuckle.—

   Oil of neroli........ 12 minims
   Oil of rose......... 10 minims
   Oil of bitter almond.. 8 minims
   Tincture of storax.... 4 ounces
   Tincture of vanilla... 6 ounces
   Essence of cassie..... 16 ounces
   Essence of rose...... 16 ounces
   Essence of tuberose.. 16 ounces
   Essence of violet..... 16 ounces

Iridia.—

   Coumarin.......... 10 grains
   Concentrated rose
     water (1 to 40) .... 2 ounces
   Neroli oil.......... 5 minims
   Vanilla bean........ 1 drachm
   Bitter almond oil..... 5 minims
   Orris root .... 1 drachm
   Alcohol............ 10 ounces

Macerate for a month.

Javanese Bouquet.—

Rose oil	15	minims
Pimento oil	20	minims
Cassia oil	3	minims
Neroli oil	3	minims
Clove oil	2	minims
Lavender oil	60	minims
Sandalwood oil	10	minims
Alcohol	10	ounces
Water	1½	ounces

Macerate for 14 days.

Lily Perfume.—

Essence of jasmine	1 ounce
Essence of orange flowers	1 ounce
Essence of rose	2 ounces
Essence of cassie	2 ounces
Essence of tuberose	8 ounces
Spirit of rose	1 ounce
Tincture of vanilla	1 ounce
Oil of bitter almond	2 minims

Lily of the Valley.—

I.—
Acacia essence	750 parts
Jasmine essence	750 parts
Orange flower essence	800 parts
Rose flower essence	800 parts
Vanilla flower essence	1,500 parts
Bitter almond oil	15 parts

II.—
Oil of bitter almond	10	minims
Tincture of vanilla	2	ounces
Essence of rose	2	ounces
Essence of orange flower	2	ounces
Essence of jasmine	2½	ounces
Essence of tuberose	2½	ounces
Spirit of rose	2½	ounces

III.—
Extract rose	200 parts
Extract vanilla	200 parts
Extract orange	800 parts
Extract jasmine	600 parts
Extract musk tincture	150 parts
Neroli oil	10 parts
Rose oil	6 parts
Bitter almond oil	4 parts
Cassia oil	5 parts
Bergamot oil	6 parts
Tonka beans essence	150 parts
Linaloa oil	12 parts
Spirit of wine (90 per cent)	3,000 parts

IV.—
Neroli extract	400 parts
Orris root extract	600 parts
Vanilla extract	400 parts
Rose extract	900 parts
Musk extract	200 parts

Orange extract	500 parts
Clove oil	6 parts
Bergamot oil	5 parts
Rose geranium oil	15 parts

Maréchal Niel Rose.—In the genus of roses, outside of the hundred-leaved or cabbage rose, the Maréchal Niel rose (Rosa Noisetteana Red), also called Noisette rose and often, erroneously, tea rose, is especially conspicuous. Its fine, piquant odor delights all lovers of precious perfumes. In order to reproduce the fine scent of this flower artificially at periods when it cannot be had without much expenditure, the following recipes will be found useful:

I.—
Infusion rose I (from pomades)	1,000	parts
Genuine rose oil	10	parts
Infusion Tolu balsam	150	parts
Infusion genuine musk I	40	parts
Neroli oil	30	parts
Clove oil	2	parts
Infusion tubereuse I (from pomades)	1,000	parts
Vanillin	1	part
Coumarin	0.5	parts

II.—
Triple rose essence	50	grams
Simple rose essence	60	grams
Neroli essence	30	grams
Civet essence	20	grams
Iris essence	30	grams
Tonka beans essence	20	grams
Rose oil	5	drops
Jasmine essence	60	grams
Violet essence	50	grams
Cassia essence	50	grams
Vanilla essence	45	grams
Clove oil	20	drops
Bergamot oil	10	drops
Rose geranium oil	20	drops

May Flowers.—

Essence of rose	10	ounces
Essence of jasmine	10	ounces
Essence of orange flowers	10	ounces
Essence of cassie	10	ounces
Tincture of vanilla	20	ounces
Oil of bitter almond	½	drachm

Narcissus.—

Caryophyllin	10	minims
Extract of tuberose	16	ounces
Extract of jasmine	4	ounces
Oil of neroli	20	minims
Oil of ylang-ylang	20	minims
Oil of clove	5	minims
Glycerine	30	minims

Almond Blossom.—

Extract of heliotrope	30 parts
Extract of orange flower	10 parts
Extract of jasmine	10 parts
Extract of rose	3 parts
Oil of lemon	1 part
Spirit of bitter almond, 10 per cent	6 parts
Deodorized alcohol	40 parts

Artificial Violet.—Ionone is an artificial perfume which smells exactly like fresh violets, and is therefore an extremely important product. Although before it was discovered compositions were known which gave fair imitations of the violet perfume, they were wanting in the characteristic tang which distinguishes all violet preparations. Ionone has even the curious property possessed by violets of losing its scent occasionally for a short time. It occasionally happens that an observer, on taking the stopper out of a bottle of ionone, perceives no special odor, but a few seconds after the stopper has been put back in the bottle, the whole room begins to smell of fresh violets. It seems to be a question of dilution. It is impossible, however, to make a usable extract by mere dilution of a 10 per cent solution of ionone.

It is advisable to make these preparations in somewhat large quantities, say 30 to 50 pounds at a time. This enables them to be stocked for some time, whereby they improve greatly. When all the ingredients are mixed, 10 days or a fortnight, with frequent shakings, should elapse before filtration. The filtered product must be kept in well-filled and well-corked bottles in a dry, dark, cool place, such as a well-ventilated cellar. After 5 or 6 weeks the preparation is ready for use.

Quadruple Extract.— By weight

Jasmine extract, 1st pomade	100 parts
Rose extract, 1st pomade	100 parts
Cassia extract, 1st pomade	200 parts
Violet extract, 1st pomade	200 parts
Oil of geranium, Spanish	2 parts
Solution of vanillin, 10 per cent	10 parts
Solution of orris, 10 per cent	100 parts
Solution of ionone, 10 per cent	20 parts

Infusion of musk	10 parts
Infusion of orris from coarsely ground root	260 parts

Triple Extract.— By weight

Cassia extract, 2d pomade	100 parts
Violet extract, 2d pomade	300 parts
Jasmine extract, 2d pomade	100 parts
Rose extract, 2d pomade	100 parts
Oil of geranium, African	1 part
Ionone, 10 per cent	15 parts
Solution of vanillin, 10 per cent	5 parts
Infusion of orris from coarse ground root	270 parts
Infusion of musk	10 parts

Double Extract.— By weight

Cassia extract, 2d pomade	100 parts
Violet extract, 2d pomade	150 parts
Jasmine extract, 2d pomade	100 parts
Rose extract, 2d pomade	100 parts
Oil of geranium, reunion	2 parts
Ionone, 10 per cent	10 parts
Solution of vanillin, 10 per cent	10 parts
Infusion of ambrette	20 parts
Infusion of orris from coarse ground root	300 parts
Spirit	210 parts

White Rose.—

Rose oil	25 minims
Rose geranium oil	20 minims
Patchouli oil	5 minims
Ionone	3 minims
Jasmine oil (synthetic)	5 minims
Alcohol	10 ounces

Ylang-Ylang Perfume.—

I.—

Ylang-ylang oil	10 minims
Neroli oil	5 minims
Rose oil	5 minims
Bergamot oil	3 minims
Alcohol	10 ounces

One grain of musk may be added.

II.—

Extract of cassie (2d)	96 ounces
Extract of jasmine (2d)	24 ounces

Extract of rose . . . . . 24 ounces
Tincture of orris . . . . 4 ounces
Oil of ylang-ylang . . 6 drachms
Glycerine . . . . 6 ounces

TOILET WATERS.

Toilet waters proper are perfumed liquids designed more especially as refreshing applications to the person—accessories to the bath and to the operations of the barber. They are used sparingly on the handkerchief also, but should not be of so persistent a character as the "extracts" commonly used for that purpose, as they would then be unsuitable as lotions.

Ammonia Water.—Fill a 6-ounce ground glass stoppered bottle with a rather wide mouth with pieces of ammonium carbonate as large as a marble, then drop in the following essential oils:

Oil of lavender . . . . . 30 drops
Oil of bergamot . . . . 30 drops
Oil of rose . . . . 10 drops
Oil of cinnamon . . . . 10 drops
Oil of clove . . . . 10 drops

Finally fill the bottle with stronger water of ammonia, put in the stopper and let stand overnight.

Birch-Bud Water.—Alcohol (96 per cent), 350 parts; water, 70 parts; soft soap, 20 parts; glycerine, 15 parts; essential oil of birch buds, 5 parts; essence of spring flowers, 10 parts; chlorophyll, quantity sufficient to tint. Mix the water with an equal volume of spirit and dissolve the soap in the mixture. Mix the oil and other ingredients with the remainder of the spirit, add the soap solution gradually, agitate well, allow to stand for 8 days and filter. For use, dilute with an equal volume of water.

Carmelite Balm Water.—

Melissa oil . . . . . . 30 minims
Sweet marjoram oil . . . . 3 minims
Cinnamon oil . . . . 10 minims
Angelica oil . . . . . 3 minims
Citron oil . . . . 30 minims
Clove oil . . . . 15 minims
Coriander oil . . . . 5 minims
Nutmeg oil . . . . . . 5 minims
Alcohol (90 per cent) . . . . 10 fluidounces

Angelica oil is obtained principally from the aromatic root of *Angelica archangelica*, L. (N. O. Umbelliferæ), which is commonly cultivated for the sake of the volatile oil which it yields.

Cypress Water.—

Essence of ambergris ½ ounce
Spirits of wine . . . . 1 gallon
Water . . . . 2 quarts

Distill a gallon.

Eau de Botot.—

Aniseed . . . . 80 parts
Clover . . . . 20 parts
Cinnamon cassia . . 20 parts
Cochineal . . . . 5 parts
Refined spirit . . . . . 800 parts
Rose water . . . . 200 parts

Digest for 8 days and add

Tincture of ambergris . . . . 1 part
Peppermint oil . . . . 10 parts

Eau de Lais.—

Eau de cologne . . . . . 1 part
Jasmine extract . . . . 0.5 parts
Lemon essence . . . . . 0.5 parts
Balm water . . . . 0.5 parts
Vetiver essence . . . . . 0.5 parts
Triple rose water . . . 0.5 parts

Eau de Merveilleuse.—

Alcohol . . . . 3 quarts
Orange flower water 4 quarts
Peru balsam . . . . 2 ounces
Clove oil . . . . 4 ounces
Civet . . . . 1½ ounces
Rose geranium oil . . ½ ounce
Rose oil . . . . 4 drachms
Neroli oil . . . . 4 drachms

Edelweiss.—

Bergamot oil . . . . . 10 grams
Tincture of ambergris . . . . 2 grams
Tincture of vetiver (1 in 10) . . . . 25 grams
Heliotropin . . . . . . 5 grams
Rose oil spirit (1 in 100) . . . . 25 grams
Tincture of musk . 5 drops
Tincture of angelica . . . . 12 drops
Neroli oil, artificial . . . . 10 drops
Hyacinth, artificial . . . . 15 drops
Jasmine, artificial . 1 gram
Spirit of wine, 80 per cent . . . . 1,000 grams

Honey Water.—

I.—Best honey . . . . 1 pound
Coriander seed . . . . . 1 pound
Cloves . . . . 1½ ounces
Nutmegs . . . . 1 ounce
Gum benjamin . . . . . 1 ounce
Vanilloes, No. 4 . . . . 1 drachm
The yellow rind of 3 large lemons.

Bruise the cloves, nutmegs, coriander seed, and benjamin, cut the vanilloes in pieces, and put all into a glass alembic with 1 gallon of clean rectified spirit, and, after digesting 48 hours, draw off the spirit by distillation. To 1 gallon of the distilled spirit add

Damask rose water.	1½ pounds
Orange flower water	1½ pounds
Musk.	5 grains
Ambergris	5 grains

Grind the musk and ambergris in a glass mortar, and afterwards put all together into a digesting vessel, and let them circulate 3 days and 3 nights in a gentle heat; then let all cool. Filter, and keep the water in bottles well stoppered.

II.—Oil of cloves	2½ drachms
Oil of bergamot	10 drachms
English oil of lavender	2½ drachms
Musk	4 grains
Yellow sandalwood.	2½ drachms
Rectified spirit	32 ounces
Rose water	8 ounces
Orange flower water	8 ounces
English honey	2 ounces

Macerate the musk and sandalwood in the spirit 7 days, filter, dissolve the oils in the filtrate, add the other ingredients, shake well, and do so occasionally, keeping as long as possible before filtering.

Lilac Water.—

Terpineol	2 drachms
Heliotropin	8 grains
Bergamot oil	1 drachm
Neroli oil	8 minims
Alcohol	12 ounces
Water	4 ounces

Orange Flower Water.—

Orange flower essence	8 ounces
Magnesium carbonate	1 ounce
Water	8 pints

Triturate the essence with the magnesium carbonate, add the water, and filter.

To Clarify Turbid Orange Flower Water.—Shake 1 quart of it with ¼ pound of sand which has previously been boiled out with hydrochloric acid, washed with water, and dried at red heat. This process doubtless would prove valuable for many other purposes.

Violet Waters.—

I.—Spirit of ionone, 10 per cent	½ drachm
Distilled water	5 ounces
Orange flower water	1 ounce

Rose water	1 ounce
Cologne spirit	8 ounces

Add the spirit of ionone to the alcohol and then add the waters. Let stand and filter.

II.—Violet extract	2 ounces
Cassie extract	1 ounce
Spirit of rose	½ ounce
Tincture of orris	½ ounce
Green coloring, a sufficiency.	
Alcohol to 20 ounces.	

PERFUMED PASTILLES.

These scent tablets consist of a compressed mixture of rice starch, magnesium carbonate, and powdered orris root, saturated with heliotrope, violet, or lilac perfume.

Violet.—

Ionone	50 parts
Ylang-ylang oil	50 parts
Tincture of musk, extra strong	200 parts
Tincture of benzoin	200 parts

Heliotrope.—

Heliotropin	200 parts
Vanillin	50 parts
Tincture of musk	100 parts
Tincture of benzoin	200 parts

Lilac.—

Terpineol	200 parts
Muguet	200 parts
Tincture of musk	200 parts
Tincture of benzoin	200 parts

Sandalwood	2 drachms
Vetivert	2 drachms
Lavender flowers	4 drachms
Oil of thyme	½ drachm
Charcoal	2 ounces
Potassium nitrate	½ ounce
Mucilage of tragacanth, a sufficient quantity.	

Perfumes for Hair Oils.—

I.—Heliotropin	8 grains
Coumarin	1 grain
Oil of orris	1 drop
Oil of rose	15 minims
Oil of bergamot	30 minims

II.—Coumarin	2 grains
Oil of cloves	4 drops
Oil of cassia	4 drops
Oil of lavender flowers	15 minims
Oil of lemon	45 minims
Oil of bergamot	75 minims

Soap Perfumes.—
See also Soap.

I.—Oil of lavender	½ ounce
Oil of cassia	30 minims
Add 5 pounds of soap stock.	

II.—Oil of caraway......
Oil of clove........
Oil of white thyme..
Oil of cassia........
Oil of orange leaf
(neroli petit grain)
Oil of lavender.....

1½ drachms of each

Add to 5 pounds of soap stock.

PERFUMES (FUMIGANTS):
See Fumigants.

PERSPIRATION REMEDY:
See Cosmetics.

Petroleum

(See also Oils.)

The Preparation of Emulsions of Crude Petroleum.—Kerosene has long been recognized as a most efficient insecticide, but its irritating action, as well as the very considerable cost involved, has prevented the use of the pure oil as a local application in the various parasitic skin diseases of animals.

In order to overcome these objections various expedients have been resorted to, all of which have for their object the dilution or emulsification of the kerosene. Probably the best known and most generally employed method for accomplishing this result is that which is based upon the use of soap as an emulsifying agent. The formula which is used almost universally for making the kerosene soap emulsion is as follows:

Kerosene........... 2 gallons
Water .... 1 gallon
Hard soap .... ½ pound

The soap is dissolved in the water with the aid of heat, and while this solution is still hot the kerosene is added and the whole agitated vigorously. The smooth white mixture which is obtained in this way is diluted before use with sufficient water to make a total volume of 20 gallons, and is usually applied to the skin of animals or to trees or other plants by means of a spray pump. This method of application is used because the diluted emulsion separates quite rapidly, and some mechanical device, such as a self-mixing spray pump, is required to keep the oil in suspension.

It will be readily understood that this emulsion would not be well adapted either for use as a dip or for application by hand, for in the one case the oil, which rapidly rises to the surface, would adhere to the animals when they emerged from the dipping tank and the irritating effect would be scarcely less than that produced by the plain oil, and in the second case the same separation of the kerosene would take place and necessarily result in an uneven distribution of the oil on the bodies of the animals which were being treated.

Within recent years it has been found that a certain crude petroleum from the Beaumont oil fields is quite effective for destroying the Texas fever cattle ticks. This crude petroleum contains from 40 to 50 per cent of oils boiling below 300° C. (572° F.), and from 1 to 1.5 per cent of sulphur. After a number of trials of different combinations of crude oil, soap, and water, the following formula was decided upon as the one best suited to the uses in view:

Crude petroleum..... 2 gallons
Water.............. ½ gallon
Hard soap.......... ½ pound

Dissolve the soap in the water with the aid of heat; to this solution add the crude petroleum, mix with a spray pump or shake vigorously, and dilute with the desired amount of water. Soft water should, of course, be used. Various forms of hard and soft soaps have been tried, but soap with an amount of free alkali equivalent to 0.9 per cent of sodium hydroxide gives the best emulsion. All the ordinary laundry soaps are quite satisfactory, but toilet soaps in the main are not suitable.

An emulsion of crude petroleum made according to this modified formula remains fluid and can be easily poured; it will stand indefinitely without any tendency toward a separation of the oil and water and can be diluted in any proportion with cold soft water. After sufficient dilution to produce a 10 per cent emulsion, a number of hours are required for all the oil to rise to the surface, but if the mixture is agitated occasionally, no separation takes place. After long standing the oil separates in the form of a creamlike layer which is easily mixed with the water again by stirring. It is therefore evident that for producing an emulsion which will hold the oil in suspension after dilution, the modified formula meets the desired requirements.

In preparing this emulsion for use in the field, a large spray pump capable of mixing 25 gallons may be used with perfect success.

In using the formula herewith given, it should be borne in mind that it is recommended especially for the crude petroleum obtained from the Beaumont oil fields, the composition of which has already been given. As crude petroleums from different sources vary greatly in their composition, it is impracticable to give a formula that can be used with all crude oils. Nevertheless, crude petroleum from other sources than the Beaumont wells may be emulsified by modifying the formula given above. In order to determine what modification of this formula is necessary for the emulsification of a given oil, the following method may be used:

Dissolve ½ pound of soap in ½ gallon of hot water; to 1 measure of this soap solution add 4 measures of the crude petroleum to be tested and shake well in a stoppered bottle or flask for several minutes.

If, after dilution, there is a separation of a layer of pure oil within half an hour the emulsion is imperfect, and a modification of the formula will be required. To accomplish this the proportion of oil should be varied until a good result is obtained.

Petroleum for Spinning.—In order to be able to wash out the petroleum or render it "saponifiable," the following process is recommended: Heat the mineral oil with 5 to 10 per cent of olein, add the proper amount of alcoholic lye and continue heating until the solvent (water alcohol) evaporates. A practical way is to introduce an aqueous lye at 230° F. in small portions and to heat until the froth disappears. For clearness it is necessary merely to evaporate all the water. In the same manner, more olein may be added as desired if the admixture of lye is kept down so that not too much soap is formed or the petroleum becomes too thick. After cooling, a uniform gelatinous mass results. This is liquefied mechanically, during or after the cooling, by passing it through fine sieves. Soap is so finely and intimately distributed in the petroleum that the finest particles of oil are isolated by soap, as it were. When a quantity of oil is intimately stirred into the water an emulsion results so that the different parts cannot be distinguished. The same process takes place in washing, the soap contained in the oil swelling between the fibers and the oil particles upon mixture with water, isolating the oil and lifting it from the fiber.

Deodorized Petroleum. — Petroleum may be deodorized by shaking it first with 100 parts of chlorinated lime for every 4,500 parts, adding a little hydro-chloric acid, then transferring the liquid to a vessel containing lime, and again shaking until all the chlorine is removed. After standing, the petroleum is decanted.

Petroleum Briquettes.—Mix with 1,000 parts of petroleum oil 150 parts of ground soap, 150 parts of rosin, and 300 parts of caustic soda lye. Heat this mixture while stirring. When solidification commences, which will be in about 40 minutes, the operation must be watched. If the mixture tends to overflow, pour into the receiver a few drops of soda, and continue to stir until the solidification is complete. When the operation is ended, flow the matter into molds for making the briquettes, and place them for 10 or 15 minutes in a stove; then they may be allowed to cool. The briquettes can be employed a few hours after they are made.

To the three elements constituting the mixture it is useful to add per 1,000 parts by weight of the briquettes to be obtained, 120 parts of sawdust and 120 parts of clay or sand, to render the briquettes more solid.

Experiments in the heating of these briquettes have demonstrated that they will furnish three times as much heat as briquettes of ordinary charcoal, without leaving any residue.

PETROLEUM EMULSION:
See Insecticides.

PETROLEUM JELLIES:
See Lubricants.

PETROLEUM SOAP:
See Soap.

PEWTER:
See Alloys.

PEWTER, TO CLEAN:
. See Cleaning Preparations and Methods.

PEWTER, AGEING:

If it is desired to impart to modern articles of pewter the appearance of antique objects, plunge the pieces for several moments into a solution of alum to which several drops of hydrochloric or sulphuric acid have been added.

PICTURES, GLOW.

These can be easily produced by drawing the outlines of a picture, writing, etc., on a piece of white paper with a solution of 40 parts of saltpeter and 20 parts of gum arabic in 40 parts of warm water, using a writing pen for this purpose. All the lines must connect and one of them must run to the edge of the paper, where it should be marked with a fine lead-pencil line. When a burning match is held to this spot, the line immediately glows on, spreading over the whole design, and the design formerly invisible finally appears entirely singed. This little trick is not dangerous.

PHOSPHATE SUBSTITUTE.

An artificial phosphate is thus prepared: Melt in an oven a mixture of 100 parts of phosphorite, ground coarsely, 70 parts of acid sulphate of soda; 20 parts of carbonate of lime; 22 parts of sand, and 607 parts of charcoal. Run the molten matter into a receiver filled with water; on cooling it will become granular. Rake out the granular mass from the water, and after drying, grind to a fine powder. The phosphate can be kept for a long time without losing its quality, for it is neither caustic nor hygroscopic. Wagner has, in collaboration with Dorsch, conducted fertilizing experiments for determining its value, as compared with superphosphate or with Thomas slag. The phosphate decomposes more rapidly in the soil than Thomas slag, and so far as the experiments have gone, it appears that the phosphoric acid of the new phosphate exercises almost as rapid an action as the phosphoric acid of the superphosphate soluble in water.

PHOSPHORESCENT MASS.

See also Luminous Bodies and Paints.

Mix 2 parts of dehydrated sodium carbonate, 0.5 parts of sodium chloride, and 0.2 parts of manganic sulphate with 100 parts of strontium carbonate and 30 parts of sulphur and heat 3 hours to a white heat with exclusion of air.

PHOSPHOR BRONZE:

See Alloys, under Bronzes.

PHOSPHORUS SUBSTITUTE.

G. Graveri recommends persul focyanic acid—$H_2(CN)_2S_3$ as meeting all the requirements of phosphorus on matches. It resists shock and friction, it is readily friable, and will mix with other substances; moreover, it is non-poisonous and cheaper than phosphorus.

Photography

DEVELOPERS AND DEVELOPING OF PLATES.

No light is perfectly safe or non-actinic, even that coming through a combined ruby and orange window or lamp. Therefore use great care in developing.

A light may be tested this way: Place a dry plate in the plate holder in total darkness, draw the slide sufficiently to expose one-half of the plate, and allow the light from the window or lamp, 12 to 18 inches distant, to fall on this exposed half for 3 or 4 minutes. Then develop the plate the usual length of time in total darkness. If the light is safe, there will be no darkening of the exposed part. If not safe, the remedy is obvious.

The developing room must be a perfectly dark room, save for the light from a ruby- or orange-colored window (or combination of these two colors). Have plenty of pure running water and good ventilation.

Plates should always be kept in a dry room. The dark room is seldom a safe place for storage, because it is apt to be damp.

Various developing agents give different results. Pyrogallic acid in combination with carbonate of sodium or carbonate of potassium gives strong, vigorous negatives. Eikonogen and metol yield soft, delicate negatives. Hydrochinon added to eikonogen or metol produces more contrast or greater strength.

It is essential to have a bottle of bromide of potassium solution, 10 per cent, in the dark room. (One ounce of bromide of potassium, water to 10 ounces.) Overtimed plates may be much improved by adding a few drops of bromide solution to the developer as soon as the overtimed condition is apparent (a plate is overtimed when the image appears almost immediately, and then blackens all over).

Undertimed plates should be taken out of the developer and placed in a tray of water where no light can reach them. If the detail in the shadows begins to appear after half an hour or so, the plate can be replaced in the developer and development brought to a finish.

Quick development, with strong solutions, means a lack of gradation or half-tones.

A developer too warm or containing too much alkali (carbonate of sodium or potassium) will yield flat, foggy negatives.

A developer too cold is retarded in its action, and causes thin negatives.

Uniform temperature is necessary for uniform results.

If development is continued too long, the negative will be too dense.

In warm weather, the developer should be diluted; in cold weather, it should be stronger.

The negative should not be exposed to white light until fixation is complete.

The negative should be left fully 5 minutes longer in the fixing bath than is necessary to dissolve out the white bromide of silver.

In hot weather a chrome alum fixing bath should be used to prevent frilling.

Always use a fresh hypo or fixing bath. Hypo is cheap.

Plates and plate holders must be kept free from dust, or pinholes will result.

After the negative is fixed, an hour's washing is none too much.

The plate should be dried quickly in warm weather else the film will become dense and coarse-grained.

Do not expect clean, faultless negatives to come out of dirty developing and fixing solutions and trays.

Pyro and Soda Developer.—

I.—Pure water......... 30 ounces
Sulphite soda, crystals .... 5 ounces
Carbonate soda, crystals .... 2½ ounces

II.—Pure water......... 24 ounces
Oxalic acid........ 15 grains
Pyrogallic acid..... 1 ounce

To develop, take of

Solution No. I...... 1 ounce
Solution No. II..... ½ ounce
Pure water......... 3 ounces

More water may be used in warm weather and less in cool weather.

If solution No. I is made by hydrometer test, use equal parts of the following:

Sulphite soda testing, 80°.
Carbonate soda testing, 40°.

One ounce of this mixture will be equivalent to 1 ounce of solution No. I.

Pyro and Potassium Developer.—

I.—Pure water......... 32 ounces
Sulphite soda, crystals .... 8 ounces
Carbonate potassium, dry........ 1 ounce

II.—Pure water......... 24 ounces
Oxalic acid........ 15 ounces
Pyrogallic acid..... 1 ounce

To develop, take of

Solution No. I...... 1 ounce
Solution No. II..... ½ ounce
Pure water......... 3 ounces

When the plate is fully developed, if the lights are too thin, use less water in the developer; if too dense, use more water.

Pyro and Metol Developer.—Good for short exposures:

I.—Pure water.......... 57 ounces
Sulphite soda, crystals............. 2½ ounces
Metol............ 1 ounce

II.—Pure water......... 57 ounces
Sulphite soda, crystals............. 2½ ounces
Pyrogallic acid..... ¼ ounce

III.—Pure water......... 57 ounces
Carbonate potassium............ 2½ ounces

To develop, take of

Pure water......... 3 ounces
Solution No. I..... 1 ounce
Solution No. II..... 1 ounce
Solution No. III.... 1 ounce

This developer may be used repeatedly by adding a little fresh developer as required.

Keep the used developer in a separate bottle.

Rodinal Developer.—One part rodinal to 30 parts pure water.

Use repeatedly, adding fresh as required.

Bromo-Hydrochinon Developer.—For producing great contrast and intensity, also for developing over-exposed plates.

I.—Distilled or ice water 25 ounces
Sulphite of soda, crystals .... 3 ounces
Hydrochinon....... ½ ounce
Bromide of potassium............ ¼ ounce

Dissolve by warming, and let cool before use.

II.—Water............. 25 ounces
Carbonate of soda, crystals.......... 6 ounces

Mix Nos. I and II, equal parts, for use.

Eikonogen Hydrochinon Developer.—

I.—Distilled or pure well water..... 32 ounces
Sodium sulphite, crystals........ 4 ounces
Eikonogen........ 240 grains
Hydrochinon...... 60 grains

II.—Water............. 32 ounces
Carbonate of potash 4 ounces

To develop, take

No. I............ 2 ounces
No. II............ 1 ounce
*Water............. 1 ounce

* For double-coated plates use 5 ounces of water.

By hydrometer:

I.—Sodium sulphite.
    solution to test 30   34 ounces
    Eikonogen........ 240 grains
    Hydrochinon...... 60 grains

II.—Carbonate of pot-
    ash solution to
    test 50........

To develop, take
    No. I............. 2 ounces
    No. II............. 1 ounce
    *Water............. 1 ounce

Hydrochinon Developer.—

I.—Hydrochinon....... 1 ounce
    Sulphite of soda,
      crystals......... 5 ounces
    Bromide of potas-
      sium............ 10 grains
    Water (ice or dis-
      tilled).......... 55 ounces

II.—Caustic potash.....180 grains
    Water............. 10 ounces

To develop:

Take of I, 4 ounces; II, ½ ounce. After use pour into a separate bottle. This can be used repeatedly, and with uniformity of results, by the addition of 1 drachm of I and 10 drops of II to every 8 ounces of old developer.

In using this developer it is important to notice the temperature of the room, as a slight variation in this respect causes a very marked difference in the time it takes to develop, much more so than with pyro. The temperature of room should be from 70° to 75° F.

Metol Developer.—

I.—Water............. 8 ounces
    Metol.............100 grains
    Sulphite of soda,
      crystals........ 1 ounce

II.—Water............. 10 ounces
    Potassium carbonate 1 ounce

Take equal parts of I and II and 6 parts of water. If more contrast is needed, take equal parts of I and II and 3 parts of water, with 5 drops to the ounce of a $\frac{1}{10}$ solution of bromide of potassium.

Metol and Hydrochinon Developer.—

I.—Pure hot water..... 80 ounces
    Metol............. 1 ounce
    Hydrochinon....... ⅛ ounce
    Sulphite soda, crys-
      tals............. 6 ounces

II.—Pure water........ 80 ounces
    Carbonate soda,
      crystals.......... 5 ounces

To develop, take of
    Pure water........ 2 ounces
    Solution No. I...... 1 ounce
    Solution No. II..... 1 ounce

Metol-Bicarbonate Developer.—Thoroughly dissolve
    Metol............. 1 ounce
    In water.......... 60 ounces

Then add
    Sulphite of soda,
      crystals.......... 6 ounces
    Bicarbonate of soda. 3 ounces

To prepare with hydrometer, mix
    Sulphite of soda so-
      lution, testing 75.. 30 ounces
    Bicarbonate of soda
      solution, testing 50 30 ounces
    Metol............. 1 ounce

Dissolved in 12 ounces water.

Ferrous-Oxalate Developer.—For transparencies and opals.

I.—Oxalate of potash... 8 ounces
    Water............. 30 ounces
    Citric acid........ 60 grains
    Citrate of ammonia
      solution......... 2 ounces

II.—Sulphate of iron.... 4 ounces
    Water............. 32 ounces
    Sulphuric acid...... 16 drops

III.—Citrate of ammonia
    solution saturated.

Dissolve 1 ounce citric acid in 5 ounces distilled water, add liquor ammonia until a slip of litmus paper just loses the red color, then add water to make the whole measure 8 ounces.

Add 1 ounce of II to 2 of I, and ½ ounce of water, and 3 to 6 drops of 10 per cent solution bromide potassium.

To develop, first rinse developing dish with water, lay film or plate down, and flow with sufficient developer to well cover. Careful attention must be given to its action, and when detail is just showing in the face, or half-tone lights in a view, pour off developer, and well wash the film before placing in the fixing bath.

Tolidol Developer.—Standard formula for dry plates and films:
    Water......... 16 ounces
    Tolidol........ 24 grains
    Sodium sul-
      phite....... 72 (144) grains
    Sodium car-
      bonate...... 96 (240) grains

The figures in parenthesis are for crystals. It will be seen that in every case

---

* For double-coated plates use 5 ounces of water.

the weight of sulphite required in crystals is double that of dry sulphite, while the weight of carbonate crystals is $2\frac{1}{2}$ times as much as dry carbonate.

For tank development Dr. John M. Nicol recommends the standard formula diluted with 6 times the amount of water, and the addition of 1 drop of retarder to every ounce after dilution.

To obtain very strong negatives:

Water........	16 ounces
Tolidol.......	50 to 65 grains
Sodium sulphite.......	80 (160) grains
Sodium carbonate.....	120 (300) grains

On some brands of plates the addition of a little retarder will be necessary.

If stock solutions are preferred, they may be made as follows:

Solution A

Water..........	32 ounces
Tolidol..........	1 ounce
Sodium sulphite..	1 (2) ounce

Solution B

Water..........	32 ounces
Sodium sulphite..	2 (4) ounces

Solution C

Water..........	32 ounces
Sodium carbonate	4 (10) ounces

If preferred, stock solutions B and C can be made by hydrometer, instead of by weight as above. The solutions will then show:

Solution B

Sodium sulphite.... 40

Solution C

Sodium carbonate.. 75

Or if potassium carbonate is preferred instead of sodium:

Solution C

Potassium carbonate 60

For standard formula for dry plates and films, mix

Solution A........	1 part
Solution B........	1 part
Solution C........	1 part
Water...........	7 parts

For strong negatives (for aristo-platino):

Solution A.....$1\frac{1}{2}$ to 2	parts
Solution B.....	1 part
Solution C.....	1 part
Water........4 to $4\frac{1}{2}$	parts

For tank development:

Solution A........	1 part
Solution B........	1 part
Solution C........	1 part
Water	35 parts

For developing paper:

Solution A........	2 parts
Solution B........	2 parts
Solution C........	1 part

The reading of the hydrometer for stock solutions is the same whether dried chemicals or crystals are used. No water is used.

Pyrocatechin-Phosphate Developer.—

Solution A

Crystallized sulphite of soda.........	386 grains
Pyrocatechin......	77 grains
Water...........	8 ounces

Solution B

Ordinary crystal phosphate of sodium..........	725 grains
Caustic soda (purified in sticks)....	77 grains
Water...........	8 ounces

Mix 1 part of A with 1 part of B and from 1 to 3 parts of water. If the exposure is not absolutely normal we recommend to add to the above developer a few drops of a solution of bromide of potassium (1.10).

Pyrocatechin Developer (One Solution).—Dissolve in the following range:

Sulphite of soda crystallized..........	$25\frac{1}{2}$ drachms
Caustic soda (purified in sticks)	$3\frac{1}{2}$ drachms
Distilled water.....	14 ounces
Pyrocatechin.......	308 grains

The pyrocatechin must not be added until the sulphite and caustic soda are entirely dissolved. For use the concentrated developer is to be diluted with from 10 to 20 times as much water. The normal proportion is 1 part of developer in 15 parts of water.

Vogel's Pyrocatechin Combined Developer and Fixing Solution.—

Sulphite of soda crystallized	468 grains
Water...........	$2\frac{5}{8}$ ounces
Caustic potash (purified in sticks)........	108 grains
Pyrocatechin......	108 grains

Mix for a formally fixing plate of 5 x 7 inches.

Developer.........	3 drachms
Fixing soda solution (1:5)	$5\frac{1}{2}$ drachms
Water...........	1 ounce

The process of developing and fixing with this solution is accomplished in a few minutes. The picture first appears usually, strengthens very quickly, and shortly after the fixing is entirely done.

Ellon's Pyrocatechin Developer.— Pyrocatechin, 2 per cent solution (2 grams pyrocatechin in 100 cubic centimeters of water).

Carbonate of potassium, 10 per cent solution (10 grams carbonate in 100 cubic centimeters of water).

For use take equal parts and add water as desired.

Imperial Standard Pyro Developer.—

I.—Metabisulphite of
potassium...... 120 grains
Pyrogallic acid.... 55 grains
Bromide of potassium........... 20 grains
Metol........... 45 grains
Water........... 20 ounces

II.—Carbonate of soda. 4 ounces
Water........... 20 ounces
For use mix equal parts I and II.

Bardwell's Pyro-Acetone Developer.—

Water........... 4 ounces
Sulphite of sodium
(saturated solution)........... 4 drachms
Acetone........... 2 drachms
Pyro........... 10 grams

Hauff's Adurol Developer.—One solution.

Water........... 10 ounces
Sulphide of sodium,
crystals........... 4 ounces
Carbonate of potassium........... 3 ounces
Adurol........... ½ ounce

For studio work and snap shots take 1 part with 3 parts water.

For time exposures out-door take 1 part with 5 parts water.

Glycin Developer.—

I.—Hot water........ 10 ounces
Sulphite of sodium,
crystals........... 1½ ounces
Carbonate of sodium ¼ ounce
Glycin........... ½ ounce
Add to water in order given.

II.—Water........... 10 ounces
Carbonate of potash 1¼ ounces

For normal exposure take I, 1 ounce; II, 2 ounces; water, 1 ounce.

Imogen Developer.—

I.—Hot water........ 9 ounces
Sulphite of sodium,
crystals........ 385 grains
Imogen......... 123 grains

II.—Hot water........ 4½ ounces
Carbonate of sodium 2 ounces

For use take 2 ounces of I and 1 ounce of II.

Diogen Developer.—

Water........... 9 ounces
Sulphite of sodium.. 3½ ounces
Diogen........... 7 drachms
Carbonate of potassium........... 4½ ounces

For normal exposure take 4 drachms of this solution; dilute with 2 ounces, 1 drachm of water, and add 2 drops bromide of potassium, 10 per cent solution.

Ortol Developer.—Formula by Pentlarge.

I.—Water........... 1 ounce
Metabisulphite of
potassium....... 4 grains
Ortol........... 8 grains

II.—Water........... 1 ounce
Sulphite of sodium.. 48 grains
Carbonate of potassium........... 16 grains
Carbonate of sodium 32 grains

For use take equal parts I and II, and an equal bulk of water.

Metacarbol Developer.—

Metacarbol....... 25 grains
Sulphite of soda,
crystals........ 100 grains
Caustic soda...... 50 grains
Water........... 10 ounces

Dissolve the metacarbol in water, then add the sulphite, and when dissolved add the caustic soda and filter.

DEVELOPING POWDERS.

	By weight
I.—Pyrogallol.........	0.3 parts
Sodium bisulphite..	1.2 parts
Sodium carbonate..	1.2 parts
II.—Eikonogen.........	1.1 parts
Sodium sulphite....	2.4 parts
Potassium carbonate	1.5 parts
III.—Hydroquinone.....	0.6 parts
Sodium sulphite....	3.4 parts
Potassium bromide.	0.3 parts
Sodium carbonate..	7.0 parts

These three formulas each yield one powder. The powders should be put up in oiled paper, and carefully inclosed, besides, in a wrapper of black paper. For use, one powder is dissolved in about 60 parts of distilled water.

DEVELOPING PAPERS.

Light.—The paper can be safely handled 8 feet from the source of light which may be Welsbach gas light, covered with post-office paper, incandescent light, ordinary gas light, kerosene light, or reduced daylight, the latter produced by covering a window with one or more thicknesses of orange post-office paper, as necessitated by strength of light.

Expose by holding the printing frame close to gas, lamp, or incandescent light, or to subdued daylight. Artificial light is recommended in preference to daylight because of uniformity, and it being in consequence easier to judge the proper length of time to expose.

Exposure.—The amount of exposure required varies with the strength of the light; it takes about the same time with an ordinary gas burner and an incandescent light; a Welsbach gas light requires only about one-half as much time as the ordinary gas burner, and a kerosene light of ordinary size about three times as much as an ordinary gas burner. If daylight is to be used the window should be covered with post-office paper, in which a sub-window about 1 foot square for making the exposure may be made. Cover this window first with a piece of white tissue paper, then with a piece of black cloth or post-office paper to exclude the white light when not wanted. Make exposure according to strength of light at from 1 to 2 feet away from the tissue paper. Keep the printing frame when artificial light is used constantly in motion during exposure.

Timing the Exposure.—The time necessary for exposing is regulated by density of negative and strength of light. The further away the negative is from the source of light at the time of exposure the weaker the light; hence, in order to secure uniformity in exposure it is desirable always to make the exposure at a given distance from the light used. With a negative of medium density exposed 1 foot from an ordinary gas burner, from 1 to 10 minutes' exposure is required.

A test to ascertain the length of exposure should be made. Once the proper amount of exposure is ascertained with a given light, the amount of exposure required can be easily approximated by making subsequent exposures at the same distance from the same light; the only difference that it would then be necessary to make would be to allow for variation in density of different negatives.

Fixing.—Allow the prints to remain in the fixing solution 10 to 20 minutes, when they should be removed to a tray containing clear water.

Washing.—Wash 1 hour in running water, or in 10 or 12 changes of clear water, allowing prints to soak 2 to 3 minutes in each change.

Pyrocatechin Formula.—

Solution A

Pyrocatechin	2	parts
Sulphite of soda, crystals	2.5	parts
Water	100	parts

Solution B

Carbonate of soda	10 parts	
Water	100 parts	

Before using mix 20 parts of Solution A, and ½ part of Solution B.

Metol Quinol.—

Water	10	ounces
Metol	7	grains
Sodium sulphite, crystals, pure	½	ounce
Hydroquinone	30	grains
Sodium carbonate, dessicated (or 400 grains of crystallized carbonate).	200	grains
Ten per cent bromide of potassium solution, about	10	drops

Amidol Formula.—

Water	4 ounces	
Sodium sulphite, crystals, pure	200 grains	
Amidol, about	20 grains	
Ten per cent bromide of potassium solution, about	5 drops	

If the blacks are greenish, add more amidol; if whites are grayish, add more bromide of potassium.

Hypo-Acid Fixing Bath.—

Hypo	16 ounces	
Water	64 ounces	

Then add the following hardening solution:

Water	5	ounces
Sodium sulphite crystals	½	ounce
Commercial acetic acid (containing 25 per cent pure acid)	3	ounces
Powdered alum	½	ounce

Amidol Developer.—

Amidol	2	grains
Sodium sulphite	30	grains
Potassium bromide	1	grain
Water	1	ounce

With a fairly correct exposure this will be found to produce prints of a rich black tone, and of good quality. The whole secret of successful bromide printing lies in correctness of exposure. It is generally taken for granted that any poor, flat negative is good enough to yield a bromide print, but this is not so. A negative of good printing quality on printing-out paper will also yield a good print on bromide paper, but considerable care and skill are necessary to obtain a good result from a poor negative. The above developer will not keep in solution, and should be freshly prepared as required. The same formula will also be found useful for the development of lantern plates, but will only yield black-toned slides.

PLATINUM PAPERS:

General Instructions.—To secure the most brilliant results the sensitized paper, before, during, and after its exposure to light, must be kept as dry as possible.

The paper is exposed to daylight, in the printing frame, for about one-third of the time necessary for ordinary silver paper.

The print is then immersed in the developer for about 30 seconds, then cleared in 3 acid baths containing 1 part of muriatic acid C. P. to 60 parts of water, washed for a short time in running water, the whole operation of printing, clearing, and washing being complete in about half an hour.

As a general rule all parts of the picture except the highest lights should be visible when the exposure is complete.

When examining the prints in the printing frames, care should be taken not to expose them unduly to light; for the degradation of the whites of the paper due to slight action of light is not visible until after development.

Ansco Platinum Paper.—Print until a trace of the detail *desired* is slightly visible in the high lights.

Development.—Best results are obtained with the temperature of the developer from 60° to 80° F. Immerse the print in the developer with a quick sweeping motion to prevent air bells. Develop in artificial or weak daylight. The development of a print from a normal negative will require 40 seconds or more.

Formula for Developer.—

Water	50 ounces
Neutral oxalate of potash	8 ounces
Potassium phosphate (monobasic)	1 ounce

Care must be used to obtain the monobasic potassium phosphate.

Immediately after prints are developed, place them face down in the first acid bath, composed of

Muriatic acid, C. P.	1 ounce
Water	60 ounces

After remaining in this bath for a period of about 5 minutes, transfer to the second acid bath of the same strength. The prints should pass through at least 3 and preferably 4 acid baths, to remove all traces of iron that may remain in the pores of the paper.

When thoroughly cleared, the print should be washed from 10 to 20 minutes in running water. If running water is not available, several changes of water in the tray will be necessary.

"Water Tone" Platinum Paper.—"Water tone" platinum paper is very easily affected by moisture; it will, therefore, be noticed when printing in warm, damp weather that the print will show quite a tendency to print out black in the deep shadows. This must not be taken into consideration, as the same amount of exposure is necessary as in dry days.

Print by direct light (sunlight preferred) until the shadows are clearly outlined in a deep canary color. At this stage the same detail will be observed in the half tones that the finished print will show. For developing, use plain water, heated to 120° F. (which will be as hot as they can bear).

The development will be practically instantaneous, and care must be taken to avoid air bubbles forming upon the surface of the prints. Place prints, after developing, directly into a clearing bath of muriatic acid, 1 drachm to 12 ounces of water, and let them remain in this bath about 10 minutes, when they are ready for the final washing of 15 minutes in running water, or 5 changes of about 3 minutes each. Lay out between blotters to dry, and mount by attaching the corners.

Bradley Platinum Paper.—Developer.

A.—For black tones:

Neutral oxalate potassium	8 ounces
Potassium phosphate	1 ounce
Water	30 ounces

B.—For sepia tones:

Of above mixed solution	8 ounces
Saturated bichloride mercury solution	1 ounce
Citrate soda	5 grains

If deep red tones are desired add to B

Nitrate uranium..... 10 grains

Then filter and use as a developer.

W. & C. Platinotype.—Development.— The whole contents of the box of the W. & C. developing salts must be dissolved at one time, as the salts are mixed; and if this be not done, too large a proportion of one of the ingredients may be used.

Development should be conducted in a feeble white light, similar to that used when cutting up the paper, or by gas light.

It may take place immediately after the print is exposed, or at the end of the day's printing.

Develop by floating the print, exposed side downwards, on the developing solution.

Development may take 30 seconds or more.

During the hot summer days it is not advisable to unduly delay the development of exposed prints. If possible develop within 1 hour after printing.

Either porcelain or agate—preferably porcelain—dishes are necessary to hold the developing solution.

To clear the developed prints: These must be washed in a series of baths (not less than three) of a weak solution of muriatic acid C. P. This solution is made by mixing 1 part of acid in 60 parts of water.

As soon as the print has been removed from the developing dish it must be immersed face downwards in the first bath of this acid, contained in a porcelain dish, in which it should remain about 5 minutes; meanwhile other prints follow until all are developed. The prints must then be removed to a second acid bath for about 10 minutes; afterwards to the third bath for about 15 minutes. While the prints remain in these acid baths they should be moved so that the solution has free access to their surfaces, but care should be taken not to abrade them by undue friction.

Pure muriatic acid must be used.

If commercial muriatic acid be used, the prints will be discolored and turn yellow.

For each batch of prints fresh acid baths must be used.

After the prints have passed through the acid baths they should be well washed in three changes of water during about a half hour. It is advisable to add a pinch of washing soda to the second washing water to neutralize any acid remaining in the print. Do not use water that contains iron, as it tends to turn paper yellow. Soft water is the best for this purpose.

W. & C. Sepia Paper.—With a few exceptions the method of carrying out the operations is the same as for the "black" kinds of platinotype paper. The following points should be attended to:

The "sepia" paper is more easily affected by faint light, and, therefore, increased care must be taken when printing.

To develop, add to each ounce of the developing solution 1½ drachms of sepia solution supplied for this purpose, and proceed as described for black paper.

The solution must be heated to a temperature of 150° to 160° F., to obtain the greatest amount of brilliance and the warmest color, but very good results can be obtained by using a cooler developer.

Variations of the Sepia Developer.— Primarily the object of the sepia solution in the developer is to increase the brightness of the prints, as, for example, when the negative is thin and flat, or pense and flat, the addition of the sepia solution to the developer clears up, to some extent, the flatness of the print by taking out traces of the finer detail in the higher lights, which is often a decided improvement. If, however, the negative be dense, with clear shadows, the sepia solution may be discarded altogether. This will prevent the loss of any of the finer detail and greatly reduce harshness in the prints. Sometimes a half, or even a quarter, of the quantity of the sepia solution recommended as an addition to the developer will be sufficient, depending altogether upon the strength of the negatives. Prints developed without the solution have less of the sepia quality but are very agreeable nevertheless. It should be remembered that the sepia paper is totally different from the black, and will develop sepia tones on a developer to which no sepia solution has been added. The sepia solution clears up and brightens the flat, muddy (to some extent, not totally) effects from the thinner class of negatives.

The Glycerine Process.—The "glycerine process," or the process of developing platinotype prints by application of the developing agent with the brush, is perhaps one of the most interesting and fascinating of photographic processes, owing to its far-reaching possibilities.

By this method of developing platino-type paper, many negatives which have been discarded on account of the dim, flat, non-contrasty results which they yield, in the hands of one possessing a little artistic skill, produce snappy, animated pictures. On the other hand, from the sharp and hard negative, soft, sketchy effects may be secured.

There are required for this process: Some glass jars; some soft brushes, varying from the fine spotter and the Japanese brush to the $1\frac{1}{2}$-inch duster, and several pieces of special blotting paper.

Manipulation.—Print the paper a trifle deeper than for the ordinary method of developing. Place the print face up on a piece of clean glass (should the print curl so that it is unmanageable, moisten the glass with glycerine), and, with the broad camel's-hair brush, thinly coat the entire print with pure glycerine, blotting same off in 3 or 4 seconds; then recoat more thickly such portions as are desired especially restrained, or the details partly or entirely eliminated. Now brush or paint such portion of the print as is first desired with solution of 1 part glycerine and 4 parts normal developer, blotting the portion being developed from time to time to avoid developing too far. Full strength developer (without glycerine) is employed where a pronounced or deep shade is wanted.

When any part of the print has reached the full development desired, blot that portion carefully with the blotter and coat with pure glycerine.

A brown effect may be obtained by using saturated solution of mercury in the developer (1 part mercury to 8 parts developer). By the use of diluted mercury the "flesh tones" are produced in portraits, etc.

When print has reached complete development, place in hydrochloric (muriatic) acid and wash as usual.

Eastman's Sepia Paper.—This paper is about 3 times as rapid as blue paper. It should be under rather than over printed, and is developed by washing in plain water. After 2 or 3 changes of water fix 5 minutes in a solution of hypo ($1\frac{1}{2}$ grains to the ounce of water), and afterwards wash thoroughly.

Short fixing gives red tones. Longer fixing produces a brown tone.

Development of Platinum Prints.—In the development of platinotype prints by the hot bath process, distinctly warmer tones are obtained by using a bath which has been several times heated, colder blacks resulting from the use of a freshly prepared solution, and colder tones still if the developing solution be faintly acidified. The repeated heating of the solution of the neutral salt apparently has the effect of rendering the bath slightly alkaline by the conversion of a minute proportion of the oxalate into potassium carbonate. If this be the case, it allows a little latitude in choice of tone which may be useful. Some photographers recommend the use of potassium phosphate with the neutral oxalate, stating that the solution should be rendered acid by the addition of a small proportion of oxalic acid. When the potassium phosphate was first recommended for this purpose, probably the acid salt, KH_2PO_4, was intended, by the use of which cold steely black tones were obtained. The use of the oxalic acid with the ordinary phosphate K_2HPO_4, is probably intended to produce the same result.

THE CARBON PROCESS.

The paper used is coated on one surface with a mixture of gelatin and some pigment (the color of which depends upon the color the required print is to be), and then allowed to dry. When required for printing it is sensitized by floating upon a solution of bichromate of potassium, and then again drying, in the dark this time. The process is based upon the action of light upon this film of chromatized gelatin; wherever the light reaches, the gelatin is rendered insoluble, even in hot water.

The paper is exposed in the usual way. But as the appearance of the paper before and after printing is precisely the same, it is impossible to tell when it is printed by examining the print. This is usually accomplished by exposing a piece of gelatino-chloride paper under a negative of about the same density, and placing it alongside of the carbon print. When the gelatino-chloride paper is printed, the carbon will be finished. The paper is then removed from the printing frame and immersed in cold water, which removes a great deal of the bichromate of potassium, and also makes the print lie out flat. It is then floated on to what is known as a support, and pressed firmly upon it, face downwards, and allowed to remain for 5 or 10 minutes. Then the support, together with the print, is placed in hot water for a short time, and when the gelatin commences to ooze out at the edges the print is removed by stripping from the support, this process leaving the greater quantity of the gelatin and pigment upon the support. The gelatin and pigment are then treated with hot water by running the hot water over the face of the support by means of a sponge. This removes the soluble gelatin, and leaves the gelatin, together with the pigment it contains, which was acted upon by light; this then constitutes the picture.

The reason for transferring the gelatin film is quite apparent, since the greater portion of the unacted-upon gelatin will be at the back of the film, and in order to get at it to remove it, it is necessary to transfer it to a support. In this condition the print can be dried and mounted, but on consideration it will be seen that the picture is in a reversed position, that is to say, that the right-hand side of the original has become the left, and vice versa.

If the picture be finished in this condition, it is said to have been done by the single transfer method. In some instances this reversal would be of no consequence, such as some portraits, but with views which are known this would never do. In order to remedy this state of affairs, the picture is transferred once more, by pressing, while wet, upon another support, and allowed to dry upon it; when separated, the picture remains upon the latter support, and is in its right position. This is what is known as the double transfer method. When the double transfer method is used, the first support consists of a specially prepared support, which has been waxed in order to prevent the pictures from adhering permanently to it; this is then known as a temporary support. The paper upon which the print is finally received is prepared with a coating of gelatin, and is known as the final support.

LANTERN SLIDES.

The making of a good slide begins with the making of the negative, the operations in both cases being closely allied, and he who has mastered the first, which is the corner stone to all successful results in any branch of photography, may well be expected to be able to make a good lantern slide. A slide is judged not by what it appears to be when held in the hand, but by its appearance when magnified two to five thousand times on the screen, where a small defect in the slide will show up as a gross fault. Patience and cleanliness are absolutely necessary. The greatest caution should be observed to keep the lantern plates free from dust, both before and after exposure and development, for small pinholes and dust spots, hardly noticeable on the slide, assume huge proportions on the screen and detract materially from the slide's beauty.

The high lights in a slide should, in rare cases only, be represented by clear glass, and the shadows should always be transparent, even in the deepest part. The balance between these extremes should be a delicate gradation of tone from one to the other. The contrast between the strongest high light and the deepest shadow should be enough to give brilliancy without hardness and delicacy or softness without being flat. This is controlled also, to some extent, by the subject summer sunshine requiring a more vigorous rendering than hazy autumn effects, and herein each individual must decide for himself what is most necessary to give the correct portrayal of the subject. It is a good idea to procure a slide, as near technically perfect as possible, from some slide-making friend, or dealer, to use it as a standard, and to make slide after slide from the same negative until a satisfactory result is reached.

A black tone of good quality is usually satisfactory for most slides, but it is very agreeable to see interspersed a variety of tone, and beautiful slides can be made, where the subject warrants, in blue, brown, purple, and even red and green, by varying the exposure and development and by using gold or uranium toning baths and other solutions for that purpose, the formulas and materials for which are easily obtainable from the magazines and from stock dealers, respectively.

It must be understood, however, that these toning solutions generally act as intensifiers, and that if toning is contemplated, it should be borne in mind at the time of developing the slide, so that it may not finally appear too dense. Toning will improve otherwise weak slides, but will not help under-exposed ones, as its tendency will be in such case to increase the contrast, which in such slides is already too great. Another method of getting a fine quality of slides is to make rather strong exposures to over-develop, and then to reduce with persulphate of ammonium.

The popular methods of making the exposure are: First, by contact in the printing frame, just as prints are made on velox or other developing paper, provided the subject on the negative is of the right size for a lantern slide; and the other and better method is the camera method, by which the subject of any negative, large or small, or any part thereof, can be reduced or enlarged, and thus brought to the proper size desired for the slide. This is quite a knack, and should be considered and studied by the slide maker very carefully.

Hard and inflexible rules cannot be laid down in this relation. Portrait studies of bust or three-fourths figures or baby figures need not be made for a larger opening than 1½ by 2 inches, and often appear to good advantage if made quite a bit smaller. Figure or group compositions, with considerable background or accessories, may, of course, have a larger opening to suit the particular circumstances. Monuments, tall buildings, and the like should have the benefit of the whole height of mat opening of 2¾ inches, and should be made of a size to fill it out properly, providing, however, for sufficient foreground and a proper sky line. Landscapes and marine views generally can be made to fill out the full length of mat opening, which, however, should not exceed 2⅞ inches, and may be of any height to suit the subject, up to 2¾ inches.

The subject should be well centered on the plate and the part intended to be shown as the picture should be well within the size of the mat opening decided upon, so that with a slight variation of the placing of the mat no part of the picture will be cut off by the carrier in the stereopticon. The horizon line in a landscape, and more particularly in a marine view, should always be in proper position, either below or above the center line of the slide, as may suit the subject, but should never divide the picture in the middle and should not appear to be running either up or down hill. And the vertical lines in the pictures should not be leaning, but should run parallel with the side lines of the mat; this refers especially to the vertical lines in architecture, except, however, the Tower of Pisa and kindred subjects, which should in every case be shown with their natural inclinations.

As to time of exposure, very little can be said. That varies with the different makes of plates, with the quality of the light, and the nature and density of each individual negative. Therefore every one must be a judge unto himself and make as good a guess as he can for the first trial from each negative and gauge further exposures from the results thus obtained; but this much may be said, that a negative strong in contrast should be given a long exposure, close to the light, if artificial light is used, or in strong daylight, and developed with a weak or very much diluted developer to make a soft slide with full tone values. And a flat, weak negative will yield better results if exposed farther from the light or to a weaker light, and developed by a normal or more aggressive developer. Over exposure and under exposure show the same results in slide plates as in negative plates, and the treatment should be similar in both kinds of plates except that, perhaps, in cases of under exposure of slide plates, the better plan would be to cast them aside and make them over, as very little can be done with them. For getting bright and clear effects it is now well understood that better and more satisfactory results are obtained by backing the slide plates as well as by backing negative plates. This is accomplished by coating the back or glass side of the plate with the following mixture:

Gum arabic	½	ounce
Caramel	1	ounce
Burnt sienna	2	ounces
Alcohol	2	ounces

Mix and apply with small sponge or wad of absorbent cotton.

It should coat thin and smooth and dry hard enough so it will not rub off when handled. If the plates are put into a light-proof grooved box as fast as backed, they can be used about half an hour after being coated. Before developing, this backing should be removed; this is best done by first wetting the film side of the plate under the tap, which will prevent staining it, and then letting the water run on the backing, and, with a little rubbing, it will disappear in a few moments, when development may proceed. Other preparations for this purpose, ready for use, may be found at the stock houses. The mat should be carefully selected or cut of a size and shape to show up the subject to best advantage, and should cover everything not wanted in the picture. The opening should not exceed 2¾ x 2⅞ inches in any case, and must not be ragged or fuzzy, but clean cut and symmetrical. The lines of the opening of square mats should be parallel with the outside lines of the plate. Oval, or round, or other variously shaped mats, should be used sparingly, and in special cases only where the nature of the subject will warrant their use.

Statuary shows up to best advantage when the background is blocked out.

This is easily done with a small camel's-hair artist's brush and opaque or india ink, in a retouching frame, a good eye and a steady hand being the only additional requirements. This treatment may also be applied to some flower studies and other botanical subjects.

Binding may be performed with the aid of a stationer's spring clamp, such as is used for holding papers together, and can be purchased for 10 cents. Cut the binding strips the length of the sides and ends of the slide, and gum them on separately, rubbing them firmly in contact with the glass with a piece of cloth or an old handkerchief, which might be kept handy for that purpose, so that the binding may not loosen or peel off after the slides are handled but half a dozen times. Before storing the slides away for future use they should be properly labeled and named. The name label should be affixed on the right end of the face of the slide as you look at it in its proper position, and should contain the maker's name and the title of the slide. The thumb label should be affixed to the lower left-hand corner of the face of the slide, and may show the number of the slide.

HOW TO UTILIZE WASTE MATERIAL.

Undoubtedly spoiled negatives form the greatest waste. The uses to which a ruined negative may be put are manifold. Cut down to $3\frac{1}{4}$ inches square and the films cleaned off, they make excellent cover glasses for lantern slides. Another use for them in the same popular branch of photography is the following: If, during development, you see that your negative is spoiled through uneven density, over exposure, or what not, expose it to the light and allow it to blacken all over. Now with sealing wax fasten a needle to a penholder, and by means of this little tool one can easily manufacture diagram slides from the darkened film (white lines on black ground).

Take a spoiled negative, dissolve out all the silver with a solution of potassium ferricyanide and hypo. Rinse, dry, rub with sandpaper, and you will have a splendid substitute for ground glass.

Remove the silver in a similar manner from another negative, but this time wash thoroughly. Squeegee down on this a print, and an opaline will be your reward. From such an opaline, by cementing on a few more glasses, a tasteful letter weight may soon be made. Another way in which very thin negatives may be used is this: Bleach them in bichloride of mercury, back them with black paper, and positives will result. Old negatives also make good trimming boards, the film preventing a rapid blunting of the knife, and they may be successfully used as mounting tables. Clean off the films, polish with French chalk, and squeegee your prints thereto. When dry they may be removed and will have a fine enameled, if hardly artistic, appearance. Many other uses for them may also be found if the amateur is at all ingenious.

Users of pyro, instead of throwing the old developer away, should keep some of it and allow it to oxidize. A thin negative, if immersed in this for a few minutes, will be stained a deep yellow all over, and its printing quality will be much improved.

Old hypo baths should be saved, and, when a sufficient quantity of silver is thought to be in solution, reduced to recover the metal.

Printing paper of any sort is another great source of waste, especially to the inexperienced photographer. Prints are too dark or not dark enough successfully to undergo the subsequent operations. Spoiled material of this kind, however, is not without its uses in photography. Those who swear by the "combined bath," will find that scraps of printing-out paper, or any silver paper, are necessary to start the toning action.

Spoiled mat surface, printing-out paper, bromide paper, or platinotype should be allowed to blacken all over. Here we have a dead-black surface useful for many purposes. A leak in the bellows when out in the field may be repaired temporarily by moistening a piece of mat printing-out paper and sticking it on the leak; the gelatin will cause it to adhere. These papers may also be used to back plates, platinotypes, of course, requiring some adhesive mixture to make them stick.

In every photographer's possession there will be found a small percentage of stained prints. Instead of throwing these away, they may often be turned to good account in the following manner: Take a large piece of cardboard, some mountant, and the prints. Now proceed to mount them tastefully so that the corners of some overlap, arranging in every case to hide the stain. If you have gone properly to work, you will have an artistic mosaic. Now wash round with india ink, or paint a border of leaves, and the whole thing will form a very neat "tit bit."

Keep the stiff bits of cardboard between which printing paper is packed. They are useful in many ways—from opaque cards in the dark slide to partitions between negatives in the storing boxes.

In reclaiming old gold solutions, all liquids containing gold, with the exception of baths of which cyanide forms a part, must be strongly acidulated with chlorhydric or sulphuric acid, if they are not already acid in their nature. They are afterwards diluted with a large proportion of ordinary water, and a solution of sulphate of ferroprotoxide (green vitriol) is poured in in excess. It is recognized that the filtered liquid no longer contains gold when the addition of a new quantity of ferric sulphate does not occasion any cloudiness. Gold precipitated in the form of a reddish or blackish powder is collected on a filter and dried in an oven with weights equal to its own of borax, saltpeter, and carbonate of potash. The mass is afterwards introduced gradually into a fireproof crucible and carried to a white-red heat in a furnace. When all the matter has been introduced, a stronger blast is given by closing the furnace, so that all the metal collects at the bottom of the crucible. On cooling, a gold ingot, chemically pure, will be obtained. This mode of reduction is also suitable for impure chloride of gold, and for the removal of gilding, but not for solutions containing cyanides, which never give up all the gold they contain; the best means of treating the latter consists in evaporating them to dryness in a cast-iron boiler, and in calcining the residue in an earthen crucible at the white red. A small quantity of borax or saltpeter may be added for facilitating the fusion, but it is not generally necessary. The gold separated collects at the bottom of the crucible. It is red, if saltpeter is employed; and green, if it is borax.

To reclaim silver place the old films, plates, paper, etc., in a porcelain dish, so arranged that they will burn readily. To facilitate combustion, a little kerosene or denatured alcohol poured over the contents will be found serviceable.

Before blowing off the burnt paper, place the residue in an agateware dish, the bottom of which is covered with a solution of saltpeter and water. Place the whole on the fire, and heat it until the silver is separated as a nitrate.

The solution being complete, add to the mass a little water and hydrochloric acid, when in a short time the serviceable silver chloride will be obtained. If the films should not give up their silver as freely as the plates, then add a little more hydrochloric acid or work them up separately. Silver reclaimed in this way is eminently suitable for silver-plating all sorts of objects.

FIXING AND CLEARING BATHS:

The Acid Fixing and Clearing Bath.— Add 2 ounces of S. P. C. clarifier (acid bisulphite of sodium) solution to 1 quart of hypo solution 1 in 5.

Combined Alum and Hypo Bath.— Add saturated solution of sulphite of sodium to saturated solution of alum till the white precipitate formed remains undissolved, and when the odor of sulphurous acid becomes perceptible.

Mix this solution with an equal bulk of freshly prepared hypo solution 1 in 5, and filter.

This bath will remain clear.

Clearing Solution (Edward's).—

Alum	1 ounce	avoirdupois
Citric acid . .	1 ounce	avoirdupois
Sulphate of iron, crystals	3 ounces	avoirdupois
Water	1 imperial pint	

This should be freshly mixed.

Clearing Solution.—

Saturated solution of alum	20 ounces
Hydrochloric acid	1 ounce

Immerse negative after fixing and washing. Wash well after removal.

Reducer for Gelatin Dry-Plate Negatives.—

I.—Saturated solution of ferricyanide of potassium . . . . 1 part
Hyposulphite of sodium solution (1 in 10) 10 parts

II.—Perchloride of iron . . 30 grains
Citric acid . . . . 60 grains
Water . . . . 1 pint

Belitski's Acid Ferric-Oxalate Reducer for Gelatin Plates.—

Water	7	ounces
Potassium ferric oxalate . . . . : . . .	2½	drachms
Crystallized neutral sulphite of sodium .	2	drachms
Powdered oxalic acid, from . . . . 30 to	45	grains
Hyposulphite of soda .	1½	ounces

The solution must be made in this order, filtered, and be kept in tightly closed bottles; and as under the influence of light the ferric salt is reduced to ferrous, the preparation must be kept in subdued light, in non-actinic glass bottles.

Orthochromatic Dry Plates—Erythrosine Bath (Mallman and Scolik).—Preliminary bath:

Water......	200 cubic centimeters
Stronger ammonia....	2 cubic centimeters

Soak a plate for 2 minutes.

Color bath:

Erythrosine solution (1 in 1,000) .	25 cubic centimeters
Stronger ammonia (0.900)...	4 cubic centimeters
Water......	175 cubic centimeters

The plate should not remain longer in the bath than $1\frac{1}{4}$ minutes.

PAPER–SENSITIZING PROCESSES:

Blueprint Paper.—I.—The ordinary blue photographic print in which white lines appear on a blue ground may be made on paper prepared as follows:

A.—Potassium ferricyanide	....	10 drachms
Distilled water	....	4 ounces

B.—Iron ammonia citrate.		15 drachms
Distilled water	....	4 ounces

Mix when wanted for use, filter, and apply to the surface of the paper.

With this mixture no developer is required. The paper after exposure is simply washed in water to remove the unaltered iron salts. The print is improved by immersion in dilute hydrochloric acid, after which it must be again well washed in water.

II.—The following process, credited to Captain Abney, yields a photographic paper giving blue lines on a white ground:

Common salt.......	3	ounces
Ferric chloride......	8	ounces
Tartaric acid.......	$3\frac{1}{4}$	ounces
Acacia.............	25	ounces
Water.............	100	ounces

Dissolve the acacia in half the water and dissolve the other ingredients in the other half; then mix.

The liquid is applied with a brush to strongly sized and well rolled paper in a subdued light. The coating should be as even as possible. The paper should be dried rapidly to prevent the solution sinking into its pores. When dry, the paper is ready for exposure.

In sunlight, 1 or 2 minutes is generally sufficient to give an image; while in a dull light as much as an hour is necessary.

To develop the print, it is floated immediately after leaving the printing frame upon a saturated solution of potassium ferrocyanide. None of the developing solution should be allowed to reach the back. The development is usually complete in less than a minute. The paper may be lifted off the solution when the face is wetted, the development proceeding with that which adheres to the print.

When the development is complete, the print is floated on clean water, and after 2 or 3 minutes is placed in a bath, made as follows:

Sulphuric acid......	3	ounces
Hydrochloric acid...	8	ounces
Water.............	100	ounces

In about 10 minutes the acid will have removed all iron salts not turned into the blue compound. It is next thoroughly washed and dried. Blue spots may be removed by a 4 per cent solution of caustic potash.

The back of the tracing must be placed in contact with the sensitive surface.

III.—Dissolve $3\frac{3}{4}$ ounces of ammonia citrate of iron in 18 ounces of water, and put in a bottle. Then dissolve $2\frac{3}{8}$ ounces of red prussiate of potash in 18 ounces of water, and put in another bottle. When ready to prepare the paper, have the sheets piled one on top of the other, coating but one at a time. Darken the room, and light a ruby lamp. Now, mix thoroughly equal parts of both solutions and apply the mixture with a sponge in long parallel sweeps, keeping the application as even as possible. Hang the paper in the dark room to dry and keep it dark until used. Any of the mixture left from sensitizing the paper should be thrown away, as it deteriorates rapidly.

Often, in making blueprints by sunlight, the exposure is too long, and when the frame is opened the white lines of the print are faint or obscure. Usually these prints are relegated to the waste basket; but if, after being washed as usual, they are sponged with a weak solution of chloride of iron, their reclamation is almost certain. When the lines reappear, the print should be thoroughly rinsed in clear water.

Often a drawing, from which prints have already been made, requires changing. The blueprints then on hand are worthless, requiring more time to correct than it would take to make a new print. An economical way of using the worthless prints is to cancel the drawing already thereon, sensitize the reverse side, and use the paper again.

How to Make Picture Postal Cards and Photographic Letter Heads.—I.— Well-sized paper is employed. If the sizing should be insufficient, resizing can be done with a 10 per cent gelatin solution, with a 2 per cent arrowroot paste, or with a 50 per cent decoction of carrageen. This size is applied on the crude paper with a brush and allowed to dry. The well-sized or resized papers are superior and the picture becomes stronger on them than on insufficiently sized paper. Coat this paper uniformly with a solution of 154 grains of ferric oxalate in $3\frac{1}{2}$ fluidounces of distilled water, using a brush, and allow to dry. Next, apply the solution of $15\frac{1}{2}$ grains of silver nitrate in $3\frac{1}{2}$ fluidounces of water with a second brush, and dry again. Coating and drying must be conducted with ruby light or in the dark.

The finished paper keeps several days. Print deep so as to obtain a strong picture and develop in the following bath:

Distilled water...	$3\frac{1}{2}$	fluidounces
Potassium oxalate (neutral)..	340	grains
Oxalic acid.....	4	grains

After developing the well-washed prints, fix them preferably in the following bath:

Distilled water..	$3\frac{1}{2}$	fluidounces
Sodium thiosulphate........	75	grains
Gold chloride solution (1 in 100)........	80	minims

Any other good bath may be employed.

II.—Starch is dissolved in water and the solution is boiled until it forms a thin paste. Carmine powder is added, and the mixture is rapidly and assiduously stirred until it is homogeneous throughout. It is now poured through muslin and spread by means of a suitable pencil on the paper to be sensitized. Let dry, then float it, prepared side down on a solution of potassium chromate, 30 parts in 520 parts of distilled water, being careful to prevent any of the liquid from getting on the back or reverse side. Dry in the dark room, and preserve in darkness. When desired for use lay the negative on the face of the paper, and expose to the full sunlight for 5 or 6 minutes (or about an hour in diffused light). Washing in plenty of water completes the process.

A Simple Emulsion for Mat or Printing-Out Paper.—One of the very best surfaces to work upon for coloring in water color is the carbon print. Apart from its absolute permanency as a base, the surface possesses the right tooth for the adhering of the pigment. It is just such a surface as this that is required upon other prints than carbon, both for finished mat surfaces and for the purposes of coloring. The way to obtain this surface upon almost any kind of paper, and to print it out so that the correct depth is ascertained on sight, will be described. Some of the crayon drawing papers can be utilized, as well as many other plain photographic papers that may meet the desires of the photographer. If a glossy paper is desired, the emulsion should be coated on a baryta-coated stock.

There will be required, in the first place, 2 half-gallon stoneware crocks with lids. The best shape to employ is a crock with the sides running straight, with no depressed ridge at the top. One of these crocks is for the preparation of the emulsion, the other to receive the emulsion when filtered. An enameled iron saucepan of about 2 gallons capacity will be required in which to stand the crock for preparing the emulsion, and also to remelt the emulsion after it has become set. The following is the formula for the emulsion, which must be prepared and mixed in the order given. Failure will be impossible if these details are scrupulously attended to.

Having procured 2 half-gallon stoneware crocks with lids, clean them out well with hot and cold water, and place into one of these the following:

Distilled water......	10 ounces
Gelatin (Heinrich's, hard)...........	4 ounces

Cut the gelatin into shreds with a clean pair of scissors. Press these shreds beneath the water with a clean strip of glass and allow to soak for 1 hour. Now proceed to melt the water-soaked gelatin by placing the crock into hot water in the enameled saucepan, the water standing about half way up on the outside of the crock. Bring the water to boiling point, and keep the gelatin occasionally stirred until it is completely dissolved. Then remove the crock to allow the contents to cool down to 120° F. Now prepare the following, which can be done while the gelatin is melting:

No. 1

Rochelle salts....... 90 grains
Distilled water....... 1 ounce

No. 2

Chloride of ammo-
nium............. 45 grains
Distilled water....... 1 ounce

No. 3

Nitrate of silver,.
1 ounce and...... 75 grains
Citric acid (crushed
crystals).......... 95 grains
Distilled water....... 10 ounces

No. 4

Powdered white alum 90 grains
Distilled water (hot).. 5 ounces

The latter solution may be made with boiling water. When these solutions are prepared, pour into the hot gelatin solution No. 1, stirring all the while with a clean glass rod. Then add No. 2. Rinse the vessel with a little distilled water, and add to the gelatin. Now, while stirring gradually, add No. 3, and lastly add No. 4, which may be very hot. This will cause a decided change in the color of the emulsion. Lastly add 2 ounces of pure alcohol (photographic). This must be added very gradually with vigorous stirring, because if added too quickly it will coagulate the gelatin and form insoluble lumps. The emulsion must, of course, be mixed under a light not stronger than an ordinary small gas-jet, or under a yellow light obtained by covering the windows with yellow paper. The cover may now be placed upon the crock, and the emulsion put aside for 2 or 3 days to ripen.

At the end of this time the contents of the crock, now formed into a stiff emulsion, may be remelted in hot water by placing the crock in the enameled saucepan over a gas stove. The emulsion may be broken up by cutting it with a clean bone or hard-rubber paper cutter to facilitate the melting. Stir the mixture occasionally until thoroughly dissolved, and add the following as soon as the emulsion has reached a temperature of about 150° F.:

Distilled water....... 4 ounces
Pure alcohol......... 1 ounce

The emulsion must now be filtered into the second crock. The filtering is best accomplished in the following manner: Take an ordinary plain-top kerosene lamp chimney, tie over the small end two thicknesses of washed cheese cloth. Invert the chimney and insert a tuft of absorbent cotton about the size of an ordinary egg. Press it carefully down upon the cheese cloth. Fix the chimney in the ring of a retort stand (or cut a hole about 3 inches in diameter in a wooden shelf), so that the crock may stand conveniently beneath. In the chimney place a strip of glass, resting upon the cotton, to prevent the cotton from lifting. Now pour in the hot emulsion and allow the whole of it to filter through the absorbent cotton. This accomplished, we are now ready for coating the paper, which is best done in the following manner:

Cut the paper into strips or sheets, say 12 inches wide and the full length of the sheet. This will be, let us suppose, 12 x 26 inches. Attach, by means of the well-known photographic clips, a strip of wood at each end of the paper upon the back. Three clips at each end will be required. Having a number of sheets thus prepared, the emulsion should be poured into a porcelain pan or tray, kept hot by standing within another tray containing hot water. The emulsion tray being, say, 11 x 14 size, the paper now is easily coated by holding the clipped ends in each hand, then holding the left end of the paper up, and the right-hand end lowered so that the curve of the paper just touches the emulsion. Then raise the right hand, at the same time lowering the left hand at the same rate. Then lower the right hand, lifting the left. Repeat this operation once more; then drain the excess of emulsion at one corner of the tray, say, the left-hand corner. Just as soon as the emulsion has drained, the coated sheet of paper may be hung up to dry, by the hooks attached to the clips, upon a piece of copper wire stretched from side to side of a spare closet or room that can be kept darkened until the paper is dry. In this way coat as much paper as may be required. When it is dry it may be rolled up tight or kept flat under pressure until needed.

If any emulsion remains it may be kept in a cool place for 2 weeks, and still be good for coating. Be sure to clean out all the vessels used before the emulsion sets, otherwise this will present a difficult task, since the emulsion sets into an almost insoluble condition.

This emulsion is so made that it does not require to be washed. If it is washed it will become spoiled. It is easy to make and easy to use. If it is desired that only small sheets of paper are to be coated, they may be floated on the emulsion, but in this case the paper must be damp, which is easily accomplished by wetting a sheet of blotting paper, then covering this with two dry sheets of blotting paper. Place the sheets to be coated upon these, and place under pressure during the night. Next day they will be in good condition for floating.

When the coated paper is dry it may be printed and toned just the same as any other printing-out paper, with any toning bath, and fixed in hyposulphite of soda as usual. Toning may be carried to a rich blue black, or if not carried too far will remain a beautiful sepia color. After well washing and drying, it will be observed that the surface corresponds with that of a carbon print; if the paper has been of a somewhat absorbent character, the surface will be entirely mat, and will give an excellent tooth for coloring or finishing in sepia, black and white, etc.

How to Sensitize Photographic Printing Papers.—I.—The older form of paper is one in which the chemicals are held by albumen. Silver is said to combine with this, forming an albuminate. Pictures printed on this would be too sharp in their contrasts, and consequently "hard"; this is avoided by introducing silver chloride.

To prepare this form of paper, beat 15 ounces of fresh egg albumen with 5 ounces of distilled water, dissolve in it 300 grains of ammonium chloride, set aside for a time, and decant or filter. Suitable paper is coated with this solution by floating, and then dried. The paper is "sensitized" by floating it on a solution of silver nitrate in distilled water, about 80 grains to the ounce, with a drop of acetic acid. The paper is dried as before, and is then ready for printing. The sensitizing must, of course, be done in the dark room.

The reaction between the ammonium chloride present in the albumen coating produces a certain quantity of silver chloride, the purpose of which is shown above. Of course, variations in the proportions of this ingredient will give different degrees of softness to the picture.

II.—The bromide and chloride papers which are now popular consist of the ordinary photographic paper sensitized by means of a thin coating of bromide or chloride emulsion. In "Photographic Printing Methods," by the Rev. W. H. Burbank, the following method is given for bromide paper:

A.—Gelatin (soft)........ 42½ grains
    Bromide of potassium 26 grains
    Distilled water....... 1 ounce
B.—Nitrate of silver...... 33½ grains
    Distilled water.,...... 1 ounce

Dissolve the bromide first, then add the gelatin and dissolve by gentle heat (95° to 100° F.). Bring the silver solution to the same temperature, and add in a small stream to the gelatin solution, stirring vigorously, of course in non-actinic light. Keep the mixed emulsion at a temperature of 105° F. for half an hour, or according to the degree of sensitiveness required, previously adding 1 drop of nitric acid to every 5 ounces of the emulsion. Allow it to set, squeeze through working canvas, and wash 2 hours in running water. In his own practice he manages the washing easily enough by breaking the emulsion up into an earthen jar filled with cold water, and placed in the dark room sink. A tall lamp chimney standing in the jar immediately under the tap conducts fresh water to the bottom of the jar, and keeps the finely divided emulsion in constant motion; a piece of muslin, laid over the top of the jar to prevent any of the emulsion running out, completes this simple, inexpensive, but efficient washing apparatus.

Next melt the emulsion and add one-tenth of the whole volume of glycerine and alcohol; the first to prevent troublesome cockling of the paper as it dries, the second to prevent air bubbles and hasten drying. Then filter.

With the emulsion the paper may be coated just as it comes from the stock dealer, plain, or, better still, given a substratum of insoluble gelatin, made as follows:

Gelatin............. 1⅜ grains
Water.............. 1 ounce

Dissolve and filter; then add 11 drops of a 1 in 50 filtered chrome alum solution. The paper is to be floated for half a minute on this solution, avoiding air bubbles, and then hung up to dry in a room free from dust. The purpose of this substratum is to secure additional brilliancy in the finished prints by keeping the emulsion isolated from the surface of the paper. The paper should now be cut to the size desired.

We do not know of these processes having been applied to postal cards, but unless there is some substance in the sizing of the card which would interfere, there is no reason why it should not be. Of course, however, a novice will not get the results by using it that an experienced hand would.

Ferro-Prussiate Paper.—The following aniline process of preparing sensitive paper is employed by the Prussian and Hessian railway administrations. The ordinary paper on reels is used for the purpose, and sensitized as follows:

Two hundred and fifty parts, by weight, of powdered potassium bichromate are dissolved in water; the solution should be completely saturated; 10 parts of concentrated sulphuric acid, 10 parts of alcohol (962), and 30 parts of phosphoric acid, are added successively, and the whole stirred together. The solution is sponged over the paper. It is not necessary to have the room absolutely dark, or to work by a red light, still the light should be obscured. The drying of the paper, in the same place, takes about 10 minutes, after which the tracing to be reproduced and the paper are placed in a frame, as usual, and exposed to daylight. On a sunny day, an exposure of 35 seconds is enough; in cloudy weather, 60 to 70 seconds; on a very dark day, as much as 5 minutes.

After exposure, the paper is fixed by suspending it for 20 minutes upon a bar in a closed wooden box, on the bottom of which are laid some sheets of blotting paper, sprinkled with 40 drops of benzine and 20 of crude aniline oil. The vapors given off will develop the design. Several impressions may be taken at the same time.

For fixing, crude aniline oil is to be used (anilinum purum), not refined (purissimum), for the reason that the former alone contains the substances necessary for the operation. The reproduced design is placed in water for a few minutes, and hung up to dry.

Pigment Paper for Immediate Use.— Pigment paper is usually sensitized in the bichromate solution on the evening before it is desired for use. If it is not then used it will spoil. By proceeding as follows the paper may be used within a quarter of an hour after treating it in the bichromate bath. Make a solution of

Ammonium bi-
chromate..... 75 grains
Water.......... 3½ fluidounces
Sodium carbonate 15 grains

Mix 0.35 ounces of this solution with 0.7 ounces alcohol, and with a broad brush apply to surface of the pigment paper, as evenly as possible. Dry this paper as quickly as possible in a pasteboard box of suitable size, 15 minutes being usually long enough for the purpose. It may then be used at once.

Photographing on Silk.—China silk is thoroughly and carefully washed to free it from dressing, and then immersed in the following solution:

Sodium chloride.... 4 parts
Arrowroot......... 4 parts
Acetic acid........ 15 parts
Distilled water...... 100 parts

Dissolve the arrowroot in the water by warming gently, then add the remaining ingredients. Dissolve 4 parts of tannin in 100 parts of distilled water and mix the solutions. Let the silk remain in the bath for 3 minutes, then hang it carefully on a cord stretched across the room to dry. The sensitizing mixture is as follows:

Silver nitrate....... 90 parts
Distilled water...... 750 parts
Nitric acid......... 1 part

Dissolve. On the surface of this solution the silk is to be floated for 1 minute, then hung up till superficially dry, then pinned out carefully on a flat board until completely dry. This must, of course, be done in the dark room. Print, wash, and tone in the usual manner.

TONING BATHS FOR PAPER.

The chief complaints made against separate baths are (1) the possibility of double tones, and (2) that the prints sometimes turn yellow and remain so. Such obstacles may easily be removed by exercising a little care. Double tones may be prevented by soaking the prints in a 10 per cent solution of common salt before the preliminary washing, and by not touching the films with the fingers; and the second objection could not be raised provided fresh solution were used, with no excess of sulphocyanide, if this be the bath adopted.

A very satisfactory solution may be made as follows:

Sodium phosphate... 20 grains
Gold chloride........ 1½ grains
Distilled (or boiled)
water............. 10 ounces

This tones very quickly and evenly, and the print will be, when fixed, exactly the color it is when removed from the bath. Good chocolate tints may be obtained, turning to purple gray on prolonged immersion.

Next to this, as regards ease of manipulation, the tungstate bath may be placed, the following being a good formula:

Sodium tungstate.... 40 grains
Gold chloride....... 2 grains
Water.............. 12 ounces

The prints should be toned a little further than required, as they change color, though only slightly, in the hypo.

Provided that ordinary care be exercised, the sulphocyanide bath cannot well be improved upon. The formulas given by the various makers for their respective papers are all satisfactory, and differ very little. One that always acts well is

Ammonium sulpho-
  cyanide........... 28 grains
Distilled water....... 16 ounces
Gold chloride........ 2½ grains

For those who care to try the various baths, and to compare their results, here is a table showing the quantities of different agents that may be used with sufficient water to make up 10 ounces:

Gold chloride, 1 gr. to 1 oz. water....	12 dr.	16 dr.	16 dr.	11 dr.	11 dr.	14 dr.
Borax....	60 gr.					
Sod. bicarbonate...			10 gr.			
Sod. carbonate..				20 gr.		
Sod. phosphate...					20 gr.	
Sod. tungstate.....						40 gr.
Amm. sulphocyanide...						17.5 gr.

We may take it that any of these substances reduce gold trichloride, $AuCl_3$ to $AuCl$; this $AuCl$ apparently acts as an electrolyte, from which gold is deposited on the silver of the image, and at the same time a small quantity of silver combines with the chlorine of the gold chloride thus:

$$AuCl + Ag = AgCl + Au$$

When toning has been completed, the prints are washed and placed in the fixing bath, when the sodium thiosulphate present dissolves any silver chloride that has not been affected by light.

Besides the well-known, every-day tones we see, which never outstep the narrow range between chocolate brown and purple, a practically infinite variety of color, from chalk red to black, may be obtained by a little careful study of toning baths instead of regarding them as mere unalterable machines. Most charming tints are produced with platinum baths, a good formula being

Strong nitric acid.... 5 drops
Water.............. 4 ounces
Chloro-platinite of po-
  tassium........... 1 grain

The final tone of a print cannot be judged from its appearance in the bath, but some idea of it may be got by holding it up to the light and looking through it. A short immersion gives various reds, while prolonged toning gives soft grays.

Results very similar to platinotype may be obtained with the following combined gold and platinum bath:

A.—Sodium acetate...... 1 drachm
    Water.............. 4 ounces
    Gold chloride....... 1 grain

B.—Chloro-platinite of po-
    tassium........... 1 grain
    Water............. 4 ounces

Mix A and B and neutralize with nitric acid. (The solution will be neutral when it just ceases to turn red litmus paper blue.)

Another toning agent is stannous chloride. Two or three grains of tin foil are dissolved in strong hydrochloric acid with the aid of heat. The whole is then made up to about 4 ounces with water.

Toning Baths for Silver Bromide Paper.—The picture, which has been exposed at a distance of 1½ feet for about 8 to 10 seconds, is developed in the customary manner and fixed in an acid fixing bath composed of

Distilled water.. 1,000 cubic centimeters
Hyposulphite of
  soda......... 100 grams
Sodium sulphite 20 grams
Sulphuric acid.. 4 to 5 grams

First dissolve the sodium sulphite, then add the sulphuric acid, and finally the hyposulphite, and dissolve.

Blue tints are obtained by laying the picture in a bath composed as follows:

A.—Uranium ni-
    trate..... 2 grams
    Water...... 200 cubic centimeters

B.—Red prussiate of
    potash... 2 grams
    Water...... 200 cubic centimeters

C.—Ammonia-
    iron-alum 10 grams
    Water...... 100 cubic centimeters
    Pure hydrochloric
    acid...... 15 cubic centimeters

Immediately before the toning, mix

Solution A.. 200 cubic centimeters
Glacial acetic acid... 20 cubic centimeters
Solution B.. 200 cubic centimeters
Solution C.. 30 to 40 cubic centimeters

Brown tints. Use the following solutions:

A.—Uranium nitrate..... 12 grams
Water......1,000 cubic centimeters

B.—Red prussiate of potash... 9 grams
Water......1,000 cubic centimeters

And mix immediately before use

Solution *A*.. 100 cubic centimeters
Solution *B*.. 100 cubic centimeters
Glacial acetic acid... 10 cubic centimeters

Pictures toned in this bath are then laid into the following solution:

Water......1,500 cubic centimeters
Pure hydrochloric acid...... 5 cubic centimeters
Citric acid.. 20 grams

To Turn Blueprints Brown.—A piece of caustic soda about the size of a bean is dissolved in 5 ounces of water and the blueprint immersed in it, on which it will take on an orange-yellow color. When the blue has entirely left the print it should be washed thoroughly and immersed in a bath composed of 8 ounces of water in which has been dissolved a heaping teaspoonful of tannic acid. The prints in this bath will assume a brown color that may be carried to almost any tone, after which they must again be thoroughly washed and allowed to dry.

COMBINED TONING AND FIXING BATHS.

The combined toning and fixing bath consists essentially of five parts—(1) water, the solvent; (2) a soluble salt of gold, such as gold chloride; (3) the fixing agent, sodium thiosulphate; (4) a compound which will readily combine with "nascent" sulphur—i. e., sulphur as it is liberated—this is usually a soluble lead salt, such as the acetate or nitrate, and (5) an auxiliary, such as a sulphocyanide.

The simplest bath was recommended by Dr. John Nicol, and is as follows:

Sodium thiosulphate. 3 ounces
Distilled water....... 16 ounces

When dissolved, add

Gold chloride.... 4 grains
Distilled water... 4 fluidrachms

A bath which contains lead is due to Dr. Vogel, whose name alone is sufficient to warrant confidence in the formula:

Sodium thiosulphate 7 ounces
Ammonium sulphocyanide......... 1 ounce
Lead acetate ......67 grains
Alum............. 1 ounce
Gold chloride......12 grains
Distilled water.....35 fluidounces

A bath which contains no lead is one which has produced excellent results and is due to the experimental research of Dr. Liesegang. It is as follows:

Ammonium sulphocyanide.... ¼ ounce
Sodium chloride.. 1 ounce
Alum........... ½ ounce
Sodium thiosulphate......... 4 ounces
Distilled water... 24 fluidounces

Allow this solution to stand for 24 hours, during which time the precipitated sulphur sinks to the bottom of the vessel; decant or filter, and add

Gold chloride.... 8 grains
Distilled water... 1 fluidounce

It is curious that, with the two baths last described, the addition to them of some old, exhausted solution makes them work all the better.

ENLARGEMENTS.

TIMES OF ENLARGEMENT AND REDUCTION

Focus of Lens. In.	1 inch	2 inches	3 inches	4 inches	5 inches	6 inches	7 inches	8 inches
2	4 4	6 3	8 2 2/3	10 2 1/2	12 2 2/5	14 2 1/3	16 2 2/7	18 2 1/4
2 1/2	5 5	7 1/2 3 3/4	10 3 1/3	12 1/2 3 1/8	15 3	17 1/2 2 9/16	20 2 6/7	22 1/2 2 3/16
3	6 6	9 4 1/2	12 4	15 3 3/4	18 3 3/5	21 3 1/2	24 3 3/7	27 3 3/8
3 1/2	7 7	10 1/2 5 1/4	14 4 2/3	17 1/2 4 3/4	21 4 1/5	24 1/2 4 1/12	28 4	31 1/2 3 9/16
4	8 8	12 6	16 5 1/3	20 5	24 4 4/5	28 4 2/3	32 4 4/7	36 4 1/2
4 1/2	9 9	13 1/2 6 3/4	18 6	22 1/2 5 3/8	27 5 5/8	31 1/2 5 1/4	36 5 1/7	40 1/2 5 1/16
5	10 10	15 7 1/2	20 6 2/3	25 6 1/4	30 6	35 5 5/6	40 5 5/7	45 5 5/8
5 1/2	11 11	16 1/2 8 1/4	22 7 1/3	27 1/2 6 1/6	33 6 1/2	38 1/2 6 5/12	44 6 2/7	49 1/2 6 3/16
6	12 12	18 9	24 8	30 7 1/2	36 7 1/5	42 7	48 6 6/7	54 6 3/4
7	14 14	21 10 1/2	28 9 1/3	35 8 3/4	42 8 2/5	49 8 1/6	56 8	63 7 7/8
8	16 16	24 12	32 10 2/3	40 10	48 9 3/5	56 9 1/3	64 9 1/7	72 9
9	18 18	27 13 1/2	36 12	45 11 1/4	54 10 4/5	63 10 1/2	72 10 2/7	81 10 1/8

The object of this table is to enable any manipulator who is about to enlarge (or reduce) a copy any given number of times to do so without troublesome calculation. It is assumed that the photographer knows exactly what the focus of his lens is, and that he is able to measure accurately from its optical center. The use of the table will be seen from the following illustration: A photographer has a *carte* to enlarge to four times its size, and the lens he intends employing is one of 6 inches equivalent focus. He must therefore look for 4 on the upper horizontal line and for 6 in the first vertical column, and carry his eye to where these two join, which will be at 30–7½. The greater of these is the distance the sensitive plate must be from the center of the lens; and the lesser, the distance of the picture to be copied. To reduce a picture any given number of times, the same method must be followed; but in this case the greater number will represent the distance between the lens and the picture to be copied, the latter that between the lens and the sensitive plate. This explanation will be sufficient for every case of enlargement or reduction.

If the focus of the lens be 12 inches, as this number is not in the column of focal lengths, look out for 6 in this column and multiply by 2, and so on with any other numbers.

To make a good enlargement five points should be kept constantly in view, viz.:

1. Most careful treatment of the original negative.

2. Making a diapositive complete in all its parts.

3. Scrupulous consideration of the size of the enlargement.

4. Correct exposure during the process of enlargement.

5. The most minute attention to the details of development, including the chemical treatment of the enlarged negative.

The original negative should not be too dense, nor, on the contrary, should it be too thin. If necessary, it should be washed off, or strengthened, as the case may be. Too strong a negative is usually weakened with ammonium persulphate, or the fixing hypo solution is quite sufficient. All spots, points, etc., should be retouched with the pencil and carmine.

The diapositive should be produced by contact in the copying apparatus. A border of black paper should be used to prevent the entry of light from the side.

The correct period of exposure depends upon the thickness of the negative, the source of the light, its distance, etc. Here there is no rule, experience alone must teach.

For developing one should use not too strong a developer. The metol-soda developer is well suited to this work, as it gives especially soft lights and half tones. Avoid too short a development. When the finger laid behind the thickest spot, and held toward the light, can no longer be detected, the negative is dense enough.

The denser negatives should be exposed longer, and the development should be quick, while with thin, light negatives the reverse is true; the exposure should be briefer and the development long, using a strong developer, and if necessary with an addition of potassium bromide.

The silver chloro-bromide diapositive plates, found in the shops, are totally unsuited for enlargements, as they give overdone, hard pictures.

To produce good artistic results in enlarging, the diapositive should be kept soft, even somewhat too thin. It should undergo, also, a thorough retouching. All improvements are easily carried out on the smaller positive or negative pictures. Later on, after the same have been enlarged, corrections are much more difficult and troublesome.

VARNISHES:

Cold Varnish.—

I.—Pyroxylin.......... 10 grains
    Amyl alcohol....... 1 ounce
    Amyl acetate....... 1 ounce

Allow to stand, shaking frequently till dissolved. Label: The negative should be thoroughly dried before this solution is applied, which may be done either by flowing it over the solution or with a flat brush. The negative should be placed in a warm place for at least 12 hours to thoroughly dry.

II.—Japanese gold size .. ⎰ Equal parts.
    Benzol............. ⎱

Label: In applying this varnish great care should be taken not to use it near a light or open fire. It can be flowed over or brushed on the negative.

Black Varnish.—

    Brunswick black... 1½ ounces
    Benzol........... 1 ounce

Label: The varnish should be applied with a brush, care being taken not to use it near a light or open fire.

Dead Black Varnish.—

Borax	30 grains
Shellac	60 grains
Glycerine	30 minims
Water	2 ounces

Boil till dissolved, filter, and add aniline black, 120 grains.

Label: Apply the solution with a brush, and repeat when dry if necessary.

Ordinary Negative Varnish.—

Gum sandarac	1 ounce
Orange shellac	½ ounce
Castor oil	90 minims
Methyl alcohol	1 pint

Allow to stand with occasional agitation till dissolved, and then filter. Label: The negative should be heated before a fire till it can be comfortably borne on the back of the hand, and then the varnish flowed over, any excess being drained off, and the negative should then be again placed near the fire to dry.

Water Varnish.—It is not only in connection with its application to a wet collodion film that water varnish forms a valuable addition to the stock of chemicals in all-round photography; it is almost invaluable in the case of gelatin as with wet collodion films. In the case of gelatin negatives the water varnish is applied in the shape of a wash directly after the negatives have been washed to free their films from all traces of hypo, or in other words, at that stage when the usual drying operation would begin. After the varnish has been applied the films are dried in the usual manner, and its application will soon convince anyone that has experienced the difficulty of retouching by reason of the want of a tooth in the film to make a lead-pencil bite, as the saying goes, that were this the only benefit accruing from its application it is well worthy of being employed.

The use of water varnish, however, does away with the necessity of employing collodion as an additional protection to a negative, and is, perhaps, the best known remedy against damage from silver staining that experienced workers are acquainted with. As a varnish it is not costly, neither is it difficult to make in reasonably small quantities, while its application is simplicity itself. The following formula is an excellent sample of water varnish:

Place in a clean, enameled pan 1 pint of water, into which insert 4 ounces of shellac in thin flakes, and place the vessel on a fire or gas stove until the water is raised to 212° F. When this temperature is reached a few drops of hot, saturated solution of borax is dropped into the boiling pan containing the shellac and water, taking care to stir vigorously with a long strip of glass until the shellac is all dissolved. Too much borax should not be added, only just sufficient to cause the shellac to dissolve, and it is better to stop short, if anything, before all the flakes dissolve out than to add too much borax. The solution is then filtered carefully and, when cold, the water varnish is ready for use.

FADED PHOTOGRAPHS AND THEIR TREATMENT:

Restoring Faded Photographs.—I.—As a precaution against a disaster first copy the old print in the same size. Soak the faded photograph for several hours in clean water and, after separating print from mount, immerse the former in nitric acid, highly dilute (1 per cent), for a few minutes. Then the print is kept in a mercury intensifier (mercuric chloride, ½ ounce; common salt, ½ ounce; hot water, 16 ounces, used cold), until bleached as much as possible. After half an hour's rinsing, a very weak ammonia solution will restore the photograph, with increased vigor, the upper tones being much improved, though the shadows will show some tendency to clog. The net result will be a decided improvement in appearance; but, at this stage, any similarly restored photographs should be recopied if their importance warrants it, as mercury intensifier results are not permanent. It may be suggested that merely rephotographing and printing in platinotype will probably answer.

II.—Carefully remove the picture from its mount, and put it in a solution of the following composition:

	By weight
Hydrochloric acid	2 parts
Sodium chloride	8 parts
Potassium bichromate	8 parts
Distilled water	250 parts

The fluid bleaches the picture, but photographs that have been toned with gold do not quite vanish. Rinse with plenty of water, and develop again with very dilute alkaline developer.

MOUNTANTS:

See also Adhesives.

I.—If buckling of the mount is to be cured, the prints must be mounted in a dry state, and the film of mountant borne by the print must be just sufficient to attach it firmly to the mount and no more. The great virtue of the method here described consists of the marvelously thin film of tenacious mountant applied to the print in its dry condition, shrinkage by this means being entirely obviated. A drawing board with a perfectly smooth surface and of fair dimensions, an ivory or bone burnisher attached to a short handle, with some common glue, are the principal requisites. Take, say, a quarter of a pound of the glue broken into small pieces and cover it with water in a clean gallipot, large enough to allow for the subsequent swelling of the glue. Place on one side until the glue has become thoroughly permeated by the water, then pour off the excess and dissolve the glue in the water it has absorbed, by placing the gallipot in a vessel of hot water. The solution tested with a piece of blue litmus paper will show a distinctly acid reaction, which must be carefully neutralized by adding some solution of carbonate of soda. The amount of water absorbed by the glue will probably be too little to give it the best working consistency, and, if this is the case, sufficient should be added to make it about the thickness of ordinary molasses. Careful filtration through a cambric handkerchief, and the addition of about 10 grains of thymol, completes the preparation of the mounting solution. As glue deteriorates by frequent and prolonged heating, it is preferable to make up a stock solution, from which sufficient for the work in hand can be taken in the form of jelly, melted, and used up at once.

The finished prints, dried and trimmed to the required size, are placed on the boards they are to occupy when mounted, and, as it is impossible to remove a print for readjustment once it is laid down for final mounting, the wisest course is to indicate by faint pencil marks on the mount the exact position the print is to occupy; then it may be laid down accurately and without any indecision. A small gas or oil stove is required on the mounting table to keep the glue liquid, but maintaining the solution in a constant state of ebullition throughout the operation is unnecessary and harmful to the glue; the flame should be regulated so that the mountant is kept just at the melting point. Place the drawing board beside the gas stove and with a house-painter's brush of good quality and size spread the glue over an area considerably exceeding the dimensions of the print to be mounted. A thin coating of glue evenly applied to the board is the end to aim at, to accomplish which the brush should be worked in horizontal strokes, crossing these with others at right angles. Have at hand a small pile of paper cut into pieces somewhat larger than the print to be mounted (old newspaper answers admirably for these pieces), lay one down on the glued patch and press it well into contact by passing the closed hand across it in all directions. Raise one corner of the paper, and slowly but firmly strip it from the board. Repeat the operations of gluing the board (in the same place) and stripping the newspaper 2 or 3 times, when a beautifully even cushion of glue will remain on the board.

Mounting the prints is the next step. The cushion of glue obtained on the board has to be coated with glue for, say, every second print, but the amount applied must be as small as possible. After applying the glue the print is laid down upon it, a square of the waste newspaper laid over the print, which has then to be rubbed well into contact with the glue. Raise a corner of the print with the point of a penknife and strip it from the board, as in the case of the newspaper. Care must be taken when handling the print in its glued condition to keep the fingers well beyond the edges of the print, in order that no glue may be abstracted from the edges. Lay the print quickly down upon its mount; with a clean, soft linen duster smooth it everywhere into contact, place upon it a square of photographic drying board, and with the bone burnisher go over it in all directions, using considerable pressure. The finished result is a mounted print that shows no signs of buckling, and which adheres to the mount with perfect tenacity.

II.—Gelatin	2 parts
Water	4 parts
Alcohol	8 parts

The alcohol is added slowly as soon as the gelatin is well dissolved in the water, and the vessel turned continually to obtain a homogeneous mixture. The solution must be kept hot during the operation on a water bath, and should be applied quickly, as it soon dries; the print must be placed exactly the first time, as it adheres at once. The solution keeps for a long time in well-corked bottles.

TRANSPARENT PHOTOGRAPHS:

I.—The following mixture may be employed at 176° F., to render photographs transparent. It consists of 4 parts paraffine and 1 part linseed oil. After immersion the photographs are at once dried between blotting paper. For fastening these photographs to glass, glue or gelatin solution alone cannot be employed. This is possible only when one-fourth of its weight of sugar has been added to the glue before dissolving. The glasses for applying the photographs must be perfect, because the slightest defects are visible afterwards.

II.—If on albumen paper, soak the print overnight in a mixture of 8 ounces of castor oil and 1 ounce of Canada balsam. Plain paper requires a much shorter time. When the print is thoroughly soaked, take it from the oil, drain well, and lay it on the glass face downward, and squeeze till all is driven out and the print adheres. If a curved glass is used, prepare a squeegee with edge parallel with the curvature of the glass. It will take several hours before the print is dry enough to apply color to it.

THE GUM - BICHROMATE PHOTO-PRINTING PROCESS.

Gum bichromate is not a universal printing method. It is not suited for all subjects or for all negatives, but where there is simplicity and breadth in sizes of 8½ x 6½ and upward, direct or enlarged prints by it have a charm altogether their own, and afford an opportunity for individuality greater than any other method.

While almost any kind of paper will do, there are certain qualities that the beginner at least should endeavor to secure. It should be tough enough to stand the necessary handling, which is considerably more than in either the printing-out or developing methods. It must not be so hard or smooth as to make coating difficult, nor so porous as to absorb or let the coating sink in too much; but a few trials will show just what surface is best. Till that experience is acquired it may be said that most of Whatman's or Michallet's drawing papers, to be had at any artist's materials store, will be found all that can be desired; or, failing these, the sizing of almost any good paper will make it almost as suitable.

For sizing, a weak solution of gelatin is generally employed, but arrowroot is better; half an ounce to a pint of water. It should be beaten into a cream with a little of the water, the rest added, and brought to the boil. When cold it may be applied with a sponge or tuft of cotton, going several times, first in one direction and then in the other, and it saves a little future trouble to pencil mark the non-sized side.

The quality of the gum is of less importance than is generally supposed, so long as it is the genuine gum arabic, and in round, clean "tears." To make the solution select an 8-ounce, wide-mouthed bottle, of the tall rather than the squat variety, and place in it 6 ounces of water. Two ounces of the gum are then tied loosely in a piece of thin muslin and suspended in the bottle so as to be about two-thirds covered by the water. Solution begins at once, as may be seen by the heavier liquid descending, and if kept at the ordinary temperature of the room may not be complete for 24 or even 48 hours; but the keeping qualities of the solution will be greater than if the time had been shortened by heat. When all that will has been dissolved, there will still be a quantity of gelatinous matter in the muslin, but on no account must it be squeezed out, as the semi-soluble matter thus added to the solution would be injurious. With the addition of a few drops of carbolic acid and a good cork the gum solution will keep for months.

The selection of the pigments is not such a serious matter as some of the writers would lead us to believe. Tube water colors are convenient and save the trouble of grinding, but the cheap colors in powder take a better grip and give richer images. The best prints are made with mixtures of common lampblack, red ocher, sienna, umber, and Vandyke brown, the only objection to their employment being the necessity of rather carefully grinding. This may be done with a stiffish spatula and a sheet of finely ground glass, the powder mixed with a little gum solution and rubbed with the spatula till smooth, but better still is a glass paper weight in the shape of a cone with a base of about 1½ inches in diameter, bought in the stationer's for 25 cents.

The sensitizer is a 10 per cent solution of potassium bichromate, and whatever be the pigment or whatever the method of preparing the coating, it may be useful to keep in mind that the right strength or proportion, or at least a strength of coating that answers very well, is equal parts of that and the gum solution.

In preparing the coating measure the gum solution in a cup from a toy tea set that holds exactly 1 ounce, it being easier to get it all out of this than out of a conical graduate. From 20 to 30 grains of the color or mixture of colors in powder is placed on the slab—the ground surface of an "opal" answers well—and enough of the gum added to moisten it, and work the paper weight "muller," aided by the spatula, as long as any grittiness remains, or till it is perfectly smooth, adding more and more gum till it is like a thick cream. It is then transferred to a squat teacup and 1 ounce of the bichromate solution gradually added, working it in with one of the brushes to perfect homogeneity. Of course, it will be understood that this mixture should be used all at once, or rather only as much as is to be used at once should be made, as notwithstanding what has been said to the contrary, it will not keep. After each operation, both or all of the brushes should be thoroughly cleaned before putting them away.

Not the least important are the brushes; one about 2 inches wide and soft for laying on the coating, the other, unless for small work, twice that breadth and of what is known as "badger" or a good imitation thereof, for softening.

The paper can be bought in sheets of about 17 x 22 inches. Cut these in two, coating pieces of about 17 x 11. The sheet is fastened to a drawing board by drawing pins, one at each corner. The coating brush—of camel's hair, but it is said that hog's is better—is filled with the creamy mixture, which has been transferred to a saucer as more convenient, and with even strokes, first one way and then the other, drawn all over the paper. It is easier to do than to describe, but all three joints, wrist, elbow, and shoulder take part, and unless the surface of the paper is too smooth, there is really no difficulty to speak of.

By the time the whole surface has been covered the paper will have expanded to an extent that makes it necessary to remove three of the pins and tighten it, and then comes the most important and the only really difficult part of the work, the softening. The softener is held exactly as one holds the pen in writing, and the motion confined altogether to the wrist, bringing only the points of the hair in contact with the coating, more like stippling than painting.

If much of the coating has been laid on, and too much is less of an evil than too little, the softener will soon have taken up so much as to require washing. This is done at the tap, drying on a soft cloth, and repeat the operation, the strokes or touches gradually becoming lighter and lighter, till the surface is as smooth and free from markings as if it had been floated.

Just how thick the coating should be is most easily learned by experience, but as, unlike ordinary carbon, development begins from the exposed surface, it must be as deep; that is, as dark on the paper as the deepest shadow on the intended print, and it should not be deeper.

While it is true that the bichromate colloid is not sensitive while wet, the coating is best done in subdued light, indeed, generally at night. Hang the sheets to dry in the dark room.

Exposure should be made with some form of actino-meter.

Development may be conducted in various ways, and is modified according to the extent of the exposure. Float the exposed sheet on water at the ordinary temperature from the tap. The exposure should admit of complete, or nearly complete, development in that position in from 5 to 10 minutes; although it should not generally be allowed to go so far. By turning up a corner from time to time one may see how it goes, and at the suitable stage depending on what one really wants to do, the otherwise plain outcome of the negative is modified, gently withdrawn from the water, and pinned up to dry.

The modifying operation may be done at once, where the exposure has been long enough to admit it, but generally, and especially when it has been such as to admit of the best result, the image is too soft, too easily washed off to make it safe. But after having been dried and again moistened by immersion in water, the desired modification may be made with safety.

The moistened print is now placed on a sheet of glass, the lower end of which rests on the bottom of the developing tray, and supported by the left hand at a suitable angle; or, better still, in some other way so as to leave both hands free. In this position, and with water at various temperatures, camel's-hair brushes of various sizes, and a rubber syringe, it is possible to do practically anything.

TABLES AND SCALES:

Comparative Exposures of Various Subjects.—

	Seconds
Open panorama, with fields and trees	1
Snow, ice, marine views	1
Panorama, with houses, etc.	2
Banks of rivers	3
Groups and portraits in open air (diffused light)	6
Underneath open trees	6
Groups under cover	10
Beneath dense trees	10
Ravines, excavations	10
Portraits in light interiors	10
Portraits taken 4 feet from a window, indoors, diffused light	30

TABLE SHOWING DISPLACEMENT ON GROUND GLASS OF OBJECTS IN MOTION

By Henry L. Tolman
From the Photographic Times

Lens 6-inch Equivalent Focus, Ground Glass at Principal Focus of Lens

Miles per Hour.	Feet per Second.	Distance on Ground Glass, in inches, with Object 30 Feet away.	Same with Object 60 Feet away.	Same with Object 120 Feet away.
1	1½	.29	.15	.073
2	3	.59	.29	.147
3	4½	.88	.41	.220
4	6	1.17	.59	.293
5	7½	1.47	.73	.367
6	9	1.76	.88	.440
7	10½	2.05	1.03	.513
8	12	2.35	1.17	.587
9	13	2.64	1.32	.660
10	14½	2.93	1.47	.733
11	16	3.23	1.61	.807
12	17½	3.52	1.76	.880
13	19	3.81	1.91	.953
14	20½	4.11	2.05	1.027
15	22	4.40	2.20	1.100
20	29	5.87	2.93	1.467
25	37	7.33	3.67	1.833
30	44	8.80	4.40	2.200
35	51	10.27	5.13	2.567
40	59	11.73	5.97	2.933

W. D. Kilbey, in the *American Annual of Photography*, gives still another table for the exposure that should be given to objects in motion.

According to his method the table is made out for a distance from the camera 100 times that of the focus of the lens; that is, for a 6-inch focus lens at 50 feet, a 7-inch at 58 feet, an 8-inch at 67 feet, a 9-inch at 75 feet, or a 12-inch at 100 feet.

	Toward the Camera.	At Right Angles to the Camera.
Man walking slowly, street scenes.......	$\frac{1}{15}$ sec.	$\frac{1}{45}$ sec.
Cattle grazing.......	$\frac{1}{15}$ "	$\frac{1}{45}$ "
Boating............	$\frac{1}{20}$ "	$\frac{1}{60}$ "
Man walking, children playing, etc........	$\frac{1}{40}$ "	$\frac{1}{120}$ "
Pony and trap, trotting............	$\frac{1}{100}$ "	$\frac{1}{300}$ "
Cycling, ordinary....	$\frac{1}{100}$ "	$\frac{1}{300}$ "
Man running a race and jumping......	$\frac{1}{150}$ "	$\frac{1}{450}$ "
Cycle racing........	$\frac{1}{200}$ "	$\frac{1}{600}$ "
Horses galloping.....	$\frac{1}{200}$ "	$\frac{1}{600}$ "

If the object is twice the distance, the length of allowable exposure is doubled, and vice versa.

To Reduce Photographs.—When one wishes to copy a drawing or photograph he is usually at a loss to know how high the plate will be when any particular base is selected. A plan which has the merit of being simple and reliable has been in use in engravers' offices for years. Here are the details:

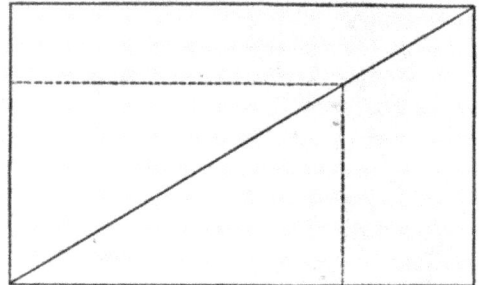

Reducing Scale for Copying Photographs.

Turn the drawing face down and rule a diagonal line from the left bottom to the right top corner. Then measure from the left, on the bottom line, the width required. Rule a vertical line from that point until it meets the diagonal. Rule from that point to the left, and the resulting figure will have the exact proportions of the reduction. If the depth wanted is known, and the width is required, the former should be measured on the left upright line, carried to the diagonal, and thence to the lower horizon. The accompanying diagram explains the matter simply.

COLOR PHOTOGRAPHY:

A Three-Color Process.—Prepare 7 solutions, 4 of which are used for color screens, the remaining 3 serving as dyes for the plates.

A.—Screen Solutions.—

Blue violet.	By weight
Methylene blue....	5 parts
Tetraethyldiamido-oxytriphenyl carbinol...........	2 parts

Or:	By weight
Methyl violet......	5 parts
Alcohol..........	200 parts
Water, distilled....	300 parts

Green.	By weight
Malachite green...	10 parts
Alcohol..........	200 parts
Water, distilled....	300 parts

Yellow.   By weight
    Acridin yellow N.
    O.............   10 parts
    Alcohol..........   200 parts
    Water, distilled.... 300 parts

Red.   By weight
    Congo rubin......   10 parts
    Alcohol..........   200 parts
    Water, distilled...   300 parts

B.—Dyes (Stock Solutions).—
                    By weight
I.—Acridin yellow or
    acridin orange,
    N. O...........   1 part
    Alcohol..........   100 parts
    Water, distilled.... 400 parts

                    By weight
II.—Congo rubin......   1 part
    Alcohol..........   100 parts
    Water, distilled.... 400 parts

                    By weight
III.— Tetraethyldiamido-
    oxytriphenyl car-
    binol...........   1 part
    Alcohol..........   100 parts
    Water, distilled.... 400 parts

The screen solutions, after being filtered through paper filters into clean dishes, are utilized to bathe 6 clean glass plates previously coated with 2 per cent raw collodion; we require 1 plate for blue violet, 2 plates for red, 2 plates for yellow, and 1 plate for green, which in order to obtain the screens are combined in the following way: Yellow and red plate, yellow and green plate. For special purposes the other red plate may be combined with the blue violet. Another method of preparing the screens is to add the saturated solutions drop by drop to a mixture of Canada balsam and 2 per cent castor oil and cement the glasses together. Those who consider the screens by the first method too transparent, coat the glass plates with a mixture of 2 to 3 per cent raw collodion and 1 per cent color solution. Others prefer gelatin screens, using

                    By weight
    Hard gelatin (Nelson's)..........   8 parts
    Water...........   100 parts
    Absolute alcohol...   10 parts
    Pigment..........   1 part

This is poured over the carefully leveled and heated plate after having been filtered through flannel.

The collodion screens are cemented together by moistening the edges with Canada balsam (containing castor oil) and pressing the plates together in a printing frame, sometimes also binding the edges with strips of Japanese paper.

On the evening before the day of work, good dry plates of about 18° to 24° W. are dyed in the following solution:

                    By weight
    Stock solution, No. 1   16 parts
    Distilled water......   100 parts
    Alcohol...........   5 parts
    Nitrate of silver
      (1.500)..........   50 parts
    Ammonia..........   1–2 parts

This bath sensitizes almost uninterruptedly to line A. The total sensitiveness is high, and the plate develops cleanly and fine. Blue sensitiveness is very much reduced, and the blue screen is used for exposure. As far as the author's recollection goes, the plate for the yellow color has never been color-sensitized, many operators using the commercial Vogel-Obernetter eosin silver plates made by Perutz, of Munich; others again only use ordinary dry plates with a blue-violet screen. This is, however, a decided mistake, necessitating an immense amount of retouching, as otherwise it produces a green shade on differently colored objects of the print.

For the red color plate the dry plate is dyed in

                    By weight
    Stock solution, No. 2   10 parts
    Distilled water......   100 parts
    Nitrate of silver
      (1.500)..........   100 parts
    Ammonia..........   2 parts

The resulting absorption band is closed until E, reaching from violet to red (over C). This red pigment was examined by Eder, who obtained very good results, using ammonia in the solution.

The corresponding screen is a combination of malachite green with acridin yellow or acridin orange N. O.

For the blue color plate the dye is made up as follows:

                    By weight
    Stock solution, No. 3 0.5–1 part
    Distilled water......   100 parts
    Nitrate of silver
      (1.500)..........   100 parts
    Ammonia..........   1–2 parts

This dye yields a strong band, commencing at B, reaching to C ¾ D; since the orange screen used herewith necessitates a long exposure, the action seems to extend into the infra-red (beyond A).

As a rule, cyanine is used instead of the tetraethyldiamidooxytriphenyl carbinol (HCl salt), but the former is apt to produce fogged plates. Methyl violet or crystal violet has also been suggested.

Exposures should be made in direct sunlight or with artificial pure white light (acetylene); electric light is too variable.

The most suitable methods of reproduction are half-tone, and the prototype methods; also Turati's Isotypie. The greatest difficulty in 3-color printing nowadays is presented by the want of accurate printing. We must use the proper paper and pure fast colors; the inking rollers should be smooth, not too soft, and free from pores or weals. The blocks must be firmly fixed typehigh, otherwise they take color irregularly. A good printing machine is, of course, most essential.

To supplement the above working directions: After having kept the plates for 2 or 3 minutes (constantly moving the dish) in the dyes, they are removed into a dish containing filtered alcohol, which extracts the superfluous pigment. Plates thus treated dry much more rapidly, develop cleaner, and show no fogging.

Most of the above dyes may be obtained from the "Berliner Actiengesellschaft für Anilinfabrikation," the acridin only from the "Farbwerk Mühlheim, a/Main, vorm. A. Leonhard & Company."

Solution for Preparing Color Sensitive Plates.—H. Vollenbruch maintains that plates sensitized with erythrosin silver citrate are not only more sensitive to color impressions, but also have better keeping qualities than ordinary erythrosin bathed plates.

For depression of the over-active blue rays he recommends the addition of picric acid to the coloring solution. The picric acid erythrosin silver citrate ammonia solution is prepared as follows:

Solution I

Citrate of potassa   1 gram
Distilled water .. 10 cubic centimeters

Solution II

Silver nitrate....   1 gram
Distilled water .. 10 cubic centimeters

Both solutions are mixed and a white precipitate is formed which is allowed to subside. The clear supernatant liquid is poured off carefully, precipitate washed with water, allowed again to subside, and the wash water again decanted. This process is repeated two or three times.

Finally a large bulk of water (20 cubic centimeters) is added to the precipitate and well shaken; 5 cubic centimeters of this is reserved, the remainder is treated to ammonia, drop by drop, until the precipitate is redissolved. Now add the 5 cubic centimeters of reserved solution and shake the whole until every particle is dissolved. Then make up the solution to 50 cubic centimeters and filter; this forms Solution III.

Solution IV

Distilled water .. 300 cubic centimeters
Pure erythrosin..   1 grain

Under lamplight the 50 cubic centimeters of Solution III are poured slowly with repeated shaking in Solution IV, by which the originally beautiful red is converted into a dirty turbid bluish red somewhat viscid fluid; add—

Solution V

Picric acid......   4 grams
Absolute alcohol. 30 cubic centimeters

Shake well, and add to the whole 33 cubic centimeters ammonia (specific gravity, 0.91), wherewith the beautiful red color is restored.

After the filtration call this Solution VI. This solution keeps well. The slight deposit formed is redissolved on shaking.

The plates are sensitized as follows: The plate to be sensitized is first laid in a tray of distilled water for 2 or 3 minutes, then bathed in a mixture of 1 cubic centimeter ammonia for 1 minute and finally for 2 minutes in a bath composed of the following:

Color Solution VI   10 cubic centimeters
Distilled water... 300 cubic centimeters

The plate is well drained and dried in a perfectly dark room. These plates keep well for several months.

MICROPHOTOGRAPHS.

The instruments used are an objective of very short focus and a small camera with a movable holder. This camera and the original negative to be reduced are fastened to the opposite ends of a long, heavy board, similar to the arrangement in use for the making of lantern slides. The camera must be movable in the direction of the objective axis, and the negative must be fastened to a vertically stationary stand. It is then uniformly lighted from the reversed side by either daylight or artificial light. Some difficulty is experienced in getting a sharp focus of the picture. The ordinary ground glass cannot be used, not being fine enough, and the best medium for this purpose is a perfectly plain piece of glass, coated with pretty strongly iodized collodion, and sensitized in the silver bath, the same way as in the wet process. The focusing is done with a small lens or even with a microscope. The plate intended for the picture has, of course, to lie in exactly the same plane as the plate used for focusing. To be certain on this point, it is best to focus upon the picture plate, inserting for this purpose a yellow glass between objective and plate. If satisfactory sharpness has been obtained, the apparatus is once for all in order for these distances. Bromide of silver gelatin plates, on account of their comparatively coarse grain, are not suitable for these small pictures, and the collodion process has to come to the rescue.

Dagron, in Paris, a prominent specialist in this branch, gives the following directions: A glass plate is well rubbed on both sides with a mixture of 1,000 parts of water, 50 parts powdered chalk, and 200 parts of alcohol, applied with a cotton tuft, after which it is gone over with a dry cotton tuft, and thereafter cleaned with a fine chamois leather. The side used for taking the picture is then finally cleaned with old collodion. The collodion must be a little thinner than ordinarily used for wet plates. Dissolve

Ether................	400 parts
Alcohol..............	100 parts
Collodion cotton....	3 parts
Iodide ammonia....	4 parts
Bromide ammonia..	1 part

The plate coated herewith is silvered in a silver bath of 7 or 8 per cent. From 12 to 15 seconds are sufficient for this.

The plate is then washed in a tray or under a faucet with distilled water, to liberate it from the free nitrate of silver and is afterwards placed upon blotting paper to drip off. The still moist plate is then coated with the albumen mixture:

Albumen....... 150 cubic centimeters

Add

Water..........	15 cubic centimeters
Iodide potassium	3 grams
Ammonia.......	5 grams
White sugar.....	2 grams

Iodine, a small cake.

With a wooden quirl this is beaten to snow (foam) for about 10 minutes, after which it must stand for 14 hours to settle. The albumen is poured on to the plate the same as collodion, and the surplus filtered back. After drying, the plate is laid for 15 seconds in a silver bath, con-

sisting of 100 parts of water, 10 parts nitrate of silver, and 10 cubic centimeters of acetic acid. The plate is then carefully washed and left to dry. If carefully kept, it will retain its properties for years. To the second silver bath, when it assumes a dirty coloration, is added 25 parts kaolin to each 100 parts, by shaking the same well, and the bath is then filtered, after which a little nitrate of silver and acetic acid is added.

After each exposure the plate holder is moved a certain length, so that 10 or more reproductions are obtained upon one and the same plate. The time of exposure depends upon the density of the negative and differs according to light. It varies between a second and a minute.

The developer is composed as follows:

Water.............	100 parts
Gallic acid........	0.3 parts
Pyro.............	0.1 part
Alcohol...........	2.5 parts

The exposed plate is immersed in this bath, and after 10 to 20 seconds, from 1 to 2 drops of a 2 per cent nitrate of silver solution are added to each 100 cubic centimeters of the solution, whereby the picture becomes visible. To follow the process exactly, the plate has to be laid—in yellow light—under a weakly enlarging microscope, and only a few drops of the developer are put upon the same. As soon as the picture has reached the desired strength, it is rinsed and fixed in a fixing soda solution, 1 to 5. Ten to 15 seconds are sufficient generally. Finally it is washed well.

After the drying of the plate, the several small pictures are cut with a diamond and fastened to the small enlarging lenses. For this purpose, the latter are laid upon a metal plate heated from underneath, a drop of Canada balsam is put to one end of the same, and, after it has become soft, the small diapositive is taken up with a pair of fine pincers, and is gradually put in contact with the fastener. Both glasses are then allowed to lie until the fastener has become hard. If bubbles appear, the whole method of fastening the picture has to be repeated.

Photographs on Brooches.—These may be produced by means of a paper (celuidin paper) whose upper layer after exposure by means of ordinary negative can be detached in lukewarm water. The picture copied on this paper is first laid in tepid water. After a few minutes it is taken out and placed on the article in question, naturally with the face upon it. The enamel surface upon which the pic ture is laid is previously coated with gelatin solution to insure a safe adhesion. When dry, the article is placed in water in which the paper is loosened and the photographic image now adheres firmly to the object. It may now be colored further and finally is coated with a good varnish.

FLASHLIGHT POWDERS AND APPARATUS.

Flash powders to be ignited by simply applying the flame of a match or laying on an oiled paper and igniting that, may be made by the following formulas:

I.—Magnesium........ 6 parts
Potassium chloride.. 12 parts

II.—Aluminum......... 4 parts
Potassium chloride.. 10 parts
Sugar............. 1 part

The ingredients in each case are to be powdered separately, and then lightly mixed with a wooden spatula, as the compound may be ignited by friction and burn with explosive violence.

It is best to make only such quantity as may be needed for use at the time, which is 10 or 15 grains.

To Prevent Smoke from Flashlight.—Support over the point where the ignition is to take place a large flat pad of damp wool lint. This may be done by tacking the lint to the underside of a board supported on legs. When ignition takes place the products of combustion for the most part will become absorbed by the wool.

A Flashlight Apparatus with Smoke Trap.—A light box, not too large to be conveniently carried out into the open air, is the first essential, and to the open front of this grooves must be fitted, in which grooves a lid will slide very easily, a large sheet of millboard being convenient as a sliding lid. The box being so placed that the sliding lid can be drawn out upward, a thread is attached to the lower edge of the lid, after which the thread is passed over a pulley fixed inside the box near the top, when the end is attached to the bottom of the box, so that the thread holds the sliding lid up. The lid will then slide down the grooves quickly, and close the box, if the thread is severed, the thread being cut at the right instant by placing the lower part across the spot where the flash is to be produced. So small is the cloud of smoke at the first instant that practically the whole of it can be caught in a drop trap of the above-mentioned kind. If the apparatus is not required again

for immediate use, the smoke may be allowed to settle down in the box; but in other cases the box may be taken out into the open air, and the smoke buffeted out with a cloth. In the event of several exposures being required in immediate succession, the required number of apparatus might be set up, as each need not cost much to construct.

INTENSIFIERS AND REDUCERS:

Intensifier (Mercuric) with Sodium Sulphite, for Gelatin Dry Plates.—Whiten the negative in the saturated solution of mercuric chloride, wash and blacken with a solution of sulphite of sodium, 1 in 5. Wash well.

The reduction is perfect, with a positive black tone.

Intensifier with Iodide of Mercury.—Dissolve 1 drachm of bichloride of mercury in 7 ounces of water and 3 drachms of iodide of potassium in 3 ounces of water, and pour the iodide solution into the mercury till the red precipitate formed is completely dissolved.

For use, dilute with water, flow over the negative till the proper density is reached, and wash, when the deposit will turn yellow. Remove the yellow color by flowing a 5 per cent solution of hypo over the plate, and give it the final washing.

Agfa Intensifier.—One part of agfa solution in 9 parts water (10 per cent solution). Immerse negative from 4 to 6 minutes.

Intensifying Negatives Without Mercury.—Dissolve 1 part of iodine and 2 parts of potassium iodide in 10 parts of water. When required for use, dilute 1 part of this solution with 100 parts of water. Wash the negative well and place in this bath, allowing it to remain until it has become entirely yellow, and the image appears purely dark yellow on a light-yellow ground. The negative should then be washed in water until the latter runs off clearly, when it is floated with the following solution until the whole of the image has become uniformly brown:

Schlippe's salt....... 60 grains
Water............. 1 ounce
Caustic soda solution,
  10 per cent........ 6 drops

Finally the negative is again thoroughly washed and dried. The addition of the small quantity of caustic soda is to prevent surface crystallization. It is claimed that with this intensifier the operation may be carried out to a greater extent than with bichloride of mercury; that it gives clear shadows, and that it possesses the special advantage of removing entirely any yellow stain the negative may have acquired during development and fixing. Furthermore, with this intensifying method it is not necessary to wash the negative, even after fixing, as carefully as in the case of the intensifying processes with mercury, because small traces of hypo which may have been left in the film will be rendered innocuous by the free iodine. The iodine solution may be employed repeatedly if its strength is kept up by the addition of concentrated stock solution.

Uranium Intensifier.—

Potassium ferricyanide (washed)	48 grains
Uranium nitrate	48 grains
Sodium acetate	48 grains
Glacial acetic acid	1 ounce
Distilled water to	10 ounces.

Label: Poison. Immerse the well-washed negative till the desired intensification is reached, rinse for 5 minutes and dry. This intensifier acts very strongly and should not therefore be allowed to act too long.

MISCELLANEOUS FORMULAS:

Renovating a Camera.—

The following formula should be applied to the mahogany of the camera by means of a soft rag, rubbing it well in, finally polishing lightly with a clean soft cloth:

Raw linseed oil	6	ounces
White wine vinegar	3	ounces
Methylated spirit	3	ounces
Butter of antimony	$\frac{1}{2}$	ounce

Mix the oil with vinegar by degrees, shaking well to prevent separation after each addition, then add the spirit and antimony, and mix thoroughly. Shake before using.

Exclusion of Air from Solutions.—

Water is free from air only when it has been maintained for several minutes in bubbling ebullition. In order to keep out the air from the bottle, when using the contents, the air-pressure contrivances are very convenient; one glass tube reaching through the rubber stopper into the bottle to the bottom, while the second tube, provided with a rubber pressing-ball, only runs into the flask above. If the long bent tube is fitted with a rubber tube, a single pressure suffices to draw off the desired quantity of the developer. It is still more convenient to pour a thin layer of good sweet oil on top of the developer besides. The developer is not injured thereby, and the exclusion of air is perfect.

Bottle Wax.—

Many ready-prepared solutions, such as developers and other preparations from which light has to be excluded, should be packed in bottles whose neck, after complete drying of the stopper, is dipped in a pot with molten sealing wax. A good recipe is the following, pigments being added if desired: For black take: Colophony, 6 parts; paraffine, 3 parts. Melt together and add 20 parts of black. For yellow, only 7 parts of chrome yellow. For blue, 7 parts of ultramarine.

Bleaching Photographic Prints White.

—To make a salt print, ink over it with waterproof ink, then bleach out white all but the black lines. Sensitize Clemon's mat surface paper on a 40-grain bath of nitrate of silver. After fuming and printing, the print is thoroughly fixed in hyposulphite of soda solution, and washed in running water until every trace of the hypo is out of the print. On this the permanency of the bleaching operation depends. The bleaching bath is:

Bichloride of mercury	1 ounce
Water	5 ounces
Alcohol	1 ounce
Hydrochloric acid	1 drachm

If the drawing has been made with non-waterproof ink, then alcohol is substituted for the water in the formula. For safety, use an alcoholic solution of mercury. The bleaching solution is poured on and off the drawing, and, when the print is bleached white, the mercury is washed off the drawing by holding it for a few moments under running water. Photographs bleached in this way will keep white for years.

To Render Negatives Permanent.—

A fine negative, one that we would like to preserve, may be rendered permanent by placing it, after it has been fixed, in a 10 per cent solution of alum, and letting it remain a few minutes. This makes the plate wonderfully clear and clean, and absolutely unalterable. The alum acts upon the gelatin, rendering it insoluble.

Stripping Photograph Films.—

This is generally done by immersing the plate in formaldehyde solution until the film has become almost insoluble and impermeable. Then it is placed in a solution of sodium carbonate until the gelatin has absorbed a sufficient quantity of it. When the negative is immersed in weak hydrochloric acid, carbon dioxide is liberated, and the little bubbles of gas which lodge themselves between the film and the glass cause a separation of the two, so that the film may be stripped off. After having hardened the film with formaldehyde, it is a lengthy process to get it saturated with sodium carbonate. It is advisable to use a combined bath of 1 part of carbonate, 3 of 40 per cent formaldehyde, and 20 of water; its tanning action is enhanced by the alkaline reaction, and two operations are superseded by one. After 10 minutes' soaking, the surface of the film must be wiped and the plate dried. A sharp knife is then used to cut all around the film a slight distance from the edge, and when this is done the negative is put into a 5 per cent solution of hydrochloric acid, when the film will probably float off unaided; but, if necessary, may be assisted by gently raising one corner.

Phosphorescent Photographs. — The necessary chemicals belong to the class of phosphorescent bodies, among others, calcium sulphite, strontium sulphite, barium sulphide, calcareous spar, fluorspar. These placed in the magnesium light or sunlight, acquire the property of giving forth, for a shorter or longer time, a light of their own. The best examples of these substances are the well-known "Balmains light colors," which yield a very clear and strong light after exposure. They consist of calcium sulphide, 10,000 parts; bismuth oxide, 13 parts; sodium hyposulphite, 1,000 parts.

According to Professor Schnauss, plates for phosphorographs are prepared as follows: Dissolve 10 parts of pure gelatin in 50 parts of hot water, add and dissolve 30 parts of "light" color (as above), and 1 part of glycerine.

If a plate or a paper, prepared as above detailed, be placed under a diapositive, in a copying apparatus, and submitted to the action of sunlight for a few minutes, when taken out in a dark room a phosphorescent picture of the diapositive will be found. It is also a known fact that duplicate negatives or positives may be made with this phosphorograph by simply bringing the latter in contact in a copying apparatus, with the ordinary silver bromide plate for 30 seconds, in the dark room, and then developing the same.

Printing Names on Photographs. — The name or other matter to be printed on the photograph is set up in type, and printed on cardboard; from this make an exposure on a transparency plate, developing it strongly. After the print has been made from the regular printing negative, it is placed under the dense transparency of the regular negative, and the name printed in. The only precaution necessary is to time the transparency negative properly, and develop strongly, so as to get good contrast. Photographers will find this a much easier and quicker method than the old one of printing on tissue paper and fastening the paper to the negative by means of varnish; moreover, the result is black instead of white, usually much more pleasing.

Spots on Photographic Plates. — Spots on photographic plates may be caused by dust or by minute bubbles in the emulsion, both of which are easily preventable, but some spots cannot be ascribed to either of these causes. On investigating this trouble, Mumford found that it is due to the presence on the surface of the film of small colonies of microorganisms which, under conditions favorable to their growth, are capable of producing large mold colonies, from which the organisms can easily be separated. Experiments were instituted in order to find whether these growths can be produced on the plate by artificial means, by inoculating the surface with a fluid culture of one of these organisms, with affirmative results, but with one slight difference, namely, that in the inoculated film, on microscopic examination, no dust particle was visible in the center of each spot, which had formerly been the case. As these microorganisms do not exist in the air as isolated units, but travel upon small or large dust particles in the case under consideration, the carrying medium most probably is the fine impalpable dust from which it is practically impossible to free the air of a building. In order that these organisms may grow into colonies of sufficient size to cause spots, they must be able to grow rapidly, there being only about 12 hours before the plate is dry in which they can grow; and they must also be capable of growing at the rather high temperature of 70° F. On testing some of the organisms causing the spots it was found that they grew best under exactly such conditions. A bacteriological examination of some of the gelatin used in the manufacture of plates, both in the raw state and in the form of emulsion, also revealed the fact that there were numerous organisms present. No means for the prevention of this troublesome defect is suggested;

most dry-plate manufacturers use the precaution to add a small quantity of a chemical antiseptic to the emulsion, but it is not possible to employ a sufficient quantity to destroy any organisms that may be present without damaging the plate for photographic purposes.

To Remove Pyro Stains from the Fingers.—Make a strong solution of chlorinated lime; dip the fingers which are stained in this, and rub the stains with a large crystal of citric acid. Apply the lime solution and acid alternately until the stain is removed; then rinse with water.

To Remove Pyro Stain from Negatives.—Immerse in a clearing bath as follows:

Protosulphate of iron.	3 ounces
Alum	1 ounce
Citric acid	1 ounce
Water	20 ounces

Prevention is better than cure, however; therefore immerse the negatives in the above directly they are taken from the fixing bath. After clearing the negatives, they should be well washed.

PHOTOGRAPHY WITHOUT LIGHT:
See Catatypy.

PIANO POLISHES:
See Polishes.

PICKLE FOR BRASS:
See Brass and Plating.

PICKLE FOR BRONZE:
See Bronze Coloring.

PICKLE FOR COPPER:
See Copper and Plating.

PICKLE VINEGAR:
See Vinegar.

PICKLING OF GERMAN-SILVER ARTICLES:
See Plating.

PICKLING IRON SCRAP BEFORE ENAMELING:
See Enameling.

PICRIC ACID STAINS, TO REMOVE:
See Cleaning Preparations and Methods.

PICTURE COPYING:
See Copying.

PICTURE FRAMES, REPAIRING:
See Adhesives and Lutes.

PICTURE POSTAL CARDS:
See Photography.

Pigments

(See also Paints.)

Nature, Source, and Manufacture of Pigments.—A pigment is a dry earthy or clayey substance that, when mixed with oil, water, etc., forms a paint. Most pigments are of mineral origin, but there are vegetable pigments, as logwood, and animal pigments, as cochineal. In modern practice the colors are produced mainly by dyeing certain clays, which excel in a large percentage of silicic acid, with aniline dyestuffs. The coloring matters best adapted for this purpose are those of a basic character. The colors obtained in this manner excel in a vivid hue, and fastness to light and water.

Following is a general outline of their manufacture: One hundred parts, by weight, of washed clay in paste form are finely suspended in 6 to 8 times the volume of water and acidulated with about $1\frac{1}{2}$ parts, by volume, of 5 per cent hydrochloric or acetic acid, and heated by means of steam almost to the boiling temperature. There is next introduced, according to the shade desired, 1 to 2 parts, by weight, of the dyestuff, such as auramin, diamond green, Victoria blue, etc., with simultaneous stirring and heating, for 1 to 2 hours, or until a sample filtered off from the liquor shows no dyestuff. Next the clay dyed in this manner is isolated by filtration and washed with hot water and dried. The colors thus obtained may be used as substitutes for mineral colors of all description.

The method of manufacture varies greatly. According to the Bennett and Mastin English patent the procedure is as follows: Grind together to a paste in water, substances of a clayey, stony, earthy, or vitreous nature, and certain metallic oxides, or "prepared oxides," such as are commonly used in the pottery trades; dry and powder the paste, and subject the powder to the heat of a furnace, of such a temperature that the requisite color is obtained, and for such length of time that the color strikes through the whole substance. For example, 8 parts of black oxide of cobalt, 12 parts of oxide of zinc, and 36 parts of alumina, when incorporated with 20 times their combined bulk of clay and treated as described, yield a rich blue pigment in the case of a white clay, and a rich green in the case of a yellow clay. Long-continued firing in this case improves the color.

Many minerals included in formulas for pigments have little or no coloring power in themselves; nevertheless they are required in producing the most beautiful shades of color when blended one with another, the color being brought out by calcination.

Mixing Oil Colors and Tints.—It must not be expected that the formulas given will produce the exact effect desired, because the strength of the various brands of colors vary to a great extent, and therefore the painter must exercise his own judgment. The table simply gives an idea of what can be produced by following the formulas given, when chemically pure material is employed in the mixing. It is also recommended that the parts mentioned be weighed out in paste form, and the white or black and each color separately thinned and strained before mixing them together, because the arriving at the proper hue of color or depth and tone of tint will be simplified by using that precaution. By thinning it is not meant that they should be quite ready for application, but of such consistency that they will pass an ordinary strainer with the aid of a brush.

Unless otherwise indicated, the materials are understood to be ground fine in paste form.

NOTE.—The majority of the following are by Joseph Griggs, in the *Painters' Magazine*:

GROUNDS FOR GRAINING COLORS:

Ash Ground.—Four hundred parts white lead; 4 parts French ocher; 1 part raw Turkey umber.

Ash.—Raw umber; raw sienna; and a little black or Vandyke brown.

Hungarian Ash.—Raw sienna and raw and burnt umber.

Bun Ash.—Raw sienna; burnt umber; and Vandyke brown.

Cherry Ground.—One hundred parts white lead; 5 parts burnt sienna; 1 part raw sienna.

Natural Cherry.—Raw and burnt sienna and raw umber.

Stained Cherry.—Burnt sienna; burnt umber; and Vandyke brown.

Chestnut.—Raw sienna; burnt umber; Vandyke brown; and a little burnt sienna.

Maple.—Raw sienna and raw umber.

Silver Maple.—Ivory black over a nearly white ground.

Light Maple Ground.—One hundred parts white lead; 1 part French ocher.

Dark Maple Ground.—One hundred parts white lead; 1 part dark golden ocher.

Oak.—Raw sienna; burnt umber; a little black.

Pollard Oak.—Raw and burnt sienna, or burnt umber and Vandyke brown.

Light Oak Ground.—Fifty parts white lead; 1 part French ocher.

Dark Oak Ground.—Fifty parts white lead; 1 part dark golden ocher.

Satinwood.—Add a little ivory black to maple color.

Mahogany.—Burnt sienna; burnt umber; and Vandyke brown.

Mahogany Ground.—Ten parts white lead; 5 parts orange chrome; and 1 part burnt sienna.

Rosewood.—Vandyke brown and a little ivory black.

Rosewood Ground.—Drop black.

Walnut Ground.—Fifty parts white lead; 3 parts dark golden ocher; 1 part dark Venetian red; and 1 part drop black.

Black Walnut.—Burnt umber with a little Vandyke brown for dark parts.

French Burl Walnut.—Same as black walnut.

Hard Pine.—Raw and burnt sienna; add a little burnt umber.

Cypress.—Raw and burnt sienna and burnt umber.

Whitewood.—Ground same as for light ash; graining color, yellow ocher, adding raw umber and black for dark streaks.

POSITIVE COLORS:

Blue.—Twelve parts borate of lime; 6 parts oxide of zinc; 10 parts litharge; 9 parts feldspar; 4 parts oxide of cobalt.

Blue Black A.—Nine parts lampblack; 1 part Chinese or Prussian blue.

Blue Black B.—Nineteen parts drop black; 1 part Prussian blue.

Bright Mineral.—Nine parts light Venetian red; 1 part red lead.

Brilliant Green.—Nine parts Paris green; 1 part C. C. chrome green, light.

Bronze Green, Light.—Three parts raw Turkey umber; 1 part medium chrome yellow.

Bronze Green, Medium.—Five parts medium chrome yellow; 3 parts burnt Turkey umber; 1 part lampblack.

Bronze Green, Dark.—Twenty parts drop black; 2 parts medium chrome yellow; and 1 part dark orange chrome.

Bottle Green.—Five parts commercial chrome green, medium, and 1 part drop black.

Brown.—Ten parts crude antimony; 12 parts litharge; 2 parts manganese; 1 part oxide of iron.

Brown Stone.—Eighteen parts burnt umber; 2 parts dark golden ocher; and 1 part burnt sienna.

Cherry Red.—Equal parts of best imitation vermilion and No. 40 carmine.

Citron A.—Three parts medium chrome yellow and 2 parts raw umber.

Citron B.—Six parts commercial chrome green, light, and 1 part medium chrome yellow.

Coffee Brown.—Six parts burnt Turkey umber; 2 parts French ocher; and 1 part burnt sienna.

Emerald Green.—Use Paris green.

Green.—Twenty parts litharge; 12 parts flint; 2 parts oxide of copper; $2\frac{1}{2}$ parts ground glass; $2\frac{1}{2}$ parts whiting; $1\frac{1}{2}$ parts oxide of chrome.

Flesh Color.—Nineteen parts French ocher; 1 part deep English vermilion.

Fern Green.—Five parts lemon chrome yellow and 1 part each of light chrome green and drop black.

Foliage Green.—Three parts medium chrome yellow and 1 part of ivory or drop black.

Foliage Brown.—Equal parts of Vandyke brown and orange chrome yellow.

Golden Ocher.—Fourteen parts French yellow ocher and 1 part medium chrome yellow for the light shade, and 9 parts Oxford ocher and 1 part orange chrome yellow for the dark shade.

Gold Russet.—Five parts lemon chrome yellow and 1 part light Venetian red.

Gold Orange.—Equal parts of dry orange mineral and light golden ocher in oil.

Indian Brown.—Equal parts of light Indian red, French ocher, and lamp black.

Mahogany, Cheap.—Three parts dark golden ocher and 1 part of dark Venetian red.

Maroon, Light.—Five parts dark Venetian red; 1 part drop black.

Maroon, Dark.—Nine parts dark Indian red; 1 part lampblack.

Olive Green.—Seven parts light golden ocher; 1 part drop black.

Ochrous Olive.—Nine parts French ocher; 1 part raw umber.

Orange-Brown.—Equal parts burnt sienna and orange chrome yellow.

Oriental Red.—Two parts Indian red, light, in oil; 1 part dry red lead.

Purple A.—Eight parts crocus martis; 2 parts red hematite; 1 part oxide of iron.

Purple B.—Two parts rose pink; 1 part ultramarine blue.

Purple Black.—Three parts lampblack and 1 part rose pink, or 9 parts drop black and 1 part rose pink.

Purple Brown.—Five parts Indian red, dark, and 1 part each of ultramarine blue and lampblack.

Roman Ocher.—Twenty-three parts French ocher and 1 part each burnt sienna and burnt umber.

Royal Blue, Dark.—Eighteen parts ultramarine blue and 2 parts Prussian blue. To lighten use as much white lead or zinc white as is required.

Royal Purple.—Two parts ultramarine blue; 1 part No. 40 carmine or carmine lake.

Russet. — Fourteen parts orange chrome yellow and 1 part C. P. chrome green, medium.

Seal Brown.—Ten parts burnt umber; 2 parts golden ocher, light; 1 part burnt sienna.

Snuff Brown.—Equal parts burnt umber and golden ocher, light.

Terra Cotta.—Two parts white lead; 1 part burnt sienna; also 2 parts French ocher to 1 part Venetian red.

Turkey Red.—Strong Venetian red or red oxide.

Tuscan Red. Ordinary.—Nine parts Indian red to 1 part rose pink.

Brilliant.—Four parts Indian red to 1 part red madder lake.

Violet.—Three parts ultramarine blue; 2 parts rose lake; 1 part best ivory black.

Yellow.—Four and one-half parts tin ashes; 1 part crude antimony; 1 part litharge; and 1 part red ocher.

Yellow, Amber.—Ten parts medium chrome yellow; 7 parts burnt umber; 3 parts burnt sienna.

Yellow, Canary.—Five parts white lead; 2 parts permanent yellow; 1 part lemon chrome yellow.

Yellow, Golden.—Ten parts lemon chrome yellow; 3 parts orange chrome, dark; 5 parts white lead.

Yellow, Brimstone. — Three parts white lead; 1 part lemon chrome yellow; 1 part permanent yellow.

Azure Blue.—Fifty parts white lead; 1 part ultramarine blue.

Blue Gray.—One hundred parts white lead; 3 parts Prussian blue; 1 part lampblack.

Bright Blue.—Twenty parts zinc white; 1 part imitation cobalt blue.

Blue Grass.—Seven parts white lead; 2 parts Paris green; 1 part Prussian blue.

Deep Blue.—Fifteen parts white lead; 1 part Prussian blue or Antwerp blue.

French Blue.—Five parts imitation cobalt blue; 2 parts French zinc white.

Green Blue.—One hundred parts white lead; 5 parts lemon chrome yellow; 3 parts ultramarine blue.

Hazy Blue.—Sixty parts white lead; 16 parts ultramarine blue; 1 part burnt sienna.

Mineral Blue.—Five parts white lead; 4 parts imitation cobalt blue; 2 parts red madder lake; 1 part best ivory or drop black.

Orient Blue.—Twenty-five parts white lead; 2 parts Prussian blue; 1 part lemon chrome yellow.

Royal Blue.—Thirty-four parts white lead; 19 parts ultramarine blue; 2 parts Prussian blue; 1 part rose madder or rose lake.

Sapphire Blue.—Two parts French zinc white and 1 part best Chinese blue.

Sky Blue.—One hundred parts white lead; 1 part Prussian blue.

Solid Blue.—Five parts white lead; 1 part ultramarine blue.

Turquoise Blue.—Twenty parts white lead; 3 parts ultramarine blue; 1 part lemon chrome yellow.

RED TINTS:

Cardinal Red.—Equal parts of white lead and scarlet lake.

Carnation Red.—Fifteen parts white lead; 1 part scarlet lake.

Claret.—Twenty-one parts oxide of zinc; 4 parts crocus martis; 4 parts oxide of chrome; 3 parts red lead; 3 parts boracic acid.

Coral Pink.—Fifteen parts white lead; 2 parts bright vermilion; 1 part deep orange chrome.

Deep Rose.—Ten parts white lead; 1 part red lake.

Deep Purple.—Five parts white lead; 1 part ultramarine blue; 1 part rose pink.

Deep Scarlet.—Fifteen parts bright vermilion; 2 parts red lake; 5 parts white lead.

Flesh Pink.—One hundred parts white lead; 1 part orange chrome yellow; 1 part red lake.

Indian Pink.—One hundred parts white lead; 1 part light Indian red.

Lavender.—Fifty parts white lead; 2 parts ultramarine blue; 1 part red lake.

Light Pink.—Fifty parts white lead; 1 part bright vermilion.

Lilac.—Fifty parts white lead; 1 part best rose pink.

Mauve.—Fifteen parts white lead; 2 parts ultramarine blue; 1 part carmine lake or red lake.

Orange Pink.—Two parts white lead; 1 part dark orange chrome or American vermilion.

Purple.—Five parts white lead; 2 parts ultramarine blue; 1 part red madder lake.

Royal Pink.—Five parts white lead; 1 part carmine lake or red madder lake.

Royal Rose.—Twenty parts white lead; 1 part rich rose lake.

Red Brick.—Ten parts white lead; 3 parts light Venetian red; 1 part yellow ocher.

Reddish Terra Cotta.—Two parts white lead; 1 part rich burnt sienna.

Salmon.—Fifty parts white lead; 5 parts deep orange chrome.

Shell Pink.—Fifty parts white lead; 2 parts bright vermilion; 1 part orange chrome; 1 part burnt sienna.

Violet.—Fifteen parts white lead; 4 parts ultramarine blue; 3 parts rose lake; 1 part drop black.

GREEN TINTS:

Apple Green.—Fifty parts white lead; 1 part chrome green, light or medium shade.

Citrine Green.—One hundred parts white lead; 2 parts medium chrome yellow; 1 part drop black.

Citron Green.—One hundred parts white lead; 3 parts medium chrome yellow; 1 part lampblack.

Emerald Green.—Ten parts white lead; 1 part Paris (emerald) green.

Grass Green A.—Five parts white lead; 7 parts Paris green.

Grass Green B.—Ten parts oxide of chrome; 2 parts tin ashes; 5 parts whiting; 1 part crocus martis; 1 part bichromate potash.

Gray Green.—Five parts white lead; 1 part Verona green.

Marine Green.—Ten parts white lead; 1 part ultramarine green.

Nile Green.—Fifty parts white lead; 6 parts medium chrome green; 1 part Prussian blue.

Olive Green.—Fifty parts white lead; 2 parts medium chrome yellow; 3 parts raw umber; 1 part drop black.

Olive Drab.—Fifty parts white lead; 8 parts raw umber; 5 parts medium chrome green; 1 part drop black.

Pea Green.—Fifty parts white lead; 1 part light chrome green.

Satin Green.—Three parts white lead; 1 part Milori green.

Sage Green.—One hundred parts white lead; 3 parts medium chrome green; 1 part raw umber.

Sea Green.—Fifty parts white lead; 1 part dark chrome green.

Stone Green.—Twenty-five parts white lead; 2 parts dark chrome green; 3 parts raw umber.

Velvet Green.—Twenty parts white lead; 7 parts medium chrome green; 2 parts burnt sienna.

Water Green.—Fifteen parts white lead; 10 parts French ocher; 1 part dark chrome green.

BROWN TINTS:

Chocolate.—Twenty-five parts white lead; 3 parts burnt umber.

Cocoanut.—Equal parts white lead and burnt umber.

Cinnamon.—Ten parts white lead; 2 parts burnt sienna; 1 part French ocher.

Dark Drab.—Forty parts white lead; 1 part burnt umber.

Dark Stone.—Twenty parts white lead; 1 part raw umber.

Fawn.—Fifty parts white lead; 3 parts burnt umber; 2 parts French ocher.

Golden Brown.—Twenty-five parts white lead; 4 parts French ocher; 1 part burnt sienna.

Hazel Nut Brown.—Twenty parts white lead; 5 parts burnt umber; 1 part medium chrome yellow.

Mulberry.—Ten parts manganese; 2 parts cobalt blue; 2 parts saltpeter.

Purple Brown.—Fifty parts white lead; 6 parts Indian red; 2 parts ultramarine blue; 1 part lampblack.

Red Brown.—Twelve parts hematite ore; 3 parts manganese; 7 parts litharge; 2 parts yellow ocher.

Seal Brown.—Thirty parts white lead; 5 parts burnt umber; 1 part medium chrome yellow.

Snuff Brown.—Twenty-five parts white lead; 1 part burnt umber; 1 part Oxford ocher.

GRAY TINTS:

Ash Gray.—Thirty parts white lead; 2 parts ultramarine blue; 1 part burnt sienna.

Cold Gray.—Five hundred parts white lead; 6 parts lampblack; 1 part Antwerp blue.

Dove Color.—Twelve parts manganese; 5 parts steel filings; 3 parts whiting; 1 part oxide of cobalt.

Dove Gray.—Two hundred parts white lead; 5 parts ultramarine blue; 2 parts drop black.

French Gray.—One hundred and fifty parts white lead; 2 parts lampblack; 1 part orange chrome yellow; 1 part chrome red (American vermilion).

Lead Color.—Fifty parts white lead; 1 part lampblack (increase proportion of white lead for light tints).

Lustrous Gray.—Ten parts white lead; 1 part graphite (plumbago).

Olive Gray.—Two hundred parts white lead; 2 parts lampblack; 1 part medium chrome green.

Pure Gray.—One hundred parts white lead; 1 part drop black.

Pearl Gray.—One hundred parts white lead; 1 part ultramarine blue; 1 part drop black.

Silver Gray.—One hundred and fifty parts white lead; 2 parts lampblack; 3 parts Oxford ocher.

Warm Gray.—One hundred parts white lead; 3 parts drop black; 2 parts French ocher; 1 part light Venetian red.

NOTE.—For inside work and whenever desirable, the white lead may be replaced by zinc white or a mixture of the two white pigments may be used. Be it also remembered that pure colors, as a rule, will produce the cleanest tints and that fineness of grinding is an important factor. It will not be amiss to call attention to the fact that the excessive use of driers, especially of dark japans or liquid driers, with delicate tints is bad practice, and liable to ruin otherwise good effects in tints or delicate solid colors.

COLOR TESTING.

Expense and trouble deter many a painter from having a color examined, although such an examination is often very necessary. For the practical man it is less important to know what percentage of foreign matter a paint contains, but whether substances are contained therein, which may act injuriously in some way or other.

If a pigment is to be tested for arsenic, pour purified hydrochloric acid into a test tube or a U-shaped glass vessel which withstands heat, add a little of the pigment or the colored fabric, wall paper, etc. (of pigment take only enough to strongly color the hydrochloric acid simply in the first moment), and finally a small quantity of stannous chloride. Now heat the test tube with its contents moderately over a common spirit lamp. If the liquid or mass has assumed a brown or brownish color after being heated, arsenic is present in the pigment or fabric, etc.

An effective but simple test for the durability of a color is to paint strips of thick paper and nail them on the wall in the strongest light possible. A strip of paper should then be nailed over one-half of the samples of color so as to protect them from the light. On removing this the difference in shade between the exposed and unexposed portions will be very apparent. Some colors, such as the vermilionettes, will show a marked difference after even a few weeks.

Testing Body Colors for Gritty Admixtures.—The fineness of the powdered pigment is not a guarantee of the absence of gritty admixtures. The latter differ from the pigment proper in their specific gravity. If consisting of metallic oxides or metallic sulphides the sandy admixtures are lighter than the pigments and rise to the surface upon a systematic shaking of the sample. In the case of other pigments, e. g., aluminas and iron varnish colors, they collect at the bottom. For carrying out the test, a smoothly bored metallic tube about $\frac{1}{2}$ to $\frac{3}{4}$ inch in diameter and 6 to 7 inches long is used. Both ends are closed with screw caps and at one side of the tube some holes about $\frac{1}{8}$ of an inch in diameter are bored, closed by pieces of a rubber hose pushed on. The tube is filled with the pigment powder, screwed up and feebly shaken for some time in a vertical position (the length of time varying according to the fineness of the powder). Samples may now be taken from all parts of the tube. Perhaps glass tubes would be preferable, but lateral apertures cannot be so readily made. After the necessary samples have been collected in this manner, they must be prepared with a standard sample, which is accomplished either by feeling the powder between the fingers or by inspecting it under a microscope, or else by means of the scratching test, which last named is the usual way. The requisites for these scratch tests consist of two soft, well-polished glass plates ($2\frac{1}{2}$ x $2\frac{1}{2}$ inches) which are fixed by means of cement in two stronger plates of hard wood suitably hollowed out. The surface of the glass must project about $\frac{1}{2}$ inch over the wooden frame. If a sample of the pigment powder is placed on such a glass plate, another plate is laid on top and both are rubbed slowly together; this motion will retain a soft, velvety character in case the pigment is free from gritty admixtures; if otherwise, the glass is injured and a corresponding sound becomes audible. Next the powder is removed from the plate, rubbing the latter with a soft rag, and examining the surface with a microscope. From the nature of the scratches on the plate the kind of gritty ingredients can be readily determined. The human finger is sufficiently sensitive to detect the presence of gritty substances, yet it is not capable of distinguishing whether they consist of imperfectly reduced or badly sifted grains of pigment or real gritty admixtures.

To Determine the Covering Power of Pigments.—To determine the covering power of white lead, or any other pigment, take equal quantities of several varieties of white lead and mix them with a darker pigment, black, blue, etc., the latter also in equal proportions. The white lead which retains the lightest color is naturally the most opaque. In a similar manner, on the other hand, the mixing power of the dark pigments can be ascertained. If experiments are made with a variety of white lead or zinc white, by the admixture of dark pigments, the color which tints the white lead or zinc white most, also possesses the greatest covering or mixing power.

To Detect the Presence of Aniline in a Pigment.—Lay a little of the color upon letter paper and pour a drop of spirit on it. If it is mixed with aniline the paper is colored right through thereby, while a pure pigment does not alter the shade of the paper and will never penetrate it.

Vehicle for Oil Colors.—Petroleum, 20 to 30 pounds; tallow, 3 to 5 pounds; cotton-seed oil, 5 to 7 pounds; colophony, 5 to 7 pounds. The pigments having been ground up with this mixture, the mixed paint can be made still better by adding to it about a sixth of its weight of the following mixture: Vegetable oil, 8 to 20 pounds; saponified rosin, 6 to 16 pounds; turpentine, 4 to 30 ounces.

Frankfort Black.—Frankfort black, also known as German black, is a name applied to a superior grade of lampblack. In some districts of Germany it is said to be made by calcining wine lees and tartar. The material is heated in large cylindrical vessels having a vent in the cover for the escape of smoke and vapors that are evolved during the process. When no more smoke is observed, the operation is finished. The residuum in the vessels is then washed several times in boiling water to extract the salts contained therein and finally is reduced to the proper degree of fineness by grinding on a porphyry.

Paris Green.—Emerald or Paris green is rather permanent to light, but must not be mixed with pigments containing sulphur, because of the tendency to blacken when so mixed. It will not resist acids, ammonia, and caustics.

PIGMENT PAPER:
See Photography.

PILE OINTMENTS.

I.—"Extract" witch-
hazel	2 fluidounces
Lanum	2 ounces
Petrolatum	6 ounces
Glycerine	4 fluidounces
Tannic acid	1 drachm
Powdered opium	1 drachm

II.—
Tannic acid	20 grains
Bismuth subnitrate	1 drachm
Powdered opium	10 grains
Lanum	3 drachms
Petrolatum	5 drachms

PINE SYRUP:
See Essences and Extracts.

PINEAPPLE ESSENCE:
See Essences and Extracts.

PINEAPPLE LEMONADE:
See Beverages.

PING PONG FRAPPÉ:
See Beverages, under Lemonades.

PINS OF WATCHES:
See Watchmakers' Formulas.

PINION ALLOY:
See Watchmakers' Formulas.

PINK SALVE:
See Ointments.

PINKEYE:
See Veterinary Formulas.

PIPE-JOINT CEMENT:
See Cement.

PIPE LEAKS:
See Leaks.

PIPES, RUST-PREVENTIVE FOR:
See Rust Preventives.

PISTACHIO ESSENCE:
See Essences and Extracts.

PLANTS:

Temperature of Water for Watering Plants.—Experiments were made several years ago at the Wisconsin Agricultural Experiment Station to determine whether cold water was detrimental to plants. Plants were grown under glass and in the open field, and in all cases the results were similar. Thus, coleus planted in lots of equal size and vigor were watered with water at 35°, 50°, 65°, and 86° F. At the end of 60 days it was impossible to note any difference, and when the experiment was repeated with water at 32°, 40°, 70°, and 100° F., the result was the same. Beans watered with water at 32°, 40°, 70°, and 100° F., were equally vigorous; in fact, water at 32° and 40° F. gave the best results. Lettuce watered with water at 32° F. yielded slightly more than the other lots. From these experiments it was concluded that for vegetable and flowering plants commonly grown under glass, ordinary well or spring water may be used freely at any time of the year without warming.

PLANT PRESERVATIVES:
See Flowers.

Plaster

(See also Gypsum.)

Therapeutic Grouping of Medicinal Plasters.—The vehicle for medicated plasters requires some other attribute than simply adhesiveness. From a study of the therapy of plasters they may be put in three groups, similarly to the ointments with reference to their general therapeutic uses, which also governs the selection of the respective vehicles.

1.—Epidermatic: Supportive, protective, antiseptic, counter-irritant, vesicant. Vehicle: Rubber or any suitable adhesive. Official plasters: Emp. adhesivum, E. capsici.

2.—Endermatic: Anodyne, astringent, alterative, resolvent, sedative, stimulant. Vehicle: Oleates or lead plaster, sometimes with rosins or gum rosins. Official plasters: Emp. Belladonnæ, E. opii, E. plumbi, E. saponis.

3.—Diadermatic: For constitutional or systemic effects. Vehicle: Lanolin or plaster-mull. Official plasters: Emp. hydrargyri.

Methods of Preparing Rubber Plasters.—Mechanic Roller Pressure Method. —This method of incorporating the rubber with certain substances to give it the necessary body to serve as a vehicle is at present the only one employed. But since it requires the use of the heaviest machinery—some of the apparatus weighing many tons—and enormous steam power, its application for pharmaceutical purposes is out of the question.

As is well known, the process consists in: 1. Purification of the rubber by mascerating and pressing it and removing foreign impurities by elutriating it with water. 2. Forming a homogeneous mass of the dried purified rubber by working it on heated revolving rollers and incorporating sufficient quantities of orris powder and oleoresins. 3. Incorporating the medicinal agent, i. e., belladonna extract, with the rubber mass by working it on warmed revolving rollers. 4. Spreading the prepared plaster.

Solution in Volatile Solvents.—This process has been recommended from time to time, the principal objection being the use of so relatively large quantities of inflammable solvents.

The German Pharmacopœia Method. —The following is the formula of "Arzneibuch für das Deutsche Reich," 1900: Emplastrum adhesivum: Lead plaster, waterfree, 40 parts; petrolatum, 2.5 parts; liquid petrolatum, 2.5 parts, are melted together, and to the mixture add rosin, 35 parts; dammar, 10 parts, previously melted. To the warm mixture is added caoutchouc, 10 parts; dissolved in benzine, 75 parts, and the mixture stirred on the water-bath until all the benzine is lost by evaporation.

The Coleplastrum adhesivum of the Austrian Society is still more complex, the formula containing the following: Rosin oil, empyreumatic, 150 parts; copaiba, 100 parts; rosin, 100 parts; lard, 50 parts; wax, 30 parts; dissolved in ether, 1,200 parts, in which caoutchouc, 250 parts, has been previously dissolved; to this

is then added orris powder, 220 parts; sandarac, 50 parts; ether, 400 parts. The mixture, when uniform, is spread on cloth.

Solution of Rubber in Fixed Solvent: Petrolatum and Incorporation with Lead Acetate.—India rubber dissolves, though with difficulty, in petrolatum. The heat required to melt the rubber being comparatively high, usually considerably more than 212° F., as stated in the U. S. P., it is necessary to melt the rubber first and then add the petrolatum, in order to avoid subjecting the latter to the higher temperature. The mixture of equal parts of rubber and petrolatum is of a soft jelly consistence, not especially adhesive, but when incorporated with the lead oleate furnishes a very adhesive plaster. While at first 5 per cent of each rubber and petrolatum was used, it has been found that the petrolatum would melt and exude around the edges of the plaster when applied to the skin, and the quantity was therefore reduced to 2 per cent of each. This mass affords a plaster which is readily adhesive to the body, does not run nor become too soft. Plasters spread on cloth have been kept for months exposed to the sun in the summer weather without losing their stability or permanency.

The lead oleate made by the interaction of hot solution of soap and lead acetate, thoroughly washed with hot water, and freed from water by working the precipitated oleate on a hot tile, is much to be preferred to the lead plaster made by the present official process. The time-honored method of boiling litharge, olive oil, and water is for the requirements of the pharmacists most tedious and unsatisfactory. Since in the beginning of the process, at least, a temperature higher than that of 212° F. is required, the water bath cannot be employed, and in the absence of this limiting device the product is usually "scorched." When the steam bath under pressure can be used this objection does not apply. But the boiling process requires from 3 to 4 hours, with more or less attention, while the precipitation method does not take over half an hour. Besides, true litharge is difficult to obtain, and any other kind will produce unsatisfactory results.

The following is the process employed:

Lead oleate (Emplastrum plumbi):
Soap, granular and
dried............ 100 parts
Lead acetate....... 60 parts
Distilled water, a sufficient quantity.

Dissolve the soap in 350 parts hot distilled water and strain the solution. Dissolve the lead acetate in 250 parts hot distilled water and filter the solution while hot into the warm soap solution, stirring constantly. When the precipitate which has formed has separated, decant the liquid and wash the precipitate thoroughly with hot water. Remove the precipitate, let it drain, free from water completely by kneading it on a warm slab, form it into rolls, wrap in paraffine paper, and preserve in tightly closed containers.

Emplastrum adhesivum:

Rubber, cut in small pieces...........	20 parts
Petrolatum.........	20 parts
Lead plaster........	960 parts

Melt the rubber at a temperature not exceeding 302° F., add the petrolatum, and continue the heat until the rubber is dissolved. Add the lead plaster to the hot mixture, continue the heat until it becomes liquid; then let it cool and stir until it stiffens.

Court Plaster or Sticking Plaster.—I. —Brush silk over with a solution of isinglass, in spirits or warm water, dry and repeat several times. For the last application apply several coats of balsam of Peru. This is used to close cuts or wounds, by warming and applying it. It does not wash off until the skin partially heals.

II.—Isinglass, 1 part; water, 10 parts; dissolve, strain the solution, and gradually add to it of tincture of benzoin, 2 parts; apply this mixture gently warmed, by means of a camel's-hair brush, to the surface of silk or sarcenet, stretched on a frame, and allow each coating to dry before applying the next one, the application being repeated as often as necessary; lastly, give the prepared surface a coating of tincture of benzoin or tincture of balsam of Peru. Some manufacturers apply this to the unprepared side of the plaster, and others add to the tincture a few drops of essence of ambergris or essence of musk.

III. (Deschamps).—A piece of fine muslin, linen, or silk is fastened to a flat board, and a thin coating of smooth, strained flour paste is given to it; over this, when dry, two coats of colorless gelatin, made into size with water, quantity sufficient, are applied warm. Said to be superior to the ordinary court plaster.

Coloring of Modeling Plaster.—I.—If burnt gypsum is stirred up with water containing formaldehyde and with a little alkali, and the quantity of water necessary for the induration of the plaster containing in solution a reducible metallic salt is added thereto, a plaster mass of perfectly uniform coloring is obtained. The hardening of the plaster is not affected thereby. According to the concentration of the metallic salt solutions and the choice of the salts, the most varying shades of color, as black, red, brown, violet, pearl gray, and bronze may be produced. The color effect may be enhanced by the addition of certain colors. For the production of a gray-colored gypsum mass, for example, the mode of procedure is as follows: Stir 15 drachms of plaster with one-fourth its weight of water, containing a few drops of formaldehyde and a little soda lye and add 10 drops of a one-tenth normal silver solution, which has previously been mixed with the amount of water necessary for hardening the gypsum. The mass will immediately upon mixing assume a pearl-gray shade, uniform throughout. In order to produce red or copper-like, black or bronze-like shades, gold salts, copper salts or silver salts, bismuth salts or lead salts, singly or mixed, are used. Naturally, these colorings admit of a large number of modifications. In lieu of formaldehyde other reducing agents may be employed, such as solutions of sulphurous acid or hydrogen peroxide with a little alkali. Metals in the elementary state may likewise be made use of, e. g., iron, which, stirred with a little copper solution and plaster, produces a brown mass excelling in special hardness, etc. This process of coloring plaster is distinguished from the former methods in that the coloration is caused by metals in the nascent state and that a very fine division is obtained. The advantage of the dyeing method consists in that colorings can be produced with slight quantities of a salt; besides, the fine contours of the figures are in no way affected by this manner of coloring, and another notable advantage lies in the mass being colored throughout, whereby a great durability of the color against outside actions is assured. Thus a peeling off of the color or other way of becoming detached, such as by rubbing off, is entirely excluded.

II.—Frequently, in order to obtain colored plaster objects, ocher or powdered colors are mixed with the plaster. This method leaves much to be desired, because the mixture is not always perfect, and instead of the expected uniform color, blotches appear. Here is a more certain recipe: Boil brazil wood, logwood, or yellow wood, in water, according to the desired color, or use extracts of the woods. When the dye is cold mix it with the plaster. The dye must be passed through a cloth before use. One may also immerse the plaster articles, medals, etc., in this dye, but in this case they must be left for some time and the operation repeated several times.

Treatment of Fresh Plaster.—Freshly plastered cement surfaces on walls may be treated as follows:

The freshly plastered surface first remains without any coating for about 14 days; then it is coated with a mixture of 50 parts water and 10 parts ammonia carbonate dissolved in hot water; leave this coat alone for a day, paint it again and wait until the cement has taken on a uniform gray color, which takes place as a rule in 12 to 14 days. Then prime the surface thus obtained with pure varnish and finish the coating, after drying, with ordinary varnish paint or turpentine paint.

Plaster for Foundry Models.—Gum lac, 1 part; wood spirit, 2 parts; lampblack in sufficient quantity to dye.

Plaster from Spent Gas Lime.—Spent lime from gas purifiers, in which the sulphur has been converted into calcium sulphate, by exposure to weather, if necessary, is mixed with clay rich in alumina. The mixture is powdered, formed into balls or blocks with water, and calcined at a temperature below that at which the setting qualities of calcium sulphate are destroyed. Slaked lime, clay, and sand are added to the calcined product, and the whole is finely powdered.

Plaster Mold.—Nearly all fine grades of metals can be cast in plaster molds, provided only a few pieces of the castings are wanted. Dental plaster should be used, with about one-half of short asbestos. Mix the two well together, and when the mold is complete let it dry in a warm place for several days, or until all the moisture is excluded. If the mold is of considerable thickness it will answer the purpose better. When ready for casting, the plaster mold should be warmed, and smoked over a gas light; then the metal should be poured in, in as cool a state as it will run.

Cleaning of Statuettes and Other Plaster Objects.—Nothing takes the dust more freely than plaster objects, more or less artistic, which are the modest ornaments of our dwellings. They rapidly contract a yellow-gray color, of unpleasant appearance. Here is a practical method for restoring the whiteness: Take finely powdered starch, quite white, and make a thick paste with hot water. Apply, when still hot, with a flexible spatula or a brush on the plaster object. The layer should be quite thick. Let it dry slowly. On drying, the starch will split and scale off. All the soiled parts of the plaster will adhere, and be drawn off with the scales. This method of cleaning does not detract from the fineness of the model.

Hardening and Toughening Plaster of Paris.—I.—Plaster of Paris at times sets too rapidly; therefore the following recipe for toughening and delaying drying will be useful. To calcined plaster of Paris add 4 per cent of its weight of powdered marshmallow root, which will keep it from setting for about an hour, and augment its hardness when set, or double the quantity of marshmallow root powder, and the plaster will become very firm, and may be worked 2 or 3 hours after mixing, and may be carved and polished when hard. It is essential that these powders, which are of different densities and specific gravities, should be thoroughly mixed, and the plaster of Paris be quite fresh, and it must be passed through fine hair sieves to ensure its being an impalpable powder. To ensure thorough mixing, pass the combined powders through the hair sieve three times. Make up with water sufficient for the required model or models. Should any of the powder be left over it may be kept by being put in an air-tight box and placed in a warm room.

The marshmallow root powder may be replaced by dextrin, gum arabic, or glue. The material treated is suitable while yet in a soft state, for rolling, glasstube developing, making plates, etc.

II.—Plaster of Paris may be caused to set more quickly if some alum be dissolved in the water used for rendering it plastic. If the gypsum is first moistened with a solution of alum and then again burned, the resulting compound sets very quickly and becomes as hard as marble. Borax may be similarly employed. The objects may also be treated with a solution of caustic baryta. But it has been found that no matter how deep this penetrates, the baryta is again drawn toward the surface when the water evaporates, a portion efflorescing on the outside, and only a thin layer remaining in the outer shell, where it is converted into carbonate. This at the same time stops up the pores, rendering it impossible to repeat the operation. It was later found that the whole mass of the cast might be hardened by applying to it with a brush made of glass bristles, a hot solution of baryta. To prevent separation of the crystallized baryta at the surface, the object must be raised to a temperature of 140° to 175° F. To produce good results, however, it is necessary to add to the plaster before casting certain substances with which the baryta can combine. These are silicic acid in some form, or the sulphates of zinc, magnesium, copper, iron, aluminum, etc. With some of these the resulting object may be colored. As it is, however, difficult to insure the production of uniform tint, it is better when employing salts producing color, to mix the plaster with about 5 per cent of quicklime, or, better, to render it plastic with milk of lime, and then to soak the object in a solution of metallic sulphate.

Preservation of Plaster Casts.—Upon complete drying, small objects are laid for a short while in celluloid varnish of 4 per cent, while large articles are painted with it, from the top downward, using a soft brush. Articles set up outside and exposed to the weather are not protected by this treatment, while others can be readily washed off and cleaned with water. To cover 100 square feet of surface, 1¾ pints of celluloid varnish are required.

To Arrest the Setting of Plaster of Paris.—Citric acid will delay the setting of plaster of Paris for several hours. One ounce of acid, at a cost of about 5 cents, will be sufficient to delay the setting of 100 pounds of plaster of Paris for 2 or 3 hours. Dissolve the acid in the water before mixing the plaster.

Weatherproofing Casts. — I. — Brethauer's method of preparing plaster of Paris casts for resisting the action of the weather is as follows: Slake 1 part of finely pulverized lime to a paste, then mix gypsum with limewater and intimately mix both. From the compound thus prepared the figures are cast. When perfectly dry they are painted with hot linseed oil, repeating the operation several times, then with linseed-oil varnish, and finally with white oil paint. Statues, etc., prepared in this way have been constantly exposed to the action of the weather for 4 years without suffering any change.

II.—Jacobsen prepares casts which retain no dust, and can be washed with lukewarm soap water by immersing them or throwing upon them in a fine spray a hot solution of a soap prepared from stearic acid and soda lye in ten times its quantity, by weight, of hot water.

Reproduction of Plaster Originals.— This new process consists in making a plaster mold over the original in the usual manner. After the solidification of the plaster the mass of the original is removed, as usual, by cutting out and rinsing out. The casting mold thus obtained is next filled out with a ceramic mass consisting of gypsum, 1 part; powdered porcelain, 5 parts; and flux, 1 part. After the mass has hardened it is baked in the mold. This renders the latter brittle and it falls apart on moistening with water while the infusion remains as a firm body, which presents all the details of the original in a true manner.

PLASTER ARTICLES, REPAIRING OF:
See Adhesives and Lutes.

PLASTER GREASE:
See Lubricants.

PLASTER, PAINTS FOR:
See Paints.

PLASTER OF PARIS, MOLDS FOR CASTING:
See Casting.

PLASTIC COMPOSITIONS:
See Celluloid and Matrix Mass.

PLASTER, IRRITATING:
See Ointments.

PLATES, CARE OF PHOTOGRAPHIC:
See Photography.

PLATINA, BIRMINGHAM:
See Alloys, under Brass.

Plating

The plating of metal surfaces is accomplished in four different ways: (1) By oxidation, usually involving dipping in an acid bath; (2) by electrodeposition, involving suspension in a metallic solution, through which an electric current is passed; (3) by applying a paste that is fixed, as by burning in; (4) by pouring on molten plating metal and rolling. For convenience the methods of plating are arbitrarily classified below under the following headings:

1. Bronzing.
2. Coloring of Metals.
3. Electrodeposition Processes.
4. Gilding and Gold-Plating.

5. Oxidizing Processes.
6. Patina Oxidizing Processes.
7. Platinizing.
8. Silvering and Silver-Plating.
9. Tinned Lead-Plating.
10. Various Recipes.

BRONZING:

Art Bronzes.—These are bronzes of different tints, showing a great variety according to the taste and fancy of the operator.

I.—After imparting to an object a coating of vert antique, it is brushed to remove the verdigris, and another coat is applied with the following mixture: Vinegar, 1,000 parts, by weight; powdered bloodstone, 125 parts, by weight; plumbago, 25 parts, by weight. Finish with a waxed brush and a coat of white varnish.

II.—Cover the object with a mixture of vinegar, 1,000 parts, by weight; powdered bloodstone, 125 parts, by weight; plumbago, 25 parts, by weight; sal ammoniac, 32 parts, by weight; ammonia, 32 parts, by weight; sea salt, 32 parts, by weight. Finish as above.

Antique Bronzes.—In order to give new bronze castings the appearance and patina of old bronze, various compositions are employed, of which the following are the principal ones:

I.— Vert Antique: Vinegar, 1,000 parts, by weight; copper sulphate, 16 parts, by weight; sea salt, 32 parts, by weight; sal ammoniac, 32 parts, by weight; mountain green (Sanders green), 70 parts, by weight; chrome yellow, 30 parts, by weight; ammonia, 32 parts, by weight.

II.—Vert Antique: Vinegar, 1,000 parts, by weight; copper sulphate, 16 parts, by weight; sea salt, 32 parts, by weight; sal ammoniac, 32 parts, by weight; mountain green, 70 parts, by weight; ammonia, 32 parts, by weight.

III.—Dark Vert Antique: To obtain darker vert antique, add a little plumbago to the preceding mixtures.

IV.—Vinegar, 1,000 parts, by weight; sal ammoniac, 8 parts, by weight; potassium bioxalate, 1 part, by weight.

Brass Bronzing.—I.—Immerse the articles, freed from dirt and grease, into a cold solution of 10 parts of potassium permanganate, 50 parts of iron sulphate, 5 parts of hydrochloric acid, in 1,000 parts of water. Let remain 30 seconds; then withdraw, rinse off, and dry in fine, soft sawdust. If the articles have be-

come too dark, or if a reddish-brown color be desired, immerse for about 1 minute into a warm (60° C. or 140° F.) solution of chromic acid, 10 parts; hydrochloric acid, 10 parts; potassium permanganate, 10 parts; iron sulphate, 50 parts; water, 1,000 parts. Treat as before. If the latter solution alone be used the product will be a brighter dark yellow or reddish-brown color. By heating in a drying oven the tone of the colors is improved.

II.—Rouge, with a little chloride of platinum and water, will form a chocolate brown of considerable depth of tone and is exceedingly applicable to brass surfaces which are to resemble a copper bronze.

Copper Bronzing.—I.—After cleaning the pieces, a mixture made as follows is passed over them with a brush: Castor oil, 20 parts; alcohol, 80 parts; soft soap, 40 parts; water, 40 parts. The day after application, the piece has become bronzed; and if the time is prolonged, the tint will change. Thus, an affinity of shades agreeable to the eye can be procured. The piece is dried in hot sawdust, and colorless varnish with large addition of alcohol is passed over it. This formula for bronzing galvanic apparatus imparts any shade desired, from Barbodienne bronze to antique green, provided the liquid remains for some time in contact with the copper.

II.—Acetate of copper, 6 parts; sal ammoniac, 7 parts; acetic acid, 1 part; distilled water, 100 parts. Dissolve all in water in an earthen or porcelain vessel. Place on the fire and heat slightly; next, with a brush give the objects to be bronzed 2 or 3 coats, according to the shade desired. It is necessary that each coat be thoroughly dry before applying another.

Bronzing of Gas Fixtures.—Gas fixtures which have become dirty or tarnished from use may be improved in appearance by painting with bronze paint and then, if a still better finish is required, varnishing after the paint is thoroughly dry with some light-colored varnish that will give a hard and brilliant coating.

If the bronze paint is made up with ordinary varnish it is liable to become discolored from acid which may be present in the varnish. One method proposed for obviating this is to mix the varnish with about 5 times its volume of spirit of turpentine, add to the mixture dried slaked lime in the proportion of about 40 grains to the pint, agitate well, repeating the agitation several times, and finally allowing the suspended matter to settle and decanting the clear liquid. The object of this is, of course, to neutralize any acid which may be present. To determine how effectively this has been done, the varnish may be chemically tested.

Iron Bronzing.—I.—The surface of a casting previously cleaned and polished is evenly painted with a vegetable oil, e. g., olive oil, and then well heated, care being taken that the temperature does not rise to a point at which the oil will burn. The cast iron absorbs oxygen at the moment when the decomposition of the oil begins, and a brown layer of oxide is formed which adheres firmly to the surface and which may be vigorously polished, giving a bronze-like appearance to the surface of the iron.

II.—To give polished iron the appearance of bronze commence by cleaning the objects, then subject them for about 5 minutes to the vapor of a mixture of concentrated hydrochloric and nitric acids; then smear them with vaseline and heat them until the vaseline begins to decompose. The result is a fine bronzing.

Liquid for Bronze Powder.—Take 2 ounces gum animi and dissolve in ½ pint linseed oil by adding gradually while the oil is being heated. Boil, strain, and dilute with turpentine.

Bronzing Metals.—I.—The following composition is recommended for bronzing metal objects exposed to the air: Mix about equal parts of siccative, rectified oil of turpentine, caoutchouc oil, and dammar varnish, and apply this composition on the objects, using a brush. This bronze has been found to resist the influences of the weather.

II.—Cover the objects with a light layer of linseed oil, and then heat over a coal fire, prolonging the heat until the desired shade is reached.

III.—Expose the objects to be bronzed for about 5 minutes to the vapors of a bath composed of 50 parts of nitric acid and 50 parts of concentrated hydrochloric acid. Then rub the articles with vaseline and heat until the vaseline is decomposed. The objects to be bronzed must always be perfectly polished.

IV.—To bronze iron articles they should be laid in highly heated coal dust; the articles must be covered up in the glowing dust, and the heat must be the same throughout. The iron turns at first yellow, then blue, and finally rather black. Withdraw the objects when they have attained the blue shade or the black color; then while they are still hot, rub them with a wad charged with tallow.

· V.—For electrolytic bronzing of metals the baths employed differ from the brass baths only in that they contain tin in solution instead of zinc. According to Elsner, dissolve 70 parts, by weight, of cupric sulphate in 1,000 parts of water and add a solution of 8 parts of stannic chloride in caustic lye. For a positive pole plate put in a bronze plate. The bath works at ordinary temperature.

VI.—A good bath consists of 10 parts of potash, 2 parts of cupric chloride, 1 part of tin salt, 1 part of cyanide of potassium dissolved in 100 parts of water.

VII.—Mix a solution of 32 parts of copper sulphate in 500 parts of water with 64 parts of cyanide of potassium. After the solution has become clear, add 4 to 5 parts of stannic chloride dissolved in potash lye.

VIII.—Precipitate all soda from a solution of blue vitriol by phosphate of sodium, wash the precipitate well, and dissolve in a concentrated solution of pyrophosphate of copper. Also, saturate a solution of the same salt with tin salt. Of both solutions add enough in such proportion to a solution of 50 parts, by weight, of pyrophosphate of sodium in 1,000 parts of water until the solution appears clear and of the desired color. A cast bronze plate serves as an anode. From time to time a little soda, or if the precipitate turns out too pale, copper solution should be added.

Tin Bronzing.—The pieces are well washed and all grease removed; next plunged into a solution of copperas (green vitriol), 1 part; sulphate, 1 part; water, 20 parts. When dry they are plunged again into a bath composed of verdigris, 4 parts; dissolved in distilled wine vinegar, 11 parts. Wash, dry, and polish with English red.

Zinc Bronzing.—The zinc article must be first electro-coppered before proceeding to the bronzing. The process used is always the same; the different shades are, however, too numerous to cover all of them in one explanation. The bronzing of zinc clocks is most frequently done on a brown ground, by mixing graphite, lampblack, and sanguine stirred in water in which a little Flanders Dutch glue is dissolved. The application is made by means of a brush. When it is dry a spirit varnish is applied; next, before the varnish is perfectly dry, a little powdered bronze or sanguine or powdered bronze mixed with sanguine or with graphite, according to the desired shades. For green bronze, mix green sanders with chrome yellow stirred with spirit in which a little varnish is put. When the bronzing is dry, put on the varnish and the powdered bronze as above described. After all has dried, pass the brush over a piece of wax, then over the bronzed article, being careful to charge the brush frequently with wax.

COLORING OF METALS:

Direct Coloration of Iron and Steel by Cupric Selenite.—Iron precipitates copper and selenium from their salts. Immersed in a solution of cupric selenite, acidulated with a few drops of nitric acid, it precipitates these two metals on its surface in the form of a dull black deposit, but slightly adherent. But, if the object is washed with water, then with alcohol, and rapidly dried over a gas burner, the deposit becomes adherent. If rubbed with a cloth, this deposit turns a blue black or a brilliant black, according to the composition of the bath.

The selenite of copper is a greenish salt insoluble in water, and but slightly soluble in water acidulated with nitric or sulphuric acid. It is preferable to mix a solution of cupric sulphate with a solution of selenious acid, and to acidulate with nitric acid, in order to prevent the precipitation of the selenite of copper.

This process, originated by Paul Malherbe, is quite convenient for blackening or bluing small objects of iron or steel, such as metallic pens or other small pieces. It does not succeed so well for objects of cast iron; and the selenious acid is costly, which is an obstacle to its employment on large metallic surfaces. The baths are quickly impoverished, for insoluble yellow selenite of iron is deposited.

Brilliant Black Coloration.—Selenious acid, 6 parts; cupric sulphate, 10 parts; water, 1,000 parts; nitric acid, 4 to 6 parts.

Blue-Black Coloration.—Selenious acid, 10 parts; cupric sulphate, 10 parts; water, 1,000 parts; nitric acid, 4 to 6 parts.

By immersing the object for a short time the surface of the metal can be colored in succession yellow, rose, purple, violet and blue.

Coloration of Copper and Brass with Cupric Selenite.—When an object of copper or brass is immersed in a solution of selenite of copper acidulated with nitric acid, the following colors are obtained, according to the time of the immersion: Yellow, orange, rose, purple, violet, and blue, which is the last color which can be obtained. In general, the solution should be slightly acid; otherwise the color is fugacious and punctate.

	a.	b.
Selenious acid	6.5	2.9 parts
Sulphate of copper	12.5	20.0 parts
Nitric acid	2.0	2.5 parts
Water	1,000.0	1,000.0 parts

Production of Rainbow Colors on Metals (iron, copper, brass, zinc, etc.)—I.—The following process of irisation is due to Puscher. It allows of covering the metals with a thick layer of metallic sulphide, similar to that met with in nature—in galena, for example.

These compounds are quite solid and are not attacked by concentrated acids and alkalies, while dilute reagents are without action. In 5 minutes thousands of objects of brass can be colored with the brightest hues. If they have been previously cleaned chemically, the colors deposited on the surface adhere with such strength that they can be worked with the burnisher.

Forty-five parts of sodium hyposulphite are dissolved in 500 parts of water; a solution of 15 parts of neutral acetate of lead in 500 parts of water is poured in. The clear mixture, which is composed of a double salt of hyposulphite of lead and of sodium, possesses, when heated to 212° F., the property of decomposing slowly and of depositing brown flakes of lead sulphide. If an article of gold, silver, copper, brass, tombac, iron, or zinc is put into this bath while the precipitation is taking place, the object will be covered with a film of lead sulphide, which will give varied and brilliant colors, according to its thickness. For a uniform coloration, it is necessary that the pieces should be heated quite uniformly. However, iron assumes under this treatment only a blue color, and zinc a bronze color. On articles of copper the first gold color which appears is defective. Lead and tin are not colored.

By substituting for the neutral acetate of lead an equal quantity of cupric sulphate and proceeding in a similar way, brass or imitation gold is covered with a very beautiful red, succeeded by an imperfect green, and finally a magnificent brown, with iridescent points of greenish red. The latter coating is fairly permanent.

Zinc is not colored in this solution, and

precipitates in it a quantity of flakes of greenish brown (cupric sulphide), but if about one-third of the preceding solution of lead acetate is added, a solid black color is developed, which, when covered with a light coating of wax, gains much in intensity and solidity. It is also useful to apply a slight coating of wax to the other colors.

II.—Beautiful designs may be obtained, imitating marble, with sheets of copper plunged into a solution of lead, thickened by the addition of gum tragacanth, and heated to 212° F. Afterwards they are treated with the ordinary lead solution. The compounds of antimony, for example the tartrate of antimony and potash, afford similar colorations, but require a longer time for their development. The solutions mentioned do not change, even after a long period, and may be employed several times.

III.—By mixing a solution of cupric sulphate with a solution of sodium hyposulphite, a double hyposulphite of sodium and of copper is obtained.

If in the solution of this double salt an article of nickel or of copper, cleaned with nitric acid, then with soda, is immersed, the following colors will appear in a few seconds: Brilliant red, green, rose, blue, and violet. To isolate a color, it is sufficient to take out the object and wash it with water. The colors obtained on nickel present a moiré appearance, similar to that of silk fabrics.

IV.—Tin sulphate affords with sodium hyposulphite a double salt, which is reduced by heat, with production of tin sulphide. The action of this double salt on metallic surfaces is the same as that of the double salts of copper and lead. Mixed with a solution of cupric sulphate, all the colors of the spectrum will be readily obtained.

V.—Coloration of Silver.—The objects of copper or brass are first covered with a layer of silver, when they are dipped in the following solution at the temperature of 205° to 212° F.: Water, 3,000 parts; sodium hyposulphite, 300 parts; lead acetate, 100 parts.

VI.—Iron precipitates bismuth from its chlorhydric solution. On heating this deposit, the colors of the rainbow are obtained.

Coloration by Electrolysis.—I.—Colored Rings by Electrolysis (Nobili, Becquerel).—In order to obtain the Nobili rings it is necessary to concentrate the current coming from one of the poles of the battery through a platinum wire, whose point alone is immersed in the liquid to be decomposed, while the other pole is connected with a plate of metal in the same liquid. This plate is placed perpendicularly to the direction of the wire, and at about 0.04 inches from the point.

Solutions of sulphate of copper, sulphate of zinc, sulphate of manganese, acetate of lead, acetate of copper, acetate of potassium, tartrate of antimony and potash, phosphoric acid, oxalic acid, carbonate of soda, chloride of manganese, and manganous acetate, may be employed.

II.—A process, due to M. O. Mathey, allows of coloring metals by precipitating on their surface a transparent metallic peroxide. The phenomenon of electrochemical coloration on metals is the same as that which takes place when an object of polished steel is exposed to heat. It first assumes a yellow color, from a very thin coating of ferric oxide formed on its surface. By continuing the heating, this coating of oxide increases in thickness, and appears red, then violet, then blue. Here, the coloration is due to the increase in the thickness of a thin coating of a metallic oxide precipitated by an alkaline solution.

The oxides of lead, tin, zinc, chromium, aluminum, molybdenum, tungsten, etc., dissolved in potash, may be employed; also protoxide of iron, zinc, cadmium, cobalt, dissolved in ammonia.

Lead Solution.—Potash, 400 parts; litharge or massicot, 125 parts. Boil 10 minutes, filter, dilute until the solution marks 25° Bé.

Iron Solution.—Dissolve ferrous sulphate in boiling water, and preserve sheltered from air. When desired for use, pour a quantity into a vessel and add ammonia until the precipitate is redissolved. This solution, oxidizing rapidly in the air, cannot be used for more than an hour.

III.—Electro-chemical coloration succeeds very well on metals which are not oxidizable, such as gold and platinum, but not well on silver. This process is employed for coloring watch hands and screws. The object is placed at the positive pole, under a thickness of $1\frac{1}{2}$ inches of the liquid, and the negative electrode is brought to the surface of the bath. In a few seconds all the colors possible are obtained. Generally, a ruby-red tint is sought for.

IV.—Coloration of Nickel.—The nickel piece is placed at the positive pole in a solution of lead acetate. A netting of copper wires is arranged at the negative pole according to the contours of the design, and at a short distance from the object. The coloration obtained is uniform if the distance of the copper wires from the object is equal at all points.

Coloring of Brass.—I.—(a) Brown bronze: Acid solution of nitrate of silver and bismuth or nitric acid. (b) Light bronze: Acid solution of nitrate of silver and of copper. (c) Black: Solution of nitrate of copper. In all cases, however, the brass is colored black, if after having been treated with the acid solution, it is placed for a very short time in a solution of potassium sulphide, of ammonium sulphydrate, or of hydrogen sulphide.

II.—The brass is immersed in a dilute solution of mercurous nitrate; the layer of mercury formed on the brass is converted into black sulphide, if washed several times in potassium sulphide. By substituting for the potassium sulphide the sulphide of antimony or that of arsenic, beautiful bronze colors are obtained, varying from light brown to dark brown.

III.—Clean the brass perfectly. Afterwards rub with sal ammoniac dissolved in vinegar. Strong vinegar, 1,000 parts; sal ammoniac, 30 parts; alum, 15 parts; arsenious anhydride, 8 parts.

IV.—A solution of chloride of platinum is employed, which leaves a very light coating of platinum on the metal, and the surface is bronzed. A steel tint or gray color is obtained, of which the shade depends on the metal. If this is burnished, it takes a blue or steel gray shade, which varies with the duration of the chemical action, the concentration, and the temperature of the bath. A dilute solution of platinum is prepared thus: Chloride of platinum, 1 part; water, 5,000 parts.

Another solution, more concentrated at the temperature of 104° F., is kept ready. The objects to be bronzed are attached to a copper wire and immersed for a few seconds in a hot solution of tartar, 30 parts to 5,000 parts of water. On coming from this bath they are washed 2 or 3 times with ordinary water, and a last time with distilled water, and then put in the solution of platinum chloride, stirring them from time to time. When a suitable change of color has been secured, the objects are passed to the concentrated solution of platinum chloride (40°). They are stirred, and taken out when the wished-for color has been reached. They are then washed 2 or 3 times, and dried in wood sawdust.

V.—To give to brass a dull black color, as that used for optical instruments, the metal is cleaned carefully at first, and covered with a very dilute mixture of neutral nitrate of tin, 1 part; chloride of gold, 2 parts. At the end of 10 minutes this covering is removed with a moist brush. If an excess of acid has not been employed, the surface of the metal will be found to be of a fine dull black.

The nitrate of tin is prepared by decomposing the chloride of this metal with ammonia and afterwards dissolving in nitric acid the oxide of tin formed.

VI.—For obtaining a deposit of bismuth the brass is immersed in a boiling bath, prepared by adding 50 to 60 parts of bismuth to nitric acid diluted with 1,000 parts of water, and containing 32 parts of tartaric acid.

VII.—The electrolysis of a cold solution of 25 to 30 parts per 1,000 parts of the double chloride of bismuth and ammonium produces on brass or on copper a brilliant adherent deposit of bismuth, whose appearance resembles that of old silver.

Production of Rainbow Hues.—Various colors.—I.—Dissolve tartrate of antimony and of potash, 30 parts; tartaric acid, 30 parts; water, 1,000 parts. Add hydrochloric acid, 90 to 120 parts; pulverized antimony, 90 to 120 parts. Immerse the object of brass in this boiling liquid, and it will be covered with a film, which, as it thickens, reflects quite a series of beautiful tints, first appearing iridescent, then the color of gold, copper, or violet, and finally of a grayish blue. These colors are adherent, and do not change in the air.

II.—The sulphide of tin may be deposited on metallic surfaces, especially on brass, communicating shades varying with the thickness of the deposit. For this purpose, Puscher prepares the following solutions: Dissolve tartaric acid, 20 parts, in water, 1,000 parts; add a salt of tin, 20 parts; water, 125 parts. Boil the mixture, allow it to repose, and filter. Afterwards pour the clear portion a little at a time, shaking continually, into a solution of hyposulphite of soda, 80 parts; water, 250 parts. On boiling, sulphide of tin is formed, with precipitation of sulphur. On plunging the pieces of brass in the liquid, they are covered, according to the period of immersion, with varied shades, passing from go'd yellow to red, to crimson, to blue, and finally to light brown.

III.—The metal is treated with the

following composition: Solution A.—Cotton, well washed, 50 parts; salicylic acid, 2 parts, dissolved in sulphuric acid, 1,000 parts, and bichromate of potash, 100 parts. Solution B.—Brass, 20 parts; nitric acid, density 1.51, 350 parts; nitrate of soda, 10 parts. Mix the two solutions, and dilute with 1,500 parts of water. These proportions may be modified according to the nature of the brass to be treated. This preparation is spread on the metal, which immediately changes color. When the desired tint is obtained, the piece is quickly plunged in an alkaline solution; a soda salt, 50 parts; water, 1,000 parts. The article is afterwards washed, and dried with a piece of cloth. Beautiful red tints are obtained by placing the objects between 2 plates, or better yet, 2 pieces of iron wire-cloth.

IV.—Put in a flask 100 parts of cupric carbonate and 750 parts of ammonia and shake. This liquid should be kept in well-stoppered bottles. When it has lost its strength, this may be renewed by pouring in a little ammonia. The objects to be colored should be well cleaned. They are suspended in the liquid and moved back and forth. After a few minutes of immersion, they are washed with water and dried in wood sawdust. Generally, a deep-blue color is obtained.

V.—Plunge a sheet of perfectly clean brass in a dilute solution of neutral acetate of copper, and at the ordinary temperature, and in a short time it will be found covered with a fine gold yellow.

VI.—Immerse the brass several times in a very dilute solution of cupric chloride, and the color will be deadened and bronzed a greenish gray.

A plate of brass heated to 302° F. is colored violet by rubbing its surface gently with cotton soaked with cupric chloride.

VII.—On heating brass, perfectly polished, until it can be no longer held in the hand, and then covering it rapidly and uniformly with a solution of antimony chloride by means of a wad of cotton, a fine violet tint is communicated.

VIII.—For greenish shades, a bath may be made use of, composed of water, 100 parts; cupric sulphate, 8 parts; sal ammoniac, 2 parts.

IX.—For orange-brown and cinnamon-brown shades: Water, 1,000 parts; potassium chlorate, 10 parts; cupric sulphate, 10 parts.

X.—For obtaining rose-colored hues, then violet, then blue: Water, 400 parts; cupric sulphate, 30 parts; sodium hyposulphite, 20 parts; cream of tartar, 10 parts.

XI.—For yellow, orange, or rose-colored shades, then blue, immerse the objects for a longer or shorter time in the following bath: Water, 400 parts, ammoniacal ferrous sulphate, 20 parts; sodium hyposulphite, 40 parts; cupric sulphite, 30 parts; cream of tartar, 10 parts. By prolonging the boiling, the blue tint gives place to yellow, and finally to a fine gray.

XII.—A yellowish brown may be obtained with water, 50 parts; potassium chlorate, 5 parts; nickel carbonate, 2 parts; sal nickel, 5 parts.

XIII.—A dark brown is obtained with water, 50 parts; sal nickel, 10 parts; potassium chlorate, 5 parts.

XIV.—A yellowish brown is obtained with water, 350 parts; a crystallized sodium salt, 10 parts; orpiment, 5 parts.

XV.—Metallic moire is obtained by mixing two liquids: (a) Cream of tartar, 5 parts; cupric sulphate, 5 parts; water, 250 parts. (b) Water, 125 parts; sodium hyposulphite, 15 parts.

XVI.—A beautiful color is formed with one of the following baths: (a) Water, 140 parts; ammonia, 5 parts; potassium sulphide, 1 part. (b) Water, 100 parts; ammonium sulphydrate, 2 parts.

Bronzing of Brass.—The object is boiled with zinc grains and water saturated with ammoniacal chlorhydrate. A little zinc chloride may be added to facilitate the operation, which is completed as above.

It may also be terminated by plunging the object in the following solution: Water, 2,000 parts; vinegar, 100 parts; sal ammoniac, 475 parts; pulverized verdigris, 500 parts.

ELECTRODEPOSITION PROCESSES.

The electrodeposition process is that used in electroplating and electrotyping. It consists in preparing a bath in which a metal salt is in solution, the articles to be plated being suspended so that they hang in the solution, but are insulated. The bath being provided with an anode and cathode for the passing of an electric current, and the article being connected with the cathode or negative pole, the salts are deposited on its surface (on the unprotected parts of its surface), and thus receive a coating or plating of the metal in solution.

When a soft metal is deposited upon a hard metal or the latter upon a metal softer than itself, the exterior metal should be polished and not burnished, and for this reason: If silver is deposited upon lead, for instance, the great pressure which is required in burnishing to produce the necessary polish would cause the softer metal to expand, and consequently a separation of the two metals would result. On the other hand, silver being softer than steel, if the burnisher is applied to silver-coated steel the exterior metal will expand and separate from the subjacent metal.

Many articles which are to receive deposits require to have portions of their surfaces topped off, to prevent the deposit spreading over those parts; for instance, in taking a copy of one side of a bronze medallion, the opposite side must be coated with some kind of varnish, wax, or fat, to prevent deposition; or, in gilding the inside of a cream jug which has been silvered on the outside, varnish must be applied all around the outer side of the edge, for the same reason. For gilding and other hot solutions, copal varnish is generally used; but for cold liquids and common work, an ordinary varnish, such as engravers use for similar purposes, will do very well. In the absence of other substances, a solution of sealing wax, dissolved in naphtha, may be employed.

Plating of Aluminum.—The light metal may be plated with almost any other metal, but copper is most commonly employed. Two formulas for coppering aluminum follow:

I.—Make a bath of cupric sulphate, 30 parts; cream of tartar, 30 parts; soda, 25 parts; water, 1,000 parts. After well scouring the objects to be coppered, immerse in the bath. The coppering may also be effected by means of the battery with the following mixture: Sodium phosphate, 50 parts; potassium cyanide, 50 parts; copper cyanide, 50 parts; distilled water, 1,000 parts.

II.—First clean the aluminum in a warm solution of an alkaline carbonate, thus making its surface rough and porous; next wash it thoroughly in running water, and dip it into a hot solution of hydrochloric acid of about 5 per cent strength. Wash it again in clean water, and then place it in a somewhat concentrated acid solution of copper sulphate, until a uniform metallic deposit is formed; it is then again thoroughly washed and returned to the copper sulphate bath, when an electric current is passed until a coating of copper of the required thickness is obtained.

Brassing.—The following recipe is recommended for the bath: Copper acetate, 50 parts, by weight; dry zinc chloride, 25 parts, by weight; crystallized sodium sulphite, 250 parts, by weight; ammonium carbonate, 35 parts, by weight; potassium cyanide, 110 parts, by weight. Dissolve in 3,000 parts of water.

Coppering.—I.—This is the Dessolle process for the galvanic application of copper. The special advantage claimed is that strong currents can be used, and a deposit obtained of 0.004 inch in 1½ hours. After having cleaned the object to be coppered, with sand or in an acid bath, a first coat is deposited in an ordinary electrolytic bath; then the object is placed in a final bath, in which the electrolyte is projected on the electrode, so as to remove all bubbles of gas or other impurities tending to attach themselves to the surface. The electrolyte employed is simply a solution of cupric sulphate in very dilute sulphuric acid. For the preliminary bath the double cyanide of potassium and copper is made use of.

II.—Those baths which contain cyanide work best, and may be used for all metals. The amount of the latter must not form too large an excess. The addition of a sulphide is very dangerous. It is of advantage that the final bath contain an excess of alkali, but only as ammonia or ammonium carbonate. For a copper salt the acetate is preferable. According to this, the solution A is prepared in the warm, and solution B is added with heating. Solution A: Neutral copper acetate, 30 parts, by weight; crystallized sodium sulphite, 30 parts, by weight; ammonium carbonate, 5 parts, by weight; water, 500 parts, by weight. Solution B: Potassium cyanide (98 to 99 per cent), 35 parts, by weight; and water, 500 parts, by weight.

Coppering Glass.—I.—Glass vessels may be coated with copper by electrolytic process, by simply varnishing the outer surface of the vessel, and when the varnish is nearly dry, brushing plumbago well over it. A conducting wire is then attached to the varnished surface, which may be conveniently done by employing a small piece of softened gutta percha or beeswax, taking care to employ the plumbago to the part which unites the wire to the plumbagoed surface.

II.—Dissolve gutta percha in essence of turpentine or benzine; apply a coat of the solution on the glass in the places to be coppered and allow to dry; next rub it with graphite and place in the electric bath. The rubber solution is spread with a brush.

Coppering Plaster Models, etc.—Busts and similar objects may be coated by saturating them with linseed oil, or better, with beeswax, then well blackleading, or treating them with phosphorous, silver and gold solutions, attaching a number of guiding wires, connected with all the most hollow and distant parts, and then immersing them in the sulphate of copper solution and causing just sufficient copper to be deposited upon them, by the battery process, to protect them, but not to obliterate the fine lines or features.

Coppering Zinc Plate.—The zinc plate should first be cleaned with highly diluted hydrochloric acid and the acid completely removed with water. Then prepare an ammoniacal copper solution from 3 parts copper sulphate, 3 parts spirits of sal ammoniac, and 50 parts water. If possible the zinc articles are dipped into this solution or else the surface is coated a few times quickly and uniformly with a flat, soft brush, leaving to dry between the coats. When sufficient copper has precipitated on the zinc, brush off the object superficially.

Cobaltizing of Metals.—Following are various processes for cobaltizing on copper or other metals previously coppered: I.—Cobalt, 50 parts, by weight; sal ammoniac, 25 parts; liquid ammonia, 15 parts; distilled water, 1,000 parts. Dissolve the cobalt and the sal ammoniac in the distilled water, and add the liquid ammonia.

II.—Pure potash in alcohol, 50 parts, by weight; cobalt chloride, 10 parts; distilled water, 1,000 parts. Dissolve the cobalt in half the distilled water and the potash in the other half and unite the two.

III.—Potassium sulphocyanide, 13 parts, by weight; cobalt chloride, 10 parts; pure potash in alcohol, 2 parts; distilled water, 1,000 parts. Proceed as described above. All these baths are used hot and require a strong current.

Nickel Plating with the Battery.—The nickel bath is prepared according to the following formula:

I.—Nickel and ammo-
nium sulphate... 10 parts
Boracic acid...... 4 parts
Distilled water.... 175 parts
A sheet of nickel is used as an
anode.

Perfect cleanliness of the surface to be coated is essential to success. With nickel especially is this the case, as traces of oxide will cause it to show dark streaks. Finger marks will in any case render the deposit liable to peel off.

Cleansing is generally accomplished either by boiling in strong solution of potassium hydrate, or, when possible, by heating to redness in a blow-pipe flame to burn off any adhesive grease, and then soaking in a pickle of dilute sulphuric acid to remove any oxide formed during the heating. In either case it is necessary to subject the article to a process of scratch brushing afterwards; that is, long-continued friction with wire brushes under water, which not only removes any still adhering oxide, but renders the surface bright.

To certain metals, as iron, nickel, and zinc, metallic deposits do not readily adhere. This difficulty is overcome by first coating them with copper in a bath composed as follows:

II.—Potassium cyanide. 2 parts
Copper acetate, in
crystals......... 2 parts
Sodium carbonate,
in crystals...... 2 parts
Sodium bisulphite .. 2 parts
Water............ 100 parts

Moisten the copper acetate with a small quantity of water and add the sodium carbonate dissolved in 20 parts of water. When reaction is complete, all the copper acetate being converted into carbonate, add the sodium bisulphite, dissolved in another 20 parts of water; lastly, add the potassium cyanide, dissolved in the remainder of the water. The finished product should be a colorless liquid.

If a dynamo is not available for the production of a current, a Daniell's battery is to be recommended, and the "tank" for a small operation may be a glass jar. The jar is crossed by copper rods in connection with the battery; the metal to be deposited is suspended from the rod in connection with the positive pole, and is called the anode. The articles to be coated are suspended by thin copper wires from the rod in connection with the negative pole; these form the cathode. The worker should bear in mind that it is very difficult to apply a thick coating of nickel without its peeling.

Replating with Battery.—It is well known to electro-metallurgists that metals deposited by electricity do not adhere so firmly to their kind as to other metals. Thus gold will adhere more tenaciously to silver, copper, or brass, than it will to gold or to a gilt surface, and silver will attach itself more closely to copper or brass than to a silver-plated surface. Consequently, it is the practice to remove, by stripping or polishing the silver from old plated articles before electroplating them. If this were not done, the deposited coating would in all probability "strip," as it is termed, when the burnisher is applied to it—that is, the newly deposited metal would peel off the underlying silver. It must be understood that these remarks apply to cases in which a good, heavy deposit of silver is required, for, of course, the mere film would not present any remarkable peculiarity.

Silver Plating.—The term silver deposit designates a coating of silver which is deposited upon glass, porcelain, china, or other substances. This deposit may be made to take the form of any desired design, and to the observer it has the appearance (in the case of glass) of having been melted on.

Practically all of the plated articles are made by painting the design upon the glass or other surface by means of a mixture of powdered silver, a flux and a liquid to make the mixture in the form of a paint so that it may be readily spread over the surface. This design is then fired in a muffle until the flux melts and causes the silver to become firmly attached to the glass. A thin silver deposit is thus produced, which is a conductor of electricity, and upon which any thickness of silver deposit may be produced by electroplating in the usual cyanide silver-plating bath.

To be successful in securing a lasting deposit a suitable flux must be used. This flux must melt at a lower temperature than the glass upon which it is put, in order to prevent the softening of the articles by the necessary heat and the accompanying distortion. Second, a suitable muffle must be had for firing the glass articles upon which the design has been painted. Not only must a muffle be used in which the heat can be absolutely controlled, but one which allows the slow cooling of the articles. If this is not done they are apt to crack while cooling.

The manufacture of the flux is the most critical part of the silver deposit process. Without a good flux the operation will not be a success. This flux is frequently called an enamel or frit. After a series of experiments it was found that the most suitable flux is a borate of lead. This is easily prepared, fuses before the glass softens, and adheres tenaciously to the glass surface.

To make it, proceed as follows: Dissolve ¼ pound of acetate of lead (sugar of lead) in 1 quart of water and heat to boiling. Dissolve ¼ pound of borax in 1 quart of hot water and add to the sugar of lead solution. Borate of lead follows as a white precipitate. This is filtered out and washed until free from impurities. It is then dried.

The precipitated borate of lead is then melted in a porcelain or clay crucible. When in the melted condition it should be poured into a basin of cold water. This serves to granulate and render it easily pulverized. After it has been poured into water it is removed and dried. Before using in the paint it is necessary that this fused borate of lead be ground in a mortar as fine as possible. Unless this is done the deposit will not be smooth.

The silver to be used should be finely powdered silver, which can be purchased in the same manner as bronze powders.

The mixture used for painting the design upon the glass is composed of 2 parts of the powdered silver, and 1 part of the fused borate of lead. Place the parts in a mortar and add just enough oil of lavender to make the mass of a paint-like consistency. The whole is then ground with the pestle until it is as fine as possible. The amount of oil of lavender which is used must not be too great, as it will then be found that a thick layer cannot be obtained upon the glass.

The glass to be treated must be cleaned by scouring with wet pumice stone and washing soda. The glass should be rinsed and dried. The design is then painted on the glass with a brush, painting as thick as possible and yet leaving a smooth, even surface. The glass should be allowed to dry for 24 hours, when it is ready for firing.

When placed in the gas muffle, the glass should be subjected to a temperature of a very low red heat. The borate of lead will melt at this temperature, and after holding this heat a short time to enable the borate of lead to melt and attach itself, the muffle is allowed to cool.

After cooling, the articles are removed and scratch brushed and placed in a silver bath for an electro deposit of silver of a thickness desired.

Before the plating the glass article is dipped into a cyanide dip, or, if found necessary, scoured lightly with pumice stone and cyanide, and then given a dip in the customary blue dip or mercury solution, so as to quickly cover all parts of the surface. It next passes to the regular cyanide silver solution, and is allowed to remain until the desired deposit is obtained.

A little potassium cyanide and some mono-basic potassium citrate in powder form is added from time to time to the bath generally used, which is prepared by dissolving freshly precipitated silver cyanide in a potassium cyanide solution. After this the glass is rinsed and dried, and may be finished by buffing.

Steel Plating.—The following is a solution for dipping steel articles before electroplating: Nitrate of silver, 1 part; nitrate of mercury, 1 part; nitric acid (specific gravity, 1.384), 4 parts; water, 120 parts. The article, free from grease, is dipped in the pickle for a second or two.

The following electroplating bath is used: Pure crystallized ferrous sulphate, 40 parts, by weight, and ammonium chloride, 100 parts, by weight, in 1,000 parts, by weight, of water. It is of advantage to add to this 100 parts, by weight, of ammonium citrate, in order to prevent the precipitation of basic iron salts, especially at the anode.

Tin Plating by Electric Bath.—Most solutions give a dead-white film of tin, and this has to be brightened by friction of some sort, either by scratch brushing, burnishing, polishing, or rubbing with whiting. The bright tin plates are made bright by rolling with polished steel rollers. Small articles may be bright-tinned by immersion in melted tin, after their surfaces have been made chemically clean and bright, all of which processes entail much time and labor. Benzoic acid, boric acid, or gelatin may be tried with a well-regulated current and the solution in good working order, but all will depend upon the exact working of the solution, the same conditions being set up as are present in the deposition of other metals. These substances may be separately tried, in the proportion of 1 ounce to each gallon of the tin solution, by boiling the latter and adding either one during the boiling, as they dissolve much easier with the tin salts than in water separately. Tin articles are usually brightened and polished with Vienna lime or whiting, the first being used with linen rags and the latter with chamois leather. Tin baths must be used hot, not below 75° F., with a suitable current according to their composition. Too strong a current produces a bad color, and the deposit does not adhere well. A current of from 2 to 6 volts will be sufficient. Small tinned articles are brightened by being shaken in a leather bag containing a quantity of bran or by revolving in a barrel with the same substance; but large objects have to be brightened by other means, such as scratch brushing and mopping to give an acceptable finish to the deposited metal.

GILDING AND GOLD PLATING:

Genuine gilding readily takes up mercury, while imitation gilding does not or only very slowly. Any coating of varnish present should, however, be removed before conducting the test. Mercurous nitrate has no action on genuine gold, but on spurious gilding a white spot will form which quickly turns dark. A solution of neutral copper chloride does not act upon genuine gold, but on alloys containing copper a black spot will result. Gold fringe, etc., retains its luster in spirit of wine, if the gilding is genuine; if not, the gilding will burn and oxidize. Imitation gilding might be termed "snuff gilding," as in Germany it consists of dissolved brass, snuff, saltpeter, hydrochloric acid, etc., and is used for tin toys. An expert will immediately see the difference, as genuine gilding has a different, more compact pore formation and a better color. There are also some gold varnishes which are just as good.

The effect of motion while an article is receiving the deposit is most clearly seen during the operation of gilding. If a watch dial, for instance, be placed in the gilding bath and allowed to remain for a few moments undisturbed and the solution of gold has been much worked, it is probable that the dial will acquire a dark fox-red color; but if it be quickly moved about, it instantly changes color and will sometimes even assume a pale straw color. In fact, the color of a deposit may be regulated greatly by motion of the article in the bath—a fact which the operator should study with much attention, when gilding.

The inside of a vessel is gilded by filling the vessel with the gilding solution, suspending a gold anode in the liquid, and passing the current. The lips of cream jugs and the upper parts of vessels of irregular outline are gilded by passing the current from a gold anode through a rag wetted with the gilding solution and laid upon the part.

Sometimes, when gilding the insides of mugs, tankards, etc., which are richly chased or embossed, it will be found that the hollow parts do not receive the deposit at all, or very partially. When this is the case, the article must be rinsed and well scratch brushed, and a little more cyanide added to the solution. The anode must be slightly kept in motion and the battery power increased until the hollow surfaces are coated. Frequent scratch brushing aids the deposit to a great extent by imparting a slight film of brass to the surface.

In gilding chains, brooches, pins, rings, and other articles which have been repaired, i. e., hard soldered, sometimes, it is found that the gold will not deposit freely upon the soldered parts; when such is the case, a little extra scratch brushing applied to the part will assist the operation greatly and it has sometimes been found that dry scratch brushing for an instant—that is, without the stream of beer usually employed—renders the surface a better and more uniform conductor and consequently it will more readily receive the deposit. In fact, dry scratch brushing is very useful in many cases in which it is desirable to impart an artificial coating of brass upon an article to which silver or gold will not readily adhere. In scratch brushing without the employment of beer or some other liquid, however, great care must be taken not to continue the operation too long, as the minute particles of metal given off by the scratch brush would be likely to prove prejudicial to the health of the operator, were he to inhale them to any great extent.

The following solutions are for gilding without a battery: I.—In 1,000 parts of distilled water dissolve in the following order:

Crystalline sodium
  pyrophosphate.... 80 parts
Twelve per cent solu-
  tion of hydrocyanic
  acid............. 8 parts
Crystalline gold chlo-
  ride............. 2 parts

Heat to a boiling temperature, and dip the article, previously thoroughly cleaned, therein.

II.—Dissolve in boiling distilled water, 1 part of chloride of gold and 4 parts of cyanide of potassium. Plunge the objects into this solution, while still hot, and leave them therein for several hours, keeping them attached to a copper wire or a very clean strip of zinc. They will become covered with a handsome gold coating.

Aluminum Gilding.—I.—Dissolve 6 parts of gold in aqua regia and dilute the solution with distilled water; on the other hand, put 30 parts of lime in 150 parts of distilled water; at the end of 2 hours add the gold solution to the lime, shake all and allow to settle for 5 to 6 hours, decant and wash the precipitate, which is lime aurate. Place this aurate of lime in 1,000 parts of distilled water, with 20 parts of hyposulphite of soda; put all on the fire for 8 to 10 minutes, without allowing to boil; remove and filter. The filtered liquor serves for gilding in the cold, by plunging into this bath the aluminum articles previously pickled by passing through caustic potash and nitric acid. This gilding is obtained without the aid of the battery.

II.—The gold bath is prepared with gold dissolved in the usual way, and the addition of salts, as follows: Gold, 20 parts, by weight; sulphate of soda, 20 parts; phosphate of soda, 660 parts; cyanuret of potassium, 40 parts; water, 1,000 parts. The bath ought to be of the temperature of 68° to 77° F.

Amalgam Gold Plating.—Gold amalgam is chiefly used as a plating for silver, copper, or brass. The article to be plated is washed over with diluted nitric acid or potash lye and prepared chalk, to remove any tarnish or rust that might prevent the amalgam from adhering. After having been polished perfectly bright, the amalgam is applied as evenly as possible, usually with a fine scratch brush. It is then set upon a grate over a charcoal fire, or placed into an oven and heated to that degree at which mercury exhales. The gold, when the mercury has evaporated, presents a dull yellow color. Cover it with a coating of pulverized niter and alum in equal parts, mixed to a paste with water, and heat again till it is melted, then plunge into water. Burnish up with a steel or bloodstone burnisher.

Brass Gilding.—On brass, which is an electropositive metal, an electromagnetic metal, such as gold, can be deposited very cheaply from the dilute solutions of its salts. The deposit is naturally very thin, but still quite adhesive. In preparing it, the proportions stated below have to be accurately observed, otherwise no uniform, coherent coating will result, but one that is uneven and spotted.

I.—In 750 parts, by weight, of water dissolve: Phosphate of soda, 5 parts, and caustic potash, 3 parts, and in 250 parts of water, gold chloride, 1 part, and potassium cyanide, 16 parts. Mix both solutions well and cause the mixture to boil, whereupon the brass articles to be gilded are immersed. The gold in the mixture can be utilized almost entirely. When the solution does not gild well any more a little potassium cyanide is added, and it is used for pre-gilding the articles, which can then be gilded again in a fresh solution. This solution is very weak. A stronger one can be prepared mechanically by dissolving 2 to 3 parts of gold chloride in very little water to which 1 part of saltpeter is added. Into this solution dip linen rags, let them dry in a dark place, and cause them to char into tinder, which is rubbed up in a porcelain dish. Into the powder so made, dip a soft, slightly charred cork, moistened with a little vinegar, or else use only the finger, and rub the gold powder upon the brass articles.

II.—To Give Brass a Golden Color, it is dipped until the desired shade is obtained into a solution of about 175° F., produced as follows: Boil 4 parts of caustic soda, 4 parts of milk sugar, and 100 parts of water for 15 minutes; next add 4 parts of blue vitriol, dissolved in as little water as possible.

Copper and Brass Gilding.—The solutions used to gild copper can be generally used also for brass articles. Copper gilding acquires importance because in order to gild iron, steel, tin, and zinc, they must first be coated with copper, if the boiling method is to be employed. Following is Langbein's bath for copper and brass:

Dissolve 1 part, by weight, of chloride of gold and 16 parts, by weight, of potassium cyanide in 250 parts, by weight, of water; dissolve also and separately, 5 parts, by weight, of sodium phosphate and 3 parts, by weight, of caustic potash in 750 parts, by weight, of cold water. Mix these solutions and bring them to a boil. If the action subsides, add from 3 to 5 parts, by weight, more potassium cyanide. The polished iron and steel objects must first be copper-plated by dipping them into a solution of 5 parts, by weight, of blue vitriol and 2 parts, by weight, of sulphuric acid in 1,000 parts, by weight, of water. They may now be dipped into a hot solution containing 6 parts, by weight, of gold chloride and 22½ parts, by weight, of soda crystals in 75 parts, by weight, of water. This coating of gold may be polished.

Cold Chemical Gilding.—The chemical gilding by the wet process is accomplished by E. E. Stahl with the aid of three baths: A gold bath, a neutralization

bath, and a reduction bath. The gold bath is prepared from pure hydrochloric acid, 200 parts; nitric acid, 100 parts; and pure gold. The gold solution evaporated to crystallization is made to contain 1½ per cent of gold by diluting with water. The neutralization bath consists of soda lye of 6°, of pure sodium hydroxide, and distilled water. The reduction bath contains a mixture of equal parts of 90 per cent alcohol and distilled water, wherein pure hydrogen has been dissolved. The gilding proper is conducted by first entering the article in the gold bath, next briskly moving it about in the neutralization bath, and finally adding the reducing bath with further strong agitation of the liquid. The residues from the gilding are melted with 3 parts each of potash, powdered borax, and potash niter, thus recovering the superfluous gold. The gilding or silvering respectively produces a deposit of gold or silver of very slight thickness and of the luster of polishing gold. Besides the metal solution an "anti-reducer" is needed, consisting of 50 grams of rectified and rosinified turpentine oil and 10 grams of powdered roll sulphur. From this is obtained, by boiling, a syrupy balsam, to which is added, before use, lavender oil, well-ground basic bismuth nitrate, and the solution for gilding or silvering. The last takes place by a hydrochloric solution of aluminum with the above balsam.

Colored Gilding.—A variety of shades of green and red gold can be obtained by the electro-chemical process, which method may be employed for the decoration of various objects of art. In order to produce red gold in the different shades, a plate of pure copper is hung into a rather concentrated gold bath (5 to 6 parts, by weight, per 1,000 parts of liquid), which is connected with the battery in such a manner that gold is deposited on the article immersed in the bath. By the action of the electric current copper is dissolved as well from the copper plate and is separated simultaneously with the gold, so that, after a certain time, a deposit containing a gold copper alloy, conforming in color to the quantities of gold and copper contained in it, is obtained by the electric process. When the desired shade of color of the deposit is reached the copper plate is taken out and replaced by another consisting of the copper gold alloy, likewise produced by electrodeposition, and the articles are now gilt in this liquid. In some large manufactories of gold articles this last coloring is used even for pure gold articles, to give them a popular color. To produce green gold (alloy of gold and silver), a silver plate is first employed, which is dipped into the gold bath and from which enough silver is dissolved until the separating alloy shows the desired shade. The silver plate is then exchanged for a gold-silver plate of the respective color, and the articles are gilt with green gold.

Gilding German Silver.—In gilding German silver the solution may be worked at a low temperature, the solution being weakened and a small surface of anode exposed. German silver has the power of reducing gold from its solution in cyanide (especially if the solution be strong) without the aid of the battery; therefore, the solution should be weaker, in fact, so weak that the German silver will not deposit the gold *per se;* otherwise the deposit will take place so rapidly that the gold will peel off when being burnished or even scratch brushed.

Gilding of Glass.—I.—In order to produce a good gilding on glass, the gold salt employed must be free from acid. Prepare three solutions, viz.:

a. 20 parts acid-free gold chloride in 150 parts of distilled water.

b. 5 parts dry sodium hydrate in 80 parts of distilled water.

c. 2½ parts of starch sugar in 30 parts distilled water; spirit of wine, 20 parts; and commercial pure 40 per cent aldehyde, 20 parts. These liquids are quickly mixed together in the proportion of 200, 50, and 5 parts, whereupon the mixture is poured on the glass previously cleaned with soda solution, and the gilding will be effected in a short time. The gold coating is said to keep intact for years.

II.—Coat the places to be gilded thinly with a saturated borax solution, lay the gold leaf on this and press down well and uniformly with cotton-wool. Heat the glass over a spirit flame, until the borax melts, and allow to cool off. If the glass is to be decorated with gilt letters or designs, paint the places to be gilded with water-glass solution of 40° Bé.; lay on the gold leaf, and press down uniformly. Then heat the object to 86° F., so that it dries a little, sketch the letters or figures on with a lead pencil, erase the superfluous gold, and allow the articles to dry completely at a higher temperature.

Green Gilding.—This can be obtained conveniently by the galvanic process, by means of anodes of sheet platinum with the following composition: Water, 10,000 parts, by weight; sodium phosphate, 200 parts; sodium sulphate, 35 parts; potassium carbonate, 10 parts; 1 ducat gold from gold chloride, potassium cyanide (100 per cent), 20 parts. Dissolve the first three salts in 10,000 parts of cold water and add, with stirring, the gold chloride and potassium cyanide. Before the first use boil down the solution thoroughly about one-half, replacing the evaporating water and filter after cooling, in case a sediment should appear. To this gold bath very carefully add some silver bath. The platinum sheets which are to serve as anodes are employed 1¾ inches long, ⅓ inch broad, and $\frac{1}{100}$ of an inch thick. With these anodes the gold tone can be somewhat regulated by hanging more or less deeply into the solution during the gilding. The current should have a tension of 3 to 4 volts. In the case of batteries three Busen elements are connected for current tension. It is difficult to produce old gold on silver, especially if the raised portions are to appear green. It is most advantageous first to lightly copper the silver goods, taking the copper off again on the high places by brushing with pumice stone. After that hang at once in the above gold bath. If the embossed portions should be too mat, brighten slightly by scratching with a very fine brass wire brush. In this manner a handsome brown shade is obtained in the deep places and a green color on the raised portions. This process requires practice. Since this method will produce only a very light gilding, a coating of white varnish will protect the articles from tarnishing.

Incrusting with Gold.—The article is first made perfectly bright, and those places which are to be gilt are covered with a matt consisting of white lead ground with gum water, made into a paste which can be applied like a thick paint by means of a pen or brush. Those places of the metal surface not covered by the paint are coated with asphalt varnish—a solution of asphaltum in benzine to which oil of turpentine is added to render it less volatile. After this is done lay the article in water, so that the white lead paint comes off, and put it into a gilding bath. By the electric current gold is precipitated on the bright parts of the metal. When the layer of gold is thick enough lift the object from the bath, wash, let dry and lay it into a vessel filled with benzol. The asphalt dissolves in the benzol, and the desired design appears in gold on the bronze or silver ground. This operation may also be performed by coating the whole article with asphalt varnish and executing the design by means of a blunt graver which only takes away the varnish covering without scratching the metal itself. On the parts thus bared gold is deposited by the electric current and the varnish coating is then removed.

Ivory Gilding.—I.—The pattern is painted with a fine camel's-hair pencil, moistened with gold chloride. Hold the ivory over the mouth of a bottle in which hydrogen gas is generated (by the action of dilute sulphuric acid on zinc waste). The hydrogen reduces the auric chloride in the painted places into metallic gold, and the gold film precipitated in this manner will quickly obtain a considerable luster. The gold film is very thin, but durable.

II.—This is especially suitable for monograms. Take gold bronze and place as much as can be taken up with the point of a knife in a color-cup, moistening with a few drops of genuine English gold paint. Coat the raised portions sparingly with gold, using a fine pencil; next, coat the outer and inner borders of the design. When the work is done, and if the staining and gilding have been unsuccessful, which occurs frequently at the outset, lay the work for 5 or 10 minutes in warmed lead water and brush off with pumice stone. By this process very fine shades are often obtained which cannot be produced by mere staining. Since the gold readily wears off on the high places of the work, it is well to lightly coat these portions with a thin shellac solution before gilding. This will cause the gilding to be more permanent.

Mat Gilding.—To obtain a handsome mat gilding the article, after having been neatly polished, is passed through a sand-blast, such as is found in glass-grinding and etching establishments; next, the object is carefully cleansed of fine sand (if possible, by annealing and decocting), whereupon it is gilt and subsequently brushed mat with the brass brush. Where there is no sand-blast, the article is deadened with the steel wire brush, which will produce a satisfactory result, after some practice. After that, treatment is as above. The above-mentioned applies in general only to silver articles. In case of articles of gold, brass, or tombac, it is better to previously silver them strongly, since they are too hard for direct treatment

with the steel wire brush, and a really correct mat cannot be attained. The brushes referred to are, of course, circular brushes for the lathe.

Dead-Gilding of an Alloy of Copper and Zinc.—The parts which are to be deadened must be isolated from those which are to be polished, and also from those which are to be concealed, and which therefore are not to be gilded. For this purpose they are coated with a paste made of Spanish white mixed with water. The articles prepared in this manner are then attached by means of iron wire to an iron rod and suspended in a furnace constructed for this process. The floor of this furnace is covered on four sides with plates of enameled earthenware for receiving the portions spattered about of the salt mixture given off later.

In the middle is an oven constructed like a cooking stove, on which is an iron tripod for carrying the deadening pan; this latter is cemented into a second pan of cast iron, the intervening space being filled up with stove cement. In the middle of the pan is the bottom or sill, provided with a thick cast-iron plate, forming the hearth. On all four sides of the latter are low brick walls, connecting with the floor of the furnace, and the whole is covered with thick sheet metal. On the side of the furnace opposite the side arranged for carrying the pans, is a boiler in which boiling water is kept. On the same side of the furnace, but outside it, is a large oval tub of a capacity of about 700 or 800 quarts, which is kept filled with water. The upper portions of the staves of this tub are covered with linen to absorb all parts that are spattered about.

Powder for Gilding Metals.—I.—In a solution of perchloride of gold soak small pieces of linen which are dried over the solution so that the drops falling therefrom are saved. When the rags are dry burn them, carefully gathering the ashes, which ashes, stirred with a little water, are used for gilding either with pumice stone or with a cork. For the hollows, use a small piece of soft wood, linden, or poplar.

II.—Dissolve the pure gold or the leaf in nitro-muriatic acid and then precipitate it by a piece of copper or by a solution of iron sulphate. The precipitate, if by copper, must be digested with distilled vinegar and then washed by pouring water over it repeatedly and dried. This precipitate will be in the form of very fine powder; it works better and is more easily burnished than gold leaf ground with honey.

Gilding Pastes.—I.—A good gilding paste is prepared as follows: Slowly melt an ounce of pure lard over the fire, add ½ a teaspoonful of juice of squills, and stir up the mixture well, subsequently adding 10 drops of spirit of sal ammoniac. If the mixture is not stiff enough after cooling, the firmness may be enhanced by an admixture of ⅓ to ½ ounce of pure melted beef-tallow. A larger addition of tallow is necessary if the white of an egg is added. After each addition the mixture should be stirred up well and the white of egg should be added, not to the warm, but almost cold, mixture.

II.—Alum, 3 parts, by weight; saltpeter, 6 parts; sulphate of zinc, 3 parts; common salt, 3 parts. Mix all into a thick paste, dip the articles into it, and heat them, until nearly black, on a piece of sheet iron over a clear coke or charcoal fire; then plunge them into cold water.

Red Gilding.—This is obtained by the use of a mixture of equal parts of verdigris and powdered tartar, with which the article is coated; subsequently burning it off on a moderate coal fire. Cool in water, dip the article in a pickle of tartar, scratch it, and a handsome red shade will be the result, which has not attacked the gilding in any way.

Regilding Mat Articles.—In order to regenerate dead gold trinkets without having to color them again—which is, as a rule, impossible, because the gold is too weak to stand a second coloring—it is advisable to copper these articles over before gilding them. After the copper has deposited all over, the object, well cleaned and scratched, is hung in the gilding. By this manipulation much time and vexation is saved, such as every jeweler will have experienced in gilding mat gold articles. The article also acquires a faultless new appearance. Here are two recipes for the preparation of copper baths:

I.—Distilled boiling water, 2,000 parts, by weight; sodium sulphate, 10 parts; potassium cyanide, 15 parts; cupric acetate, 15 parts; sodium carbonate, 20 parts; ammonia, 12 parts.

II.—Dissolve crystallized verdigris, 20 parts, by weight, and potassium cyanide, 42 parts, in 1,000 parts of boiling water.

Silk Gilding.—This can only be accomplished by the electric process. The fiber is first rendered conductive by impregnation with silver nitrate solution and reduction of same with grape sugar and diluted alkali, or, best of all, with Raschig's reduction salt. In place of the silver nitrate, a solution of lead acetate or copper acetate may be employed. The silk thus impregnated is treated in the solution of an alkaline sulphide, e. g., sodium sulphide, ammonium sulphide, or else with hydrogen sulphide, thus producing a conductive coating of metallic sulphide. Upon this gold can be precipitated by electrodeposition in the usual way.

Spot Gilding.—Gilding in spots, producing a very fine appearance, is done by putting a thin coat of oil on those parts of the metal where the gilding is not to appear; the gold will then be deposited in those spots only where there is no oil, and the oil is easily removed when the work is finished.

Gilding Steel.—Pure gold is dissolved in aqua regia; the solution is allowed to evaporate until the acid in excess has gone. The precipitate is placed in clean water, 3 times the quantity of sulphuric acid is added and the whole left to stand for 24 hours in a well-closed flask, until the ethereal gold solution floats on top. By moistening polished steel with the solution a very handsome gilding is obtained. By the application of designs with any desired varnish the appearance of a mixture of gold and steel may be imparted to the article.

Wood Gilding. — I. — The moldings, ledges, etc., to be gilded are painted with a strong solution of joiners' glue, which is left to harden well, whereupon 8 to 10 coatings of glue mixed with whitening are given. Each coat must, of course, be thoroughly dry, before commencing the next. After this has been done, paint with a strong mixture of glue and minium, and while this is still wet, put on the gold leaflets and press them down with cotton. To impart the fine gloss, polish with a burnishing agate after the superfluous gold has been removed.

II.—Proceed as above, but take silver leaf instead of gold leaf, and after all is thoroughly dry and the superfluous silver has been removed, apply a coating of good gold lacquer. The effect will be equally satisfactory.

Zinc Gilding.—I.—Gilding by means of zinc contact may be accomplished with the following formula: Two parts, by weight, of gold chloride; 5 parts, by weight, of potassium cyanide; 10 parts, by weight, of sulphite of soda; and 60 parts, by weight, of sodium phosphate are dissolved in 1,000 parts of water. When used the bath must be hot. A cold bath without the addition of potassium cyanide may also be used for gilding, and this consists of 7 parts, by weight, of gold chloride; 30 parts, by weight, of yellow prussiate of potash; 30 parts, by weight, of potash; 30 parts, by weight, of common salt in 1,000 parts of water.

II.—To gild zinc articles, dissolve 20 parts of gold chloride in 20 parts of distilled water, and 80 parts of potassium cyanide in 80 parts of water, mix the solutions, stir a few times, filter, and add tartar, 5 parts, and fine chalk, 100 parts. The resulting paste is applied with a brush. Objects of copper and brass are previously coated with zinc. This is done in the following manner: Heat a concentrated sal ammoniac solution to the boiling point with addition of zinc dust and immerse the thoroughly cleaned objects until a uniform zinc coating has formed. Or boil the articles in a concentrated caustic soda solution with zinc dust.

OXIDIZING PROCESSES:

Aluminum Plating.—I.—To plate iron and other metals with pure aluminum, deoxidize the pieces with a solution of borax and place them in an enameling oven, fitted for receiving metallic vapors. Raise the temperature to 1,832° to 2,732° F. Introduce the aluminum vapors generated by heating a quantity of the metal on the sand bath. When the vapors come in contact with the metallic surfaces, the aluminum is deposited. The vapors that have not been used or are exhausted may be conducted into a vessel of water.

To Copper Aluminum,
take

II.—Sulphate of copper.	30 parts
Cream of tartar....	30 parts
Soda............	25 parts
Water	1,000 parts

The articles to be coppered are merely dipped in this bath, but they must be well cleaned previously.

Antimony Baths.—I.—By dissolving 15 parts, by weight, of tartar emetic and 15 parts of prepared tartar in 500 parts of hot water and adding 45–60 parts of hydrochloric acid and 45–60 parts of powdered antimony, brass becomes coated in the boiling liquid with beautiful antimony colors. In this manner it is possible to impart to brass golden, copper-red, violet, or bluish-gray shades, according to a shorter or longer stay of the objects in the liquid. These antimony colors possess a handsome luster, are permanent, and never change in the air.

II.—Carbonate of soda, 200 parts, by weight; sulphide of antimony, 50 parts; water, 1,000 parts. Heat the whole in a porcelain capsule for 1 hour, keeping constantly in ebullition; next, filter the solution, which, on cooling, leaves a precipitate, which boil again with the liquid for one-half hour, whereupon the bath is ready for use.

To Coat Brass Articles with Antimony Colors.—Dissolve 15 parts, by weight, of tartar emetic and 15 parts, by weight, of powdered tartar in 500 parts, by weight, of hot water and add 50 parts, by weight, of hydrochloric acid, and 50 parts, by weight, of powdered antimony. Into this mixture, heated to a boil, the immersed articles become covered with luster colors, a golden shade appearing at first, which is succeeded by one of copper red. If the objects remain longer in the liquid, the color passes into violet and finally into bluish gray.

Brassing.—I.—To brass small articles of iron or steel drop them into a quart of water and ½ ounce each of sulphate of copper and protochloride of tin. Stir the articles in this solution until desired color is obtained.

II.—Brassing Zinc, Steel, Cast Iron, etc.—Acetate of copper, 100 parts, by weight; cyanide of potassium, 250 parts; bisulphite of soda, 200 parts; liquid ammonia, 100 parts; protochloride of zinc, 80 parts; distilled water, 10,000 parts. Dissolve the cyanide of potassium and the bisulphite of soda. On the other hand, dissolve the ammonia in three-fourths of the water and the protochloride of zinc in the remaining water; next, mix the two solutions. This bath is excellent for brassing zinc and is used cold.

III.—Acetate of copper, 125 parts, by weight; cyanide of potassium, 400 parts; protochloride of zinc, 100 parts; liquid ammonia, 100 parts; distilled water, 8,000 to 10,000 parts. Proceed as above described.

IV.—Acetate of copper, 150 parts, by weight; carbonate of soda, 1,000 parts; cyanide of potassium, 550 parts; bisulphite of soda, 200 parts; protochloride of zinc, 100 parts. Proceed as above. This bath serves for iron, cast iron, and steel, and is used cold.

Colored Rings on Metal.—Dissolve 200 parts, by weight, of caustic potash in 2,000 parts of water and add 50 parts of litharge. Boil this solution for half an hour, taking care that a little of the litharge remains undissolved. When cold, pour off the clear fluid; it is then ready for use. Move the object to and fro in the solution; a yellow-brown color appears, becoming in turn white, yellow, red, and finally a beautiful violet and blue. As soon as the desired color is obtained, remove the article quickly from the solution, rinse in clean water, and dry in sawdust.

Green or Gold Color for Brass.—French articles of brass, both cast and made of sheet brass, mostly exhibit a golden color, which is produced by a copper coating. This color is prepared as follows: Dissolve 50 parts, by weight, of caustic soda and 40 parts of milk sugar in 1,000 parts of water and boil a quarter of an hour. The solution finally acquires a dark-yellow color. Now add to the mixture, which is removed from the fire, 40 parts of concentrated cold blue vitriol solution. A red precipitate is obtained from the vitriol, which falls to the bottom at 167° F. Next a wooden sieve, fitted to the vessel, is put into the liquid with the polished brass articles. Toward the end of the second minute the golden color is usually dark enough. The sieve with the articles is taken out and the latter are washed and dried in sawdust. If they remain in the copper solution they soon assume a green color, which in a short time passes into yellow and bluish green, and finally into the iridescent colors. These shades must be produced slowly at a temperature of 133° to 135° F.

To Give a Green Color to Gold Jewelry.—Take verdigris, 120 parts, by weight; sal ammoniac, 120 parts; nitrate of potassium, 45 parts; sulphate of zinc, 16 parts. Grind the whole and mix with strong vinegar. Place on the fire and boil in it the articles to be colored.

Nickeling by Oxidation.—I.—Nickeling may be performed on all metals cold, by means of nickelene by the Mitressey process, without employing electrical apparatus, and any desired thickness deposited. It is said to be more solid than nickel.

First Bath.—Clean the objects and take 5 parts, by weight, of American potash per 25 parts, by weight, of water. If the pieces are quite rusted, take 2 parts, by weight, of chlorhydric acid per 1 part, by weight, of water. The bath is employed cold.

Second Bath.—Put 250 parts, by weight, of sulphate of copper in 25,000 parts, by weight, of water. After dissolution add a few drops of sulphuric acid, drop by drop, stirring the liquid with a wooden stick until it becomes as clear as spring water.

Take out the pieces thus cleaned and place them in what is called the copper bath, attaching to them leaves of zinc; they will assume a red tint. Then pass them into the nickeling bath, which is thus composed:

	By weight
Cream of tartar	20 parts
Sal ammoniac, in powder	10 parts
Kitchen salt	5 parts
Oxychlorhydrate of tin	20 parts
Sulphate of nickel, single	30 parts
Sulphate of nickel, double	50 parts

Remove the pieces from the bath in a few minutes and rub them with fine sand on a moist rag. Brilliancy will thus be obtained. To improve the appearance, apply a brass wire brush. The nickeling is said to be more solid and beautiful than that obtained by the electrical method.

Brilliancy may be also imparted by means of a piece of buff glued on a wooden wheel and smeared with English red stuff. This will give a glazed appearance.

II.—Prepare a bath of neutral zinc chloride and a neutral solution of a nickel salt. The objects are immersed in the bath with small pieces of zinc and kept boiling for some time. This process has given satisfactory results. It is easy to prepare the zinc chloride by dissolving it in hydrochloric acid, as well as a saturated solution of ammoniacal nickel sulphate in the proportion of two volumes of the latter to one of the zinc chloride. The objects should be boiled for 15 minutes in the bath. Nickel salt may also be employed, preferably in the state of chloride.

Pickling Solutions.—Oxidized copper, brass, and German silver articles must be cleansed by acid solutions. In the case of brass alloys, this process, through which the object acquires a dull yellow surface, is known as dipping or yellowing. The treatment consists of several successive operations. The article is first boiled in a lye composed of 1 part caustic soda and 10 parts water, or in a solution of potash or soda or in limewater; small objects may be placed in alcohol or benzine. When all the grease has been removed, the article is well rinsed with water, and is then ready for the next pickling. It is first plunged into a mixture of 1 part sulphuric acid and 10 parts water, and allowed to remain in it till it acquires a reddish tinge. It is then immersed in 40° Bé. nitric acid, for the purpose of removing the red tinge, and then for a few seconds into a bath of 1 part nitric acid, 1.25 parts sulphuric acid of 66° Bé., 0.01 part common salt, and 0.02 parts lampblack. The article must then be immediately and carefully washed with water till no trace of acid remains. It is then ready for galvanizing or drying in bran or beech sawdust. When articles united with soft solder are pickled in nitric acid, the solder receives a gray-black color.

Palladiumizing Watch Movements.— Palladium is successfully employed for coating parts of timepieces and other pieces of metals to preserve them against oxidation. To prepare a palladium bath use the following ingredients: Chloride of palladium, 10 parts, by weight; phosphate of ammonia, 100 parts; phosphate of soda, 300 parts; benzoic acid, 8 parts; water, 2,000 parts.

Metal Browning by Oxidation.— The article ought first to be cleaned with either nitric acid or muriatic acid, then immersed in an acid affecting the metal and dried in a warm place. A light coating is thus formed. For a second coating acetic or formic acid is used preferably for aluminum, nickel, and copper; but for iron and steel, muriatic or nitric acid. After cleaning, the article is placed in a solution of tannin or gallic acid, and is then dried in a warm place as before. The second coating is of a yellowish-brown color. On placing it near the fire, the color can be deepened until it becomes completely black; care must be taken to withdraw it when the desired shade is produced. Instead of the acids employed for the first coating, ammonia may be used.

Silvering by Oxidation.— The oxidizing of silver darkens it, and gives an antique appearance that is highly prized.

I.—The salts of silver are colorless when the acids, the elements of which enter into their composition, are not colored, but they generally blacken on exposure to light. It is easy, therefore, to blacken silver and obtain its oxide; it is sufficient to place it in contact with a sulphide, vapor of sulphur, sulphohydric acids, such as the sulphides or polysulphides of potash, soda, dissolved in water and called *eau de barège*. The chlorides play the same part, and the chloride of lime in solution or simply Javelle water may be used. It is used hot in order to accelerate its action. The bath must be prepared new for each operation for two reasons: (1) It is of little value; (2) the sulphides precipitate rapidly and give best effects only at the time of their direct precipitations. The quantity of the reagent in solution, forming the bath, depends upon the thickness of the deposit of silver. When this is trifling, the oxidation penetrates the entire deposit and the silver exfoliates in smaller scales, leaving the copper bare. It is necessary, therefore, in this case to operate with dilute baths inclosing only about 45 grains of oxidizant at most per quart. The operation is simple: Heat the necessary quantity of water, add the sulphide or chloride and agitate to effect the solution of the mixture, and then at once plunge in the silver-plated articles, leaving them immersed only for a few seconds, which exposure is sufficient to cover it with a pellicle of deep black-blue silver. After withdrawing they are plunged in clean cold water, rinsed and dried, and either left mat or else polished, according to the nature of the articles.

Should the result not be satisfactory, the articles are brightened by immersing them in a lukewarm solution of cyanide of potassium. The oxide, the true name of which would be the sulphuret or chloruret, can be raised only on an object either entirely of silver or silver plated.

II.—Rub the article with a mixture of graphite, 6 parts, and powdered bloodstone, 1 part, moistened with oil of turpentine. Allow to dry and brush with soft brushes passed over wax. Or else, brush with a soft brush wet with alcoholic or aqueous platinic chloride solution of 1 in 20.

III.—Sulphurizing is effected with the following methods: Dip in a solution heated to about 175° F., of potassium sulphide, 5 parts, by weight; ammonium carbonate, 10 parts; water, 1,000 parts; or, calcium sulphide, 1 to 2 parts; sal ammoniac, 4 parts; water, 1.000 parts.

IV.—In the following solution articles of silver obtain a warm brown tone: Copper sulphate, 20 parts, by weight; potassium nitrate, 10 parts; ammonium chloride, 20 parts. By means of bromine, silver and silver alloys receive a black coloring. On engraved surfaces a niello-like effect may be produced thereby.

Oxidized Steel.—I.—Mix together bismuth chloride, 1 part; mercury bichloride, 2 parts; copper chloride, 1 part; hydrochloric acid, 6 parts; alcohol, 5 parts; and water, 5 parts. To use this mixture successfully the articles to be oxidized must be cleaned perfectly and freed from all grease, which is best accomplished by boiling them in a soda solution or by washing in spirit of wine. Care should be taken not to touch the article with the fingers again after this cleaning. However clean the hand may be, it always has grease on it and leaves spots after touching, especially on steel. Next the object is dipped into the liquid, or if this is not possible the solution is applied thin but evenly with a brush, pencil, or rabbit's foot. When the liquid has dried, the article is placed for a half hour in simple boiling water. If a very dark shade is desired the process is repeated until the required color is attained.

II.—Apply, by means of a sponge, a solution of crystallized iron chloride, 2 parts; solid butter of antimony, 2 parts; and gallic acid, 1 part in 5 parts of water. Dry the article in the air and repeat the treatment until the desired shade is reached. Finally rinse with water, dry, and rub with linseed-oil varnish.

Tinning by Oxidation.—A dipping bath for tinning iron is prepared by dissolving 300 parts, by weight, ammonia alum (sulphate of alumina and sulphate of ammonia) and 10 parts of melted stannous chloride (tin salt) in 20,000 parts of warm water. As soon as the solution boils, the iron articles, previously pickled and rinsed in fresh water, are plunged into the fluid; they are immediately covered with a layer of tin of a beautiful dull-white color, which can be made bright by treatment in a tub or sack. Small quantities of tin salt are added from time to time as may be required to replace the tin deposited on the iron. This bath is also well adapted for tinning zinc, but here also, as with iron, the deposit is not sufficient to prevent oxidation of the metal below. Larger articles tinned in this way are polished by scratch brushing. In tinning zinc by this process, the ammonia alum may be replaced by any other kind of alum, or aluminum sulphate may be used alone; experience has shown, however, that this cannot be done with iron, cast iron, or steel. If it is desired to tin other metals besides iron and zinc in the solution which we have described, the battery must be resorted to; if the latter is used, the above solution should be applied in preference to any other.

PATINA OXIDIZING PROCESSES:

Patina of Art Bronzes. — For all patinas, whether the ordinary brown of commerce, the green of the Barye bronzes, or the dark-orange tint of the Florentine bronzes, a brush is used with pigments varying according to the shade desired and applied to the metal after it is warmed. Recipes are to be met with on every hand that have not been patented. But the details of the operation are the important thing, and often the effect is produced by a handicraft which it is difficult to penetrate.

I.—A dark tint may be obtained by cleaning the object and applying a coat of hydrosulphate of ammonia; then, after drying it, by rubbing with a brush smeared with red chalk and plumbago. The copper may also be moistened with a dilute solution of chloride of platina and warmed slightly, or still by plunging it in a warm solution of the hydrochlorate of antimony. For the verde antique a solution is recommended composed of 200 grams of acetic acid of 8° strength, the same quantity of common vinegar, 30 parts, by weight, of carbonate of ammonia; 10 parts, by weight, of sea salt; with the same quantities of cream of tartar and acetate of copper and a little water. To obtain the bronze of medals several processes afford a selection: For example, the piece may be dipped in a bath consisting of equal parts of the perchloride and the sesquiazotate of iron, warming to the evaporation of the liquid, and rubbing with a waxed brush.

II.—Dissolve copper nitrate, 10 parts, by weight, and kitchen salt, 2 parts, in 500 parts of water and add a solution of ammonium acetate obtained by neutralization of 10 parts of officinal spirit of sal ammoniac with acetic acid to a faintly acid reaction, and filling up with water to 500 parts. Immerse the bronze, allow to dry, brush off superficially and repeat this until the desired shade of color has been obtained.

A Permanent Patina for Copper.—
Green.—

I.—Sodium chloride. 37 parts
   Ammonia water.. 75 parts
   Ammonium chloride.......... 37 parts
   Strong wine vinegar.......... 5,000 parts

Mix and dissolve. Apply to object to be treated, with a camel's-hair pencil. Repeat the operation until the desired shade of green is reached.

Yellow Green.—

II.—Oxalic acid....... 5 parts
   Ammonium chloride........... 10 parts
   Acetic acid, 30 per cent dilution.... 500 parts

Mix and dissolve. Use as above indicated. The following will produce the same result:

III.—Potassium oxalate, acid.......... 4 parts
   Ammonium chloride..........16–17 parts
   Vinegar containing 6 per cent of acetic acid..... 1,000 parts

IV.—Bluish Green.—After using the first formula (for green) pencil over with the following solution:

   Ammonium chloride.......... 40 parts
   Ammonium carbonate........ 120 parts
   Water.......... 1,000 parts

Mix and dissolve.

Greenish Brown.—

V.—Potassium sulphuret........ 5 parts
   Water........... 1,000 parts

Mix and dissolve. With this, pencil over object to be treated, let dry, then pencil over with 10 parts a mixture of a saturated solution of ammonia water and acetic acid and 5 parts of ammonium chloride thinned with 1,000 parts of water. Let dry again, then brush off well. Repeat, if necessary, until the desired hue is attained.

Another Blue Green.—

VI.—Corrosive sublimate. 25 parts
   Potassium nitrate.. 86 parts
   Borax............. 56 parts
   Zinc oxide........ 113 parts
   Copper acetate ...220–225 parts

Mix and heat together on the surface of the object under treatment.

VII.—Brown.—The following is a Parisian method of producing a beautiful deep brown:

   Potassium oxalate, acid.......... 3 parts
   Ammonium chloride........... 15 parts
   Water, distilled.... 280 parts

Mix and dissolve. The object is penciled over with this several times, each time allowing the solution to dry before putting on any more. The process is slow, but makes an elegant finish.

Green Patina Upon Copper.—To produce a green patina upon copper take tartaric acid, dilute it half and half with boiling water; coat the copper with this; allow to dry for one day and rub the applied layer off again the next day with oakum. The coating must be done in dry weather, else no success will be obtained. Take hydrochloric acid and dilute it half and half with boiling water, but the hydrochloric acid should be poured in the water, not vice-versa, which is dangerous. In this hydrochloric acid water dissolve as much zinc as it can solve and allow to settle. The clear liquid is again diluted half with boiling water and the copper is coated with this a few times.

Black Patina.—Black patina is obtained by coating with tallow the pieces to be oxidized and lighting with a rosin torch. Finally, wipe the reliefs and let dry.

Blue-Black Patina.—Use a dilute solution of chloride of antimony in water and add a little free hydrochloric acid. Apply with a soft brush, allow the article to dry and rub with a flannel. If expense is no object, employ a solution of chloride of palladium, which gives a magnificent blue black. It is necessary, however, to previously clean the articles thoroughly in a hot solution of carbonate of soda, in order to remove the dirt and greasy matter, which would prevent the patina from becoming fixed.

Red Patina.—The following is a new method of making a red patina, the so-called blood bronze, on copper and copper alloys. The metallic object is first made red hot, whereby it becomes covered with a coating consisting of cupric oxide on the surface and cuprous oxide beneath. After cooling, it is worked upon with a polishing plate until the black cupric oxide coating is removed and the cuprous oxide appears. The metal now shows an intense red color, with a considerable degree of luster, both of which are so permanent that it can be treated with chemicals, such as blue vitriol, for instance, without being in the least affected.

If it is desired to produce a marbled surface, instead of an even red color, borax or some chemical having a similar action is sprinkled upon the metal during the process of heating. On the places covered by the borax, oxidation is prevented, and after polishing, spots of the original metallic color will appear in the red surface. These can be colored by well-known processes, so as to give the desired marbled appearance.

PLATINIZING:

Platinizing Aluminum. — Aluminum vessels coated with a layer of platinum are recommended in place of platinum vessels, when not exposed to very high temperatures. The process of platinizing is simple, consisting in rubbing the aluminum surface, previously polished, with platinic chloride, rendered slightly alkaline. The layer of platinum is made thicker by repeated application. Potash lye is carefully added to a solution of 5 to 10 per cent of platinic chloride in water till a slightly alkaline reaction is produced on filtering paper or a porcelain plate by means of phenolphthalein. This solution must always be freshly prepared, and is the best for the purpose. Neither galvanizing nor amalgamating will produce the desired result. Special care must be taken that the aluminum is free from iron, otherwise black patches will arise which cannot be removed. Vessels platinized in this way must not be cleaned with substances such as seasand, but with a 5 to 10 per cent solution of oxalic acid in water, followed by thorough rinsing in water. These vessels are said to be specially suitable for evaporating purposes.

Platinizing Copper and Brass.—I.—The articles are coated with a thin layer of platinum in a boiling solution of platinum sal ammoniac, 1 part; sal ammoniac, 8 parts; and water, 40 parts, and next polished with chalk. A mixture of equal parts of platinum sal ammoniac and tartar may also be rubbed on the objects. Steel and iron articles can be platinized with an ethereal solution of platinic chloride. For small jewelry the boiling solution of platinic chloride, 10 parts; cooking salt, 200 parts; and water, 1,000 parts, is employed, which is rendered alkaline with soda lye. In this, one may also work with zinc contact.

II.—Heat 800 parts of sal ammoniac and 10 parts of platinum sal ammoniac to the boiling point with 400 parts of water, in a porcelain dish, and place the articles to be platinized into this, whereby they soon become covered with a coating of platinum. They are then removed from the liquid, dried and polished with whiting.

Platinizing on Glass or Porcelain.— First dissolve the platinum at a moderate temperature in aqua regia, and next evaporate the solution to dryness, observing the following rules: When the solution commences to turn thick it is necessary to diminish the fire, while carrying the evaporation so far that the salt becomes dry, but the solution should not be allowed to acquire a brown color, which occurs if the heat is too strong. The result of this first operation is chloride of platina. When the latter has cooled off it should be dissolved in alcohol (95 per cent). The dissolution accomplished, which takes place at the end of 1 or 2 hours, throw the solution gradually into four times its weight of essence of lavender, then put into a well-closed flask.

For use, dip a brush into the solution and apply it upon the objects to be platinized, let dry and place in the muffle, leaving them in the oven for about one-half hour. In this operation one should be guided as regards the duration of the baking by the hardness or fusibility of the objects treated. The platinization accomplished, take a cotton cloth, dipped into whiting in the state of pulp, and rub the platinated articles with this, rinsing with water afterwards.

Platinizing Metals.—Following are several processes of platinizing on metals:

It is understood that the metals to be covered with platinum must be copper or coppered. All these baths require strong batteries.

I.—Take borate of potash, 300 parts, by weight; chloride of platina, 12 parts; distilled water, 1,000 parts.

II.—Carbonate of soda, 250 parts, by weight; chloride of platina, 10 parts; distilled water, 1,000 parts.

III.—Sulphocyanide of potash, 12 parts, by weight; chloride of platina, 12 parts; carbonate of soda, 12 parts; distilled water, 1,000 parts.

IV.—Borate of soda, 500 parts, by weight; chloride of platina, 12 parts; distilled water, 1,000 parts.

SILVERING, SILVER-PLATING, AND DESILVERING:

See also Silvering by Oxidation, under Oxidation Processes, under Plating.

Antique Silver—There are various processes for producing antique silver, either fat or oxidized:

To a little copal varnish add some finely powdered ivory black or graphite. Thin with spirits of turpentine and rub with a brush dipped into this varnish the objects to be treated. Allow to dry for an hour and wipe off the top of the articles with some rag, so that the black remains only in the hollows. If a softer tint is desired, apply again with a dry brush and wipe as the first time. The coating of black will be weaker and the shade handsomer.

Britannia Silver-Plating. — I. — The article should first be cleaned and then rubbed by means of a wet cloth with a pinch of powder obtained by mixing together: Nitrate of silver, 1 part; cyanide of potassium, 2 parts; chalk, 5 parts. Then wipe with a dry cloth, and polish well with rouge to give brilliancy.

II.—By the electric method the metal is simply plunged into a hot saturated solution of crude potassium carbonate, and the plating is then done directly, using a strong electrical current. The potassium carbonate solution dissolves the surface of the britannia metal and thus enables the silver to take a strong hold on the article.

To Silver Brass, Bronze, Copper, etc. —I.—In order to silver copper, brass, bronze, or coppered metallic articles, dissolve 10 parts of lunar caustic in 500 parts of distilled water, and 35 parts of potassium cyanide (98 per cent) in 500 parts of distilled water; mix both solutions with stirring, heat to 176° to 194° F. in an enameled vessel, and enter the articles, well cleansed of fat and impurities, until a uniform coating has formed.

II.—Zinc, brass, and copper are silvered by applying a paste of the following composition: Ten parts of silver nitrate dissolved in 50 parts of distilled water, and 25 parts of potassium cyanide dissolved in distilled water; mix, stir, and filter. Moisten 100 parts of whiting and 400 parts of powdered tartar with enough of the above solution to make a paste-like mass, which is applied by means of a brush on the well-cleaned objects. After the drying of this coating, rinse off, and dry in sawdust.

III.—To silver brass and copper by friction, rub on the articles, previously cleaned of grease, a paste of silver chloride, 10 parts; cooking salt, 20 parts; powdered tartar, 20 parts; and the necessary water, using a rag.

Desilvering.—I.—It often happens in plating that, notwithstanding all precautions, some pieces have failed and it is necessary to commence the work again. For removing the silver that has been applied, a rapid method is to take sulphuric acid, 100 parts, and nitrate of potash, 10 parts. Put the sulphuric acid and the nitrate of potash (saltpeter) in a vessel of stoneware or porcelain, heated on the water bath. When the silver has been removed from the copper, rinse the object several times and recommence the silvering. This bath may be used repeatedly, taking care each time to put it in a stoppered bottle. When it has been saturated with silver and has no more strength, decant the deposit, boil the liquor to dryness, add the residue to the deposit, and melt in a crucible to regenerate the metal.

II.—To dissolve the silver covering of a metallic object, a bath is made use of, composed of 66 per cent sulphuric acid, 3 parts, and 40 per cent nitric acid, 1 part. This mixture is heated to about 176° F., and the objects to be desilvered are suspended in it by means of a copper wire. The operation is accomplished in a few seconds. The objects are washed and then dried in sawdust.

To Silver Glass Balls and Plate Glass. —The following is a method for silvering the glass balls which are used as ornaments in gardens, glass panes, and concave mirrors: Dissolve 300 parts of nitrate of silver and 200 parts of ammonia in 1,300 parts of distilled water. Add 35 parts of tartaric acid dissolved in 4 times its weight of water. Dilute the whole with 15,000 to 17,000 parts of distilled water. Prepare a second solution containing twice the amount of tartaric acid as the preceding one. Apply each of these solutions successively for 15 to 20 minutes on the glass to be silvered, which must previously have been cleaned and dried. When the silvering is sufficient, wash the object with hot water, let dry, and cover with a brown varnish.

Iron Silver-Plating.—I.—Iron articles are plated with quicksilver in a solution of nitrate of mercury before being silvered. The quicksilver is then removed by heating to 572° F. The articles may also be first tinned to economize the silver. Steel is dipped in a mixture of nitrate of silver and mercury, each dissolved separately in the proportion of 5 parts, by weight, to 300 parts, by weight, of water, then wiped to remove the black film of carbon, and silvered till a sample dipped in a solution of blue vitriol ceases to turn red. According to H. Krupp, articles made of an alloy of nickel, copper, and zinc, such as knives, forks, spoons, etc., should be coated electrically with nickel, put into a solution of copper like that used for galvanic coppering, and then electroplated.

II.—A brilliant silver color may be imparted to iron (from which all grease has been previously removed) by treating it with the following solution: Forty parts, by weight, chloride of antimony; 10 parts, by weight, powdered arsenious acid; and 80 parts levigated hematite are mixed with 1,000 parts of 90 per cent alcohol and gently heated for half an hour on a water bath. A partial solution takes place, and a small cotton pad is then dipped in the liquid and applied with a gentle pressure to the iron. A thin film consisting of arsenic and antimony is precipitated, as described by Dr. Langbein, in his "Handbuch der galv. Metallniederschläge." The brilliancy of the effect depends upon the care with which the iron has previously been polished.

To Silver-Plate Metals.—I.—Nitrate of silver, 30 parts, by weight; caustic potash, 30 parts; distilled water, 100 parts. Put the nitrate of silver into the water; one-quarter hour afterwards add the potash, and, when the solution is done, filter. It is sufficient to dip the objects to be silvered into this bath, moving them about in it for 1 or 2 minutes at most; then rinsing and drying in sawdust. It is necessary to pickle the pieces before using the bath. To make the nitrate of silver one's self, take 30 parts of pure silver and 60 parts of nitric acid, and when the metal is dissolved add the caustic potash and the water.

II.—Kayser's silvering liquid, which is excellent for all kinds of metals, is prepared from lunar caustic, 11 parts; sodium hyposulphite, 20 parts; sal ammoniac, 12 parts; whiting, 20 parts; and distilled water, 200 parts. The articles must be cleaned well.

Mosaic Silver.—This compound consists of tin, 3 parts, by weight; bismuth, 3 parts; and mercury, 1½ parts. The alloy of these metals is powdered finely, thus forming a silvery mass used for imitation silvering of metals, paper, wood, etc. In order to impart to metals,

especially articles of copper and brass, an appearance similar to silver, they are made perfectly bright. The powder of the mosaic silver is mixed with six times the volume of bone ashes, adding enough water to cause a paste and rubbing this on the metallic surface by means of a cork of suitable shape. In order to silver paper by means of this preparation it is ground with white of egg, diluted mucilage, or varnish, and treated like a paint.

Pastes for Silvering.—I.—Carbonate of lime, 65 parts; sea salt, 60 parts; cream of tartar, 35 parts; nitrate of silver, 20 parts. Bray all in a mortar, not adding the carbonate of lime until the other substances are reduced to a fine powder. Next, add a little water to form a homogeneous paste, which is preserved in blue bottles away from the light. For use, put a little of this paste on a small pad and rub the article with it.

II.—Articles of zinc, brass, or copper may also be silver-plated by applying to them a pasty mass of the following composition: First dissolve 10 parts, by weight, of nitrate of silver in 50 parts, by weight, of distilled water; also 25 parts, by weight, of potassium cyanide in sufficient distilled water to dissolve it. Pour the two together, stir well, and filter. Now 100 parts, by weight, of whiting or levigated chalk and 400 parts, by weight, of potassium bitartrate, finely powdered, are moistened with the above solution sufficiently to form a soft paste, which may be applied to the objects, previously well cleansed, with a brush. After this coating has dried well, rinse it off, and dry the object in clean sawdust.

Resilvering.—I.—Take 100 parts, by weight, of distilled water and divide it into two equal portions. In the one dissolve 10 parts of silver nitrate and in the other 25 parts of potassium cyanide. The two solutions are reunited in a single vessel as soon as completed. Next prepare a mixture of 100 parts of Spanish white, passed through a fine sieve, 10 parts of cream of tartar, pulverized, and 1 part of mercury. This powder is stirred in a portion of the above liquid so as to form a rather thick paste. The composition is applied by means of the finger, covered with a rag, on the object to be silvered. The application must be as even as possible. Let the object dry and wash in pure water. The excess of powder is removed with a brush.

II.—The following is a process used when the jeweler has to repair certain pieces from which silvering has come off in places, and which he would like to repair without having recourse to the battery, and specially without having to take out the stones or pearls: Take nitrate of silver, 25 parts, by weight; cyanide of potassium, 50 parts; cream of tartar, 20 parts; Paris white, 200 parts; distilled water, 200 parts; mercury, 2 parts. Dissolve the nitrate of silver in half of the distilled water and the cyanide in the other half; mix the two liquids; next bray well in a mortar the mercury, Paris white, and cream of tartar. Preserve the products of these two operations separately, and when you wish to use them make a rather soft paste of the two, which apply with a little cotton or a brush on the portion to be silvered. Let dry and subsequently rub with a soft brush.

Tin Silver-Plating.—Prepare a solution of 3 parts, by weight, of bismuth subnitrate in 10 parts of nitric acid of 1.4 specific gravity, to which add a solution of 10 parts of tartar and 40 parts of hydrochloric acid in 1,000 parts of water. In the mixture of these solutions immerse the tin articles freed from grease and oxide. The pulverous bismuth precipitated on the surface is rubbed off, whereupon the objects appear dark steel gray. For silvering prepare a mixture of 10 parts of silver chloride; 30 parts of cooking salt; 20 parts of tartar, and 100 parts of powdered chalk, which is rubbed in a slightly moist state on the bismuth surface of the tin articles, using a flannel rag. The silver separates only in a very thin layer, and must be protected against power and light before tarnishing by a coating of preservative or celluloid varnish.

Zinc Contact Silver-Plating.—According to Buchner, 10 parts, by weight, of silver nitrate is dissolved in water and precipitated by the addition of hydrochloric acid in the form of silver chloride, which is washed several times in clean water; now dissolve 70 parts, by weight, of spirit of sal ammoniac in water, and add to it 40 parts, by weight, of soda crystals, 40 parts, by weight, of pure potassium cyanide, and 15 parts, by weight, of common salt. Now thin down the compound with sufficient distilled water to make a total of 1,000 parts.

Tin Plating of Lead.—Lead plates are best tinned by plating. For this purpose a table with a perfectly even iron surface and provided with vertical raised edges to prevent the melted metal from flowing away, is employed. The lead is poured on this table, and covered with grease to prevent oxidation of the surface. As soon as the lead is congealed, melted tin is poured over it, care being taken that the tin is sufficiently heated to remelt the surface of the lead and combine thoroughly with it. When the plate is sufficiently cooled, it is turned over, and the lower surface treated in the same way. The plate, thus tinned on both sides, is then placed between rollers, and can be rolled into very thin sheets without injury to the tin coating. These sheets, doubly coated with tin by this process, are specially adapted for lining cases intended for the transport of biscuits, chocolate, candies, tea, snuff, etc. If lead plates are only to be tinned superficially, they are heated to a tolerably high temperature, and sprinkled with powdered rosin; melted tin is then rubbed on the surface of the plate with a ball of tow. It is advisable to give the lead a fairly thick coating of tin, as the latter is rendered thinner by the subsequent rolling.

VARIOUS RECIPES:

To Ascertain whether an Article is Nickeled, Tinned, or Silvered.—When necessary to ascertain quickly and accurately the nature of the white metal covering an object, the following process will be found to give excellent results:

Nickeled Surface.—If the article has a nickel coating, a drop of hydrochloric acid, deposited on a spot clean and free from grease, will quickly develop a greenish tint. If the object is kept for 5 or 10 minutes in a solution composed of 60 parts of sea salt and 110 parts of water, it will receive a very characteristic reddish tint. A drop of sulphuret of sodium does not change a nickeled surface.

Tinned Surface.—A tinned object may be recognized readily by applying hydrochloric acid, which, even diluted, will remove the tin. The salt solution, used as previously described, produces a gray tint, faint in certain cases. The sulphuret of sodium dissolves tin.

Silvered Surface.—In the case of a silvered article a drop of nitric acid will remove the silver, while hydrochloric acid will scarcely attack it. The salt solution will produce no effect. The sulphuret of sodium will blacken it rapidly.

PLATINIZING:
See Plating.

PLATINOTYPE PAPER:
See Photography.

PLATINUM PAPERS AND THEIR DE-VELOPMENT:

See Photography, under Developing Papers.

PLATINUM WASTE, TO SEPARATE SILVER FROM:

See Silver.

PLUMBAGO:

See Lubricants.

PLUMES:

See Feathers.

PLUSH:

To Make Plush Adhere to Metal.— Wash off with ordinary soda water the bottom of a tin box, wiping it dry with cloth. Coat the tin with the juice of onion and press on this space a piece of strong paper, smoothing it out so that there will be no blisters. When this has dried, the paper will adhere, so that it can be removed only by scraping with a sharp instrument. Then give a coat of hot glue to the paper and press the plush down into the glue, and when dry and hard, the plush can be removed only by placing the tin box in boiling water.

PLUSH, TO REMOVE GREASE SPOTS FROM:

See Cleaning Preparations and Methods.

POISONS, ANTIDOTES FOR:

See Antidotes.

Polishes

Polishes for Aluminum. — I. — M. Mouray recommends the use of an emulsion of equal parts of rum and olive oil, made by shaking these liquids together in a bottle. When a burnishing stone is used, the peculiar black streaks first appearing should not cause vexation, since they do not injure the metal in the least, and may be removed with a woolen rag. The object in question may also be brightened in potash lye, in which case, however, care must be taken not to have the lye too strong. For cleaning purposes benzol has been found best.

II.—Aluminum is susceptible of taking a beautiful polish, but it is not white like that of silver or nickel, rather slightly bluish, like tin. The shade can be improved. First, the grease is to be removed from the object with pumice stone. Then, for polishing, use is made of an emery paste mingled with tallow, forming cakes which are rubbed on the polishing brushes. Finally, rouge powder is employed with oil of turpentine.

Polishes for Bars, Counters, etc.

I.—Linseed oil......... 8 ounces
    Stale ale........... 8 ounces
    Hydrochloric acid.. 1 ounce
    Alcohol, 95 per cent. 1 ounce
    White of 1 egg.

Mix. Shake before using. Clean out the dust, dirt, etc., using an appropriate brush, or a bit of cloth wrapped around a stick, then apply the above, with a soft brush, or a bit of cotton wrapped in a bit of silk—or, in fact, any convenient method of applying it.

II.—Japan wax...... 1 av. ounce
    Oil of turpentine 3 fluidounces
    Linseed oil...... 16 fluidounces
    Alcohol........ 3 fluidounces
    Solution of pot-
    ash.......... 1½ fluidounces
    Water to make 32 fluidounces.

Dissolve the wax in the turpentine, add the other ingredients, diluting the potash solution with the water before adding to the other ingredients, and stir briskly until well mixed.

POLISHES FOR BRASS, BRONZE, COPPER, ETC.

Objects of polished copper, bronze, brass, and other alloys of copper tarnish through water and it is sometimes necessary to give them again their bright appearance. Pickle the articles in an acid bath; wash them next in a neutral bath; dry them, and subsequently rub them with a polishing powder. Such is the general formula; the processes indicated below are but variants adapted to divers cases and recommended by disinterested experimenters:

Sharp Polishes.—The following three may be used on dirty brasses, copper articles, etc., where scratching is not objectionable:

I.—Quartz sand, pow-
    dered and levigat-
    ed............... 20 parts
    Paris red.......... 30 parts
    Vaseline........... 50 parts

Mix intimately and make a pomade.

II.—Emery flour, finest
    levigated........ 50 parts
    Paris red.......... 50 parts
    Mutton suet........ 40 parts
    Oleic acid......... 40 parts

III.—Levigated emery
    powder .... 100 parts
    Anhydrous sodium
    carbonate...... 5 parts
    Tallow soap....... 20 parts
    Water............ 100 parts

Copper Articles.—Make a mixture of powdered charcoal, very fine, 4 parts; spirit of wine, 3 parts; and essence of turpentine, 2 parts. To this add water in which one-third of its weight of sorrel salt or oxalic acid has been stirred, and rub the objects with this mixture.

Bronze Articles.—Boil the objects in soap lye, wash in plenty of water, and dry in sawdust.

Highly Oxidized Bronzes.—First dip in strong soda lye, then in a bath containing 1 part of sulphuric acid to 12 parts of water. Rinse in clean water, and next in water containing a little ammonia. Dry and rub with a polishing powder or paste.

POLISHES FOR FLOORS.

I.—Throw a handful of permanganate potash crystals into a pail of boiling water, and apply the mixture as hot as possible to the floor with a large flat brush. If the stain produced is not dark enough, apply one or two more coats as desired, leaving each wash to dry thoroughly before applying another. If it is desired to polish the surface with beeswax, a coat of size should be applied to the boards before staining, as this gives depth and richness to the color. After 3 or 4 days, polish well with a mixture of turpentine and beeswax. A few cents will cover the cost of both size and permanganate of potash.

II.—Potash............ 1 part
Water............. 4 parts
Yellow beeswax.... 5 parts
Hot water, a sufficient quantity.

Emulsify the wax by boiling it in the water in which the potash has been dissolved; stir the whole time. The exact amount of boiling is determined by the absence of any free water in the mass. Then remove the vessel from the fire, and gently pour in a little boiling water, and stir the mixture carefully. If a fat-like mass appears without traces of watery particles, one may know the mass is in a fit condition to be liquefied by the addition of more hot water without the water separating. Then put in the water to the extent of 200 to 225 parts, and reheat the compound for 5 to 10 minutes, without allowing it to reach the boiling point. Stir constantly until the mixture is cool, so as to prevent the separation of the wax, when a cream-like mass results which gives a quick and brilliant polish on woodwork, if applied in the usual way, on a piece of flannel rag, and polished by rubbing with another piece of flannel.

Colored Floor Polishes.—Yellow: Caustic soda solution, $7\frac{1}{2}$ parts, mixed with $1\frac{1}{2}$ to 2 parts of finely powdered ocher, heated with $2\frac{1}{4}$ parts of yellow wax, and stirred until uniformly mixed. A reddish-brown color may be obtained by adding 2 parts of powdered umber to the above mixture.

Nut Brown.—I.—Natural umber, $\frac{1}{2}$ part; burnt umber, 1 part; and yellow ocher, 1 part, gives a fine red-brown color when incorporated with the same wax and soda mixture.

II.—Treat 5 pounds of wax with 15 pounds of caustic soda lye of 3° Bé. so that a uniform wax milk results; boil with $\frac{1}{2}$ pound of annatto, 3 pounds of yellow ocher, and 2 pounds of burnt umber.

Mahogany Brown.—Boil 5 pounds of wax with 15 pounds of caustic soda lye as above. Then add 7 pounds of burnt umber very finely powdered, making it into a uniform mass by boiling again.

Yellow Ocher.—The wax milk obtained as above is boiled with 5 pounds of yellow ocher.

The mass on cooling has the consistency of a salve. If it is to be used for rubbing the floor it is stirred with sufficient boiling water so as to form a fluid of the consistency of thin syrup or oil. This is applied very thin on the floor, using a brush; then it is allowed to dry only half way, and is rubbed with a stiff floor brush. The polishing is continued with a woolen rag until a mirror-like gloss is obtained. It is best not to paint the whole room and then brush, but the deals should be taken one after the other, otherwise the coating would become too dry and give too dull a luster. The floors thus treated with gloss paste are very beautiful. To keep them in this condition they should be once in a while rubbed with a woolen rag, and if necessary the color has to be renewed in places. If there are parquet floors whose patterns are not to be covered up, the ocher (yellow) paste or, better still, the pure wax milk is used.

French Polish.—The wood to be polished must be made perfectly smooth and all irregularities removed from the surface with glass paper; next oil the work with linseed oil, taking care to rub off all superfluous oil. (If the wood is white no oil should be used, as it imparts a slight color.) Then prepare a wad or rubber of wadding, taking care there are no hard lumps in it. After the rubber is prepared pour on it a small quantity of polish. Then cover it with a piece of old cotton rag (new will not answer). Put a small drop of oil with the finger on the surface of the rubber, and then proceed to polish, moving the rubber in lines, making a kind of figure of eight over the work. Be very careful that the rubber is not allowed to stick or the work will be spoilt. A little linseed oil facilitates the process. When the rubber requires more polish, turn back the rag cover, pour on the polish, replace the cover, oil and work as before. After this rubbing has proceeded for a little time and the whole surface has been gone over, the work must be allowed to stand for a few hours to harden, and then be rubbed down smooth with very fine emery paper. Then give another coat of polish. If not smooth enough, emery paper again. This process must continue until the grain is filled up. Finish off with a clean rubber with only spirit on it (no polish), when a clear bright surface should be the result. Great care must be taken not to put the polish on too freely, or you will get a rough surface. After a little practice all difficulties will vanish. The best French polish will be found to be one made only from good pale orange shellac and spirit, using 3 pounds of shellac for each gallon of spirit. The latter should be of 63 to 64° over-proof. A weak spirit is not suitable and does not make a good polish. A few drops of pure linseed oil make the polish work more freely.

POLISHES FOR FURNITURE.

First make a paste to fill cracks as follows: Whiting, plaster of Paris, pumice stone, litharge, equal parts; japan dryer, boiled linseed oil, turpentine, coloring matter of sufficient quantity. Rub the solids intimately with a mixture of 1 part of the japan, 2 parts of the linseed oil, and 3 parts of turpentine, coloring to suit with Vandyke brown or sienna. Lay the filling on with a brush, let it set for about 20 minutes, and then rub off clean except where it is to remain. In 2 or 3 days it will be hard enough to polish.

After the surface has been thus prepared, the application of a coat of first-class copal varnish is in order. It is recommended that the varnish be applied in a moderately warm room, as it is injured by becoming chilled in drying. To get the best results in varnishing, some skill and experience are required. The varnish must be kept in an evenly warm temperature, and put on neither too plentifully nor too gingerly. After a satisfactorily smooth and reg-

ular surface has been obtained, the polishing proper may be done. This may be accomplished by manual labor and dexterity, or consist in the application of a very thin, even coat of a very fine, transparent varnish.

If the hand-polishing method be preferred, it may be pursued by rubbing briskly and thoroughly with the following finishing polish:

I.—Alcohol........... 8 ounces
Shellac............ 2 drachms
Gum benzoin...... 2 drachms
Best poppy oil...... 2 drachms

Dissolve the shellac and gum in the alcohol in a warm place, with frequent agitation, and, when cold, add the poppy oil. This may be applied on the end of a cylindrical rubber made by tightly rolling a piece of flannel which has been torn, not cut, into strips 4 to 6 inches wide.

A certain "oily sweating" of articles of polished wood occurs which has been ascribed to the oil used in polishing, but has been found to be due to a waxy substance present in shellac, which is often used in polishing. During the operation of polishing, this wax enters into close combination with the oil, forming a soft, greasy mass, which prevents the varnish from ever becoming really hard. This greasy matter exudes in the course of time. The remedy is to use only shellac from which the vegetable wax has been completely removed. This is accomplished by making a strong solution of the shellac in alcohol and then shaking it up with fresh seed lac or filtering it through seed lac. In this way the readily soluble rosins in the seed lac are dissolved, and with them traces of coloring matter. At the same time the vegetable wax, which is only slightly soluble, is deposited. The shellac solution which has exchanged its vegetable wax for rosin is not yet suitable for fine furniture polishing. It is not sufficiently taken up by the wood, and an essential oil must be added to give it the necessary properties, one of the best oils to employ for this purpose being that of rosemary. The following recipe is given:

II.—Twenty pounds of shellac and 4 pounds of benzoin are dissolved in the smallest possible quantity of alcohol, together with 1 pound of rosemary oil. The solution then obtained is filtered through seed lac so as to remove whatever vegetable wax may be present.

Red Furniture Paste.—

Soft water........... 6 pints
Turpentine......... 6 pints

Beeswax	3	pounds
White wax	1½	ounces
White soap	18	ounces
Red lead	12	ounces

Cut up soap and dissolve in water by aid of heat; then evaporate to 6 pounds. Melt the waxes and add turpentine in which red lead has been stirred, pour into this the soap solution, and stir until it is nearly cold. If a darker color is wanted add more red lead, 4 to 6 ounces.

Beechwood Furniture.—The wood of the red beech is known to acquire, by the use of ordinary shellac polish, a dirty yellow color, and by the use of white polish, prepared from bleached shellac, an unsightly gray-white color. Therefore, where light colors are desired, only filtered shellac polish should be employed, and in order to impart some fire to the naturally dull color of the beechwood the admixture of a solution of dragon's blood in alcohol for a red shade, or turmeric in alcohol for yellow may be used. A compound of the red and yellow liquids gives a good orange shade. A few trials will soon show how much coloring matter may be added to the polish.

Polishes for Glass.—I.—Mix calcined magnesia with purified benzine to a semi-liquid paste. Rub the glass with this mixture by means of a cotton wad, until it is bright.

II.—Crush to powder cologne chalk, 60 parts, by weight; tripoli, 30 parts, by weight; bole, 15 parts, by weight. For use moisten the glass a little, dip a linen rag into the powder and rub the glass until it is clean.

III.—Tin ashes may be employed with advantage. The glass is rubbed with this substance and then washed off with a piece of soft felt. In this manner a very handsome polish is obtained.

Polishes for Ivory, Bone, etc.—I.—First rub with a piece of linen soaked with a paste made of Armenian bole and oleic acid. Wash with Marseilles soap, dry, rub with a chamois skin, and finally render it bright with an old piece of silk. If the ivory is scratched, it may be smoothed by means of English red stuff on a cloth, or even with a piece of glass if the scratches are rather deep. In the hollow parts of ivory objects the paste can be made to penetrate by means of an old toothbrush.

II.—Tortoise-shell articles have a way of getting dull and dingy looking. To repolish dip the finger in linseed oil and rub over the whole surface. Very little oil should be used, and if the article is a patterned one it may be necessary to use a soft brush to get it into the crevices. Then rub with the palm of the hand until all oil has disappeared, and the shell feels hot and looks bright and shiny.

Marble Polishing.—Polishing includes five operations. Smoothing the roughness left on the surface is done by rubbing the marble with a piece of moist sandstone; for moldings either wooden or iron mullers are used, crushed, and wet sandstone, or sand, more or less fine, according to the degree of polish required, being thrown under them. The second process is continued rubbing with pieces of pottery without enamel, which have only been baked once, also wet. If a brilliant polish is required, Gothland stone instead of pottery is used, and potter's clay or fuller's earth is placed beneath the muller. This operation is performed upon granites and porphyry with emery and a lead muller, the upper part of which is incrusted with the mixture until reduced by friction to clay or impalpable powder. As the polish depends almost entirely upon these two operations, care must be taken that they are performed with a regular and steady movement. When the marble has received the first polish, the flaws, cavities, and soft spots are sought out and filled with mastic of a suitable color.

This mastic is usually composed of a mixture of yellow wax, rosin, and Burgundy pitch, mixed with a little sulphur and plaster passed through a fine sieve, which gives it the consistency of a thick paste; to color this paste to a tone analogous to the ground tints or natural cement of the material upon which it is placed, lampblack and rouge, with a little of the prevailing color of the material, are added. For green and red marbles, this mastic is sometimes made of gum lac, mixed with Spanish sealing wax of the color of the marble. It is applied with pincers, and these parts are polished with the rest. Sometimes crushed fragments of marble are introduced into the cement, but for fine marbles the same colors are employed which are used in painting, and which will produce the same tone as the ground; the gum lac is added to give it body and brilliancy.

The third operation in polishing consists in rubbing it again with a hard pumice stone, under which water is being constantly poured, unmixed with sand. For the fourth process, called softening the ground, lead filings are mixed with the emery mud produced by the polishing of mirrors or the working of precious stones, and the marble is rubbed by a compact linen cushion well saturated with this mixture; rouge is also used for this polish. For some outside works, and for hearths and paving tiles, marble workers confine themselves to this polish. When the marbles have holes or grains, a lead muller is substituted for the linen cushion. In order to give a perfect brilliancy to the polish, the gloss is applied. Wash well the prepared surfaces and leave them until perfectly dry, then take a linen cushion, moistened only with water, and a little powder of calcined tin of the first quality. After rubbing with this for some time take another cushion of dry rags, rub with it lightly, brush away any foreign substance which might scratch the marble, and a perfect polish will be obtained. A little alum mixed with the water used penetrates the pores of the marble, and gives it a speedier polish. This polish spots very easily and is soon tarnished and destroyed by dampness. It is necessary when purchasing articles of polished marbles to subject them to the test of water; if there is too much alum, the marble absorbs the water and a whitish spot is left.

POLISHING POWDERS.

Polishing powders are advantageously prepared according to the following recipes:

I.—Four pounds magnesium carbonate, 4 pounds chalk, and 4 pounds rouge are intimately mixed.

II.—Four pounds magnesium carbonate are mixed with $\frac{1}{4}$ pound fine rouge.

III.—Five pounds fine levigated whiting and 2 pounds Venetian red are ground together.

IV.—Kieselguhr......... 42 pounds
Putty powder...... 14 pounds
Pipe clay.......... 14 pounds
Tartaric acid....... 1$\frac{1}{2}$ pounds

Powder the acid, mix well with the others. This is styled "free from mercury, poisonous mineral acids, alkalies, or grit." It may be tinted with 12 ounces of oxide of iron if desired.

Liquid Polishes.—

I.—Malt vinegar....... 4 gallons
Lemon juice....... 1 gallon
Paraffine oil........ 1 gallon
Kieselguhr......... 7 pounds
Powdered bath brick 3 pounds
Oil lemon.......... 2 ounces

II.—Kieselguhr......... 56 pounds
Paraffine oil........ 3 gallons
Methylated spirit... 1$\frac{1}{2}$ gallons
Camphorated spirit. $\frac{1}{2}$ gallon
Turpentine oil...... $\frac{1}{4}$ gallon
Liquid ammonia fort............. 3 pints

III.—Rotten stone..... 16 av. ounces
Paraffine........ 8 av. ounces
Kerosene (coal oil) 16 fluidounces
Oil of mirbane enough to perfume.

Melt the paraffine, incorporate the rotten stone, add the kerosene, and the oil of mirbane when cold.

IV.—Oxalic acid ...... $\frac{1}{2}$ av. ounce
Rotten stone..... 10 av. ounces
Kerosene (coal oil) 30 fluidounces
Paraffine....... 2 av. ounces

Pulverize the oxalic acid and mix it with rotten stone; melt the paraffine, add to it the kerosene, and incorporate the powder; when cool, add oil of mirbane or lavender to perfume.

Pour the ammonia into the oil, methylated spirits, and turpentine, add the camphorated spirit and mix with the kieselguhr. To prevent setting, keep well agitated during filling. The color may be turned red by using a little sesquioxide of iron and less kieselguhr. Apply with a cloth, and when dry use another clean cloth or a brush.

Polishing Soaps.—

I.—Powdered pipe clay 112 pounds
Tallow soap....... 16 pounds
Tartaric acid...... 1$\frac{1}{4}$ pounds

Grind until pasty, afterwards press into blocks by the machine.

II.—Levigated flint..... 60 pounds
Whiting........... 52 pounds
Tallow............ 20 pounds
Caustic soda....... 5 pounds
Water............. 2 gallons

Dissolve the soda in water and add to the tallow; when saponified, stir in the others, pressing as before.

III.—Saponified cocoanut oil.............. 56 pounds
Kieselguhr......... 12 pounds
Alum............. 5$\frac{1}{2}$ pounds
Flake white........ 5$\frac{1}{2}$ pounds
Tartaric acid....... 1$\frac{3}{4}$ pounds

Make as before.

IV.—Tallow soap........ 98 pounds
Liquid glycerine soap............. 14 pounds
Whiting............ 18 pounds
Levigated flint..... 14 pounds
Powdered pipe clay. 14 pounds

METAL POLISHES:

Polishing Pastes.—

I.—White petroleum
jelly............ 90 pounds
Kieselguhr......... 30 pounds
Refined paraffine
wax............. 10 pounds
Refined chalk or
whiting.......... 10 pounds
Sodium hyposulphite 8 pounds

Melt wax and jelly, stir in others and grind.

It is an undecided point as to whether a scented paste is better than one without perfume. The latter is added merely to hide the nasty smell of some of the greases used, and it is not very nice to have spoons, etc., smelling, even tasting, of mirbane, so perhaps citronelle is best for this purpose. It is likely to be more pure. The dose of scent is usually at the rate of 4 ounces to the hundredweight.

II.—Dehydrated soda.. 5 parts
Curd soap........ 20 parts
Emery flour...... 100 parts

To be stirred together on a water bath with water, 100 parts, until soft.

III.—Turpentine....... 1 part
Emery flour...... 1 part
Paris red......... 2 parts
Vaseline......... 2 parts

Mix well and perfume.

IV.—Stearine........ 8 to 9 parts
Mutton suet..... 32 to 38 parts
Stearine oil...... 2 to 2.5 parts

Melt together and mix with Vienna chalk, in fine powder, 48 to 60 parts; Paris red, 20 parts.

V.—Rotten stone....... 1 part
Iron subcarbonate.. 3 parts
Lard oil, a sufficient quantity.

VI.—Iron oxide........ 10 parts
Pumice stone...... 32 parts
Oleic acid, a sufficient quantity.

VII.—Soap, cut fine...... 16 parts
Precipitated chalk.. 2 parts
Jewelers' rouge.... 1 part
Cream of tartar.... 1 part
Magnesium carbonate............ 1 part
Water, a sufficient quantity.

Dissolve the soap in the smallest quantity of water over a water bath. Add the other ingredients to the solution while still hot, stirring all the time to make sure of complete homogeneity. Pour the mass into a box with shallow sides, and afterwards cut into cubes.

Non-Explosive Liquid Metal Polish.—Although in a liquid form, it does not necessarily follow that a liquid polish is less economical than pastes, because the efficiency of both is dependent upon the amount of stearic or oleic acid they contain, and a liquid such as that given below is as rich in this respect as most of the pastes, especially those containing much mineral jelly and earthy matters which are practically inert, and can only be considered as filling material. Thus it is a fact that an ounce of fluid polish may possess more polishing potency than an equal weight of the paste. Proportions are: Sixteen pounds crude oleic acid; 4 pounds tasteless mineral oil; 5 pounds kieselguhr; $1\frac{1}{2}$ ounces lemon oil. Make the earthy matter into a paste with the mixed fluids and gradually thin out, avoiding lumps. Apply with one rag, and finish with another.

Miscellaneous Metal Polishes.—I.—Articles of polished copper, such as clocks, stove ornaments, etc., become tarnished very quickly. To restore their brilliancy dip a brush in strong vinegar and brush the objects to be cleaned. Next pass through water and dry in sawdust. A soap water, in which some carbonate of soda has been dissolved, will do the same service.

II.—This is recommended for machinery by the chemical laboratory of the industrial museum of Batavia:

Oil of turpentine..... 15 parts
Oil of stearine....... 25 parts
Jewelers' red........ 25 parts
Animal charcoal, of
superior quality.... 45 parts

Alcohol is added to that mixture in such a quantity as to render it almost liquid, then by means of a brush it is put on those parts that are to be polished. When the alcohol has dried, the remaining cover is rubbed with a mixture of 45 parts of animal charcoal and 25 parts jewelers' red. The rubbed parts will become quite clean and bright.

III.—The ugly spots which frequently show themselves on nickel-plated objects may be easily removed with a mixture of 1 part sulphuric acid and 50 parts alcohol. Coat the spots with this solution, wipe off after a few seconds, rinse off thoroughly with clean water, and rub dry with sawdust.

IV.—Crocus, dried and powdered, when applied with chamois leather to nickel-plated goods, will restore their brilliancy without injuring their surface.

V.—Articles of tin should be ground

and polished with Vienna lime or Spanish white. The former may be spread on linen rags, the latter on wash leather. Good results may be obtained by a mixture of about equal parts of Vienna lime, chalk, and tripoli. It should be moistened with alcohol, and applied with a brush. Subsequent rubbing with roe skin (chamois) will produce a first-rate polish. Tin being a soft metal, the above polishing substances may be very fine.

VI.—To polish watch cases, take two glasses with large openings, preferably two preserving jars with ground glass covers. Into one of the glass vessels pour 1 part of spirit of sal ammoniac and 3 parts water, adding a little ordinary barrel soap and stirring everything well. Fill the other glass one half with alcohol. Now lay the case to be cleaned, with springs and all, into the first-named liquid and allow to remain therein for about 10 to 20 seconds. After protracted use this time may be extended to several minutes. Now remove the case, quickly brush it with water and soap and lay for a moment into the alcohol in the second vessel. After drying off with a clean cloth heat over a soldering flame for quick drying and the case will now look almost as clean and neat as a new one. The only thing that may occur is that a polished metal dome may become tarnished, but this will only happen if either the mixture is too strong or the case remains in it too long, both of which can be easily avoided with a little practice. Shake before using.

VII.—This is a cleanser as well as polisher:

    Prepared chalk......... 2 parts
    Water of ammonia...... 2 parts
    Water sufficient to make. 8 parts

The ammonia saponifies the grease usually present.

It must be pointed out that the alkali present makes this preparation somewhat undesirable to handle, as it will affect the skin if allowed too free contact.

The density of the liquid might be increased by the addition of soap; the solid would, of course, then remain longer in suspension.

VIII.—Serviettes Magiques.—These fabrics for polishing articles of metal consist of pure wool saturated with soap and tripoli, and dyed with a little coralline. They are produced by dissolving 4 parts of Marseilles soap in 20 parts of water, adding 2 parts of tripoli and saturating a piece of cloth 3 inches long and 4 inches wide with it, allowing to dry.

IX.—In order to easily produce a mat polish on small steel articles use fine powdered oil stone, ground with turpentine.

Polishes for Pianos.—

I.—Alcohol, 95 per cent.. 300 parts
    Benzol............. 700 parts
    Gum benzoin....... 8 parts
    Sandarac........... 16 parts

Mix and dissolve. Use as French polish.

II.—Beeswax.......... 2,500 parts
    Potassium carbon-
        ate............. 25 parts
    Oil of turpentine.... 4,000 parts
    Water, rain or dis-
        tilled............ 4,500 parts

Dissolve the potassium carbonate in 1,500 parts of the water and in the solution boil the wax, shaved up, until the latter is partially saponified, replacing the water as it is driven off by evaporation. When this occurs remove from the fire and stir until cold. Now add the turpentine little by little, and under constant agitation, stirring until a smooth, homogeneous emulsion is formed. When this occurs add the remainder of the water under constant stirring. If a color is wanted use alkanet root, letting it macerate in the oil of turpentine before using the latter (about an ounce to the quart is sufficient). This preparation is said to be one of the best polishes known. The directions are very simple: First wash the surface to be polished, rinse, and dry. Apply the paste as evenly and thinly as possible over a portion of the surface, then rub off well with a soft woolen cloth.

Polishes for Silverware.—The best polish for silverware—that is, the polish that, while it cleans, does not too rapidly abrade the surface—is levigated chalk, either alone or with some vegetable acid, like tartaric, or with alum. The usual metal polishes, such as tripoli (diatomaceous earth), finely ground pumice stone, etc., cut away the surface so rapidly that a few cleanings wear through ordinary plating.

I.—White lead........ 5 parts
    Chalk, levigated.... 20 parts
    Magnesium carbon-
        ate .... 2 parts
    Aluminum oxide.... 5 parts
    Silica............. 3 parts
    Jewelers' rouge..... 2 parts

Each of the ingredients must be reduced to an impalpable powder, mixed carefully, and sifted through silk several times to secure a perfect mixture, and to avoid any possibility of leaving in the powder anything that might scratch the silver or gold surface. This may be left in the powder form, or incorporated with soap, made into a paste with glycerine, or other similar material. The objection to mixtures with vaseline or greasy substances is that after cleaning the object must be scrubbed with soap and water, while with glycerine simple rinsing and running water instantly cleans the object. The following is also a good formula:

II.—Chalk, levigated.... 2 parts
Oil of turpentine.... 4 parts
Stronger ammonia water........... 4 parts
Water............. 10 parts

Mix the ammonia and oil of turpentine by agitation, and rub up the chalk in the mixture. Finally rub in the water gradually or mix by agitation. Three parts each of powdered tartaric acid and chalk with 1 part of powdered alum make a cheap and quick silver cleaning powder.

III.—Mix 2 parts of beechwood ashes with $\frac{1}{100}$ of a part of Venetian soap and 2 parts of common salt in 8 parts of rain water. Brush the silver with this, using a pretty stiff brush. A solution of crystallized permanganate of potash is often recommended, or even the spirits of hartshorn, for removing the grayish violet film which forms upon the surface of the silver. Finally, when there are well-determined blemishes upon the surface of the silver, they may be soaked 4 hours in soapmakers' lye, then cover them with finely powdered gypsum which has been previously moistened with vinegar, drying well before a fire; now rub them with something to remove the powder. Finally, they are to be rubbed again with very dry bran.

POLISHES FOR STEEL AND IRON.

The polishing of steel must always be preceded by a thorough smoothing, either with oilstone dust, fine emery, or coarse rouge. If any lines are left to be erased by means of fine rouge, the operation becomes tedious and is rarely successful. The oilstone dust is applied on an iron or copper polisher. When it is desired to preserve the angles sharp, at a shoulder, for instance, the polisher should be of steel. When using diamantine an iron polisher, drawn out and flattened with a hammer, answers very well. With fine rouge, a bronze or bellmetal polisher is preferable for shoulders; and for flat surfaces, discs or large

zinc or tin polishers, although glass is preferable to either of these. After each operation with oilstone dust, coarse rouge, etc., the polisher, cork, etc., must be changed, and the object should be cleaned well, preferably by soaping, perfect cleanliness being essential to success. Fine rouge or diamantine should be made into a thick paste with oil; a little is then taken on the polisher or glass and worked until quite dry. As the object is thus not smeared over, a black polish is more readily obtained, and the process gets on better if the surface be cleaned from time to time.

For Fine Steel.—Take equal parts (by weight) of ferrous sulphate—green vitriol—and sodium chloride—cooking salt—mix both well together by grinding in a mortar and subject the mixture to red heat in a mortar or a dish. Strong fumes will develop, and the mass begin to flow. When no more fumes arise, the vessel is removed from the fire and allowed to cool. A brown substance is obtained with shimmering scales, resembling mica. The mass is now treated with water, partly in order to remove the soluble salt, partly in order to wash out the lighter portions of the non-crystallized oxide, which yield an excellent polishing powder. The fire must be neither too strong nor too long continued, otherwise the powder turns black and very hard, losing its good qualities. The more distinct the violet-brown color, the better is the powder.

For polishing and cleaning fenders, fireirons, horses' bits, and similar articles: Fifty-six pounds Bridgewater stone; 28 pounds flour emery; 20 pounds rotten stone; 8 pounds whiting. Grind and mix well.

To make iron take a bright polish like steel, pulverize and dissolve in 1 quart of hot water, 1 ounce of blue vitriol; 1 ounce of borax; 1 ounce of prussiate of potash; 1 ounce of charcoal; $\frac{1}{2}$ pint of salt, all of which is to be added to one gallon of linseed oil and thoroughly mixed. To apply, bring the iron or steel to the proper heat and cool in the solution.

Stove Polish.—The following makes an excellent graphite polish:

I.—Ceresine.......... 12 parts
Japan wax........ 10 parts
Turpentine oil..... 100 parts
Lampblack, best... 12 parts
Graphite, levigated 10 parts

Melt the ceresine and wax together, remove from the fire, and when half cooled off add and stir in the graphite and lampblack, previously mixed with the turpentine.

II.—Ceresine......... 23 parts
Carnauba wax.... 5 parts
Turpentine oil..... 220 parts
Lampblack....... 300 parts
Graphite, finest
levigated....... 25 parts

Mix as above.

III.—Make a mixture of water glass and lampblack of about the consistency of thin syrup, and another of finely levigated plumbago and mucilage of Soudan gum (or other cheap substitute for gum arabic), of a similar consistency. After getting rid of dust, etc., go over the stove with mixture No. I and let it dry on, which it will do in about 24 hours. Now go over the stove with the second mixture, a portion of the surface at a time, and as this dries, with an old blacking brush give it a polish. If carefully done the stove will have a polish resembling closely that of new Russian iron. A variant of this formula is as follows: Mix the graphite with the water glass to a smooth paste; add, for each pound of paste, 1 ounce of glycerine and a few grains of aniline black. Apply to the stove with a stiff brush.

POLISHES FOR WOOD:

See also Polishes for Furniture, Floors and Pianos.

In the usual method of French polishing, the pad must be applied along curved lines, and with very slight pressure, if the result is to be uniform. To do this requires much practice and the work is necessarily slow. Another disadvantage is that the oil is apt to sweat out afterwards, necessitating further treatment. According to a German patent all difficulty can be avoided by placing between the rubber and its covering a powder composed of clay or loam, or better, the powder obtained by grinding fragments of terra cotta or of yellow bricks. The powder is moistened with oil for use. The rubber will then give a fine polish, without any special delicacy of manipulation and with mere backward and forward rubbing in straight lines, and the oil will not sweat out subsequently. Another advantage is that no priming is wanted, as the powder fills up the pores. The presence of the powder also makes the polish adhere more firmly to the wood.

Oak Wood Polish.—The wood is first carefully smoothed, then painted with the following rather thickly liquid mass, using a brush, viz.: Mix 1½ parts, by weight, of finely washed chalk (whiting), ½ part of dryer, and 1 part of boiled linseed oil with benzine and tint (umber with a little lampblack, burnt sienna). After the applied mixture has become dry, rub it down, polish with glass powder, and once more coat with the same mixture. After this filling and after rubbing off with stickwood chips or fine sea grass, one or two coats of shellac are put on (white shellac with wood alcohol for oak, brown shellac for cherry and walnut). This coating is cut down with sandpaper and given a coat of varnish, either polishing varnish, which is polished off with the ball of the hand or a soft brush, or with interior varnish, which is rubbed down with oil and pumice stone. This polish is glass hard, transparent, of finer luster, and resistive.

Hard Wood Polish.—In finishing hard wood with a wax polish the wood is first coated with a "filler," which is omitted in the case of soft wood. The filler is made from some hard substance, very finely ground; sand is used by some manufactures.

The polish is the same as for soft wood. The simplest method of applying wax is by a heated iron, scraping off the surplus, and then rubbing with a cloth. It is evident that this method is especially laborious; and for that reason solution of the wax is desirable. It may be dissolved rather freely in turpentine spirit, and is said to be soluble also in kerosene oil.

The following recipes give varnish-like polishes:

I.—Dissolve 15 parts of shellac and 15 parts of sandarac in 180 parts of spirit of wine. Of this liquid put some on a ball of cloth waste and cover with white linen moistened with raw linseed oil. The wood to be polished is rubbed with this by the well-known circular motion. When the wood has absorbed sufficient polish, a little spirit of wine is added to the polish, and the rubbing is continued. The polished articles are said to sustain no damage by water, nor show spots or cracks.

II.—Orange shellac, 3 parts; sandarac, 1 part; dissolved in 30 parts of alcohol. For mahogany add a little dragon's blood.

III.—Fifteen parts of oil of turpentine, dyed with anchusine, or undyed, and 4 parts of scraped yellow wax are stirred into a uniform mass by heating on the water bath.

IV.—Melt 1 part of white wax on the water bath, and add 8 parts of petroleum. The mixture is applied hot. The petroleum evaporates and leaves behind a thin layer of wax, which is subsequently rubbed out lightly with a dry cloth rag.

V.—
Stearine	100 parts
Yellow wax	25 parts
Caustic potash	60 parts
Yellow laundry soap	10 parts
Water, a sufficient quantity.	

Heat together until a homogeneous mixture is formed.

VI.—
Yellow wax	25 parts
Yellow laundry soap	6 parts
Glue	12 parts
Soda ash	25 parts
Water, a sufficient quantity.	

Dissolve the soda in 400 parts of water, add the wax, and boil down to 250 parts, then add the soap. Dissolve the glue in 100 parts of hot water, and mix the whole with the saponified wax.

VII.—This is waterproof. Put into a stoppered bottle 1 pint alcohol; 2 ounces gum benzoin; ¼ ounce gum sandarac, and ¼ ounce gum anime. Put the bottle in a sand bath or in hot water till the solids are dissolved, then strain the solution, and add ¼ gill best clear poppy oil. Shake well and the polish is ready for use.

VIII.—A white polish for wood is made as follows:

White lac	1½ pounds
Powdered borax	1 ounce
Alcohol	3 pints

The lac should be thoroughly dried, especially if it has been kept under water, and, in any case, after being crushed, it should be left in a warm place for a few hours, in order to remove every trace of moisture. The crushed lac and borax are then added to the spirit, and the mixture is stirred frequently until solution is effected, after which the polish should be strained through muslin.

IX.—To restore the gloss of polished wood which has sweated, prepare a mixture of 100 parts of linseed oil, 750 parts of ether, 1,000 parts of rectified oil of turpentine, and 1,000 parts of petroleum benzine, perfumed, if desired, with a strongly odorous essential oil, and colored, if required, with cuicuma, orlean, or alkanna. The objects to be treated are rubbed thoroughly with this mixture, using a woolen rag.

MISCELLANEOUS POLISHING AGENTS:

Polishing Agent which may also be used for Gilding and Silvering.—The following mediums hitherto known as possessing the aforenamed properties, lose these qualities upon having been kept for some time, as the metal salt is partly reduced. Furthermore, it has not been possible to admix reducing substances such as zinc to these former polishing agents, since moisture causes the metal to precipitate. The present invention obviates these evils. The silver or gold salt is mixed with chalk, for instance, in a dry form. To this mixture, fine dry powders of one or more salts (e. g., ammonia compounds) in whose solutions the metal salt can enter are added; if required, a reducing body, such as zinc, may be added at the same time. The composition is pressed firmly together and forms briquettes, in which condition the mass keeps well. For use, all that is necessary is to scrape off a little of the substance and to prepare it with water.

Silver Polishing Balls.—This polishing agent is a powder made into balls by means of a binding medium and enjoys much popularity in Germany. It is prepared by adding 5 parts of levigated chalk to 2 parts of yellow tripoli, mixing the two powders well and making into a stiff paste with very weak gum water —1 part gum arabic to 12 parts of water. This dough is finally shaped by hand into balls of the size of a pigeon's egg. The balls are put aside to dry on boards in a moderately warm room, and when completely hard are wrapped in tin-foil paper.

To Prepare Polishing Cloths.—The stuff must be pure woolen, colored with aniline red, and then put in the following:

Castile soap, white	4 parts
Jewelers' red	2 parts
Water	20 parts

Mix. One ounce of this mixture will answer for a cloth 12 inches square, where several of them are saturated at the same time. For the workshop, a bit of chamois skin of the same size (a foot square), is preferable to wool, on account of its durability. After impregnation with the soap solution, it should be dried in the air, being manipulated while drying to preserve its softness and suppleness.

To Polish Delicate Objects.—Rub the objects with a sponge charged with a mixture of 28 parts of alcohol, 14 parts of water, and 4 parts of lavender oil.

Polish for Gilt Frames.—Mix and beat the whites of 3 eggs with one-third, by weight, of javelle water, and apply to the gilt work.

Steel Dust as a Polishing Agent.—Steel dust is well adapted for polishing precious stones and can replace emery with advantage. It is obtained by spraying water on a bar of steel brought to a high temperature. The metal becomes friable and can be readily reduced to powder in a mortar. This powder is distinguished from emery by its mordanting properties and its lower price. Besides, it produces a finer, and consequently, a more durable polish.

Polishing Bricks.—Stir into a thick pulp with water 10 parts of finely powdered and washed chalk, 1 part of English red, and 2 parts of powdered gypsum; give it a square shape and dry.

Polishing Cream.—

Denaturized alcohol	400 parts
Spirit of sal ammoniac	75 parts
Water	150 parts
Petroleum ether	80 parts
Infusorial earth	100 parts
Red bole or white bole	50 parts
Calcium carbonate	100 parts

Add as much of the powders as desired. Mirbane oil may be used for scenting.

Polishing Paste.—

Infusorial earth (Kieselguhr)	8 ounces
Paraffine	2 ounces
Lubricating oil	6 fluidounces
Oleic acid	1 fluidounce
Oil mirbane	30 minims

Melt the paraffine with the lubricating oil, and mix with the infusorial earth, then add the oleic acid and oil of mirbane.

To Polish Paintings on Wood.—According to the statements of able cabinet makers who frequently had occasion to cover decorations on wood, especially aquarelle painting, with a polish, a good coating of fine white varnish is the first necessity, dammar varnish being employed for this purpose. This coat is primarily necessary as a protective layer so as to preserve the painted work from destructive attacks during the rubbing for the production of a smooth surface and the subsequent polishing. At all events, the purest white polishing varnish must be used for the polish so as to prevent a perceptible subsequent darkening of the white painting colors. Naturally the success here is also dependent upon the skill of the polisher. To polish painting executed on wood it is necessary to choose a white, dense, fine grained wood, which must present a well-smoothed surface before the painting. After the painting the surface is faintly coated with a fine, quickly drying, limpid varnish. When the coating has dried well, it is carefully rubbed down with finely pulverized pumice stone, with tallow or white lard, and now this surface is polished in the usual manner with a good solution prepared from the best white shellac.

Polishing Mediums.—For iron and steel, stannic oxide or Vienna lime or iron oxide and sometimes steel powder is employed. In using the burnisher, first oil is taken, then soap water, and next Vienna lime.

For copper, brass, German silver, and tombac, stearine oil and Vienna lime are used. Articles of brass can be polished, after the pickling, in the lathe with employment of a polish consisting of shellac, dissolved in alcohol, 1,000 parts; powdered turmeric, 1,000 parts; tartar, 2,000 parts; ox gall, 50 parts; water, 3,000 parts.

Gold is polished with ferric oxide (red stuff), which, moistened with alcohol, is applied to leather.

For polishing silver, the burnisher or bloodstone is employed, using soap water, thin beer, or a decoction of soap wort. Silver-plated articles are also polished with Vienna lime.

To produce a dull luster on gold and silver ware, glass brushes, i. e., scratch brushes of finely spun glass threads, are made use of.

Pewter articles are polished with Vienna lime or whiting; the former on a linen rag, the latter on leather.

If embossed articles are to be polished, use the burnisher, and for polish, soap water, soap-wort decoction, ox gall with water.

Antimony-lead alloys are polished with burnt magnesia on soft leather or with fine jewelers' red.

Zinc is brightened with Vienna lime or powdered charcoal.

Vienna lime gives a light-colored polish on brass, while ferric oxide imparts a dark luster.

Rouge or Paris Red.—This appears in commerce in many shades, varying from brick red to chocolate brown. The color, however, is in no wise indicative of its purity or good quality, but it can be accepted as a criterion by which to determine the hardness of the powder. The darker the powder, the greater is its degree of hardness; the red or reddish is always very soft, wherefore the former is used for polishing steel and the latter for softer metals.

For the most part, Paris red consists of ferric oxide or ferrous oxide. In its production advantage is taken of a peculiarity common to most salts of iron, that when heated to a red heat they separate the iron oxide from the acid combination. In its manufacture it is usual to take commercial green vitriol, copperas crystals, and subject them to a moderate heat to drive off the water of crystallization. When this is nearly accomplished they will settle down in a white powder, which is now placed in a crucible and raised to a glowing red heat till no more vapor arises, when the residue will be found a soft smooth red powder. As the temperature is raised in the crucible, the darker will become the color of the powder and the harder the abrasive.

Should an especially pure rouge be desired, it may be made so by boiling the powder we have just made in a weak solution of soda and afterwards washing it out repeatedly and thoroughly with clean water. If treated in this way, all the impurities that may chance to stick to the iron oxide will be separated from it.

Should a rouge be needed to put a specially brilliant polish upon any object its manufacture ought to be conducted according to the following formula: Dissolve commercial green vitriol in water; dissolve also a like weight of sorrel salt in water; filter both solutions; mix them well, and warm to 140° F.; a yellow precipitate, which on account of its weight, will settle immediately; decant the fluid, dry out the residue, and afterwards heat it as before in an iron dish in a moderately hot furnace till it glows red.

By this process an exceptionally smooth, deep-red powder is obtained, which, if proper care has been exercised in the various steps, will need no elutriation, but can be used for polishing at once. With powders prepared in this wise our optical glasses and lenses of finest quality are polished.

POLISHES FOR THE LAUNDRY:
See Laundry Preparations.

POMADE, PUTZ:
See Cleaning Preparations and Methods.

POMADES:
See Cosmetics.

POMEGRANATE ESSENCE:
See Essences and Extracts.

PORCELAIN:
See also Ceramics.

Mending Porcelain by Riveting (see Adhesives for methods of mending Porcelain by means of cements).—Porcelain and glass can be readily pierced with steel tools. Best suited are hardened drills of ordinary shape, moistened with oil of turpentine, if the glazed or vitreous body is to be pierced. In the case of majolica and glass without enamel the purpose is best reached if the drilling is done under water. Thus, the vessel should previously be filled with water, and placed in a receptacle containing water, so that the drill is used under water, and, after piercing the clay body, reaches the water again. In the case of objects glazed on the inside, instead of filling them with water, the spot where the drill must come through may be underlaid with cork. The pressure with which the drill is worked is determined by the hardness of the material, but when the tool is about to reach the other side it should gradually decrease and finally cease almost altogether, so as to avoid chipping. In order to enlarge small bore holes already existing, three-cornered or four-square broaches, ground and polished, are best adapted. These are likewise employed under water or, if the material is too hard (glass or enamel), moistened with oil of turpentine. The simultaneous use of oil of turpentine and water is most advisable in all cases, even where the nature of the article to be pierced does not admit the use of oil alone, as in the case of majolica and non-glazed porcelain, which absorb the oil, without the use of water.

Porcelain Decoration.—A brilliant yellow color, known as "gold luster," may be produced on porcelain by the use of paint prepared as follows: Melt over a sand bath 30 parts of rosin, add 10 parts of uranic nitrate, and, while constantly stirring, incorporate with the liquid 35 to 40 parts of oil of lavender. After the mixture has become entirely homogeneous, remove the source of heat, and add 30 to 40 parts more of oil of lavender. Intimately mix the mass thus obtained with a like quantity of bismuth glass prepared by fusing together equal parts of oxide of bismuth and crystallized boric acid. The paint is to be burned in in the usual manner.

PORCELAIN, HOW TO TELL POTTERY AND PORCELAIN:
See Ceramics.

PORTLAND CEMENT:
See Cement.

PORTLAND CEMENT, SIZE OVER:
See Adhesives.

POSTAL CARDS, HOW TO MAKE SENSITIZED:
See Photography, under Paper-Sensitizing Processes.

POTASSIUM SILICATE AS A CEMENT:
See Adhesives, under Water-Glass Cements.

POTATO STARCH:
See Starch.

POTTERY:
See Ceramics.

POULTRY APPLICATIONS:
See Insecticides.

POULTRY FOODS AND POULTRY DISEASES AND THEIR REMEDIES:
See Veterinary Formulas.

POULTRY WINE:
See Wines and Liquors.

POUNCE:
See Cleaning Preparations and Methods, under Ink Eradicators.

POWDER FOR COLORED FIRES:
See Pyrotechnics.

POWDER, FACE:
See Cosmetics.

POWDER, ROUP:
See Roup Powder.

POWDERS FOR STAMPING:
See Stamping.

POWDERS FOR THE TOILET:
See Cosmetics.

Preservatives

(See also Foods.)

Preservative Fluid for Museums.—

Formaldehyde solution	6 parts
Glycerine	12 parts
Alcohol	3 parts
Water	100 parts

The addition of glycerine becomes necessary only if it is desired to keep the pieces in a soft state. Filtering through animal charcoal renders the liquid perfectly colorless. For dense objects, such as lungs and liver, it is best to make incisions so as to facilitate the penetration of the fluid. In the case of very thick pieces, it is best to take 80 to 100 parts of formaldehyde solution for above quantities.

Preservative for Stone, etc.—A new composition, or paint, for protecting stone, wood, cement, etc., from the effects of damp or other deleterious influences consists of quicklime, chalk, mineral colors, turpentine, boiled oil, galipot, rosin, and benzine. The lime, chalk, colors, and turpentine are first fixed and then made into a paste with the boiled oil. The paste is finely ground and mixed with the rosins previously dissolved in the benzine.

Preservative for Stuffed Animals.—
For the exterior preservation use

Arsenic	0.7 parts
Alum	15.0 parts
Water	100.0 parts

For sprinkling the inside skin as well as filling bones, the following is employed:

Camphor	2 parts
Insect powder	2 parts
Black pepper	1 part
Flowers of sulphur	4 parts
Alum	3 parts
Calcined soda	3 parts
Tobacco powder	3 parts

Preservatives for Zoological and Anatomical Specimens.—The preparations are first placed in a solution or mixture of

Sodium fluoride	5 parts
Formaldehyde (40 per cent)	2 parts
Water	100 parts

After leaving this fixing liquid they are put in the following preservative solution:

Glycerine (28° Bé.)	5	parts
Water	10	parts
Magnesium chloride	1	part
Sodium fluoride	0.2	parts

In this liquid zoological preparations, especially reptiles, retain their natural coloring. Most anatomical preparations likewise remain unchanged therein.

PRESERVATIVES FOR WOOD:
See Wood.

Preserving

Canning.—There should be no trouble in having canned fruit keep well if perfect or "chemical cleanliness" is observed in regard to jars, lids, etc., and if the fruit or vegetables are in good order, not overripe or beginning to ferment where bruised or crushed. Fruit will never come out of jars better than it goes in. It is better to put up a little fruit at a time when it is just ripe than to wait for a large amount to ripen, when the first may be overripe and fermenting and likely to spoil the whole lot. Use only the finest flavored fruit.

Have everything ready before beginning canning. Put water in each jar, fit on rubbers and tops, and invert the jar on the table. If any water oozes out try another top and rubber until sure the jar is air-tight. Wash jars and tops, put them in cold water and bring to a boil. When the fruit is cooked ready take a jar from the boiling water, set it on a damp cloth laid in a soup plate, dip a rubber in boiling water, and fit it on firmly. Fill the jar to overflowing, wipe the brim, screw on the top, and turn it upside down on a table. If any syrup oozes out empty the jar back into the kettle and fit on a tighter rubber. Let it stand upside down till cold, wipe clean, wrap in thick paper, and keep in a cool, dry place.

These general directions are for all fruits and vegetables that are cooked before putting in the jars. Fruit keeps its shape better if cooked in the jars, which should be prepared as above, the fruit carefully looked over and filled into the jars. If a juicy fruit, like blackberries or raspberries, put the sugar in with it in alternate layers. For cherries the amount of sugar depends on the acidity of the fruit and is best made into a syrup with a little water and poured down through them. Peaches and pears after paring, are packed into the jars and a syrup of about a quarter of a pound of sugar to a pound of fruit poured over them. Most fruits need to be cooked from 10 to 15 minutes after the water around them begins to boil.

Red raspberries ought not to be boiled. Put them into jars as gently as possible; they are the tenderest of all fruits and will bear the slightest handling. Drop them in loosely, fold a saucer into a clean cloth, and lay over the top, set on a perforated board in a boiler, pour water to two-thirds, cover and set over a slow fire. As the fruit settles add more until full. When it is cooked soft lift the jar out and fill to the top with boiling syrup of equal parts of sugar and water, and seal.

Do not can all the fruit, for jams and jellies are a welcome change and also easier to keep. Raspberries and currants mixed make delicious jam. Use the juice of a third as many currants and ¾ of a pound of sugar to a pound of fruit.

The flavor of all kinds of fruit is injured by cooking it long with the sugar, so heat the latter in the oven and add when the fruit is nearly done.

Jelly is best made on a clear day, for small fruits absorb moisture, and if picked after a rain require longer boiling, and every minute of unnecessary boiling gives jelly a less delicate color and flavor. When jelly is syrupy, it has been boiled too long; if it drops from the spoon with a spring, or wrinkles as you push it with the spoon in a saucer while cooling, it is done enough. Try it after 5 minutes' boil. Cook the fruit only until the skin is broken and pulp softened. Strain without squeezing for jelly, and use the last juice you squeeze out for jam. Measure the juice and boil uncovered, skimming off. For sweet fruits ¾ of a pound of sugar is enough to a pint of juice. Heat the sugar in the oven, add to the boiling juice; stir till dissolved. When it boils up, draw to the back of the stove. Scald the jelly glasses, fill and let stand in a clean, cool place till next day; then cover. Blackberries make jelly of a delicious flavor and jelly easily when a little underripe. Currants should be barely ripe; the ends of the bunches may be rather green.

A highly prized way of canning cherries: Stone and let them stand overnight. In the morning pour off the juice, add sugar to taste, and some water if there is not much juice, and boil and skim till it is a rich syrup. If the cherries are sweet a pint of juice and ¾ of a pint of sugar will be right. Heat the jars, put in the uncooked cherries till they are nearly full; then pour over them the boiling syrup and fasten on the covers. Set the jars in a washboiler, fill it with very hot water and let it stand all night. The heat of the syrup and of the water will cook the fruit, but the flavor and color will be that of fresh and uncooked cherries.

Canning without Sugar.—I.—In order to preserve the juices of fruit merely by sterilization, put the juice into the bottles in which it is to be kept, filling them very nearly full; place the bottles, unstoppered, in a kettle filled with cold water, so arranging them on a wooden perforated "false bottom," or other like contrivance, as to prevent their immediate contact with the metal, thus preventing unequal heating and possible fracture. Now heat the water, gradually raising the temperature to the boiling point, and maintain at that until the juice attains a boiling temperature; then close the bottles with perfectly fitting corks, which have been kept immersed in boiling water for a short time before use. The corks should not be fastened in any way, for if the sterilization is not complete, fermentation and consequent explosion of the bottle might occur, unless the cork should be forced out. The addition of sugar is not necessary to secure the success of the operation; in fact a small proportion would have no antiseptic effect. If the juice is to be used for syrup as for use at the soda fountain, the best method is to make a concentrated syrup at once, using about 2 pounds of refined sugar to 1 pint of juice, dissolving by a gentle heat. The syrup may be made by simple agitation without heat and a finer flavor thus results, but its keeping quality would be uncertain.

II.—Fruit juices may be preserved by gentle heating and after protection from the air in sterilized containers. The heat required is much below the boiling point. Professor Müller finds that a temperature of from 140° to 158° F., maintained for 15 minutes, is sufficient to render the fermenting agents present inactive. The bottles must also be heated to destroy any adherent germs. The juices may be placed in them as expressed and the container then placed in a water bath. As soon as the heating is finished the bottles must be securely closed. The heating process will, in consequence of coagulating certain substances, produce turbidity, and if clear liquid is required, filtration is, of course, necessary. In this case it is better to heat the juice in bulk in a kettle, filter through felt, fill the bottles, and then heat again in the containers as in the first instance. It is said that grape juice prepared in this manner has been found unaltered after keeping for many years. Various antiseptics have been proposed as preservatives for fruit juices and other articles of food, but all such agents are objectionable both on account of their direct action on the system and their effect in rendering food less digestible. While small quantities of such drugs occasionally taken may exert no appreciable effect, continuous use is liable to be more or less harmful.

CRUSHED FRUIT PRESERVING:

Crushed Pineapples.—Secure a good brand of canned grated pineapple and drain off about one-half of the liquor by placing on a strainer. Add to each pound of pineapple 1 pound of granulated sugar. Place on the fire and bring to boiling point, stirring constantly. Just before removing from the fire, add to each gallon of pulp 1 ounce saturated alcoholic solution salicylic acid. Put into air-tight jars until wanted for use.

Crushed Peach.—Take a good brand of canned yellow peaches, drain off liquor, and rub through a No. 8 sieve. Add sugar, bring to the boiling point, and when ready to remove from fire add to each gallon 1 ounce saturated alcoholic solution of salicylic acid. Put into jars and seal hermetically.

Crushed Apricots.—Prepared in similar manner to crushed peach, using canned apricots.

Crushed Orange.—Secure oranges with a thin peel and containing plenty of juice. Remove the outer or yellow peel first, taking care not to include any of bitter peel. The outer peel may be used in making orange phosphate or tincture sweet orange peel. After removing the outer peel, remove the inner, bitter peel, quarter and remove the seeds. Extract part of the juice and grind the pulp through an ordinary meat grinder. Add sugar, place on the fire, and bring to the boiling point. When ready to remove, add to each gallon 1 ounce saturated alcoholic solution of salicylic acid and 1 ounce glycerine. Put into jars and seal.

Crushed Cherries.—If obtainable, the large, dark California cherry should be used. Stone the cherries, and grind to a pulp. Add sugar, and place on the fire, stirring constantly. Before removing, add to each gallon 1 ounce of the saturated solution of salicylic acid. Put into jars and seal.

Dry Sugar Preserving.—The fruits are embedded in a thick layer of dry, powdered sugar to which they give up the greater part of the water contained in them. At the same time, a quantity of sugar passes through the skins into the interior of the fruits. Afterwards, the fruits are washed once, wiped, and completely dried.

Fruit Preserving.—Express the juice and filter at once, through two thicknesses of best white Swedish paper, into a container that has been sterilized immediately before letting the juice run into it, by boiling water. The better plan is to take out of water in active ebullition at the moment you desire to use it. Have ready some long-necked, 8-ounce vials, which should also be kept in boiling water until needed. Pour the juice into these, leaving room in the upper part of the body of the vial to receive a teaspoonful of the best olive oil. Pour the latter in so that it will trickle down the neck and form a layer on top of the juice, and close the neck with a wad of antiseptic cotton thrust into it in such manner that it does not touch the oil, and leaves room for the cork to be put in without touching it. Cork and cap or seal the vial, and put in a cool, dark place, and keep standing upright. If carried out faithfully with due attention to cleanliness, this process will keep the juice in a perfectly natural condition for a very long time. The two essentials are the careful and rapid filtration, and the complete aseptization of the containers. Another process, in use in the French Navy, depends upon the rapid and careful filtering of the juice, and the addition of from 8 to 10 per cent of alcohol.

Raspberry Juice.—A dark juice is obtained by adding to the crushed raspberries, before the fermentation, slight quantities of sugar in layers. The ethyl-alcohol forming during the fermentation is said to cause a better extraction of the raspberry red. Furthermore, the boiling should not be conducted on a naked fire, but by means of superheated steam, so as to avoid formation of caramel. Finally, the sugar used should be perfectly free from ultramarine and lime, since both impurities detract from the red color of the raspberries.

Spice for Fruit Compote.—This is greatly in demand in neighborhoods where many plums and pears are preserved.

	Parts		Parts
Lemon peel	15	or	...
Cinnamon, ordinary	15	or	50
Star aniseed	10	or	15
Coriander	3	or	100
Carob pods	5	or	..
Ginger root, peeled	2	or	200
Pimento	..	or	100
Licorice	..	or	100
Cloves, without stems	..	or	30
Spanish peppers	..	or	2
Oil of lemon	..	or	4
Oil of cinnamon	..	or	2
Oil of cloves	..	or	2

All the solid constituents are powdered moderately fine and thoroughly mixed; the oils dropped in last, and rubbed into the powder.

Strawberries.—Carefully remove the stems and calyxes, place the strawberries on a sieve, and move the latter about in a tub of water for a few moments, to remove any dirt clinging to them. Drain and partially dry spontaneously, then remove from the sieve and put into a porcelain-lined kettle provided with a tight cover. To every pound of berries take a half pound of sugar and 2 ounces of water and put the same in a kettle over the fire. Let remain until the sugar has dissolved or become liquid, and then pour the same, while still hot, over the berries, cover the kettle tightly and let it stand overnight. The next morning put the kettle over the fire, removing the cover when the berries begin to boil, and let boil gently for 6 to 8 minutes (according to the mass), removing all scum as it arises. Remove from the fire, and with a perforated spoon or dipper take the fruit from the syrup, and fill into any suitable vessel. Replace the syrup on the fire and boil for about the same length of time as before, then pour, all hot, over the berries. The next day empty out the contents of the vessel on a sieve, and let the berries drain off; remove the syrup that drains off, add water, put on the fire, and boil until you obtain a syrup which flows but slowly from the stirring spoon. At this point add the berries, and let boil gently for a few moments. Have your preserve jars as hot as possible, by putting them into a pot of cold water and bringing the latter to a boil, and into them fill the berries, hot from the kettle. Cool down, cover with buttered paper, and immediately close the jars hermetically. If corks are used, they should be protected below with parchment paper, and afterwards covered with wet bladder stretched over the top, securely tied and waxed. The process seems very troublesome and tedious, but all of the care expended is repaid by the richness and pureness of the flavor of the preserve, which maintains the odor and taste of the fresh berry in perfection.

Hydrogen Peroxide as a Preservative.—Hydrogen peroxide is one of the best, least harmful, and most convenient agents for preserving syrups, wine, beer, cider, and vinegar. For this purpose $2\frac{1}{2}$ fluidrachms of the commercial peroxide of hydrogen may be added to each quart of the article to be preserved. Hydrogen peroxide also affords an easy test for bacteria in water. When hydrogen peroxide is added to water that contains bacteria, these organisms decompose it, and consequently oxygen gas is given off. If the water be much contaminated the disengagement of gas may be quite brisk.

To Preserve Milk (which should be as fresh as possible) there should be added enough hydrogen peroxide to cause it to be completely decomposed by the enzymes of the milk. For this purpose 1.3 per cent, by volume, of a 3 per cent hydrogen peroxide solution is required. The milk is well shaken and kept for 5 hours at 122° to 125° F. in well-closed vessels. Upon cooling, it may keep fresh for about a month and also to retain its natural fresh taste. With this process, if pure milk is used, the ordinary disease germs are killed off soon after milking and the milk sterilized.

Powdered Cork as a Preservative.—Tests have shown that powdered cork is very efficacious for packing and preserving fruits and vegetables. A bed of cork is placed at the bottom of the case, and the fruits or vegetables and the cork are then disposed in alternate layers, with a final one of cork at the top. Care should be taken to fill up the interstices, in order to prevent friction. Fruit may thus be kept fresh a year, provided any unsound parts have been removed preliminarily. When unpacking for sale, it suffices to plunge the fruit into water. Generally speaking, 50 pounds of cork go with 1,000 or 1,200 pounds of fruit. The cork serves as a protection against cold, heat, and humidity. Various fruits, such as grapes, mandarines, tomatoes, and early vegetables, are successfully packed in this way.

PRESSURE TABLE:
See Tables.

PRINT COPYING:
See Copying.

PRINTERS' OIL:
See Oil.

PRINTING ON PHOTOGRAPHS:
See Photography.

PRINTS, RESTORATION OF:
See Engravings.

PRINTS, THEIR PRESERVATION:
See Engravings.

PRINTING OILCLOTH AND LEATHER IN GOLD:
See Gold.

PRINTING-OUT PAPER, HOW TO SENSITIZE:
See Photography, under Paper-Sensitizing Processes.

PRINTING-ROLLER COMPOSITIONS:
See Roller Compositions for Printers.

PRUSSIC ACID:
See Poisons.

PUMICE STONE.

While emery is used for polishing tools, polishing sand for stones and glass, ferric oxide for fine glassware, and lime and felt for metals, pumice stone is more frequently employed for polishing softer objects. Natural pumice stone presents but little firmness, and the search has therefore been made to replace the natural product with an artificial one. An artificial stone has been produced by means of sandstone and clay, designed to be used for a variety of purposes. No. 1, hard or soft, with coarse grain, is designed for leather and waterproof garments, and for the industries of felt and wool; No. 2, hard and soft, of average grain, is designed for work in stucco and sculptors' use, and for rubbing down wood before painting; No. 3, soft, with fine grain, is used for polishing wood and tin articles; No. 4, of average hardness, with fine grain, is used for giving to wood a surface previous to polishing with oil; No. 5, hard, with fine grain, is employed for metal work and stones, especially lithographic stones. These artificial products are utilized in the same manner as the volcanic products. For giving a smooth surface to wood, the operation is dry; but for finishing, the product is diluted with oil.

PUMICE-STONE SOAP:
See Soaps.

PUNCHES:
See Ice Creams.

PUNCTURE CEMENT:
See Cement.

PURPLE OF CASSIUS:
See Gold.

Putty

(See also Lutes, under Adhesives and Cements.)

Common putty, as used by carpenters, painters, and glaziers, is whiting mixed with linseed oil to the consistency of dough. Plasterers use a fine lime mortar that is called putty. Jewelers use a tin oxide for polishing, called putty powder or putz powder. (See Putz Powder, under Jewelers' Polishes, under Polishes.)

Acid-Proof Putty.—I.—Melt 1 part of gum elastic with 2 parts of linseed oil and mix with the necessary quantity of white bole by continued kneading to the desired consistency. Hydrochloric acid and nitric acid do not attack this putty, it softens somewhat in the warm and does not dry readily on the surface. The drying and hardening is effected by an admixture of ½ part of litharge or red lead.

II.—A putty which will even resist boiling sulphuric acid is prepared by melting caoutchouc at a moderate heat, then adding 8 per cent of tallow, stirring constantly, whereupon sufficiently slaked lime is added until the whole has the consistency of soft dough. Finally about 20 per cent of red lead is still added, which causes the mass to set immediately and to harden and dry. A solution of caoutchouc in double its weight of linseed oil, added by means of heat and with the like quantity (weight) of pipe clay, gives a plastic mass which likewise resists most acids.

Black Putty.—Mix whiting and antimony sulphide, the latter finely powdered, with soluble glass. This putty, it is claimed, can be polished, after hardening, by means of a burnishing agate.

Durable Putty.—According to the "Gewerbeschau," mix a handful of burnt lime with 4¼ ounces of linseed oil; allow this mixture to boil down to the consistency of common putty, and dry the extensible mass received, in a place not accessible to the rays of the sun. When the putty, which has become very hard through the drying, is to be used, it is warmed. Over the flame it will become soft and pliable, but after having been applied and become cold, it binds the various materials very firmly.

Glaziers' Putty. — I. — For puttying panes or looking glasses into picture frames a mixture prepared as follows is well adapted: Make a solution of gum elastic in benzine, strong enough so that a syrup-like fluid results. If the solution be too thin, wait until the benzine evaporates. Then grind white lead in linseed-oil varnish to a stiff paste and add the gum solution. This putty may be used, besides the above purposes, for the tight puttying-in of window panes into their frames. The putty is applied on the glass lap of the frames and the panes are firmly pressed into it. The glass plates thereby obtain a good, firm support and stick to the wood, as the putty adheres both to the glass and to the wood.

II.—A useful putty for mirrors, etc., is prepared by dissolving gummi elasticum (caoutchouc) in benzol to a syrupy solution, and incorporating this latter with a mixture of white lead and linseed oil to make a stiff pulp. The putty adheres strongly to both glass and wood, and may therefore be applied to the framework of the window, mirror, etc., to be glazed, the glass being then pressed firmly on the cementing layer thus formed.

Hard Putty.—This is used by carriage painters and jewelers. Boil 4 pounds brown umber and 7 pounds linseed oil for 2 hours; stir in 2 ounces beeswax; take from the fire and mix in 5½ pounds chalk and 11 pounds white lead; the mixing must be done very thoroughly.

Painters' Putty and Rough Stuff.—Gradually knead sifted dry chalk (whiting) or else rye flour, powdered white lead, zinc white, or lithopone white with good linseed-oil varnish. The best putty is produced from varnish with plenty of chalk and some zinc white. This mixture can be tinted with earth colors. These oil putties must be well kneaded together and rather compact (like glaziers' putty).

If flour paste is boiled (this is best produced by scalding with hot water, pouring in, gradually, the rye flour which has been previously dissolved in a little cold water and stirring constantly until the proper consistency is attained) and dry sifted chalk and a little varnish are added, a good rough stuff for wood or iron is obtained, which can be rubbed. This may also be produced from glaziers' oil putty by gradually kneading into it flour paste and a little more sifted dry chalk.

To Soften Glaziers' Putty.—I.—Glaziers' putty which has become hard can be softened with the following mixture: Mix carefully equal parts of crude powdered potash and freshly burnt lime and make it into a paste with a little water. This dough, to which about ¼ part of soft soap is still added, is applied on the putty to be softened, but care has to be taken not to cover other paint, as it would be surely destroyed thereby. After a few hours the hardest putty will be softened by this caustic mass and can be removed from glass and wood.

II.—A good way to make the putty soft and plastic enough in a few hours so that it can be taken off like fresh putty, is by the use of kerosene, which entirely dissolves the linseed oil of the putty, transformed into rosin, and quickly penetrates it.

Substitute for Putty.—A cheap and effective substitute for putty to stop cracks in woodwork is made by soaking newspapers in a paste made by boiling a pound of flour in 3 quarts of water, and adding a teaspoonful of alum. This mixture should be of about the same consistency as putty, and should be forced into the cracks with a blunt knife. It will harden, like papier maché, and when dry may be painted or stained to match the boards, when it will be almost imperceptible.

Waterproof Putties.—I.—Grind powdered white lead or minium (red lead) with thick linseed-oil varnish to a stiff paste. This putty is used extensively for tightening wrought-iron gas pipes, for tightening rivet seams on gas meters, hot-water furnaces, cast-iron flange pipes for hot-water heating, etc. The putty made with minium dries very slowly, but becomes tight even before it is quite hard, and holds very firmly after solidification. Sometimes a little ground gypsum is added to it.

The two following putties are cheaper than the above-mentioned red lead putty: **II.—**One part white lead, 1 part manganese, one part white pipe clay, prepared with linseed-oil varnish.

III.—Two parts red lead, 5 parts white lead, 4 parts clay, ground in or prepared with linseed-oil varnish.

IV.—Excellent putty, which has been found invaluable where waterproof closing and permanent adhesion are desired, is made from litharge and glycerine. The litharge must be finely pulverized and the glycerine very concentrated, thickly liquid, and clear as water. Both substances are mixed into a viscid, thickly liquid pulp. The pegs of kerosene lamps, for instance, can be fixed in so firmly with this putty that they can only be removed by chiseling it out. For putting in the glass panes of aquariums it is equally valuable. As it can withstand higher temperatures it may be successfully used for fixing tools, curling irons, forks, etc., in the wooden handles. The thickish putty mass is rubbed into the hole, and the part to be fixed is inserted. As this putty hardens very quickly it cannot be prepared in large quantities, and only enough for immediate use must be compounded in each case.

V.—Five parts of hydraulic lime, 0.3 parts of tar, 0.3 parts of rosin, 1 part of horn water (the decoction resulting from boiling horn in water and decanting the latter). The materials are to be mixed and boiled. After cooling, the putty is ready for use. This is an excellent cement for glass, and may be used also for reservoirs and any vessels for holding water, to cement the cracks; also for many other purposes. It will not give way, and is equally good for glass, wood, and metal.

VI.—This is especially recommended for boiler leaks: Mix well together 6 parts of powdered graphite, 3 parts of slaked lime, 8 parts of heavy spar (barytes), and 8 parts of thick linseed-oil varnish, and apply in the ordinary way to the spots.

PUTTY FOR ATTACHING SIGN-LETTERS TO GLASS:
See Adhesives, under Sign-Letter Cements.

PUTTY, TO REMOVE:
See Cleaning Preparations and Methods.

PUTZ POMADE:
See Cleaning Preparations and Methods.

PYROGALLIC ACID:
See Photography.

PYROGALLIC ACID STAINS, TO REMOVE, FROM THE SKIN:
See Cleaning Preparations and Methods and Photography.

PYROCATECHIN DEVELOPER:
See Photography.

Pyrotechnics

FIREWORKS.

The chief chemical process is, of course, oxidation. Oxidation may be produced by the atmosphere, but in many cases this is not enough, and then the pyrotechnist must employ his knowledge of chemistry in selecting oxidizing agents.

The chief of these oxidizing agents are chlorates and nitrates, the effect of which is to promote the continuance of combustion when it is once started. They are specially useful, owing to their solid non-hygroscopic nature. Then ingredients are needed to prevent the too speedy action of the oxidizing agents, to regulate the process of combustion, such as calomel, sand, and sulphate of potash. Thirdly, there are the active ingredients that produce the desired effect, prominent among which are substances that in contact with flame impart some special color to it. Brilliancy and brightness are imparted by steel, zinc, and copper filings. Other substances employed are lampblack with gunpowder, and, for theatre purposes, lycopodium.

Fireworks may be classified under four heads, viz.:

1. Single fireworks.
2. Terrestrial fireworks, which are placed upon the ground and the fire issues direct from the surface.
3. Atmospheric fireworks, which begin their display in the air.
4. Aquatic fireworks, in which oxidation is so intense that they produce a flame under water.

Rockets.—First and foremost among atmospheric fireworks are rockets, made in different sizes, each requiring a slightly different percentage composition. A good formula is

Sulphur	1 part
Carbon, wood	2 parts
Niter	4 parts
Meal powder	1 part

Meal powder is a fine black or brown dust, which acts as a diluent.

Roman Candles.—Roman candles are somewhat after the same principle. An average formula is:

Sulphur	4 parts
Carbon	3 parts
Niter	8 parts

Pin Wheels.—These are also similar in composition to the preceding. The formula for the basis is

Sulphur	5 parts
Niter	9 parts
Meal powder	15 parts
Color as desired.	

Bengal Lights.—Bengal lights have the disadvantage of being poisonous. A typical preparation can be made according to this formula:

Realgar	1 part
Black antimony	5 parts
Red lead	1 part
Sulphur	3 parts
Niter	14 parts

COLORED FIRES.

The compounds should be ignited in a small pill box resting on a plate. All the ingredients must be dried and powdered separately, and then lightly mixed on a sheet of paper. Always bear in mind that sulphur and chlorate of potassium explode violently if rubbed together.

Smokeless Vari-Colored Fire.—First take barytes or strontium, and bring to a glowing heat in a suitable dish, remove from the fire, and add the shellac. The latter (unpowdered) will melt at once, and can then be intimately mixed with the barytes or strontium by means of a spatula. After cooling, pulverize. One may also add about 2½ per cent of powdered magnesium to increase the effect. Take for instance 4 parts of barytes or strontium and 1 part of shellac.

The following salts, if finely powdered and burned in an iron ladle with a little spirits, will communicate to the flame their peculiar colors.

Potassium nitrate or sodium chlorate, yellow.

Potassium chlorate, violet.
Calcium chloride, orange.
Strontium nitrate, red.
Barium nitrate, apple green.
Copper nitrate, emerald green.
Borax, green.
Lithium chloride, purple.

The colored fires are used largely in the production of various theatrical effects.

Blue Fire.—

I.—Ter-sulphuret of antimony	1 part	
Sulphur	2 parts	
Nitrate of potassium	6 parts	
II.—Sulphur	15 parts	
Potassium sulphate	15 parts	
Ammonio-cupric sulphate	15 parts	
Potassium nitrate	27 parts	
Potassium chlorate	28 parts	
III.—Chlorate of potash	8 parts	
Calomel	4 parts	
Copper sulphate	5 parts	
Shellac	3 parts	
IV.—Ore pigment	2 parts	
Charcoal	3 parts	
Potassium chloride	5 parts	
Sulphur	13 parts	
Potassium nitrate	77 parts	
V.—Potassium chlorate	10 parts	
Copper chlorate	20 parts	
Alcohol	20 parts	
Water	100 parts	
VI.—Copper chlorate	100 parts	
Copper nitrate	50 parts	
Barium chlorate	25 parts	
Potassium chlorate	100 parts	
Alcohol	500 parts	
Water	1,000 parts	

Green.—

I.—Barium chlorate	20 parts	
Alcohol	20 parts	
Water	100 parts	
II.—Barium nitrate	10 parts	
Potassium chlorate	10 parts	
Alcohol	20 parts	
Water	100 parts	

III.—Shellac.......... 5 parts
    Barium nitrate.... 1¼ parts
Pound after cooling, and add
    Barium chlorate, 2 to 5 per cent.

Red.—

I.—Shellac.......... 5 parts
    Strontium nitrate 1 to 1.2 parts

Preparation as in green fire. In damp weather add 2 to 4 per cent of potassium chlorate to the red flame; the latter causes a little more smoke.

II.—Strontium nitrate.. 20 parts
    Potassium chlorate 10 parts
    Alcohol.......... 20 parts
    Water........... 100 parts

Yellow.—

I.—Sulphur.......... 16 parts
    Dried carbonate of
      soda.......... 23 parts
    Chlorate of potas-
      sium........... 61 parts

II.—Sodium chlorate... 20 parts
    Potassium oxalate. 10 parts
    Alcohol.......... 20 parts
    Water........... 100 parts

Violet.—

I.—Strontium chlorate. 15 parts
    Copper chlorate... 15 parts
    Potassium chlorate 15 parts
    Alcohol.......... 50 parts
    Water........... 100 parts

II.—Potassium chlorate 20 parts
    Strontium chlorate. 20 parts
    Copper chlorate... 10 parts
    Alcohol.......... 50 parts
    Water........... 100 parts

Lilac.—

    Potassium chlorate 20 parts
    Copper chlorate... 10 parts
    Strontium chloride. 10 parts
    Alcohol.......... 50 parts
    Water........... 100 parts

Mauve.—

    Chlorate of potash. 28 parts
    Calomel.......... 12 parts
    Shellac.......... 4 parts
    Strontium nitrate.. 4 parts
    Cupric sulphate... 2 parts
    Fat.............. 1 part

Purple.—

    Copper sulphide... 8 parts
    Calomel.......... 7 parts
    Sulphur.......... 2 parts
    Chlorate of potash. 16 parts

White.—

I.—Gunpowder....... 15 parts
    Sulphur.......... 22 parts
    Nitrate of potassium 64 parts

II.—Potassium nitrate... 30 parts
    Sulphur.......... 10 parts
    Antimony sulphide
      (black)......... 5 parts
    Flour............ 3 parts
    Powdered camphor. 2 parts

III.—Charcoal.......... 1 part
    Sulphur.......... 11 parts
    Potassium sulphide. 38 parts

IV.—Stearine,.......... 1 part
    Barium carbonate.. 1 part
    Milk sugar........ 4 parts
    Potassium nitrate.... 4 parts
    Potassium chlorate. 12 parts

As a general rule, a corresponding quantity of shellac may be taken instead of the sulphur for inside fireworks.

The directions for using these solutions are simply to imbibe bibulous papers in them, then carefully dry and roll tightly into rolls of suitable length, according to the length of time they are to burn.

Fuses.—For fuses or igniting papers, the following is used:

    Potassium nitrate... 2 parts
    Lead acetate....... 40 parts
    Water............100 parts

Mix and dissolve, and in the solution place unsized paper; raise to nearly a boil and keep at this temperature for 20 minutes. If the paper is to be "slow," it may now be taken out, dried, cut into strips, and rolled. If to be "faster," the heat is to be continued longer, according to the quickness desired. Care must be taken to avoid boiling, which might disintegrate the paper.

In preparing these papers, every precaution against fire should be taken, and their preparation in the shop or house should not be thought of. In making the solutions, etc., where heat is necessary, the water bath should invariably be used.

PYROTECHNIC MAGIC.

[Caution.—When about to place any lighted material in the mouth be sure that the mouth is well coated with saliva, and that you are exhaling *the breath continuously*, with greater or less force, *according to the amount of heat you can bear.*

If the lighted material shows a tendency to burn the mouth, *do not attempt to drag it out quickly*, but simply shut the lips tight, and breathe through the nose, and the fire must go out instantly.

In the Human Gas Trick, where a flame 10 to 15 inches long is blown from the mouth, be careful after lighting the gas, *to continue to exhale the breath.* When you desire the gas to go out, simply shut the lips tight and hold the breath for a few seconds. In this trick, until the gas is well out, any inhalation is likely to be attended with the most serious results.

The several cautions above given may be examined with a lighted match, first removing, after lighting the match, any brimstone or phosphorus from its end.]

To Fire Paper, etc., by Breathing on it.—This secret seems little known to conjurers. Pay particular attention to the caution concerning phosphorus at the head of this article, and the caution respecting the dangerous nature of the prepared fluid given.

Half fill a half-ounce bottle with carbon disulphide, and drop in 1 or 2 fragments of phosphorus, each the size of a pea, which will quickly dissolve. Shake up the liquid, and pour out a small teaspoonful onto a piece of blotting paper. The carbon disulphide will quickly evaporate, leaving a film of phosphorus on the paper, which will quickly emit fumes and burst into flame. The once-popular term Fenian fire was derived from the supposed use of this liquid by the Fenians for the purpose of setting fire to houses by throwing a bottle down a chimney or through a window, the bottle to break and its contents to speedily set fire to the place.

For the purpose of experiment this liquid should only be prepared in small quantities as above, and any left over should be poured away onto the soil in the open air, so as to obviate the risk of fire. Thin paper may be fired in a similar manner with the acid bulbs and powder already mentioned. The powder should be formed into a paste, laid on the paper, and allowed to dry. Then the acid bulb is pasted over the powder.

Burning Brimstone.—Wrap cotton around two small pieces of brimstone and wet it with gasoline; take between the fingers, squeezing the surplus liquid out, light it with a candle, throw back the head well, and put it on the tongue blazing. Blow fire from mouth, and observe that a freshly blown-out candle may be lighted from the flame, which makes it more effective. After lighting candle chew up brimstone and pretend to swallow.

Blazing Sponge Trick.—Take 2 or 3 small sponges, place them in a ladle; pour just enough oil or gasoline over them to wet them. Be very careful not to have enough oil on them to cause them to drip. Set fire to the sponges and take one of them up with the tongs, and throw the head back and drop the blazing sponge in the mouth, expelling the breath all the time. Now close your mouth quickly; this cuts off the air from the flame and it immediately goes out. Be careful not to drop the sponge on the face or chin. Remove sponge under cover of a handkerchief before placing the second one in the mouth.

Burning Sealing Wax.—Take a stick of common sealing wax in one hand and a candle in the other, melt the wax over the candle, and put on your tongue while blazing. The moisture of the mouth cools it almost instantly. Care should be taken not to get any on the lips, chin, or hands.

Demon Bowls of Fire.—The performer has three $6\frac{1}{2}$-inch brass bowls on a table, and openly pours ordinary clean water (may be drunk) into bowls, until each is about half full. Then by simply passing the hand over bowls they each take fire and produce a flame 12 to 20 inches high.

Each bowl contains about 2 teaspoonfuls of ether, upon which is placed a small piece of the metal potassium, about the size of a pea. If the ether be pure the potassium will not be acted upon. When the water is poured into the bowl the ether and potassium float up, the latter acting vigorously on the water, evolving hydrogen and setting fire thereto, and to the ether as well.

The water may be poured into the bowl and lighted at command. In this case the potassium and ether are kept separated in the bowl, the former in a little cup on one side, and the latter in the body of the bowl. The water is poured in, and on rocking the bowl it is caused to wash into the little cup, the potassium floats up, and the fire is produced.

N. B.—The above tricks are not safe in any but specially made bowls, i. e., bowls with the wide flange round edge to prevent the accidental spilling of any portion of the burning ether.

The Burning Banana.—Place some alcohol in a ladle and set fire to it. Dip a banana in the blazing alcohol and eat it while it is blazing. As soon as it is placed in the mouth the fire goes out.

Sparks from the Finger Tips.—Take a small piece of tin about $\frac{1}{2}$ inch wide and $1\frac{1}{2}$ inches long. Bend this in the shape of a ring. To the center of this piece solder another small piece of tin bent in the shape of a letter U; between the ends of this U place a small piece of wax tape about ½ inch long. Take a piece of small rubber tubing about 2 feet in length and to one end of this attach a hollow rubber ball, which you must partly fill with iron filings. Place the rubber ball containing the iron filings under the arm and pass the rubber tube down through the sleeve of the coat to the palm of the hand; now place the tin ring upon the middle finger, with the wax taper inside of the hand. Light this taper. By pressing the arm down sharply on the rubber ball, the force of the air will drive some of the iron filings through the rubber tube and out through the flame of the burning taper, when they will ignite and cause a beautiful shower of sparks to appear to rain from the finger tips.

To Take Boiling Lead in the Mouth.— The metal used, while not unlike lead in appearance, is not the ordinary metal, but is really an alloy composed of the following substances:

Bismuth	8 parts
Lead	5 parts
Tin	2 parts

To prepare it, first melt the lead in a crucible, then add the bismuth and finally the tin, and stir well together with a piece of tobacco pipe stem. This "fusible metal" will melt in boiling water, and a teaspoon cast from the alloy will melt if very hot water be poured into it, or if boiling water be stirred with it. If the water be not quite boiling, as is pretty sure to be the case if tea from a teapot is used, in all probability the heat will be insufficient to melt the spoon. But by melting the alloy and adding to it a small quantity of quicksilver a compound will be produced, which, though solid at the ordinary temperature, will melt in water *very much below the boiling point.* Another variety of easily fusible alloy is made by melting together

Bismuth	7 to 8 parts
Lead	4 parts
Tin	2 parts
Cadmium	1 to 2 parts

This mixture melts at 158°, that given above at 208° F.

Either one of the several alloys above given will contain considerably less heat than lead, and in consequence be the more suitable for the purposes of a "Fire King."

When a body is melted it is raised to a certain temperature and then gets no hotter, not even if the fire be increased— all the extra heat goes to melt the remainder of the substance.

Second Method.—This is done with a ladle constructed similarly to the tin cup in a previous trick. The lead, genuine in this case, is, apparently, drunk from the ladle, which is then tilted, that it may be seen to be empty. The lead is concealed in the secret interior of the ladle, and a solid piece of lead is in conclusion dropped from the mouth, as congealed metal.

To Eat Burning Coals.—In the first place make a good charcoal fire in the furnace. Just before commencing the act throw in three or four pieces of soft pine. When burnt to a coal one cannot tell the difference between this and charcoal, except by sticking a fork into it. This will not burn in the least, while the genuine charcoal will. You can stick your fork into these coals without any difficulty, but the charcoal is brittle and hard; it breaks before the fork goes into it.

Chain of Fire.—Take a piece of candle wick 8 or 10 inches long, saturated with kerosene oil, squeeze out surplus oil. Take hold of one end with your fire tongs, light by furnace, throw back your head, and lower it into your mouth *while exhaling the breath freely.* When all in, close your lips and remove in handkerchief.

Note.—Have a good hold of the end with the tongs, for if it should fall it would probably inflict a serious burn; for this reason also no burning oil must drop from the cotton.

Biting Off Red-Hot Iron.—Take a piece of hoop iron about 2 feet long, place it in a vise and bend it backwards and forwards, about an inch from the end, until it is nearly broken off. Put this in a furnace until it becomes red hot, then take it in your right hand, grasp the broken end in your teeth, being careful not to let it touch your lips or your tongue, make a "face" as though it was terribly hard to bite off, and let the broken end drop from between your teeth into a pail of water (which you should always have at hand in case of fire), when the hissing will induce the belief that the portion bitten off is still "red hot"—it may be, for that matter, if the iron be nearly broken off in the first place and if you have good teeth and are not afraid to injure them.

Water Stirred Yellow, Scarlet, and Colorless.—Obtain a glass tube with one end hermetically sealed and drawn into a fine point that will break easily. Into an ale glass put a solution of mercury bichloride (corrosive sublimate, a deadly poison) and into the tube a strong solution of potassium iodide so adjusted in strength that it will redissolve the scarlet precipitate formed by the union of the two liquids. While stirring the solution in the glass the bottom of the tube (apparently a glass rod) is broken and a small portion of its contents allowed to escape, which produces a beautiful scarlet. The balance of the fluid in the tube is retained there by simply keeping the thumb on the open top end. Continue the stirring, allowing the balance of the contents of the tube to escape, and the scarlet fluid again becomes colorless. Before the scarlet appears the liquid is yellow.

To heighten the effect, another ale glass, containing only clean water and a solid glass stirring-rod, may be handed to one of the company, with instructions to do the same as the performer; the result is amusing.

QUICK-WATER:
See Alloys.

QUILTS, TO CLEAN:
See Cleaning Preparations and Methods.

QUINCE EXTRACT:
See Essences and Extracts.

RAGS FOR CLEANING AND POLISHING:
See Cleaning Preparations and Methods.

RASPBERRYADE POWDER:
See Salts, Effervescent.

RASPBERRY SYRUP:
See Essences and Extracts.

Rat Poisons

(See also Turpentine.)

Poisons for rats may be divided into two classes, quick and slow. Potassium cyanide and strychnine belong to the first, and phosphorus and arsenic to the second. Both should be kept away from children, dogs, and cats, and this is best done by putting them in places too narrow for anything larger than a rat to squeeze into. If the poison is too quick, the effect of it is visible to the same rats which saw the cause, and those which have not eaten of the bait will leave it alone. On the other hand, if it is too slow, the poisoned rat may spread it to edible things in the pantry, by vomiting. Slow poisons generally cause the rat to seek water, and when they are used water should not be left about promiscuously.

The substances most useful as rat poisons, and which are without danger to the larger domestic animals, are plaster of Paris and fresh squills. Less dangerous than strychnine and arsenic are the baryta preparations, of which the most valuable is barium carbonate. Like plaster of Paris, this substance, when used for the purpose, must be mixed with sugar and meal, or flour, and as a decoy some strong-smelling cheese should be added. In closed places there should be left vessels containing water easily accessible to the creatures.

One advantage over these substances possessed by the squill is that it is greedily eaten by rats and mice. When it is used, however, the same precaution as to water, noted above, is necessary, a circumstance too frequently forgotten. In preparing the squill for this purpose, by the addition of bacon, or fat meat of any kind, the use of a decoy like cheese is unnecessary, as the fats are sufficiently appetizing to the rodents. It is to be noted that only fresh squills should be used for this purpose, as in keeping the bulb the poisonous principle is destroyed, or, at least, is so modified as to seriously injure its value.

Squill Poisons.—The preparation of the squill as a rat poison can be effected in several different ways. Usually, after the removal of the outer peel, the bulb is cut up into little slices and mixed with milk and flour; these are stirred into a dough or paste, which, with bits of bacon rind, is put into the oven and baked. Another plan is to grate the squill on a grater and mingle the gratings with mashed, boiled, or roasted potato. This method of preparing them necessitates the immediate use of the poison. The following is, however, a stable preparation that keeps well:

I.—Hog's lard	500 grams
Acid salicylic	5 grams
Squill	1 bulb
Beef suet	50 to 100 grams
Barium carbonate	500 grams
Solution of ammonium copper acetate, 20 per cent	50 grams

Cut or grate the squill into very small pieces, and fry it in the lard and suet until it has acquired a dark-brown color and the fats have taken up the characteristic squill odor; then to the mess add the other substances, and stir well together.

II.—Squill, bruised......   4 ounces
     Bacon, chopped fine   6 ounces
     Flour or meal, enough.
     Water, enough.

Make into a stiff mass, divide into small cakes, and bake.

Phosphorus Poisons.—Next to the squill in value as a poison comes phosphorus in the shape of an electuary, or in pills. For readily preparing the electuary, when needed or ordered, it is a good plan to keep on hand a phosphorated syrup made as follows:

To 200 parts of simple syrup, in a strong flask, add 50 parts of phosphorus and 10 parts of talc powder; place the container in a suitable vessel and surround it with water heated to 120° to 130° F., and let it stand until the phosphorus is melted. Now, cork the flask well, tie down the cork, and agitate until the mixture is completely cold. As a measure of precaution, the flask should be wrapped with a cloth.

To make the poison take 50 parts of rye flour and mix with it 10 parts of powdered sugar. To the mixture add about 40 parts of water and from 30 to 40 parts of the phosphorated syrup, and mix the mass thoroughly.

While it is best to make the phosphorated syrup fresh every time that it is required, a stable syrup can be made as follows:

Heat together very carefully in a water bath 5 parts of phosphorus, 3 parts of sublimed sulphur, and 30 parts of water, until the phosphorus is completely melted and taken up; then add 30 parts of wheat flour and 6 parts of ground mustard seed, and work up, with the addition of warm water from time to time, if necessary, into a stiff paste, finally adding and working in from 1 to 2 parts of oil of anise.

Borax in powder, it may be noticed, is also useful as a preservative of phosphorated paste or the electuary.

Mühsam gives the following formula for an electuary of phosphorus for this purpose:

I.—Phosphorus, granu-
     lated............   1 part
     Rye flour..........   30 parts
     Simple syrup.......   10 parts
     Mustard seed, pow-
     dered............   1 part
     Sublimed sulphur...   1 part
     Water............   10 parts

Proceed as indicated above.

Hager's formula for "Phosphorus globules" is as follows:

II.—Phosphorus, amor-
     phous..........   10 parts
     Glycerine.........   20 parts
     Linseed, powdered 100 parts
     Meat extract......   15 parts
     Quark, recently coagulated, quantity sufficient.

Mix, and make a mass, and divide into 200 globules, weighing about 15 grains each. Roll in wheat flour, in which a little powdered sugar has been mixed.

Phosphorus electuary, made as indicated above, may be smeared upon bits of fried bacon, which should be tacked firmly to a bit of board or to the floor. It is essential that either flour or sugar, or both, be strewn over the surface of the phosphorus.

The most convenient in practice, on the whole, are the phosphorus globules, either made after Hager's formula, or, more readily, by adding rye flour and sugar to the electuary and working up to a pill mass, or barium carbonate and plaster may be added.

Arsenical Poisons.—The following are some of the formulas given by Hager for preparing globules, or pills, of arsenic:

I.—Arsenic, white, pow-
     dered...........100 parts
     Soot from the kitch-
     en..............   5 parts
     Oil of anise.......   1 part
     Lard, sufficient.
     Wheat flour, sufficient.

Make into 400 globules.

II.—Beef suet..........500 parts
     Rye flour..........500 parts
     Arsenic, white, pow-
     dered...........   50 parts
     Ultramarine.......   10 parts
     Oil of anise.......   1 part

Melt the suet, and add to the flour, mix in the other ingredients, and work up while hot, beating the mass with a roller. Make 1,000 globules.

Strychnine Poisons.—The strychnine preparations are also valuable in the destruction of rats and mice. The first of these in point of usefulness is strychnine-wheat, or strychnine-oats (Strychninweizen or Strychninhafer), in the proportion of 1 part of strychnine to 100 or 150 parts of wheat or oat flour, prepared by dissolving 1 part of strychnine in 40 to 50 parts of hot water, mixing well up with the flour, and drying in the water bath. Strychnine may also be used on fresh or salted meat, sausage, etc., by insertion of the powder, or the heads of fried fish are opened and the powder strewn on the inside. The latter is an especially deadly method, since the odor of the fish acts as a powerful lure, as also do the bits of bacon or other fats used in frying fish. Strong cheese is also a good vehicle for strychnine, acting as a powerful lure for the rodents.

Strychnine sulph	1	drachm
Sugar milk	3	drachms
Prussian blue	5	grains
Sugar	½	ounce
Oat flour	½	ounce

Nux Vomica Poison.—

Oatmeal	1	pound
Powdered nux vomica	1	ounce
Oil of anise	5	drops
Tincture of asafetida	5	drops

Barium Poison.—

Barium carbonate	4	ounces
Sugar	6	ounces
Oatmeal	6	ounces
Oil of anise	4	drops
Oil of caraway	4	drops

RAZOR PAPER:
See Paper.

RAZOR PASTES:
See also Pastes.

The razor pastes, razor creams, etc., on the market, have for their cutting, or sharpening, agent jewelers' rouge, or rouge and emery. When emery is used it should be ground to an impalpable powder and levigated.

I.—The simplest formula is a mixture in equal parts of rouge and emery powder, rubbed up with spermaceti ointment. Coke is also used as a cutting agent. Suet, prepared lard, in fact, any greasy or soapy substance, will answer for the vehicle.

II.—Melt 1,000 parts of beef tallow and pour 250 parts of oil to it. To this mixture, which is uniformly combined by thorough stirring, add in the same manner 150 parts of washed emery, 100 parts of tin ashes, and 50 parts of iron oxide. The stirring of these ingredients must be continued until the mass is cool, as otherwise they would be unevenly distributed. The leather of the strop should be rubbed with this grease, applying only small quantities at a time. This renders it possible to produce a very uniform coating, since little quantities penetrate the fibers of the leather more easily.

III.—Tin putty (tin ashes)	2	parts
Colcothar	2	parts
Forged iron scales or filings	1	part
Pure levantine honing stone finely powdered	7	parts
Beef suet	3	parts

All the ingredients with the exception of the suet should be finely powdered. The suet is melted, the ingredients poured in, and the whole thoroughly mixed to form a doughy mass.

IV.—Colcothar	1½	parts
Pumice stone	1½	parts
Graphite	4½	parts
Bloodstone (red hematite)	2	parts
Iron filings	1	part

These ingredients are finely powdered, washed, and mixed with the following:

Grafting wax	2	parts
Soap	2	parts
Lard	2	parts
Olive oil	2	parts

Naturally the fatty ingredients are to be heated before the solid substances are commingled with them.

The side of the blade to be polished should be treated with the following compositions:

a. Tin ashes (tin putty) rubbed down to a fine powder on a honing stone and mixed with axle grease.

b. Washed graphite mingled with olive oil.

REDUCERS:
See Photography.

REDUCING PHOTOGRAPHS, SCALE FOR:
See Photography.

REFLECTOR METAL:
See Alloys.

REFRIGERANTS.

I.—Potassium nitrate	2	pounds
Ammonium chloride	2	pounds
Water	5	pints

II.—Potassium nitrate	2½	pounds
Ammonium chloride	2½	pounds
Sodium sulphate	4	pounds
Water	9	pints

III.—Ammonia nitrate	4	pounds
Water	4	pints

IV.—Sodium sulphate	8	parts
Dilute hydrochloric acid	5	parts

V.—Snow............ 1 part
Water............ 1 part
Sulphuric acid...... 4 parts

VI.—Snow............ 3 parts
Calcium chloride... 4 parts

Refrigeration

If water to be frozen is placed in a tin bucket or other receptacle it can be readily congealed by putting it in a pail containing a weak dilution of sulphuric acid and water. Into this throw a handful of common Glauber salts, and the resulting cold is so great that water immersed in the mixture will be frozen solid in a few minutes, and ice cream or ices may be quickly and easily prepared. The cost is only a few cents. The same process in an ice-cream freezer will do the trick for ice cream.

Home-Made Refrigerators.—I.—Partly fill with water a shallow granite-ware pan. Place it in an open, shady window where there is a good draught of air. In this put bottles of water, milk, and cream (sealed), wrapped with wet cloths reaching into the water. Put butter in an earthen dish deep enough to prevent water getting in. Over this turn an earthen flower-pot wrapped with a wet cloth reaching into the water. The pan should be fixed every morning and evening. With several of these pans one can keep house very comfortably without ice.

II.—Procure a wire meat-safe—that is, a box covered by wire netting on three sides, with a fly-proof door. On top place a deep pan filled with water. Take a piece of burlap the height of the pan and safe, and of sufficient length to reach around the entire safe. Tack it fast where the door opens and closes. Tuck the upper edge in the water. Place it where there is a draught and where the dripping will do no damage. This constitutes a well-ventilated refrigerator that costs nothing but water to maintain.

III.—Take a store box, any convenient size, and place in this a smaller box, having the bottom and space around the sides packed with sawdust. Have a galvanized iron pan made, the size of the inside box and half as deep, to hold the ice. Have the pan made with a spout 6 inches long to drain off the water as the ice melts. Bore a hole the size of the spout through the double bottom and sawdust packing to admit the spout. Short legs may be nailed on the sides of the box and a vessel set underneath to catch the drippings. Put on a tight board cover. A shelf may be placed in the box above the ice. This box will keep ice for three days.

IV.—Select a large cracker box with a hinged cover. Knock out the bottom and cut windows in each side, leaving a 3-inch frame, over which tack wire gauze. In the coolest part of the cellar dig away the earth to a level depth of 3 inches and fit the box into the space.

Mix plaster of Paris to a consistency of thick cream and pour into the box for a ½-inch thick bottom. Twenty-four hours will harden it sufficiently. Put a hook and catch on the lid. A box of this sort can be cleaned easily, and insects cannot penetrate it.

To Drain a Refrigerator.—I.—Have a stout tin funnel made, 7 inches in diameter at the top. The tube portion should be at least 8 inches long and of uniform diameter. Bore a hole through the floor directly under the drain-pipe of the refrigerator; insert the funnel, then force a piece of rubber tubing (a tight fit) over the funnel from the cellar side. Pass the tubing through a hole cut in the screen frame of a cellar window, and drain into any convenient place. This avoids the necessity of continually emptying the drain-pan, and prevents the overflow that frequently occurs when it is forgotten.

II.—This simple device saves the inconvenience of having a drip-pan under the refrigerator: If the refrigerator is placed near the outer wall get a piece of rubber hose long enough to reach from the waste pipe to the outside of the wall. Bore a hole through the wall under the refrigerator, where baseboard and floor meet. Attach the hose to the waste-pipe and pass through the hole in the wall. A small trough outside should carry the water away from the house.

REFRIGERATORS, THEIR CARE:
See Household Formulas.

REPLATING:
See Plating.

RESILVERING OF MIRRORS:
See Mirrors.

REVOLVER LUBRICANTS:
See Lubricants.

RHUBARB AS A REMEDY FOR CHOLERA:
See Cholera Remedies.

RIBBONS FOR TYPEWRITERS:
See Typewriter Ribbons.

RICE PASTE:
See Adhesives.

RICE POWDER:
See Cosmetics.

RIFLE LUBRICANTS:
See Lubricants.

RING, HOW TO SOLDER A JEWELED:
See Solders.

RINGS ON METAL, PRODUCING COLORED:
See Plating.

ROACH EXTERMINATORS:
See Insecticides.

ROBURITE:
See Explosives.

RODINAL DEVELOPER:
See Photography.

ROLLER COMPOSITIONS FOR PRINTERS.

Rollers for transferring ink to types have to possess special properties, which have reference both to the nature of the ink and that of the types to which it is to be transferred. They must be as little liable as possible to changes of temperature. They must be sticky, but only just sticky enough, and must have elasticity enough to exert a uniform pressure over the varying surface with which they meet in the form. Originally, the composition was one of glue and molasses in varying proportions, and the only practical improvement that has been made is the addition of glycerine. This being slightly hygroscopic, helps to keep the roller at the right degree of softness, and being practically unfreezable, it is a great assistance in keeping the rollers from hardening in cold weather.

The recipes given in technical works for printing roller compositions are numerous and very different. All contain glue and molasses, and it is the practice to put a larger proportion of glue in rollers to be used in the summer than in those intended for winter use. The following is a selection of recipes:

I.—Soak 8 pounds of glue in as much water as it will absorb. When there is no visible water, treat the glue till melted, and add 7 pounds of hot molasses.

II.—Glue (summer).... 8 pounds
Glue (winter)..... 4 pounds
Molasses......... 1 gallon

III.—Molasses......... 12 pounds
Glue............. 4 pounds

IV.—Molasses......... 24 pounds
Glue............. 16 pounds
Paris white....... 2 pounds

V.—Glue or gelatin.... 64 pounds
Water............ 48 pounds
Linseed oil........ 96 pounds
Molasses or sugar.
64 to 96 pounds
Chloride of calcium 3 pounds
Powdered rosin ... 8 pounds

Soak the glue in the water and then liquefy by heat. Then stir in the oil, first heated to 150° F. Then add the molasses and the chloride of calcium, and finally the fused rosin. The latter ingredient is only to be added when very tough rollers are required. This recipe is interesting from the inclusion in it of the hygroscopic salt, chloride of calcium, the object of which is obviously to keep the rollers moist.

ROOFS, HOW TO LAY GALVANIZED.
See Household Formulas.

ROOFS, PREVENTION OF LEAKAGE:
See Household Formulas.

ROOF PAINTS:
See Paint.

ROOM DEODORIZER:
See Household Formulas.

ROPES.

To protect ropes, cordage, and cloths made of flax and hemp against rot, it has been recommended to leave them for 4 days in a solution of copper sulphate, 20 parts by weight to a liter, then allow them to dry, and then, to prevent the copper sulphate being washed away by the water, place in tar or a solution of soap—1 to 10. In the latter case an insoluble copper soap is formed. To secure the same result with twine, the following process has been recommended: Place the string for an hour in a solution of glue, then allow to dry, and place in a solution of tannin. After removal from the tannin, again dry, and soak in oil. The process first described has been shown by experience to be very effective; but to prevent the washing away of the copper sulphate, it is advisable to use the solution of soap in preference to the tar, as articles steeped in the latter substance are apt to become stiff, and consequently brittle. The treatment with glue and tannin in the second process has the drawback that it tends to make the string too stiff and inflexible, and thus impair its usefulness.

ROPE LUBRICANTS:
See Lubricant.

ROPES, WATERPROOFING:
See Waterproofing.

ROSE CORDIAL:
See Wines and Liquors.

ROSEWOOD:
See Wood.

ROSE POWDERS:
See Cosmetics.

ROSIN, TESTS FOR, IN EXTRACTS:
See Foods.

ROSIN OIL:
See Oil.

ROSIN STICKS:
See Depilatories.

ROT:

Remedies for Dry Rot.—A good remedy for dry rot is petroleum. The sick parts of the wood are painted with it, which causes the fungi to die, turn black, and finally drop off. The best preventive of dry rot is plenty of draught. If the portions are already affected so badly that they must be removed and renewed, the freshly inserted wood is coated with "carbolineum" to prevent a fresh appearance of dry rot. Another remedy is ordinary salt, which is known to have a highly hygroscopic action. It absorbs the moisture of the wood, whereby it is itself dissolved, thus gradually impregnating the planks, etc. In order to combat dry rot with salt, proceed as follows: Throw salt into boiling water until a perfectly saturated solution is obtained. With this repeatedly wash the wood and masonry afflicted with dry rot. Wherever practicable the salt may be sprinkled direct upon the affected place.

ROUGE:
See Cosmetics.

ROUGE FOR BUFF WHEELS.

The rouge employed by machinists, watchmakers, and jewelers, is obtained by directly subjecting crystals of sulphate of iron or copperas to a high heat by which the sulphuric acid is expelled and the oxide of iron remains. Those portions least calcined, when ground, are used for polishing gold and silver. These are of bright crimson color. The darker and more calcined portions are known as "crocus," and are used for

polishing brass and steel. Others prefer for the production of rouge the peroxide of iron precipitated by ammonia from a dilute solution of sulphate of iron, which is washed, compressed until dry, then exposed to a low red heat and ground to powder. Of course, there are other substances besides rouge which are employed in polishing, as powdered emery, kieselguhr, carborundum, rotten stone, etc.

ROUGE POWDER:
See Polishes.

ROUGH STUFF:
See Wood.

ROUP CURES:
See Veterinary Formulas.

Rubber

ARTIFICIAL RUBBER.

Austin G. Day tried hundreds of experiments and took out many patents for rubber substitutes. He was in a measure successful, his "Kerite" compound proving of great value and being a result of his seeking for something that would wholly supplant rubber. As far back as 1866 he made public the results of some of his work, giving as formulas for rubber substitutes the following compounds:

I.—Linseed oil......... 2 pounds
    Cottonseed oil...... 1 pound
    Petroleum......... 2 pounds
    Raw turpentine.... 2 pounds
    Sulphur.......... 2 pounds

Boil 2 hours.

II.—Linseed oil......... 2 pounds
    Cottonseed oil...... 1 pound
    Petroleum......... 1 pound
    Raw turpentine..... 2 pounds
    Castor oil.......... 1 pound
    Sulphur.......... 2 pounds

Boil ½ hour.

III.—Linseed oil......... 2 pounds
    Cottonseed oil...... 1 pound
    Petroleum......... 1 pound
    Raw turpentine.... ½ pound
    Liquid coal tar..... 3 pounds
    Peanut oil......... 1 pound
    Spirits turpentine... 1 pound
    Sulphur.......... 4 pounds

Boil 35 minutes.

IV.—Linseed oil......... 2 pounds
    Cottonseed oil...... 1 pound
    Petroleum......... 2 pounds
    Raw turpentine.... ½ pound
    Liquid coal tar..... 2 pounds

Spirits turpentine... 1 pound
Rubber ....   pound
Sulphur..........   2 pounds

Boil 1 hour.

In 1871 Mr. Day had brought his experimenting down to the following formula:

V.—Cottonseed oil...... 14 pounds
Linseed oil......... 14 pounds
Asphaltum........   8 pounds
Coal tar..........   8 pounds
Sulphur.......... 10 pounds
Camphor.........   ½ pound

In this the tar and asphaltum were first mixed with the cottonseed oil, after which was added the linseed oil and camphor, and, last of all, the sulphur, when the temperature was about 270° F.

A substitute designed to be used in rubber compounding in place, say, of reclaimed rubber, was made as follows:

VI.—Cottonseed oil...... 27 pounds
Coal tar........... 30 pounds
Earthy matter......   5 pounds

To be mixed and heated to 300° F., and then strained and cooled to 200° F. Then were added 27 pounds linseed oil, the heat raised to 220° F., and 15 to 18 pounds of sulphur added, the heat being continually raised until the mass was sulphurized. When the heat reached 240° F., 1 to 1½ ounces of nitric acid were added, and at 270° to 280° F., from 1 to 3 ounces camphor were added to help the sulphurization. The resultant compound was used on the following basis:

VII.—Para rubber....... 20 pounds
Litharge..........   5 pounds
Sulphur...........   1 pound
A b o v e   c o m -
pound...... 20 to 40 pounds

Mr. Day did not insist on the compound quoted, but advised that the proportions be varied as widely as the exigencies of the case might demand. Whiting, barytes, infusorial earth, white lead, blacks, in fact almost any of the oxides, carbonates, or earthy materials commonly used in compounding, were used in connection with his substitute, as also were any grades of crude rubber. Among other ingredients that he found of use in making his substitutes were vegetable and animal waxes, together with ozokerite and paraffine. These were only used in small quantities, and always in connection with the linseed and cottonseed oils, and generally asphaltum or coal tar. One of his compounds also called for a quantity of golden sulphuret of antimony, presumably to assist in the sulphurization, and a small amount of tannic acid.

Another line of experimenting that is interesting, and that will yet produce good results, although so far it has not amounted to much, is in the use of cellulose. A very simple formula is of French origin and calls for the treating of cellulose with sulphuric acid, washing, drying, granulating, treating with resinate of soda—which is afterwards precipitated by sulphate of alumina—then drying and molding under pressure. As a matter of fact, the resultant mass would not be mistaken for rubber. An English formula is more like it. This consists of

VIII.—Cellulose......... 15 pounds
Pitch............ 25 pounds
Asphalt.......... 20 pounds
Silica............ 20 pounds
Mastic............   5 pounds
Bitumen..........   5 pounds
Rosin............ 10 pounds
Coal tar.......... 12 pounds

This makes a thick gummy varnish which is of little use except as for its waterproof qualities. Allen's formula for a cellulose substitute might have a value if it were carried further. It is made up of 100 pounds of rosinous wood pulp treated with animal gelatin, 100 pounds asphalt, and 10 pounds asphalt oil, all heated and molded.

The Greening process, which is English, is more elaborate than Allen's, but seems a bit laborious and costly. This process calls for the treatment of the cellulose by a mixture of sulphuric acid and nitrate of potash, and, after drying, a treatment to a bath of liquid carbonic acid. When dry again, it is mixed in a retort with refined rosin, gum benzoin, castor oil, and methylated alcohol. The distillate from this is dried by redistilling over anhydrous lime.

Another curious line of substitutes is that based upon the use of glue and glycerine. Some of these have uses, while others, that look very attractive, are of no use at all, for the simple reason that they will absorb water almost as readily as a dry sponge. The first of these is more than 30 years old and is said to be of French origin. The formula is:

IX.—Glue............   4 pounds
Glycerine........   8 ounces
Nutgall ....   3 ounces
Acetic acid, 1 pound in 5 pounds
of water.

Ten years later this was approached by an English formula in which in place of the nutgall and acetic acid, chromic and tannic acids were substituted, and a modicum of ground cork was added as a cheapener probably. Some four years later an ingenious Prussian gave out a formula in which to the glue and glycerine and tannic acid were added Marseilles soap and linseed oil. None of the above have ever had a commercial value, the nearest approach being the glue and glycerine compound used as a cover for gas tubing.

The substitutes that have really come into use generally are made either from linseed, cottonseed, or maize oil. Scores of these have been produced and thousands of dollars have been spent by promoters and owners in trying to make these gums do just what crude rubber will. A German formula which was partially successful is

X.—Linseed oil, in solu-
tion........... 80 pounds
L i m e - hardened
rosin, in solution 50 pounds
Add to above
Sulphur.......... 8 pounds
Linseed oil....... 42 pounds

Add 20 pounds sulphur and heat to 375° F.

Rubber and Rubber Articles.—As regards the action of coal gas on rubber tubes, it has been observed that it is weakest on ordinary gray rubber which withstands it the longest, and gives off no odor. Red rubber is more readily affected, and the black kind still more so.

To prevent rubber tubes from drying up and becoming brittle, they should be coated with a 3 per cent aqueous solution of carbolic acid, which preserves them. If they have already turned stiff and brittle, they can be rendered soft and pliant again by being placed in ammonia which has been made liquid with double the amount of water.

In France rubber tubes are used as a core for casting pipes from cement and sand. In order to construct a connected pipe conduit in the ground, a groove is dug and a layer of cement mortar spread out. Upon this the rubber tube is laid, which is wrapped up in canvas and inflated. The remaining portion of the channel is then filled up with cement mortar, and as soon as it has set, the air is let out of the rubber hose and the latter is pulled out and used as before.

To cover cloth with rubber, there are chiefly employed for dissolving the rubber, naphtha, alcohol, and benzol. They are mixed with purified solid paraffine, and ground together.

Rubber boots and shoes are rendered waterproof by melting 4 parts of spermaceti and 1 part of rubber on a moderate fire, adding tallow or fat, 10 parts, and lastly 5 parts of copal varnish or amber varnish. This mixture is applied on the shoes with a brush. It should be stated that the rubber used for this purpose must be cut up very small and allowed 4 to 5 hours to dissolve.

To rid rubber articles of unpleasant odor, cover both sides with a layer of animal charcoal and heat to about 140° F.

To prevent gas from escaping through rubber hose, cover it with a mixture prepared as follows: Dissolve 5 parts of gum arabic and 3 parts of molasses in 15 parts of white wine and add, with constant stirring, 6 parts of alcohol in small quantities. Stirring is necessary to prevent the alcohol from precipitating the gum arabic.

Repairing Rubber Goods.—First, clean off all adherent matter, and dry thoroughly. Varnish or lacquer, as for instance on rubber shoes, may be removed with sand or emery paper, or even with a file, in the absence of one of these. The surface thus produced is then rubbed with benzine. A solution of Para rubber in benzine is then painted over the surface around the break or tear, and a strip of natural rubber fitted over it. Then prepare a vulcanizing solution as follows:

Sulphur chloride.... 18 parts
Benzine........... 400 parts
Carbon disulphide.. 300 parts

This is applied to the edges of the joint by means of a pledget of cotton wrapped on the end of a little stick, and press the jointed parts well together.

One may repair rubber bulbs by the following method: Put some pure gum in three times its bulk of benzine, and cork tightly. Let stand several days. Get some rubber in sheet form; it will be better if it is backed with cloth. To make a patch, dampen some little distance around the hole to be mended with benzine. After a moment, scrape with a knife; repeat the process several times till the site to be patched is thoroughly clean. Cut a patch from sheet of rubber a little larger than the hole to be mended, and apply to its surface several coats of the benzine solution. Then apply a good coat of the solution to both patch and about the hole, and press the patch firmly in place. Again apply the solution to make coating over the patch, and allow to dry till it will not stick to the finger. Do not use for several days.

Cracked rubber goods may be successfully mended in the following manner: Before patching, the cracked surfaces to unite well must be dried, entirely freed from all dirt and dust and greased well, otherwise the surfaces will not combine. In case of a cover, waterproof coat, or rubber boots, etc., take a moderately thick piece of india rubber, suited to size of the object, cut off the edges obliquely with a sharp knife moistened in water, coat the defective places as well as the cut pieces of rubber with oil of turpentine, lay the coated parts together and subject them for 24 hours to a moderate pressure. The mended portions will be just as waterproof as the whole one. Rubber cushions or articles containing air are repaired in a very simple manner, after being cleaned as aforesaid. Then take colophony, dissolve it in alcohol (90 per cent) so that a thick paste forms, smear up the holes, allow all to harden well, and the rubber article, pillow, ball, knee caps, etc., may be used again.

Softening Rubber.—The hardening of gum articles is generally referable to these having been kept for a long time in some warm, dry place, though keeping them in the cold will produce the same effect. Hardness and brittleness, under any reasonable care and conditions, are usually signs of an inferior article of goods. Articles of Para rubber, of good workmanship, usually maintain their elasticity for a very long time. Before attempting to soften hollow rubber ware, such as flasks, water bags, or bottles, etc., they should be well scrubbed with a wire brush (bottle cleaner) and warm water, so as to remove all dirt and dust. This scrubbing should be continued until the wash water comes away clean and bright. For softening, the best agent is dilute water of ammonia, prepared by mixing pharmacopœial ammonia water, 1 part, and water, 2 parts. There should be enough of this to cover the articles, inside and out. Let them remain in the mixture until the ammonia has evaporated. Warm water works better than cold. From 1 to 2 hours will be long enough, as a usual thing. Thick and massive articles such as large rubber tubing, require more energetic treatment, and the journal recommends for the treatment of these that they be filled nearly full with the ammonia mixture, corked at both ends, and coiled up in a kettle, or other vessel, of sufficient size, warm water poured in sufficient to cover the coil completely, and lightly boiled for from 1 to 2 hours. The water lost by evaporation should be replaced from time to time, and the vessel should never be allowed to boil violently. When the proper time has arrived (and this must be learned, it appears, by experience, as the article quoted gives no directions save those translated), remove from the fire, and allow to cool gradually.

Glycerine has been also recommended, and it may be used with advantage in certain cases. The articles must first be cleaned with the brush and warm water, as above detailed. Heat them in water and rub them with a wad of cotton soaked in glycerine, drawing the wad over them, backwards and forwards. This wad should be wrapped with good stout wire, the ends of which are prolonged, to serve as a handle. Where possible the articles should be stricken with the glycerine inside and out, the article being, naturally, held out of the boiling water, sufficiently, at least, to make bare the part being rubbed at the time. Let rest for 24 hours, and repeat this process. With goods kept in stock, that show a tendency to grow brittle, this treatment should be repeated every 6 months or oftener. Never put away tubing, etc., treated in this manner until every particle of moisture has drained off or evaporated.

Another authority, Zeigler, has the following on this subject: Tubing, bands, and other articles of vulcanized caoutchouc that have become brittle and useless, may be restored to usefulness, indeed, to their pristine elasticity, by treating them as follows: First, put them in a hot aqueous solution of tannic acid and tartar emetic. Next, transfer them to a cold aqueous solution of tannic acid and calcium sulphate. Mix the two solutions and heat to about the boiling point, and transfer the articles to the hot solution. This treatment should be maintained from 1 day to 3 or 4, according to the nature and condition of the articles.

To restore rubber stoppers that have become too hard for usefulness, digest them in 5 per cent soda lye for about 10 days at 86° to 104° F., replacing the lye repeatedly. Next, wash the stoppers in water and scrape off the softened outer layer with a knife, until no more can be removed. The stoppers (which have become quite soft and elastic again) are next rinsed in warm water to remove the caustic soda. If it is desired to trim them it should be done with a knife moistened with soap spirit.

Treatment and Utilization of Rubber Scraps.—The scraps, assorted according to their composition, are first cleaned by boiling to remove the adhering dirt, absorbed and adhering acids, salts, etc., as well as to eliminate the free sulphur. Next, the waste is ground between rollers and reduced to powder in emery grinders with automatic feeding. In many cases the material obtained may be added at once dry to the mixture, but generally it first receives a chemical treatment. This is carried out by boiling in caustic soda solution, or sulphuric or hydrochloric acid respectively, and steaming for about 20 hours with 4 atmospheres pressure.

According to another method, the ground scraps are steamed with soda lye under pressure, washed twice thoroughly for the elimination of the lye, and dried in the vacuum. Subsequently mix between cold rollers with 5 to 10 per cent of benzol or mineral oil and steam for some hours under hydraulic pressure at 4 atmospheres. The product thus obtained is rolled in plates and added to the mixture. The finely ground dry waste must not be stored for a long time in large quantities, as it hardens very easily and takes fire.

Old articles of vulcarized rubber are first "devulcanized" by grinding, boiling with caustic soda, and washing thoroughly. After drying, the scraps are heated to 302° F. with linseed oil in a kettle provided with stirring mechanism which is kept in continual motion. When the rubber has dissolved, a quantity of natural or coal-tar asphalt is added, and as soon as the contents of the kettle have become well mixed, the temperature is raised so high that dense fumes begin to rise and air is forced through the mass until a cooled sample shows the desired consistence. This composition being very tough and flexible, forms an excellent covering for electric cables. It finds many other uses, the proportions of rubber, asphalt, and oil being varied in accordance with the purpose for which it is designed.

Vulcanization.—Besides the Goodyear, Mason, and other patented processes, the process now usually followed in vulcanizing rubber stamps and similar small objects of rubber, is as follows:

Sulphur chloride is dissolved in carbon disulphide in various proportions, according to the degree of hardness the vulcanized object is to receive; the rubber cast is plunged in the solution and left there from 60 to 70 seconds. On removing, it is placed in a box or space warmed to 80° F., and left long enough for the carbon disulphide to evaporate, or about 90 to 100 seconds. It is then washed in a weakly alkaline bath of water, and dried.

Another method (recommended by Gerard) depends upon letting the rubber lie in a solution of potassium *ter* or *penta* sulphide, of 25° Bé., heated to about 280° F. for 3 hours.

Testing Rubber Gloves.—In testing rubber gloves it is best to inflate them with air, and then put them under water. Thus one may discover many small holes in new ones which otherwise would have been impossible to find.

Dissolving Old Rubber.—The material is shredded finely and then heated, under pressure, for several hours, with a strong solution of caustic soda. All cloth, paint, glue, fillers, etc., in the rubber are disintegrated, but the rubber is not affected. The mass is then washed repeatedly with water, to remove all alkali, and the resultant pure rubber may then be formed into sheets.

Rubber Stamps.—Set up the desired name and address in common type, oil the type and place a guard about ½ inch high around the form. Mix plaster of Paris to the proper consistence, pour in and allow it to set. Have the vulcanized rubber all ready, as made in long strips 3 inches wide and ⅛ of an inch thick, cut off the size of the intended stamp, remove the plaster cast from the type, and place both the cast and the rubber in a screw press, applying sufficient heat to thoroughly soften the rubber. Then turn down the screw hard and let it remain until the rubber receives the exact impression of the cast and becomes cold, when it is removed, neatly trimmed with a sharp knife, and cemented to the handle ready for use.

RUBBER CEMENTS:
See Adhesives.

RUBBER GLOVES, SUBSTITUTE FOR:
See Antiseptics.

RUBBER, ITS PROPERTIES AND USES IN WATERPROOFING:
See Waterproofing.

RUBBER VARNISHES:
See Varnishes.

RUBY SETTINGS:
See Watchmakers' Formulas.

RUOLTZ METAL:
See Alloys.

RUM, BAY:
See Bay Rum.

Rust Preventives

(See also Enamels, Glazes, Paints, Varnishes, Waterproofing.)

In spite of the numerous endeavors to protect metal objects from oxidation, a thoroughly satisfactory process has not yet been found, and we still have to resort to coatings and embrocations.

By covering the metals with a pale, colorless linseed-oil varnish, a fat or spirit lacquer, an unfailing protection against oxidation is obtained. This method, though frequently employed, however, is too laborious and expensive to admit of general use, and instead we frequently see employed ordinary or specially composed greases, especially for scythes, straw-knives, and many other bright iron goods. These greases are not suited to retard oxidation, for they are without exception acid-reacting bodies, which absorb oxygen in the air and under the action of light, thus rather assisting oxidation than retarding it. A covering of wax dissolved in oil of turpentine would be more recommendable, because wax is an impervious body, and a firm and rather hard layer remains after evaporation of the oil of turpentine, which excludes the air. If the treatment with the wax salve is carefully attended to no other objection can be urged against this preserving agent than that it is likewise comparatively expensive if used in large quantities. As regards the greases, and treatment with petroleum or vaseline, the easy attrition of these substances is another drawback, which makes a lasting protection impossible.

According to Shedlok, cast-iron articles are treated with acids, then exposed to the action of steam, hot or cold water, and dried. The receptacle is exhausted of air and a solution of pitch, rosin, rubber, or caoutchouc, applied under pressure. Objects prepared in this manner are said to be impervious even to weak acids.

The inoxidizing process of Ward is founded on the simultaneous employment of silicates and heat. The cast iron or wrought iron are coated with a siliceous mass by means of a brush or by immersion. This covering dries quickly, becomes liquid when the articles are exposed to a suitable heat, and soaks into the pores of the metal, forming a dense and uniform coat of dull black color after cooling, which is not changed by long-continued influence of the atmosphere, and which neither scales nor peels from the object. By the admixture of glass coloring matters to the siliceous mass, decorated surfaces may be produced.

Another inoxidation process for cast iron is the following: The cast-iron objects, such as whole gas chandeliers, water pipes, ornaments, balcony railings, cooking vessels, etc., are laid upon an iron sliding carriage 3.5 meters long and are exposed in a flame furnace of special construction first 15 minutes to the influence of gas generators with oxidizing action, then 20 minutes to such with reducing action. After being drawn out and cooled off the inoxidized pieces take on a uniform slate-blue shade of color, but can be enameled and ornamented in any manner desired. In applying the enamel the corroding with acid is obviated, for which reason the enamel stands exceedingly well.

A bronze-colored oxide coating which withstands outward influences fairly well, is produced as follows: The brightly polished and degreased objects are exposed from 2 to 5 minutes to the vapors of a heated mixture of concentrated hydrochloric acid and nitric acid (1:1) until the bronze color becomes visible on the articles. After these have been rubbed well with vaseline, heat once more until the vaseline commences to decompose. After cooling, the object is smeared well with vaseline. If vapors of a mixture of concentrated hydrochloric acid and nitric acid are allowed to act on the iron object, light reddish-brown shades are obtained, but if acetic acid is added to the above named two acids, oxide coatings of a bronze-yellow color can be obtained by the means of the vapors. By the use of different mixtures of acids any number of different colorings can be produced.

"Emaille de fer contre-oxide" is the name of an enamel which is said to protect iron pipes cheaply. The enamel is composed as follows: One hundred and thirty parts powdered crystal glass, 20.5 parts soda, 12 parts boracic acid. These substances mixed in the most careful manner are melted together in crucibles, the mass is chilled and transformed into a fine powder by crushing and grinding. The iron pipes and other objects of iron are first cleaned in the usual manner by corroding, dried and then coated with a very dilute gum arabic solution or any other gluing agent, and the powdered mass is spread over them by means of a sieve. The objects thus powdered are put in a room which is heated to 160° C. to drive out all moisture and are heated to dark redness, at which temperature the oxide coating melts.

Those processes, which produce a black protoxide layer on the iron by heating iron objects in supersaturated aqueous vapor, have not stood the test, as the layer formed will drop off or peel off after a short time, thus opening the way for rust after all.

The anti-rust composition called rubber oil is prepared as follows, according to the specification of the patent: The crude oil obtained by the dry distillation of brown oil, peat and other earthy substances are subjected to a further distillation. Thinly rolled India rubber, cut in narrow strips, is saturated with four times the bulk of the oil and left alone for a week or so. The mass thus composed is then subjected to the action of mineral sperm oil or a similar substance, until an entirely uniform clear substance has formed. This substance, which is applied on the metallic surfaces in as thin a layer as possible, forms a sort of film after slowly drying, which is perfectly proof against atmospheric influences.

The rust-preventive composition of Jones & Co., Sheffield, is a composition of wax, fat, turpentine, and small quantities of iron oxide.

According to a process patented by A. Buchner in Germany, the iron objects are first painted with a mixture of an alkaline glue solution and rosin soap. The alkaline mass enters all the pores and fissures and prevents the rust from extending under the coating. After the first coat is dry a second one is applied of the following composition: Five parts linseed oil boiled with peroxide of manganese; 2.25 parts turpentine; 0.25 parts benzol; 20 parts zinc dust, carbonate of calcium, lead oxide, or peroxide of manganese. The mixing of the liquid with the powders must be done immediately before use, as the mass solidifies after 10 hours, and is then no longer of working consistency. The second coating, which should only be thin, hardens quickly. The paint is weatherproof, does not peel off or blister, and adheres so firmly that it can only be removed with mechanical means.

A patented process to prevent rusting of wrought or cast iron consists in applying with a brush a strong solution of potassium dichromate and drying in a stove or over an open fire. Drying at ordinary temperature is not sufficient. To ascertain if the heat is strong enough the iron is moistened with a little water. So long as this takes up any color the heat must be increased. When the proper degree of heat is reached a fine deep black layer results, which is not acted upon by water, and protects the surface from the action of the atmosphere.

A permanent lustrous rust preventive is secured as follows: The well-cleaned iron parts are suspended for a few minutes in a blue vitriol solution, so that a delicate skin of copper forms on the surface; if the pieces rinsed off with water are then moved about for a few minutes in a solution of sodium hyposulphite faintly acidulated with hydrochloric acid, they assume a blue-black coating of copper sulphide, which is equally permanent in air and in water. The black surface may be immediately rinsed with water, dried with a rag or blotting paper, and polished at once. It possesses a steel-blue luster, adheres well to the iron, will stand treatment with the scratch brush, and protects against rust in a most satisfactory manner.

Black Sheet Rust Preventive.—Before black plate is ready to receive a rust protective coating, it is necessary to render the surface free from grease and scales, for which purpose the sheet iron is placed for some time into a warmed solution of 10 parts of sulphuric acid in 100 parts of water, whereby the impurities become detached, a process which may be assisted and accelerated by scouring with sand. Then rinse in clean water and rub dry in sawdust. The sheets thus prepared are placed for a short while into a feeble solution of blue vitriol, where they assume a reddish coloring. Next, they are rinsed in water, and after that moved to and fro, for a short time, in a feeble solution of hyposulphite of soda acidulated with a little hydrochloric acid. The result is a dark-blue coating on the sheets, which prevents all oxidation.

To Keep Machinery Bright.—I.—In order to keep machinery from rusting take 1 ounce of camphor, dissolve it in 1 pound of melted lard; take off the scum, and mix as much fine black lead as will give it iron color. Clean the machinery and smear it with this mixture. After 24 hours, rub clean with soft linen cloth. It will keep clean for months under ordinary circumstances.

II.—Mastic, transparent
grains.......... 10 parts
Camphor.......... 5 parts
Sandarac.......... 5 parts
Gum elemi......... 5 parts
Alcohol, wood, quantity sufficient to dissolve.

Mix and cover the articles with the solution. The latter will take the lacquer better if warmed slightly, but may be easily covered in the cold, if necessary.

Magnetic Oxide.—A layer of magnetic oxide is a good preservative from rust. To obtain it the objects are placed in the furnace at a temperature sufficient for decomposing steam. Steam superheated to 1,040° F. is then injected for from 4 to 6 hours. The thickness of the layer of oxide formed varies with the duration of the operation. This process can replace zincing, enameling, and tinning.

The deposit of magnetic oxide may also be obtained by electrolysis. The iron object is placed at the anode in a bath of distilled water heated to 176° F. The cathode is a copper plate, or the vessel itself, if it is of iron or copper. By electrolysis a layer of magnetic oxide is formed. Other peroxides may be deposited in the same manner. With an alkaline solution of litharge, a very adherent, brilliant, black deposit of peroxide of lead is secured. Too energetic a current must be avoided, as it would cause a pulverulent deposit. To obtain a good coating it is necessary, after putting the objects for a moment at the positive pole, to place them at the other pole until the oxide is completely reduced, and then bring them back to their first position.

Paper as Protection for Iron and Steel.—That paraffine paper is a very good protector of iron and steel has been proven by tests conducted by Louis H. Barker for the Pennsylvania Railroad. The mode of applying the paraffine paper is as follows: After the rust is carefully cleaned off by means of stiff wire brushes, a tacky paint is applied. The paper is then covered over and tightly pressed upon the painted surface, the joints of the paper slightly lapping. As soon as the paper is in place it is ready for the outside coat of paint. Iron and steel girders and beams subjected to the action of smoke and gases may thus be admirably protected from decomposition.

Anti-Rust Paper for Needles.—This is paper covered with logwood, and prepared from a material to which fine graphite powder has been added, and which has been sized with glue and alum. It is used for wrapping around steel goods, such as sewing needles, etc., and protecting them against rust. According to Lake, the paper is treated with sulphuric acid, like vegetable parchment, the graphite being sprinkled on before the paper is put into the water.

Rust Paper.—Rust paper is produced by coating strong packing paper with linseed-oil varnish, size, or any other binder, and sprinkling on the powder given in previous formula. For use the paper must be moistened with petroleum.

Anti-Rust Pastes.—I.—This preparation serves for removing rust already present, as well as for preventing same, by greasing the article with it: Melt 5 parts of crude vaseline on the water bath, and mix with 5 parts of finely levigated powdered pumice stone into a uniform mass. To the half-way cooled mass add ½ part of crude acid oxalate of potassium (sorrel salt) in a finely powdered state and grind into complete homogeneity.

II.—Dry tallow, 25 parts; white wax, 23 parts; olive oil, 22 parts; oil of turpentine, 25 parts; mineral oil, 10 parts. Apply with a brush at the fusing temperature of the mixture.

Rust Prevention for Iron Pipes.—The pieces of pipe are coated with tar and filled with light wood sawdust, which is set afire. This method will fully protect the iron from rust for an unlimited period, rendering a subsequent coat altogether superfluous.

Rust Preventive for Tools, etc.—I.—To preserve tools, dies, etc., from rust, they should be greased well with yellow vaseline. To use oil is not advisable, since all oils, except the dear ones, which are too expensive for this purpose, contain a certain percentage of acid that has an injurious effect upon the steel and iron articles. For greasing the cavities use a hard brush.

II.—Carefully heat benzine and add half its weight of white wax, which dissolves completely in this ratio. This solution is applied to the tools by means of a brush. It is also said to protect against the action of acidiferous fumes.

III.—Take a pound of vaseline and melt with it 2 ounces of blue ointment— what druggists call one-third—and add, to give it a pleasant odor, a few drops of oil of wintergreen, cinnamon, or sassafras. When thoroughly mixed pour into a tin can—an old baking-powder can will do. Keep a rag saturated with the preventive to wipe tools that are liable to rust.

To Separate Rusty Pieces.—By boiling the objects in petroleum, success is certain. It is necessary to treat them with alcohol or spirit to avoid subsequent oxidation, petroleum being in itself an oxidant.

To Protect Zinc Roofing from Rust.— Zinc sheets for roofing can easily be protected against rust by the following simple process. Clean the plates by immersing them in water to which 5 per cent of sulphuric acid has been added, then wash with pure water, allow to dry and coat with asphalt varnish. Asphalt varnish is prepared by dissolving 1 to 2 parts asphalt in 10 parts benzine; the solution should be poured evenly over the plates, and the latter placed in an upright position to dry.

RUST SPOT REMOVER:
See Cleaning Preparations and Methods.

SACCHARINE IN FOOD:
See Food.

SADDLE GALLS:
See Veterinary Formulas.

SADDLE SOAP:
See Soap.

SALAMANDRINE DESSERT:
See Pyrotechnics.

SALICYL (SWEET):
See Dentifrices.

SALICYLIC ACID IN FOOD:
See Foods.

SALICYLIC SOAP:
See Soap.

Salts, Effervescent

Granulated effervescent salts are produced by heating mixtures of powdered citric acid, tartaric acid, sodium bicarbonate, and sugar to a certain temperature, until they assume the consistency of a paste, which is then granulated and dried.

If effervescent caffeine citrate, antipyrin, lithium citrate, etc., are to be prepared, the powder need not be dried before effecting the mixture, but if sodium phosphate, sodium sulphate, or magnesium sulphate are to be granulated, the water of crystallization must first be removed by drying, otherwise a hard, insoluble and absolutely non-granulable mass will be obtained. Sodium phosphate must lose 60 per cent of its weight in drying, sodium sulphate 56 per cent, and magnesium sulphate 23 per cent.

Naturally, water and carbonic acid escape on heating, and the loss will increase with the rise of temperature. For the production of the granulation mass it must not exceed 158° F., and for drying the grains a temperature of 122° F. is sufficient.

The fineness of the mesh should vary according to the necessary admixture of sugar and the size of the grains.

If the ingredients should have a tendency to cling to the warm bottom, an effort should be made immediately upon the commencement of the reaction to cause a new portion of the surface to come in contact with the hot walls.

When the mass is of the consistency of paste it is pressed through a wire sieve, paper or a fabric being placed underneath. Afterwards dry at sufficient heat. For wholesale manufacture, surfaces of large size are employed, which are heated by steam.

In the production of substances containing alkaloids, antipyrin, etc., care must be taken that they do not become colored. It is well, therefore, not to use heat, but to allow the mixture to stand in a moist condition for 12 hours, adding the medicinal substances afterwards and kneading the whole in a clay receptacle. After another 12 hours the mass will have become sufficiently paste-like, so that it can be granulated as above.

According to another much employed method, the mass is crushed with alcohol, then rubbed through a sieve, and dried rapidly. This process is somewhat dearer, owing to the great loss of alcohol, but presents the advantage of furnishing a better product than any other recipe.

Effervescent magnesium citrate cannot be very well made; for this reason the sulphate was used in lieu of the citrate. A part of the customary admixture of sulphate is replaced by sugar and aromatized with lemon or similar substances.

An excellent granulation mass is obtained from the following mixture by addition of alcohol:

	Parts by weight
Sodium bicarbonate......	30
Tartaric acid............	15
Citric acid.............	13
Sugar.................	30

The total loss of this mass through granulation amounts to from 10 to 15 per cent.

To this mass, medicinal substances, such as antipyrin, caffeine citrate, lithium citrate, lithium salicylate, phenacetin, piperacin, ferric carbonate, and pepsin may be added, as desired.

In order to produce a quinine preparation, use tincture of quinine instead of alcohol for moistening; the quinine tincture is prepared with alcohol of 96 per cent.

Basis for Effervescent Salts.—

Sodium bicarbonate,
   dried and powdered 53 parts
Tartaric acid, dried
   and powdered..... 28 parts
Citric acid, unefflor-
   esced crystals...... 18 parts

Powder the citric acid and add the tartaric acid and sodium bicarbonate. This basis may be mixed with many of the medicaments commonly used in the form of granular effervescent salts, in the proportion which will properly represent their doses and such substances as sodium phosphate, magnesium sulphate, citrated caffeine, potassium bromide, lithium citrate, potassium citrate, and others, will produce satisfactory products. A typical formula for effervescent sodium phosphate would be as follows:

Sodium phosphate,
   uneffloresced crys-
   tals............. 500 parts
Sodium bicarbonate,
   dried and pow-
   dered............ 477 parts
Tartaric acid, dried
   and powdered.... 252 parts
Citric acid, unefflor-
   esced crystals..... 162 parts

Dry the sodium phosphate on a water bath until it ceases to lose weight; after powdering the dried salt, mix it intimately with the citric acid and tartaric acid, then thoroughly incorporate the sodium bicarbonate. The mixed powders are now ready for granulation. The change in manipulation which is suggested to replace that usually followed, requires either a gas stove or a blue-flame coal-oil stove, and one of the small tin or sheet-iron ovens which are so largely used with these stoves. The stove itself will be found in almost every drug store; the oven costs from $1 to $2.

The oven is heated to about 200° F. (the use of a thermometer is desirable at first, but one will quickly learn how to regulate the flame to produce the desired temperature), and the previously mixed powders are placed on, preferably, a glass plate, which has been heated with the oven, about ½ pound being taken at a time, dependent upon the size of the oven. The door of the oven is now closed for about one minute, and, when opened, the whole mass will be found to be uniformly moist and ready to pass through a suitable sieve, the best kind and size being a tinned iron, No. 6. This moist, granular powder may then be placed upon the top of the oven, where the heat is quite sufficient to thoroughly dry the granules, and the operator may proceed immediately with the next lot of mixed powder, easily granulating 10 or more pounds within an hour.

Sugar has often been proposed as an addition to these salts, but experience has shown that the slight improvement in taste, which is sometimes questioned, does not offset the likelihood of darkening, which is apt to occur when the salt is being heated, or the change in color after it has been made several months. It should be remembered that in making a granular effervescent salt by the method which depends upon the liberation of water of crystallization, a loss in weight, amounting to about 10 per cent, will be experienced. This is due, in part, to the loss of water which is driven off, and also to a trifling loss of carbon dioxide when the powder is moistened.

EFFERVESCENT POWDERS:

Magnesian Lemonade Powder.—

Fine white sugar.....	2 pounds
Magnesium carbonate	6 ounces
Citric acid..........	4 ounces
Essence of lemon....	2 drachms

Rub the essence into the dry ingredients, work well together, sift, and bottle.

Magnesian Orgeat Powder.—

Fine sugar..........	1 pound
Carbonate of magne-	
sia...............	3 ounces
Citric acid..........	1 ounce
Oil of bitter almonds.	3 drops
Vanilla flavoring, quantity sufficient.	

Thoroughly amalgamate the dry ingredients. Rub in the oil of almonds and sufficient essence of vanilla to give a slight flavor. Work all well together, sift, and bottle.

Raspberryade Powder.—

Fine sugar......... ..	2 pounds
Carbonate of soda....	2 ounces
Tartaric acid........	2 ounces
Essence of raspberry.	4 drachms
Carmine coloring, quantity sufficient.	

Rub the essence well into the sugar, and mix this with the soda and acid. Then work in sufficient liquid carmine to make the powder pale red, sift through a fine sieve, and pack in air-tight bottles.

Ambrosia Powder.—

Fine sugar	2 pounds
Carbonate of soda	12 drachms
Citric acid	10 drachms
Essence of ambrosia	20 drops

Amalgamate the whole of the above, and afterwards sift and bottle in the usual manner.

Noyeau Powder.—

Fine sugar	2 pounds
Carbonate of soda	12 drachms
Tartaric acid	10 drachms
Essence of Noyeau	6 drops

After the dry ingredients have been mixed, and the essence rubbed into them, sift and bottle the powder.

Lemon Sherbet.—

Fine sugar	9 pounds
Tartaric acid	40 ounces
Carbonate of soda	36 ounces
Oil of lemon	2 drachms

Having thoroughly mixed the dry ingredients, add the lemon, rubbing it well in between the hands; then sift the whole thrice through a fine sieve, and cork down tight.

As oil of lemon is used in this recipe, the blending must be quite perfect, otherwise when the powder is put in water the oil of lemon will float.

Any other flavoring may be substituted for lemon, and the sherbet named accordingly.

Cream Soda Powder.—

Fine sugar	30 parts
Tartaric acid	7 parts
Carbonate of soda	6 parts
Finely powdered gum arabic	1 part
Vanilla flavoring, quantity sufficient.	

Proceed exactly as for lemon sherbet.

Kissingen Salt.—

Potassium chloride	17 parts
Sodium chloride	367 parts
Magnesium sulphate (dry)	59 parts
Sodium bicarbonate	107 parts

For the preparation of Kissingen water, dissolve 1.5 grams in 180 grams of water.

Vichy Salt.—

Sodium bicarbonate	846 parts
Potassium carbonate	38 parts
Magnesium sulphate (dry)	38 parts
Sodium chloride	77 parts

For making Vichy water dissolve 1 part in 200 parts of water.

Seidlitz Salt.—This is one of the many old names for magnesium sulphate. It has at various times been known as Seidlitz salt, Egra salt, canal salt, bitter salt, cathartic salt, English salt, and Epsom salt. Its earliest source was from the salt springs of Epsom in England and from this fact it took its last two names. For a long time sea-salt makers supplied the markets of the world. They procured it as a by-product in the making of salt. The bitter water that remained after the table salt had been crystallized out was found to contain it. Now it is chiefly procured from such minerals as dolomite, siliceous magnesium hydrate, and schistose rock containing the sulphide of magnesia. Many medical men deem it our best saline cathartic.

SALTS, SMELLING.

I.—Moisten coarsely powdered ammonium carbonate with a mixture of

Strong tincture of orris root	2½ ounces
Extract of violet	3 drachms
Spirit of ammonia	1 drachm

II.—Fill suitable bottles with coarsely powdered ammonium carbonate, and add to the salt as much of the following solution as it will absorb:

Oil of orris	5 minims
Oil of lavender flowers	10 minims
Extract of violet	30 minims
Stronger water of ammonia	2 ounces

SALVES:

See Ointments.

SAND:

Colored Sand.—Sift fine white sand from the coarser particles and color it as follows:

I.—Blue.—Boil 106 parts of sand and 4 of Berlin blue with a small quantity of water, stirring constantly, and dry as soon as the sand is thoroughly colored.

II.—Black Sand.—Heat very fine quartz sand, previously freed from dust by sifting, and add to every ¼ pound of it 6 to 8 spoonfuls of fat. Continue the heating as long as smoke or a flame is observed on stirring. The sand is finally washed and dried. This black sand will not rub off.

III.—Dark-Brown Sand.—Boil white sand in a decoction of brazil wood and dry it over a fire.

IV.—Rose-colored sand is obtained by mixing 100 parts of white sand with 4 parts of vermilion.

Lawn Sand.—Lawn sand may be prepared by mixing crude ammonium sulphate, 65 parts, with fine sand, 35 parts. This mixture will kill daisies and plantains, but does not permanently injure the grass of lawns. A most effective method of killing plantains is to put, during dry weather, a full teaspoonful of common salt in the head of each.

SAND HOLES IN BRASS:
See Castings.

SAND SOAP:
See Soap.

SANDSTONE CEMENTS:
See Adhesives.

SANDSTONE COATING:
See Acid-Proofing.

SANDSTONES, TO REMOVE OIL SPOTS FROM:
See Cleaning Preparations and Methods.

SAND, TO PREVENT ADHESION OF SAND TO CASTINGS:
See Castings.

SARSAPARILLA.

Each fluidounce of Ayer's sarsaparilla represents

Sarsaparilla root.....	10 parts
Yellow dock root.....	8 parts
Licorice root........	8 parts
Buckthorn bark.....	4 parts
Burdock root........	3 parts
Senna leaves........	2 parts
Black cohosh root....	2 parts
Stillingia root.......	4 parts
Poke root...........	1 part
Cinchona red bark...	2 parts
Potassium iodide....	4 parts

Solvent.—Alcohol, $10\frac{1}{2}$ minims to each fluidrachm; glycerin, syrup, water.

This is the formula as given by Dr. Charles H. Stowell, of the Ayer Company, to the daily papers, for advertising purposes.

Sarsaparilla Flavoring.—

Oil wintergreen......	6 parts
Oil sassafras........	2 parts
Oil cassia..........	$1\frac{1}{2}$ parts
Oil clove...........	$1\frac{1}{2}$ parts
Oil anise..........	$1\frac{1}{2}$ parts
Alcohol............	60 parts

Sarsaparilla Syrup.—

Simple syrup........	40 ounces
Sarsaparilla flavoring.	1 drachm
Caramel to color.	

SARSAPARILLA EXTRACT:
See Essences and Extracts.

SAUCES, TABLE:
See Condiments.

SATINWOOD:
See Wood.

SAUSAGE COLOR:
See Foods.

SAWDUST IN BRAN:
See Bran.

SAWDUST FOR JEWELERS AND WATCHMAKERS:
See Watchmakers' Formulas.

SCALD HEAD, SOAP FOR:
See Soap.

SCALD REMEDIES:
See Cosmetics.

SCALE FOR PHOTOGRAPHIC REDUCTION:
See Photography.

SCALE PAN CLEANER:
See Cleaning Preparations and Methods.

SCALE IN BOILERS:
See Boiler Compounds.

SCALE INSECTS, EXTERMINATION OF:
See Insecticides.

SCALP WASHES:
See Hair Preparations.

SCISSORS HARDENING:
See Steel.

SCOURING LIQUIDS:
See Laundry Preparations.

SCRATCH BRUSHING:
See Plating, under Gilding.

SCREWS:

To Prevent Screws from Rusting and Becoming Fast.—Screws will sometimes rust in their seats, even when carefully oiled before driving them to their seats, but if they are anointed with a mixture of graphite and soft tallow they will remain unrusted and unaltered for years.

A screw rusted in may also be removed by placing the flat extremity of a red-hot rod of iron on it for 2 or 3 minutes. When the screw is heated, it will be found to turn quite easily.

SCREWS, BLUEING:
See Steel.

SCREWS IN WATCHES:
See Watchmakers' Formulas.

SEALING (BURNING) TRICK:
See Pyrotechnics.

SEALING WAX
See Waxes.

SEA SICKNESS.

I.—To prevent sea sickness, take 2 or 3 grams of potassium bromide dissolved in plain or carbonated water every evening either with supper or just before retiring for several weeks before going on the voyage. During the voyage, breathing should be deep and a tight bandage should be worn around the abdomen.

II.—Menthol............ 0.1 part
Cocaine hydrochloride.......... 0.2 parts
Alcohol............... 60.0 parts
Syrup............... 30.0 parts

A dessertspoonful to be taken at intervals of half an hour.

SEASONINGS:
See Condiments.

SEED, BIRD:
See Bird Foods.

SEEDS, TESTS FOR FOREIGN:
See Foods.

SEIDLITZ POWDERS:
See Salts (Effervescent).

SELTZER WATER:
See Water.

SERPENTS, PHARAOH'S.

An old form consisted of pellets of a very poisonous mercurial compound which gave off dangerous fumes when heated. The "eggs" may be made of comparatively safe material by the following formula:

Potassium bichromate. 2 parts
Potassium nitrate..... 1 part
White sugar.......... 2 parts

Powder each ingredient separately, mix, and press into small paper cones. These must be kept from light and moisture.

Of course, neither this nor other chemical toys containing substances in the slightest degree harmful if swallowed should be placed in the hands of children not old enough fully to understand the danger of eating or even tasting unknown things.

SERVIETTES MAGIQUES:
See Polishes.

SETTING OF TOOLS:
See Tool Setting.

SEWING-MACHINE OIL:
See Lubricants.

SHAMPOO LOTIONS AND PASTES:
See Hair Restorers and Soaps.

SHARPENING PASTES:
See Razor Pastes.

SHARPENING STONES:
See Whetstones.

SHAVING PASTE.

An emulsion of paraffine wax, melting at 131° F., should be used. This is prepared with 25 per cent of wax and 2 per cent of tragacanth, the wax being melted and mixed with the tragacanth previously made into a mucilage with some of the water. The addition of a little stearine or lard renders the emulsification of the wax easier, while about 10 per cent of alcohol makes the preparation more agreeable to use. The fatty odor of the preparation may be covered by the addition of ½ to 1 per cent of lavender oil, and the finished product then appears as a thick white cream. In use a small quantity is rubbed over the area to be shaved and the razor immediately applied. As the water in the emulsion evaporates, the particles of wax previously distributed in the emulsion become coherent and fill up the depressions in the surface of the skin from which the hairs arise, thus forming a mechanical support during the passage of the razor. The quantity required is very small, 1 ounce being sufficient for shaving the face about 6 times.

SHAVING SOAP:
See Soap.

SHEEP-DIPS:
See Disinfectants.

SHEEP DISEASES:
See Veterinary Formulas.

SHELL CAMEOS.

If shell cameos and corals have become too hot in cementing and cracks have appeared in consequence, olive oil is applied and allowed to soak in by heating. The same process is employed for shell cameos which have developed white fissures, owing to being filed smaller.

SHELL, IMITATION OF:
See Casein Compounds.

SHELLS, LUBRICANTS FOR RE-DRAWING:
See Lubricants.

SHELL POLISHES:
See Polishes.

SHELLAC:
See Varnishes.

SHELLAC BLEACHING.

In bleaching, shellac is brought into contact with an acidified solution of chloride of lime for some time, then washed, kneaded in hot water, placed back into the chloride of lime solution, and brushed. Through this treatment with the chloride of lime solution the bleached shellac sometimes loses its solubility in alcohol, which, however, can be restored if the shellac is melted in boiling water, or if it is moistened with a little ether in a well-closed vessel. A quantity of ether in the proportion of 1 part to 20 parts shellac is sufficient. Great caution is recommended in the handling of ether. The ether vapors easily ignite when in proximity to a burning light and a mixture of ether vapor and atmospheric air may cause most vehement explosions. After an action of the ether upon the shellac for several hours, the alcohol necessary to dissolve it may either be added directly or the shellac moistened with ether is placed in the open air for half an hour in a dish, after which time the ether will have evaporated and the shellac can then be dissolved by the use of alcohol.

Bleached shellac is known to lose its solubility in alcohol, especially if treated with chlorine in bleaching. This solubility can be readily restored, however, by first moistening the rosin with $\frac{1}{20}$ its weight of ether, placing it in a closed vessel and allowing it to swell there. Shellac thus treated becomes perfectly soluble again.

SHIMS IN ENGINE BRASSES.

In taking up the wear of engine brasses on wrist pin or crosshead pin when the key is driven clear down, back out the key and instead of putting in sheet-iron shims, put in a small piece of pine wood of just the right thickness to allow the key to come even with the under side of the strap, then pour in melted babbitt. A hole must be drilled through the flange of the brasses to allow for pouring the babbitt.

Every engineer knows the trouble it is to put several shims between the brass box and the end of the strap, especially if the box is a round-end one, as many are. By using the method described, brasses may be worn up much closer, even if worn through; the babbitt will form part of the bearing.

Shoe Dressings
(See also Leather.)

Acid-Free Blacking.—

Lampblack......	27–36 parts
Bone black......	3 parts
Syrup..........	60–70 parts

Put in a kettle and under gentle heat stir together until a smooth, homogeneous mass has been attained. In another kettle put 3 parts of finely shredded gutta percha and warm over an open fire until it begins to run, then add, with constant stirring, 5 parts of olive oil, continuing the heat until the gum is completely dissolved. When this occurs dissolve in 1 part of stearine, and add the whole while still hot in a slow stream, and under diligent and constant stirring, to the mixture of syrup and blacks. Continue the agitation of the mass until it is completely homogeneous. Now dissolve 4 parts of Senegal gum in 12 parts of water, and add the solution to the foregoing mass. Stir well in and finally add sufficient mirbane (about $\frac{1}{4}$ part) to perfume.

Blacking Pastes.—While shellac is not soluble in water alone, it is soluble in water carrying borax, the alkaline carbonates, etc. In paste blacking the object of the sulphuric acid is to remove from the bone black the residual calcium phosphate. The ordinary bone black of commerce consists of only about 10 per cent of carbon, the residue being chiefly calcium phosphate. This is the reason that we cannot obtain a pure black color from it, but a dirty brown. To make a good blacking, one that is of a black in color, either use purified bone black, or a mineral acid (sulphuric or hydrochloric acid) with crude bone black. The residual acid is entirely neutralized by the sodium carbonate and has no bad effect on the leather. The following formula contains no acid and makes a good paste:

I.—Marseilles soap...	122 parts
Potassium carbonate........	61 parts
Beeswax.........	500 parts
Water..........2,000 parts	

Mix and boil together with occasional stirring until a smooth, homogeneous paste is obtained, then add, a little at a time, and under constant stirring, the following:

Rock candy, pow-
dered......... 153 parts
Gum arabic, pow-
dered......... 61 parts
Ivory black......1,000 parts

Stir until homogeneous, then pour, while still hot, into boxes.

The following makes a very brilliant and durable black polish for shoes:

II.—Bone black...... 40 parts
Sulphuric acid... 10 parts
Fish oil......... 10 parts
Sodium carbonate
crystal....... 18 parts
Sugar, common
brown, or mo-
lasses .... 20 parts
Liquid glue, pre-
pared as below. 20 parts
Water, sufficient.

Soak 10 parts of good white glue in 40 parts of cold water for 4 hours, then dissolve by the application of gentle heat, and add 1.8 parts of glycerine (commercial). Set aside. Dissolve the sodium carbonate in sufficient water to make a cold saturated solution (about 3 parts of water at 60° F.), and set aside. In an earthenware vessel moisten the bone black with a very little water, and stirring it about with a stick, add the sulphuric acid, slowly. Agitate until a thick dough-like mass is obtained, then add and incorporate the fish oil. Any sort of animal oil, or even colza will answer, but it is best to avoid high-smelling oils. Add a little at a time, and under vigorous stirring, sufficient of the saturated sodium carbonate solution to cause effervescence. Be careful not to add so freely as to liquefy the mass. Stir until effervescence ceases, then add the molasses or sugar, the first, if a soft, damp paste is desired, and the latter if a dryer one is wanted. Finally, add, a little at a time, and under constant stirring, sufficient of the solution of glue to make a paste of the desired consistency. The exact amount of this last ingredient that is necessary must be learned by experience. It is a very important factor, as it gives the finished product a depth and brilliancy that it could not otherwise have, as well as a certain durability, in which most of the blackings now on the market are deficient.

III.—Soap............ 122 parts
Potassium car-
bonate....... 61 parts
Beeswax......... 500 parts
Water..........2,000 parts

Mix and boil together until a smooth, homogeneous paste is obtained, then add

Bone black......1,000 parts
Powdered sugar.. 153 parts
Powdered gum
arabic........ 61 parts

Mix thoroughly, remove from the fire, and pour while still hot into boxes.

Boot-Top Liquid.—

Solution of muriate of
tin................ 3 drachms
French chalk (in pow-
der)............. 1 ounce
Salt of sorrel .... ½ ounce
Flake white......... 1 ounce
Burnt alum......... ½ ounce
Cuttle-fish bones
(powdered)....... 1 ounce
White arsenic....... 1 ounce
Boiling water........ 1 quart

Brown Dressing for Untanned Shoes.—

Yellow wax......... 30 parts
Soap.............. 12 parts
Nankin yellow....... 15 parts
Oil of turpentine..... 100 parts
Alcohol............. 12 parts
Water............. 100 parts

Dissolve in the water bath the wax in the oil of turpentine; dissolve, also by the aid of heat, the soap in the water, and the Nankin yellow (or in place of that any of the yellow coal-tar colors) in the alcohol. Mix the solutions while hot, and stir constantly until cold. The preparation is smeared over the shoes in the usual way, rubbed with a brush until evenly distributed, and finally polished with an old silk or linen cloth.

Heel Polish.—

I.—Carnauba wax.... 5 parts
Japanese wax..... 5 parts
Paraffine......... 5 parts
Oil of turpentine.. 50 parts
Lampblack ...... 1 part
Wine black....... 2 parts

Melt the wax and the paraffine, and when this has become lukewarm, add the turpentine oil, and finally the lampblack and the wine black. When the black color has become evenly distributed, pour, while still lukewarm, into tin cans.

II.—Melt together Japanese wax, 100 parts; carnauba wax, 100 parts; paraffine, 100 parts; and mix with turpentine oil, 500 parts, as well as a trituration of lampblack, 10 parts; wine black, 20 parts; turpentine oil, 70 parts.

LIQUID BLACKINGS.

The following formulas make a product of excellent quality:

I.—Ivory black....... 120 parts
Brown sugar...... 90 parts
Olive oil......... 15 parts
Stale beer........ 500 parts

Mix the black, sugar and olive oil into a smooth paste, adding the beer, a little at a time, under constant stirring. Let stand for 24 hours, then put into flasks, lightly stoppered.

II.—Ivory black....... 200 parts
Molasses......... 200 parts
Gallnuts, bruised. 12 parts
Iron sulphate..... 12 parts
Sulphuric acid.... 40 parts
Boiling water..... 700 parts

Mix the molasses and ivory black in an earthen vessel. In an iron vessel let the gallnuts infuse in 100 parts of boiling water for 1 hour, then strain and set aside. In another vessel dissolve the iron sulphate; in another, 100 parts of the boiling water. One-half of this solution is added at once to the molasses mixture. To the remaining half add the sulphuric acid, and pour the mixture, a little at a time, under constant stirring, into the earthen vessel containing the molasses mixture. The mass will swell up and thicken, but as soon as it commences to subside, add the infusion of gallnuts, also under vigorous stirring. If a paste blacking is desired the preparation is now complete. For a liquid black add the remaining portion of the boiling water (500 parts), stir thoroughly and bottle.

Patent-Leather Polish.—

Yellow wax or ceresine 3 ounces
Spermaceti......... 1 ounce
Oil of turpentine..... 11 ounces
Asphaltum varnish... 1 ounce
Borax.............. 80 grains
Frankfort black...... 1 ounce
Prussian blue........ 150 grains

Melt the wax, add the borax, and stir until an emulsion has been formed. In another pan melt the spermaceti; add the varnish, previously mixed with the turpentine; stir well and add to the wax; lastly add the colors.

Preservatives for Shoe Soles.—I.—This preparation, destined for impregnating leather shoe soles, is produced as follows: Grind 50 parts of linseed oil with 1 part of litharge; next heat for 2 hours to the boiling point with ½ part of zinc vitriol, which is previously calcined (dehydrated). The composition obtained in this manner, when perfectly cold, is mixed with 8 parts of benzine and filled in bottles or other receptacles. To render this preservative effective, the soles must be coated with it until the leather absorbs it.

II.—Dissolve ordinary household soap in water; on the other hand, dissolve an aluminum salt—the cheapest is the commercial aluminum sulphate—in water and allow both solutions to cool. Now pour the aluminum salt solution, with constant stirring, into the soap solution, thereby obtaining a very fine precipitate of aluminum oleate. The washed-out residue is dried with moderate heat. By adding 10 to 30 per cent to petroleum with slight heating, a solid petroleum of vaseline-like consistency is received, which may be still further solidified by additional admixture. A 10 per cent solution of aluminum oleate in petroleum is a very excellent agent for preserving the soles, a single saturation of the soles sufficing forever. The sole will last about 1 year.

III.—The following mixture is prepared by melting together over the fire in an enameled iron vessel: Vaseline, 400 parts; ceresine, 100 parts. The melted mass, which is used as a grease, is filled in wooden boxes or tin cans.

IV.—The oleic acid of the stearine factories is heated with strong alcohol and sulphuric acid. Take 16 parts of oleic acid, 2 parts of alcohol (90 per cent), and 1 part of concentrated sulphuric acid. The oleic-acid ether formed separates as a thin brownish oil. It is liberated from free sulphuric acid and the alcohol in excess by agitation with warm water and allowing to settle. This oleic-acid ether is mixed with the same weight of fish oil, and 4 to 8 parts of nitro-benzol are added per 1,000 parts to disguise the odor.

TAN AND RUSSET SHOE POLISHES:

To Renovate and Brighten Russet and Yellow Shoes.—First, clean off all dirt and dust with a good stiff brush, then with a sponge dipped in benzine go over the leather, repeating the process as soon as the benzine evaporates. A few wipings will bring back the original color. Then use a light-yellow dressing and brush well.

The liquid application consists usually of a solution of yellow wax and soap in oil of turpentine, and it should be a matter of no difficulty whatever to compound a mixture of this character at least equal to the preparations on the market. As a type of the mixture occasionally recommended we may quote the following:

I.—Yellow wax........ 4 ounces
    Pearl ash.......... 4 drachms
    Yellow soap........ 1 drachm
    Spirit of turpentine. 7 ounces
    Phosphine (aniline). 4 grains
    Alcohol........... 4 drachms
    Water, a sufficient quantity.

Scrape the wax fine and add it, together with the ash and soap, to 12 ounces of water. Boil all together until a smooth, creamy mass is obtained; remove the heat and add the turpentine and the aniline (previously dissolved in the alcohol). Mix thoroughly, and add sufficient water to bring the finished product up to 1½ pints.

II.—Water............. 18 parts
    Rosin oil.......... 4½ parts
    Spirit of sal ammo-
      niac, concentrated 1⅛ parts
    White grain soap... 1.93 parts
    Russian glue....... 1.59 parts
    Brown rock candy.. 0.57 parts
    Bismarck brown.... 0.07 parts

Boil all the ingredients together, excepting the pigment; after all has been dissolved, add the Bismarck brown and filter. The dressing is applied with a sponge.

III.—Beeswax, yellow.... 2 ounces
    Linseed oil......... 3 ounces
    Oil turpentine...... 10 ounces

Dissolve by heat of a water bath, and add 1¼ ounces soap shavings, hard yellow. Dissolve this in 14 ounces of hot water.

IV.—A simpler form of liquid mixture consists of equal parts of yellow wax and palm oil dissolved with the aid of heat in 3 parts of oil of turpentine.

V.—Soft or green soap... 1 ounce
    Linseed oil, raw.... 2 ounces
    Annatto solution (in
      oil)............ 7 ounces
    Yellow wax........ 2 ounces
    Gum turpentine.... 7 ounces
    Water............. 7 ounces

Dissolve the soap in the water and add the solution of annatto; melt the wax in the oil of turpentine, and gradually stir in the soap solution, stirring until cold.

The paste to accompany the foregoing mixtures is composed of yellow wax and rosin thinned with petrolatum, say 4 parts of wax, 1 part of rosin, and 12 parts of petrolatum.

Paste Dressings for Russet Shoes.— The paste dressings used on russet leather consist of mixtures of wax with oil and other vehicles which give a mixture of proper working quality.

A simple formula is:

I.—Yellow wax........ 9 parts
    Oil of turpentine.... 20 parts
    Soap.............. 1 part
    Boiling water...... 20 parts

Dissolve the wax in the turpentine on a water bath and the soap in the water and stir the two liquids together until the mixture becomes sufficiently cold to remain homogeneous.

Another formula in which stearine is used is appended:

II.—Wax.............. 1 part
    Stearine........... 2 parts
    Linseed oil......... 1 part
    Oil of turpentine.... 6 parts
    Soap.............. 1 part
    Water............. 10 parts

Proceed as above.

Carnauba wax is often used by manufacturers of such dressings instead of beeswax, as it is harder and takes a higher polish. These dressings are sometimes colored with finely ground yellow ocher or burnt umber. If the leather be badly worn, however, it is best to apply a stain first, and afterwards the waxy dressing.

Suitable stains are made by boiling safflower in water, and annatto is also used in the same way, the two being sometimes mixed together. Oxalic acid darkens the color of the safflower. Aniline colors would also doubtless yield good results with less trouble and expense. By adding finely ground lampblack to the waxy mixture instead of ocher, it would answer as a dressing for black leather.

WATERPROOF SHOE DRESSINGS.

I.—Caoutchouc....... 10 parts
    Petroleum........ 10 parts
    Carbon disulphide. 10 parts
    Shellac........... 40 parts
    Lampblack........ 20 parts
    Oil lavender...... 1 part
    Alcohol.......... 200 parts

Upon the caoutchouc in a bottle pour the carbon disulphide, cork well, and let stand a few days, or until the caoutchouc has become thoroughly gelatinized or partly dissolved. Then add the petroleum, oil of lavender, and alcohol, next the shellac in fine powder, and heat it to about 120° F., taking care that as little as possible is lost by evaporation. When the substances are all dissolved and the liquid is tolerably clear, add the lampblack, mix thoroughly, and fill at once into small bottles.

II.—A waterproof blacking which will give a fine polish without rubbing, and will not injure the leather:

Beeswax	18 parts
Spermaceti	6 parts
Turpentine oil	66 parts
Asphalt varnish	5 parts
Powdered borax	1 part
Frankfort black	5 parts
Prussian blue	2 parts
Nitro-benzol	1 part

Melt the wax, add the powdered borax and stir till a kind of jelly has formed. In another pan melt the spermaceti, add the asphalt varnish, previously mixed with the oil of turpentine, stir well, and add to the wax. Lastly add the color previously rubbed smooth with a little of the mass. The nitro-benzol gives fragrance.

Waterproof Varnish for Beach Shoes.—
Yellow.—

Water	150 parts
Borax	5 parts
Glycerine	3 parts
Spirit of ammonia	1 part
White shellac	25 parts
Yellow pigment, water soluble	1 part
Formalin, a few drops.	

Orange.—

Water	150	parts
Borax	5	parts
Glycerine	2	parts
Spirit of ammonia	1	part
Ruby shellac	22	parts
Orange, water soluble	1	part
Brown	0.3	parts
Formalin	0.1	part

Pale Brown.—

Water	150	parts
Borax	5	parts
Glycerine	2	parts
Spirit of ammonia	0.25	parts
White shellac	25	parts
Yellow, water soluble	8	parts
Orange	0.3	parts
Formalin	0.1	part

Stir the glycerine and the spirit of ammonia together in a special vessel before putting both into the kettle. It is also advisable, before the water boils, to pour a little of the nearly boiling water into a clean vessel and to dissolve the colors therein with good stirring, adding this solution to the kettle after the shellac has been dissolved.

White Shoe Dressing.—

I.—Cream of tartar	3	ounces
Oxalic acid	1	ounce
Alum	1	ounce
Milk	3	pints

Mix and rub on the shoes. When they are thoroughly dry, rub them with a mixture of prepared chalk and magnesium carbonate.

II.—Water	136	parts
Fine pipe clay	454	parts
Shellac, bleached	136	parts
Borax, powdered	68	parts
Soft soap	8	parts
Ultramarine blue	5	parts

Boil the shellac in the water, adding the borax, and keeping up the boiling until a perfect solution is obtained, then stir in the soap (5 or 6 parts of "ivory" soap, shaved up, and melted with 2 or 3 parts of water, is better than common soft soap), pipe clay, and ultramarine. Finally strain through a hair-cloth sieve. This preparation, it is said, leaves absolutely nothing to be desired. A good deal of stiffness may be imparted to the leather by it. The addition of a little glycerine would remedy this. The old application should be wiped away before a new one is put on. This preparation is suitable for military shoes, gloves, belts, and uniforms requiring a white dressing.

SHOES, WATERPROOFING:
See Waterproofing.

SHIO LIAO:
See Adhesives, under Cements.

SHIP COMPOSITIONS AND PAINTS:
See Paints.

SHOW BOTTLES FOR DRUGGISTS:
See Bottles.

SHOW CASES.

Dents in show cases and counters, and, indeed, almost all forms of "bruises" on shop and other furniture, may be removed by the exercise of a little patience, and proceeding as follows: Sponge the place with water as warm as can be borne by the hand. Take a piece of filtering or other bibulous paper large enough to fold 6 or 8 times and yet cover the bruise, wet in warm water and place over the spot. Take a warm (not hot) smoothing iron and hold it on the paper until the moisture is evaporated (renewing its heat, if necessary). If the bruise does not yield to the first trial, repeat the process. A dent as large as a dollar and ¼ inch deep in the center, in black walnut of tolerably close texture, was brought up smooth and level with the surrounding surface by two applications of the paper and iron as described. If the bruise be small, a sponge dipped in warm water placed upon it, renewing the warmth from time to time, will be all-sufficient. When the dent is removed and the wood dry, the polish can be restored by any of the usual processes. If the wood was originally finished in oil, rub with a little boiled linseed cut with acetic acid (oil, 8 parts; acid, 1 part). If it was "French polished," apply an alcoholic solution of shellac, and let dry; repeat if necessary, and when completely dry proceed as follows: Rub the part covered with shellac, first with crocus cloth and a few drops of olive oil, until the ridges, where the new and old polish come together, disappear; wipe with a slightly greased but otherwise clean rag and finish with putz pomade.

SHOW-CASE SIGNS:
See Lettering.

SHOW-CASES, TO PREVENT DIMMING OF:
See Glass.

Siccatives

The oldest drier is probably litharge, a reddish-yellow powder, consisting of lead and oxygen. Formerly it was ground finely in oil, either pure or with admixture of white vitriol and added to the dark oil paints. Litharge and sugar of lead are used to-day only rarely as drying agents, having been displaced by the liquid manganese siccatives, which are easy to handle. E. Ebelin, however, is of the opinion that the neglect of the lead compounds has not been beneficial to decorative painting. Where these mediums were used in suitable quantities hard-drying coatings were almost always obtained. Ebelin believes that formerly there used to be less lamentation on account of tacky floors, pews, etc., than at the present time.

Doubtless a proposition to grind litharge into the oil again will not be favorably received, although some old master painters have by no means discarded this method.

Sugar of lead (lead acetate) is likewise used as a drier for oil paint. While we may presume in general that a siccative acts by imparting its oxygen to the linseed oil or else prepares the linseed oil in such a manner as to render it capable of readily absorbing the oxygen of the air,

it is especially sugar of lead which strengthens us in this belief. If, according to Leuchs, a piece of charcoal is saturated with lead acetate, the charcoal can be ignited even with a burning sponge, and burns entirely to ashes. (Whoever desires to make the experiment should take 2 to 3 parts, by weight, of sugar of lead per 100 parts of charcoal.) This demonstrates that the sugar of lead readily parts with its oxygen, which though not burning itself, supports the combustion. Hence, it may be assumed that it will also as a siccative freely give off its oxygen.

Tormin reports on a siccative, of which he says that it has been found valuable for floor coatings. Its production is as follows: Pour 1 part of white lead and 1½ parts each of litharge, sugar of lead and red lead to 12½ parts of linseed oil, and allow this mixture to boil for 8 to 10 hours. Then remove the kettle from the fire and add to the mixture 20 parts of oil of turpentine. During the boiling, as well as during and after the pouring in of the oil turpentine, diligent stirring is necessary, partly to prevent anything from sticking to the kettle (which would render the drier impure) and partly to cause the liquid mass to cool off sooner. After that, it is allowed to stand for a few days, whereby the whole will clarify. The upper layer is then poured off and added to the light tints, while the sediment may be used for the darker shades.

If white vitriol (zinc sulphate or zinc vitriol) has been introduced among the drying agents, this is done in the endeavor to create a non-coloring admixture for the white pigments and also not to be compelled to add lead compounds, which, as experience has shown, cause a yellowing of white coatings to zinc white. For ordinary purposes, Dr. Koller recommends to add to the linseed oil 2 per cent (by weight) of litharge and ½ per cent of zinc vitriol, whereupon the mixture is freely boiled. If the white vitriol is to be added in powder form, it must be deprived of its constitutional water. This is done in the simplest manner by calcining. The powder, which feels moist, is subjected to the action of fire on a sheet-iron plate, whereby the white vitriol is transformed into a vesicular, crumbly mass. At one time it was ground in oil for pure zinc white coatings only, while for the other pigments litharge is added besides, as stated above.

As regards the manganese preparations which are employed for siccatives, it must be stated that they do not possess certain disadvantages of the lead preparations as, for instance, that of being acted upon by hydrogen sulphide gas. The ordinary brown manganese driers, however, are very liable to render the paint yellowish, which, of course, is not desirable for pure white coatings. In case of too large an addition of the said siccative, a strong subsequent yellowing is perceptible, even if, for instance, zinc white has been considerably "broken" by blue or black. But there are also manganese siccatives or drying preparations offered for sale which are colorless or white, and therefore may unhesitatingly be used in comparatively large quantities for white coatings. A pulverulent drying material of this kind consists, for example, of equal parts of calcined (i. e., anhydrous) manganese vitriol, manganous acetate, and calcined zinc vitriol.

Of this mixture 3 per cent is added to the zinc white. Of the other manganese compounds, especially that containing most oxygen, viz., manganic peroxide, is extensively employed. This body is treated as follows: It is first coarsely powdered, feebly calcined, and sifted. Next, the substance is put into wire gauze and suspended in linseed oil, which should be boiled slightly. The weight of the linseed oil should be 10 times that of the manganese peroxide.

According to another recipe a pure pulverous preparation may be produced by treating the manganic peroxide with hydrochloric acid, next filtering, precipitating with hot borax solution, allowing to deposit, washing out and finally drying. Further recipes will probably be unnecessary, since the painter will hardly prepare his own driers.

Unless for special cases driers should be used but sparingly. As a rule 3 to 5 per cent of siccative suffices; in other words, 3 to 5 pounds of siccative should be added to 100 pounds of ground oil paint ready. for use. As a standard it may be proposed to endeavor to have the coating dry in 24 hours. For lead colors a slight addition of drier is advisable; for red lead, it may be omitted altogether. Where non-tacky coatings are desired, as for floors, chairs, etc., as well as a priming for wood imitations, lead color should always be employed as foundation, and as a drier also a lead preparation. On the other hand, no lead compounds should be used for pure zinc-white coats and white lacquering.

Testing Siccatives.—Since it was discovered that the lead and manganese compounds of rosin acids had a better and more rapid action on linseed oil than the older form of driers, such as red lead, litharge, manganese dioxide, etc., the number of preparations of the former class has increased enormously. Manufacturers are continually at work endeavoring to improve the quality of these compounds, and to obtain a preparation which will be peculiarly their own. Consequently, with such a large variety of substances to deal with, it becomes a matter of some difficulty to distinguish the good from the bad. In addition to the general appearance, color, hardness, and a few other such physical properties, there is no means of ascertaining the quality of these substances except practical testing of their drying properties, that is, one must mix the driers with oil and prove their value for oneself. Even the discovery of an apparently satisfactory variety does not end the matter, for experience has shown that such preparations, even when they appear the same, do not give similar results. A great deal depends upon their preparation; for example, manganese resinate obtained from successive consignments, and containing the same percentage of manganese, does not always give identical results with oil. In fact, variation is the greatest drawback to these compounds. With one preparation the oil darkens, with another it remains pale, or sometimes decomposition of the oil takes place in part. The addition of a small proportion of drier has been known to cause the separation of 50 per cent of the oil as a dark viscous mass. One drier will act well, and the oil will remain thin, while with another, the same oil will in the course of a few months thicken to the consistency of stand oil. These various actions may all be obtained from the same compound of rosin with a metal, the source only of the drier varying.

The liquid siccatives derived from these compounds by solution in turpentine or benzine also give widely divergent results. Sometimes a slight foot will separate, or as much as 50 per cent may go to the bottom of the pan, and at times the whole contents of the pan will settle to a thick, jelly-like mass. By increasing the temperature, this mass will become thin and clear once more, and distillation will drive over pure unaltered turpentine or benzine, leaving behind the metallic compound of rosin in its original state.

The compounds of metals with fatty acids which, in solution in turpentine, have been used for many years by varnish-makers, show even greater variation. At the same time, a greater drying power is obtained from them than from rosin acids, quantities being equal. As these compounds leave the factory, they are often in solution in linseed oil or turpentine, and undoubtedly many of the products of this nature on the market are of very inferior quality.

The examination of these bodies may be set about in two ways:

A.—By dissolving in linseed oil with or without heat.

B.—By first dissolving the drier in turpentine and mixing the cooled solution (liquid siccatives) with linseed oil.

Before proceeding to describe the method of carrying out the foregoing tests, it is necessary to emphasize the important part which the linseed oil plays in the examination of the driers. As part of the information to be gained by these tests depends upon the amount of solid matter which separates out, it is essential that the linseed oil should be uniform. To attain this end, the oil used must always be freed from mucilage before being used for the test. If this cannot readily be obtained, ordinary linseed oil should be heated to a temperature of from 518° to 572° F., so that it breaks, and should then be cooled and filtered. With the ordinary market linseed oil, the amount of solid matter which separates varies within wide limits, so that if this were not removed, no idea of the separation of foot caused by the driers would be obtained. It is not to be understood from this that unbroken linseed oil is never to be used for ordinary paint or varnish, the warning being only given for the sake of arriving at reliable values for the quality of the driers to be tested.

A.—*Solution of Drier in Linseed Oil.*—The precipitated metallic compounds of rosin (lead resinate, manganese resinate and lead manganese resinate) dissolve readily in linseed oil of ordinary temperature (60° to 70° F.). The oil is mixed with 1½ per cent of the drier and subjected to stirring or shaking for 24 hours, the agitation being applied at intervals of an hour. Fused metallic resinates are not soluble in linseed oil at ordinary temperatures, so different treatment is required for them. The oil is heated in an enameled pan together with the finely powdered drier, until the latter is completely in solution, care being taken not to allow the temperature to rise above 390° F. The pan is then removed from the fire and its contents allowed to settle. The quantity of drier used should not exceed 1½ to 3 per cent. In the case of metallic linoleates (lead linoleate, manganese linoleate and lead-manganese linoleate), the temperature must be raised above 290° F. before they will go into solution. In their case also the addition should not be greater than 3 per cent. Note, after all the tests have settled, the amount of undissolved matter which is left at the bottom, as this is one of the data upon which an idea of the value of the drier must be formed.

B.—*Solution of Drier in Turpentine or Benzine.*—For the preparation of these liquid siccatives 1 to 1.4 parts of the metallic resinate or linoleate are added to the benzine or turpentine and dissolved at a gentle heat, or the drier may first be melted over a fire and added to the solvent while in the liquid state. The proportion of matter which does not go into solution must be carefully noted as a factor in the valuation of the drier. From 5 to 10 per cent of the liquid siccative is now added to the linseed oil, and the mixture shaken well, at intervals during 24 hours.

Samples of all the oils prepared as above should be placed in small clear bottles, which are very narrow inside, so that a thin layer of the oil may be observed. The bottles are allowed to stand for 3 or 4 days in a temperate room, without being touched. When sufficient time has been allowed for thorough settling, the color, transparency, and consistency of the samples are carefully observed, and also the quantity and nature of any precipitate which may have settled out. A note should also be made of the date for future reference. Naturally the drier which has colored the oil least and left it most clean and thin, and which shows the smallest precipitate, is the most suitable for general use. The next important test is that of drying power, and is carried out as follows: A few drops of the sample are placed on a clear, clean glass plate, 4 x 6 inches, and rubbed evenly over with the fingers. The plate is then placed, clean side up, in a sloping position with the upper edge resting against a wall. In this way any excess of oil is run off and a very thin equal layer is obtained. It is best to start the test early in the morning as it can then be watched throughout the day. It should be remarked that the time from the "tacky" stage to complete dryness is usually very short, so that the observer must be constantly on the watch. If a good drier has been used, the time may be from 4 to 5 hours, and should not be more than 12 or at the very highest

15. The bleaching of the layer should also be noted. Many of the layers, even after they have become as dry as they seem capable of becoming, show a slight stickiness. These tests should be set aside in a dust-free place for about 8 days, and then tested with the finger.

SIGN LETTERS:

To Remove Black Letters from White Enameled Signs.—It frequently happens that a change has to be made on such signs, one name having to be taken off and another substituted. Priming with white lead followed by dull and glossy zinc white paint always looks like a daub and stands out like a pad. Lye, glass paper or steel chips will not attack the burned-in metallic enamel. The quickest plan is to grind down carefully with a good grindstone.

SIGN-LETTER CEMENTS:
See Adhesives, under Cements.

SIGNS, TO REPAIR ENAMELED:
See Enamels.

SILK:

Artificial "Rubbered" Silk.—A solution of caoutchouc or similar gum in acetone is added, in any desired proportion, to a solution of nitro-cellulose in acetone, and the mixture is made into threads by passing it into water or other suitable liquid. The resulting threads are stated to be very brilliant in appearance, extremely elastic, and very resistant to the atmosphere and to water. The product is not more inflammable than natural silk.

Artificial Ageing of Silk Fabrics.—To give silk goods the appearance of age, exposure to the sun is the simplest way, but as this requires time it cannot always be employed. A quicker method consists in preparing a dirty-greenish liquor of weak soap water, with addition of a little blacking and gamboge solution. Wash the silk fabric in this liquor and dry as usual, without rinsing in clean water, and calender.

Bleaching Silk.—The Lyons process of bleaching skeins of silk is to draw them rapidly through a sort of aqua regia bath. This bath is prepared by mixing 5 parts of hydrochloric acid with 1 of nitric, leaving the mixture for 4 or 5 days at a gentle heat of about 77° F., and then diluting with about 15 times its volume of water. This dilution is effected in large tanks cut from stone. The temperature of the bath should be from 68° to 85° F., and the skeins should not be in it over 15 minutes, and frequently not so long as that; they must be kept in motion during all that time. When taken out, the silk is immediately immersed successively in 2 troughs of water, to remove every trace of the acid, after which they are dried.

Hydrogen peroxide is used as a silk bleach, the silk being first thoroughly washed with an alkaline soap and ammonium carbonate to free it of its gummy matter. After repeated washings in the peroxide (preferably rendered alkaline with ammonia and soda), the silk is "blued" with a solution of blue aniline in alcohol.

Washing of Light Silk Goods.—The best soap may change delicate tints. The following method is therefore preferable: First wash the silk tissue in warm milk. Prepare a light bran infusion, which is to be decanted, and after resting for a time, passed over the fabric. It is then rinsed in this water, almost cold. It is moved about in all directions, and afterwards dried on a napkin.

SILK SENSITIZERS FOR PHOTOGRAPHIC PURPOSES:
See Photography, under Paper-Sensitizing Processes.

Silver

Antique Silver (see also Plating).—Coat the polished silver articles with a thin paste of powdered graphite, 6 parts; powdered bloodstone, 1 part; and oil of turpentine. After the drying take off the superfluous powder with a soft brush and rub the raised portions bright with a linen rag dipped in spirit. By treatment with various sulphides an old appearance is likewise imparted to silver. If, for example, a solution of 5 parts of liver of sulphur and 10 parts of ammonium carbonate are heated in 1 quart of distilled water to 180° F., placing the silver articles therein, the latter first turn pale gray, then dark gray, and finally assume a deep black-blue. In the case of plated ware, the silvering must not be too thin; in the case of thick silver plating or solid silver 1 quart of water is sufficient. The colors will then appear more quickly. If the coloring is spotted or otherwise imperfect dip the objects into a warm potassium cyanide solution, whereby the silver sulphide formed is immediately dissolved. The bath must be renewed after a while. Silver containing much copper is subjected, previous to the coloring, to a blanching process, which is accomplished in a boiling solution of 15 parts of powdered tartar and 30 parts of cooking salt in 2 pints of water. Objects which are to be mat are coated with a paste of potash and water after the blanching, then dry, anneal, cool in water, and boil again.

Imitation of Antique Silver.—Plated articles may be colored to resemble old objects of art made of solid silver. For this purpose the deep-lying parts, those not exposed to friction, are provided with a blackish, earthy coating, the prominent parts retaining a leaden but bright color. The process is simple. A thin paste is made of finely powdered graphite and oil of turpentine (a little bloodstone or red ocher may be added, to imitate the copper tinge in articles of old silver) and spread over the whole of the previously plated article. It is then allowed to dry, and the particles not adhering to the surface removed with a soft brush. The black coating should then be carefully wiped off the exposed parts by means of a linen rag dipped in alcohol. This process is very effective in making imitations of objects of antique art, such as goblets, candlesticks, vessels of every description, statues, etc. If it is desired to restore the original brightness to the object, this can be done by washing with caustic soda or a solution of cyanide of potassium. Benzine can also be used for this purpose.

Blanching Silver.—I.—Mix powdered charcoal, 3 parts, and calcined borax, 1 part, and stir with water so as to make a homogeneous paste. Apply this paste on the pieces to be blanched. Put the pieces on a charcoal fire, taking care to cover them up well; when they have acquired a cherry red, withdraw them from the fire and leave to cool off. Next place them in a hot bath composed of 9 parts of water and 1 part of sulphuric acid, without causing the bath to boil. Leave the articles in for about 1 hour. Remove them, rinse in clean water, and dry.

II.—If the coat of tarnish on the surface of the silver is but light and superficial, it suffices to rub the piece well with green soap to wash it thoroughly in hot water; then dry it in hot sawdust and pass it through alcohol, finally rubbing with a fine cloth or brush. Should the coat resist this treatment, brush with Spanish white, then wash, dry, and

pass through alcohol. The employment of Spanish white has the drawback of shining the silver if the application is strong and prolonged. If the oxidation has withstood these means and if it is desired to impart to the chain the handsome mat appearance of new goods, it should be annealed in charcoal dust and passed through vitriol, but this operation, for those unused to it, is very dangerous to the soldering and consequently may spoil the piece.

Coloring Silver.—A rich gold tint may be imparted to silver articles by plunging them into dilute sulphuric acid, saturated with iron rust.

Frosting Polished Silver.—Articles of polished silver may be frosted by putting them into a bath of nitric acid diluted with an equal volume of distilled water and letting them remain a few minutes. A better effect may be given by dipping the article frequently into the bath until the requisite degree of frosting has been attained. Then rinse and place for a few moments in a strong bath of potassium cyanide; remove and rinse. The fingers must not be allowed to touch the article during either process. It should be held with wooden forceps or clamps.

Fulminating Silver.—Dissolve 1 part of fine silver in 10 parts of nitric acid of 1.36 specific gravity at a moderate heat; pour the solution into 20 parts of spirit of wine (85 to 90 per cent) and heat the liquid. As soon as the mixture begins to boil, it is removed from the fire and left alone until cooled off. The fulminic silver crystallizes on cooling in very fine needles of dazzling whiteness, which are edulcorated with water and dried carefully in the air.

Hollow Silverware.—A good process for making hollow figures consists in covering models of the figures, made of a base or easily soluble metal, with a thin and uniform coating of a nobler metal, by means of the electric current in such a way that this coating takes approximately the shape of the model, the latter being then removed by dissolving it with acid. The model is cast from zinc in one or more pieces, a well-chased brass mold being used for this purpose, and the separate parts are then soldered together with an easily fusible solder. The figure is then covered with a galvanized coating of silver, copper, or other metal. Before receiving the coating of silver, the figure is first covered with a thin deposit of copper, the silver being added afterwards in the required thickness. But in order that the deposit of silver may be of the same thickness throughout (this is essential if the figure is to keep the right shape), silver anodes, so constructed and arranged as to correspond as closely as possible to the outlines of the figure, should be suspended in the solution of silver and cyanide of potassium on both sides of the figure, and at equal distances from it. As soon as the deposit is sufficiently thick, the figure is removed from the bath, washed, and put into a bath of dilute sulphuric or hydrochloric acid, where it is allowed to remain till the zinc core is dissolved. The decomposition of the zinc can be accelerated by adding a pin of copper. The figure now requires only boiling in soda and potassic tartrate to acquire a white color. If the figure is to be made of copper, the zinc model must be covered first with a thin layer of silver, then with the copper coating, and then once more with a thin layer of silver, so that while the zinc is being dissolved, the copper may be protected on either side by the silver. Similar precautions must be taken with other metals, regard being paid to their peculiar properties. Another method is to cast the figures, entire or in separate parts, out of some easily fusible alloy in chased metal molds. The separate portions are soldered with the same solder, and the figure is then provided with a coating of copper, silver, etc., by means of the galvanic current. It is then placed in boiling water or steam, and the inner alloys melted by the introduction of the water or steam through holes bored for this purpose.

Lustrous Oxide on Silver (see also Plating and Silver, under Polishes).—Some experience is necessary to reproduce a handsome black luster. Into a cup filled with water throw a little liver of sulphur and mix well. Scratch the silver article as bright as possible with the scratch brush and dip into the warm liquid. Remove the object after 2 minutes and rinse off in water. Then scratch it up again and return it into the liquid. The process should be repeated 2 or 3 times, whereby a wonderful glossy black is obtained.

Ornamental Designs on Silver.—Select a smooth part of the silver, and sketch on it a monogram or any other design with a sharp lead pencil. Place the article in a gold solution, with the battery in good working order, and in a short time all the parts not sketched with the lead pencil will be covered with a coat of gold. After cleaning the article the black lead is easily removed with the finger, whereupon the silver ornament is disclosed. A gold ornament may be produced by reversing the process.

Separating Silver from Platinum Waste.—Cut the waste into small pieces, make red hot to destroy grease and organic substances, and dissolve in aqua regia (hydrochloric acid, 3 parts, and nitric acid, 1 part). Platinum and all other metals combined with it are thus dissolved, while silver settles on the bottom as chloride in the shape of a gray, spongy powder. The solution is then drawn off and tested by oxalic acid for gold, which is precipitated as a fine yellowish powder. The other metals remain untouched thereby. The platinum still present in the solution is now obtained by a gradual addition of sal ammoniac as a yellowish-gray powder. These different precipitates are washed with warm water, dried, and transformed into the metallic state by suitable fluxes. Platinum filings, however, have to be previously refined. They are also first annealed. All steel or iron filings are removed with a magnet and the rest is dipped into concentrated sulphuric acid and heated with this to the boiling point. This process is continued as long as an action of the acid is noticeable. The remaining powder is pure platinum. Hot sulphuric acid dissolves silver without touching the platinum. The liquid used for the separation of the platinum is now diluted with an equal quantity of water and the silver expelled from it by means of a saturated cooking salt solution. The latter is added gradually until no more action, j. e., separation, is perceptible. The liquid is carefully drawn off, the residue washed in warm water, dried and melted with a little soda ashes as flux, which yields pure metallic silver.

The old process for separating silver from waste was as follows: The refuse was mixed with an equal quantity of charcoal, placed in a crucible, and subjected to a bright-red heat, and in a short time a silver button formed at the bottom. Carbonate of soda is another good flux.

Silvering Glass Globes.—Take ⅓ ounce of clean lead, and melt it with an equal weight of pure tin; then immediately add ½ ounce of bismuth, and carefully skim off the dross; remove the alloy from the fire and before it grows cold add 5 ounces of mercury, and stir the whole well together; then put the fluid amalgam into a clean glass, and it is fit for use. When this amalgam is used for silvering let it be first strained through a linen rag; then gently pour some ounces thereof into the globe intended to be silvered; the alloy should be poured into the globe by means of a paper or glass funnel reaching almost to the bottom of the globe, to prevent it splashing the sides; the globe should be turned every way very slowly, to fasten the silvering.

Silvering Powder for Metals.—Copper, brass, and some other metals may be silvered by rubbing well with the following powder: Potassium cyanide, 12 parts; silver nitrate, 6 parts; calcium carbonate, 30 parts. Mix and keep in a well-closed bottle. It must be applied with hard rubbing, the bright surface being afterwards rinsed with water, dried, and polished. Great care must be exercised in the use of the powder on account of its poisonous nature. It should not be allowed to come in contact with the hands.

Silver Testing.—For this purpose a cold saturated solution of potassium bichromate in pure nitric acid of 1.2 specific gravity is employed. After the article to be tested has been treated with spirit of wine for the removal of any varnish coating which might be present, a drop of the above test liquor is applied by means of a glass rod and the resultant spot rubbed off with a little water.

A testing solution of potassium bichromate, 1 ounce, pure nitric acid, 6 ounces, and water, 2 ounces, gives the following results on surfaces of the metals named:

Metal.	Color in one minute.	Color of mark left.
Pure silver	Bright blood-red	Grayish white
.925 silver	Dark red	Dark brown
.800 silver	Chocolate	Dark brown
.500 silver	Green	Dark brown
German silver	Dark blue	Light gray
Nickel	Turquoise blue	Scarcely any
Copper	Very dark blue	Cleaned copper
Brass	Dark brown	Light brown
Lead	Nut brown	Leaden
Tin	Reddish brown	Dark
Zinc	Light chocolate	Steel gray
Aluminum	Yellow	No stain
Platinum	Vandyke brown	No stain
Iron	Various	Black
9-carat gold	Unchanged	No stain

The second column in the table shows such change of color as the liquid—not the metal—undergoes during its action for the period of 1 minute. The test liquid being then washed off with cold water, the third column shows the nature of the stain that is left.

In the case of faintly silvered goods, such as buttons, this test fails, since the slight quantity of resulting silver chromate does not become visible or dissolves in the nitric acid present. But even such a thin coat of silver can be recognized with the above test liquor, if the bichromate solution is used, diluted with the equal volume of water, or if a small drop of water is first put on the article and afterwards a little drop of the undiluted solution is applied by means of a capillary tube. In this manner a distinct red spot was obtained in the case of very slight silvering.

A simpler method is as follows: Rub the piece to be tested on the touchstone and moisten the mark with nitric acid, whereupon it disappears. Add a little hydrochloric acid with a glass rod. If a white turbidness (silver chloride) appears which does not vanish upon addition of water, or, in case of faint silvering or an alloy poor in silver, a weak opalescence, the presence of silver is certain. Even alloys containing very little silver give this reaction quite distinctly.

Pink Color on Silver.—To produce a beautiful pink color upon silver, dip the clean article for a few seconds into a hot and strong solution of cupric chloride, swill it in water and then dry it or dip it into spirit of wine and ignite the spirit.

SILVER, IMITATION:
See Alloys.

SILVERING:
See Plating.

SILVERING OF MIRRORS:
See Mirrors.

SILVERING, TEST FOR:
See Plating.

SILVER FOIL SUBSTITUTE:
See Metal Foil.

SILVER NITRATE SPOTS, TO REMOVE:
See Cleaning Preparations and Methods.

SILVER-PLATING:
See Plating.

SILVER, RECOVERY OF PHOTOGRAPHIC:
See Photography.

SILVER SOLDERS:
See Solders.

SILVER, TO CLEAN:
See Cleaning Preparations and Methods.

SILVER, TO RECOVER GOLD FROM:
See Gold.

SILVERWARE POLISHES:
See Polishes.

SIMILOR:
See Alloys.

SINEWS, TREATMENT OF, IN MANU-FACTURING GLUE:
See Adhesives.

SYRUP (RASPBERRY):
See Raspberry.

SYRUPS:
See Essences and Extracts.

SIZING:
See Adhesives.

SIZING WALLS FOR KALSOMINE:
See Kalsomine.

SKIN-CLEANING PREPARATIONS:
See Cleaning Preparations and Methods.

SKIN OINTMENTS:
See Ointments.

SKIN FOODS:
See Cosmetics.

SKIN TROUBLES:
See Soap.

SLATE:
Artificial Slate.—The artificial slate coating on tin consists of a mixture of finely ground slate, lampblack, and a water-glass solution of equal parts of potash and soda water glass (1.25 specific gravity). The process is as follows:

I.—First prepare the water-glass solution by finely crushing equal parts of solid potash and soda water glass and pouring over this 6 to 8 times the quantity of soft river water, which is kept boiling about 1½ hours, whereby the water glass is completely dissolved. Add 7 parts finely crushed slate finely ground with a little water into impalpable dust, 1 part lampblack, which is ground with it, and grind enough of this mass with the previously prepared water-glass solution as is necessary for a thick or thin coating. With this compound the roughened tin plates are painted as uniformly as possible. For roofing, zinc plate may be colored in the same manner. The coating protects the zinc from oxidation and consequently from destruction. For painting zinc plate, however, only pure potash water glass must be added to the mixture, as the paint would loosen or peel off from the zinc if soda water glass were used.

II.—Good heavy paper or other substance is saturated with linseed-oil varnish and then painted, several coats, one after another with the following mixture:

Copal varnish.......	1 part
Oil of turpentine.....	2 parts
Fine, dry sand, powdered............	1 part
Powdered glass......	1 part
Ground slate........	2 parts
Lampblack..........	1 part

SLIDES FOR LANTERNS:
See Photography.

SLUGS ON ROSES:
See Insecticides.

SMARAGDINE:
See Alcohol (Solid).

SMUT, TREATMENT FOR:
See Grain.

SNAKE BITES.
About 25 years ago, Dr. S. Weir Mitchell and Dr. Reichert published results of their investigations of snake venom which indicated that permanganate of potassium may prove of material value as an antidote to this lethal substance. Since that time permanganate has been largely used all over the world as a remedy when men and animals were bitten by poisonous snakes, and Sir Lauder Brunton devised an instrument by means of which the permanganate may be readily carried in the pocket, and immediately injected into, or into the neighborhood of, the wound. Captain Rodgers, of the Indian Medical Service, recently reported several cases treated by this method, the wounds being due to the bites of the cobra. After making free crucial incisions of the bitten part, the wound was thoroughly flushed with a hot solution of permanganate of potassium, and then bandaged. Recovery occurred in each instance, although the cauterant action of the hot solution of permanganate of potassium delayed healing so long that the part was not well for about 3 weeks. About 12 or 13 years ago, Dr. Amos Barber, of Cheyenne, Wyoming, reported cases in which excellent results had followed this method of treatment.

Soaps

(See also Cleaning Compounds and Polishes.)

ANTISEPTIC SOAP.

I.—Various attempts have been made to incorporate antiseptics and cosmetics with soap, but for the most part unsuccessfully, owing to the unfavorable action of the added components, a good instance of this kind being sodium peroxide, which, though a powerful antiseptic, soon decomposes in the soap and loses its properties, while the caustic character of the oxide renders its use precarious, even when the soap is fresh, unless great care is taken. However, according to a German patent, zinc peroxide is free from these defects, since it retains its stability and has no corrosive action on the skin, while possessing powerful antiseptic and cosmetic properties, and has a direct curative influence when applied to cuts or wounds.

II.—The soap is prepared by melting 80 parts of household soap in a jacketed pan, and gradually adding 20 parts of moist zinc peroxide (50 per cent strength), the whole being kept well stirred all the time. The finished mixture will be about as stiff as dough, and is easily shaped into tablets of convenient size.

III.—Take 50 parts, by weight, of caustic soda of 70 per cent, and free from carbonic acid, if possible; 200 parts, by weight, of sweet almond oil; 160 parts, by weight, of glycerine of 30° Bé.; and sufficient distilled water to make up 1,000 parts by weight. First, dissolve the alkali in double its weight of water, then add the glycerine and oil and stir together. Afterwards, add the remainder of the water and keep the whole on the water bath at a temperature of 140° to 158° F., for 24 to 36 hours; remove the oil not saponified, which gives a gelatinous mass. Mix 900 parts, by weight, of it with 70 parts, by weight, of 90 per cent alcohol and 10 parts, by weight, of lemon oil, and as much of the oil of bergamot and the oil of vervain. Heat for some hours at 140° F., then allow to cool and filter on wadding to eliminate the needles of stearate of potash. The liquid after filtering remains clear.

Carpet Soap.—

Fuller's earth	4 ounces
Spirits of turpentine	1 ounce
Pearlash	8 ounces

Rub smooth and make into a stiff paste with a sufficiency of soft soap.

To Cut Castile Soap.

—A thin spatula must be used. To cut straight, a trough with open ends made with ½-inch boards should be taken, the inside dimensions being 2⅞ inches wide, 3¾ inches deep, and about 14 inches long. Near the end a perpendicular slit is sawed through the side pieces. Passing the spatula down through this slit the bar is cut neatly and straight. For trimming off the corners a carpenter's small iron plane works well.

COLORING SOAP.

The first point to be observed is to select the proper shade of flower corresponding with the perfume used, for instance, an almond soap is left white; rose soap is colored pink or red; mignonette, green, etc.

The colors from which the soapmaker may select are numerous; not only are most of the coal-tar colors adapted for his purpose, but also a very great number of mineral colors. Until recently, the latter were almost exclusively employed, but the great advance in the tar-color industry has brought about a change. A prominent advantage of the mineral colors is their stability; they are not changed or in any way affected by exposure to light. This advantage, however, is offset in many cases by the more difficult method of application, the difficulty of getting uniform shades. The coal-tar colors give brilliant shades and tints, are easy to use, and produce uniform tints. The specific gravity of mineral colors being rather high, in most cases they will naturally tend to settle toward the bottom of soap, and their use necessitates crutching of the soap until it is too thick to allow the color to settle. For mottled soap, however, vermilion, red oxide, and ultramarine are still largely employed.

For transparent soap mineral colors are not applicable, as they would detract from their transparency; for milled toilet soap, on the other hand, they are very well adapted, as also for cold-made soaps which require crutching anyway until a sufficient consistency is obtained to keep the coloring material suspended.

A notable disadvantage in the use of aniline colors, besides their sensitiveness to the action of light, is the fact that many of them are affected and partly destroyed by the action of alkali. A few of them are proof against a small excess of lye, and these may be used with good effect. Certain firms have made a specialty of manufacturing colors answering the peculiar requirements of soap, being very easy of application, as they are simply dissolved in boiling water and the solution stirred into the soap. To some colors a little weak lye is added; others are mixed with a little oil before they are added to the soap.

For a soluble red color there were formerly used alkanet and cochineal; at present these have been displaced to a great extent, on account of their high cost, by magenta, which is very cheap and of remarkable beauty. A very small amount suffices for an intense color, nor is a large proportion desirable, as the soap would then stain. Delicate tints are also produced by the eosine colors, of which rose bengal, phloxine, rhodamine, and eosine are most commonly used. These colors, when dissolved, have a brilliant fluorescence which heightens their beautiful effect.

The following minerals, after being ground and washed several times in boiling water, will produce the colors stated:

Hematite produces deep red.
Purple oxide iron produces purple.
Oxide of manganese produces brown.
Yellow ocher produces yellow.
Yellow ocher calcined produces orange.
Umber produces fawn.
Cinnabar produces medium red.

There are also a number of the azo dyes, which are suitable for soaps, and these, as well as the eosine colors, are used principally for transparent soaps. For opaque soaps both aniline and mineral reds are used, among the latter being vermilion, chrome red, and iron oxide. Chrome red is a basic chromate of lead, which is now much used in place of vermilion, but, as it becomes black on exposure to an atmosphere containing even traces only of sulphureted hydrogen, it is not essentially adapted for soap. Vermilion gives a bright color, but its price is high. Iron oxide, known in the trade as colcothar, rouge, etc., is used for cheap soaps only.

Among the natural colors for yellow are saffron, gamboge, turmeric, and caramel (sugar color); the first named of these is now hardly used, owing to its high cost. Of the yellow aniline colors special mention must be made of picric acid (trinitrophenol), martius yellow, naphthol yellow, acid yellow, and auramine. If an orange tint is wanted, a trace of magenta or safranine may be added to the yellow colors named. The use of some unbleached palm oil with the stock answers a similar purpose, but the color fades on exposure. A mineral yellow is chrome yellow (chromate of

lead), which has the same advantages and disadvantages as chrome red.

Of the blue aniline colors, there may be used alkali blue, patent blue, and indigo extract. Alkali or aniline blue is soluble only in alkaline liquids; while patent blue is soluble in water and in alcohol. Both blues can be had in different brands, producing from green blues to violet blues. Indigo extract, which should be classed among the natural colors rather than among the tar colors, is added to the soap in aqueous solution.

Of ultramarine there are two modifications, the sulphate and the soda. Both of these are proof against the action of alkali, but are decomposed by acids or salts having an acid reaction. The former is much paler than the latter; the soda ultramarine is best adapted for coloring soda soaps blue. The ultramarine is added to the soap in the form of a fine powder. Smalt is unsuitable, although it gives soap a color of wonderful beauty because a considerable quantity of it is required to produce a deep color, and, furthermore, it makes the soap rough, owing to the gritty nature which smalt has even when in the finest powder. By mixing the blue and yellow colors named, a great variety of greens are obtained. Both component colors must be entirely free from any reddish tint, for the latter would cause the mixture to form a dirty-green color.

Of the colors producing green directly the two tar colors, Victoria and brilliant green, are to be noted; these give a bright color, but fade rapidly; thereby the soap acquires an unsightly appearance. For opaque soap of the better grades, green ultramarine or chrome green are used. Gray and black are produced by lampblack. For brown, there is Bismarck brown among the aniline colors and umber among the earthy pigments.

Garment-Cleaning Soap.—The following is excellent:

I.—White soap, rasped
    or shaved........ 12 parts
Ammonia water.... 3 parts
Boiling water...... 18 parts

Dissolve the soap in the water and when it cools down somewhat, add to the solution the ammonia water. Pour the solution into a flask of sufficient capacity (or holding about three times as much as the mixture) and add enough water to fill it about three-quarters full. Shake and add, a little at a time, under active agitation, enough benzine to make 100 parts. This constitutes the stock bottle. To make up the mass or paste put a teaspoonful in an 8-ounce bottle and add, a little at a time, with constant agitation, benzine to about fill the bottle. This preparation is a rapid cleaner and does not injure the most delicate colors.

II.—Good bar soap,
    shaved up.....   165 parts
    Ammonia water..   45 parts
    Benzine.........   190 parts
    Water sufficient
      to make.......   1,000 parts

Dissolve the soap in 600 parts of water by heating on the water bath, remove, and add the ammonia under constant stirring. Finally add the benzine, and stir until homogeneous, and quite cold. The directions to go with this paste are: Rub the soap well into the spot and lay the garment aside for a half hour. Then using a stiff brush, rub with warm water and rinse. This is especially useful in spots made by rosins, oils, grease, etc. Should the spot be only partially removed by the first application, repeat.

Glycerine Soaps.—Dr. Sarg's liquid glycerine soap consists of 334 parts of potash soda soap, and 666 parts of glycerine free from lime, the mixture being scented with Turkish rose oil and orange blossom oil in equal proportions, the actual amount used being varied according to taste. The soap should be perfectly free from alkali; but as this is a condition difficult of attainment in the case of ordinary potash soaps, it is presupposed that the soap used has been salted out with potassium chloride, this being the only way to obtain a soap free from alkali.

Another variety of liquid glycerine soap is prepared from purified medicinal soft soap, 300 parts; glycerine free from lime, 300 parts; white sugar syrup, 300 parts; doubly rectified spirit (96 per cent), 300 parts. The mixture is scented with oil of cinnamon, 1 part; oil of sassafras, 2 parts; oil of citronella, ½ part; oil of wintergreen, 1 part; African geranium oil, 1 part; clove oil, ¼ part; oil of bergamot, 3 parts; pure tincture of musk, ½ part. These oils are dissolved in spirit, and shaken up with the other ingredients; then left for 8 days with frequent shaking, and 3 days in absolute quiet, after which the whole is filtered, and is then ready for packing.

Iodine Soaps.—In British hospitals, preference is given to oleic acid over alcoholic preparations for iodine soaps, as the former do not stain and can be washed off with soap and water. The following formula is given:

I.—Iodine..........   1 av. ounce
    Oleic acid.......   1 fluidounce
    Alcohol..........   6 fluidrachms
    Stronger water of
      ammonia......   2 fluidrachms

This makes a soapy paste soluble in all liquids, except fixed oils.

II.—Iodine..........   1 av. ounce
    Oleic acid.......   2 fluidounces
    Stronger water of
      ammonia......   3 fluidrachms
    Paraffine oil, col-
      orless, to make   20 fluidounces

III.—Iodine..........   1 av. ounce
    Alcohol..........   5 fluidounces
    Solution of am-
      monium oleate.   1 fluidounce
    Glycerine to make   20 fluidounces

The solution of ammonium oleate is made from oleic acid and spirit of ammonia.

Liquid Soaps.—Liquid soaps, or, as they are sometimes called, soap essences, are made from pure olive-oil soap by dissolving it in alcohol and adding some potassium carbonate. Tallow or lard soaps cannot be used, as they will not make a transparent preparation. The soap is finely shaved and placed with the alcohol and potassium carbonate in a vessel over a water bath, the temperature slowly and gradually raised, while the mixture is kept in constant agitation by stirring. The soap should be of a pure white color and the alcohol gives the best product when it is about 80 per cent strength. After about three-quarters of an hour to one hour, solution will be complete and a perfectly transparent article obtained. This can be scented as desired by adding the proper essential oil as soon as the mixture is removed from the water bath.

If an antiseptic soap is wanted the addition of a small amount of benzoic acid, formaldehyde, or corrosive sublimate will give the desired product. Liquid soaps should contain from 20 to 40 per cent of genuine white castile soap and about 2 to 2¼ per cent of potassium carbonate.

This is a common formula:

                                By weight
I.—Olive or cottonseed
      oil..............   60 parts
    Caustic potash, U.
      S. P.............   15 parts
    Alcohol and water,
      sufficient of each.

Dissolve the potash in 1 ounce of water, heat the oil on a water bath, add the solution of potash previously warmed, and stir briskly. Continue the heat until saponification is complete. If oil globules separate out and refuse to saponify, the potash is not of proper strength, and more must be added—1 or 2 parts dissolved in water. If desired transparent add a little alcohol, and continue the heat without stirring until a drop placed in cold water first solidifies and then dissolves.

Commercial potash may be used, but the strength must be ascertained and adjusted by experiment. The soap thus made will be like jelly; it is dissolved in alcohol, 4 to 6 ounces of soap to 2 of alcohol, and after standing a day or two is filtered and perfumed as desired. A rancid oil would be easier to saponify, but the soap would likely be rancid or not as good.

II.—Ammonium sulphoichthyolate, 10 parts; distilled water, 15 parts; hebra's soap spirit (a solution of potash soap, 120 parts, in 90 per cent spirit, 60 parts; and spirit of lavender, 5 parts), 75 parts.

MEDICATED SOAPS.

First make up a suitable soap body and afterwards add the medicament. For instance, carbolic soaps may be made as follows:

```
I.—Cocoanut oil........  20 pounds
    Tallow.............   4 pounds
    Soda lye (38° to 40°
      B.).............   12 pounds
    Phenol.............   1 pound
```

Prepare the body soap by stirring the liquefied fat into the lye at 113° F., and when combination has set in, incorporate the phenol and quickly pour into molds. Cover the latter well. Instead of the phenol 2 pounds of sulphur may be used, and a sulphur soap made.

	Parts by weight
II.—Cotton oil...............	200
Alcohol, 91 per cent.....	300
Water..................	325
Caustic soda............	45
Potassium carbonate.....	10
Ether..................	15
Carbolic acid..........	25

The oil is mixed in a large bottle with water, 100 parts; alcohol, 200 parts; and caustic soda, 45 parts, and after saponification the remaining alcohol and the potassium carbonate dissolved in the rest of the water, and finally the carbolic acid and the ether are added and the whole well shaken. The mixture is filled in tightly closed bottles and stored at medium temperature. The preparation may be scented as desired, and the carbolic acid replaced with other antiseptics.

Liquid Tar Soap.—Mix 200 parts of tar with 400 parts of oleic acid, warm lightly and filter. In this way the aqueous content produces no trouble. Now warm the filtrate on the water bath, neutralize by stirring in an alcoholic potash solution. To the soap thus produced, add 100 parts of alcohol, and further a little olive oil, in order to avoid a separation of any overplus of alkaline matter. Finally, bring up to 1,000 parts with glycerine. This soap, containing 22 per cent of tar, answers all possible demands that may be made upon it. Mixed with 2 parts of distilled water it leaves no deposit on the walls of the container.

Liquid Styrax Soap.—The process is identical with the foregoing. For digestion with oleic acid, the crude balsam will answer, since filtration deprives the product of all contaminating substances. While this soap will separate, it is easily again rendered homogeneous with a vigorous shake. Preparations made with it should be accompanied with a "shake" label.

Superfatted Liquid Lanolin-Glycerine Soap.—Dissolve about 10 per cent of lanolin in oleic acid, saponify as in the tar soap, and perfume (for which a solution of coumarin in geranium oil is probably the most suitable agent). The prepared soap is improved by the addition of a little tincture of benzoin.

Massage Soaps.—I.—An excellent recipe for a massage soap is: Special cocoanut oil ground soap, 2,500 pounds; lanolin, 50 pounds; pine-needle oil, 20 pounds; spike oil, 3 pounds. Other massage soaps are made from olive oil ground soap, to which in special cases, as in the treatment of certain rheumatic affections, ichthyol is added. Massage soaps are always wanted white, so that Cochin cocoanut oil should be preferred to other kinds.

II.—Cocoanut oil, 1,000 pounds; caustic soda lye, 37° B., 500 pounds; pine-needle oil, 4 pounds; artificial bitter almond oil, 2 pounds. There is also a "massage cream," which differs from the ordinary massage soaps in being made with a soft potash soap as a ground soap. The oils, etc., incorporated with the ground mass are exactly the same in the "cream" as in the soap.

Metallic Soaps.—Metallic soaps are obtained by means of double decomposition. First a soap solution is produced which is brought to a boil. On the other hand, an equally strong solution of the metallic salt of which the combination is to be made (chlorides and sulphides are employed with preference) is prepared, the boiling solutions are mixed together, and the metallic soap obtained is gathered on a linen cloth. This is then put on enameled plates and dried, first at 104° F., later at 140° F.

Aluminum soap is the most important. Dissolved in benzine or oil of turpentine, it furnishes an excellent varnish. It has been proposed to use these solutions for the varnishing of leather; they furthermore serve for the production of waterproof linen and cloths, paper, etc. Jarry recommended this compound for impregnating railroad ties to render them weatherproof.

Manganese soap is used as a siccative in the preparation of linseed-oil varnish, as well as for a drier to be added to paints. Zinc soap is used in the same manner.

Copper soap enters into the composition of gilding wax, and is also employed for bronzing plaster of Paris articles. For the same purpose, a mixture is made use of consisting of copper soap and iron soap melted in white lead varnish and wax. Iron soap is used with aluminum soap for waterproofing purposes and for the production of a waterproof varnish. By using wax instead of a soap, insoluble metallic soaps are obtained, which, melted in oils or wax, impart brilliant colorings to them; but colored waterproof and weather-resisting varnishes may also be produced with them. Metallic rosin soaps may be produced by double decomposition of potash rosin soaps and a soluble metal salt. From these, good varnishes are obtained to render paper carriage covers, etc., waterproof; they may also be employed for floor wax or lacquers.

Petroleum Soap.—

I.—Beeswax, refined... 4 parts
  Alcohol........... 5 parts
  Castile soap, finely
    grated.......... 10 parts
  Petroleum........ 5 parts

Put the petroleum into a suitable vessel along with the wax and alcohol and cautiously heat on the water bath, with an occasional agitation, until complete solution is effected. Add the soap and continue the heat until it is dissolved. When this occurs remove from the bath and stir until the soap begins to set, then pour into molds.

II.—The hydrocarbons (as petroleum, vaseline, etc.) are boiled with a sufficient quantity of alkali to form a soap, during which process they absorb oxygen and unite with the alkali to form fatty acid salts. The resulting soap is dissolved in water containing alkali, and the solution is heated along with alkali and salt. The mass of soap separates out in three layers, the central one being the purest; and from this product the fatty acids may be recovered by treatment with sulphuric acid.

Perfumes for Soap.—From 1 to 2 ounces of the following mixtures are to be used to 10 pounds of soap:

I.—Oil of rose geranium 2   ounces
  Oil of patchouli..... ½ ounce
  Oil of cloves....... ½ ounce
  Oil of lavender
    flowers.......... 1   ounce
  Oil of bergamot.... 1   ounce
  Oil of sandalwood.. 1   ounce

II.—Oil of bergamot.... 2   ounces
  Oil of orange flow-
    ers.............. 2   ounces
  Oil of sassafras..... 2   ounces
  Oil of white thyme.. 3   ounces
  Oil of cassia....... 3   ounces
  Oil of cloves....... 3   ounces

III.—Oil of citronella.... 1   ounce
  Oil of cloves....... 1   ounce
  Oil of bitter al-
    monds.......... 2   ounces

Pumice-Stone Soaps.—These soaps are always produced by the cold process, either from cocoanut oil alone or in conjunction with tallow, cotton oil, bleached palm oil, etc. The oil is melted and the lye stirred in at about 90° F.; next, the powdered pumice stone is sifted into the soap and the latter is scented. Following are some recipes:

I.—Cocoanut oil...... 40,000 parts
  Cotton oil........ 10,000 parts
  Caustic soda lye,
    38° Bé......... 24,000 parts
  Caustic potash lye,
    30° Bé......... 1,000 parts
  Powdered pumice
    stone.......... 25,000 parts
  Cassia oil........ 150 parts
  Rosemary oil...... 100 parts
  Lavender oil...... 50 parts
  Safrol........... 50 parts
  Clove oil........ 10 parts

II.—Cocoanut oil...... 50,000 parts
  Caustic soda lye,
    40° Bé......... 25,000 parts

Powdered pumice
stone.......... 50,000 parts
Lavender oil....... 250 parts
Caraway oil...... 80 parts

Shaving Soaps.

I.—Palm oil soap....... 5 pounds
Oil of cinnamon.... 10 drachms
Oil of caraway..... 2 drachms
Oil of lavender..... 2 drachms
Oil of thyme....... 1 drachm
Oil of peppermint.. 45 minims
Oil of bergamot.... $2\frac{1}{2}$ drachms

Melt the soap, color if desired, and incorporate the oils.

II.—Soap............... 10 pounds
Alcohol........... 1 ounce
Oil of bitter almonds $1\frac{1}{4}$ ounces
Oil of bergamot.... $\frac{3}{4}$ ounce
Oil of mace........ 3 drachms
Oil of cloves........ $\frac{1}{2}$ ounce

Melt the soap with just enough water to convert it into a soft paste when cold; dissolve the oils in the alcohol, mix with the paste, and rub up in a mortar, or pass several times through a kneading machine.

III.—White castile soap.. 5 parts
Alcohol........... 15 parts
Rose water........ 15 parts

SOAP POWDERS.

The raw materials of which soap powder is made are soap and soda, to which ingredients an addition of talcum or water glass can be made, if desired, these materials proving very useful as a filling. An excellent soap powder has been made of 20 parts of crystallized soda, 5 parts of dark-yellow soap (rosin curd), and 1 part of ordinary soft soap. At first the two last mentioned are placed in a pan, then half the required quantity of soda is added, and the whole is treated. Here it must be mentioned that the dark-yellow curd soap, which is very rosinous, has to be cut in small pieces before placing the quantity into the pan. The heating process must continue very slowly, and the material has to be crutched continually until the whole of the substance has been thoroughly melted. Care must be taken that the heating process does not reach the boiling point. The fire underneath the pan must now be extinguished, and then the remaining half of the crystallized soda is added to be crutched with the molten ingredients, until the whole substance has become liquid. The liquefaction is assisted by the residual heat of the first heated material and the pan. The slow cooling facilitates the productive process by thickening the mass, and when the soda has been absorbed, the whole has become fairly thick. With occasional stirring of the thickened liquid the mass is left for a little while longer, and when the proper moment has arrived the material contained in the pan is spread on sheets of thin iron, and these are removed to a cool room, where, after the first cooling, they must be turned over by means of a shovel, and the turning process has to be repeated at short intervals until the material has quite cooled down and the mixture is thoroughly broken. The soap is now in a very friable condition, and the time has now come to make it into powder, for which purpose it is rubbed through the wire netting or the perforated sieves. Generally the soap is first rubbed through a coarse sieve, and then through finer ones, until it has reached the required conditions of the powder. Some of the best soap powders are coarse, but other manufacturers making an equally good article prefer the finer powder, which requires a little more work, since it has to go through three sieves, whereas the coarse powder can do with one or at most two treatments. But this is, after all, a matter of local requirements or personal taste.

The powder obtained from the above-mentioned ingredients is fine and yellow colored, and it has all the qualities needed for a good sale. Instead of the dark-yellow soap, white stock soap can also be used, and this makes only a little difference in the coloring. But again white stock soap can be used, and the same color obtained by the use of palm oil, or other coloring ingredients, as these materials are used for giving the toilet soaps their manifold different hues. Many makers state that this process is too expensive, and not only swallows up all the profit, but some of the color materials influence the soap and not to its advantage.

Soft soap is used only to make the powder softer and easier soluble, and for this reason the quantity to be used varies a little and different manufacturers believe to have a secret by adding different quantities of this material. As a general statement it may be given that the quantity of soft soap for the making of soap powder should not overstep the proportion of one to three, compared with the quantity of hard soap; any excess in this direction would frustrate the desires of the maker, and land him with a product which has become smeary and moist, forming into balls and lumping together in bags or cases, to become discolored and useless. It is best to stick to the proportion as given, 5 parts of hard and 1 part of soft soap, when the produced powder will be reliable and stable and not form into balls even if the material is kept for a long while.

This point is of special importance, since soap powder is sold mostly in weighed-out packages of one and a half pounds. Most manufacturers will admit that loose soap powder forms only a small part of the quantities produced, as only big laundries and institutions purchase same in bags or cases. The retail trade requires the soap powder wrapped up in paper, and if this has to be done the powder must not be too moist, as the paper otherwise will fall to pieces. This spoils the appearance of the package, and likely a part of the quantity may be lost. When the powder is too moist or absorbs easily external moisture, the paper packages swell very easily and burst open.

The best filling material to be employed when it is desired to produce a cheaper article is talcum, and in most cases this is preferred to water glass. The superiority of the former over the latter is that water glass hardens the powder, and this is sometimes done to such an extent, when a large quantity of filling material is needed, that it becomes very difficult to rub the soap through the sieves. In case this difficulty arises, only one thing can be done to lighten the task, and that is to powderize the soap when the mixed materials are still warm, and this facilitates the work very much. It is self-evident that friction under these conditions leaves a quantity of the soap powder material on the sieves, and this cannot be lost. Generally it is scraped together and returned to the pan to be included in the next batch, when it is worked up, and so becomes useful, a need which does not arise when talcum has been used as a filling material. Again, the soap powder made with the addition of water glass is not so soluble, and at the same time much denser than when the preparation has been made without this material. It is thus that the purchaser receives by equal weight a smaller-looking quantity, and as the eye has generally a great influence when the consumer determines a purchase, the small-sized parcels will impress him unfavorably. This second quality of soap powder is made of the same ingredients as the other, except that an addition of about 6 parts of talcum is made, and this is stirred up with the other material after all the soda has been dis-

solved. Some makers cheapen the products also by reducing the quantity of hard soap from 5 to 3 parts and they avoid the filling; the same quantity of soda is used in all cases. On the same principle a better quality is made by altering the proportions of soda and soap the other way. Experiments will soon show which proportions are most suitable for the purpose.

So-called ammonia-turpentine soap powder has been made by crutching oil of turpentine and ammonia with the materials just about the time before the whole is taken out of the heating pan. Some of the powder is also scented, and the perfume is added at the same time and not before. In most of the latter cases mirbane oil is used for the purpose.

These powders are adaptable to hard water, as their excess of alkali neutralizes the lime that they contain:

I.—Curd (hard) soap,
    powdered........ 4 parts
    Sal soda........... 3 parts
    Silicate of soda..... 2 parts

Make as dry as possible, and mix intimately.

Borax Soap Powder.—

II.—Curd (hard) soap, in
    powder.......... 5 parts
    Soda ash.......... 3 parts
    Silicate of soda..... 2 parts
    Borax (crude)...... 1 part

Each ingredient is thoroughly dried, and all mixed together by sieving.

London Soap Powder.—

III.—Yellow soap........ 6 parts
    Soda crystals....... 3 parts
    Pearl ash........... $1\frac{1}{2}$ parts
    Sulphate of soda.... $1\frac{1}{2}$ parts
    Palm oil........... 1 part

TOILET SOAPS.

The question as to the qualities of toilet soaps has a high therapeutical significance. Impurity of complexion and morbid anomalies of the skin are produced by the use of poor and unsuitable soaps. The latter, chemically regarded, are salts of fatty acids, and are prepared from fats and a lye, the two substances being mixed in a vessel and brought to a boil, soda lye being used in the preparation of toilet soaps. In boiling together a fat and a lye, the former is resolved into its component parts, a fatty acid and glycerine. The acid unites with the soda lye, forming a salt, which is regarded as soap. By the addition of sodium chloride, this (the soap) is separated and swims on the residual liquid as "kern," or granulated soap. Good soaps were formerly made only from animal fats, but some of the vegetable oils or fats have been found to also make excellent soap. Among them the best is cacao butter.

From a hygienic standpoint it must be accepted as a law that a good toilet soap must contain no free (uncombined) alkali, every particle of it must be chemically bound up with fatty acid to the condition of a salt, and the resultant soap should be neutral in reaction. Many of the soaps found in commerce to-day contain free alkali, and exert a harmful effect upon the skin of those who use them. Such soaps may readily be detected by bringing them into contact with the tongue. If free alkali be present it will make itself known by causing a burning sensation—something that a good toilet soap should never do.

The efficiency of soap depends upon the fact that in the presence of an abundance of water the saponified fat is decomposed into acid and basic salts, in which the impurities of the skin are dissolved and are washed away by the further application of water. Good soap exerts its effects on the outer layer of the skin, the so-called horny (epithelial) layer, which in soapy water swells up and is, in fact, partially dissolved in the medium and washed away. This fact, however, is unimportant, since the superficial skin cells are reproduced with extraordinary rapidity and ease. When a soap contains or carries free alkali, the caustic effects of the latter are carried further and deeper, reaching below the epithelial cells and attacking the true skin, in which it causes minute rifts and splits and renders it sore and painful. Good soap, on the contrary, makes the skin smooth and soft.

Since the employment of poor soaps works so injuriously upon the skin, many persons never, or rarely ever, use soap, but wash the face in water alone, or with a little almond bran added. Their skins cannot bear the regular application of poor soap. This, however, applies only to poor, free-alkali containing soaps. Any skin can bear without injury any amount of a good toilet soap, free from uncombined alkali and other impurities. The habit of washing the face with water only, without the use of soap, must be regarded as one altogether bad, since

the deposits on the skin, mostly dust-particles and dead epithelial cells, mingling with the oily or greasy matter exuded from the fat glands of the skin—excellent nutrient media for colonies of bacteria—cannot be got rid of by water alone. Rubbing only forces the mass into the openings in the skin (the sweat glands, fat glands, etc.), and stops them up. In this way are produced the so-called "black heads" and other spots and blotches on the skin usually referred to by the uneducated, or partially educated, as "parasites." The complexion is in this manner injured quite as much by the failure to use good soap as by the use of a poor or bad article.

All of the skin troubles referred to may be totally avoided by the daily use of a neutral, alkali-free soap, and the complexion thus kept fresh and pure. Completely neutral soaps, however, are more difficult to manufacture—requiring more skill and care than those in which no attention is paid to excess of alkali—and consequently cost more than the general public are accustomed, or, in fact, care to pay for soaps. While this is true, one must not judge the quality of a soap by the price demanded for it. Some of the manufacturers of miserable soaps charge the public some of the most outrageous prices. Neither can a soap be judged by its odor or its style of package and putting on the market.

To give a soap an agreeable odor the manufacturers add to it, just when it commences to cool off, an etheric oil (such as attar of rose, oil of violets, bergamot oil, etc.), or some balsamic material (such as tincture of benzoin, for instance). It should be known, however, that while grateful to the olfactory nerves, these substances do not add one particle to the value of the soap, either as a detergent or as a preserver of the skin or complexion.

Especially harmful to the skin are soaps containing foreign substances, such, for instance, as the starches, gelatin, clay, chalk, gums, or rosins, potato flour, etc., which are generally added to increase the weight of soap. Such soaps are designated, very significantly, "filled soaps," and, as a class, are to be avoided, if for no other reason, on account of their lack of true soap content. The use of these fillers should be regarded as a criminal falsification under the laws regarding articles of domestic use, since they are sold at a relatively high price, yet contain foreign matter, harmful to health.

RECIPES FOR COLD-STIRRED TOILET SOAPS.

	Parts by weight
I.—Cocoanut oil	30
Castor oil	3
Caustic soda lye (38° Bé).	17½

Pink Soap.—

	Parts by weight
II.—Pink No. 114	10
Lemon oil	60
Cedar-wood oil	60
Citronella oil	50
Wintergreen oil	15

Pale-Yellow Soap.—

	Parts by weight
III.—Orange No. 410	10
Citronella oil	60
Sassafras oil	60
Lavender oil	45
Wintergreen oil	15
Aniseed oil	25

Toilet Soap Powder.—

Marseilles soap, powdered	100 parts
Bran of almonds	50 parts
Lavender oil	5 parts
Thyme oil	3 parts
Spike oil	2 parts
Citronella oil	2 parts

Soft Toilet Soaps.—Soft toilet soaps or creams may be prepared from fresh lard with a small addition of cocoanut oil and caustic potash solution, by the cold process or by boiling. For the cold process, 23 parts of fresh lard and 2 parts of Cochin cocoanut oil are warmed in a jacketed pan, and when the temperature reaches 113° F. are treated with 9 parts of caustic potash and 2½ parts of caustic soda solution, both of 38° Bé. strength, the whole being stirred until saponification is complete. The soap is transferred to a large marble mortar and pounded along with the following scenting ingredients: 0.15 parts of oil of bitter almonds and 0.02 parts of oil of geranium rose, or 0.1 part of the latter, and 0.05 parts of lemon oil. The warm process is preferable, experience having shown that boiling is essential to the proper saponification of the fats. In this method, 80 parts of lard and 20 parts of Cochin cocoanut oil are melted together in a large pan, 100 parts of potash lye (20° Bé.) being then crutched in by degrees, and the mass raised to boiling point. The combined influence of the heat and crutching vaporizes part of the water in the lye, and the soap thickens. When the soap has combined, the fire is made up, and another 80 parts of the same potash lye

are crutched in gradually. The soap gets thicker and thicker as the water is expelled and finally throws up "roses" on the surface, indicating that it is nearly finished. At this stage it must be crutched vigorously, to prevent scorching against the bottom of the pan and the resulting more or less dark coloration. The evaporation period may be shortened by using only 50 to 60 parts of lye at first, and fitting with lye of 25° to 30° strength. For working on the large scale iron pans heated by steam are used, a few makers employing silver-lined vessels, which have the advantage that they are not attacked by the alkali. Tinned copper pans are also useful. The process takes from 7 to 8 hours, and when the soap is finished it is transferred into stoneware vessels for storage. Clear vegetable oils (castor oil) may be used, but the soaps lack the requisite nacreous luster required.

TRANSPARENT SOAPS.

The mode of production is the same for all. The fats are melted together, sifted into a double boiler, and the lye is stirred in at 111° F. Cover up for an hour, steam being allowed to enter slowly. There is now a clear, grain-like soap in the kettle, into which the sugar solution and the alcohol are crutched, whereupon the kettle is covered up. If cuttings are to be used, they are now added. When same are melted, the kettle will contain a thin, clear soap, which is colored and scented as per directions, and subsequently filled into little iron molds and cooled.

Rose-Glycerine Soap.—

I.—Cochin cocoanut oil	70,000 parts
Compressed tallow	40,000 parts
Castor oil	30,000 parts
Caustic soda lye, 38° Bé.	79,000 parts
Sugar	54,000 parts

Dissolved in

Water	60,000 parts
Alcohol	40,000 parts
Geranium oil (African)	250 parts
Lemon oil	200 parts
Palmarosa oil	1,200 parts
Bergamot oil	80 parts

Benzoin-Glycerine Soap.—

II.—Cochin cocoanut oil	66,000 parts
Compressed tallow	31,000 parts

Castor oil........ 35,000 parts
Caustic soda lye,
    38° Bé........ 66,000 parts
Sugar........... 35,000 parts

Dissolved in

Water........... 40,000 parts
Alcohol......... 35,000 parts
Brown, No. 120... 200 parts
Powdered benzoin
    (Siam)........ 4,200 parts
Styrax liquid..... 1,750 parts
Tincture of ben-
    zoin.......... 1,400 parts
Peru balsam..... 700 parts
Lemon oil....... 200 parts
Clove oil........ 70 parts

Sunflower-Glycerine Soap.—

III.—Cochin cocoanut
    oil............ 70,000 parts
Compressed tal-
    low........... 50,000 parts
Castor oil........ 23,000 parts
Caustic soda lye,
    39° Bé........ 71,000 parts
Sugar........... 40,000 parts

Dissolved in

Water........... 30,000 parts
Alcohol......... 40,000 parts
Brown, No. 55... 250 parts
Geranium oil..... 720 parts
Bergamot oil..... 300 parts
Cedar-wood oil... 120 parts
Palmarosa oil.... 400 parts
Vanillin......... 10 parts
Tonka tincture... 400 parts

MISCELLANEOUS FORMULAS:

Szegedin Soap.—Tallow, 120 parts; palm kernel oil, 80 parts. Saponify well with about 200 parts of lye of 24° Bé. and add, with constant stirring, the following fillings in rotation, viz., potash solution, 20° Bé., 150 parts, and cooling salt solution 20° Bé., 380 parts.

Instrument Soap.—A soap for cleaning surgical instruments, and other articles of polished steel, which have become specked with rust by exposure, is made by adding precipitated chalk to a strong solution of cyanide of potassium in water, until a cream-like paste is obtained. Add to this white castile soap in fine shavings, and rub the whole together in a mortar, until thoroughly incorporated. The article to be cleaned should be first immersed, if possible, in a solution of 1 part of cyanide of potash in 4 parts of water, and kept there until the surface dirt and rust disappears. It should then be polished with the soap, made as above directed.

Stain-Removing Soaps.—These are prepared in two ways, either by making a special soap, or by mixing ordinary soap with special detergents. A good recipe is as follows:

I.—Ceylon cocoanut
    or palm seed oil 320 pounds
Caustic soda lye,
    38° Bé........ 160 pounds
Carbonate of pot-
    ash, 20° Bé.... 56 pounds
Oil of turpentine. 9 pounds
Finely powdered
    kieselguhr..... 280 pounds
Brilliant green.... 2 pounds

The oil having been fused, the dye is mixed with some of it and stirred into the contents of the pan. The kieselguhr is then crutched in from a sieve, then the lye, and then the carbonate of potash. These liquids are poured in in a thin stream. When the soap begins to thicken, add the turpentine, mold, and cover up the molds.

II.—Rosin grain soap, 1,000 pounds
    Talc (made to a
    paste with weak
    carbonate of
    potash)....... 100 pounds
Oil of turpentine. 4 pounds
Benzine......... 3 pounds

Mix the talc and soap by heat, and when cool enough add the turpentine and benzine, and mold.

III.—Cocoanut oil..... 600 pounds
    Tallow.......... 400 pounds
    Caustic soda lye.. 500 pounds
    Fresh ox gall..... 200 pounds
    Oil of turpentine. 12 pounds
    Ammonia (sp. gr.,
    0.91)......... 6 pounds
    Benzine......... 5 pounds

Saponify by heat, cool, add the gall and the volatile liquids, and mold.

Soap Substitutes.—

I.—Linseed oil......... 28 pounds
    Sulphur.......... 8 pounds
    Aluminum soap.... 28 pounds
    Oil of turpentine.... 4 pounds

II.—Aluminum soap.... 15 pounds
    Almadina......... 25 pounds
    Caoutchouc....... 50 pounds
    Sulphur.......... 6 pounds
    Oleum succini...... 4 pounds

Shampoo Soap.—

Linseed oil.......... 20 parts
Malaga olive oil..... 20 parts
Caustic potash....... 9½ parts
Alcohol .... 1 part
Water.............. 30 parts

Warm the mixed oils on a large water bath, then the potash and water in another vessel, heating both to 158° F., and adding the latter hot solution to the hot oil while stirring briskly. Now add and thoroughly mix the alcohol. Stop stirring, keep the heat at 158° F. until the mass becomes clear and a small quantity dissolves in boiling water without globules of oil separating. Set aside for a few days before using to make the liquid soap.

The alcohol may be omitted if a transparent product is immaterial.

Sapo Durus.—

Olive oil	100 parts
Soda lye, sp. gr., 1.33.	50 parts
Alcohol (90 per cent).	30 parts

Heat on a steam bath until saponification is complete. The soap thus formed is dissolved in 300 parts of hot distilled water, and salted out by adding a filtered solution of 25 parts of sodium chloride and 5 parts of crystallized sodium carbonate in 80 parts of water.

Sapo Mollis.—

Olive oil	100 parts
Solid potassium hydroxide	21 parts
Water	100 parts
Alcohol (90 per cent).	20 parts

Boil by means of a steam bath until the oil is saponified, adding, if necessary, a little more spirit to assist the saponification.

Sand Soap.—Cocoa oil, 24 parts; soda lye, 38° Bé., 12 parts; sand, finely sifted, 28 parts; cassia oil, .0100 parts; sassafras oil, .0100 parts.

Salicylic Soap.—When salicylic acid is used in soap it decomposes, as a rule, and an alkali salicylate is formed which the skin does not absorb. A German chemist claims to have overcome this defect by thoroughly eliminating all water from potash or soda soap, then mixing it with vaseline, heating the mixture, and incorporating free salicylic acid with the resulting mass. The absence of moisture prevents any decomposition of the salicylic acid.

Olein Soap Substitute.—Fish oil or other animal oil is stirred up with sulphuric acid, and then treated with water. After another stirring, the whole is left to settle, and separate into layers, whereupon the acid and water are drawn off, and caustic soda solution is stirred in with the oil. The finishing stage consists in stirring in refined mineral oil, magnesium chloride, borium chloride, and pure seal or whale oil, in succession.

Mottled Soap.—Tallow, 30 parts; palm kernel oil, 270 parts; lye, 20°, 347½ parts; potassium chloride solution, 20°, 37½ parts. After everything has been boiled into a soap, crutch the following dye solution into it: Water, 5½ parts; blue, red, or black, .0315 parts; water glass, 38°, 10 parts; and lye, 38°, 1½ parts.

Laundry Soap.—A good, common hard soap may be made from clean tallow or lard and caustic soda, without any very special skill in manipulation. The caustic soda indicated is a crude article which may now be obtained from wholesale druggists in quantities to suit, at a very moderate price. A lye of average strength is made by dissolving it in water in the proportion of about 2 pounds to the gallon. For the saponification of lard, a given quantity of the grease is melted at a low heat, and ¼ its weight of lye is then added in small portions with constant stirring; when incorporation has been thoroughly effected, another portion of lye equal to the first is added, as before, and the mixture kept at a gentle heat until saponification appears to be complete. If the soap does not readily separate from the liquid, more lye should be added, the soap being insoluble in strong lye. When separation has occurred, pour off the lye, add water to the mass, heat until dissolved, and again separate by the use of more strong lye or a strong solution of common salt. The latter part of the process is designed to purify the soap and may be omitted where only a cruder article is required. The soap is finally remelted on a water bath, kept at a gentle heat until as much water as possible is expelled, and then poured into frames or molds to set.

Dog Soap.—

Petroleum	5	
Wax	4	Parts
Alcohol	5	by
Good laundry soap.	15	weight

Heat the petroleum, wax, and alcohol on a water bath until they are well mixed, and dissolve in the mixture the soap cut in fine shavings. This may be used on man or beast for driving away vermin.

Liquid Tar Soap (Sapo Picis liquidus).—

Wood tar	25 parts
Hebra's soap spirit	75 parts

Ox-Gall Soap for Cleansing Silk Stuffs.—To wash fine silk stuffs, such as piece goods, ribbons, etc., employ a soap containing a certain amount of ox gall, a product that is not surpassed for the purpose. In making this soap the following directions will be found of advantage: Heat 1 pound of cocoanut oil to 100° F. in a copper kettle. While stirring vigorously add ½ pound of caustic soda lye of 30° Baumé. In a separate vessel heat ½ pound of white Venice turpentine, and stir this in the soap in the copper kettle. Cover the kettle well, and let it stand, mildly warmed for 4 hours, when the temperature can be again raised until the mass is quite hot and flows clear; then add the pound of ox gall to it. Now pulverize some good, perfectly dry grain soap, and stir in as much of it as will make the contents of the copper kettle so hard that it will yield slightly to the pressure of the fingers. From 1 to 2 pounds is all the grain soap required for the above quantity of gall soap. When cooled, cut out the soap and shape into bars. This is an indispensable adjunct to the dyer and cleaner, as it will not injure the most delicate color.

SOAP-BUBBLE LIQUIDS.

I.—White hard soap...   25 parts
      Glycerine.........   15 parts
      Water........... 1,000 parts

II.—Dry castile soap...   2 parts
      Glycerine.........   30 parts
      Water...........   40 parts

SOAP POLISHES:
See Polishes.

SOAP, TOOTH:
See Dentifrice.

SODA PAINT:
See Paint.

SODA WATER:
See Beverages.

SODIUM HYPOSULPHITE:
See Photography.

SODIUM SILICATE AS A CEMENT:
See Adhesives, under Water-Glass Cements.

SODIUM SALTS, EFFERVESCENT:
See Salts.

Solders

SOLDERING OF METALS AND THE PREPARATION OF SOLDERS.

The object of soldering is to unite two portions of the same metal or of different metals by means of a more fusible metal or metallic alloy, applied when melted, and known by the name of solder. As the strength of the soldering depends on the nature of the solder used, the degree of strength required for the joint must be kept in view in choosing a solder. The parts to be joined must be free from oxide and thoroughly clean; this can be secured by filing, scouring, scraping, or pickling with acids. The edges must fit exactly, and be heated to the melting point of the solder. The latter must have a lower melting point than either of the portions of metal that require to be joined, and if possible only those metals should be chosen for solder which form alloys with them. The solder should also as far as possible have the same color and approximately the same strength as the article whose edges are to be united.

To remove the layers of oxide which form during the process of soldering, various so-called "fluxes" are employed. These fluxes are melted and applied to the joint, and act partly by keeping off the air, thus preventing oxidation, and partly by reducing and dissolving the oxides themselves. The choice of a flux depends on the quantity of heat required for soldering.

Solders are classed as soft and hard solders. Soft solders, also called tin solders or white solders, consist of soft, readily fusible metals or alloys, and do not possess much strength; they are easy to handle on account of their great fusibility. Tin, lead-tin, and alloys of tin, lead, and bismuth are used for soft solders, pure tin being employed only for articles made of the same metal (pure tin).

The addition of some lead makes the solder less fusible but cheaper, while that of bismuth lowers the melting point. Soft solders are used for soldering easily fusible metals such as Britannia metal, etc., also for soldering tin plate. To prepare solder, the metals are melted together in a graphite crucible at as low a temperature as possible, well stirred with an iron rod, and cast into ingots in an iron mold. To melt the solder when required for soldering, the soldering iron is used; the latter should be kept as free from oxidation as possible, and the part applied should be tinned over.

To make so-called "Sicker" solder, equal parts of lead and tin are melted together, well mixed, and allowed to stand till the mixture begins to set, the part still in a liquid condition being then poured off. This mixture can, however, be more easily made by melting together 37 parts of lead and 63 parts of tin (exactly measured).

Soldering irons are usually made of copper, as copper is easily heated and easily gives up its heat to the solder. The point of the iron must be "tinned." To do this properly, the iron should be heated hot enough easily to melt the solder; the point should then be quickly dressed with a smooth flat file to remove the oxide, and rubbed on a piece of tin through solder and sal ammoniac. The latter causes the solder to adhere in a thin, even coat to the point of the iron. A gas or gasoline blow torch or a charcoal furnace is best for heating the iron, but a good, clean coal fire, well coked, will answer the purpose.

When in use, the iron should be hot enough to melt the solder readily. A cold iron produces rough work. This is where the beginner usually fails. If possible, it is well to warm the pieces before applying the iron. The iron must not be heated too hot, however, or the tin on the point will be oxidized. The surfaces to be soldered must be clean. Polish them with sandpaper, emery cloth, a file, or a scraper. Grease or oil will prevent solder from sticking.

Some good soldering fluid should be used. A very good fluid is made by dissolving granulated zinc in muriatic acid. Dissolve as much zinc as possible in the acid. The gas given off will explode if ignited. To granulate the zinc, melt it in a ladle, and pour it slowly into a barrel of water. A brush or swab should be used to spread the fluid on the surfaces to be soldered. If the point of the soldering iron becomes dirty, it should be wiped on a cloth or piece of waste that has been dampened with the soldering fluid.

Soldering of Metallic Articles.—In a recently invented process the parts to be united are covered, on the surfaces not to be soldered, with a protective mass, which prevents an immediate contact of the solder with the surfaces in question, and must be brushed off only after the soldered pieces have cooled perfectly, whereby the possibility of a change of position of these pieces seems precluded.

For the execution of this process the objects to be soldered, after the surfaces to be united have been provided with a water-glass solution as the soldering agent and placed together as closely as possible or united by wires or rivets, are coated in the places where no solder is desired with a protective mass, consisting essentially of carbon (graphite, coke, or charcoal), powdered talc or asbestos, ferric hydrate (with or without ferrous hydrate), and, if desired, a little aluminum oxide, together with a binding agent of the customary kind (glue solution, beer).

Following are some examples of the composition of these preparations:

I.—Graphite, 50 parts; powdered coke, 5 parts; powdered charcoal, 5 parts; powdered talc, 10 parts; glue solution, 2.5 parts; drop beer, 2.5 parts; ferric hydrate, 10 parts; aluminum oxide, 5 parts.

II.—Graphite, burnt, 4 parts; graphite, unburnt, 6 parts; powdered charcoal, 3 parts; powdered asbestos, 1 part; ferric hydrate, 3 parts; ferrous hydrate, 2 parts; glue solution, 1 part.

The article thus prepared is plunged, after the drying of the protective layer applied, in the metal bath serving as solder (molten brass, copper, etc.), and left to remain therein until the part to be soldered has become red hot, which generally requires about 50 to 60 seconds, according to the size of the object. In order to avoid, in introducing the article into the metal bath, the scattering of the molten metal, it is well previously to warm the article and to dip it warm. After withdrawal from the metal bath the soldered articles are allowed to cool, and are cleaned with wire brushes, so as to cause the bright surfaces to reappear.

The process is especially useful for uniting iron or steel parts, such as machinery, arms, and bicycle parts in a durable manner.

Soldering Acid.—A very satisfactory soldering acid may be made by the use of the ordinary soldering acid for the base and introducing a certain proportion of chloride of tin and sal ammoniac. This gives an acid which is superior in every way to the old form. To make 1 gallon of this soldering fluid take 3 quarts of common muriatic acid and allow it to dissolve as much zinc as it will take up. This method, of course, is the usual one followed in the manufacture of ordinary soldering acid. The acid, as is well known, must be placed in an earthenware or glass vessel. The zinc may be sheet clippings or common plate spelter broken into small pieces. Place the acid in the vessel and add the zinc in small portions so as to prevent the whole from boiling over. When all the zinc has been added and the action has stopped, it indicates that enough has been taken up. Care must be taken to see that there is a little zinc left in the bottom, as otherwise the acid will be in excess. The idea is to have the acid take up as much zinc as it can.

After this has been done there will remain some residue in the form of a black precipitate. This is the lead which all zinc contains, and which is not dissolved by the muriatic acid. This lead may be removed by filtering through a funnel in the bottom of which there is a little absorbent cotton, or the solution may be allowed to remain overnight until the lead has settled and the clear solution can then be poured off. This lead precipitate is not particularly injurious to the soldering fluid, but it is better to get rid of it so that a good, clear solution may be obtained. Next, dissolve 6 ounces of sal ammoniac in a pint of warm water. In another pint dissolve 4 ounces of chloride of tin. The chloride of tin solution will usually be cloudy, but this will not matter. Now mix the 3 solutions together. The solution will be slightly cloudy when the 3 have been mixed, and the addition of a few drops of muriatic acid will render it perfectly clear. Do not add any more acid than is necessary to do this, as the solution would then contain too much of this ingredient and the results would be injurious.

This soldering acid will not spatter when the iron is applied to it. It has also been found that a poorer grade of solder may be used with it than with the usual soldering acid.

ALUMINUM SOLDERS.

To solder aluminum it is necessary previously to tin the parts to be soldered. This tinning is done with the iron, using a composition of aluminum and tin. Replace the ordinary soldering iron by an iron of pure aluminum. Preparation of aluminum solder: Commence by fusing the copper; then add the aluminum in several installments, stir the mixture well with a piece of iron; next add the zinc and a little tallow or benzine at the same time. Once the zinc is added do not heat too strongly, to avoid the volatilization of the zinc.

I.—Take 5 parts of tin and 1 part of aluminum. Solder with the iron or with the blowpipe, according to the article in question.

II.—The pieces to be soldered are to be tinned, but instead of using pure tin, alloys of tin with other metals are employed, preferably those of tin and aluminum. For articles to be worked after soldering, 45 parts of tin and 10 parts of aluminum afford a good alloy, malleable enough to be hammered, cut, or turned. If they are not to be worked, the alloy requires less aluminum and may be applied in the usual manner as in soldering iron.

Aluminum Bronze.—I.—Strong solder: Gold, 89 parts; fine silver, 5 parts; copper, 6 parts.

II.—Medium solder: Gold, 54 parts; fine silver, 27 parts; copper, 19 parts.

III.—Weak solder: Gold, 14 parts; silver, 57 parts; copper, 15 parts; brass, 14 parts.

BRASS SOLDERS.

Brass solder consists of brass fusible at a low temperature, and is made by melting together copper and zinc, the latter being in excess. A small quantity of tin is often added to render the solder more fusible. Hard solders are usually sold in the form of granules. Although many workers in metals make their own solder, it is advisable to use hard solder made in factories, as complete uniformity of quality is more easily secured where large quantities are manufactured.

In making hard solder the melted metal is poured through birch twigs in order to granulate it. The granules are afterwards sorted by passing them through sieves.

When brass articles are soft-soldered, the white color of the solder contrasts unpleasantly with the brass. If this is objected to, the soldered part can be colored yellow in the following manner:

Dissolve 10 parts of copper sulphate in 35 parts of water; apply the solution to the solder, and stir with a clean iron wire. This gives the part the appearance of copper. To produce the yellow color, paint the part with a mixture consisting of 1 part of a solution of equal parts of zinc and water (1 part each) and 2 parts of a solution of 10 to 35 parts respectively of copper sulphate and water and rub on with a zinc rod. The resulting yellow color can, if desired, be improved by careful polishing.

The quality of soft solder is always judged in the trade from the appearance of the surface of the castings, and it is considered important that this surface should be radiant and crystalline, showing the so-called "flowers." These should be more brilliant than the dull background, the latter being like mat silver in appearance. If the casting has a uniform whitish-gray color, this is an indication that the alloy contains an insufficient quantity of tin. In this case the alloy should be remelted and tin added, solder too poor in tin being extremely viscid.

Most of the varieties of brass used in the arts are composed of from 68 to 70 per cent copper and from 32 to 30 per cent zinc. Furthermore, there are some kinds of brass which contain from 24 to 40 per cent zinc. The greater the quantity of zinc the greater will be the resemblance of the alloy to copper. Consequently, the more crystalline will the structure become. For hard soldering only alloys can be employed which, as a general rule, contain no more than 34 per cent of zinc. With an increase in copper there follows a rise in the melting point of the brass. An alloy containing 90 per cent of copper will meet at 1,940° F.; 80 per cent copper, at 1,868° F.; 70 per cent copper, at 1,796° F.; 60 per cent copper, at 1,742° F. Because an increase in zinc causes a change in color, it is sometimes advisable to use tin for zinc, at least in part, so that the alloy becomes more bronze-like in its properties. The durability of the solder is not seriously affected, but its fusibility is lowered. If more than a certain proportion of tin be added, thin and very fluid solders are obtained of grayish-white color, and very brittle—indeed, so brittle that the soldering joints are apt to open if the object is bent. Because too great an addition of tin is injurious, the utmost caution must be exercised. If very refractory metals are to be soldered, brass alone can be used. In some cases, a solder can be produced merely by melting brass and adding copper. The following hard solders have been practically tested and found of value.

YELLOW HARD SOLDERS:
Applebaum's Compositions.—

I.—Copper	58	parts
Zinc	42	parts

II.—Sheet brass	85.42	parts
Zinc	13.58	parts

Karmarsch's Composition.—

III.—Brass	7	parts
Zinc	1	part

IV.—Zinc	49	parts
Copper	44	parts
Tin	4	parts
Lead	2	parts

Prechtl's Composition.—

V.—Copper	53.3	parts
Zinc	43.1	parts
Tin	1.3	parts
Lead	0.3	parts

All these hard-solder compositions have the fine yellow color of brass, are very hard, and can be fused only at high temperatures. They are well adapted for all kinds of iron, steel, copper, and bronze.

Solders which fuse at somewhat lower temperatures and, therefore, well adapted for the working of brass, are the following:

VI.—Sheet brass	81.12	parts
Zinc	18.88	parts

VII.—Copper	54.08	parts
Zinc	45.29	parts

VIII.—Brass	3 to 4	parts
Zinc	1	part

A solder which is valuable because it can be wrought with the hammer, rolled out, or drawn into wire, and because it is tough and ductile, is the following:

IX.—Brass	78.26	parts
Zinc	17.41	parts
Silver	4.33	parts

Fusible White Solder.—

X.—Copper	57.4	parts
Zinc	28	parts
Tin	14.6	parts

Easily Fusible Solders.—

XI.—Brass	5	parts
Zinc	2.5	parts

XII.—Brass	5	parts
Zinc	5	parts

Semi-White Hard Solders.—

XIII.—Copper	53.3	parts
Zinc	46.7	parts

XIV.—Brass	12	parts
Zinc	4 to 7	parts
Tin	1	part

XV.—Brass	22	parts
Zinc	10	parts
Tin	1	part

XVI.—Copper	44	parts
Zinc	49	parts
Tin	3.20	parts
Lead	1.20	parts

Formulas XIII and XVI are fairly fusible.

White Hard Solders.—

XVII.—Brass	20	parts
Zinc	1	part
Tin	4	parts

XVIII.—Copper	58	parts
Zinc	17	parts
Tin	15	parts

XIX.—Brass	11	parts
Zinc	1	part
Tin	2	parts

XX.—Brass........ 6 parts
Zinc......... 4 parts
Tin .... 10 parts

XXI.—Copper...... 57.44 parts
Zinc......... 27.98 parts
Tin.......... 14.58 parts

For Brass Tubes.—I.—Copper, 100 parts; lead, 25 parts.

II.—A very strong solder for soldering brass tubes to be drawn, etc., is composed of 18 parts brass, 4 parts zinc, and 1 part fine silver.

For Fastening Brass to Tin.—To 20 parts of fine, reduced copper, add sufficient sulphuric acid to make a stiff paste. To this add 70 parts of metallic mercury, and work in, at the same time applying heat until the mass assumes a wax-like consistency. Warm or heat the plates to be united, to about the same temperature, apply the mixture, hot, to each, then press together, and let cool.

COPPER SOLDERS.

The copper solders which are used for soldering copper as well as bronze are mixtures of copper and lead. By increasing the quantity of lead the fusibility is increased, but the mixture departs from the color and toughness of copper. The most commonly employed copper solder is the following:

I.—Copper........... 5 parts
Lead .... 1 part

II.—Copper........... 80 parts
Lead .... 15 parts
Tin .... 5 parts

For Red Copper.—I.—Copper, 3 parts; zinc, 1 part.

II.—Copper, 7 parts; zinc, 3 parts; tin, 2 parts.

FATS FOR SOLDERING.

I.—Soldering fat or grease is commonly a mixture of rosin and tallow with the addition of a small quantity of sal ammoniac. It is particularly adapted to the soldering of tinned ware, because it is easily wiped off the surface after the joint is made, whereas if rosin were used alone, the scraping away might remove some of the tin and spoil the object.

II.—The following is a well-tried recipe for a soldering grease: In a pot of sufficient size and over a slow fire melt together 500 parts of olive oil and 400 parts of tallow; then stir in slowly 250 parts of rosin in powder, and let the whole boil up once. Now let it cool

down, and add 125 parts of saturated solution of sal ammoniac, stirring the while. When cold, this preparation will be ready for use.

FLUIDS FOR SOLDERING.

I.—To the ordinary zinc chloride, prepared by digesting chips of zinc in strong hydrochloric acid to saturation, add $\frac{1}{3}$ spirits of sal ammoniac and $\frac{1}{3}$ part rain water, and filter the mixture. This soldering liquid is especially adapted to the soft soldering of iron and steel, because it does not make rust spots.

To solder zinc, the zinc chloride may be used without any spirit sal ammoniac.

II.—Mix phosphoric acid with strong spirits of wine in the following proportions:

Phosphoric acid solution.............. 1 quart
Spirits of wine (80 per cent)............. $1\frac{1}{2}$ quarts

More or less of the spirits of wine is used depending upon the concentration of the phosphoric acid solution. When this soldering liquid is applied to the metal to be soldered, the phosphoric acid immediately dissolves the oxide. The hot soldering iron vaporizes the spirits of wine very quickly and causes the oxide released by the phosphoric acid to form a glazed mass with the surplus phosphoric acid, which mass can be easily removed.

III.—Dissolve in hydrochloric acid: Zinc, 50 parts (by weight); sal ammoniac, 50 parts.

IV.—Hydrochloric acid, 600 parts (by weight); sal ammoniac, 100 parts. Put zinc chips into the acid to saturation, next add the sal ammoniac. Filter when dissolved and preserve in flasks.

V.—Eight hundred parts of water with 100 parts of lactic acid and 100 parts of glycerine. This dispenses with the use of chloride of zinc.

Acid-Free Soldering Fluid.—I.—Five parts of zinc chloride dissolved in 25 parts of boiling water. Or, 20 parts of zinc chloride, 10 parts of ammonia chloride, dissolved in 100 parts of boiling water and put into glass carboys.

II.—Chloride zinc...... 1 drachm
Alcohol.......... 1 ounce

Substitute for Soldering Fluid.—As a substitute for the customary soldering fluid and soldering mediums an ammonia soap is recommended, which is obtained by the mixture of a finely powdered rosin with strong ammonia solution. Of this soap only the finely divided rosin remains on the soldered place after the soldering. This soldering process is well adapted for soldering together copper wires for electrical conduits, since the rosin at the same time serves as an insulator.

FLUXES FOR SOLDERING.

The fluxes generally used in the soft-soldering of metals are powdered rosin or a solution of chloride of zinc, alone or combined with sal ammoniac. A neutral soldering liquid can be prepared by mixing 27 parts neutral zinc chloride, 11 parts sal ammoniac, and 62 parts water; or, 1 part sugar of milk, 1 part glycerine, and 8 parts water.

A soldering fat for tin-plate, preferable to ordinary rosin, as it can be more easily removed after soldering, is prepared as follows: One hundred and fifty parts beef tallow, 250 parts rosin, and 150 parts olive oil are melted together in a crucible and well stirred, 50 parts powdered sal ammoniac dissolved in as little water as possible being added.

Soldering fat for iron is composed of 50 parts olive oil and 50 parts powdered sal ammoniac. Soldering fat for aluminum is made by melting together equal parts of rosin and tallow, half the quantity of zinc chloride being added to the mixture.

Soldering paste consists of neutral soldering liquid thickened with starch paste. This paste must be applied more lightly than the soldering liquid.

Soldering salt is prepared by mixing equal parts of neutral zinc chloride, free from iron, and powdered sal ammoniac. When required for use, 1 part of the salt should be dissolved in 3 or 4 parts water.

Borax is the flux most frequently used for hard-soldering; it should be applied to the soldering seam either dry or stirred to a paste with water. It is advisable to use calcined borax, i. e., borax from which the water of crystallization has been driven out by heat, as it does not become so inflated as ordinary borax. Borax dissolves the metallic oxides forming on the joint.

Finely powdered cryolite, or a mixture of 2 parts powdered cryolite and 1 part phosphoric acid, is also used for hard-soldering copper and copper alloys.

Muller's hard-soldering liquid consists of equal parts of phosphoric acid and alcohol (80 per cent).

A mixture of equal parts of cryolite and barium chloride is used as a flux in hard-soldering aluminum bronze.

A very good dry-soldering preparation consists of two vials, one of which is filled with zinc chloride, and the other with ammonium chloride. To use, dissolve a little of each salt in water, apply the ammonium chloride to the object to be soldered and heat the latter until it begins to give off vapor of ammonium, then apply the other, and immediately thereafter the solder, maintaining the heat in the meantime. This answers for very soft solder. For a harder solder dissolve the zinc in a very small portion of the ammonium chloride solution (from $\frac{1}{4}$ to $\frac{1}{2}$ pint).

When steel is to be soldered on steel, or iron on steel, it is necessary to remove every trace of oxide of iron between the surfaces in contact. Melt in an earthen vessel: Borax, 3 parts; colophony, 2 parts; pulverized glass, 3 parts; steel filings, 2 parts; carbonate of potash, 1 part; hard soap, powdered, 1 part. Flow the melted mass on a cold plate of sheet iron, and after cooling break up the pieces and pulverize them. This powder is thrown on the surfaces a few minutes before the pieces to be soldered are brought together. The borax and glass contained in the composition dissolve, and consequently liquefy all of the impurities, which, if they were shut up between the pieces soldered, might form scales, at times dangerous, or interfering with the resistance of the piece.

To prepare rosin for soldering bright tin, mix $1\frac{1}{2}$ pounds of olive oil, $1\frac{1}{2}$ pounds of tallow, and 12 ounces of pulverized rosin, and let them boil up. When this mixture has become cool, add $1\frac{3}{4}$ pints of water saturated with pulverized sal ammoniac, stirring constantly

GAS SOLDERING.

The soldering of small metallic articles where the production is a wholesale one, is almost exclusively done by the use of gas, a pointed flame being produced by air pressure. The air pressure is obtained by the workman who does the soldering setting in motion a treadle with his foot, which, resting on rubber bellows, drives by pressure on the same the aspirated air into wind bellows. From here it is sent into the soldering pipe, where it is connected with the gas and a pointed flame is produced. In order to obtain a rather uniform heat the workman has to tread continually, which, however, renders it almost impossible to hold the article to be soldered steady, although this is necessary if the work is to proceed quickly. Hence, absolutely skillful and expensive hands are required, on whom the employer is often entirely dependent. To improve this method of soldering and obviate its drawbacks, the soldering may be conducted with good success in the following manner: For the production of the air current a small ventilator is set up. The wind is conducted through two main conduits to the work tables. Four or six tables may, for instance, be placed together, the wind and the gas pipe ending in the center. The gas is admitted as formerly. the wind is conducted into wind bellows by means of joint and hose to obtain a constant pressure and from here into the soldering pipe. In this manner any desired flame may be produced, the workman operates quietly and without exertion, which admits of employing youthful hands and consequently of a saving in wages. The equipment is considerably cheaper, since the rubber bellows under the treadle are done away with.

GERMAN-SILVER SOLDERS.

Because of its peculiar composition German-silver solder is related to the ordinary hard solders. Just as hard solders may be regarded as varieties of brass to which zinc has been added, German-silver solders may be regarded as German silver to which zinc has been added. The German-silver solder becomes more easily fused with an increase in zinc, and vice versa. If the quantity of zinc be increased beyond a certain proportion, the resultant solder becomes too brittle. German-silver solders are characterized by remarkable strength, and are therefore used not only in soldering German silver, but in many cases where special strength is required. As German silver can be made of the color of steel, it is frequently used for soldering fine steel articles.

Solder for ordinary German silver can be made of 1,000 parts German-silver chips, 125 parts sheet-brass chips, 142 parts zinc, and 33 parts tin; or, of 8 parts German silver and 2 to 3 parts zinc.

Soft German-Silver Solder.—

I.—Copper........... 4.5 parts
    Zinc............. 7   parts
    Nickel.......... 1   part

II.—Copper...........35   parts
    Zinc.............56.5 parts
    Nickel........... 8.5 parts

III.—German silver..... 5   parts
     Zinc............. 5   parts

Compositions I and II have analogous properties. In composition III "German silver" is to be considered as a mixture of copper, zinc, and nickel, for which reason it is necessary to know the exact composition of the German silver to be used. Otherwise it is advisable to experiment first with small quantities in order to ascertain how much zinc is to be added. The proper proportion of German silver to zinc is reached when the mixture reveals a brilliancy and condition which renders it possible to barely pulverize it while hot. A small quantity when brought in contact with the soldering iron should just fuse.

Hard German-Silver or Steel Solder.—

I.—Copper...........35   parts
    Zinc.............56.5 parts
    Nickel.......... 9.5 parts

II.—Copper...........38   parts
    Zinc.............50   parts
    Nickel..........12   parts

Composition I requires a fairly high temperature in order to be melted. Composition II requires a blow pipe.

GOLD SOLDERS:

Hard Solder for Gold.—The hard solder or gold solder which the jeweler frequently requires for the execution of various works, not only serves for soldering gold ware, but is also often employed for soldering fine steel goods, such as spectacles, etc. Fine gold is only used for soldering articles of platinum. The stronger the alloy of the gold, the more fusible must be the solder. Generally the gold solder is a composition of gold, silver, and copper. If it is to be very easily fusible, a little zinc may be added, but, on the other hand, even the copper is sometimes left out and a mixture consisting only of gold and silver (e. g., equal parts of both) is used. The shade of the solder also requires attention, which must be regulated by varying proportions of silver and copper, so that it may be as nearly as possible the same as that of the gold to be soldered.

I.—For 24-carat gold: Twenty-two parts gold (24 carat), 2 parts silver, and 1 part copper; refractory.

II.—For 18-carat gold: Nine parts gold (18 carat), 2 parts silver, and 1 part copper; refractory.

III.—For 16-carat gold: Twenty-four parts gold (16 carat), 10 parts silver, and 8 parts copper; refractory.

IV.—For 14-carat gold: Three parts gold (14 carat), 2 parts silver, and 1 part copper; more fusible.

V.—Gold solder for alloys containing smaller quantities of gold is composed of 8 parts gold, 10.5 parts silver, and 5.5 parts copper, or,

VI.—Ten parts gold (13.5 carat), 5 parts silver, and 1 part zinc.

VII.—The following easily fusible solder is used for ordinary gold articles: Two parts gold, 9 parts silver, 1 part copper, and 1 part zinc. Articles soldered with this solder cannot be subjected to the usual process of coloring the gold, as the solder would become black.

VIII.—A refractory enamel solder for articles made of 20-carat and finer gold, which can bear the high temperature required in enameling, consists of 37 parts gold and 9 parts silver, or 16 parts gold (18 carat), 3 parts silver, and 1 part copper.

Which of these compositions should be employed depends upon the degree of the fusibility of the enamel to be applied. If it is very difficult of fusion only the first named can be used; otherwise it may happen that during the melting on of the enamel the soldering spots are so strongly heated that the solder itself melts. For ordinary articles, as a rule, only readily fusible enamels are employed, and consequently the readily fusible enameling solder may here be made use of. Soldering with the latter is readily accomplished with the aid of the soldering pipe. Although the more hardly fusible gold solders may also be melted by the use of the ordinary soldering pipe, the employment of a special small blowing apparatus is recommended on account of the resulting ease and rapidity of the work.

SOLDERS FOR GLASS.

I.—Melt tin, and add to the melted mass enough copper, with constant stirring, until the melted metal consists of 95 per cent of tin and 5 per cent of copper. In order to render the mixture more or less hard, add ½ to 1 per cent of zinc or lead.

II.—A compound of tin (95 parts) and zinc (5 parts) melts at 392° F., and can then be firmly united to glass. An alloy of 90 parts of tin and 10 parts of aluminum melts at 734° F., adheres, like the preceding, to glass, and is equally brilliant. With either of these alloys glass may be soldered as easily as metal, in two ways. In one, heat the pieces of glass in a furnace and rub a stick of soldering alloy over their surfaces. The alloy will melt, and can be easily spread by means of a roll of paper or a slip of aluminum. Press the pieces firmly together, and keep so until cool. In the other method a common soldering iron, or a rod of aluminum, is heated over a coal fire, a gas jet, or a flame supplied by petroleum. The hot iron is passed over the alloy and then over the pieces to be soldered, without the use of a dissolvent. Care should be taken that neither the soldering irons nor the glass be brought to a temperature above the melting point of the alloy, lest the latter should be oxidized, and prevented from adhering.

HARD SOLDERS.

Hard solders are distinguished as brass, German silver, copper, gold, silver, etc., according to the alloys used (see Brass Solders, Copper Solders, etc., for other hard solders).

The designation "hard solder" is used to distinguish it from the easily running and softer solder used by tinsmiths, and it applies solely to a composition that will not flow under a red heat. For the purposes of the jeweler solder may be classified according to its composition and purpose, into gold or silver solder, which means a solder consisting of an alloy of gold with silver, copper, tin, or zinc-like metal or an alloy of silver with copper, tin, or zinc-like metal. According to the uses, the solder is made hard or soft; thus in gold solders there is added a greater amount of silver, whereas for silver solders there is added more tin or zinc-like metal.

In the production of solder for the enameler's use, that is for combining gold with gold, gold with silver, or gold with copper, which must be enameled afterwards, it is necessary always to keep in mind that no solder can be used effectually that contains any tin, zinc, zinc alloys, or tin or zinc-like metals in any great quantities, since it is these very metals that contribute to the cracking of the enamel. Yet it is not possible to do without such an addition entirely, otherwise the solder would not flow under the melting point of the precious metals themselves and we should be unable to effect a union of the parts. It is therefore absolutely necessary to confine these additions to the lowest possible percentage, so that only a trace is apparent. Moreover, care must be taken to use for enameling purposes no base alloy, because the tenacity or durability of the compound will be affected thereby; in other words, it must come up to the standard.

In hard soldering with borax, direct, several obstacles are encountered that make the process somewhat difficult. In the first place the salt forms great bubbles in contact with the soldering iron, and easily scales away from the surface of the parts to be soldered. Besides this, the parts must be carefully cleaned each time prior to applying the salt. All these difficulties vanish if instead of borax we use its component parts, boric acid and sodium carbonate. The heat of the soldering iron acting on these causes them to combine in such a way as to produce an excellent flux, free from the difficulties mentioned.

Composition of Various Hard Solders. —Yellow solders for brass, bronze, copper, and iron:

I.—Sheet-brass chips, 5 parts, and zinc, 3 to 5 parts, easily fusible.

II.—Sheet brass chips, 3 parts, and zinc, 1 part; refractory.

III.—Sheet-brass chips, 7 parts, and zinc, 1 part; very refractory and firm.

Semi-white solders, containing tin and consequently harder:

I.—Sheet brass, 12 parts; zinc, 4 to 7 parts, and tin, 1 part.

II.—Copper, 16 parts; zinc, 16 parts, and tin, 1 part.

III.—Yellow solder, 20 to 30 parts, and tin, 1 part.

White solders:

I.—Sheet brass, 20 parts; zinc, 1 part, and tin, 4 parts.

II.—Copper, 3 parts; zinc, 1 part, and tin, 1 part.

To Hard-Solder Parts Formerly Soldered with Tin Solder.—To repair gold or silver articles which have been spoiled with tin solder proceed as follows: Heating the object carefully by means a of small spirit lamp, brush the tin off as much as possible with a chalk brush; place the article in a diluted solution of hydrochloric acid for about 8 to 10 hours, as required. If much tin remains, perhaps 12 hours may be necessary. Next withdraw it, rinse off and dry; whereupon it is carefully annealed and finally put in a pickle of dilute sulphuric acid, to remove the annealing film. When the article has been dipped, it may be hard soldered again.

SILVER SOLDERS.

Silver solder is cast in the form of ingots, which are hammered or rolled into thin sheets. From these small chips or "links," as they are called, are cut off. The melted solder can also be poured, when slightly cooled, into a dry iron mortar and pulverized while still warm. The solder can also be filed and the filings used for soldering.

Silver solders are used not only for soldering silver objects, but also for soldering metals of which great resistance is expected. A distinction must be drawn between silver solder consisting either of copper and silver alone, and silver solder to which tin has been added.

Very Hard Silver Solder for Fine Silverware.—

I.—Copper	1 part
Silver	4 parts
Hard silver solder.	

II.—Copper	1 part
Silver	20 parts
Brass	9 parts

III.—Copper	2 parts
Silver	28 parts
Brass	10 parts
Soft silver solder.	

IV.—Silver	2 parts
Brass	1 part

V.—Silver	3 parts
Copper	2 parts
Zinc	1 part

VI.—Silver	10 parts
Brass	10 parts
Tin	1 part

These solders are preferably to be employed for the completion of work begun with hard silver solders, defective parts alone being treated. For this purpose it is sometimes advisable to use copper-silver alloys mixed with zinc, as for example:

VII.—Silver	12 parts
Copper	4 parts
Zinc	1 part

VIII.—Silver	5 parts
Brass	6 parts
Zinc	2 parts

This last formula (VIII) is most commonly used for ordinary silverware.

Silver Solders for Soldering Iron, Steel, Cast Iron, and Copper.—

I.—Silver	10	parts
Brass	10	parts

II.—Silver	20	parts
Copper	30	parts
Zinc	10	parts

III.—Silver	30	parts
Copper	10	parts
Tin	0.5	parts

IV.—Silver	60	parts
Brass	60	parts
Zinc	5	parts

In those solders in which brass is used care should be taken that none of the metals employed contains iron. Even an inappreciable amount of iron deleteriously affects the solder.

V.—Copper, 30 parts; zinc, 12.85 parts; silver, 57.15 parts.

VI.—Copper, 23.33 parts; zinc, 10 parts; silver, 66.67 parts.

VII.—Copper, 26.66 parts; zinc, 10 parts; silver, 63.34 parts.

VIII.—Silver, 66 parts; copper, 24 parts, and zinc, 10 parts. This very strong solder is frequently used for soldering silver articles, but can also be used for soldering other metals, such as brass, copper, iron, steel band-saw blades, etc.

IX.—Silver, 4 parts, and brass, 3 parts.

X.—A very refractory silver solder, which, unlike the silver solder containing zinc, is of great ductility and does not break when hammered, is composed of 3 parts silver and 1 part copper.

Soft Silver Solders.—I.—A soft silver solder for resoldering parts already soldered is made of silver, 3 parts; copper, 2 parts, and zinc, 1 part.

II.—Silver, 1 part, and brass, 1 part; or, silver, 7 parts; copper, 3 parts, and zinc, 2 parts.

III.—A readily fusible silver solder for ordinary work: Silver, 5 parts; copper, 6 parts, and zinc, 2 parts.

IV.—(Soft.) Copper, 14.75 parts; zinc, 8.20 parts; silver, 77.05 parts.

V.—Copper, 22.34 parts; zinc, 10.48 parts; silver, 67.18 parts.

VI.—Tin, 63 parts; lead, 37 parts.

French Solders for Silver.—I.—For fine silver work: Fine silver, 87 parts; brass, 13 parts.

II.—For work 792 fine: Fine silver, 83 parts; brass, 17 parts.

III.—For work 712 fine: Fine silver, 75 parts; brass, 25 parts.

IV.—For work 633 fine: Fine silver, 66 parts; brass, 34 parts.

V.—For work 572 fine: Fine silver, 55 parts; brass, 45 parts.

Solder for Silversmiths, etc.—Gold, 10 parts; silver, 55 parts; copper, 29 parts; zinc, 6 parts.

Hard Solder.—Silver, 60 parts; bronze, 39 parts; arsenic, 1 part.

Soft Solder. — Powdered copper, 30 parts; sulphate of zinc, 10 parts; mercury, 60 parts; sulphuric acid. Put the copper and the zinc sulphate in a porcelain mortar, and then the sulphuric acid. Enough acid is required to cover the composition; next add the mercury while stirring constantly. When the amalgamation is effected, wash several times with hot water to remove the acid, then allow to cool. For use, it is sufficient to heat the amalgam until it takes the consistency of wax. Apply on the parts to be soldered and let cool.

Solder for Silver-Plated Work.—I.—Fine silver, 2 parts; bronze, 1 part.

II.—Silver, 68 parts; copper, 24 parts; zinc, 17 parts.

Solder for Silver Chains.—I.—Fine silver, 74 parts; copper, 24 parts; orpiment, 2 parts.

II.—Fine silver, 40 parts; orpiment, 20 parts; copper, 40 parts.

SOFT SOLDERS:

See also Brass Solders, Copper Solders, Gold Solders.

I.—Fifty parts bismuth, 25 parts tin, and 25 parts lead. This mixture melts at 392° F.

II.—Fifty parts bismuth, 30 parts lead, and 20 parts tin. This will melt at 374° F.

III.—The solder that is used in soldering Britannia metal and block tin pipes is composed of 2 parts tin and 1 part lead. This melts in the blow-pipe flame at many degrees lower temperature than either tin or Britannia metal, and it is nearly of the same color. Care must be taken in mixing these solders to keep them well stirred when pouring into molds. Care should also be taken that the metal which melts at a higher temperature be melted first and then allowed to cool to the melting temperature of the next metal to be added, and so on. Articles to be soldered with these solders should be joined with a blow pipe to get the best results, but if a copper is used it must be drawn out to a long, thin point. For a flux use powdered rosin or sweet oil.

Tin solders for soldering lead, zinc, tin, tin-plate, also copper and brass when special strength is not required, are prepared as follows:

I.—Tin, 10 parts; lead, 4 parts; melting point, 356° F.

II.—Tin, 10 parts; lead, 5 parts; melting point, 365° F.

III.—Tin, 10 parts; lead, 6 parts; melting point, 374° F.

IV.—Tin, 10 parts; lead, 10 parts; melting point, 392° F.

V.—Tin, 10 parts; lead, 15 parts; melting point, 432° F.

VI.—Tin, 10 parts; lead, 20 parts; melting point, 464° F.

The last of the above mixtures is the cheapest, on account of the large quantity of lead.

Bismuth solder or pewterer's solder fusible at a low temperature is prepared by melting together:

I.—Tin, 2 parts; lead, 1 part; bismuth, 1 part; melting point, 266° F.

II.—Tin, 3 parts; lead, 4 parts; bismuth, 2 parts; melting point, 297° F.

III.—Tin, 2 parts; lead, 2 parts; bismuth, 1 part; melting point, 320° F.

STEEL SOLDERING.

Dissolve scraps of cast steel in as small a quantity as possible of nitric acid, add finely pulverized borax and stir vigorously until a fluid paste is formed, then dilute by means of sal ammoniac and put in a bottle. When soldering is to be done, apply a thin layer of the solution to the two parts to be soldered, and when these have been carried to ordinary redness, and the mass is consequently plastic, beat lightly on the anvil with a flat hammer. This recipe is useful for cases when the steel is not to be soldered at an elevation of temperature to the bright red.

To Solder a Piece of Hardened Steel.— To hard-solder a piece of hardened steel such as index (regulator), stop spring (in the part which is not elastic), click, etc., take a very flat charcoal if the piece is difficult to attach; hard-solder and as soon as the soldering has been done, plunge the piece into oil. All that remains to be done is to blue it again and to polish.

Soldering Powder for Steel.— Melt in an earthen pot 3 parts of borax, 2 of colophony, 1 of potassium carbonate, as much powdered hard soap, to which must be added 3 parts of finely powdered glass and 2 parts of steel filings. The melted mass is run out upon a cold plate of sheet iron, and when it is completely chilled it is broken into small bits or finely powdered. To solder, it is necessary to sprinkle the powder on the surfaces to be joined several minutes before bringing them together.

Soldering Solution for Steel.— A soldering solution for steel that will not rust or blacken the work is made of 6 ounces alcohol, 2 ounces glycerine, and 1 ounce oxide of zinc.

PLATINUM SOLDERS.

There are many platinum solders in existence, but the main principle to be borne in mind in jewelry work is that the soldering seam should be as little perceptible as possible; the solder, therefore, should have the same color as the alloy.

I.—A platinum solder which meets these requirements very satisfactorily is composed of 9 parts gold and 1 part palladium; or, 8 parts gold and 2 parts palladium.

II.—The following is a readily fusible platinum solder: Fine silver, 1.555 parts, and pure platinum, 0.583 parts. This melts easily in the ordinary draught furnace, as well as before the soldering pipe on a piece of charcoal. Of similar action is a solder of the following composition, which is very useful for places not exposed to the view:

III.—Fine gold, 1.555 parts; fine silver, 0.65 parts; and pure copper, 0.324 parts.

SOLDER FOR IRON:

See also under Silver Solders.

Copper, 67 parts; zinc, 33 parts; or, copper, 60 parts; zinc, 40 parts.

TIN SOLDERS:

See also Soft Solders.

Gold jewelry which has been rendered unsightly by tin solder may be freed from tin entirely by dipping the article for a few minutes into the following solution and then brushing off the tin: Pulverize 2 parts of green vitriol and 1 part of saltpeter and boil in a cast-iron pot with 10 parts of water until the larger part of the latter has evaporated. The crystals forming upon cooling are dissolved in hydrochloric acid (8 parts of hydrochloric acid to 1 part of crystals). If the articles in question have to be left in the liquid for some time, it is well to dilute it with 3 or 4 parts of water. The tin solder is dissolved by this solution without attacking or damaging the article in the least.

VARIOUS RECIPES FOR SOLDERING:

To Conceal Soldering.— Visible soldering may be obviated by the following methods: For copper goods a concentrated solution of blue vitriol is prepared and applied to the places by means of an iron rod or iron wire. The thickness of the layer may be increased by a repetition of the process. In order to give the places thus coppered the appearance of the others, use a saturated solution of zinc vitriol, 1 part, and blue vitriol, 2 parts, and finish rubbing with a piece of zinc. By sprinkling on gold powder and subsequently polishing, the color is rendered deeper. In the case of gold articles the places are first coppered over, then covered with a thin layer of fish glue, after which bronze filings are thrown on. When the glue is dry rub off quickly to produce a fine polish. The places can, of course, also be electro-gilt, whereby a greater uniformity of the shade is obtained. In silver objects, the soldering seams, etc., are likewise coppered in the above-described manner; next they are rubbed with a brush dipped into silver powder and freshly polished.

Solder for Articles which will not Bear a High Temperature.—Take powdered copper, the precipitate of a solution of the sulphate by means of zinc, and mix it with concentrated sulphuric acid. According to the degree of hardness required, take from 20 to 30 or 36 parts of copper. Add, while constantly shaking, 70 parts of quicksilver, and when the amalgam is complete, wash with warm water to remove the acid; then allow it to cool. In 10 or 12 hours the composition will be hard enough to scratch tin. For use, warm it until it reaches the consistency of wax, and spread it where needed. When cold it will adhere with great tenacity.

Soldering a Ring Containing a Jewel. —I.—Fill a small crucible with wet sand and bury the part with the jewel in the sand. Now solder with soft gold solder, holding the crucible in the hand. The stone will remain uninjured.

II. Take tissue paper, tear it into strips about 3 inches in width, and make them into ropes; wet them thoroughly and wrap the stone in them, passing around the stone and through the ring until the center of the latter is slightly more than half filled with paper, closely wound around. Now fix on charcoal, permitting the stone to protrude over the edge of the charcoal, and solder rapidly. The paper will not only protect the stone, but also prevent oxidation of the portion of the ring which is covered.

Soldering without Heat.—For soldering objects without heating, take a large copper wire filed to a point; dip into soldering water and rub the parts to be soldered. Then heat the copper wire

and apply the solder, which melts on contact. It may then be applied to the desired spot without heating the object.

COLD SOLDERING:
See also Adhesives and Cements.

For soldering articles which cannot stand a high temperature, the following process may be employed:

I.—Take powdered copper precipitated from a solution of sulphate by means of zinc and mix it in a cast-iron or porcelain mortar with concentrated sulphuric acid. The number of parts of copper varies according to the degree of hardness which it is wished to obtain. Next add, stirring constantly, 70 parts of mercury, and when the amalgam is finished, allow to cool. At the end of 10 to 12 hours the composition is sufficiently hard. For use, heat until it acquires the consistency of wax. Apply to the surface. When cool it will adhere with great tenacity.

II.—Crush and mix 6 parts of sulphur, 6 parts of white lead, and 1 part of borax. Make a rather thick cement of this powder by triturating it with sulphuric acid. The paste is spread on the surfaces to be welded, and the articles pressed firmly together. In 6 or 7 days the soldering is so strong that the two pieces cannot be separated, even by striking them with a hammer.

Cast-Iron Soldering.—A new process consists in decarbonizing the surfaces of the cast iron to be soldered, the molten hard solder being at the same time brought into contact with the red-hot metallic surfaces. The admission of air, however, should be carefully guarded against. First pickle the surfaces of the pieces to be soldered, as usual, with acid and fasten the two pieces together. The place to be soldered is now covered with a metallic oxygen compound and any one of the customary fluxes and heated until red hot. The preparation best suited for this purpose is a paste made by intimately mingling together cuprous oxide and borax. The latter melts in soldering and protects the pickled surfaces as well as the cuprous oxide from oxidation through the action of the air. During the heating the cuprous oxide imparts its oxygen to the carbon contained in the cast iron and burns it. Metallic copper separates in fine subdivision. Now apply hard solder to the place to be united, which in melting forms an alloy with the eliminated copper, the alloy combining with the decarburized surfaces of the cast iron.

Soldering Block.—This name is given to a very useful support for hard soldering and can be readily made. The ingredients are: Charcoal, asbestos, and plaster of Paris. These are powdered in equal parts, made into a thick paste with water, and poured into a suitable mold. Thus a sort of thick plate is obtained. When this mass has dried it is removed from the mold and a very thin cork plate is affixed on one surface by means of thin glue. The mission of this plate is to receive the points of the wire clamps with which the articles to be soldered are attached to the soldering block, the asbestos not affording sufficient hold for them.

SOLDERS FOR JEWELERS:
See Jewelers' Formulas.

SOLDER FROM GOLD, TO REMOVE:
See Gold.

SOLDERING PASTE.

The semi-liquid mass termed soldering paste is produced by mixing zinc chloride solution or that of ammonia-zinc chloride with starch paste. For preparing this composition, ordinary potato starch is made with water into a milky liquid, the latter is heated to a boil with constant stirring, and enough of this mass, which becomes gelatinous after cooling, is added to the above-mentioned solutions as to cause a liquid resembling thin syrup to result. The use of all zinc preparations for soldering presents the drawback that vapors of a strongly acid odor are generated by the heat of the soldering iron, but this evil is offset by the extraordinary convenience afforded when working with these preparations. It is not necessary to subject the places to be soldered to any special cleaning or preparation. All that is required is to coat them with the soldering medium, to apply the solder to the seam, etc., and to wipe the places with a sponge or moistened rag after the solder has cooled. Since the solder adheres readily with the use of these substances, a skillful workman can soon reach such perfection that he has no, or very little, subsequent polishing to do on the soldering seams.

Soft Soldering Paste.—Small articles of any metals that would be very delicate to solder with a stick of solder, especially where parts fit into another and only require a little solder to hold them together, can best be joined with a soldering paste. This paste contains the solder and flux combined, and is easily applied to seams, or a little applied before the parts are put together. The soldering flame will cause the tin in the paste to amalgamate quickly. The paste is made out of starch paste mixed with a solution of chloride of tin to the consistency of syrup.

SOLUTIONS, PERCENTAGE:
See Tables.

SOOTHING SYRUP:
See Pain Killers.

SOUP HERB EXTRACT:
See Condiments.

SOZODONT:
See Dentifrices.

SPARKS FROM THE FINGER TIPS:
See Pyrotechnics.

SPATTER WORK:
See Lettering.

SPAVIN CURES:
See Veterinary Formulas.

SPECULUM METAL:
See Alloys.

SPICES, ADULTERATED:
See Foods.

SPICES FOR FLAVORING:
See Condiments.

Spirit

INDUSTRIAL AND POTABLE ALCOHOL: SOURCES AND MANUFACTURE.

Abstract of a Farmers' Bulletin prepared for the United States Department of Agriculture by Dr. Harvey W. Wiley.

The term "industrial alcohol," or spirit, is used for brevity, and also because it differentiates sharply between alcohol used for beverages or for medicine and alcohol used for technical purposes in the arts.

Alcohol Defined.—The term "alcohol" as here used and as generally used means that particular product which is obtained by the fermentation of a sugar, or a starch converted into sugar, and which, from a chemical point of view, is a compound of the hypothetical substance "ethyl" with water, or with that part of water remaining after the separation of one of the atoms of hydrogen. This is a rather technical expression, but it is very difficult, without using technical language, to give a definition of alcohol from the chemical point of view. There are three elementary substances represented in alcohol: Carbon, the chemical symbol of which is C; hydrogen, symbol

H; and oxygen, symbol O. These atoms are put together to form common alcohol, or, as it is called, ethyl alcohol, in which preparation 2 atoms of carbon and 5 atoms of hydrogen form the hypothetical substance "ethyl," and 1 atom of oxygen and 1 atom of hydrogen form the hydroxyl derived from water. The chemical symbol of alcohol therefore is C_2H_5OH. Absolutely pure ethyl alcohol is made only with great difficulty, and the purest commercial forms still have associated with them traces of other volatile products formed at the time of the distillation, chief among which is that group of alcohols to which the name "fused oil" is applied. So far as industrial purposes are concerned, however, ethyl alcohol is the only component of any consequence, just as in regard to the character of beverages the ethyl alcohol is the component of least consequence.

Sources of Potable Alcohol. — The raw materials from which alcohol is made consist of those crops which contain sugar, starch, gum, and cellulose (woody fiber) capable of being easily converted into a fermentable sugar. Alcohol as such is not used as a beverage. The alcohol occurring in distilled beverages is principally derived from Indian corn, rye, barley, and molasses. Alcohol is also produced for drinking purposes from fermented fruit juices such as the juice of grapes, apples, peaches, etc. In the production of alcoholic beverages a careful selection of the materials is required in order that the desired character of drink may be secured. For instance, in the production of rum, the molasses derived from the manufacture of sugar from sugar cane is the principal raw material. In the fermentation of molasses a particular product is formed which by distillation gives the alcohol compound possessing the aroma and flavor of rum. In the making of brandy, only sound wine can be used as the raw material, and this sound wine, when subjected to distillation, gives a product containing the same kind of alcohol as that found in rum, but associated with the products of fermentation which give to the distillate a character entirely distinct and separate from that of rum. Again, when barley malt or a mixture of barley malt and rye is properly mashed, fermented, and subjected to distillation, a product is obtained which, when properly concentrated and aged, becomes potable malt or rye whisky. In a similar manner, if Indian corn and bar-

ley malt are properly mashed, with a small portion of rye, the mash fermented and subjected to distillation, and the distillate properly prepared and aged, the product is known as Bourbon whisky. Thus, every kind of alcoholic beverage gets its real character, taste, and aroma, not from the alcohol which it contains but from the products of fermentation which are obtained at the same time the alcohol is made and which are carried over with the alcohol at the time of distillation.

Agricultural Sources of Industrial Alcohol.—The chief alcohol-yielding material produced in farm crops is starch, the second important material is sugar, and the third and least important raw material is cellulose, or woody fiber. The quantity of alcohol produced from cellulose is so small as to be of no importance at the present time, and therefore this source of alcohol will only be discussed under the headings "Utilization of Waste Material or By-Products" and "Wood Pulp and Sawdust."

Starch-Producing Plants.—Starch is a compound which, from the chemical point of view, belongs to the class known as carbohydrates, that is, compounds in which the element carbon is associated by a chemical union with water. Starch is therefore a compound made of carbon, hydrogen, and oxygen, existing in the proportion of 2 atoms of hydrogen to 1 atom of oxygen. Each molecule of starch contains at least 6 atoms of carbon, 10 atoms of hydrogen, and 5 atoms of oxygen. The simplest expression for starch is therefore $C_6H_{10}O_5$. Inasmuch as this is the simplest expression for what the chemist knows as a molecule of starch, and it is very probable that very many, perhaps a hundred or more, of these molecules exist together, the proper expression for starch from a chemical point of view would be $(C_6H_{10}O_5)x$.

The principal starch-producing plants are the cereals, the potato, and cassava. With the potato may be classed, though not botanically related thereto, the sweet potato and the yam. Among cereals rice has the largest percentage of starch and oats the smallest. The potato, as grown for the table, has an average content of about 15 per cent of starch. When a potato is grown specifically for the production of alcohol it contains a larger quantity, or nearly 20 per cent. Cassava contains a larger percentage of starch than the potato, varying from 20 to 30 per cent.

Sugar-Producing Plants.—*Sugar cane,*

etc. While sugar is present in some degree in all vegetable growths, there are some plants which produce it in larger quantities than are required for immediate needs, and this sugar is stored in some part of the plant. Two plants are preëminently known for their richness in sugar, namely, the sugar cane and the sugar beet. In Louisiana the sugar canes contain from 9 to 14 per cent of sugar, and tropical canes contain a still larger amount.

The juices of the sugar beet contain from 12 to 18 per cent of sugar. There are other plants which produce large quantities of sugar, but which are less available for sugar-making purposes than those just mentioned. Among these, the sorghum must be first mentioned, containing in the stalk at the time the seed is just mature and the starch hardened from 9 to 15 per cent of sugar. Sorghum seed will also yield as much alcohol as equal weights of Indian corn. The juices of the stalks of Indian corn contain at the time the grain is hardening and for some time thereafter large quantities of sugar, varying from 8 to 15 per cent.

In the case of the sorghum and the Indian-corn stalk a large part of the sugar present is not cane sugar or sucrose as it is commonly known, but the invert sugar derived therefrom. For the purposes of making alcohol the invert sugar is even more suitable than cane sugar. Many other plants contain notable quantities of sugar, but, with the exception of fruits, discussed under the following caption, not in sufficient quantities to be able to compete with those just mentioned for making either sugar or alcohol.

Cane sugar is not directly susceptible to fermentation. Chemically considered, it has the formula expressed by the symbols: $C_{12}H_{22}O_{11}$. When cane sugar having the above composition becomes inverted, it is due to a process known as hydrolysis, which consists in the molecule of cane sugar taking up 1 molecule of water and splitting off into 2 molecules of sugar having the same formula but different physical and chemical properties. Thus the process may be represented as follows: $C_{12}H_{22}O_{11}$ (cane sugar) $+ H_2O$ (water) $= C_6H_{12}O_6$ (dextrose) $+ C_6H_{12}O_6$ (levulose). These two sugars (dextrose and levulose) taken together are known as invert sugar and are directly susceptible to fermentation. All cane sugar assumes the form of invert sugar before it becomes fermented.

Fruits.—Nearly all fruit juices are rich in sugar, varying in content from 5 to 30 per cent. The sugar in fruits is composed of both cane sugar and its invert products (dextrose and levulose), in some fruits principally the latter. Of the common fruits the grape yields the largest percentage of sugar. The normal grape used for wine making contains from 16 to 30 per cent of sugar, the usual amount being about 20 per cent. Fruit juices are not usually employed in any country for making industrial alcohol, because of their very much greater value for the production of beverages.

Composition and Yield of Alcohol-Producing Crops.—The weight of alcohol that may be produced from a given crop is estimated at a little less than one-half of the amount of fermentable substance present, it being understood that the fermentable substance is expressed in terms of sugar. Pasteur was the first to point out the fact that when sugar was fermented it yielded theoretically a little over one-half of its weight of alcohol. It must be remembered, however, that in the production of alcohol a process of hydrolysis is taking place which adds a certain quantity of alcohol to the products which are formed. For this reason 100 parts of sugar yield more than 100 parts of fermentable products. The distribution of the weights produced, as theoretically calculated by Pasteur, is as follows:

One hundred parts of sugar yield the following quantities of the products of fermentation:

Alcohol	51.10 parts
Carbonic acid	49.20 parts
Glycerine	3.40 parts
Organic acids, chiefly succinic	.65 parts
Ethers, aldehydes, furfural, fat, etc.	1.30 parts
Total weight fermentation products produced	105.65 parts

Artichokes.—The artichoke has been highly recommended for the manufacture of alcohol. The fermentable material in the artichoke is neither starch nor sugar, but consists of a mixture of a number of carbohydrates of which inulin and levulin are the principal constituents. When these carbohydrate materials are hydrolized into sugars they produce levulose instead of dextrose. The levulose is equally as valuable as dextrose for the production of alcohol. Artichokes may be harvested either in the autumn or in the spring. As they keep well during the winter, and in a few places may be kept in hot weather, they form a raw material which can be stored for a long period and still be valuable for fermentation purposes.

Under the term "inulin" are included all the fermentable carbohydrates. The above data show, in round numbers, 17 per cent of fermentable matter. Theoretically, therefore, 100 pounds of artichokes would yield approximately $8\frac{1}{2}$ pounds of industrial alcohol, or about $1\frac{1}{4}$ gallons.

Bananas.—The banana is a crop which grows in luxurious abundance in tropical countries, especially Guatemala and Nicaragua. The fruit contains large quantities of starch and sugar suitable for alcohol making. From 20 to 25 per cent of the weight of the banana consists of fermentable material. It is evident that in the countries where the banana grows in such luxuriance it would be a cheap source of industrial alcohol.

Barley and the Manufacture of Malt.— A very important cereal in connection with the manufacture of alcohol is barley which is quite universally employed for making malt, the malt in its turn being used for the conversion of the starch of other cereals into sugar in their preparation for fermentation.

Malt is made by the sprouting of barley at a low temperature (from 50° to 60° F.) until the small roots are formed and the germ has grown to the length of $\frac{1}{2}$ an inch or more. The best malts are made at a low temperature requiring from 10 to 14 days for the growth of the barley. The barley is moistened and spread upon a floor, usually of cement, to the depth of 1 foot or 18 inches. As the barley becomes warm by the process of germination, it is turned from time to time and the room is kept well ventilated and cool. It is better at this point in the manufacture of malt to keep the temperature below 60° F. After the sprouting has been continued as above noted for the proper length of time, the barley is transferred to a drier, where it is subjected to a low temperature at first and finally to a temperature not to exceed 140° or 158° F., until all the water is driven off, except 2 or 3 per cent. Great care must be exercised in drying the barley not to raise the temperature too high, lest the diastase which is formed be deprived of its active qualities. The malt has a sweetish taste, the principal portion of the starch having been converted into sugar, which is known chemically as "maltose." This sugar is, of course, utilized in the fermentation for the production of alcohol. Malt is

chiefly valuable, however, not because of the amount of alcohol that may be produced therefrom, but from the fact that in quantities of about 10 per cent it is capable of converting the starch of the whole of the unmalted grains, whatever their origin may be, into maltose, thus preparing the starch for fermentation. Barley is not itself used in this country as a source of industrial alcohol, but it is employed for producing the highest grades of whisky, made of pure barley malt, which, after fermentation, is distilled in a pot still, concentrated in another pot still to the proper strength, placed in wood, and stored for a number of years. Barley malt is too expensive a source of alcohol to justify its use for industrial purposes. It is, however, one of the cheapest and best methods of converting the starch of other cereals into sugar preparatory to fermentation.

Barley has, in round numbers, about 68 per cent of fermentable matter. The weight of a bushel of barley (48 pounds) multiplied by 0.68 gives 32 pounds of fermentable matter in a bushel of barley.

Cassava.—Cassava is grown over a large area of the South Atlantic and Gulf States of this country. Of all the substances which have been mentioned, except the cereals, cassava contains the largest amount of alcoholic or fermentable substances. The root, deprived of its outer envelope, contains a little over 30 per cent of starch, while the undetermined matter in the analyses is principally sugar. If this be added to the starch, it is seen that approximately 35 per cent of the fresh root is fermentable. This of course represents a very high grade of cassava, the ordinary roots containing very much less fermentable matter. If, however, it is assumed that the fermentable matter of cassava root will average 25 per cent, this amount is much greater than the average of the potato, or even of the sweet potato and the yam. Twenty-five per cent is undoubtedly a low average content of fermentable matter. In the dry root there is found nearly 72 per cent of starch and 17 per cent of extract, principally sugar. Assuming that 15 per cent of this is fermentable, and adding this to the 72 per cent, it is seen that 87 per cent of the dry matter of the cassava is fermentable. This appears to be a very high figure, but it doubtless represents almost exactly the conditions which exist. It would be perfectly safe to say, discounting any exceptional qualities of the samples examined, that 80 per cent of the dry matter of the cassava root is capable of being converted into alcohol. It thus becomes in a dry state a source of alcohol almost as valuable, pound for pound, as rice.

Careful examinations, however, of actual conditions show that if 5 tons per acre of roots are obtained it is an average yield. In very many cases, where no fertilizer is used and where the roots are grown in the ordinary manner, the yield is far less than this, while with improved methods of agriculture it is greater. The bark of the root, has very little fermentable matter in it. If the whole root be considered, the percentage of starch is less than it would be for the peeled root. If cassava yields 4 tons, or 8,000 pounds, per acre and contains 25 per cent of fermentable matter, the total weight of fermentable matter is 2,000 pounds, yielding approximately 1,000 pounds of 95 per cent alcohol, or 143 gallons of 95 per cent alcohol per acre.

Corn (Indian Corn or Maize).—The crop which at the present time is the source of almost all of the alcohol made in the United States is Indian corn.

The fermentable matter in Indian corn—that is, the part which is capable of being converted into alcohol—amounts to nearly 70 per cent of the total weight, since the unfermentable cellulose and pentosans included in carbohydrates do not exceed 2 per cent. Inasmuch as a bushel of Indian corn weighs 56 pounds, the total weight of fermentable matter therein, in round numbers, is 39 pounds. The weight of the alcohol which is produced under the best conditions is little less than one-half of the fermentable matter. Therefore the total weight of alcohol which would be yielded by a bushel of average Indian corn would be, in round numbers, about 19 pounds. The weight of a gallon of 95 per cent alcohol is nearly 7 pounds. Hence 1 bushel of corn would produce 2.7 gallons.

If the average price of Indian corn be placed, in round numbers, at 40 cents a bushel, the cost of the raw material—that is, of the Indian corn—for manufacturing 95 per cent industrial alcohol is about 15 cents a gallon. To this must be added the cost of manufacture, storage, etc., which is perhaps as much more, making the estimated actual cost of industrial alcohol of 95 per cent strength made from Indian corn about 30 cents per gallon. If to this be added the profits of the manufacturer and dealer, it appears that under the conditions cited, industrial alcohol, untaxed, should be sold for about 40 cents per gallon.

Potatoes.—The weight of a bushel of potatoes is 60 pounds. As the average amount of fermentable matter in potatoes grown in the United States is 20 per cent, the total weight of fermentable matter in a bushel of potatoes is 12 pounds, which would yield approximately 6 pounds or 3.6 quarts of alcohol.

The quantity of starch in American-grown potatoes varies from 15 to 20 per cent. Probably 18 per cent might be stated as the general average of the best grades of potatoes.

Under the microscope the granules of potato starch have a distinctive appearance. They appear as egg-shaped bodies on which, especially the larger ones, various ring-like lines are seen. With a modified light under certain conditions of observation a black cross is developed upon the granule. It is not difficult for an expert microscopist to distinguish potato from other forms of starch by this appearance.

The potato contains very little material which is capable of fermentation aside from starch and sugars.

Although the potato is not sweet to the taste in a fresh state, it contains notable quantities of sugar. This sugar is lost whenever the potato is used for starch-making purposes, but is utilized when it is used for the manufacture of industrial alcohol. The percentage of sugar of all kinds in the potato rarely goes above 1 per cent. The average quantity is probably not far from 0.35 per cent, including sugar, reducing sugar, and dextrin, all of which are soluble in water. In the treatment of potatoes for starch making, therefore, it may be estimated that 0.35 per cent of fermentable matter is lost in the wash water.

Average Composition.—The average composition of potatoes is:

Water	75.00 per cent
Starch	19.87 per cent
Sugars and dextrin	.77 per cent
Fat	.08 per cent
Cellulose	.33 per cent
Ash	1.00 per cent

According to Maercker, the sugar content, including all forms of sugar, varies greatly. Perfectly ripe potatoes contain generally no sugar or only a fractional per cent. When potatoes are stored under unfavorable conditions, large quantities of sugar may be developed, amounting to as high as 5 per cent altogether. In general, it may be stated that the content of sugar of all kinds will vary from 0.4 per cent to 3.4 per cent, according to conditions.

The liberal application of nitrogenous fertilizers increases the yield per acre of tubers and of starch to a very marked extent, although the average percentage of starch present is increased very little.

Of all the common root crops, the potatoes, including the yam and the sweet potato, are the most valuable for the production of alcohol, meaning by this term that they contain more fermentable matter per 100 pounds than other root crops.

While sugar beets, carrots, and parsnips contain relatively large amounts of fermentable matter, these roots could not compete with potatoes even if they could all be produced at the same price per 100 pounds.

A general review of all the data indicates that under the most favorable circumstances and with potatoes which have been grown especially for the purpose an average content of fermentable matter of about 20 per cent may be reasonably expected. It is thus seen that approximately 10 pounds of industrial alcohol can be made from 100 pounds of potatoes. If 60 pounds be taken as the average weight of a bushel of potatoes, there are found therein 12 pounds of fermentable matter, from which 6 pounds of industrial alcohol can be produced, or ⅘ of a gallon. It has also been shown that the amount of Indian corn necessary for the production of a gallon of industrial alcohol costs not less than 15 cents. From this it is evident that the potatoes for alcohol making will have to be produced at a cost not to exceed 15 cents per bushel, before they can compete with Indian corn for the manufacture of industrial alcohol.

Rice.—Rice is not used to any great extent in this country for making alcohol, but it is extensively used for this purpose in Japan and some other countries, and has the largest percentage of fermentable matter of all the cereals. The percentage of fermentable matter in rice is nearly 78 per cent. A bushel of rice weighs, unhulled, 45 pounds, hulled, 56 pounds, and it therefore has about 34 and 43 pounds, respectively, of fermentable matter for the unhulled and the hulled rice. It is not probable that rice will ever be used to any extent in this country as a source of industrial alcohol, although it is used to a large extent in the manufacture of beverages, as for instance in beers, which are often made partly of rice.

Rye.—Large quantities of alcohol, chiefly in the form of alcoholic beverages, are manufactured from rye. It is, in connection with Indian corn, the principal source of the whiskies made in the United States. Rye, however, is not used to any extent in this or other countries for making industrial alcohol.

Rye contains almost as much fermentable matter as Indian corn. A bushel of rye weighs 56 pounds. Wheat and other cereals, not mentioned above, are not used in this country to any appreciable extent in the manufacture of alcohol.

Spelt.—This grain, which is botanically a variety of wheat, more closely resembles barley. Under favorable conditions as much as 73 bushels per acre have been reported, and analyses show 70 per cent of fermentable carbohydrates. The weight per bushel is about the same as that of oats. It would appear that this crop might be worthy of consideration as a profitable source of industrial alcohol.

Sugar Beets.—The sugar beet is often used directly as a source of alcohol. Working on a practical scale in France, it has been found that from 10,430 tons of beets there were produced 183,624 gallons of crude alcohol of 100 per cent strength. The beets contain 11.33 per cent of sugar. From 220 pounds of sugar 15.64 gallons of alcohol were produced. The weight of pure alcohol obtained is a little less than one-half the weight of the dry fermentable matter calculated as sugar subjected to fermentation. About 18 gallons of alcohol are produced for each ton of sugar beets employed.

Sweet Potatoes.—Experiments show that as much as 11,000 pounds of sweet potatoes can be grown per acre. The average yield of sweet potatoes, of course, is very much less. On plots to which no fertilizer is added the yield is about 8,000 pounds of sweet potatoes per acre, yielding in round numbers 1,900 pounds of starch. The quantity of sugar in the 8,000 pounds is about 350 pounds, which added to the starch, makes 2,250 pounds of fermentable matter per acre. This will yield 1,125 pounds of industrial alcohol of 95 per cent strength, or approximately 160 gallons per acre. The percentage of starch is markedly greater than in the white or Irish potato. In all cases over 20 per cent of starch was obtained in the South Carolina sweet potatoes, and in one instance over 24 per cent. As much as 2,600 pounds of starch were produced per acre.

In addition to starch, the sweet potato contains notable quantities of sugar, sometimes as high as 6 per cent being present, so that the total fermentable matter in the sweet potato may be reckoned at the minimum at 25 per cent. A bushel of sweet potatoes weighs 55 pounds, and one-quarter of this is fermentable matter, or nearly 14 pounds. This would yield, approximately, 7 pounds, or a little over 1 gallon of 95 per cent alcohol. It may be fairly stated, therefore, in a general way, that a bushel of sweet potatoes will yield 1 gallon of industrial alcohol.

Experiments have shown that the quantity of starch diminishes and the quantity of sugar increases on storing. Further, it may be stated that in the varieties of sweet potatoes which are most esteemed for table use there is less starch and perhaps more sugar than stated above. The total quantity of fermentable matter, however, does not greatly change, although there is probably a slight loss.

Utilization of Waste Material or By-Products. — *Molasses.* **—** The utilization of the waste materials from the sugar factories and sugar refineries for the purpose of making alcohol is a well-established industry. The use of these sources of supply depends, of course, upon the cost of the molasses. When the sugar has been exhausted as fully as possible from the molasses the latter consists of a saccharine product, containing a considerable quantity of unfermentable carbohydrate matter, large quantities of mineral salts, and water. In molasses of this kind there is probably not more than 50 pounds of fermentable matter to 100 pounds of the product. Assuming that a gallon of such molasses weighs 11 pounds, it is seen that it contains $5\frac{1}{2}$ pounds of fermentable matter, yielding $2\frac{1}{4}$ pounds of industrial alcohol of 95 per cent strength. It requires about 3 gallons of such molasses to make 1 gallon of industrial alcohol.

When the price of molasses delivered to the refineries falls as low as 5 or 6 cents a gallon it may be considered a profitable source of alcohol.

Wood Pulp and Sawdust. — Many attempts have been made to produce alcohol for industrial purposes from sawdust, wood pulp, or waste wood material. The principle of the process rests upon the fact that the woody substance is composed of cellulose and kindred matters which, under the action of dilute acid (preferably sulphuric or sulphurous) and heat, with or without pressure, undergo hydrolysis and are changed into sugars. A large part of the sugar which is formed is nonfermentable, consisting of a substance known as xylose. Another part of the sugar produced is dextrose, made from the true cellulose which the wood contains.

The yield of alcohol in many of the experiments which have been made has not been very satisfactory. It is claimed, however, by some authors that paying quantities of alcohol are secured. In Simmonsen's process for the manufacture of alcohol $\frac{1}{2}$ per cent sulphuric acid is employed and from 4 to 5 parts of the liquid heated with 1 part of the finely comminuted wood for a quarter of an hour under a pressure of 9 atmospheres. It is claimed by Simmonsen that he obtained a yield of 6 quarts of alcohol from 110 pounds of air-dried shavings. Another process which has been tried in this and other countries for converting comminuted wood into alcohol is known as Classen's. The comminuted wood is heated for 15 minutes in a closed apparatus at a temperature of from 248° to 293° F. in the presence of sulphurous acid (fumes of burning sulphur) instead of sulphuric acid. It is claimed by the inventor that he has made as much as 12 quarts of alcohol from 110 pounds of the air-dried shavings. There is reason to doubt the possibility of securing such high yields in actual practice as are claimed in the above processes. That alcohol can be made from sawdust and wood shavings is undoubtedly true, but whether or not it can be made profitably must be determined by actual manufacturing operations.

Waste Products of Canneries, etc. — The principal waste materials which may be considered in this connection are the refuse of wine making, fruit evaporating, and canning industries, especially the waste of factories devoted to the canning of tomatoes and Indian corn. In addition to this, the waste fruit products themselves, which are not utilized at all, as, for instance, the imperfect and rotten apples, tomatoes, grapes, etc., may be favorably considered. The quantity of waste products varies greatly in different materials.

The quantities of waste material in grapes and apples, as shown by Lazenby, are as follows: About 25 per cent of the total weight in grapes, with the exception of the wild grape, where it is about 60 per cent; with apples the average percentage of waste was found to be 23.8 per cent from 25 varieties. This included the waste in the core, skin, and the defective apples caused by insects, fungi, bruises, etc. In general it may be said that in the preparation of fruits for preserving purposes about 25 per cent of their weight is waste, and this, it is evident, could be utilized for the manufacture of alcohol. If apples be taken as a type of fruits, we may assume that the waste portions contain 10 per cent of fermentable matters, which, however, is perhaps rather a high estimate. Five per cent of this might be recovered as industrial alcohol. Thus, each 100 pounds of fruit waste in the most favorable circumstances might be expected to produce 5 pounds of industrial alcohol. The quantity of waste which could be utilized for this purpose would hardly

established it might be profitable to devote them to this purpose.

Manufacture of Alcohol.—The three principal steps in the manufacture of alcohol are (1) the preparation of the mash or wort, (2) the fermentation of the mash or wort drawn off from the mash tun, and (3) the distillation of the dilute alcohol formed in the beer or wash from the fermentation tanks. The preparation of the mash includes (1) the treatment of the material used with hot water to form a paste of the starch or the sugar, and (2) the action of the malt or ferment

FIG. 1.—MASH TUN IN AN IRISH DISTILLERY.

render it profitable to engage in the manufacture. A smaller percentage could be expected from the waste of the tomato, where the quantity of sugar is not so great. In the waste of the sweet-corn factory the amount of fermentable matter would depend largely on the care with which the grain was removed. There is usually a considerable quantity of starchy material left on the cobs, and this, with the natural sugars which the grown cobs contain, might yield quite large quantities of fermentable matter. It would not be profitable to erect distilleries simply for the utilization of waste of this kind, but if these wastes could be utilized in distilleries already

on the paste to convert the starch into fermentable sugar.

Mashing.—Figs. 1 and 2 show two views of the mashing tun or tank, the first figure giving the general appearance, and the second a view of the interior of the tun, showing the machinery by which the stirring is effected and the series of pipes for cooling the finished product down to the proper temperature for the application of the malt.

The object of the mash tun is to reduce the starch in the ground grain to a pasty, gummy mass, in order that the ferment of the malt may act upon it vigorously and convert it into sugar. If the mashing be done before the addition

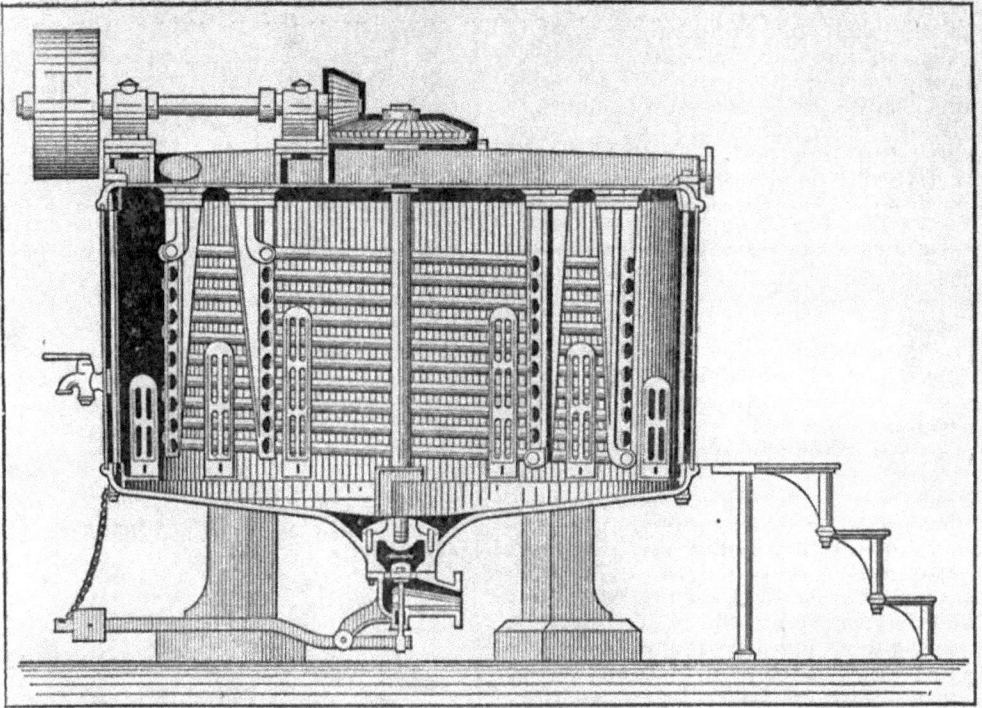

FIG. 2.—MASHING AND COOLING APPARATUS, CROSS SECTION.

FIG. 3.—FERMENTATION TANKS IN AN IRISH DISTILLERY.

of the malt the temperature may be raised to that of boiling water. If, however, the malt be added before the mashing begins, the temperature should not rise much, if any, above 140° F., since the fermenting power is retarded and disturbed at higher temperatures. The mashing is simply a mechanical process by means of which the starch is reduced to a form of paste and the temperature maintained at that point which is best suited to the conversion of the starch into sugar.

Fermentation.—The mash, after the starch has all been converted into sugar, goes into fermenting tanks, which in Scotland are called "wash backs," when the yeast is added. A view of the typical wash back is shown in Fig. 3. They often have a stirring apparatus, as indicated in the figure; whereby the contents can be thoroughly mixed with the yeast and kept in motion. This is not necessary after the fermentation is once well established, but it is advisable, especially in the early stages, to keep the yeast well distributed throughout the mass. In these tanks the fermentations are conducted, the temperature being varied according to the nature of the product to be made. For industrial alcohol the sole purpose should be to secure the largest possible percentage of alcohol without reference to its palatable properties.

An organism belonging to the vegetable family and to which the name "yeast" has been given is the active agent in fermentation. The organism itself does not take a direct part in the process, but it secretes another ferment of an unorganized character known as an "enzym" or a "diastase." This enzym has the property, under proper conditions of food, temperature, and dilution, of acting upon sugar and converting it into alcohol and carbonic acid. Anyone who has ever seen a fermenting vat in full operation and noticed the violent boiling or ebullition of the liquor, can understand how rapidly the gas "carbon dioxide" or "carbonic acid," as it is usually called, may be formed, as it is the escape of this gas which gives the appearance to the tank of being in a violent state of ebullition. The yeast which produces the fermentation belongs to the same general family as the ordinary yeast which is used in the leavening of bread. The leavening of bread under the action of yeast is due to the conversion of the sugar in the dough into alcohol and carbon dioxide or carbonic acid. The gas thus formed becomes

entangled in the particles of the gluten, and these expanding cause the whole mass to swell or "rise," as it is commonly expressed. Starch cannot be directly fermented, but must be first converted into sugar, either by the action of a chemical like an acid, or a ferment or enzym, known as diastase, which is one of the abundant constituents of malt, especially of barley malt. In the preparation of a cereal, for instance, for fermentation, it is properly softened and ground, and then usually heated with water to the boiling point or above in order that the starch may be diffused throughout the water. After cooling, it is treated with barley malt, the diastase of which acts vigorously upon the starch, converting it into a form of sugar, namely, maltose, which lends itself readily to the activities of the yeast fermentation. (Fig. 4.)

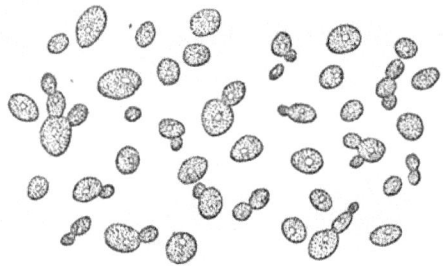

When ordinary sugar (cane sugar, beet sugar, and sucrose) is subjected to fermentation it is necessary that the yeast, which also exerts an activity similar to that of malt, should first convert the cane sugar into invert sugar (equal mixtures of dextrose and levulose) before the alcoholic fermentation is set up. The cane sugar is also easily inverted by heating with an acid.

When different kinds of sugars and starches are fermented for the purpose of making a beverage it is important that the temperature of fermentation be carefully controlled, since the character of the product depends largely upon the temperature at which the fermentation takes place. On the contrary, when industrial alcohol is made, the sole object is to get as large a yield as possible, and for this reason that temperature should be employed which produces the most alcohol and the least by-products, irrespective of the flavor or character of the product made. Also, in the making of alcoholic beverages, it is important that the malt be of the very best quality in order that the resulting product may have the proper flavor. In the production of alcohol for industrial purposes this is of no consequence, and the sole purpose here should be to produce the largest possible yield. For this reason there is no objection to the use of acids for converting the starch, cane sugar, and cellulose into fermentable sugars. Therefore, the heating of the raw materials under pressure with dilute acids in order to procure the largest quantity of sugar is a perfectly legitimate method of procedure in the manufacture of industrial alcohols.

Sugars and starches are usually associated in nature with another variety of carbohydrates known as cellulose, and this cellulose itself, when acted upon by an acid, is converted very largely into sugars, which, on fermentation, yield alcohol. For industrial purposes, the alcohol produced in this manner is just as valuable as that made from sugar and starch. Whether the diastatic method of converting the starch and sugar into fermentable sugars be used, or the acid method, is simply a question of economy and yield. On the other hand, when alcoholic beverages are to be made, those processes must be employed, irrespective of the magnitude of the yield, which give the finest and best flavors to the products.

Distillation.—The object of distillation is to separate the alcohol which has been formed from the non-volatile substances with which it is mixed. A typical form of distilling apparatus for the concentration of the dilute alcohol which is formed in the beer or wash from the fermentation tanks, is represented in Fig. 5.

This apparatus is of the continuous type common to Europe and America. It consists of a "beer still" provided with a number of chambers fitted with perforated plates and suitable overflow pipes. It is operated as follows:

The syrup and alcohol are pumped into the top of the beer still through a pipe *G*; the tank *G* may also be placed above the center of the still and the contents allowed to flow into the still by gravity; steam is admitted through an open pipe into the kettle *A* at the bottom of the column or is produced by heating the spent liquor by means of a coil. The steam ascends through the perforations in the plates, becoming richer and richer in alcohol as it passes through each layer of liquor, while the latter gradually descends by means of the overflow pipes to the bottom of the column *B* and finally reaches the kettle completely exhausted of alcohol, whence it is removed by

means of a pump connected with the pipe line *H*. On reaching the top of the beer still *B* the vapors of the alcohol and the steam continue to rise and pass into the alcohol column *C*. This column is also divided into chambers, but by solid instead of perforated plates, as shown at

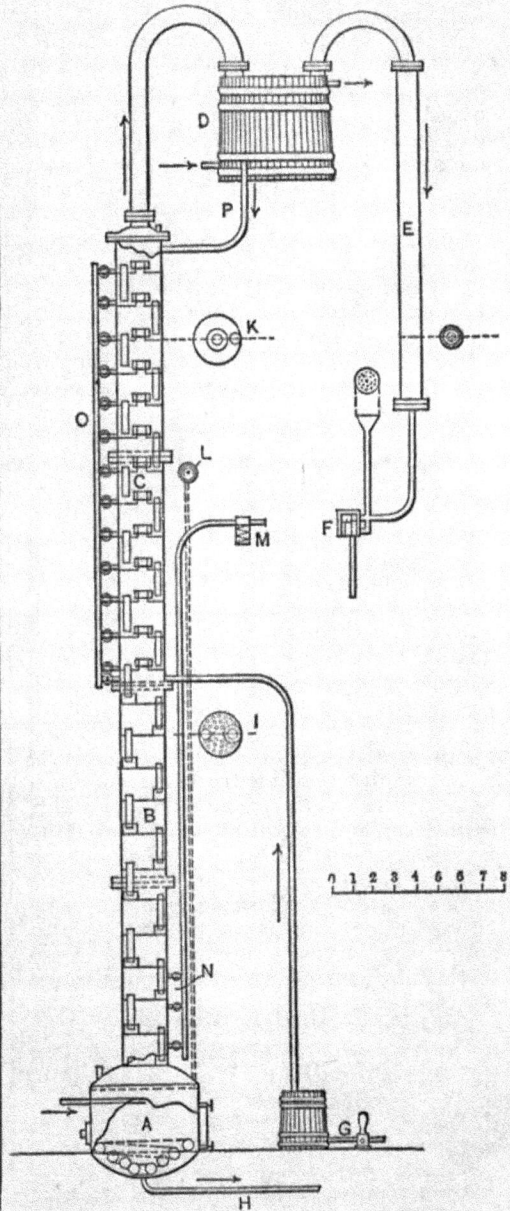

Fig. 5.—CONTINUOUS DISTILLING APPARATUS.

K. Each chamber is provided with a return or overflow pipe and an opening through which the vapors ascend. In the alcohol column the vapors are so directed as to pass through a layer of liquid more or less rich in alcohol which is retained by the plate separating the compartments. An excess of liquids in these compartments overflows through the down pipes, gradually works its way into the beer still, and thence to the kettle. On reaching the top of the column the vapors, which have now become quite rich in alcohol, are passed into a coil provided with an outlet at the lowest part of each bend. These outlets lead into the return pipe P, which connects with the top chamber of the alcohol column. This coil is technically termed the "goose" and is immersed in a tank called the "goose tub." A suitable arrangement is provided for controlling the temperature of the water in the tub by means of outlet and inlet water pipes. When the still is in operation the temperature of the "goose" is regulated according to the required density of the alcohol. The object of the "goose" is the return to the column of all low products which condense at a temperature below the boiling point of ethyl alcohol of the desired strength. On leaving the "goose" the vapors enter a condenser E, whence the liquid alcohol is conducted into a separator F. This separator consists simply of a glass box provided with a cylinder through which a current of alcohol is constantly flowing. An alcohol spindle is inserted in this cylinder and shows the density of the spirit at all times. A pipe, with a funnel-shaped opening at its upper extremity, connects with the pipe leading from the condenser and gives vent to any objectionable fumes. The separator is connected by means of a pipe with the alcohol storage tank. The pipe O is for emptying the upper chambers when necessary. The valves N, communicating by means of a small pipe with a condenser M, are for testing the vapors in the lower chambers for alcohol.

Substances Used for Denaturing Alcohol.—The process of rendering alcohol unsuitable for drinking is called "denaturing," and consists, essentially, in adding to the alcohol a substance soluble therein of a bad taste or odor, or both, of an intensity which would render it impossible or impracticable to use the mixture as a drink. Among the denaturing substances which have been proposed are the following:

Gum shellac (with or without the addition of camphor, turpentine, wood spirit, etc.), colophonium, copal rosin, Manila gum, camphor, turpentine, acetic acid, acetic ether, ethylic ether, methyl alcohol (wood alcohol), pyridine, acetone, methyl acetate, methyl violet, methylene blue, aniline blue, eosin, fluorescein, naphthalene, castor oil, benzine, carbolic acid, caustic soda, musk, animal oils, etc.

Methyl (wood) alcohol and benzine are the denaturing agents authorized in the United States, in the following proportions: To 100 parts, by volume, of ethyl alcohol (not less than 90 per cent strength) add 10 parts of approved methyl (wood) alcohol and $\frac{1}{2}$ of 1 part of approved benzine. Such alcohol is classed as completely denatured. Formulas for special denaturation may be submitted for approval by manufacturers to the Commissioner of Internal Revenue, who will determine whether they may be used or not, and only one special denaturant will be authorized for the same class of industries unless it shall be shown that there is good reason for additional special denaturants. Not less than 300 wine gallons can be withdrawn from a bonded warehouse at one time for denaturing purposes.

Spirit.—Proof spirit is a term used by the revenue department in assessing the tax on alcoholic liquors. It means a liquid in which there is 50 per cent (by volume) of absolute alcohol. As it is the actual alcohol in the whisky, brandy, dilute alcohol, etc., which is taxed, and as this varies so widely, it is necessary that the actual wine gallons be converted into proof gallons before the tax rate can be fixed. A sample that is half alcohol and half water (let us say for convenience) is "100 proof." A sample that is $\frac{3}{4}$ alcohol and $\frac{1}{4}$ water is 150 proof, and the tax on every gallon of it is $1\frac{1}{2}$ times the regular government rate per proof gallon. Absolute alcohol is 200 proof and has to pay a double tax.

The legal definition of proof spirit is, "that alcoholic liquor which contains one-half its volume of alcohol of a specific gravity of 0.7939 at 60° F."

SPONGES:

Bleaching Sponges. — I. — Soak in dilute hydrochloric acid to remove the lime, then wash in water, and place for 10 minutes in a 2 per cent solution of potassium permanganate. The brown color on removal from this solution is due to the deposition of manganous oxide, and this may be removed by steeping for a few minutes in very dilute sulphuric acid. As soon as the sponges appear white, they are washed out in water to remove the acid.

II.—A sponge that has been used in

surgical operations or for other purposes, should first be washed in warm water, to every quart of which 20 drops of liquor of soda have been added; afterwards washed in pure water, wrung or pressed out and put into a jar of bromine water, where it is left until bleached. Bleaching is accelerated by exposing the vessel containing the bromine water to the direct rays of the sun. When the sponge is bleached it is removed from the bromine water, and put for a few minutes in the water containing soda lye. Finally it is rinsed in running water until the odor of bromine disappears. It should be dried as rapidly as possible by hanging it in the direct sunlight.

Sterilization of Sponges.—I.—Allow the sponges to lie for 24 hours in an 8 per cent hydrochloric acid solution, to eliminate lime and coarse impurities; wash in clean water, and place the sponges in a solution of caustic potash, 10 parts; tannin, 10 parts; and water, 1,000 parts. After they have been saturated for 5 to 20 minutes with this liquid, they are washed out in sterilized water or a solution of carbolic acid or corrosive sublimate, until they have entirely lost the brown coloring acquired by the treatment with tannin. The sponges thus sterilized are kept in a 2 per cent or 15 per cent carbolic solution.

Sponge Window Display.—Soak a large piece of coarse sponge in water, squeeze half dry, then sprinkle in the openings red clover seed, millet, barley, lawn grass, oats, rice, etc. Hang this in the window, where the sun shines a portion of the day, and sprinkle lightly with water daily. It will soon form a mass of living green vegetation very refreshing to the eyes. While the windows are kept warm this may be done at any season. The seeds used may be varied, according to fancy.

SPONGES AS FILTERS:
See Filters.

SPONGE CLEANERS:
See Cleaning Preparations and Methods, under Miscellaneous Methods.

SPONGE-TRICK, BURNING:
See Pyrotechnics.

SPOT ERADICATORS:
See Cleaning Preparations and Methods and Soaps.

SPOT GILDING:
See Plating.

SPRAY SOLUTION:
See Balsams.

SPEARMINT CORDIAL:
See Wines and Liquors.

SPRAIN WASHES:
See Veterinary Formulas.

SPRING CLEANING:
See Cleaning Preparations and Methods.

SPRING HARDENING:
See Steel.

SPRINGS OF WATCHES:
See Watchmakers' Formulas.

SPRUCE BEER:
See Beverages.

STAIN REMOVERS:
See Cleaning Preparations and Methods.

STAINS:
See Paints, Varnishes and Wood Stains.

STAINS FOR LACQUERS:
See Lacquers.

Stamping

(See also Dyes.)

Stamping Colors for Use with Rubber Stamps.—Blue: 0.3 parts of water-blue 1 B, 1.5 parts of dextrin, 1.5 parts of distilled water. Dissolve the aniline dye and the dextrin in the distilled water, over a water bath, and add 7 parts of refined glycerine, 28° Bé.

Other colors may be made according to the same formula, substituting the following quantities of dyes for the water-blue: Methyl violet 3 B, 0.02 parts; diamond fuchsine I, 0.02 parts; aniline green D, 0.04 parts; vesuvine B, 0.05 parts; phenol black, 0.03 parts. Oleaginous colors are mostly used for metallic stamps, but glycerine colors can be used in case of necessity.

Oleaginous Stamping Colors.—Mix 0.8 parts of indigo, ground fine with 2.5 parts of linseed-oil varnish, and 0.5 parts of olein. Add 2 parts of castor oil and 5 parts of linseed oil. For other colors according to the same formula, use the following quantities: Cinnabar, 2½ parts; verdigris, 2½ parts; lampblack, 1.2 parts; oil-soluble aniline blue A, 0.35 parts; oil-soluble aniline scarlet B, 0.3 parts; aniline yellow (oil-soluble), 0.45 parts; oil-soluble aniline black L, 0.6 parts.

Stamping Liquids and Powders.—Dissolve 1 drachm each of rosin and copal in 4 fluidounces of benzine and with a little of this liquid triturate ½ drachm of Prussian blue and finally mix thoroughly with the remainder.

Ultramarine, to which has been added a small proportion of powdered rosin, is generally used for stamping embroidery patterns on white goods. The powder is dusted through the perforated pattern, which is then covered with a paper and a hot iron passed over it to melt the rosin and cause the powder to adhere to the cloth. The following are said to be excellent powders:

I.—White.—One part each of rosin, copal, damar, mastic, sandarac, borax, and bronze powder, and 2 parts white lead.

II.—Black.—Equal parts of rosin, damar, copal, sandarac, Prussian blue, ivory black, and bronze powder.

III. — Blue. — Equal parts of rosin, damar, copal, sandarac, Prussian blue, ultramarine, and bronze powder.

In all these powders the gums are first to be thoroughly triturated and mixed by passing through a sieve, and the other ingredients carefully added. Other colors may be made by using chrome yellow, burnt or raw sienna, raw or burnt umber, Vandyke brown, etc. For stamping fabrics liable to be injured by heat, the stamping is done by moistening a suitable powder with alcohol and using it like a stencil ink.

Stamping Powder for Embroideries.— "Stamping powders" used for outlining embroidery patterns are made by mixing a little finely powdered rosin with a suitable pigment. After dusting the powder through the perforated pattern it is fixed on the fabric by laying over it a piece of paper and then passing a hot iron carefully over the paper. By this means the rosin is melted and the mixture adheres. When white goods are to be "stamped," ultramarine is commonly used as the pigment; for dark goods, zinc white may be substituted. Especial care should be taken to avoid lead compounds and other poisonous pigments, as they may do mischief by dusting off. On velvets or other materials likely to be injured by heat, stamping is said to be done by moistening a suitable powder with alcohol and using it as stencil paint. A small addition of rosinous matter would seem required here also.

Starch

Black Starch.—Add to the starch a certain amount of logwood extract before the starch mixture is boiled. The quantity varies according to the depth of the black and the amount of starch. A small quantity of potassium bichromate dissolved in hot water is used to bring out the proper shade of black. In place of bichromate, black iron liquor may be used. This comes ready prepared.

Starch Gloss.—I. — Melt 2½ pounds of the best paraffine wax over a slow fire. When liquefied remove from the fire to stir in 100 drops of oil of citronella. Place several new pie tins on a level table, coat them slightly with sweet oil, and pour about 6 tablespoonfuls of the melted paraffine wax into each tin. The pan may be floated in water sufficiently to permit the mixture to be cut or stamped out with a tin cutter into small cakes about the size of a peppermint lozenge. Two of these cakes added to each pint of starch will cause the smoothing iron to impart the finest possible finish to muslin or linen, besides perfuming the clothes.

II.—

Gum arabic, powdered	3 parts
Spermaceti wax	6 parts
Borax, powdered	4 parts
White cornstarch	8 parts

All these are to be intimately mixed in the powder form by sifting through a sieve several times. As the wax is in a solid form and does not readily become reduced to powder by pounding in a mortar, the best method of reducing it to such a condition is to put the wax into a bottle with some sulphuric or rectified ether and then allow the fluid to evaporate. After it has dissolved the wax, as the evaporation proceeds, the wax will be deposited again in the solid form, but in fine thin flakes, which will easily break down to a powder form when rubbed up with the other ingredients in a cold mortar. Pack in paper or in cardboard boxes. To use, 4 teaspoonfuls per pound of dry starch are to be added to all dry starch, and then the starch made in the usual way as boiled starch.

Refining of Potato Starch.—A suitable quantity of chloride of lime, fluctuating according to its quality between ½ to 1 part per 100 parts of starch, is made with little water into a thick paste. To this paste add gradually with constant stirring 10 to 15 times the quantity of water, and filter.

The filtrate is now added to the starch stirred up with water; ½ part of ordinary hydrochloric acid of 20° Bé. previously diluted with four times the quantity of water is mixed in, for every part of chloride of lime, the whole is stirred thoroughly, and the starch allowed to stand.

When the starch has settled, the supernatant water is let off and the starch is washed with fresh water until all odor of chlorine has entirely disappeared. The starch now obtained is the resulting final product.

If the starch thus treated is to be worked up into dextrin, it is treated in the usual manner with hydrochloric acid or nitric acid and will then furnish a dextrin perfectly free from taste and smell.

In case the starch is to be turned into "soluble" starch proceed as usual, in a similar manner as in the production of dextrin, with the single difference that the starch treated with hydrochloric or nitric acid remains exposed to a temperature of 212° F., only until a test with tincture of iodine gives a bluish-violet reaction. The soluble starch thus produced, which is clearly soluble in boiling water, is odorless and tasteless.

Starch Powder. — Finely powdered starch is a very desirable absorbent, according to Snively, who says that for toilet preparations it is usually scented by a little otto or sachet powder. Frangipanin powder, used in the proportion of 1 part to 30 of the starch, he adds, gives a satisfactory odor.

STARCHES:
See Laundry Preparations.

STARCH IN JELLY, TESTS FOR:
See Foods.

STARCH PASTE:
See Adhesives.

STATUE CLEANING:
See Cleaning Preparations and Methods.

STATUETTES, CLEANING OF:
See Plaster.

STATUETTES OF LIPOWITZ METAL:
See Alloys.

Steel

(See also Iron and Metals.)

ANNEALING STEEL:
See also Hardening Steel and Tempering Steel.

This work requires the use of substances which yield their carbon readily and quickly to the tools on contact at a high temperature. Experience has shown that the best results are obtained by the use of yellow blood-lye salt (yellow prussiate of potash), which, when brought in contact with the tool at a cherry-red heat, becomes fluid, and in this condition has a strong cementing effect. The annealing process is as follows: The tool is heated to a cherry red and the blood-lye salt sprinkled over the surface which is to be annealed. A fine sieve should be used, to secure an even distribution of the substance. The tool is then put back into the fire, heated to the proper temperature for tempering, and tempered. If it is desired to give a higher or more thorough tempering to iron or soft steel, the annealing process is repeated 2 or 3 times. The surface of the tool must, of course, be entirely free from scale. Small tools to which it is desired to impart a considerable degree of hardness by annealing with blood-lye salt are tempered as follows: Blood-lye salt is melted in an iron vessel over a moderate fire, and the tool, heated to a brown-red heat, placed in the melted salt, where it is allowed to remain for about 15 minutes. It is then heated to the hardening temperature and hardened. A similar but milder effect is produced in small, thin tools by making them repeatedly red hot, immersing them slowly in oil or grease, reheating them, and finally tempering them in water. To increase the effect, soot or powdered charcoal is added to the oil or grease (train oil) till a thick paste is formed, into which the red-hot tool is plunged. By this means the tool is covered with a thick, not very combustible, coating, which produces a powerful cementation at the next heating. By mixing flour, yellow blood-lye salt, saltpeter, horn shavings, or ground hoofs, grease, and wax, a paste is formed which serves the same purpose. A choice may be made of any of the preparations sold as a "hardening paste"; they are all more or less of the same composition. This is a sample: Melt 500 grains of wax, 500 grains tallow, 100 grains rosin, add a mixture of leather-coal, horn shavings, and ground hoofs in equal parts till a paste is formed, then add 10 grains saltpeter and 50 to 100 grains powdered yellow blood-lye salt, and stir well. The tools are put into this paste while red hot, allowed to cool in it, then reheated and tempered.

More steel is injured, and sometimes spoiled, by over-annealing than in any other way. Steel heated too hot in annealing will shrink badly when being hardened; besides, it takes the life out of it. It should never be heated above a low cherry red, and it should be a lower heat than it is when being hardened. It should be heated slowly and given a uniform heat all over and through the piece.

This is difficult to do in long bars and in an ordinary furnace. The best way to heat a piece of steel, either for annealing or hardening, is in red-hot, pure lead. By this method it is done uniformly, and one can see the color all the time. Some heating for annealing is done in this way: Simply cover up the piece in sawdust, and let it cool there, and good results will be obtained.

Good screw threads cannot be cut in steel that is too soft. Soft annealing produces a much greater shrinkage and spoils the lead of the thread.

This mixture protects the appearance of polished or matted steel objects on heating to redness: Mix 1 part of white soap, 6 parts of chemically pure boracic acid, and 4 parts of phosphate of soda, after pulverizing, and make with water into a paste. For use, apply this to the article before the annealing.

COLORING STEEL:

Black.—I.—Oil or wax may be employed on hard steel tools; with both methods the tool loses more or less of its hardness and the blacking process therefore is suited only for tools which are used for working wood or at least need not be very hard, at any rate not for tools which are employed for working steel or cast iron. The handsomest glossy black color is obtained by first polishing the tool neatly again after it has been hardened in water, next causing it to assume on a grate or a hot plate the necessary tempering color, yellow, violet blue, etc., then dipping it in molten, not too hot, yellow wax and burning off the adhering wax, after withdrawal, at a fire, without, however, further heating the tool. Finally dip the tool again into the wax and repeat the burning off at the flame until the shade is a nice lustrous black, whereupon the tool may be cooled off in water. The wax is supposed to impart greater toughness to the tool. It is advisable for all tools to have a trough of fat ready, which has been heated to the necessary tempering degree, and the tools after hardening in water are suspended in the fat until they have acquired the temperature of the fat bath. When the parts are taken out and slowly allowed to cool, they will be a nice, but not lustrous, black.

II.—The following has been suggested for either steel or iron:

Bismuth chloride...	1 part
Mercury bichloride.	2 parts
Copper chloride....	1 part
Hydrochloric acid..	6 parts
Alcohol..........	5 parts

Water sufficient to make 64 parts.

Mix. As in all such processes a great deal depends upon having the article to be treated absolutely clean and free from grease. Unless this is the case uniform results are impossible. The liquid may be applied with a swab, or a brush, but if the object is small enough to dip into the liquid better results may thus be obtained than in any other way. The covering thus put on is said to be very lasting, and a sure protection against oxidation.

Blue.—I.—Heat an iron bar to redness and lay it on a receptacle filled with water. On this bar place the objects to be blued, with the polished side up. As soon as the article has acquired the desired color cause it to fall quickly into the water. The pieces to be blued must always previously be polished with pumice stone or fine emery.

II.—For screws: Take an old watch barrel and drill as many holes into the head of it as the number of screws to be blued. Fill it about one-fourth full of brass or iron filings, put in the head, and then fit a wire long enough to bend over for a handle, into the arbor holes—head of the barrel upward. · Brighten the heads of the screws, set them, point downward, into the holes already drilled, and expose the bottom of the barrel to the lamp until the screws assume the color you wish.

III.—To blue gun-barrels, etc., dissolve 2 parts of crystallized chloride of iron; 2 parts solid chloride of antimony; 1 part gallic acid in 4 or 5 parts of water; apply with a small sponge, and let dry in the air. Repeat this two or three times, then wash with water, and dry. Rub with boiled linseed oil to deepen the shade. Repeat this until satisfied with the result.

IV.—The bluing of gun barrels is effected by heating evenly in a muffle until the desired blue color is raised, the barrel being first made clean and bright with emery cloth, leaving no marks of grease or dirt upon the metal when the bluing takes place, and then allow to cool in the air. It requires considerable experience to obtain an even clear blue.

Brown.—I.—The following recipe for browning is from the United States Ordnance Manual: Spirits of wine, 1½ ounces; tincture of iron, 1½ ounces; corrosive sublimate, 1½ ounces; sweet spirits of niter, 1½ ounces; blue vitriol, 1 ounce; nitric acid, ¾ ounce. Mix and dissolve in 1 quart of warm water and keep in a glass jar. Clean the barrel well with caustic soda water to remove grease or oil. Then clean the surface of all stains and marks with emery paper or cloth, so as to produce an even, bright surface for the acid to act upon, and one without finger marks. Stop the bore and vent with wooden plugs. Then apply the mixture to every part with a sponge or rag, and expose to the air for 24 hours, when the loose rust should be rubbed off with a steel scratch brush. Use the mixture and the scratch brush twice, and more if necessary, and finally wash in boiling water, dry quickly, and wipe with linseed oil or varnish with shellac.

II.—Apply four coats of the following solution, allowing each several hours to dry. Brush after each coat if necessary. After the last coat is dry, rub down hard.

Sulphate of copper... 1 ounce
Sweet spirits of niter.. 1 ounce
Distilled water....... 1 pint

Niello.—This is a brightly polished metal, which is provided with a black or blue-black foundation by heating, is covered with a design by the use of a suitable matrix and then treated with hydrochloric acid in such a manner that only the black ground is attacked, the metal underneath remaining untouched. Next, the acid is rinsed off and the reserve is removed with suitable solvents. The parts of the metal bared by the acid may also be provided with a galvanic coating of silver or other metal.

Another method is to plunge the articles for a few minutes into a solution of oxalic acid and to clean them by passing them through alcohol. In this way the polish can even be brought back without the use of rouge or diamantine.

Whitening or Blanching.—If dissatisfied with the color acquired in tempering, dip the article into an acid bath, which whitens it, after which the bluing operation is repeated. This method is of great service, but it is important to remember always thoroughly to wash after the use of acid and then allow the object to remain for a few minutes in alcohol. Sulphuric acid does not whiten well, often leaving dark shades on the surface. Hydrochloric acid gives better results. Small pieces of steel are also whitened with a piece of pith moistened with dilute sulphuric acid, else the fine steel work, such as a watch hand, is fixed with lacquer on a plate and whitened by means of pith and polishing rouge, or a small stiff brush is charged with the same material. It is then detached by heating and cleaned in hot alcohol.

TEMPERING STEEL.

The best temperature at which to quench in the tempering of tool steel is the one just above the transformation point of the steel, and this temperature may be accurately determined in the following manner, without the use of a pyrometer. The pieces of steel are introduced successively at equal intervals of time into a muffle heated to a temperature a little above the transformation point of the steel. If, after a certain time, the pieces be taken out in the reverse order they will at first show progressively increasing degrees of brightness, these pieces being at the transformation point. When this point is passed the pieces again rapidly acquire a brightness superior to that of their neighbors, and should then be immediately quenched.

I.—Heat red hot and dip in an unguent made of mercury and the fat of bacon. This produces a remarkable degree of hardness and the steel preserves its tenacity and an elasticity which cannot be obtained by other means.

II.—Heat to the red white and thrust quickly into a stick of sealing wax. Leave it a second, and then change it to another place, and so continue until the metal is too cool to penetrate the wax. To pierce with drills hardened in this way, moisten them with essence of turpentine.

To Temper Small Coil Springs and Tools.—To temper small coil springs in a furnace burning wood the springs are exposed to the heat of the flame and are quenched in a composition of the following preparation: To a barrel of fish oil, 10 quarts of rosin and 12 quarts of tallow are added. If the springs tempered in this mixture break, more tallow is added, but if the break indicates brittleness of the steel rather than excessive hardness, a ball of yellow beeswax about 6 inches in diameter is added. The springs are drawn to a reddish purple by being placed on a frame having horizontally radiating arms like a star which is mounted on the end of a vertical rod. The springs are laid on the star and are lowered into a pot of melted lead, being held there for such time as is required to draw to the desired color.

It is well known that the addition of

certain soluble substances powerfully affects the action of tempering water. This action is strengthened if the heat-conducting power of the water is raised by means of these substances; it is retarded if this power is reduced, or the boiling point substantially lowered. The substance most frequently used for the purpose of increasing the heat-conducting power of tempering water is common salt. This is dissolved in varying proportions of weight, a saturated solution being generally used as a quenching mixture. The use of this solution is always advisable when tools of complicated shape, for which a considerable degree of hardness is necessary, are to be tempered in large quantities or in frequent succession. In using these cooling fluids, care must be taken that a sufficient quantity is added to the water to prevent any great rise of temperature when the tempering process is protracted. For this reason the largest possible vessels should be used, wide and shallow, rather than narrow and deep, vessels being selected. Carbonate of soda and sal ammoniac do not increase the tempering action to the same extent as common salt, and are therefore not so frequently employed, though they form excellent additions to tempering water in certain cases. Tools of very complicated construction, such as fraises, where the danger of fracture of superficial parts has always to be kept in view, can with advantage be tempered in a solution of soda or sal ammoniac. Acids increase the action of tempering water considerably, and to a far greater extent than common salt. They are added in quantities up to 2 per cent, and frequently in combination with salts. Organic acids (e. g., acetic or citric) have a milder action than mineral acids (e. g., hydrochloric, nitric, or sulphuric). Acidulous water is employed in tempering tools for which the utmost degree of hardness is necessary, such as instruments for cutting exceptionally hard objects, or when a sufficiently hard surface has to be given to a kind of steel not capable of much hardening. Alcohol lowers the boiling point of water, and causes so vigorous an evaporation when the water comes in contact with the red-hot metal, that the tempering is greatly retarded (in proportion to the amount of alcohol in the mixture). Water containing a large quantity of alcohol will not temper. Soap and soap suds will not temper steel; this property is made use of in the rapid cooling of steel for which a great degree of hardness is not

desirable. When certain parts of completely tempered steel have to be rendered soft, these parts are heated to a red heat and then cooled in soap suds. This is done with the tangs of files, knives, swords, saws, etc. Soluble organic substances retard the tempering process in proportion to the quantity used, and thus lessen the effect of pure water. Such substances (e. g., milk, sour beer, etc.) are employed only to a limited extent.

To Caseharden Locally.—In case-hardening certain articles it is sometimes necessary, or desirable, to leave spots or sections in the original soft uncarbonized condition while the remainder is carbonized and hardened. This may be effected by first covering the parts to be hardened with a protecting coat of japan, and allowing it to dry. Then put the piece in an electroplating bath and deposit a heavy coat of nickel over the parts not protected by the japan. The piece thus prepared may be treated in the usual manner in casehardening. The coat of nickel prevents the metal beneath being carbonized, so it does not harden when dipped in the bath.

A plating of copper answers the same purpose as nickel and is often used. A simpler plan, where the shape of the piece permits, is to protect it from the action of the carbonizing material with an iron pipe or plate closely fitted or luted with clay. Another scheme is to machine the parts wanted soft after carbonizing but before hardening. By this procedure the carbonized material is removed where the metal is desired soft, and when heated and dipped these parts do not harden.

To Harden a Hammer.—To avoid the danger of "checking" a hammer at the eye, heat the hammer to a good uniform hardening heat and then dip the small end almost up to the eye and cool as quickly as possible by moving about in the hardening bath; then dip the large end. To harden a hammer successfully by this method one must work quickly and cool the end dipped first enough to harden before the heat is lost on the other end. Draw the temper from the heat left about the eye. The result is a hammer hard only where it should be and free from "checks."

Hardening Steel Wire.—Pass the steel wire through a lead bath heated to a temperature of 1,200° to 1,500° F. after it has previously been coated with a paste of chalk, so as to prevent the formation of oxides. The wire is thus heated in a uniform manner and, according to whether it is desired hard or elastic, it is cooled in water or in oil.

Hardening of Springs.—A variety of steel must be chosen which is suitable for the production of springs, a very tough quality with about 0.8 per cent of carbon being probably the best. Any steel works of good reputation would no doubt recommend a certain kind of steel. In shaping a spring, forging and hammering should be avoided if possible. In forging, an uneven treatment can scarcely be avoided; one portion is worked more than the other, causing tensions which, especially in springs, must be guarded against. It is most advantageous if a material of the thickness and shape of the spring can be obtained, which, by bending and pressing through, is shaped into the desired spring. Since this also entails slight tension, a careful annealing is advisable, so as to prevent cracking or distorting in hardening. The annealing is best conducted with exclusion of the air, by placing the springs in a sheet-iron box provided with a cover, smearing all the joints well up with loam. The heating may be done in a muffled furnace; the box, with contents, is, not too slowly, heated to cherry red and then allowed to cool gradually, together with the stove. The springs must only be taken out when they have cooled off enough that they will give off no hissing sound when touched by water. In order to uniformly heat the springs for hardening, a muffle furnace is likewise employed, wherein they are heated to cherry-red heat. For cooling liquid, a mixture of oil, tallow, and petroleum is employed. A mass consisting of fish oil, tallow, and wax also renders good service, but one should see to it that there is a sufficient quantity of these cooling liquids, so that the springs may be moved about, same as when cooled in water, without causing an appreciable increase in the temperature of the liquid. In most cases too small a quantity of the liquid is responsible for the many failures in hardening. When the springs have cooled in the hardening liquid, they are taken out, dried off superficially, and the oil still adhering is burned off over a charcoal fire. This enables one to moderate the temper according to the duration of the burning off and to produce the desired elasticity. An even heating being of great importance in hardening springs, the electric current has of late been successfully employed for this purpose.

To Temper a Tap.—After the tap has been cut and finished heat it in a pair of tongs to a blood-red heat over a charcoal fire or the blue flame of a Bunsen burner or blow pipe, turning it around so that one point does not get heated before another. Have ready a pail of clean, cold water, into which a handful of common salt has been put. Stir the water in the pail so that a whirlpool is set up. Then plunge the tap, point first and vertically, into the vortex to cool. The turning of the tap during heating, as well as the swirl of the quenching water, prevents distortion. In tempering, the temper of the tap requires to be drawn to a light straw color, and this may be done as follows: Get a piece of cast-iron tube about 3 inches in diameter and heat it to a dull-red heat for about 4 inches of its length. Then hold the tap, with the tongs, up the center of the tube, meanwhile turning the tap around until the straw color appears all over it. Then dip the tap in the water, when it will be found perfectly hard. The depth of the color, whether light or dark straw, must be determined by the nature of the cast steel being used, which can be gained only from experience of the steel.

Scissors Hardening.—The united legs of the scissors are uniformly heated to a dark cherry red, extending from the point to the screw or rivet hole. This may be done in the naked fire, a feeble current of air being admitted until the steel commences to glow. Then the fire is left to itself and the scissor parts are drawn to and fro in the fire, until all the parts to be hardened show a uniform dark cherry red. The two legs are hardened together in water and then tempered purple red to violet.

The simultaneous heating, hardening, and tempering of the parts belonging together is necessary, so that the degree of heat is the same and the harder part does not cut the softer one.

In accordance with well-known rules, the immersion in the hardening bath should be done with the point first, slowly and vertically up to above the riveting hole.

Hardening without Scaling.—Articles made of tool steel and polished may be hardened without raising a scale, thereby destroying the polish, by the following method: Prepare equal parts in bulk of common salt and (fine) corn meal, well mixed. Dip the article to be hardened first into water, then into the mixture and place it carefully into the fire. When hot enough to melt the mixture, take from the fire and dip or roll in the salt and meal, replace in the fire and bring to the required heat for hardening. Watch the piece closely and if any part of it shows signs of getting dry, sprinkle some of the mixture on it. The mixture, when exposed to heat, forms a flux over the surface of the steel which excludes the air and prevents oxidation, and when cooled in water or oil comes off easily, leaving the surface as smooth as before heating. Borax would possibly give the same result, but is sometimes difficult to remove when cold.

Hardening with Glycerine.—I.—The glycerine employed must be of the density of 1.08 to 1.26 taken at the temperature of 302° F. Its weight must be equal to about 6 times the weight of the pieces to be tempered. For hard temper add to the glycerine ¼ to 4 per cent of sulphate of potash or of manganese, and for soft temper 1 to 10 per cent of chloride of manganese, or 1 to 4 per cent of chloride of potassium. The temperature of the tempering bath is varied according to the results desired.

II.—Glycerine, 8,000 parts, by weight; cooking salt, 500 parts, by weight; sal ammoniac, 100 parts, by weight; concentrated hydrochloric acid, 50 parts; and water, 10,000 parts, by weight. Into this liquid the steel, heated, for example, to a cherry red, is dipped. A reheating of the steel is not necessary.

To Remove Burnt Oil from Hardened Steel.—To remove excess oil from parts that have been hardened in oil, place the articles in a small tank of gasoline, which, when exposed to the air, will dry off immediately, allowing the part to be polished and tempered without the confusing and unsightly marks of burnt oil.

VARIOUS RECIPES:

To Put an Edge on Steel Tools.— Aluminum will put an edge on fine cutting instruments such as surgical knives, razors, etc. It acts exactly like a razor-hone of the finest quality. When steel is rubbed on the aluminum, as, for instance, in honing a knife blade, the metal disintegrates, forming an infinitely minute powder of a greasy unctuous quality that clings to steel with great tenacity and thus assists in cutting away the surface of the harder metal. So fine is the edge produced that it can in no wise be made finer by the strop, which used in the ordinary way merely tends to round the edge.

To Restore Burnt Steel.—To restore burnt cast steel heat the piece to a red heat and sprinkle over it a mixture of 8 parts red chromate of potassium; 4 parts saltpeter; ⅛ part aloes; ⅛ part gum arabic; and ¼ part rosin.

To Remove Strains in Metal by Heating.—In making springs of piano wire, or, in fact, any wire, if the metal is heated to a moderate degree the spring will be improved. Piano or any steel wire should be heated to a blue, brass wire to a degree sufficient to cause tallow to smoke. Heating makes the metal homogeneous; before heating, it is full of strains.

If a piece of metal of any kind is straightened cold and then put into a lathe and a chip turned off, it will be far from true. Before turning, it was held true by the strain of the particles on the outside, they having changed position, while the particles near the axis are only sprung. The outside particles being removed by the lathe tool, the sprung particles at the center return to their old positions. If, after straightening, the metal is heated to a temperature of 400° F., the particles settle together and the strains are removed.

This is the case in the manufacture of saws. The saw is first hardened and tempered and then straightened on an anvil by means of a hammer. After it is hammered true, it is ground and polished a little, then blued to stiffen it and then is subjected to the grinding process. Before bluing, the metal is full of strains; these are entirely removed by the heat required to produce the blue color. Often a piano-wire spring will not stand long wear if used without heating, while if heated it will last for years.

To Render Fine Cracks in Tools Visible.—It is often of importance to recognize small cracks which appear in the metal of the tools. For this purpose it is recommended to moisten the fissured surface with petroleum; next rub and dry with a rag and rub again, but this time with chalk. The petroleum which has entered the cracks soon comes out again and the trace is plainly shown by the chalk.

To Utilize Drill Chips.—There is one modern machining process that produces a shaving that has more value than that of mere scrap, and that is drilling rifle barrels with the oil-tube drill. The cutting edge of this drill is broken up into steps and the chips produced are literally shavings, being long hair-like threads of steel. These shavings are considerably used in woodworking factories for smoothing purposes.

To Remove Fragments of Steel from Other Metals.—The removal of broken spiral drills and taps is an operation which even the most skillful machinist has to perform at times. A practical process for removing such broken steel pieces consists in preparing in a suitable kettle (not iron) a solution of 1 part, by weight, of commercial alum in 4 to 5 parts, by weight, of water and boiling the object in this solution until the piece which is stuck works itself out. Care must be taken to place the piece in such a position that the evolving gas bubbles may rise and not adhere to the steel to protect it from the action of the alum solution.

Testing Steel.—A bar of the steel to be tested is provided with about nine notches running around it in distances of about ⅝ of an inch. Next, the foremost notched piece is heated in a forge in such a manner that the remaining portion of the bar is heated less by the fire proper than by the transmitted heat. When the foremost piece is heated to burning, i. e., to combustion, and the color of the succeeding pieces gradually passes to dark-brownish redness, the whole rod is hardened. A test with the file will now show that the foremost burned piece possesses the greatest hardness, that several softer pieces will follow, and that again a piece ordinarily situated in the second third, whose temperature was the right one for hardening, is almost as hard as the first one. If the different pieces are knocked off, the fracture of the piece hardened at the correct temperature exhibits the finest grain. This will give one an idea of the temperature to be employed for hardening the steel in question and its behavior in general. Very hard steel will readily crack in this process.

Welding Compound.—Boracic acid, 41½ parts; common salt 35 parts; ferrocyanide of potassium, 20 parts; rosin, 7½ parts; carbonate of sodium, 4 parts. Heat the pieces to be welded to a light-red heat and apply the compound; then heat to a strong yellow heat and the welding may be accomplished in the usual manner.

The precaution should be observed, the same as with any of the cyanides, to avoid breathing the poisonous fumes.

Softening Steel.—Heat the steel to a brown red and plunge into soft water, river water being the best. Care should be taken, however, not to heat over brown red, otherwise it will be hard when immersed. The steel will be soft enough to be cut with ease if it is plunged in the water as soon as it turns red.

Draw-Tempering Cast Steel.—First heat the steel lightly by means of charcoal until of a cherry-red shade, whereupon it is withdrawn to be put quickly into ashes or dry charcoal dust until completely cooled. The steel may also be heated in the forge to a red cherry color, then hammered until it turns blue and then plunged into water.

Drilling Hard Steel.—To accomplish the object quickly, a drill of cast steel should be made, the point gradually heated to the red, the scales taken off, and the extremity of the point immersed at once in quicksilver; then the whole quenched in cold water. Thus prepared, the drill is equal to any emergency; it will bore through the hardest pieces. The quantity of quicksilver needed is trifling.

Engraving or Etching on Steel.—Dissolve in 150 parts of vinegar, sulphate of copper, 30 parts; alum, 8 parts; kitchen salt, 11 parts. Add a few drops of nitric acid. According to whether this liquid is allowed to act a longer or shorter time, the steel may be engraved upon deeply or the surface may be given a very ornamental, frosted appearance.

To Distinguish Steel from Iron.—Take a very clean file and file over the flame of an alcohol lamp. If the filed piece is made of steel, little burning and crackling sparks will be seen. If it consists of iron, the sparks will not crackle.

STEEL, BROWNING OF:
See Plating.

STEEL, DISTINGUISHING IRON FROM:
See Iron.

STEEL ETCHING:
See Etching.

STEEL-HARDENING POWDER:
See Iron.

STEEL, OXIDIZED:
See Plating.

STEEL PLATING:
See Plating.

STEEL POLISHES:
See Polishes.

STEEL, TO CLEAN:
See Cleaning Preparations and Methods.

STENCILS FOR PLOTTING LETTERS OF SIGN PLATES:

See Enameling.

STENCIL INKS:

See Inks.

STEREOCHROMY.

Stereochromatic colors can be bought ground in a thickly liquid water-glass solution. They are only diluted with water-glass solution before application on the walls. The two solutions are generally slightly dissimilar in their composition, the former containing less silicic acid, but more alkali, than the latter, which is necessary for the better preservation of the paint. Suitable pigments are zinc white, ocher with its different shades of light yellow, red, and dark brown, black consisting of a mixture of manganese and lampblack, etc., etc. White lead cannot be used, as it coagulates with the water glass, nor vermilion, because it fades greatly under the action of the light. The plastering to be coated must be porous, not fresh, but somewhat hardened. Otherwise the caustic lime of the plaster will quickly decompose the water glass. This circumstance may account for the unsatisfactory results which have frequently been obtained with water-glass coatings. Before applying the paint the wall should first be impregnated with a water-glass solution. The colors may be kept on hand ground, but must be protected from contact with the air. If air is admitted a partial separation of silica in the form of a jelly takes place. Only pure potash water glass, or, at least, such as only contains little soda, should be used, as soda will cause efflorescence.

STEREOPTICON SLIDES:

See Photography.

STEREOTYPE METAL:

See Alloys.

STONE, ARTIFICIAL.

The following is a process of manufacture in which the alkaline silicates prepared industrially are employed.

The function of the alkaline silicates, or soluble glass, as constituents of artificial stone, is to act as a cement, forming with the alkaline earths, alumina, and oxide of lead, insoluble silicates, which weld together the materials (quartz sand, pebbles, granite, fluorspar, and the waste of clay bricks). The mass may be colored black by the addition of a quantity of charcoal or graphite to the extent of 10 per cent at the maximum, binoxide of manganese, or ocher; red, by 6 per cent of colcothar; brick red, by 4 to 7 per cent of cinnabar; orange, by 6 to 8 per cent of red lead; yellow, by 6 per cent of yellow ocher, or 5 per cent of chrome yellow; green, by 8 per cent of chrome green; blue, by 6 to 10 per cent of Neuwied blue, Bremen blue, Cassel blue, or Napoleon blue; and white, by 20 per cent, at the maximum, of zinc white.

Chrome green and zinc oxide produce an imitation of malachite. An imitation of lapis lazuli is obtained by the simultaneous employment of Cassel blue and pyrites in grains. The metallic oxides yield the corresponding silicates, and zinc oxide, mixed with cleansed chalk, yields a brilliant marble. The ingredients are mixed in a kind of mechanical kneading trough, furnished with stirrers, in variable proportions, according to the percentage of the solution of alkaline silicate. The whole is afterwards molded or compressed by the ordinary processes.

The imitation of granite is obtained by mixing lime, 100 parts; sodium silicate (42° Bé.), 35 parts; fine quartz sand, 120 to 180 parts; and coarse sand, 180 to 250 parts.

Artificial basalt may be prepared by adding potassium sulphite and lead acetate, or equal parts of antimony ore and iron filings.

To obtain artificial marble, 100 pounds of marble dust or levigated chalk are mixed with 20 parts of ground glass and 8 parts of fine lime and sodium silicate. The coloring matter is mixed in proportion depending on the effect to be produced.

A fine product for molding is obtained by mixing alkaline silicate, 100 parts; washed chalk, 100 parts; slaked lime, 40 parts; quick lime, 40 parts; fine quartz sand, 200 parts; pounded glass, 80 parts; infusorial earths, 80 parts; fluorspar, 150 parts. On hardening, there is much contraction.

Other kinds of artificial stone are prepared by mixing hydraulic lime or cement, 50 parts; sand, 200 parts; sodium silicate, in dry powder, 50 parts; the whole is moistened with 10 per cent of water and molded.

A hydraulic cement may be employed, to which an alkaline silicate is added. The stone or object molded ought to be covered with a layer of fluosilicate.

A weather-proof water-resisting stone is manufactured from sea mud, to which 5 per cent of calcic hydrate is added. The mass is then dried, lixiviated, and dried once more at 212° F., whereupon the stones are burned. By an admixture of crystallized iron sulphate the firmness of these stones is still increased.

Sand-Lime Brick.—In a French patent for making bricks from pitch and coal tar, powdered coke and sea sand are gently heated in a suitable vessel, and 20 per cent of pitch and 10 per cent of coal tar added, with stirring. The pasty mass obtained is then molded under pressure. The product obtained may be employed alone, or together with a framework of iron, or with hydraulic lime or cement.

According to a French patent for veining marble, etc., in one or more colors, coloring matters of all kinds are mixed with a sticky liquid, which is then spread in a very thin layer on the surface of another immiscible and heavier liquid. By agitating the surface, colored veins, etc., are obtained, which are then transferred to the object to be decorated (which may be of most varied kind) by applying it to the surface of the heavy liquid. A suitable composition with which the colors may be mixed consists of: Oil of turpentine, 100 parts; colophony, 10 parts; linseed oil, 10 parts; *siccatif soleil*, 5 parts. The heavy liquid may be water, mercury, etc.; and any colors, organic or mineral, may be used.

CONCRETE.

Concrete is the name applied to an artificial combination of various mineral substances which under chemical action become incorporated into a solid mass. There are one or two compositions of comparatively trifling importance which receive the same name, though differing fundamentally from true concrete, their solidification being independent of chemical influence. These compositions only call for passing mention; they are: *Tar concrete*, made of broken stones (macadam) and tar; *iron concrete*, composed of iron turnings, asphalt, bitumen, and pitch; and *lead concrete*, consisting of broken bricks set in molten lead. The last two varieties, with rare exceptions, are only used in connection with military engineering, such as for fortifications.

Concrete proper consists essentially of two groups or classes of ingredients. The first, termed the *aggregate*, is a heterogeneous mass, in itself inactive, of mineral material, such as shingle, broken stone, broken brick, gravel, and sand. These are the substances most commonly in evidence, but other ingredients are also occasionally employed, such as slag from iron furnaces. Burnt clay, in any form, and earthenware, make admirable material for incorporation. The second class constitutes the active agency which produces adhesion and solidification. It is termed the matrix, and consists of hydraulic lime or cement, combined with water.

One of the essential features in good concrete is cleanliness and an entire absence of dirt, dust, greasy matter, and impurities of any description. The material will preferably be sharp and angular, with a rough, porous surface, to which the matrix will more readily adhere than to smooth, vitreous substances. The specific gravity of the aggregate will depend upon the purpose for which the concrete is to be used. For beams and lintels, a light aggregate, such as coke breeze from gasworks, is permissible, especially when the work is designed to receive nails. On the other hand, for retaining walls, the heaviest possible aggregate is desirable on the ground of stability.

The aggregate by no means should be uniform in size. Fragments of different dimensions are most essential, so that the smaller material may fill up the interstices of the larger. It is not infrequently stipulated by engineers that no individual fragment shall be more than 4 inches across, and the material is often specified to pass through a ring $1\frac{1}{2}$ to 2 inches in diameter. The absolute limits to size for the aggregate, however, are determinable by a number of considerations, not the least important of which is the magnitude and bulk of the work in which it is to be employed. The particles of sand should also be of varying degrees of coarseness. A fine, dust-like sand is objectionable; its minute subdivision prevents complete contact with the cement on all its faces. Another desideratum is that the particles should not be too spherical, a condition brought about by continued attrition. Hence, pit sand is better in many cases than river sand or shore sand.

The matrix is almost universally Portland cement. It should not be used in too hot a condition, to which end it is usually spread over a wooden floor to a depth of a few inches, for a few days prior to use. By this means, the aluminate of lime becomes partially hydrated, and its activity is thereby modified.

Roman cement and hydraulic lime may also be used as matrices.

Portland cement will take a larger proportion of sand than either Roman cement or hydraulic lime; but with the larger ratios of sand, its tenacity is, of course, correspondingly reduced. One part of cement to 4 parts of sand should therefore be looked upon as the upper limit, while for the strongest mortar the proportion need hardly exceed 1 part of cement to $1\frac{1}{2}$ or 2 parts of sand. In the ensuing calculations there is assumed a ratio of 1 to 3. For impermeability, the proportion of 1 to 2 should be observed, and for Roman cement this proportion should never be exceeded. The ratio will even advantageously be limited to 2 to 3. For hydraulic lime equal parts of sand and cement are suitable, though 2 parts of sand to 1 part of cement may be used.

The quantity of mortar required in reference to the aggregate is based on the vacuities in the latter. For any particular aggregate the amount of empty space may be determined by filling a tank of known volume with the minerals and then adding sufficient water to bring to a level surface. The volume of water added (provided, of course, the aggregate be impervious or previously saturated) gives the net volume of mortar required. To this it is necessary to make some addition (say 10 per cent of the whole), in order to insure the thorough flushing of every part of the work.

Assuming that the proportion of interstices is 30 per cent and adding 10 for the reason just stated, we derive 40 parts as the quantity of mortar to 100 — 10 = 90 parts of the aggregate. An allowance of $\frac{1}{4}$ volume for shrinkage brings the volume of the dry materials (sand and cement) of the mortar to $40 + 40/3 = 53\frac{1}{3}$ parts, which, divided in the ratio of 1 to 3, yields:

Cement $\dfrac{53\frac{1}{3}}{4}$ = ... $13\frac{1}{3}$ parts

Sand, $\frac{3}{4} \times 53\frac{1}{3}$ = ... 40 parts

Aggregate.......... 90 parts

Total.......... $143\frac{1}{3}$ parts

As the resultant concrete is 100 parts, the total shrinkage is 30 per cent. Expressed in terms of the cement, the concrete would have a composition of 1 part cement, 3 parts sand, 7 parts gravel and broken stone, and it would form, approximately, what is commonly known as 7 to 1 concrete.

There are other ratios depending on the proportion of sand. Thus we have:

Cement	Sand	Aggregate
1	$1\frac{1}{2}$	$4\frac{1}{3}$
1	2	5
1	$2\frac{1}{2}$	6
1	3	7
1	$3\frac{1}{2}$	$7\frac{1}{2}$
1	4	$8\frac{1}{4}$

The cost of concrete may be materially reduced without affecting the strength or efficacy of the work, by a plentiful use of stone "plums" or "burrs." These are bedded in the fluid concrete during its deposition *in situ*, but care must be taken to see that they are thoroughly surrounded by mortar and not in contact with each other. Furthermore, if they are of a porous nature, they should be well wetted before use.

The mixing of concrete is important. If done by hand, the materials forming the aggregate will be laid out on a platform and covered by the cement in a thin layer. The whole should be turned over thrice in the dry state, and as many times wet, before depositing, in order to bring about thorough and complete amalgamation. Once mixed, the concrete is to be deposited immediately and allowed to remain undisturbed until the action of setting is finished. Deposition should be effected, wherever possible, without tipping from a height of more than about 6 feet, as in greater falls there is a likelihood of the heavier portions of the aggregate separating from the lighter. In extensive undertakings, concrete is more economically mixed by mechanical appliances.

The water used for mixing may be either salt or fresh, so far as the strength of the concrete is concerned. For surface work above the ground level, salinity in any of the ingredients is objectionable, since it tends to produce efflorescence— an unsightly, floury deposit, difficult to get rid of. The quantity of water required cannot be stated with exactitude; it will depend upon the proportion of the aggregate and its porosity. It is best determined by experiment in each particular case. Without being profuse enough to "drown" the concrete, it should be plentiful enough to act as an efficient intermediary between every particle of the aggregate and every particle of the matrix. Insufficient moisture is, in fact, as deleterious as an excess.

Voids.—The strength of concrete depends greatly upon its density, and this is secured by using coarse material which contains the smallest amount of voids or empty spaces. Different kinds of sand, gravel, and stone vary greatly in the amount of voids they contain, and by judiciously mixing coarse and fine material the voids may be much reduced and the density increased. The density and percentage of voids in concrete material may be determined by filling a box of 1 cubic foot capacity and weighing it. One cubic foot of solid quartz or limestone, entirely free from voids, would weigh 165 pounds, and the amount by which a cubic foot of any loose material falls short of this weight represents the proportion of voids contained in it. For example, if a cubic foot of sand weighs 115½ pounds, the voids would be 49½-165ths of the total volume, or 30 per cent.

The following table gives the per cent of voids and weight per cubic foot of some common concrete materials:

	Per Cent Voids	Wt. per Cu. Ft.
Sandusky Bay sand.	32.3	111.7 pounds
Same through 20-mesh screen	38.5	101.5 pounds
Gravel, ⅛ to ¼ inch	42.4	95.0 pounds
Broken limestone, egg-size	47.0	87.4 pounds
Limestone screenings, dust to ½ inch	26.0	122.2 pounds

It will be noted that screening the sand through a 20-mesh sieve, and thus taking out the coarse grains, considerably increased the voids and reduced the weight; thus decidedly injuring the sand for making concrete.

The following figures show how weight can be increased and voids reduced by mixing fine and coarse material:

	Per Cent Voids	Wt. per Cu. Ft.
Pebbles, about 1 inch	38.7	101.2 pounds
Sand, 30 to 40 mesh	35.9	105.8 pounds
Pebbles plus 38.7 per cent sand, by vol.	19.2	133.5 pounds

Experiments have shown that the strength of concrete increases greatly with its density; in fact, a slight increase in weight per cubic foot adds very decidedly to the strength.

The gain in strength obtained by adding coarse material to mixtures of cement and sand is shown in the following table of results of experiments made in Germany by R. Dykerhoff. The blocks tested were 2½-inch cubes, 1 day in air and 27 days in water,

Proportions by Measure.			Per Cent. Cement.	Compression Strength.
Cement.	Sand.	Gravel.	By Volume.	Lbs. per Sq. In.
1	2	...	33.0	2,125
1	2	5	12.5	2,387
1	3	...	25.0	1,383
1	3	6½	9.5	1,515
1	4	...	20.0	1,053
1	4	8½	7.4	1,204

These figures show how greatly the strength is improved by adding coarse material, even though the proportion of cement is thereby reduced. A mixture of 1 to 12½ of properly proportioned sand and gravel is, in fact, stronger than 1 to 4, and nearly as strong as 1 to 3, of cement and sand only.

In selecting materials for concrete, those should be chosen which give the greatest density. If it is practicable to mix two materials, as sand and gravel, the proportion which gives the greatest density should be determined by experiment, and rigidly adhered to in making concrete, whatever proportion of cement it is decided to use. Well-proportioned dry sand and gravel or sand and broken stone, well shaken down, should weigh at least 125 pounds per cubic foot. Limestone screenings, owing to minute pores in the stone itself, are somewhat lighter, though giving equally strong concrete. They should weigh at least 120 pounds per cubic foot. If the weight is less, there is probably too much fine dust in the mixture.

The density and strength of concrete are also greatly improved by use of a liberal amount of water. Enough water must be used to make the concrete thoroughly soft and plastic, so as to quake strongly when rammed. If mixed too dry it will never harden properly, and will be light, porous, and crumbling.

Thorough mixing of concrete materials is essential, to increase the density and give the cement used a chance to produce its full strength. The cement, sand, and gravel should be intimately mixed dry, then the water added and the mixing continued. If stone or coarse gravel is added, this should be well wetted and thoroughly mixed with the mortar.

Materials for Concrete Building Blocks. —In the making of building blocks the spaces to be filled with concrete are generally too narrow to permit the use of very coarse material, and the blockmaker is limited to gravel or stone not exceeding $\frac{1}{2}$ or $\frac{3}{4}$ inch in size. A considerable proportion of coarse material is, however, just as necessary as in other kinds of concrete work, and gravel cr screenings should be chosen which will give the greatest possible density. For good results, at least one-third of the material, by weight, should be coarser than $\frac{1}{8}$ inch. Blocks made from such gravel or screenings, 1 to 5, will be found as good as 1 to 3 with sand only. It is a mistake to suppose that the coarse fragments will show on the surface; if the mixing is thorough this will not be the case. A moderate degree of roughness or variety in the surface of blocks is, in fact, desirable, and would go far to overcome the prejudice which many architects hold against the smooth, lifeless surface of cement work. Sand and gravel are, in most cases, the cheapest material to use for block work. The presence of a few per cent of clay or loam is not harmful provided the mixing is thorough. Stone screenings, if of good quality, give fully as strong concrete as sand and gravel, and usually yield blocks of somewhat lighter color. Screenings from soft stone should be avoided, also such as contain too much dust. This can be determined from the weight per cubic foot, and by a sifting test. If more than two-thirds pass $\frac{1}{8}$ inch, and the weight (well jarred down) is less than 120 pounds, the material is not the best.

Cinders are sometimes used for block work; they vary greatly in quality, but if clean and of medium coarseness will give fair results. Cinder concrete never develops great strength, owing to the porous character and crushability of the cinders themselves. Cinder blocks may, however, be strong enough for many purposes, and suitable for work in which great strength is not required.

Lime.—It is well known that slaked lime is a valuable addition to cement mortar, especially for use in air. In sand mixtures, 1 to 4 or 1 to 5, at least one-third of the cement may be replaced by slaked lime without loss of strength. The most convenient form of lime for use in block-making is the dry-slaked or hydrate lime, now a common article of commerce. This is, however, about as expensive as Portland cement, and there is no great saving in its use. Added to block concrete, in the proportion of $\frac{1}{4}$ to $\frac{1}{2}$ the cement used, it will be found to make the blocks lighter in color, denser, and decidedly less permeable by water.

Cement.—Portland cement is the only hydraulic material to be seriously considered by the blockmaker. Natural and slag cements and hydraulic lime are useful for work which remains constantly wet, but greatly inferior in strength and durability when exposed to dry air. A further advantage of Portland cement is the promptness with which it hardens and develops its full strength; this quality alone is sufficient to put all other cements out of consideration for block work.

Proportions.—There are three important considerations to be kept in view in adjusting the proportions of materials for block concrete—strength, permeability, and cost. So far as strength goes, it may easily be shown that concretes very poor in cement, as 1 to 8 or 1 to 10, will have a crushing resistance far beyond any load that they may be called upon to sustain. Such concretes are, however, extremely porous, and absorb water like a sponge. The blocks must bear a certain amount of rough handling at the factory and while being carted to work and set up in the wall. Safety in this respect calls for a much greater degree of hardness than would be needed to bear the weight of the building. Again, strength and hardness, with a given proportion of cement, depend greatly on the character of the other materials used; blocks made of cement and sand, 1 to 3, will not be so strong or so impermeable to water as those made from a good mixed sand and gravel, 1 to 5. On the whole, it is doubtful whether blocks of satisfactory quality can be made, by hand mixing and tamping, under ordinary factory conditions, from a poorer mixture than 1 to 5. Even this proportion requires for good results the use of properly graded sand and gravel or screenings, a liberal amount of water, and thorough mixing and tamping. When suitable gravel is not obtainable, and coarse mixed sand only is used, the proportion should not be less than 1 to 4. Fine sand alone is a very bad material, and good blocks cannot be made from it except by the use of an amount of cement which would make the cost very high.

The mixtures above recommended, 1 to 4 and 1 to 5, will necessarily be somewhat porous, and may be decidedly so if the gravel or screenings used is not properly graded. The water-resisting qualities may be greatly improved, without loss of strength, by replacing a part of the cement by hydrate lime. This is a light, extremely fine material, and a given weight of it goes much further than the same amount of cement in filling the pores of the concrete. It has also the effect of making the wet mixture more plastic and more easily compacted by ramming, and gives the finished blocks a lighter color.

The following mixtures, then, are to be recommended for concrete blocks. By "gravel" is meant a suitable mixture of sand and gravel, or stone screenings, containing grains of all sizes, from fine to $\frac{1}{2}$ inch.

1 to 4 Mixtures, by Weight.

Cement, 150 parts; gravel, 600 parts.
Cement, 125 parts; hydrated lime, 25 parts; gravel, 600 parts.
Cement, 100 parts; hydrated lime, 50 parts; gravel, 600 parts.

1 to 5 Mixtures, by Weight.

Cement, 120 parts; gravel, 600 parts.
Cement, 100 parts; hydrated lime, 20 parts; gravel, 600 parts.

Proportion of Water.—This is a matter of the utmost consequence, and has more effect on the quality of the work than is generally supposed. Blocks made from too dry concrete will always remain soft and weak, no matter how thoroughly sprinkled afterwards. On the other hand, if blocks are to be removed from the machine as soon as made, too much water will cause them to stick to the plates and sag out of shape. It is perfectly possible, however, to give the concrete enough water for maximum density and first-class hardening properties, and still to remove the blocks at once from the mold. A good proportion of coarse material allows the mixture to be made wetter without sticking or sagging. Use of plenty of water vastly improves the strength, hardness, and waterproof qualities of blocks, and makes them decidedly lighter in color. The rule should be:

Use as much water as possible without causing the blocks to stick to the plates or to sag out of shape on removing from the machine.

The amount of water required to produce this result varies with the materials used, but is generally from 8 to 9 per cent of the weight of the dry mixture. A practiced blockmaker can judge closely when the right amount of water has been added, by squeezing some of the mixture in the hand. Very slight variations in proportion of water make such a marked difference in the quality and color of the blocks that the water, when the proper quantity for the materials used has been determined, should always be accurately measured out for each batch. In this way much time is saved and uncertainty avoided.

Facing.—Some blockmakers put on a facing of richer and finer mixture, making the body of the block of poorer and coarser material. As will be explained later, the advantage of the practice is, in most cases, questionable, but facings may serve a good purpose in case a colored or specially waterproof surface is required. Facings are generally made of cement and sand, or fine screenings, passing a $\frac{1}{8}$-inch sieve. To get the same hardness and strength as a 1 to 5 gravel mixture, at least as rich a facing as 1 to 3 will be found necessary. Probably 1 to 2 will be found better, and if one-third the cement be replaced by hydrate lime the waterproof qualities and appearance of the blocks will be improved. A richer facing than 1 to 2 is liable to show greater shrinkage than the body of the block, and to adhere imperfectly or develop hair-cracks in consequence.

Poured Work.—The above suggestions on the question of proportions of cement, sand, and gravel for tamped blocks apply equally to concrete made very wet, poured into the mold, and allowed to harden a day or longer before removing. Castings in a sand mold are made by the use of very liquid concrete; sand and gravel settle out too rapidly from such thin mixtures, and rather fine limestone screenings are generally used.

Mixing.—To get the full benefit of the cement used it is necessary that all the materials shall be very thoroughly mixed together. The strength of the block as a whole will be only as great as that of its weakest part, and it is the height of folly, after putting a liberal measure of cement, to so slight the mixing as to get no better result than half as much cement, properly mixed, would have given. The poor, shoddy, and crumbly blocks turned out by many small-scale makers owe their faults chiefly to careless mixing and use of too little water, rather than to too small proportion of cement.

The materials should be mixed dry, until the cement is uniformly distributed and perfectly mingled with the sand and gravel or screenings; then the water is to be added and the mixing continued until all parts of the mass are equally moist and every particle is coated with the cement paste.

Concrete Mixers.—Hand mixing is always imperfect, laborious, and slow, and it is impossible by this method to secure the thorough stirring and kneading action which a good mixing machine gives. If a machine taking 5 or 10 horse-power requires 5 minutes to mix one-third of a yard of concrete, it is of course absurd to expect that two men will do the same work by hand in the same time. And the machine never gets tired or shirks if not constantly urged, as it is the nature of men to do. It is hard to see how the manufacture of concrete blocks can be successfully carried on without a concrete mixer. Even for a small business it will pay well in economy of labor and excellence of work to install such a machine, which may be driven by a small electric motor or gasoline engine. In work necessarily so exact as this, requiring perfectly uniform mixtures and use of a constant percentage of water, batch mixers, which take a measured quantity of material, mix it, and discharge it, at each operation, are the only satisfactory type, and continuous mixers are unsuitable. Those of the pug-mill type, consisting of an open trough with revolving paddles and bottom discharge, are positive and thorough in their action, and permit the whole operation to be watched and controlled. They should be provided with extensible arms of chilled iron, which can be lengthened as the ends become worn.

Concrete Block Systems.—For smaller and less costly buildings, *separate blocks*, made at the factory and built up into the walls in the same manner as brick or blocks of stone, are simpler, less expensive, and much more rapid in construction than monolithic work. They also avoid some of the faults to which solid concrete work, unless skillfully done, is subject, such as the formation of shrinkage cracks.

There are two systems of block making, differing in the consistency of the concrete used:

1. Blocks tamped or pressed from semi-wet concrete, and removed at once from the mold.

2. Blocks poured or tamped from wet concrete, and allowed to remain in the mold until hardened.

Tamped Blocks from Semi-Wet Mixture.—These are practically always made on a block machine, so arranged that as soon as a block is formed the cores and side plates are removed and the block lifted from the machine. By far the larger part of the blocks on the market are made in this way. Usually these are of the one-piece type, in which a single block, provided with hollow cores, makes the whole thickness of the wall. Another plan is the *two-piece* system, in which the face and back of the wall are made up of different blocks, so lapping over each other as to give a bond and hold the wall together. Blocks of the two-piece type are generally formed in a hand or hydraulic press.

Various shapes and sizes of blocks are commonly made; the builders of the most popular machines have, however, adopted the standard length of 32 inches and height of 9 inches for the full-sized block, with thickness of 8, 10, and 12 inches. Lengths of 24, 16, and 8 inches are also obtained on the same machines by the use of parting plates and suitably divided face plates; any intermediate lengths and any desired heights may be produced by simple adjustments or blocking off.

Blocks are commonly made plain, rock-faced, tool-faced, paneled, and of various ornamental patterns. New designs of face plates are constantly being added by the most progressive machine makers.

Block Machines.—There are many good machines on the market, most of which are of the same general type, and differ only in mechanical details. They may be divided into two classes: those with vertical and those with horizontal face. In the former the face plate stands vertically, and the block is simply lifted from the machine on its base plate as soon as tamped. In the other type the face plate forms the bottom of the mold; the cores are withdrawn horizontally, and by the motion of a lever the block with its face plate is tipped up into a vertical position for removal. In case it is desired to put a facing on the blocks, machines of the horizontal-face type are considered the more convenient, though a facing may easily be put on with the vertical-face machine by the use of a parting plate.

Blocks Poured from Wet Concrete. —As already stated, concrete made too dry is practically worthless, and an excess of water is better than a deficiency. The above-described machine process, in which blocks are tamped from damp concrete and at once removed, gives blocks of admirable hardness and quality if the maximum of water is used. A method of making blocks from very wet concrete, by the use of a large number of separable molds of sheet steel, into which the wet concrete is poured and in which the blocks are left to harden for 24 hours or longer, has come into considerable use. By this method blocks of excellent hardening and resistance to water are certainly obtained. Whether the process is the equal of the ordinary machine method in respect of economy and beauty of product must be left to the decision of those who have had actual experience with it.

The well-known cast-stone process consists in pouring liquid concrete mixture into a sand mold made from a pattern in a manner similar to that in which molds for iron castings are produced. The sand absorbs the surplus water from the liquid mixture, and the casting is left in the mold for 24 hours or longer until thoroughly set. This process necessitates the making of a new sand mold for every casting, and is necessarily much less rapid than the machine method. It is less extensively used for building blocks than for special ornamental architectural work, sills, lintels, columns, capitals, etc., and for purposes of this kind it turns out products of the highest quality and beauty.

Tamping of Concrete Blocks. — This is generally done by means of hand rammers. Pneumatic tampers, operated by an air compressor, are in use at a few plants, apparently with considerable saving in time and labor and improvements in quality of work. Hand tamping must be conscientious and thorough, or poor work will result. It is important that the mold should be filled a little at a time, tamping after each addition; at least four fillings and tampings should be given to each block. If the mixture is wet enough no noticeable layers will be formed by this process.

Hardening and Storage. — Tripledecked cars to receive the blocks from the machines will be found a great saving of labor, and are essential in factories of considerable size. Blocks will generally require to be left on the plates for at least 24 hours, and must then be kept under roof, in a well-warmed room, with frequent sprinkling, for not less than 5 days more. They may then be piled up out of doors, and in dry weather should be wetted daily with a hose. Alternate wetting and drying is especially favorable for the hardening of cement, and concrete so treated gains much greater strength than if kept continuously in water or dry air.

Blocks should not be used in building until at least 4 weeks from the time they are made. During this period of seasoning, blocks will be found to shrink at least $\frac{1}{16}$ inch in length, and if built up in a wall when freshly made, shrinkage cracks in the joints or across the blocks will surely appear.

Efflorescence, or the appearance of a white coating on the surfaces, sometimes takes place when blocks are repeatedly saturated with water and then dried out; blocks laid on the ground are more liable to show this defect. It results from diffusion of soluble sulphates of lime and alkalies to the surface. It tends to disappear in time, and rarely is sufficient in amount to cause any complaint.

Properties of Concrete Blocks — Strength. — In the use of concrete blocks for the walls of buildings, the stress to which they are subjected is almost entirely one of compression. In compressive strength well-made concrete does not differ greatly from ordinary building stone. It is difficult to find reliable records of tests of sand and gravel concrete, 1 to 4 and 1 to 5, such as is used in making blocks; the following figures show strength of concrete of approximately this richness, also the average of several samples each of well-known building stones, as stated by the authorities named:

Limestone, Bedford, Ind.
(Indiana Geographical
Survey)............ 7,792 pounds
Limestone, Marblehead,
Ohio (Q. A. Gillmore)
7,393 pounds
Sandstone, N. Amherst,
Ohio (Q. A. Gillmore)............ 5,831 pounds
Gravel concrete, 1:1.6-
:2.8, at 1 year (Candlot)............ 5,500 pounds
Gravel concrete, 1:1.6-
:3.7, at 1 year (Candlot)............ 5,050 pounds
Stone concrete, 1:2:4 at
1 year (Boston El.
R. R.)............ 3,904 pounds

Actual tests of compression strength of hollow concrete blocks are difficult to make, because it is almost impossible to apply the load uniformly over the whole surface, and also because a block 16 inches long and 8 inches wide will bear a load of 150,000 to 200,000 pounds, or more than the capacity of any but the largest testing machines. Three one-quarter blocks, 8 inches long, 8 inches wide, and 9 inches high, with hollow space equal to one-third of the surface, tested at the Case School of Science, showed strengths of 1,805, 2,000, and

1,530 pounds per square inch, respectively, when 10 weeks old.

Two blocks 6 × 8 × 9 inches, 22 months old, showed crushing strength of 2,530 and 2,610 pounds per square inch. These blocks were made of cement 1¼ parts, lime ½ part, sand and gravel 6 parts, and were tamped from damp mixture. It is probably safe to assume that the minimum crushing strength of well-made blocks, 1 to 5, is 1,000 pounds per square inch at 1 month and 2,000 pounds at 1 year.

A block 12 inches wide and 24 inches long has a total surface of 288 square inches, or, deducting ⅓ for openings, a net area of 192 inches. Such a block, 9 inches high, weighs 130 pounds. Assuming a strength of 1,000 pounds and a factor of safety of 5, the safe load would be 200 pounds per square inch, or 200 × 192 = 38,400 pounds for the whole surface of the block. Dividing this by the weight of the block, 130 pounds, we find that 295 such blocks could be placed one upon another, making a total height of wall of 222 feet, and still the pressure on the lowest block would be less than one-fifth of what it would actually bear. This shows how greatly the strength of concrete blocks exceeds any demands that are ever made upon it in ordinary building construction.

The safe load above assumed, 200 pounds. seems low enough to guard against any possible failure. In Taylor and Thompson's work on concrete, a safe load of 450 pounds for concrete 1 to 2 to 4 is recommended; this allows a factor of safety of 5½. On the other hand, the Building Code of the city of Cleveland permits concrete to be loaded only to 156 pounds per square inch, and limits the height of walls of 12-inch blocks to 44 feet. The pressure of such a wall would be only 40 pounds per square inch; adding the weight of two floors at 25 pounds per square foot each, and roof with snow and wind pressure, 40 pounds per square foot, we find that with a span of 25 feet the total weight on the lowest blocks would be only 52 pounds per square inch, or about one-twentieth of their minimum compression strength.

Blocks with openings equal to only one-third the surface, as required in many city regulations, are heavy to handle, especially for walls 12 inches and more in thickness, and, as the above figures show, are enormously stronger than there is any need of. Blocks with openings of 50 per cent would be far more acceptable to the building trade,

and if used in walls not over 44 feet high, with floors and roof calculated as above for 25 feet span, would be loaded only to 56 pounds per square inch of actual surface. This would give a factor of safety of 18, assuming a minimum compression strength of 1,000 pounds.

There is no doubt that blocks with one-third opening are inconveniently and unnecessarily heavy. Such a block, 32 inches long, 12 inches wide, and 9 inches high, has walls about 3½ inches thick, and weighs 180 pounds. A block with 50 per cent open space would have walls and partitions 2 inches in thickness, and would weigh about 130 pounds. With proper care in manufacture, especially by using as much water as possible, blocks with this thickness of walls may be made thoroughly strong, sound, and durable. It is certainly better for strength and water-resisting qualities to make thin-walled blocks of rich mixture, rather than heavy blocks of poor and porous material.

Filling the voids with cement is a rather expensive method of securing waterproof qualities, and gives stronger concretes than are needed. The same may be accomplished more cheaply by replacing part of the cement by slaked lime, which is an extremely fine-grained material, and therefore very effective in closing pores. Hydrate lime is the most convenient material to use, but nearly as costly as Portland cement at present prices. A 1 to 4 mixture in which one-third the cement is replaced by hydrate lime will be found equal to a 1 to 3 mixture without the lime. A 1 to 4 concrete made from cement, 1; hydrate lime, ½; sand and gravel, 6 (by weight), will be found fairly water-tight, and much superior in this respect to one of the same richness consisting of cement, 1½; sand and gravel, 6.

The cost of lime may be greatly reduced by using ordinary lump lime slaked to a paste. The lime must, however, be very thoroughly hydrated, so that no unslaked fragments may remain to make trouble by subsequent expansion. Lime paste is also very difficult to mix, and can be used successfully only in a concrete mixer of the pug-mill type. Ordinary stiff lime paste contains about 50 per cent water; twice as much of it, by weight, should therefore be used as of dry hydrate lime.

Waterproof Qualities.—The chief fault of concrete building blocks, as ordinarily made, is their tendency to absorb water. In this respect they are generally no worse than sandstone or common brick; it is well known that stone or brick walls are too permeable to allow plastering directly on the inside surface, and must be furred and lathed before plastering, to avoid dampness. This practice is generally followed with concrete blocks, but their use and popularity would be greatly increased if they were made sufficiently waterproof to allow plastering directly on the inside surface.

For this purpose it is not necessary that blocks should be perfectly waterproof, but only that the absorption of water shall be *slow*, so that it may penetrate only part way through the wall during a long-continued rain. Walls made entirely water-tight are, in fact, objectionable, owing to their tendency to "sweat" from condensation of moisture on the inside surface. For health and comfort, walls must be slightly porous, so that any moisture formed on the inside may be gradually absorbed and carried away.

Excessive water absorption may be avoided in the following ways:

1. Use of Properly Graded Materials. —It has been shown by Feret and others that porosity and permeability are two different things; porosity is the total proportion of voids or open spaces in the mass, while permeability is the rate at which water, under a given pressure, will pass through it. Permeability depends on the *size* of the openings as well as on their total amount. In two masses of the same porosity or percentage of voids, one consisting of coarse and the other of fine particles, the permeability will be greater in the case of the coarse material. The least permeability, and also the least porosity, are, however, obtained by use of a suitable mixture of coarse and fine particles. Properly graded gravel or screenings, containing plenty of coarse fragments and also enough fine material to fill up the pores, will be found to give a much less permeable concrete than fine or coarse sand used alone.

2. Use of Rich Mixtures.—All concretes are somewhat permeable by water under sufficient pressure. Mixtures rich in cement are of course much less permeable than poorer mixtures. If the amount of cement used is more than sufficient to fill the voids in the sand and gravel, a very dense concrete is obtained, into which the penetration of water is extremely slow. The permeability also decreases considerably with age, owing to the gradual crystallization of the cement in the pores, so that concrete

which is at first quite absorbent may become practically impermeable after exposure to weather for a few weeks or months. There appears to be a very decided increase in permeability when the cement is reduced below the amount necessary to fill the voids. For example, a well-mixed sand and gravel weighing 123 pounds per cubic foot, and therefore containing 25 per cent voids, will give a fairly impermeable concrete in mixtures up to 1 to 4, but with less cement will be found quite absorbent. A gravel with only 20 per cent voids would give about equally good results with a 1 to 5 mixture; such gravel is, however, rarely met with in practice. On the other hand, the best sand, mixed fine and coarse, seldom contains less than 33 per cent voids, and concrete made from such material will prove permeable if poorer than 1 to 3.

3. Use of a Facing.—Penetration of water may be effectively prevented by giving the blocks a facing of richer mixture than the body. For the sake of smooth appearance, facings are generally made of cement and fine sand, and it is often noticed that these do not harden well. It should be remembered that a 1 to 3 sand mixture is no stronger and little if any better in water absorption than a 1 to 5 mixture of well-graded sand and gravel. To secure good hardness and resistance to moisture a facing as rich as 1 to 2 should be used.

4. Use of an Impervious Partition.— When blocks are made on a horizontal-face machine, it is a simple matter, after the face is tamped and cores pushed into place, to throw into each opening a small amount of rich and rather wet mortar, spread this fairly evenly, and then go on tamping in the ordinary mixture until the mold is filled. A dense layer across each of the cross walls is thus obtained, which effectually prevents moisture from passing beyond it. A method of accomplishing the same result with vertical-face machines, by inserting tapered wooden blocks in the middle of the cross walls, withdrawing these blocks after tamping, and filling the spaces with rich mortar, has been patented. In the two-piece system the penetration of moisture through the wall is prevented by leaving an empty space between the web of the block and the inside face, or by filling this space with rich mortar.

5. Use of Waterproof Compounds.— There are compounds on the market, of a fatty or waxy nature, which, when mixed with cement to the amount of

only 1 or 2 per cent of its weight, increase its water-resisting qualities in a remarkable degree. By thoroughly mixing 1 to 2 pounds of suitable compound with each sack of cement used, blocks which are practically waterproof may be made, at very small additional cost, from 1 to 4 or 1 to 5 mixtures. In purchasing waterproof compound, however, care should be taken to select such as has been proved to be *permanent* in its effect, and some of the materials used for this purpose lose their effect after a few days' exposure to weather, and are entirely worthless.

6. Application to Surface after Erecting.—Various washes, to make concrete and stone impervious to water, have been used with some success. Among these the best known is the Sylvester wash of alum and soap solution. It is stated that this requires frequent renewal, and it is hardly likely to prove of any value in the concrete industry. The writer's experience has been that the most effective remedy, in case a concrete building proves damp, is to give the outside walls a very thin wash of cement suspended in water. One or two coats will be found sufficient. If too thick a coating is formed it will show hair cracks. The effect of the cement wash is to make the walls appear lighter in color, and if the coating is thin the appearance is in no way injured.

General Hints on Waterproof Qualities.—To obtain good water-resisting properties the first precaution is to make the concrete sufficiently wet. Dry-tamped backs, even from rich mixture, will always be porous and absorbent, while the same mixture in plastic condition will give blocks which are dense, strong, and water-tight. The difference in this respect is shown by the following tests of small concrete blocks, made by the writer. The concrete used was made of 1 part cement and 5 parts mixed fine and coarse sand, by weight.

No. 1. With 8 per cent water, rather dryer than ordinary block concrete, tamped in mold.

No. 2. With 10 per cent water, tamped in the mold, and the mold removed at once.

No. 3. With 25 per cent water, poured into a mold resting on a flat surface of dry sand; after 1 hour the surface was troweled smooth; mold not removed until set.

These blocks were allowed to harden a week in moist air, then dried. The weights, voids, and water absorption were as follows:

	1 Damp-tamped	2 Wet-tamped	3 Poured
Weight, per cubic foot, pounds.....	122.2	123.9	110.0
Voids, calculated, per cent of volume	25.9	24.9	33.3
Water required to fill voids, per cent of weight.......	9.8	9.4	12.5
Water absorbed, after 2 hours, per cent of weight...	8.8	6.4	10.5

The rate at which these blocks absorbed water was then determined by drying them thoroughly, then placing them in a tray containing water $\frac{1}{4}$ inch in depth, and weighing them at intervals.

	1 Damp-tamped	2 Wet-tamped	3 Poured
$\frac{1}{2}$ hour..........	2.0	0.9	1.8
1 hour..........	3.2	1.1	2.5
2 hours.........	4.1	1.6	3.2
4 hours.........	5.2	2.0	3.8
24 hours.........	6.1	3.4	7.0
48 hours.........	6.4	4.3	7.5

These figures show that concrete which is sufficiently wet to be thoroughly plastic absorbs water much more slowly than dryer concrete, and prove the importance of using as much water as possible in the damp-tamping process.

Cost.—Concrete blocks can be sold and laid up at a good profit at 25 cents per cubic foot of wall. Common red brick costs (at this writing) generally about $12 per thousand, laid. At 24 to the cubic foot, a thousand brick are equal to 41.7 cubic foot of wall; or, $12, 29 cents per cubic foot. Brick walls with pressed brick facing cost from 40 cents to 50 cents per cubic foot, and dressed stone from $1 to $1.50 per foot.

The factory cost of concrete blocks varies according to the cost of materials. Let us assume cement to be $1.50 per barrel of 380 pounds, and sand and gravel 25 cents per ton. With a 1 to 4 mixture, 1 barrel cement will make 1,900 pounds of solid concrete, or at 130 pounds per cubic foot, 14.6 cubic feet. The cost of materials will then be:

Cement, 380 pounds........... $1.50
Sand and gravel, 1,500 pounds... 0.19

Total................... $1.69

or 11.5 cents per cubic foot solid concrete. Now, blocks 9 inches high and 32 inches long make 2 square feet of face of wall, each. Blocks of this height and length, 8 inches thick, make $1\frac{1}{3}$ cubic feet of wall; and blocks 12 inches thick make 2 cubic feet of wall. From these figures we may calculate the cost of materials for these blocks, with cores or openings equal to $\frac{1}{3}$ or $\frac{1}{2}$ the total volume, as follows:

Per cubic foot of block, $\frac{1}{3}$ opening	7.7 cts.
Per cubic foot of block, $\frac{1}{2}$ opening	5.8 cts.
Block 8 x 9 x 32 inches, $\frac{1}{3}$ opening	10.3 cts.
Block 8 x 9 x 32 inches, $\frac{1}{2}$ opening	7.7 cts.
Block 12 x 9 x 32 inches, $\frac{1}{3}$ opening	15.4 cts.
Block 12 x 9 x 32 inches, $\frac{1}{2}$ opening	11.6 cts.

If one-third of the cement is replaced by hydrate lime the quality of the blocks will be improved, and the cost of material reduced about 10 per cent. The cost of labor required in manufacturing, handling, and delivering blocks will vary with the locality and the size and equipment of factory. With hand mixing, 3 men at an average of $1.75 each will easily make 75 8-inch or 50 12-inch blocks, with $\frac{1}{3}$ openings, per day. The labor cost for these sizes of blocks will therefore be 7 cents and $10\frac{1}{2}$ cents respectively. At a factory equipped with power concrete mixer and cars for transporting blocks, in which a number of machines are kept busy, the labor cost will be considerably less. An extensive industry located in a large city is, however, subject to many expenses which are avoided in a small country plant, such as high wages, management, office rent, advertising, etc., so that the total cost of production is likely to be about the same in both cases. A fair estimate of total factory cost is as follows:

	Material	Labor	Total
8 x 32 inch, $\frac{1}{3}$ space	10.3	7	17.3 cts.
8 x 32 inch, $\frac{1}{2}$ space	7.7	6	13.7 cts.
12 x 32 inch, $\frac{1}{3}$ space	15.4	10.5	25.9 cts.
12 x 32 inch, $\frac{1}{2}$ space	11.6	9	20.6 cts.

With fair allowance for outside expenses and profit, 8-inch blocks may be sold at 30 cents and 12-inch at 40 cents each. For laying 12-inch blocks in the wall, contractors generally figure about 10 cents each. Adding 5 cents for teaming, the blocks will cost 55 cents each, erected, or $27\frac{1}{2}$ cents per cubic foot of wall. This is less than the cost of common brick, and the above figures show that this price could be shaded somewhat, if necessary, to meet competition.—*S. B. Newberry in a monograph issued by the American Association of Portland Cement Manufacturers.*

Artificial Marbles.—I.—The mass used by Beaumel consists of alum and heavy spar (barium sulphate) with addition of water and the requisite pigments. The following proportions have been found to be serviceable: Alum, 1,000 parts; heavy spar, 10 to 100 parts; water, 100 parts; the amount of heavy spar being governed by the degree of translucence desired. The alum is dissolved in water with the use of heat. As soon as the solution boils the heavy spar is mixed in, stirred with water and the pigment; this is then boiled down until the mixture has lost about 3 per cent of its weight, at which moment the mass exhibits a density of 34° Bé. at a temperature of 212° F. The mixture is allowed to cool with constant stirring until the substance is semi-liquid. The resultant mass is poured into a mold covered on the inside with several layers of collodion and the cast permitted to cool completely in the mold, whereupon it is taken out and dried entirely in an airy room. Subsequently the object may be polished, patinized, or finished in some other way.

II.—**Imitation Black Marble.**—A black marble of similar character to that exported from Belgium—the latter product being simply prepared slate—may be produced in the following manner: The slate suitable for the purpose is first smoothly polished with a sandstone, so that no visible impression is made on it with a chisel—this being rough—after which it is polished finely with artificial pumice stone, and lastly finished with extremely light natural pumice stone, the surface then presenting a soft, velvet-like appearance. After drying and thoroughly heating the finely polished surface is impregnated with a heated mixture of oil and fine lampblack. This is allowed to remain 12 hours; and, according to whether the slate used is more or less gray, the process is repeated until the gray appearance is lost. Polishing thoroughly with emery on a linen rag follows, and the finishing polish is done with tin ashes, to which is added some lampblack. A finish being made thus, wax dissolved in turpentine, with some lampblack, is spread on the polished plate and warmed again, which after a while is rubbed off vigorously with a clean linen rag. Treated thus, the slate has the appearance of black marble.

STONE CEMENTS:
See Adhesives.

STONE CLEANING:
See Cleaning Preparations and Methods.

STONES FOR SHARPENING:
See Tool Setting and Whetstones.

STONES (PRECIOUS), IMITATION OF:
See Gems, Artificial.

STONEWARE:
See Ceramics.

STONEWARE CEMENTS:
See Adhesives and Lutes.

STOPPERS.

I.—To make an anti-leak and lubricating mixture for plug-cocks use 2 parts of tried suet and 1 part of beeswax melted together; stir thoroughly, strain, and cool.

II.—A mixture for making glass stoppers tight is made by melting together equal parts of glycerine and paraffine.

To Loosen a Glass Stopper.—I.—Make a mixture of

Alcohol	2 drachms
Glycerine	1 drachm
Sodium chloride	1 drachm

Let a portion of this stand in the space above the stopper for a few hours, when a slight tap will loosen the stopper.

II.—A circular adjustable clamp, to which is attached a strip of asbestos in which coils of platinum wire are imbedded, is obtained. By placing this on the neck of the bottle, and passing a current of electricity through the coils of wire, sufficient heat will be generated to expand the neck and liberate the stopper. Heat may also be generated by passing a yard of cord once around the bottle neck and, by taking one end of the cord in each hand, drawing it rapidly back and forth. Care should be taken that the contents of the bottle are not spilled on the hand or thrown into the face when the stopper does come out—or when the bottle breaks.

STOPPER LUBRICANTS:
See Lubricants.

STOVE POLISH:
See also Polishes.

The following formula gives a liquid stove blacking:

Graphite, in fine powder	1 pound
Lampblack	1 ounce
Rosin	4 ounces
Turpentine	1 gallon

The mixture must be well shaken when used, and must not be applied when there is a fire or light near on account of the inflammability of the vapor.

This form may be esteemed a convenience by some, but the rosin and turpentine will, of course, give rise to some disagreeable odor on first heating the stove, after the liquid is applied.

Graphite is the foundation ingredient in many stove polishes; lampblack, which is sometimes added, as in the foregoing formula, deepens the color, but the latter form of carbon is of course much more readily burned off than the former. Graphite may be applied by merely mixing with water, and then no odor follows the heating of the iron. The coating must be well rubbed with a brush to obtain a good luster.

The solid cakes of stove polish found in the market are made by subjecting the powdered graphite, mixed with spirit of turpentine, to great pressure. They have to be reduced to powder and mixed with water before being applied.

Any of them must be well rubbed with a brush after application to give a handsome finish.

STOVE CEMENT:
See Cement.

STOVE CLEANERS:
See Cleaning Compounds.

STOVE LACQUER:
See Lacquers.

STOVE VARNISHES:
See Varnishes.

STRAMONIUM, ANTIDOTE FOR:
See Atropine.

STRAP LUBRICANT:
See Lubricant.

STRAW FIREPROOFING:
See Fireproofing.

STRAWBERRIES, FROZEN:
See Ice Creams.

STRAWBERRY JUICE:
See Essences and Extracts.

STRAW-HAT CLEANERS:
See Cleaning Preparations and Methods.

STRAW-HAT DYES:
See Hats.

STROPPING PASTES:
See Razor Pastes.

STYPTICS.

Styptics are substances which arrest local bleeding. Creosote, tannic acid, alcohol, alum, and most of the astringent salts belong to this class.

Brocchieri's Styptic.—A nostrum consisting of the water distilled from pine tops.

Helvetius's Styptic.—Iron filings (fine) and cream of tartar mixed to a proper consistence with French brandy.

Eaton's Styptic.—A solution of sulphate disguised by the addition of some unimportant substances. Helvetius's styptic was for a long time employed under this title.

Styptic Paste of Gutta Percha.—Gutta percha, 1 ounce; Stockholm tar, 1½ or 2 ounces; creosote, 1 drachm; shellac, 1 ounce; or quantity sufficient to render it sufficiently hard. To be boiled together with constant stirring, till it forms a homogeneous mass. For alveolar hemorrhage, and as a styptic in toothache. To be softened by molding with the fingers.

SULPHATE STAINS, TO REMOVE:
See Cleaning Preparations and Methods.

SULTANA ROLL:
See Ice Creams.

SUNBURN REMEDIES:
See Cosmetics.

SUTURES OF CATGUT, THEIR PREPARATION:
See Catgut.

SYNDETICON:
See Adhesives.

Syrups

(See also Essences and Extracts.)

The syrups should either be made from the best granulated sugar, free from ultramarine, or else rock-candy syrup. If the former, pure distilled water should be used in making the syrup, as only in this manner can a syrup be obtained that will be free from impurities and odor. There are two methods by which syrup can be made, namely, by the cold process, or by boiling. The advantage of the former is its convenience; of the latter, that it has better keeping qualities. In the cold process, the sugar is either stirred up in the water until it is dissolved, or water is percolated or filtered through the sugar, thus forming a solution. In the hot process, the sugar is simply dissolved in the water by the aid of heat, stirring until solution is effected. The strength of the syrup for fountain use should be about 6 pounds in the gallon of finished syrup; it is best, however, to make the stock syrup heavier, as it will keep much better, using 15 pounds of granulated sugar, and 1 gallon of water. When wanted for use it can be diluted to the proper density with water. The syrups of the market are of this concentrated variety. Unless the apartments of the dispenser are larger than is usual, it is often best to buy the syrup, the difference in cost being so small that when the time is taken into consideration the profit is entirely lost. Foamed syrups should, however, never be purchased; they are either contaminated with foreign flavor, or are more prone to fermentation than plain syrup.

Fruit Syrups.—These may be prepared from fruit juices, and the desired quantity of syrup, then adding soda foam, color, and generally a small amount of fruit-acid solution. They may also be made by reducing the concentrated fruit syrups of the market with syrup, otherwise proceeding as above. As the fruit juices and concentrated syrups always have a tried formula attached, it is needless to use space for this purpose.

When a flavor is weak it may be fortified by adding a small amount of flavoring extract, but under no condition should a syrup flavored entirely with an essence be handed out to the consumer as a fruit syrup, for there is really no great resemblance between the two. Fruit syrups may be dispensed solid by adding the syrup to the soda water and stirring with a spoon. Use nothing but the best ingredients in making syrups.

Preservation of Syrups.—The preservation of syrups is purely a pharmaceutical question. They must be made right in order to keep right. Syrups, particularly fruit syrups, must be kept aseptic, especially when made without heat. The containers should be made of glass, porcelain, or pure block tin, so that they may be sterilized, and should be easily and quickly removed, so that the operation may be effected with promptness and facility. As is well known, the operation of sterilization is very simple, consisting in scalding the article with boiling water. No syrup should ever be filled into a container without first sterilizing the container. The fruit acids, in the presence of sugar, serve as a media for the growth and development of germ life upon exposure to the air. Hence the employment of heat as pasteurization and sterilization in the preserving of fruits, etc.

A pure fruit syrup, filled into a glass bottle, porcelain jar, or block-tin can, which has been rendered sterile with boiling water, maintained at a cool temperature, will keep for any reasonable length of time. All danger of fracturing the glass, by pouring water into it, may be obviated by first wetting the interior of the bottle with cold water.

The fruits for syrups must not only be fully ripe, but they must be used immediately after gathering. The fruit must be freed from stems, seeds, etc., filled into lightly tied linen sacks, and thus subjected to pressure, to obtain their juices. Immediately after pressure the juice should be heated quickly to 167° F., and filtered through a felt bag. The filtrate should fall directly upon the sugar necessary to make it a syrup. The heating serves the purpose of coagulating the albuminous bodies present in the juices, and thus to purify the latter.

Syrups thus prepared have not only a most agreeable, fresh taste, but are very stable, remaining in a good condition for years.

Hints on Preparation of Syrups.— Keep the extracts in a cool, dark place. Never add flavoring extracts to hot syrup. It will cause them to evaporate, and weaken the flavor. Keep all the mixing utensils scrupulously clean. Never mix fruit syrups, nor let them stand in the same vessels in which sarsaparilla, ginger, and similar extract flavors are mixed and kept. If possible, always use distilled water in making syrup. Never allow a syrup containing acid to come in contact with any metal except pure block tin. Clean the syrup jars each time before refilling. Keep all packages of concentrated syrups and crushed fruits tightly corked. Mix only a small quantity of crushed fruit in the bowl at a time, so as to have it always fresh.

How to Make Simple Syrups—Hot Process. — Put 25 pounds granulated sugar in a large pail, or kettle, and pour on and stir hot water enough to make 4 gallons, more or less depending on how thick the syrup is desired. Then strain while hot through fine cheese cloth.

Cold Process.—By agitation. Sugar, 25 pounds; water, 2 gallons. Put the sugar in a container, add the water, and agitate with a wooden paddle until the sugar is dissolved. An earthenware jar with a cover and a faucet at the bottom makes a very convenient container.

Cold Process. — By percolation. A good, easy way to keep syrup on hand all the time: Have made a galvanized iron percolator, 2 feet long, 8 inches across top, and 4 inches at base, with a 4-inch wire sieve in bottom. Finish the bottom in shape of a funnel. Put a syrup faucet in a barrel, and set on a box, so that the syrup can be drawn into a gallon measure. Bore a hole in the barrel head, and insert the percolator. Fill three-fourths full of sugar, and fill with water. As fast as the syrup runs into the barrel fill the percolator, always putting in plenty of sugar. By this method 20 to 25 gallons heavy syrup can be made in a day.

Rock-Candy Syrup.—Sugar, 32 pounds; water, 2 gallons. Put the sugar and water in a suitable container, set on stove, and keep stirring until the mixture boils up once. Strain and allow to cool. When cool there will be on top a crust, or film, of crystallized sugar. Strain again to remove this film, and the product will be what is commonly known as rock-candy syrup. This may be reduced with one-fifth of its bulk of water when wanted for use.

COLORS FOR SYRUPS:

Caramel.—Place 3 pounds of crushed sugar in a kettle with 1 pint of water, and heat. The sugar will at first dissolve, but as the water evaporates a solid mass will be formed. This must be broken up.

Continue to heat, with constant stirring, until the mass has again become liquefied. Keep on a slow fire until the mass becomes very dark; then remove the kettle from the fire and pour in slowly 3 pints of boiling water. Set the kettle back on the fire and permit contents to boil for a short time, then remove, and cool. Add simple syrup to produce any required consistency.

Blue.—

I.—Indigo carmine......   1 part
      Water.............. 20 parts

Indigo carmine may usually be obtained commercially;

II.—Tincture of indigo also makes a harmless blue.

Sap Blue.—

    Dark blue.......... 3 parts
    Grape sugar........ 1 part
    Water............. 6 parts

Green.—The addition of indigo-carmine solution to any yellow solution will give various shades of green. Indigo carmine added to a mixture of tincture of crocus and glycerine will give a fine green color. A solution of commercial chlorophyll yields grass-green shades.

Pink.—

I.—Carmine............ 1 part
    Liquor potassæ...... 6 parts
    Rose water to make.. 48 parts

Mix. If the color is too high, dilute with distilled water until the required tint is obtained.

II.—Soak red-apple parings in California brandy. The addition of rose leaves makes a fine flavoring as well as coloring agent.

Red.—

    Carmine, No. 40.... 1 part
    Strong ammonia
        water............ 4 parts
    Distilled water to make 24 parts

Rub up the carmine and ammonia water and to the solution add the water under trituration. If, in standing, this shows a tendency to separate, a drop or two of water of ammonia will correct the trouble. This statement should be put on the label of the bottle as the volatile ammonia soon escapes even in glass-stoppered vials. Various shades of red may be obtained by using fruit juices, such as black cherry, raspberry, etc., and also the tinctures of sudbear, alkanet, red saunders, erythroxylon, etc.

Orange.—

    Tincture of red sandal-
        wood............. 1 part
    Ethereal tincture of Orlean, q. s.

Add the orlean tincture to the sandalwood gradually until the desired tint is obtained. A red color added to a yellow one gives an orange color.

Purple.—A mixture of tincture of indigo, or a solution of indigo carmine, added to cochineal red gives a fine purple.

Yellow.—Various shades of yellow may be obtained by the maceration of saffron or turmeric in alcohol until a strong tincture is obtained. Dilute with water until the desired tint is reached.

SYRUP, TABLE:
See Tables.

Tables

ALCOHOL DILUTION.

The following table gives the percentage, by weight, of alcohol of 95 per cent and of distilled water to make 1 liter (about 1 quart), or 1 kilogram (2.2 pounds), of alcohol of various dilutions.

TABLE FOR THE DILUTION OF ALCOHOL.

Percentage by Volume.	1 Liter contains		Specific Gravity at 60° F.	1 Kilogram contains		Percentage by Weight.
	Alcohol 95%.	Distilled Water.		Alcohol 95%.	Distilled Water.	
	Gms.	Gms.		Gms.	Gms.	
5	42.87	950.13	0.993	43.17	956.83	3.99
10	85.89	900.11	0.986	87.11	912.89	8.05
15	128.87	852.13	0.981	131.37	868.63	12.14
20	171.83	804.17	0.976	176.06	823.94	16.27
25	214.77	756.23	0.971	221.18	778.82	20.44
30	257.93	707.07	0.965	267.28	732.72	24.70
35	300.74	658.26	0.959	313.60	686.40	28.98
40	343.77	608.23	0.952	361.10	638.90	33.37
45	386.75	557.25	0.944	409.69	590.31	37.86
50	429.65	504.35	0.934	460.01	539.99	42.51
55	472.64	451.36	0.924	511.52	488.48	47.27
60	515.60	398.40	0.914	564.11	435.89	52.13
65	558.61	343.39	0.902	619.30	380.70	57.23
70	601.55	288.45	0.890	675.90	324.10	62.46
75	644.58	232.42	0.877	734.98	265.02	67.92
80	687.57	176.43	0.864	795.80	204.20	73.54
85	730.51	119.49	0.850	859.43	140.57	79.42
90	773.53	0.47	0.834	927.49	72.51	85.71

Capacities of Common Utensils.—For ordinary measuring purposes a wineglass may be said to hold 2 ounces.

A tablespoon, ½ ounce.
A dessertspoon, ¼ ounce.
A teaspoon, ⅛ ounce, or 1 drachm.
A teacupful of sugar weighs ½ pound.
Three tablespoonfuls weigh ¼ pound.

Cook's Table.—Two teacupfuls (well heaped) of coffee and of sugar weigh 1 pound.

Two teacupfuls (level) of granulated sugar weigh 1 pound.

Two teacupfuls soft butter (well packed) weigh 1 pound.

One and one-third pints of powdered sugar weigh 1 pound.

Two tablespoonfuls of powdered sugar or flour weigh 1 pound.

Four teaspoonfuls are equal to 1 tablespoon.

Two and one-half teacupfuls (level) of the best brown sugar weigh 1 pound.

Two and three-fourths teacupfuls (level) of powdered sugar weigh 1 pound.

One tablespoonful (well heaped) of granulated or best brown sugar equals 1 ounce.

One generous pint of liquid, or 1 pint finely chopped meat, packed solidly, weighs 1 pound.

Table of Drops.—Used in estimating the amount of a flavoring extract necessary to flavor a gallon of syrup. Based on the assumption of 450 drops being equal to 1 ounce.

One drop of extract to an ounce of syrup is equal to 2 drachms to a gallon.

Two drops of extract to an ounce of syrup are equal to $4\frac{1}{2}$ drachms to a gallon.

Three drops of extract to an ounce of syrup are equal to $6\frac{1}{2}$ drachms to a gallon.

Four drops of extract to an ounce of syrup are equal to 1 ounce and 1 drachm to a gallon.

Five drops of extract to an ounce of syrup are equal to 1 ounce and $3\frac{1}{8}$ drachms to a gallon.

Six drops of extract to an ounce of syrup are equal to 1 ounce and $5\frac{1}{4}$ drachms to a gallon.

Seven drops of extract to an ounce of syrup are equal to 2 ounces to the gallon.

Eight drops of extract to an ounce of syrup are equal to 2 ounces and $2\frac{1}{2}$ drachms to a gallon.

Nine drops of extract to an ounce of syrup are equal to 2 ounces and $4\frac{1}{2}$ drachms to a gallon.

Ten drops of extract to an ounce of syrup are equal to 2 ounces and $6\frac{3}{4}$ drachms to a gallon.

Twelve drops of extract to an ounce of syrup are equal to 3 ounces and $3\frac{1}{4}$ drachms to a gallon.

Fourteen drops of extract to an ounce of syrup are equal to 4 ounces to a gallon.

Sixteen drops of extract to an ounce of syrup are equal to 4 ounces and $4\frac{1}{8}$ drachms to a gallon.

Eighteen drops of extract to an ounce of syrup are equal to 5 ounces and 1 drachm to a gallon.

NOTE.—The estimate 450 drops to the ounce, while accurate and reliable enough in this particular relation, must not be relied upon for very exact purposes, in which, as has frequently been demonstrated, the drop varies within a very wide range, according to the nature of the liquid, its consistency, specific gravity, temperature; the size and shape of the aperture from which it is allowed to escape, etc.

Fluid Measure.—U. S. Standard, or **Wine Measure.**—Sixty minims are equal to 1 fluidrachm.

Eight fluidrachms are equal to 1 fluidounce.

Sixteen fluidounces are equal to 1 pint.

Two pints are equal to 1 quart.

Four quarts are equal to 1 gallon.

One pint of distilled water weighs about 1 pound.

Percentage Solutions.—To prepare the following approximately correct solutions, dissolve the amount of medicament indicated in sufficient water to make one imperial pint.

For $\frac{1}{50}$ per cent, or 1 in 5,000 solution, use $1\frac{3}{4}$ grains of the medicament.

For $\frac{1}{20}$ per cent, or 1 in 2,000 solution, use $4\frac{3}{8}$ grains of the medicament.

For $\frac{1}{10}$ per cent, or 1 in 1,000 solution, use $8\frac{3}{4}$ grains of the medicament.

For $\frac{1}{4}$ per cent, or 1 in 400 solution, use $21\frac{7}{8}$ grains of the medicament.

For $\frac{1}{2}$ per cent, or 1 in 200 solution, use $43\frac{3}{4}$ grains of the medicament.

For 1 per cent, or 1 in 100 solution, use $87\frac{1}{2}$ grains of the medicament.

For 2 per cent, or 1 in 50 solution, use 175 grains of the medicament.

For 4 per cent, or 1 in 25 solution, use 350 grains of the medicament.

For 5 per cent, or 1 in 20 solution, use $437\frac{1}{2}$ grains of the medicament.

For 10 per cent, or 1 in 10 solution, use 875 grains of the medicament.

To make smaller quantities of any solution, use less water and reduce the medicament in proportion to the amount of water employed; thus $\frac{1}{2}$ imperial pint of a 1 per cent solution will require $43\frac{3}{4}$ grains of the medicament.

Pressure Table.—This table shows the amount of commercial sulphuric acid (H_2SO_4) and sodium bicarbonate necessary to produce a given pressure:

120 Pounds Pressure.

Water, gallons	Soda Bicar., Av. ounces	Acid Sulph., Av. ounces
10	86	50
20	123	71
30	161	93
40	198	118
50	236	138

135 Pounds Pressure.

Water, gallons	Soda Bicar., Av. ounces	Acid Sulph., Av. ounces.
10	96	56
20	134	73
30	171	100
40	209	122
50	246	144

If marble dust be used, reckon at the rate of 18 ounces hot water for use.

Syrup Table.—The following table shows the amount of syrup obtained from

1. The addition of pounds of sugar to 1 gallon of water; and the

2. Amount of sugar in each gallon of syrup resulting therefrom:

Pounds of sugar added to one gallon of cold water.	Quantity of syrup actually obtained.			Pounds of sugar in one gallon of syrup.
	Gallons.	Pints.	Fluid-ounces.	
1	1	—	10	.93
2	1	1	4	1.73
3	1	1	14	2.43
4	1	2	3	3.05
5	1	3	2	3.6
6	1	3	12	4.09
7	1	4	6	4.52
8	1	5	—	4.92
9	1	5	10	5.28
10	1	6	4	5.62
11	1	6	14	5.92
12	1	7	8	6.18
13	2	—	2	6.38
14	2	—	12	6.7
15	2	1	6	6.91

TABLE-TOPS, ACID-PROOF:
See Acid-Proofing.

TABLES FOR PHOTOGRAPHERS:
See Photography.

TAFFY:
See Confectionery.

TALCUM POWDER:
See Cosmetics.

TALLOW:
See Fats.

TALMI GOLD:
See Alloys.

TAMPRING:
See Tampring, under Steel.

TAN REMEDY:
See Cosmetics.

TANK:

To Estimate Contents of a Circular Tank.—The capacity of a circular tank may be determined by multiplying the diameter in inches by itself and by .7854 and by the length (or depth) in inches, which gives the capacity of the tank in inches, and then dividing by 231, the number of cubic inches in a United States gallon.

TANNING:
See Leather.

TAPS, TO REMOVE BROKEN.

First clean the hole by means of a small squirt gun filled with kerosene. All broken pieces of the tap can be removed with a pair of tweezers, which should be as large as possible. Then insert the tweezers between the hole and flutes of the tap. By slowly working back and forth and occasionally blowing out with kerosene, the broken piece is easily released.

TAR PAINTS:
See Wood.

TAR-SPOTS ON WOODWORK:
See Paint.

TAR-SULPHUR SOAP:
See Soap.

TAR SYRUP:
See Essences and Extracts.

TATTOO MARKS, REMOVAL OF.

Apply a highly concentrated tannin solution on the tattooed places and treat them with the tattooing needle as the tattooer does. Next vigorously rub the places with a lunar caustic stick and allow the silver nitrate to act for some time, until the tattooed portions have turned entirely black. Then take off by dabbing. At first a silver tannate forms on the upper layers of the skin, which dyes the tattooing black; with slight symptoms of inflammation a scurf ensues which comes off after 14 to 16 days, leaving behind a reddish scar. The latter assumes the natural color of the skin after some time. The process is said to have given good results.

TAWING:
See Leather.

TEA EXTRACT:
See Essences and Extracts.

TEETH, TO WHITEN DISCOLORED.

Moisten the corner of a linen handkerchief with hydrogen peroxide, and with it rub the teeth, repeating the rubbing occasionally. Use some exceedingly finely pulverized infusorial earth, or pumice ground to an impalpable powder, in connection with the hydrogen peroxide, and the job will be quicker than with the peroxide alone.

TEMPERING OF STEEL:
See Steel.

TERRA COTTA SUBSTITUTE.

A substance, under this name, designed to take the place of terra cotta and plaster of Paris in the manufacture of small ornamental objects, consists of

Albumen............. 10 parts
Magnesium sulphate. 4 parts
Alum............... 9 parts
Calcium sulphate, cal-
   cined............. 45 parts
Borax.............. 2 parts
Water.............. 30 parts

The albumen and alum are dissolved in the water and with the solution so obtained the other ingredients are made into a paste. This paste is molded at once in the usual way and when set the articles are exposed in an oven to a heat of 140° F.

TERRA COTTA CLEANING:
See Cleaning Preparations and Methods.

TEXTILE CLEANING:
See Cleaning Preparations and Methods and Household Formulas.

Thermometers

Table Showing the Comparison of the Readings of Thermometers.

Celsius, or Centigrade (C). Réaumur (R). Fahrenheit (F).

C.	R.	F.	C.	R.	F.
−30	−24.0	−22.0	23	18.4	73.4
−25	−20.0	−13.0	24	19.2	75.2
−20	−16.0	− 4.0	25	20.0	77.0
−15	−12.0	+ 5.0	26	20.8	78.8
−10	− 8.0	14.0	27	21.6	80.6
− 5	− 4.0	23.0	28	22.4	82.4
− 4	− 3.2	24.8	29	23.2	84.2
− 3	− 2.4	26.6	30	24.0	86.0
− 2	− 1.6	28.4	31	24.8	87.8
− 1	− 0.8	30.2	32	25.6	89.6
			33	26.4	91.4
Freezing point of water.			34	27.2	93.2
			35	28.0	95.0
0	0.0	32.0	36	28.8	96.8
1	0.8	33.8	37	29.6	98.6
2	1.6	35.6	38	30.4	100.4
3	2.4	37.4	39	31.2	102.2
4	3.2	39.2	40	32.0	104.0
5	4.0	41.0	41	32.8	105.8
6	4.8	42.8	42	33.6	107.6
7	5.6	44.6	43	34.4	109.4
8	6.4	46.4	44	35.2	111.2
9	7.2	48.2	45	36.0	113.0
10	8.0	50.0	50	40.0	122.0
11	8.8	51.8	55	44.0	131.0
12	9.6	53.6	60	48.0	140.0
13	10.4	55.4	65	52.0	149.0
14	11.2	57.2	70	56.0	158.0
15	12.0	59.0	75	60.0	167.0
16	12.8	60.8	80	64.0	176.0
17	13.6	62.6	85	68.0	185.0
18	14.4	64.4	90	72.0	194.0
19	15.2	66.2	95	76.0	203.0
20	16.0	68.0	100	80.0	212.0
21	16.8	69.8			
22	17.6	71.6	Boiling point of water.		

Readings on one scale can be changed into another by the following formulas,

in which $t°$ indicates degrees of temperature:

Réau. to Fahr.	Cent. to Fahr.
$\frac{9}{4}t°\,R+32°=t°\,F$	$\frac{9}{5}t°\,C+32°=t°\,F$
Réau to Cent.	Cent. to Réau.
$\frac{5}{4}t°\,R=t°\,C$	$\frac{4}{5}t°\,C=t°\,R$

Fahr. to Cent.
$$\frac{5}{9}\left(t°\,F-32°\right)=t°\,C$$

Fahr. to Réau.
$$\frac{4}{9}\left(t°\,F-32°\right)=t°\,R$$

THREAD:
See also Cordage.

Dressing for Sewing Thread.—For colored thread: Irish moss, 3 pounds; gum arabic, 2½ pounds; Japan wax, ½ pound; stearine, 185 grams; borax, 95 grams; boil together for ¼ hour.

For white thread: Irish moss, 2 pounds; tapioca, 1½ pounds; spermaceti, ¾ pound; stearine, 110 grams; borax, 95 grams; boil together for 20 minutes.

For black thread: Irish moss, 3 pounds; gum Senegal, 2½ pounds; ceresin, 1 pound; borax, 95 grams; logwood extract, 95 grams; blue vitriol, 30 grams; boil together for 20 minutes. Soak the Irish moss in each case overnight in 45 liters of water, then boil for 1 hour, strain and add the other ingredients to the resulting solution. It is of advantage to add the borax to the Irish moss before the boiling.

THROAT LOZENGES:
See Confectionery.

THYMOL:
See Antiseptics.

TICKS, CATTLE DIP FOR:
See Insecticides.

TIERCES:
See Disinfectants.

TILEMAKERS' NOTES:
See Ceramics.

Tin

Etching Bath for Tin.—The design is either freely drawn upon the metal with a needle or a lead pencil, or pricked into the metal through tracing paper with a needle. The outlines are filled with a varnish (wax, colophony, asphalt). The varnish is rendered fluid with turpentine and applied with a brush. The article after having dried is laid in a ½ solution of nitric acid for 1½ to 2 hours. It is then washed and dried with blotting paper. The protective coating of asphalt is removed by heating. The zinc oxide in the deeper portions is cleaned away with a silver soap and brush.

Recovery of Tin and Iron in Tinned-Plate Clippings.—The process of utilizing tinned-plate scrap consists essentially in the removal of the tin. This must be very completely carried out if the remaining iron is to be available for casting. The removal of the outer layer of pure tin from the tinned plate is an easy matter. Beneath this, however, is another crystalline layer consisting of an alloy of tin and iron, which is more difficult of treatment. It renders the iron unavailable for casting, as even 0.2 per cent of tin causes brittleness. Its removal is best accomplished by electrolysis. If dilute sulphuric acid is used as an electrolyte, the deposit is spongy at first, and afterwards, when the acid has been partly neutralized, crystalline. After 6 hours the clippings are taken out and the iron completely dissolved in dilute sulphuric acid; the residue of tin is then combined with the tin obtained by the electrolysis. Green vitriol is therefore a by-product in this process.

Gutensohn's process has two objects: To obtain tin and to render the iron fit for use. The tin is obtained by treating the tinned plate repeatedly with hydrochloric acid. The tin is then removed from the solution by means of the electric current. The tinned plate as the positive pole is placed in a tank made of some insulating material impervious to the action of acids, such as slate. A copper plate forms the cathode. The bichloride of tin solution, freed from acid, is put round the carbon cylinder in the Bunsen element. Another innovation in this process is that the tank with the tinned-plate clippings is itself turned into an electric battery with the aid of the tin. A still better source of electricity is, however, obtained during the treatment of the untinned iron which will be described presently. The final elimination of the tin takes place in the clay cup of the Bunsen elements. Besides the chloride of tin solution (free from acid), another tin solution, preferably chromate of tin, nitrate of tin, or sulphate of tin, according to the strength of the current desired, may be used. To render the iron of the tinned plate serviceable the acid is drawn off as long as the iron is covered with a thin layer of an alloy of iron and tin. The latter makes the iron unfit for use in rolling mills or for the precipitation of copper. Fresh hydrochloric acid or sulphuric acid is therefore poured over the plate to remove the alloy, after the treatment with the bichloride of tin solution. This acid is also systematically used in different vats to the point of approximate saturation. This solution forms the most suitable source of electricity, a zinc-iron element being formed by means of a clay cell and a zinc cylinder. The electrical force developed serves to accelerate the solution in the next tank, which contains tinned plate, either fresh or treated with hydrochloric acid. Ferrous oxide, or spongy metallic iron if the current is very strong, is liberated in the iron battery. Both substances are easily oxidized, and form red oxide of iron when heated. The remaining solution can be crystallized by evaporation, so that ferrous sulphate (green vitriol) or ferric chloride can be obtained, or it can be treated to form red oxide of iron.

Tin in Powder Form.—To obtain tin in powder form the metal is first melted; next pour it into a box whose sides, etc., are coated with powdered chalk. Agitate the box vigorously and without discontinuing, until the metal is entirely cold. Now pass this powder through a sieve and keep in a closed flask. This tin powder is eligible for various uses and makes a handsome effect, especially in bronzing. It can be browned.

TINFOIL:
See also Metal Foil.

By pouring tin from a funnel with a very long and narrow mouth upon a linen surface, the latter being tightly stretched, covered with a mixture of chalk and white of egg, and placed in a sloping position, very thin sheets can be produced, and capable of being easily transformed into thin foil. Pure tin should never be used in the preparation of foil intended for packing tobacco, chocolate, etc., but an alloy containing 5 to 40 per cent of lead. Lead has also been recently plated on both sides with tin by the following method: A lead sheet from 0.64 to .80 inches thick is poured on a casting table as long as it is hot, a layer of tin from 0.16 to 0.20 inches in thickness added, the sheet then turned over and coated on the other side with tin in the same manner. The sheet is then stretched between rollers. Very thin sheet tin can also be made in the same way as sheet lead, by cutting up a tin cylinder into spiral sections. Colored tinfoil is prepared by making the foil thoroughly bright by rubbing with purified chalk and cotton, then adding a coat of gelatin, colored as required, and covering the whole finally with a transparent spirit varnish. In place of this somewhat troublesome process, the following much simpler method has lately been introduced: Aniline dyes dissolved in alcohol are applied on the purified foil, and the coat, when dry, covered with a very thin layer of a colorless varnish. This is done by pouring the varnish on the surface and then inclining the latter so that the varnish may reach every part and flow off.

TIN, SILVER-PLATING:
See Plating.

TIN VARNISHES:
See Varnishes.

TINNING:
See Plating.

TIRE:
Anti-Leak Rubber Tire.—Pneumatic tires can be made quite safe from punctures by using a liberal amount of the following cheap mixture: One pound of sheet glue dissolved in hot water in the usual manner, and 3 pints of molasses. This mixture injected into the tire through the valve stem, semi-hardens into an elastic jelly, being, in fact, about the same as the well-known ink roller composition used for the rollers of printing presses. This treatment will usually be found to effectually stop leaks in punctured or porous tires.

TIRE CEMENTS:
See Adhesives, under Rubber Cements.

TISSIER'S METAL:
See Alloys.

TITANIUM STEEL:
See Steel.

TODDY, HOT SODA:
See Beverages.

TOILET CREAMS, MILKS, POWDERS, ETC.:
See Cosmetics.

TOLIDOL DEVELOPER:
See Photography.

TOMATO BOUILLON EXTRACT:
See Condiments.

TOMBACK:
See Jewelers' Formulas.

TONING BATHS:
See Photography.

TONKA EXTRACT:
See Essences and Extracts.

TONKA, ITS DETECTION IN VANILLA EXTRACTS:
See Vanilla.

TOOL SETTING.

The term "setting" (grinding) is applied to the operation of giving an edge to the tools designed for cutting, scraping, or sawing. Cutting tools are rubbed either on flat sandstones or on rapidly turned grindstones. The wear on the faces of the tools diminishes their thickness and renders the cutting angle sharper. Good edges cannot be obtained except with the aid of the grindstone; it is therefore important to select this instrument with care. It should be soft, rather than hard, of fine, smooth grain, perfectly free from seams or flaws. The last condition is essential, for it often happens that, under the influence of the revolving motion, a defective stone suddenly yields to the centrifugal force, bursts and scatters its pieces with such violence as to wound the operator. This accident may also happen with perfectly formed stones. On this account artificial stones have been substituted, more homogeneous and coherent than the natural ones.

Whatever may be the stone selected, it ought to be kept constantly moist during the operation. If not, the tools will soon get heated and their temper will be impaired. When a tool has for a certain time undergone the erosive action of the stone, the cutting angle becomes too acute, too thin, and bends over on itself, constituting what is called "the feather edge." This condition renders a new setting necessary, which is usually effected by bending back the feather edge, if it is long, and whetting the blade on a stone called a "setter." There are several varieties of stones used for this purpose, though they are mostly composed of calcareous or argilaceous matter, mixed with a certain proportion of silica.

The scythestone, of very fine grain, serves for grinding off the feather edge of scythes, knives, and other large tools. The Lorraine stone, of chocolate color and fine grain, is employed with oil for carpenters' tools. American carborundum is very erosive. It is used with water and with oil to obtain a fine edge. The lancet stone is not inferior to any of the preceding. As its name indicates, it is used for sharpening surgical instruments, and only with oil. The Levant stone (Turkish sandstone) is the best of all for whetting. It is gray and semitransparent; when of inferior quality, it is somewhat spotted with red. It is usually quite soft.

To restore stones and efface the inequalities and hollows caused by the friction of the tools, they are laid flat on a marble or level stone, spread over with fine, well-pulverized sandstone, and rubbed briskly. When tools have a curved edge, they are subjected to a composition formed of pulverized stone, molded into a form convenient for the concavity or convexity. Tools are also whetted with slabs of walnut or aspen wood coated with emery of different numbers, which produces an excellent setting.

TOOL LUBRICANT:
See Lubricant.

Toothache

TOOTHACHE GUMS:
See also Pain Killers.

I.—Paraffine......... 94 grains
Burgundy pitch...800 grains
Oil of cloves...... $\frac{1}{2}$ fluidrachm
Creosote......... $\frac{1}{2}$ fluidrachm

Melt the first two ingredients, and, when nearly cool, add the rest, stirring well. May be made into small pills or turned out in form of small cones or cylinders.

II.—Melt white wax or spermaceti, 2 parts, and when melted add carbolic-acid crystals, 1 part, and chloral-hydrate crystals, 2 parts; stir well until dissolved. While still liquid, immerse thin layers of carbolized absorbent cotton wool and allow them to dry. When required for use a small piece may be snipped off and slightly warmed, when it can be inserted into the hollow tooth, where it will solidify.

Toothache Remedy.—

Camphor........ 4 drachms
Chloral hydrate.. 4 drachms
Oil of cloves..... 2 drachms
Oil of cajeput.... 2 drachms
Chloroform...... 12 drachms
Tincture of capsicum......... 24 drachms

TOOTH CEMENTS:
See Cements.

TOOTH PASTES, POWDERS, SOAPS, AND WASHES:
See Dentifrices.

TORTOISE-SHELL POLISHES:
See Polishes.

TOOTH STRAIGHTENING:
See Watchmakers' Formulas.

TOUCHSTONE, AQUAFORTIS FOR THE:
See Aquafortis.

TOY PAINT:
See Paint.

TRACING-CLOTH CLEANERS:
See Cleaning Preparations and Methods.

TRAGACANTH, MUCILAGE OF:
See Adhesives, under Mucilages.

TRANSPARENCIES:
See also Photography.

A good method of preparing handsome London transparencies is as follows:

White paper is coated with a liquid whose chief constituent is Iceland moss strongly boiled down in water to which a slight quantity of previously dissolved gelatin is added. In applying the mass, which should always be kept in a hot condition, the paper should be covered uniformly throughout. After it has been dried well it is smoothed on the coated side and used for a proof. The transparent colors to be used must be ground in stronger varnish than the opaque ones. In order to produce a handsome red, yellow lake and red sienna are used; the tone of the latter is considerably warmer than that of the yellow lake. Where the cost is no consideration, aurosolin may also be employed. For pale red, madder lakes should be employed, but for darker shades, crimson lakes and scarlet cochineal lakes. The vivid geranium lake gives a magnificent shade, which, however, is not at all fast in sunlight. The most translucent blue will always be Berlin blue. For purple, madder purple is the most reliable color, but possesses little gloss. Luminous effects can be obtained with the assistance of aniline colors, but these are only of little permanence in transparencies. Light, transparent green is hardly available. Recourse has to be taken to mixing Berlin blue with yellow lake, or red sienna. Green chromic oxide may be used if its sober, cool tone has no disturbing influence. Almost all brown coloring bodies give transparent colors, but the most useful are madder lakes and burnt umber. Gray is produced by mixing purple tone colors with suitable brown, but a gray color hardly ever occurs in transparent prints. Liquid siccative must always be added to the colors, otherwise the drying will occupy too much time. After the drying, the prints are varnished on both sides. For this purpose, a well-covering, quickly drying, colorless, not too thick varnish must be used, which is elastic enough not to crack nor to break in bending.

Frequently the varnishing of the placards is done with gelatin. This imparts to the picture an especially handsome, luminous luster. After an equal quantity of alcohol has been added to a readily flowing solution of gelatin, kept for use in a zinc vessel, the gelatin solution is poured on the glass plates destined for the transparencies. After a quarter of an hour, take the placard, moisten its back uniformly, and lay it upon a gelatin film which has meanwhile formed on the glass plate, where it remains 2 to 3 days. When it is to be removed from the plate, the edge of the gelatin film protruding over the edge of the placard is lifted up with a dull knife, and it is thus drawn off. A fine, transparent gloss remains on the placard proper. In order to render the covering waterproof and pliable, it is given a coating of collodion, which does not detract from the transparence. The glass plates and their frames must be cleaned of adhering gelatin particles before renewed use.

TRANSFER PROCESSES:

To Transfer Designs.—Designs can be transferred on painted surfaces, cloth, leather, velvet, oil cloth, and linen sharply and in all the details with little trouble. Take the original design, lay it on a layer of paper, and trace the lines of design accurately with a packing needle, the eye of which is held by a piece of wood for a handle. It is necessary to press down well. The design becomes visible on the back by an elevation. When everything has been accurately pressed through, take, e. g., for dark objects, whiting (formed in pieces), lay the design face downward on the knee and pass mildly with the whiting over the elevations; on every elevation a chalk line will appear. Then dust off the superfluous whiting with the fingers, lay the whiting side on the cloth to hold it so that it cannot slide, and pass over it with a soft brush. For light articles take powdered lead pencil, which is rubbed on with the finger, or limewood charcoal. For tracing use oil paint on cloth and India ink on linen.

To Copy Engravings.—To make a facsimile of an engraving expose it in a warm, closed box to the vapor of iodine, then place it, inkside downward, on a smooth, dry sheet of clean white paper, which has been brushed with starch water. After the two prepared surfaces have been in contact for a short time a facsimile of the engraving will be reproduced more or less accurately, according to the skill of the operator.

To Transfer Engravings.—The best way to transfer engraving from one piece to another is to rub transfer wax into the engraved letters. This wax is made of beeswax, 3 parts; tallow, 3 parts; Canada balsam, 1 part; olive oil, 1 part. If the wax becomes too hard, add a few drops of olive oil, and if too soft, a little more beeswax. Care should be taken that the wax does not remain on the surface about the engraving, otherwise the impression would be blurred. Then moisten a piece of paper by drawing it over the tongue and lay it on the engraving. Upon this is laid another piece of dry paper, and securing both with the thumb and forefinger of the left hand, so they will not be moved, go over the entire surface with a burnisher made of steel or bone, with a pointed end. This will press the lower paper into the engraving and cause the wax to adhere to it. Then the top paper is removed and the corner of the lower one gently raised. The whole is then carefully peeled off, and underneath will be found a reversed, sharp impression of the engraving. The edges of the paper are then cut so it can be fitted in a position on the other articles similar to that on the original one. When this is done lay the paper in the proper position and rub the index finger lightly over it, which will transfer a clear likeness of the original engraving. If due care is taken two dozen or more transfers can be made from a single impression.

TRICKS WITH FIRE:
See Pyrotechnics.

TUBERS, THEIR PRESERVATION:
See Roots.

TUBS: TO RENDER SHRUNKEN TUBS WATER-TIGHT:
See Casks.

TUNGSTEN STEEL:
See Steel.

TURMERIC IN FOOD:
See Foods.

TURPENTINE STAINS:
See Wood.

TURTLE (MOCK) EXTRACT:
See Condiments.

TWINE:
See also Thread and Cordage.

Tough twine may be greatly strengthened by dissolving plenty of alum in water and laying the twine in this solution. After drying, the twine will have much increased tensile strength.

Typewriter Ribbons

(See also Inks.)

The constituents of an ink for typewriter ribbons may be broadly divided into four elements: 1, the pigment; 2, the vehicle; 3, the corrigent; 4, the solvent. The elements will differ with the kind of ink desired, whether permanent or copying.

Permanent (Record) Ink.—Any finely divided, non-fading color may be used as the pigment; vaseline is the best vehicle and wax the best corrigent. In order to make the ribbon last a long time with one inking, as much pigment as feasible should be used. To make black record ink: Take some vaseline, melt it on a slow fire or water bath, and incorporate by constant stirring as much lampblack as it will take up without becoming granular. Take from the fire and allow it to cool. The ink is now practically finished, except, if not entirely suitable on trial, it may be improved by adding the corrigent wax in small quantity. The ribbon should be charged with a very thin, evenly divided amount of ink. Hence the necessity of a solvent—in this instance a mixture of equal parts of petroleum benzine and rectified spirit of turpentine. In this mixture dissolve a sufficient amount of the solid ink by vigorous agitation to make a thin paint. Try the ink on one extremity of the ribbon; if too soft, add a little wax to make it harder; if too pale, add more coloring matter; if too hard, add more vaseline. If carefully applied to the ribbon, and the excess brushed off, the result will be satisfactory.

On the same principle, other colors may be made into ink; but for delicate colors, albolene and bleached wax should be the vehicle and corrigent, respectively.

The various printing inks may be used if properly corrected. They require the addition of vaseline to make them non-drying on the ribbon, and of some wax if found too soft. Where printing inks are available, they will be found to give excellent results if thus modified, as the pigment is well milled and finely divided. Even black cosmetic may be made to answer, by the addition of some lampblack to the solution in the mixture of benzine and turpentine.

After thus having explained the principles underlying the manufacture of permanent inks, we can pass more rapidly over the subject of copying inks, which is governed by the same general rules.

For copying inks, aniline colors form the pigment; a mixture of about 3 parts of water and 1 part of glycerine, the vehicle; transparent soap (about $\frac{1}{4}$ part), the corrigent; stronger alcohol (about 6 parts), the solvent. The desired aniline color will easily dissolve in the hot vehicle, soap will give the ink the necessary body and counteract the hygroscopic tendency of the glycerine, and in the stronger alcohol the ink will readily dissolve, so that it can be applied in a finely divided state to the ribbon, where the evaporation of the alcohol will leave it in a thin film. There is little more to add. After the ink is made and tried—if too soft, add a little more soap; if too hard, a little more glycerine; if too pale, a little more pigment. Printer's copying ink can be utilized here likewise.

Users of the typewriter should so set a fresh ribbon as to start at the edge nearest the operator, allowing it to run back and forth with the same adjustment until exhausted along that strip; then shift the ribbon forward the width of one letter, running until exhausted, and so on. Finally, when the whole ribbon is exhausted, the color will have been equably used up, and on reinking, the work will appear even in color, while it will look patchy if some of the old ink has been left here and there and fresh ink applied over it.

UDDER INFLAMMATION:
See Veterinary Formulas.

VALVES.

The manufacturers of valves test each valve under hydraulic pressure before it is sent out from the factory, yet they frequently leak when erected in the pipe lines. This is due to the misuse of the erector in most cases. The following are the most noteworthy bad practices to be avoided when fitting in valves:

I.—Screwing a valve on a pipe very tightly, without first closing the valve. Closing the valve makes the body much more rigid and able to withstand greater strains and also keeps the iron chips from lodging under the seats, or in the working parts of the valves. This, of course, does not apply to check valves.

II.—Screwing a long mill thread into a valve. The threads on commercial pipes are very long and should never be screwed into a valve. An elbow or tee will stand the length of thread very well, but a suitable length thread should be cut in every case on the pipe, when used to screw into a valve. If not, the end of pipe will shoulder against the seat of valve and so distort it that the valve will leak very badly.

III.—The application of a pipe wrench on the opposite end of the valve from the end which is being screwed on the pipe. This should never be done, as it invariably springs or forces the valve seats from their true original bearing with the disks.

IV.—Never place the body of a valve in the vise to remove the bonnet or center-piece from a valve, as it will squeeze together the soft brass body and throw all parts out of alignment. Properly to remove the bonnet or centerpiece from a valve, either screw into each end of the valve a short piece of pipe and place one piece of the pipe in the vise, using a wrench on the square of bonnet; or if the vise is properly constructed, place the square of the bonnet in same and use the short piece of pipe screwed in each end as a lever. When using a wrench on square of bonnet or centerpiece, use a Stillson or Trimo wrench with a piece of tin between the teeth of the jaws and the finished brass. It may mark the brass slightly, but this is preferable to rounding off all the corners with an old monkey wrench which is worn out and sprung. As the threads on all bonnets or center-pieces are doped with litharge or cement, a sharp jerk or jar on the wrench will start the bonnet much more quickly than a steady pull. Under no circumstances try to replace or remove the bonnet or centerpiece of a valve without first opening it wide. This will prevent the bending of the stem, forcing the disk down through the seat or stripping the threads on bonnet where it screws into body. If it is impossible to remove bonnet or centerpiece by ordinary methods, heat the body of the valve just outside the thread. Then tap lightly all around the thread with a soft hammer. This method never fails, as the heat expands the body ring and breaks the joint made by the litharge or cement.

V.—The application of a large monkey wrench to the stuffing box of valve. Many valves are returned with the stuffing boxes split, or the threads in same stripped. This is due to the fact that the fitter or engineer has used a large-sized monkey wrench on this small part.

VI.—The screwing into a valve of a long length of unsupported pipe. For example, if the fitter is doing some repair work and starts out with a run of 2-inch horizontal pipe from a 2-inch valve connected to main steam header, the pipe being about 18 feet long, after he has screwed the pipe tightly into the valve, he leaves the helper to support the pipe at the other end, while he gets the hanger ready. The helper in the meantime has become tired and drops his shoulder on which the pipe rests about 3 inches and in consequence the full weight of this 18-foot length of pipe bears on the valve. The valve is badly sprung and when the engineer raises steam the next morning the valve leaks. When a valve is placed in the center of a long run of pipe, the pipe on each side, and close to the valve, should be well supported.

VII.—The use of pipe cement in valves. When it is necessary to use pipe cement in joints, this mixture should always be placed on the pipe thread which screws into the valve, and never in the valve itself. If the cement is placed in the valve, as the pipe is screwed into the valve it forces the cement between the seats and disks, where it will soon harden and thus prevent the valve from seating properly.

VIII.—Thread chips and scale in pipe. Before a pipe is screwed into a valve it should be stood in a vertical position and struck sharply with a hammer. This will release the chips from the thread cutting, and loosen the scale inside of pipe. When a pipe line containing valves is connected up, the valves should all be opened wide and the pipe well blown out before they are again closed. This will remove foreign substances which are liable to cut and scratch the seats and disks.

IX. — Expansion and contraction. Ample allowance must be provided for expansion and contraction in all steam lines, especially when brass valves are included. The pipe and fittings are much more rigid and stiff than the brass valves and in consequence the expansion strains will relieve themselves at the weakest point, unless otherwise provided for.

X.—The use of wrenches or bars on valve wheels to close the valves tightly. This should never be done, as it springs the entire valve and throws all parts out of alignment, thus making the valve leak. The manufacturer furnishes a wheel sufficiently large properly to close against any pressure for which it is suitable. If the valves cannot be closed tightly by this means, there is something between the disks and seats or they have been cut or scratched by foreign substances.

Vanilla

(See also Essences and Extracts.)

The best Mexican vanilla yields only in the neighborhood of 1.7 per cent of vanillin; that from Reunion and Guadeloupe about 2.5 per cent; and that from Java 2.75 per cent. There seems to be but little connection between the quantity of vanillin contained in vanilla pods and their quality as a flavor producer. Mexican beans are esteemed the best and yet they contain far less than the Java. Those from Brazil and Peru contain much less than those from Mexico, and yet they are considered inferior in quality to most others. The vanillin of the market is chiefly, if not entirely, artificial and is made from the coniferin of such pines and firs as abies excelsa, a. pectinata, pinus cembra, and p. strobus, as well as from the eugenol of cloves and allspice. Vanillin also exists in asparagus, lupine seeds, the seeds of the common wild rose, asafetida, and gum benzoin.

A good formula for a vanilla extract is the following:

Vanilla..........	1 ounce
Tonka...........	2 ounces
Alcohol, deodorized..........	32 fluidounces
Syrup...........	8 fluidounces

Cut and bruise the vanilla, afterwards adding and bruising the Tonka; macerate for 14 days in 16 fluidounces of the alcohol, with occasional agitation; pour off the clear liquid and set aside; pour the remaining alcohol on the magma, and heat by means of a water bath to about 168° F., in a closely covered vessel. Keep it at that temperature for 2 or 3 hours, then strain through flannel with slight pressure; mix the two portions of liquid and filter through felt. Lastly, add the syrup. To render this tincture perfectly clear it may be treated with pulverized magnesium carbonate, using from ½ to 1 drachm to each pint.

To Detect Artificial Vanillin in Vanilla Extracts (see also Foods).—There is no well-defined test for vanillin, but one can get at it in a negative way. The artificial vanillin contains vanillin identical with the vanillin contained in the vanilla bean; but the vanilla bean, as the vanilla extract, contains among its many "extractive matters" which enter into the food and fragrant value of vanilla extract, certain rosins which can be identified with certainty in analysis by a number of determining reactions. Extract made without true vanilla can be detected by negative results in all these reactions.

Vanilla beans contain 4 to 11 per cent of this rosin. It is of a dark red to brown color and furnishes about one-half the color of the extract of vanilla. This rosin is soluble in 50 per cent alcohol, so that in extracts of high grade, where sufficient alcohol is used, all rosin is kept in solution. In cheap extracts, where as little as 20 per cent of alcohol by volume is sometimes used, an alkali—usually potassium bicarbonate—is added to aid in getting rosin, gums, etc., in solution, and to prevent subsequent turpidity. This treatment deepens the color very materially.

Place some of the extract to be examined in a glass evaporating dish and evaporate the alcohol on the water bath. When alcohol is removed, make up about the original volume with hot water. If alkali has not been used in the manufacture of the extract, the rosin will appear as a flocculent red to brown residue. Acidify with acetic acid to free rosin from bases, separating the whole of the rosin and leaving a partly decolorized, clear supernatant liquid after standing a short time. Collect the rosin on a filter, wash with water, and reserve the filtrate for further tests.

Place a portion of the filter with the attached rosin in a few cubic centimeters of dilute caustic potash. The rosin is dissolved to a deep-red solution. Acidify. The rosin is thereby precipitated. Dissolve a portion of the rosin in alcohol; to one fraction add a few drops of ferric chloride; no striking coloration is produced. To another portion add hydrochloric acid; again there is little change in color. In alcoholic solution most rosins give color reactions with ferric chloride or hydrochloric acid. To a portion of the filtrate obtained above add a few drops of basic lead acetate. The precipitate is so bulky as to almost solidify, due to the excessive amount of organic acids, gums, and other extractive matter. The filtrate from this precipitate is nearly, but not quite, colorless. Test another portion of the filtrate from the rosin for tannin with a solution of gelatin. Tannin is present in varying but small quantities. It should not be present in great excess.

To Detect Tonka in Vanilla Extract.— The following test depends on the chemical difference between coumarin and vanillin, the odorous principles of the two beans. Coumarin is the anhydride of coumaric acid, and on fusion with a caustic alkali yields acetic and salicylic acids, while vanillin is methyl protocatechin aldehyde, and when treated similarly yields protocatechuic acid. The test is performed by evaporating a small quantity of the extract to dryness, and melting the residue with caustic potash. Transfer the fused mass to a test tube, neutralize with hydrochloric acid, and add a few drops of ferric chloride solution. If Tonka be present in the extract, the beautiful violet coloration characteristic of salicylic acid will at once become evident.

Vanilla Substitute.— A substitute for vanilla extract is made from synthetic vanillin. The vanillin is simply dissolved in diluted alcohol and the solution colored with a little caramel and sweetened perhaps with syrup. The following is a typical formula:

Vanillin...........	1 ounce
Alcohol...........	6 quarts
Water.............	5 quarts
Syrup.............	1 quart
Caramel sufficient to color.	

An extract so made does not wholly represent the flavor of the bean; while vanillin is the chief flavoring constituent of the bean, there are present other substances which contribute to the flavor; and connoisseurs prefer this combination, the remaining members of which have not yet been made artificially.

VANILLIN:
See Vanilla.

Varnishes

(See also Enamels, Glazes, Oils, Paints, Rust Preventives, Stains, and Waterproofing.)

Varnish is a solution of resinous matter forming a clear, limpid fluid capable of hardening without losing its transparency. It is used to give a shining, transparent, hard, and preservative covering to the finished surface of woodwork, capable of resisting in a greater or less degree the influence of the air and moisture. This coating, when applied to metal or mineral surfaces, takes the name of lacquer, and must be prepared from rosins at once more adhesive and tenacious than those entering into varnish.

The rosins, commonly called gums, suitable for varnish are of two kinds— the hard and the soft. The hard varieties are copal, amber, and the lac rosins. The dry soft rosins are juniper gum (commonly called sandarac), mastic, and dammar. The elastic soft rosins are benzoin, elemi, anime, and turpentine. The science of preparing varnish consists in combining these classes of rosins in a suitable solvent, so that each conveys its good qualities and counteracts the bad ones of the others, and in giving the desired color to this solution without affecting the suspension of the rosins, or detracting from the drying and hardening properties of the varnish.

In spirit varnish (that made with alcohol) the hard and the elastic gums must be mixed to insure tenderness and solidity, as the alcohol evaporates at once after applying, leaving the varnish wholly dependent on the gums for the tenacious and adhesive properties; and if the soft rosins predominate, the varnish will remain "tacky" for a long time. Spirit varnish, however good and convenient to work with, must always be inferior to oil varnish, as the latter is at the same time more tender and more solid, for the oil in oxidizing and evaporating thickens and forms rosin which continues its softening and binding presence, whereas in a spirit varnish the alcohol is promptly dissipated, and leaves the gums on the surface of the work in a more or less granular and brittle precipitate which chips readily and peels off.

Varnish must be tender and in a manner soft. It must yield to the movements of the wood in expanding or contracting with the heat or cold, and must not inclose the wood like a sheet of glass. This is why oil varnish is superior to spirit varnish. To obtain this suppleness the gums must be dissolved in some liquid not highly volatile like spirit, but one which mixes with them in substance permanently to counteract their extreme friability. Such solvents are the oils of lavender, spike, rosemary, and turpentine, combined with linseed oil. The vehicle in which the rosins are dissolved must be soft and remain so in order to keep the rosins soft which are of themselves naturally hard. Any varnish from which the solvent has completely dried out must of necessity become hard and glassy and chip off. But, on the other hand, if the varnish remains too soft and "tacky," it will "cake" in time and destroy the effect desired.

Aside from this, close observers if not chemists will agree that for this work it is much more desirable to dissolve these rosins in a liquid closely related to them in chemical composition, rather than in a liquid of no chemical relation and which no doubt changes certain properties of the rosins, and cuts them into solution more sharply than does turpentine or linseed oil. It is a well-known fact that each time glue is liquefied it loses some of its adhesive properties. On this same principle it is not desirable to dissolve varnish rosins in a liquid very unlike them, nor in one in which they are quickly and highly soluble. Modern effort has been bent on inventing a cheap varnish, easily prepared, that will take the place of oil varnish, and the market is flooded with benzine, carbon bisulphide, and various ether products which are next to worthless where wearing and durable properties are desired.

Alcohol will hold in solution only about one-third of its weight in rosins. Turpentine must be added always last to spirit varnish. Turpentine in its clear recently distilled state will not mix with alcohol, but must first be oxidized by exposing it to the air in an uncorked bottle until a small quantity taken therefrom mixes perfectly with alcohol. This usually takes from a month to six weeks. Mastic must be added last of all to the ingredients of spirit varnish, as it is not wholly soluble in alcohol but entirely so in a solution of rosins in alcohol. Spirit varnishes that prove too hard and brittle may be improved by the addition of either of the oils of turpentine, castor seed, lavender, rosemary, or spike, in the proportion required to bring the varnish to the proper temper.

Coloring "Spirit" Varnishes. — In modern works the following coloring substances are used, separately and in blends: Saffron (brilliant golden yellow), dragon's blood (deep reddish brown), gamboge (bright yellow), Socotrine or Bombay aloes (liver brown), asphalt, ivory, and bone black (black), sandalwood, pterocarpus santalinus, the heartwood (dark red), Indian sandalwood, pterocarpus indica, the heartwood (orange red), brazil wood (dark

yellow), myrrh (yellowish to reddish brown; darkens on exposure), madder (reddish brown), logwood (brown), red scammony rosin (light red), turmeric (orange yellow), and many others according to the various shades desired.

Manufacturing Hints.—Glass, coarsely powdered, is often added to varnish when mixed in large quantities for the purpose of cutting the rosins and preventing them from adhering to the bottom and sides of the container. When possible, varnish should always be compounded without the use of heat, as this carbonizes and otherwise changes the constituents, and, besides, danger always ensues from the highly inflammable nature of the material employed. However, when heat is necessary, a water bath should always be used; the varnish should never fill the vessel over a half to three-fourths of its capacity.

The Gums Used in Making Varnish.— Juniper gum or true sandarac comes in long, yellowish, dusty tears, and requires a high temperature for its manipulation in oil. The oil must be so hot as to scorch a feather dipped into it, before this gum is added; otherwise the gum is burned. Because of this, juniper gum is usually displaced in oil varnish by gum dammar. Both of these gums, by their dryness, counteract the elasticity of oil as well as of other gums. The usual sandarac of commerce is a brittle, yellow, transparent rosin from Africa, more soluble in turpentine than in alcohol. Its excess renders varnish hard and brittle. Commercial sandarac is also often a mixture of the African rosin with dammar or hard Indian copal, the place of the African rosin being sometimes taken by true juniper gum. This mixture is the pounce of the shops, and is almost insoluble in alcohol or turpentine. Dammar also largely takes the place of tender copal, gum anime, white amber, white incense, and white rosin. The latter three names are also often applied to a mixture of oil and Grecian wax, sometimes used in varnish. When gum dammar is used as the main rosin in a varnish, it should be first fused and brought to a boiling point, but not thawed. This eliminates the property that renders dammar varnish soft and "tacky" if not treated as above.

Venetian turpentine has a tendency to render varnish "tacky" and must be skillfully counteracted if this effect is to be avoided. Benzoin in varnish exposed to any degree of dampness has a tendency to swell, and must in such cases be avoided. Elemi, a fragrant rosin from Egypt, in time grows hard and brittle, and is not so soluble in alcohol as anime, which is highly esteemed for its more tender qualities. Copal is a name given rather indiscriminately to various gums and rosins. The East Indian or African is the tender copal, and is softer and more transparent than the other varieties; when pure it is freely soluble in oil of turpentine or rosemary. Hard copal comes in its best form from Mexico, and is not readily soluble in oil unless first fused. The brilliant, deep-red color of old varnish is said to be based on dragon's blood, but not the kind that comes in sticks, cones, etc. (which is always adulterated), but the clear, pure tear, deeper in color than a carbuncle, and as crystal as a ruby. This is seldom seen in the market, as is also the tear of gamboge, which, mixed with the tear of dragon's blood, is said to be the basis of the brilliant orange and gold varnish of the ancients.

Of all applications used to adorn and protect the surface of objects, oil varnishes or lacquers containing hard rosins are the best, as they furnish a hard, glossy coating which does not crack and is very durable even when exposed to wind and rain.

To obtain a varnish of these desirable qualities the best old linseed oil, or varnish made from it, must be combined with the residue left by the dry distillation of amber or very hard copal. This distillation removes a quantity of volatile oil amounting to one-fourth or one-fifth of the original weight. The residue is pulverized and dissolved in hot linseed-oil varnish, forming a thick, viscous, yellow-brown liquid, which, as a rule, must be thinned with oil of turpentine before being applied.

Hard rosin oil varnish of this sort may conveniently be mixed with the solution of asphalt in the oil of turpentine with the aid of the simple apparatus described below, as the stiffness of the two liquids makes hand stirring slow and laborious. A cask is mounted on an axle which projects through both heads, but is inclined to the axis of the cask, so that when the ends of the axle are set in bearings and the cask is revolved, each end of the cask will rise and fall alternately, and any liquid which only partly fills the cask will be thoroughly mixed and churned in a short time. The cask is two-thirds filled with the two thick varnishes (hard rosin in linseed oil and asphalt in the oil of turpentine) in the desired proportion, and after these have been intimately mixed by turning the cask, a sufficient quantity of rectified oil of turpentine to give proper consistence is added and the rotation is continued until the mixture is perfectly uniform.

To obtain the best and most durable result with this mixed oil, rosin, and asphalt varnish it is advisable to dilute it freely with oil of turpentine and to apply 2 or 3 coats, allowing each coat to dry before the next is put on. In this way a deep black and very glossy surface is obtained which cannot be distinguished from genuine Japanese lacquer.

Many formulas for making these mixed asphalt varnishes contain rosin—usually American rosin. The result is the production of a cheaper but inferior varnish. The addition of such soft rosins as elemi and copaiba, however, is made for another reason, and it improves the quality of the varnish for certain purposes. Though these rosins soften the lacquer, they also make it more elastic, and therefore more suitable for coating leather and textile fabrics, as it does not crack in consequence of repeated bending, rolling, and folding.

In coloring spirit varnish the alcohol should always be colored first to the desired shade before mixing with the rosin, except where ivory or bone black is used. If the color is taken from a gum, due allowance for the same must be made in the rosins of the varnish. For instance, in a varnish based on mastic, 10 parts, and tender copal, 5 parts, in 100 parts, if this is to be colored with, say, 8 parts of dragon's blood (or any other color gum), the rosins must be reduced to mastic, 8 parts, and tender copal, 4 parts. Eight parts of color gum are here equivalent to 3 parts of varnish rosin. This holds true with gamboge, aloes, myrrh, and the other gum rosins used for their color. This seeming disproportion is due to the inert matter and gum insoluble in alcohol, always present in these gum rosins.

Shellac Varnish.—This is made in the general proportion of 3 pounds of shellac to a gallon of alcohol, the color, temper, etc., to be determined by the requirements of the purchaser, and the nature of the wood to which the varnish is to be applied. Shellac varnish is usually tempered with sandarac, elemi, dammar, and the oil of linseed, turpentine, spike, or rosemary.

Various impurities held in suspension in shellac varnish may be entirely precipitated by the gradual addition of some crystals of oxalic acid, stirring the varnish to aid their solution, and then setting it aside overnight to permit the impurities to settle. No more acid should be used than is really necessary.

Rules for Varnishing.—1. Avoid as far as possible all manipulations with the varnishes; do not dilute them with oil of turpentine, and least of all with siccative, to expedite the drying. If the varnish has become too thick in consequence of faulty storing, it should be heated and receive an addition of hot, well-boiled linseed-oil varnish and oil of turpentine. Linseed-oil varnish or oil of turpentine added to the varnish at a common temperature renders it streaky (flacculent) and dim and has an unfavorable influence on the drying; oil of turpentine takes away the gloss of varnish.

2. Varnishing must be done only on smooth, clean surfaces, if a fine, mirror-like gloss is desired.

3. Varnish must be poured only into clean vessels, and from these never back into the stationary vessels, if it has been in contact with the brush. Use only dry brushes for varnishing, which are not moist with oil of turpentine or linseed oil or varnish.

4. Apply varnishes of all kinds as uniformly as possible; spread them out evenly on the surfaces so that they form neither too thick nor too thin a layer. If the varnish is put on too thin the coating shows no gloss; if applied too thick it does not get even and does not form a smooth surface, but a wavy one.

5. Like all oil-paint coatings, every coat of varnish must be perfectly dry before a new one is put on; otherwise it is likely that the whole work will show cracks. The consumer of varnish is only too apt to blame the varnish for all defects which appear in his work or develop after some time, although this can only be proven in rare cases. As a rule, the ground was not prepared right and the different layers of paint were not sufficiently dry, if the surfaces crack after a comparatively short time and have the appearance of maps. The cracking of paint must not be confounded with the cracking of the varnish, for the cracking of the paint will cause the varnish to crack prematurely. The varnish has to stand more than the paint; it protects the latter, and as it is transparent, the defects of the paint are visible through the varnish, which frequently causes one to form the erroneous conclusion that the varnish has cracked,

6. All varnish coatings must dry slowly, and during the drying must be absolutely protected from dust, flies, etc., until they have reached that stage when we can pass the back of the hand or a finger over them without sticking to it.

The production of faultless varnishing in most cases depends on the accuracy of the varnisher, on the treatment of his brush, his varnish pot, and all the other accessories. A brush which still holds the split points of the bristles never varnishes clear; they are rubbed off easily and spoil the varnished work. A brush which has never been used does not produce clean work; it should be tried several times, and when it is found that the varnishing accomplished by its use is neat and satisfactory it should be kept very carefully.

The preservation of the brush is thus accomplished: First of all do not place it in oil or varnish, for this would form a skin, parts of which would adhere to it, rendering the varnished surface unclean and grainy; besides these skins there are other particles which accumulate in the corners and cannot be removed by dusting off; these will also injure the work. In order to preserve the brush properly, insert it in a glass of suitable size through a cork in the middle of which a hole has been bored exactly fitting the handle. Into the glass pour a mixture of equal parts of alcohol and oil of turpentine, and allow only the point of the brush to touch the mixture, if at all. If the cork is air-tight the brush cannot dry in the vapor of oil of turpentine and spirit. From time to time the liquids in the glass should be replenished.

If the varnish remains in the varnish receptacle, a little alcohol may be poured on, which can do the varnish no harm. At all events the varnish will be prevented from drying on the walls of the vessel and from becoming covered by a skin which is produced by the linseed oil, and which indicates that the varnish is both fat and permanent. No skin forms on a meager varnish, even when it drys thick.

After complete drying of the coat of varnish it sometimes happens that the varnish becomes white, blue, dim, or blind. If varnish turns white on exposure to the air the quality is at fault. The varnish is either not fat enough or it contains a rosin unsuitable for exterior work (copal). The whitening occurs a few days after the drying of the varnish and can be removed only by rubbing off the varnish.

Preventing Varnish from Crawling.—Rub down the surface to be varnished with sharp vinegar. Coating with strongly diluted ox gall is also of advantage.

Amber Varnish.—This varnish is capable of giving a very superior polish or surface, and is especially valuable for coach and other high-class work. The amber is first bleached by placing a quantity—say about 7 pounds—of yellow amber in a suitable receptacle, such as an earthenware crucible, of sufficient strength, adding 14 pounds of sal gemmæ (rock or fossil salt), and then pouring in as much spring water as will dissolve the sal gemmæ. When the latter is dissolved more water is added, and the crucible is placed over a fire until the color of the amber is changed to a perfect white. The bleached amber is then placed in an iron pot and heated over a common fire until it is completely dissolved, after which the melting pot is removed from the fire, and when sufficiently cool the amber is taken from the pot and immersed in spring water to eliminate the sal gemmæ, after which the amber is put back into the pot and is again heated over the fire till the amber is dissolved. When the operation is finished the amber is removed from the pot and spread out upon a clean marble slab to dry until all the water has evaporated, and is afterwards exposed to a gentle heat to entirely deprive it of humidity.

Asphalt Varnishes.—Natural asphalt is not entirely soluble in any liquid. Alcohol dissolves only a small percentage of it, ether a much larger proportion. The best solvents are benzol, benzine, rectified petroleum, the essential oils, and chloroform, which leave only a small residue undissolved. The employment of ether as a solvent is impracticable because of its low boiling point, 97° F., and great volatility. The varnish would dry almost under the brush. Chloroform is not open to this objection, but it is too expensive for ordinary use. Rectified petroleum is a good solvent of asphalt, but it is not a desirable ingredient of varnish because, though the greater part of it soon evaporates, a small quantity of less volatile substances, which is usually present in even the most thoroughly rectified petroleum, causes the varnish to remain "tacky" for a considerable time and to retain a disagreeable odor much longer. Common coal-tar benzine is also a good solvent and has the merit of cheapness, but its great volatility makes the varnish dry too quickly for convenient use, especially in summer.

The best solvent, probably, is oil of turpentine, which dissolves asphalt almost completely, producing a varnish which dries quickly and forms a perfect coating if the turpentine has been well rectified. The turpentine should be a "water white," or entirely colorless, liquid of strong optical refractive power and agreeable odor, without a trace of smokiness. A layer $\frac{1}{8}$ of an inch in depth should evaporate in a short time so completely as to leave no stain on a glass dish.

But even solutions of the best Syrian asphalt in the purest oil of turpentine, if they are allowed to stand undisturbed for a long time in large vessels, deposit a thick, semi-fluid precipitate which a large addition of oil of turpentine fails to convert into a uniform thin liquid. It may be assumed that this deposit consists of an insoluble or nearly insoluble part of the asphalt which, perhaps, has been deprived of solubility by the action of light. Hence, in order to obtain a uniform solution, this thick part must be removed. This can be done, though imperfectly, by carefully decanting the solution after it has stood for a long time in large vessels. This tedious and troublesome process may be avoided by filtering the solution as it is made, by the following simple and quite satisfactory method: The solution is made in a large cask, lying on its side, with a round hole about 8 inches in diameter in its upper bilge. This opening is provided with a well-fitting cover, to the bottom of which a hook is attached. The asphalt is placed in a bag of closely woven canvas, which is inclosed in a second bag of the same material. The diameter of the double bag, when filled, should be such as to allow it to pass easily through the opening in the cask, and its length such that, when it is hung on the hook, its lower end is about 8 inches above the bottom of the cask. The cask is then filled with rectified oil of turpentine, closed, and left undisturbed for several days. The oil of turpentine penetrates into the bag and dissolves the asphalt, and the solution, which is heavier than pure oil of turpentine, exudes through the canvas and sinks to the bottom of the cask. Those parts of the asphalt which are quite insoluble, or merely swell in the oil of turpentine, cannot pass through the canvas, and are removed with the bag, leaving a perfect solution. When all soluble portions have been dissolved, the bag, with the cover, is raised and hung over the opening to drain. If pulverized asphalt has been used the bag is found to contain only a small quantity of semi-fluid residue. This, thinned with oil of turpentine and applied with a stiff brush and considerable force, forms a thick, weather-resisting, and very durable coating for planks, etc.

The proportion of asphalt to oil of turpentine is so chosen as to produce, in the cask, a pretty thick varnish, which may be thinned to any desired degree by adding more turpentine. For use, it should be just thick enough to cover bright tin and entirely conceal the metal with a single coat. When dry, this coat is very thin, but it adheres very firmly, and continually increases in hardness, probably because of the effect of light. This supposition is supported by the difficulty of removing an old coat of asphalt varnish, which will not dissolve in turpentine even after long immersion, and usually must be removed by mechanical means.

For a perfect, quick-drying asphalt varnish the purest asphalt must be used, such as Syrian, or the best Trinidad. Trinidad seconds, though better than some other asphalts, yield an inferior varnish, owing to the presence of impurities.

Of artificial asphalt, the best for this purpose is the sort known as "mineral caoutchouc," which is especially suitable for the manufacture of elastic dressings for leather and other flexible substances. For wood and metal it is less desirable, as it never becomes as hard as natural asphalt.

FORMULAS:

I.—A solution of 1 part of caoutchouc in 16 parts of oil of turpentine or kerosene is mixed with a solution of 16 parts of copal in 8 parts of linseed-oil varnish. To the mixture is added a solution of 2 parts of asphalt in 3 or 4 parts of linseed-oil varnish diluted with 8 or 10 parts of oil of turpentine, and the whole is filtered. This is a fine elastic varnish.

II.—Coal-tar asphalt, American asphalt, rosin, benzine, each 20 parts; linseed-oil varnish, oil of turpentine, coal-tar oil, each 10 parts; binoxide of manganese, roasted lampblack, each 2 parts. The solid ingredients are melted together and mixed with the linseed-oil varnish, into which the lampblack has been stirred, and, finally, the other liquids are added. The varnish is strained through tow.

Bicycle Varnish.—This is a spirit varnish, preferably made by a cold process, and requires less technical knowledge than the preparation of fatty varnishes. The chief dependence is upon the choice of the raw materials. These raw materials, copal, shellac, etc., are first broken up small and placed in a barrel adapted for turning upon an axis, with a hand crank, or with a belt and pulley from a power shaft. The barrel is of course simply mounted in a frame of wood or iron, whichever is the most convenient. After the barrel has received its raw material, it may be started and kept revolving for 24 hours. Long interruptions in the turning must be carefully avoided, particularly in summer, for the material in the barrel, when at rest, will, at this season, soon form a large lump, to dissolve which will consume much time and labor. To prevent the formation of a semi-solid mass, as well as to facilitate the dissolving of the gum, it would be well to put some hard, smooth stones into the barrel with the varnish ingredients.

Bicycle Dipping Varnish (Baking Varnish).—Take 50 parts, by weight, of Syrian asphalt; 50 parts, by weight, of copal oil; 50 parts, by weight, of thick varnish oil, and 105 parts, by weight, turpentine oil, to which add 7 parts, by weight, of drier. When the asphalt is melted through and through, add the copal oil and heat it until the water is driven off, as copal oil is seldom free from water. Now take it off the fire and allow it to cool; add first the siccative, then the turpentine and linseed oil, which have been previously thoroughly mixed together. This bicycle varnish does not get completely black until it is baked.

Black Varnishes.—Black spirit lacquers are employed in the wood and metal industries. Different kinds are produced according to their use. They are called black Japanese varnishes, or black brilliant varnishes.

Black Japanese Varnish.—I.—Sculpture varnish, 5 parts; red acaroid varnish, 2 parts; aniline black, ¼ part; Lyons blue, .0015 parts. If a sculpture varnish prepared with heated copal is employed, a black lacquer of especially good quality is obtained. Usually 1 per cent of oil of lavender is added.

II.—

Shellac	4 parts
Borax	2 parts
Glycerine	2 parts
Aniline black	5 parts
Water	50 parts

Dissolve the borax in the water, add

the shellac, and heat until solution is effected; then add the other ingredients. This is a mat-black varnish.

For Blackboards.—For blackening these boards mix ½ liter (1.05 pints) good alcohol, 70 grams (1,080 grains) shellac, 6 grams (92 grains) fine lampblack, 3 grams (46 grains) fine chalk free from sand. If red lines are to be drawn, mix the necessary quantity of red lead in alcohol and shellac.

Bookbinders' Varnishes.—

	I Per Cent	II Per Cent	III Per Cent	IV Per Cent	V Per Cent
Shellac	14.5	6.5	13.5	6.3	8.3
Mastic	6.0	2.0	..	..	1.1
Sandarac	6.0	13.0	..	1.3	1.1
Camphor	1.0	..	0.5	1.5	..
Benzoin	..	..	..	..	13.7
Alcohol	72.5	78.5	86.0	79.2	75.8

Scent with oil of benzoin, of lavender, or of rosemary. Other authors give the following recipes:

	VI Per Cent	VII Per Cent	VIII Per Cent	IX Per Cent
Blond shellac	11.5	13.0	9.0	..
White shellac	11.5	..	..	..
Camphor	..	0.7	..	..
Powdered sugar	..	0.7	..	..
Sandarac	..	..	18.0	6.6
Mastic	..	..	..	13.0
Venice turpentine	..	..	2.0	6.6
Alcohol	77.0	85.6	71.0	73.8

All solutions may be prepared in the cold, but the fact that mastic does not dissolve entirely, must not be lost sight of.

Bottle Varnish.—Bottles may be made to exclude light pretty well by coating them with asphaltum lacquer or varnish. A formula recommended for this purpose is as follows: Dissolve asphaltum, 1 part, in light coal-tar oil, 2 parts, and add to the solution about 1 per cent of castor oil. This lacquer dries somewhat slowly, but adheres very firmly to the glass. Asphaltum lacquer may also be rendered less brittle by the addition of elemi. Melt together asphaltum, 10 parts, and elemi, 1 part, and dissolve the cold fused mass in light coal-tar oil, 12 parts.

Amber-colored bottles for substances acted upon by the actinic rays of light may be obtained from almost any manufacturer of bottles.

Can Varnish.—Dissolve shellac, 15 parts, by weight; Venice turpentine, 2 parts, by weight; and sandarac, 8 parts, by weight, in spirit, 75 parts, by weight.

Copal Varnish.—Very fine copal varnish for those parts of carriages which require the highest polish, is prepared as follows:

I.—Melt 8 pounds best copal and mix with 20 pounds very clear matured oil. Then boil 4 to 5 hours at moderate heat until it draws threads; now mix with 35 pounds oil of turpentine, strain and keep for use. This varnish dries rather slowly, therefore varnishers generally mix it one-half with another varnish, which is prepared by boiling for 4 hours, 20 pounds clear linseed oil and 8 pounds very pure, white anime rosin, to which is subsequently added 35 pounds oil of turpentine.

II.—Mix the following two varnishes:

(a) Eight pounds copal, 10 pounds linseed oil, ½ pound dried sugar of lead, 35 pounds oil of turpentine.

(b) Eight pounds good anime rosin, 10 pounds linseed oil, ¼ pound zinc vitriol, 35 pounds oil of turpentine. Each of these two sets is boiled separately into varnish and strained, and then both are mixed. This varnish dries in 6 hours in winter, and in 4 hours in summer. For old articles which are to be re-varnished black, it is very suitable.

Elastic Limpid Gum Varnishes.—I.—In order to obtain a limpid rubber varnish, it is essential to have the rubber entirely free from water. This can be obtained by cutting the rubber into thin strips, or better, into shreds as fine as possible, and drying them, at a temperature of from 104° to 122° F., for several days or until they are water free, then proceed as follows:

II.—Dissolve 1 part of the desiccated rubber in 8 parts of petroleum ether (benzine) and add 2 parts of fat copal varnish and stir in. Or, cover 2 parts of dried rubber with 1 part of ether; let stand for several days, or until the rubber has taken up as much of the ether as it will, then liquefy by standing in a vessel of moderately warm water. While still warm, stir in 2 parts of linseed oil, cut with 2 parts of turpentine oil.

ENAMEL VARNISHES:

Antiseptic Enamel.—This consists of a solution of spirituous gum lac, rosin, and copal, with addition of salicylic acid, etc. Its purpose is mainly the prevention or removal of mold or fungous formation. The salicylic acid contained in the mass acts as an antiseptic during the painting, and destroys all fungi present.

Bath-Tub Enamel Unaffected by Hot Water.—I.—In order to make paint hold on the zinc or tinned copper lining of a bath tub, a wash must be used to produce a film to which oil paint will adhere. First remove all grease, etc., with a solution of soda or ammonia and dry the surface thoroughly; then apply with a wide, soft brush equal parts, by weight, of chloride of copper, nitrate of copper, and sal ammoniac, dissolved in 64 parts, by weight, of water. When dissolved add 1 part, by weight, of commercial muriatic acid. This solution must be kept in glass or earthenware. It will dry in about 12 hours, giving a grayish-black coating to which paint will firmly adhere.

The priming coat should be white lead thinned with turpentine, with only just sufficient linseed oil to bind it. After this is thoroughly dry, apply one or more coats of special bath-tub enamel, or a gloss paint made by mixing coach colors ground in Japan with hard-drying varnish of the best quality. Most first-class manufacturers have special grades that will stand hot water.

II.—The following preparation produces a brilliant surface on metals and is very durable, resisting the effect of blows without scaling or chipping off, and being therefore highly suitable for cycles and any other articles exposed to shock:

For the manufacture of 44 gallons, 11 pounds of red copper, 8.8 pounds of yellow copper, 4.4 pounds of hard steel, and 4.4 pounds of soft steel, all in a comminuted condition, are well washed in petroleum or mineral spirit, and are then treated with concentrated sulphuric acid in a lead-lined vessel, with continued stirring for 2 hours. After 12 hours' rest the sulphuric acid is neutralized with Javel extract, and the fine powder left in the vessel is passed through a silk sieve to remove any fragments of metal, then ground along with linseed oil, ivory black, and petroleum, the finely divided mass being afterwards filtered through flannel and incorporated with a mixture of Bombay gum, 22 pounds; Damascus gum, 11 pounds; Judea bitumen, 22 pounds; Norwegian tar rosin, 11 pounds; and 11 pounds of ivory black ground very fine in refined petroleum. When perfectly homogeneous the mass is again filtered, and is then ready for use. It is laid on with a brush, and then fixed by exposure to a temperature of between 400° and 800° F. The ivory black may be replaced by other coloring matters, according to requirements.

A Color Enamel.—On the piece to be enameled apply oil varnish or white lead, and add a powder giving brilliant reflections, such as diamantine, brilliantine, or argentine. Dry in a stove. Apply a new coat of varnish. Apply the powder again, and finally heat in the oven. Afterwards, apply several layers of varnish; dry each layer in the oven. Apply pumice stone in powder or tripoli, and finally apply a layer of Swedish varnish, drying in the oven. This enamel does not crack. It adheres perfectly, and is advantageous for the pieces of cycles and other mobiles.

Cold Enameling.—This style of enameling is generally employed for repairing purposes. The various colors are either prepared with copal varnish and a little oil of turpentine, or else they are melted together with mastic and a trifle of oil of spike. In using the former, the surface usually settles down on drying, and ordinarily the latter is preferred, which is run on the cracked-off spot by warming the article. After the cooling, file the cold enamel off uniformly, and restore the gloss by quickly drawing it through the flame. For black cold enamel melt mastic together with lampblack, which is easily obtained by causing the flame of a wick dipped into linseed oil to touch a piece of tin.

White.—White lead or flake white.

Red.—Carmine or cinnabar (vermilion).

Blue.—Ultramarine or Prussian blue.

Green.—Scheele's green or Schweinfurt green.

Brown.—Umber.

Yellow.—Ocher or chrome yellow.

The different shades are produced by mixing the colors.

Enamel for Vats, etc.—Two different enamels are usually employed, viz., one for the ground and one for the top, the latter being somewhat harder than the former. Ground enamel is prepared by melting in an enameled iron kettle 625 parts brown shellac, 125 parts French oil of turpentine, with 80 parts colophony, and warming in another vessel 4,500 parts of spirit (90 per cent). As soon as the rosins are melted, remove the pot from the fire and add the spirit in portions of 250 parts at a time, seeing to it that the spirit added is completely combined with the rosins by stirring before adding any more. When all the spirit is added, warm the whole again for several minutes on the water bath (free fire should be avoided, on account of danger of fire), and allow to settle. If a yellow color is desired, add yellow ocher, in which case the mixture may also be used as floor varnish.

The top enamel (hard) consists of 500 parts shellac, 125 parts French oil of turpentine, and 3,500 parts spirit (90 per cent). Boiling in the water bath until the solution appears clear can only be of advantage. According to the thickness desired, one may still dilute in the cold with high-strength spirit. Tinting may be done, as desired, with earth colors, viz., coffee brown with umber, red with English red, yellow with ocher, silver gray with earthy cerussite, and some lampblack. Before painting, dry out the vats and putty up the joints with a strip of dough which is prepared from ground enamel and finely sifted charcoal or brown coal ashes, and apply the enamel after the putty is dry. The varnish dries quickly, is odorless and tasteless, and extraordinarily durable. If a little annealed soot black is added to this vat enamel, a fine iron varnish is obtained which adheres very firmly. Leather •(spattering leather on carriages) can also be nicely varnished with it.

Finishing Enamel for White Furniture.—Various methods are practiced in finishing furniture in white enamel, and while numerous preparations intended for the purpose named are generally purchasable of local dealers in paint supplies, it is often really difficult, and frequently impossible, to obtain a first-class ready-made enamel. To prepare such an article take ½ pint of white lead and add to it ¼ pint of pure turpentine, ¼ gill of pale coach Japan, and ½ gill of white dammar varnish. Mix all the ingredients together thoroughly. Apply with a camel's-hair brush, and for large surfaces use a 2-inch double thick brush. There should be at least three coats for good work, applied after an interval of 24 hours between coats; and for strictly high-class work four coats will be necessary. Each coat should be put on thin and entirely free from brush marks, sandpapering being carefully done upon each coat of pigment. Work that has been already painted or varnished needs to be cut down with, say, No. ½ sandpaper, and then smoothed fine with No. ½ paper. Then thin white lead to a free working consistency with turpentine, retaining only a weak binder of oil in the pigment, and apply two coats of it to the surface. Give each coat plenty of time

to harden (36 hours should suffice), after which sandpapering with No. 1½ paper had best be done. Ordinarily, upon two coats of white lead, the enamel finish, as above detailed, may be successfully produced. For the fine, rich enamel finish adapted to rare specimens of furniture and developed in the mansions of the multimillionaires, a more elaborate and complex process becomes necessary.

Quick-Drying Enamel Colors. — Enamel colors which dry quickly, but remain elastic so that applied on tin they will stand stamping without cracking off, can be produced as follows:

In a closed stirrer or rolling cask place 21.5 parts, by weight, of finely powdered pale French rosin, 24½ parts, by weight, of Manila copal, as well as 35 parts, by weight, of denaturized spirit (95 per cent), causing the cask or the stirrer to rotate until all the gum has completely dissolved, which, according to the temperature of the room in which the stirrer is and the hardness of the gums, requires 24 to 48 hours. When the gums are entirely dissolved add to the mixture a solution of 21½ parts, by weight, of Venice oil turpentine in 0.025 parts, by weight, of denaturized spirit of 95 per cent, allowing the stirrer to run another 2 to 3 hours. For the purpose of removing any impurities present or any undissolved rosin from the varnish, it is poured through a hair sieve or through a threefold layer of fine muslin (organdie) into suitable tin vessels or zinc-lined barrels for further clarification. After 10 to 14 days the varnish is ready for use. By grinding this varnish with the corresponding dry pigments the desired shades of color may be obtained; but it is well to remark that chemically pure zinc white cannot be used with advantage because it thickens and loses its covering power. The grinding is best carried out twice on an ordinary funnel mill. Following are some recipes:

I. — **Enamel White.** — Lithopone, 2 parts, by weight; white lead, purest, ½ part, by weight; varnish, 20 parts, by weight.

II.—**Enamel Black.**—Ivory black, 2 parts, by weight; Paris blue, 0.01 part, by weight; varnish, 23 parts, by weight.

III.—**Pale Gray.**—Graphite, 2 parts, by weight; ultramarine, 0.01 part, by weight; lithopone, 40 parts, by weight; varnish, 100 parts, by weight.

IV.—**Dark Gray.**—Graphite, 3 parts, by weight; ivory black, 2 parts, by weight; lithopone, 40 parts, by weight; varnish, 110 parts, by weight.

V.—Chrome Yellow, Pale.—Chrome yellow, 2 parts, by weight; lithopone, 2 parts, by weight; varnish, 40 parts, by weight; benzine, 1½ parts, by weight.

VI.—Chrome Yellow, Dark.—Chrome yellow, dark, 2 parts, by weight; chrome orange, ⅛ part, by weight; lithopone, 1 part, by weight; varnish, 35 parts, by weight; benzine, 1 part, by weight.

VII.—Pink, Pale.—Carmine, ½ part, by weight; lithopone, 15 parts, by weight; varnish, 40 parts, by weight; benzine, 1½ parts, by weight.

VIII.—Pink, Dark.—Carmine, ½ part, by weight; Turkey red, 1 part, by weight; lithopone, 15 parts, by weight; varnish, 40 parts, by weight.

IX.—Turkey Red.—Turkey red, pale, 2 parts, by weight; lithopone, 1 part, by weight; Turkey red, dark, 1 part, by weight; white lead, pure, ½ part, by weight; varnish, 18 parts, by weight; benzine, ½ part, by weight.

X.—Flesh Tint.—Chrome yellow, pale, 1½ parts, by weight; graphite, ⅛ part, by weight; lithopone, 15 parts, by weight; Turkey red, pale, 2 parts, by weight; varnish, 42 parts, by weight; benzine, ½ part, by weight.

XI.—Carmine Red.—Lead sulphate, 5 parts, by weight; Turkey red, pale, 6 parts, by weight; carmine, 1½ parts, by weight; orange minium, 3 parts, by weight; vermilion, 2 parts, by weight; varnish, 50 parts, by weight; benzine, 1½ parts, by weight.

XII.—Sky Blue.—Ultramarine, 5 parts, by weight; lithopone, 5 parts, by weight; ultramarine green, 0.05 parts, by weight; varnish, 30 parts, by weight; benzine, 1 part, by weight.

XIII.—Ultramarine.—Ultra blue, 5 parts, by weight; varnish, 12 parts, by weight; benzine, ½ part, by weight.

XIV.—Violet.—Ultramarine, with red tinge, 10 parts, by weight; carmine, 0.5 parts, by weight; varnish, 25 parts, by weight.

XV.—Azure.—Paris blue, 10 parts, by weight; lithopone, 100 parts, by weight; varnish, 300 parts, by weight.

XVI.—Leaf Green.—Chrome green, pale, 5 parts, by weight; varnish, 25 parts, by weight; benzine, ½ part, by weight.

XVII.—Silk Green.—Silk green, 10 parts, by weight; chrome yellow, pale, ½ part, by weight; lead sulphate, 5 parts, by weight; varnish, 30 parts, by weight; benzine, ½ part, by weight.

XVIII. — Brown. — English red, 10 parts, by weight; ocher, light, 3 parts, by weight; varnish, 30 parts, by weight; benzine, ½ part, by weight.

XIX.—Ocher.—French ocher, 10 parts, by weight; chrome yellow, dark, ½ part, by weight; varnish, 30 parts, by weight; benzine, ½ part, by weight.

XX.—Chocolate.—Umber, 10 parts, by weight; Florentine lake, ⅛ part, by weight; varnish, 25 parts, by weight; benzine, ½ part, by weight.

XXI.—Terra Cotta.—Chrome yellow, pale, 10 parts, by weight; Turkey red, dark, 3 parts, by weight; varnish, 35 parts, by weight.

XXII. — Olive, Greenish. — French ocher, 5 parts, by weight; Paris blue, ½ part, by weight; graphite, ½ part, by weight; varnish, 25 parts, by weight; lithopone, 5 parts, by weight.

XXIII.—Olive, Brownish.—Chrome orange, 5 parts, by weight; Paris blue, 2 parts, by weight; lead sulphate, 10 parts, by weight; English red, 1 part, by weight; varnish, 40 parts, by weight; benzine, 1½ parts, by weight.

XXIV.—Olive, Reddish.—Turkey red, dark, 75 parts, by weight; sap green, 75 parts, by weight; ocher, pale, 5 parts, by weight; varnish, 300 parts, by weight; benzine, 1½ parts, by weight.

ENGRAVERS' VARNISHES.

In copper-plate engraving the plate must be covered with a dark-colored coating which, though entirely unaffected by the etching fluid, must be soft enough to allow the finest lines to be drawn with the needle and must also be susceptible of complete and easy removal when the etching is finished. Varnishes which possess these properties are called "etching grounds." They are made according to various formulas, but in all cases the principal ingredient is asphalt, of which only the best natural varieties are suitable for this purpose. Another common ingredient is beeswax, or tallow.

Etching grounds are usually made in small quantities, at a single operation, by melting and stirring the solid ingredients together and allowing the mass to cool in thin sheets, which are then dissolved in oil of turpentine. The plate is coated uniformly with this varnish through which the engraver's tool readily penetrates, laying bare the metal beneath. After the lines thus drawn have been etched by immersing the plate in acid, the varnish is washed off with oil of turpentine.

The following formulas for etching grounds have been extensively used by engravers:

	I	II	III	IV
Yellow wax	50	30	110	40 parts
Syrian asphalt	20	20	25	40 parts
Rosin	..	..	..	20 parts
Amber	..	..	20	.. parts
Mastic	25	25	25	.. parts
Tallow	..	..	..	2 parts
Bergundy pitch	..	..	..	10 parts

FLOOR VARNISHES.

I.—Manila copal, spirit-
soluble............... 12 parts
Ruby shellac, pow-
dered................ 62 parts
Venice, turpentine.... 12 parts
Spirit, 96 per cent.... 250 parts

The materials are dissolved cold in a covered vat with constant stirring, or better still, in a stirring machine, and filtered. For the pale shades take light ocher; for dark ones, Amberg earth, which are well ground with the varnish in a paint mill.

II.—Shellac, A C leaf, 1.2 parts; sandarac, 8 parts; Manila copal, 2 parts; rosin, 5 parts; castor or linoleic acid or wood oil acid, 1.50 parts; spirit (96 per cent), 65 parts.

French Varnish.—So-called French varnish is made by dissolving 1 part of bleached or orange shellac in 5 parts of alcohol, the solution being allowed to stand and the clear portion then being decanted. The varnish may be colored by materials which are soluble in alcohol.

For red, use 1 part of eosin to 49 parts of the bleached shellac solution. For blue, use 1 part of aniline blue to 24 parts of the bleached shellac solution, as the orange shellac solution would impart a greenish cast. For green, use 1 part of aniline green (brilliant green) to 49 parts of the orange shellac solution. For yellow, use either 2 parts of extract of turmeric or 1 part of gamboge to 24 parts of the solution, or 1 part of aniline yellow to 49 parts of the solution. For golden yellow, use 2 parts of gamboge and 1 part of dragon's blood to 47 parts of the orange shellac solution. The gamboge and dragon's blood should be dissolved first in a little alcohol.

Golden Varnishes.—

I.—Powdered benzoin.. 1 part
Alcohol enough to make 10 parts.
Pure saffron, roughly broken up,
about 6 threads to the ounce.

Macerate 3 days and filter. Vary the quantity of saffron according to the shade desired. Mastic and juniper gum may be added to this varnish if a heavier body is desired.

II.—Benzoin, juniper gum, gum mastic, equal parts.

Dissolve the gums in 9 times their weight of alcohol (varied more or less according to the consistency wanted), and color to the desired shade with threads of pure saffron. This varnish is very brilliant and dries at once.

India-Rubber Varnishes.—I.—Dissolve 10 pounds of India rubber in a mixture of 10 pounds of turpentine and 20 pounds of petroleum by treating same on a water bath. When the solution is completed add 45 pounds of drying oil and 5 pounds of lampblack and mix thoroughly.

II.—Dissolve 7 pounds of India rubber in 25 pounds of oil of turpentine. By continued heating dissolve 14 pounds of rosin in the mixture. Color while hot with 3 pounds of lampblack.

Inlay Varnish.—

Ozokerite.......... 17 parts
Carnauba wax....... 3 parts
Turpentine oil....... 15 parts

Melt the ozokerite and Carnauba wax, then stir in the turpentine oil. This varnish is applied like a polish and imparts to the wood a dark natural color and a dull luster.

Japanning Tin.—The first thing to be done when a vessel is to be japanned, is to free it from all grease and oil, by rubbing it with turpentine. Should the oil, however, be linseed, it may be allowed to remain on the vessel, which must in that case be put in an oven and heated till the oil becomes quite hard.

After these preliminaries, a paint of the shade desired, ground in linseed oil, is applied. For brown, umber may be used.

When the paint has been satisfactorily applied it should be hardened by heating, and then smoothed down by rubbing with ground pumice stone applied gently by means of a piece of felt moistened with water. To be done well, this requires care and patience, and, it might be added, some experience.

The vessel is next coated with a varnish, made by the following formula:

Turpentine spirit.... 8 ounces
Oil of lavender...... 6 ounces
Camphor........... 1 drachm
Bruised copal........ 2 ounces

Perhaps some other good varnish would give equally satisfactory results.

After this the vessel is put in an oven and heated to as high a temperature as it will bear without causing the varnish to blister or run. When the varnish has become hard, the vessel is taken out and another coat is put on, which is submitted to heat as before. This process may be repeated till the judgment of the operator tells him that it is no longer advisable.

Some operators mix the coloring matter directly with the varnish; when this is done, care should be taken that the pigment is first reduced to an impalpable powder, and then thoroughly mixed with the liquid.

LABEL VARNISHES.

I.—Sandarac........ 3 ounces av.
Mastic.......... ¾ ounce av.
Venice turpentine 150 grains
Alcohol........ 16 fluidounces

Macerate with repeated stirring until solution is effected, and then filter.

The paper labels are first sized with diluted mucilage, then dried, and then coated with this varnish. If the labels have been written with water-soluble inks or color, they are first coated with 2 coats of collodion, and then varnished.

II.—The varnished labels of stock vessels often suffer damage from the spilling of the contents and the dripping after much pouring.

Formalin gelatin is capable of withstanding the baneful influence of ether, benzine, water, spirit of wine, oil, and most substances. The following method of applying the preservative is recommended: Having thoroughly cleaned the surface of the vessel, paste the label on and allow it to dry well. Give it a coat of thin collodion to protect the letters from being dissolved out or caused to run, then after a few minutes paint over it a coat of gelatin warmed to fluidity—5 to 25—being careful to cover in all the edges. Just before it solidifies go over it with a tuft of cotton dipped into a 40 per cent formalin solution. It soon dries and becomes as glossy as varnish, and may be coated again and again without danger of impairing the clear white of the label or decreasing its transparency.

Leather Varnishes.—I.—An excellent varnish for leather can be made from the following recipe: Heat 400 pounds of boiled oil to 212° F., and add little by little 2 pounds of bichromate of potash, keeping the same temperature. The addition of the bichromate should take about 15 minutes. Raise to 310° F., and add gradually during 1 hour at that temperature, 40 pounds Prussian blue. Heat for 3 hours more, gradually raising to 482° to 572° F., with constant stirring.

In the meantime, heat together at 392° F., for ½ an hour, 25 pounds linseed oil, 35 pounds copal, 75 pounds turpentine, and 7 pounds ceresine. Mix the two varnishes, and dilute, if necessary, when cold with turpentine. The varnish should require to be warmed for easy application with the brush.

II.—Caoutchouc, 1 part; petroleum, 1 part; carbon bisulphide, 1 part; shellac, 4 parts; bone black, 2 parts; alcohol, 20 parts. First the caoutchouc is brought together with carbon bisulphide in a well-closed bottle and stood aside for a few days. As soon as the caoutchouc is soaked add the petroleum and the alcohol, then the finely powdered shellac, and heat to about 125° F. When the liquid appears pretty clear, which indicates the solution of all substances, the bone black is added by shaking thoroughly and the varnish is at once filled in bottles which are well closed. This pouch composition excels in drying quickly and produces upon the leather a smooth, deep black coating, which possesses a certain elasticity.

METAL VARNISHES.

The purpose of these varnishes is to protect the metals from oxidation and to render them glossy.

Aluminum Varnish.—The following is a process giving a special varnish for aluminum, but it may also be employed for other metals, giving a coating unalterable and indestructible by water or atmospheric influences: Dissolve, preferably in an enameled vessel, 10 parts, by weight, of gum lac in 30 parts of liquid ammonia. Heat on the water bath for about 1 hour and cool. The aluminum to be covered with this varnish is carefully cleaned in potash, and, having applied the varnish, the article is placed in a stove, where it is heated, during a certain time, at a suitable temperature (about 1062° F.).

Brass Varnishes Imitating Gold.—I.—An excellent gold varnish for brass objects, surgical or optical instruments, etc., is prepared as follows: Gum lac, in grains, pulverized, 30 parts; dragon's blood, 1 part; red sanders wood, 1 part; pounded glass, 10 parts; strong alcohol, 600 parts; after sufficient maceration, filter. The powdered glass simply serves for accelerating the dissolving, by interposing between the particles of gum lac and opal.

II.—Reduce to powder, 160 parts, by weight, of turmeric of best quality, and pour over it 2 parts, by weight, of saffron, and 1,700 parts, by weight, of spirit; digest in a warm place 24 hours, and filter. Next dissolve 80 parts, by weight, of dragon's blood; 80 parts, by weight, of sandarac; 80 parts, by weight, of elemi gum; 50 parts, by weight, of gamboge; 70 parts, by weight, of seedlac. Mix these substances with 250 parts, by weight, of crushed glass, place them in a flask, and pour over this mixture the alcohol colored as above described. Assist the solution by means of a sand or water bath, and filter at the close of the operation. This is a fine varnish for brass scientific instruments.

Bronze Varnishes.—I.—The following process yields a top varnish for bronze goods and other metallic ware in the most varying shades, the varnish excelling, besides, in high gloss and durability. Fill in a bottle, pale shellac, best quality, 40 parts, by weight; powdered Florentine lake, 12 parts, by weight; gamboge, 30 parts, by weight; dragon's blood, also powdered, 6 parts, by weight; and add 400 parts, by weight, of spirit of wine. This mixture is allowed to dissolve, the best way being to heat the bottle on the water bath until the boiling point of water is almost reached, shaking from time to time until all is dissolved. Upon cooling, decant the liquid, which constitutes a varnish of dark-red color, from any sediment that may be present. In a second bottle dissolve in the same manner 24 parts, by weight, of gamboge in 400 parts, by weight, of spirit of wine, from which will result a varnish of golden-yellow tint. According to the hue desired, mix the red varnish with the yellow variety, producing in this way any shade from the deepest red to the color of gold. If required, dilute with spirit of wine. The application of the varnish should be conducted as usual, that is, the article should be slightly warm, it being necessary to adhere strictly to a certain temperature, which can be easily determined by trials and maintained by experience. In order to give this varnish a pale-yellow to greenish-yellow tone, mix 10 drops of picric acid with about 3 parts, by weight, of spirit of wine, and add to a small quantity of the varnish some of this mixture until the desired shade has been reached. Picric acid is poisonous, and the keeping of varnish mixed with this acid in a closed bottle is not advisable, because there is danger of an explosion. Therefore, it is best to prepare only so much varnish at one time as is necessary for the immediate purpose.

Brown Varnish.—An excellent and quickly drying brown varnish for metals is made by dissolving 20 ounces of gum kino and 5 ounces of gum benjamin in 60 ounces of the best cold alcohol; 20 ounces of common shellac and 2 ounces of thick turpentine in 36 ounces of alcohol also give a very good varnish. If the brown is to have a reddish tint, dissolve 50 ounces of ruby shellac, 5 ounces balsam of copaiba, and 2 to 5 ounces of aniline brown, with or without $\frac{1}{2}$ to 1 ounce of aniline violet, in 150 ounces of alcohol.

Copper Varnishes.—These two are for polished objects:

I.—One hundred and ten parts of sandarac and 30 parts of rosin, dissolved in sufficient quantity of alcohol; 5 parts of glycerine are to be added.

II.—Sandarac......... 10 parts
Rosin............. 3 parts
Glycerine......... $\frac{1}{2}$ part
Alcohol, a sufficient quantity.

Dissolve the two rosins in sufficient alcohol and add the glycerine.

Decorative Metal Varnishes.—

	I Per Cent	II Per Cent	III Per Cent	IV Per Cent
Seed lac.....	11.5	..	..	..
Amber......	7.6	..	..	13.5
Gamboge....	7.6	..	..	..
Dragon's blood.....	0.18	..	..	..
Saffron......	0.16	..	..	..
Sandarac....	..	11.2	15.9	16.6
Mastic......	..	6.5	14.0	3.4
Elemi.......	..	3.3	..	..
Venice turpentine....	..	..	1.0	3.4
Camphor....	..	1.5	..	..
Aloe........	..	..	7.0	..
Alcohol.....	72.96	77.5	66.1	63.2

As will be seen, only natural colors are used. The so-called "gold lacquer" is composed as follows: Sandarac, 6.25 parts; mastic, 3 parts; shellac, 12.5 parts; Venice turpentine, 2.5 parts; aloe, 0.75 parts; gamboge, 3 parts; alcohol, 72 parts. The solution is filtered. Applied in a thin coating this varnish shows a handsome golden shade. Other metal varnishes have the following composition:

	V Per Cent	VI Per Cent	VII Per Cent
Shellac.............	17.5	..	18.0
Yellow acaroid gum..	13.1	25.0	..
Manila	..	8.0	9.0
Alcohol.............	69.4	67.0	63.0

Gold Varnish.—I.—A good gold varnish for coating moldings which produces great brilliancy is prepared as follows: Dissolve 3 pounds of shellac in 30 quarts of alcohol, 5 pounds of mastic in 5 quarts of alcohol, 3 pounds of sandarac in 5 quarts of alcohol, 5 pounds of gamboge in 5 quarts of alcohol, 1 pound of dragon's blood in 1 quart of alcohol, 3 pounds of saunders in 5 quarts of alcohol, 3 pounds of turpentine in 3 quarts of alcohol. After all the ingredients have been dissolved separately in the given quantity of absolute alcohol and filtered, the solutions are mixed at a moderate heat.

II.—A varnish which will give a splendid luster, and any gold color from deep red to golden yellow, is prepared by taking 50 ounces pale shellac, 15 pounds Florentine lake (precipitated from cochineal or redwood decoction by alum onto strach, kaolin, or gypsum), 25 ounces of sandalwood, and 8 ounces of dragon's blood. These in fine powder are dissolved on the water bath, in 500 ounces rectified spirit. The spirit must boil and remain, with occasional shaking, for 2 to 3 hours on the bath. Then cool and decant. In the meantime heat in another flask on the bath 30 ounces of gamboge in 500 ounces of the same spirit. The two liquids are mixed until the right color needed for the particular purpose in hand is obtained. Dilute with spirit if too thick. The addition of a little picric acid gives a greenish-yellow bronze but makes the varnish very liable to explode. These varnishes are applied to gently warmed surfaces with a soft bristle brush.

Gold Varnish for Tin.—This is obtained in the following manner: Spread out 5 parts, by weight, of finely powdered crystallized copper acetate in a warm spot, allowing it to lie for some time; then grind the powder, which will have acquired a light-brown shade, with oil of turpentine and add, with stirring, 15 parts, by weight, of fat copal varnish heated to 140° F. When the copper acetate has dissolved (in about ¼ hour), the mass is filled in a bottle and allowed to stand warm, for several days, shaking frequently. The gold varnish is then ready for use. Coat the articles uniformly with it, and heat in a drying chamber, whereupon, according to the degree of temperature, varying colorations are obtained, changing from green to yellow, then golden yellow, and finally orange to brown. When good copal varnish is employed, the varnish will adhere very firmly, so that the article can be pressed without damage.

Iron Varnishes.—I.—A varnish obtained by dissolving wax in turpentine is useful. It gives a fairly hard coat, but has the drawback of filling up fine grooves, and so injuring the appearance of many metal ornaments.

II.—Shellac, 15 pounds; Siam benjamin, 13 pounds; alcohol, 80 pounds; formylchloride, 20 pounds.

III.—Sierra Leone copal, 6 pounds; dammar, 18 pounds; oleic acid, 3 pounds; alcohol, 40 pounds; oil of turpentine, 20 pounds; formylchloride, 15 pounds. The formylchloride not only effects the rapid drying necessary to prevent the varnish gravitating into hollows, but enables the alcohol to make a perfect solution of the rosin. The varnishes are excessively volatile, and must be stored accordingly.

Stove Varnishes.—

Shellac...............	12	parts
Manila copal........	14	parts
Rosin...............	12	parts
Gallipot............	2	parts
Benzoin............	1	part
Lampblack.........	5	parts
Nigrosin, spirit-soluble.............	1½	parts
Alcohol.............	250	parts

Tin Varnishes.—I.—For Tin Boxes.—In 75 parts of alcohol dissolve 15 parts of shellac, 2 parts of Venice turpentine, and 8 parts of sandarac.

II.—For Trays and Other Tinware.—The ground is prepared by adding to the white lead the tinting colors ground in good rubbing varnish and half oil of turpentine. For drier an admixture of "terebine" is recommended. With this lean and dull paint, coat the tins 2 or 3 times and blend. Next, grain with water or vinegar glaze, and varnish with pure Zanzibar copal varnish, or finest amber table-top varnish. There are other tried methods for varnishing tin, which are applicable for new goods, manufactured in large quantities, while they are less advantageous for the restoration of old, repeatedly used articles.

VARNISH SUBSTITUTES.

A substitute for varnish is produced by adding to 100 parts of casein 10 to 25 parts of a 1 to 10 per cent soap solution and then 20 to 25 parts of slaked lime. The mixture is carefully kneaded until a perfectly homogeneous mass results. Then gradually add 25 to 40 parts of turpentine oil and sufficient water for the mass to assume the consistency of varnish. If it is desired to preserve it for some time a little ammonia is added so that the casein lime does not separate. The surrogate is considerably cheaper than varnish and dries so quickly that paint ground with it may be applied twice in quick succession.

Zapon Varnishes.—In the case of many articles which have been colored mechanically or by the battery, particularly with large pieces, an opaque varnish is used as a protection against atmospheric influences. The so-called brassoline, of a brown color, negroline, black, and zapon, which is colorless, are employed, according to the color of the article. The last-named varnish is most commonly used, and gives a fine and durable coating, insoluble in almost all liquids which would come into consideration here, except that it will wash off in soap and water. Zapon varnish is a solution of collodion cotton and camphor in amyl acetate and amyl alcohol, and was formerly used to preserve old manuscripts and legal documents. In the process of zaponizing, the article is slightly warmed and immersed in the varnish, or the latter is applied with a brush. The solution is very durable, and has the advantage that after drying it will not show edges, rings, or spots. Zapon varnish which has become too thick must be diluted, and the brushes must be kept from becoming dry. If it is desired to give an especially warm tone, the article is treated with brushes which have been drawn over beeswax or mineral wax.

For the production of zapon or celluloid varnish, pour 20 parts of acetone over 2 parts of colorless celluloid waste, allowing it to stand for several days in a closed vessel, stirring frequently until the whole has dissolved into a clear, thick mass. Admix 78 parts of amyl acetate and clarify the zapon varnish by allowing it to settle for weeks.

VARNISH, HOW TO POUR OUT:
See Castor Oil.

VARNISHES, INSULATING:
See Insulation.

VARNISHES, PHOTOGRAPHIC RETOUCHING:
See Photography.

VARNISH REMOVERS:
See Cleaning Preparations and Methods.

VASELINE STAINS, TO REMOVE FROM CLOTHING:
See Cleaning Preparations and Methods.

VASOLIMENTUM.

This unguent is of two kinds, liquid and semi-solid. The former is prepared by mixing 500 parts of olein, 250 parts of alcoholic ammonia, and 1,000 parts of liquid paraffine, the whole being warmed until completely dissolved, and any loss in weight made up by addition of spirit. The semi-solid preparation is made of the same ingredients, except the paraffine salve is substituted for the liquid. The product is used as a basis for ointments in place of vasogene, and can be incorporated with a number of medicaments, such as 10 per cent of naphthol, 20 per cent of guaiacol, 25 per cent of juniper tar, 5 per cent of thiol, 6 per cent of iodine, 5 per cent of creosote, 10 per cent of ichthyol, 5 per cent of creolin, 2 per cent of menthol, etc.

VAT ENAMELS AND VARNISHES:
See Varnishes.

VEGETABLES, TESTS FOR CANNED:
See Foods.

VEGETABLE PARCHMENT:
See Parchment.

VICHY:
See Waters.

VICHY SALT:
See Salts (Effervescent).

Veterinary Formulas

FOR BIRDS:

Asthma in Canaries.—

Tincture capsicum...	5	drachms
Spirits chloroform...	90	minims
Iron citrate, soluble..	45	grains
Fennel water........,.	3½	ounces

Give a few drops on lump of sugar in the cage once daily.

Colas.—

Tincture ferri perchloride.........	1	drachm
Acid hydrochloric, dil.	½	drachm
Glycerine..........	1½	drachms
Aqua camphor, q. s..	1	ounce

Use 3 to 6 drops in drinking water.

Ointment for Healing.—

Peru balsam........	60	grains
Cola cream........	1	ounce

Apply.

Constipation in Birds.—

F. E. senna	2 drachms
Syrup manna	1 ounce
Fennel water, q. s.	4 ounces

Give a few drops on sugar in cage once daily.

Diarrhœa.—

Tincture iron chloride	2 drachms
Paregoric	2 drachms
Caraway water	3½ ounces

Give few drops on lump of sugar once daily.

Mocking-Bird Food.—

Crackers	8 ounces
Corn	9 ounces
Rice	2 ounces
Hemp seed	1 ounce
Capsicum	10 grains

Mix and reduce to a coarse powder.

Foods for Red Birds.—

Sunflower seed	8 ounces
Hemp seed	16 ounces
Canary seed	10 ounces
Cracked wheat	8 ounces
Unshelled rice	6 ounces

Mix and grind to a coarse powder.

Canary-Bird Food.—

Yolk of egg (dry)	2 ounces
Poppy heads (powdered)	1 ounce
Cuttlefish bone (powdered)	1 ounce
Sugar	2 ounces
Powdered crackers	8 ounces

Bird Tonic.—

Powdered capsicum	20 grains
Powdered gentian	1 drachm
Ferri peroxide	½ ounce
Powdered sugar	½ ounce
Syrup, q. s.	

Put a piece size of pea in cage daily.

Tonic.—

I.—

Tincture cinchona	½ drachm
Tincture iron	2 drops
Glycerine	1 drachm
Caraway water	1 ounce

Put a few drops on lump of sugar in cage daily.

II.—

Compound tincture cinchona	2 drachms
Compound tincture gentian	2 drachms
Syrup orange	1 ounce
Simple elixir	2½ ounces

Put a few drops on lump of sugar in the cage daily.

Antiseptic Wash for Cage Birds.—

Chinosol, F	2 drachms
Sugar (burnt)	20 minims
Aqua cinnamon	4 ounces
Aqua	20 ounces

Add 1 or 2 teaspoonfuls to the bath water and allow the birds to use it, when it will quickly destroy all parasites or germs in the feathers. To wash out the cages, use a mixture of 1 tablespoonful in a pint of hot water.

Mixed Bird Seed.—

Sicily canary	10 ounces
German rape	2 ounces
Russian hemp	1 ounce
German millet	3 ounces

FOR HORSES AND CATTLE:

Blistering. — Tincture cantharides, 1 ounce; camphorated oil, ½ ounce. Apply a portion with friction 3 times a day until a blister shows. As it subsides apply again.

Horse-Colic Remedy.—I.—In making a horse-colic remedy containing tincture of opium, ether and chloroform, to be given in tablespoonful doses, apportion the ingredients about equally, and mix the dose with a pint of water.

Other formulas are:

II.—

Chloroform anodyne	1 ounce
Spirit of nitrous ether	2 ounces
Linseed oil	13 ounces

Give in one dose and repeat in an hour if necessary.

Condition Powders.—I.—Sulphur, 2 pounds; Glauber salts, 1 pound; black antimony, ½ pound; powdered bloodroot, 4 ounces; copperas, ½ pound; rosin, ½ pound; asafetida, 2 ounces; saltpeter ½ pound. Powder and mix well.

II.—Gentian, 4 ounces; potassium nitrate, 1 ounce; sulphur, 4 ounces; ginger (African), 4 ounces; antimony, 4 ounces; rosin, 2 ounces; Fœnugreek, 2 ounces; capsicum, 2 ounces; serpentaria, 2 ounces; sodium sulphate, 9 ounces; flaxseed meal, 16 ounces. All ingredients in fine powder. Dose: 1 tablespoonful in feed twice a day.

Veterinary Dose Table.—For a colt 1 month old give $\frac{1}{14}$ of the full dose; 3 months old, $\frac{1}{12}$; 6 months old, ⅓; 1 year old, ¼; 2 years old, ½; 3 years old, ¾. Fluids for cattle usually the same dose as for the horse. Solids for cattle usually 1½ times the dose for the horse.

Drug.	Horses.	Cattle.
Aloes	1 to 8 dr.	½ to 2 oz.
Alum	1 to 3 dr.	1 to 3 dr.
Aqua ammonia	3 to 5 dr.	3 to 5 dr.
Ammonia bromide	¼ to 2 oz.	¼ to 2 oz.
Ammonia carbonate	1 to 3 dr.	2 to 5 dr.
Ammonia iodide	½ to 3 dr.	1 to 5 dr.
Antimony black	15 to 50 gr.	....
Areca nut	3 to 5 dr.	....
Arsenic	5 to 12 gr.	5 to 12 gr.
Asafetida	1 to 4 dr.	½ to 2 oz.
Belladonna leaves	½ to 2 oz.	½ to 2 oz.
Buchu leaves	½ to 3 oz.	½ to 4 oz.
Calaber bean	4 to 12 gr.	4 to 12 gr.
Camphor	½ to 2 dr.	2 to 3 dr.
Cantharides	5 to 25 gr.	12 to 30 gr.
Capsicum	1 to 2 dr.	1 to 3 dr.
Catechu	1 to 2 dr.	2 to 4 dr.
Chalk preparation	2 to 3 oz.	2 to 4 oz.
Chloral hydrate	½ to 1½ oz.	½ to 1½ oz.
Chloroform	½ to 1 dr.	½ to 2 dr.
Cinchona	1 to 3 dr.	½ to 2 oz.
Copper sulphate	½ to 2 dr.	½ to 3 dr.
Creolin	1 to 5 dr.	2 to 5 dr.
Creosote	15 to 30 min.	1 to 2 dr.
Digitalis leaves	10 to 20 gr.	20 to 50 gr.
Dover powder	½ to 2 dr.	½ to 2 dr.
Ergot	¼ to 1 oz.	¼ to 1 oz.
Ether	½ to 2½ oz.	1 to 3 oz.
Ex. belladonna fluid	½ to 2 dr.	2 to 4 dr.
Extract buchu fluid	1 to 5 dr.	....
Extract cannabis indica	¼ to ½ dr.	¼ to 1 dr.
Fœnugreek	½ to 3 oz.	1 to 3 oz.
Gallnuts	2 to 4 dr.	½ to 1 oz.
Gentian	2 to 6 dr.	½ to 1 oz.
Ginger	3 to 5 dr.	½ to 2 oz.
Ipecac	½ to 2 dr.	½ to 3 dr.
Iron carbonate	1 to 2 dr.	....
Iron sulphate	½ to 2 dr.	1 to 3 dr.
Juniper berries	1 to 2 oz.	1 to 3 oz.
Limewater	3 to 6 oz.	3 to 6 oz.
Magnesia sulphate	½ to 3 lb.	½ to 3 lb.
Mustard	2 to 4 dr.	2 to 6 dr.
Nux vomica	½ to 1 dr.	2 to 3 dr.
Oil castor	½ to 1 pt.	½ to 1 pt.
Oil Croton	10 to 20 min.	1 to 2 dr.
Oil juniper	½ to 2 dr.	½ to 2 dr.
Oil linseed	½ to 1 pt.	½ to 2 pt.
Oil olive	½ to 2 pt.	1 to 2 pt.
Oil savin	1 to 3 dr.	1 to 3 dr.
Oil turpentine	½ to 2 oz.	½ to 2 oz.
Opium	½ to 2 dr.	½ to 2 dr.
Potassium iodide	2 to 4 dr.	2 to 6 dr.
Potassium nitrate	1 to 2 oz.	1 to 2 oz.
Potassium sulphide	1 to 2 dr.	1 to 2 dr.
Quinine	10 to 30 gr.	20 to 40 gr.
Rhubarb	½ to 1 oz.	1 to 2 oz.
Santonine	15 to 40 gr.	½ to 1 dr.
Sodium hyposulphite	½ to 1 oz.	1 to 3 oz.
Sodium sulphate	½ to 2 lb.	1 to 2 lb.
Sodium sulphite	½ to 1 oz.	1 to 3 oz.
Spirits ammonia, aromatic	½ to 2 oz.	1 to 3 oz.
Spirits chloroform	½ to 1 oz.	1 to 2 oz.
Spirits nitrous ether	1 to 3 oz.	1 to 3 oz.
Spirits peppermint	1 to 2 oz.	1 to 2 oz.
Strychnine sulphite	½ to 1 gr.	1 to 3 gr.
Sulphur	2 to 4 oz.	2 to 4 oz.
Tincture aconite	5 to 30 min.	5 to 20 min.
Tincture asafetida	1 to 4 dr.	....
Tincture belladonna	1 to 3 dr.	2 to 4 dr.
Tincture cantharides	1 to 2 oz.	½ to 1 oz.
Tincture columbo	½ to 2 oz.	1 to 2 oz.
Tincture digitalis	1 to 3 dr.	2 to 4 dr.
Tincture iron	1 to 2 oz.	1 to 2 oz.
Tincture ginger	½ to 2 oz.	1 to 2 oz.
Tincture nux vomica	2 to 4 dr.	½ to 1 oz.
Tincture opium	½ to 3 oz.	1 to 3 oz.
Tobacco	½ to 1 dr.	½ to 1 dr.
Vinegar	1 to 3 oz.	2 to 6 oz.
Whisky	2 to 10 oz.	....
White vitriol	5 to 15 gr.	5 to 15 gr.

Astringent.—

I.—Opium............. 12 grains
Camphor........... ½ drachm
Catechu............ 1 drachm

One dose.

II.—Opium............. 12 grains
Camphor........... 1 drachm
Ginger............... 2 drachms
Castile soap.......... 2 drachms
Anise............... 3 drachms
Licorice............ 2 drachms

Contracted Hoof or Sore Feet.—

I.—Lard...............
Yellow wax.........
Linseed oil.........
Venice turpentine...
Tar................
} Equal parts.

Apply to the edge of the hair once a day.

II.—Rosin............... 4 ounces
Lard............... 8 ounces

Melt and add
Powdered vertigris... 1 ounce

Stir well; when partly cool add
Turpentine......... 2 ounces

Apply to hoof about 1 inch down from the hair.

Cough.—

I.—Sodii bromide....... 180 grains
Creosote water....... 2 ounces
Fennel water........ 4 ounces

Half tablespoonful 4 times daily.

II.—Ammonia bromide... 180 grains
Fennel water........ 4 ounces
Syrup licorice....... 4 ounces

Teaspoonful 4 times daily.

Cow Powder.—

Powdered catechu... 60 grains
Powdered ginger..... 240 grains
Powdered gentian.... 240 grains
Powdered opium..... 30 grains

CUTS, WOUNDS, SORES.

I.—Tincture opium, 2 ounces; tannin, ¼ ounce.

II.—Tincture aloes, 1 ounce; tincture of myrrh, ½ ounce; tincture of opium, ½ ounce; water, 4 ounces. Apply night and morning.

III.—Lard, 4 ounces; beeswax, 4 ounces; rosin, 2 ounces; carbolic acid, ¼ ounce.

Diarrhœa.—

I.—Opium........... 15 grains
Peppermint........ ¼ ounce
Linseed meal...... 1 ounce

Give half in morning and remainder in evening in a pint of warm water.

II.—Prepared chalk..... 6 ounces
    Catechu.......... 3 ounces
    Opium............ 1½ ounces
    Ginger........... 3 ounces
    Gentian.......... 3 ounces

One powder 3 times a day in half a pint of warm water. One-sixth of dose for calves.

Diuretic Ball.—

I.—Oil juniper....... ½ drachm
    Rosin............. 2 drachms
    Saltpeter......... 2 drachms
    Camphor.......... ½ drachm
    Castile soap....... 1 ounce
    Flaxseed meal..... 1 ounce

Make 1 pill.

II.—Rosin............. 90 grains
    Potassium nitrate... 90 grains
    Po buchu leaves.... 45 grains

Dose: 1 twice a day.

Drying Drink.—

    Powdered alum...... 6 ounces
    Armenian bole....... 2 ounces
    Powdered juniper berries .... ½ ounce

Once daily in 1 quart of warm gruel.

Epizooty or Pinkeye.—

    Sublimed sulphur .... ½ ounce
    Epsom salt.......... 1 ounce
    Charcoal........... ½ ounce
    Extract licorice...... 1 ounce

Fever.—

I.—Salicylic acid....... ¾ ounce
    Sodium bicarbonate.. ½ ounce
    Magnesium sulphate. 10 ounces

Give half in quart of warm bran water at night.

II.—Spirits niter......... 3 ounces
    Tincture aconite..... 2 drachms
    Fluid extract belladonna............. ½ ounce
    Nitrate potash...... 2 ounces
    Muriate ammonia.... 2 ounces
    Water, q. s.......... 1 quart

Dose: Teaspoonful every 2 or 3 hours till better.

Heaves. — I. — Balsam copaiba, 1 ounce; spirits of turpentine, 2 ounces; balsam fir, 1 ounce; cider vinegar, 16 ounces.

Tablespoonful once a day.

II.—Saltpeter, 1 ounce; indigo, ½ ounce; rain or distilled water, 4 pints.
Dose: 1 pint twice a day.

Hide Bound.—

    Elecampane......... 2 ounces
    Licorice root........ 2 ounces
    Fœnugreek.......... 2 ounces
    Rosin............... 2 ounces
    Copperas........... ½ ounce
    Ginger............. 2 drachms
    Gentian............ 1 drachm
    Saltpeter........... 1 drachm
    Valerian............ 1 drachm
    Linseed meal........ 3 ounces
    Sublimed sulphur.... 1 ounce
    Black antimony...... 4 drachms

Tablespoonful twice a day.

HORSE EMBROCATIONS AND LINIMENTS.

I.—Camphor.......... 1 ounce
    Acetic acid......... 15 ounces
    Alcohol............. 18 ounces
    Oil turpentine....... 51 ounces
    Eggs............... 6
    Distilled witch hazel. 45 ounces

II.—Iodine............ 50 grains
    Pot iodide..........125 grains
    Soap liniment....... 6 ounces

INFLUENZA.

I.—Ammonia muriate... 1½ ounces
    Gum camphor...... ½ ounce
    Pot chloride........ 1 ounce
    Extract licorice, powdered............. 2 ounces
    Molasses, q. s.

Make a mass. Dose: Tablespoonful in form of pill night and morning.

II.—Ammonium chloride. 30 parts
    Potassium nitrate.... 30 parts
    Potassium sulphate in little crystals...... 100 parts
    Licorice powder..... 65 parts

Mix. Dose: A tablespoonful, in a warm mash, 3 times daily.

INFLAMMATION OF THE UDDER.

I.—Salicylic acid........ 40 grains
    Mercurial ointment.. 1 ounce
    Liniment of camphor 3¼ ounces

Apply and rub the udder carefully twice a day.

II.—Belladonna root....... 1 drachm
    Oil turpentine........ 1 ounce
    Camphor............ 1 drachm
    Solution green soap, q.s. 6 ounces

Mix and make a liniment. Bathe the udder several times with hot water. Dry and apply above liniment.

MANGE.

Sulphur is a specific for mange; the trouble consists in its application. The old-fashioned lotion of train oil and black sulphur serves well enough, but for stabled animals something is wanted which will effectually destroy the parasites in harness and saddlery without injury to those expensive materials. The creosote emulsions and coal-tar derivatives generally are fatal to the sarcopts if brought into actual contact, but a harness pad with ridges of accumulated grease is a sufficient retreat for a few pregnant females during a perfunctory disinfection, and but a few days will be needed to reproduce a new and vigorous stock. A cheap and efficient application can be made by boiling together flowers of sulphur and calcis hydras in the proportion of 4 parts of the former to 1 of the latter, and 100 of water, for half an hour. It should be applied warm, or immediately after washing with soft soap.

Milk Powder for Cows.—For increasing the flow of milk, in cows, Hager recommends the following mixture:

Potassium nitrate....	1 part
Alum................	1 part
Sublimed sulphur....	1 part
Prepared chalk......	1 part
White bole..........	2 parts
Red clover..........	5 parts
Anise...............	10 parts
Fennel..............	10 parts
Salt................	10 parts

All should be in tolerably fine powder and should be well mixed. The directions are to give 1 or 2 handfuls with the morning feed.

LAXATIVES.

I.—Aloes..............	1 drachm
Soap.................	12 drachms
Caraway.............	4 drachms
Ginger..............	4 drachms
Treacle, q. s.	

Make 4 balls. Dose: 1 daily.

II.—Rochelle salts.......	2 ounces
Aloes, powdered.....	150 grains
Linseed meal.......	150 grains

One dose, given in warm water.

Lice.—

Crude oil...........	1 ounce
Oil tar.............	1 ounce
Oil cedar...........	1 drachm
Cottonseed oil........	5 ounces

Apply to parts.

DOMESTIC PETS.

The sarcoptic itch of the dog, as well as that of the cat, is transmissible to man. The *Tinea tonsurans*, the so-called barbers' itch, due to a trychophyton, and affecting both the dog and cat, is highly contagious to man. Favus, *Tinea favos*, caused by *achorion schoenleini*, of both animals, is readily transmissible to human beings. The dog carries in his intestines many kinds of *tænia* (tapeworm), among them *Tænia echinococcus*, the eggs of which cause hydatic cysts. Hydatic cysts occur in persons who are always surrounded with dogs, or in constant contact with them.

Aviar diphtheria (i. e., the diphtheria of birds), caused by at least two microbes (bacillus of Klebs-Loeffler and bacillus coli), may easily be transmitted to man and cause in him symptoms analogous to those of true diphtheritic angina.

Parrots are subject to an infectious enteritis which may be communicated to human beings, giving rise to the so-called psittacosis (from the Greek, *psitta*, a parrot), of which there have been a number of epidemics in France. It is determined by the bacillus of Nocard.

Human tuberculosis is certainly transmitted to dogs, cats, and birds. Cadiot, Gibert, Roger, Benjamin, Petit, and Basset, as well as other observers, cite cases where dogs, cats, and parrots. presenting all the lesions of tuberculosis, were shown to have contracted it from contact with human beings; while there are no recorded cases, there can scarcely be a natural doubt that man may, in a similar manner, become attainted through them, and that their tuberculosis constitutes an actual danger to man.

Need we recall here the extraordinary facility with which hydrophobia is communicated to man through the dog, cat, etc.?

We may, therefore, conclude that we should not permit these animals to take up so much space in our apartments, nor should they be petted and caressed either by adults or children in the reckless manner common in many households. The disgusting habit of teaching animals to take bits of food, lumps of sugar, etc., from between the lips of members of the family is also to be shunned.

Finally, any or all of them should be banished from the house the moment that they display certain morbid symptoms. Besides, in certain cases, there should be a rigid prophylaxis against certain diseases—as echinococcus, for instance.

Worms.—In cats and dogs, round worms, of which ascaris mystax is the most common in cats, are found chiefly in young animals. This worm has hirsute appendages somewhat resembling a mustache. To treat an animal infected with such "guests," the patient should be made to fast for 24 hours. For a small kitten ½ grain of santonin, up to a grain or two for large cats, followed in an hour by a dose of castor oil, is recommended. To avoid spilling the oil on the animal's coat the "doctor" should have it heated and whipped with warm milk. Another way to get cats to take it is to smear it on the bottoms of their front feet, when they will lick it off.

Areca nut, freshly ground by the druggist himself and administered in liberal doses, say 30 to 60 grains, will usually drive out any worms in the alimentary canal.

It is important that animals successfully treated for worms once should undergo the treatment a second or third time, as all the parasites may not have been killed or removed the first time, or their progeny may have developed in the field vacated by the parents.

The following is an effective formula:

German wormseed, powdered......... 1 drachm
Fluid extract of spigelia............. 3 drachms
Fluid extract of senna. 1 drachm
Fluid extract of valerian.............. 1 drachm
Syrup of buckthorn .. 2 ounces

Dose: From ½ to 1 teaspoonful night and morning.

Foot Itch.—The itch that affects the feet of poultry is contagious in a most insidious way. The various birds of a poultry yard in which the disease is prevalent, rarely contract it until after a comparatively long period of exposure, but sooner or later every bird will contract it. One infected bird is enough to infect a whole yard full, and once infected, it is exceedingly difficult to get rid of. The disease, however, affects birds only.

The treatment is simple. Having softened the feet by keeping them for some minutes in tepid water, the scabs that cover them are carefully detached, avoiding, as far as possible, causing them to bleed, and taking the precaution of throwing every scab into the fire. The feet are then carefully dried, with a bit of soft cotton material, which should afterwards be burned; then the entire surface is covered with ointment (*Unguentum sulphuris kalinum*). An alcoholic solution of Canada balsam is preferred by some.

Protect the ointment by a proper appliance, and allow it to remain in contact 2 or 3 days. At the end of this time remove the applications and wash off with tepid suds. The bird will generally be found cured, but if not, repeat the treatment—removing the remaining scabs, which will be found soft enough without resorting to soaking in tepid water, and apply the ointment directly.

There is another method of treatment that has been found successful, which not only cures the infected birds but prevents the infection of others. It is simply providing a sand bath for the birds, under a little shed, where they can indulge themselves in rolling and scratching, the bath being composed of equal parts fine sand, charcoal in fine powder, ashes, and flowers of sulphur, sifted together. The bath should be renewed every week. In the course of a few weeks the cure is complete.

Foods.—

I.—Powdered egg shell or phosphate of lime. 4 ounces
Iron sulphate...... 4 ounces
Powdered capsicum.. 4 ounces
Powdered Fœnugreek 2 ounces
Powdered black pepper............. 1 ounce
Silver sand......... 2 ounces
Powdered lentils .... 6 ounces

A tablespoonful to be mixed with sufficient feed for 20 hens.

II.—Oyster shell, ground. 5 ounces
Magnesia........... 1 ounce
Calcium carbonate.. 3 ounces
Bone, ground....... 1½ ounces
Mustard bran....... 1½ ounces
Capsicum.......... 1 ounce

Powders.—

I.—Cayenne pepper..... 2 parts
Allspice............. 4 parts
Ginger.............. 6 parts

Powder and mix well together. A teaspoonful to be mixed with every pound of food, and fed 2 or 3 times a week. Also feed fresh meat, finely chopped.

II.—Powdered egg shells.. 4 parts
Powdered capsicum.. 4 parts
Sulphate of iron..... 4 parts
Powdered Fœnugreek 2 parts
Powdered black pepper.............. 1 part
Sand.:............. 2 parts
Powdered dog biscuit 6 parts

A tablespoonful to be mixed with sufficient meal or porridge to feed 20 hens.

Lice Powders.—

I.—Sulphur............ 4 ounces
Tobacco dust....... 6 ounces
Cedar oil.......... ¼ ounce
White hellebore..... 4 ounces
Crude naphthol..... 1 ounce
Powdered chalk, q. s. 2 pounds

II.—Sulphur............ 1 ounce
Carbolic acid....... ¼ ounce
Crude naphthol..... 1 ounce
Powdered chalk..... 1 pound

Roup or Gapes.—Roup in poultry is caused by the presence of parasites or entozoa in the windpipe. Young birds are most commonly affected. The best method of treatment is to expose the affected bird to the fumes of heated carbolic acid until on the point of suffocation. The bird may be placed in a box with a hot brick, and carbolic acid placed thereon. The fowls soon recover from the incipient suffocation, and are almost always freed from the disease. Care must be taken to burn the parasites coughed out, and the bodies of any birds which may die of the disease. The following powders for the treatment of "roup" in poultry have been recommended:

I.—Potassium chlorate.. 1 ounce
Powdered cubebs.... 1 ounce
Powdered anise..... ½ ounce
Powdered licorice.... 1½ ounces

Mix a teaspoonful with the food for 20 hens.

II.—Ammonium chloride. 1 ounce
Black antimony..... ¼ ounce
Powdered anise..... ½ ounce
Powdered squill..... ¼ ounce
Powdered licorice.... 2 ounces

Mix and use in the foregoing.

FOR SHEEP:

Dips.—For the prevention of "scab" in sheep, which results from the burrowing of an acarus or the destruction of the parasite when present, various preparations of a somewhat similar character are used. The following formulas for sheep dips are recommended by the United States Department of Agriculture:

I.—Soap.............. 1 pound
Crude carbolic acid.. 1 pint
Water.............. 50 gallons

Dissolve the soap in a gallon or more of boiling water, add the acid, and stir thoroughly.

II.—Fresh skimmed milk.. 1 gallon
Kerosene........... 2 gallons

Churn together until emulsified, or mix and put into the mixture a force pump and direct the stream from the pump back into the mixture. The emulsification will take place more rapidly if the milk be added while boiling hot.

Use 1 gallon of this emulsion to each 10 gallons of water required.

Constipation.—

I.—Green soap........ 150 grains
Linseed oil........ 1½ ounces
Water............. 15 ounces

Give ⅛ every ½ hour till action takes place.

II.—Calomel........... 1½ grains
Sugar............. 15 grains

One dose.

Loss of Appetite.—

Sodium sulphate,
dried............. 90 grains
Sodium bicarbonate.. 30 grains
Rhubarb........... 30 grains
Calamus........... 90 grains

Form the mass into 6 pills. Give one twice daily.

Inflammation of the Eyes.—

Zinc sulphate........ 20 grains
Mucilage quince seed. 4 ounces
Distilled water....... 4 ounces

Bathe eyes twice daily.

Vinegar

I.—Into a hogshead with a large bung-hole put 1,500 parts, by weight, of honey, 125 parts of carob-pods, cut into pieces, 50 parts of powdered red or white potassium bitartrate, 125 parts of powdered tartaric acid, 2,000 parts of raisin stems, 400 parts of the best brewers' yeast, or 500 of leaven rubbed up in water; add 16,000 parts of triple vinegar and 34,000 parts of 40 per cent spirit, containing no fusel oil. Stir all vigorously together; fill up the hogshead with hot water (100° F.), close the bunghole with gauze to keep out insects, and let the contents of the cask stand for from 4 to 6 weeks or until they have turned to vinegar. The temperature of the room should be from 77° to 88° F.

Draw off half the vinegar, and fill the hogshead up again with 15 parts of soft water and 1 part of spirit (40 per cent). Do this 4 times, then draw off all the vinegar and begin the first process over again. This method of making vinegar is suitable for households and small dealers, but would not suffice for wholesale manufacturers, since it would take too long to produce any large amount.

II.—Put into an upright wine cask open at the top, 14,000 parts, by weight, of lukewarm water, 2,333 parts of 60 per cent alcohol, 500 parts of brown sugar, 125 parts of powdered red or white potassium bitartrate, 250 parts of good brewers' yeast, or 125 parts of leaven, 1,125 parts of triple vinegar, and stir until the substances are dissolved. Lay a cloth and a perforated cover over the cask and let it stand in a temperature of 72° to 77° F. from 4 to 6 weeks; then draw off the vinegar. The thick deposit at the bottom, the "mother of vinegar," so called, can be used in making more vinegar. Pour over it the same quantities of water and alcohol used at first; but after the vinegar has been drawn off twice, half the first quantity of sugar and potassium bitartrate, and the whole quantity of yeast, must be added. This makes excellent vinegar.

III.—A good strong vinegar for household use may be made from apple or pear peelings. Put the peelings in a stone jar (not glazed with lead) or in a cask, and pour over them water and a little vinegar, fermented beer, soured wine, or beet juice. Stir well, cover with a linen cloth and leave in a warm room. The vinegar will be ready in 2 or 3 weeks.

IV.—Two wooden casks of any desired size, with light covers, are provided. They may be called A and B. A is filled with vinegar, a tenth part of this is poured off into B, and an equal amount of fermented beer, wine, or any other sweet or vinous liquid, or a mixture of 1,125 parts, by weight, of alcohol, 11,500 to 14,000 parts of water, and 1,125 parts of beet juice, put into A.

When vinegar is needed, it is drawn out of B, an equal quantity is poured from A into B and the same quantity of vinegar-making liquids put into A. In this way vinegar is constantly being made and the process may go on for years, provided that the casks are large enough so that not more than a tenth of the contents of A is used in a week. If too much is used, so that the vinegar in the first cask becomes weak, the course of the vinegar making is disturbed for a long time, and this fact, whose importance has not been understood, prevents this method—in its essential principles the best—from being employed on a large scale. The surplus in A acts as a fermentative.

Aromatic Vinegar.—I.—Sixteen ounces glacial acetic acid, 40 drops oil of cloves, 40 drops oil of rosemary, 40 drops oil of bergamot, 16 drops oil of neroli, 30 drops oil of lavender, 1 drachm benzoic acid, ½ ounce camphor, 30 to 40 drops compound tincture of lavender, 3 ounces spirit of wine. Dissolve the oils, the benzoic acid, and the camphor in the spirit of wine, mix with acetic acid and shake until bright, lastly adding the tincture of lavender to color.

II.—Dried leaves of rosemary, rue, wormwood, sage, mint, and lavender flowers, each ½ ounce; bruised nutmegs, cloves, angelica root, and camphor, each ¼ of an ounce; rectified alcohol, 4 ounces; concentrated acetic acid, 16 ounces. Macerate the materials for a day in the alcohol; then add the acid and digest for 1 week longer at a temperature of 490° F. Finally press out the now aromatised acid and filter it.

Cider Vinegar.—By "artificial vinegar" is meant vinegar made by the quick method with beechwood shavings. This cannot be carried out with any economy on a small scale, and requires a plant. A modification of the regular plan is as follows: Remove the head from a good tight whisky barrel, and put in a wooden faucet near the bottom. Fill the barrel with corn cobs and lay an empty coffee sack over them. Moisten the cobs by sprinkling them with some good, strong, natural vinegar, and let them soak for a few hours. After the lapse of 2 or 3 hours draw off the vinegar and again moisten the cobs, repeating this until they are rendered sour throughout, adding each time 1 quart of high wines to the vinegar before throwing it back on the cobs. This prevents the vinegar from becoming flat, by the absorption of its acetic acid by the cobs. Mix a gallon of molasses with a gallon of high wine and 14 gallons of water and pour it on the cobs. Soak for 8 hours, then draw off and pour on the cobs again. Repeat this twice daily, until the vinegar becomes sour enough to suit. By having a battery of barrels, say 4 barrels prepared as above, the manufacture may be made remunerative, especially if the residue of sugar casks in place of molasses, and the remnants of ale, etc., from the bar-rooms around town are used. All sugar-containing fruit may be utilized for vinegar making.

VINEGAR, TESTS FOR:
See Foods.

VINEGAR, TOILET:
See Cosmetics.

VIOLET AMMONIA:
See Cosmetics.

VIOLET WATER:
See Perfumes.

VIOLIN ROSIN:
See Rosin.

VIOLIN VARNISH:
See Varnishes.

VISCOSE:
See Celluloid.

VOICE LOZENGES:
See Confectionery.

VULCANIZATION OF RUBBER:
See Rubber.

WAGON GREASE:
See Lubricants.

WALLS, DAMP:
See Household Formulas.

WALL AND WALL-PAPER CLEAN-ERS:
See Cleaning Preparations and Methods, also Household Formulas.

WALL-PAPER DYES:
See Dyes.

WALL-PAPER PASTE:
See Adhesives.

WALL PAPER, REMOVAL OF:
See Household Formulas.

WALL WATERPROOFING:
See Waterproofing and Household Formulas.

WALL PRIMING:
See Paints.

WALNUT:
See Wood.

WARMING BOTTLE:
See Bottles.

WARPING, PREVENTION OF:
See Wood.

Warts

Wart Cure.—The following is especially useful in cases where the warts are very numerous:

I.—Chloral hydrate	1 part
Acetic acid	1 part
Salicylic acid	4 parts
Sulphuric ether	4 parts
Collodion	15 parts

Mix. Directions: Every morning apply the foregoing to the warts, painting one coat on another. Should the mass fall off without taking the warts with it, repeat the operation. Take, internally 10 grains of burnt magnesia daily.

II.—Sulphur	10 parts
Acetic acid	5 parts
Glycerine	25 parts

Keep the warts covered with this mixture.

WASHING FLUIDS AND POWDERS:
See Laundry Preparations.

WASTE, PHOTOGRAPHIC, ITS DISPOSITION:
See Photography.

WATCH-DIAL CEMENTS:
See Adhesives, under Jewelers' Cements.

WATCH GILDING:
See Plating.

Watchmakers' Formulas

WATCH MANUFACTURERS' ALLOYS.

Some very tenacious and hard alloys, for making the parts of watches which are not sensitive to magnetism, are as follows:

	I	II	III	IV	V	VI	VII
Platinum	62.75	62.75	62.75	54.32	0.5	0.5	—
Copper	18	16.20	16.20	16	18.5	18.5	25
Nickel	18	18	16.50	24.70	—	2	1
Cadmium	1.25	1.25	1.25	1.25	—	—	—
Cobalt	—	—	1.50	1.96	—	—	—
Tungsten	—	1.80	1.80	1.77	—	—	—
Palladium	—	—	—	—	72	72	70
Silver	—	—	—	—	6.5	7	4
Rhodium	—	—	—	—	1	—	—
Gold	—	—	—	—	1.5	—	—

A non-magnetic alloy for watch-springs, wheels, etc.: Gold, 30 to 40 parts; palladium, 30 to 40 parts; copper, 10 to 20 parts; silver, 1 to 5 per cent; cobalt, 0.1 to 2.5 per cent; tungsten, 0.1 to 5 per cent; rhodium, 0.1 to 5 per cent; platinum, 0.1 to 5 per cent.

An Alloy for Watch Pinion Sockets.—Gold, 31 parts; silver, 19 parts; copper, 39 parts; palladium, 1 part.

Replacing Rubies whose Settings have Deteriorated.—Enlarge, with the squarer (steel brooch for enlarging holes), the hole of the old setting, and adjust it, with hard rubbing, to the extremity of a stem of pierced brass wire. Take the stem in an American nippers, and set the ruby at the extremity (the setting may be driven back by using a flat burnishing tool, very gently). Then take off with a cleaving file the part of the stem where the ruby is set, and diminish it to the thickness desired, by filing on the finger, or on cork. These operations finished, a set stopper is obtained which now needs only to be solidly fixed at the suitable height, in the hole prepared.

To Straighten Bent Teeth.—Bent teeth are straightened by means of the screwdriver used as a lever against the root of the adjacent teeth, and bent pivots may be held in the jaws of the pliers and the pinion bent with the fingers in the direction and to the extent required. For such a purpose, pliers having the jaws lined with brass are used so that the pivot is not bruised, and the bending has to be done with great care.

To Renew a Broken Barrel Tooth.—Frequently, in consequence of the breaking of a spring, a tooth of a barrel is broken. Sometimes it may only be bent, in which case the blade of a penknife may be used with care. If 2 or 3 successive teeth are lacking, the best way is to change the barrel, but a single tooth may be easily renewed in this way: Drill a hole through the thickness of the tooth, taking care not to penetrate the drum; then fit in a piece of metal tightly and give it, as well as possible, the correct form of the tooth. To assure solidity, solder it; then clean and round the edges. Properly executed the repair will scarcely be noticed.

Heated Sawdust.—Sawdust is known to have been employed from time immemorial by watchmakers and goldsmiths for the purpose of drying rinsed articles. The process of drying can be accelerated four-fold if the sawdust is heated before use. This must, however, be done with great caution and constant stirring.

To Repair a Dial, etc., with Enamel Applied Cold.—There are two kinds of false enamel for application, when cold, to damaged dials. The first, a mixture of white rosin and white lead, melts like sealing wax, which it closely resembles. It is advisable when about to apply it to gently heat the dial and the blade of a knife, and with the knife cut the piece of enamel of the requisite size and lay it on the dial. The new enamel must project somewhat above the old. When cold the surface is leveled by scraping, and a shining surface is at once produced by holding at a little distance from the flame of a spirit lamp. It is necessary to be very careful in conducting this operation, as the least excess of heat will burn the enamel and turn it yellow. It is, however, preferable to the following although more difficult to apply, as it is harder and does not become dirty so soon. The second false enamel contains white lead mixed with melted white wax. It is applied like cement, neatly filling up the space and afterwards rubbing with tissue paper to produce a shining surface. If rubbed with a knife blade or other steel implement its surface will be discolored.

Lettering a Clock Dial.—Painting Roman characters on a clock dial is not such a difficult task as might at first be imagined. If one has a set of drawing instruments and properly proportions the letters, it is really simple. The letters should be proportioned as follows: The breadth of an "I" and a space should equal $\frac{1}{2}$ the breadth of an "X," that is, if the "X" is $\frac{1}{2}$ inch broad, the "I" will be $\frac{3}{16}$ inch broad and the space between letters $\frac{1}{16}$ inch, thus making the "I" plus one space equal to $\frac{1}{4}$ inch or half the breadth of an "X." The "V's" should be the same breadth as the "X's." After the letters have been laid off in pencil, outline them with a ruling pen and fill in with a small camel's-hair brush, using gloss black paint thinned to the proper consistency to work well in the ruling pen. Using the ruling pen to outline the letters gives sharp straight edges, which it would be impossible to obtain with a brush in the hands of an inexperienced person.

Verification of the Depthings.—In the verge watches, the English watches, and those of analogous caliber, it is often difficult to verify the depthings, except by the touch. For this reason we often find the upper plate pierced over each depth. In the jeweled places, instead of perforating the upper plate, it suffices to deposit a drop of very limpid oil on the ruby, taking care that it does not scatter. In this manner a lens is formed and one may readily distinguish the depthing.

To Make or Enlarge a Dial Hole.—By wetting the graver or the file with spirit of turpentine, cracks may be avoided and the work will be accomplished much quicker.

To Repair a Repeating Clock-Bell.—When the bell is broken, whether short off or at a distance, file it away and pierce it, and after having sharpened a little the stem of the spring which remains, push by force, in the hole just made, a thin piece of solder (pewter). The sound will not have changed in any appreciable manner.

A seconds pendulum of a regulator, which has no compensation for temperature will cause the clock to lose about

1 second per day for each 3 degrees of increase in heat. A watch without a compensation balance will lose 6.11 seconds in 24 hours for each increase of 1° F. in heat.

To Remedy Worn Pinions.—Turn the leaves or rollers so that the worn places upon them will be toward the arbor or shaft and fasten them in that position. If they are "rolling pinions," and they cannot be secured otherwise, a little soft solder should be used.

Watchmakers' Oil. — I. — Put some lead shavings into neat's foot oil, and allow to stand for some time, the longer the better. The lead neutralizes the acid, and the result is an oil that never corrodes or thickens.

II.—Stir up for some time best olive oil with water kept at the boiling point; then after the two fluids have separated, decant the oil and shake up with a little freshly burned lime. Let the mixture stand for some weeks in a bottle exposed to the sunlight and air, but protected from wet and dirt. When filtered, the oil will be nearly colorless, perfectly limpid, and will never thicken or become rancid.

To Weaken a Balance Spring.—A balance spring may need weakening; this is effected by grinding the spring thinner. Remove the spring from the collet and place it upon a piece of pegwood cut to fit the center coil. A piece of soft iron wire, flattened so as to pass freely between the coils and charged with a little powdered oilstone, will serve as a grinder, and with it the strength of the spring may soon be reduced. Operations will be confined to the center coil, for no other part of the spring will rest sufficiently against the wood to enable it to be ground, but this will generally suffice. The effect will be rather rapid; therefore care should be taken or the spring may be made too weak.

To Make a Clock Strike Correctly.— Pry the plates apart on the striking side, slip the pivots of the upper wheels out, and having disconnected them from the train, turn them partly around and put them back. If still incorrect, repeat the experiment. A few efforts at most will get them to work properly. The sound in cuckoo clocks is caused by a wire acting on a small bellows which is connected with two small pipes like organ pipes.

To Reblack Clock Hands.—One coat of asphaltum varnish will make old rusty hands look as good as new, and will dry in a few minutes.

To Tighten a Ruby Pin.—Set the ruby pin in asphaltum varnish. It will become hard in a few minutes and be much firmer and better than the gum shellac, generally used.

To Loosen a Rusty Screw in a Watch Movement.—Put a little oil around the screw; heat the head lightly by means of a red-hot iron rod, applying the same for 2 or 3 minutes. The rusty screw may then be removed as easily as though it had just been put in.

Gilding Watch Movements. (See also Gilding.)—In gilding watch movements, the greatest care must be observed with regard to cleanliness. The work is first to be placed into a weak solution of caustic potash for a few minutes, and then rinsed in cold water. The movements are now to be dipped into pickling acid (nitrous acid) for an instant, and then plunged immediately into cold water. After being finally rinsed in hot water, they may be placed in the gilding bath and allowed to remain therein until they have received the required coating. A few seconds will generally be sufficient, as this class of work does not require to be very strongly gilt. When gilt, the movements are to be rinsed in warm water, and scratch-brushed; they may then be returned to the bath, for an instant, to give them a good color. Lastly, rinse in hot water and place the movements in clean box sawdust. An economical mode of gilding watch movements is to employ a copper anode—working from the solution, add 10 parts of cream of tartar and a corresponding quantity of elutriated chalk to obtain a pulp that can be put on with the brush. The gilding or silvering obtained in this manner is pretty, but of slight durability. At the present time this method is only seldom employed, since the electroplating affords a means of producing gilding and silvering in a handsome and comparatively cheap manner, the metallic coating having to be but very thin. Gold and silver for this kind of work are used in the form of potassium cyanide of gold or potassium cyanide of silver solutions, it being a custom to copper the zinc articles previously by the aid of a battery, since the appearance will then be much handsomer than on zinc alone. Gilding or silvering with leaf metal is done by polishing the surface of the zinc bright and coating it with a very tough linseed-oil varnish diluted with 10 times the quantity of benzol. The metallic leaf is then laid on and polished with an agate.

WATCHMAKERS' CLEANING PREP-
ARATIONS:
See Cleaning Preparations and Meth-
ods.

WATCH MOVEMENTS, PALLADIUM
PLATING OF:
See Plating.

Water, Natural and Artificial

In making an artificial mineral water
it must be remembered that it is seldom possible to reproduce the water
by merely combining its chemical components. In other words, the analysis
of the water cannot serve as a basis from
which to prepare it, because even though
all of the components were put together,
many would be found insoluble, and
others would form new chemical combinations, so that the result would differ
widely from the mineral water imitated.

For example, carbonate of magnesia
and carbonate of lime, which are important ingredients in most mineral
waters, will not make a clear solution
unless freshly precipitated. Hence,
when these are to be reproduced in a
mineral water it is customary to employ
other substances which will dissolve at
once, and which will, upon combining,
produce these salts. The order in which
the salts are added is also a very important matter, for by dissolving the
salts separately and then carefully combining them, solutions may be effected
which would be impossible were all the
salts added together to the water in the
portable fountain.

In this connection the following table
will be found useful:

Group 1

Ammonium carbon-ate.	Sodium carbonate.
Ammonium chloride.	Sodium chloride.
Sodium borate (bo-rax).	Sodium fluoride.
	Sodium iodide.
Potassium carbon-ate.	Sodium nitrate.
	Sodium phosphate.
Potassium chloride.	Sodium pyrophos-phate.
Potassium nitrate.	Sodium silicate.
Potassium sulphate.	Sodium sulphate.
Sodium bromide.	

Group 2

Lithium carbonate.

Group 3

Aluminum chloride.	Magnesium chlo-ride.
Barium chloride.	
Calcium bromide.	Magnesium nitrate.
Calcium chloride.	Strontium chloride.
Calcium nitrate.	Lithium chloride.

Group 4

Magnesium sul-phate.	Alum (potassa or soda alum).

Group 5

Lime carbonate.	Lime sulphate pre-cipitate.
Magnesium carbon-ate hydrate.	

Group 6

Lithium carbonate.	Iron pyrophosphate.
Acid hydrochloric.	Iron sulphate.
Acid sulphuric.	Manganese chloride.
Iron chloride.	Manganese sulphate.

Group 7

Sodium arseniate, or sodium sulphide,
or acid hydrosulphuric.

Explanation of Groups.—The explanation of the use of these groups is simple.
When about to prepare an artificial
mineral water, first ascertain from the
formula which of the ingredients belong
to group 1. These should be dissolved in
water, and then be filtered and added to
distilled water, and thoroughly agitated.
Next the substance or substances belonging to group 2 should be dissolved
in water, then filtered and added to the
water, which should again be agitated.
And so the operation should proceed;
whatever ingredients are required from
each group should be taken in turn, a
solution made, and this solution, after
being filtered, should be separately added to the fountain, and the latter be well
agitated before the following solution is
added.

For groups 1, 3, and 4, the salts should
be dissolved in 5 times their weight of
boiling, or 10 times their weight of cold,
water. For group 2 (lithium carbonate)
the proportions should be 1 part of
lithium carbonate to about 130 parts of
cold or boiling water. The substances
mentioned in group 5 are added to the
portable fountain in their solid state, and
dissolve best when freshly precipitated.
As carbonic acid gas aids their solution,
it is best to charge the fountain after they
are added, and agitate thoroughly, blowing off the charge afterwards if necessary.

In group 5 the lithium carbonate is
dissolved in the acids (see also group 2),
the iron and manganese salts are dissolved in 5 parts of boiling, or 10 parts of
cold, water, the solution quickly filtered,
the acids added to it, and the whole
mixture added to the fountain already
charged with gas, the cap being quickly
taken off, and the solution poured in.
The iron and manganese salts easily
oxidize and produce turbidity, therefore
the atmospheric air should be carefully

blown off under high pressure several times while charging fountains. The substances mentioned in group 7 are never put into the fountain, except the arseniate of sodium in the case of Vichy water, which contains but a trifling amount of this compound.

Most of the solutions may be prepared beforehand and be used when required, thus saving considerable time.

Formulas for various waters will be given at the end of this article.

A question which arises in preparing mineral waters is: What is the best charging pressure? As a general rule, they are charged to a lower pressure than plain soda; good authorities even recommend charging certain mineral waters as low as 30 pounds pressure to the square inch, but this seems much too low a pressure for the dispensing counter. From 50 to 120 pounds pressure would be a good limit, while plain soda may be served out as high as 180 pounds. There must be enough pressure completely to empty the fountain, while enabling sufficient gas to be retained by the water to give it a thorough pungency. Moreover, a high pressure to the mineral water enables a druggist at a pinch, when he runs out of plain soda, to use his Vichy water, instead, with the syruped drinks. The taste of the Vichy is not very perceptible when covered by the syrup, and most customers will not notice it.

Apollinaris Water.—

Sodium carbonate....	2,835 grains
Sodium sulphate.....	335 grains
Sodium silicate......	10 grains
Magnesium chloride.	198 grains
Calcium chloride.....	40 grains
Potassa alum........	57 grains
Magnesium carbonate hydrate........	158 grains
Iron sulphate........	21 grains

Hunyadi Water.—

Magnesium sulphate.	400 parts
Sodium sulphate.....	400 parts
Potassium sulphate..	2 parts
Sodium chloride.....	31 parts
Sodium bicarbonate..	12 parts
Water..............	1 quart

Lithia Water.—

Lithium carbonate...	120 grains
Sodium bicarbonate.	1,100 grains
Carbonated water....	10 gallons

For "still" lithia water, substitute lithium citrate for the carbonate in the above formula.

Seltzer Water. — Hydrochloric acid (chemically pure), 2,520 grains; pure water, 40 ounces. Mix and add marble dust, 240 grains; carbonate of magnesium, 420 grains. Dissolve, and after 1 hour add bicarbonate of sodium, 2,540 grains. Dissolve, then add sufficient pure water to make 10 gallons. Filter and charge to 100 pounds pressure.

Vichy Water.—The following formula, based on the analysis of Bauer-Struve, yields an imitation of

Vichy (Grande Grille).

Sodium iodide......	0.016	parts
Sodium bromide....	0.08	parts
Sodium phosphate..	2	parts
Sodium silicate.....	80	parts
Potassium sulphate .	125	parts
Sodium chloride....	139	parts
Sodium carbonate...	6,792	parts
Aluminum chloride.	1	part
Strontium chloride..	1	part
Ammonium chloride	3	parts
Magnesium chloride	24	parts
Calcium chloride...	170	parts
Manganese sulphate	0.46	parts
Iron sulphate......	1	part
Sulphuric acid......	40	parts
Water to make......	10	gallons

Mix the first 7 ingredients with about 10 times their weight of water and filter. In the same manner, mix the next 5 ingredients with water and filter; and then the last 3 ingredients. Pour these solutions into sufficient water contained in a fountain to make 10 gallons, and charge at once with carbon dioxide gas.

Waters like the above are more correctly named "imitation" than "artificial," as the acidic and basic radicals may bear different relations to one another in the natural and the other.

PURIFYING WATER.

See also Filters.

If an emulsion of clay is poured into a soap solution, the clay gradually separates out without clarifying the liquid. When a few drops of hydrochloric acid, however, are added to a soap solution and a small quantity—about 1.5 per cent—of a clay emulsion poured in, the liquid clarifies at once, with formation of a plentiful sediment. Exactly the same process takes place when the waste waters from the combing process in spinning are treated with clay. The waters which remain turbid for several days contain 500 to 800 grams of fatty substances per cubic meter. If to 1 liter of this liquid 1 gram of clay is added, with 15 to 20 per cent of water, the liquid clarifies with separation of a sediment and assumes a golden-brown color. Besides the fatty substances, this deposit also contains a certain quantity of nitrogenous bodies. Dried at (100°C.) 212° F., it weighs about 1.6 grams and contains 30 per cent of fat. The grease obtained from it is clear, of good quality, and deliquesces at 95° F. After removal of this fat, the mass still contains 1.19 per cent of nitrogen.

Sterilization of Water with Lime Chloride.—In order to disinfect and sterilize 1,000 parts of drinking water, 0.15 parts of dry chloride of lime are sufficient. The lime is stirred with a little water into a thin paste and introduced, with stirring, into the water to be disinfected and a few drops of officinal hydrochloric acid are added. After ¼ hour the clarification and disinfection is accomplished, whereupon 0.3 parts of calcium sulphite are added, in order to kill the unpleasant smell and taste of the chlorine.

Clarifying Muddy Water.—The water supply from rivers is so muddy at times that it will not go through the filter. When this happens agitate each barrel of water with 2 pounds of phosphate of lime and allow it to settle. This will take but a few minutes, and it will be found that most of the impurities have been carried down to the bottom. The water can then be drawn off carefully and filtered.

Removal of Iron from Drinking Water.—The simplest method for removing the taste of iron in spring water is to pass the water through a filter containing a layer of tricalcic phosphate either in connection with other filtering materials or alone. The phosphate is first recovered in a gelatinous form, then dried and powdered.

For Hardness.—A solution perfectly adapted to this purpose, and one which may be kept a long time, is prepared as follows:

Thirty-five parts of almond oil are mixed with 50 parts of glycerine of 1.26 specific gravity and 8.5 parts of 50 per cent soda lye, and boiled to saponification. To this mixture, when it has cooled to from 85° to 90° C. (185° to 194° F.), are added 100 to 125 parts of boiling water. After cooling again, 500 parts of water are added, and the solution is poured into a quart flask, with 94 per cent alcohol to make up a quart. After standing 2 months it is filtered. Twenty hydrolimeter degrees of this solution make, with 40 parts of a solution of 0.55 grams of barium chloride in 1 quart of water, a dense lather 1 centimeter high.

WATER (COPPER):
See Copper.

WATER ICES:
See Ice Creams.

WATER, TO FREEZE:
See Refrigeration.

WATER JACKETS, ANTI-FREEZING SOLUTIONS FOR:
See Freezing Preventives.

WATER SPOTS, PRIMING FOR:
See Paint.

WATER STAINS:
See Wood.

WATER-LILY ROOTS:
See Pyrotechnics.

WATER, STIRRED YELLOW, SCARLET AND COLORLESS:
See Pyrotechnics.

WATERS (TOILET):
See Cosmetics.

WATER-GLASS CEMENTS:
See Adhesives.

WATER GLASS IN STEREOCHROMATIC PAINTING:
See Stereochromy.

Waterproofing

(See also Enamels, Glazes, Paints, Preservatives, Varnishes.)

Waterproofing Brick Arches.—Waterproofing of brick arches is done in the following manner: The masonry is first smoothed over with cement mortar. This is then covered with a special compound on which a layer of Hydrex felt is laid so as to lap at least 12 inches on the transverse seams. Five layers of compound and 5 of felt are used, and special attention is paid to securing tightness around the drain pipes and at the spandrel walls. In fact the belt is carried up the back of the latter and turned into the joint under the coping about 2 inches, where it is held with cement mortar. The waterproofing on the arches is protected with 1 inch of cement mortar and that on the walls with a single course of brickwork.

Waterproofing Blue Prints.—Use refined paraffine, and apply by immersing the print in the melted wax, or more conveniently as follows: Immerse in melted paraffine until saturated, a number of pieces of an absorbent cloth a foot or more square. When withdrawn and cooled they are ready for use at any time.

To apply to a blue print, spread one of the saturated cloths on a smooth surface, place the dry print on it with a second waxed cloth on top, and iron with a moderately hot flatiron. The paper immediately absorbs paraffine until saturated, and becomes translucent and highly waterproofed. The lines of the print are intensified by the process, and there is no shrinking or distortion. As the wax is withdrawn from the cloths, more can be added by melting small pieces directly under the iron.

By immersing the print in a bath of melted paraffine the process is hastened, but the ironing is necessary to remove the surplus wax from the surface, unless the paper is to be directly exposed to the weather and not to be handled. The irons can be heated in most offices by gas or over a lamp, and a supply of saturated cloths obviates the necessity of the bath. This process, which was originally applied to blue prints to be carried by the engineer corps in wet mines, is equally applicable to any kind of paper, and is convenient for waterproofing typewritten or other notices to be posted up and exposed to the weather.

Waterproof Coatings.—I.—Rosin oil, 50 parts; rosin, 30 parts; white soap, 9 parts. Apply hot on the surfaces to be protected.

II.—It has been observed that when gluten dried at an ordinary temperature, hence capable of absorbing water, is mixed with glycerine and heated, it becomes water-repelling and suitable for a waterproof paint. One part of gluten is mixed with $1\frac{1}{2}$ parts of glycerine, whereby a slimy mass is obtained which is applied on fabrics subsequently subjected to a heat of 248° F. The heating should not last until all glycerine has evaporated, otherwise the coating becomes brittle and peels off.

Waterproofing Canvas.—I.—The canvas is coated with a mixture of the three solutions named below:

1. Gelatin, 50 parts, by weight, boiled in 3,000 parts of water free from lime. 2. Alum, 100 parts, dissolved in 3,000 parts of water. 3. Soda soap dissolved in 2,000 parts of water.

II.—Prepare a zinc soap by entirely dissolving 56 parts of soft soap in 125 to 150 parts of water. To the boiling liquid add, with constant stirring, 28 to 33 parts of zinc vitriol (white vitriol). The zinc soap floats on top and forms, after cooling, a hard white mass, which is taken out. In order to clean it of admixed carbonic alkali, it must be remelted in boiling fresh water. Next place 232.5 parts of raw linseed oil (free from mucus) in a kettle with 2.5 parts of best potash, and 5 parts of water. This mass is boiled until it has become white and opaque and forms a liquid, soap-like compound. Now, add sugar of lead, 1.25 parts; litharge, 1 part; red lead, 2 parts; and brown rosin, 10.5 parts. The whole is boiled together about 1 hour, the temperature not being allowed to exceed 212° F., and stirring well from time to time. After this add 15 parts of zinc soap and stir the whole until the metal soap has combined with the oil, the temperature not exceeding 212° F. When the mixture is complete, add a solution of caoutchouc, 1.2 parts, and oil of turpentine, 8.56 parts, which must be well incorporated by stirring. The material is first coated on one side by means of a brush with this composition, which must have a temperature of 158° F. Thereupon hang it up to dry, then apply a second layer of composition possessing the same temperature, which is likewise allowed to dry. The fiber is now filled out, so that the canvas is waterproof.

Waterproofing Corks.—For the purpose of making corks as impervious as possible, while at the same time keeping them elastic, saturate them with caoutchouc solution. Dissolve caoutchouc in benzine in the ratio of 1 part of caoutchouc to 19 parts of benzine. Into this liquid lay the corks to be impregnated and subject them to a pressure of 150 to 180 pounds by means of a force pump, so that the liquid can thoroughly enter. The corks thus treated must next be exposed to a strong draught of air until all trace of benzine has entirely evaporated and no more smell is noticeable.

WATERPROOFING FABRICS.

It will be convenient to divide waterproof fabrics into two classes, viz., those which are *impervious* to water, and those which are *water-repellent*. It is important to make this distinction, for, although all waterproof material is made for the purpose of resisting water, there is a vast difference between the two classes. The physical difference between them can be briefly summed up as follows: Fabrics which are completely impervious to water comprise oil-skins, mackintoshes, and all materials having a water-resisting film on one or both sides, or in the interior of the fabric. Those coming under the second heading of water-repellent materials do not possess this film, but have their fibers so treated as to offer less attraction to the water than the water molecules have for themselves.

The principal members of the first group are the rubber-proofed goods; in these the agent employed is rubber in greater or less quantity, together with other bodies of varying properties. Before enlarging on this class, it will be necessary to give a short description of the chemical and physical properties of rubber.

Rubber, or caoutchouc, is a natural gum exuding from a large number of plants, those of the *Euphorbiaceæ* being the chief source for the commercial variety. The raw material appears on the market in the shape of blocks, cakes, or bottle-shaped masses, according to the manner in which it has been collected. It possesses a dark-brown — sometimes nearly black—exterior; the interior of the mass is of a lighter shade, and varies from a dingy brown to a dirty white, the color depending on the different brands and sources. In the raw state its properties are very different from what they are after going through the various manufacturing processes, and it has only a few of the characteristics which are generally associated with India rubber. Chemically it is a complex hydrocarbon with the formula $C_{45}H_{36}$, and appears to consist of a highly porous network of cells having several different rosins in their interstices. It is perfectly soluble in no single solvent, but will yield some of its constituents to many different solvents. At a temperature of $10°$ C. ($50°$ F.) raw caoutchouc is a solid body and possesses very little elasticity. At $36°$ C. ($97°$ F.) it is soft and elastic to a high degree, and is capable of being stretched 16 times its length. Further increase of temperature lessens its elastic properties, and at $120°$ C. ($248°$ F.) it melts. While in the raw condition it has several peculiar properties, one of which is: After stretching, and cooling suddenly while stretched, it retains its new form, and only regains its former shape on being warmed. Another striking feature is its strong adhesive capacity; this property is so powerful that the rubber cannot be cut with a knife unless the blade is wet; and freshly cut portions, if pressed together, will adhere and form a homogeneous mass. From these facts it will be seen how it differs from rubber in the shape of a cycle tire or other manufactured form.

The most valuable property possessed by raw caoutchouc is that of entering into chemical combination with sulphur, after which its elasticity is much increased; it will then bear far greater gradations of heat and cold. This chemical treatment of caoutchouc with sulphur is known as "vulcanizing," and, if properly carried out, will yield either soft vulcanized rubber or the hard variety known as vulcanite. On the other hand, caoutchouc, after vulcanizing, has lost its plastic nature, and can no longer be molded into various shapes, so that in the production of stamped or molded objects, the customary method is to form them in unvulcanized rubber and then to vulcanize them.

Raw caoutchouc contains a number of natural impurities, such as sand, twigs, soil, etc.; these require removing before the manufacturing processes can be carried out. The first operation, after rough washing, is to shred the raw material into small strips, so as to enable the impurities to be washed out. This process is carried out by pressing the rubber against the surface of a revolving drum (A, Fig. 1), carrying a

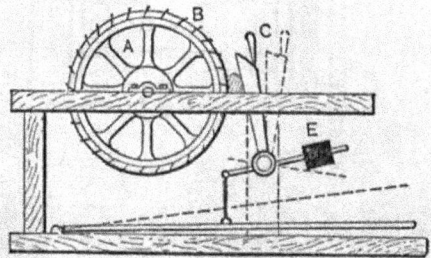

number of diagonally arranged knives, B, on its surface. A lever, C, presses the rubber against the knives; D is the fulcrum on which C works, E being a weight which throws back the lever on the pressure being removed. During

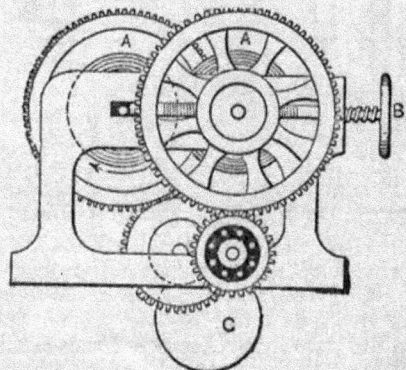

this operation a jet of water is kept playing onto the knives to cool and enable them to cut.

Following this comes the passage between a pair of corrugated steel rollers (as shown in Fig. 2). These rollers have each a different speed, so that the rubber gets stretched and squeezed at the same time. Immediately over the rollers a water pipe is fixed, so that a steady stream of water washes out all the sand and other extraneous matter. In Fig. 2, AA are the steel rollers, while B is a screw working springs which regulate the pressure between the rollers. The power is transmitted from below from the pulley, C, and thence to the gearing.

The next operation, after well drying, is to thoroughly masticate the shredded rubber between hot steel rollers, which resemble those already described, but usually have a screw-thread cut on their surfaces. Fig. 3 shows the front view

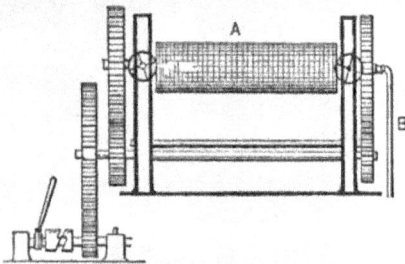

FIG. 3.

of this masticating machine, A being the rollers, while the steam pipe for heating is shown at B. Fig. 3a gives a top view

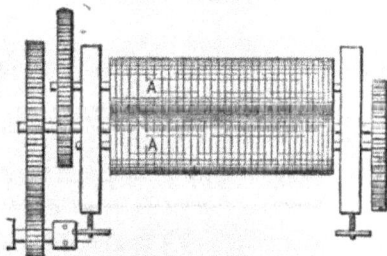

FIG. 3A.

of the same machine, showing the two rollers.

After passing several times through these, the rubber will be in the form of homogeneous strips, and is then ready either for molding or dissolving. As we are dealing solely with waterproofed textiles, the next process which concerns us is the dissolving of the rubber in a suitable solvent. Benzol, carbon bisulphide, oil of turpentine, ether, and absolute alcohol, will each dissolve a

certain amount of rubber, but no one of them used alone gives a thorough solution. The agent commonly employed is carbon bisulphide, together with 10 per cent of absolute alcohol. Whatever solvent is used, after being steeped in it for some hours the caoutchouc swells out enormously, and then requires the addition of some other solvent to effect a complete solution. A general method is to place the finely shredded rubber in a closed vessel, to cover it with carbon bisulphide, and allow to stand for some hours. Toward the end of the time the vessel is warmed by means of a steam coil or jacket, and 10 parts absolute alcohol are added for every 100 parts of carbon bisulphide. The whole is then kept gently stirred for a few hours. Fig. 4 shows a common type of the vessel

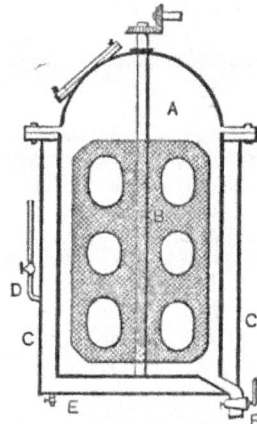

FIG. 4.

used for dissolving rubber. In this diagram A is the interior of the vessel, and B a revolving mixer in the same. The whole vessel is surrounded by a steam jacket, C, with a steam inlet at D and a tap for condensed water at E. F is the cock by which the solution is drawn off.

After the rubber is dissolved, about 12 to 24 per cent of sulphur is added, and thoroughly incorporated with the solution. The sulphur may be in the form of chloride of sulphur, or as sulphur pure and simple. A very small quantity of sulphur is required to give the necessary result, 2 to 3 per cent being sufficient to effect vulcanization; but a large quantity is always added to hasten the operation.

Even after prolonged treatment with the two solvents, a solution of uniform consistency is never obtained: clots of a thicker nature will be found floating in the solution, and the next operation is to knead it up so as to obtain equal density throughout. Fig. 5 will give an idea of how this mixing is done.

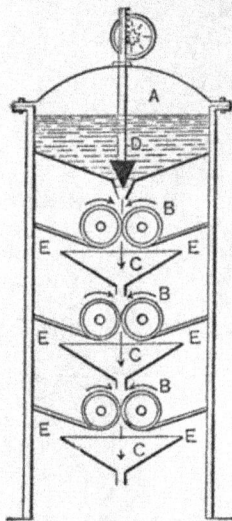

FIG. 5.

At the top of a closed wooden chamber is a covered reservoir, A, containing the solution of rubber. A long slit at the base of this reservoir allows the solution to fall between sets of metal rollers, BBB below. Neighboring rollers are revolving in opposite directions, and at different speeds, so that, after passing all three sets of rollers, and emerging at the bottom, the solution should be of uniform consistency. CCC are the guiding funnels, and EE are scrapers to clear the solution from the rollers. D is a wedge-shaped plug worked by a rack and pinion, and regulates the flow of the solution.

It now remains to apply the rubber to the fabric and vulcanize it. Up to this stage the sulphur has only been mechanically mixed with the rubber; the aid of heat is now required to bring about chemical combination between the two. This process, which is known as "burning," consists in subjecting the rubber-covered fabric to a temperature of about 248° F. Sulphur itself melts at 239° F.,

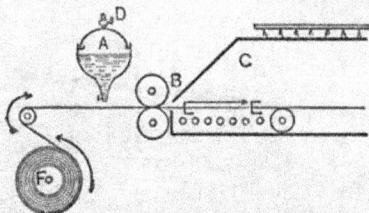

FIG. 6.

and the temperature at which combination takes place must be above this. Fig. 6 shows one of the methods of spreading the rubber on the cloth. A is the tank containing the solution with an outlet at the bottom arranged so as to regulate the flow of solution. The fabric passes slowly underneath this, receiving as it travels a thin coating of the waterproofing. The two rollers at B press the solution into the fabric and distribute the proofing evenly over the entire surface.

After leaving the two squeezing rollers, the cloth travels slowly through a covered chamber, C, having a series of steam pipes, EE, underneath, to evaporate the solvent; this condenses on the upper portion of the chamber, which is kept cooled, and flows down the sides into suitable receptacles. After this the proofed cloth is vulcanized by passing round metal cylinders heated to the necessary temperature, or by passing through a heated chamber. Fig. 7 shows the spreading of

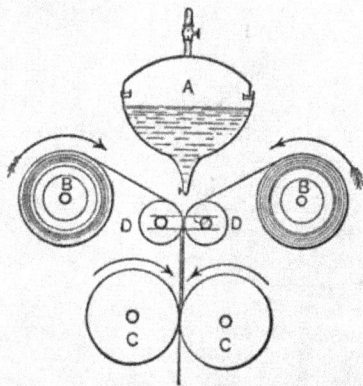

FIG. 7.

rubber between two fabrics. The two cloths are wound evenly on the rollers, BB; from this they are drawn conjointly through the rollers, D, the stream of proofing solution flowing down between the rollers, which then press the two fabrics together with the rubber inside. The lower rollers marked CC are heated to the necessary degree, and cause the rubber and sulphur to combine in chemical union.

So far the operation of proofing has been described as though pure rubber only was used; in practice the rubber forms only a small percentage of the proofing material, its place being taken by cheaper bodies. One of the common ingredients of proofing mixtures is boiled linseed oil, together with a small quantity of litharge; this dries very quickly, and forms a glassy flexible film. Coal tar, shellac, colophony, etc., are all used, together with India-rubber varnish, to make different waterproof compositions. Oil of turpentine and benzol form good solvents for rubber, but it is absolutely essential that both rubber and solvent be perfectly anhydrous before mixing. Oil of turpentine, alcohol, etc., can be best deprived of water by mixing with either sulphuric acid or dehydrated copper sulphate, and allowing to stand. The acid or the copper salt will absorb the water and sink to the bottom, leaving a supernatant layer of dehydrated turpentine or whatever solvent is used. All the sulphur in a rubber-proofed cloth is not in combination with the rubber; it is frequently found that, after a lapse of time, rubber-proofed material shows an efflorescence of sulphur on the surface, due to excess of sulphur, and occasionally the fabric becomes stiff and the proofing scales off. Whenever a large proportion of sulphur is present, there is always the danger of the rubbers forming slowly into the hard vulcanite state, as the substance commonly called vulcanite consists only of ordinary vulcanized rubber carried a stage further by more sulphur being used and extra heat applied. If after vulcanizing, rubber is treated with caustic soda, all this superfluous sulphur can be extracted; if it is then well washed the rubber will retain its elasticity for a long period. With the old methods of proofing, a sheet of vulcanized rubber was cemented to a fabric with rubber varnish, and frequently this desulphurizing was performed before cementing together. The result was a flexible and durable cloth, but of great weight and thickness, and expensive to produce.

The chemistry of rubber is very little understood; as mentioned previously, rubber is a highly complex body, liable to go through many changes. These changes are likely to be greater in rubber varnish, consisting of half a dozen or more ingredients, than in the case of rubber alone. The action of sunlight has a powerful effect on rubber, much to its detriment, and appears to increase its tendency to oxidize. Vulcanized rubber keeps its properties better under water than when exposed to the air, and changes more slowly if kept away from the light. It appears as though a slight decomposition always takes place even with pure rubber; but the presence of so many differently constituted substances as sometimes occur in rubber solutions no doubt makes things worse. Whenever a number of different bodies with varying properties are consolidated together by heat, as in the case of rubber compositions, it is only reasonable to expect there will be some molecular rearrangement going on in the mass; and this can be assigned as the reason why some proofings last as long again as others. Some metallic salts have a very injurious action on rubber, one of the worst being copper sulphate. Dyers are frequently warned that goods for rubber-proofing must be free from this metal, as its action on rubber is very powerful, though but little understood. As is generally known, grease in any form is exceedingly destructive to rubber, and it should never be allowed in contact in the smallest proportion. Some compositions are made up by dissolving rubber in turpentine and coal tar; but in this case some of the rubber's most valuable properties are destroyed, and it is doubtful if it can be properly vulcanized. Owing to rubber being a bad conductor of heat, it requires considerable care to vulcanize it in any thickness. A high degree of heat applied during a short period would tend to form a layer of hard vulcanite on the surface, while that immediately below would be softer and would gradually merge into raw rubber in the center.

The different brands of rubber vary so much, especially with regard to solubility, that it is always advisable to treat each brand by itself, and not to make a solution of two or more kinds. Oilskins and tarpaulins, etc., are mostly proofed by boiled linseed oil, with or without thickening bodies added. They are not of sufficient interest to enlarge upon in this article, so the second, or "water-repellent," class has now to be dealt with.

All the shower-proof fabrics come under this heading, as well as every cloth which is pervious to air and repulsive to water. The most time-honored recipe for proofing woollen goods is a mixture of sugar of lead and alum, and dates back hundreds of years. The system of using this is as follows: The two ingredients are dissolved separately, and the solutions mixed together. A mutual decomposition results, the base of the lead salt uniting with the sulphuric acid out of the alum to form lead sulphate, which precipitates to the bottom. The clear solution contains alumina in the form of acetate, and this supplies the proofing quality to the fabric. It is applied in a form of machine shown in Fig. 8, which will be seen to consist of a trough containing the proofing solution, C, with a pair of squeezing rollers, A, over the top. The fabric is drawn down through the solution and up through the squeezers in the direction of the arrows. At the back of the machine the cloth automatically winds itself onto a roll, *B*, and then only requires drying to develop the water-

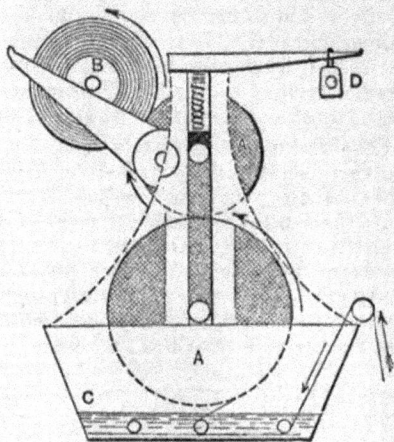

FIG. 8.

resisting power. *D* is a weight acting on a lever which presses the two rollers, *A*, together. The water-repelling property is gained as follows:

Drying the fabric, which is impregnated with acetate of alumina, drives off some of the volatile acetic acid, leaving a film of basic acetate of alumina on each wool fiber. This basic salt is very difficult to wet, and has so little attraction for moisture that in a shower of rain the drops remain in a spheroidal state, and fall off. In a strong wind, or under pressure, water eventually penetrates through fabrics proofed in this manner; but they will effectually resist a sharp shower. Unfortunately, shower-proofed goods, with wear, gradually lose this property of repelling water. The equation representing the change between alum and sugar of lead is given below. In the case of common alum there would, of course, be potassium acetate in solution besides the alumina.

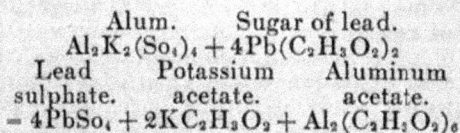

Alum.   Sugar of lead.
$$Al_2K_2(So_4)_4 + 4Pb(C_2H_3O_2)_2$$
Lead   Potassium   Aluminum
sulphate.   acetate.   acetate.
$$= 4PbSo_4 + 2KC_2H_3O_2 + Al_2(C_2H_3O_2)_6$$

Now that sulphate of alumina is in common use, alum need not be used, as the potash in it serves no purpose in proofing.

There are many compositions conferring water: resisting powers upon textiles, but unfortunately they either affect the general handle of the material and make it stiff, or they stain and discolor it, which is equally bad. A large range of waterproof compositions can be got by using stearates of the metals: these, in nearly every case, are insoluble bodies, and when deposited in the interior of a fabric form a water-resisting "filling" which is very effective. As a rule these stearates are deposited on the material by means of double baths; for example, by passing the fabric through (say) a bath of aluminum acetate, and then, after squeezing out the excess of liquid, passing it through a bath of soap. The aluminum salt on the fabric decomposes the soap, resulting in a deposit of insoluble stearate of alumina. This system of proofing in two baths is cleaner and more economical than adding all the ingredients together, as the stearate formed is just where it is required "on the fibers," and not at the bottom of the bath.

One of the most important patents now worked for waterproofing purposes is on the lines of the old alumina process. In this case the factor used is rosin, dissolved in a very large bulk of petroleum spirit. The fabrics to be proofed (usually dress materials) are passed through a bath of this solution, and carefully dried to drive off the solvent. Following this, the goods are treated by pressing with hot polished metal rollers. This last process melts the small quantity of rosin, which is deposited on the cloth, and leaves each single fiber with an exceedingly thin film of rosin on it. It will be understood that only a very attenuated solution of rosin is permissible, so that the fibers of the threads and not the threads themselves are coated with it. If the solution contains too much rosin the fabric is stiffened, and the threads cemented together; whereas if used at the correct strength (or, rather, weakness) neither fabric nor dye suffers, and there is no evidence of stickiness of any description.

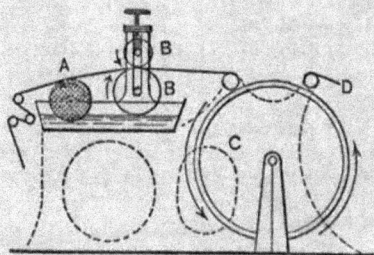

FIG. 9.

Fig. 9 shows a machine used for spreading a coat of either proofing or any other fluid on one side of the fabric.

This is done by means of a roller, A, running in the proofing solution, the material to be coated traveling slowly over the top and just in contact with the roller, A, which transfers the proofing to it. Should the solution used be of a thick nature, then a smooth metal roller will transfer sufficient to the fabric. If the reverse is the case, and the liquid used is very thin, then the roller is covered with felt, which very materially adds to its carrying power. As shown in Fig. 9, after leaving the two squeezing rollers, BB, the fabric passes slowly round a large steam-heated cylinder, C, with the coated side uppermost. This dries the proofing and fastens it, and the cloth is taken off at D.

Besides stearates of the metals, glues and gelatins have been used for proofing purposes, but owing to their stiffening effect, they are only of use in some few isolated cases. With glue and gelatin the fixing agent is either tannic acid or some metallic salt. Tannic acid converts gelatin into an insoluble leather-like body; this can be deposited in the interstices of the fabric by passing the latter through a gelatin bath first, and then squeezing and passing through the tannic acid. Bichromate of potash also possesses the property of fixing the proteid bodies and rendering them insoluble.

The following are special processes used to advantage in the manufacture of waterproof fabrics:

I.—Ordinary Fabrics, Dressing Apparel, etc.—Immerse in a vat of acetate of alumina (5° Bé.) for 12 hours, lift, dry, and let evaporate at a temperature of from 140° to 149° F.

II.—Sailcloth, Awnings, Thick Blankets, etc.—Soak in a 7 per cent solution of gelatin at 104° F., dry, pass through a 4 per cent solution of alum, dry again, rinse in water, and dry.

III.—Fabrics of Cotton, Linen, Jute, and Hemp.—Put into a bath of ammoniacal cupric sulphate of 10° Bé. at a temperature of 87° F.; let steep thoroughly, then put in a bath of caustic soda (20° Bé.) and dry. To increase the impermeability, a bath of sulphate of alumina may be substituted for the caustic-soda bath.

IV.—Saturate the fabrics with the following odorless compound, subjecting them several times to a brushing machine having several rollers, where the warp threads will be well smoothed, and a waterproof product of fine sheen and scarcely fading will be the result. The compound is made with 30 parts, by weight, of Japan wax, 22½ parts, by weight, of paraffine, 12 parts, by weight, of rosin soap, 35 parts, by weight, of starch, and 5 parts, by weight, of a 5 per cent solution of alum. Fabrics thus prepared are particularly adapted to the manufacture of haversacks, shoes, etc.

V.—White or Light Fabrics.—Pass first through a bath of acetate of alumina of 4° to 5° Bé. at a temperature of 104° F., then through the rollers to rid of all liquid; put into a warm solution of soap (5 parts, by weight, of olive-oil soap to 100 parts, by weight, of fresh water) and finally pass through a 2 per cent solution of alum, dry for 2 or 3 days on the dropping horse, and brush off all particles of soap.

VI.—Dissolve 1½ parts, by weight, of gelatin in 50 parts, by weight, of boiling water, add 1½ parts, by weight, of scraped tallow soap and 2½ parts, by weight, of alum, the latter being put in gradually; lower the temperature of the bath to 122° F., lift out the fabric, dry, and calender.

VII.—Tent Cloth.—Soak in a warm solution of 1 part, by weight, of gelatin, 1 part, by weight, of glycerine, and 1 part, by weight, of tannin in 12 parts, by weight, of wood vinegar (pyroligneous acid) of 12° Bé. The whole is melted in a kettle and carefully mixed. The mass is poured into the receiver of the brushing machine, care being taken to keep it liquid. For a piece of 500 feet in length and 20 inches in width, 50 to 80 parts, by weight, of this compound are needed.

VIII.—To freshen worn waterproof material, cover with the following: Fifty-five thousand parts, by weight, of gelatin; 100 parts, by weight, of bichromate of potash; 100 parts, by weight, of acetic acid (to keep glue from congealing), and from 3,000 to 5,000 parts, by weight, of water; to this add 500 parts, by weight, of peroxide of ammoniacal copper, 100° Bé. This compound is put on the fabric with a brush and then exposed to air and light.

IX.—Soft Hats.—The hats are stiffened as usual, then put through the following three baths: Dissolve ½ part, by weight, of tallow soap in from 40 to 50 parts, by weight, of warm water (140° F.). Put 3 to 4 dozen hats into this solution, leave them in it for half an hour, then take out and put them as they are into another bath prepared with 40 to 50 parts, by weight, of water and ½ part, by weight, of alum and heated to 86° to 104° F. After having been left in the second bath for $\frac{1}{4}$ or $\frac{1}{2}$ hour, take out as before, put into the third bath of 40 to 50 parts, by weight, of water, $\frac{1}{2}$ part, by weight, of alum, and about 13 parts, by weight, of fish glue. In this cold bath the hats are left for another $\frac{1}{2}$ hour or more until they are completely saturated with the liquid, then dried and the other operations continued.

X.—Woolen cloth may be soaked in a vat filled with aluminum acetate, of 5° Bé., for 12 hours, then removed, dried, and dried again at a temperature of 140° F.

XI.—Wagon covers, awnings, and sails are saturated with a 7 per cent gelatin solution, at a temperature of 104° F., dried in the air, put through a 4 per cent solution of alum, dried again in the air, carried through water, and dried a third time.

XII.—Cotton, linen, jute, and hemp fabrics are first thoroughly saturated in a bath of ammonio-cupric sulphate, of 10° Bé., at a temperature of 77° F., then put into a solution of caustic soda, 2° Bé., and dried. They may be made still more impervious to water by substituting a solution of aluminum sulphate for the caustic soda.

XIII.—White and light-colored fabrics are first put into a bath of aluminum acetate, 4° to 5° Bé., at a temperature of 102° F., the superfluous liquid being removed from the fabric by press rollers. The fabric is put into a soap solution (5 parts of good Marseilles soap in 100 parts of soft water). Finally it is put through a 2 per cent alum solution, and left to dry for 2 or 3 days on racks. The adhering particles of soap are removed by brushing with machinery.

XIV.—Dissolve 1.5 parts of gelatin in 50 parts of boiling water, add 1.5 parts of shavings of tallow grain soap, and gradually, 2.5 parts of alum. Let this cool to 122° F., draw the fabric through it, dry and calender.

XV.—Cellular tissues are made waterproof by impregnating them with a warm solution of 1 part, by weight, of gelatin, 1 part, by weight, of glycerine, and 1 part, by weight, of tannin, in 12 parts, by weight, of wood vinegar, 12° Bé.

XVI.—Linen, hemp, jute, cotton, and other fabrics can be given a good odorless waterproof finish by impregnating them, and afterwards subjecting them to the action of several mechanical brush rollers. By this process the fabric is brushed dry, the fibers are laid smooth, the threads of the warp brought out, and a glossy, odorless, unfading waterproof stuff results. Fabrics manufactured in the usual way from rough and colored yarns are put through a bath of this waterproof finish, whose composition is as follows: Thirty parts, by weight, of Japanese wax; 22.5 parts, by weight, of paraffine; 15 parts, by weight, of rosin soap; 35 parts, by weight, of starch, and 5 parts, by weight, of a 5 per cent alum solution. The first three components are melted in a kettle, the starch and, lastly, the alum added, and the whole stirred vigorously.

XVII.—One hundred parts, by weight, of castor oil are heated to nearly 204° F., with 50 parts, by weight, of caustic potash, of 50° Bé., to which 50 parts, by weight, of water have previously been added. Forty parts, by weight, of cooler water are then added slowly, care being taken to keep the temperature of the mixture constant. As soon as the liquor begins to rise, 40 parts, by weight, of cooler water are again added, with the same precaution to keep the temperature from falling below 204° F. At the same time care must be taken to prevent the liquor boiling, as this would produce too great saponification. By the prolonged action of heat below the boiling point, the oil absorbs water and caustic potash without being changed, and the whole finally forms a perfectly limpid, nearly black liquid. This is diluted with 5 times its weight of hot or cold water, and is then ready for use without any further preparation. Other vegetable oils may be employed besides castor oil, and the quantity of unsaponified oil present may be increased by stirring the prepared liquid with a fresh quantity of castor or other vegetable oil. The product is slightly alkaline, but wool fiber is not injured, as the oiling may be done in the cold. The solution is clear and limpid, and will not separate out on standing like an emulsion. This product in spinning gives a 10 per cent better utilization of the raw material owing to the greater evenness and regularity with which the fibers are oiled; in weaving less oiling is required.

The product can be completely removed by water, preferably by cold water, and scouring of the goods subsequently with soap, soda, or fuller's earth can thus be dispensed with.

XVIII.—Cloth may be rendered waterproof by rubbing the under side with a lump of beeswax until the surface presents a uniform white or grayish appearance. This method it is said renders the cloth practically waterproof, although still leaving it porous to air.

XIX.—Coating the under side of the cloth with a solution of isinglass and then applying an infusion of galls is another method, a compound being thus formed which is a variety of leather.

XX.—An easy method is the formation of aluminum stearate in the fiber of the cloth, which may readily be done by immersing it in a solution of aluminum sulphate in water (1 in 10) and without allowing it to dry passing through a solution of soap made from soda and tallow or similar fat, in hot water. Reaction between the aluminum sulphate and the soap produces aluminum stearate and sodium sulphate. The former is insoluble and remains in the fiber; the latter is removed by subsequently rinsing the fabric in water.

XXI.—A favorite method for cloth is as follows: Dissolve in a receptacle, preferably of copper, over a bright coal fire, 1 liter (1.76 pints) of pure linseed oil, 1 liter (1.76 pints) of petroleum, ½ liter (0.88 pints) of oil turpentine, and 125 grams (4.37 ounces) of yellow wax, the last named in small bits. As there is danger of fire, boiling of this mass should be avoided. With this hot solution removed from the fire, of course the felt material is impregnated; next it is hung up in a warm, dry room or spread out, but in such a manner that the uniform temperature can act upon all parts.

Waterproofing Leather. — I. — Tenning's process is as follows: Melt together equal parts of zinc and linseed oil, at a temperature not above 225° F. Put the leather in the molten mixture and let it remain until saturated. The "zinc soap" is made by dissolving 6 parts of white soap in 16 parts of water, and stirring into the solution 6 parts of zinc sulphate. To make sure of a homogeneous mixture remelt the whole and stir until it begins to cool. The process, including the saturation of the leather, requires about 48 hours. Instead of zinc sulphate, copper or iron sulphate may be used. The philosophy of the process is that the moisture and air contained in the pores of the leather are driven out by the heat of the soap mixture, and their place is taken, on cooling, by the mixture. The surface of the leather is scraped after cooling, and the article is dried, either by heating over an open fire or by hanging in a drying room, strongly heated.

II.—Prideaux' process consists in submitting the leather to treatment with a solution of caoutchouc until it is thoroughly saturated with the liquid. The latter consists of 30 parts of caoutchouc in 500 parts of oil of turpentine. Complete impregnation of the leather requires several days, during which the solution must be frequently applied to the surface of the leather and rubbed in.

III.—Villon's process consists in applying a soap solution to the leather, about as follows: The leather is first treated to a solution of 62 parts of soap, 124 parts of glue, and 2,000 parts of water. When it has become saturated with the solution, it is treated to rubbing with a mixture of 460 parts of common salt and 400 parts of alum, in sufficient water to dissolve the same. After this it is washed with tepid water and dried. This process is much the quickest. The application of the soap requires about 2 hours, and the subsequent treatment about as much more, or 4 or 5 hours in all.

Oilskins.—The art of painting over textile fabrics with oily preparations to make them waterproof is probably nearly as old as textile manufacture itself, an industry of prehistoric, nay, geologic, origin. It is certainly more ancient than the craft of the artistic painter in oils, whose canvases are nothing more nor less than art oilskins, and when out of their frames, have served the usual purpose of those things in protecting goods or the human body before now. The art of waterproofing has been extended beyond the domain of the oilskin by chemical processes, especially those in which alum or lead salts, or tannin, are used, as well as by the discovery of India rubber and gutta percha. These two have revolutionized the waterproofing industry in quite a special manner, and the oilskin manufacture, although it still exists and is in a fairly flourishing condition, has found its products to a very large extent replaced by rubber goods. The natural result has been that the processes used in the former industry have remained now unchanged for a good many years. They had already been brought to a very perfect state when the rubber-waterproofing business sprang up, so that improvements were even then difficult to hit upon in oilskin making, and the check put upon the trade by India rubber made people less willing to spend time and money in experimenting with a view to improving what many years had already made it difficult to better. Hence the three cardinal defects of the oilskin: its weight, its stiffness, and the liability of its folds to stick together when it is wrapped up, or in the other extreme to crack, still remains. The weight, of course, is inevitable. An oilskin must be heavy, comparatively, from the very essence of the process by which it is made, but there seems no reason why it should not in time be made much more pliable (an old-time oilskin coat could often stand up on end when empty) and free from the danger of cracking or being compacted into a solid block when it has been stored folded on a shelf.

Probably the best oilskins ever made are those prepared by combining Dr. Stenhouse's process (patented in 1864) with the ordinary method, which consists in the main of painting over the fabric with two or more coats of boiled linseed oil, allowing each coat to dry before the next is applied. This, with a few variations in detail, is the whole method of making oilskins. Dr. Stenhouse's waterproofing method is to impregnate the fabric with a mixture of hard paraffine and boiled oil in proportions varying according to circumstances from 95 per cent of paraffine and 5 of oil to 70 per cent of the former and 30 of the latter. The most usual percentages are 80 and 20. The mixture is made with the aid of heat, and is then cast into blocks for storage. It is applied to the cloth stretched on a hot plate by rubbing the fabric thoroughly all over with a block of the composition, which may be applied on one or both sides as may be wished. The saturation is then made complete, and excess of composition is removed by passing the cloth between hot rollers. When the cloth is quite cold the process is complete. The paraffine and the drying oil combine their waterproofing powers, and the paraffine prevents the oil from exerting any injurious action upon the material. Drying oil, partly on account of the metallic compounds in it, and partly on account of its absorbing oxygen from the atmosphere, has a decided slow weakening effect upon textile fibers. Dr. Stenhouse points out that the inflammability of oilskins may be much lessened by the use of the ordinary fireproofing salts, such as tungstate of soda, or alum, either before or after the waterproofing process is carried out.

The following are some of the best recommended recipes for making oilskins:

I.—Dissolve 1 ounce of yellow soap in 1½ pints of boiling water. Then stir in 1 quart of boiled oil. When cold, add ¼ pint of gold size.

II.—Take fine twilled calico. Soak it in bullock's blood and dry it. Then give it 2 or 3 coats of boiled oil, mixed with a little litharge, or with an ounce of gold size to every pint of the oil.

III.—Make ordinary paint ready to be applied thin with a strong solution of soap.

IV.—Make 96 pounds of ocher to a thin paste with boiled oil, and then add 16 pounds of ordinary black paint mixed ready for use. Apply the first coat of this with soap, the subsequent coats without soap.

V.—Dissolve rosin in hot boiled oil till it begins to thicken.

VI.—Mix chalk or pipe clay in the finest powder, and in the purest state obtainable to a thin paste with boiled oil.

VII.—Melt together boiled oil, 1 pint; beeswax and rosin, each, 2 ounces.

VIII.—Dissolve soft soap in hot water and add solution of protosulphate of iron till no further precipitate is produced. Filter off, wash, and dry, and form the mass into a thin paste with boiled oil.

All these compositions are painted on with an ordinary painter's brush. The fabric should be slightly stretched, both to avoid folds and to facilitate the penetration of the waterproofing mixture. To aid the penetration still further, the mixture should be applied hot. It is of the greatest importance that the fabric should not be damp when the composition is applied to it. It is best to have it warm as well as the composition. If more than one coat is applied, which is practically always the case, three being the usual number, it is essential that the last coat should be perfectly dry before the next is applied. Neglect of this precaution is the chief cause of stickiness, which frequently results in serious damage to the oilskins when they have to be unfolded. In fact, it is advisable to avoid folding an oilskin when it can be avoided. They should be hung up when not in use, whenever practicable, and be allowed plenty of room. It goes without saying that no attempt should be made to sell or use the oilskin, whether garment or tarpaulin, until the final coat of composition is perfectly dry and set. It is unadvisable to use artificial heat in the drying at any stage in the manufacture.

Waterproofing Paper.—Any convenient and appropriate machinery or apparatus may be employed; but the best method for waterproofing paper is as follows: The treatment may be applied while the pulp is being formed into paper, or the finished paper may be treated. If the material is to be treated while being formed into paper, then the better method is to begin the treatment when the web of pulpy material leaves the Foudrinier wire or the cylinders, it then being in a damp condition, but with the larger percentage of moisture removed. From this point the treatment of the paper is the same whether it be pulp in a sheet, as above stated, or finished paper.

The treatment consists, first, in saturating the paper with glutinous material, preferably animal glue, and by preference the bath of glutinous material should be hot, to effect the more rapid absorption and more perfect permeation, impregnation, and deposit of the glutinous material within all the microscopic interstices throughout the body of the paper being treated. By preference a suitable tank is provided in which the glutinous material is deposited, and in which it may be kept heated to a constant temperature, the paper being passed through the tank and saturated during its passage. The material being treated should pass in a continuous sheet—that is, be fed from a roll and the finished product be wound in a roll after final treatment. This saves time and the patentee finds that the requisite permeation or incorporation of glutinous matter in the fiber will with some papers —for instance, lightly sized manila hemp—require but a few seconds. As the paper passes from the glutin tank the surplus of the glutinous matter is removed from the surface by mechanical means, as contradistinguished from simply allowing it to pass off by gravity, and in most instances it is preferred to pass the paper between suitable pressure rolls to remove such surplus. The strength and consistency of the glutinous bath may be varied, depending upon the material being treated and the uses for which such material is designed. It may, however, be stated that, in a majority of cases, a hot solution of about 1 part of animal glue to about 10 parts of water, by weight, gives the best results. After leaving the bath of glutinous material and having the surplus adhering to the surfaces removed, the paper before drying is passed into or through a solution of formaldehyde and water to "set" the glutinous material. The strength of this solution may also be variable, depending, as heretofore stated, upon the paper and uses for which it is designed. In the majority of cases, however, a solution of 1 part of formaldehyde (35 per cent solution) to 5 parts of water, by weight, gives good results, and the best result is attained if this bath is cold instead of hot, though any particular temperature is not essentially necessary. The effect of the formaldehyde solution upon the glutin-saturated paper is to precipitate the glutinous matter and render it insoluble.

As the material comes from the formaldehyde bath, the surplus adhering to the surfaces is removed by mechanical means, pressure rolls being probably most convenient. The paper is then dried in any convenient manner. The best result in drying is attained by the air-blast, i. e., projecting blasts of air against both surfaces of the paper. This drying removes all the watery constituents and leaves the paper in a toughened or greatly strengthened condition, but not in practical condition for commercial uses, as it is brittle, horny, and stiff, and has an objectionable odor and taste on account of the presence of the aldehydes, paraldehydes, formic acid, and other products, the result of oxidation. Hence it needs to be "tempered." Now while the glutinous material is rendered insoluble—that is, it is so acted upon by formaldehyde and the chemical action which takes place while the united solutions are giving off their watery constituents that it will not fully dissolve—it is, however, in a condition to be acted on by moisture, as it will swell and absorb, or take up permanently by either chemical or mechanical action a percentage of water, and will also become improved in many respects, so that to temper and render the paper soft and pliable and adapt it for most commercial uses it is subjected to moisture, which penetrates the paper, causing a welling in all directions, filling the interstices perfectly and resulting in "hydration" throughout the entire cellular structure. Two actions, mechanical and chemical, appear to take place, the mechanical action being the temporary absorption of water analogous to the absorption of water by a dry sponge, the chemical action being the permanent union of water with the treated paper, analogous to the union of water and tapioca, causing swelling, or like the chemical combination of water with lime or cement. For this purpose it is preferred to pass the paper into a bath of hot water, saturated steam or equivalent heat-and-moisture medium, thus causing the fibers and the non-soluble glutinous material filling the interstices to expand in all directions and forcing the glutinous material into all the microscopic pores or openings and into the masses of fiber, causing a commingling or thorough incorporation of the fibers and the glutinous compound. At the same time, as heretofore indicated, a change (hydration) takes place, whereby the hardened mass of fiber, glutinous material, and formaldehyde become tempered and softened and the strength imparted by the previous treatment increased. To heighten the tempering and softening effect, glycerine may, in some instances, be introduced in the tempering bath, and in most cases one two-hundredths in volume of glycerine gives the best results.

The paper may be dried in any convenient manner and is in condition for most commercial uses, it being greatly strengthened, more flexible, more impervious to moisture, acids, grease, or alkalies, and is suitable for the manufacture of binding-twine, carpets, and many novelties, for dry wrappings and lining packing cases, etc., but is liable to have a disagreeable taste and may carry traces of acids, rendering it impracticable for some uses—for instance, wrapping butter, meats, cheese, etc., after receiving the alkali treatment. The paper is also valuable as a packing for joints in steam, water, and other pipes or connections. For the purpose, therefore, of rendering the material absolutely free from all traces of acidity and all taste and odors and, in fact, to render it absolutely hygienic, it is passed through a bath of water and a volatile alkali (ammonium hydrate), the proportion by preference in a majority of cases being one-hundredth of ammonium hydrate to ninety-nine one-hundredths of water by volume. A small percentage of wood alcohol may be added. This bath is preferably cool, but a variation in its temperature will not interfere to a serious extent with the results. The effect of this bath followed by drying is to complete the chemical reaction and destroy all taste or odor, removing all traces of acids and rendering the paper hygienic in all respects. The material may be calendered or cut and used for any of the purposes desired. If the material is to be subjected to the volatile alkali bath, it is not necessary to dry it between the tempering and volatile alkali baths.

The paper made in accordance with the foregoing will, it is claimed, be found to be greatly strengthened, some materials being increased in strength from 100 to 700 per cent. It will be non-absorbent to acids, greases, and alkalies, and substantially waterproof, and owing to its component integrate structure will be practically non-conductive to electricity, adapting it as a superior insulating material. It may with perfect safety be employed for wrapping butter, meats, spices, groceries, and all materials, whether unctuous or otherwise.

The term "hydration" means the subjecting of the material (after treatment with glutinous material and formaldehyde and drying) to moisture, whereby the action described takes place.

A sheet or web of paper can be treated by the process as rapidly as it is manufactured, as the time for exposure to the action of the glutinous material need not be longer than the time required for it to become saturated, this, of course, varying with different thicknesses and densities, and the length of time of exposure may be fixed without checking the speed by making the tank of such length that the requisite time will elapse while the sheet is passing through it and the guides so arranged as to maintain the sheet in position to be acted on by such solution the requisite length of time. Four seconds' exposure to the action of formaldehyde is found sufficient in most cases.

Waterproof Ropes.—For making ropes and lines impervious to weather, the process of tarring is recommended, which can be done either in the separate strands or after the rope is twisted. An addition of tallow gives greater pliability.

Waterproof Wood. — I. — Soak in a mixture of boracic acid, 6 parts; ammonium chloride, 5 parts; sodium borate, 3 parts, and water, 100 parts.

II.—Saturate in a solution of zinc chloride.

Wax

Adulteration of Wax.—Wax is adulterated with the following among other substances: Rosins, pitch, flowers of sulphur, starch, fecula, stearine, paraffine, tallow, palm oil, calcined bones, yellow ocher, water, and wood sawdust.

Rosins are detected by cold alcohol, which dissolves all rosinous substances and exercises no action on the wax. The rosins having been extracted from the alcoholic solution by the evaporation of the alcohol, the various kinds may be distinguished by the odors disengaged by burning the mass several times on a plate of heated iron.

All earthy substances may be readily

separated from wax by means of oil of turpentine, which dissolves the wax, while the earthy matters form a residue.

Oil of turpentine also completely separates wax from starchy substances, which, like earthy matters, do not dissolve, but form a residue. A simpler method consists in heating the wax with boiling water; the gelatinous consistency assumed by the water, and the blue coloration in presence of iodine, indicate that the wax contains starchy substances. Adulteration by means of starch and fecula is quite frequent. These substances are sometimes added to the wax in a proportion of nearly 60 per cent. To separate either, the suspected product is treated hot with very dilute sulphuric acid (2 parts of acid per 100 parts of water). All amylaceous substances, converted into dextrin, remain dissolved in the liquid, while the wax, in cooling, forms a crust on the surface. It is taken off and weighed; the difference between its weight and that of the product analyzed will give the quantity of the amylaceous substances.

Flowers of sulphur are recognized readily from the odor of sulphurous acid during combustion on red-hot iron.

Tallow may be detected by the taste and odor. Pure wax has an aromatic, agreeable taste, while that mixed with tallow is repulsive both in taste and smell. Pure wax, worked between the fingers, grows soft, preserving a certain cohesion in all parts. It divides into lumps, which adhere to the fingers, if it is mixed with tallow. The adulteration may also be detected by the thick and nauseating fumes produced when it is burned on heated iron.

Stearic acid may be recognized by means of boiling alcohol, which dissolves it in nearly all proportions and causes it to deposit crystals on cooling, while it is without action on the wax. Blue litmus paper, immersed in alcohol solution, reddens on drying in air, and thus serves for detecting the presence of stearic acid.

Ocher is found by treating the wax with boiling water. A lemon-yellow deposit results, which, taken up with chlorhydric acid, yields with ammonia a lemon-yellow precipitate of ferric oxide.

The powder of burnt bones separates and forms a residue, when the wax is heated with oil of turpentine.

Artificial Beeswax.—This is obtained by mixing the following substances, in approximately the proportions stated: Paraffine, 45 parts, by weight; white Japan vegetable wax, 30 parts, by weight; rosins, or colophonies, 10 parts, by weight; white pitch, 10 parts, by weight; tallow, 5 parts, by weight; ceresine, colorant, 0.030 parts, by weight; wax perfume, 0.100 parts, by weight. If desired, the paraffine may be replaced with ozokerite, or by a mixture of vaseline and ozokerite, for the purpose of varying the fusing temperature, or rendering it more advantageous for the various applications designed. The following is the method of preparation: Melt on the boiling water bath, shaking constantly, the paraffine, the Japan wax, the rosins, the pitch, and the tallow. When the fusion is complete, add the colorant and the perfume. When these products are perfectly mingled, remove from the fire, allow the mixture to cool, and run it into suitable molds. The wax thus obtained may be employed specially for encaustics for furniture and floors, or for purposes where varnish is employed.

Waxes for Floors, Furniture, etc.—
I.—White beeswax..... 16 parts
Colophony........ 4 parts
Venice turpentine.. 1 part

Melt the articles together over a gentle fire, and when completely melted and homogeneous, pour into a sizable earthenware vessel, and stir in, while still warm, υ parts of the best French turpentine. Cool for 24 hours, by which time the mass has acquired the consistence of soft butter, and is ready for use. Its method of use is very simple. It is smeared, in small quantities, on woolen cloths, and with these is rubbed into the wood.

This is the best preparation, but one in which the beeswax is merely dissolved in the turpentine in such a way as to have the consistence of a not too thin oil color, will answer. The wood is treated with this, taking care that the surface is evenly covered with the mixture, and that it does not sink too deeply in the ornaments, corners, etc., of the woodwork. This is best achieved by taking care to scrape off from the cloths all excess of the wax.

If, in the course of 24 hours, the surface is hard, then with a stiff brush go over it, much after the way of polishing a boot. For the corners and angles smaller brushes are used; when necessary, stiff pencils may be employed. Finally, the whole is polished with plush, or velvet rags, in order not to injure the original polish. Give the article a good coat of linseed oil or a washing with petroleum before beginning work.

II.—Articles that are always exposed to the water, floors, doors, especially of oak, should, from time to time, be satu rated with oil or wax. A house door, plentifully decorated with wood carving, will not shrink or warp, even where the sun shines hottest on it, when it is frequently treated to saturation with wax and oil. Here a plain dosage with linseed oil is sufficient. Varnish, without the addition of turpentine, should never be used, or if used it should be followed by a coat of wax.

III.—A good floor wax is composed of 2 parts of wax and 3 parts of Venice turpentine, melted on the water bath, and the mixture applied while still hot, using a pencil, or brush, for the application, and when it has become solid and dry, diligently rubbed, or polished down with a woolen cloth, or with a floor brush, especially made for the purpose.

IV.—An emulsion of 5 parts of yellow wax, 2 parts of crude potassium carbonate, and 12 parts of water, boiled together until they assume a milky color and the solids are dissolved, used cold, makes an excellent composition for floors. Any desired color may be given this dressing by stirring in the powdered coloring matter. Use it exactly as described for the first mass.

Gilders' Wax.—For the production of various colorings of gold in fire gilding, the respective places are frequently covered with so-called gilders' wax. These consist of mixtures of various chemicals which have an etching action in the red heat upon the bronze mass, thus causing roughness of unequal depth, as well as through the fact that the composition of the bronze is changed somewhat on the surface, a relief of the gold color being effected in consequence of these two circumstances. The gilding wax is prepared by melting together the finely powdered chemicals with wax according to the following recipes:

	I	II	III	IV	V
Yellow wax	32	32	32	96	36
Red chalk	3	24	18	48	18
Verdigris	2	4	18	32	18
Burnt alum	2	4	—	—	—
Burnt borax	—	—	2	1	3
Copper ash	—	4	6	20	8
Zinc vitriol	—	—	—	32	18
Green vitriol	—	—	—	1	6

Grafting Wax.—

I.—
Beeswax	7	parts
Purified rosin	12	parts
Turpentine	3	parts
Rape oil	1	part
Venice turpentine	2.5	parts
Zinc white	2.5	parts

Color yellow with turmeric.

II.—
Japan wax	1	part
Yellow wax	3	parts
Rosin	8	parts
Turpentine	4	parts
Hard paraffine	1	part
Suet	3	parts
Venice turpentine	6	parts

Harness Wax.—
Oil of turpentine	90	parts
Wax, yellow	9	parts
Prussian blue	1	part
Indigo	0.5	parts
Bone black	5	parts

Dissolve the wax in the oil by aid of a low heat, on a water bath. Mix the remaining ingredients, which must be well powdered, and work up with a portion of the solution of wax. Finally, add the mixture to the solution, and mix thoroughly on the bath. When a homogeneous liquid is obtained, pour into earthen boxes.

Modeling Wax.—I.—Yellow wax, 1,000 parts; Venice turpentine, 130 parts; lard, 65 parts; bole, 725 parts. The mixture when still liquid is poured into tepid water and kneaded until a plastic mass is obtained.

II.—Summer Modeling Wax.—White wax, 20 parts; ordinary turpentine, 4 parts; sesame oil, 1 part; vermilion, 2 parts.

III.—Winter Modeling Wax.—White wax, 20 parts; ordinary turpentine, 6 parts; sesame oil, 2 parts; vermilion, 2 parts. Preparation same as for Formula I.

Sealing Waxes.—The following formulas may be followed for making sealing wax: Take 4 pounds of shellac, 1 pound of Venice turpentine, and 3 pounds of vermilion. Melt the lac in a copper pan suspended over a clear charcoal fire, then add the turpentine slowly to it, and soon afterwards add the vermilion, stirring briskly all the time with a rod in either hand. In forming the round sticks of sealing wax, a certain portion of the mass should be weighed while it is ductile, divided into the desired number of pieces, and then rolled out upon a warm marble slab by means of a smooth wooden block like that used by apothecaries for rolling a mass of pills.

The oval and square sticks of sealing wax are cast in molds, with the above compound, in a state of fusion. The marks of the lines of junction of the mold box may be afterwards removed by holding the sticks over a clear fire, or passing them over a blue gas flame. Marbled sealing wax is made by mixing two, three, or more colored kinds together while they are in a semi-fluid state. From the viscidity of the several portions their incorporation is left incomplete, so as to produce the appearance of marbling. Gold sealing wax is made simply by adding gold chrome instead of vermilion into the melted rosins. Wax may be scented by introducing a little essential oil, essence of musk, or other perfume. If 1 part of balsam of Peru be melted along with 99 parts of the sealing-wax composition, an agreeable fragrance will be exhaled in the act of sealing with it. Either lampblack or ivory black serves for the coloring matter of black wax. Sealing wax is often adulterated with rosin, in which case it runs into thin drops at the flame of a candle.

The following mistakes are sometimes made in the manufacture of sealing wax:

I.—Use of filling agents which are too coarsely ground.

II.—Excessive use of filling agents.

III.—Insufficient binding of the pigments and fillings with a suitable adhesive agent, which causes these bodies to absorb the adhesive power of the gums.

IV.—Excessive heating of the mass, caused by improper melting or faulty admixture of the gummy bodies. Turpentine and rosin must be heated before entering the shellac. If this rule is inverted, as is often the case, the shellac sticks to the bottom and burns partly.

Great care must be taken to mix the coloring matter to a paste with spirit or oil of turpentine before adding to the other ingredients. Unless this is done the wax will not be of a regular tint.

Dark Blue Wax.—Three ounces Venetian turpentine, 4 ounces shellac, 1 ounce rosin, 1 ounce Prussian blue, ½ ounce magnesia.

Green Wax.—Two ounces Venetian turpentine, 4 ounces shellac, 1½ ounces rosin, ½ ounce chrome yellow, ¼ ounce Prussian blue, 1 ounce magnesia.

Carmine Red Wax.—One ounce Venetian turpentine, 4 ounces shellac, 1 ounce rosin, colophony, 1¼ ounces Chinese red, 1 drachm magnesia, with oil of turpentine.

Gold Wax.—Four ounces Venetian turpentine, 8 ounces shellac, 14 sheets of genuine leaf gold, ½ ounce bronze, ¼ ounce magnesia, with oil of turpentine.

White Wax.—I.—The wax is bleached by exposing to moist air and to the sun, but it must first be prepared in thin sheets or ribbons or in grains. For this purpose it is first washed, to free it from the honey which may adhere, melted, and poured into a tin vessel, whose bottom is perforated with narrow slits. The melted wax falls in a thin stream on a wooden cylinder arranged below and half immersed in cold water. This cylinder is turned, and the wax, rolling round in thin leaves, afterwards falls into the water. To melt it in grains, a vessel is made use of, perforated with small openings, which can be rotated. The wax is projected in grains into the cold water. It is spread on frames of muslin, moistened with water several times a day, and exposed to the sun until the wax assumes a fine white. This whiteness, however, is not perfect. The operation of melting and separating into ribbons or grains must be renewed. Finally, it is melted and flowed into molds. The duration of the bleaching may be abridged by adding to the wax, treated as above, from 1.25 to 1.75 per cent of rectified oil of turpentine, free from rosin. In 6 or 8 days a result will be secured which would otherwise require 5 or 6 weeks.

II.—Bleached shellac..... 28 parts
Venetian turpentine.. 13 parts
Plaster of Paris..... 30 parts

WAX FOR BOTTLES:
See Photography.

WAX, BURNING, TRICK:
See Pyrotechnics.

WAXES, DECOMPOSITION OF:
See Oil.

WAX FOR IRONING:
See Laundry Preparations.

WAX FOR LINOLEUM:
See Linoleum.

Weather Forecasters

(See also Hygrometers and Hygroscopes.)

I.—It is known that a leaf of blotting paper or a strip of fabric made to change color according to the hygrometric state of the atmosphere has been employed for weather indications in place of a barometer. The following compound is recommended for this purpose: One part of cobalt chloride, 75 parts of nickel oxide, 20 parts of gelatin, and 200 parts of water. A strip of calico, soaked in this solution, will appear green in fine weather, but when moisture intervenes the color disappears.

II.—Copper chloride.... 1 part
    Gelatin........... 10 parts
    Water............ 100 parts

III.—This is a method of making old-fashioned weather glasses containing a liquid that clouds or solidifies under certain atmospheric conditions:

    Camphor........ 2½ drachms
    Alcohol......... 11 drachms
    Water.......... 9 drachms
    Saltpeter........ 38 grains
    Sal ammoniac.... 38 grains

Dissolve the camphor in the alcohol and the salts in the water and mix the solutions together. Pour in test tubes, cover with wax after corking and make a hole through the cork with a red-hot needle, or draw out the tube until only a pin hole remains. When the camphor, etc., appear soft and powdery, and almost filling the tube, rain with south or southwest winds may be expected; when crystalline, north, northeast, or northwest winds, with fine weather, may be expected; when a portion crystallizes on one side of the tube, wind may be expected from that direction. Fine weather: The substance remains entirely at bottom of tube and the liquid perfectly clear. Coming rain: Substance will rise gradually, liquid will be very clear, with a small star in motion. A coming storm or very high wind: Substance partly at top of tube, and of a leaflike form, liquid very heavy and in a fermenting state. These effects are noticeable 24 hours before the change sets in. In winter: Generally the substance lies higher in the tube. Snow or white frost: Substance very white and small stars in motion. Summer weather: The substance will lie quite low. The substance will lie closer to the tube on the opposite side to the quarter from which the storm is coming. The instrument is nothing more than a scientific toy.

WEATHERPROOFING:
See Paints.

WEED KILLERS:
See Disinfectants.

Weights and Measures

INTERNATIONAL ATOMIC WEIGHTS.

The International Committee on Atomic Weights have presented this table as corrected:

		O=16	H=1
Aluminum.....	Al	27.1	26.9
Antimony......	Sb	120.2	119.3
Argon.........	A	39.9	39.6
Arsenic........	As	75	74.4
Barium........	Ba	137.4	136.4
Bismuth.......	Bi	208.5	206.9
Boron.........	B	11	10.9
Bromine.......	Br	79.96	79.36
Cadmium......	Cd	112.4	111.6
Cæsium	Cs	132.9	131.9
Calcium.......	Ca	40.1	39.7
Carbon........	C	12	11.91
Cerium	Ce	140.25	139.2
Chlorine.......	Cl	35.45	35.18
Chromium.....	Cr	52.1	51.7
Cobalt........	Co	59	58.55
Columbium....	Cb	94	93.3
Copper........	Cu	63.6	63.1
Erbium........	Er	166	164.8
Fluorine.......	F	19	18.9
Gadolinium....	Gd	156	154.8
Gallium........	Ga	70	69.5
Germanium....	Ge	72.5	72
Glucinum......	Gl	9.1	9.03
Gold..........	Au	197.2	195.7
Helium........	He	4	4
Hydrogen......	H	1.008	1
Indium........	In	115	114.1
Iodine.........	I	126.97	126.01
Iridium........	Ir	193	191.5
Iron	Fe	55.9	55.5
Krypton.......	Kr	81.8	81.2
Lanthanum....	La	138.9	137.9
Lead..........	Pb	206.9	205.35
Lithium.......	Li	7.03	6.98
Magnesium....	Mg	24.36	24.18
Manganese	Mn	55	54.6
Mercury.......	Hg	200	198.5
Molybdenum ..	Mo	96	95.3
Neodymium....	Nd	143.6	142.5
Neon..........	Ne	20	19.9
Nickel.........	Ni	58.7	58.3
Nitrogen.......	N	14.04	13.93
Osmium.......	Os	191	189.6
Oxygen........	O	16	15.88
Palladium.....	Pd	106.5	105.7
Phosphorus....	P	31	30.77
Platinum......	Pt	194.8	193.3
Potassium	K	39.15	38.85
Praseodymium.	Pr	140.5	139.4
Radium.......	Ra	225	223.3
Rhodium......	Rh	103	102.2
Rubidium.....	Rb	85.5	84.9
Ruthenium....	Ru	101.7	100.9
Samarium.....	Sm	150.3	149.2
Scandium......	Sc	44.1	43.8
Selenium......	Se	79.2	78.6
Silicon........	Si	28.4	28.2
Silver.........	Ag	107.93	107.11
Sodium........	Na	23.05	22.88
Strontium......	Sr	87.6	86.94
Sulphur.......	S	32.06	31.82
Tantalum.....	Ta	183	181.6
Tellurium.....	Te	127.6	126.6
Terbium.......	Tb	160	158.8
Thallium......	Tl	204.1	202.6

INTERNATIONAL ATOMIC WEIGHTS—Continued.

		O=16	H=1				O=16	H=1
Thorium	Th	232.5	230.8	Vanadium	V		51.2	50.8
Thulium	Tm	171	169.7	Xenon	Xe		128	127
Tin	Sn	119	118.1	Ytterbium	Yb		173	171.7
Titanium	Ti	48.1	47.7	Yttrium	Yt		89	88.3
Tungsten	W	184	182.6	Zinc	Zn		65.4	64.9
Uranium	U	238.5	236.7	Zirconium	Zr		90.6	89.9

UNITED STATES WEIGHTS AND MEASURES

(According to existing standards)

LINEAL

	Inches.	Feet.	Yards.	Rods.	Fur's.	Mile.
12 inches = 1 foot.	12 =	1				
3 feet = 1 yard.	36 =	3 =	1			
5.5 yards = 1 rod.	198 =	16.5 =	5.5 =	1		
40 rods = 1 furlong.	7,920 =	660 =	220 =	40 =	1	
8 furlongs = 1 mile.	63,360 =	5,280 =	1,760 =	320 =	8 =	1

SURFACE—LAND

	Feet.	Yards.	Rods.	Roods.	Acres.
144 sq. inches = 1 square foot.	9 =	1			
9 square feet = 1 square yard.	272.25 =	30.25 =	1		
30.25 square yards = 1 square rod.	10,890 =	1,210 =	40 =	1	
40 square rods = 1 square rood.	43,560 =	4,840 =	160 =	4 =	1
4 square roods = 1 acre.	27,878,400 =	3,097,600 =	102,400 =	2,560 =	640
640 acres = 1 square mile.					

VOLUME—LIQUID

	Gills.	Pints.	Gallon.	Cub. In.
4 gills = 1 pint.	32 =	8 =	1 =	231
2 pints = 1 quart.				
4 quarts = 1 gallon.				

FLUID MEASURE

Gallon.	Pints.	Ounces.	Drachms.	Minims.	Cubic Centimeters.
1 =	8 =	128 =	1,024 =	61,440 =	3,785.435
	1 =	16 =	128 =	7,680 =	473.179
		1 =	8 =	480 =	29.574
			1 =	60 =	3.697

16 ounces, or a pint, is sometimes called a fluidpound.

TROY WEIGHT

Pound.	Ounces.	Pennyweights.	Grains.	Grams.
1 =	12 =	240 =	5,760 =	373.24
	1 =	20 =	480 =	31.10
		1 =	24 =	1.56

APOTHECARIES' WEIGHT

℔.	℥	ℨ	℈	gr.	
Pound.	Ounces.	Drachms.	Scruples.	Grains.	Grams.
1 =	12 =	96 =	288 =	5,760 =	373.24
	1 =	8 =	24 =	480 =	31.10
		1 =	3 =	60 =	3.89
			1 =	20 =	1.30
				1 =	.06

The pound, ounce, and grain are the same as in Troy weight.

AVOIRDUPOIS WEIGHT

Pound.	Ounces.	Drachms.	Grains (Troy)	Grams.
1 =	16 =	256 =	7,000 =	453.60
	1 =	16 =	437.5 =	28.35
		1 =	27.34 =	1.77

ENGLISH WEIGHTS AND MEASURES

APOTHECARIES' WEIGHT

20 grains	= 1 scruple =	20 grains
3 scruples	= 1 drachm =	60 grains
8 drachms	= 1 ounce =	480 grains
12 ounces	= 1 pound =	5,760 grains

FLUID MEASURE

60 minims	= 1 fluidrachm
8 drachms	= 1 fluidounce
20 ounces	= 1 pint
8 pints	= 1 gallon

The above weights are usually adopted in formulas.

All chemicals are usually sold by

AVOIRDUPOIS WEIGHT

27½ grains	= 1 drachm =	27½ grains
16 drachms	= 1 ounce =	437½ grains
16 ounces	= 1 pound =	7,000 grains

Precious metals are usually sold by

TROY WEIGHT

24 grains	= 1 pennyweight =	24 grains
20 pennyweights	= 1 ounce =	480 grains
12 ounces	= 1 pound =	5,760 grains

NOTE.—An ounce of metallic silver contains 480 grains, but an ounce of nitrate of silver contains only 437½ grains.

METRIC SYSTEM OF WEIGHTS AND MEASURES

MEASURES OF LENGTH

Denominations and Values.		Equivalents in Use.	
Myriameter	10,000 meters	6.2137	miles
Kilometer	1,000 meters	.62137	miles, or 3,280 feet, 10 inches
Hectometer	100 meters	328	feet and 1 inch
Dekameter	10 meters	393.7	inches
Meter	1 meter	39.37	inches
Decimeter	1-10th of a meter	3.937	inches
Centimeter	1-100th of a meter	.3937	inches
Millimeter	1-1,000th of a meter	.0394	inches

MEASURES OF SURFACE

Denominations and Values.		Equivalents in Use.	
Hectare	10,000 square meters	2.471	acres
Are	100 square meters	119.6	square yards
Centare	1 square meter	1,550	square inches

MEASURES OF VOLUME

			Equivalents in Use.			
Names.	No. of Liters.	Cubic Measures.	Dry Measure.		Wine Measure.	
Kiloliter or stere.	1,000	1 cubic meter	1.308	cubic yards	264.17	gallons
Hectoliter	100	1-10th cubic meter	2	bushels and 3.35 pecks	26.417	gallons
Dekaliter	10	10 cubic decimeters	9.08	quarts	2.6417	gallons
Liter	1	1 cubic decimeter	.908	quarts	1.0567	quarts
Deciliter	1-10	1-10th cubic decimeter	6.1023	cubic inches	.845	gills
Centiliter	1-100	10 cubic centimeters	.6102	cubic inches	.338	fluidounces
Milliliter	1-1,000	1 cubic centimeter	.061	cubic inches	.27	fluidrachms

WEIGHTS

			Equivalents in Use.
Names.	Number of Grams.	Weight of Volume of Water at its Maximum Density.	Avoirdupois Weight.
Millier or Tonneau	1,000,000	1 cubic meter	2,204.6 pounds
Quintal	100,000	1 hectoliter	220.46 pounds
Myriagram	10,000	10 liters	22.046 pounds
Kilogram or Kilo	1,000	1 liter	2.2046 pounds
Hectogram	100	1 deciliter	3.5274 ounces
Dekagram	10	10 cubic centimeters	.3527 ounces
Gram	1	1 cubic centimeter	15.432 grains
Decigram	1-10	1-10th of a cubic centimeter	1.5432 grains
Centigram	1-100	10 cubic millimeters	.1543 grains
Milligram	1-1,000	1 cubic millimeter	.0154 grains

For measuring surfaces, the square dekameter is used under the term of ARE; the hectare, or 100 ares, is equal to about 2½ acres. *The unit of capacity* is the cubic decimeter or LITER, and the series of measures is formed in the same way as in the case of the table of lengths. The cubic meter is the unit of measure for solid bodies, and is termed STERE. *The unit of weight* is the GRAM, which is the weight of one cubic centimeter of pure water weighed in a vacuum at the temperature of 4° C. or 39.2° F., which is about its temperature of maximum density. In practice, the term cubic centimeter, abbreviated c.c., is generally used instead of milliliter, and cubic meter instead of kiloliter.

THE CONVERSION OF METRIC INTO ENGLISH WEIGHT

The following table, which contains no error greater than one-tenth of a grain, will suffice for most practical purposes:

1 gram =	15⅖ grains	
2 grams =	30⅘ grains	
3 grams =	46⅕ grains	
4 grams =	61⅘ grains, or	1 drachm, 1⅘ grains
5 grams =	77⅕ grains, or	1 drachm, 17⅕ grains
6 grams =	92⅗ grains, or	1 drachm, 32⅗ grains
7 grams =	108 grains, or	1 drachm, 48 grains
8 grams =	123⅖ grains, or	2 drachms, 3⅖ grains
9 grams =	138⅘ grains, or	2 drachms, 18⅘ grains
10 grams =	154⅖ grains, or	2 drachms, 34⅖ grains
11 grams =	169⅘ grains, or	2 drachms, 49⅘ grains
12 grams =	185⅕ grains, or	3 drachms, 5⅕ grains
13 grams =	200⅗ grains, or	3 drachms, 20⅗ grains
14 grams =	216 grains, or	3 drachms, 36 grains
15 grams =	231⅖ grains, or	3 drachms, 51⅖ grains
16 grams =	247 grains, or	4 drachms, 7 grains
17 grams =	262⅖ grains, or	4 drachms, 22⅖ grains
18 grams =	277⅘ grains, or	4 drachms, 37⅘ grains
19 grams =	293⅕ grains, or	4 drachms, 53⅕ grains
20 grams =	308⅗ grains, or	5 drachms, 8⅗ grains
30 grams =	463 grains, or	7 drachms, 43 grains
40 grams =	617⅕ grains, or	10 drachms, 17⅕ grains
50 grams =	771⅗ grains, or	12 drachms, 51⅗ grains
60 grams =	926 grains, or	15 drachms, 26 grains
70 grams =	1,080⅕ grains, or	18 drachms, 0⅕ grains
80 grams =	1,234⅗ grains, or	20 drachms, 34⅗ grains
90 grams =	1,389 grains, or	23 drachms, 9 grains
100 grams =	1,543⅕ grains, or	25 drachms, 43⅕ grains
1,000 grams =	1 kilogram = 32 ounces, 1 drachm, 12⅖ grains	

THE CONVERSION OF METRIC INTO ENGLISH MEASURE

1 cubic centimeter =	17 minims	
2 cubic centimeters =	34 minims	
3 cubic centimeters =	51 minims	
4 cubic centimeters =	68 minims, or	1 drachm, 8 minims
5 cubic centimeters =	85 minims, or	1 drachm, 25 minims
6 cubic centimeters =	101 minims, or	1 drachm, 41 minims
7 cubic centimeters =	118 minims, or	1 drachm, 58 minims
8 cubic centimeters =	135 minims, or	2 drachms, 15 minims
9 cubic centimeters =	152 minims, or	2 drachms, 32 minims
10 cubic centimeters =	169 minims, or	2 drachms, 49 minims
20 cubic centimeters =	338 minims, or	5 drachms, 38 minims
30 cubic centimeters =	507 minims, or	1 ounce, 0 drachm, 27 minims
40 cubic centimeters =	676 minims, or	1 ounce, 3 drachms, 16 minims
50 cubic centimeters =	845 minims, or	1 ounce, 6 drachms, 5 minims
60 cubic centimeters =	1,014 minims, or	2 ounces, 0 drachms, 54 minims
70 cubic centimeters =	1,183 minims, or	2 ounces, 3 drachms, 43 minims
80 cubic centimeters =	1,352 minims, or	2 ounces, 6 drachms, 32 minims
90 cubic centimeters =	1,521 minims, or	3 ounces, 1 drachm, 21 minims
100 cubic centimeters =	1,690 minims, or	3 ounces, 4 drachms, 10 minims
1,000 cubic centimeters =	1 liter = 34 fluidounces nearly, or 2⅛ pints.	

WELDING POWDERS.

See also Steel.

Powder to Weld Wrought Iron at Pale-red Heat with Wrought Iron.—I.—Borax, 1 part (by weight); sal ammoniac, $\frac{1}{2}$ part; water, $\frac{1}{2}$ part. These ingredients are boiled with constant stirring until the mass is stiff; then it is allowed to harden over the fire. Upon cooling, the mass is rubbed up into a powder and mixed with one-third wrought-iron filings free from rust. When the iron has reached red heat, this powder is sprinkled on the parts to be welded, and after it has liquefied, a few blows are sufficient to unite the pieces.

II. — Borax, 2 parts; wrought-iron filings, free from rust, 2 parts; sal ammoniac, 1 part. These pulverized parts are moistened with copaiba balsam and made into a paste, then slowly dried over a fire and again powdered. The application is the same as for Formula I.

Welding Powder to Weld Steel on Wrought Iron at Pale-red Heat.—Borax, 3 parts; potassium cyanide, 2 parts; Berlin blue, 1-100 part. These substances are powdered well, moistened with water; next they are boiled with constant stirring until stiff; then dry over a fire. Upon cooling, the mass is finely pulverized and mixed with 1 part of wrought-iron filings, free from rust. This powder is sprinkled repeatedly upon the hot pieces, and after it has burned in the welding is taken in hand.

WHEEL GREASE:

See Lubricants.

WHETSTONES.

To make artificial whetstones, take gelatin of good quality, dissolve it in equal weight of water, operating in almost complete darkness, and add $1\frac{1}{2}$ per cent of bichromate of potash, previously dissolved. Next take about 9 times the weight of the gelatin employed of very fine emery or fine powdered gun stone, which is mixed intimately with the gelatinized solution. The paste thus obtained is molded into the desired shape, taking care to exercise an energetic pressure in order to consolidate the mass. Finally dry by exposure to the sun.

WHITING:

To Form Masses of Whiting.—Mix the whiting into a stiff paste with water, and the mass will retain its coherence when dry.

Whitewash

(See also Paint.)

Wash the ceiling by wetting it twice with water, laying on as much as can well be floated on, then rub the old color up with a stumpy brush and wipe off with a large sponge. Stop all cracks with whiting and plaster of Paris. When dry, claricole with size and a little of the whitewash when this is dry. If very much stained, paint those parts with turps, color, and, if necessary, claricole again. To make the whitewash, take a dozen pounds of whiting (in large balls), break them up in a pail, and cover with water to soak. During this time melt over a slow fire 4 pounds common size, and at the same time, with a palette knife or small trowel, rub up fine about a dessertspoonful of blue-black with water to a fine paste; then pour the water off the top of the whiting and with a stick stir in the black; when well mixed, stir in the melted size and strain. When cold, it is fit for use. If the jelly is too stiff for use, beat it up well and add a little cold water. Commence whitewashing over the window and so work from the light. Distemper color of any tint may be made by using any other color instead of the blue-black—as ocher, chrome, Dutch pink, raw sienna for yellows and buff; Venetian red, burnt sienna, Indian red or purple brown for reds; celestial blue, ultramarine, indigo for blues; red and blue for purple, gray or lavender; red lead and chrome for orange; Brunswick green for greens.

Ox blood in lime paint is an excellent binding agent for the lime, as it is chiefly composed of albumin, which, like casein or milk, is capable of transforming the lime into casein paint. But the ox blood must be mixed in the lime paint; to use it separately is useless, if not harmful. Whitewashing rough mortar-plastering to saturation is very practical, as it closes all the pores and small holes.

A formula used by the United States Government in making whitewash for light-houses and other public buildings is as follows:

Unslaked lime	2	pecks
Common salt	1	peck
Rice flour	3	pounds
Spanish whiting	$\frac{1}{2}$	pound
Glue (clean and white)	1	pound
Water, a sufficient quantity.		

Slake the lime in a vessel of about 10 gallons capacity; cover it, strain, and add the salt previously dissolved in warm water. Boil the rice flour in water; soak the glue in water and dissolve on a water bath, and add both, together with the whiting and 5 gallons of hot water to the mixture, stirring all well together. Cover to protect from dirt, and let it stand for a few days, when it will be ready for use. It is to be applied hot, and for that reason should be used from a kettle over a portable furnace.

To Soften Old Whitewash.—Wet the whitewash thoroughly with a wash made of 1 pound of potash dissolved in 10 quarts of water.

WHITEWASH, TO REMOVE:
See Cleaning Preparations and Methods.

WHITE METAL:
See Alloys.

WINDOW-CLEANING COMPOUND:
See Cleaning Compounds.

WINDOW DISPLAY:
See also Sponges.

An attractive window display for stores can be prepared as follows:

In a wide-mouth jar put some sand, say, about 6 inches in depth. Make a mixture of equal parts of aluminum sulphate, copper sulphate, and iron sulphate, coarsely powdered, and strew it over the surface of the sand. Over this layer gently pour a solution of sodium silicate, dissolved in 3 parts of hot water, taking care not to disturb the layer of sulphates. In about a week or 10 days the surface will be covered with crystals of different colors, being silicates of different metals employed. Now take some pure water and let it run into the vessel by a small tube, using a little more of it than you used of the water-glass solution. This will displace the water-glass solution, and a fresh crop of crystals will come in the silicates, and makes, when properly done, a pretty scene. Take care in pouring in the water to let the point of the tube be so arranged as not to disturb the crop of silicates.

WINDOW PERFUME.

In Paris an apparatus has been introduced consisting of a small tube which is attached lengthwise on the exterior of the shop windows. Through numerous little holes a warm, lightly perfumed current of air is passed, which pleasantly tickles the olfactory nerves of the looker-on and at the same time keeps the panes clear and clean, so that the goods exhibited present the best possible appearance.

WINDOW POLISHES:
See Polishes.

WINDOWS, FROSTED:
See Glass.

WINDOWS, TO PREVENT DIMMING OF:
See Glass.

Wines and Liquors

BITTERS.

Bitters, as the name indicates, are merely tinctures of bitter roots and barks, with the addition of spices to flavor, and depend for their effect upon their tonic action on the stomach. Taken too frequently, however, they may do harm, by overstimulating the digestive organs.

The recipes for some of these preparations run to great lengths, one for Angostura bitters containing no fewer than 28 ingredients. A very good article, however, may be made without all this elaboration. The following, for instance, make a very good preparation:

Gentian root (sliced)..	12 ounces
Cinnamon bark......	10 ounces
Caraway seeds.......	10 ounces
Juniper berries......	2 ounces
Cloves..............	1 ounce
Alcohol, 90 per cent..	7 pints

Macerate for a week; strain, press out, and filter, then add

Capillaire...........	1¼ pints
Water to make up....	2½ gallons

Strength about 45 u. p.

Still another formula calls for Angostura bark, 2½ ounces; gentian root, 1 ounce; cardamom seeds, ½ ounce; Turkey rhubarb, ¼ ounce; orange peel, 4 ounces; caraways, ½ ounce; cinnamon bark, ½ ounce; cloves, ¼ ounce.

Brandy Bitters.—

Sliced gentian root...	3	pounds
Dried orange peel....	2	pounds
Cardamom seed......	1	pound
Bruised cinnamon....	½	pound
Cochineal...........	2	ounces
Brandy.............	10	pints

Macerate for 14 days and strain.

Hostetter's Bitters.—

Calamus root........	1 pound
Orange peel.........	1 pound
Peruvian bark.......	1 pound
Gentian root........	1 pound

Calumba root....... 1 pound
Rhubarb root........ 4 ounces
Cinnamon bark...... 2 ounces
Cloves............. 1 ounce
Diluted alcohol..... 2 gallons
Water............. 1 gallon
Sugar............. 1 pound

Macerate together for 2 weeks.

CORDIALS.

Cordials, according to the *Spatula*, are flavored liquors containing from 40 to 50 per cent of alcohol (from 52 to 64 fluidounces to each gallon) and from 20 to 25 per cent of sugar (from 25 to 32 ounces avoirdupois to each gallon).

Cordials, while used in this country to some degree, have their greatest consumption in foreign lands, especially in France and Germany.

Usually such mixtures as these are clarified or "fined" only with considerable difficulty, as the finally divided particles of oil pass easily through the pores of the filter paper. Purified talcum will be found to be an excellent clarifying medium; it should be agitated with the liquid and the liquid then passed through a thoroughly wetted filter. The filtrate should be returned again and again to the filter until it filters perfectly bright. Purified talcum being chemically inert is superior to magnesium carbonate and other substances which are recommended for this purpose.

When the filtering process is completed the liquids should at once be put into suitable bottles which should be filled and tightly corked and sealed. Wrap the bottles in paper and store away, laying the bottles on their sides in a moderately warm place. A shelf near the ceiling is a good place. Warmth and age improve the beverages, as it appears to more perfectly blend the flavors, so that the older the liquor becomes the better it is. These liquids must never be kept in a cold place, as the cold might cause the volatile oils to separate.

The following formulas are for the production of cordials of the best quality, and therefore only the very best of materials should be used; the essential oils should be of unquestionable quality and strictly fresh, while the alcohol must be free from fusel oil, the water distilled, and the sugar white, free from bluing, and if liquors of any kind should be called for in any formula only the very best should be used. The oils and other flavoring substances should be dissolved in the alcohol and the sugar in the water. Then mix the two solutions and filter clear.

Alkermes Cordial.—

Mace.......... 1½ avoirdupois ounces
Ceylon cinnamon 1⅛ avoirdupois ounces
Cloves.......... ¾ avoirdupois ounce
Rose water
(best)........ 6 fluidounces
Sugar..........28 avoirdupois ounces
Deodorized alcohol........52 fluidounces
Distilled water,
q. s.......... 1 gallon

Reduce the mace, cinnamon, and cloves to a coarse powder macerate with the alcohol for several days, agitating occasionally, then add the remaining ingredients, and filter clear.

Anise Cordial.—

Anethol.......... 7 fluidrachms
Oil of fennel seed.. 80 minims
Oil of bitter
almonds........ 16 drops
Deodorized alcohol 8 pints
Simple syrup...... 5 pints
Distilled water, q. s. 16 pints

Mix the oils and anethol with the alcohol and the syrup with the water; mix the two and filter clear, as directed.

Blackberry Cordial.—This beverage is usually misnamed "blackberry brandy" or "blackberry wine." This latter belongs only to wines obtained by the fermentation of the blackberry juice. When this is distilled then a true blackberry brandy is obtained, just as ordinary brandy is obtained by distilling ordinary wines.

The name is frequently applied to a preparation containing blackberry root often combined with other astringents, but the true blackberry cordial is made according to the formulas given herewith. Most of these mention brandy, and this article should be good and fusel free, or it may be replaced by good whisky, or even by diluted alcohol, depending on whether a high-priced or cheap cordial is desired.

I.—Fresh blackberry juice, 3 pints; sugar, 7½ ounces; water, 30 fluidounces; brandy, 7½ pints; oil of cloves, 3 drops; oil of cinnamon, 3 drops; alcohol, 6 fluidrachms. Dissolve the sugar in the water and juice, then add the liquor. Dissolve the oils in the alcohol and add ½ to the first solution, and if not sufficiently flavored add more of the second solution. Then filter.

II.—Fresh blackberry juice, 4 pints; powdered nutmeg (fresh), 1 ounce; powdered cinnamon (fresh), 1 ounce; powdered pimento (fresh), ½ ounce; powdered cloves (fresh), ½ ounce; brandy, 2½ pints; sugar, 2½ pounds. Macerate the spices in the brandy for several days. Dissolve the sugar in the juice and mix and filter clear.

Cherry Cordials.—

I.—
Oil of bitter almonds	8 drops
Oil of cinnamon	1 drop
Oil of cloves	1 drop
Acetic ether	12 drops
Ceuanthic ether	1 drop
Vanilla extract	1 drachm
Alcohol	3 pints
Sugar	3 pounds
Cherry juice	20 ounces
Distilled water, q. s.	1 gallon

The oils, ethers, and extracts must be dissolved in the alcohol, the sugar in part of the water, then mix, add the juice and filter clear. When the juice is not sufficiently sour, add a small amount of solution of citric acid. To color, use caramel.

II.—
Vanilla extract	10	drops
Oil of cinnamon	10	drops
Oil of bitter almonds	10	drops
Oil of cloves	3	drops
Oil of nutmeg	3	drops
Alcohol	2½	pints
Cherry juice	2½	pints
Simple syrup	3	pints

Dissolve the oils in the alcohol, then add the other ingredients and filter clear. It is better to make this cordial during the cherry season so as to obtain the fresh expressed juice of the cherry.

Curacoa Cordials.—

I.—
Curacoa orange peel	6	ounces
Cinnamon	¾	ounce
Mace	2½	drachms
Alcohol	3½	pints
Water	4½	pints
Sugar	12	ounces

Mix the first three ingredients and reduce them to a coarse powder, then mix with the alcohol and 4 pints of water and macerate for 8 days with an occasional agitation, express, add the sugar and enough water to make a gallon of finished product. Filter clear.

II.—
Curacoa or bitter orange peel	2	ounces
Cloves	80	grains
Cinnamon	80	grains
Cochineal	60	grains
Oil of orange (best)	1	drachm
Orange-flower water	½	pint
Holland gin	1	pint
Alcohol	2	pints
Sugar	3	pints
Water, q. s.	1	gallon

Reduce the solids to a coarse powder, add the alcohol and macerate 3 days. Then add the oil, gin, and 3 pints of water and continue the maceration for 8 days more, agitating once a day, strain and add sugar dissolved in balance of the water. Then add the orange-flower water and filter.

Kola Cordial.—

Kola nuts, roasted and powdered	7	ounces
Cochineal powder	30	grains
Extract of vanilla	3	drachms
Arrac	3	ounces
Sugar	7	pounds
Alcohol	6	pints
Water, distilled	6	pints

Macerate kola and cochineal with alcohol for 10 days, agitate daily, add arrac, vanilla, and sugar dissolved in water. Filter.

Kümmel Cordials.—

I.—
Oil of caraway	30	drops
Oil of peppermint	3	drops
Oil of lemon	3	drops
Acetic ether	30	drops
Spirit of nitrous ether	30	drops
Sugar	72	ounces
Alcohol	96	ounces
Water	96	ounces

Dissolve the oils and ethers in the alcohol, and the sugar in the water. Mix and filter.

II.—
Oil of caraway	20	drops
Oil of sweet fennel	2	drops
Oil of cinnamon	1	drop
Sugar	14	ounces
Alcohol	2	pints
Water	4	pints

Prepare as in Formula I.

Orange Cordials.—Many of the preparations sold under this name are not really orange cordials, but are varying mixtures of uncertain composition, possibly flavored with orange. The following are made by the use of oranges:

I.—
Sugar	8	avoirdupois pounds
Water	2¾	gallons
Oranges	15	

Dissolve the sugar in the water by the aid of a gentle heat, express the oranges, add the juice and rinds to the syrup, put the mixture into a cask, keep the whole in a warm place for 3 or 4 days, stirring frequently, then close the cask, set aside in a cool cellar and draw off the clear liquid.

II.—Express the juice from sweet oranges, add water equal to the volume of juice obtained, and macerate the expressed oranges with the juice and water for about 12 hours. For each gallon of juice, add 1 pound of granulated sugar, grape sugar, or glucose, put the whole into a suitable vessel, covering to exclude the dust, place in a warm location until fermentation is completed, draw off the clear liquid, and preserve in well-stoppered stout bottles in a cool place.

III.—Orange wine suitable for "soda" purposes may be prepared by mixing 3 fluidounces of orange essence with 13 fluidounces of sweet Catawba or other mild wine. Some syrup may be added to this if desired.

Rose Cordial.—

Oil of rose, very best..	3 drops
Palmarosa oil........	3 drops
Sugar..............	28 ounces
Alcohol.............	52 ounces
Distilled water, q. s..	8 pints

Dissolve the sugar in the water and the oils in the alcohol; mix the solutions, color a rose tint, and filter clear.

Spearmint Cordial.—

Oil of spearmint.....	30 drops
Sugar..............	28 ounces
Alcohol.............	52 ounces
Distilled water, q. s..	8 pints

Dissolve the sugar in the water and the oil in the alcohol; mix the two solutions, color green, and filter clear.

Absinthe.—

I.—
Oil of wormwood...	96 drops
Oil of star anise...	72 drops
Oil of aniseed.....	48 drops
Oil of coriander...	48 drops
Oil of fennel, pure.	48 drops
Oil of angelica root...........	24 drops
Oil of thyme......	24 drops
Alcohol (pure)....162 fluidounces
Distilled water....	30 fluidounces

Dissolve the oils in the alcohol, add the water, color green, and filter clear.

II.—
Oil of wormwood..	36 drops
Oil of orange peel.	30 drops
Oil of star anise...	12 drops
Oil of neroli petate.	5 drops
Fresh oil of lemon.	9 drops
Acetic ether......	24 drops
Sugar.............	30 avoirdupois ounces
Alcohol, deodorized 90 fluidounces
Distilled water....	78 fluidounces

Dissolve the oils and ether in the alcohol and the sugar in the water; then mix thoroughly, color green, and filter clear.

DETANNATING WINE.

According to Caspari, the presence of appreciable quantities of tannin in wine is decidedly objectionable if the wine is to be used in connection with iron and other metallic salts; moreover, tannin is incompatible with alkaloids, and hence wine not deprived of its tannin should never be used as a menstruum for alkaloidal drugs. The process of freeing wines from tannin is termed detannation, and is a very simple operation. The easiest plan is to add $\frac{1}{2}$ ounce of gelatin in number 40 or number 60 powder to 1 gallon of the wine, to agitate occasionally during 24 or 48 hours, and then to filter. The operation is preferably carried out during cold weather or in a cold apartment, as heat will cause the gelatin to dissolve, and the maceration must be continued until a small portion of the wine mixed with a few drops of ferric chloride solution shows no darkening of color. Gelatin in large pieces is not suitable, especially with wines containing much tannin, since the newly formed tannate of gelatin will be deposited on the surface and prevent further intimate contact of the gelatin with the wine. Formerly freshly prepared ferric hydroxide was much employed for detannating wine, but the chief objection to its use was due to the fact that some iron invariably was taken up by the acid present in the wine; moreover, the process was more tedious than in the case of gelatin. As the removal of tannin from wine in no way interferes with its quality—alcoholic strength and aroma remaining the same, and only coloring matter being lost—a supply of detannated wine should be kept on hand, for it requires very little more labor to detannate a gallon than a pint.

If ferric hydroxide is to be used, it must be freshly prepared, and a convenient quantity then be added to the wine—about 8 ounces of the expressed, but moist, precipitate to a gallon.

PREVENTION OF FERMENTATION.

Fermentation may be prevented in either of two ways:

(1) By chemical methods, which consist in the addition of germ poisons or antiseptics, which either kill the germs or prevent their growth. Of these the principal ones used are salicylic, sulphurous, boracic, and benzoic acids, formalin, fluorides, and saccharine. As these substances are generally regarded as adulterants and injurious, their use is not recommended.

(2) The germs are either removed by

some mechanical means such as a filtering or a centrifugal apparatus, or they are destroyed by heat or electricity. Heat has so far been found the most practical.

When a liquid is heated to a sufficiently high temperature all organisms in it are killed. The degree of heat required, however, differs not only with the particular kind of organism, but also with the liquid in which it is held. Time is also a factor. An organism may not be killed if heated to a high temperature and quickly cooled. If, however, the temperature is kept at the same high degree for some time, it will be killed. It must also be borne in mind that fungi, including yeasts, exist in the growing and the resting states, the latter being much more resistant than the former. One characteristic of the fungi and their spores is their great resistance to heat when dry. In this state they can be heated to 212° F. without being killed. The spores of the common mold are even more resistant. This should be well considered in sterilizing bottles and corks, which should be steamed to 240° F. for at least 15 minutes.

Practical tests so far made indicate that grape juice can be safely sterilized at from 165° to 176° F. At this temperature the flavor is hardly changed, while at a temperature much above 200° F. it is. This is an important point, as the flavor and quality of the product depend on it.

Use only clean, sound, well-ripened, but not over-ripe grapes. If an ordinary cider mill is at hand, it may be used for crushing and pressing, or the grapes may be crushed and pressed with the hands. If a light-colored juice is desired, put the crushed grapes in a cleanly washed cloth sack and tie up. Then either hang up securely and twist it or let two persons take hold, one on each end of the sack and twist until the greater part of the juice is expressed. Next gradually heat the juice in a double boiler or a large stone jar in a pan of hot water, so that the juice does not come in direct contact with the fire at a temperature of 180° to 200° F., never above 200° F. It is best to use a thermometer, but if there be none at hand heat the juice until it steams, but do not allow it to boil. Put it in a glass or enameled vessel to settle for 24 hours; carefully drain the juice from the sediment, and run it through several thicknesses of clean flannel, or a conic filter made from woolen cloth or felt may be used. This filter is fixed to a hoop of iron, which

can be suspended wherever necessary. After this fill into clean bottles. Do not fill entirely, but leave room for the liquid to expand when again heated. Fit a thin board over the bottom of an ordinary wash boiler, set the filled bottles (ordinary glass fruit jars are just as good) in it, fill in with water around the bottles to within about an inch of the tops, and gradually heat until it is about to simmer. Then take the bottles out and cork or seal immediately. It is a good idea to take the further precaution of sealing the corks over with sealing wax or paraffine to prevent mold germs from entering through the corks. Should it be desired to make red juice, heat the crushed grapes to not above 200° F., strain through a clean cloth or drip bag (no pressure should be used), set away to cool and settle, and proceed the same as with light-colored juice. Many people do not even go to the trouble of letting the juice settle after straining it, but reheat and seal it up immediately, simply setting the vessel away in a cool place in an upright position where they will be undisturbed. The juice is thus allowed to settle, and when wanted for use the clear juice is simply taken off the sediment. Any person familiar with the process of canning fruit can also preserve grape juice, for the principles involved are identical.

One of the leading defects so far found in unfermented juice is that much of it is not clear, a condition which very much detracts from its otherwise attractive appearance, and due to two causes already alluded to. Either the final sterilization in bottles has been at a higher temperature than the preceding one, or the juice has not been properly filtered or has not been filtered at all. In other cases the juice has been sterilized at such a high temperature that it has a disagreeable scorched taste. It should be remembered that attempts to sterilize at a temperature above 195° F. are dangerous so far as the flavor of the finished product is concerned.

Another serious mistake is sometimes made by putting the juice into bottles so large that much of it becomes spoiled before it is used after the bottles are opened. Unfermented grape juice properly made and bottled will keep indefinitely, if it is not exposed to the atmosphere or mold germs; but when a bottle is once opened it should, like canned goods, be used as soon as possible to keep from spoiling.

Another method of making unfermented grape juice, which is often resorted to where a sufficiently large quantity is made at one time, consists in this:

Take a clean keg or barrel (one that has previously been made sweet). Lay this upon a skid consisting of two scantlings or pieces of timber of perhaps 20 feet long, in such a manner as to make a runway. Then take a sulphur match, made by dipping strips of clean muslin about 1 inch wide and 10 inches long into melted brimstone, cool it and attach it to a piece of wire fastened in the lower end of a bung and bent over at the end, so as to form a hook. Light the match and by means of the wire suspend it in the barrel, bung the barrel up tight, and allow it to burn as long as it will. Repeat this until fresh sulphur matches will no longer burn in the barrel.

Then take enough fresh grape juice to fill the barrel one-third full, bung up tight, roll and agitate violently on the skid for a few minutes. Next burn more sulphur matches in it until no more will burn, fill in more juice until the barrel is about two-thirds full; agitate and roll again. Repeat the burning process as before, after which fill the barrel completely with grape juice and roll. The barrel should then be bunged tightly and stored in a cool place with the bung up, and so secured that the package cannot be shaken. In the course of a few weeks the juice will have become clear and can then be racked off and filled into bottles or jars direct, sterilized, and corked or sealed up ready for use. By this method, however, unless skillfully handled, the juice is apt to have a slight taste of the sulphur.

The following are the component parts of a California and a Concord unfermented grape juice:

	Concord Per Cent	California Per Cent
Solid contents	20.37	20.60
Total acids (as tartaric)	.663	.53
Volatile acids	.023	.03
Grape sugar	18.54	19.15
Free tartaric acids	.025	.07
Ash	.255	.19
Phosphoric acids	.027	.04
Cream of tartar	.55	.59

This table is interesting in so far that the California unfermented grape juice was made from Viniferas or foreign varieties, whereas the Concord was a Labruska or one of the American sorts. The difference in taste and smell is even more pronounced than the analysis would indicate.

Small quantities of grape juice may be preserved in bottles. Fruit is likely to be dusty and to be soiled in other ways, and grapes, like other fruits, should be well washed before using. Leaves or other extraneous matter should also be removed. The juice is obtained by moderate pressure in an ordinary screw press, and strained through felt. By gently heating, the albuminous matter is coagulated and may be skimmed off, and further clarification may be effected by filtering through paper, but such filtration must be done as rapidly as possible, using a number of filters and excluding the air as much as possible.

The juice so obtained may be preserved by sterilization, in the following manner: Put the juice in the bottles in which it is to be kept, filling them very nearly full; place the bottles, unstoppered, in a kettle filled with cold water, so arranging them on a wooden perforated "false bottom" or other like contrivance as to prevent their immediate contact with the metal, this preventing unequal heating and possible fracture. Now heat the water, gradually raising the temperature to the boiling point, and maintain at that until the juice attains a boiling temperature; then close the bottles with perfectly fitting corks, which have been kept immersed in boiling water for a short time before use.

The corks should not be fastened in any way, for, if the sterilization is not complete, fermentation and consequent explosion of the bottle may occur unless the cork should be forced out.

If the juice is to be used for syrup, as for use at the soda fountain, the best method is to make a concentrated syrup at once, using about 2 pounds of refined sugar to 1 pint of juice, dissolving by a gentle heat. This syrup may be made by simple agitation without heat; and a finer flavor thus results, but its keeping quality would be uncertain.

The juices found in the market are frequently preserved by means of antiseptics, but so far none have been proposed for this purpose which can be considered entirely wholesome. Physiological experiments have shown that while bodies suited for this purpose may be apparently without bad effect at first, their repeated ingestion is likely to cause gastric disturbance.

SPARKLING WINES.

An apparatus for converting still into foaming wines, and doing this efficiently, simply, and rapidly, consists of a vertical steel tube, which turns on an axis, and bears several adjustable glass globes that are in connection with each other by means of distributing valves, the latter being of silver-plated bronze. The glass globes serve as containers for carbonic acid, and are kept supplied with this gas from a cylinder connected therewith.

The wine to be impregnated with the acid is taken from a cask, through a special tube, which also produces a light pressure of carbonic acid on the cask, the object of which is to prevent the access of atmospheric air to the wine within, and, besides, to cause the liquid to pass into the bottle without jar or stroke. The bottles stand under the distributing valves, or levers, placed above and below them. Now, if the cock, by means of which the glass bulbs and the bottles are brought into connection, is slightly opened, and the desired lever is put in action, the carbonic acid at once forces the air out of the bottles, and sterilizes them. The upper bottles are now gradually filled. The whole apparatus, including the filled bottles, is now tilted over, and the wine, of its own weight, flows through collectors filled with carbonic acid, and passes, impregnated with the gas, into other bottles placed below. Each bottle is filled in course, the time required for each being some 45 seconds. The saturation of the liquid with carbonic acid is so complete and plentiful that there is no need of hurry in corking.

By means of this apparatus any desired still wine is at once converted into a sparkling one, preserving at the same time its own peculiarities of taste, bouquet, etc. The apparatus may be used equally well upon fruit juices, milk, and, in fact, any kind of liquid, its extreme simplicity permitting of easy and rapid cleansing.

ARTIFICIAL FRENCH BRANDY.

I.—The following is Eugene Dieterich's formula for *Spiritus vini Gallici artificialis:*

Tincture of gall- apples..........	10 parts
Aromatic tincture...	5 parts
Purified wood vinegar.............	5 parts
Spirit of nitrous ether	10 parts
Acetic ether........	1 part
Alcohol, 68 per cent.	570 parts
Distilled water.....	400 parts

Mix, adding the water last, let stand for several days, then filter.

II.—The *Münchener Apotheker Verein* has adopted the following formula for the same thing:

Acetic acid, dilute, 90 per cent.......	4 parts
Acetic ether........	4 parts
Tincture aromatic..	40 parts
Cognac essence.....	40 parts
Spirit of nitrous ether............	20 parts
Alcohol, 90 per cent.	5,000 parts
Water, distilled.....	2,500 parts

Add the acids, ethers, etc., to the alcohol, and finally add the water. Let stand several days, and, if necessary, filter.

III.—The Berlin Apothecaries have adopted the following as a magistral formula:

Aromatic tincture...	4 parts
Spirit of nitrous ether............	5 parts
Alcohol, 90 per cent.	1,000 parts
Distilled water, quantity sufficient to make.........	2,000 parts

Mix the tincture and ether with the alcohol, add the water and for every ounce add one drop of tincture of rhatany.

Of these formulas the first is to be preferred as a close imitation of the taste of the genuine article. To imitate the color use burnt sugar.

LIQUEURS.

Many are familiar with the properties of liqueurs but believe them to be very complex and even mysterious compounds. This is, of course, due to the fact that the formulas are of foreign origin and many of them have been kept more or less secret for some time. Owing to the peculiar combination of the bouquet oils and flavors, it is impossible to make accurate analyses of them. But by the use of formulas now given, these products seem to be very nearly duplicated.

It is necessary to use the best sugar and oils obtainable in the preparation of the liqueurs. As there are so many grades of essential oils on the market, it is difficult to obtain the best indirectly. The value of the cordials is enhanced by the richness and odor and flavor of the oils, so only the best qualities should be used.

For filtering, flannel or felt is valuable. Flannel is cheaper and more easily washed. It is necessary to return filtrate several times with any of the filtering media.

As a clarifying agent talcum allowed to stand several days acts well. These rules are common to all.

The operations are all simple:

First: Heat all mixtures. Second: Keep the product in the dark. Third: Keep in warm place.

The liqueurs are heated to ripen the bouquet flavor, it having effect similar to age. To protect the ethereal oils, air and light are excluded; hence it is recommended that the bottles be filled to the stopper. The liqueurs taste best at a temperature not exceeding 55° F. They are all improved with age, especially many of the bouquet oils.

Bénédictine.—

I.—Bitter almonds..	40	grams
Powdered nutmeg........	4.500	grams
Extract vanilla..	120	grams
Powdered cloves........	2	grams
Lemons, sliced..	2	grams
True saffron....	.600	grams
Sugar.........	2,000	grams
Boiling milk....	1,000	c.c.
Alcohol, 95 per cent........	2,000	c.c.
Distilled water..	2,500	c.c.

Mix. Let stand 9 days with occasional agitation. Filter sufficiently.

II.—Essence Bénédictine.......	75	c.c
Alcohol, 95 per cent........	1,700	c.c.

Mix.

Sugar..........	1,750	grams
Water, distilled.	1,600	c.c.

Mix together, when clear solution of sugar is obtained. Color with caramel. Filter sufficiently.

NOTE.—This liqueur should be at least 1 year old before used.

Essence Bénédictine for Bénédictine No. II.—

I.—Myrrh..........	1	part
Decorticated cardamom..........	1	part
Mace...........	1	part
Ginger..........	10	parts
Galanga root......	10	parts
Orange peel (cut)..	10	parts
Extract aloe.......	4	parts
Alcohol..........	160	parts
Water..........	80	parts

Mix, macerate 10 days and filter.

II.—Extract licorice....	20	parts
Sweet spirits niter..	200	parts
Acetic ether.......	30	parts
Spirits ammonia...	1	part
Coumarin........	.12	parts
Vanillin..........	1	part

III.—Oil lemon.........	3	drops
Oil orange peel....	3	drops
Oil wormwood.....	2.5	drops
Oil galanga.......	2	drops
Oil ginger........	1	drop
Oil anise........	15	drops
Oil cascarilla......	15	drops
Oil bitter almond..	12	drops
Oil milfoil........	10	drops
Oil sassafras......	7	drops
Oil angelica.......	6	drops
Oil hyssop........	4	drops
Oil cardamom.....	2	drops
Oil hops..........	2	drops
Oil juniper........	1	drop
Oil rosemary......	1	drop

Mix A, B, and C.

NOTE.—This essence should stand 2 years before being used for liqueurs.

Chartreuse.—I.—Elixir végétal de la Grande Chartreuse.

Fresh balm mint herbs............	64 parts
Fresh hyssop herbs..	64 parts
Angelica herbs and root, fresh, together	32 parts
Cinnamon..........	16 parts
Saffron.............	4 parts
Mace.............	4 parts

Subject the above ingredients to maceration for a week with alcohol (96 per cent), 1,000 parts, then squeeze off and distill the liquid obtained over a certain quantity of fresh herbs of balm and hyssop. After 125 parts of sugar have been added to the resultant liqueur, filter.

The genuine Chartreuse comes in three different colors, viz., green, white, and yellow. The coloration, however, is not artificial, but is determined by the addition of varying quantities of fresh herbs in the distillation. But since it would require long and tedious trials to produce the right color in a small manufacture, the yellow shade is best imparted by a little tincture of saffron, and the green one by the addition of a few drops of indigo solution.

II.—Eau des Carmes.....	3½	ounces
Alcohol............	1	quart
Distilled water......	1	quart
Sugar.............	1½	pounds
Tincture of saffron...	1	ounce

Mix. Dissolve sugar in warm water, cool, strain, add remainder of ingredients, and filter. This is known as yellow Chartreuse.

Curaçao Liqueur.—

A.—

Oil lemon, q. s.......	10 drops
Oil bitter almond, q. s.	5 drops
Oil curaçoa orange...	15 parts
Oil sweet orange.....	1 part
Oil bitter orange.....	1 part
Cochineal...........	1 part
French brandy......	50 parts

B.—Alcohol.............4,500 parts

C.—Sugar..............3,500 parts
Water (distilled).....4,000 parts

Mix A, B, and C. Filter. Color with caramel.

May Bowl or May Wine.—The principal ingredient of May bowl, or that which gives it its flavor and bouquet, is fresh *Waldmeisterkraut* (*Asperula odorata*), the "woodruff" or "sweet grass," "star grass," and a dozen other aliases, of a plant growing wild all over Europe, both continental and insular, and cultivated by some gardeners in this country. It is accredited with being a diuretic, deobstruent and hepatic stimulant, of no mean order, though it has long been banished from the pharmacopœia.

In Baden and in Bavaria in preparing *Maitrank* the practice was formerly to first make an essence—*Maitrankessenz*, for the preparation of which every housewife had a formula of her own. The following was that generally used in the south of Germany:

I.—

Fresh, budding woodruff, cut fine	500 parts
Alcohol, commercial (90 per cent).	1,000 parts

Digest together for 14 days, then filter and press off. Many add to this some flavoring oil. As coumarin has been found to be the principle to which the Waldmeister owes its odor, many add to the above Tonka bean, chopped fine, 1 part to the thousand. From about 12 to 15 drachms of this essence is added to make a gallon of the wine, which has about the following formula:

French brandy, say	4 drachms
Oil of unripe oranges.........	80 drops
Sugar............4 to 8 ounces
Essence..........	12 drachms
Wine to make.....	1 gallon

II.—Take enough good woodruff (*Waldmeister*) of fine aroma and flavor. Remove all parts that will not add to the excellence of the product, such as wilted, dead, or imperfect leaves, stems, etc., and wash the residue thoroughly in cold water, and with as little pressure as possible. Now choose a flask with a neck sufficiently wide to receive the stems without pressing or bruising them, and let the pieces fall into it. Pour in sufficient strong alcohol (96 per cent) to cover the herbs completely. In from 30 to 40 minutes the entire aroma is taken up by the alcohol, which takes on a beautiful green color, which, unfortunately, does not last, disappearing in a few days, but without affecting the aroma in the least. The alcohol should now be poured off, for if left to macerate longer, while it would gain in aroma, it will also take up a certain bitter principle that detracts from the delicacy of flavor and aroma. The extract is now poured on a fresh quantity of the herb, and continue proceeding in this manner until a sufficiently concentrated extract is obtained to give aroma to 100 times its weight of wine or cider.

III.—Fresh woodruff, in bloom or flower, is freed from the lower part of its stem and leaves, and also of all foreign or inert matter. The herb is then lightly stuck into a wide-mouth bottle, and covered with strong alcohol. After 30 minutes pour off the liquor on fresh woodruff. In another half hour the essence is ready, though it should not be used immediately. It should be kept at cellar heat (about 60° F.) for a few days, or until the green color vanishes. Any addition to the essence of aromatics, such as orange peel, lemons, spices, etc., is to be avoided. To prepare the Maitrank, add the essence to any good white wine, tasting and testing, until the flavor suits.

The following are other formulas for the drink:

IV.—

Good white wine or cider............	65 parts
Alcohol, dilute......	20 parts
Sugar.............	10 parts
Maitrankessenz.....	1 part

Mix.

Maraschino Liqueur.—

Oil bitter almonds....	15 minims
Essence vanilla......	1 drachm
Jasmine extract......	2 drops
Raspberry essence....	10 drops
Oil neroli...........	10 drops
Oil lemon...........	15 minims
Spirits nitrous ether..	2 drachms
Alcohol.............	6 pints
Sugar..............	8 pounds
Rose water.........	10 ounces
Water sufficient to make.............	2 gallons

Make a liquor in the usual manner.

To Clarify Liqueurs.—For the clarification of turbid liqueurs, burnt powdered alum is frequently employed. Make a trial with 200 parts of the dim liqueur, to which 1.5 parts of burnt powdered alum is added; shake well and let stand until the liquid is clear. Then decant and filter the last portion. If the trial is successful, the whole stock may be clarified in this manner.

MEDICINAL WINES:

Beef and Iron.—The following formula is recommended by the American Pharmaceutical Association:

I.—Extract of beef.... 35 grams
Tincture of citro-
 chloride of iron.. 35 c.c.
Compound spirit
 of orange....... 1 c.c.
Hot water........ 60 c.c.
Alcohol.......... 125 c.c.
Syrup............ 125 c.c.
Sherry wine suffi-
 cient to make....1,000 c.c.

Rub the extract of beef with the hot water, and add, while stirring, the alcohol. Allow to stand 3 days or more, then filter and distill off the alcohol. Add to the residue 750 cubic centimeters of the wine, to which the compound spirit of orange has been previously added. Finally add the tincture of citro-chloride of iron, syrup, and enough wine to make 1,000 cubic centimeters. Filter if nece..sary.

II.—For Poultry and Stock.—A good formula for wine of beef and iron is as follows:

Beef extract.......256 grains
Tincture of iron
 citro-chloride. ..256 minims
Hot water........ 1 fluidounce
Sherry wine enough
 to make........ 1 pint

Pour the hot water in the beef extract and triturate until a smooth mixture is made. To · this add, gradually and under constant stirring, 12 ounces of the wine. Add now, under same conditions, the iron, stir in well, and finally add the remainder of the wine.

Cinchona.—I.—Macerate 100 parts of cinchona succirubra in coarse powder for 30 minutes in 100 parts of boiling water. Strain off the liquor and set aside. Macerate the residuum in 1,000 parts of California Malaga for 24 hours, strain off the liquid and set aside. Finally macerate the magma in 500 parts of alcohol, of 50 per cent, for 1 hour, strain off and set aside. Wash the residue with a little water to recover all the alcoholic tincture; then unite all the liquids, let stand for 24 hours, and filter. To the filtrate add 800 parts loaf sugar and dissolve by the aid of gentle heat and again filter. The product is all that could be asked of a wine of cinchona. To make a ferrated wine of this, dissolve 1 part of citro-ammoniacal pyrophosphate of iron to every 1,000 parts of wine.

II.—Yvon recommends the following formula:

Red cinchona, coarse
 powder........... 5 parts
Alcohol, 60 per cent.. 10 parts
Diluted hydrochloric
 acid............. 1 part
Bordeaux wine......100 parts

Macerate the bark with the acid and alcohol for 6 days, shaking from time to time, add the wine, macerate for 24 hours, agitating frequently, then filter.

Removal of Musty Taste and Smell from Wine.—For the removal of this unpleasant quality, Kulisch recommends the use of a piece of charcoal of about the size of a hazel nut—5 to 10 parts per 1,000 parts of wine. After this has remained in the cask for 6 to 8 weeks, and during this time has been treated once a week with a chain or with a stirring rod, the wine can be racked off. Obstinate turbidness, as well as stalk taste and pot flavor, can also be obviated by the use of the remedy.

WINTERGREEN, TO DISTINGUISH METHYL SALICYLATE FROM OIL OF.

A quantity of the sample is mixed in a test tube with an equal volume of pure concentrated sulphuric acid. Under these conditions the artificial compound shows no rise in temperature and acquires only a slight yellowish tint, while with the natural oil there is a marked rise in temperature and the mixture assumes a rose-red color, gradually passing into darker shades.

WIRE ROPE.

See also Steel.

A valuable anti-friction and preservative compound for mine cables is as follows: Seven parts soft tallow and 3 parts plumbago, mixed thoroughly; make a long, hollow box or trough, gouge out a 4 by 6 piece of scantling about 2 feet long, sawing it down lengthwise and hollowing out the box or trough enough to hold several pounds of the compound, making also a hole lengthwise of the trough for the cable to run through; then affix to rope and clamp securely, having the box or trough so fixed that it cannot play, and letting the cable pass through it while going up or down, so that it will get a thorough coating. This, it is found, will preserve a round cable very well, and can be used at least once a week. For a flat steel cable raw linseed oil can be used instead of the tallow, in about the proportion of 6 parts oil and 3 plumbago. If tar is used, linseed oil is to be added to keep the tar from adhering, both ingredients to be mixed while warm.

To preserve wire rope laid under ground, or under water, coat it with a mixture of mineral tar and fresh slaked lime in the proportion of 1 bushel of lime to 1 barrel of tar. The mixture is to be boiled, and the rope saturated with it while hot; sawdust is sometimes added to give the mixture body. Wire rope exposed to the weather is coated with raw linseed oil, or with a paint composed of equal parts of Spanish brown or lamp-black with linseed oil.

WIRE HARDENING:
See Steel.

WITCH-HAZEL JELLY:
See Cosmetics.

Wood

DECORATIVE WOOD-FINISH.

Paint or stencil wood with white-lime paint. When it has dried slowly in the shade, brush it off and a handsome dark-brown tone will be imparted to the oak-wood. Some portions which may be desired darker and redder are stained again with lime, whereby these places become deeper. It is essential that the lime be applied in even thickness and dried slowly, for only then the staining will be red and uniform.

After the staining saturate the wood with a mixture of varnish, 2 parts; oil of turpentine, 1 part; turpentine, ½ part. When the oil ground is dry apply 2 coatings of pale amber varnish.

Colored decorations on pinewood can be produced as follows:

The most difficult part of the work is to remove the rosin accumulations without causing a spot to appear. Burn out the places carefully with a red-hot iron. Great care is necessary to prevent the iron from setting the rosin on fire, thus causing black smoke clouds.

The resulting holes are filled up with plaster to which a little light ocher is added to imitate the shade of the wood as perfectly as possible. Plaster up no more than is necessary.

Rub the wood down with very fine sandpaper, taking especial care to rub only with the grain of the wood, since all cross scratches will remain permanently visible.

After this preliminary work cover the wood with a solution of white shellac, in order not to injure the handsome golden portions of the wood and to preserve the pure light tone of the wood in general.

On this shellac ground paint and stencil with glazing colors, ground with isinglass solution. The smaller, more delicate portions, such as flowers and figures, are simply worked out in wash style with water colors, using the tone of the wood to remain as high lights, surrounding the whole with a black contour.

After this treatment the panels and decorated parts are twice varnished with dammar varnish. The friezes and pilaster strips are glazed darker and set off with stripes; to varnish them use amber varnish.

The style just mentioned does not exclude any other. Thus, for instance, a very good effect is produced by decorating the panels only with a black covering color or with black and transparent red (burnt sienna and a little carmine) after the fashion of boule work in rich ornaments, in such a way that the natural wood forms the main part and yet quite a considerable portion of the ornament.

Intarsia imitation is likewise well adapted, since the use of variegated covering colors is in perfect keeping with the decoration of natural wood. How it should be applied, and how much of it, depends upon one's taste, as well as the purpose and kind of the object.

It is a well-known fact that the large pores of oak always look rather smeary, according to whether the workshop is more or less dusty. If this is to be avoided, which is essential for neat work, take good wheat starch, pound it fine with a hammer and stir by means of a wooden spatula good strong polish with the wheat starch to a paste and work the paste into the pores by passing it cross-wise over the wood. After about ½ hour, rub down the wood thus treated in such a manner that the pores are filled. In case any open pores remain, repeat the process as before. After that, rub down, polish or deaden. If this operation is not performed, the pores will always look somewhat dirty, despite all care. Every cabinetmaker will readily perceive that this filling of the pores will save both time and polish in the subsequent finishing.

WOOD FILLERS.

The novice in coach painting is quite as likely to get bewildered as to be aided by much of the information given about roughstuff, the more so as the methods differ so widely. One authority tells us to use a large proportion of lead ground in oil with the coarser pigment, while another says use dry lead and but a small percentage, and still another insists that lead must be tabooed altogether. There are withal a good many moss-grown superstitions associated with the subject. Not the least of these is the remarkably absorbent nature which the surface that has been roughstuffed and "scoured" is supposed to possess. By many this power of absorption is believed to be equal to swallowing up, not only all the color applied, but at least 3 coats of varnish, and none of these would think of applying a coat of color to a roughstuffed surface without first giving it a coat of liquid filler as a sort of sacrificial oblation in recognition of this absorbing propensity. Another authority on the subject has laid down the rule that in the process of scouring, the block of pumice stone must always be moved in one direction, presumably for the reason that some trace of the stone is likely to be visible after the surface is finished.

If the block of stone is scratching, perhaps the appearance of the finished panel may be less objectionable with the furrows in parallel lines than in what engravers call "cross-hatching," but if the rubbing is properly done it is not easy to discover what difference it could make whether the stone is moved in a straight line or a circle. As to absorption, it cannot be distinguished in the finished panel between the surface that was coated with liquid filler and that to which the color was applied directly, except that cracking always occurs much sooner in the former, and this will be found to be the case with surfaces that have been coated with liquid filler and finished without roughstuff. Among the pigments that may be used for roughstuff, and there are half a dozen or more, any of which may be used with success, there is no doubt but that known as "English filler" is best, but it is not always to be had without delay and inconveniences.

Yellow ocher, Reno umber and Key-stone filler are all suitable for roughstuff, the ocher having been used many years for the purpose, but, as already remarked, the English filler is best. This is the rule for mixing given by Nobles and Hoare: Four pounds filler, 1 pound ground white lead, 1 pint gold size, 1 pint varnish and $1\frac{1}{4}$ pints turpentine, or $\frac{3}{4}$ pint good size and $\frac{1}{2}$ pint boiled oil in lieu of the varnish. In regard to the use of white lead ground in oil, it makes the rubbing more laborious, increases the liability to scratching, and requires a much longer time to harden before the scouring can be done, without in any appreciable manner improving the quality of the surface when finished.

It may be remarked here that the addition of white lead, whether ground in oil or added dry to the coarser pigment, increases the labor of scouring just in proportion as it is used until sufficient may be used to render the scouring process impossible; hence, it follows that the mixing should be governed by the character of the job in hand. If the job is of a cheap class the use of very little or no lead at all is advisable, and the proportion of Japan and turpentine may also be increased, with the result that a fairly good surface may be obtained with much less labor than in the formula given.

The number of coats of filler required to effect the purpose in any given case must depend upon how well the builder has done his part of the work. If he has left the surface very uneven it follows, as a matter of course, that more coats will be required to make it level, and more of the roughstuff will remain after the leveling process than if the woodwork had been more perfectly done. While the merits of a system or method are not to be judged by its antiquity, there should be a good reason to justify the substitution of a new method for one that has given perfect satisfaction for generations and been used by the best coach painters who ever handled a brush.

A well-known writer on paints says that the effect of a varnish is usually attributed to the manner of its application and the quantity of thinners used for diluting the melted gums, with the prepared oils and the oxidizing agents used in its manufacture. While this has undoubtedly much to do with the successful application of varnish, there are other facts in this connection that should not be overlooked. For example, varnish is sometimes acted on by the breaking up, or the disintegration of the filling coats; which in turn is evidently acted on by the wood itself, according to its nature.

With the aid of the microscope in examining the component parts of wood a cellular tissue is observed which varies in form according to the species and the parts which are inspected. This cellular tissue is made up of small cavities called pores or cells, which are filled with a widely diversified matter and are covered with a hard and usually brittle substance called *lignin*.

This diversified matter consists of mineral salts and various organic substances, gelatinous in their nature and held in solution by a viscous liquid and containing nitrogenous matter in different combinations, the whole being designated by the general name of albuminous substances. The older the wood the more viscous is the matter; while wood of recent growth (sapwood) contains less viscous matter holding these substances in solution. This albumen in wood acts on substances like filler and varnish in one way or the other, good or bad. The seasoning of wood does not dispose of these substances. The water evaporates, leaving them adhering to the sides of the cells. The drier these substances are the less action they exert on the filler or whatever substance is coated on the surface. If the filler disintegrates, it affects the varnish.

All albuminous substances, be they dry or in liquid form, are subject, more or less, according to the protein they contain—which seems, or rather is, the essential principle of all albuminous matter —to the influence of caustic potash and soda. Thus, the albumen of an egg is exactly like that contained in the composition of wood. As albumen in wood becomes solid by drying, it is easily dissolved again, and will then be acted on chemically by any extraneous substance with which it comes in contact.

Some of the shellacs, substitutes for shellacs, and some of the liquid fillers are manufactured from some of the following substances: Old linseed oil, old varnish, old and hard driers, turpentine, benzine, often gasoline, rosin, whiting, cornstarch flour, hulls, paint skins, silica, and so on. The list is long. To these must be added a large volume of potash, to bring it to and hold it in solution. There must be an excess of potash which is not combined into a chemical compound, which if it did, might mitigate its influence on the albumen of the wood. But as there is potash in its pure state remaining in the solution it necessarily attacks the albumen of the wood, causing disintegration, which releases it from the wood, causing white, grayish flakes, and the formation of a powder. This is not a conclusion drawn from an inference but an established scientific fact resulting from experiments with fillers the various compositions of which were known. All alkalies act on albumen. No one would knowingly varnish over a surface such as it would be were the white of an egg applied to it and then washed with an alkali solution; but that is just what is done when varnish is put over a wood surface filled with a filler which contains an alkali.

Most of the combinations of material used in the painting trade are mixtures; that is, each part remains the same— exerting the same chemical action on another substance, or any other substance coming in contact with a paint mixture will exert the same chemical action on any part, or on any ingredient it contains, the same as if that part was by itself.

We can now account for some of the numerous peculiarities of varnish. We know that any alkali when coming in contact with albumen forms a compound, which on drying is a white, brittle substance easily disintegrated. This is why potash, sal soda, and kindred substances will remove paint. The alkali attacks the albumen in the oil, softening it, causing easy removal, whereas if it were allowed to dry, the albumen in the oil would take on a grayish color quite brittle. Potash or other alkalies in filler not only attack the albumen in the wood, but also attack the albumen in the oil by forming a compound with it. Probably this compound is very slight, only forming a compound in part, enough, nevertheless, to start a destroying influence, which is demonstrated by the following results of experiments. The reader has, perhaps, some time in his career applied a rosin varnish over a potash filler and has been surprised by the good results, a more permanent effect being obtained than in other instances where the best of varnish was used. This is accounted for by the rosin of the potash. Again, the reader may have had occasion to remove varnish with potash and found that potash would not touch it. This is because of its being a rosin varnish. Potash in filler may be rendered somewhat inert, by reason of its compounding with other parts of the filler, but owing to the quantity used in some of the commercial fillers it is not possible that all the alkali is rendered inert. Hence it will attack the albumen wherever found, as all albumen is identical in its chemical composition.

Alkalies have but little effect on the

higher classes of gums, because of their effect on the albumen in the wood and oil. All alcohol varnishes or varnishes made by the aid of heat stand well over an alkali filler. Varnishes which contain little oil seem to stand well. This is accounted for by the fact that alcohol renders albumen insoluble. Alkalies of all kinds readily attack shellac and several other of the cheap gums, forming unstable compounds on which oil has but little effect.

Close-grained wood contains less albumen and more lignin than open-grained varieties, and consequently does not take so much filler, which accounts for the finish invariably lasting longer than the same kind used on an open-grained wood. Open-grained wood contains more sap than close grained; consequently there is more albumen to adhere to the sides of the cells. The more albumen, the more readily it is attacked by the potash, and the more readily decomposed, or rather destroyed.

Alcohol renders albumen insoluble immediately on application. It prevents it from compounding with any other substance, or any other substance compounding with it. Hence, we must conclude that an application of alcohol to wood before the filler is applied is valuable, which is proven to be a fact by experiment. Wash one half of a board with alcohol, then apply the potash filler over all. Again, wash the portion of the board on which is the filler and apply a heavy-bodied oil varnish. Expose to sunlight and air the same as a finished door or the like, and wait for the result. At the end of a few months a vast difference will be found in the two parts of the surface. The one on which there is no alcohol will show the ravages of time and the elements much sooner than the one on which it is.

Wood finishers demand a difference in the composition of fillers, paste and liquid, for open- and close-grained wood, respectively; but unfortunately they do not demand a difference between either kind in themselves, according to the kind of wood. Paste fillers are used indiscriminately for open-grained wood and liquid for close-grained wood.

To find the fillers best adapted for a certain wood, and to classify them in this respect will require a large amount of chemical work and practical experiments; but that it should be done is evidenced by the fact that both success and failure result from the use of the same filler on different varieties of wood. After once being classified (owing to the large number now on the market), they will not number nearly so many in the aggregate as might be supposed; as it will be found in many instances that two entirely different varieties of wood resemble each other more closely in their vascular formation and cell characteristics than do two other specimens of the same variety. It is a recognized fact that paste fillers whose base is starch or the like work better and give better results in certain instances, while those whose base is mineral matter seem to do better in other cases.

It is noticed that rosewood as a finishing veneer is obsolete. This is not because of its scarcity, but because it is so hard to finish without having been seasoned for a long time. In these days, manufacturers cannot wait. It takes longer for the sap of rosewood to become inactive, or in trade parlance to "die," than any other wood. This is because it takes so long for the albumen in the sap to coagulate. Rosewood has always been a source of trouble to piano makers, on account of the action of the sap on the varnish. However, if this wood, previously to filling, was washed with a weak solution of phosphoric acid, and then with wood spirit, it might be more easily finished. The phosphoric acid would coagulate the albumen on the surface of the wood immediately, while alcohol would reduce it to an insoluble state. The idea here is to destroy the activity of the sap, on the same principle as sappy places and knot sap are destroyed by alcohol-shellac before being painted.

Oak is another wood which gives the painter trouble to finish. This may be accounted for as follows: Oak contains a sour acid principle called tannic acid. It is a very active property. Wood during the growing season contains more albumen; thus in the circulation of the sap a large quantity of soft matter is deposited on the lignin which lines the cells, which lignin, if it contains any acid matter, acts on the material of the filler. Tannic acid has a deleterious effect on some of the material of which a number of fillers are made. Starch and many gums are susceptible to its influence, making some of them quite soft. Oak, like most other timber cut at the season when the least sap is in circulation, is the more easily finished.

The vascular formation may, and no doubt has, something to do with wood finishing. Different species of wood differ materially in their vascular and cellular formation. Wood finishers recognize a difference in treatment of French burl walnut and the common American variety. Circassian and Italian walnut, although of the same species, demand widely different treatment in finishing to get the best results.

The only way to find the best materials to use in certain cases is to study and experiment with that end in view. If, by aid of a microscope, a certain piece of wood shows the same cellular formation that another piece did which was successfully finished by a certain process, it may be regarded as safe to treat both alike. If observation on this line is indulged in, it will not take the finisher very long to learn just what treatment is best for the work in hand. How often it has been noticed in something of two parts, like a door, that the panels when finished will pit, run, or sag, while the sides will present a surface in every way desirable and *vice versa*. This is due to the difference in the cellular construction of the wood and to the cellulose, and cannot be otherwise for the parts have been seasoned the same time and treated exactly alike. The physiology of wood is imperfectly understood, but enough is known to warrant us in saying with a certainty that the chemicals in fillers do act upon the principles embodied in its formation.

Some tried formulas follow:

I.—Make a paste to fill the cracks as follows: Old furniture polish: Whiting, plaster of Paris, pumice stone, litharge, equal parts, Japan drier, boiled linseed oil, turpentine, coloring matter, of each a sufficient quantity.

Rub the solids intimately with a mixture of 1 part of the Japan, 2 parts of the linseed oil, and 3 parts of turpentine, coloring to suit with vandyke brown or sienna. Lay the filling on with a brush, let it set for about 20 minutes, and then rub off clean except where it is to remain. In 2 days it will be hard enough to polish. After the surface has been thus prepared, the application of a coat of first-class copal varnish is in order. It is recommended that the varnish be applied in a moderately warm room, as it is injured by becoming chilled in drying. To get the best results in varnishing, some skill and experience are required. The varnish must be kept in an evenly warm temperature, and put on neither too plentifully nor too gingerly. After a satisfactorily smooth and regular surface has been obtained, the polishing proper may be done. This may be accomplished by manual labor and dexterity, or by the application of a very thin, even coat of a very fine, transparent varnish.

If the hand-polishing method be preferred, it may be pursued by rubbing briskly and thoroughly with the following finishing polish:

Alcohol	8 ounces
Shellac	2 drachms
Gum benzoin	2 drachms
Best poppy oil	2 drachms

Dissolve the shellac and gum in the alcohol in a warm place, with frequent agitation, and, when cold, add the poppy oil. This may be applied on the end of a cylindrical rubber made by tightly rolling a piece of flannel, which has been torn, not cut, into strips 4 to 6 inches wide. It should be borne in mind that the surface of the cabinet work of a piano is generally veneered, and this being so, necessitates the exercise of much skill and caution in polishing.

II.—Prepare a paste from fine starch flour and a thick solution of brown shellac, with the spatula upon a grinding stone, and rub the wooden object with this. After the drying, rub off with sandpaper and polish lightly with a rag moistened with a thin shellac solution and a few drops of oil. The ground thus prepared varnish once or twice and a fine luster will be obtained. This method is well adapted for any wood with large pores, such as oak.

Removal of Heat Stains from Polished Wood.—Fold a sheet of blotting paper a couple of times (making 4 thicknesses of the paper), cover the place with it, and put a hot smoothing iron thereon. Have ready at hand some bits of flannel, also folded and made quite hot. As soon as the iron has made the surface of the wood quite warm, remove the paper, etc., and go over the spot with a piece of paraffine, rubbing it hard enough to leave a coating of the substance. Now with one of the hot pieces of flannel rub the injured surface. Continue the rubbing, using freshly warmed cloths until the whiteness leaves the varnish or polish. The operation may have to be repeated.

PRESERVATION OF WOOD.

I.—An excellent way of preserving wood is to cut it between August and October. The branches are removed, leaving only the leaves at the top. The trunks, carefully cut or sawn (so that their pores remain open), are immediately placed upright, with the lower part immersed in tanks three-quarters filled with water, into which 3 or 4 kilograms of powdered cupric sulphate per hectoliter have been introduced. The mass of leaves left at the extremity of each trunk is sufficient to cause the ascent of the liquid by means of the capillary force and a reserve of energy in the sap.

II.—Wood which can be well preserved may be obtained by making a circular incision in the bark of the trees a certain time before cutting them down. The woodcutters employed in the immense teak forests of Siam have adopted in an empirical way a similar process, which has been productive of good results. The tree is bled, making around the trunk, at the height of 4 feet above ground, a circular incision 8 inches wide and 4 inches deep, at the time when it is in bloom and the sap rising. Sometimes the tree is left standing for 3 years after this operation. Frequently, also, a deep incision reaching the heart is made on two opposite sides, and then it takes sometimes only 6 months to extract the sap.

It is probable that it is partly in consequence of this method that the teak-wood acquires its exceptional resistance to various destructive agents.

III.—A good preservation of piles, stakes, and palisades is obtained by leaving the wood in a bath of cupric sulphate of 4° of the ordinary acidimeter for a time which may vary from 8 to 15 days, according to greater or less dryness of the wood and its size. After they are half dried they are immersed in a bath of limewater; this forms with the sulphate an insoluble compound, preventing the rain from dissolving the sulphate which has penetrated the wood. This process is particularly useful for vine props and the wood of white poplars.

A good way to prevent the decay of stakes would be to plant them upside down; that is, to bury the upper extremity of the branch in the ground. In this way, the capillary tubes do not so easily absorb the moisture which is the cause of decay. It frequently happens that for one or another reason, the impregnation of woods designed to be planted in the ground, such as masts, posts, and supports has been neglected. It would be impracticable, after they are placed, to take up these pieces in order to coat them with carbolineum or tar, especially if they are fixed in a wall, masonry, or other structure. Recourse must be had to other means. Near the point where the piece rises from the ground, a hole about one centimeter in width is made in a downward slanting direction, filled with carbolineum, and closed with a wooden plug.

It depends upon the consistency of the wood whether the liquid will be absorbed in 1 or 2 days. The hole is filled again for a week. The carbolineum replaces by degrees the water contained in the wood. When it is well impregnated, the hole is definitely closed with a plug of wood, which is sawn level with the opening. The wood will thus be preserved quite as well as if it had been previously coated with carbolineum.

IV.—Wooden objects remaining in the open air may be effectually protected against the inclemency of the weather by means of the following coating: Finely powdered zinc oxide is worked into a paste with water and serves for white-washing walls, garden fences, benches, and other wooden objects. After drying, probably at the end of 2 or 3 hours, the objects must be whitewashed again with a very dilute solution of zinc chloride in glue or water. Zinc oxide and zinc chloride form a brilliant, solid compound, which resists the inclemency of the weather.

As a paint for boards, planks for covering greenhouses, garden-frames, etc., Inspector Lucas, of Reutlingen (Würtemberg), has recommended the following coating: Take fresh cement of the best quality, which has been kept in a cool place, work it up with milk on a stone until it is of the consistency of oil paint. The wood designed to receive it must not be smooth, but left rough after sawing. Two or 3 coats are also a protection from fire. Wood to be thus treated must be very dry.

V.—Wood treated with creosote resists the attacks of marine animals, such as the teredo. Elm, beech, and fir absorb creosote very readily, provided the wood is sound and dry. Beechwood absorbs it the best. In fir the penetration is complete, when the wood is of a species of rapid growth, and of rather compact grain. Besides, with the aid of pressure it is always possible to force the creosote into the wood. Pieces of wood treated with creosote have resisted for 10 or 11 years under conditions in which oak wood not treated in this way would have been completely destroyed.

The prepared wood must remain in store at least 6 months before use. The creosote becomes denser during this time and causes a greater cohesion in the fibers. In certain woods, as pitch pine, the injection is impossible, even under pressure, on account of the presence of rosin in the capillary vessels.

VI.—M. Zironi advises heating the wood

in vacuo. The sap is eliminated in this way. Then the receiver is filled with rosin in solution with a hydrocarbide. The saturation takes place in two hours, when the liquid is allowed to run off, and a jet of vapor is introduced, which carries off the solvent, whole the rosin remains in the pores of the wood, increasing its weight considerably.

VII.—Wood can be well preserved by impregnating it with a solution of tannate of ferric protoxide. This method is due to Hazfeld.

VIII.—The Hasselmann process (xylolized wood), which consists in immersing the wood in a saline solution kept boiling under moderate pressure, the liquid containing copper and iron sulphates (20 per cent of the first and 80 per cent of the second), as well as aluminum and kainit, a substance until recently used only as a fertilizer, is now much employed on the railways in Germany.

IX.—Recently the discovery has been made that wood may be preserved with dissolved betuline, a vegetable product of the consistency of paste, called also birchwood rosin. Betuline must first be dissolved. It is procurable in the crude state at a low price. The wood is immersed for about 12 hours in the solution, at a temperature of from 57° to 60° F.

After the first bath the wood is plunged into a second, formed of a solution of pectic acid of 40° to 45° Bé., and with a certain percentage of an alkaline carbonate—for instance, potassium carbonate of commerce—in the proportion of 1 part of carbonate to about 4 parts of the solution. The wood remains immersed in this composition for 12 hours; then it is taken out and drained from 8 to 15 hours, the time varying according to the nature of the wood and the temperature. In consequence of this second bath, the betulin which was introduced through the first immersion, is fixed in the interior of the mass. If it is desirable to make the wood more durable and to give it special qualities of density, hardness, and elasticity, it must be submitted to strong pressure. In thus supplementing the chemical with mechanical treatment, the best results are obtained.

X.—A receiver of any form or dimensions is filled with a fluid whose boiling point is above 212° F., such as heavy tar oil, saline solutions, etc. This is kept at an intermediate temperature varying between 212° F. and the boiling point; the latter will not be reached, but if into this liquid a piece of wood is plunged, an agitation analogous to boiling is manifested, produced by the water and sap contained in the pores of the wood. These, under the action of a temperature above 212° F., are dissolved into vapor and traverse the bath.

If the wood is left immersed and a constant temperature maintained until every trace of agitation has disappeared, the water in the pores of the wood will be expelled, with the exception of a slight quantity, which, being in the form of vapor, represents only the seventeen-hundredth part of the original weight of the water contained; the air which was present in the pores having been likewise expelled.

If the liquid is left to cool, this vapor is condensed, forming a vacuum, which is immediately filled under the action of the atmospheric pressure. In this way the wood is completely saturated by the contents of the bath, whatever may be its form, proportions or condensation.

To attain the desired effect it is not necessary to employ heavy oils. The latter have, however, the advantage of leaving on the surface of the prepared pieces a kind of varnish, which contributes to protect them against mold, worms, moisture, and dry rot. The same phenomenon of penetration is produced when, without letting the wood grow cold in the bath, it is taken out and plunged immediately into a cold bath of the same or of a different fluid. This point is important, because it is possible to employ as fluids to be absorbed matters having a boiling point below 212° F., and differing in this respect from the first bath, which must be composed of a liquid having a boiling point above 212° F.

If, instead of a cold bath of a homogeneous nature, two liquids of different density separated in two layers, are employed, the wood can, with necessary precautions, be immersed successively in them, so that it can be penetrated with given quantities of each. Such liquids are heavy tar oil and a solution of zinc chloride of 2° to 4° Bé. The first, which is denser, remains at the bottom of the vessel, and the second above. If the wood is first immersed in a saline solution, it penetrates deep into the pores, and when finally the heavy oil is absorbed, the latter forms a superficial layer, which prevents the washing out of the saline solution in the interior, as well as the penetration of moisture from the outside.

XI.—Numerous experiments have been made with all kinds of wood, even with hard oak. In the preparation of oak railway ties it was discovered that pieces subjected to a temperature of 212° F. in a bath of heavy tar oil for 4 hours lost from 6 to 7 per cent of their weight, represented by water and albuminous substances, and that they absorbed in heavy oil and zinc chloride enough to represent an increase of from 2 to 3 per cent on their natural original weight. The oak wood in question had been cut for more than a year and was of a density of 1.04 to 1.07.

This system offers the advantage of allowing the absorption of antiseptic liquids without any deformation of the constituent elements of the wood, the more as the operation is performed altogether in open vessels. Another advantage is the greater resistance of the wood to warping and bending, and to the extraction of metallic pieces, such as nails, cramp irons, etc.

XII.—In the Kyanizing process seasoned timber is soaked in a solution of bichloride of mercury (corrosive sublimate) which coagulates the albumen. The solution is very poisonous and corrodes iron and steel, hence is unsuited for structural purposes in which metallic fastenings are used. The process is effective, but dangerous to the health of the workers employed.

XIII.—The Wellhouse process also uses zinc chloride, but adds a small percentage of glue. After the timber has been treated under pressure the zinc chloride solution is drawn off and one of tannin is substituted. The tannin combines with the glue and forms an insoluble substance that effectually seals the pores.

XIV.—The Allardyce process makes use of zinc chloride and dead oil of tar, the latter being applied last, and the manner of application being essentially the same for both as explained in the other processes.

XV.—The timber is boiled in a solution of copper, iron, and aluminum sulphate, to which a small quantity of kainit is added.

XVI.—In the creo-rosinate process the timber is first subjected to a steaming process at 200° F. to evaporate the moisture in the cells; the temperature is then gradually increased to 320° F. and a pressure of 80 pounds per square inch. The pressure is slowly reduced to 26 inches vacuum, and then a solution of dead oil of tar, melted rosin, and formaldehyde is injected. After this process the timber is placed in another cylinder where a solution of milk of lime is applied at a temperature of 150° F. and a pressure of 200 pounds per square inch.

XVII.—The vulcanizing process of treating timber consists essentially in subjecting it to a baking process in hot air which is heated to a temperature of about 500° F. by passing over steam coils. The heat coagulates the albumen, expels the water from the cells, kills the organisms therein, and seals the cells by transforming the sap into a preservative compound. This method is used with success by the elevated railway systems of several cities.

XVIII.—A durable coating for wood is obtained by extracting petroleum asphalt, with light petroleum, benzine, or gasoline. For this purpose the asphalt, coarsely powdered, is digested for 1 to 2 days with benzine in well-closed vessels, at a moderately warm spot. Petroleum asphalt results when the distillation of petroleum continued until a glossy, firm, pulverizable mass of conchoidal fracture and resembling colophony in consistency remains. The benzine dissolves from this asphalt only a yellowish-brown dyestuff, which deeply enters the wood and protects it from the action of the weather, worms, dry rot, etc. The paint is not opaque, hence the wood retains its natural fiber. It is very pleasant to look at, because the wood treated with it keeps its natural appearance. The wood can be washed off with soap, and is especially suited for country and summer houses.

XIX.—A liquid to preserve wood from mold and dry rot which destroys the albuminous matter of the wood and the organisms which feed on it, so there are neither germs nor food for them if there were any, is sold under the name of carbolineum. The specific gravity of a carbolineum should exceed 1.105, and should give the wood a fine brown color. It should, too, be perfectly waterproof. The three following recipes can be absolutely relied on: a. Heat together and mix thoroughly 95 pounds of coal-tar oil and 5 pounds of asphalt from coal tar. b. Amalgamate together 30 pounds of heavy coal-tar oil, 60 pounds of crude wood-tar oil, and 25 pounds of heavy rosin oil. c. Mix thoroughly 3 pounds of asphalt, 25 pounds of heavy coal-tar oil, and 40 pounds of heavy rosin oil.

XX.—Often the wooden portions of machines are so damaged by dampness prevailing in the shops that the following compound will be found useful for their protection: Melt 375 parts of colophony in an iron vessel, and add 10,000 parts of tar, and 500 parts of sulphur. Color with brown ocher or any other coloring matter diluted with linseed oil. Make a first light application of this mixture while warm, and after drying apply a second coat.

XXI.—For enameling vats, etc., 1,000 parts of brown shellac and 125 parts of colophony are melted in a spacious kettle. After the mass has cooled somewhat, but is still thinly liquid, 6.1 parts of alcohol (90 per cent) is gradually added. In order to prevent the ignition of the spirit vapor, the admixture of spirit is made at a distance from the stove. By this addition the shellac swells up into a semiliquid mass, and a larger amount of enamel is obtained than by dissolving it cold. The enamel may be used for wood or iron.

The wood must be well dried; only then will the enamel penetrate into the pores. Two or three coats suffice to close up the pores of the wood thoroughly and to render the surface smooth and glossy. Each coating will harden perfectly in several hours. The covering endures a heat of 140° to 150° F. without injury. This glaze can also be mixed with earth colors. Drying quickly and being tasteless, its applications are manifold. Mixed with ocher, for instance, it gives an elegant and durable floor varnish, which may safely be washed off with weak soda solution. If it is not essential that the objects be provided with a smooth and glossy coating, only a preservation being aimed at the following coat is recommended by the same source: Thin, soluble glass (water glass) as it is found in commerce, with about 24 per cent of water, and paint the dry vessel rather hot with this solution. When this has been absorbed, repeat the application, allow to dry, and coat with a solution of about 1 part of sodium bicarbonate in 8 parts of water. In this coating silicic acid is separated by the carbonic acid of the bicarbonate; from the water glass (sodium silicate) absorbed by the pores of the wood, which, as it were, silicifies the wooden surfaces, rendering them resistive against the penetration of liquids. The advantages claimed for both processes are increased durability and facilitated cleaning.

XXII.—Tar paints, called also mineral or metallic paints, are sold in barrels or boxes, at varying prices. Some dealers color them—yellow ocher, red ocher, brown, gray, etc. They are prepared by mixing equal parts of coal tar and oil of turpentine or mineral essence (gasoline). The product, if it is not colored artificially, is of a brilliant black, even when cold. It dries in a few hours, especially when prepared with oil of turpentine. The paints with mineral essence are, however, generally preferred, on account of their lower cost. Either should be spread on with a hard brush, in coats as thin as possible. They penetrate soft woods, and even semi-hard woods sufficiently deep, and preserve them completely. They adhere perfectly to metals. Their employment can, therefore, be confidently advised, so far as concerns the preservation directly of iron cables, reservoirs, the interior surface of generators, etc. However, it has been shown that atmospheric influence or variations of temperature cause the formation of ammoniacal solutions, which corrode the metals. Several companies for the care and insurance of steam engines have for some time recommended the abandonment of tar products for applications of this kind and the substitution of hot linseed oil.

XXIII.—Coal-tar paints are prepared according to various formulas. One in current use has coal tar for a base, with the addition of gum rosin. It is very black. Two thin coats give a fine brilliancy. It is employed on metals, iron, sheet iron, etc., as well as on wood. It dries much quicker than the tars used separately. Its preserving influence against rust is very strong.

The following Tissandier formula has afforded excellent results. Its facility of preparation and its low cost are among its advantages. Mix 10 parts of coal tar, 1 to 1.6 parts of slaked lime, 4,000 parts of oil of turpentine, and 400 parts of strong vinegar, in which $\frac{1}{2}$ part of cupric sulphate has been previously boiled. The addition of 2 or 3 cloves of garlic in the solution of cupric sulphate aids in producing a varnish, brilliant as well as permanent. The compound can be colored like ordinary paints.

XXIV. — Rectified rosinous oil for painting must not be confounded with oils used in the preparation of lubricants for metallic surfaces exposed to friction. It contains a certain quantity of rosin in solution, which, on drying, fills the pores of the wood completely, and prevents decomposition from the action of various saprophytic fungi. It is well adapted to the preservation of pieces to be buried in the ground or exposed to the inclemency of the weather. Paints can also be prepared with it by the addition of coloring powders, yellow, brown, red, green, blue, etc., in the proportion of 1 kilo to 5 liters of oil. The addition ought to take place slowly, while shaking, in order to obtain quite a homogeneous mixture. Paints of this kind are economical, in consequence of the low price of rosin, but they cannot be used in the interior of dwellings by reason of the strong and disagreeable odor disengaged, even a long time after their application. As an offset, they can be used like tar and carbonyl, for stalls, stables, etc.

To Prevent Warping.—Immerse the wood to be worked upon in a concentrated solution of sea salt for a week or so. The wood thus prepared, after having been worked upon, will resist all changes of temperature.

STAINS FOR WOOD.

In the staining of wood it is not enough to know merely how to prepare and how to apply the various staining solutions; a rational exercise of the art of wood staining demands rather a certain acquaintance with the varieties of wood to be operated upon, a knowledge of their separate relations to the individual stains themselves; for with one and the same stain very different effects are obtained when applied to the varying species of wood.

Such a diversity of effects arises from the varying chemical composition of wood. No unimportant rôle is played by the presence in greater or lesser quantities of tannin, which acts chemically upon many of the stains and forms with them various colored varnishes in the fibers. Two examples will suffice to make this clear. (1) Let us take pine or fir, in which but little of the tanning principle is found, and stain it with a solution of 50 parts of potassium chromate in 1,000 parts of pure water; the result will be a plain pale yellow color, corresponding with the potassium chromate, which is not fast and as a consequence is of no value. If, with the same solution, on the contrary, we stain oak, in which the tanning principle is very abundant, we obtain a beautiful yellowish-brown color which is capable of withstanding the effects of both light and air for some time; for the tannin of the oak combines with the penetrating potassium chromate to form a brown dyestuff which deposits in the woody cells. A similar procedure occurs in the staining of mahogany and walnut with

the chromate because these varieties of wood are very rich in tannin.

(2) Take some of the same pine or fir and stain it with a solution of 20 parts of sulphate of iron in 1,000 parts of water and there will be no perceptible color. Apply this stain, however, to the oak and we get a beautiful light gray, and if the stain be painted with a brush on the smoother oaken board, in a short time a strong bluish-gray tint will appear. This effect of the stain is the result of the combination of the green vitriol with the tannin; the more tannin present, the darker the stain becomes. The hardness or density of the wood, too, exerts a marked influence upon the resulting stain. In a soft wood, having large pores, the stain not only sinks further in, but much more of it is required than in a hard dense wood; hence in the first place a stronger, greasier stain will be obtained with the same solution than in the latter.

From this we learn that in soft woods it is more advisable to use a thinner stain to arrive at a certain tone; while the solution may be made thicker or stronger for hard woods.

The same formula or the same staining solution cannot be relied upon to give the same results at all times even when applied to the same kinds of wood. A greater or lesser amount of rosin or sap in the wood at the time the tree is felled, will offer more or less resistance to the permeating tendencies of the stain, so that the color may be at one time much lighter, at another darker. Much after the same manner we find that the amount of the tanning principle is not always equal in the same species of wood.

Here much depends upon the age of the tree as well as upon the climatic conditions surrounding the place where it grew. Moreover, the fundamental color of the wood itself may vary greatly in examples of the same species and thus, particularly in light, delicate shades, cause an important delay in the realization of the final color tone. Because of this diversification, not only in the different species of wood, but even in separate specimens of the same species, it is almost impossible always, and at the first attempt, to match a certain predetermined color.

It is desirable that trials at staining should first be made upon pieces of board from the same wood as the object to be stained; the results of such experiments furnishing exact data concerning the strength and composition of the stain to be employed for the exact reproduction of a prescribed color.

Many cases occur in which the color tone obtained by staining cannot always be judged directly after applying the stain. Especially is this the case when stain is employed which slowly develops under the action of the air or when the dye-stuff penetrates only slowly into the pores of the wood. In such cases the effect of the staining may only be fully and completely appreciated after the lapse of 24 or 48 hours.

Wood that has been stained should always be allowed 24 or 48 hours to dry in ordinary temperatures, before a coat of varnish, polish, or wax is applied. If any dampness be left in the wood this will make itself apparent upon the varnish or polish. It will become dull, lose its glossy appearance, and exhibit white spots which can only be removed with difficulty. If a certain effect demand the application of two or more stains one upon the other, this may only be done by affording each distinct coat time to dry, which requires at least 24 hours.

Not all the dyes, which are applicable to wood staining, can be profitably used together, either when separately applied or mixed. This injunction is to be carefully noted in the application of coal tar or aniline colors.

Among the aniline dyes suitable for staining woods are two groups—the so-called acid dyes and the basic dyes. If a solution of an acid dye be mixed with a basic dye the effect of their antagonistic dispositions is shown in the clouding up of the stain, a fine precipitate is visible and often a rosin-like separation is noticeable.

It is needless to say that such a staining solution is useless for any practical purpose. It cannot penetrate the wood fibers and would present but an unseemly and for the most part a flaky appearance. In preparing the stains it is therefore of the greatest importance that they remain lastingly clear. It would be considerably of advantage, before mixing aniline solutions of which the acid or basic characteristics are unknown, to make a test on a small scale in a champagne glass and after standing a short time carefully examine the solution. If it has become cloudy or wanting in transparency it is a sign that a separation of the coloring matter has taken place.

The mixing of acid or basic dyestuffs even in dry powdered form is attended with the same disadvantages as in the state of solubility, for just as soon as they are dissolved in water the reactions commence and the natural process of precipitation takes place with all its attending disagreeable consequences.

COLOR STAINS:

Bronze.—I.—Prepare first a thin glue size by soaking good animal glue over night in cold water and melting it next morning in the usual water bath. Strain it, before using, through old linen or cheese cloth into a clean vessel. Sandpaper smooth and dust the articles, then apply with a soft bristle brush 2 or 3 coats of the size, allowing sufficient time for each coat to harden before applying the next. Now, a ground coat made by thoroughly mixing finely bolted gilders' whiting and glue size is applied, and when this has become hard it is rubbed to a smooth, even surface with selected fine pumice, and then given 1 coat of thin copal varnish. When this is nearly but not quite dry, the bronze powder is applied with a suitable brush or wad of cotton, and when dry the surplus bronze is removed with the same tool. If collected on clean paper, the dusted-off bronze powder may be used again.

II. — Diluted water-glass solution makes a good ground for bronze. Bronze powder is sprinkled on from a wide-necked glass tied up with gauze, and the excess removed by gently knocking. The bronze powder adheres so firmly after drying that a polish may be put on by means of an agate. The process is especially useful for repairing worn-off picture frames, book ornamentations, etc. The following bronze ground also yields good results: Boil 11,000 parts of linseed oil with 25 parts of impure zinc carbonate, 100 parts of red lead, 25 parts of litharge, and 0.3 parts of mercuric chloride, until a drop taken out will stand like a pea upon a glass surface. Before complete cooling, the mass is diluted with oil of turpentine to a thick syrup.

Ebony Stains.—I.—To 1 pint of boiling water add ¾ ounce of copperas and 1 ounce logwood chips. Apply this to the wood hot. When the surface has dried thoroughly wet it with a solution composed of 7 ounces steel filings dissolved in ¼ pint of vinegar.

II.—Give the wood several applications of a stout decoction of logwood chips, finishing off with a free smear of vinegar in which rusty nails have been for some time submerged.

III.—In 1 quart of water boil ½ pound of logwood chips, subsequently adding ½ ounce pearl ash, applying the mixture hot. Then again boil the same quantity of logwood in the same quantity of water, adding ¼ ounce of verdigris and ¼ ounce of copperas, after which strain and put in ¼ pound of rusty steel filings. With this latter mixture coat the work, and, should the wood not be sufficiently black, repeat the application.

Metallic Luster.—A valuable process to impart the luster of metal to ordinary wood, without injuring its natural qualities, is as follows: The wood is laid, according to its weight, for 3 or 4 days in a caustic alkaline solution, such as, for instance, of calcined soda, at a temperature of 170° F. Then it is at once placed in a bath of calcium hydrosulphite, to which, after 24 to 36 hours, a saturated solution of sulphur in caustic potash is added. In this mixture the wood is left for 48 hours at 100° to 120° F. The wood thus prepared, after having been dried at a moderate temperature, is polished by means of a smoothing iron, and the surface assumes a very handsome metallic luster. The effect of this metallic gloss is still more pleasing if the wood is rubbed with a piece of lead, zinc, or tin. If it is subsequently polished with a burnisher of glass of porcelain, the wood gains the brilliancy of a metallic mirror.

Nutwood.—One part permanganate of potassium is dissolved in 30 parts clear water; with this the wood to be stained is coated twice. After an action of 5 minutes, rinse off with water, dry, oil, and polish. It is best to prepare a fresh solution each time.

Oak.—I.—Water-color stains do not penetrate deep enough into wood to make the effect strong enough, hence solutions of other material than color are being employed for the purpose. Aqua ammonia alone, applied with a rag or brush repeatedly, will darken the color of oak to a weathered effect, but it is not very desirable, because of its tendency to raise the grain. Bichromate of potash, dissolved in cold water, applied in a like manner, until the desired depth is obtained, will serve the purpose. These washes or solutions, however, do not give the dark, almost black, effect that is at the present time expected for weathered oak, and in order to produce this, 4 ounces of logwood chips and 3 ounces of green copperas should be boiled together in 2 quarts of water for 40 minutes and the solution applied hot. When this has dried it should be gone over with a wash made from 4 ounces steel filings and 1 pint of strong vinegar. The steel filings are previously put into the vinegar and allowed to stand for several days. This will penetrate into the wood deeply, and the stain will be permanent. Picture-frame manufacturers use a quick-drying stain, made from aniline blacks.

II.—Dissolve ¼ part of permanganate of potassium in 1,000 parts of cold water and paint the wood with the violet solution obtained. As soon as the solution comes in contact with the wood it decomposes in consequence of chemical action, and a handsome light-brown precipitate is produced in the wood. The brushes used must be washed out immediately, as the permanganate of potassium destroys animal bristles, but it is preferable to use sponges or brushes of glass threads for staining. Boil 2 parts of cutch in 6 parts of water for 1 hour, stir while boiling, so that the rosiniferous catechu cannot burn on the bottom of the vessel; strain the liquid as soon as the cutch is dissolved, through linen, and bring again to a boil. Now dissolve therein ⅛ part of alum, free from iron; apply the stain while hot, and cover after the drying, with a solution of 1 part of bichromate of potassium in 25 parts of water.

Rosewoood.—First procure ½ pound logwood, boiling it in 3 pints water. Continue the boiling until the liquid assumes a very dark color, at which point add 1 ounce salt of tartar. When at the boiling point stain your wood with 2 or 3 coats, but not in quick succession, as the latest coat must be nearly dry before the succeeding one is applied. The use of a flat graining brush, deftly handled, will produce a very excellent imitation of dark rosewood.

Silver Gray.—This stain is prepared by dissolving 1 part of pyrogallic acid in 25 parts of warm water and the wood is coated with this. Allow this coating to dry and prepare, meanwhile, a solution of 2 parts of green vitriol in 50 parts of boiling water, with which the first coating is covered again to obtain the silver-gray shade.

Walnut.—I.—Prepare a solution of 6 ounces of a solution of permanganate of potassium, and 6 ounces of sulphate of magnesia in 2 quarts of hot water. The solution is applied on the wood with a brush and the application should be repeated once. In contact with the wood the permanganate decomposes, and a handsome, lasting walnut color results. If small pieces of wood are to be thus stained, a very dilute bath is prepared according to the above description, then the wooden pieces are immersed and left therein from 1 to 5 minutes, according to whether a lighter or darker coloring is desired.

II.—One hundredweight Vandyke brown, ground fine in water, and 28 pounds of soda, dissolved in hot water, are mixed while the solutions are hot in a revolving mixer. The mixture is then dried in sheet-iron trays.

Yellow.—The wood is coated with a hot concentrated solution of picric acid, dried, and polished. (Picric acid is poisonous.)

IMITATION STAINS.

Yellow, green, blue, or gray staining on wood can be easily imitated with a little glazing color in oil or vinegar, which will prove better and more permanent than the staining. If the pores of the wood are opened by a lye or a salt, almost any diluted color can be worked into it. With most stains the surface is thus prepared previously.

Light-Fast Stains.—Stains fast to light are obtained by saturating wood in a vacuum chamber, first with dilute sulphuric acid, then with dilute alkali to neutralize the acid, and finally with a solution with or without the addition of a mordant. The action of the acid is to increase the affinity of the wood for dye very materially. As wood consists largely of cellulose, mercerization, which always increases the affinity of that substance for dyes, may be caused to some extent by the acid.

SPIRIT STAINS:

Black.—

I.—White shellac....... 12 ounces
Vegetable black,..... 6 ounces
Methylated spirit.... 3 pints

II.—Lampblack......... 1 pound
Ground iron scale.... 5 pounds
Vinegar............ 1 gallon

Mahogany Brown.—Put into a vessel, say 4 pounds of bichromate of potash, and as many ounces of burnt umber, let it stand a day or two, then strain or lawn for use.

Vandyke Brown.—

Spirit of wine........ 2 pints
Burnt umber........ 3 ounces
Vandyke brown color 1 ounce
Carbonate of soda... 1 ounce
Potash............ ½ ounce

Mahogany.—Rub the wood with a solution of nitrous acid, and then apply with a brush the following:

I.—Dragon's blood...... 1 ounce
Sodium carbonate... 6 drachms
Alcohol........... 20 ounces

Filter just before use.

II.—Rub the wood with a solution of potassium carbonate, 1 drachm to a pint of water, and then apply a dye made by boiling together:

Madder........... 2 ounces
Logwood chips...... ½ ounce
Water............ 1 quart

Maple.—

I.—Pale button lac..... 3 pounds
Bismarck brown.... ⅛ ounce
Vandyke brown.... ½ ounce
Gamboge......... 4 ounces
Methylated spirit... 1 gallon

II.—Use 1 gallon of methylated spirit, 4 ounces gamboge (powdered), ½ ounce Vandyke brown, 1 drachm Bismarck brown, 3 pounds shellac.

Maroon.—To produce a rich maroon or ruby, steep red Janders wood in rectified naphtha and stir into the solution a little cochineal; strain or lawn for use.

Turpentine Stains.—Turpentine stains are chiefly solutions of oil-soluble coaltar dyes in turpentine oil, with small quantities of wax also in solution. They do not roughen the wood, making a final polishing unnecessary. They enter the wood slowly, so that an even stain, especially on large surfaces, is secured. The disadvantages of turpentine stains are the lack of permanence of the coloring, when exposed to light and air, and their high price.

Varnish Stains.—Shellac is the chief article forming the basis of varnish stains the coloring matter being usually coal tar or aniline dyes, as they give better results than dye wood tincture. To prevent the varnish stain being too brittle, the addition of elemi rosin is a much better one than common rosin, as the latter retards the drying quality, and if too much be used, renders the stain sticky.

Water Stains.—Water stains are solutions of chemicals, dye extracts, astringent substances, and coal-tar dyes in water. They roughen the wood, a disadvantage, however, which can be remedied to a large extent by previous treatment, as follows: The wood is moistened with a wet sponge, allowed to dry, and then rubbed with sandpaper, or made smooth by other agencies. This almost entirely prevents roughening of the surface by the stain. Another disadvantage of these stains is that they are rapidly absorbed by the wood, which makes an even staining of large surfaces difficult. For this too there is a remedy. The surface of the wood is rubbed all over evenly with raw linseed oil, applied with a woolen cloth, allowed to dry, and then thoroughly smoothed with sandpaper. The water stain, applied with a sponge, now spreads evenly, and is but slightly absorbed by the wood.

Among good water stains are the long-known Cassel brown and nut brown, in granules. Catechine is recommended for brown shades, with tannin or pyrogallic acid and green vitriol for gray. For bright-colored stains the tar-dyes azine green, croceine scarlet, Parisian red, tartrazine, water-soluble nigrosin, walnut, and oak brown are very suitable. With proper mixing of these dyes, all colors except blue and violet can be produced, and prove very fast to light and air, and superior to turpentine stains. Only the blue and violet dyes, methyl blue, naphthol blue, and pure violet, do not come up to the standard, and require a second staining with tannin.

A very simple method of preparing water stains is as follows: Solutions are made of the dyes most used, by dissolving 500 parts of the dye in 10,000 parts of hot water, and these are kept in bottles or casks. Any desired stain can be prepared by mixing proper quantities of the solutions, which can be diluted with water to make lighter stains.

Stains for Wood Attacked by Alkalies or Acids.—

Solution A

Copper sulphate... 125 grams
Potassium chlorate. 125 grams
Water............. 1,000 cu. cm.

Boil until all is dissolved.

Solution B

Aniline hydro-
chloride........ 150 grams
Water............ 1,000 cu. cm.

Apply Solution A twice by means of a brush, allowing time to dry after each coat; next, put on Solution B and let dry again. On the day following, rub on a little oil with a cloth and repeat this once a month.

SUBSTITUTES FOR WOOD.

I.—Acetic paraldehyde or acetic aldehyde respectively, or polymerized formaldehyde is mixed with methylic alcohol and carbolic acid, as well as fusel oil saturated with hydrochloric acid gas or sulphuric acid gas or methylic alcohol, respectively, are added to the mixture. The mass thus obtained is treated with paraffine. The final product is useful as a substitute for ebonite and wood as well as for insulating purposes.

II.—"Carton Pierre" is the name of a mass which is used as a substitute for carved wood. It is prepared in the following manner: Glue is dissolved and boiled; to this, tissue paper in suitable quantity is added, which will readily go to pieces. Then linseed oil is added, and finally chalk is stirred in. The hot mass forms a thick dough which crumbles in the cold, but softens between the fingers and becomes kneadable, so that it can be pressed into molds (of glue, gypsum, and sulphur). After a few days the mass will become dry and almost as hard as stone. The paper imparts to it a high degree of firmness, and it is less apt to be injured than wood. It binds well and readily adheres to wood.

III.—Wood Pulp.—The boards for painters' utensils are manufactured in the following manner: The ordinary wood fiber (not the chemical wood cellulose) is well mixed with soluble glass of 33° Bé., then spread like cake upon an even surface, and beaten or rolled until smooth. Before completely dry, the cake is removed, faintly satined (for various other purposes it is embossed) and finally dried thoroughly at a temperature of about 133° F., whereupon the mass may be sawed, carved, polished, etc., like wood.

Any desired wood color can be obtained by the admixture of the corresponding pulverized pigment to the mass. The wood veining is produced by placing a board of the species of timber to be imitated, in vinegar, which causes the soft parts of the wood to deepen, and making an impression with the original board thus treated upon the wood pulp when the latter is not quite hard. By means of one of these original boards (with the veins embossed), impressions can be made upon a large number of artificial wood plates. The veins will show to a greater advantage if the artificial wood is subsequently saturated and treated with colored oil, colored stain and colored polish, as is done with palettes.

WOOD, ACID-PROOF:
See Acid-Proofing.

WOOD CEMENTS:
See Adhesives.

WOOD, CHLORINE-PROOFING:
See Acid-Proofing.

WOOD, FIREPROOFING:
See Fireproofing.

WOOD GILDING:
See Plating.

WOOD, IMITATION:
See Plaster.

WOOD POLISHES:
See Polishes.

WOOD RENOVATORS:
See Cleaning Preparations and Methods under Paint, Varnish, and Enamel Removers.

WOOD, SECURING METALS TO:
See Adhesives.

WOOD, WATERPROOFING:
See Waterproofing.

WOOD'S METAL:
See Alloys.

WOOL FAT:
See Fats.

WORM POWDER FOR STOCK:
See Veterinary Formulas.

WRITING, RESTORING FADED:

Writing on old manuscripts, parchments, and old letters that has faded into nearly or complete invisibility can be restored by rubbing over it a solution of ammonium sulphide, hydrogen sulphide or of "liver of sulphur." On parchment the restored color is fairly permanent but on paper it does not last long. The letters however could be easily retraced, after such treatment, by the use of India ink and thus made permanent. This treatment will not restore faded aniline ink. It only works with ink containing a metal-like iron that forms a black sulphide.

WRINKLES, REMOVAL OF:
See Cosmetics.

Yeast

DRY YEAST.

Boil together for $\frac{1}{2}$ hour, 95 parts of the finest, grated hops and 4,000 parts of water. Strain. Add to the warm liquor 1,750 parts of rye meal or flour. When the temperature has fallen to that of the room add 167 parts of good yeast. On the following day the mass will be in a state of fermentation. While it is in this condition add 4,000 parts of barley flour, so as to form a dough. This dough is cut up into thin disks, which are dried

as rapidly as possible in the open air or sun. For use, the disks are broken into small pieces and soaked overnight in warm water. The yeast can be used on the following day as if it were ordinary brewers' yeast.

PRESERVATION OF YEAST.

I.—The yeast is laid in a vessel of cold water which is thereupon placed in a well-ventilated, cool spot. In this manner the yeast can be preserved for several weeks. In order to preserve the yeast for several months a different process must be followed. The yeast, after having been pressed, is thoroughly dried. For this purpose the yeast is cut up into small pieces which are rolled out, placed on blotting paper, and allowed to dry in a place which is not reached by the sun. These rolls are then grated, again dried, and finally placed in glass bottles. For use, the yeast is dissolved, whereupon it immediately regains its freshness. This process is particularly to be recommended because it preserves the yeast for a long period.

II.—For liquid yeast add one-eighth of its volume in glycerine. In the case of compressed yeast, the cakes are to be covered with glycerine and kept in closed vessels. Another method of preserving compressed yeast is to mix it intimately with animal charcoal to a dough, which is to be dried by exposure to sunlight. When it is to be used, it is treated with water, which will take up the ferment matter, while the charcoal will be deposited. Liquid and compressed yeast have been kept for a considerable time, without alteration, by saturating the former with chloroform and keeping the latter under chloroform water.

YEAST TESTS.

I.—Pour a few drops of yeast into boiling water. If the yeast sinks, it is spoiled; if it floats, it is good.

II.—To 1 pound yeast add $\frac{1}{2}$ tablespoonful of corn whisky or brandy, a pinch of sugar, and 2 tablespoonfuls of wheat flour. Mix thoroughly and allow the resultant compound to stand in a warm place. If the yeast is good it will rise in about an hour.

YEAST AND FERTILIZERS:
See Fertilizers.

YELLOW (CHROME), TEST FOR:
See Pigments.

INDEX

A

Absinthe, 765
Absolute Alcohol, 45
Abrasion Remedy, 225, 486
Acacia, Mucilage of, 43
Acid-free Soldering Fluid, 659
Acid-proof Alloy, 62
  Cement, 26
  Corks, 10
  Glass, 374
Acid-proofing, 9
Acid-proof Pastes, 38
  Putty, 607
  Table Top, 9
Acid Receptacles, Lining for, 10
Acid-resisting Paint, 499
Acids, Soldering, 656
Acid Stains Removed, 184
  Test for Gold, 432
  for Vinegar, 358
Aconite-Monkshood Poison, 93
Adhesion, 105
  Belt Pastes for Increasing, 105
Adhesive Paste, 37, 39
Adhesives, 10
Advertising Matter, to Scent, 510
Adulterants in Foods, 348
Adulteration of Linseed Oil, 460
  of Wax, 753
Adurol Developer, 527
Affixing Labels to Glass, 42
Agar Agar Paste, 37
Agate, Buttons of Artificial, 44
Agate (Imitation), 370
Age of Eggs, 283
Aging of Silk, 639
Agricultural Sources of Industrial
  Alcohol, 668
Air Bath, 44
  Bubbles in Gelatine, 370
  Exclusion of, 553
Air-purifying, 44
Albata Metal, 63
Albumen, 34
  in Urine, Detection of, 44
  Paste, 37
Alcohol, 44
  Absolute, 45
  Defined, 667
  Deodorized, 45, 514
  Dilution of, 45, 703
  in Beer, 45
  Manufacture, 667, 674
  Solid, 45
  Tests for Absolute, 45
Ale, 46
  Ginger, 107
Alfenide Metal, 63
Alkali Blue and Nicholson's Blue
  Dye, 267
Alkalis and Their Salts Poison,
  93
Alkaline Glycerine of Thymol,
  100
Alkaloids, Antidotes to, 102
Alkermes Cordial, 763
Alloy, Acid-proof, 62
  for Caliper and Gage-rod Cast-
    ings, 80
  for Watch Pinion Sockets, 736
  Lipowitz's, 61
  Moussets', 76

Alloys, 47
  Copper, Silver, Cadmium, 76
  for Casting Coins, etc., 62
  for Cementing Glass, 52
  for Drawing Colors on Steel, 80
  for Metal Foil, 474
  for Small Casting Molds, 80
  having a Density, 48
  Silver, Nickel, Zinc, 76
  Tin, 77
  Unclassified, 80
Almond Blossom Perfumery, 518
  Cold Cream, 235
  Extracts, 312
  Powders for the Toilet, 242
Altars, to Clean, 185
Alum, 80
  Baking Powder, 102
  Bath, 535
  Process of Water Purification,
    340
Aluminum Alloys, 48
  Electrical Conductivity of, 50
Aluminum-brass, 50
Aluminum Bronze, 56, 657
  Castings, 150
Aluminum-Copper, 50
Aluminum Gilding, 576
  Gold, 68
  Etching Fluid for, 324
  How to Color, 80
  Lacquer for, 438
  Paper, 507
  Plating, 572, 581
  Polishes, 590
Aluminum-Silver, 50, 75
Aluminum Solders, 657
Aluminum-Tin, 50
Aluminum, to Clean, 204
  Toughness, Density and Te-
    nacity, 83
Aluminum-Tungsten, 50
Aluminum Varnish, 725
  Working of Sheet, 83
Aluminum-Zinc, 50
Amalgam for Cementing Glass,
  etc., 90
  for Plaster, 65
  for Silvering Glass Balls, 90
  for the Rubber of Electric
    Machines, 90
  Gold Plating, 576
Amalgams, 64, 85
  for Mirrors, 72
Amber, 90
  Cements, 26
  Varnish, 718
Ambrosia Powder, 628
American Champagne, 118
  Factory Cheese, 176
  Lemonade, 110
  Soda Fountain Company's
    Whipped Cream, 248
Amethyst (Imitation), 370
Amidol Developer, 528
Ammon-carbonite, 331
Ammonia, 91
  for Fixing Prints, 536
  Household, 91
  Poison, 93
  Violet Color for, 91
  Water, 245, 519
    Perfumed, 91
Anchovies, Essence of, 98

Anchovy Paste, 98
  Preparations, 98
  Sauce, Extemporaneous, 98
Angostura Bitters, 762
Anise Cordial, 763
Aniline, 266
  Black Dye, 266, 279
    Substitutes, 279
  Black Lake Dye, 278
  Blue Dye, 268
  Green Dye for Wool, 269
    for Silk, 269
  in Pigments, Tests for, 560
  Scarlet Dye, 271
  Stains, to Remove, 185
  Yellow Dye, 271
Animals, Fly Protection for, 419
Ankara, 142
Annealing Bronze, 56
  Copper, 219
Annealing of Steel, Wire, etc.,
  681
Anodynes, 486
Ansco Platinum Paper, 529
Ant Destroyers, 420
Anti-corrosive or Asiatic Ink, 414
Antidotes for Belladonna, 93
  for Poisons, 92
Anti-ferments, 97
Anti-fouling Compositions, 498
Anti-freezing Solution, 362, 363
  for Automobilists, 363
Anti-friction Bearing or Babbitt
  Metals, 50
  Metal, 58
Anti-frost Solution, 363
Anti-leak Rubber Tire, 708
Antimony Poison, 93
  Baths, 581
Antique Bronzes, 566
  Silver, 587, 639
    Imitation of, 640
Antiques, to Preserve, 98
Anti-rust Compositions, 625
  Paper for Needles, 625
  Pastes, 625
Antiseptic Bromine Solution, 100
  Enamel, 720
  Nervine Ointment, 487
  Oil of Cinnamon, 100
  Paste (Poison), 99
  Pencils, 99
  Powders, 98
  Soap, 644
  Solution, Coloring for, 100
  Tooth Powder, 253
Antiseptics, 98
  for Caged Birds, 729
  Mouth, 99
Aphtite, 70
Apollinaris Lemonade, 110
  Water, 740
Apple Extract, 312
  Syrup, 312
Applications for Prickly Heat, 398
  of Barium Amalgams, 86
  of Bismuth Amalgams, 88
  of Cadmium Amalgams, 87
  of Copper Amalgams, 87
  of Gold Amalgams, 89
  of Lead Amalgams, 88
  of Manganese Amalgams, 87

Applications of Potassium Amal-
  gams, 86
  of Silver Amalgams, 88
  of Sodium Amalgams, 86
  of Strontium Amalgams, 86
  of Tin Amalgams, 87
  of Zinc Amalgams, 87
Applying Decalcomania Pictures, 250
Apricot Extract, 312
Aquarium Putty, 608
Argentan, 69
Arguzoid, 70
Armenian Cement, 20
Arms, Oil for, 460
Arnica Salve, 486
Aromatic Cod-Liver Oil, 482
  Cotton, 246
  Rhubarb Remedy, 180
  Vinegar, 735
Arsenic Alloys, 63, 75
Arsenic Poison, 93, 614
Art Bronzes, 57, 556
  of Lacquering, 437
Artificial Aging of Fabrics, 639
  Beeswax, 754
  Butter, 142
  Ciders, 181
  Coloring of Flowers, 346
  Egg Oil, 284
  Fertilizers for Pot Plants, 336
  Flowers, Dyes for, 272
  Flower Fertilizer, 337
  Horn, 396
  Leather, 447
  Marbles, 699
  Rubber, 618
  "Rubbered" Silk, 639
  Slate, 643
  Violet Perfumery, 518
  Water, 739
Asbestos Cement, 30
  Fabric, 342
Asphalt and Pitch, 33
  as Ingredient of Rubber, 619
  in Painting, 718
  Varnishes, 718
Assaying of Gold, 381
Asthma Cures, 101
  Fumigating Powders, 101
  in Canaries, 728
  Papers, 101
Astringent for Horses, 730
  Wash for Flabby Skin, 234
Atomic Weights, 758
Atomizer Liquid for Sick Rooms, 264
Attaching Enamel Letters to Glass, 19
  by Cement, 17
Atropine, Antidote to, 102
Aqua Aromatica, 102
  Fortis for the Touchstone, 383
  Poison, 92
  Regia, 102
Aquarium Cements, 31
Automobile Engines, Cooling, 363
Automobiles, Anti-freezing Solution, 363
Axle Grease, 462

B

Babbitt-Metals, 50
Baking Powders, 102
Balance Spring, 738
Baldness, 392
Balkan Paste, 38
Ball Blue, 281, 444
Ball-Room Floor Powder, 345
Balsam, Birch, 103
  of Sulphur, 380
  Spray Solution, 103

Balsam, Stains, to Remove, 194
  Wild-cherry, 103
Balsams, 102
Balsamic Cough Syrup, 211
Banana Bronzing Solution, 489
  Cream, 115
  Trick, the Burning, 611
  Syrup, 312
Banjo Sour, 110
Barbers' Itch, 486
  Powder, 243
Barium Amalgams, 86
  Poison, 615
Barometers (Paper), 402
Bath, Air, 44
  Metal, 63
  Powder, 242
  Tablets, Effervescent, 103
Bath-tub Enamel, 721
  Paint, 501
Batteries, Solution for, 104
Basis for Effervescent Salts, 627
Baudoin Metal, 63
Bavaroise au Cognac, 118
Bay Rum, 104, 513
Bear Fat, 333
Bearing Lubricant, 461
  Metal, 50
Beauty Cream, 231
  Water, 244
Bedbug Destroyers, 420
Beechwood Furniture Polish, 593
Beef and Iron, 771
  Iron, and Wine, 104
Beef-marrow Pomade, 227
Beef Peptonoids, 509
  Preservatives, 360
  Tea, 112
Beer, 118
  Ginger, 108
  Lemon, 108
  Restoration of Spoiled, 105
  Spruce, 119
  Treacle, 119
  Weiss, 119
Beers, Alcohol in, 45
Beetle Powder, 425
Bees, Foul Brood in, 105
Beeswax, Artificial, 754
Belladonna, Antidotes to, 93
Bell Metal, 51
Belt Cement, 31
  Glue, 15
  Lubricant, 462
  Pastes for Increasing Adhesion, 105
Bénédictine, 769
Bengal Lights, 609
Bent Glass, 371
Benzine, 106
  Cleaning with, 209
  Purification of, 106
  to Color Green, 106
Benzoic Acid, Detection of, 350
  in Food, 350
Benzoic-acid Pastilles, 211
Benzoin-Glycerine Soap, 652
Benzoparal, 107
Berge's Blasting Powder, 330
Beverages, 107
  Yellow Coloring for, 119
Bibra Alloy, 71
Bicycle Dipping Varnish, 719
Bicycle-tire Cement, 23
Bicycle Varnishes, 719
Bicycles, Black Paint for, 495
Bidery Metal, 80
Billiard Balls, 148, 428
Birch Balsam, 103
Birch-Bud Water, 519
Birch Water, 244, 389
Bird Diseases, Remedies, 728
  Foods, 120, 729

Bird Lime, 458
  Paste, 145
  Tonic, 729
Birds, Antiseptic Wash for, 729
  Constipation in, 729
  Diarrhœa in, 729
Biscuit, Dog, 265
Bismarck Brown Dye, 267
Bismuth, 49
  Alloys, 52
  Amalgams, Applications of, 88
  Bronze, 70
  Purification of, 380
  to Purify, 380
Biting Off Red-hot Iron, 612
Bitter Almond Oil Poison, 93
Bitters, 762
Blackberry Cholera Mixture, 180
  Cordial, 763
Blackboard Paint and Varnish, 489
  Varnish, 720
Black Color on Brass, 129
  Dye for Tanned Leather, 447
  on Cotton, 266
  on Wool, for Mixtures, 267
Blackening Iron, 495
"Black Eye" Lotion, 333
Black Finish for Brass, 129
  Grease Paints, 229
  Hair Dye without Silver, 390
Blackhead Remedies, 232
Blacking Copper, 221
  for Harness, 450
  for Shoes, 631
  Stove, 700
Black Japanese Varnish, 719
  Lake Dyes for Wall-paper, 278
  Marble, Imitation, 699
  Marking Inks, 407
  Paint for Polished Iron, 495
  Patina, 585
  Putty, 607
  Ruling Ink, 403
  Sheet Rust Preventive, 624
  Starch, 680
  Straw Hat Varnish, 266
  Varnish, 543, 544, 719
  Wash for Casting Molds, 150
Blanching Silver, 640
Blanket Washing, 399
Blasting Powder, 330
Blazing Sponge Trick, 611
Bleach for Hands, 233
Bleaches, Bone, 430
Bleaching, 120
  and Coloring Feathers, 335
  Bone Fat, 333
  Cotton by Steaming, 245
  Cotton, 245
  Feathers, 121, 335
  Linen, 120
  of Linseed Oil, 459
  of Vegetable Fibers with Hydrogen Peroxide, 245
  Oils, 484
  Photographic Prints White, 553
  Silk, 120, 639
  Skin Salves, 234
  Solution, 121
  for Photographs, 553
  Solutions for the Laundry, 446
  Sponges, 678
  Straw, 120
  Tallows and Fats, 334
  Wool, 120
Bleeding, Local, 701
Blight Remedies, 121
Blisters, for Horses, 729
Block for Soldering, 667
  Hollow Concrete Building, 691
  Machines, 694
Blocks Poured from Wet Concrete, 694
Bood-red Brick Stain, 166

Blotting Paper, 503
Blue, Ball, 281
Blue-black Ink, 414
  Patina, 585
Blue Bronze, 138
  Dye for Hosiery, 268
  from Green at Night, 121
  Indelible Ink, 406
  Paving Bricks, 166
Blueprint Inks, 403
  Paper Making, 536
Blueprints, to Change, 121
  to Turn Brown, 542
  Waterproofing, 741
Blue Ruling Ink, 403
  Sanitary Powder, 263
  Vitriol Poison, 94
Bluing, 443
  Compounds, 443
  of Steel, 682
Bluish-black Lake Dye, 278
Blush Pink Dye on Cotton Textile, 279
Board-sizing, 38
Boiled Oil, 484
Boiler Compounds, 121
  Plates, Protecting from Scales, 122
  Pressure, 123
  Scales, Prevention of, 122
Boiling the Linseed Oil, 409
Boil Remedy, 121
Bone Black, 123
  Bleaches, 430
  Fat, 333
  Fertilizers, 338
  or Ivory Black, 123
  Polishes, 395
  Uniting Glass With, 17
Bones, A Test for Broken, 124
  Treatment of, in Manufacturing Glue, 10
Bookbinders' Varnish, 720
Book Disinfectant, 263
  How to Open, 125
Bookworms, 425
Books, their Preservation, 124
  to Remove Marks from, 186
Boot Dressings, 631
  Lubricant, 460
Boot-top Liquid, 632
Boots, Waterproofing, 750
Borated Apple Blossom Powder, 243
  Talcum, 510
Borax in Food, 350
  for Sprinkling, 125
  Soap Powder, 650
Boric Acid, Detection of, 350
Borotonic, 258
Bottling Sweet Cider, 181
Bottle-cap Lacquer, 440
Bottle-Capping Mixtures, 126
Bottle Cleaners, 210
  Deodorizer, 127
  Stoppers, 700
  Varnish, 720
  Wax, 553
Bottles, 126
  White Glass for, 373
Bouillon, 113
  Chicken, 112
  Clam, 113
  Hot Egg, 112
  Tomato Extract, 212
Bowls of Fire Trick, 611
Box Glue, 15
Bradley Platinum Paper, 529
"Braga," 117
Bran, Sawdust in, 126
Brandy, Artificial French, 768
  and Brandy Bitters, 762
Brass, 127, 435
  A Bronze for, 136

Brass and Bronze Protective Paint, 495
  Articles, Restoration of, 132
  Black Color on, 129
  Black Finish for, 129
  Bronzing, 566
  Brown Color to, 130
  Cleaners, 202, 203
  Coloring, 129, 473
  Colors for Polished, 127
  Etching Bath for, 324
  Fluid for, 323
  Fastening Porcelain to, 17
  Gilding, 576
  Graining of, 130
Brass-Iron (Aich's Metal), 53
Brass Parts, Improved, 132
  Pickle for, 132
  Platinizing, 566
  Polishes, 590
  Sand Holes in, 150
  Solders, 657
  to Cast Yellow, 54
  Tombac Color on, 130
  Unpolished Coloring, 128
  Varnishes Imitating Gold, 725
Brassing, 572, 581
  Zinc, Steel, Cast Iron, 581
Brassware, Gold Lacquers for, 440
Bread, Dog, 265
Breath, Fetid, Remedies for, 133
  Perfumes, 258
Brewers' Yeast, 339
Brick and Tilemakers' Glazed Bricks, 164
  Arches, Waterproofing, 741
Brickbat, Cheese, 176
Brick, Blood-red Stain, 166
  Colors, 165
Brickmakers' Notes, 167
Brick Polishes, 600
  Stain, 133, 166
  Walls to Clean, 197
  to Renovate, 190
  Waterproofing, 134
Bricks, 164
  Glaze for, 377
  of Sand-lime, 689
  Polish for, 600
Brie, Cheese, 176
Brightening Pickle, 469
Bright Red Rouge, 229
Brilliantine, 390
  Florician, 483
Brimstone (Burning), 611
Bristol Brass (Prince's Metal), 53
Britannia Metal, 55
  to Clean, 201
  Silver-plating, 587
British Champagne, 118
  Oil, 484
Brocchieri's Styptic, 701
Brocq's Pomade for Itching, 228
Broken Bones, A Test for, 124
Bromine, Antiseptic, 100
Bromoform, 134
  Rum, 134
Bronze, Aluminum, 56
  Annealing, 56
  Articles, Polish for, 591
  Casting, 150
  Cleaning, 202, 205
  Coloring, 138
  Dye, 272
  for Brass, 136
  Gilding, 137
  Leather, 447
  Lettering, 456
  Machine, 58
  Phosphor, 58
  Polishes, 591
  Powder, Liquid for, 567
Bronze Powders, 134, 139
  Preparations, 135

Bronze, Renovation of, 205
  Silicon, 61
  Steel, 61
  Substitutes, 137
  Tincture, 135, 137
  to Renovate, 201
  Varnishes, 726
Bronzes, 55
  Art, 57
  Pickle for, 138
  Statuary, 57
Bronzing, 566
  and Patinizing of Articles, 136
  Engraved Ornaments, 137
  General Directions for, 135
  Liquid, 136
  Metals, 567
  of Brass, 571
  of Gas Fixtures, 566
  of Wood, 782
  of Zinc, 137
  Solutions for Paints, 489
  with Soluble Glass, 139
Brooches, Photographing on, 551
Brown Dye for Cotton, 267
  for Silk, 267
  for Wool, 267
  and Silk, 267
  Hair Dye, 390
Browning of Steel, 583
Brown Ink, 414
  Ointment, 486
  Oxidation on Bronze, 139
  Shoe Dressing, 632
Brownstone, Imitation, 133
Brown Tints, 559
  Varnish, 726
Brunette or Rachelle Powder, 242
Brushes, 140
Bubble (Soap), Liquid, 655
Bubbles, 141
  in Gelatine, 370
Buff Terra-Cotta Slip, 166
  Wheels, Rouge for, 618
Bug Killers, 420
Building Blocks, Concrete, 691
Bunions, 224
Burning Banana Trick, 103
  Brimstone, 611
  Sealing Wax, 611
Burns, 486
  Carbolic Acid, 147
  Mixture for, 142
Burnt Alum, 80
  Steel, to Restore, 686
Butter, 142, 354
  Artificial, Tests for, 354
  Color, 142, 359
Buttermilk, Artificial, 143
Buttons of Artificial Agate, 44
  Platine for, 80

C

Cadmium Alloy, about the Hardness of Zinc, 77
  Alloys, 61, 64
  with Gold, Silver, and Copper, 62
  Amalgams, Applications of, 87
Calcium Carbide, 144
  Sulphide (Luminous), 494
Camera, Renovating a, 553
Campchello, 117
Camphor for Cholera, 180
Camphorated and Carbolated Powders, 252
  Cold Cream, 226
  Ice, 145
  Pomade, 145
  Preparations, 144

Camphorated Substitutes in the Preparation of Celluloid, 157
Canary-Bird Food, 729
  Paste, 145
Canary Birds, Their Diseases, 729
Concrete, 689
Candles, 145
  Coloring, 145, 146
  Fumigating, 365
  Transparent, 145
Candy, 216
  Colors and Flavors, 218
  Orange Drops, 216
Canned Vegetables, 352
Canning, 602
  without Sugar, 603
Cantharides and Modern Potato Bug Poison, 94
  Pomade, 392
Can Varnish, 720
Canvas Waterproofing, 742
Caoutchouc, 618
  Solution for Paints, 719
Capacities of Utensils, 703
Capsule Varnish, 720
Capping Mixtures for Bottles, 126
Caramels, 146, 216
Caramel in Food, 352
12-Carat, 433
4-Carat Gold, 433
18-Carat Gold for Rings, 433
22-Carat Solder, 433
Carats, to Find the Number of, 432
Carbolic Acid, 147
Carbolic-acid Burns, 147
  Decolorization of, 147
  Disguising Odor of, 147
Carbolic Powder, 263
  Soap, 647
Carbolineum, 497
Carbonated Pineapple Champagne, 118
Carbon Ink, 403
  Paper, 503
  Printing, 531
  Process in Photography, 531
Carbuncle Remedies, 121
Cardboard or Leather Glue, 15
  Waterproofing, 751
Cards (Playing), to Clean, 209
Care of Refrigerators, 401
Carmelite Balm Water, 519
Carmine, 403
  Lake Dye for Wall Paper, 278
Carnation Lake Dye, 277
Carpet Preservation, 399
  Soap, 644
Carpets, How to Preserve, 399
Carriage-top Dressing, 448
Carron Oil, 242
Case Hardening, 648
Casein, 34, 148
  Albumen, and Glue, 34
  Cements, 20
  Massage Cream, 233
  Paste, 38
  Varnish, 34
Cashmere Perfumery, 516
Casket Trimmings, 150
Casks, 149
  Watertight, 149
Cassius, Purple of, 383
Cast Brass, 53
Cast-brass Work, Sand Holes in, 150
Castile Soap, to Cut, 644
Casting, 149
  Copper, 63
  in Wax, 755
  Molds, Alloys for, 80
  of Soft Metal Castings, 151
Castings, Making in Aluminum, 81

Castings Out of Various Metals, 149
  to Soften Iron, 427
Cast-iron Soldering, 666
Castor Oil, 153
Castor-oil Chocolate Lozenges, 154
Castor Oil, How to Take, 154
  Tasteless, 153
Casts from Wax Models, 755
  (Plaster), Preservation of, 565
  Repairing of Broken, 26
  Waterproofing, 565
Catatypy, 154
Cat Diseases and Remedies, 732
Caterpillar Destroyers, 423
Catgut, 155
  Sutures, Preparation of, 155
Catsup, Adulterated, 353
Cattle Dips and Applications, 264
Caustic Potash Poison, 93, 94
Ceiling Cleaners, 400
Celery Clam Punch, 112
  Compound, 155
Cellars, Waterproof, 400
Celloidin Paper, 504
Cells, Solutions and Fillers for Battery, 104
Celluloid, 155
  Cements and Glues, 17
  Glue for, 12
  Lacquer, 439
  of Reduced Inflammability, 159
  Putty, 161
Cement, 692
  Armenian, 20
  Asbestos, 30
  Cheap and Excellent, 30
  Colors, 688
  Diamond Glass, 29
  for Belts, 31
  for Chemical Apparatus, 31
  for Cracks in Stoves, 162
  for Enameled Dials, 20
  for General Use, 31
  for Glass, 21, 25, 28
  for Iron and Marble, 17
  for Ivory, 31
  for Leather and Iron, 25
  for Metals, 21, 25
  for Metal on Hard Rubber, 22
  for Pallet Stones, 162
  for Pasteboard and Paper, 21
  for Patching Boots, 23
  for Pipe Joints, 162
  for Porcelain Letters, 19
  for Sandstones, 17
  for Steam and Water Pipes, 161
  for Watch-lid, 20
  for Waterpipe, 162
  Hydraulic, 33
Cementing Celluloid and Hard-rubber Articles, 18
Cement Jewelers, 20
  Mordant for, 479
  on Marble Slabs, 16
  Paints for, 499
  Parisian, 30
  Protection of, Against Acid, 9
  Rubber for Cloth, 24
  to Paint Over Fresh, 499
  Transparent for Glass, 29
  Strong, 30, 32
  Universal, 31
  Work, Protection for, 162
Cements, 16, 161
  Amber, 26
  Aquarium, 31
  Casein, 20
  Celluloid, 17
  for Attaching Letters on Glass, 19
  for Fastening Porcelain to Metal, 25

Cements, for Iron, 24
  for Leather, 22, 23
  for Metals, 24
  for Rubber, 22
  for Stone, 16
  for Tires, 23
  for Water-glass, 19
  Meerschaum, 30
  Sign-letters, 18
  Silicate of Oxychloride, 35
Ceramics, 164
Chain of Fire, 612
Chains (Watch), to Clean, 206
Chalk for Tailors, 164
Chamois Skin, to Clean, 186
Champagne, 118
  Cider, 181
Chapped Skin, 232
Chappine Cream, 237
Charta Sinapis, 480
Chartreuse, 769
Cheddar Cheese, 176
Cheese, 174
  Color, 359
  Wrapping, Tin Foil for, 474
Chemical Apparatus, Cement for, 31
  Gardens, 368
  Reagents, 349
Cherry Balsam, 103
  Cordial, 764
  Phosphate, 112
  Tooth Paste, 257
Chewing Candy, 217
  Gums, 178
Cheshire Cheese, 176
Chestnut Brown Dye for Straw Bonnets, 267
  Hair Dye, 391
Chicken Bouillon, 112
Chicken-coop Application, 419
Chicken Diseases, 734
Chicory, Tests for, 353
Chilblains, 486
Children, Doses for, 265
Children's Tooth Powder, 255
China, 173
  Pomade, 227
  Repairing, 601
  Riveting, 179
  Silver Alloy, 75
  to Toughen, 173
Chinese Tooth Paste, 257
Chlorides, Platt's, 264
Chloriding Mineral Lubricating Oils, 462
Chlorine-proofing, 9
Chocolate, 179
  and Milk, 114
  Castor-oil Lozenges, 154
  Extracts, 312
  Frappé, 114
  Hot, 111
  Soda Water, 111
Cholera Remedies, 179
Chowchow, 212
Chrome Black Dye for Wool, 267
Chromium Glue, 15
Chromo Making, 180
Cider, 180
  Preservative, 181
  Vinegar, 735
Cigarettes, Asthma, 101
Cigar Flavoring, 183
  Sizes and Colors, 182
  Spots, 183
Cigars, 182
Cinnamon Essence, 312
  Oil as an Antiseptic, 100
  or Brown Dye for Cotton and Silk, 267
Cinchona, 771
  Pomade, 392
Citrate of Magnesium, 464
Clam Bouillon, 113

Claret Lemonade, 110
  Punch, 110, 112
Clarification of Gelatin and Glue, 370
Clarifying, 184
  Muddy Water, 741
Clay, 33, 184
Claying Mixture for Forges, 184
Clean Bronze, 202
Cleaner, Universal, 209
Cleaning Linoleum, 398
  Marble, 196
  Polished Woodwork, 194
  Brass on Clock, 206
  Bronze Objects, 205
  Clocks, 207
  Copper, 200
  Copper Sinks, 202
  Electro-plate Goods, 205
  Funnels and Measures, 204
  Gilded Work on Altars, 185
  Gilded Articles, 185
  Gilded Bronzes, 205
  Gilt Bronze Ware, 201
  Glass, Paste for, 208
  Inferior Gold Articles, 207
  Lamp Globes, 209
  Marble, Furniture, etc., 197
  Methods and Processes, 209
  of Copperplate Engravings, 309
  of Statuettes and Plaster Objects, 564
  of Walls, Ceilings, and Paper, 190, 397
  Oil Stains on Wall Paper, 190
  Optical Lenses, 208
  Paint Brushes, 140
  Painted and Varnished Surfaces, 194
  Painted Doors, Walls, etc., 190
  Pearls, 208
  Preparations, 184, 397, 590, 644
  Preparation for Glass with Metal Decorations, 208
  Pewter Articles, 205
  Powder, 194
  Skins and Leather, 186
  Silver-plated Ware, 200
  Terra Cotta, 197
  Tracings, 194
  Varnish Brushes, 141
  Wall Paper, 191
  Whitewashed Walls, 190
  Window Panes, 208
Cleansing Fluids, 185
Clearing Baths, 535
Cleary's Asthma Fumigating Powder, 101
Cliché Metal, 52
Clock-bell Repairing, 737
Clock Cleaning, 207
Clock-dial Lettering, 737
Clock Hands, to Reblack, 738
Clockmakers' Cleaning Processes, 206
Clock Oil, 482
  Repairing, 738
Clothes and Fabric Cleaners, 191
  Cleaners, 191
Clothes-Cleaning Fluids, 192
Cloth Paper, 504
  Strips Attached to Iron, 14
  to Iron, Gluing, 37
  Waterproofing, 748
Cloths for Polishing, 599
Clouding of Mouth Mirrors, 477
Cloudless Caramel Coloring, 146
Clove Pink Perfumery, 516
Coal Oil, 484
Coals, to Eat Burning, 612
Coating for Bathrooms, 498
  for Damp Walls, 499
  for Name Plates, 501
  Metallic Surfaces with Glass, 377
  Tablets with Chocolate, 179

Cobaltizing of Metals, 573
Cobalt, or Fly Powder Poison, 94
Cochineal Insect Remedy, 422
Cocoa Mint, 115
  Syrup, 112
Cocoas, 112
Cod Liver Oil and Its Emulsion, 482
Coffee, 353
  Cocktail, 114
  Cordial, 763
  Cream Soda, 113
  Essence, 314
  Extracts, 313
  for the Soda Fountain, 111
  Frappé, 114
  Hot, 111
  Iced, 114
  Nogg, 114, 115
  Substitutes for, 210
  Syrups, 313
Coil Spring, 683
  Springs, to Temper, 683
Coin Cleaning, 200
  Metal, 62
Coins, Impressions of, 467
  Matrix for, 467
Colas, 728
Cold and Cough Mixtures, 211
  Chemical Gilding, 577
  Cream, 225
  Enameling, 721
  Soldering, 666
  Varnish, 543
Colic in Cattle, 729
Collapsible Tubes, Skin Cream, 239
  Tooth Paste for, 257
Collodion, 212
Cologne, 514
  for Headaches, 394
  Spirits or Deodorized Alcohol, 514
Coloration of Copper and Brass with Cupric Selenite, 568
Colored Alloys for Aluminum, 50
  Celluloid, 161
  Fireproofing, 344
  Fires, 609
  Floor Polishes, 591
  Gilding, 577
  Glass, 165, 371
  Gold Alloys, 66
  Hygroscopes, 402
  Inks, 414
  Lacquer, 439
  Marking Inks, 407
  Rings on Metal, 582
  Sand, 628
Coloring Benedine Green, 106
  Brass, 473
  Ceresine Candles for the Christmas Tree, 145
  Common Gold, 431
  Copper, 473
  Electric-light Bulbs and Globes, 371
  Fluid for Brass, 129
  Gold Jewelry, 430
  Incandescent Lamps, 442
  Matter in Fats, 334
  Metals, 471, 568
  of Brass, 128, 570
  of Modeling Plaster, 563
  Perfumes, 511
  Silver, 640
  Soap, 644
  "Spirit" Varnishes, 715
  Steel, 682
  Unpolished Brass, 128
Colorings for Jewelers' Work, 433
Color Enamel, 721
  Photography, 548
  Stains, for Wood, 782

Color Stamps for Rough Paper, 411
  Testing, 559
Colors, 266
  and Sizes of Cigars, 182
  for Confectionery, 218
  for Paints, 555
  for Polished Brass, 127
  for Pomade, 228
  for Syrups, 702
  Fusible Enamel, 306
Combined Alum and Hypo Bath, 535
  Toning and Fixing Baths, 542
Comfortable, Washing, 399
Commercial Enameling, 290
  Formaldehyde, 362
  Mucilage, 43
Common Silver for Chains, 434
  Silver Solder, 434
Composition Files, 339
  for Cleaning Copper, Nickel, and other Metals, 203
  for Linoleum, Oilcloth, etc., 459
  for Writing on Glass, 376
  of Various Hard Solders, 663
Compositions for Ships' Bottoms, 498
Compost for Indoor Plants, 337
Compound for Cleaning Brass, 203
  Salicylated Collodion Corn Cure, 224
  Solution of Thymol, 100
Concentrated Lye Poison, 93
Concrete, 689
  Blocks, Properties of, 695
  Tamping of, 695
Concrete Block Systems, 694
  Building Block, 691
  Mixers, 693
Condimental Sauces, 353
Condiments, 212
  Tests for Adulterated, 349
Condition Powders, 729
  for Cattle, 729
Conductivity of Aluminum Alloys, 48
Confectionery, 216
  Colors, 218
Constipation in Birds, 729
Contracted Hoof or Sore Feet in Cattle, 730
Conversion of Metric into English Measure, 760
Cooling Screen, 616
Cooking Vessels, Glazes for, 377
Cook's Table, 703
Cooper's Pen Metal, 74
Copal Varnish, 720
Copper, 219
  Alloys, 51, 76
  Amalgam, 90
  Amalgams, Applications of, 87
  and Brass Gilding, 577
    Platinizing, 586
  A Permanent Patina for, 585
  Arsenic, 63
  Articles, Polish for, 591
  Bronzing, 566
  Cleaning, 200
  Coloring, 221, 473
  Enameling, 294
  Etching, 324
  in Food, 351
  Iron, 63
  Lacquers, 439
  Nickel, 63
  Paint for, 495
  Paper, 507
  Patinizing and Plating, 586
  Polishes, 590
  Separation of Gold from, 382
Copper-Silver Alloy, 75

Copper, Silver, and Cadmium Alloys, 76
Solder for Plating, 434
Solders, 659
to Bronze, 136
Varnishes, 726
Coppering, 572
Glass, 572
Plaster Models, etc., 573
Zinc Plate, 573
Copying Ink, 415
Printed Pictures, 222
Process on Wood, 222
Cordage, 223
Lubricant, 463
Waterproofing, 753
Cordials, 763
Cork as a Preservative, 606
Cleaner, 210
to Metal, Fastening, 36
Corks, 223
Impermeable and Acid-proof, 10
to Clean, 210
Waterproofing, 742
Corn Plaster, 224
Cures, 224
Corrosive Sublimate Poison, 94
Cosmetic Jelly, 232
Cosmetics, 225
Cottenham Cheese, 176
Cotton, 245
Belts, Lubrication, 462
Degreasing, 246
Cottonseed, Extracting Oil, 482
Hulls as Stock Food, 246
Oil, 482
Compress Cough Balsam with Iceland Moss, 211
Drops, 217
Mixtures and Remedies, 211
for Cattle, 730
Syrup, 211
Counter Polishes, 590
Court Plasters, 247, 563
Cow Diseases — Remedies, 730
Powder, 730
Cow's Milk, Powder for, 732
Cracked Leather, 448
Cracks in Tools, to Render Visible, 686
Crayons, 374
for Graining and Marbling, 247
for Writing on Glass, 374
Cream, 247
Beef Tea, 112
Bonbons for Hoarseness, 216
Cheese, 176
How to Determine, 474
Soda Powder, 628
Creams for the Face and Skin, 225
Creosote-carbolic Acid Poison, 94
Cresol Emulsion, 248
Crimson Dye for Silk, 271
Indelible Ink, 406
Crystal Cements, 248
Crystalline Coatings or Frostwork on Glass or Paper, 376
Honey Pomade, 227
Crystallization, Ornamental, 368
Crockery, 167
Plaster and Meerschaum Repairing, 27
Crocus, 248
Crude Petroleum, Emulsion of, 521
Crushed Apricot, 365, 604
Cherries, 365, 604
Fruit Preserving, 604
Orange, 365, 604
Peach, 365, 604
Pineapples, 364, 604
Raspberry, 364
Strawberry, 364
Cucumber Creams, 237

Cucumber Essence, 314
Jelly, Juice, and Milk, 228
Juice, 239
Milk, 239
Pomade, 228
Cummins's Whipped Cream, 248
Curaçoa Cordial, 764
Liqueur, 770
Cure for Barber's Itch, 486
for Snake Bites, 96
for Tan, 242
for Warts, 736
Currant Cream, 115
Curry Powder, 213
Curtains, Coloring of, 446
Cutlers' Cements for Fixing Knife Blades into Handles, 16
Cutlery Cements, 16
Cutting, Drilling, Grinding, and Shaping Glass, 371
Cuspidor Powder, 263
Custard Powder, 249
Cyanide of Potassium Poison, 93
Cylinder Oil, 464
Cymbal Metal, 64
Cypress Water, 519

D

Dairy Products, 354
Damaskeening, 249
by Electrolysis, 249
on Enamel Dials, 250
Damp Walls, Coating for, 400, 499
Damson Cheese, 176
Dandruff Cures, 388
Darcet Alloy, 64
Dark-blue Dye, 268
Dark Gold Purple, 383
Dark-Green Blackboard Paint, 489
Dark Red Grease Paint, 229
Snuff-Brown Dye for Wool, 267
Steel Dye, 269
Deadening Paint, 491
Dead-gilding of an Alloy of Copper and Zinc, 579
Dead, or Matt, Dip for Brass, 131
Deadly Nightshade Poison, 94
Decalcomania Processes, 250
Decolorization of Carbolic Acid, 147
Decolorizing and Deodorizing Oils, 484
or Bleaching Linseed Oil, 483
Decomposition of Oils, Fats, 484
Decorating Aluminum, 81
Decorative Metal Varnishes, 726
Wood-finish, 772
Deep Red Grease Paint, 229
Red Raspberry Syrup, 318
Dehorners or Horn Destroyers, 397
Delta Metal, 63
Demon Bowls of Fire, 611
Denaturized Alcohol, 45, 678
Dental Cements, 163
Platinum, 74
Dentifrices, 251
Deodorants for Water-closets, 263
Deodorization of Calcium Carbide, 144
Deodorized Alcohol, 514
Cod Liver Oil, 482
Petroleum, 522
Deodorizing Benzine, 106
Depilatory Cream, 259
Depthings, Verification of, 737
Derbyshire Cheese, 176
Desilvering, 587
Detannating Wine, 765
Detecting Dyed Honey, 396

Detection of Albumen in Urine, 44
of Formaldehyde in Food, 351
in Milk, 474
of Copper in Food, 351
of Cottonseed Oil in Lard, 442
of Glucose in Food, 357
of Saccharine in Food, 351
of Salicylic Acid in Food, 349
of Starch in Food, 357
Detergent for Skin Stains, 235
Detergents, 186
Determination of Artificial Colors in Food, 351
of Preservatives, 349
Determining Cream, 474
Developers for Photographic Purposes, 523
Development of Platinum Prints, 531
Dextrine Pastes, 35
Diabetics, Lemonade for, 109
Dial Cements, 20
Cleaners, 207
Repairing, 737
Diamalt, 475
Diamantine, 432
Diamond Cement, 20
Glass Cement, 29
Tests, 260
Diarrhœa in Birds, 729
Remedies, 179
Die Venting, 261
Digestive Powders, 261
Relish, 213
Diogen Developer, 527
Dip for Brass, 131
Dipping Metals, Danger of, 470
Dips, 469
for Cattle, 264
Direct Coloration of Iron and Steel by Cupric Selenite, 568
Directions for Bronzing, 135
for Making Perfumes, 512
Disinfectants, 264
Disguising Odor of Carbolic Acid, 147
Dish Washing, 399
Disinfectant for Books, 125
Disinfectants, 262
for Sick Room, 264
Disinfecting Coating, 265
Fluids, 262
or Weed-killers, 262
Powders, 262
Dissolving Old Rubber, 622
Distemper in Cattle, 729
Distinguishing Blue from Green, 121
Diuretic Ball, 731
Dog Applications, 419
Biscuit, 265
Soap, 654
Domestic Ointments, 486
Pets, 732
Donarite, 330
Doors, to Clean, 190
Doses for Adults and Children, 265
Dose Table for Veterinary Purposes, 729
Double Extract Perfumery, 518
Drawing Inks, 403
Paper, 504
Temper from Brass, 133
Drawings, Preservation of, 266
to Clean, 206
Draw-tempering Cast Steel, 687
Dressing for Carriage Tops, 448
for Sewing Thread, 706
Dressings for Harness, 451
for Leather, 448
for Linoleum, 459
for the Hair, 389

Dried Casein, its Mfg., 148
  Yolk of Egg, 284
Driers, 636
Driffield Oils, 485
Drill Chips, to Utilize, 686
Drilling Hard Steel, 687
  Lubricant for, 463
  Shaping, and Filing Glass, 372
Drinking Water, Removal of Iron from, 741
Drinks for Summer and Winter, 107
  Soda Water, 111
Drops of Lime in the Eye, 333
  Table of, 704
Drosses, 151
Dry Bases for Paints, 489
  Perfumes, 509
  Powder Fire Extinguishers, 341
  Rot, Remedies for, 618
  Sugar Preserving, 604
  Yeast, 786
Drying Oils, 485
Druggists' Label Paste, 41
Dubbing for Leather, 460
Duesseldorff Mustard, 215
Dunlop Cheese, 176
Durable Bronze on Banners, 137
  Putty, 607
Dust-laying, 485
Dust Preventers and Cloths, 401
Dutch (Holland) Cheese, 176
  Pink Dye, 278
Dyeing Feathers, 335
  Leather, 450
  Silk or Cotton Fabrics, 280
  Straw Hats, 394
Dyes, 266
  and Dyestuffs, 274
  Colors, etc., for Textile Goods, 279
  for Artificial Flowers, 272
  for Feathers, 272
  for Food, 359
  for Furs, 272
  for Hats, 273
  for Leather, 450
Dye Stains, Removal from Skin, 184
Dynamite, 329

E

Earthenware, 168
Easily Fusible Alloys, 64
Eastman's Sepia Paper, 531
Eaton's Styptic, 701
Eau de Botot Water, 519
  de Lais Water, 519
  de Merveilleuse Water, 519
  de Quinine, 392
Eberle's Whipped Cream, 248
Ebony, 783
  Lacquer, 439
  Stains, 782
Eczema Dusting Powder, 282
Edible Oils, 355
Effervescent Bath Tablets, 103
  Powders, 627
Eggs, 282, 355
Egg Chocolate, 114
  Claret, 115
  Coffee, 115
  Crème de Menthe, 115
  Dyes, 275
  Lemonade, 111, 115
  Malted Milk Coffee, 114
  Oil, 284
  Orgeat, 115
  Phosphate, 113
  Powder, 284
  Shampoo, 393
  Sherbet, 115
  Sour, 115
  Wine, 118

Egg-stain Remover, 201
Eikonogen Developer, 524
Ektogan, 98
Elaine Substitute, 286
Elastic Glue, 14
  Limpid Gum Varnishes, 720
  or Pliable Paste, 39
  Substitute for Celluloid, 158
Electrical Conductivity of Aluminum Alloys, 50
Electric Installations, Fusible Alloys for, 64
  Insulation, 425
  Light Bulbs, Coloring, 371
Electrodeposition Processes, 571
Electro-etching, 324
Electrolysis in Boilers, 123
Electroplating and Electrotyping, 286
Elm Tea, 288
Embalming Fluids, 288
Embroideries, Stamping Powder for, 680
Embroidery, Ink for, 411
Emerald, Imitation, 370
Emery, 289
  Grinder, 289
  Substitute, 289
Emmenthaler Cheese, 176
Emollient Skin Balm, 234
Emulgen, 290
Emulsifiers, 289
Emulsion, Cresol, 248
  of Bromoform, 134
Emulsions of Petroleum, 521
Enamel Colors, 727
  for Copper Cooking Vessels, 305
  for Vats, 721
  How to Remove, 189
  Letters Attaching to Glass, 19
  Mixing, 302
  Removers, 187
  Solder, 434
  Varnishes, 720
Enameled Dials, Cement for, 20
  Iron Recipes, 305
Enameling, 290
  Alloys, 67
Enamels, Metallic Glazes on, 173
  Unaffected by Hot Water, 721
Engines (Gasoline), Anti-freezing Solution for, 363
English Margarine, 143
  Pink Dye, 278
  Weights and Measures, 758
Engravers' Varnishes, 723
Engraving, Matting, and Frosting Glass, 375
  on Steel, 687
  or Etching on Steel, 687
  Spoon Handles, 309
Engravings, their Preservation, 309
  to Reduce, 310
  to Transfer, 710
Enlargements, 542
Envelope Gum, 43
Epicure's Sauce, 213
Epizooty, 731
Eradicators, 205
Erasing Powder or Pounce, 189
Essence Bénédictine, 769
  of Anchovies, 98
  of Cinnamon, 312
  of Extract of Soup Herbs, 212
  of Savory Spices, 214
Essences and Extracts of Fruits, 310, 312
Etching, 322
  Bath for Brass, 324
  for Tin, 706
  Copper, Brass, and Tombac, 323
  Fluids, 322
  Fluid for Aluminum, 324

Etching, Fluid, for Brass, 323
  to Make Stencils, 323
  for Copper, Zinc, and Steel, 324
  for Gold, 324
  for Lead, Antimony, and Britannica Metal, 324
  for Tin or Pewter, 324
  for Zinc, 323
  Fluids for Copper, 325
  for Iron and Steel, 322
  for Silver, 324
  Glass by Means of Glue, 326
  -ground for Copper Engraving, 322
  on Copper, 324
  on Glass, 325
  on Ivory, 327, 428
  on Marble, 327
  on Steel, 687
  Powder for Iron and Steel, 323
  for Metals, 324
  Steel, Liquids for, 327
  with Wax, 326
Eucalyptus Bonbons, 212
  Paste, 257
Examination of Foods, 352
Expectorant Mixtures, 212
Explosives, 328, 330
Exposures in Photographing, 528
Extemporaneous Anchovy Sauce, 98
Extract, Ginger-ale, 107
  of Meat Containing Albumen, 361
  of Milk, 474
Extracting Oil from Cottonseed, 482
Extracts, 312
  Coffee, 313
Eye, Foreign Matter in, 333
Eyeglasses, 376
Eye Lotions, 333

F

Fabric Cleaners, 191
Fabrics, Waterproofing of, 742
Façade Paint, 499
Face Black and Face Powder, 230
  Bleach or Beautifier, 231
  Cream without Grease, 239
  Powder, Fatty, 230
Faded Photographs, 544
Fairthorne's Dental Cement, 163
Falling Hair, 392
Fancy Soda Drinks, 113
Fastening Cork to Metal, 36
Fats, 333, 334, 335
  Decomposition of, 484
  for Soldering, 659
Fatty Acid Fermentation Process, 334
  Face Powders, 230
Feather Bleaching and Coloring, 121, 335
  Dyes, 272, 335
Felt Waterproofing, 749
Fermentation, Prevention of, 765
  Process, Fatty Acid, 334
Ferro-argentan, 71
Ferro-prussiate Paper, 539
Ferrous-oxalate Developer, 525
Fertilizer with Organic Matter, for Pot Flowers, 337
Fertilizers, 336
  Bone, 338
Fever in Cattle, 731
Fig Squares, 216
File Alloys, 64
  Metal, 64
Files, 339
  Geneva Composition, 64
  to Clean, 205, 339
  Vogel's Composition, 64

Filigree Gilding, 576
Fillers for Letters, 457
  for Wood, 773
Film-stripping, 553
Filter Paper, 504
Filters for Water, 339
Finger-marks, to Remove, 125
Fingers, Pyrogallic-acid Stains on, 185
Finger-tips, Sparks from, 611
Finishing Enamel for White Furniture, 722
Firearm Lubricants, 460
Firearms, Oil for, 460
Fire, Chain of, 612
  Colored, 609
  Grenades, Substitutes for, 341
  Trick, 611
  Extinguishers, 340
Fireproof and Waterproof Paints, 491
  Coating, 344
  Compositions, 344
  Glue, 16
  Paints, 490
  Papers, 344, 504
Fireproofing, 341, 344
  Celluloid, 159
  Clothing, 342
  for Wood, Straw, Textiles, 343
  Light Woven Fabrics, 342
  Mosquito Netting, 342
  Rope and Straw Matting, 342
  Stage Decorations, 342
  Tents, 342
Fireworks, 608
Fish Bait, 344
Fishing Net, Preservation of, 223
Fixing and Clearing Baths, 535
  Agents in Perfumes, 512
  Baths for Paper, 542
Fixatives for Crayon Drawings, etc., 344
Flabby Skin, Wash for, 234
Flashlight Apparatus, 552
  Apparatus with Smoke Trap, 552
Flannels, Whitening of, 446
Flavoring Cigars, 183
  Extracts, 355
  Peppermint as a, 252
  Sarsaparilla, 629
Flavorings, 213
  for Dentifrice, 255
  Spices, 213
Flea Destroyers, 423
Flesh Face Powder, 243
Flexible Ivory, 428
Flies and Paint, 501
  in the House, 399
Floor Coating, 500
  Dressings, 344
  Oils, 485
  Paper, 506
  Polish, 591
  Varnishes, 724
  Waterproofing, 753
  Wax, 754
Floral Hair Oil, 483
  Hair Pomade, 483
Florentine Bronzes, 136
Floricin Brilliantine, 483
  Oil, 483
Florida Waters, 514
Flower Preservatives, 345
Flowers, Coloring for, 346
Flour and Starch Compositions, 35
  Paste, 39
Fluid Measure, U. S. Standard, 704

Fluid Measures, 758
Fluids, Clothes-cleaning, 192
  Disinfecting, 262
  for Embalming, 288
  for Soldering, 659
Fluorescent Liquids, 347
Fluxes for Soldering, 660
  Used in Enameling, 305
Flux for Enameled Iron, 305
Fly Essences, 421
Fly-papers and Fly-poisons, 347
Fly-killers, 421
Fly Protectives for Animals, 419
Foam Preparations, 348
Foamy Scalp Wash, 389
Foreign Matter in the Eye, 333
Food Adulterants, Tests for, 348
  Benzoic Acid in, 107
  Colorants, 358
  Cooked in Copper Vessels, 94
Foods, Bird, 120, 729
  for Pets, 733
  for Red Birds, 729
Foot Itch, 733
Foot-powders and Solutions, 361
Footsores on Cattle, 730
Formaldehyde, 362
  for Disinfecting Books, 263
  in Milk, Detection of, 474
Formalin for Grain Smut, 384
  Treatment of Seed Grain for Smut, 384
Formol Albumen for Preparation of Celluloid, 156
Formulas for Bronzing Preparations, 135
  for Cements for Repairing Porcelain, Glassware, Crockery, Plaster, and Meerschaum, 27
  to Drive Ants Away, 420
Foul Brood in Bees, 105
Fowler's Solution Poison, 93
Foxglove, or Digitalis Poison, 94
Fcy's Whipped Cream, 248
Fragrant Naphthalene Camphor, 14
Frames, Protection from Flies, 363
Frame Cleaning, 185
  Polishes, 600
Framing, Passe-partout, 508
Frangipanni Perfumery, 516
Frankfort Black, 561
Freckle Lotions, 240
Freckles and Liver Spots, 241
Freezing Mixtures, 615, 616
  Preventives, 363
French Brandy, 768
  Bronze, Preparation of, 136
  Dentrifice, 256
  Floor Polish, 591
  Gelatin, 369
  Hide Tanning Process, 453
  Solders for Silver, 664
  Varnish, 724
Fresh Crushed Fruits, 365
Frost Bite, 363
  Preventive, 363
  Removers, 376
Frosted Glass, 374
  Mirrors, 375
Frosting Polished Silver, 640
Fruit Essences and Extracts, 310
  Frappe, 116
  Jelly Extract, 314
  Preserving, 364, 604
  Products, 357
  Syrups, 701
  Vinegar, 735
Fuel, 152
Fuller's Purifier for Cloths, 274

Fulminates, 332
Fulminating Antimony, 332
  Bismuth, 332
  Copper, 332
  Mercury, 333
  Powder, 333
  Silver, 640
Fumigants, 365
Fumigating Candles, 365
Funnels, to Clean, 204
Furnace Jacket, 368
Furniture Cleaners, 206
  Enamel, 722
  Its Decoration, 772
  Polishes, 592
  Wax, 754
Fuses, 610
  for Electrical Circuits, 64
Fusible Alloys for Electric Installations, 64
  Enamel Colors, 306
  Safety Alloys for Steam Boilers, 65
Fusion Point of Metals, 473

G

Galvanized Iron, 496
  Roofing, 397
  Paper, 507
Gamboge Stain, 439
Gapes in Poultry, 734
Garancine Process, 277
Gardens, Chemical, 368
Garment-cleaning Soap, 645
Gas Fixtures, 130
  Bronzing of, 566
Gasoline Pumps, Packing for, 488
Gas Soldering, 660
  Stove, to Clean, 202
  Trick, 610
Gear Lubricant, 463
Gelatin, 369
  Air Bubbles in, 370
Gems, Artificial, 370
Gem Cements, 20
Geneva Composition Files, 64
Genuine Silver Bronze, 140
German Matches, 467
  Method of Preserving Meat, 361
  Silver or Argentan, 69
German-silver Solders, 661
German Table Mustard, 215
Gilders' Sheet Brass, 55
  Wax, 755
Gilding, 493
  and Gold Plating, 575
  German Silver, 578
  Glass, 373, 578
  in Size, 493
  Metals, Powder for, 579
  Pastes, 580
  Plating and Electrotyping, 288
  Renovation of, 185
  Steel, 580
  Substitute, 575
  to Clean, 185
  Watch Movements, 738
Gilt Frames, Polish for, 600
  Test for, 383
  Work, to Burnish, 384
Ginger, 112
Ginger-Ale Extract, 107
Ginger Ale, Flavoring for, 108
  Soluble Extract, 108
  Beer, 107, 108
  Extracts, 314
Gold-leaf Alloys, 67
  Striping, 383
Gold Varnish for Tin, 727
Glass, 371
  Acid-proof, 374

Glass and Porcelain Cement, 28
and Glassware Cement, 25
Balls, Amalgam for, 90
Silvering, 587
Celluloid, and Metal Inks, 403
Cement for, 21
Cleaning, 208
Coppering, Gilding, and Plating, 572
Etching, 325
Fastening Metals on, 25
Gilding, 373, 578
Globe, Silvering, 641
How to Affix Sign-letters on, 18
Lettering, 457
Lubricants, 372
Manufacturing, 373
Polishes for, 593
Porcelain Repairing, 26
Refractory to Heat, 373
Stop Cock Lubricant, 462
Stopper, to Loosen, 700
Silvering of, 476
Solders for, 662
Soluble, as a Cement, 28
to Affix Paper on, 19
to Cut, 371
to Fasten Brass Upon, 17
to Fix Gold Letters to, 18
to Remove Glue from, 208
to Silver, 641
Waterproof Cements for, 21
Globes, How to Color, 371
Silvering, 476
Glossy Paint for Bicycles, 495
Gloucester Cheese, 176
Glove Cleaners, 195
Gloves, Substitute for Rubber, 100
Testing, 622
Glaziers' Putty, 607
Glazing on Size Colors, 377
Glaze for Bricks, 377
Glazes, 377
and Pottery Bodies, 167
for Cooking Vessels, 377
for Laundry, 444
Glucose in Jelly, 357
Glue, Box, 15
Chromium for Wood, Paper and Cloth, 15
Clarifier, 370
Elastic, 14
Fireproof, 16
for Articles of a Metallic or Mineral Character, 15
for Attaching Cloth Strips to Iron, 14
for Attaching Gloss to Precious Metals, 14
for Belts, 15
for Cardboard, 15
for Celluloid, 12
for Glass, 15
for Leather or Cardboard, 15
for Paper and Metal, 14
for Tablets, 13
for Uniting Metals with Fabrics, 15
for Wood, 15
Manufacture, 10
Marine, 13
or Paste for Making Paper Boxes, 15
Prevented from Cracking, 10
Test, 10
to Fasten Linoleum on Iron Stairs, 14
to Form Paper Pads, 12
Glues, 10, 34, 378
Liquid, 11
Waterproof, 13

Glycerine, 378
and Cucumber Jelly, 228
Applications, 228, 236, 237, 239
as a Detergent, 186
Creams, 237
Developer, 530
Lotion, 379
Milk, 239
Process, 531
Soap, 646, 652
Goats' Milk Cheese, 178
Gold, 379
Acid Test for, 432
Alloys, 66, 435
Amalgams, 89
and Silver Bronze Powders, 139
Assaying of, 381
Enameling Alloys, 67
Enamel Paints, 493
Etching Fluid for, 324
Extraction of, by Amalgamation, 89
Foil Substitutes and Gold Leaf, 747
from Acid Coloring Baths, 381
Imitations of, 433
Indelible Ink, 406
Ink, 405, 415
Jewelry, to Give a Green Color to, 582
Lacquers, 440
Leaf and its Applications, 492
Gold-leaf Alloys, 67
Gold-leaf Waste, to Recover, 381
Gold Lettering, 456
Letters on Glass, Cements for Affixing, 18
Oil Suitable for Use, 485
Paints, 492
Gold-plate Alloys, 67
Gold Plating, 575
Printing on Oilcloth, 379
Purple, 383
Recovery of Waste, 381
Reduction of Old Photographic, 535
Renovator, 199
Solders, 434, 661
Testing, 432
Varnish, 726, 727
Ware Cleaner, 200
Welding, 381
Goldenade, 114
Golden Fizz, 115
Varnishes, 724
"Golf Goblet," 114
Gong Metal, 64
Grafting Wax, 755
Grain, 384
Graining and Marbling, 247
Colors, 556
Crayons, 247
of Brass, 130
with Paint, 494
Granola, 110
Grape Glace, 114
Juice, Preservation of, 767
Graphite Lubricating Compound, 463
Gravel Walks, 385
Gravers, 385
Gray Dyes, 269
Tints, 559
Grease Eradicators, 205
for Locomotive Axles, 462
Greasless Face Cream, 239
Grease Paints, 228
Greases, 462
Wagon and Axle, 462
Green Bronze on Iron, 138
Coloring for Antiseptic Solutions, 100
Dyes, 269

Green Dye for Cotton, 269
for Silk, 269
for Wool and Silk, 269
Fustic Dye, 269
Gilding, 578
Ginger Extract, 315
Ink, 415
or Gold Color for Brass, 582
or Sage Cheese, 176
Patina Upon Copper, 585
Salve, 486
to Distinguish Blue from, 121
Grenades, 341
Grinder Disk Cement, Substitute for, 31
Grinding, 708
Glass, 372
Grindstone Oil, 386
Grindstones, 386
Ground Ceramics, Laying Oil for, 485
for Relief Etching, 322
Grounds for Graining Colors, 556
Grosser's Washing Brick, 445
Gruyere Cheese, 176
Gum Arabic, Substitute, 43, 386
Bichromate Process, 546
Drops, 216
for Envelopes, 43
Gums, 386
their Solubility in Alcohol, 386
Used in Making Varnish, 715
Gun Barrels, to Blue, 682
Bronze, 59
Cotton, 331
Lubricants, 460
Gunpowder, 328
Stains, 387
Gutta-percha, 387
Gutter Cement, 162
Gypsum, 387
Flowers, 346
Paint for, 293

H

Haenkel's Bleaching Solution, 445
Hair-curling Liquids, 389
Hair Dressings and Washes, 389
Dyes, 390
Embrocation, 389
for Mounting, 388
Oil, 390
Oils, Perfumes for, 520
Preparations, 388
Removers, 259
Restorers and Tonics, 389, 391
Shampoo, 392
Hammer, to Harden, 684
Hand Bleach, 233
Creams and Lotions, 232
Hand-cleaning Paste, 232
Handkerchief Perfumes, 516
Hand Stamps, Ink for, 411
Hands, Remove Stains from, 184, 185
Perspiring, 233
Hard-finished Walls, 499
Hard German-silver or Steel Solder, 661
Glaze Bricks, 164
Lead, 71
Metal Drilling Lubricant, 463
Putty, 607
Solders, 662, 664
Solder for Gold, 661
Wood Polish, 598
Hardened Ivory, 429
Steel, to Solder, 665
Hardening Plaster of Paris, 564
of Springs, 685
Steel without Scaling, 685
Steel Wire, 684

Hare-lip Operation, 99
Harmless Butter Color, 143
  Colors for Use in Syrups, 321
Harness Dressings, 450
  Grease, 451
  Oils, 451
  Preparations, 450
  Pastes, 451
  Wax, 755
Hartshorn Poison, 93
Hat-cleaning Compounds, 187
Hat Waterproofing, 748
Hats, 394
  to Dye, 273
Headache Cologne, 394
  Remedies, 394
Head Lice in Children, 422
Heat-indicating Paint, 501
Heat Insulation, 426
  Prickly, 398
Heat-resistant Lacquers, 441
Heaves, 731
Hectograph Pads and Inks, 395, 416
Hedge Mustard, 394
Heel Polish, 632
Helleborе Poison, 94
Helvetius's Styptic, 701
Hemlock Poison, 94
Hemorrhoids, 561
Henbane Poison, 94
Herbarium Specimens, Mounting, 394
  Pomade, 227
Herb Vinegar, 735
Hide Bound, 731
Hide-cleaning Processes, 186
Hides, 454
Hoarfrost Glass, 375
Hoarseness, Bonbons for, 216
  Remedy for, 211
Holland Cheese, 176
Hollow Concrete Blocks, 691
  Silverware, 640
Home-made Outfit for Grinding Glass, 372
  Refrigerators, 616
Honey, 396
  Clarifier, 396
  Water, 519
  Wine, 468
Honeysuckle Perfumery, 516
Honing, 761
Hoof Sores, 730
Hop Beer, 108
  Bitter Beer, 118
  Syrup, 315
Horehound Candy, 217
Horn, 396
  Bleaches, 430
  Uniting Glass with, 17
Horns, Staining, 397
Horse Blistering, 730
Horse-colic Remedy, 729
Horse Embrocations and Liniments, 731
Horses and Cattle, 729
  Treatment of Diseases, 729
Horticultural Ink, 405
Hosiery, Dye for, 268
Hostetter's Bitters, 762
Hot Beef Tea, 112
  Bouillon, 113
  Celery Punch, 112
  Chocolate and Milk, 111
  Egg Bouillon, 112
    Chocolate, 111, 113
    Coffee, 113
    Drinks, 113
    Lemonade, 113
    Milk, 113
    Nogg, 113
    Orangeade, 111
    Phosphate, 113
  Lemonades, 110, 111
  Malt, 112

Hot Malted Milk Coffee (or Chocolate), 112
  Orange Phosphate, 112
  Soda Toddy, 112
  Soda-water Drinks, 111
  Tea, 113
Household Ammonia, 91
  Formulas, 397
House Paint, 500
How to Bronze Metals, 136
  to Clean a Panama Hat, 187
    Brass and Steel, 202
    Tarnished Silver, 204
  to Color Aluminum, 80
  to Keep Cigars, 187
    Fruit, 364
    Lamp Burners in Order, 399
  to Lay Galvanized Roofing, 397
  to Make Castings of Insects, 151
    a Cellar Waterproof, 400
    a Plaster Cast of a Coin or Medal, 150
    Picture Postal Cards and Photographic Letter Head, 537
    Simple Syrups; Hot Process, 702
  to Open a Book, 125
  to Paste Labels on Tin, 40
  to Pour Out Castor Oil, 153
  to Renovate Bronzes, 201
  to Reproduce Old Prints, 223
  to Sensitize Photographic Printing Papers, 539
  to Take Care of Paint Brushes, 140
    Castor Oil, 154
  to Tell Pottery, 173
  to Unite Rubber and Leather, 22
  to Tell the Character of Enamel 304
Huebner's Dental Cement, 163
Hunyadi Water, 740
Huyler's Lemonade, 110
Hydraulic Cement, 33
Hydrochinon Developer, 525
Hydrocyanic Acid Gas for Exterminating Household Insects, 418
Hydrofluoric Formulas, 326
Hydrographic Paper, 504
Hydrogen Peroxide as a Preservative, 605
Hygrometer and Its Use, 401
Hydrometers and Hygroscopes, 402
Hyoscyamus, Antidote to, 102

I

Ice, 402
  Flowers, 402
Iced Coffee, 114
Iceland Moss, Cough Mixture, 211
Ideal Cosmetic Powder, 243
Igniting Composition, 403
Imitation Black Marble, 699
  Cider, 182
  Diamonds, 432
  Egg Shampoos, 393
  Gold, 67, 433
    Foils, 474
  Japanese Bronze, 138
  of Antique Silver, 640
  Ivory, 429
  Platinum, 74
  Silver Alloys, 77
    Bronze, 140
    Foil, 474
  Stains for Wood, 784
Imogen Developer, 527

Impervious Corks, 223
Impregnation of Papers with Zapon Varnish, 506
Improved Celluloid, 156
Incandescent Lamps, 442
Incense, 366
Incombustible Bronze Tincture, 135, 137
Increasing the Toughness, Density and Tenacity of Aluminum, 83
Incrustation, Prevention of, 122
Indelible Hand-stamp Ink, 411
  Inks, 405
  for Glass or Metal, 404
  Labels on Bottles, 327
  Stencil Inks, 412
India, China or Japan Ink, 406
India-rubber Varnishes, 724
Indigo, 268, 281
Indoor Plants, Compost for, 337
Industrial and Potable Alcohol: Sources and Mfg., 667
Infant Foods, 359
Infants, Milk for, 475
Inflammable Explosive with Chlorate of Potash, 331
Inflammability of Celluloid Reduced, 159
Inflammation of the Udder, 731
Influenza in Cattle, 731
  in Horses, 731
Ink Eradicators, 189
  Erasers, 189
  for Laundry, 446
  for Leather Finishers, 453
  for Steel Tools, 404
  for Writing on Glass, 325, 376
    on Glazed Cardboard, 404
    on Marble, 404
  Powders and Lozenges, 407
  Stains, Removing, 189
Inks, 403
  for Hand Stamps, 411
  for Shading Pen, 416
  for Stamp Pads, 408
  for Typewriters, 711
  Hectograph, 395
Inlay Varnish, 724
Inlaying by Electrolysis, 324
Insect Bites, 417
  Casting, 151
  Powders, 419, 424
  Trap, 425
Insecticides, 418
  for Animals, 419
  for Plants, 422
Instructions for Etching, 322
Instrument Alloys, 71
  Cleaning, 199
  Lacquer, 440
  Soap, 653
Instruments, to Remove Rust, 199
Insulating Varnishes, 425
Insulation, 425
  Against Heat, 426
  Moisture, Weather, etc., 426
Intensifiers and Reducers, 552
International Atomic Weights, 757
Iodine Poison, 94
  Soap, 646
  Solvent, 427
Iodoform Deodorizer, 427
Iridescent Paper, 504
Iridia Perfumery, 516
Iron, 427
  and Marble, Cement for, 17
  and Steel, Etching Fluids for, 322
  Polishes, 597
  Powder for Hardening, 427
Biting Off Red Hot, 612
Black Paint for, 495

Iron, Bronzing, 567
Castings, to Soften, 427
Cements for, 17, 25
How to Attach Rubber to, 22
Pipes, Rust Prevention for, 625
Silver-plating, 587
Solders, 665
to Cement Glass to, 17
to Clean, 204
to Cloth, Gluing, 14
to Color Blue, 427
to Whiten, 427
Varnishes, 727
Ironing Wax, 444
Irritating Plaster, 486
Itch, Barbers', 486
Ivory, 428
and Bone Bleaches, 430
Black, 123
Cement, 31
Coating for Wood, 500
Etching on, 428
Gilding, 579
Polishes, 593
Tests, 430

J

Jaborandi Scalp Waters, 392
Jackson's Mouth Wash, 259
Jandrier's Test for Cotton, 246
Japan Black, 495
Paint, 495
Japanese Alloys, 69
Bronze, 138
(Gray), Silver, 76
Japanning and Japan Tinning, 724
Jasmine Milk, 240
Jelly (Fruit) Extract, 314
Jet Jewelry, to Clean, 431
Jewelers' Alloys, 433
Cements, 20
Cleaning Processes, 206
Enamels, 308
Formulas, 430
Glue Cement, 20
Jewelry, to Clean, 206

K

Kalsomine, 436
Karats, to Find Number of, 432
Keeping Flies Out of a House, 399
Keramics, 164
Kerit, 619
Kerosene-cleaning Compounds, 193
Kerosene Deodorizer, 484
Emulsions, 521
Ketchup (Adulterated), 353
Khaki Color Dyeing, 276
Kid, 449
Leather Dressings, 449
Reviver, 453
Kirschner Wine Mustard, 214
Kissingen Salts, 628
Knife-blade Cement, 16
Knife-sharpening Pastes, 615
Knockenplombe, 31
Kola Cordial, 764
Tincture, 321
Koumiss, 116
Substitute, 437
Krems Mustard, Sour, 215
Krems Mustard, Sweet, 215
Kümmel, 764
Kwass, 117

L

Label Pastes, 39
Varnishes, 725

Labels on Tin, How to Paste, 40
Lac and the Art of Lacquering, 437
Lace Leather, 454
to Clean Gold and Silver, 193
Laces, Washing and Coloring of, 446
Lacquer for Aluminum, 438
for Brass, 438
for Bronze, 438
for Copper, 439
for Oil Paintings, 440
for Microscopes, etc., 440
for Stoves and other Articles, 441
Lacquered Ware, to Clean, 195
Lacquers, 437
for Papers, 441
Lakes, 277
Lampblack, 441
Lamp Burners, to Clean, 200, 399
Lamps, 442
Lanoline Creams, 238
Hair Wash, 389
Soap, 647
Toilet Milk, 239
Lantern Slides, 532
Lard, 442
Lathe Lubricant, 461
Laudanum Poison, 95
Laundry Blue, 443
Tablets, 444
Gloss Dressing, 444
Inks, 399
Preparations, 443
Soap, 654
Laundrying Laces, 446
Laurel Water, Poison, 93
Lavatory Deodorant, 398
Lavender Sachets, 510
Water, 514
Lawn Sand, 629
Laxatives for Cattle, etc., 732
Lead, 48, 446
Alloys, 48, 71
Amalgams, Application of, 88
Paper, 507
Plate, Tinned, 589
Poison, 95
to Take Boiling, in the Mouth, 612
Leaf Brass, 54
Leaks, 446
in Boilers, Stopping, 608
Leather, 447
and Rubber Cements, 22
as an Insulator, 426
Cements for, 23
Leather-cleaning Processes, 186
Leather Dyeing, 450
Lac, 441
Lubricants, 460
or Cardboard Glue, 15
Painting on, 455
Polish Lac, 441
Removing Spots from, 206
Russian, 454
Varnish, 725
Waste Insulation, 426
Waterproofing, 750
Leguminous Cheese, 176
Lemon Beer, 108
Essences, 315
Extract (Adulterated), 356
Juice, Plain, 112
Sherbet, 628
Sour, 116
Lemons, 456
Lemonade, 109, 112
for Diabetics, 109
Powder, 627
Preparations for the Sick, 109
Lemonades and Sour Drinks, 110
Lenses and their Care, 456

Letter-head Sensitizers, 537
Lettering, 456
a Clock Dial, 737
on Glass, 457
on Mirrors, 457
Ley Pewter, 75
Lice Killers, 422
Powders, 734
Lichen Removers, 4
Licorice, 458
Syrup, 321
Liebermann's Bleaching Test, 246
Light, Inactinic, 154
Lilac Dye for Silk, 270
Water Perfumery, 520
Limburger Cheese, 176
Lime, 33, 692
Limeade, 110
Lime as a Fertilizer, 339
Bird, 458
Juice, 112, 316
Lime-juice Cordial, 118
Limewater for Dyers' Use, 274
Lincoln Cheese, 176
Lincolnshire Relish, 213
Linen Bleaching, 120
Dressing, 444
to Distinguish Cotton from, 246
Linoleum, 459
Cleaning and Polishing, 206, 398
Glue to Fasten, 14
Liniments, 459
for Horses, 731
Lining for Acid Receptacles, 10
Linseed Oil, 34, 459
Adulteration of, 460
Bleaching of, 459
for Varnish Making, 483
or Poppy Oil, 484
Refining, 484
Solid, 483
Lipol, 226
Lipowitz Metal, 61, 65
Lip, Pomades, 226
Salves and Lipol, 226
Liqueurs, 768
to Clarify, 770
Liquid Bedbug Preparations, 421
Bottle Lac, 440
Bronzes, 135
Cloth and Glove Cleaner, 195
Court Plaster, 247
Dentifrices, 256
Dye Colors, 273
for Bronze Powder, 567
for Cooling Automobile Engines, 363
Liquids for Etching Steel, 327
Liquid Gold, 380
Glues, 11
Headache Remedies, 394
Indelible Drawing Ink, 403
Laundry Blue, 444
Metal Polish, Non-explosive, 595
Perfumes, 511, 515
Polishes, 594
Porcelain Cement, 28
Rouge, 230
Shampoos, 393
Shoe Blackings, 633
Soaps, 646
Styrax Soap, 647
Tar Soap, 647, 654
Liquor Ammonii Anisatus, 91
Liquors, 762
Lithia Water, 740
Lithographic Inks, 407
Lacquer, 440
Paper, 505
Liver-spot Remedies, 241, 242
Lobelia-Indian Poke Poison, 95

Locomotive Axles, Grease for, 462
Lubricants, 462
Locust Killer, 422
Logwood and Indigo Blue Dye, 268
London Soap Powder, 650
Lotion for the Hands, 232
Louse Wash, 423
Lozenges, Voice and Throat, 219
Lubricants, 460, 462
for Cutting Tools, 461
for Heavy Bearings, 461
for Highspeed Bearings, 461
for Lathe Centers, 461
for Redrawing Shells, 463
for Watchmakers, 738
Luhn's Washing Extract, 445
Luminous Paints, 494
Lunar Blend, 114
Lustrous Oxide on Silver, 641
Luster Paste, 464
Lutes, 32

M

Machine Bronze, 58
Oil, 460
Machinery, to Clean, 200, 201, 203
to Keep it Bright, 624
Macht's Yellow Metal, 63
Madder Lake Dye, 277
Magic, 610
Bottles, 126
Mirrors, 478
Magnesian Lemonade Powder, 627
Orgeat Powder, 627
Magnesium, 49
Citrate, 464
Flash-light Powders, 552
Magnetic Alloys, 71
Curves of Iron Filings, their Fixation, 464
Oxide, 625
Magnolia Metal, 51
Mahogany, 784
Make Extract of Indigo Blue Dye, 268
Making Castings in Aluminum, 81
Malleable Brass, 54
Malt, Hot, 112
Malted Food, 359
Milk, 112, 474
Manganese Alloys, 72
Amalgams, Applications of, 87
Argentan, 70
Copper, 72
Manganin, 72
Mange Cures, 731
Manicure Preparations, 226
Mannheim Gold or Similor, 68
Mantles, 465
Manufacture of Alcohol, 674
of Cheese, 174
of Chewing Gum, 178
of Compounds Imitating Ivory, Shell, etc., 429
of Composite Paraffine Candles, 145
of Glue, 10
of Matches, 465
of Pigments, 555
Manufacturing Varnish Hints, 715
Manures, 337
Manuscript Copying, 223
Maple, 784
Maraschino Liqueur, 770
Marble, Artificial, 699
Cements, 16
Cleaning, 196
Colors, 699
Etching, 327
Painting on, 488

Marble, Polishing, 593
Slabs, Cement for, 16
Marbling Crayons, 247
Paper for Books, 505
Margerine, 143
Marine Glue, 13
Paint to Resist Sea Water, 498
Marking Fluid, 465
or Labeling Inks, 407
Maroon Dye for Woolens, 280
Lake Dye, 277
Massage Application, 233
Balls, 233
Creams, 233
Skin Foods, 233
Soaps, 647
Mastic Lac, 441
Mat Aluminum, 81
Gilding, 579
Mats for Metals, 470
Matches, 465
Match Marks on Paint, 195
Phosphorus, Substitute for, 523
Materials, 172
for Concrete Building Blocks, 691
Matrix for Medals, Coins, etc., 467
Matt Etching of Copper, 323
Matzoon, 468
May Bowl or May Wine, 770
Mead, 468
Meadow Saffron Poison, 95
Measures, 760
to Clean, 204
Measuring the Weight of Ice, 402
Meat Extract Containing Albumen, 361
Preservatives, 359, 360
Products (Adulterated), 357
Medallion Metal, 62
Medal Impressions, 467
Medals, to Clean, 199
Medical Paste, 37
Medicated Cough Drops, 217
Massage Balls, 233
Soaps, 647
Medicinal Wines, 771
Medicine Doses, 265
Meerschaum, 469
Cements, 30
Repairing, 27
Mending Celluloid, 161
Porcelain by Riveting, 601
Menthol Cough Drops, 217
Tooth Powder, 253
Mercury, Poison, 95
Salves, 487
Stains, to Remove, 186
Metacarbol Developer, 527
Metal and Paper Glue, 14
Browning by Oxidation, 583
Cements, 25
Cleaning, 199
Foil, 474
Glass and Porcelain Cement, 25
Inlaying, 249
Lipowitz, 65
Polishes, 595
Protectives, 624
Temperature of, 152
Type, 78
Varnishes, 725, 727
Waterproof Cements for, 21
Metallic Articles, Soldering of, 656
Cement, 163
Coffins, 71
Glazes on Enamels, 173
Luster on Pottery, 173
Stain, 783
Paper, 507
Soaps, 648

Metals and Their Treatment, 469
Brightening and Deadening, by Dipping, 469
Bronzing, 567
Cements for, 21, 24
Coloring, 471
Etching Powder for, 324
Fusion Point of, 473
How to Attach to Rubber, 22
How to Bronze, 136
Securing Wood to, 37
Solution for Cleaning, 200
to Silver-plate, 588
Metric System of Weights and Measures, 759
Weights, 759
Meth, 468
Metheglin, 468
Method of Hardening Gypsum and Rendering it Weatherproof, 387
of Purifying Glue, 378
Methods of Preparing Rubber Plasters, 562
Methyl Salicylate, to Distinguish from Oil of Wintergreen, 771
Metol and Hydrochinon Developer, 525
Metol-bicarbonate Developer, 525
Metol Developer, 524, 525
Mice Poison, 613
Microphotographs, 550
Milk, 354, 474
Milk as a Substitute for Celluloid, Bone, and Ivory, 148
Cucumber, 239
Extracts, 474
Powder for Cows, 732
Substitute, 475
to Preserve, 475, 606
Minargent, 64
Mineral Acids, Poison, 92
Oil, 484
Waters, 739
Minofor Metal, 64
Mint Cordial, 765
Julep, 114
Mirror Alloys, 72
Mirror-lettering, 457
Mirror Polishes, 593
Silvering, 476
Mirrors, 476
Frosted, 375
to Clean, 209
to Prevent Dimming of, 374
Miscellaneous Tin Alloys, 78
Mite Killer, 422
Mixed Birdseed, 120, 729
Mixers, Concrete, 693
Mixing Castor Oil with Mineral Oils, 484
Mixture for Burns, 142
Mocking-bird Food, 120, 729
Mock Turtle Extract, 212
Modeling Wax, 755
Modification of Milk for Infants, 473
Moisture, 426
Molding Sand, 478
Molds, 152
of Plaster, 564
Moles, 479
Montpelier Cough Drops, 217
Mordant for Cement Surfaces, 479
for Gold Size, 479
Morphine Poison, 95
Mortar, Asbestos, 479
Mosaic Gold, 68, 140
Silver, 140, 588
Mosquitoes, Remedies, 425
Moss Removers, 209
Moth Exterminators, 425
Paper, 507
Moths and Caterpillars, 423

Motors, Anti-freezing Solution for, 363
Mottled Soap, 654
Mountants, 479, 544
Mounting Drawings, etc., 479
  Prints on Glass, 480
Mousset's Alloy, 76
Moutarde aux Epices, 215
  des Jesuittes, 214
Mouth Antiseptics, 99
  Washes, 258
  Wash-tablets, 259
Moving Objects, How to Photograph Them, 548
Mucilage, 42
  Commercial, 43
  Creams, 238
  of Acacia, 43
  to Make Wood and Pasteboard Adhere to Metals, 43
Mulberry Dye for Silk, 272
Muriatic Acid Poison, 92
Mushroom Poison, 96
Music Boxes, 480
Muslin, Painting on, 488
Mustache Fixing Fluid, 480
Mustard, 214
  Cakes, 214
  Paper, 480
  Vinegar, 215
Myrrh Mouth Wash, 258
  Tooth Paste, 257

N

Nadjy, 115
Nail-cleaning Washes, 227
Nail, Ingrowing, 481
  Polishes, 226
  Varnish, 227
Name Plates, Coating for, 501
Natural Glue for Cementing Porcelain, Crystal Glass, etc., 15
  Lemon Juice, 316
  Water, 739
Nature, Source and Manufacture of Pigments, 555
Neatsfoot Oil, 481
Needles, Anti-rust Paper for, 625
Negatives, How to Use Spoiled, 534
Nervine Ointment, 487
Nerve Paste, 481
Nets, 223
Neufchatel Cheese, 177
Neutral Tooth Powder, 255
New Celluloid, 155
  Mordant for Aniline Colors, 273
  Production of Indigo, 281
Nickel Alloys, 76
  Bronze, 70
Nickel-plating, 573
  with the Battery, 573
Nickel-testing, 481
Nickel, to Clean, 200
  to Remove Rust from, 199
Nickeled Surface, 589
Nickeling by Oxidation, 587
  Test for, 589
Niello, 683
Nitrate of Silver Poison, 95
  Spots, 198
Nitric Acid Poison, 92
  Stains to Remove, 185
Nitroglycerine, 329
Non-explosive Liquid Metal Polish, 595
Non-masticating Insects, 423
Non-Poisonous Textile and Egg Dyes for Household Use, 275
  Fly-papers, 347
Non-porous Corks, 224
Norfolk Cheese, 177

Normona, 115
Nose Putty, 230
Notes for Potters, Glass-, and Brick-makers, 164
Noyeau Powder, 628
Nut Candy Sticks, 216
Nutmeg Essence, 316
Nutwood Stain, 783
Nux Vomica Poison, 615

O

Oak, 775, 783
  Graining, 494
  Leather, Stains for, 455
  Stain, 783
  Wood Polish, 598
Odorless Disinfectants, 264
Odonter, 259
Œnanthic Ether as a Flavoring for Ginger Ale, 108
Oil, Carron, 242
  Castor, 153
  Clock, 482
Oilcloth, 459
  Adhesives, 36
Oil Extinguisher, 341
  for Firearms, 460
  Grease-, Paint-spot Eradicators, 205
  How to Pour Out, 153
  Lubricating, 460
  Neatsfoot, 481
  of Cinnamon as an Antiseptic, 100
  of Vitriol Poison, 92
  Paintings, Lacquer for, 440
    Protection for, 488
  Prints, Reproduced, 223
  Removers, 205
  Solidified, 461
  Stains for Hard Floors, 344
  Suitable for Use with Gold, 485
Oils, 482
  (Edible), Tests for, 355
  for Harness, 451
  Purification of, 335
Oilskins, 750
Oily Bottles, to Clean, 210
Ointments, 486
  for Veterinary Purposes, 731
Oleaginous Stamping Colors, 679
Olein Soap, 654
Oleomargarine, 142
Old-fashioned Ginger Beer, 107
  Lemonade, 110
Olive-oil Paste, 143
Onyx Cements, 16
Opium and All Its Compounds, Poison, 95
Optical Lenses, Cleaning, 208
Orangeade, 110
Orange Bitters and Cordial, 762, 764
  Drops, 216
  Dye, 271
  Extract, 316
  Flower Water, 520
  Frappé, 110
  Peel, Soluble Extract, 316
  Phosphate, 112
Ordinary Drab Dye, 281
  Green Glass for Dispensing Bottles, 373
  Negative Varnish, 544
Oreïde (French Gold), 68
Orgeat Punch, 110
Ornamental Designs on Silver, 641
Ornaments of Iron, Blackening, 495
Orris and Rose Mouth Wash, 258
Ortol Developer, 527
Ox-gall Soap for Cleansing Silk, 654

Oxide, Magnetic, 625
  of Chrome, 172
  of Tin, 172
  of Zinc Poison, 97
Oxidized Steel, 584
Oxidizing, 139
  Processes, 581
Ozonatine, 44

P

Package Pop, 107
  Wax, 755
Packing for Gasoline Pumps, 488
  for Stuffing Boxes, 488
Packings, 488
Pads of Paper, 488, 502
Pain-subduing Ointment, 487
Paint, Acid-resisting, 10
  Bases, 489
  Brushes, 490
    at Rest, 141
    Cleaning, 140
    Deadening, 491
  Dryers, 492
  for Bathtubs, 501
  for Blackboards, 489
  for Copper, 495
  for Iron, 496
  for Protecting Cement Against Acid, 9
  Grease, 229
  Peeling of, 501
  Removed from Clothes, 192
  Removers, 187
  to Prevent Crawling of, 490
  Varnish, and Enamel Removers, 187
Painters' Putty, 607
Painting on Leather, 455
  on Marble, 488
  on Muslin, 488
  Ornaments or Letters on Cloth and Paper, 488
  Over Fresh Cement, 499
  Processes, 488
Paintings, 488
  to Clean, 195
Paints, 489
  Dry Base for, 489
  for Gold and Gilding, 492
  for Metal Surfaces, 495
  for Roofs and Roof Paper, 497
  for Walls of Cement, Plaster, Hard Finish, etc., 498
  for Wood, 500
  Stains, etc., for Ships, 498
  Waterproof and Weatherproof, 499
Pale Purple Gold, 383
Pale-yellow Soap, 652
Palladium Alloys, 73
  Bearing Metal, 73
  Gold, 69
  Silver Alloy, 73
Palladiumizing, 583
Palms, their Care, 502
Panama Hat, How to Clean, 187
Paper, 502
  and Metal Glue, 14
  (Anti-rust) for Needles, 625
  as Protection for Iron, 625
  Blotting, 503
  Box Glue, 15
  Celloidin, 504
  Cements, 21
  Disinfectant, 263
  Fireproof, 344
  Floor Covering, 506
  Frosted, 374
Paperhangers' Pastes, 39
Paper Hygrometers, 402
  Making, Blue Print, 536
  on Glass, to Affix, 19
  Pads, 502

Paper Pads, Glue for, 12
  Photographic, 527
  -sensitizing Processes, 536
  Tickets Fastening to Glass, 19
  Varnishes, 725
  Waterproofing, 505, 751
Papers, Igniting, 611
Papier-mâché, 502
Paraffine, 507
  Scented Cakes, 508
Paraffining of Floors, 345
Parchment and Paper, 502
  Cement, 21
  Paste, 37
Paris Green, 561
  Red, 600
  Salts, 264
Parisian Cement, 30
Parmesan Cheese, 177
Parquet Floors, Renovating, 345
  Polishes, 591
Passe-partout Framing, 508
Paste, Agar-agar, 37
  Albumen, 37
  Antiseptic, 99
  Balkan, 38
Pasteboard Cement, 21
  Deodorizers, 399
Paste, Elastic or Pliable, 39
  for Affixing Cloth to Metal, 37·
  for Cleaning Glass, 208
  for Fastening Leather to Desk Tops, etc., 36
  for Making Paper Boxes, 15
  for Paper, 37
  for Parchment Paper, 37
  for Removing Old Paint or Varnish Coats, 188
  for Tissue Paper, 37
  for Wall Paper, 39
  Flour, 39
  Ink to Write with Water, 416
  Permanent, 38
  that will not Mold, 37
  Venetian, 39
Pastes, 35
  for Paperhangers, 39
  for Polishing Metals, 595
  for Silvering, 588
  to Affix Labels to Tin, 39
Pastilles, Fumigating, 367
Pasting Celluloid on Wood, 36
  Paper Signs on Metal 36
  Wood and Cardboard on Metal, 37
Pattern Letters and Figures, Alloys for, 80
Paving Brick, Stain for, 166
Patent Leather, 451
  Leather Dressings, 449
  Polish, 633
  Preserver, 453
  Stains for, 452
Patina of Art Bronzes, 584
  Oxidizing Processes, 584
Patinas, 584
Peach Extract, 317
  Tint Rouge, 231
Pearls, to Clean, 208
Peeling of Paints, 501
Pegamoid, 509
Pencils, Antiseptic, 99
  for Marking Glass, 374
Pen Metal, 74
Pens, Gold, 383
Peppermint as a Flavor, 252
Pepsin Phosphate, 112
Percentage Solution, 509, 704
Perfumed Ammonia Water, 91
  Fumigating Pastilles, 367
  Pastilles, 520

Perfumes, 366, 509
  Coloring, 511
  Directions for Making, 512
  Fumigating, 366
  for Hair Oils, 520
  for Soap, 648
Permanent Patina for Copper, 585
  Paste, 38
Perpetual Ink, 404
Perspiration Remedy, 233
Perspiring Hands, 233
Petrolatum Cold Cream, 226
Petroleum, 521
  Briquettes, 522
  Emulsion, 423
  for Spinning, 522
  Hair Washes, 390
  Jellies and Solidified Lubricants, 461
  Soap, 648
Pewter, 75
  Aging, 522
  to Clean, 205
Phosphate Dental Cement, 163
  of Casein and its Production, 149
Phosphor Bronze, 58
Phosphorescent Mass, 523
Photographers' Ointment, 487
  Photographs, 554
Phosphorus Poison, 96, 614
  Substitute, 523
Photographic Developing Papers, 527
  Mountants, 41
Photographing on Silk, 540
Photographs Enlarged, 542
  on Brooches, 551
  Transparent, 545
Photography, 523
  without Light, 154
Piano Polishes, 598
Piccalilli Sauce, 213
Pickle for Brass, 132
  for Bronze, 138
  for Copper, 221
  for Dipping Brass, 132
Pickling Brass like Gold, 132
  Iron Scrap before Enameling, 305
  of German-silver Articles, 582
  Process, 453
  Spice, 214
Picric Acid Stains, 186
Picture Copying, 222
  Postal Cards, 537
  Transferrer, 251
Pictures, Glow, 522
Pigment Paper, 540
Pigments, 555
Pile Ointments, 561
Pinaud Eau de Quinine, 392
Pinchbeck Gold, 69
Pineapple Essence, 317
  Lemonade, 110
Pine Syrup, 320
Pine-tar Dandruff Shampoo, 389
Ping-pong Frappé, 110
Pinion Alloy, 737
Pink Carbolized Sanitary Powder, 263
  Color on Silver, 642
  Dye for Cotton, 271
  for Wool, 271
Pinkeye, 731
Pink Grease Paint, 229
  Purple Gold, 383
  Salve, 487
  Soap, 652
Pins of Watches, 738
Pin Wheels, 609
Pipe-joint Cement, 162

Pipe Leaks, 446
  to Color a Meerschaum, 469
Pipes, Rust-preventive for, 625
Pistachio Essence, 317
Plain Rubber Cement, 34
Plant Fertilizers, 336
  Preservatives, 345
Plants, 561
Plaster, 561
  Articles, Repairing of, 27
  Cast of Coins, 150
  Casts, Preservation of, 565
  for Foundry Models, 564
  from Spent Gas Lime, 564
  Grease, 463
  Irritating, 486
  Model Lubricant, 463
  Mold, 152, 564
  Objects, Cleaning of, 564
  of Paris, Hardening, 32, 150, 564
  Repairing, 27
Plastic Alloys, 64
  and Elastic Composition, 158
  Metal Composition, 65
  Modeling Clay, 184
  Substances of Nitro-cellulose Base, 156
Polishing Paste, 600
Platina, Birmingham, 55
Plate Glass, Removing Putty, 206
  Pewter, 75
Plates, Care of Photographic, 523
  for Engraving, 71
Platine for Dress Buttons, 80
Plating, 565
  Gilding and Electrotyping, 288
  of Aluminum, 572
Platinizing, 586
  Aluminum, 586
  Copper and Brass, 586
  Metals, 586
  on Glass or Porcelain, 586
Platinotype Paper, 530
Platinum Alloys, 73
  -gold Alloys for Dental Purposes, 74
  Papers and Their Development, 529
  Silver, 74
  Solders, 665
  Waste, to Separate Silver from, 641
Platt's Chlorides, 264
Playing Cards, to Clean, 209
Plumbago, 460
Plumbers' Cement, 161
Plumes, 335
Plush, 590
  to Remove Grease Spots from, 193
Poison Ivy, 96
Poisonous Fly-papers, 347
  Mushrooms, 96
Poisons, Antidotes for, 92
Polish for Beechwood Furniture, 593
  for Bronze Articles, 591
  for Copper Articles, 591
  for Fine Steel, 597
  for Gilt Frames, 600
  for Varnished Work, 195
Polishes, 590
  Bone, 395
  for Aluminum, 590
  for Bars, Counters, etc., 590
  for Brass, Bronze, Copper, etc., 590
  for Floors, 591
  for Furniture, 592
  for Glass, 593
  for Ivory, Bone, etc., 593
  for Pianos, 596
  for Silverware, 596

Polishes, for Steel and Iron, 597
for the Laundry, 444
for Wood, 598
or Glazes for Laundry Work, 444
Polishing Agent, 599
Bricks, 600
Cloths, to Prepare, 599
Cream, 600
Mediums, 600
Pastes, 595
for the Nails, 227
Powders, 594
Soaps, 594
Polychroming of Figures, 501
Pomade, Putz, 203
Pomades, 277, 392
Colors for, 228
for the Lips, 226
Pomegranate Essence, 317
Poppy Oil, 484
-seed Oil, Bleaching of, 459
Porcelain, 601
How to Tell Pottery, 173
Letters, Cement for, 19
Production of Luster Colors, 172
Portland Cement, 162
Size Over, 30
Positive Colors, 556
Postal Cards, How to Make, 537
How to Make Sensitized, 539
Potassium Amalgams, Applications of, 86
Silicate as a Cement, 19
Potato Starch, 680
Pottery, 173
and Porcelain, How to Tell, 173
Bodies and Glazes, 167
Metallic Luster on, 173
to Cut, 164
Poultry Applications, 419
Foods and Poultry Diseases and Their Remedies, 733
Lice Destroyer, 419
Wine, 771
Pounce, 189
Powdered Camphor in Permanent Form, 144
Cork as a Preservative, 606
Nail Polishes, 226
Powder, Blasting, 330
Face, 243
for Cleaning Gloves, 195
for Colored Fires, 609
for Gilding Metals, 579
for Hardening Iron, 427
Roup, 734
to Keep Moths Away, 425
to Weld Wrought Iron at Pale-red Heat with Wrought Iron, 761
Powders for Stamping, 679
for the Toilet, 242
Preservation and Use of Calcium Carbide, 144
of Belts, 105
of Carpets, 399
of Drawings, 266
of Eggs, 284
of Fats, 335
of Fishing Nets, 223
of Fresh Lemon Juice, 456
of Fruit Juices, 310
of Gum Solution, 44
of Meats, 359
of Milk, 475
of Plaster Casts, 565
of Syrups, 701
of Wood, 776
of Yeast, 786
Preservative Fluid for Museums, 602
for Stuffed Animals, 602
Preservatives, 602

Preservatives, for Leather, 452
Prairie Oyster, 116
Preparation of Amalgams, 85
of Brick Colors, 165
of Carbolineum, 147
of Catgut Sutures, 155
of Celluloid, 156
of Emulsions of Crude Petroleum, 521
of Enamels, 308
of French Bronze, 136
of Syrups, 702
of Uninflammable Celluloid, 157
Preparations of Copper Water, 221
Prepared Mustards of Commerce, 214
Preparing Bone for Fertilizer, 338
Preparing Emery for Lapping, 289
Preservative for Stone, 602
Preservatives for Paste, 38
for Shoe Soles, 633
for Zoological and Anatomical Specimens, 602
Preserved Strawberries, 605
Preserving Antiques, 98
Eggs with Lime, 285
Meat, a German Method, 361
Pressure Table, 704
Preventing the Peeling of Coatings for Iron, 427
the Putrefaction of Strong Glues, 11
Varnish from Crawling, 717
Prevention of Boiler Scale, 122
of Electrolysis, 123
of Fermentation, 765
of Foaming and Partial Caramelization of Fruit Juices, 311
of Fogging, Dimming and Clouding, 374
Prickly Heat, Applications for, 398
Priming Coat for Water Spots, 501
Iron, 495
Print Copying, 222
Printing Ink, Savages, 409
Inks, 408
Oilcloth and Leather in Gold, 379
on Celluloid, 161
on Photographs, 554
Printing-out Paper, How to Sensitize, 539
Printing-roller Compositions, 617
Prints, their Preservation, 309
Process for Colored Glazes, 165
for Dyeing in Khaki Colors, 276
of Electroplating, 286
of Impregnating Fabrics with Celluloid, 161
Production of Consistent Mineral Oils, 484
of Lampblack, 441
of Luster Colors on Porcelain and Glazed Pottery, 172
of Minargent, 64
of Rainbow Colors on Metals, 568
of Substances Resembling Celluloid, 158
Properties of Amalgams, 85
of Concrete Blocks, Strength, 695
Protecting Boiler Plates from Scale, 122
Cement Against Acid, 9
Stuffed Furniture from Moths, 425
Protection for Cement Work, 162
for Oil Paintings, 488

Protection of Acetylene Apparatus from Frost, 363
Protective Coating for Bright Iron Articles, 496
Prussic Acid, 93
Pumice Stone, 606
Pumice-stone Soap, 648
Pumillo Toilet Vinegar, 244
Punch, Claret, 112
Puncture Cement, 162
Purification of Benzine, 106
Purifying-air, 44
Purifying Oils and Fats, 335
Rancid Castor Oil, 153
Water, 740
Purple and Violet Dyes, 269
Dye, 269
for Cotton, 270
for Silk, 270
Ink, 416
of Cassius, 383
Putty, 606
Acid-proof, 607
for Attaching Sign-letters to Glass, 19
for Celluloid, 161
Nose, 230
Substitute for, 608
to Remove, 206
Putz Pomade, 203
Pyrocatechin Developer, 526
Pyrogallic Acid Stains, 185
Pyrotechnics, 608, 610

Q

Quadruple Extract Perfumery, 518
Quince Extract, 317
Flip, 115
Quick Dryer for Inks Used on Bookbinders' Cases, 410
Quick-drying Enamel Colors, 722
Quick-water, 66
Quilts, to Clean, 194

R

Rags for Cleaning, 194
Raspberryade Powder, 627
Raspberry Essences, 318
Lemonade, 110
Sour, 116
Syrup, 317, 318
Rat Poisons, 96, 613
Ratsbane Poison, 93
Ravigotte Mustard, 215
Razor Paper, 503
Pastes, 509, 615
Recipes for Cold-stirred Toilet Soaps, 652
for Pottery and Brick Work, 167
for Soldering, 665
Recovering Glycerine from Soap Boiler's Lye, 378
Recovery of Tin and Iron in Tinned-plate Clippings, 707
Recutting Old Files, 339
Red Birds, Food for, 729
Coloring of Copper, 221
Crimson and Pink Dyes, 270
Dye for Wool, 271
Furniture Paste, 592
Gilding, 580
Gold Enamel, 67
Grease Paint, 229
Indelible Inks, 406
Ink, 416
Patina, 585
Russia Leather Varnish, 449
Reducer for Gelatin Dry-plate Negatives, 535

Reducers, 552
Reducing Photographs, 542
Refining Linseed Oil, 484
  of Potato Starch, 680
Refinishing Gas Fixtures, 130
Reflector Metal, 72
Refrigerants, 615
Refrigeration, 616
Refrigerators, Home-made, 616
  their Care, 401
Regilding Mat Articles, 580
Reinking Typewriter Ribbons,
  413
Relief Etching of Copper, Steel,
    and Brass, 323
  Ground for, 322
  of Zinc, 323
Relishes, 213
Remedies Against Human Para-
    sites, 422
  Mosquitoes, 425
  for Dry Rot, 618
  for Fetid Breath, 133
  for Insect Bites, 417
Removable Binding, 141
Removal of Aniline-dye Stains
    from the Skin, 184
  of Corns, 224
  of Dirt from Paraffine, 508
  of Heat Stains from Polished
    Wood, 776
  of Iron from Drinking Water,
    741
  of Musty Taste and Smell from
    Wine, 771
  of Odors from Wooden Boxes,
    Chests, Drawers, etc., 398
  of Paint from Clothing, 192
  of Peruvian-balsam Stains, 194
  of Picric-acid Stains, 186
  of Rust, 199
Removing Acid Stains, 184
  and Preventing Match Marks,
    195
  Egg Stains, 201
  Glaze from Emery Wheels, 289
  Grease Spots from Plush, 193
  Ingrown Dirt, 235
  Ink Stains, 189
  Iron Rust from Muslin, 193
  Odor from Pasteboard, 399
  Oil Spots from Leather, 206
  Oil Stains from Marble, 197
  Old Wall Paper, 400
  Paint from Wood, 188
  Silver Stains, 209
  Spots from Furniture, 206
  the Gum of Sticky Fly-paper,
    348
  Varnish, etc., 188
  Window Frost, 376
  Woody Odor, 399
Rendering Paraffine Transparent,
  507
Renovating a Camera, 553
  Old Parquet Floors, 345
Renovation of Polished Surfaces
    of Wood, etc., 197
Repairing Broken Glass, 26
  Hectographs, 396
  Rubber Goods, 620
Replacing Rubies whose Settings
  have Deteriorated, 736
Replating, 588
  with Battery, 573
Reproduction of Plaster Origi-
  nals, 565
Resilvering, 588
  of Mirrors, 476
Restoring Photographs, 544
  Tarnished Gold, 199
Restoration of Brass Articles, 132
  of Old Prints, 309

Restoration of Spoiled Beer, 105
  of the Color of Turquoises, 432
Retz Alloy, 64
Revolver Lubricants, 460
Rhubarb for Cholera, 180
Ribbon, Fumigating, 366
Ribbons for Typewriters, 711
Rice Paste, 38
Rifle Lubricants, 460
Ring, How to Solder, 666
Rings on Metal, Producing Col-
  ored, 582
Riveting China, 179
Roach Exterminators, 425
Rock-candy Syrup, 702
Rockets, 609
Rockingham Glazes, 171
Rodinal Developer, 524
Roller Compositions for Printers,
  617
Roman Candles, 609
Roof Paints, 497
Roofs, How to Lay, 397
  Prevention of Leakage, 397
Room Deodorizer, 400
Rope Lubricants, 463
Ropes, 617
  Waterproofing, 753
Roquefort Cheese, 177
Rose's Alloy, 64
Rose Cordial, 765
  Cream, 115
Rose-Glycerine Soap, 652
Rosemary Water for the Hair, 389
Rose Mint, 115
  Pink Dye, 278
  Pomade, 227
  Poudre de Riz Powder, 243
  Powders, 230
  Talc, 510
Rose-tint Glass, 371
Rosewood, 783
  Stain, 783
Rosin, Shellac, and Wax Cement,
  34

Soap as an Emulsifier, 289
  Sticks, 260
  Tests for, in Extracts, 356
Rottmanner's Beauty Water, 244
Rouge, 228, 229, 230
  for Buff Wheels, 618
  or Paris Red, 600
  Palettes, 230
  Powder, 600
  Tablets, 230
  Theater, 231
Roup Cures, 734
Royal Frappé, 114
  Mist, 115
Rubber, 618
  and Rubber Articles, 620
    Wood Fastened, 22
  Boots and Shoe Cement, 23
  Cement for Cloth, 24
  Cements, 22, 34
  Gloves, Substitute for, 100
    Testing, 622
  Goods, Repairing, 620
  Its Properties and Uses in
    Waterproofing, 743
  Scraps, Treatment of, 621
  Softening, 621
  Stamps, 622
  Varnishes, 724
Ruby Settings, 737
Rules for Varnishing, 717
Rum, Bay, 104
Ruoltz Metal, 64
Russet Leather Dressing, 449
Russian Leather, 454
  Polishing Lac, 411
Rust Paints, 497
  Paper, 625

Rust, Prevention for Iron Pipes,
  625
  Preventive for Tools, etc., 625
  Removers, 193, 198
  Preventives, 623
Rusty Pieces, to Separate, 625

S

Saccharine in Food, 351
Sachet Powders, 509
Safety in Explosives, 330
  Paper, 503
  Paste for Matches, 467
Sage Cheese, 176
Salicyl, Sweet, 258
Salicylic Acid in Food, 349
  Soap, 654
Saltpeter (Nitrate of Potash), 96
Salts, Effervescent, 626
  Smelling, 628
Salve, 486
Sand, 628
  Holes in Brass, 150
  in Cast-brass Work, 150
Sand-lime Brick, 689
Sand Soap, 654
  to Prevent Adhesion of Sand to
    Castings, 150
Sandstone Cements, 17
  Coating, 10
  to Remove Oil Spots from, 198
Sapo Durus, 654
Saponaceous Tooth Pastes, 257
Sarsaparilla, 629
  Beer, 118
  Extract, 318
  Soluble Extract, 318
Sauces, Table, 213
Sausage Color, 358
Savage's Printing Ink, 409
Savine Poison, 96
Sawdust for Jewelers, 737
  in Bran, 126
Saxon Blue Dye, 268
Scald Head, Soap for, 653
Scale for Photographic Reduc-
    tion, 542
  in Boilers, 122
  Insects, Extermination of, 423
    on Orange Trees, 423
  Pan Cleaner, 205
Scales and Tables, 547
Scalp Wash, 389
Scarlet Lake Dyes, 277
  with Lac Dye, 271
Schiffmann's Asthma Powder, 101
Scissors Hardening, 685
Scotch Beer, 118
Scratch Brushing, 576
Screws, 629
  Bluing, 682
  in Watches, 738
Sealing (Burning) Trick, 611
  Waxes, 755
Sea Sickness, 630
Seasonings, 213
Seed, Bird, 120
Seidlitz Salt, 628
Self-igniting Mantles, 465
Seltzer and Lemon, 110
  Lemonade, 110
  Water, 740
Separating Silver from Platinum
    Waste, 641
Serpents, Pharaoh's, 630
Serviettes Magiques, 596
Setting of Tools, 708
  the Paint-brush Bristles, 141
Sewing-machine Oil, 461
Sewing Thread, Dressing for, 706
Shades of Red, etc., on Matt Gold
    Bijouterie, 431
Shading Pen, Ink for, 416

Shampoo Lotions and Pastes, 392
Soap, 653
Sharpening Pastes, 509
Stones, 761
Shaving Paste, 630
Soaps, 649
Sheep, 734
Sheet Brass, 54
Sheet-dips, 264
Sheet Metal Alloy, 71
Lubricant, 463
Shellac, 716
Bleaching, 631
Shell Cameos, 630
Imitation of, 429
Polishes, 593
Shells, Lubricants for Redrawing, 463
Sherbet, Egg, 115
Shims in Engine Brasses, 631
"Shio Liao," 32
Ship Compositions and Paints, 498
Shoe Dressings, 631
Leather Dressing, 450
Shoes, Blacking for, 631
Waterproofing, 750
Show Bottles, 127
Show-case Signs, 457
Show Cases, 635
to Prevent Dimming of, 374
Siberian Flip, 115
Siccatives, 636
Sign Letters, 639
Sign-letter Cements, 18
Signs on Show Cases, 457
to Repair Enameled, 304
Silicate of Oxychloride Cements, 35
Silicon Bronze, 61
Silk, 639
Gilding, 580
Sensitizers for Photographic Purposes, 540
Silver, 639
Alloys, 75
Amalgam, 88, 90
Bromide Paper, Toning Baths for, 541
Bronze, 71
Silver-coin Cleaner, 200
Silver, Copper, Nickel, and Zinc Alloys, 76
Etching Fluid for, 324
Fizz, 115
Foil Substitute, 474
Gray Dye for Straw, 269
Stain, 783
Imitation, 77
Ink, 416
Nitrate Spots, to Remove, 194
Test for Cottonseed Oil, 482
Ornamental Designs on, 641
Silver-plating, 574, 587
Silver Polishing Balls, 599
Solder for Enameling, 434
for Plated Metal, 434
Solders, 663
for Soldering Iron, Steel, Cast Iron, and Copper, 663
Testing, 642
to Clean, 204
to Color Pink, 642
to Recover Gold from, 382
Silvering by Oxidation, 583
Bronze, 587
Copper, 587
Glass Balls, Amalgam for, 90
Globes, 641
Globes, 476
of Mirrors, 476
Powder for Metals, 642
Silver-plating, and Desilvering, 587
Test for, 642

Silverware Cleaner, 200
Polishes, 596
Wrapping Paper for, 506
Silver-zinc, 76
Similor, 68
Simple Coloring of Bronze Powder, 134
Test for Red Lead and Orange Lead, 446
Way to Clean a Clock, 207
Sinews, Treatment of, 11
Sinks, to Clean, 202
Size Over Portland Cement, 31
Sizing, 38
Walls for Kalsomine, 436
Skin Bleaches, Balms, etc., 234
Chapped, 232
Skin-cleaning Preparations, 184
Skin Cream, 239
Discoloration, 235
Foods, 231, 234
Lotion, 234
Ointments, 487
Troubles, 644
Slate, 643
Dye for Silk, 269
for Straw Hats, 269
Parchment, 506
Slides for Lanterns, 532
Slipcoat or Soft Cheese, 177
Slugs on Roses, 423
Smaragdine, 45
Smelling Salts, 510, 628
Smokeless Powder, 329
Vari-colored Fire, 609
Smut, Treatment for, 384
Snake Bites, 96, 643
Soap, Benzoin, 652
Soap-bubble Liquids, 655
Soap, Coloring, 644
for Surgical Instruments, 653
for Garment Cleaning, 645
Perfumes, 520
Polishes, 594
Powder, Borax, 649, 650
Substitutes, 653
Tooth, 257
Soaps, 644
and Pastes for Gloves, 195
for Clothing and Fabrics, 191
Soda, Coffee Cream, 113
Water, 111
Soda-water Fountain Drinks, 110
Sodium Amalgams, Applications of, 86
Salts, Effervescent, 627
Silicate as a Cement, 19
Soft Enamels for Iron, White, 305
German-silver Solder, 661
Glaze Brick, 165
Gold Solder, 434
Metal Castings, 151
Silver Solders, 664
Soldering Paste, 667
Solder, 664
Toilet Soaps, 652
Softening Celluloid, 160
Rubber, 621
Steel, 687
Solder, Copper, 659
for Articles which will not Bear a High Temperature, 666
for Brass Tubes, 659
for Fastening Brass to Tin, 659
for Gold, 434
for Iron, 665
for Silver Chains, 664
for Silver-plated Work, 664
for Silversmiths, 664
from Gold, to Remove, 383
Soldering, Acids, 656
a Ring Containing a Jewel, 436, 666
Block, 667

Soldering, Concealed, 665
of Metallic Articles, 656
of Metals, 655
Fluxes for, 660
Paste, 667
Powder for Steel, 665
Recipes, 665
Solution for Steel, 665
without Heat, 666
Solders, 655
for Glass, 662
for Gold, 434
for Jewelers, 436
for Silver, 434
Solid Alcohol, 45
Cleansing Compound, 209
Linseed Oil, 483
Solidified Lubricants, 462
Soluble Blue, 443
Essence of Ginger, 314
Extract of Ginger Ale, 108
Glass, Bronzing with, 139
Gun Cotton, 332
Solution for Removing Nitrate of Silver Spots, 194
Solutions for Batteries, 104
for Cleaning Metals, 200
Percentage, 704
Solvent for Iron Rust, 201
Solvents for Celluloid, 160
Sorel's Dental Cement, 163
Soup Herb Extract, 212
Sources of Potable Alcohol, 668
Sozodont, 256
Sparkling Wines, 767
Sparks from the Finger Tips, 611
Spatter Work, 457
Spavin Cures, 730
Spearmint Cordial, 765
Special Glazes for Bricks, 167
Specific Gravity Test, 382
Speculum Metal, 73
Spice for Fruit Compote, 605
Pickling, 214
Spices, Adulterated, 358
for Flavoring, 213
Spirit, 667, 678
Stains for Wood, 784
Spirits of Salts Poison, 92
Sponge Trick, Blazing, 611
Window Display, 679
Sponges, 678
as Filters, 339
Sterilization of, 679
to Clean, 210
Spot and Stain Removers, 185
Gilding, 580
Spots on Photographic Plates, 554
Sprain Washes, 730
Spray Solution, 103
Spring Cleaning, 207
Hardening, 685
Springs of Watches, 737
to Clean, 207
Sprinkling Powders for Flies, 421
Spruce Beer, 118, 119
Squibb's Diarrhœa Mixture, 179
Squill Poisons, 613
Stage Decorations, Fireproofing, 342
Stain, Brick, 133
for Blue Paving Bricks, 166
Stain-removing Soaps, 653
Stained Ceilings, 400
Staining Horns, 397
Stains, 781
for Lacquers, 438
for Oak Leather, 455
for Patent Leather, 452
for Wood, 781
Attacked by Alkalies or Acids, 785
Stamping, 679
Colors for Use with Rubber Stamps, 679

Stamping Liquids and Powders, 679
Powder for Embroideries, 680
Starch, 445, 680
in Jelly, Tests for, 357
Luster, 399
Paste, 35
Powder, 681
Starch-producing Plants, 668
Statuary Bronze, 57
Statue Cleaning, 197
Statuettes, Cleaning of, 564
of Lipowitz Metal, 64
Steam Cylinder Lubricant, 463
Steel, 681
Alloys, 77
for Drawing Colors on, 80
for Locomotive Cylinders, 77
and Iron Polishes, 597
Blue and Old Silver on Brass, 130
Bluing, 682
Bronze, 61
Browning of, 682
Cleaner, 199
Coloring, 682
Distinguing Iron from, 427
Dust as a Polishing Agent, 600
Etching, 323
on, 687
Fragments, 687
Steel-hardening Powder, 427
Steel, Oxidized, 584
Paint for, 497
Plating, 575
Polishes, 597
Soldering, 665
Testing, 687
to Clean, 199
Tools, to Put an Edge on, 686
Wire Hardening, 684
Stencil Inks, 411
Marking Ink that will Wash Out, 399
Stencils for Plotting Letters of Sign Plates, 296
Stereochromy, 688
Stereopticon Slides, 532
Stereotype Metal, 77
Sterilization of Sponges, 679
of Water with Lime Chloride, 741
Sterling Silver, 434
Stick Pomade, 228
Sticky Fly-papers, 347
Fly Preparations, 421
Stilton Cheese, 177
Stone, Artificial, 688
Cements, 16
Cleaning, 196
Preservative for, 602
Stones for Sharpening, 708, 761
(Precious), Imitation of, 370
Stoneware, 167
and Glass Cements, 26
Waterproof Cements for, 21
Stopper Lubricants, 462, 700
Store Windows, to Clean, 209
Stove, Blacking, 700
Cement, 162
Cleaners, 202
Lacquer, 441
Polish, 597, 700
Varnishes, 727
Stramonium, Antidote for, 102
Strap Lubricant, 460
Strawberries, Preserved, 605
Strawberry Essence, 318
Juice, 318
Pomade, 227
Straw, Bleaching, 120
Fireproofing, 343
Straw-hat Cleaners, 187
Dyes, 394
Strengthened Filter Paper, 503

Stripping Gilt Articles, 205
Photograph Films, 553
Strong Adhesive Paste, 37, 39
Cement, 32
Twine, 223
Strontium Amalgams, 86
Stropping Pastes, 615
Strychnine or Nux Vomica, 96
Poisons, 614
Stuffed Animals, Preserved, 602
Styptic Paste of Gutta Percha, 701
Styptics, 701
Substances Used for Denaturing Alcohol, 678
Substitute for Benzine, 106
for Camphor in the Preparation of Celluloid and Applicable to Other Purposes, 157
for Cement on Grinder Disks, 31
for Cork, 224
for Fire Grenades, 341
for Gum Arabic, 386
for Putty, 608
for Rubber Gloves, 100
for Soldering Fluid, 659
Substitutes for Coffee, 210
for German Silver, 70
for Wood, 785
Suffolk Cheese, 177
Sugar-producing Plants, 668
Sulphate of Zinc Poison, 97
Stains, to Remove, 186
Sulphuric Acid Poison, 92
Summer Drink, 118
Taffy, 217
Sun Bronze, 61
Cholera Mixture, 179
Sunburn Remedies, 240, 241
Sunflower-glycerine Soap, 653
Superfatted Liquid Lanolin-glycerine Soap, 647
Sutures of Catgut, 155
Swiss Cheese, 177
Sympathetic Inks, 412
Syndeticon, 32
Syrup of Bromoform, 134
(Raspberry), 317
Table, 704
Syrups, 321, 701
Szegedin Soap, 653

T

Table of Drops, 704
Sauces, 213
Showing Displacement on Ground Glass of Objects in Motion, 548
Top, Acid-proof, 9
Tables, 703
and Scales, 547
for Photographers, 547
Tablet Enameling, 293
Tablets, Chocolate Coated, 179
for Mouth Wash, 259
Glue for, 13
Taffy, 217
Tailor's Chalk, 164
Talc Powder, 243
Talcum Powder, 243
Tallow, 334
Talmi Gold, 69
Tamping of Concrete Blocks, 695
Tan and Freckle Lotion, 241
and Russet Shoe Polishes, 633
Tank, 705
Tanned Leather, Dye for, 447
Tanning, 453
Hides, 454
Taps, to Remove Broken, 705
Tar Paints, 780
Tarragon Mustard, 215

Tar Syrup, 320
Tasteless Castor Oil, 153
Tattoo Marks, Removal of, 705
Tawing, 448
Tea Extract, 319
Hot, 113
Tea-rose Talc Powder, 243
Teeth, to Whiten Discolored, 705
Telescope Metal, 71
Temperature for Brushes, 140
of Metal, 152
of Water for Plants, 561
Tempered Copper, 221
Tempering Brass, 132
Steel, 683
Terra Cotta Cleaning, 197
Substitute, 705
Test for Glue, 10
Testing Nickel, 481
Rubber Gloves, 622
Siccatives, 637
Silver, 642
Steel, 687
Tests for Absolute Alcohol, 45
for Aniline in Pigments, 560
for Cotton, 245
for Lubricants, 463
for Yeast, 786
Textile Cleaning, 191
Theater Rouge, 231
The Burning Banana, 611
Gum-bichromate Photoprinting Process, 546
Preservation of Books, 124
Prevention of the Inflammability of Benzine, 106
Therapeutic Grouping of Medicinal Plasters, 561
Thermometers, 706
Thread, 706
Three-color Process, 548
Throat Lozenges, 218
Thymol, 100
Ticks, Cattle Dip for, 419
Tiers-Argent Alloy, 75
Tilemakers' Notes, 164
Tin, 49, 706
Alloys, 77
Amalgams, Applications of, 87
Ash, 172
Bismuth, and Magnesium, 49
Bronzing, 567
Chloride of Tin, Poison, 97
Tinctures for Perfumes, 513
Tin, Etching Fluid for, 324
Tinfoil, 707
Tin Foils for Capsules, 474
for Wrapping Cheese, 474
Tin in Powder Form, 707
Tin-lead, 77
Alloys, 78
Tinned Surface, 589
Tinning, 584
by Oxidation, 584
Tin Plating by Electric Bath, 575
of Lead, 589
Tinseled Letters, or Chinese Painting on Glass, 458
Tin Silver-Plating, 589
Solders, 665
Statuettes, Buttons, etc., 78
Varnishes, 727
Tipping Gold Pens, 383
Tire, 708
Cements, 23
Tissier's Metal, 64
Tissue Paper, Paste for, 37
To Ascertain whether an Article is Nickeled, Tinned, or Silvered, 589
Attach Glass Labels to Bottles, 41
Gold Leaf Permanently, 474

Tobin Bronze, 61
To Blacken Aluminum, 81
  Bleach Glue, 378
Tobacco Poison, 97
To Bronze Copper, 136
  Burnish Gilt Work, 384
  Caseharden Locally, 684
  Cast Yellow Brass, 54
  Cement Glass to Iron, 17
  Clarify Liqueurs, 770
    Solutions of Gelatin, Glues, etc., 370
    Turbid Orange Flower Water, 512
  Clean a Gas Stove, 202
    Aluminum, 204
    Articles of Nickel, 201
    Brushes of Dry Paint, 188
    Colored Leather, 186
    Dull Gold, 204
    Files, 205
    Fire-gilt Articles, 185
    Furs, 368
    Gilt Frames, etc., 185
    Gilt Objects, 203
    Gold and Silver Lace, 193
    Gummed Parts of Machinery, 203
    Gummed-up Springs, 207
    Jet Jewelry, 431
    Lacquered Goods, 195
    Linoleum, 206
    Milk Glass, 209
    Mirrors, 209
    Oily Bottles, 210
    Old Medals, 199
    Painted Walls, 190
    Paintings, 195
    Petroleum Lamp Burners, 200
    Playing Cards, 209
    Polished Parts of Machines, 201
    Quilts, 194
    Silver Ornaments, 201
    Skins Used for Polishing Purposes, 186
    Soldered Watch Cases, 207
    Sponges, 210
    Store Windows, 209
    Tarnished Zinc, 205
    the Tops of Clocks in Repairing, 20
    Very Soiled Hands, 185
    Watch Chains, 206
    Wool, 273
    Zinc Articles, 203
  Coat Brass Articles with Antimony Colors, 581
  Color a Meerschaum Pipe, 469
    Billiard Balls Red, 428
    Bronze, 138
    Butter, 359
    Cheese, 359
    Gold, 383
    Iron Blue, 427
    Ivory, 428
  Conceal Soldering, 665
  Copper Aluminum, 581
  Copy Old Letters, etc., 223
  Cut Castile Soap, 644
    Glass, 371
To Cut Glass under Water, 372
  Pottery, 164
Toddy, Hot Soda, 112
To Detect Artificial Vanillin in Vanilla Extracts, 713
    the Presence of Aniline in a Pigment, 560
    Tonka in Vanilla Extract, 714
  Determine the Covering Power of Pigments, 560
  Dissolve Copper from Gold Articles, 382

To Distinguish Cotton from Linen, 246
    Genuine Diamonds, 260
    Glue and Other Adhesives, 378
    Iron from Steel, 427
    Steel from Iron, 687
  Do Away with Wiping Dishes, 399
  Drain a Refrigerator, 616
  Drill Optical Glass, 372
  Dye Copper Parts Violet and Orange, 221
    Cotton Dark Brown, 280
    Feathers, 282
    Felt Goods, 281
    Silk a Delicate Greenish Yellow, 280
    Silk Peacock Blue, 281
    Stiffen, and Bleach Felt Hats, 273
    Woolen Yarns, etc., Various Shades of Magenta, 280
    Woolens with Blue de Lyons, 280
  Eat Burning Coals, 612
  Estimate Contents of a Circular Tank, 705
  Extract Oil Spots from Finished Goods, 273
    Shellac from Fur Hats, 394
  Fasten Brass upon Glass, 17
    Paper Tickets to Glass, 19
    Rubber to Wood, 22
  Fill Engraved Letters on Metal Signs, 457
  Find the Number of Carats, 432
  Fire Paper, etc., by Breathing on it, 611
  Fix Alcoholic Lacquers on Metallic Surfaces, 440
    Dyes, 274
    Gold Letters, etc., upon Glass, 18
    Paper upon Polished Metal, 37
    Iron in Stone, 162
  Fuse Gold Dust, 384
  Give a Brown Color to Brass, 130
    a Green Color to Gold Jewelry, 582
    Brass a Golden Color, 577
    Dark Inks a Bronze or Changeable Hue, 409
  Grind Glass, 372
  Harden a Hammer, 684
  Hard-solder Parts Formerly Soldered with Tin Solder, 663
  Impart the Aroma and Taste of Natural Butter to Margarine, 143
  Improve Deadened Brass Parts 132
  Increase the Toughness, Density, and Tenacity of Aluminum, 83
Toilet Creams, 235
  Milks, 239
  Powders, 242
  Soap Powder, 652
Toilet Soaps, 650
  Vinegars, 244
  Waters, 244, 519
To Keep Files Clean, 339
    Flaxseed Free from Bugs, 424
    Flies from Fresh Paint, 501
    Ice in Small Quantities, 402
    India Ink Liquid, 407
    Liquid Paint in Workable Condition, 501
  Keep Machinery Bright, 624
Tolidol Developer, 52

To Loosen a Glass Stopper, 700
  a Rusty Screw in a Watch Movement, 738
Tomato Bouillon Extract, 212
Tombac Volor on Brass, 130
To Make a Belt Pull, 106
    a Clock Strike Correctly, 738
    a Transparent Cement for Glass, 29
    Cider, 180
    Corks Impermeable and Acid-proof, 10
    Fat Oil Gold Size, 382
    Holes in Thin Glass, 372
    Loose Nails in Walls Rigid, 399
    or Enlarge a Dial Hole, 737
    Plush Adhere to Metal, 590
    Matt Gilt Articles, 432
    Mend Grindstones, 386
    Wedgwood Mortars, 29
  Toning Baths, 540
    for Silver Bromide Paper, 541
    Black Inks, 409
  Tonka Extract, 319
    Its Detection in Vanilla Extracts, 714
  Tool Lubricant, 461
    Setting, 708
  Tools, Rust Prevention, 625
  Toothache, 709
  Tooth Cements, 163
    Paste to be put in Collapsible Tubes, 257
    Pastes, Powders, and Washes, 251
    Powder for Children, 255
    Powders and Pastes, 253
    Soaps and Pastes, 257
    Straightening, 737
To Overcome Odors in Freshly Prepared Rooms, 400
    Paint Wrought Iron with Graphite, 496
    Paste Paper on Smooth Iron, 37
    Pickle Black Iron-plate Scrap Before Enameling, 305
    Polish Delicate Objects, 599
    Paintings on Wood, 600
    Prepare Polishing Cloths, 599
    Preserve Beef, 360
    Furs, 368
    Milk, 606
    Steel from Rust, 199
    Prevent Crawling of Paints, 490
    Dimming of Eyeglasses, etc., 376
    Glue from Cracking, 10
    Screws from Rusting and Becoming Fast, 629
    Smoke from Flashlight, 552
    the Adhesion of Modeling Sand to Castings, 150
    the Trickling of Burning Candles, 145
    Wood Warping, 781
    Wooden Vessels from Leaking, 446
    Produce Fine Leaves of Metal, 473
    Protect Papered Walls from Vermin, 401
    Zinc Roofing from Rust, 626
    Purify Bismuth, 380
    Put an Edge on Steel Tools, 686
    Quickly Remove a Ring from a Swollen Finger, 431
    Reblack Clock Hands, 738
    Recognize Whether an Article is Gilt, 383
    Recover Gold-leaf Waste, 381
    Reduce Engravings, 310

To Reduce Photographs, 548
Refine Board Sweepings, 432
Remedy Worn Pinions from Watches, 738
Remove a Name from a Dial, 207
Aniline Stains, 185
from Ceilings, etc., 190
Balsam Stains, 194
Black Letters from White Enameled Signs, 639
Burnt Oil from Hardened Steel, 686
Enamel and Tin Solder, 188
Fragments of Steel from Other Metals, 687
Finger Marks from Books, etc., 186
Glue from Glass, 208
Gold from Silver, 382
Grease Spots from Marble, 197
Hard Grease, Paint, etc., from Machinery, 200
Ink Stains on Silver, 201
Nitric-acid Stains, 185
Oil-paint Spots from Glass, 209
Oil-paint Spots from Sandstones, 198
Old Enamel, 189
Old Oil, Paint, or Varnish Coats, 187
Paint, Varnish, etc., from Wood, 188
Putty, Grease, etc., from Plate Glass, 206
Pyro Stains from the Fingers, 555
Red (Aniline) Ink, 190
Rust from Instruments, 199
Rust from Iron Utensils, 198
Rust from Nickel, 199, 203
Silver Plating, 203
Silver Stains from White Fabrics, 193
Soft Solder from Gold, 383
Spots from Drawings, 206
Spots from Tracing Cloth, 192
Stains from the Hands, 184
Stains of Sulphate, 186
Strains in Metal by Heating, 686
Varnish from Metal, 188
Vegetable Growth from Buildings, 209
Water Stains from Varnished Furniture, 188
Vaseline Stains from Clothing, 192
Render Aniline Colors Soluble in Water, 274
Fine Cracks in Tools Visible, 686
Gum Arabic More Adhesive, 43
Negatives Permanent, 553
Pale Gold Darker, 383
Shrunken Wooden Casks Watertight, 149
Window Panes Opaque, 375
Renew Old Silks, 274
Renovate and Brighten Russet and Yellow Shoes, 633
Brick Walls, 190
Old Oil Paintings, 488
Straw Hats, 187
Repair a Dial, etc., with Enamel Applied Cold, 737
a Repeating Clock-bell, 737
Enameled Signs, 304
Meerschaum Pipes, 469
Restore Brushes, 141
Patent Leather Dash, 452

To Restore Reddened Carbolic Acid, 147
the Color of a Gold or Gilt Dial, 207
Burnt Steel, 686
Tortoise-shell Polishes, 593
To Scale Cast Iron, 204
Scent Advertising Matter, 510
Separate Rusty Pieces, 625
Silver Brass, Bronze, Copper, 587
Glass Balls and Plate Glass, 587
Silver-plate Metals, 588
Soften Glaziers' Putty, 607
Horn, 397
Iron Castings, 427
Old Whitewash, 762
Solder a Piece of Hardened Steel, 665
Stop Leakage in Iron Hot-Water Pipes, 446
Sweeten Rancid Butter, 143
Take Boiling Lead in the Mouth, 612
Tell Genuine Meerschaum, 469
Temper Small Coil Springs and Tools, 683
Test Extract of Licorice, 458
Fruit Juices and Syrups for Aniline Colors, 321
Fruit Juices for Salicylic Acid, 321
the Color to See if it is Precipitating, 277
Tighten a Ruby Pin, 738
Toughen China, 173
Transfer Designs, 710
Engravings, 710
Turn Blueprints Brown, 542
Utilize Drill Chips, 686
Touchstone, Aquafortis for the, 383
Toughening Leather, 455
To Weaken a Balance Spring, 733
Whiten Flannels, 446
Iron, 427
Widen a Jewel Hole, 431
Tracing-cloth Cleaners, 194
Tracing Cloth, Removing Spots from, 192
Tracing, How to Clean, 194
Paper, 503
Tragacanth, Mucilage of, 42
Transfer Processes, 710
Transparencies, 709
Transparent Candles, 145
Brick Glaze, 167
Ground Glass, 373
Photographs, 545
Soaps, 652
Trays, Varnish for, 727
Treacle Beer, 119
Treatment and Utilization of Rubber Scraps, 621
of Bunions, 224
of Carbolic-acid Burns, 147
of Cast-iron Grave Crosses, 202
of Corns, 225
of Damp Walls, 400
of Fresh Plaster, 564
of Newly Laid Linoleum, 459
of the Grindstone, 386
Tricks with Fire, 608
Triple Extract Perfumery, 513
Pewter, 75
Tubs: to Render Shrunken Tubs Water-tight, 149
Turmeric in Food, 352
Turpentine Stains, 784
Turquoises, Restoration of the Color of, 432
Turtle (Mock) Extract, 212
Twine, 711
Strong, 223

Two-solution Ink Remover, 189
Type Metal, 78
Typewriter Ribbon Inks, 413
Ribbons, 711

U

Udder Inflammation, 731
Unclassified Alloys, 80
Dyers' Recipes, 273
Unclean Lenses, 456
Uninflammable Celluloid, 157
United States Weights and Measures, 758
Uniting Glass with Horn, 17
Rubber and Leather, 22
Universal Cement, 31
Cleaner, 209
Urine, Detection of Albumen, 44
Utensils, Capacities of, 703
to Remove Rust, 198
Utilization of Waste Material or By-products, 673

V

Valves, 711
Vanilla, 713
Extract, 319, 355
Substitute, 714
Vanillin, 713
Vaseline Pomade, 228
Stains, to Remove, 192
Vasolimentum, 728
Varnish and Paint Remover, 188
Bookbinders', 720
Brushes at Rest, 141
for Bicycles, 719
for Blackboards, 720
for Floors, 724
for Trays and Tinware, 727
Gums Used in Making, 715
How to Pour Out, 153
Making, Linseed Oil for, 483
Manufacturing Hints, 715
Removers, 187
Substitutes, 727
Varnished Paper, 506
Varnishes, 543, 714
Engravers', 723
Insulating, 426
Photographic Retouching, 543
Varnishing, Rules for, 717
Vat Enamels and Varnishes, 721
Vegetable Acids, Poison, 92
Vegetables, Canned, 352
Vehicle for Oil Colors, 560
Venetian Paste, 39
Vermilion Grease Paint, 229
Vermin Killer, 422
Very Hard Silver Solder, 663
Veterinary Dose Table, 729
Formulas, 728
Vichy, 740
Salt, 628
Violet Ammonia, 244, 245
Color for Ammonia, 91
Cream, 115
Dye for Silk or Wool, 270
for Straw Bonnets, 270
Flavor for Candy, 217
Ink, 417
Poudre de Riz Powder, 242
Sachet, 510
Smelling Salts, 510
Talc, 510
Powder, 243
Tooth Powder, 252
Water, 520
Witch Hazel, 245
Vinaigre Rouge, 244
Vinegar, 358, 734
Toilet, 244
Viscose, 159

Vogel's Composition Files, 64
Voice Lozenges, 219
Vulcanization of Rubber, 622

W

Wagon and Axle Greases, 462
Wall Cleaners, 190
Wall-paper Dyes, 278
    Removal of, 400
Wall-paper Paste, 39
Wall Priming, 501
    Waterproofing, 741
Walls, Damp, 400
    Hard-finished, 499
Walnut, 783
Warming Bottle, 127
Warping, Prevention of, 781
Warts, 736
Washes, Nail-cleaning, 227
Washing Blankets, 399
    Brushes, 141
    Fluids and Powders, 445
    of Light Silk Goods, 639
Waste, Photographic, Its Disposition, 534
Watch Chains, to Clean, 206
Watch-dial Cements, 20
Watch Gilding, 738
Watch-lid Cement, 20
Watchmakers' Alloys, 736
    and Jewelers' Cleaning Preparations, 206
    Formulas, 736
    Oil, 738
Watch Manufacturers' Alloys, 736
    Movements, Palladium Plating of, 583
Waterproof and Acid-proof Pastes, 38
    Cements for Glass, Stoneware, and Metal, 21
    Coatings, 742
    Glues, 13
    Harness Composition, 451
    Ink, 417
    Paints, 491
    Papers, 505
    Putties, 608
    Ropes, 753
    Shoe Dressings, 634
    Stiffening for Straw Hats, 187
    Varnish for Beach Shoes, 635
    Wood, 753
Waterproofing, 741
    Blue Prints, 741
    Brick Arches, 741
    Canvas, 742
    Cellars, 400
    Corks, 742
    Fabrics, 742
    Leather, 750
    Paper, 751
Water- and Acid-resisting Paint, 499
Water-closets, Deodorants for, 263
Water, Copper, 221
    Filters for, 339
Water-glass Cements, 19
Water Glass in Stereochromatic Painting, 688
    Jackets, Anti-freezing Solutions for, 363
    Natural and Artificial, 739
    Purification, Alum Process of, 340
    Spots, Priming for, 501
    Stains, 784

Water Stirred Yellow, Scarlet and Colorless, 612
Water-tight Casks, 149
    Glass, 373
    Roofs, 373
"Water Tone" Platinum Paper, 529
    to Freeze, 616
    Varnish, 544
Waters, Toilet, 244
Wax, 753
    Burning, Trick, 611
    for Bottles, 553
    for Ironing, 444
    for Linoleum, 459
    Paper, 505
Waxes for Floors, Furniture, etc., 754
Weather Forecasters, 756
Weatherproofing, 499
    Casts, 565
Weed Killers, 262
Weights and Measures, 757
    of Eggs, 284
Weiss Beer, 119
Welding Compound, 687
    Powder to Weld Steel on Wrought Iron at Pale-red Heat, 761
    Powders, 761
Westphalian Cheese, 177
Wheel Grease, 462
Whetstones, 761
Whipped Cream, 247, 248
White Brass, 55
    Bricks, 164
    Coating for Signs, etc., 490
    Cosmetique, 228
    Face Powder, 243
    Flint Glass Containing Lead, 373
    Furniture, Enamel for, 722
    Glass for Ordinary Molded Bottles, 373
    Glazes, 167
White-gold Plates Without Solder, 384
White Grease Paints, 229
    Ink, 417
    Metals, 78
White-metal Alloys, 79
White Metals Based on Copper, 79
    Based on Platinum, 79
    Pine and Tar Syrup, 320
    Petroleum Jelly, 462
    Portland Cement, 162
    Rose Perfumery, 518
    Shoe Dressing, 635
    Solder for Silver, 434
    Stamping Ink, 417
        for Embroidery, 411
    Vitriol, Poison, 97
Whitewash, 761
    to Remove, 190
Whiting, 761
Whooping-cough Remedies, 211
Wild-cherry Balsam, 103
    Extract, 321
Wiltshire Cheese, 177
Window-cleaning Compound, 208
Window Display, 762
    Panes, Cleaning, 208
        Opaque, to Render, 375
    Perfume, 762
    Polishes, 593
Windows, Frosted, 376
    to Prevent Dimming of, 376
Wine Color Dye, 270
Wines and Liquors, 762
    Medicinal, 771
    Removal of Musty Taste, 771

Winter Beverages, 117
Wintergreen, to Distinguish Methyl Salicylate from Oil of, 771
Wire Hardening, 684
    Rope, 771
Witch-hazel Creams, 238
    Jelly, 228
    Violet, 245
Wood, 772
    Acid-proof, 9
    Cements, 26
    Chlorine-proofing, 9
    Fillers, 773
    Fireproofing, 342
Wooden Gears, 463
Wood Gilding, 580
    Polishes, 598
    Pulp, Fireproofing, 343
    Renovators, 194, 197
    Securing Metals to, 37
    Stain for, 781
    Substitutes for, 785
    Warping, to Prevent, 781
    Waterproofing, 753
Wood's Metal, 64
Woodwork, Cleaning, 194
Wool Oil, 485
    Silk, or Straw Bleaching, 120
    to Clean, 273
Woorara Poison, 97
Worcestershire Sauce, 213
Working of Sheet Aluminum, 83
Worm Powder for Stock, 732
Wrapping Paper for Silverware, 506
Wrinkles, Removal of, 231, 233
Writing Inks, 414
    on Glass, 376, 405
    on Ivory, Glass, etc., 405
    on Zinc, 405
    Restoring Faded, 786

Y

Yama, 116
Yeast, 786
    and Fertilizers, 339
Yellow Coloring for Beverages, 119
    Dye for Cotton, 271
        for Silk, 271
    Hard Solders, 658
    Ink, 417
    Orange and Bronze Dyes, 271
    Stain for Wood, 784
Ylang-Ylang Perfume, 518
Yolk of Egg as an Emulsifier, 290
York Cheese, 177

Z

Zapon, 728
    for Impregnating Paper, 506
    Varnishes, 728
Zinc, 49
    Alloys, 80
    Amalgam for Electric Batteries, 89
        for Dentists' Zinc, 163
    Amalgams, Applications of, 87
    Articles, Bronzing, 136
        to Clean, 203
    Bronzing, 137, 567
    Contact Silver-plating, 589
    Etching, 323
    Gilding, 580
Zinc-Nickel, 80
Zinc Plates, Coppering, 573
    Poison, 97
    to Clean, 205

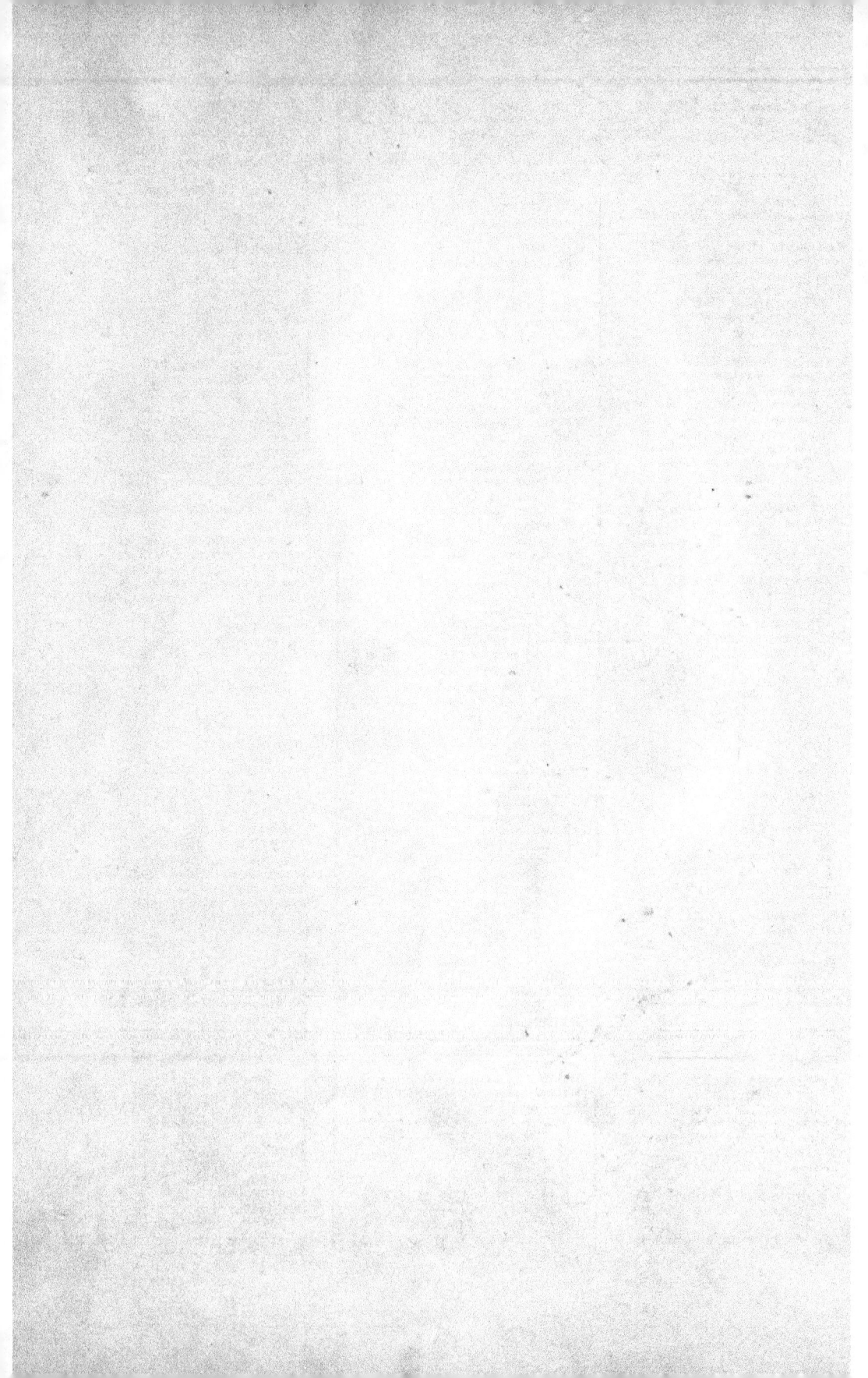

Also from Benediction Books ...

Wandering Between Two Worlds: Essays on Faith and Art
Anita Mathias
Benediction Books, 2007
152 pages
ISBN: 0955373700

Available from www.amazon.com, www.amazon.co.uk
www.wanderingbetweentwoworlds.com

In these wide-ranging lyrical essays, Anita Mathias writes, in lush, lovely prose, of her naughty Catholic childhood in Jamshedpur, India; her large, eccentric family in Mangalore, a sea-coast town converted by the Portuguese in the sixteenth century; her rebellion and atheism as a teenager in her Himalayan boarding school, run by German missionary nuns, St. Mary's Convent, Nainital; and her abrupt religious conversion after which she entered Mother Teresa's convent in Calcutta as a novice. Later rich, elegant essays explore the dualities of her life as a writer, mother, and Christian in the United States-- Domesticity and Art, Writing and Prayer, and the experience of being "an alien and stranger" as an immigrant in America, sensing the need for roots.

About the Author

Anita Mathias was born in India, has a B.A. and M.A. in English from Somerville College, Oxford University and an M.A. in Creative Writing from the Ohio State University. Her essays have been published in The Washington Post, The London Magazine, The Virginia Quarterly Review, Commonweal, Notre Dame Magazine, America, The Christian Century, Religion Online, The Southwest Review, Contemporary Literary Criticism, New Letters, The Journal, and two of HarperSanFrancisco's The Best Spiritual Writing anthologies. Her non-fiction has won fellowships from The National Endowment for the Arts; The Minnesota State Arts Board; The Jerome Foundation, The Vermont Studio Center; The Virginia Centre for the Creative Arts, and the First Prize for the Best General Interest Article from the Catholic Press Association of the United States and Canada. Anita has taught Creative Writing at the College of William and Mary, and now lives and writes in Oxford, England.